AF870178

University of New Mexico

MATH 180: Elements of Calculus I, 8e

Applied Calculus for the Managerial, Life and Social Sciences

Tan

CENGAGE Learning

Australia • Brazil • Japan • Korea • Mexico • Singapore • Spain • United Kingdom • United States

CENGAGE Learning

MATH 180: Elements of Calculus I: Applied Calculus for the Managerial, Life and Social Sciences, 8th Edition

Executive Editors:
Maureen Staudt
Michael Stranz

Senior Project Development Manager:
Linda deStefano

Marketing Specialist:
Courtney Sheldon

Senior Production/Manufacturing Manager:
Donna M. Brown

Production Editorial Manager:
Kim Fry

Sr. Rights Acquisition Account Manager:
Todd Osborne

Printed in the United States of America

Applied Calculus for the Managerial, Life, and Social Sciences, 8th Edition
Soo T. Tan

ISBN-13: 978-1-285-10828-5

ISBN-10: 1-285-10828-0

Cengage Learning
5191 Natorp Boulevard
Mason, Ohio 45040
USA
Cengage Learning is a leading provider of customized learning solutions with office locations around the globe, including Singapore, the United Kingdom, Australia, Mexico, Brazil, and Japan. Locate your local office at:
international.cengage.com/region.

Cengage Learning products are represented in Canada by Nelson Education, Ltd.
For your lifelong learning solutions, visit **www.cengage.com/custom.**
Visit our corporate website at **www.cengage.com.**

TO PAT, BILL, AND MICHAEL

Brief Contents

PREFACE

Math is an integral part of our daily life. *Applied Calculus for the Managerial, Life, and Social Sciences,* Eighth Edition, attempts to illustrate this point with its applied approach to mathematics. This text is appropriate for use in a two-semester or a three-quarter introductory calculus course for students in the managerial, life, and social sciences. My objective for this Eighth Edition is twofold: (1) To write an applied text that motivates students and (2) to make the book a useful teaching tool for instructors. I hope that with the present edition I have come one step closer to realizing my goal.

THE APPROACH

Level of Presentation

My approach is intuitive, and I state the results informally. However, I have taken special care to ensure that this approach does not compromise the mathematical content and accuracy.

Problem-Solving Approach

A problem-solving approach is stressed throughout the book. Numerous examples and applications illustrate each new concept and result. Special emphasis is placed on helping students formulate, solve, and interpret the results of the problems involving applications. Because students often have difficulty setting up and solving word problems, extra care has been taken to help students master these skills:

- Very early on in the text, students are given practice in setting up word problems (see Section 2.3).
- Guidelines are given to help formulate and solve related-rates problems in Section 3.6.
- In Chapter 4, optimization problems are covered in two sections. First, the techniques of calculus are used to solve problems in which the function to be optimized is given (Section 4.4); second, in Section 4.5, optimization problems that require the additional step of formulating the problem are solved.

Intuitive Introduction to Concepts

Mathematical concepts are introduced with concrete, real-life examples wherever appropriate. An illustrative list of some of the topics introduced in this manner follows:

- **Limits:** The Motion of a Maglev
- **The algebra of functions:** The U.S. Budget Deficit
- **The chain rule:** The Population of Americans Aged 55 Years and Older
- **Differentials:** Calculating Mortgage Payments
- **Increasing and decreasing functions:** The Fuel Economy of a Car
- **Concavity:** U.S. and World Population Growth
- **Inflection points:** The Point of Diminishing Returns
- **Curve sketching:** The Dow Jones Industrial Average on "Black Monday"
- **Exponential functions:** Income Distribution of American Families
- **Area between two curves:** Petroleum Saved with Conservation Measures
- **Approximating definite integrals:** The Cardiac Output of a Heart

Connections

One example (the maglev) is used as a common thread throughout the development of calculus—from limits through integration. The goal here is to show students the connections between the concepts presented—limits, continuity, rates of change, the derivative, the definite integral, and so on.

Motivation

Illustrating the practical value of mathematics in applied areas is an important objective of my approach. Many of the applications are based on mathematical models (functions) that I have constructed using data drawn from various sources, including current newspapers, magazines, and the Internet. Sources are given in the text for these applied problems.

Modeling

I believe that one of the important skills that a student should acquire is the ability to translate a real problem into a mathematical model that can provide insight into the problem. In Section 2.3, the modeling process is discussed, and students are asked to use models (functions) constructed from real-life data to answer questions. Students get hands-on experience constructing these models in the Using Technology sections.

NEW TO THIS EDITION

Algebra Review Gives Students a Plan of Action

A Diagnostic Test now precedes the precalculus review. Each question is referenced by the section and example in the text where the relevant topic can be reviewed. Students can now use this test to diagnose their weaknesses and review the material on an as needed basis.

Diagnostic Test

1. a. Evaluate the expression:

(i) $\left(\frac{16}{9}\right)^{3/2}$ **(ii)** $\sqrt[3]{\frac{27}{125}}$

b. Rewrite the expression using positive exponents only: $(x^{-2}y^{-1})^3$

(Exponents and radicals, Examples 1 and 2, pages 6–7)

2. Rationalize the numerator: $\sqrt[3]{\frac{x^2}{yz^3}}$

(Rationalization, Example 4, page 7)

Algebra Review Where Students Need It Most

Well-placed algebra review notes, keyed to the review chapter, appear where students need them most throughout the text. These are indicated by the icon. See this feature in action on pages 105 and 533.

EXAMPLE 6 Evaluate:

$$\lim_{h \to 0} \frac{\sqrt{1+h}-1}{h}$$

Solution Letting h approach zero, we obtain the indeterminate form 0/0. Next, we rationalize the numerator of the quotient by multiplying both the numerator and the denominator by the expression $(\sqrt{1+h}+1)$, obtaining

$$\frac{\sqrt{1+h}-1}{h} = \frac{(\sqrt{1+h}-1)(\sqrt{1+h}+1)}{h(\sqrt{1+h}+1)}$$

* See page 19.

$$= \frac{1+h-1}{h(\sqrt{1+h}+1)} \qquad (\sqrt{a}-\sqrt{b})(\sqrt{a}+\sqrt{b}) = a-b$$

$$= \frac{h}{h(\sqrt{1+h}+1)}$$

Motivating Real-World Applications

More than 160 new applications have been added to the Applied Examples and Exercises. Among these applications are world population growth, hedge-fund investments, mobile instant messaging accounts, consumption of bottled water, global warming, depletion of Social Security trust fund assets, driving costs for a medium-sized sedan, hiring lobbyists, web conferencing, the autistic brain, the revenue of Polo Ralph Lauren, and U.S. health care IT Spending.

APPLIED EXAMPLE 2 Global Warming The increase in carbon dioxide (CO_2) in the atmosphere is a major cause of global warming. The Keeling curve, named after Charles David Keeling, a professor at Scripps Institution of Oceanography, gives the average amount of CO_2, measured in parts per million volume (ppmv), in the atmosphere from 1958 through 2007. Even though data were available for every year in this time interval, we'll construct the curve based only on the following randomly selected data points.

Year	1958	1970	1974	1978	1985	1991	1998	2003	2007
Amount	315	325	330	335	345	355	365	375	380

The **scatter plot** associated with these data is shown in Figure 18a. A mathematical model giving the approximate amount of CO_2 in the atmosphere during this period is given by

$$A(t) = 0.010716t^2 + 0.8212t + 313.4 \qquad (1 \le t \le 50)$$

Modeling with Data

These new exercises are found throughout the text in many of the *Using Technology* sections. Students actually can see how some of the functions found in the end-of-section exercises are constructed. (See Internet Users in China, Exercise 44, page 335, and the corresponding exercise where the model is derived in Exercise 14, page 337.)

44. **INTERNET USERS IN CHINA** The number of Internet users in China is projected to be

$$N(t) = 94.5e^{0.2t} \qquad (1 \le t \le 6)$$

where $N(t)$ is measured in mi
years, with $t = 1$ correspondi
a. How many Internet users w
b. How many Internet users 2010?
c. Sketch the graph of N.
Source: C. E. Unterberg

14. **MODELING WITH DATA** The number of Internet users in China (in millions) from 2005 through 2010 are shown in the following table:

Year	2005	2006	2007	2008	2009	2010
Number	116.1	141.9	169.0	209.0	258.1	314.8

(*Note:* The number of users for the years 2007 through 2010 are projections.)

a. Use **ExpReg** to find an exponential regression model for the data. Let $t = 1$ correspond to 2005.
Hint: $a^x = e^{x \ln a}$.
b. Plot the scatter diagram and the graph of the function f found in part (a).

Making Connections with Technology

Many Using Technology sections have been updated. A new example—TV mobile phones—has been added to Using Technology 4.3. A new Using Technology section has been added to Section 5.3 ("Compound Interest"). Using Technology 5.6 includes a new example in which an exponential model is constructed—Internet gaming sales—using the logistic function of a graphing utility. Additional graphing calculator screens have been added in some sections.

USING TECHNOLOGY

Analyzing Mathematical Models

We can use a graphing utility to analyze the mathematical models encountered in this section.

APPLIED EXAMPLE 1 Internet Gaming Sales The estimated growth in global Internet-gaming revenue (in billions of dollars), as predicted by industry analysts, is given in the following table:

Year	2001	2002	2003	2004	2005	2006	2007	2008	2009	2010
Revenue	3.1	3.9	5.6	8.0	11.8	15.2	18.2	20.4	22.7	24.5

a. Use **Logistic** to find a regression model for the data. Let $t = 0$ correspond to 2001.
b. Plot the scatter diagram and the graph of the function f found in part (a) using the viewing window $[0, 9] \times [0, 30]$.
c. How fast was the revenue from global gaming on the Internet changing in 2001? In 2007?

Variety of Problem Types

Additional rote questions, true or false questions, and concept questions have been added throughout the text to enhance the exercise sets. (See, for example, the graphical questions added to Concept Questions 10.3, page 681.)

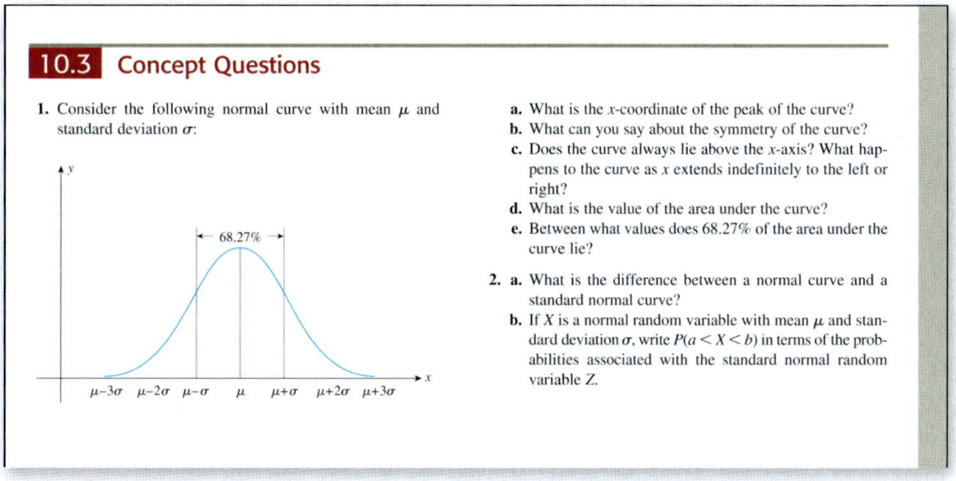

10.3 Concept Questions

1. Consider the following normal curve with mean μ and standard deviation σ:

a. What is the x-coordinate of the peak of the curve?
b. What can you say about the symmetry of the curve?
c. Does the curve always lie above the x-axis? What happens to the curve as x extends indefinitely to the left or right?
d. What is the value of the area under the curve?
e. Between what values does 68.27% of the area under the curve lie?

2. **a.** What is the difference between a normal curve and a standard normal curve?
b. If X is a normal random variable with mean μ and standard deviation σ, write $P(a < X < b)$ in terms of the probabilities associated with the standard normal random variable Z.

Action-Oriented Study Tabs

Convenient color-coded study tabs, similar to Post-it® flags, make it easy for students to tab pages that they want to return to later, whether it be for additional review, exam preparation, online exploration, or identifying a topic to be discussed with the instructor.

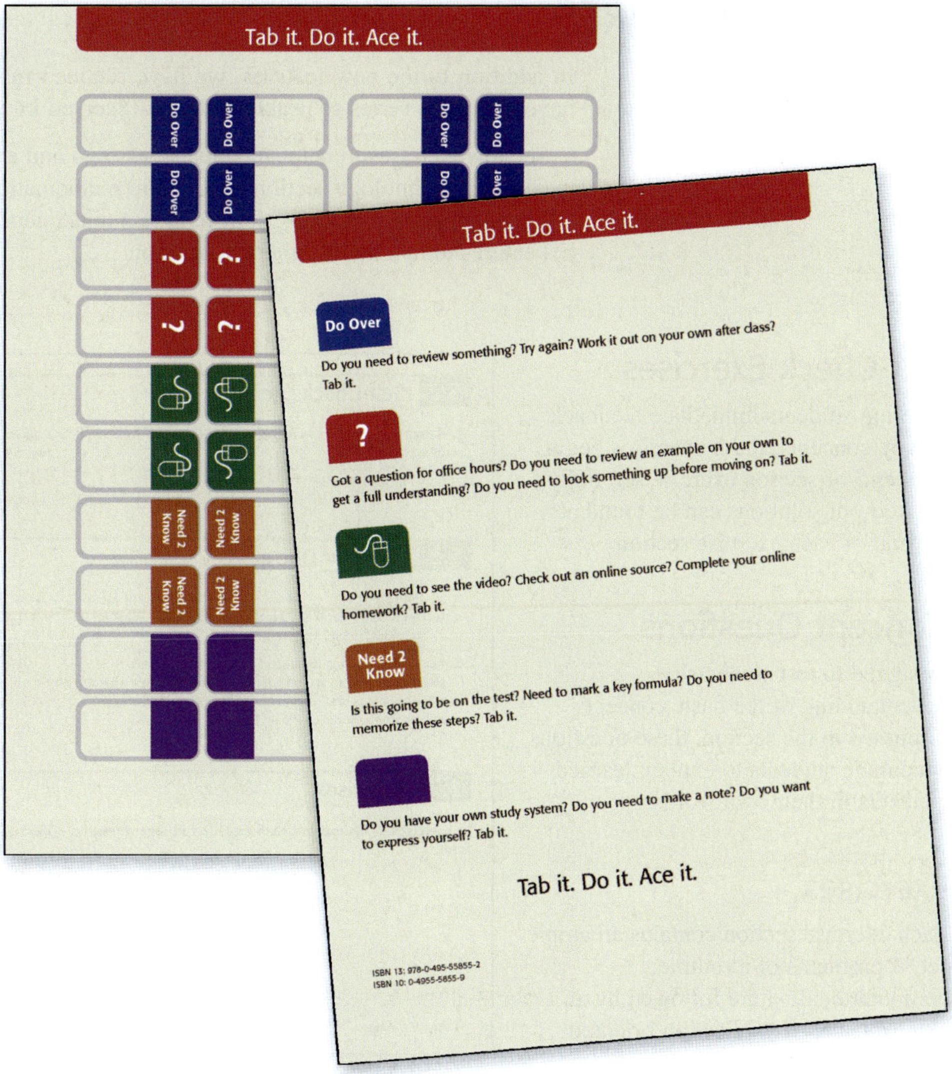

Specific Content Changes

- *The precalculus review in Sections 1.1 and 1.2 has been reorganized.* Operations with algebraic expressions and factoring are now covered in Section 1.1, and inequalities and absolute value are covered in Section 1.2. Two examples illustrating how nonlinear inequalities are solved have been added.
- *Section 2.3 on functions and mathematical models has been reorganized, and new models have been introduced.* Here, students are now asked to use a model describing global warming to predict the amount of carbon dioxide (CO_2) that will be present in the atmosphere in 2010 and a model describing the assets of the Social Security trust fund to determine when those assets are expected to be depleted.
- *The chain rule in Section 3.3 is now introduced with an application*—the population of Americans aged 55 years and older.
- *The section on Volumes of Solids of Revolution has been moved from Section 6.8 to Section 7.5.*
- *A new Exploring with Technology subsection on Applications Involving Normal Random Variables has been added to Section 10.3.*
- *Material on Indeterminate Forms and l'Hôpital's Rule has been added to the appendices.*
- *A How-To Technology Index has been added for easy reference.*

TRUSTED FEATURES

In addition to the new features, we have retained many of the following hallmarks that have made this series so usable and well-received in past editions:

- Section exercises to help students understand and apply concepts
- Optional technology sections to explore mathematical ideas and solve problems
- End-of-chapter review sections to assess understanding and problem-solving skills
- Features to motivate further exploration

Self-Check Exercises

Offering students immediate feedback on key concepts, these exercises begin each end-of-section exercise set. Fully worked-out solutions can be found at the end of each exercise section.

Concept Questions

Designed to test students' understanding of the basic concepts discussed in the section, these questions encourage students to explain learned concepts in their own words.

Exercises

Each exercise section contains an ample set of problems of a routine computational nature followed by an extensive set of application-oriented problems.

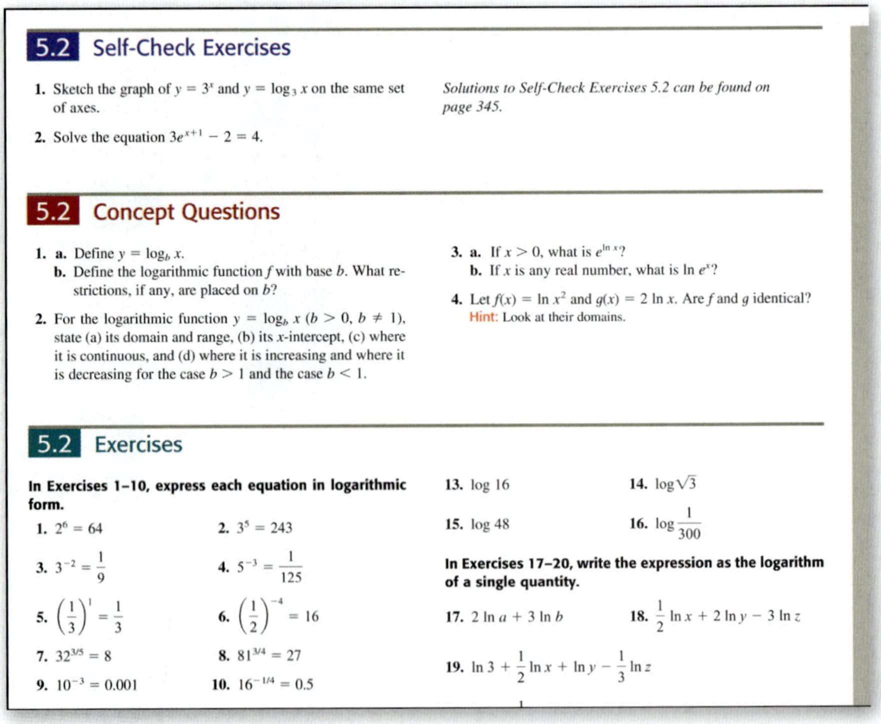

5.2 Self-Check Exercises

1. Sketch the graph of $y = 3^x$ and $y = \log_3 x$ on the same set of axes.
2. Solve the equation $3e^{x+1} - 2 = 4$.

Solutions to Self-Check Exercises 5.2 can be found on page 345.

5.2 Concept Questions

1. **a.** Define $y = \log_b x$.
   b. Define the logarithmic function f with base b. What restrictions, if any, are placed on b?
2. For the logarithmic function $y = \log_b x$ ($b > 0$, $b \neq 1$), state (a) its domain and range, (b) its x-intercept, (c) where it is continuous, and (d) where it is increasing and where it is decreasing for the case $b > 1$ and the case $b < 1$.
3. **a.** If $x > 0$, what is $e^{\ln x}$?
   b. If x is any real number, what is $\ln e^x$?
4. Let $f(x) = \ln x^2$ and $g(x) = 2 \ln x$. Are f and g identical?
   Hint: Look at their domains.

5.2 Exercises

In Exercises 1–10, express each equation in logarithmic form.

1. $2^6 = 64$
2. $3^5 = 243$
3. $3^{-2} = \frac{1}{9}$
4. $5^{-3} = \frac{1}{125}$
5. $\left(\frac{1}{3}\right)^1 = \frac{1}{3}$
6. $\left(\frac{1}{2}\right)^{-4} = 16$
7. $32^{3/5} = 8$
8. $81^{3/4} = 27$
9. $10^{-3} = 0.001$
10. $16^{-1/4} = 0.5$

13. $\log 16$
14. $\log \sqrt{3}$
15. $\log 48$
16. $\log \frac{1}{300}$

In Exercises 17–20, write the expression as the logarithm of a single quantity.

17. $2 \ln a + 3 \ln b$
18. $\frac{1}{2} \ln x + 2 \ln y - 3 \ln z$
19. $\ln 3 + \frac{1}{2} \ln x + \ln y - \frac{1}{3} \ln z$

Using Technology

These optional features appear after the section exercises. They can be used in the classroom if desired or as material for self-study by the student. Here, the graphing calculator is used as a tool to solve problems. These sections are written in the traditional example–exercise format, with answers given at the back of the book. Illustrations showing graphing calculator screens are extensively used. In keeping with the theme of motivation through real-life examples, many sourced applications are again included. Students can construct their own models using real-life data in many of the Using Technology sections. These include models for the growth of the Indian gaming industry, health-care spending, TIVO owners, nicotine content of cigarettes, computer security, and online gaming, among others.

USING TECHNOLOGY

Finding the Accumulated Amount of an Investment, the Effective Rate of Interest, and the Present Value of an Investment

Graphing Utility

Some graphing utilities have built-in routines for solving problems involving the mathematics of finance. For example, the TI-83/84 Solver function incorporates several functions that can be used to solve the problems that are encountered in this section. To access the TVM SOLVER on the TI-83 press [2ND], press [FINANCE], and then select [1: TVM Solver]. To access the TVM SOLVER on the TI-83 plus and the TI-84, press [APPS], press [1: Finance], and then select [1: TVM Solver]. Step-by-step procedures for using these functions can be found on our Companion Web site.

```
N    = 120
I%   = 10
PV   = -5000
PMT  = 0
■ FV = 13535.20745
P/Y  = 12
C/Y  = 12
PMT : END    BEGIN
```

FIGURE T1
The TI-83/84 screen showing the future value (FV) of an investment

EXAMPLE 1 Finding the Accumulated Amount of an Investment Find the accumulated amount after 10 years if \$5000 is invested at a rate of 10% per year compounded monthly.

Solution Using the TI-83/84 TVM SOLVER with the following inputs,

N = 120 (10)(12)
I% = 10
PV = −5000 Recall that an investment is an outflow.
PMT = 0
FV = 0
P/Y = 12 The number of payments each year
C/Y = 12 The number of conversion periods each year
PMT:END BEGIN

we obtain the display shown in Figure T1. We conclude that the required accumulated amount is \$13,535.21. ■

```
► Eff (10, 4)
     10.38128906
```

FIGURE T2
The TI-83/84 screen showing the effective rate of interest (Eff)

EXAMPLE 2 Finding the Effective Rate of Interest Find the effective rate of interest corresponding to a nominal rate of 10% per year compounded quarterly.

Solution Here we use the **Eff** function of the TI-83/84 calculator to obtain the result shown in Figure T2. The required effective rate is approximately 10.38% per year. ■

(continued)

Exploring with Technology

Designed to explore mathematical concepts and to shed further light on examples in the text, these optional questions appear throughout the main body of the text and serve to enhance the student's understanding of the concepts and theory presented. Often the solution of an example in the text is augmented with a graphical or numerical solution. Complete solutions to these exercises are given in the *Instructor's Solutions Manual.*

Exploring with TECHNOLOGY

In the opening paragraph of Section 5.1, we pointed out that the accumulated amount of an account earning interest *compounded continuously* will eventually outgrow by far the accumulated amount of an account earning interest at the same nominal rate but earning simple interest. Illustrate this fact using the following example.

Suppose you deposit \$1000 in account I, earning interest at the rate of 10% per year compounded continuously so that the accumulated amount at the end of t years is $A_1(t) = 1000e^{0.1t}$. Suppose you also deposit \$1000 in account II, earning simple interest at the rate of 10% per year so that the accumulated amount at the end of t years is $A_2(t) = 1000(1 + 0.1t)$. Use a graphing utility to sketch the graphs of the functions A_1 and A_2 in the viewing window $[0, 20] \times [0, 10{,}000]$ to see the accumulated amounts $A_1(t)$ and $A_2(t)$ over a 20-year period.

Summary of Principal Formulas and Terms

Each review section begins with the Summary highlighting important equations and terms with page numbers given for quick review.

CHAPTER 5 Summary of Principal Formulas and Terms

FORMULAS

1. Exponential function with base b	$y = b^x$
2. The number e	$e = \lim_{m\to\infty}\left(1 + \frac{1}{m}\right)^m = 2.71828\ldots$
3. Exponential function with base e	$y = e^x$
4. Logarithmic function with base b	$y = \log_b x$
5. Logarithmic function with base e	$y = \ln x$
6. Inverse properties of $\ln x$ and e^x	$\ln e^x = x$ and $e^{\ln x} = x$
7. Compound interest (accumulated amount)	$A = P\left(1 + \frac{r}{m}\right)^{mt}$
8. Effective rate of interest	$r_{\text{eff}} = \left(1 + \frac{r}{m}\right)^m - 1$
9. Compound interest (present value)	$P = A\left(1 + \frac{r}{m}\right)^{-mt}$
10. Continuous compound interest	$A = Pe^{rt}$
11. Derivative of the exponential function	$\frac{d}{dx}(e^x) = e^x$
12. Chain rule for exponential functions	$\frac{d}{dx}(e^u) = e^u\frac{du}{dx}$
13. Derivative of the logarithmic function	$\frac{d}{dx}\ln\|x\| = \frac{1}{x}$
14. Chain rule for logarithmic functions	$\frac{d}{dx}\ln\|u\| = \frac{1}{u}\frac{du}{dx}$

TERMS

common logarithm (338)
natural logarithm (338)
compound interest (346)
logarithmic differentiation (374)
exponential growth (380)
growth constant (380)
exponential decay (381)
decay constant (381)
half-life of a radioactive substance (382)
logistic growth function (385)

Concept Review Questions

These questions give students a chance to check their knowledge of the basic definitions and concepts given in each chapter.

CHAPTER 5 Concept Review Questions

Fill in the blanks.

1. The function $f(x) = x^b$ (b, a real number) is called a/an _____ function, whereas the function $g(x) = b^x$, where $b >$ _____ and $b \neq$ _____, is called a/an _____ function.
2. a. The domain of the function $y = 3^x$ is _____, and its range is _____.
   b. The graph of the function $y = 0.3^x$ passes through the point _____ and is decreasing on _____.
3. a. If $b > 0$ and $b \neq 1$, then the logarithmic function $y = \log_b x$ has domain _____ and range _____; its graph passes through the point _____.
   b. The graph of $y = \log_b x$ is decreasing if b _____ and increasing if b _____.
4. a. If $x > 0$, then $e^{\ln x} =$ _____.
   b. If x is any real number, then $\ln e^x =$ _____.
5. In the compound interest formula $A = P(1 + \frac{r}{m})^{mt}$, A stands for the _____ _____, P stands for the _____, r stands for the _____ _____ _____ per year, m stands for the _____ _____ _____ per year, and t stands for the _____.
6. The effective rate r_{eff} is related to the nominal interest rate r per year and the number of conversion periods per year by $r_{\text{eff}} =$ _____.
7. If interest earned at the rate of r per year is compounded continuously over t years, then a principal of P dollars will have an accumulated value of $A =$ _____ dollars.

Review Exercises

Offering a solid review of the chapter material, the Review Exercises contain routine computational exercises followed by applied problems.

CHAPTER 5 Review Exercises

1. Sketch on the same set of coordinate axes the graphs of the exponential functions defined by the equations.
   a. $y = 2^{-x}$ b. $y = \left(\frac{1}{2}\right)^x$

In Exercises 2 and 3, express each equation in logarithmic form.

2. $\left(\frac{2}{3}\right)^{-3} = \frac{27}{8}$ 3. $16^{-3/4} = 0.125$

In Exercises 4 and 5, solve each equation for *x*.

4. $\log_4(2x + 1) = 2$
5. $\ln(x - 1) + \ln 4 = \ln(2x + 4) - \ln 2$

In Exercises 6–8, given that ln 2 = *x*, ln 3 = *y*, and ln 5 = *z*, express each of the given logarithmic values in terms of *x*, *y*, an

19. $y = \frac{e^{2x}}{1 + e^{-2x}}$ 20. $f(x) = e^{2x^2-1}$
21. $f(x) = xe^{-x^2}$ 22. $g(x) = (1 + e^{2x})^{3/2}$
23. $f(x) = x^2e^x + e^x$ 24. $g(t) = t\ln t$
25. $f(x) = \ln(e^{x^2} + 1)$ 26. $f(x) = \frac{x}{\ln x}$
27. $f(x) = \frac{\ln x}{x + 1}$ 28. $y = (x + 1)e^x$
29. $y = \ln(e^{4x} + 3)$ 30. $f(r) = \frac{re^r}{1 + r^2}$
31. $f(x) = \frac{\ln x}{1 + e^x}$ 32. $g(x) = \frac{e^{x^2}}{1 + \ln x}$

Before Moving On . . .

Found at the end of each chapter review, these exercises give students a chance to see if they have mastered the basic computational skills developed in each chapter.

CHAPTER 5 Before Moving On . . .

1. Solve the equation $\frac{100}{1 + 2e^{0.3t}} = 40$ for t.
2. Find the accumulated amount after 4 yr if \$3000 is invested at 8%/year compounded weekly.
3. Find the slope of the tangent line to the graph of $f(x) = e^{\sqrt{x}}$.
4. Find the rate at which $y = x\ln(x^2 + 1)$ is changing at $x = 1$.
5. Find the second derivative of $y = e^{2x}\ln 3x$.
6. The temperature of a cup of coffee at time t (in minutes) is
   $$T(t) = 70 + ce^{-kt}$$
   Initially, the temperature of the coffee was 200°F. Three minutes later, it was 180°. When will the temperature of the coffee be 150°F?

Explore & Discuss

These optional questions can be discussed in class or assigned as homework. These questions generally require more thought and effort than the usual exercises. They may also be used to add a writing component to the class or as team projects. Complete solutions to these exercises are given in the *Instructor's Solutions Manual.*

Explore & Discuss

Consider the graph of the function h whose graph is depicted in the following figure.

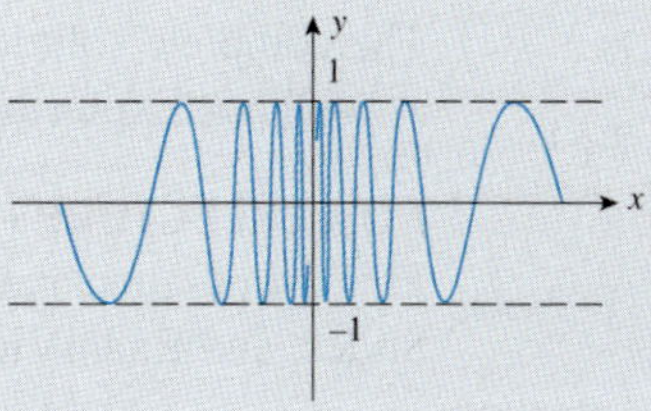

It has the property that as x approaches zero from either the right or the left, the curve oscillates more and more frequently between the lines $y = -1$ and $y = 1$.

1. Explain why $\lim_{x \to 0} h(x)$ does not exist.
2. Compare this function with those in Example 3. More specifically, discuss the different ways the functions fail to have a limit at $x = 0$.

Portfolios

The real-life experiences of a variety of professionals who use mathematics in the workplace are related in these interviews. Among those interviewed are a senior account manager at PepsiCo and an associate on Wall Street who uses statistics and calculus in writing options.

PORTFOLIO Deb Farace

TITLE Sr. National Accounts Manager
INSTITUTION PepsiCo Beverages & Foods

Working for the national accounts division for PepsiCo Beverages & Foods, I need to understand applied mathematics in order to control the variables associated with making a profit, manufacturing, production, and most importantly selling our products to mass club channels. Examples of these large, "quality product at great value" outlets are Wal*Mart, Costco and Target. The types of products I handle include Gatorade, Tropicana, and Quaker foods.

Our studies show that the grocery store channels' sales are flattening or declining as a whole in lieu of large, national outlets like the above. So in order to maximize growth in this segment of our business, I meet with regional buying offices of these chains and discuss various packaging, pricing, product, promotional and shipping options so that we can successfully compete in the market.

A number of factors must be taken into consideration in order to meet my company's financial forecasts. Precision using mathematical models is key here, since so many variables can impact last-minute decision-making. Extended variables of supply-and-demand include time of year, competitive landscape, special coupon distribution and other promotions, selling cycles and holidays, size of the outlets, and yes—even the weather.

For example, it's natural to assume that when it's hot outside people will buy more thirst-quenching products like Gatorade. But since our business is so precise, we need to understand mathematically how the weather affects sales. A mathematical model developed by Gatorade analyzes longterm data that impacts sales by geographic market due to the weather. Its findings include exponentially increased sales of Gatorade for each degree above 90 degrees. I share our mathematical analysis like this study with buyers and negotiate larger orders based on up-to-the-minute weather forecasts. The result: increased sales of product based on math.

© kolvenbach/Alamy

Example Videos

Available through the Online Resource Center and Enhanced WebAssign, these video examples offer hours of instruction from award-winning teacher Deborah Upton of Stonehill College. Watch as she walks students through key examples from the text, step by step—giving them a foundation in the skills that they need to know. Each example available online is identified by the video icon located in the margin.

VIDEO

APPLIED EXAMPLE 6 Rate of Change of DVD Sales The sales (in millions of dollars) of a DVD recording of a hit movie t years from the date of release is given by

$$S(t) = \frac{5t}{t^2 + 1}$$

a. Find the rate at which the sales are changing at time t.
b. How fast are the sales changing at the time the DVDs are released ($t = 0$)? Two years from the date of release?

TEACHING AIDS

INSTRUCTOR'S SOLUTIONS MANUAL (ISBN 0-538-73528-7) by Soo T. Tan
The complete solutions manual provides worked out solutions to all problems in the text, as well as "Exploring with Technology" and "Explore & Discuss" questions.

POWERLECTURE (ISBN 0-538-73527-9)
This comprehensive CD-ROM includes the *Instructor's Solutions Manual,* PowerPoint Slides, and ExamView® Computerized Testing to create, deliver, and customize tests.

ENHANCED WEBASSIGN Enhanced WebAssign
Instant feedback and ease of use are just two reasons why WebAssign is the most widely used homework system in higher education. WebAssign allows you to assign, collect, grade, and record homework assignments via the Web. Now this proven homework system has been enhanced to include links to textbook sections, video examples, and problem-specific tutorials. Enhanced WebAssign is more than a homework system—it is a complete learning system for math students.

LEARNING AIDS

STUDENT SOLUTIONS MANUAL (ISBN 0-538-73526-0) by Soo T. Tan
Giving you more in-depth explanations, this insightful resource includes fully worked-out solutions for the answers to select exercises included at the back of the textbook, as well as problem-solving strategies, additional algebra steps, and review for selected problems.

PREMIUM WEBSITE
Sign in, save time, and get the grade you want! One code will give you access to great tools for *Applied Calculus for the Managerial, Life, and Social Sciences,* Eighth Edition, including quizzes, solution videos, and interactive tutorials.

ACKNOWLEDGMENTS

I wish to express my personal appreciation to each of the following reviewers of the Eighth Edition, whose many suggestions have helped make a much improved book.

Paul Abraham
Kent State University—Stark

James Adair
Missouri Valley College

Jill Britton
Camosun College

Debra D. Bryant
Tennessee Technological University

Micelle Dedeo
University of North Florida

Scott L. Dennison
University of Wisconsin—Oshkosh

Christine Devena
Miles Community College

Andrew Diener
Christian Brothers University

Mike Everett
Santa Ana College

Kevin Ferland
Bloomsburg University

Tao Guo
Rock Valley College

Mark Jacobson
Montana State University—Billings

Sarah Kilby
North Country Community College

Murray Lieb
New Jersey Institute of Technology

Lia Liu
University of Illinois at Chicago

Rebecca Lynn
Colorado State University

Mary T. McMahon
North Central College

Daniela Mihai
University of Pittsburgh

Kathy Nickell
College of DuPage

Carol Overdeep
Saint Martin's University

Mohammed Rajah
Miracosta College

Dennis H. Risher
Loras College

Brian Rodas
Santa Monica College

Dr. Arthur Rosenthal
Salem State College

Abdelrida Saleh
Miami Dade College

Stephanie Anne Salomone
University of Portland

Mohammad Siddique
Virginia Union University

Jennifer Strehler
Oakton Community College

Ray Toland
Clarkson University

Justin Wyss-Gallifent
University of Maryland at College Park

I also thank the previous edition reviewers whose comments and suggestions have greatly shaped the current edition.

Faiz Al-Rubaee
University of North Florida

Jose F. Alvarado
University of Texas, Pan American

Ronald Barnes
University of Houston

Larry Blaine
Plymouth State College

Albert Bronstein
Purdue University

Kimberly Jordan Burch
Montclair State University

Peter Casazza
University of Missouri—Columbia

J. Curtis Chipman
Oakland University

Lisa Cox
Texas A&M University

Candy Giovanni
Michigan State University

Harvey Greenwald
California Polytechnic State University—San Luis Obispo

Mohammed Kazemi
University of North Carolina—Charlotte

Rasul Khan
Cleveland State University

Daniel Lee
Southwest Texas State University

Jeffrey Lee
Texas Tech University

Joyce Longman
Villanova University

C. D. Luning
Sam Houston State University

Joseph Macaluso
DeSales University

Donald K. Mason
Elmhurst College

Dean Moore
Florida Community College at Jacksonville

Marna Mozeff
Drexel University

James Olsen
North Dakota State University

Wayne B. Powell
Oklahoma State University

Deborah Primm
Jacksonville State University

Virginia Puckett
Miami Dade College

Geetha Ramachandran
California State University, Sacramento

Mary E. Rerick
University of North Dakota

Allan Silberger
Cleveland State University

Mehrdad Simkani
University of Michigan, Flint

Anne Siswanto
East Los Angeles College

Laurence Small
L. A. Pierce College

Ray Stanton
Fresno City College

Michael Sterner
University of Montevallo

Giovanni Viglino
Ramapo College of New Jersey

Joyce Vilseck
Texas A&M University, Kingston

Hiroko K. Warshauer
Texas State University—San Marcos

Jennifer Whitfield
Texas A&M University

Hing-Sing Yu
University of Texas, San Antonio

I wish to thank Jerry Grossman and Tao Guo for their many helpful suggestions for improving the text. I am also grateful to Kevin Charlwood and Jerry Grossman for helping with the accuracy checking of this text. I also thank the editorial, production, and marketing staffs of Cengage Learning: Liz Covello, Jay Campbell, Jeannine Lawless, Lauren Hamel, Cheryll Linthicum, Ashley Pickering, Mandee Eckersley, Erica O'Connell, and Sam Subity for all of their help and support during the development and production of this edition. Thanks to Martha Emry and Betty Duncan who both did an excellent job of ensuring the accuracy and readability of this edition. Simply stated, the team I have been working with is outstanding, and I truly appreciate all their hard work and efforts. Finally, a special thanks to the mathematicians—Chris Shannon and Mark van der Lann at UC Berkeley, Peter Blair Henry at Stanford, Jonathan D. Farley at Cal Tech, and Navin Khaneja at Harvard—for taking time off from their busy schedules to describe how mathematics is used in their research. Their photos and applications of their research appear on the covers of my applied mathematics series.

S. T. Tan

ABOUT THE AUTHOR

SOO T. TAN received his S.B. degree from Massachusetts Institute of Technology, his M.S. degree from the University of Wisconsin–Madison, and his Ph.D. from the University of California at Los Angeles. He has published numerous papers in Optimal Control Theory, Numerical Analysis, and Mathematics of Finance. He is currently a Professor of Mathematics at Stonehill College.

By the time I started writing the first of what turned out to be a series of textbooks in mathematics for students in the managerial, life, and social sciences, I had quite a few years of experience teaching mathematics to non-mathematics majors. One of the most important lessons I learned from my early experience teaching these courses is that many of the students come into these courses with some degree of apprehension. This awareness led to the intuitive approach I have adopted in all of my texts. As you will see, I try to introduce each abstract mathematical concept through an example drawn from a common, real-life experience. Once the idea has been conveyed, I then proceed to make it precise, thereby assuring that no mathematical rigor is lost in this intuitive treatment of the subject. Another lesson I learned from my students is that they have a much greater appreciation of the material if the applications are drawn from their fields of interest and from situations that occur in the real world. This is one reason you will see so many exercises in my texts that are modeled on data gathered from newspapers, magazines, journals, and other media. Whether it be the market for cholesterol-reducing drugs, financing a home, bidding for cable rights, broadband Internet households, or Starbucks' annual sales, I weave topics of current interest into my examples and exercises to keep the book relevant to all of my readers.

PRELIMINARIES

1

THE FIRST TWO sections of this chapter contain a brief review of algebra. We then introduce the Cartesian coordinate system, which allows us to represent points in the plane in terms of ordered pairs of real numbers. This in turn enables us to compute the distance between two points algebraically. This chapter also covers straight lines. The slope of a straight line plays an important role in the study of calculus.

How much money is needed to purchase at least 100,000 shares of the Starr Communications Company? Corbyco, a giant conglomerate, wishes to purchase a minimum of 100,000 shares of the company. In Example 11, page 21, you will see how Corbyco's management determines how much money they will need for the acquisition.

Use this test to diagnose any weaknesses that you might have in the algebra that you will need for the calculus material that follows. The review section and examples that will help you brush up on the skills necessary to work the problem are indicated after each exercise. The answers follow the test.

Diagnostic Test

1. a. Evaluate the expression:

(i) $\left(\frac{16}{9}\right)^{3/2}$ **(ii)** $\sqrt[3]{\frac{27}{125}}$

b. Rewrite the expression using positive exponents only: $(x^{-2}y^{-1})^3$

(Exponents and radicals, Examples 1 and 2, pages 6–7)

2. Rationalize the numerator: $\sqrt[3]{\frac{x^2}{yz^3}}$

(Rationalization, Example 4, page 7)

3. Simplify the following expressions:

a. $(3x^4 + 10x^3 + 6x^2 + 10x + 3) + (2x^4 + 10x^3 + 6x^2 + 4x)$

b. $(3x - 4)(3x^2 - 2x + 3)$

(Operations with algebraic expressions, Examples 5 and 6, pages 8–9)

4. Factor completely:

a. $6a^4b^4c - 3a^3b^2c - 9a^2b^2$ **b.** $6x^2 - xy - y^2$

(Factoring, Examples 7–9, pages 9–11)

5. Use the quadratic formula to solve the following equation: $9x^2 - 12x = 4$

(The quadratic formula, Example 10, pages 12–13)

6. Simplify the following expressions:

a. $\dfrac{2x^2 + 3x - 2}{2x^2 + 5x - 3}$ **b.** $\dfrac{[(t^2 + 4)(2t - 4)] - (t^2 - 4t + 4)(2t)}{(t^2 + 4)^2}$

(Rational expressions, Example 1, page 16)

7. Perform the indicated operations and simplify:

a. $\dfrac{2x - 6}{x + 3} \cdot \dfrac{x^2 + 6x + 9}{x^2 - 9}$ **b.** $\dfrac{3x}{x^2 + 2} + \dfrac{3x^2}{x^3 + 1}$

(Rational expressions, Examples 2 and 3, pages 16–18)

8. Perform the indicated operations and simplify:

a. $\dfrac{1 + \dfrac{1}{x + 2}}{x - \dfrac{9}{x}}$ **b.** $\dfrac{x(3x^2 + 1)}{x - 1} \cdot \dfrac{3x^3 - 5x^2 + x}{x(x - 1)(3x^2 + 1)^{1/2}}$

(Rational expressions, Examples 4 and 5, pages 18–19)

9. Rationalize the denominator: $\dfrac{3}{1 + 2\sqrt{x}}$

(Rationalizing algebraic fractions, Example 6, page 19)

10. Solve the inequalities:

a. $x^2 + x - 12 \le 0$

(Inequalities, Example 9, page 20)

b. $|3x - 4| \le 2$

(Absolute value, Example 14, page 22)

ANSWERS:

1. a (i). $\dfrac{64}{27}$ **(ii).** $\dfrac{3}{5}$ **b.** $\dfrac{1}{x^6y^3}$ **2.** $\dfrac{x}{z\sqrt[3]{xy}}$

3. a. $5x^4 + 20x^3 + 12x^2 + 14x + 3$ **b.** $9x^3 - 18x^2 + 17x - 12$

4. a. $3a^2b^2(2a^2b^2c - ac - 3)$ **b.** $(2x - y)(3x + y)$

5. $\dfrac{2}{3}(1 - \sqrt{2}); \dfrac{2}{3}(1 + \sqrt{2})$ **6. a.** $\dfrac{(x + 2)}{(x + 3)}$ **b.** $\dfrac{4(t^2 - 4)}{(t^2 + 4)^2}$

7. a. 2 **b.** $\dfrac{3x(2x^3 + 2x + 1)}{(x^2 + 2)(x^3 + 1)}$

8. a. $\dfrac{x}{(x + 2)(x - 3)}$ **b.** $\dfrac{x\sqrt{1 + 3x^2}(3x^2 - 5x + 1)}{(x - 1)^2}$

9. $\dfrac{3(1 - 2\sqrt{x})}{1 - 4x}$ **10. a.** $[-4, 3]$ **b.** $\left[\dfrac{2}{3}, 2\right]$

1.1 Precalculus Review I

Sections 1.1 and 1.2 review some basic concepts and techniques of algebra that are essential in the study of calculus. The material in this review will help you work through the examples and exercises in this book. You can read through this material now and do the exercises in areas where you feel a little "rusty," or you can review the material on an as-needed basis as you study the text. The self-diagnostic test that precedes this section will help you pinpoint the areas where you might have any weaknesses.

The Real Number Line

The real number system is made up of the set of real numbers together with the usual operations of addition, subtraction, multiplication, and division.

We can represent real numbers geometrically by points on a **real number, or coordinate, line.** This line can be constructed as follows. Arbitrarily select a point on a straight line to represent the number 0. This point is called the **origin.** If the line is horizontal, then a point at a convenient distance to the right of the origin is chosen to represent the number 1. This determines the scale for the number line. Each positive real number lies at an appropriate distance to the right of the origin, and each negative real number lies at an appropriate distance to the left of the origin (Figure 1).

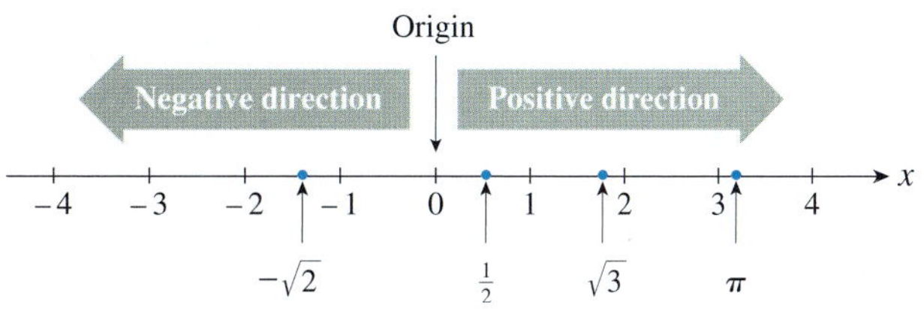

FIGURE 1
The real number line

A *one-to-one correspondence* is set up between the set of all real numbers and the set of points on the number line; that is, exactly one point on the line is associated with each real number. Conversely, exactly one real number is associated with each point on the line. The real number that is associated with a point on the real number line is called the **coordinate** of that point.

Intervals

Throughout this book, we will often restrict our attention to subsets of the set of real numbers. For example, if x denotes the number of cars rolling off a plant assembly line each day, then x must be nonnegative—that is, $x \geq 0$. Further, suppose management decides that the daily production must not exceed 200 cars. Then, x must satisfy the inequality $0 \leq x \leq 200$.

More generally, we will be interested in the following subsets of real numbers: open intervals, closed intervals, and half-open intervals. The set of all real numbers that lie *strictly* between two fixed numbers a and b is called an **open interval** (a, b). It consists of all real numbers x that satisfy the inequalities $a < x < b$, and it is called "open" because neither of its endpoints is included in the interval. A **closed interval** contains *both* of its endpoints. Thus, the set of all real numbers x that satisfy the inequalities $a \leq x \leq b$ is the closed interval $[a, b]$. Notice that square brackets are used to indicate that the endpoints are included in this interval. **Half-open intervals** contain only *one* of their endpoints. Thus, the interval $[a, b)$ is the set of all real numbers x that satisfy $a \leq x < b$, whereas the interval $(a, b]$ is described by the inequalities $a < x \leq b$. Examples of these **finite intervals** are illustrated in Table 1.

TABLE 1
Finite Intervals

Interval	Graph	Example	
Open: (a, b)	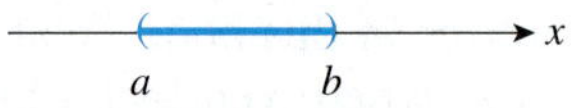	$(-2, 1)$	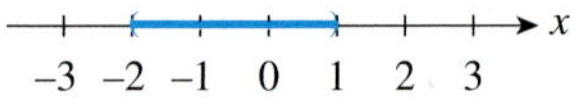
Closed: $[a, b]$	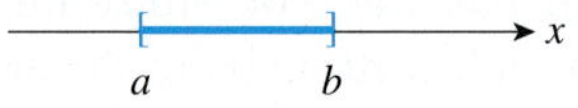	$[-1, 2]$	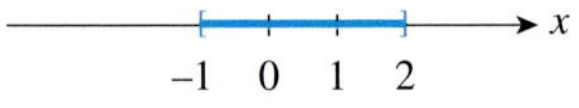
Half-open: $(a, b]$	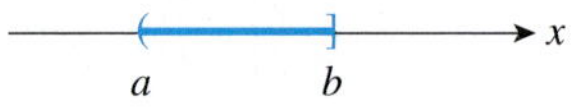	$(\frac{1}{2}, 3]$	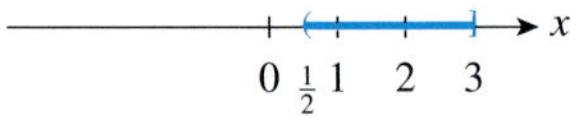
Half-open: $[a, b)$	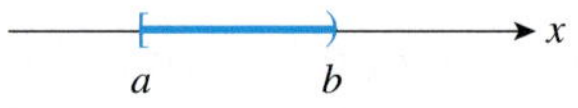	$[-\frac{1}{2}, 3)$	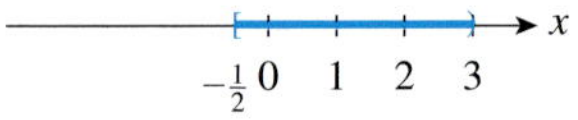

In addition to finite intervals, we will encounter **infinite intervals.** Examples of infinite intervals are the half lines (a, ∞), $[a, \infty)$, $(-\infty, a)$, and $(-\infty, a]$ defined by the set of all real numbers that satisfy $x > a$, $x \geq a$, $x < a$, and $x \leq a$, respectively. The symbol ∞, called *infinity*, is not a real number. It is used here only for notational purposes. The notation $(-\infty, \infty)$ is used for the set of all real numbers x since, by definition, the inequalities $-\infty < x < \infty$ hold for any real number x. Infinite intervals are illustrated in Table 2.

TABLE 2
Infinite Intervals

Interval	Graph	Example	
(a, ∞)		$(2, \infty)$	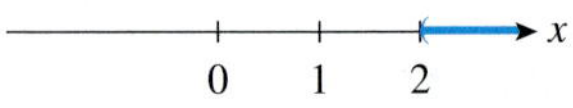
$[a, \infty)$	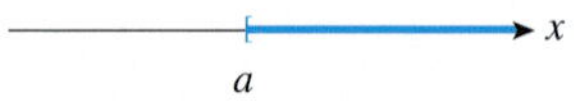	$[-1, \infty)$	
$(-\infty, a)$	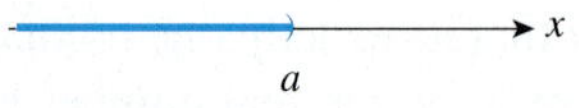	$(-\infty, 1)$	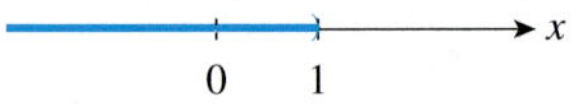
$(-\infty, a]$	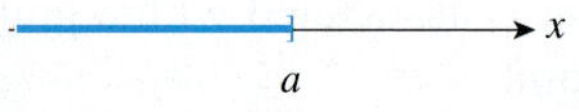	$(-\infty, -\frac{1}{2}]$	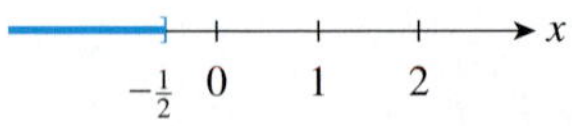

Exponents and Radicals

Recall that if b is any real number and n is a positive integer, then the expression b^n (read "b to the power n") is defined as the number

$$b^n = \underbrace{b \cdot b \cdot b \cdot \cdots \cdot b}_{n \text{ factors}}$$

The number b is called the **base,** and the superscript n is called the **power** of the exponential expression b^n. For example,

$$2^5 = 2 \cdot 2 \cdot 2 \cdot 2 \cdot 2 = 32 \qquad \text{and} \qquad \left(\frac{2}{3}\right)^3 = \left(\frac{2}{3}\right)\left(\frac{2}{3}\right)\left(\frac{2}{3}\right) = \frac{8}{27}$$

If $b \neq 0$, we define

$$b^0 = 1$$

For example, $2^0 = 1$ and $(-\pi)^0 = 1$, but the expression 0^0 is undefined.

Next, recall that if n is a positive integer, then the expression $b^{1/n}$ is defined to be the number that, when raised to the nth power, is equal to b. Thus,

$$(b^{1/n})^n = b$$

Such a number, if it exists, is called the ***n*th root of *b*,** also written $\sqrt[n]{b}$.

If n is even, the nth root of a negative number is not defined. For example, the square root of -2 ($n = 2$) is not defined because there is no real number b such that $b^2 = -2$. Also, given a number b, more than one number might satisfy our definition of the nth root. For example, both 3 and -3 squared equal 9, and each is a square root of 9. So, to avoid ambiguity, we define $b^{1/n}$ to be the positive nth root of b whenever it exists. Thus, $\sqrt{9} = 9^{1/2} = 3$. That's why your calculator will give the answer 3 when you use it to evaluate $\sqrt{9}$.

Next, recall that if p/q (p, q, positive integers with $q \neq 0$) is a rational number in lowest terms, then the expression $b^{p/q}$ is defined as the number $(b^{1/q})^p$ or, equivalently, $\sqrt[q]{b^p}$, whenever it exists. For example,

$$2^{3/2} = (2^{1/2})^3 \approx (1.4142)^3 \approx 2.8283$$

Expressions involving negative rational exponents are taken care of by the definition

$$b^{-p/q} = \frac{1}{b^{p/q}}$$

Thus,

$$4^{-5/2} = \frac{1}{4^{5/2}} = \frac{1}{(4^{1/2})^5} = \frac{1}{2^5} = \frac{1}{32}$$

The rules defining the exponential expression a^n, where $a > 0$ for all rational values of n, are given in Table 3.

The first three definitions in Table 3 are also valid for negative values of a. The fourth definition holds for all values of a if n is odd, but only for nonnegative values of a if n is even. Thus,

$$(-8)^{1/3} = \sqrt[3]{-8} = -2 \qquad n \text{ is odd.}$$

$$(-8)^{1/2} \text{ has no real value} \qquad n \text{ is even.}$$

Finally, note that it can be shown that a^n has meaning for *all* real numbers n. For example, using a pocket calculator with a $\boxed{y^x}$ key, we see that $2^{\sqrt{2}} \approx 2.665144$.

TABLE 3
Rules for Defining a^n

Definition of a^n $(a > 0)$	Example	Definition of a^n $(a > 0)$	Example
Integer exponent: If n is a positive integer, then $a^n = a \cdot a \cdot a \cdot \cdots \cdot a$ (n factors of a)	$2^5 = 2 \cdot 2 \cdot 2 \cdot 2 \cdot 2$ (5 factors) $= 32$	**Fractional exponent:** **a.** If n is a positive integer, then $a^{1/n}$ or $\sqrt[n]{a}$ denotes the nth root of a.	$16^{1/2} = \sqrt{16}$ $= 4$
Zero exponent: If n is equal to zero, then $a^0 = 1$ (0^0 is not defined.)	$7^0 = 1$	**b.** If m and n are positive integers, then $a^{m/n} = \sqrt[n]{a^m} = (\sqrt[n]{a})^m$	$8^{2/3} = (\sqrt[3]{8})^2$ $= 4$
Negative exponent: If n is a positive integer, then $a^{-n} = \dfrac{1}{a^n} \quad (a \neq 0)$	$6^{-2} = \dfrac{1}{6^2}$ $= \dfrac{1}{36}$	**c.** If m and n are positive integers, then $a^{-m/n} = \dfrac{1}{a^{m/n}} \quad (a \neq 0)$	$9^{-3/2} = \dfrac{1}{9^{3/2}}$ $= \dfrac{1}{27}$

The five laws of exponents are listed in Table 4.

TABLE 4
Laws of Exponents

Law	Example
1. $a^m \cdot a^n = a^{m+n}$	$x^2 \cdot x^3 = x^{2+3} = x^5$
2. $\dfrac{a^m}{a^n} = a^{m-n} \quad (a \neq 0)$	$\dfrac{x^7}{x^4} = x^{7-4} = x^3$
3. $(a^m)^n = a^{m \cdot n}$	$(x^4)^3 = x^{4 \cdot 3} = x^{12}$
4. $(ab)^n = a^n \cdot b^n$	$(2x)^4 = 2^4 \cdot x^4 = 16x^4$
5. $\left(\dfrac{a}{b}\right)^n = \dfrac{a^n}{b^n} \quad (b \neq 0)$	$\left(\dfrac{x}{2}\right)^3 = \dfrac{x^3}{2^3} = \dfrac{x^3}{8}$

These laws are valid for any real numbers a, b, m, and n whenever the quantities are defined.

Remember, $(x^2)^3 \neq x^5$. The correct equation is $(x^2)^3 = x^{2 \cdot 3} = x^6$.

The next several examples illustrate the use of the laws of exponents.

VIDEO

EXAMPLE 1 Simplify the expressions:

a. $(3x^2)(4x^3)$ **b.** $\dfrac{16^{5/4}}{16^{1/2}}$ **c.** $(6^{2/3})^3$ **d.** $(x^3y^{-2})^{-2}$ **e.** $\left(\dfrac{y^{3/2}}{x^{1/4}}\right)^{-2}$

Solution

a. $(3x^2)(4x^3) = 12x^{2+3} = 12x^5$ Law 1

b. $\dfrac{16^{5/4}}{16^{1/2}} = 16^{5/4-1/2} = 16^{3/4} = (\sqrt[4]{16})^3 = 2^3 = 8$ Law 2

c. $(6^{2/3})^3 = 6^{(2/3)(3)} = 6^{6/3} = 6^2 = 36$ Law 3

d. $(x^3y^{-2})^{-2} = (x^3)^{-2}(y^{-2})^{-2} = x^{(3)(-2)}y^{(-2)(-2)} = x^{-6}y^4 = \dfrac{y^4}{x^6}$ Law 4

e. $\left(\dfrac{y^{3/2}}{x^{1/4}}\right)^{-2} = \dfrac{y^{(3/2)(-2)}}{x^{(1/4)(-2)}} = \dfrac{y^{-3}}{x^{-1/2}} = \dfrac{x^{1/2}}{y^3}$ Law 5

We can also use the laws of exponents to simplify expressions involving radicals, as illustrated in the next example.

EXAMPLE 2 Simplify the expressions. (Assume that x, y, m, and n are positive.)

a. $\sqrt[4]{16x^4y^8}$ **b.** $\sqrt{12m^3n}\cdot\sqrt{3m^5n}$ **c.** $\dfrac{\sqrt[3]{-27x^6}}{\sqrt[3]{8y^3}}$

Solution

a. $\sqrt[4]{16x^4y^8} = (16x^4y^8)^{1/4} = 16^{1/4}\cdot x^{4/4}y^{8/4} = 2xy^2$

b. $\sqrt{12m^3n}\cdot\sqrt{3m^5n} = \sqrt{36m^8n^2} = (36m^8n^2)^{1/2} = 36^{1/2}\cdot m^{8/2}n^{2/2} = 6m^4n$

c. $\dfrac{\sqrt[3]{-27x^6}}{\sqrt[3]{8y^3}} = \dfrac{(-27x^6)^{1/3}}{(8y^3)^{1/3}} = \dfrac{-27^{1/3}x^{6/3}}{8^{1/3}y^{3/3}} = -\dfrac{3x^2}{2y}$

If a radical appears in the numerator or denominator of an algebraic expression, we often try to simplify the expression by eliminating the radical from the numerator or denominator. This process, called **rationalization,** is illustrated in the next two examples.

EXAMPLE 3 Rationalize the denominator of the expression $\dfrac{3x}{2\sqrt{x}}$.

Solution

$$\frac{3x}{2\sqrt{x}} = \frac{3x}{2\sqrt{x}}\cdot\frac{\sqrt{x}}{\sqrt{x}} = \frac{3x\sqrt{x}}{2\sqrt{x^2}} = \frac{3x\sqrt{x}}{2x} = \frac{3}{2}\sqrt{x}$$

EXAMPLE 4 Rationalize the numerator of the expression $\dfrac{3\sqrt{x}}{2x}$.

Solution

$$\frac{3\sqrt{x}}{2x} = \frac{3\sqrt{x}}{2x}\cdot\frac{\sqrt{x}}{\sqrt{x}} = \frac{3\sqrt{x^2}}{2x\sqrt{x}} = \frac{3x}{2x\sqrt{x}} = \frac{3}{2\sqrt{x}}$$

Operations with Algebraic Expressions

In calculus, we often work with algebraic expressions such as

$$2x^{4/3} - x^{1/3} + 1 \qquad 2x^2 - x - \frac{2}{\sqrt{x}} \qquad \frac{3xy + 2}{x + 1} \qquad 2x^3 + 2x + 1$$

An algebraic expression of the form ax^n, where the coefficient a is a real number and n is a nonnegative integer, is called a **monomial,** meaning it consists of one term. For example, $7x^2$ is a monomial. A **polynomial** is a monomial or the sum of two or more monomials. For example,

$$x^2 + 4x + 4 \qquad x^3 + 5 \qquad x^4 + 3x^2 + 3 \qquad x^2y + xy + y$$

are all polynomials.

Constant terms and terms containing the same variable factor are called **like,** or **similar, terms.** Like terms may be combined by adding or subtracting their numerical coefficients. For example,

$$3x + 7x = 10x \quad \text{and} \quad \frac{1}{2}xy + 3xy = \frac{7}{2}xy$$

The distributive property of the real number system,

$$ab + ac = a(b + c)$$

is used to justify this procedure.

To add or subtract two or more algebraic expressions, first remove the parentheses and then combine like terms. The resulting expression is written in order of decreasing degree from left to right.

EXAMPLE 5

a. $(2x^4 + 3x^3 + 4x + 6) - (3x^4 + 9x^3 + 3x^2)$

$= 2x^4 + 3x^3 + 4x + 6 - 3x^4 - 9x^3 - 3x^2$ Remove parentheses.

$= 2x^4 - 3x^4 + 3x^3 - 9x^3 - 3x^2 + 4x + 6$

$= -x^4 - 6x^3 - 3x^2 + 4x + 6$ Combine like terms.

b. $2t^3 - \{t^2 - [t - (2t - 1)] + 4\}$

$= 2t^3 - \{t^2 - [t - 2t + 1] + 4\}$

$= 2t^3 - \{t^2 - [-t + 1] + 4\}$ Remove parentheses and combine like terms within brackets.

$= 2t^3 - \{t^2 + t - 1 + 4\}$ Remove brackets.

$= 2t^3 - \{t^2 + t + 3\}$ Combine like terms within braces.

$= 2t^3 - t^2 - t - 3$ Remove braces. ■

An algebraic expression is said to be **simplified** if none of its terms are similar. Observe that when the algebraic expression in Example 5b was simplified, the innermost grouping symbols were removed first; that is, the parentheses ( ) were removed first, the brackets [ ] second, and the braces { } third.

When algebraic expressions are multiplied, each term of one algebraic expression is multiplied by each term of the other. The resulting algebraic expression is then simplified.

EXAMPLE 6 Perform the indicated operations:

a. $(x^2 + 1)(3x^2 + 10x + 3)$

b. $(e^t + e^{-t})e^t - e^t(e^t - e^{-t})$

Solution

a. $(x^2 + 1)(3x^2 + 10x + 3) = x^2(3x^2 + 10x + 3) + 1(3x^2 + 10x + 3)$

$= 3x^4 + 10x^3 + 3x^2 + 3x^2 + 10x + 3$

$= 3x^4 + 10x^3 + 6x^2 + 10x + 3$

b. $(e^t + e^{-t})e^t - e^t(e^t - e^{-t}) = e^{2t} + e^0 - e^{2t} + e^0$
$= e^{2t} - e^{2t} + e^0 + e^0$
$= 1 + 1$ Recall that $e^0 = 1$.
$= 2$ ■

Certain product formulas that are frequently used in algebraic computations are given in Table 5.

TABLE 5

Some Useful Product Formulas

Formula	Example
$(a + b)^2 = a^2 + 2ab + b^2$	$(2x + 3y)^2 = (2x)^2 + 2(2x)(3y) + (3y)^2$ $= 4x^2 + 12xy + 9y^2$
$(a - b)^2 = a^2 - 2ab + b^2$	$(4x - 2y)^2 = (4x)^2 - 2(4x)(2y) + (2y)^2$ $= 16x^2 - 16xy + 4y^2$
$(a + b)(a - b) = a^2 - b^2$	$(2x + y)(2x - y) = (2x)^2 - (y)^2$ $= 4x^2 - y^2$

Factoring

Factoring is the process of expressing an algebraic expression as a product of other algebraic expressions. For example, by applying the distributive property, we may write

$$3x^2 - x = x(3x - 1)$$

To factor an algebraic expression, first check to see if it contains any common terms. If it does, then factor out the greatest common term. For example, the common factor of the algebraic expression $2a^2x + 4ax + 6a$ is $2a$ because

$$2a^2x + 4ax + 6a = 2a \cdot ax + 2a \cdot 2x + 2a \cdot 3 = 2a(ax + 2x + 3)$$

EXAMPLE 7 Factor out the greatest common factor in each expression:

a. $-0.3t^2 + 3t$ **b.** $2x^{3/2} - 3x^{1/2}$ **c.** $2ye^{xy^2} + 2xy^3e^{xy^2}$

d. $4x(x + 1)^{1/2} - 2x^2\left(\frac{1}{2}\right)(x + 1)^{-1/2}$

Solution

a. $-0.3t^2 + 3t = -0.3t(t - 10)$

b. $2x^{3/2} - 3x^{1/2} = x^{1/2}(2x - 3)$

c. $2ye^{xy^2} + 2xy^3e^{xy^2} = 2ye^{xy^2}(1 + xy^2)$

d. $4x(x + 1)^{1/2} - 2x^2\left(\frac{1}{2}\right)(x + 1)^{-1/2} = 4x(x + 1)^{1/2} - x^2(x + 1)^{-1/2}$
$= x(x + 1)^{-1/2}[4(x + 1)^{1/2}(x + 1)^{1/2} - x]$
$= x(x + 1)^{-1/2}[4(x + 1) - x]$
$= x(x + 1)^{-1/2}(4x + 4 - x) = x(x + 1)^{-1/2}(3x + 4)$

Here we select $(x + 1)^{-1/2}$ as the greatest common factor because it is the lowest power of $(x + 1)$ in each algebraic term. In particular, observe that

$$(x + 1)^{-1/2}(x + 1)^{1/2}(x + 1)^{1/2} = (x + 1)^{-1/2+1/2+1/2} = (x + 1)^{1/2}$$ ■

Sometimes an algebraic expression may be factored by regrouping and rearranging its terms so that a common term can be factored out. This technique is illustrated in Example 8.

EXAMPLE 8 Factor:

a. $2ax + 2ay + bx + by$ **b.** $3x\sqrt{y} - 4 - 2\sqrt{y} + 6x$

Solution

a. First, factor the common term $2a$ from the first two terms and the common term b from the last two terms. Thus,

$$2ax + 2ay + bx + by = 2a(x + y) + b(x + y)$$

Since $(x + y)$ is common to both terms of the polynomial, we may factor it out. Hence,

$$2a(x + y) + b(x + y) = (2a + b)(x + y)$$

b.
$$\begin{aligned} 3x\sqrt{y} - 4 - 2\sqrt{y} + 6x &= 3x\sqrt{y} - 2\sqrt{y} + 6x - 4 && \text{Regroup terms.} \\ &= \sqrt{y}(3x - 2) + 2(3x - 2) && \text{Factor out common terms.} \\ &= (3x - 2)(\sqrt{y} + 2) \end{aligned}$$

■

As we have seen, the first step in factoring a polynomial is to find the common factors. The next step is to express the polynomial as the product of a constant and/or one or more prime polynomials.

Certain product formulas that are useful in factoring binomials and trinomials are listed in Table 6.

TABLE 6

Product Formulas Used in Factoring

Formula	Example
Difference of two squares	
$x^2 - y^2 = (x + y)(x - y)$	$x^2 - 36 = (x + 6)(x - 6)$
	$8x^2 - 2y^2 = 2(4x^2 - y^2)$
	$= 2(2x + y)(2x - y)$
	$9 - a^6 = (3 + a^3)(3 - a^3)$
Perfect-square trinomial	
$x^2 + 2xy + y^2 = (x + y)^2$	$x^2 + 8x + 16 = (x + 4)^2$
$x^2 - 2xy + y^2 = (x - y)^2$	$4x^2 - 4xy + y^2 = (2x - y)^2$
Sum of two cubes	
$x^3 + y^3 = (x + y)(x^2 - xy + y^2)$	$z^3 + 27 = z^3 + (3)^3$
	$= (z + 3)(z^2 - 3z + 9)$
Difference of two cubes	
$x^3 - y^3 = (x - y)(x^2 + xy + y^2)$	$8x^3 - y^6 = (2x)^3 - (y^2)^3$
	$= (2x - y^2)(4x^2 + 2xy^2 + y^4)$

The factors of the second-degree polynomial with integral coefficients

$$px^2 + qx + r$$

are $(ax + b)(cx + d)$, where $ac = p$, $ad + bc = q$, and $bd = r$. Since only a limited number of choices are possible, we use a trial-and-error method to factor polynomials having this form.

For example, to factor $x^2 - 2x - 3$, we first observe that the only possible first-degree terms are

$$(x \qquad)(x \qquad) \quad \text{Since the coefficient of } x^2 \text{ is 1}$$

Next, we observe that the product of the constant term is (-3). This gives us the following possible factors:

$$(x - 1)(x + 3)$$
$$(x + 1)(x - 3)$$

Looking once again at the polynomial $x^2 - 2x - 3$, we see that the coefficient of x is -2. Checking to see which set of factors yields -2 for the coefficient of x, we find that

Factors

Coefficients of inner terms ↓ Coefficients of outer terms ↓

$$(-1)(1) + (1)(3) = 2$$

Outer terms: $(x - 1)(x + 3)$ Inner terms

Coefficients of inner terms ↓ Coefficients of outer terms ↓

$$(1)(1) + (1)(-3) = -2$$

Outer terms: $(x + 1)(x - 3)$ Inner terms

and we conclude that the correct factorization is

$$x^2 - 2x - 3 = (x + 1)(x - 3)$$

With practice, you will soon find that you can perform many of these steps mentally and the need to write out each step will be eliminated.

EXAMPLE 9 Factor:

a. $3x^2 + 4x - 4$ **b.** $3x^2 - 6x - 24$

Solution

a. Using trial and error, we find that the correct factorization is

$$3x^2 + 4x - 4 = (3x - 2)(x + 2)$$

b. Since each term has the common factor 3, we have

$$3x^2 - 6x - 24 = 3(x^2 - 2x - 8)$$

Using the trial-and-error method of factorization, we find that

$$x^2 - 2x - 8 = (x - 4)(x + 2)$$

Thus, we have

$$3x^2 - 6x - 24 = 3(x - 4)(x + 2)$$

■

Roots of Polynomial Equations

A polynomial equation of degree n in the variable x is an equation of the form

$$a_n x^n + a_{n-1} x^{n-1} + \cdots + a_0 = 0$$

where n is a nonnegative integer and $a_0, a_1, \ldots, a_n$ are real numbers with $a_n \neq 0$. For example, the equation

$$-2x^5 + 8x^3 - 6x^2 + 3x + 1 = 0$$

is a polynomial equation of degree 5 in x.

The **roots of a polynomial equation** are precisely the values of x that satisfy the given equation.* One way to find the roots of a polynomial equation is to factor the polynomial and then solve the resulting equation. For example, the polynomial equation

$$x^3 - 3x^2 + 2x = 0$$

may be rewritten in the form

$$x(x^2 - 3x + 2) = 0 \quad \text{or} \quad x(x - 1)(x - 2) = 0$$

Since the product of two real numbers can be equal to zero if and only if one (or both) of the factors is equal to zero, we have

$$x = 0 \qquad x - 1 = 0 \qquad \text{or} \qquad x - 2 = 0$$

from which we see that the desired roots are $x = 0$, 1, and 2.

The Quadratic Formula

In general, the problem of finding the roots of a polynomial equation is a difficult one. But the roots of a quadratic equation (a polynomial equation of degree 2) are easily found either by factoring or by using the following quadratic formula.

Quadratic Formula

The solutions of the equation $ax^2 + bx + c = 0$ $(a \neq 0)$ are given by

$$x = \frac{-b \pm \sqrt{b^2 - 4ac}}{2a}$$

Note If you use the quadratic formula to solve a quadratic equation, first make sure that the equation is in the *standard form* $ax^2 + bx + c = 0$. ■

EXAMPLE 10 Solve each of the following quadratic equations:

a. $2x^2 + 5x - 12 = 0$ **b.** $x^2 = -3x + 8$

Solution

a. The equation is in standard form, with $a = 2$, $b = 5$, and $c = -12$. Using the quadratic formula, we find

$$\begin{aligned} x &= \frac{-b \pm \sqrt{b^2 - 4ac}}{2a} = \frac{-5 \pm \sqrt{5^2 - 4(2)(-12)}}{2(2)} \\ &= \frac{-5 \pm \sqrt{121}}{4} = \frac{-5 \pm 11}{4} \\ &= -4 \quad \text{or} \quad \frac{3}{2} \end{aligned}$$

This equation can also be solved by factoring. Thus,

$$2x^2 + 5x - 12 = (2x - 3)(x + 4) = 0$$

from which we see that the desired roots are $x = \frac{3}{2}$ or $x = -4$, as obtained earlier.

*In this book, we are interested only in the *real* roots of an equation.

b. We first rewrite the given equation in the standard form $x^2 + 3x - 8 = 0$, from which we see that $a = 1$, $b = 3$, and $c = -8$. Using the quadratic formula, we find

$$x = \frac{-b \pm \sqrt{b^2 - 4ac}}{2a} = \frac{-3 \pm \sqrt{3^2 - 4(1)(-8)}}{2(1)}$$
$$= \frac{-3 \pm \sqrt{41}}{2}$$

That is, the solutions are

$$\frac{-3 + \sqrt{41}}{2} \approx 1.7 \quad \text{and} \quad \frac{-3 - \sqrt{41}}{2} \approx -4.7$$

In this case, the quadratic formula proves quite handy!

1.1 Exercises

In Exercises 1–6, show the interval on a number line.

1. $(3, 6)$
2. $(-2, 5]$
3. $[-1, 4)$
4. $\left[-\frac{6}{5}, -\frac{1}{2}\right]$
5. $(0, \infty)$
6. $(-\infty, 5]$

In Exercises 7–22, evaluate the expression.

7. $27^{2/3}$
8. $8^{-4/3}$
9. $\left(\frac{1}{\sqrt{3}}\right)^0$
10. $(7^{1/2})^4$
11. $\left[\left(\frac{1}{8}\right)^{1/3}\right]^{-2}$
12. $\left[\left(-\frac{1}{3}\right)^2\right]^{-3}$
13. $\left(\frac{7^{-5} \cdot 7^2}{7^{-2}}\right)^{-1}$
14. $\left(\frac{9}{16}\right)^{-1/2}$
15. $(125^{2/3})^{-1/2}$
16. $\sqrt[3]{2^6}$
17. $\frac{\sqrt{32}}{\sqrt{8}}$
18. $\sqrt[3]{\frac{-8}{27}}$
19. $\frac{16^{5/8}16^{1/2}}{16^{7/8}}$
20. $\left(\frac{9^{-3} \cdot 9^5}{9^{-2}}\right)^{-1/2}$
21. $16^{1/4} \cdot 8^{-1/3}$
22. $\frac{6^{2.5} \cdot 6^{-1.9}}{6^{-1.4}}$

In Exercises 23–32, determine whether the statement is true or false. Give a reason for your choice.

23. $x^4 + 2x^4 = 3x^4$
24. $3^2 \cdot 2^2 = 6^2$
25. $x^3 \cdot 2x^2 = 2x^6$
26. $3^3 + 3 = 3^4$
27. $\frac{2^{4x}}{1^{3x}} = 2^{4x-3x}$
28. $(2^2 \cdot 3^2)^2 = 6^4$
29. $\frac{1}{4^{-3}} = \frac{1}{64}$
30. $\frac{4^{3/2}}{2^4} = \frac{1}{2}$
31. $(1.2^{1/2})^{-1/2} = 1$
32. $5^{2/3} \cdot (25)^{2/3} = 25$

In Exercises 33–38, rewrite the expression using positive exponents only.

33. $(xy)^{-2}$
34. $3s^{1/3} \cdot s^{-7/3}$
35. $\frac{x^{-1/3}}{x^{1/2}}$
36. $\sqrt{x^{-1}} \cdot \sqrt{9x^{-3}}$
37. $12^0(s + t)^{-3}$
38. $(x - y)(x^{-1} + y^{-1})$

In Exercises 39–54, simplify the expression. (Assume that *x*, *y*, *r*, *s*, and *t* are positive.)

39. $\frac{x^{7/3}}{x^{-2}}$
40. $(49x^{-2})^{-1/2}$
41. $(x^2y^{-3})(x^{-5}y^3)$
42. $\frac{5x^6y^3}{2x^2y^7}$
43. $\frac{x^{3/4}}{x^{-1/4}}$
44. $\left(\frac{x^3y^2}{z^2}\right)^2$
45. $\left(\frac{x^3}{-27y^{-6}}\right)^{-2/3}$
46. $\left(\frac{e^x}{e^{x-2}}\right)^{-1/2}$
47. $\left(\frac{x^{-3}}{y^{-2}}\right)^2\left(\frac{y}{x}\right)^4$
48. $\frac{(r^n)^4}{r^{5-2n}}$
49. $\sqrt[3]{x^{-2}} \cdot \sqrt{4x^5}$
50. $\sqrt{81x^6y^{-4}}$
51. $-\sqrt[4]{16x^4y^8}$
52. $\sqrt[3]{x^{3a+b}}$
53. $\sqrt[6]{64x^8y^3}$
54. $\sqrt[3]{27r^6} \cdot \sqrt{s^2t^4}$

In Exercises 55–58, use the fact that $2^{1/2} \approx 1.414$ and $3^{1/2} \approx 1.732$ to evaluate the expression without using a calculator.

55. $2^{3/2}$
56. $8^{1/2}$
57. $9^{3/4}$
58. $6^{1/2}$

In Exercises 59–62, use the fact that $10^{1/2} \approx 3.162$ and $10^{1/3} \approx 2.154$ to evaluate the expression without using a calculator.

59. $10^{3/2}$
60. $1000^{3/2}$
61. $10^{2.5}$
62. $(0.0001)^{-1/3}$

In Exercises 63–68, rationalize the denominator of the expression.

63. $\dfrac{3}{2\sqrt{x}}$ **64.** $\dfrac{3}{\sqrt{xy}}$ **65.** $\dfrac{2y}{\sqrt{3y}}$

66. $\dfrac{5x^2}{\sqrt{3x}}$ **67.** $\dfrac{1}{\sqrt[3]{x}}$ **68.** $\sqrt{\dfrac{2x}{y}}$

In Exercises 69–74, rationalize the numerator of the expression.

69. $\dfrac{2\sqrt{x}}{3}$ **70.** $\dfrac{\sqrt[3]{x}}{24}$ **71.** $\sqrt{\dfrac{2y}{x}}$

72. $\sqrt[3]{\dfrac{2x}{3y}}$ **73.** $\dfrac{\sqrt[3]{x^2z}}{y}$ **74.** $\dfrac{\sqrt[3]{x^2y}}{2x}$

In Exercises 75–96, perform the indicated operations and simplify each expression.

75. $(7x^2 - 2x + 5) + (2x^2 + 5x - 4)$

76. $(3x^2 + 5xy + 2y) + (4 - 3xy - 2x^2)$

77. $(5y^2 - 2y + 1) - (y^2 - 3y - 7)$

78. $3(2a - b) - 4(b - 2a)$

79. $x - \{2x - [-x - (1 - x)]\}$

80. $3x^2 - \{x^2 + 1 - x[x - (2x - 1)]\} + 2$

81. $\left(\dfrac{1}{3} - 1 + e\right) - \left(-\dfrac{1}{3} - 1 + e^{-1}\right)$

82. $-\dfrac{3}{4}y - \dfrac{1}{4}x + 100 + \dfrac{1}{2}x + \dfrac{1}{4}y - 120$

83. $3\sqrt{8} + 8 - 2\sqrt{y} + \dfrac{1}{2}\sqrt{x} - \dfrac{3}{4}\sqrt{y}$

84. $\dfrac{8}{9}x^2 + \dfrac{2}{3}x + \dfrac{16}{3}x^2 - \dfrac{16}{3}x - 2x + 2$

85. $(x + 8)(x - 2)$ **86.** $(5x + 2)(3x - 4)$

87. $(a + 5)^2$ **88.** $(3a - 4b)^2$

89. $(x + 2y)^2$ **90.** $(6 - 3x)^2$

91. $(2x + y)(2x - y)$ **92.** $(3x + 2)(2 - 3x)$

93. $(x^2 - 1)(2x) - x^2(2x)$

94. $(x^{1/2} + 1)\left(\dfrac{1}{2}x^{-1/2}\right) - (x^{1/2} - 1)\left(\dfrac{1}{2}x^{-1/2}\right)$

95. $2(t + \sqrt{t})^2 - 2t^2$

96. $2x^2 + (-x + 1)^2$

In Exercises 97–104, factor out the greatest common factor from each expression.

97. $4x^5 - 12x^4 - 6x^3$

98. $4x^2y^2z - 2x^5y^2 + 6x^3y^2z^2$

99. $7a^4 - 42a^2b^2 + 49a^3b$

100. $3x^{2/3} - 2x^{1/3}$

101. $e^{-x} - xe^{-x}$

102. $2ye^{xy^2} + 2xy^3e^{xy^2}$

103. $2x^{-5/2} - \dfrac{3}{2}x^{-3/2}$

104. $\dfrac{1}{2}\left(\dfrac{2}{3}u^{3/2} - 2u^{1/2}\right)$

In Exercises 105–118, factor each expression completely.

105. $6ac + 3bc - 4ad - 2bd$

106. $3x^3 - x^2 + 3x - 1$

107. $4a^2 - b^2$ **108.** $12x^2 - 3y^2$

109. $10 - 14x - 12x^2$ **110.** $x^2 - 2x - 15$

111. $3x^2 - 6x - 24$ **112.** $3x^2 - 4x - 4$

113. $12x^2 - 2x - 30$ **114.** $(x + y)^2 - 1$

115. $9x^2 - 16y^2$ **116.** $8a^2 - 2ab - 6b^2$

117. $x^6 + 125$ **118.** $x^3 - 27$

In Exercises 119–126, perform the indicated operations and simplify each expression.

119. $(x^2 + y^2)x - xy(2y)$

120. $2kr(R - r) - kr^2$

121. $2(x - 1)(2x + 2)^3[4(x - 1) + (2x + 2)]$

122. $5x^2(3x^2 + 1)^4(6x) + (3x^2 + 1)^5(2x)$

123. $4(x - 1)^2(2x + 2)^3(2) + (2x + 2)^4(2)(x - 1)$

124. $(x^2 + 1)(4x^3 - 3x^2 + 2x) - (x^4 - x^3 + x^2)(2x)$

125. $(x^2 + 2)^2[5(x^2 + 2)^2 - 3](2x)$

126. $(x^2 - 4)(x^2 + 4)(2x + 8) - (x^2 + 8x - 4)(4x^3)$

In Exercises 127–132, find the real roots of each equation by factoring.

127. $x^2 + x - 12 = 0$ **128.** $3x^2 - x - 4 = 0$

129. $4t^2 + 2t - 2 = 0$ **130.** $-6x^2 + x + 12 = 0$

131. $\dfrac{1}{4}x^2 - x + 1 = 0$ **132.** $\dfrac{1}{2}a^2 + a - 12 = 0$

In Exercises 133–138, solve the equation by using the quadratic formula.

133. $4x^2 + 5x - 6 = 0$

134. $3x^2 - 4x + 1 = 0$

135. $8x^2 - 8x - 3 = 0$

136. $x^2 - 6x + 6 = 0$

137. $2x^2 + 4x - 3 = 0$

138. $2x^2 + 7x - 15 = 0$

139. DISTRIBUTION OF INCOMES The distribution of income in a certain city can be described by the mathematical model $y = (2.8 \cdot 10^{11})(x)^{-1.5}$, where y is the number of families with an income of x or more dollars.

a. How many families in this city have an income of \$30,000 or more?

b. How many families have an income of \$60,000 or more?

c. How many families have an income of \$150,000 or more?

In Exercises 140–142, determine whether the statement is true or false. If it is true, explain why it is true. If it is false, give an example to show why it is false.

140. If $b^2 - 4ac > 0$, then $ax^2 + bx + c = 0$ $(a \neq 0)$ has two real roots.

141. If $b^2 - 4ac < 0$, then $ax^2 + bx + c = 0$ $(a \neq 0)$ has no real roots.

142. $\sqrt{(a + b)(b - a)} = \sqrt{b^2 - a^2}$ for all real numbers a and b.

1.2 Precalculus Review II

Rational Expressions

Quotients of polynomials are called **rational expressions.** Examples of rational expressions are

$$\frac{6x - 1}{2x + 3} \qquad \frac{3x^2y^3 - 2xy}{4x} \qquad \frac{2}{5ab}$$

Since rational expressions are quotients in which the variables represent real numbers, the properties of real numbers apply to rational expressions as well, and operations with rational fractions are performed in the same manner as operations with arithmetic fractions. For example, using the properties of the real number system, we may write

$$\frac{ac}{bc} = \frac{a}{b} \cdot \frac{c}{c} = \frac{a}{b} \cdot 1 = \frac{a}{b}$$

where a, b, and c are any real numbers and b and c are not zero.

Similarly, using the same properties of real numbers, we may write

$$\frac{(x + 2)(x - 3)}{(x - 2)(x - 3)} = \frac{x + 2}{x - 2} \qquad (x \neq 2, 3)$$

after "canceling" the common factors.

An example of incorrect cancellation is

$$\frac{\not{3} + 4x}{\not{3}} \neq 1 + 4x$$

because 3 is not a factor of the numerator. Instead, we need to write

$$\frac{3 + 4x}{3} = \frac{3}{3} + \frac{4x}{3} = 1 + \frac{4x}{3}$$

A rational expression is simplified, or in lowest terms, when the numerator and denominator have no common factors other than 1 and -1 and the expression contains no negative exponents.

EXAMPLE 1 Simplify the following expressions:

a. $\dfrac{x^2 + 2x - 3}{x^2 + 4x + 3}$ **b.** $\dfrac{(x^2 + 1)^2(-2) + (2x)(2)(x^2 + 1)(2x)}{(x^2 + 1)^4}$

Solution

a. $\dfrac{x^2 + 2x - 3}{x^2 + 4x + 3} = \dfrac{(x + 3)(x - 1)}{(x + 3)(x + 1)} = \dfrac{x - 1}{x + 1}$

b. $\dfrac{(x^2 + 1)^2(-2) + (2x)(2)(x^2 + 1)(2x)}{(x^2 + 1)^4}$

$= \dfrac{(x^2 + 1)[(x^2 + 1)(-2) + (2x)(2)(2x)]}{(x^2 + 1)^4}$ Factor out $(x^2 + 1)$.

$= \dfrac{(x^2 + 1)(-2x^2 - 2 + 8x^2)}{(x^2 + 1)^4}$ Carry out indicated multiplication.

$= \dfrac{(x^2 + 1)(6x^2 - 2)}{(x^2 + 1)^4}$ Combine like terms.

$= \dfrac{(6x^2 - 2)}{(x^2 + 1)^3}$ Cancel the common factors.

$= \dfrac{2(3x^2 - 1)}{(x^2 + 1)^3}$ Factor out 2 from the numerator. ■

The operations of multiplication and division are performed with algebraic fractions in the same manner as with arithmetic fractions (Table 7).

TABLE 7

Rules of Multiplication and Division: Algebraic Fractions

Operation	Example
If P, Q, R, and S are polynomials, then	
Multiplication	
$\dfrac{P}{Q} \cdot \dfrac{R}{S} = \dfrac{PR}{QS}$ $(Q, S \neq 0)$	$\dfrac{2x}{y} \cdot \dfrac{(x + 1)}{(y - 1)} = \dfrac{2x(x + 1)}{y(y - 1)} = \dfrac{2x^2 + 2x}{y^2 - y}$
Division	
$\dfrac{P}{Q} \div \dfrac{R}{S} = \dfrac{P}{Q} \cdot \dfrac{S}{R} = \dfrac{PS}{QR}$ $(Q, R, S \neq 0)$	$\dfrac{x^2 + 3}{y} \div \dfrac{y^2 + 1}{x} = \dfrac{x^2 + 3}{y} \cdot \dfrac{x}{y^2 + 1} = \dfrac{x^3 + 3x}{y^3 + y}$

When rational expressions are multiplied and divided, the resulting expressions should be simplified if possible.

VIDEO

EXAMPLE 2 Perform the indicated operations and simplify:

$$\frac{2x - 8}{x + 2} \cdot \frac{x^2 + 4x + 4}{x^2 - 16}$$

Solution

$$\frac{2x-8}{x+2}\cdot\frac{x^2+4x+4}{x^2-16}=\frac{2(x-4)}{x+2}\cdot\frac{(x+2)^2}{(x+4)(x-4)}$$

$$=\frac{2(x-4)(x+2)(x+2)}{(x+2)(x+4)(x-4)}$$

$$=\frac{2(x+2)}{x+4}$$ Cancel the common factors $(x+2)(x-4)$.

For rational expressions, the operations of addition and subtraction are performed by finding a common denominator of the fractions and then adding or subtracting the fractions. Table 8 shows the rules for fractions with equal denominators.

TABLE 8

Rules of Addition and Subtraction: Fractions with Equal Denominators

Operation	Example
If P, Q, and R are polynomials, then	
Addition	
$\frac{P}{R}+\frac{Q}{R}=\frac{P+Q}{R}\quad (R\neq 0)$	$\frac{2x}{x+2}+\frac{6x}{x+2}=\frac{2x+6x}{x+2}=\frac{8x}{x+2}$
Subtraction	
$\frac{P}{R}-\frac{Q}{R}=\frac{P-Q}{R}\quad (R\neq 0)$	$\frac{3y}{y-x}-\frac{y}{y-x}=\frac{3y-y}{y-x}=\frac{2y}{y-x}$

To add or subtract fractions that have different denominators, first find a common denominator, preferably the least common denominator (LCD). Then carry out the indicated operations following the procedure described in Table 8.

To find the LCD of two or more rational expressions:

1. *Find the prime factors of each denominator.*
2. *Form the product of the different prime factors that occur in the denominators.* Each prime factor in this product should be raised to the highest power of that factor appearing in the denominators.

$$\frac{x}{2+y}\neq\frac{x}{2}+\frac{x}{y}$$

EXAMPLE 3 Simplify:

a. $\frac{2x}{x^2+1}+\frac{6(3x^2)}{x^3+2}$ **b.** $\frac{1}{x+h}-\frac{1}{x}$

Solution

a. $$\frac{2x}{x^2+1}+\frac{6(3x^2)}{x^3+2}=\frac{2x(x^3+2)+6(3x^2)(x^2+1)}{(x^2+1)(x^3+2)}$$ LCD $=(x^2+1)(x^3+2)$

$$=\frac{2x^4+4x+18x^4+18x^2}{(x^2+1)(x^3+2)}$$ Carry out the indicated multiplication.

$$=\frac{20x^4+18x^2+4x}{(x^2+1)(x^3+2)}$$ Combine like terms.

$$=\frac{2x(10x^3+9x+2)}{(x^2+1)(x^3+2)}$$ Factor.

b. $\dfrac{1}{x+h} - \dfrac{1}{x} = \dfrac{x-(x+h)}{x(x+h)}$ LCD $= x(x+h)$

$= \dfrac{x-x-h}{x(x+h)}$ Remove parentheses.

$= \dfrac{-h}{x(x+h)}$ Combine like terms.

■

Other Algebraic Fractions

The techniques used to simplify rational expressions may also be used to simplify algebraic fractions in which the numerator and denominator are not polynomials, as illustrated in Example 4.

EXAMPLE 4 Simplify:

a. $\dfrac{1+\dfrac{1}{x+1}}{x-\dfrac{4}{x}}$ **b.** $\dfrac{x^{-1}+y^{-1}}{x^{-2}-y^{-2}}$

Solution

a. $\dfrac{1+\dfrac{1}{x+1}}{x-\dfrac{4}{x}} = \dfrac{\dfrac{x+1+1}{x+1}}{\dfrac{x^2-4}{x}}$ LCD for numerator is $x+1$ and LCD for denominator is x.

$= \dfrac{x+2}{x+1}\cdot\dfrac{x}{x^2-4} = \dfrac{x+2}{x+1}\cdot\dfrac{x}{(x+2)(x-2)}$

$= \dfrac{x}{(x+1)(x-2)}$

b. $\dfrac{x^{-1}+y^{-1}}{x^{-2}-y^{-2}} = \dfrac{\dfrac{1}{x}+\dfrac{1}{y}}{\dfrac{1}{x^2}-\dfrac{1}{y^2}} = \dfrac{\dfrac{y+x}{xy}}{\dfrac{y^2-x^2}{x^2y^2}}$ $x^{-n} = \dfrac{1}{x^n}$

$= \dfrac{y+x}{xy}\cdot\dfrac{x^2y^2}{y^2-x^2} = \dfrac{y+x}{xy}\cdot\dfrac{(xy)^2}{(y+x)(y-x)}$

$= \dfrac{xy}{y-x}$

■

EXAMPLE 5 Perform the given operations and simplify:

a. $\dfrac{x^2(2x^2+1)^{1/2}}{x-1}\cdot\dfrac{4x^3-6x^2+x-2}{x(x-1)(2x^2+1)}$ **b.** $\dfrac{12x^2}{\sqrt{2x^2+3}} + 6\sqrt{2x^2+3}$

Solution

a. $\dfrac{x^2(2x^2+1)^{1/2}}{x-1}\cdot\dfrac{4x^3-6x^2+x-2}{x(x-1)(2x^2+1)} = \dfrac{x(4x^3-6x^2+x-2)}{(x-1)^2(2x^2+1)^{1-1/2}}$

$= \dfrac{x(4x^3-6x^2+x-2)}{(x-1)^2(2x^2+1)^{1/2}}$

b. $$\frac{12x^2}{\sqrt{2x^2+3}} + 6\sqrt{2x^2+3} = \frac{12x^2}{(2x^2+3)^{1/2}} + 6(2x^2+3)^{1/2}$$

Write radicals in exponential form.

$$= \frac{12x^2 + 6(2x^2+3)^{1/2}(2x^2+3)^{1/2}}{(2x^2+3)^{1/2}}$$

LCD is $(2x^2+3)^{1/2}$.

$$= \frac{12x^2 + 6(2x^2+3)}{(2x^2+3)^{1/2}}$$

$$= \frac{24x^2+18}{(2x^2+3)^{1/2}} = \frac{6(4x^2+3)}{\sqrt{2x^2+3}}$$

■

Rationalizing Algebraic Fractions

When the denominator of an algebraic fraction contains sums or differences involving radicals, we may **rationalize the denominator**—that is, transform the fraction into an equivalent one with a denominator that does not contain radicals. In doing so, we make use of the fact that

$$(\sqrt{a}+\sqrt{b})(\sqrt{a}-\sqrt{b}) = (\sqrt{a})^2 - (\sqrt{b})^2$$
$$= a - b$$

This procedure is illustrated in Example 6.

EXAMPLE 6 Rationalize the denominator: $\dfrac{1}{1+\sqrt{x}}$.

Solution Upon multiplying the numerator and the denominator by $(1-\sqrt{x})$, we obtain

$$\frac{1}{1+\sqrt{x}} = \frac{1}{1+\sqrt{x}} \cdot \frac{1-\sqrt{x}}{1-\sqrt{x}}$$

$\dfrac{1-\sqrt{x}}{1-\sqrt{x}} = 1$

$$= \frac{1-\sqrt{x}}{1-(\sqrt{x})^2}$$

$$= \frac{1-\sqrt{x}}{1-x}$$

■

In other situations, it may be necessary to rationalize the numerator of an algebraic expression. In calculus, for example, one encounters the following problem.

VIDEO

EXAMPLE 7 Rationalize the numerator: $\dfrac{\sqrt{1+h}-1}{h}$.

Solution

$$\frac{\sqrt{1+h}-1}{h} = \frac{\sqrt{1+h}-1}{h} \cdot \frac{\sqrt{1+h}+1}{\sqrt{1+h}+1}$$

$$= \frac{(\sqrt{1+h})^2 - (1)^2}{h(\sqrt{1+h}+1)}$$

$$= \frac{1+h-1}{h(\sqrt{1+h}+1)}$$

$(\sqrt{1+h})^2 = \sqrt{1+h} \cdot \sqrt{1+h}$
$= 1 + h$

$$= \frac{h}{h(\sqrt{1+h}+1)}$$

$$= \frac{1}{\sqrt{1+h}+1}$$

■

Inequalities

The following properties may be used to solve one or more inequalities involving a variable.

Properties of Inequalities

If a, b, and c, are any real numbers, then

		Example
Property 1	If $a < b$ and $b < c$, then $a < c$.	$2 < 3$ and $3 < 8$, so $2 < 8$.
Property 2	If $a < b$, then $a + c < b + c$.	$-5 < -3$, so $-5 + 2 < -3 + 2$; that is, $-3 < -1$.
Property 3	If $a < b$ and $c > 0$, then $ac < bc$.	$-5 < -3$, and since $2 > 0$, we have $(-5)(2) < (-3)(2)$; that is, $-10 < -6$.
Property 4	If $a < b$ and $c < 0$, then $ac > bc$.	$-2 < 4$, and since $-3 < 0$, we have $(-2)(-3) > (4)(-3)$; that is, $6 > -12$.

Similar properties hold if each inequality sign, $<$, between a and b and between b and c is replaced by $\geq$, $>$, or $\leq$. Note that Property 4 says that an inequality sign is reversed if the inequality is multiplied by a negative number.

A real number is a *solution of an inequality* involving a variable if a true statement is obtained when the variable is replaced by that number. The set of all real numbers satisfying the inequality is called the *solution set*. We often use interval notation to describe the solution set.

EXAMPLE 8 Find the set of real numbers that satisfy $-1 \leq 2x - 5 < 7$.

Solution Add 5 to each member of the given double inequality, obtaining

$$4 \leq 2x < 12$$

Next, multiply each member of the resulting double inequality by $\frac{1}{2}$, yielding

$$2 \leq x < 6$$

Thus, the solution is the set of all values of x lying in the interval $[2, 6)$.

EXAMPLE 9 Solve the inequality $x^2 + 2x - 8 < 0$.

Solution Observe that $x^2 + 2x - 8 = (x + 4)(x - 2)$, so the given inequality is equivalent to the inequality $(x + 4)(x - 2) < 0$. Since the product of two real numbers is negative if and only if the two numbers have opposite signs, we solve the inequality $(x + 4)(x - 2) < 0$ by studying the signs of the two factors $x + 4$ and $x - 2$. Now, $x + 4 > 0$ when $x > -4$, and $x + 4 < 0$ when $x < -4$. Similarly, $x - 2 > 0$ when $x > 2$, and $x - 2 < 0$ when $x < 2$. These results are summarized graphically in Figure 2.

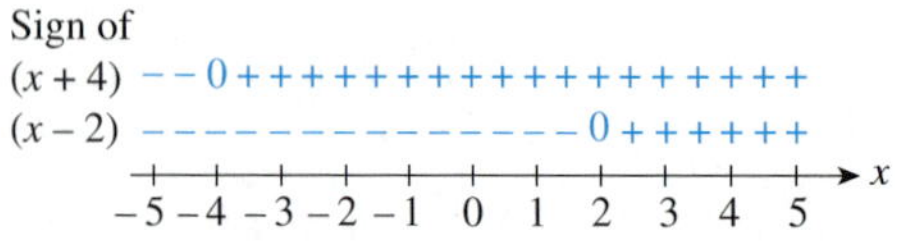

FIGURE 2
Sign diagram for $(x + 4)(x - 2)$

From Figure 2 we see that the two factors $x + 4$ and $x - 2$ have opposite signs when and only when x lies strictly between -4 and 2. Therefore, the required solution is the interval $(-4, 2)$. ■

EXAMPLE 10 Solve the inequality $\dfrac{x+1}{x-1} \geq 0$.

Solution The quotient $(x + 1)/(x - 1)$ is strictly positive if and only if both the numerator and the denominator have the same sign. The signs of $x + 1$ and $x - 1$ are shown in Figure 3.

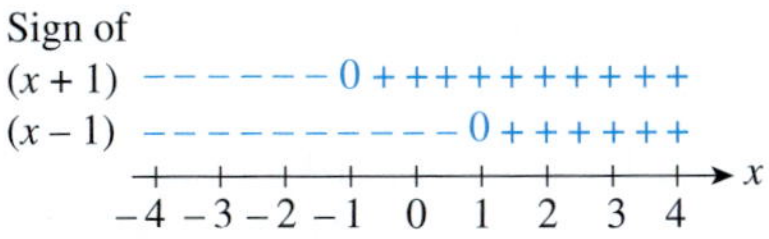

FIGURE 3 Sign diagram for $\dfrac{x+1}{x-1}$

From Figure 3, we see that $x + 1$ and $x - 1$ have the same sign when and only when $x < -1$ or $x > 1$. The quotient $(x + 1)/(x - 1)$ is equal to zero when and only when $x = -1$. Therefore, the required solution is the set of all x in the intervals $(-\infty, -1]$ and $(1, \infty)$. ■

APPLIED EXAMPLE 11 Stock Purchase The management of Corbyco, a giant conglomerate, has estimated that x thousand dollars is needed to purchase

$$100{,}000(-1 + \sqrt{1 + 0.001x})$$

shares of common stock of Starr Communications. Determine how much money Corbyco needs to purchase at least 100,000 shares of Starr's stock.

Solution The amount of cash Corbyco needs to purchase at least 100,000 shares is found by solving the inequality

$$100{,}000(-1 + \sqrt{1 + 0.001x}) \geq 100{,}000$$

Proceeding, we find

$$\begin{aligned} -1 + \sqrt{1 + 0.001x} &\geq 1 \\ \sqrt{1 + 0.001x} &\geq 2 \\ 1 + 0.001x &\geq 4 \qquad \text{Square both sides.} \\ 0.001x &\geq 3 \\ x &\geq 3000 \end{aligned}$$

so Corbyco needs at least \$3,000,000. (Recall that x is measured in thousands of dollars.) ■

Absolute Value

Absolute Value

The **absolute value** of a number a is denoted by $|a|$ and is defined by

$$|a| = \begin{cases} a & \text{if } a \geq 0 \\ -a & \text{if } a < 0 \end{cases}$$

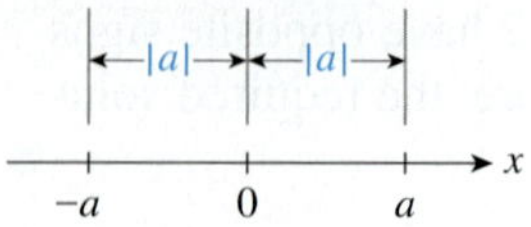

FIGURE 4
The absolute value of a number

Since $-a$ is a positive number when a is negative, it follows that the absolute value of a number is always nonnegative. For example, $|5| = 5$ and $|-5| = -(-5) = 5$. Geometrically, $|a|$ is the distance between the origin and the point on the number line that represents the number a (Figure 4).

Absolute Value Properties

If a and b are any real numbers, then

		Example
Property 5	$\lvert -a\rvert = \lvert a\rvert$	$\lvert -3\rvert = -(-3) = 3 = \lvert 3\rvert$
Property 6	$\lvert ab\rvert = \lvert a\rvert\lvert b\rvert$	$\lvert (2)(-3)\rvert = \lvert -6\rvert = 6 = (2)(3) = \lvert 2\rvert\lvert -3\rvert$
Property 7	$\left\lvert \dfrac{a}{b}\right\rvert = \dfrac{\lvert a\rvert}{\lvert b\rvert} \quad (b \neq 0)$	$\left\lvert \dfrac{(-3)}{(-4)}\right\rvert = \left\lvert \dfrac{3}{4}\right\rvert = \dfrac{3}{4} = \dfrac{\lvert -3\rvert}{\lvert -4\rvert}$
Property 8	$\lvert a + b\rvert \leq \lvert a\rvert + \lvert b\rvert$	$\lvert 8 + (-5)\rvert = \lvert 3\rvert = 3 \leq \lvert 8\rvert + \lvert -5\rvert = 13$

Property 8 is called the **triangle inequality**.

EXAMPLE 12 Evaluate each of the following expressions:

a. $|\pi - 5| + 3$ **b.** $|\sqrt{3} - 2| + |2 - \sqrt{3}|$

Solution

a. Since $\pi - 5 < 0$, we see that $|\pi - 5| = -(\pi - 5)$. Therefore,

$$|\pi - 5| + 3 = -(\pi - 5) + 3 = 8 - \pi$$

b. Since $\sqrt{3} - 2 < 0$, we see that $|\sqrt{3} - 2| = -(\sqrt{3} - 2)$. Next, observe that $2 - \sqrt{3} > 0$, so $|2 - \sqrt{3}| = 2 - \sqrt{3}$. Therefore,

$$\begin{aligned}|\sqrt{3} - 2| + |2 - \sqrt{3}| &= -(\sqrt{3} - 2) + (2 - \sqrt{3})\\ &= 4 - 2\sqrt{3} = 2(2 - \sqrt{3})\end{aligned}$$

■

EXAMPLE 13 Solve the inequalities $|x| \leq 5$ and $|x| \geq 5$.

Solution First, we consider the inequality $|x| \leq 5$. If $x \geq 0$ then $|x| = x$, so $|x| \leq 5$ implies $x \leq 5$ in this case. On the other hand, if $x < 0$ then $|x| = -x$, so $|x| \leq 5$ implies $-x \leq 5$ or $x \geq -5$. Thus, $|x| \leq 5$ means $-5 \leq x \leq 5$ (Figure 5a). To obtain an alternative solution, observe that $|x|$ is the distance from the point x to zero, so the inequality $|x| \leq 5$ implies immediately that $-5 \leq x \leq 5$.

FIGURE 5

Next, the inequality $|x| \geq 5$ states that the distance from x to zero is greater than or equal to 5. This observation yields the result $x \geq 5$ or $x \leq -5$ (Figure 5b). ■

EXAMPLE 14 Solve the inequality $|2x - 3| \leq 1$.

Solution The inequality $|2x - 3| \leq 1$ is equivalent to the inequalities $-1 \leq 2x - 3 \leq 1$ (see Example 13). Thus, $2 \leq 2x \leq 4$ and $1 \leq x \leq 2$. The solution is therefore given by the set of all x in the interval $[1, 2]$ (Figure 6).

FIGURE 6
$|2x - 3| \leq 1$

■

1.2 Exercises

In Exercises 1–6, simplify the expression.

1. $\dfrac{x^2 + x - 2}{x^2 - 4}$

2. $\dfrac{2a^2 - 3ab - 9b^2}{2ab^2 + 3b^3}$

3. $\dfrac{12t^2 + 12t + 3}{4t^2 - 1}$

4. $\dfrac{x^3 + 2x^2 - 3x}{-2x^2 - x + 3}$

5. $\dfrac{(4x - 1)(3) - (3x + 1)(4)}{(4x - 1)^2}$

6. $\dfrac{(1 + x^2)^2(2) - 2x(2)(1 + x^2)(2x)}{(1 + x^2)^4}$

In Exercises 7–24, perform the indicated operations and simplify each expression.

7. $\dfrac{2a^2 - 2b^2}{b - a} \cdot \dfrac{4a + 4b}{a^2 + 2ab + b^2}$

8. $\dfrac{x^2 - 6x + 9}{x^2 - x - 6} \cdot \dfrac{3x + 6}{2x^2 - 7x + 3}$

9. $\dfrac{3x^2 + 2x - 1}{2x + 6} \div \dfrac{x^2 - 1}{x^2 + 2x - 3}$

10. $\dfrac{3x^2 - 4xy - 4y^2}{x^2y} \div \dfrac{(2y - x)^2}{x^3y}$

11. $\dfrac{58}{3(3t + 2)} + \dfrac{1}{3}$

12. $\dfrac{a + 1}{3a} + \dfrac{b - 2}{5b}$

13. $\dfrac{2x}{2x - 1} - \dfrac{3x}{2x + 5}$

14. $\dfrac{-xe^x}{x + 1} + e^x$

15. $\dfrac{4}{x^2 - 9} - \dfrac{5}{x^2 - 6x + 9}$

16. $\dfrac{x}{1 - x} + \dfrac{2x + 3}{x^2 - 1}$

17. $\dfrac{1 + \dfrac{1}{x}}{1 - \dfrac{1}{x}}$

18. $\dfrac{\dfrac{1}{x} + \dfrac{1}{y}}{1 - \dfrac{1}{xy}}$

19. $\dfrac{4x^2}{2\sqrt{2x^2 + 7}} + \sqrt{2x^2 + 7}$

20. $6(2x + 1)^2\sqrt{x^2 + x} + \dfrac{(2x + 1)^4}{2\sqrt{x^2 + x}}$

21. $\dfrac{2x(x + 1)^{-1/2} - (x + 1)^{1/2}}{x^2}$

22. $\dfrac{(x^2 + 1)^{1/2} - 2x^2(x^2 + 1)^{-1/2}}{1 - x^2}$

23. $\dfrac{(2x + 1)^{1/2} - (x + 2)(2x + 1)^{-1/2}}{2x + 1}$

24. $\dfrac{2(2x - 3)^{1/3} - (x - 1)(2x - 3)^{-2/3}}{(2x - 3)^{2/3}}$

In Exercises 25–30, rationalize the denominator of each expression.

25. $\dfrac{1}{\sqrt{3} - 1}$

26. $\dfrac{1}{\sqrt{x} + 5}$

27. $\dfrac{1}{\sqrt{x} - \sqrt{y}}$

28. $\dfrac{a}{1 - \sqrt{a}}$

29. $\dfrac{\sqrt{a} + \sqrt{b}}{\sqrt{a} - \sqrt{b}}$

30. $\dfrac{2\sqrt{a} + \sqrt{b}}{2\sqrt{a} - \sqrt{b}}$

In Exercises 31–36, rationalize the numerator of each expression.

31. $\dfrac{\sqrt{x}}{3}$

32. $\dfrac{\sqrt[3]{y}}{x}$

33. $\dfrac{1 - \sqrt{3}}{3}$

34. $\dfrac{\sqrt{x} - 1}{x}$

35. $\dfrac{1 + \sqrt{x + 2}}{\sqrt{x + 2}}$

36. $\dfrac{\sqrt{x + 3} - \sqrt{x}}{3}$

In Exercises 37–40, determine whether the statement is true or false.

37. $-3 < -20$

38. $-5 \le -5$

39. $\dfrac{2}{3} > \dfrac{5}{6}$

40. $-\dfrac{5}{6} < -\dfrac{11}{12}$

In Exercises 41–58, find the values of *x* that satisfy the inequality (inequalities).

41. $2x + 4 < 8$

42. $-6 > 4 + 5x$

43. $-4x \ge 20$

44. $-12 \le -3x$

45. $-6 < x - 2 < 4$

46. $0 \le x + 1 \le 4$

47. $x + 1 > 4$ or $x + 2 < -1$

48. $x + 1 > 2$ or $x - 1 < -2$

49. $x + 3 > 1$ and $x - 2 < 1$

50. $x - 4 \le 1$ and $x + 3 > 2$

51. $(x + 3)(x - 5) \le 0$

52. $(2x - 4)(x + 2) \ge 0$

53. $(2x - 3)(x - 1) \ge 0$

54. $(3x - 4)(2x + 2) \le 0$

55. $\dfrac{x + 3}{x - 2} \ge 0$

56. $\dfrac{2x - 3}{x + 1} \ge 4$

57. $\dfrac{x - 2}{x - 1} \le 2$

58. $\dfrac{2x - 1}{x + 2} \le 4$

In Exercises 59–68, evaluate the expression.

59. $|-6 + 2|$

60. $4 + |-4|$

61. $\dfrac{|-12 + 4|}{|16 - 12|}$

62. $\left|\dfrac{0.2 - 1.4}{1.6 - 2.4}\right|$

63. $\sqrt{3}|-2| + 3|-\sqrt{3}|$

64. $|-1| + \sqrt{2}|-2|$

65. $|\pi - 1| + 2$

66. $|\pi - 6| - 3$

67. $|\sqrt{2} - 1| + |3 - \sqrt{2}|$

68. $|2\sqrt{3} - 3| - |\sqrt{3} - 4|$

In Exercises 69–74, suppose a and b are real numbers other than zero and that $a > b$. State whether the inequality is true or false.

69. $b - a > 0$

70. $\dfrac{a}{b} > 1$

71. $a^2 > b^2$

72. $\dfrac{1}{a} > \dfrac{1}{b}$

73. $a^3 > b^3$

74. $-a < -b$

In Exercises 75–80, determine whether the statement is true for all real numbers a and b.

75. $|-a| = a$

76. $|b^2| = b^2$

77. $|a - 4| = |4 - a|$

78. $|a + 1| = |a| + 1$

79. $|a + b| = |a| + |b|$

80. $|a - b| = |a| - |b|$

81. **Driving Range of a Car** An advertisement for a certain car states that the EPA fuel economy is 20 mpg city and 27 mpg highway and that the car's fuel-tank capacity is 18.1 gal. Assuming ideal driving conditions, determine the driving range for the car from the foregoing data.

82. Find the minimum cost C (in dollars), given that

$$5(C - 25) \geq 1.75 + 2.5C$$

83. Find the maximum profit P (in dollars) given that

$$6(P - 2500) \leq 4(P + 2400)$$

84. **Celsius and Fahrenheit Temperatures** The relationship between Celsius (°C) and Fahrenheit (°F) temperatures is given by the formula

$$C = \frac{5}{9}(F - 32)$$

a. If the temperature range for Montreal during the month of January is $-15° < °C < -5°$, find the range in degrees Fahrenheit in Montreal for the same period.

b. If the temperature range for New York City during the month of June is $63° < °F < 80°$, find the range in degrees Celsius in New York City for the same period.

85. **Meeting Sales Targets** A salesman's monthly commission is 15% on all sales over \$12,000. If his goal is to make a commission of at least \$3000/month, what minimum monthly sales figures must he attain?

86. **Markup on a Car** The markup on a used car was at least 30% of its current wholesale price. If the car was sold for \$5600, what was the maximum wholesale price?

87. **Quality Control** PAR Manufacturing manufactures steel rods. Suppose the rods ordered by a customer are manufactured to a specification of 0.5 in. and are acceptable only if they are within the *tolerance limits* of 0.49 in. and 0.51 in. Letting x denote the diameter of a rod, write an inequality using absolute value signs to express a criterion involving x that must be satisfied in order for a rod to be acceptable.

88. **Quality Control** The diameter x (in inches) of a batch of ball bearings manufactured by PAR Manufacturing satisfies the inequality

$$|x - 0.1| \leq 0.01$$

What is the smallest diameter a ball bearing in the batch can have? The largest diameter?

89. **Meeting Profit Goals** A manufacturer of a certain commodity has estimated that her profit in thousands of dollars is given by the expression

$$-6x^2 + 30x - 10$$

where x (in thousands) is the number of units produced. What production range will enable the manufacturer to realize a profit of at least \$14,000 on the commodity?

90. **Concentration of a Drug in the Bloodstream** The concentration (in milligrams/cubic centimeter) of a certain drug in a patient's bloodstream t hr after injection is given by

$$\frac{0.2t}{t^2 + 1}$$

Find the interval of time when the concentration of the drug is greater than or equal to 0.08 mg/cc.

91. **Cost of Removing Toxic Pollutants** A city's main well was recently found to be contaminated with trichloroethylene (a cancer-causing chemical) as a result of an abandoned chemical dump that leached chemicals into the water. A proposal submitted to the city council indicated that the cost, in millions of dollars, of removing x% of the toxic pollutants is

$$\frac{0.5x}{100 - x}$$

If the city could raise between \$25 and \$30 million for the purpose of removing the toxic pollutants, what is the range of pollutants that could be expected to be removed?

92. **Average Speed of a Vehicle** The average speed of a vehicle in miles per hour on a stretch of route 134 between 6 a.m. and 10 a. m. on a typical weekday is approximated by the expression

$$20t - 40\sqrt{t} + 50 \qquad (0 \leq t \leq 4)$$

where t is measured in hours, with $t = 0$ corresponding to

6 a.m. Over what interval of time is the average speed of a vehicle less than or equal to 35 mph?

93. **AIR POLLUTION** Nitrogen dioxide is a brown gas that impairs breathing. The amount of nitrogen dioxide present in the atmosphere on a certain May day in the city of Long Beach measured in PSI (pollutant standard index) at time t, where t is measured in hours, and $t = 0$ corresponds to 7 a.m., is approximated by

$$\frac{136}{1 + 0.25(t - 4.5)^2} + 28 \qquad (0 \le t \le 11)$$

Find the time of the day when the amount of nitrogen dioxide is greater than or equal to 128 PSI.

Source: Los Angeles Times

In Exercises 94–98, determine whether the statement is true or false. If it is true, explain why it is true. If it is false, give an example to show why it is false.

94. $\dfrac{a}{a+b} = \dfrac{a}{b} + \dfrac{a}{c}$

95. If $a < b$, then $a - c > b - c$.

96. $|a - b| = |b - a|$

97. $|a - b| \le |b| + |a|$

98. $\sqrt{a^2 - b^2} = |a| - |b|$

1.3 The Cartesian Coordinate System

The Cartesian Coordinate System

In Section 1.1, we saw how a one-to-one correspondence between the set of real numbers and the points on a straight line leads to a coordinate system on a line (a one-dimensional space).

A similar representation for points in a plane (a two-dimensional space) is realized through the **Cartesian coordinate system,** which is constructed as follows: Take two perpendicular lines, one of which is normally chosen to be horizontal. These lines intersect at a point O, called the **origin** (Figure 7). The horizontal line is called the ***x*-axis,** and the vertical line is called the ***y*-axis.** A number scale is set up along the x-axis, with the positive numbers lying to the right of the origin and the negative numbers lying to the left of it. Similarly, a number scale is set up along the y-axis, with the positive numbers lying above the origin and the negative numbers lying below it.

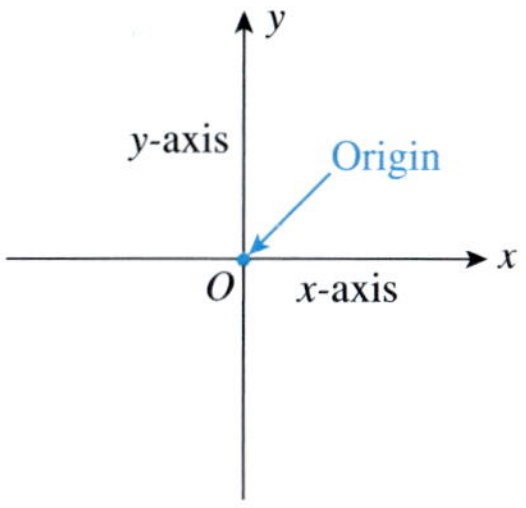

FIGURE 7
The Cartesian coordinate system

The number scales on the two axes need not be the same. Indeed, in many applications different quantities are represented by x and y. For example, x may represent the number of typewriters sold and y the total revenue resulting from the sales. In such cases, it is often desirable to choose different number scales to represent the different quantities. Note, however, that the zeros of both number scales coincide at the origin of the two-dimensional coordinate system.

A point in the plane can now be represented uniquely in this coordinate system by an **ordered pair** of numbers—that is, a pair (x, y), where x is the first number and y the second. To see this, let P be any point in the plane (Figure 8). Draw perpendiculars from P to the x-axis and y-axis, respectively. Then the number x is precisely the number that corresponds to the point on the x-axis at which the perpendicular through P hits the x-axis. Similarly, y is the number that corresponds to the point on the y-axis at which the perpendicular through P crosses the y-axis.

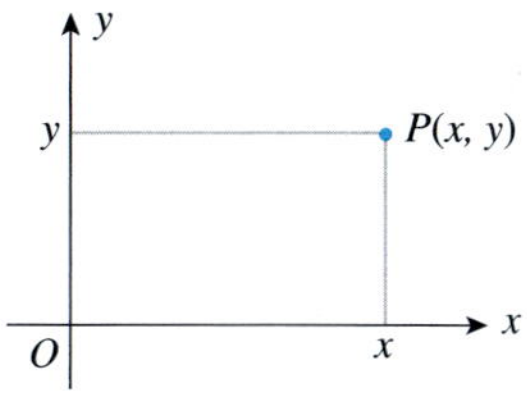

FIGURE 8
An ordered pair (x, y)

Conversely, given an ordered pair (x, y) with x as the first number and y the second, a point P in the plane is uniquely determined as follows: Locate the point on the x-axis represented by the number x and draw a line through that point parallel to the y-axis. Next, locate the point on the y-axis represented by the number y and draw a line through that point parallel to the x-axis. The point of intersection of these two lines is the point P (see Figure 8).

In the ordered pair (x, y), x is called the **abscissa,** or ***x*-coordinate;** y is called the **ordinate,** or ***y*-coordinate;** and x and y together are referred to as the **coordinates** of the point P.

Letting $P(a, b)$ denote the point with x-coordinate a and y-coordinate b, the points $A(2, 3)$, $B(-2, 3)$, $C(-2, -3)$, $D(2, -3)$, $E(3, 2)$, $F(4, 0)$, and $G(0, -5)$ are plotted in Figure 9. The fact that, in general, $P(x, y) \neq P(y, x)$ is clearly illustrated by points A and E.

The axes divide the plane into four quadrants. Quadrant I consists of the points $P(x, y)$ that satisfy $x > 0$ and $y > 0$; Quadrant II, the points $P(x, y)$, where $x < 0$ and $y > 0$; Quadrant III, the points $P(x, y)$, where $x < 0$ and $y < 0$; and Quadrant IV, the points $P(x, y)$, where $x > 0$ and $y < 0$ (Figure 10).

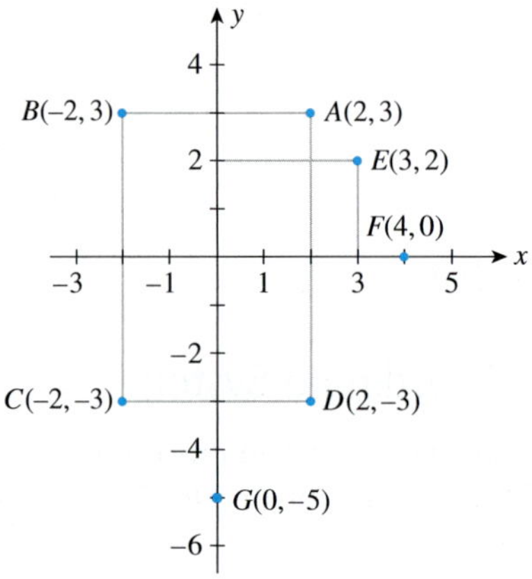

FIGURE 9
Several points in the Cartesian plane

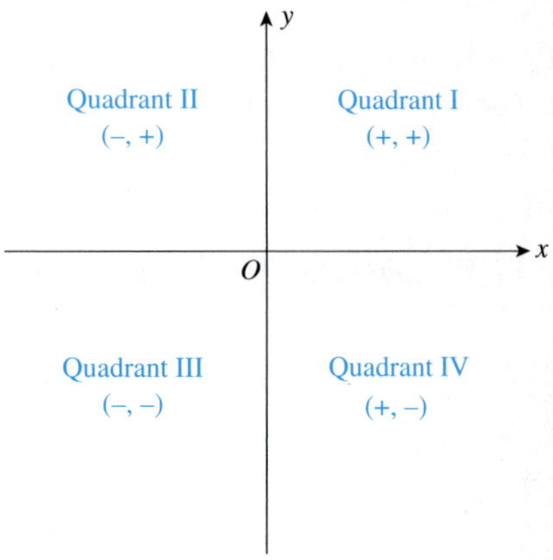

FIGURE 10
The four quadrants in the Cartesian plane

The Distance Formula

One immediate benefit that arises from using the Cartesian coordinate system is that the distance between any two points in the plane may be expressed solely in terms of their coordinates. Suppose, for example, (x_1, y_1) and (x_2, y_2) are any two points in the plane (Figure 11). Then the distance between these two points can be computed using the following formula.

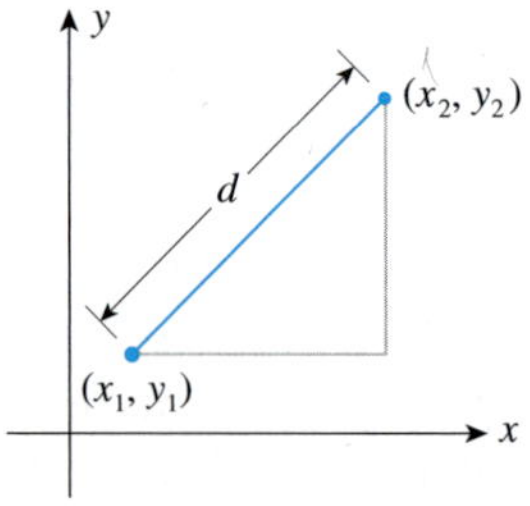

FIGURE 11
The distance d between the points (x_1, y_1) and (x_2, y_2)

Distance Formula

The distance d between two points $P_1(x_1, y_1)$ and $P_2(x_2, y_2)$ in the plane is given by

$$d = \sqrt{(x_2 - x_1)^2 + (y_2 - y_1)^2} \quad \textbf{(1)}$$

For a proof of this result, see Exercise 48, page 32.

In what follows, we give several applications of the distance formula.

EXAMPLE 1 Find the distance between the points $(-4, 3)$ and $(2, 6)$.

Solution Let $P_1(-4, 3)$ and $P_2(2, 6)$ be points in the plane. Then, we have

$$x_1 = -4 \qquad y_1 = 3 \qquad x_2 = 2 \qquad y_2 = 6$$

Using Formula (1), we have

$$\begin{aligned} d &= \sqrt{[2 - (-4)]^2 + (6 - 3)^2} \\ &= \sqrt{6^2 + 3^2} \\ &= \sqrt{45} = 3\sqrt{5} \end{aligned}$$

■

Explore & Discuss

Refer to Example 1. Suppose we label the point (2, 6) as P_1 and the point $(-4, 3)$ as P_2.
(1) Show that the distance d between the two points is the same as that obtained earlier.
(2) Prove that, in general, the distance d in Formula (1) is independent of the way we label the two points.

VIDEO

EXAMPLE 2 Let $P(x, y)$ denote a point lying on the circle with radius r and center $C(h, k)$ (Figure 12). Find a relationship between x and y.

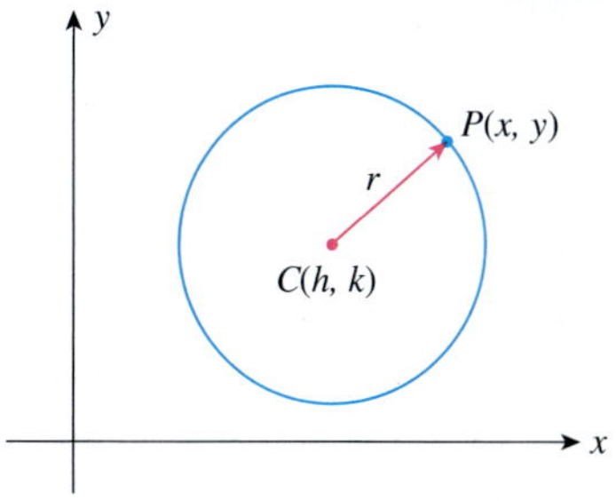

FIGURE 12
A circle with radius r and center $C(h, k)$

Solution By the definition of a circle, the distance between $C(h, k)$ and $P(x, y)$ is r. Using Formula (1), we have

$$\sqrt{(x-h)^2 + (y-k)^2} = r$$

which, upon squaring both sides, gives the equation

$$(x-h)^2 + (y-k)^2 = r^2$$

that must be satisfied by the variables x and y. ■

A summary of the result obtained in Example 2 follows.

Equation of a Circle

An equation of the circle with center $C(h, k)$ and radius r is given by

$$(x-h)^2 + (y-k)^2 = r^2 \qquad \textbf{(2)}$$

EXAMPLE 3 Find an equation of the circle with

a. Radius 2 and center $(-1, 3)$.
b. Radius 3 and center located at the origin.

Solution

a. We use Formula (2) with $r = 2$, $h = -1$, and $k = 3$, obtaining

$$[x - (-1)]^2 + (y-3)^2 = 2^2 \quad \text{or} \quad (x+1)^2 + (y-3)^2 = 4$$

(Figure 13a).

b. Using Formula (2) with $r = 3$ and $h = k = 0$, we obtain

$$x^2 + y^2 = 3^2 \quad \text{or} \quad x^2 + y^2 = 9$$

(Figure 13b).

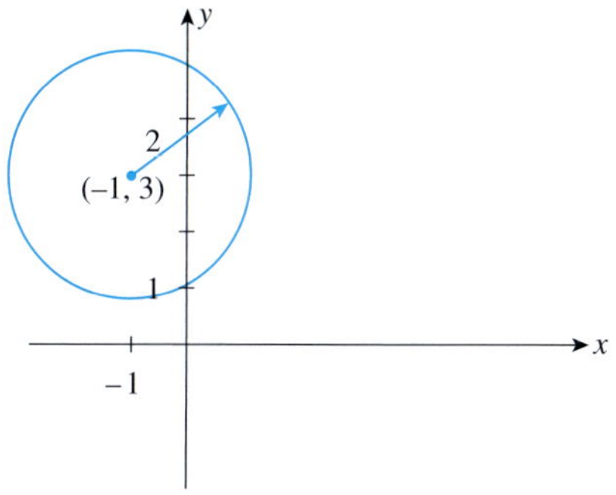

(a) The circle with radius 2 and center $(-1, 3)$

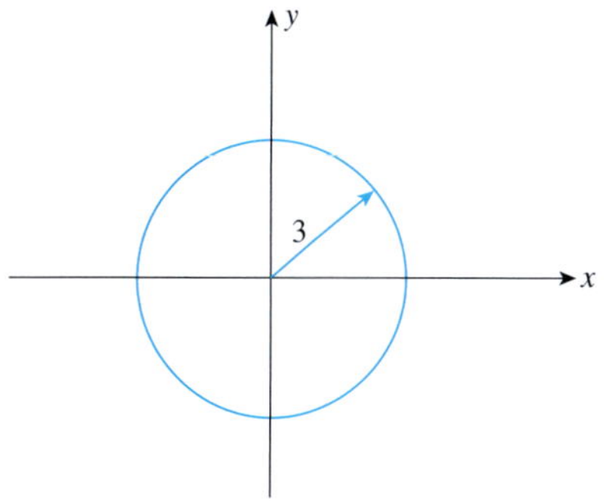

(b) The circle with radius 3 and center (0, 0)

FIGURE 13 ■

Explore & Discuss

1. Use the distance formula to help you describe the set of points in the xy-plane satisfying each of the following inequalities.
   a. $(x - h)^2 + (y - k)^2 \leq r^2$
   b. $(x - h)^2 + (y - k)^2 < r^2$
   c. $(x - h)^2 + (y - k)^2 \geq r^2$
   d. $(x - h)^2 + (y - k)^2 > r^2$
2. Consider the equation $x^2 + y^2 = 4$.
   a. Show that $y = \pm\sqrt{4 - x^2}$.
   b. Describe the set of points (x, y) in the xy-plane satisfying the following equations:
   (i) $y = \sqrt{4 - x^2}$
   (ii) $y = -\sqrt{4 - x^2}$

VIDEO

APPLIED EXAMPLE 4 Cost of Laying Cable In Figure 14, S represents the position of a power relay station located on a straight coastal highway, and M shows the location of a marine biology experimental station on an island. A cable is to be laid connecting the relay station with the experimental station. If the cost of running the cable on land is \$3.00 per running foot and the cost of running the cable under water is \$5.00 per running foot, find the total cost for laying the cable.

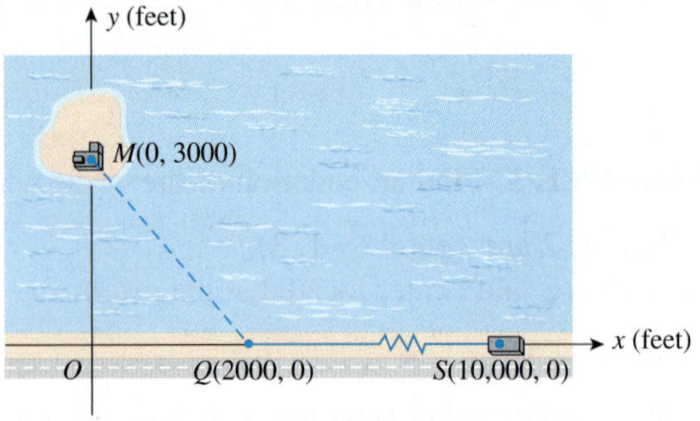

FIGURE **14**
Cable connecting relay station S to experimental station M

Solution The length of cable required on land is given by the distance from S to Q. This distance is $(10{,}000 - 2000)$, or 8000 feet. Next, we see that the length of cable required underwater is given by the distance from M to Q. This distance is

$$\begin{aligned}\sqrt{(0 - 2000)^2 + (3000 - 0)^2} &= \sqrt{2000^2 + 3000^2}\\ &= \sqrt{13{,}000{,}000}\\ &\approx 3605.55\end{aligned}$$

or approximately 3605.55 feet. Therefore, the total cost for laying the cable is

$$3(8000) + 5(3605.55) = 42{,}027.75$$

or approximately \$42,027.75.

Explore & Discuss

In the Cartesian coordinate system, the two axes are perpendicular to each other. Consider a coordinate system in which the x- and y-axes are not collinear and are not perpendicular to each other (see the accompanying figure).

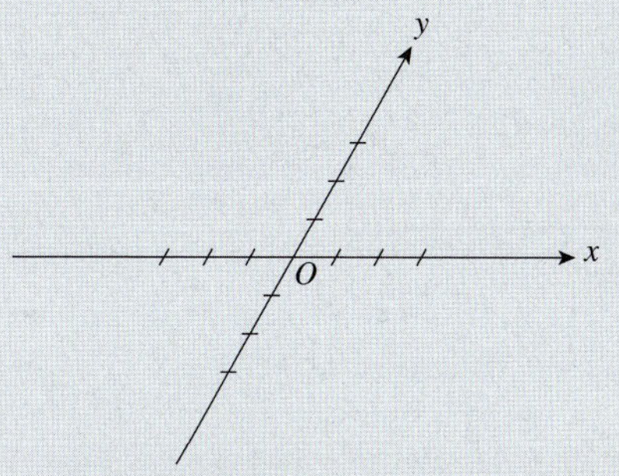

1. Describe how a point is represented in this coordinate system by an ordered pair (x, y) of real numbers. Conversely, show how an ordered pair (x, y) of real numbers uniquely determines a point in the plane.
2. Suppose you want to find a formula for the distance between two points $P_1(x_1, y_1)$ and $P_2(x_2, y_2)$ in the plane. What is the advantage that the Cartesian coordinate system has over the coordinate system under consideration? Comment on your answer.

1.3 Self-Check Exercises

1. **a.** Plot the points $A(4, -2)$, $B(2, 3)$, and $C(-3, 1)$.
   b. Find the distance between the points A and B; between B and C; between A and C.
   c. Use the Pythagorean theorem to show that the triangle with vertices A, B, and C is a right triangle.

2. The following figure shows the location of cities A, B, and C. Suppose a pilot wishes to fly from city A to city C but must make a mandatory stopover in city B. If the single-engine light plane has a range of 650 miles, can she make the trip without refueling in city B?

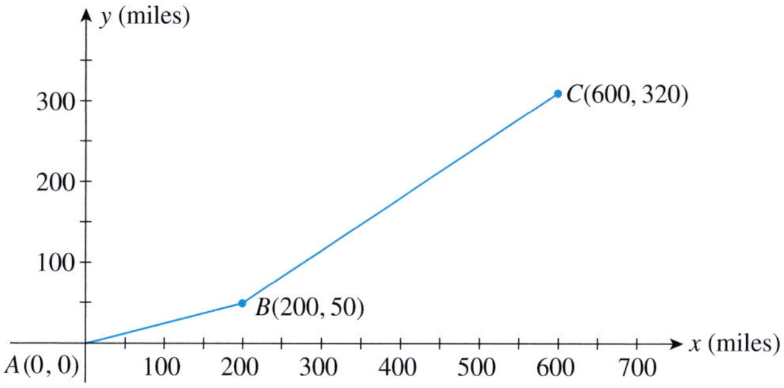

Solutions to Self-Check Exercises 1.3 can be found on page 32.

1.3 Concept Questions

1. What can you say about the signs of a and b if the point $P(a, b)$ lies in (a) the second quadrant? (b) The third quadrant? (c) The fourth quadrant?

2. **a.** What is the distance between $P_1(x_1, y_1)$ and $P_2(x_2, y_2)$?
   b. When you use the distance formula, does it matter which point is labeled P_1 and which point is labeled P_2? Explain.

1.3 Exercises

In Exercises 1–6, refer to the following figure and determine the coordinates of each point and the quadrant in which it is located.

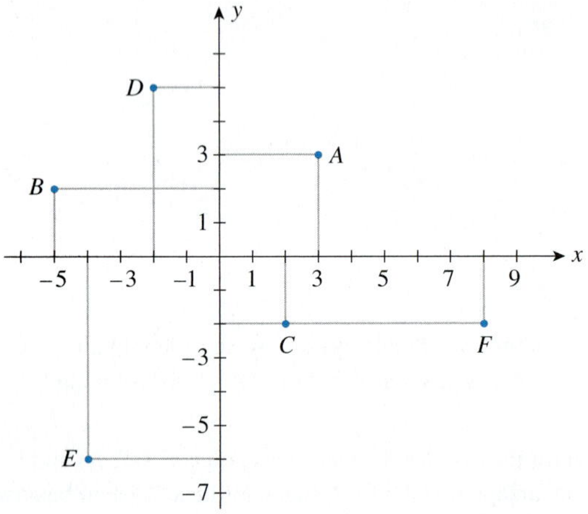

1. A
2. B
3. C
4. D
5. E
6. F

In Exercises 7–12, refer to the following figure.

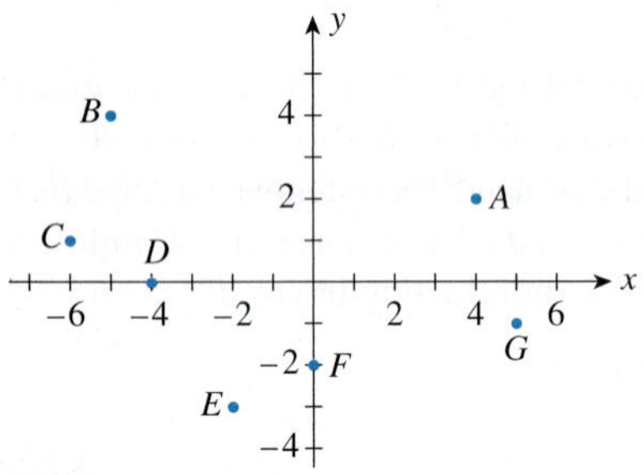

7. Which point has coordinates (4, 2)?
8. What are the coordinates of point B?
9. Which points have negative y-coordinates?
10. Which point has a negative x-coordinate and a negative y-coordinate?
11. Which point has an x-coordinate that is equal to zero?
12. Which point has a y-coordinate that is equal to zero?

In Exercises 13–20, sketch a set of coordinate axes and plot each point.

13. $(-2, 5)$
14. $(1, 3)$
15. $(3, -1)$
16. $(3, -4)$
17. $\left(8, -\frac{7}{2}\right)$
18. $\left(-\frac{5}{2}, \frac{3}{2}\right)$
19. $(4.5, -4.5)$
20. $(1.2, -3.4)$

In Exercises 21–24, find the distance between the given points.

21. $(1, 3)$ and $(4, 7)$
22. $(1, 0)$ and $(4, 4)$
23. $(-1, 3)$ and $(4, 9)$
24. $(-2, 1)$ and $(10, 6)$
25. Find the coordinates of the points that are 10 units away from the origin and have a y-coordinate equal to -6.
26. Find the coordinates of the points that are 5 units away from the origin and have an x-coordinate equal to 3.
27. Show that the points $(3, 4)$, $(-3, 7)$, $(-6, 1)$, and $(0, -2)$ form the vertices of a square.
28. Show that the triangle with vertices $(-5, 2)$, $(-2, 5)$, and $(5, -2)$ is a right triangle.

In Exercises 29–34, find an equation of the circle that satisfies the given conditions.

29. Radius 5 and center $(2, -3)$
30. Radius 3 and center $(-2, -4)$
31. Radius 5 and center at the origin
32. Center at the origin and passes through $(2, 3)$
33. Center $(2, -3)$ and passes through $(5, 2)$
34. Center $(-a, a)$ and radius $2a$
35. **Distance Traveled** A grand tour of four cities begins at city A and makes successive stops at cities B, C, and D before returning to city A. If the cities are located as shown in the following figure, find the total distance covered on the tour.

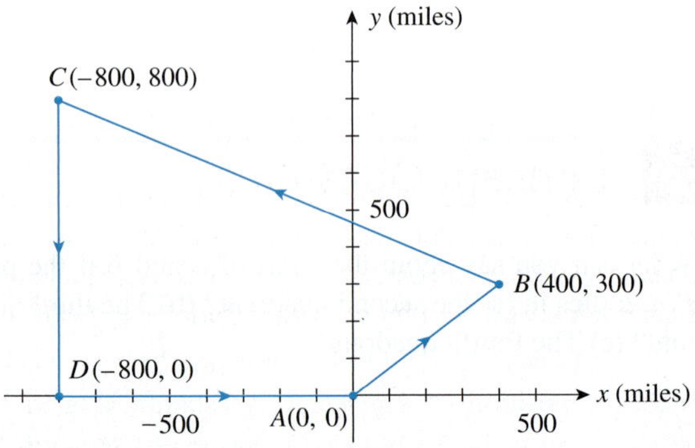

36. **DELIVERY CHARGES** A furniture store offers free setup and delivery services to all points within a 25-mi radius of its warehouse distribution center. If you live 20 mi east and 14 mi south of the warehouse, will you incur a delivery charge? Justify your answer.

37. **OPTIMIZING TRAVEL TIME** Towns A, B, C, and D are located as shown in the following figure. Two highways link town A to town D. Route 1 runs from town A to town D via town B, and Route 2 runs from town A to town D via town C. If a salesman wishes to drive from town A to town D and traffic conditions are such that he could expect to average the same speed on either route, which highway should he take to arrive in the shortest time?

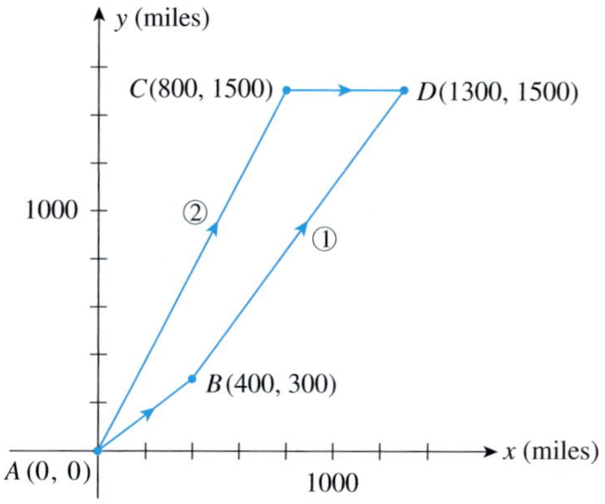

38. **MINIMIZING SHIPPING COSTS** Refer to the figure for Exercise 37. Suppose a fleet of 100 automobiles are to be shipped from an assembly plant in town A to town D. They may be shipped either by freight train along Route 1 at a cost of 22¢/mile per automobile or by truck along Route 2 at a cost of 21¢/mile per automobile. Which means of transportation minimizes the shipping cost? What is the net savings?

39. **CONSUMER DECISIONS** Ivan wishes to determine which antenna he should purchase for his home. The TV store has supplied him with the following information:

Range in Miles			
VHF	**UHF**	**Model**	**Price**
30	20	A	$40
45	35	B	$50
60	40	C	$60
75	55	D	$70

Ivan wishes to receive Channel 17 (VHF) that is located 25 mi east and 35 mi north of his home and Channel 38 (UHF) that is located 20 mi south and 32 mi west of his home. Which model will allow him to receive both channels at the least cost? (Assume that the terrain between Ivan's home and both broadcasting stations is flat.)

40. **COST OF LAYING CABLE** In the following diagram, S represents the position of a power relay station located on a straight coastal highway, and M shows the location of a marine biology experimental station on an island. A cable is to be laid connecting the relay station with the experimental station. If the cost of running the cable on land is \$3.00/running foot and the cost of running cable under water is \$5.00/running foot, find an expression in terms of x that gives the total cost for laying the cable. What is the total cost when $x = 2500$? when $x = 3000$?

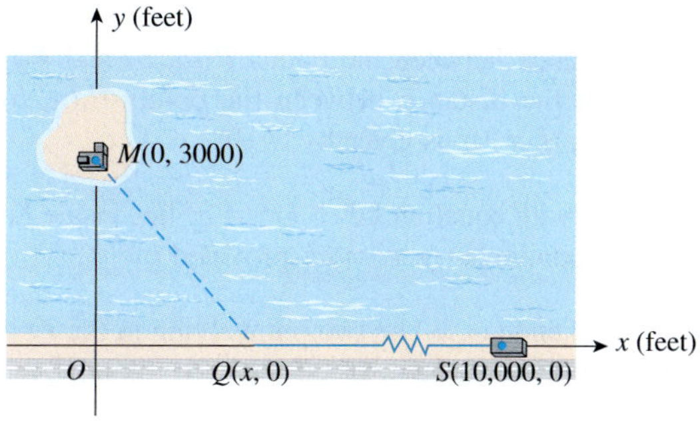

41. Two ships leave port at the same time. Ship A sails north at a speed of 20 mph while ship B sails east at a speed of 30 mph.
   a. Find an expression in terms of the time t (in hours) giving the distance between the two ships.
   b. Using the expression obtained in part (a), find the distance between the two ships 2 hr after leaving port.

42. Ship A leaves port sailing north at a speed of 25 mph. A half hour later, ship B leaves the same port sailing east at a speed of 20 mph. Let t (in hours) denote the time ship B has been at sea.
   a. Find an expression in terms of t giving the distance between the two ships.
   b. Use the expression obtained in part (a) to find the distance between the two ships 2 hr after ship A has left port.

43. **WATCHING A ROCKET LAUNCH** At a distance of 4000 ft from the launch site, a spectator is observing a rocket being launched. Suppose the rocket lifts off vertically and reaches an altitude of x ft (see the accompanying figure).

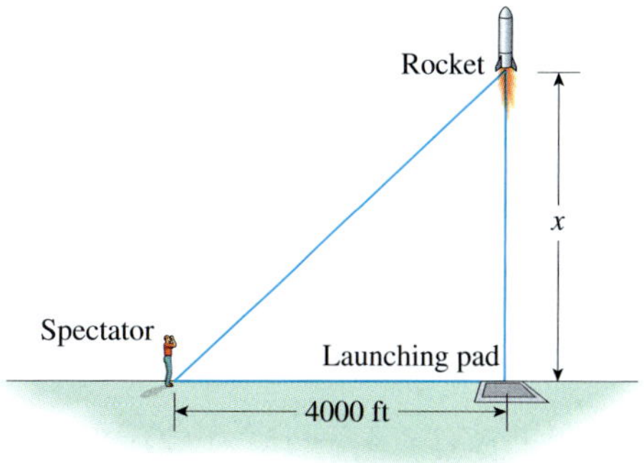

   a. Find an expression giving the distance between the spectator and the rocket.
   b. What is the distance between the spectator and the rocket when the rocket reaches an altitude of 20,000 ft?

In Exercises 44–47, determine whether the statement is true or false. If it is true, explain why it is true. If it is false, give an example to show why it is false.

44. The point $(-a, b)$ is symmetric to the point (a, b) with respect to the y-axis.

45. The point $(-a, -b)$ is symmetric to the point (a, b) with respect to the origin.

46. If the distance between the points $P_1(a, b)$ and $P_2(c, d)$ is D, then the distance between the points $P_1(a, b)$ and $P_3(kc, kd)$, $(k \neq 0)$, is given by $|k|D$.

47. The circle with equation $kx^2 + ky^2 = a^2$ lies inside the circle with equation $x^2 + y^2 = a^2$, provided $k > 1$.

48. Let (x_1, y_1) and (x_2, y_2) be two points lying in the xy-plane. Show that the distance between the two points is given by

$$d = \sqrt{(x_2 - x_1)^2 + (y_2 - y_1)^2}$$

Hint: Refer to the accompanying figure and use the Pythagorean theorem.

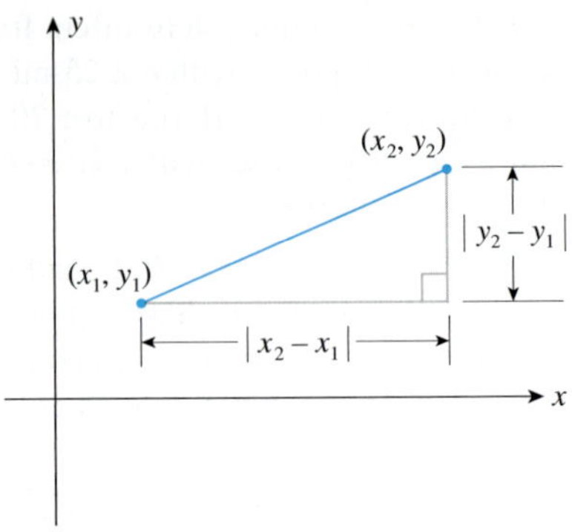

49. a. Show that the midpoint of the line segment joining the points $P_1(x_1, y_1)$ and $P_2(x_2, y_2)$ is

$$\left(\frac{x_1 + x_2}{2}, \frac{y_1 + y_2}{2}\right)$$

b. Use the result of part (a) to find the midpoint of the line segment joining the points $(-3, 2)$ and $(4, -5)$.

50. Show that an equation of a circle can be written in the form

$$x^2 + y^2 + Cx + Dy + E = 0$$

where C, D, and E are constants. This is called the general form of an equation of a circle.

1.3 Solutions to Self-Check Exercises

1. a. The points are plotted in the following figure:

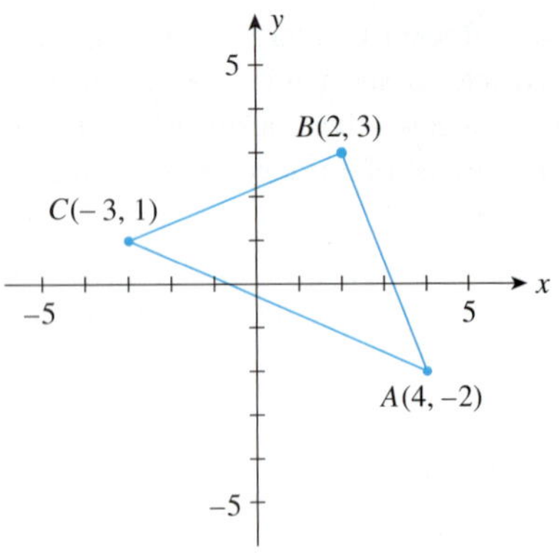

b. The distance between A and B is

$$\begin{aligned} d(A, B) &= \sqrt{(2 - 4)^2 + [3 - (-2)]^2} \\ &= \sqrt{(-2)^2 + 5^2} = \sqrt{4 + 25} = \sqrt{29} \end{aligned}$$

The distance between B and C is

$$\begin{aligned} d(B, C) &= \sqrt{(-3 - 2)^2 + (1 - 3)^2} \\ &= \sqrt{(-5)^2 + (-2)^2} = \sqrt{25 + 4} = \sqrt{29} \end{aligned}$$

The distance between A and C is

$$\begin{aligned} d(A, C) &= \sqrt{(-3 - 4)^2 + [1 - (-2)]^2} \\ &= \sqrt{(-7)^2 + 3^2} = \sqrt{49 + 9} = \sqrt{58} \end{aligned}$$

c. We will show that

$$[d(A, C)]^2 = [d(A, B)]^2 + [d(B, C)]^2$$

From part (b), we see that $[d(A, B)]^2 = 29$, $[d(B, C)]^2 = 29$, and $[d(A, C)]^2 = 58$, and the desired result follows.

2. The distance between city A and city B is

$$d(A, B) = \sqrt{200^2 + 50^2} \approx 206$$

or 206 mi. The distance between city B and city C is

$$\begin{aligned} d(B, C) &= \sqrt{[600 - 200]^2 + [320 - 50]^2} \\ &= \sqrt{400^2 + 270^2} \approx 483 \end{aligned}$$

or 483 mi. Therefore, the total distance the pilot would have to cover is 689 mi, so she must refuel in city B.

1.4 Straight Lines

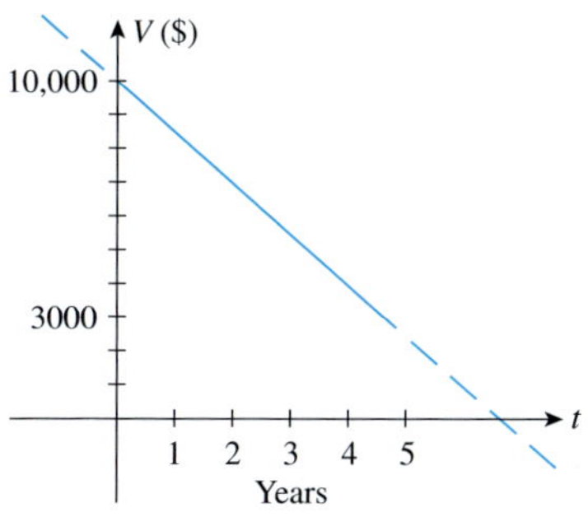

FIGURE 15
Linear depreciation of an asset

In computing income tax, business firms are allowed by law to depreciate certain assets such as buildings, machines, furniture, automobiles, and so on, over a period of time. Linear depreciation, or the straight-line method, is often used for this purpose. The graph of the straight line shown in Figure 15 describes the book value V of a server that has an initial value of \$10,000 and that is being depreciated linearly over 5 years with a scrap value of \$3,000. Note that only the solid portion of the straight line is of interest here.

The book value of the server at the end of year t, where t lies between 0 and 5, can be read directly from the graph. But there is one shortcoming in this approach: The result depends on how accurately you draw and read the graph. A better and more accurate method is based on finding an *algebraic* representation of the depreciation line.

Slope of a Line

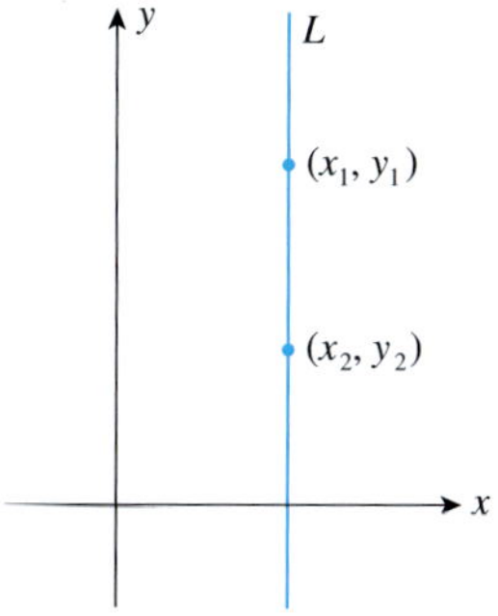

FIGURE 16
m is undefined.

To see how a straight line in the xy-plane may be described algebraically, we need to first recall certain properties of straight lines. Let L denote the unique straight line that passes through the two distinct points (x_1, y_1) and (x_2, y_2). If $x_1 = x_2$, then L is a vertical line, and the slope is undefined (Figure 16).

If $x_1 \neq x_2$, we define the slope of L as follows:

Slope of a Nonvertical Line

If (x_1, y_1) and (x_2, y_2) are any two distinct points on a nonvertical line L, then the slope m of L is given by

$$m = \frac{\Delta y}{\Delta x} = \frac{y_2 - y_1}{x_2 - x_1} \quad (3)$$

See Figure 17.

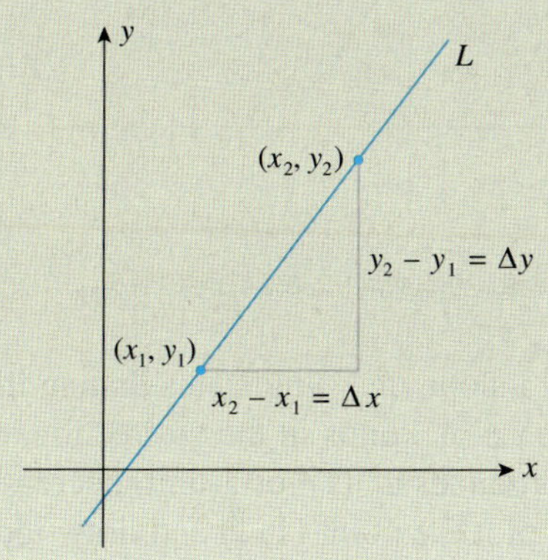

FIGURE 17

Observe that the slope of a straight line is a constant whenever it is defined. The number $\Delta y = y_2 - y_1$ (Δy is read "delta y") is a measure of the vertical change in y, and $\Delta x = x_2 - x_1$ is a measure of the horizontal change in x, as shown in Figure 17. From this figure, we can see that the slope m of a straight line L is a measure of the *rate of change of y with respect to x*.

Figure 18a shows a straight line L_1 with slope 2. Observe that L_1 has the property that a 1-unit increase in x results in a 2-unit increase in y. To see this, let $\Delta x = 1$ in

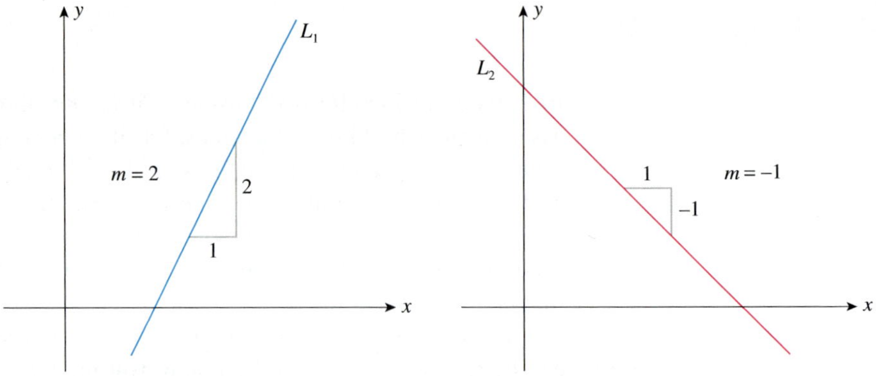

FIGURE 18

(a) The line rises ($m > 0$).

(b) The line falls ($m < 0$).

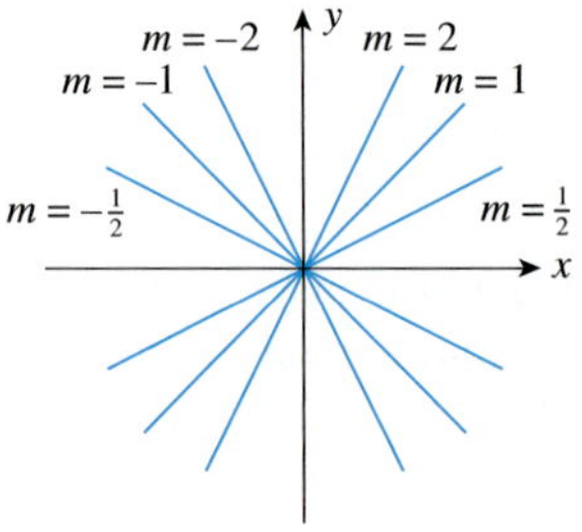

FIGURE 19
A family of straight lines

Formula (3) so that $m = \Delta y$. Since $m = 2$, we conclude that $\Delta y = 2$. Similarly, Figure 18b shows a line L_2 with slope -1. Observe that a straight line with positive slope slants upward from left to right (y increases as x increases), whereas a line with negative slope slants downward from left to right (y decreases as x increases). Finally, Figure 19 shows a family of straight lines passing through the origin with indicated slopes.

EXAMPLE 1 Sketch the straight line that passes through the point $(-2, 5)$ and has slope $-\frac{4}{3}$.

Solution First, plot the point $(-2, 5)$ (Figure 20).

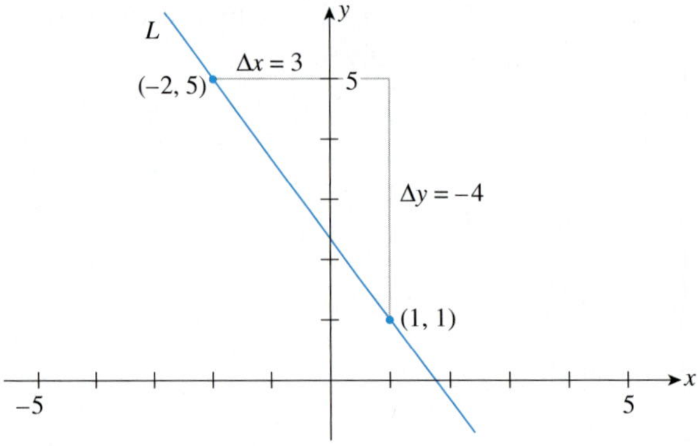

FIGURE 20
L has slope $-\frac{4}{3}$ and passes through $(-2, 5)$.

Next, recall that a slope of $-\frac{4}{3}$ indicates that an increase of 1 unit in the x-direction produces a decrease of $\frac{4}{3}$ units in the y-direction, or equivalently, a 3-unit increase in the x-direction produces a $3(\frac{4}{3})$, or 4-unit, decrease in the y-direction. Using this information, we plot the point $(1, 1)$ and draw the line through the two points. ■

Explore & Discuss

Show that the slope of a nonvertical line is independent of the two distinct points $P_1(x_1, y_1)$ and $P_2(x_2, y_2)$ used to compute it.

Hint: Suppose we pick two other distinct points, $P_3(x_3, y_3)$ and $P_4(x_4, y_4)$ lying on L. Draw a picture and use similar triangles to demonstrate that using P_3 and P_4 gives the same value as that obtained using P_1 and P_2.

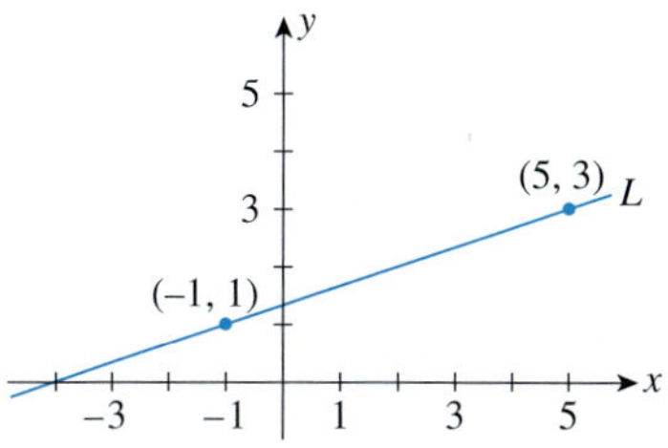

FIGURE 21
L passes through (5, 3) and (−1, 1).

EXAMPLE 2 Find the slope m of the line that passes through the points $(-1, 1)$ and $(5, 3)$.

Solution Choose (x_1, y_1) to be the point $(-1, 1)$ and (x_2, y_2) to be the point $(5, 3)$. Then, with $x_1 = -1$, $y_1 = 1$, $x_2 = 5$, and $y_2 = 3$, we find

$$m = \frac{y_2 - y_1}{x_2 - x_1} = \frac{3 - 1}{5 - (-1)} = \frac{1}{3}$$ Use Formula (3).

(Figure 21). Try to verify that the result obtained would have been the same had we chosen the point $(-1, 1)$ to be (x_2, y_2) and the point $(5, 3)$ to be (x_1, y_1). ■

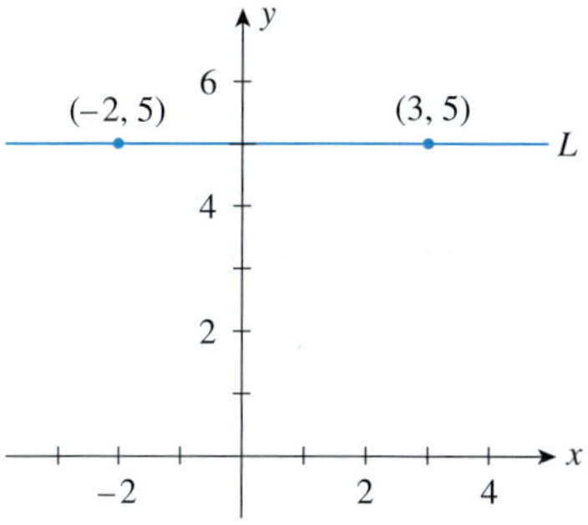

FIGURE 22
The slope of the horizontal line L is 0.

EXAMPLE 3 Find the slope of the line that passes through the points $(-2, 5)$ and $(3, 5)$.

Solution The slope of the required line is given by

$$m = \frac{5 - 5}{3 - (-2)} = \frac{0}{5} = 0$$ Use Formula (3).

(Figure 22). ■

Note In general, the slope of a horizontal line is zero. ■

We can use the slope of a straight line to determine whether a line is parallel to another line.

> **Parallel Lines**
>
> Two distinct lines are **parallel** if and only if their slopes are equal or their slopes are undefined.

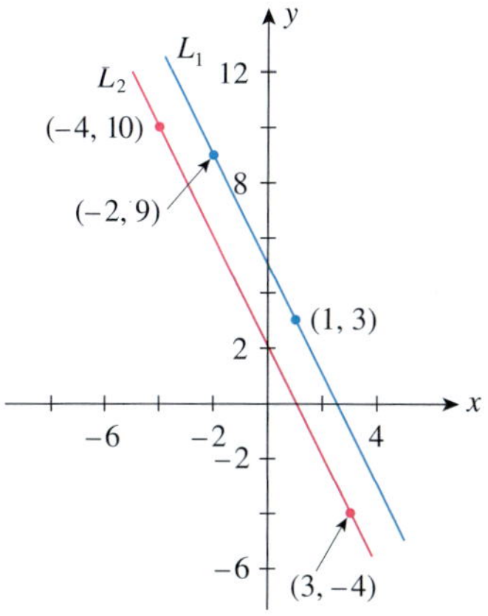

FIGURE 23
L_1 and L_2 have the same slope and hence are parallel.

EXAMPLE 4 Let L_1 be a line that passes through the points $(-2, 9)$ and $(1, 3)$ and let L_2 be the line that passes through the points $(-4, 10)$ and $(3, -4)$. Determine whether L_1 and L_2 are parallel.

Solution The slope m_1 of L_1 is given by

$$m_1 = \frac{3 - 9}{1 - (-2)} = -2$$

The slope m_2 of L_2 is given by

$$m_2 = \frac{-4 - 10}{3 - (-4)} = -2$$

Since $m_1 = m_2$, the lines L_1 and L_2 are in fact parallel (Figure 23). ■

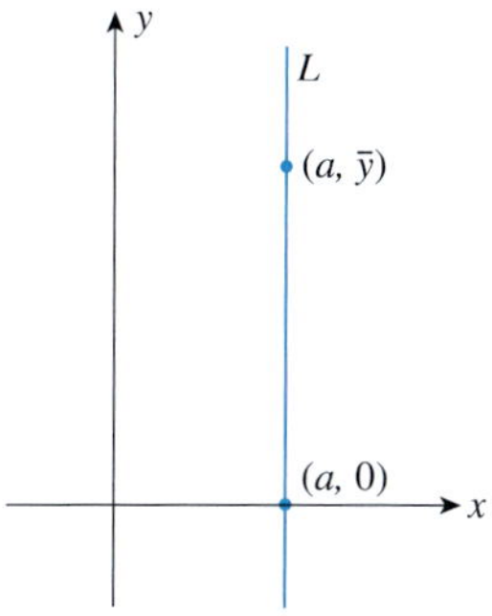

FIGURE 24
The vertical line $x = a$

Equations of Lines

We will now show that every straight line lying in the xy-plane may be represented by an equation involving the variables x and y. One immediate benefit of this is that problems involving straight lines may be solved algebraically.

Let L be a straight line parallel to the y-axis (perpendicular to the x-axis) (Figure 24). Then, L crosses the x-axis at some point $(a, 0)$ with the x-coordinate given by $x = a$, where a is some real number. Any other point on L has the form $(a, \bar{y})$, where $\bar{y}$ is an appropriate number. Therefore, the vertical line L is described by the sole condition

$$x = a$$

and this is, accordingly, the equation of L. For example, the equation $x = -2$ represents a vertical line 2 units to the left of the y-axis, and the equation $x = 3$ represents a vertical line 3 units to the right of the y-axis (Figure 25).

Next, suppose L is a nonvertical line so that it has a well-defined slope m. Suppose (x_1, y_1) is a fixed point lying on L and (x, y) is a variable point on L distinct from (x_1, y_1) (Figure 26).

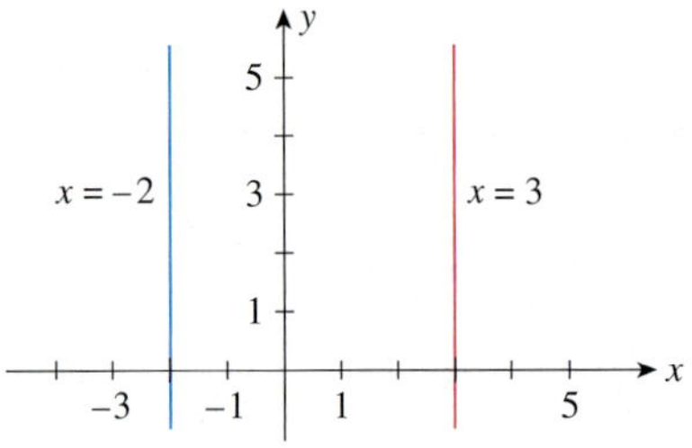

FIGURE 25
The vertical lines $x = -2$ and $x = 3$

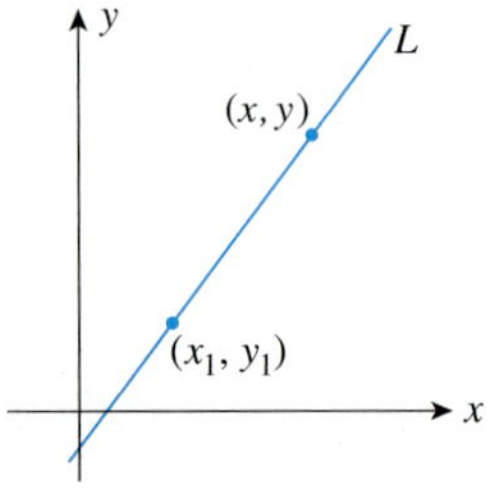

FIGURE 26
L passes through (x_1, y_1) and has slope m.

Using Formula (3) with the point $(x_2, y_2) = (x, y)$, we find that the slope of L is given by

$$m = \frac{y - y_1}{x - x_1}$$

Upon multiplying both sides of the equation by $x - x_1$, we obtain Formula (4).

Point-Slope Form

An equation of the line that has slope m and passes through the point (x_1, y_1) is given by

$$y - y_1 = m(x - x_1) \qquad \textbf{(4)}$$

Equation (4) is called the **point-slope form of the equation of a line** since it utilizes a given point (x_1, y_1) on a line and the slope m of the line.

EXAMPLE 5 Find an equation of the line that passes through the point (1, 3) and has slope 2.

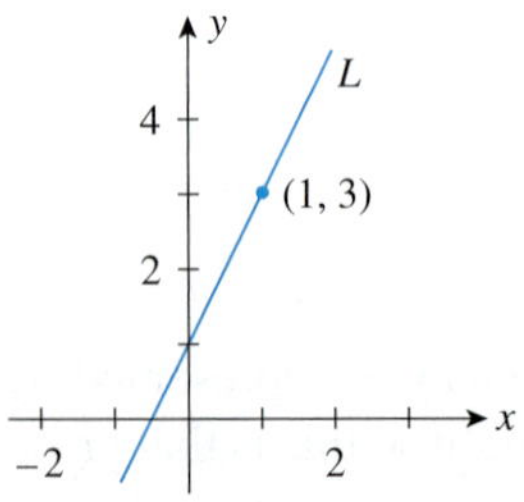

FIGURE 27
L passes through (1, 3) and has slope 2.

Solution Using the point-slope form of the equation of a line with the point (1, 3) and $m = 2$, we obtain

$$y - 3 = 2(x - 1) \qquad y - y_1 = m(x - x_1)$$

which, when simplified, becomes

$$2x - y + 1 = 0$$

(Figure 27). ■

VIDEO

EXAMPLE 6 Find an equation of the line that passes through the points $(-3, 2)$ and $(4, -1)$.

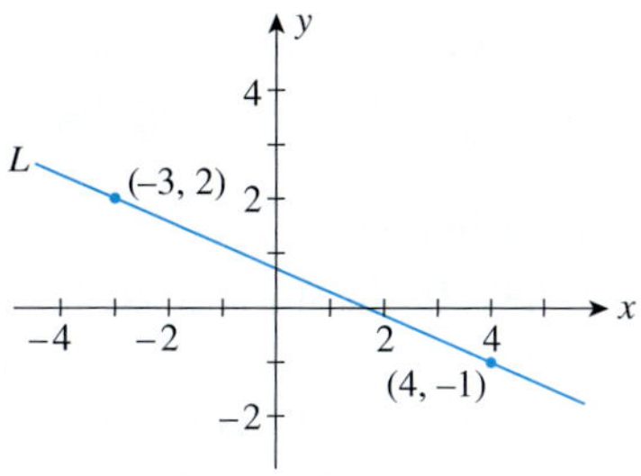

FIGURE 28
L passes through $(-3, 2)$ and $(4, -1)$.

Solution The slope of the line is given by

$$m = \frac{-1 - 2}{4 - (-3)} = -\frac{3}{7}$$

Using the point-slope form of an equation of a line with the point $(4, -1)$ and the slope $m = -\frac{3}{7}$, we have

$$y + 1 = -\frac{3}{7}(x - 4) \qquad y - y_1 = m(x - x_1)$$

$$7y + 7 = -3x + 12$$

$$3x + 7y - 5 = 0$$

(Figure 28). ■

We can use the slope of a straight line to determine whether a line is perpendicular to another line.

Perpendicular Lines

If L_1 and L_2 are two distinct nonvertical lines that have slopes m_1 and m_2, respectively, then L_1 is **perpendicular** to L_2 (written $L_1 \perp L_2$) if and only if

$$m_1 = -\frac{1}{m_2}$$

If the line L_1 is vertical (so that its slope is undefined), then L_1 is perpendicular to another line, L_2, if and only if L_2 is horizontal (so that its slope is zero). For a proof of these results, see Exercise 84, page 45.

EXAMPLE 7 Find an equation of the line that passes through the point (3, 1) and is perpendicular to the line of Example 5.

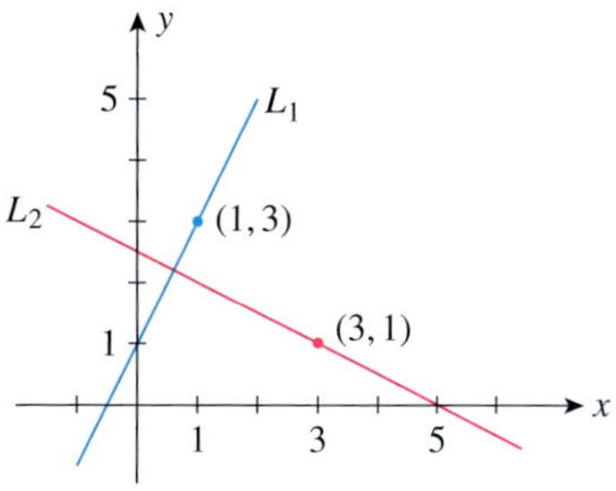

FIGURE 29
L_2 is perpendicular to L_1 and passes through (3, 1).

Solution Since the slope of the line in Example 5 is 2, the slope of the required line is given by $m = -\frac{1}{2}$, the negative reciprocal of 2. Using the point-slope form of the equation of a line, we obtain

$$y - 1 = -\frac{1}{2}(x - 3) \qquad y - y_1 = m(x - x_1)$$

$$2y - 2 = -x + 3$$

$$x + 2y - 5 = 0$$

(Figure 29). ■

A straight line L that is neither horizontal nor vertical cuts the x-axis and the y-axis at, say, points $(a, 0)$ and $(0, b)$, respectively (Figure 30). The numbers a and b are called the **x-intercept** and **y-intercept,** respectively, of L.

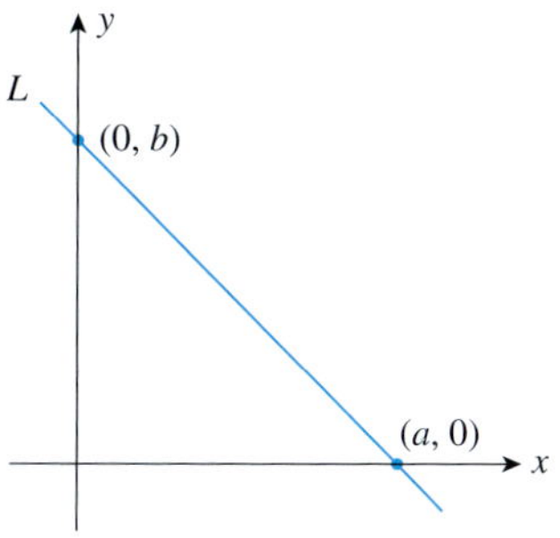

FIGURE 30
The line L has x-intercept a and y-intercept b.

Exploring with TECHNOLOGY

1. Use a graphing utility to plot the straight lines L_1 and L_2 with equations $2x + y - 5 = 0$ and $41x + 20y - 11 = 0$ on the same set of axes, using the standard viewing window.
   a. Can you tell if the lines L_1 and L_2 are parallel to each other?
   b. Verify your observations by computing the slopes of L_1 and L_2 algebraically.
2. Use a graphing utility to plot the straight lines L_1 and L_2 with equations $x + 2y - 5 = 0$ and $5x - y + 5 = 0$ on the same set of axes, using the standard viewing window.
   a. Can you tell if the lines L_1 and L_2 are perpendicular to each other?
   b. Verify your observation by computing the slopes of L_1 and L_2 algebraically.

Now, let L be a line with slope m and y-intercept b. Using Formula (4), the point-slope form of the equation of a line, with the point $(0, b)$ and slope m, we have

$$y - b = m(x - 0)$$
$$y = mx + b$$

Slope-Intercept Form

An equation of the line that has slope m and intersects the y-axis at the point $(0, b)$ is given by

$$y = mx + b \tag{5}$$

EXAMPLE 8 Find an equation of the line that has slope 3 and y-intercept -4.

Solution Using Equation (5) with $m = 3$ and $b = -4$, we obtain the required equation

$$y = 3x - 4$$ ■

EXAMPLE 9 Determine the slope and y-intercept of the line whose equation is $3x - 4y = 8$.

Solution Rewrite the given equation in the slope-intercept form. Thus,

$$3x - 4y = 8$$
$$-4y = 8 - 3x$$
$$y = \frac{3}{4}x - 2$$

Comparing this result with Equation (5), we find $m = \frac{3}{4}$ and $b = -2$, and we conclude that the slope and y-intercept of the given line are $\frac{3}{4}$ and -2, respectively. ■

Explore & Discuss

Consider the slope-intercept form of an equation of a straight line $y = mx + b$. Describe the family of straight lines obtained by keeping

1. The value of m fixed and allowing the value of b to vary.
2. The value of b fixed and allowing the value of m to vary.

Exploring with TECHNOLOGY

1. Use a graphing utility to plot the straight lines with equations $y = -2x + 3$, $y = -x + 3$, $y = x + 3$, and $y = 2.5x + 3$ on the same set of axes, using the standard viewing window. What effect does changing the coefficient m of x in the equation $y = mx + b$ have on its graph?
2. Use a graphing utility to plot the straight lines with equations $y = 2x - 2$, $y = 2x - 1$, $y = 2x$, $y = 2x + 1$, and $y = 2x + 4$ on the same set of axes, using the standard viewing window. What effect does changing the constant b in the equation $y = mx + b$ have on its graph?
3. Describe in words the effect of changing both m and b in the equation $y = mx + b$.

VIDEO

APPLIED EXAMPLE 10 Sales of a Sporting Goods Store The sales manager of a local sporting goods store plotted sales versus time for the last 5 years and found the points to lie approximately along a straight line (Figure 31). By using the points corresponding to the first and fifth years, find an equation of the trend line. What sales figure can be predicted for the sixth year?

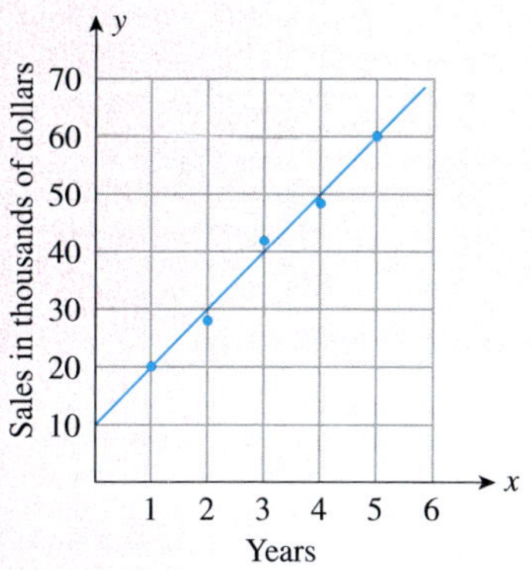

FIGURE 31
Sales of a sporting goods store

Solution Using Formula (3) with the points (1, 20) and (5, 60), we find that the slope of the required line is given by

$$m = \frac{60 - 20}{5 - 1} = 10$$

Next, using the point-slope form of the equation of a line with the point (1, 20) and $m = 10$, we obtain

$$y - 20 = 10(x - 1) \qquad y - y_1 = m(x - x_1)$$
$$y = 10x + 10$$

as the required equation.

The sales figure for the sixth year is obtained by letting $x = 6$ in the last equation, giving

$$y = 10(6) + 10 = 70$$

or $70,000. ■

APPLIED EXAMPLE 11 Appreciation in Value of an Art Object Suppose an art object purchased for $50,000 is expected to appreciate in value at a constant rate of $5000 per year for the next 5 years. Use Formula (5) to write an equation predicting the value of the art object in the next several years. What will be its value 3 years from the date of purchase?

Solution Let x denote the time (in years) that has elapsed since the date the object was purchased and let y denote the object's value (in dollars). Then, $y = 50{,}000$ when $x = 0$. Furthermore, the slope of the required equation is given by $m = 5000$, since each unit increase in x (1 year) implies an increase of 5000 units (dollars) in y. Using (5) with $m = 5000$ and $b = 50{,}000$, we obtain

$$y = 5000x + 50{,}000 \qquad y = mx + b$$

Three years from the date of purchase, the value of the object will be given by

$$y = 5000(3) + 50{,}000$$

or $65,000. ■

Explore & Discuss

Refer to Example 11. Can the equation predicting the value of the art object be used to predict long-term growth?

General Form of an Equation of a Line

We have considered several forms of an equation of a straight line in the plane. These different forms of the equation are equivalent to each other. In fact, each is a special case of the following equation.

General Form of a Linear Equation

The equation

$$Ax + By + C = 0 \tag{6}$$

where A, B, and C are constants and A and B are not both zero, is called the general form of a linear equation in the variables x and y.

We will now state (without proof) an important result concerning the algebraic representation of straight lines in the plane.

THEOREM 1

An equation of a straight line is a linear equation; conversely, every linear equation represents a straight line.

This result justifies the use of the adjective *linear* describing Equation (6).

EXAMPLE 12 Sketch the straight line represented by the equation

$$3x - 4y - 12 = 0$$

Solution Since every straight line is uniquely determined by two distinct points, we need find only two such points through which the line passes in order to sketch it. For convenience, let's compute the x- and y-intercepts. Setting $y = 0$, we find $x = 4$; thus, the x-intercept is 4. Setting $x = 0$ gives $y = -3$, and the y-intercept is -3. A sketch of the line appears in Figure 32. ■

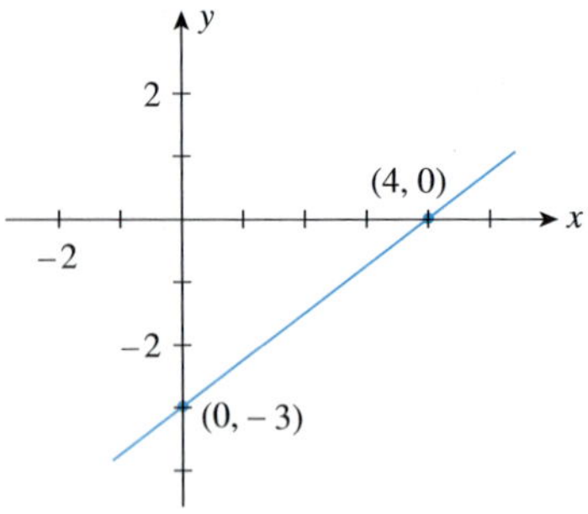

FIGURE 32
To sketch $3x - 4y - 12 = 0$, first find the x-intercept, 4, and the y-intercept, -3.

Following is a summary of the common forms of the equations of straight lines discussed in this section.

Equations of Straight Lines

Vertical line:	$x = a$
Horizontal line:	$y = b$
Point-slope form:	$y - y_1 = m(x - x_1)$
Slope-intercept form:	$y = mx + b$
General form:	$Ax + By + C = 0$

1.4 Self-Check Exercises

1. Determine the number a so that the line passing through the points $(a, 2)$ and $(3, 6)$ is parallel to a line with slope 4.

2. Find an equation of the line that passes through the point $(3, -1)$ and is perpendicular to a line with slope $-\frac{1}{2}$.

3. Does the point $(3, -3)$ lie on the line with equation $2x - 3y - 12 = 0$? Sketch the graph of the line.

4. **Satellite TV Subscribers** The following table gives the number of satellite TV subscribers in the United States (in millions) from 1998 through 2005 ($x = 0$ corresponds to 1998).

Year, x	0	1	2	3	4	5	6	7
Number, y	8.5	11.1	15.0	17.0	18.9	21.5	24.8	27.4

a. Plot the number of satellite TV subscribers in the United State (y) versus the year (x).
b. Draw the line L through the points $(0, 8.5)$ and $(7, 27.4)$.
c. Find an equation of the line L.
d. Assuming that this trend continues, estimate the number of satellite TV subscribers in the United States in 2006.

Sources: National Cable & Telecommunications Association, Federal Communications Commission.

Solutions to Self-Check Exercises 1.4 can be found on page 45.

1.4 Concept Questions

1. What is the slope of a nonvertical line? What can you say about the slope of a vertical line?

2. Give (a) the point-slope form, (b) the slope-intercept form, and (c) the general form of the equation of a line.

3. Let L_1 have slope m_1 and L_2 have slope m_2. State the conditions on m_1 and m_2 if (a) L_1 is parallel to L_2 and (b) L_1 is perpendicular to L_2.

1.4 Exercises

In Exercises 1–6, match the statement with one of the graphs (a)–(f).

1. The slope of the line is zero.
2. The slope of the line is undefined.
3. The slope of the line is positive, and its y-intercept is positive.
4. The slope of the line is positive, and its y-intercept is negative.
5. The slope of the line is negative, and its x-intercept is negative.
6. The slope of the line is negative, and its x-intercept is positive.

(a)

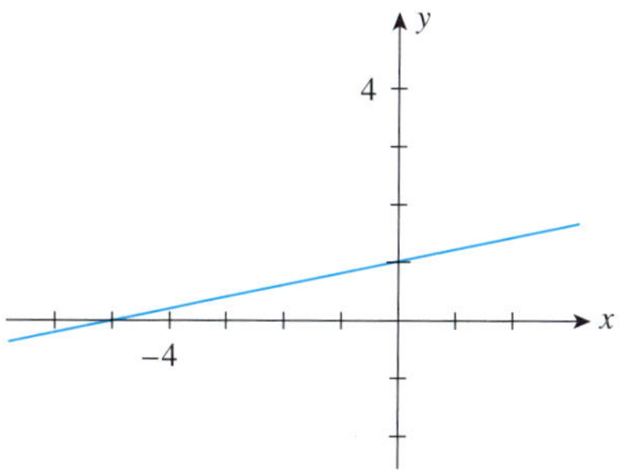

(b)

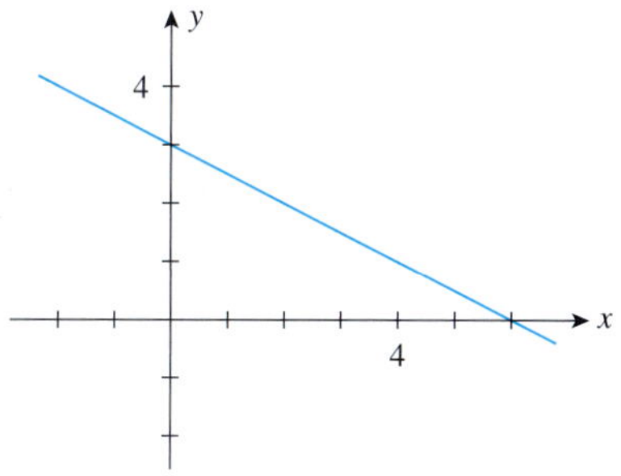

(c)

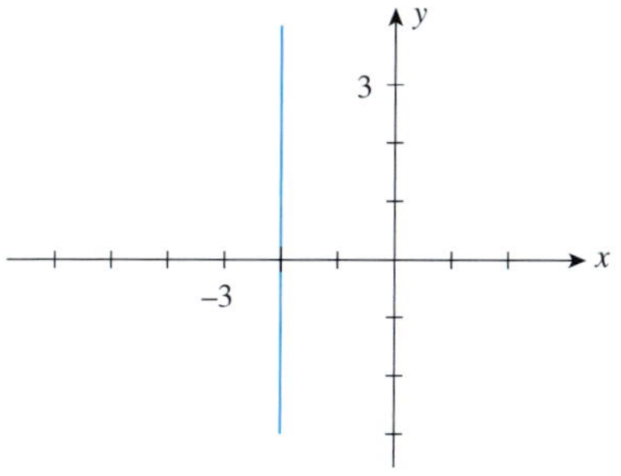

(d)

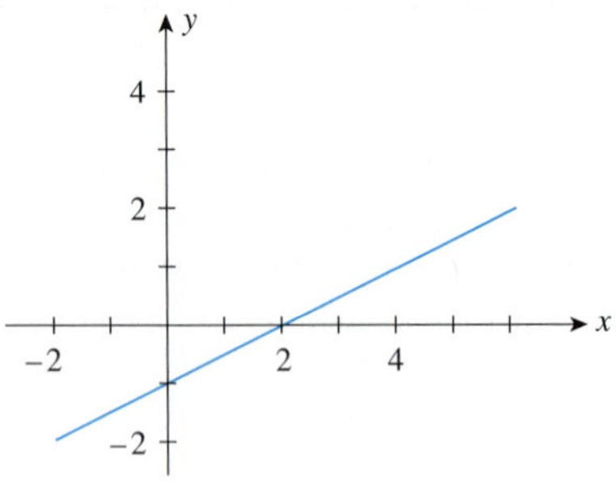

(e)

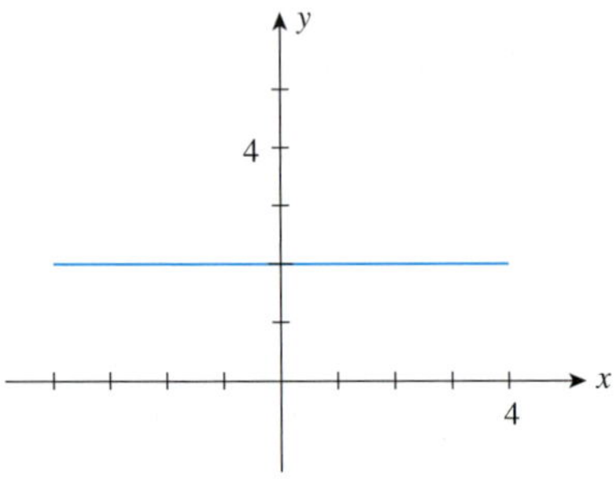

(f)

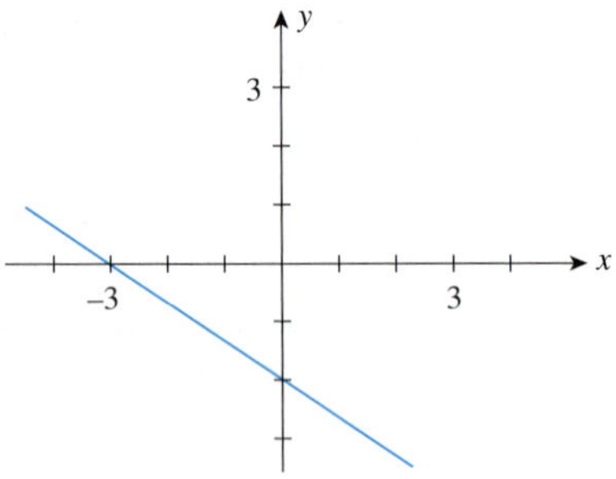

In Exercises 7–10, find the slope of the line shown in each figure.

7.

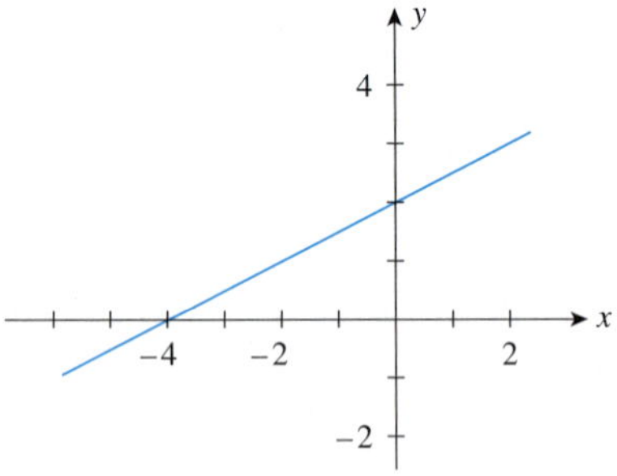

8.

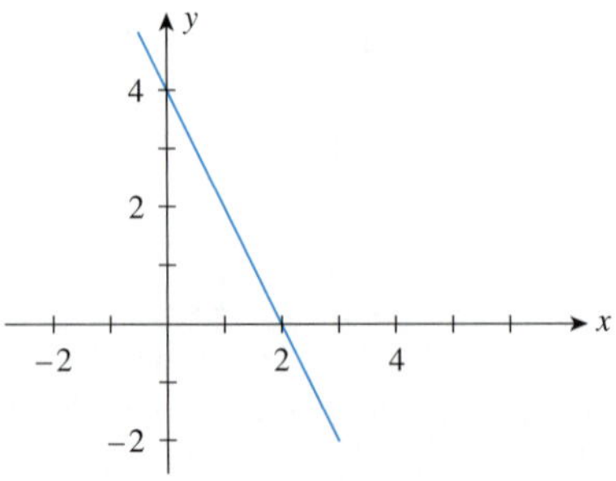

9.

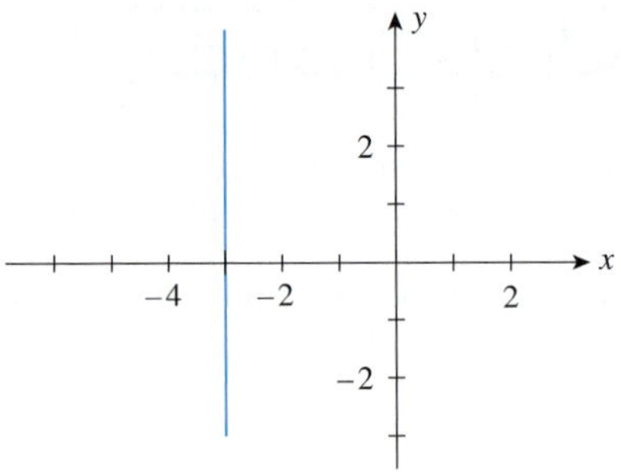

10.

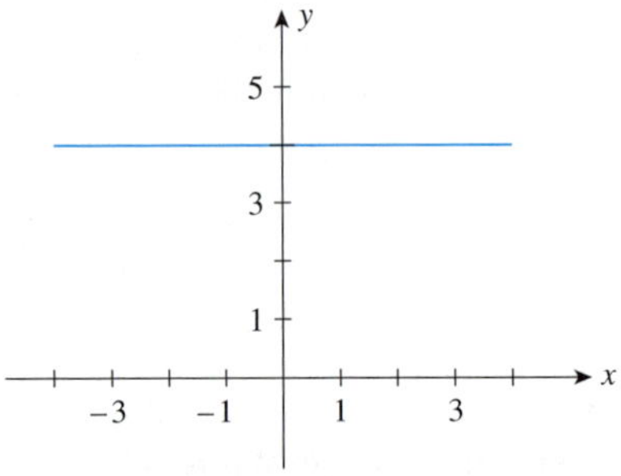

In Exercises 11–16, find the slope of the line that passes through each pair of points.

11. (4, 3) and (5, 8)

12. (4, 5) and (3, 8)

13. (−2, 3) and (4, 8)

14. (−2, −2) and (4, −4)

15. (a, b) and (c, d)

16. $(-a + 1, b - 1)$ and $(a + 1, -b)$

17. Given the equation $y = 4x - 3$, answer the following questions:

a. If x increases by 1 unit, what is the corresponding change in y?

b. If x decreases by 2 units, what is the corresponding change in y?

18. Given the equation $2x + 3y = 4$, answer the following questions:

a. Is the slope of the line described by this equation positive or negative?

b. As x increases in value, does y increase or decrease?

c. If x decreases by 2 units, what is the corresponding change in y?

In Exercises 19 and 20, determine whether the line through each pair of points is parallel.

19. $A(1, -2)$, $B(-3, -10)$ and $C(1, 5)$, $D(-1, 1)$

20. $A(2, 3)$, $B(2, -2)$ and $C(-2, 4)$, $D(-2, 5)$

In Exercises 21 and 22, determine whether the lines through each pair of points are perpendicular.

21. $A(-2, 5)$, $B(4, 2)$ and $C(-1, -2)$, $D(3, 6)$

22. $A(2, 0)$, $B(1, -2)$ and $C(4, 2)$, $D(-8, 4)$

23. If the line passing through the points $(1, a)$ and $(4, -2)$ is parallel to the line passing through the points $(2, 8)$ and $(-7, a + 4)$, what is the value of a?

24. If the line passing through the points $(a, 1)$ and $(5, 8)$ is parallel to the line passing through the points $(4, 9)$ and $(a + 2, 1)$, what is the value of a?

25. Find an equation of the horizontal line that passes through $(-4, -3)$.

26. Find an equation of the vertical line that passes through $(0, 5)$.

In Exercises 27–30, find an equation of the line that passes through the point and has the indicated slope m.

27. $(3, -4)$; $m = 2$

28. $(2, 4)$; $m = -1$

29. $(-3, 2)$; $m = 0$

30. $(1, 2)$; $m = -\frac{1}{2}$

In Exercises 31–34, find an equation of the line that passes through the points.

31. $(2, 4)$ and $(3, 7)$

32. $(2, 1)$ and $(2, 5)$

33. $(1, 2)$ and $(-3, -2)$

34. $(-1, -2)$ and $(3, -4)$

In Exercises 35–38, find an equation of the line that has slope m and y-intercept b.

35. $m = 3$; $b = 4$

36. $m = -2$; $b = -1$

37. $m = 0$; $b = 5$

38. $m = -\frac{1}{2}$; $b = \frac{3}{4}$

In Exercises 39–44, write the equation in the slope-intercept form and then find the slope and y-intercept of the corresponding line.

39. $x - 2y = 0$

40. $y - 2 = 0$

41. $2x - 3y - 9 = 0$

42. $3x - 4y + 8 = 0$

43. $2x + 4y = 14$

44. $5x + 8y - 24 = 0$

45. Find an equation of the line that passes through the point $(-2, 2)$ and is parallel to the line $2x - 4y - 8 = 0$.

46. Find an equation of the line that passes through the point $(2, 4)$ and is perpendicular to the line $3x + 4y - 22 = 0$.

In Exercises 47–52, find an equation of the line that satisfies the given condition.

47. The line parallel to the x-axis and 6 units below it

48. The line passing through the origin and parallel to the line joining the points $(2, 4)$ and $(4, 7)$

49. The line passing through the point (a, b) with slope equal to zero

50. The line passing through $(-3, 4)$ and parallel to the x-axis

51. The line passing through $(-5, -4)$ and parallel to the line joining $(-3, 2)$ and $(6, 8)$

52. The line passing through (a, b) with undefined slope

53. Given that the point $P(-3, 5)$ lies on the line $kx + 3y + 9 = 0$, find k.

54. Given that the point $P(2, -3)$ lies on the line $-2x + ky + 10 = 0$, find k.

In Exercises 55–60, sketch the straight line defined by the given linear equation by finding the x- and y-intercepts.

Hint: See Example 12, page 40.

55. $3x - 2y + 6 = 0$

56. $2x - 5y + 10 = 0$

57. $x + 2y - 4 = 0$

58. $2x + 3y - 15 = 0$

59. $y + 5 = 0$

60. $-2x - 8y + 24 = 0$

61. Show that an equation of a line through the points $(a, 0)$ and $(0, b)$ with $a \neq 0$ and $b \neq 0$ can be written in the form

$$\frac{x}{a} + \frac{y}{b} = 1$$

(Recall that the numbers a and b are the x- and y-intercepts, respectively, of thc line. This form of an equation of a line is called the *intercept form*.)

In Exercises 62–65, use the results of Exercise 61 to find an equation of a line with the given x- and y-intercepts.

62. x-intercept 3; y-intercept 4

63. x-intercept -2; y-intercept -4

64. x-intercept $-\frac{1}{2}$; y-intercept $\frac{3}{4}$

65. x-intercept 4; y-intercept $-\frac{1}{2}$

In Exercises 66 and 67, determine whether the given points lie on a straight line.

66. $A(-1, 7)$, $B(2, -2)$, and $C(5, -9)$

67. $A(-2, 1)$, $B(1, 7)$, and $C(4, 13)$

68. **Temperature Conversion** The relationship between the temperature in degrees Fahrenheit (°F) and the temperature in degrees Celsius (°C) is

$$F = \frac{9}{5}C + 32$$

a. Sketch the line with the given equation.
b. What is the slope of the line? What does it represent?
c. What is the F-intercept of the line? What does it represent?

69. **Nuclear Plant Utilization** The United States is not building many nuclear plants, but the ones it has are running full tilt. The output (as a percent of total capacity) of nuclear plants is described by the equation

$$y = 1.9467t + 70.082$$

where t is measured in years, with $t = 0$ corresponding to the beginning of 1990.
  a. Sketch the line with the given equation.
  b. What are the slope and the y-intercept of the line found in part (a)?
  c. Give an interpretation of the slope and the y-intercept of the line found in part (a).
  d. If the utilization of nuclear power continues to grow at the same rate and the total capacity of nuclear plants in the United States remains constant, by what year can the plants be expected to be generating at maximum capacity?

Source: Nuclear Energy Institute

70. **Social Security Contributions** For wages less than the maximum taxable wage base, Social Security contributions by employees are 7.65% of the employee's wages.
  a. Find an equation that expresses the relationship between the wages earned (x) and the Social Security taxes paid (y) by an employee who earns less than the maximum taxable wage base.
  b. For each additional dollar that an employee earns, by how much is his or her Social Security contribution increased? (Assume that the employee's wages are less than the maximum taxable wage base.)
  c. What Social Security contributions will an employee who earns \$65,000 (which is less than the maximum taxable wage base) be required to make?

Source: Social Security Administration

71. **College Admissions** Using data compiled by the Admissions Office at Faber University, college admissions officers estimate that 55% of the students who are offered admission to the freshman class at the university will actually enroll.
  a. Find an equation that expresses the relationship between the number of students who actually enroll (y) and the number of students who are offered admission to the university (x).
  b. If the desired freshman class size for the upcoming academic year is 1100 students, how many students should be admitted?

72. **Weight of Whales** The equation $W = 3.51L - 192$, expressing the relationship between the length L (in feet) and the expected weight W (in British tons) of adult blue whales, was adopted in the late 1960s by the International Whaling Commission.
  a. What is the expected weight of an 80-ft blue whale?
  b. Sketch the straight line that represents the equation.

73. **The Narrowing Gender Gap** Since the founding of the Equal Employment Opportunity Commission and the passage of equal-pay laws, the gulf between men's and women's earnings has continued to close gradually. At the beginning of 1990 ($t = 0$), women's wages were 68% of men's wages, and by the beginning of 2000 ($t = 10$), women's wages were projected to be 80% of men's wages. If this gap between women's and men's wages continued to narrow *linearly*, what percentage of men's wages are women's wages expected to be at the beginning of 2008?

Source: Journal of Economic Perspectives

74. **Sales Growth** Metro Department Store's annual sales (in millions of dollars) during the past 5 yr were

Annual Sales, y	5.8	6.2	7.2	8.4	9.0
Year, x	1	2	3	4	5

  a. Plot the annual sales (y) versus the year (x).
  b. Draw a straight line L through the points corresponding to the first and fifth years.
  c. Derive an equation of the line L.
  d. Using the equation found in part (c), estimate Metro's annual sales 4 yr from now ($x = 9$).

75. **Sales of GPS Equipment** The annual sales (in billions of dollars) of global positioning system (GPS) equipment from the year 2000 through 2006 follow. (Sales for 2004–2006 were projections.) Here $x = 0$ corresponds to 2000.

Annual Sales, y	7.9	9.6	11.5	13.3	15.2	17	18.8
Year, x	0	1	2	3	4	5	6

  a. Plot the annual sales (y) versus the year (x).
  b. Draw a straight line L through the points corresponding to 2000 and 2006.
  c. Derive an equation of the line L.
  d. Use the equation found in part (c) to estimate the annual sales of GPS equipment for 2005. Compare this figure with the projected sales for that year.

Source: ABI Research

76. **Digital TV Services** The percentage of homes with digital TV services, which stood at 5% at the beginning of 1999 ($t = 0$) was projected to grow linearly so that at the beginning of 2003 ($t = 4$) the percentage of such homes was projected to be 25%.
  a. Derive an equation of the line passing through the points $A(0, 5)$ and $B(4, 25)$.
  b. Plot the line with the equation found in part (a).
  c. Using the equation found in part (a), find the percentage of homes with digital TV services at the beginning of 2001.

Source: Paul Kagan Associates

In Exercises 77–81, determine whether the statement is true or false. If it is true, explain why it is true. If it is false, give an example to show why it is false.

77. Suppose the slope of a line L is $-\frac{1}{2}$ and P is a given point on L. If Q is the point on L lying 4 units to the left of P, then Q is situated 2 units above P.

78. The line with equation $Ax + By + C = 0$, $(B \neq 0)$, and the line with equation $ax + by + c = 0$, $(b \neq 0)$, are parallel if $Ab - aB = 0$.

79. If the slope of the line L_1 is positive, then the slope of a line L_2 perpendicular to L_1 may be positive or negative.

80. The lines with equations $ax + by + c_1 = 0$ and $bx - ay + c_2 = 0$, where $a \neq 0$ and $b \neq 0$, are perpendicular to each other.

81. If L is the line with equation $Ax + By + C = 0$, where $A \neq 0$, then L crosses the x-axis at the point $(-C/A, 0)$.

82. Is there a difference between the statements "The slope of a straight line is zero" and "The slope of a straight line does not exist (is not defined)"? Explain your answer.

83. Show that two distinct lines with equations $a_1x + b_1y + c_1 = 0$ and $a_2x + b_2y + c_2 = 0$, respectively, are parallel if and only if $a_1b_2 - b_1a_2 = 0$.
Hint: Write each equation in the slope-intercept form and compare.

84. Prove that if a line L_1 with slope m_1 is perpendicular to a line L_2 with slope m_2, then $m_1m_2 = -1$.
Hint: Refer to the following figure. Show that $m_1 = b$ and $m_2 = c$. Next, apply the Pythagorean theorem to triangles OAC, OCB, and OBA to show that $1 = -bc$.

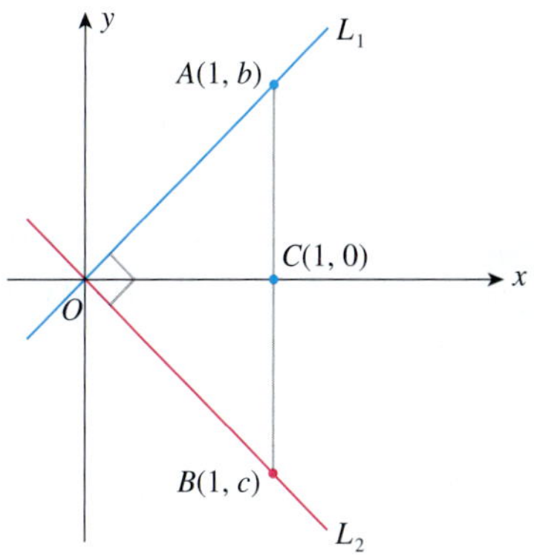

1.4 Solutions to Self-Check Exercises

1. The slope of the line that passes through the points $(a, 2)$ and $(3, 6)$ is

$$m = \frac{6 - 2}{3 - a} = \frac{4}{3 - a}$$

Since this line is parallel to a line with slope 4, m must be equal to 4; that is,

$$\frac{4}{3 - a} = 4$$

or, upon multiplying both sides of the equation by $3 - a$,

$$\begin{aligned} 4 &= 4(3 - a) \\ 4 &= 12 - 4a \\ 4a &= 8 \\ a &= 2 \end{aligned}$$

2. Since the required line L is perpendicular to a line with slope $-\frac{1}{2}$, the slope of L is

$$m = \frac{-1}{-\frac{1}{2}} = 2$$

Next, using the point-slope form of the equation of a line, we have

$$\begin{aligned} y - (-1) &= 2(x - 3) \\ y + 1 &= 2x - 6 \\ y &= 2x - 7 \end{aligned}$$

3. Substituting $x = 3$ and $y = -3$ into the left-hand side of the given equation, we find

$$2(3) - 3(-3) - 12 = 3$$

which is not equal to zero (the right-hand side). Therefore, $(3, -3)$ does not lie on the line with equation $2x - 3y - 12 = 0$. (See the accompanying figure.)

Setting $x = 0$, we find $y = -4$, the y-intercept. Next, setting $y = 0$ gives $x = 6$, the x-intercept. We now draw the line passing through the points $(0, -4)$ and $(6, 0)$ as shown.

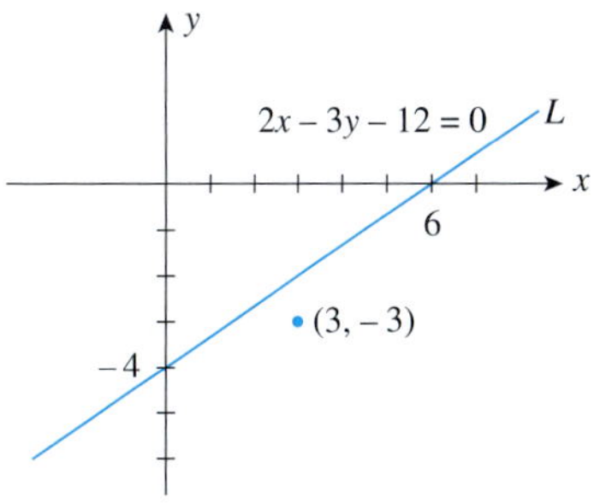

4. **a** and **b.** See the accompanying figure.

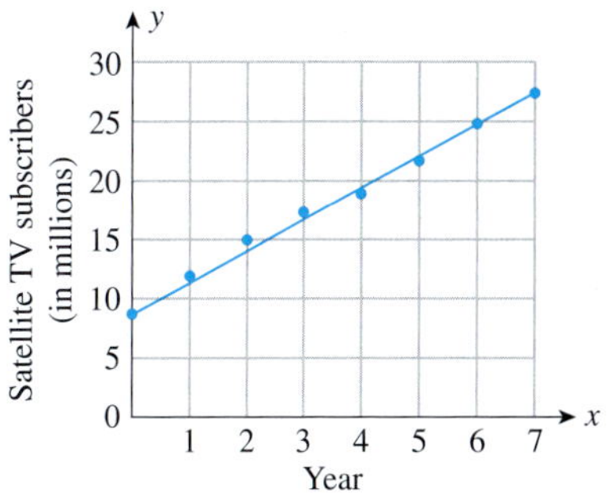

c. The slope of L is

$$m = \frac{27.4 - 8.5}{7 - 0} = 2.7$$

Using the point-slope form of the equation of a line with the point (0, 8.5), we find

$$y - 8.5 = 2.7(x - 0)$$
$$y = 2.7x + 8.5$$

d. The estimated number of satellite TV subscribers in the United States in 2006 is

$$y = 2.7(8) + 8.5 = 30.1$$

or 30.1 million.

CHAPTER 1 Summary of Principal Formulas and Terms

FORMULAS

1. Quadratic formula	$x = \dfrac{-b \pm \sqrt{b^2 - 4ac}}{2a}$
2. Distance between two points	$d = \sqrt{(x_2 - x_1)^2 + (y_2 - y_1)^2}$
3. Equation of a circle with center $C(h, k)$ and radius r	$(x - h)^2 + (y - k)^2 = r^2$
4. Slope of a line	$m = \dfrac{y_2 - y_1}{x_2 - x_1}$
5. Equation of a vertical line	$x = a$
6. Equation of a horizontal line	$y = b$
7. Point-slope form of the equation of a line	$y - y_1 = m(x - x_1)$
8. Slope-intercept form of the equation of a line	$y = mx + b$
9. General equation of a line	$Ax + By + C = 0$

TERMS

real number (coordinate) line (3)
open interval (4)
closed interval (4)
half-open interval (4)
finite interval (4)
infinite interval (4)
polynomial (8)
roots of a polynomial equation (12)
absolute value (21)
triangle inequality (22)
Cartesian coordinate system (25)
ordered pair (25)
parallel lines (35)
perpendicular lines (37)

CHAPTER 1 Concept Review Questions

Fill in the blanks.

1. A point in the plane can be represented uniquely by a/an _____ pair of numbers. The first number of the pair is called the _____, and the second number of the pair is called the _____.

2. **a.** The point $P(a, 0)$ lies on the _____-axis, and the point $P(0, b)$ lies on the _____-axis.
   b. If the point $P(a, b)$ lies in the fourth quadrant, then the point $P(-a, b)$ lies in the _____ quadrant.

3. The distance between two points $P(a, b)$ and $P(c, d)$ is _____.

4. An equation of a circle with center $C(a, b)$ and radius r is given by _____.

5. **a.** If $P_1(x_1, y_1)$ and $P_2(x_2, y_2)$ are any two distinct points on a nonvertical line L, then the slope of L is $m =$ _____.
   b. The slope of a vertical line is _____.
   c. The slope of a horizontal line is _____.
   d. The slope of a line that slants upward is _____.

6. If L_1 and L_2 are nonvertical lines with slopes m_1 and m_2, respectively, then L_1 is parallel to L_2 if and only if _____ and L_1 is perpendicular to L_2 if and only if _____.

7. **a.** An equation of the line passing through the point $P(x_1, y_1)$ and having slope m is _____. This form of the equation of a line is called the _____ _____.
   b. An equation of the line that has slope m and y-intercept b is _____. It is called the _____ form of an equation of a line.

8. **a.** The general form of an equation of a line is _____.
   b. If a line has equation $ax + by + c = 0$ $(b \neq 0)$, then its slope is _____.

CHAPTER 1 Review Exercises

In Exercises 1–4, find the values of *x* that satisfy the inequality (inequalities).

1. $-x + 3 \le 2x + 9$

2. $-2 \le 3x + 1 \le 7$

3. $x - 3 > 2$ or $x + 3 < -1$

4. $2x^2 > 50$

In Exercises 5–8, evaluate the expression.

5. $|-5 + 7| + |-2|$

6. $\left|\dfrac{5 - 12}{-4 - 3}\right|$

7. $|2\pi - 6| - \pi$

8. $|\sqrt{3} - 4| + |4 - 2\sqrt{3}|$

In Exercises 9–14, evaluate the expression.

9. $\left(\dfrac{9}{4}\right)^{3/2}$

10. $\dfrac{5^6}{5^4}$

11. $(3 \cdot 4)^{-2}$

12. $(-8)^{5/3}$

13. $\dfrac{(3 \cdot 2^{-3})(4 \cdot 3^5)}{2 \cdot 9^3}$

14. $\dfrac{3\sqrt[3]{54}}{\sqrt[3]{18}}$

In Exercises 15–20, simplify the expression.

15. $\dfrac{4(x^2 + y)^3}{x^2 + y}$

16. $\dfrac{a^6 b^{-5}}{(a^3 b^{-2})^{-3}}$

17. $\dfrac{\sqrt[4]{16x^5yz}}{\sqrt[4]{81xyz^5}}$

18. $(2x^3)(-3x^{-2})\left(\dfrac{1}{6}x^{-1/2}\right)$

19. $\left(\dfrac{3xy^2}{4x^3y}\right)^{-2}\left(\dfrac{3xy^3}{2x^2}\right)^3$

20. $\sqrt[3]{81x^5y^{10}}\sqrt[3]{9xy^2}$

In Exercises 21–24, factor each expression completely.

21. $-2\pi^2r^3 + 100\pi r^2$

22. $2v^3w + 2vw^3 + 2u^2vw$

23. $16 - x^2$

24. $12t^3 - 6t^2 - 18t$

In Exercises 25–28, solve the equation by factoring.

25. $8x^2 + 2x - 3 = 0$

26. $-6x^2 - 10x + 4 = 0$

27. $-x^3 - 2x^2 + 3x = 0$

28. $2x^4 + x^2 = 1$

In Exercises 29–32, find the value(s) of *x* that satisfy the expression.

29. $2x^2 + 3x - 2 \le 0$

30. $\dfrac{1}{x + 2} > 2$

31. $|2x - 3| < 5$

32. $\left|\dfrac{x + 1}{x - 1}\right| = 5$

In Exercises 33 and 34, use the quadratic formula to solve the quadratic equation.

33. $x^2 - 2x - 5 = 0$

34. $2x^2 + 8x + 7 = 0$

In Exercises 35–38, perform the indicated operations and simplify the expression.

35. $\dfrac{(t + 6)(60) - (60t + 180)}{(t + 6)^2}$

36. $\dfrac{6x}{2(3x^2 + 2)} + \dfrac{1}{4(x + 2)}$

37. $\dfrac{2}{3}\left(\dfrac{4x}{2x^2 - 1}\right) + 3\left(\dfrac{3}{3x - 1}\right)$

38. $\dfrac{-2x}{\sqrt{x + 1}} + 4\sqrt{x + 1}$

39. Rationalize the numerator: $\dfrac{\sqrt{x} - 1}{x - 1}$.

40. Rationalize the denominator: $\dfrac{\sqrt{x} - 1}{2\sqrt{x}}$.

In Exercises 41 and 42, find the distance between the two points.

41. $(-2, -3)$ and $(1, -7)$

42. $\left(\dfrac{1}{2}, \sqrt{3}\right)$ and $\left(-\dfrac{1}{2}, 2\sqrt{3}\right)$

In Exercises 43–48, find an equation of the line *L* that passes through the point (−2, 4) and satisfies the condition.

43. L is a vertical line.

44. L is a horizontal line.

45. L passes through the point $\left(3, \dfrac{7}{2}\right)$.

46. The x-intercept of L is 3.

47. L is parallel to the line $5x - 2y = 6$.

48. L is perpendicular to the line $4x + 3y = 6$.

49. Find an equation of the straight line that passes through the point $(2, 3)$ and is parallel to the line with equation $3x + 4y - 8 = 0$.

50. Find an equation of the straight line that passes through the point $(-1, 3)$ and is parallel to the line passing through the points $(-3, 4)$ and $(2, 1)$.

51. Find an equation of the line that passes through the point $(-3, -2)$ and is parallel to the line passing through the points $(-2, -4)$ and $(1, 5)$.

52. Find an equation of the line that passes through the point $(-2, -4)$ and is perpendicular to the line with equation $2x - 3y - 24 = 0$.

53. Sketch the graph of the equation $3x - 4y = 24$.

54. Sketch the graph of the line that passes through the point $(3, 2)$ and has slope $-2/3$.

55. Find the minimum cost C (in dollars) given that

$$2(1.5C + 80) \leq 2(2.5C - 20)$$

56. Find the maximum revenue R (in dollars) given that

$$12(2R - 320) \leq 4(3R + 240)$$

57. **A Falling Stone** A stone is thrown straight up from the roof of an 80-ft building, and the height (in feet) of the stone any time t (in seconds), measured from the ground, is given by

$$-16t^2 + 64t + 80$$

Find the interval of time when the stone is at or greater than a height of 128 ft from the ground.

58. **Sales of Navigation Systems** The projected number of navigation systems (in millions) installed in vehicles in North America, Europe, and Japan from 2002 through 2006 follow. Here $x = 0$ corresponds to 2002.

Systems Installed, y	3.9	4.7	5.8	6.8	7.8
Year, x	0	1	2	3	4

a. Plot the annual sales (y) versus the year (x).
b. Draw a straight line L through the points corresponding to 2002 and 2006.
c. Derive an equation of the line L.
d. Use the equation found in part (c) to estimate the number of navigation systems installed for 2005. Compare this figure with the projected sales for that year.

Source: ABI Research

The problem-solving skills that you learn in each chapter are building blocks for the rest of the course. Therefore, it is a good idea to make sure that you have mastered these skills before moving on to the next chapter. The Before Moving On exercises that follow are designed for that purpose. After taking this test, you can see where your weaknesses, if any, are. Then you can log in at http://academic.cengage.com/login where you will find a link to our Companion Web site. Here, you can click on the Before Moving On button, which will lead you to other versions of these tests. There you can retest yourself on those exercises that you solved incorrectly. (You can also test yourself on these basic skills before taking your course quizzes and exams.)

If you feel that you need additional help with these exercises, at this Web site you can also use the *CengageNOW* tutorials, and get live online tutoring help with *Personal Tutor*.

CHAPTER 1 Before Moving On . . .

1. Evaluate:

a. $|\pi - 2\sqrt{3}| - |\sqrt{3} - \sqrt{2}|$ b. $\left[\left(-\frac{1}{3}\right)^{-3}\right]^{1/3}$

2. Simplify:

a. $\sqrt[3]{64x^6} \cdot \sqrt{9y^2x^6}$ b. $\left(\frac{a^{-3}}{b^{-4}}\right)^2\left(\frac{b}{a}\right)^{-3}$

3. Rationalize the denominator:

a. $\dfrac{2x}{3\sqrt{y}}$ b. $\dfrac{x}{\sqrt{x} - 4}$

4. Perform each operation and simplify:

a. $\dfrac{(x^2 + 1)(\frac{1}{2}x^{-1/2}) - x^{1/2}(2x)}{(x^2 + 1)^2}$

b. $-\dfrac{3x}{\sqrt{x + 2}} + 3\sqrt{x + 2}$

5. Rationalize the numerator: $\dfrac{\sqrt{x} + \sqrt{y}}{\sqrt{x} - \sqrt{y}}$.

6. Factor completely:

a. $12x^3 - 10x^2 - 12x$ b. $2bx - 2by + 3cx - 3cy$

7. Solve each equation:

a. $12x^2 - 9x - 3 = 0$ b. $3x^2 - 5x + 1 = 0$

8. Find the distance between $(-2, 4)$ and $(6, 8)$.

9. Find an equation of the line that passes through $(-1, -2)$ and $(4, 5)$.

10. Find an equation of the line that has slope $-\frac{1}{3}$ and y-intercept $\frac{4}{3}$.

FUNCTIONS, LIMITS, AND THE DERIVATIVE

2

IN THIS CHAPTER, we define a *function,* a special relationship between two variables. The concept of a function enables us to describe many relationships that exist in applications. We also begin the study of differential calculus. Historically, differential calculus was developed in response to the problem of finding the tangent line to an arbitrary curve. But it quickly became apparent that solving this problem provided mathematicians with a method for solving many practical problems involving the rate of change of one quantity with respect to another. The basic tool used in differential calculus is the *derivative* of a function. The concept of the derivative is based, in turn, on a more fundamental notion—that of the *limit* of a function.

Unless changes are made, when is the current Social Security system expected to go broke? In Example 3, page 79, we use a mathematical model constructed from data from the Social Security Administration to predict the year in which the assets of the current system will be depleted.

2.1 Functions and Their Graphs

Functions

A manufacturer would like to know how his company's profit is related to its production level; a biologist would like to know how the size of the population of a certain culture of bacteria will change over time; a psychologist would like to know the relationship between the learning time of an individual and the length of a vocabulary list; and a chemist would like to know how the initial speed of a chemical reaction is related to the amount of substrate used. In each instance, we are concerned with the same question: How does one quantity depend upon another? The relationship between two quantities is conveniently described in mathematics by using the concept of a function.

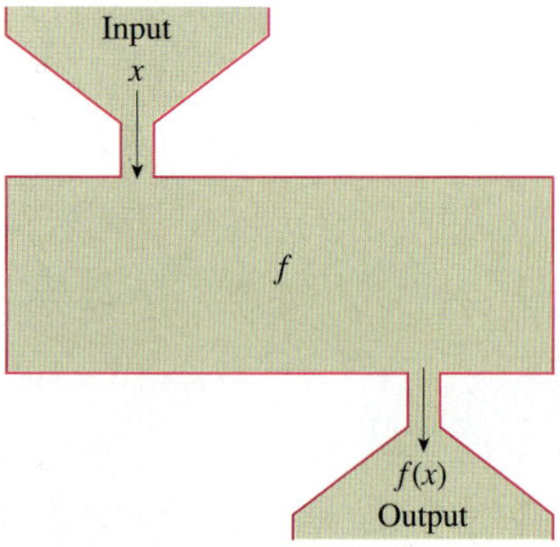

FIGURE 1
A function machine

Function

A **function** is a rule that assigns to each element in a set A one and only one element in a set B.

The set A is called the **domain** of the function. It is customary to denote a function by a letter of the alphabet, such as the letter f. If x is an element in the domain of a function f, then the element in B that f associates with x is written $f(x)$ (read "f of x") and is called the value of f at x. The set comprising all the values assumed by $y = f(x)$ as x takes on all possible values in its domain is called the **range** of the function f.

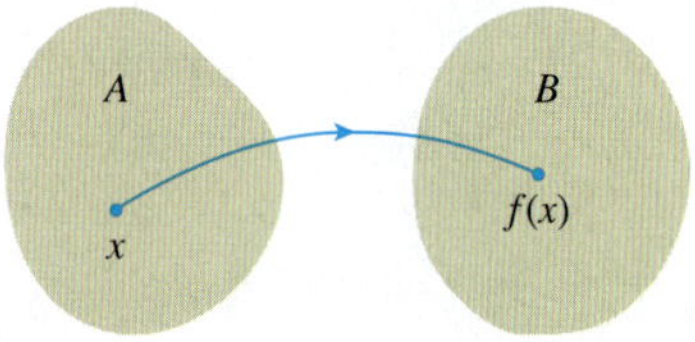

FIGURE 2
The function f viewed as a mapping

We can think of a function f as a machine. The domain is the set of inputs (raw material) for the machine, the rule describes how the input is to be processed, and the value(s) of the function are the outputs of the machine (Figure 1).

We can also think of a function f as a mapping in which an element x in the domain of f is mapped onto a unique element $f(x)$ in B (Figure 2).

Notes

1. The output $f(x)$ associated with an input x is unique. To appreciate the importance of this uniqueness property, consider a rule that associates with each item x in a department store its selling price y. Then, each x must correspond to *one and only one* y. Notice, however, that different x's may be associated with the same y. In the context of the present example, this says that different items may have the same price.
2. Although the sets A and B that appear in the definition of a function may be quite arbitrary, in this book they will denote sets of real numbers. ■

An example of a function may be taken from the familiar relationship between the area of a circle and its radius. Letting x and y denote the radius and area of a circle, respectively, we have, from elementary geometry,

$$y = \pi x^2 \tag{1}$$

Equation (1) defines y as a function of x since for each admissible value of x (that is, for each nonnegative number representing the radius of a certain circle) there corresponds precisely one number $y = \pi x^2$ that gives the area of the circle. The rule defining this "area function" may be written as

$$f(x) = \pi x^2 \tag{2}$$

To compute the area of a circle of radius 5 inches, we simply replace x in Equation (2) with the number 5. Thus, the area of the circle is

$$f(5) = \pi 5^2 = 25\pi$$

or 25π square inches.

In general, to evaluate a function at a specific value of x, we replace x with that value, as illustrated in Examples 1 and 2.

EXAMPLE 1 Let the function f be defined by the rule $f(x) = 2x^2 - x + 1$. Find:

a. $f(1)$ **b.** $f(-2)$ **c.** $f(a)$ **d.** $f(a + h)$

Solution

a. $f(1) = 2(1)^2 - (1) + 1 = 2 - 1 + 1 = 2$
b. $f(-2) = 2(-2)^2 - (-2) + 1 = 8 + 2 + 1 = 11$
c. $f(a) = 2(a)^2 - (a) + 1 = 2a^2 - a + 1$
d. $f(a + h) = 2(a + h)^2 - (a + h) + 1 = 2a^2 + 4ah + 2h^2 - a - h + 1$

APPLIED EXAMPLE 2 Profit Functions ThermoMaster manufactures an indoor–outdoor thermometer at its Mexican subsidiary. Management estimates that the profit (in dollars) realizable by ThermoMaster in the manufacture and sale of x thermometers per week is

$$P(x) = -0.001x^2 + 8x - 5000$$

Find ThermoMaster's weekly profit if its level of production is (a) 1000 thermometers per week and (b) 2000 thermometers per week.

Solution

a. The weekly profit when the level of production is 1000 units per week is found by evaluating the profit function P at $x = 1000$. Thus,

$$P(1000) = -0.001(1000)^2 + 8(1000) - 5000 = 2000$$

or \$2000.

b. When the level of production is 2000 units per week, the weekly profit is given by

$$P(2000) = -0.001(2000)^2 + 8(2000) - 5000 = 7000$$

or \$7000.

Determining the Domain of a Function

Suppose we are given the function $y = f(x)$.* Then, the variable x is called the **independent variable**. The variable y, whose value depends on x, is called the **dependent variable**.

To determine the domain of a function, we need to find what restrictions, if any, are to be placed on the independent variable x. In many practical applications, the domain of a function is dictated by the nature of the problem, as illustrated in Example 3.

*It is customary to refer to a function f as $f(x)$ or by the equation $y = f(x)$ defining it.

APPLIED EXAMPLE 3 Packaging An open box is to be made from a rectangular piece of cardboard 16 inches long and 10 inches wide by cutting away identical squares (x inches by x inches) from each corner and folding up the resulting flaps (Figure 3). Find an expression that gives the volume V of the box as a function of x. What is the domain of the function?

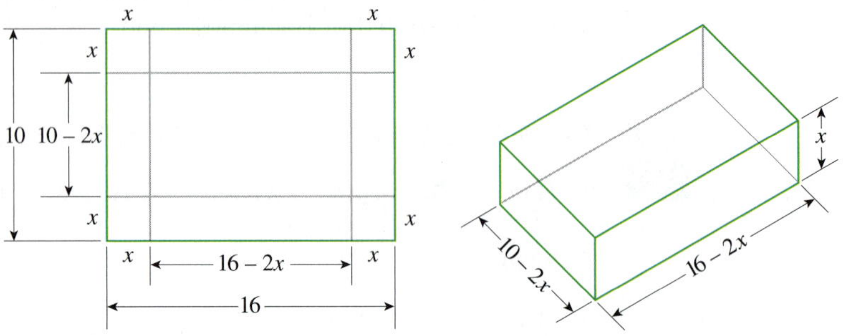

(a) The box is constructed by cutting x'' by x'' squares from each corner.

(b) The dimensions of the resulting box are $(10 - 2x)''$ by $(16 - 2x)''$ by x''.

FIGURE **3**

Solution The dimensions of the box are $(16 - 2x)$ inches by $(10 - 2x)$ inches by x inches, so its volume (in cubic inches) is given by

$$\begin{aligned} V = f(x) &= (16 - 2x)(10 - 2x)x \qquad \text{Length} \cdot \text{width} \cdot \text{height} \\ &= (160 - 52x + 4x^2)x \\ &= 4x^3 - 52x^2 + 160x \end{aligned}$$

Since the length of each side of the box must be greater than or equal to zero, we see that

$$16 - 2x \geq 0 \qquad 10 - 2x \geq 0 \qquad x \geq 0$$

simultaneously; that is,

$$x \leq 8 \qquad x \leq 5 \qquad x \geq 0$$

All three inequalities are satisfied simultaneously provided that $0 \leq x \leq 5$. Thus, the domain of the function f is the interval $[0, 5]$. ■

In general, if a function is defined by a rule relating x to $f(x)$ without specific mention of its domain, it is understood that the domain will consist of all values of x for which $f(x)$ is a real number. In this connection, you should keep in mind that (1) division by zero is not permitted and (2) the even root of a negative number is not a real number.

EXAMPLE 4 Find the domain of each function.

a. $f(x) = \sqrt{x - 1}$ **b.** $f(x) = \dfrac{1}{x^2 - 4}$ **c.** $f(x) = x^2 + 3$

Solution

a. Since the square root of a negative number is not a real number, it is necessary that $x - 1 \geq 0$. The inequality is satisfied by the set of real numbers $x \geq 1$. Thus, the domain of f is the interval $[1, \infty)$.

b. The only restriction on x is that $x^2 - 4$ be different from zero since division by zero is not allowed. But $(x^2 - 4) = (x + 2)(x - 2) = 0$ if $x = -2$ or $x = 2$. Thus, the domain of f in this case consists of the intervals $(-\infty, -2)$, $(-2, 2)$, and $(2, \infty)$.

c. Here, any real number satisfies the equation, so the domain of f is the set of all real numbers. ■

Graphs of Functions

If f is a function with domain A, then corresponding to each real number x in A there is precisely one real number $f(x)$. We can also express this fact by using **ordered pairs** of real numbers. Write each number x in A as the first member of an ordered pair and each number $f(x)$ corresponding to x as the second member of the ordered pair. This gives exactly one ordered pair $(x, f(x))$ for each x in A. This observation leads to an **alternative definition of a function** f:

Function (Alternative Definition)

A function f with domain A is the set of all ordered pairs $(x, f(x))$ where x belongs to A.

Observe that the condition that there be one and only one number $f(x)$ corresponding to each number x in A translates into the requirement that *no two ordered pairs have the same first number*.

Since ordered pairs of real numbers correspond to points in the plane, we have found a way to exhibit a function graphically.

Graph of a Function of One Variable

The **graph of a function** f is the set of all points (x, y) in the xy-plane such that x is in the domain of f and $y = f(x)$.

Figure 4 shows the graph of a function f. Observe that the y-coordinate of the point (x, y) on the graph of f gives the height of that point (the distance above the x-axis), if $f(x)$ is positive. If $f(x)$ is negative, then $-f(x)$ gives the depth of the point (x, y) (the distance below the x-axis). Also, observe that the domain of f is a set of real numbers lying on the x-axis, whereas the range of f lies on the y-axis.

EXAMPLE 5 The graph of a function f is shown in Figure 5.

a. What is the value of $f(3)$? The value of $f(5)$?
b. What is the height or depth of the point $(3, f(3))$ from the x-axis? The point $(5, f(5))$ from the x-axis?
c. What is the domain of f? The range of f?

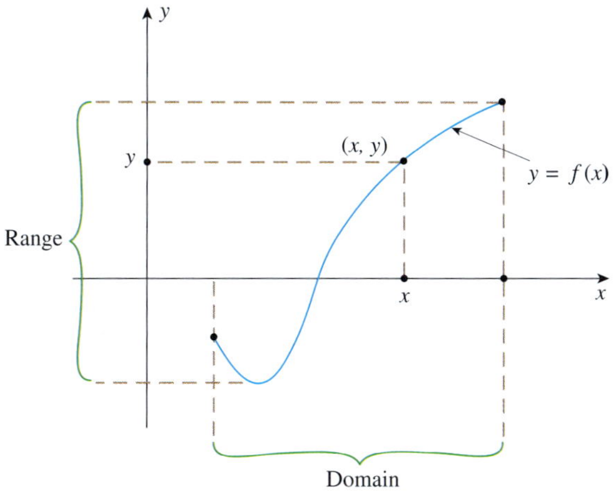

FIGURE 4
The graph of f

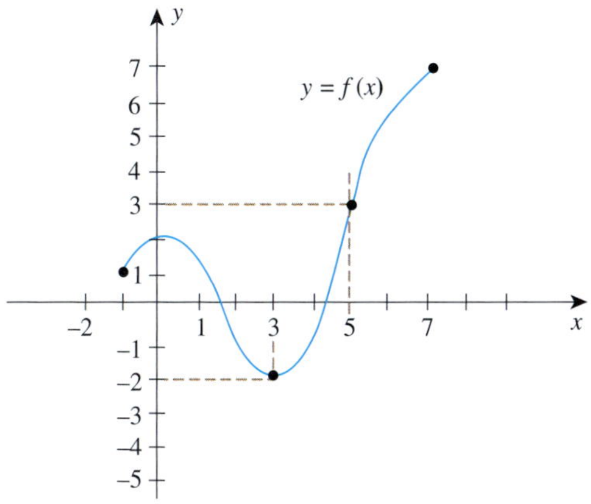

FIGURE 5
The graph of f

Solution

a. From the graph of f, we see that $y = -2$ when $x = 3$ and conclude that $f(3) = -2$. Similarly, we see that $f(5) = 3$.

b. Since the point $(3, -2)$ lies below the x-axis, we see that the depth of the point $(3, f(3))$ is $-f(3) = -(-2) = 2$ units below the x-axis. The point $(5, f(5))$ lies above the x-axis and is located at a height of $f(5)$, or 3 units above the x-axis.

c. Observe that x may take on all values between $x = -1$ and $x = 7$, inclusive, and so the domain of f is $[-1, 7]$. Next, observe that as x takes on all values in the domain of f, $f(x)$ takes on all values between -2 and 7, inclusive. (You can easily see this by running your index finger along the x-axis from $x = -1$ to $x = 7$ and observing the corresponding values assumed by the y-coordinate of each point of the graph of f.) Therefore, the range of f is $[-2, 7]$. ■

Much information about the graph of a function can be gained by plotting a few points on its graph. Later on we will develop more systematic and sophisticated techniques for graphing functions.

EXAMPLE 6 Sketch the graph of the function defined by the equation $y = x^2 + 1$. What is the range of f?

Solution The domain of the function is the set of all real numbers. By assigning several values to the variable x and computing the corresponding values for y, we obtain the following solutions to the equation $y = x^2 + 1$:

x	-3	-2	-1	0	1	2	3
y	10	5	2	1	2	5	10

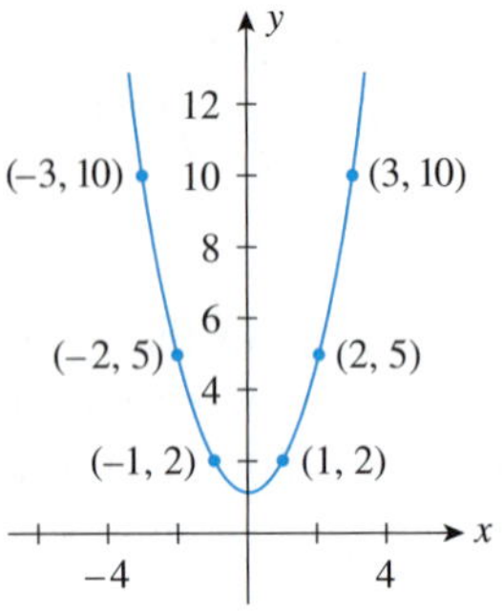

FIGURE 6
The graph of $y = x^2 + 1$ is a parabola.

By plotting these points and then connecting them with a smooth curve, we obtain the graph of $y = f(x)$, which is a parabola (Figure 6). To determine the range of f, we observe that $x^2 \geq 0$ if x is any real number, and so $x^2 + 1 \geq 1$ for all real numbers x. We conclude that the range of f is $[1, \infty)$. The graph of f confirms this result visually. ■

Exploring with TECHNOLOGY

Let $f(x) = x^2$.

1. Plot the graphs of $F(x) = x^2 + c$ on the same set of axes for $c = -2, -1, -\frac{1}{2}, 0, \frac{1}{2}, 1, 2$.
2. Plot the graphs of $G(x) = (x + c)^2$ on the same set of axes for $c = -2, -1, -\frac{1}{2}, 0, \frac{1}{2}, 1, 2$.
3. Plot the graphs of $H(x) = cx^2$ on the same set of axes for $c = -2, -1, -\frac{1}{2}, -\frac{1}{4}, 0, \frac{1}{4}, \frac{1}{2}, 1, 2$.
4. Study the family of graphs in parts 1–3 and describe the relationship between the graph of a function f and the graphs of the functions defined by (a) $y = f(x) + c$, (b) $y = f(x + c)$, and (c) $y = cf(x)$, where c is a constant.

A function that is defined by more than one rule is called a **piecewise-defined function.**

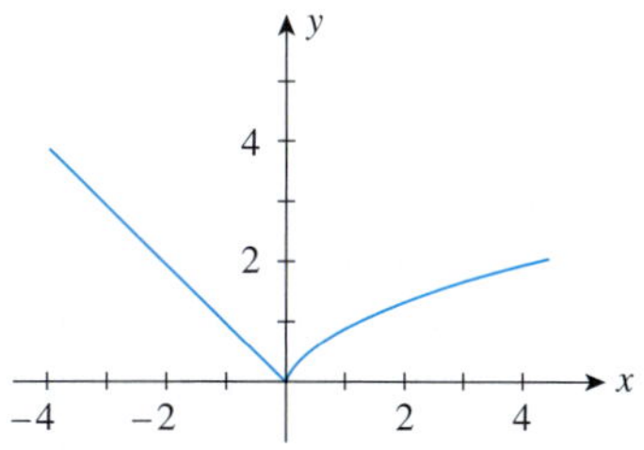

FIGURE 7
The graph of $y = f(x)$ is obtained by graphing $y = -x$ over $(-\infty, 0)$ and $y = \sqrt{x}$ over $[0, \infty)$.

EXAMPLE 7 Sketch the graph of the function f defined by

$$f(x) = \begin{cases} -x & \text{if } x < 0 \\ \sqrt{x} & \text{if } x \geq 0 \end{cases}$$

Solution The function f is defined in a piecewise fashion on the set of all real numbers. In the subdomain $(-\infty, 0)$, the rule for f is given by $f(x) = -x$. The equation $y = -x$ is a linear equation in the slope-intercept form (with slope -1 and intercept 0). Therefore, the graph of f corresponding to the subdomain $(-\infty, 0)$ is the half line shown in Figure 7. Next, in the subdomain $[0, \infty)$, the rule for f is given by $f(x) = \sqrt{x}$. The values of $f(x)$ corresponding to $x = 0, 1, 2, 3, 4, 9$, and 16 are shown in the following table:

x	0	1	2	3	4	9	16
$f(x)$	0	1	$\sqrt{2}$	$\sqrt{3}$	2	3	4

Using these values, we sketch the graph of the function f as shown in Figure 7. ■

VIDEO

APPLIED EXAMPLE 8 Bank Deposits Madison Finance Company plans to open two branch offices 2 years from now in two separate locations: an industrial complex and a newly developed commercial center in the city. As a result of these expansion plans, Madison's total deposits during the next 5 years are expected to grow in accordance with the rule

$$f(x) = \begin{cases} \sqrt{2x} + 20 & \text{if } 0 \leq x \leq 2 \\ \frac{1}{2}x^2 + 20 & \text{if } 2 < x \leq 5 \end{cases}$$

where $y = f(x)$ gives the total amount of money (in millions of dollars) on deposit with Madison in year x ($x = 0$ corresponds to the present). Sketch the graph of the function f.

Solution The function f is defined in a piecewise fashion on the interval [0, 5]. In the subdomain [0, 2], the rule for f is given by $f(x) = \sqrt{2x} + 20$. The values of $f(x)$ corresponding to $x = 0$, 1, and 2 may be tabulated as follows:

x	0	1	2
$f(x)$	20	21.4	22

Next, in the subdomain (2, 5], the rule for f is given by $f(x) = \frac{1}{2}x^2 + 20$. The values of $f(x)$ corresponding to $x = 3$, 4, and 5 are shown in the following table:

x	3	4	5
$f(x)$	24.5	28	32.5

Using the values of $f(x)$ in this table, we sketch the graph of the function f as shown in Figure 8. ■

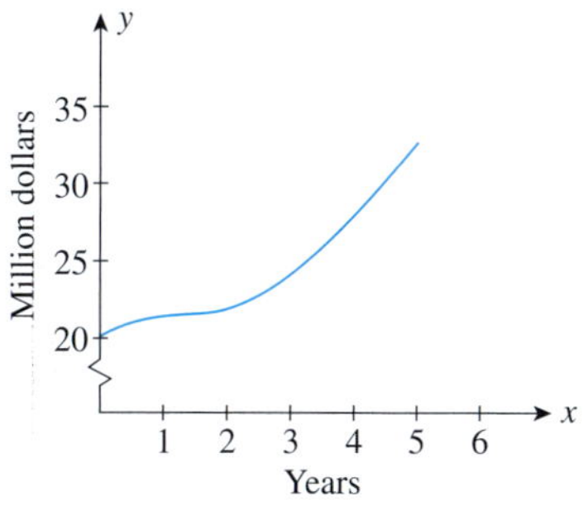

FIGURE 8
We obtain the graph of the function $y = f(x)$ by graphing $y = \sqrt{2x} + 20$ over [0, 2] and $y = \frac{1}{2}x^2 + 20$ over (2, 5].

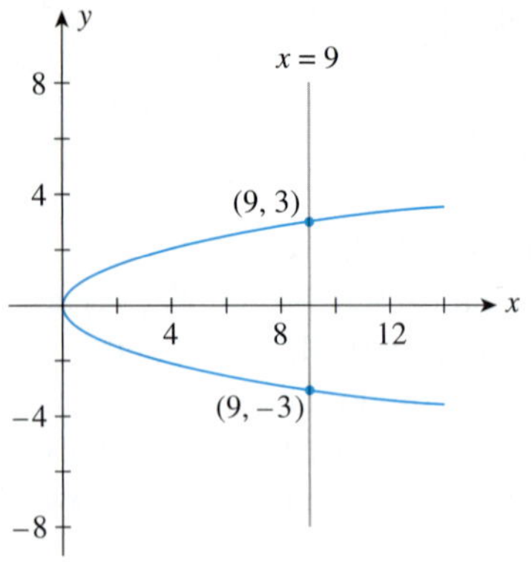

FIGURE 9
Since a vertical line passes through the curve at more than one point, we deduce that it is *not* the graph of a function.

The Vertical-Line Test

Although it is true that every function f of a variable x has a graph in the xy-plane, it is not true that every curve in the xy-plane is the graph of a function. For example, consider the curve depicted in Figure 9. This is the graph of the equation $y^2 = x$. In general, the **graph of an equation** is the set of all ordered pairs (x, y) that satisfy the given equation. Observe that the points $(9, -3)$ and $(9, 3)$ both lie on the curve. This implies that the number $x = 9$ is associated with *two* numbers: $y = -3$ and $y = 3$. But this clearly violates the uniqueness property of a function. Thus, we conclude that the curve under consideration cannot be the graph of a function.

This example suggests the following **vertical-line test** for determining whether a curve is the graph of a function.

Vertical-Line Test

A curve in the xy-plane is the graph of a function $y = f(x)$ if and only if each vertical line intersects it in at most one point.

EXAMPLE 9 Determine which of the curves shown in Figure 10 are the graphs of functions of x.

Solution The curves depicted in Figure 10a, c, and d are graphs of functions because each curve satisfies the requirement that each vertical line intersects the curve in at most one point. Note that the vertical line shown in Figure 10c does *not* intersect the graph because the point on the x-axis through which this line passes does not lie in the domain of the function. The curve depicted in Figure 10b is *not* the graph of a function of x because the vertical line shown there intersects the graph at three points. ■

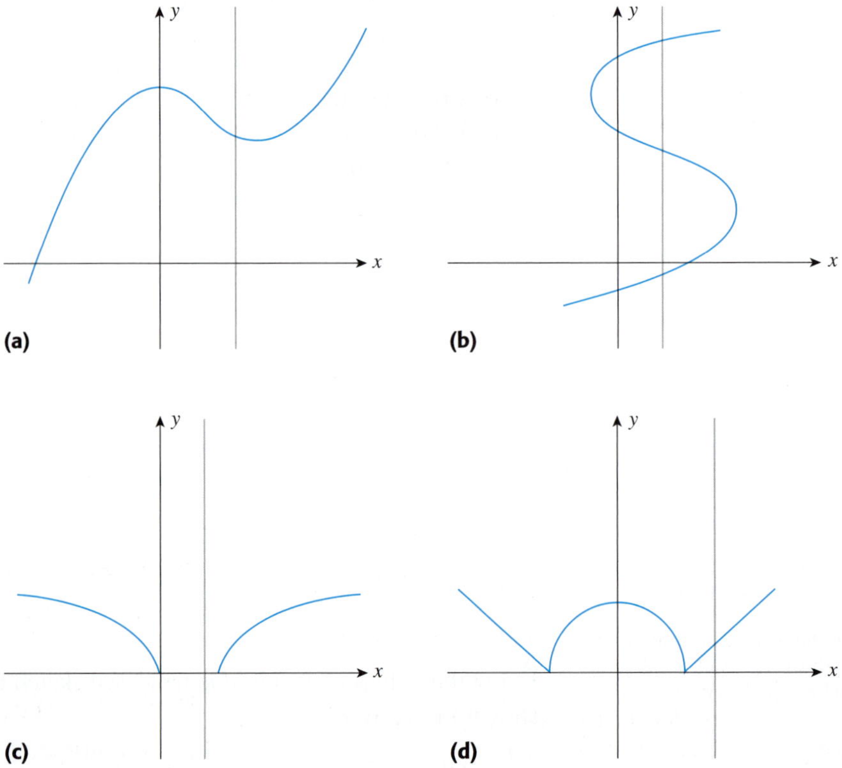

FIGURE 10
The vertical-line test can be used to determine which of these curves are graphs of functions.

2.1 Self-Check Exercises

1. Let f be the function defined by

$$f(x) = \frac{\sqrt{x+1}}{x}$$

   a. Find the domain of f. b. Compute $f(3)$.
   c. Compute $f(a + h)$.

2. Statistics show that more and more motorists are pumping their own gas. The following function gives self-serve sales as a percentage of all U.S. gas sales:

$$f(t) = \begin{cases} 6t + 17 & \text{if } 0 \le t \le 6 \\ 15.98(t - 6)^{1/4} + 53 & \text{if } 6 < t \le 20 \end{cases}$$

   Here t is measured in years, with $t = 0$ corresponding to the beginning of 1974.
   a. Sketch the graph of the function f.
   b. What percentage of all gas sales at the beginning of 1978 were self-serve? At the beginning of 1994?
   Source: Amoco Corporation

3. Let $f(x) = \sqrt{2x + 1} + 2$. Determine whether the point (4, 6) lies on the graph of f.

Solutions to Self-Check Exercises 2.1 can be found on page 63.

2.1 Concept Questions

1. **a.** What is a function?
   b. What is the domain of a function? The range of a function?
   c. What is an independent variable? A dependent variable?

2. **a.** What is the graph of a function? Use a drawing to illustrate the graph, the domain, and the range of a function.
   b. If you are given a curve in the xy-plane, how can you tell if the graph is that of a function f defined by $y = f(x)$?

3. Are the following graphs of functions? Explain.

a.

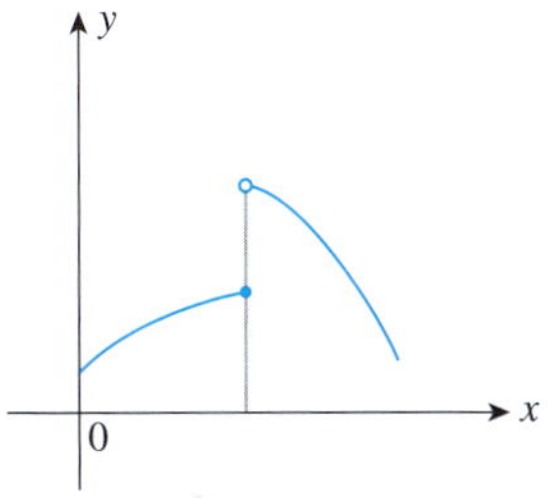

b.

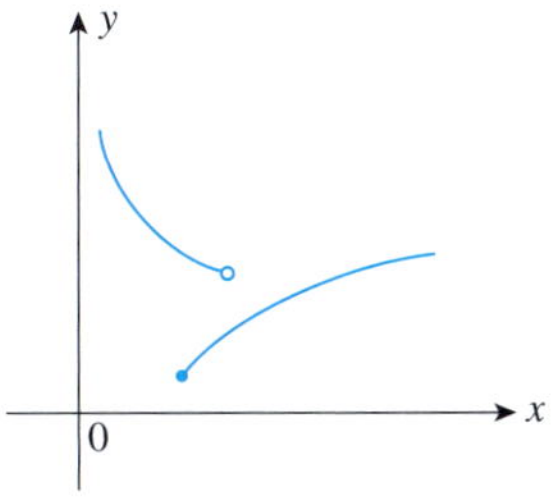

c.

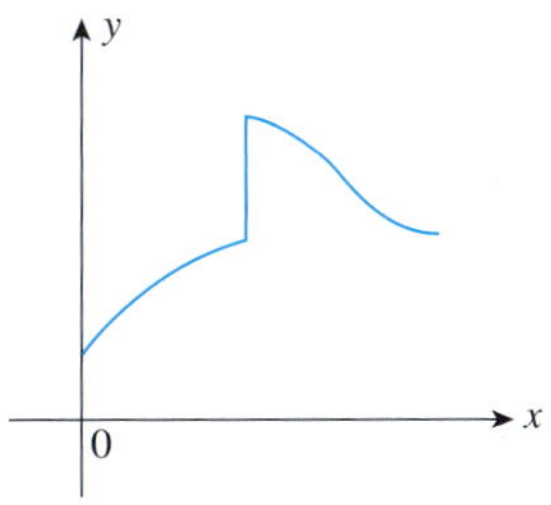

d.

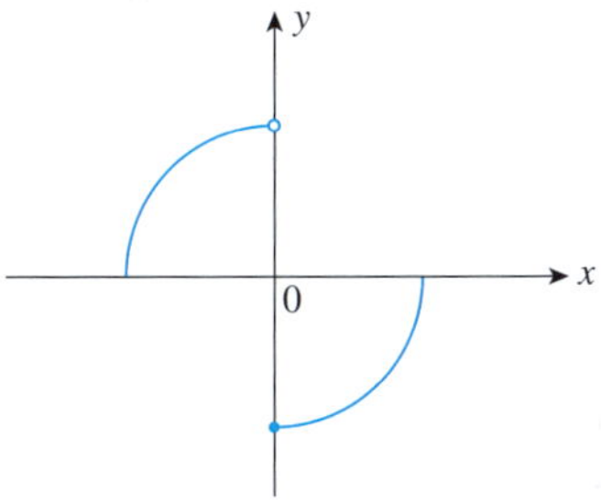

4. What are the domain and range of the function f with the following graph?

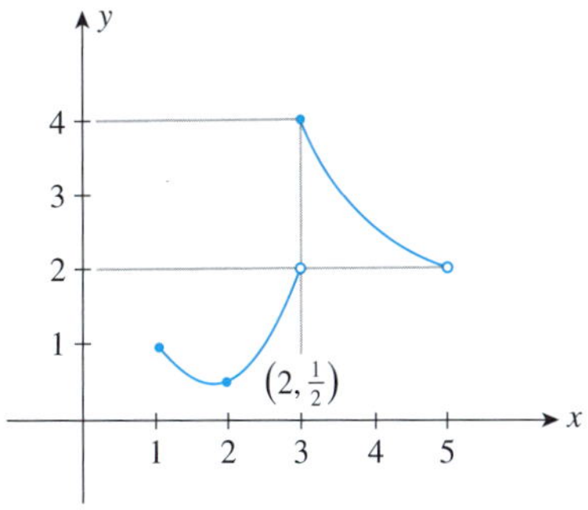

2.1 Exercises

1. Let f be the function defined by $f(x) = 5x + 6$. Find $f(3)$, $f(-3)$, $f(a)$, $f(-a)$, and $f(a + 3)$.

2. Let f be the function defined by $f(x) = 4x - 3$. Find $f(4)$, $f(\frac{1}{4})$, $f(0)$, $f(a)$, and $f(a + 1)$.

3. Let g be the function defined by $g(x) = 3x^2 - 6x - 3$. Find $g(0)$, $g(-1)$, $g(a)$, $g(-a)$, and $g(x + 1)$.

4. Let h be the function defined by $h(x) = x^3 - x^2 + x + 1$. Find $h(-5)$, $h(0)$, $h(a)$, and $h(-a)$.

5. Let f be the function defined by $f(x) = 2x + 5$. Find $f(a + h)$, $f(-a)$, $f(a^2)$, $f(a - 2h)$, and $f(2a - h)$.

6. Let g be the function defined by $g(x) = -x^2 + 2x$. Find $g(a + h)$, $g(-a)$, $g(\sqrt{a})$, $a + g(a)$, and $\dfrac{1}{g(a)}$.

7. Let s be the function defined by $s(t) = \dfrac{2t}{t^2 - 1}$. Find $s(4)$, $s(0)$, $s(a)$, $s(2 + a)$, and $s(t + 1)$.

8. Let g be the function defined by $g(u) = (3u - 2)^{3/2}$. Find $g(1)$, $g(6)$, $g(\frac{11}{3})$, and $g(u + 1)$.

9. Let f be the function defined by $f(t) = \dfrac{2t^2}{\sqrt{t - 1}}$. Find $f(2)$, $f(a)$, $f(x + 1)$, and $f(x - 1)$.

10. Let f be the function defined by $f(x) = 2 + 2\sqrt{5 - x}$. Find $f(-4)$, $f(1)$, $f(\frac{11}{4})$, and $f(x + 5)$.

11. Let f be the function defined by

$$f(x) = \begin{cases} x^2 + 1 & \text{if } x \leq 0 \\ \sqrt{x} & \text{if } x > 0 \end{cases}$$

Find $f(-2)$, $f(0)$, and $f(1)$.

12. Let g be the function defined by

$$g(x) = \begin{cases} -\frac{1}{2}x + 1 & \text{if } x < 2 \\ \sqrt{x - 2} & \text{if } x \geq 2 \end{cases}$$

Find $g(-2)$, $g(0)$, $g(2)$, and $g(4)$.

13. Let f be the function defined by

$$f(x) = \begin{cases} -\frac{1}{2}x^2 + 3 & \text{if } x < 1 \\ 2x^2 + 1 & \text{if } x \geq 1 \end{cases}$$

Find $f(-1)$, $f(0)$, $f(1)$, and $f(2)$.

14. Let f be the function defined by

$$f(x) = \begin{cases} 2 + \sqrt{1 - x} & \text{if } x \leq 1 \\ \dfrac{1}{1 - x} & \text{if } x > 1 \end{cases}$$

Find $f(0)$, $f(1)$, and $f(2)$.

15. Refer to the graph of the function f in the following figure.

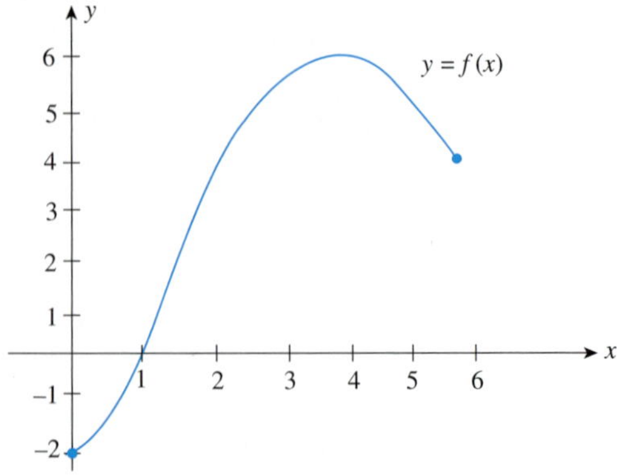

a. Find the value of $f(0)$.
b. Find the value of x for which (i) $f(x) = 3$ and (ii) $f(x) = 0$.
c. Find the domain of f.
d. Find the range of f.

16. Refer to the graph of the function f in the following figure.

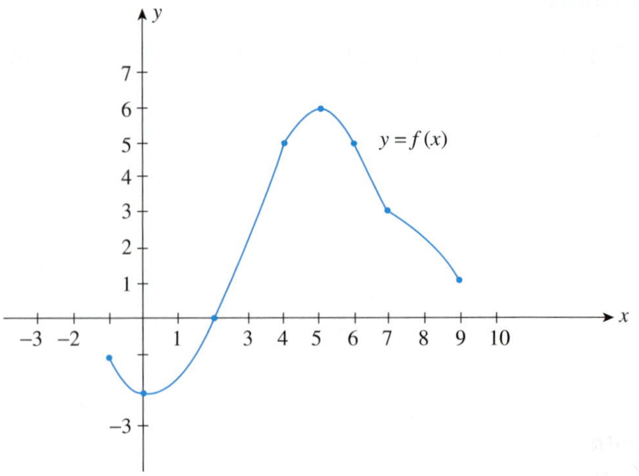

a. Find the value of $f(7)$.
b. Find the values of x corresponding to the point(s) on the graph of f located at a height of 5 units from the x-axis.
c. Find the point on the x-axis at which the graph of f crosses it. What is the value of $f(x)$ at this point?
d. Find the domain and range of f.

In Exercises 17–20, determine whether the point lies on the graph of the function.

17. $(2, \sqrt{3})$; $g(x) = \sqrt{x^2 - 1}$

18. $(3, 3)$; $f(x) = \dfrac{x + 1}{\sqrt{x^2 + 7}} + 2$

19. $(-2, -3)$; $f(t) = \dfrac{|t - 1|}{t + 1}$

20. $\left(-3, -\dfrac{1}{13}\right)$; $h(t) = \dfrac{|t + 1|}{t^3 + 1}$

In Exercises 21 and 22, find the value of *c* such that the point *P*(*a*, *b*) lies on the graph of the function *f*.

21. $f(x) = 2x^2 - 4x + c$; $P(1, 5)$

22. $f(x) = x\sqrt{9 - x^2} + c$; $P(2, 4)$

In Exercises 23–36, find the domain of the function.

23. $f(x) = x^2 + 3$

24. $f(x) = 7 - x^2$

25. $f(x) = \dfrac{3x + 1}{x^2}$

26. $g(x) = \dfrac{2x + 1}{x - 1}$

27. $f(x) = \sqrt{x^2 + 1}$

28. $f(x) = \sqrt{x - 5}$

29. $f(x) = \sqrt{5 - x}$

30. $g(x) = \sqrt{2x^2 + 3}$

31. $f(x) = \dfrac{x}{x^2 - 1}$

32. $f(x) = \dfrac{1}{x^2 + x - 2}$

33. $f(x) = (x + 3)^{3/2}$

34. $g(x) = 2(x - 1)^{5/2}$

35. $f(x) = \dfrac{\sqrt{1 - x}}{x^2 - 4}$

36. $f(x) = \dfrac{\sqrt{x - 1}}{(x + 2)(x - 3)}$

37. Let f be a function defined by the rule $f(x) = x^2 - x - 6$.
- **a.** Find the domain of f.
- **b.** Compute $f(x)$ for $x = -3, -2, -1, 0, \frac{1}{2}, 1, 2, 3$.
- **c.** Use the results obtained in parts (a) and (b) to sketch the graph of f.

38. Let f be a function defined by the rule $f(x) = 2x^2 + x - 3$.
- **a.** Find the domain of f.
- **b.** Compute $f(x)$ for $x = -3, -2, -1, -\frac{1}{2}, 0, 1, 2, 3$.
- **c.** Use the results obtained in parts (a) and (b) to sketch the graph of f.

In Exercises 39–50, sketch the graph of the function with the given rule. Find the domain and range of the function.

39. $f(x) = 2x^2 + 1$

40. $f(x) = 9 - x^2$

41. $f(x) = 2 + \sqrt{x}$

42. $g(x) = 4 - \sqrt{x}$

43. $f(x) = \sqrt{1 - x}$

44. $f(x) = \sqrt{x - 1}$

45. $f(x) = |x| - 1$

46. $f(x) = |x| + 1$

47. $f(x) = \begin{cases} x & \text{if } x < 0 \\ 2x + 1 & \text{if } x \geq 0 \end{cases}$

48. $f(x) = \begin{cases} 4 - x & \text{if } x < 2 \\ 2x - 2 & \text{if } x \geq 2 \end{cases}$

49. $f(x) = \begin{cases} -x + 1 & \text{if } x \leq 1 \\ x^2 - 1 & \text{if } x > 1 \end{cases}$

50. $f(x) = \begin{cases} -x - 1 & \text{if } x < -1 \\ 0 & \text{if } -1 \leq x \leq 1 \\ x + 1 & \text{if } x > 1 \end{cases}$

In Exercises 51–58, use the vertical-line test to determine whether the graph represents y as a function of x.

51.

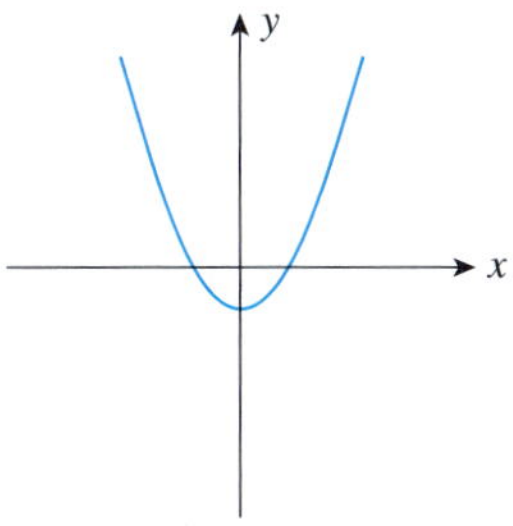

52.

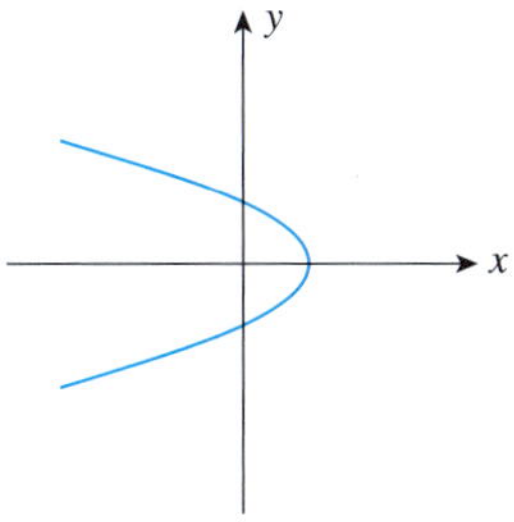

53.

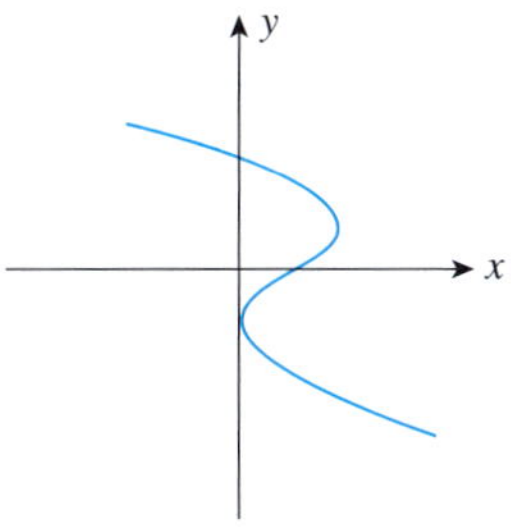

54.

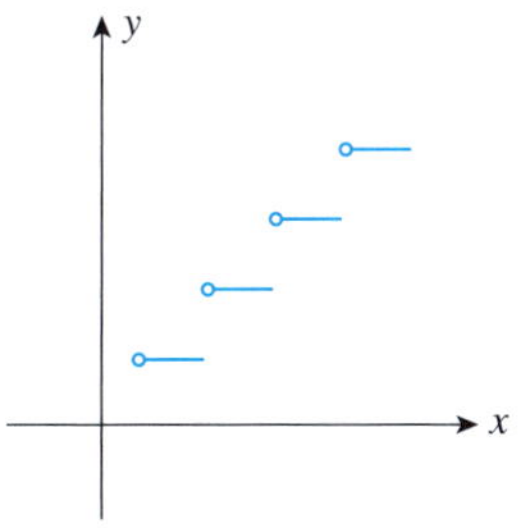

55.

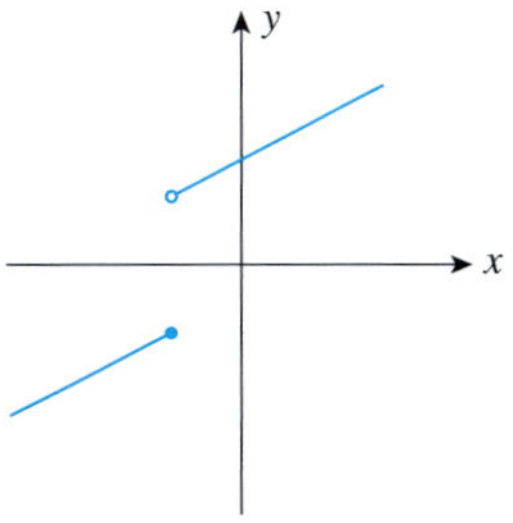

56.

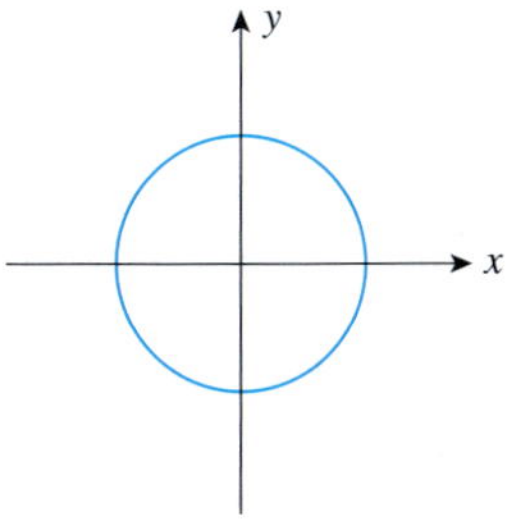

57.

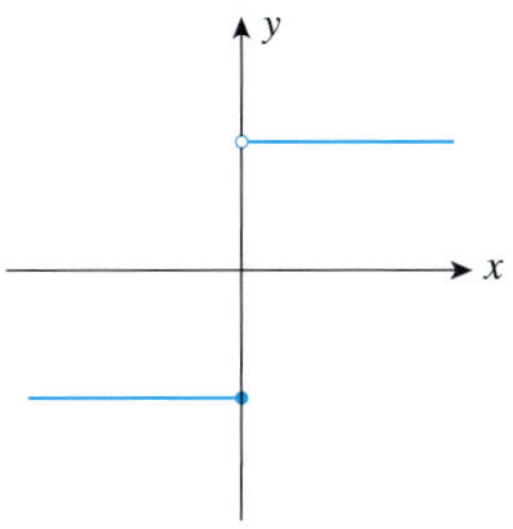

58.

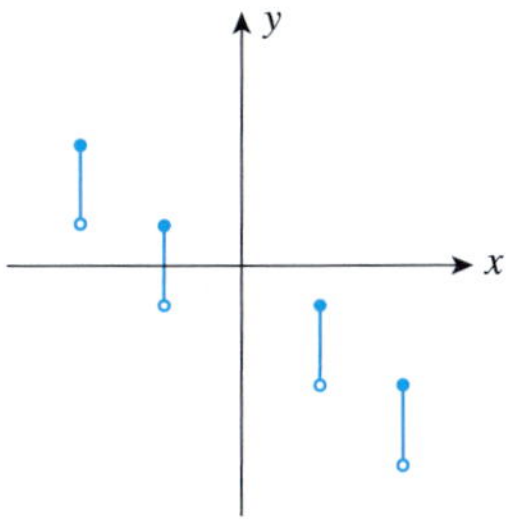

59. The circumference of a circle is given by $C(r) = 2\pi r$, where r is the radius of the circle. What is the circumference of a circle with a 5-in. radius?

60. The volume of a sphere of radius r is given by $V(r) = \frac{4}{3}\pi r^3$. Compute $V(2.1)$ and $V(2)$. What does the quantity $V(2.1) - V(2)$ measure?

61. Growth of a Cancerous Tumor The volume of a spherical cancerous tumor is given by the function

$$V(r) = \frac{4}{3}\pi r^3$$

where r is the radius of the tumor in centimeters. By what factor is the volume of the tumor increased if its radius is doubled?

62. Life Expectancy after Age 65 The average life expectancy after age 65 is soaring, putting pressure on the Social Security Administration's resources. According to the Social Security Trustees, the average life expectancy after age 65 is given by

$$L(t) = 0.056t + 18.1 \qquad (0 \leq t \leq 7)$$

where t is measured in years, with $t = 0$ corresponding to 2003.
- **a.** How fast is the average life expectancy after age 65 changing at any time during the period under consideration?
- **b.** What will the average life expectancy be after age 65 in 2010?

Source: Social Security Trustees

63. Sales of Prerecorded Music The following graphs show the sales y of prerecorded music (in billions of dollars) by format as a function of time t (in years), with $t = 0$ corresponding to 1985.

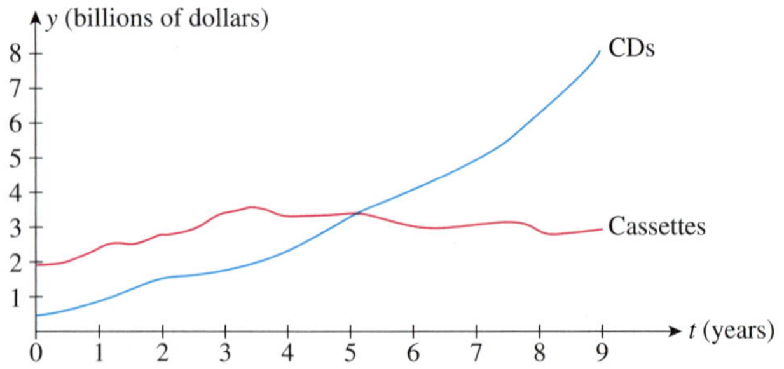

a. In what years were the sales of prerecorded cassettes greater than those of prerecorded CDs?
b. In what years were the sales of prerecorded CDs greater than those of prerecorded cassettes?
c. In what year were the sales of prerecorded cassettes the same as those of prerecorded CDs? Estimate the level of sales in each format at that time.

Source: Recording Industry Association of America

64. The Gender Gap The following graph shows the ratio of women's earnings to men's from 1960 through 2000.

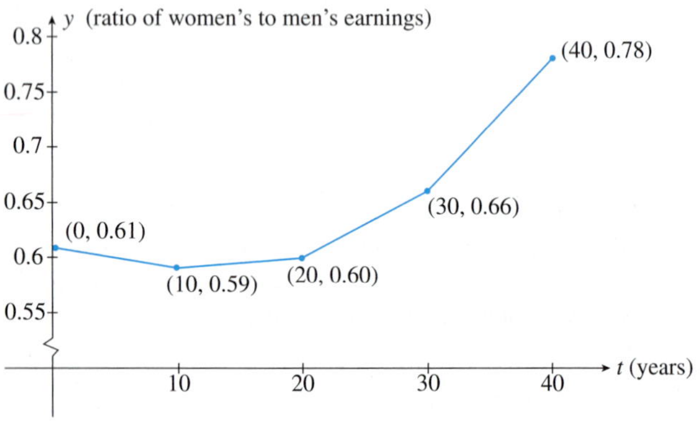

a. Write the rule for the function f giving the ratio of women's earnings to men's in year t, with $t = 0$ corresponding to 1960.
Hint: The function f is defined piecewise and is linear over each of four subintervals.
b. In what decade(s) was the gender gap expanding? Shrinking?
c. Refer to part (b). How fast was the gender gap (the ratio/year) expanding or shrinking in each of these decades?

Source: U.S. Bureau of Labor Statistics

65. Closing the Gender Gap in Education The following graph shows the ratio of the number of bachelor's degrees earned by women to that of men from 1960 through 1990.

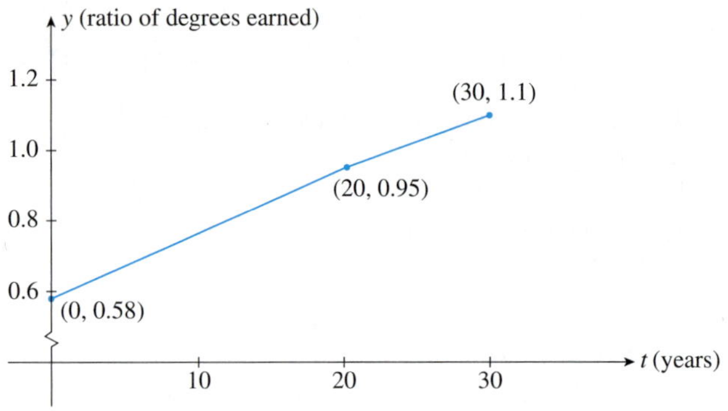

a. Write the rule for the function f giving the ratio of the number of bachelor's degrees earned by women to that of men in year t, with $t = 0$ corresponding to 1960.
Hint: The function f is defined piecewise and is linear over each of two subintervals.
b. How fast was the ratio changing in the period from 1960 to 1980? From 1980 to 1990?
c. In what year (approximately) was the number of bachelor's degrees earned by women equal for the first time to that earned by men?

Source: Department of Education

66. Consumption Function The consumption function in a certain economy is given by the equation

$$C(y) = 0.75y + 6$$

where $C(y)$ is the personal consumption expenditure, y is the disposable personal income, and both $C(y)$ and y are measured in billions of dollars. Find $C(0)$, $C(50)$, and $C(100)$.

67. Sales Taxes In a certain state, the sales tax T on the amount of taxable goods is 6% of the value of the goods purchased x, where both T and x are measured in dollars.
a. Express T as a function of x.
b. Find $T(200)$ and $T(5.65)$.

68. Surface Area of a Single-Celled Organism The surface area S of a single-celled organism may be found by multiplying 4π times the square of the radius r of the cell. Express S as a function of r.

69. Friend's Rule Friend's rule, a method for calculating pediatric drug dosages, is based on a child's age. If a denotes the adult dosage (in milligrams) and if t is the age of the child (in years), then the child's dosage is given by

$$D(t) = \frac{2}{25}ta$$

If the adult dose of a substance is 500 mg, how much should a 4-yr-old child receive?

70. COLAs Social Security recipients receive an automatic cost-of-living adjustment (COLA) once each year. Their monthly benefit is increased by the amount that consumer prices increased during the preceding year. Suppose that consumer prices increased by 5.3% during the preceding year.

a. Express the adjusted monthly benefit of a Social Security recipient as a function of his or her current monthly benefit.

b. If Harrington's monthly Social Security benefit is now \$1020, what will be his adjusted monthly benefit?

71. **Broadband Internet Households** The number of U.S. broadband Internet households stood at 20 million at the beginning of 2002 and is projected to grow at the rate of 7.5 million households per year for the next 6 yr.

a. Find a function $f(t)$ giving the projected number of U.S. broadband Internet households (in millions) in year t, where $t = 0$ corresponds to the beginning of 2002.
Hint: The graph of f is a straight line.

b. What was the projected number of U.S. broadband Internet households at the beginning of 2008?

Source: Strategy Analytics Inc.

72. **Cost of Renting a Truck** Ace Truck leases its 10-ft box truck at \$30/day and \$.45/mi, whereas Acme Truck leases a similar truck at \$25/day and \$.50/mi.

a. Find the daily cost of leasing from each company as a function of the number of miles driven.

b. Sketch the graphs of the two functions on the same set of axes.

c. Which company should a customer rent a truck from for 1 day if she plans to drive at most 70 mi and wishes to minimize her cost?

73. **Linear Depreciation** A new machine was purchased by National Textile for \$120,000. For income tax purposes, the machine is depreciated linearly over 10 yr; that is, the book value of the machine decreases at a constant rate, so that at the end of 10 yr the book value is zero.

a. Express the book value of the machine V as a function of the age, in years, of the machine n.

b. Sketch the graph of the function in part (a).

c. Find the book value of the machine at the end of the sixth year.

d. Find the rate at which the machine is being depreciated each year.

74. **Linear Depreciation** Refer to Exercise 73. An office building worth \$1 million when completed in 1990 was depreciated linearly over 50 yr. What was the book value of the building in 2005? What will be the book value in 2009? In 2013? (Assume that the book value of the building will be zero at the end of the 50th year.)

75. **Boyle's Law** As a consequence of Boyle's law, the pressure P of a fixed sample of gas held at a constant temperature is related to the volume V of the gas by the rule

$$P = f(V) = \frac{k}{V}$$

where k is a constant. What is the domain of the function f? Sketch the graph of the function f.

76. **Poiseuille's Law** According to a law discovered by the 19th-century physician Poiseuille, the velocity (in centimeters/second) of blood r cm from the central axis of an artery is given by

$$v(r) = k(R^2 - r^2)$$

where k is a constant and R is the radius of the artery. Suppose that for a certain artery, $k = 1000$ and $R = 0.2$ so that $v(r) = 1000(0.04 - r^2)$.

a. What is the domain of the function $v(r)$?

b. Compute $v(0)$, $v(0.1)$, and $v(0.2)$ and interpret your results.

c. Sketch the graph of the function v on the interval $[0, 0.2]$.

d. What can you say about the velocity of blood as we move away from the central axis toward the artery wall?

77. **Cancer Survivors** The number of living Americans who have had a cancer diagnosis has increased drastically since 1971. In part, this is due to more testing for cancer and better treatment for some cancers. In part, it is because the population is older, and cancer is largely a disease of the elderly. The number of cancer survivors (in millions) between 1975 ($t = 0$) and 2000 ($t = 25$) is approximately

$$N(t) = 0.0031t^2 + 0.16t + 3.6 \qquad (0 \le t \le 25)$$

a. How many living Americans had a cancer diagnosis in 1975? In 2000?

b. Assuming the trend continued, how many cancer survivors were there in 2005?

Source: National Cancer Institute

78. **Prevalence of Alzheimer's Patients** Based on a study conducted in 1997, the percentage of the U.S. population by age afflicted with Alzheimer's disease is given by the function

$$P(x) = 0.0726x^2 + 0.7902x + 4.9623 \qquad (0 \le x \le 25)$$

where x is measured in years, with $x = 0$ corresponding to age 65. What percentage of the U.S. population at age 65 is expected to have Alzheimer's disease? At age 90?

Source: Alzheimer's Association

79. **Worker Efficiency** An efficiency study conducted for Elektra Electronics showed that the number of "Space Commander" walkie-talkies assembled by the average worker t hr after starting work at 8:00 a.m. is given by

$$N(t) = -t^3 + 6t^2 + 15t \qquad (0 \le t \le 4)$$

How many walkie-talkies can an average worker be expected to assemble between 8:00 and 9:00 a.m.? Between 9:00 and 10:00 a.m.?

80. **Politics** Political scientists have discovered the following empirical rule, known as the "cube rule," which gives the relationship between the proportion of seats in the House of Representatives won by Democratic candidates $s(x)$ and the proportion of popular votes x received by the Democratic presidential candidate:

$$s(x) = \frac{x^3}{x^3 + (1 - x)^3} \qquad (0 \le x \le 1)$$

Compute $s(0.6)$ and interpret your result.

81. U.S. Health-Care Information Technology Spending As health-care costs increase, payers are turning to technology and outsourced services to keep a lid on expenses. The amount of health-care IT spending by payer is projected to be

$$S(t) = -0.03t^3 + 0.2t^2 + 0.23t + 5.6 \qquad (0 \le t \le 4)$$

where $S(t)$ is measured in billions of dollars and t is measured in years, with $t = 0$ corresponding to 2004. What was the amount spent by payers on health-care IT in 2004? Assuming the projection held true, what amount was spent by payers in 2008?

Source: U.S. Department of Commerce

82. Hotel Rates The average daily rate of U.S. hotels from 2001 through 2006 is approximated by the function

$$f(t) = \begin{cases} 82.95 & \text{if } 1 \le t \le 3 \\ 0.95t^2 - 3.95t + 86.25 & \text{if } 3 < t \le 6 \end{cases}$$

where $f(t)$ is measured in dollars, with $t = 1$ corresponding to 2001.

a. What was the average daily rate of U.S. hotels from 2001 through 2003?

b. What was the average daily rate of U.S. hotels in 2004? In 2005? In 2006?

c. Sketch the graph of f.

Source: Smith Travel Research

83. Investments in Hedge Funds Investments in hedge funds have increased along with their popularity. The assets of hedge funds (in trillions of dollars) from 2002 through 2007 are modeled by the function

$$f(t) = \begin{cases} 0.6 & \text{if } 0 \le t < 1 \\ 0.6t^{0.43} & \text{if } 1 \le t \le 5 \end{cases}$$

where t is measured in years, with $t = 0$ corresponding to the beginning of 2002.

a. What were the assets in hedge funds at the beginning of 2002? At the beginning of 2003?

b. What were the assets in hedge funds at the beginning of 2005? At the beginning of 2007?

Source: Hennessee Group

84. Postal Regulations In 2007 the postage for packages sent by first-class mail was raised to \$1.13 for the first ounce or fraction thereof and 17¢ for each additional ounce or fraction thereof. Any parcel not exceeding 13 oz may be sent by first-class mail. Letting x denote the weight of a parcel in ounces and $f(x)$ the postage in dollars, complete the following description of the "postage function" f:

$$f(x) = \begin{cases} \$1.13 & \text{if } 0 < x \le 1 \\ \$1.30 & \text{if } 1 < x \le 2 \\ \vdots & \\ ? & \text{if } 12 < x \le 13 \end{cases}$$

a. What is the domain of f?

b. Sketch the graph of f.

85. Harbor Cleanup The amount of solids discharged from the MWRA (Massachusetts Water Resources Authority) sewage treatment plant on Deer Island (near Boston Harbor) is given by the function

$$f(t) = \begin{cases} 130 & \text{if } 0 \le t \le 1 \\ -30t + 160 & \text{if } 1 < t \le 2 \\ 100 & \text{if } 2 < t \le 4 \\ -5t^2 + 25t + 80 & \text{if } 4 < t \le 6 \\ 1.25t^2 - 26.25t + 162.5 & \text{if } 6 < t \le 10 \end{cases}$$

where $f(t)$ is measured in tons/day and t is measured in years, with $t = 0$ corresponding to 1989.

a. What amount of solids were discharged per day in 1989? In 1992? In 1996?

b. Sketch the graph of f.

Source: Metropolitan District Commission

86. Rising Median Age Increased longevity and the aging of the baby boom generation—those born between 1946 and 1965— are the primary reasons for a rising median age. The median age (in years) of the U.S. population from 1900 through 2000 is approximated by the function

$$f(t) = \begin{cases} 1.3t + 22.9 & \text{if } 0 \le t \le 3 \\ -0.7t^2 + 7.2t + 11.5 & \text{if } 3 < t \le 7 \\ 2.6t + 9.4 & \text{if } 7 < t \le 10 \end{cases}$$

where t is measured in decades, with $t = 0$ corresponding to the beginning of 1900.

a. What was the median age of the U.S. population at the beginning of 1900? At the beginning of 1950? At the beginning of 1990?

b. Sketch the graph of f.

Source: U.S. Census Bureau

87. Distance between Ships A passenger ship leaves port sailing east at 14 mph. Two hours later, a cargo ship leaves the same port heading north at 10 mph.

a. Find a function giving the distance between the two ships t hr after the passenger ship leaves port.

b. How far apart are the two ships 3 hr after the cargo ship leaves port?

In Exercises 88–94, determine whether the statement is true or false. If it is true, explain why it is true. If it is false, give an example to show why it is false.

88. If $a = b$, then $f(a) = f(b)$.

89. If $f(a) = f(b)$, then $a = b$.

90. If f is a function, then $f(a + b) = f(a) + f(b)$.

91. A vertical line must intersect the graph of $y = f(x)$ at exactly one point.

92. The domain of $f(x) = \sqrt{x + 2} + \sqrt{2 - x}$ is $[-2, 2]$.

93. If f is a function defined on $(-\infty, \infty)$ and k is a real number, then $f(kx) = kf(x)$.

94. If f is a linear function, then $f(cx + y) = cf(x) + f(y)$, where c is a real number.

2.1 Solutions to Self-Check Exercises

1. a. The expression under the radical sign must be nonnegative, so $x + 1 \geq 0$ or $x \geq -1$. Also, $x \neq 0$ because division by zero is not permitted. Therefore, the domain of f is $[-1, 0) \cup (0, \infty)$.

   b. $f(3) = \dfrac{\sqrt{3+1}}{3} = \dfrac{\sqrt{4}}{3} = \dfrac{2}{3}$

   c. $f(a+h) = \dfrac{\sqrt{(a+h)+1}}{a+h} = \dfrac{\sqrt{a+h+1}}{a+h}$

2. a. For t in the subdomain $[0, 6]$, the rule for f is given by $f(t) = 6t + 17$. The equation $y = 6t + 17$ is a linear equation, so that portion of the graph of f is the line segment joining the points (0, 17) and (6, 53). Next, in the subdomain $(6, 20]$, the rule for f is given by $f(t) = 15.98(t - 6)^{1/4} + 53$. Using a calculator, we construct the following table of values of $f(t)$ for selected values of t.

t	6	8	10	12	14	16	18	20
$f(t)$	53	72	75.6	78	79.9	81.4	82.7	83.9

We have included $t = 6$ in the table, although it does not lie in the subdomain of the function under consideration, in order to help us obtain a better sketch of that portion of the graph of f in the subdomain $(6, 20]$. The graph of f follows:

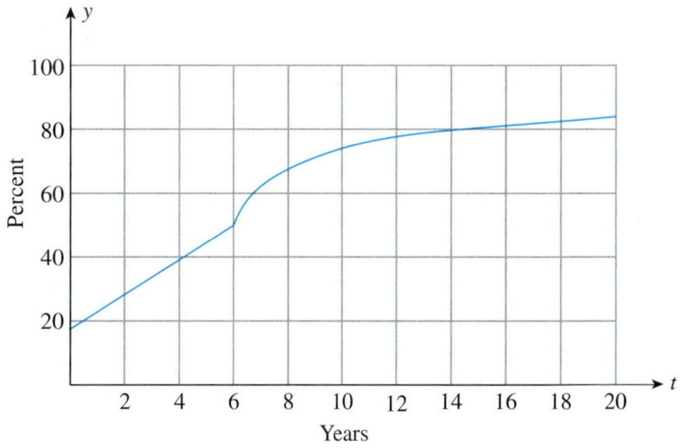

   b. The percentage of all self-serve gas sales at the beginning of 1978 is found by evaluating f at $t = 4$. Since this point lies in the interval [0, 6], we use the rule $f(t) = 6t + 17$ and find

$$f(4) = 6(4) + 17 = 41$$

giving 41% as the required figure. The percentage of all self-serve gas sales at the beginning of 1994 is given by

$$f(20) = 15.98(20 - 6)^{1/4} + 53 \approx 83.9$$

or approximately 83.9%.

3. A point (x, y) lies on the graph of the function f if and only if the coordinates satisfy the equation $y = f(x)$. Now,

$$f(4) = \sqrt{2(4) + 1} + 2 = \sqrt{9} + 2 = 5 \neq 6$$

and we conclude that the given point does *not* lie on the graph of f.

USING TECHNOLOGY

Graphing a Function

Most of the graphs of functions in this book can be plotted with the help of a graphing utility. Furthermore, a graphing utility can be used to analyze the nature of a function. However, the amount and accuracy of the information obtained using a graphing utility depend on the experience and sophistication of the user. As you progress through this book, you will see that the more knowledge of calculus you gain, the more effective the graphing utility will prove to be as a tool in problem solving.

Finding a Suitable Viewing Window

The first step in plotting the graph of a function with a graphing utility is to select a suitable viewing window. We usually do this by experimenting. For example, you might first plot the graph using the *standard viewing window* $[-10, 10]$ by $[-10, 10]$. If necessary, you then might adjust the viewing window by enlarging it or reducing it to obtain a sufficiently complete view of the graph or at least the portion of the graph that is of interest.

(continued)

EXAMPLE 1 Plot the graph of $f(x) = 2x^2 - 4x - 5$ in the standard viewing window.

Solution The graph of f, shown in Figure T1a, is a parabola. From our previous work (Example 6, Section 2.1), we know that the figure does give a good view of the graph.

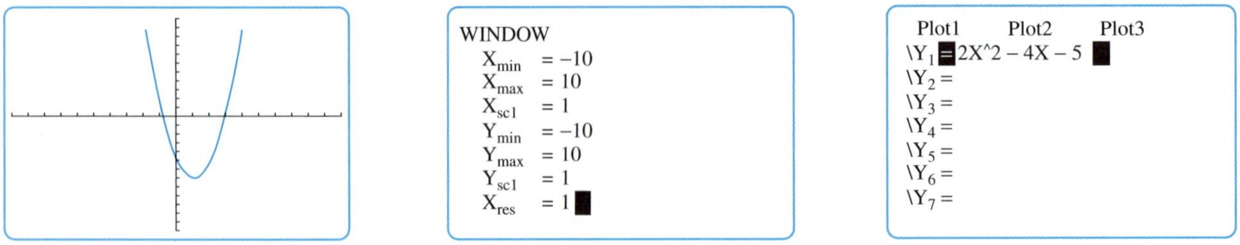

(a) **(b)** **(c)**

FIGURE T1
(a) The graph of $f(x) = 2x^2 - 4x - 5$ on $[-10, 10] \times [-10, 10]$; (b) the TI-83/84 window screen for (a); (c) the TI-83/84 equation screen

EXAMPLE 2 Let $f(x) = x^3(x - 3)^4$.

a. Plot the graph of f in the standard viewing window.
b. Plot the graph of f in the window $[-1, 5] \times [-40, 40]$.

Solution

a. The graph of f in the standard viewing window is shown in Figure T2a. Since the graph does not appear to be complete, we need to adjust the viewing window.

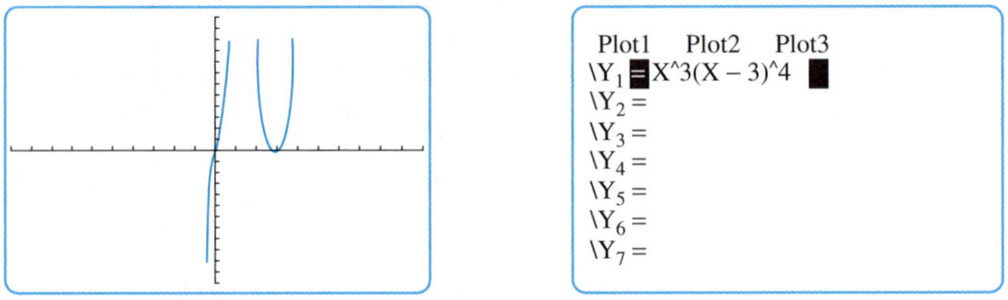

(a) **(b)**

FIGURE T2
(a) An incomplete sketch of $f(x) = x^3(x - 3)^4$ on $[-10, 10] \times [-10, 10]$; (b) the TI-83/84 equation screen

b. The graph of f in the window $[-1, 5] \times [-40, 40]$, shown in Figure T3a, is an improvement over the previous graph. (Later we will be able to show that the figure does in fact give a rather complete view of the graph of f.)

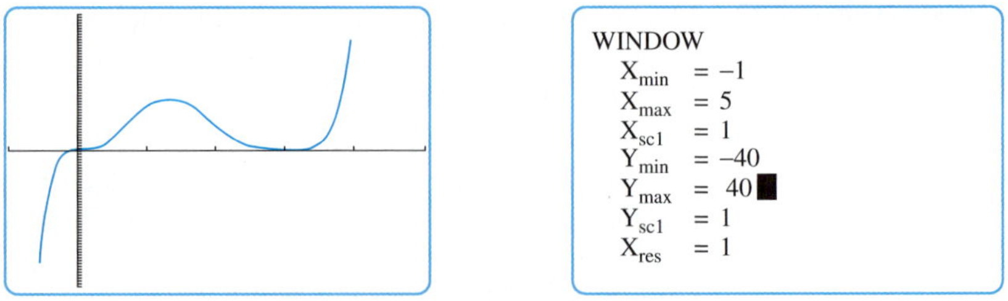

(a) **(b)**

FIGURE T3
(a) A complete sketch of $f(x) = x^3(x - 3)^4$ is shown using the window $[-1, 5] \times [-40, 40]$; (b) the TI-83/84 window screen

Evaluating a Function

A graphing utility can be used to find the value of a function with minimal effort, as the next example shows.

EXAMPLE 3 Let $f(x) = x^3 - 4x^2 + 4x + 2$.

a. Plot the graph of f in the standard viewing window.
b. Find $f(3)$ and verify your result by direct computation.
c. Find $f(4.215)$.

Solution

a. The graph of f is shown in Figure T4a.

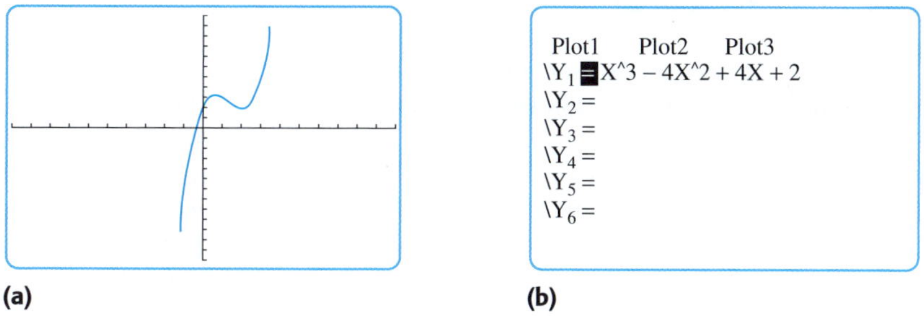

FIGURE **T4**
(a) The graph of $f(x) = x^3 - 4x^2 + 4x + 2$ in the standard viewing window; (b) the TI-83/84 equation screen

b. Using the evaluation function of the graphing utility and the value 3 for x, we find $y = 5$. This result is verified by computing

$$f(3) = 3^3 - 4(3^2) + 4(3) + 2 = 27 - 36 + 12 + 2 = 5$$

c. Using the evaluation function of the graphing utility and the value 4.215 for x, we find $y = 22.679738375$. Thus, $f(4.215) = 22.679738375$. The efficacy of the graphing utility is clearly demonstrated here! ■

APPLIED EXAMPLE 4 Number of Alzheimer's Patients The number of Alzheimer's patients in the United States is approximated by

$$f(t) = -0.0277t^4 + 0.3346t^3 - 1.1261t^2 + 1.7575t + 3.7745 \qquad (0 \le t \le 6)$$

where $f(t)$ is measured in millions and t is measured in decades, with $t = 0$ corresponding to the beginning of 1990.

a. Use a graphing utility to plot the graph of f in the viewing window $[0, 7] \times [0, 12]$.
b. What is the anticipated number of Alzheimer's patients in the United States at the beginning of 2010 ($t = 2$)? At the beginning of 2030 ($t = 4$)?

Source: Alzheimer's Association

Solution

a. The graph of f in the viewing window $[0, 7] \times [0, 12]$ is shown in Figure T5a.

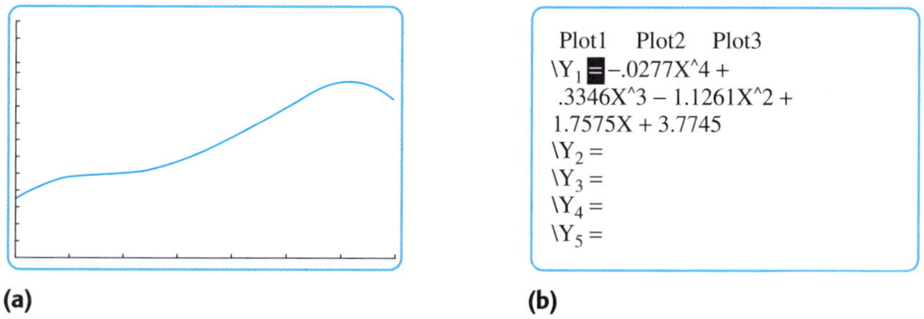

FIGURE **T5**
(a) The graph of f in the viewing window $[0, 7] \times [0, 12]$; (b) the TI-83/84 equation screen

(continued)

b. Using the evaluation function of the graphing utility and the value 2 for x, we see that the anticipated number of Alzheimer's patients at the beginning of 2010 is given by $f(2) = 5.0187$, or approximately 5 million. The anticipated number of Alzheimer's patients at the beginning of 2030 is given by $f(4) = 7.1101$ or approximately 7.1 million. ■

TECHNOLOGY EXERCISES

In Exercises 1–4, plot the graph of the function *f* in (a) the standard viewing window and (b) the indicated window.

1. $f(x) = x^4 - 2x^2 + 8; [-2, 2] \times [6, 10]$

2. $f(x) = x^3 - 20x^2 + 8x - 10; [-20, 20] \times [-1200, 100]$

3. $f(x) = x\sqrt{4 - x^2}; [-3, 3] \times [-2, 2]$

4. $f(x) = \dfrac{4}{x^2 - 8}; [-5, 5] \times [-5, 5]$

In Exercises 5–8, plot the graph of the function *f* in an appropriate viewing window. (*Note:* The answer is *not* unique.)

5. $f(x) = 2x^4 - 3x^3 + 5x^2 - 20x + 40$

6. $f(x) = -2x^4 + 5x^2 - 4$

7. $f(x) = \dfrac{x^3}{x^3 + 1}$

8. $f(x) = \dfrac{2x^4 - 3x}{x^2 - 1}$

In Exercises 9–12, use the evaluation function of your graphing utility to find the value of *f* at the indicated value of *x*. Express your answer accurate to four decimal places.

9. $f(x) = 3x^3 - 2x^2 + x - 4; x = 2.145$

10. $f(x) = 5x^4 - 2x^2 + 8x - 3; x = 1.28$

11. $f(x) = \dfrac{2x^3 - 3x + 1}{3x - 2}; x = 2.41$

12. $f(x) = \sqrt{2x^2 + 1} + \sqrt{3x^2 - 1}; x = 0.62$

13. **Hiring Lobbyists** Many public entities like cities, counties, states, utilities, and Indian tribes are hiring firms to lobby Congress. One goal of such lobbying is to place earmarks—money directed at a specific project—into appropriation bills. The amount (in millions of dollars) spent by public entities on lobbying from 1998 through 2004, where $t = 0$ corresponds to 1998, is given by

$$f(t) = -0.425t^3 + 3.6571t^2 + 4.018t + 43.7 \qquad (0 \le t \le 6)$$

a. Plot the graph of f in the viewing window $[0, 6] \times [0, 110]$.

b. What amount was spent by public entities on lobbying in the year 2000? In 2004?

Source: Center for Public Integrity

14. **Surveillance Cameras** Research reports indicate that surveillance cameras at major intersections dramatically reduce the number of drivers who barrel through red lights. The cameras automatically photograph vehicles that drive into intersections after the light turns red. Vehicle owners are then mailed citations instructing them to pay a fine or sign an affidavit that they weren't driving at the time. The function

$$N(t) = 6.08t^3 - 26.79t^2 + 53.06t + 69.5 \qquad (0 \le t \le 4)$$

gives the number, $N(t)$, of U.S. communities using surveillance cameras at intersections in year t, with $t = 0$ corresponding to 2003.

a. Plot the graph of N in the viewing window $[0, 4] \times [0, 250]$.

b. How many communities used surveillance cameras at intersections in 2004? In 2006?

Source: Insurance Institute for Highway Safety

15. **Keeping with the Traffic Flow** By driving at a speed to match the prevailing traffic speed, you decrease the chances of an accident. According to data obtained in a university study, the number of accidents/100 million vehicle miles, y, is related to the deviation from the mean speed, x, in mph by

$$y = 1.05x^3 - 21.95x^2 + 155.9x - 327.3 \qquad (6 \le x \le 11)$$

a. Plot the graph of y in the viewing window $[6, 11] \times [20, 150]$.

b. What is the number of accidents/100 million vehicle miles if the deviation from the mean speed is 6 mph, 8 mph, and 11 mph?

Source: University of Virginia School of Engineering and Applied Science

16. **Safe Drivers** The fatality rate in the United States (per 100 million miles traveled) by age of driver (in years) is given by the function

$$f(x) = 0.00000304x^4 - 0.0005764x^3 + 0.04105x^2 - 1.30366x + 16.579 \qquad (18 \le x \le 82)$$

a. Plot the graph of f in the viewing window $[18, 82] \times [0, 8]$.

b. What is the fatality rate for 18-yr-old drivers? For 50-yr-old drivers? For 80-yr-old drivers?

Source: National Highway Traffic Safety Administration

2.2 The Algebra of Functions

The Sum, Difference, Product, and Quotient of Functions

Let $S(t)$ and $R(t)$ denote, respectively, the federal government's spending and revenue at any time t, measured in billions of dollars. The graphs of these functions for the period between 1990 and 2000 are shown in Figure 11.

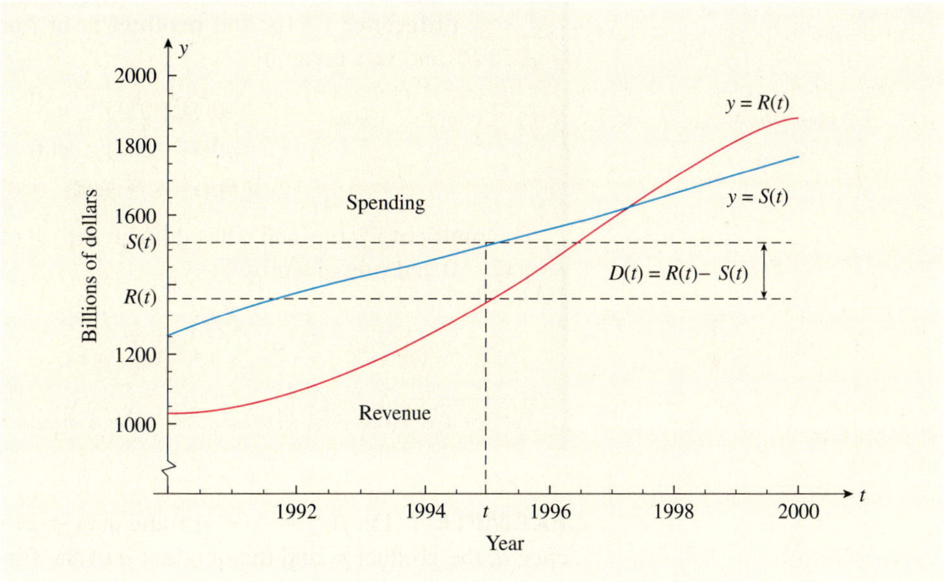

Source: Office of Management and Budget

FIGURE 11
$R(t) - S(t)$ gives the federal budget deficit (surplus) at any time t.

The difference $R(t) - S(t)$ gives the deficit (surplus) in billions of dollars at any time t if $R(t) - S(t)$ is negative (positive). This observation suggests that we can define a function D whose value at any time t is given by $R(t) - S(t)$. The function D, the *difference* of the two functions R and S, is written $D = R - S$ and may be called the "deficit (surplus) function" since it gives the budget deficit or surplus at any time t. It has the same domain as the functions S and R. The graph of the function D is shown in Figure 12.

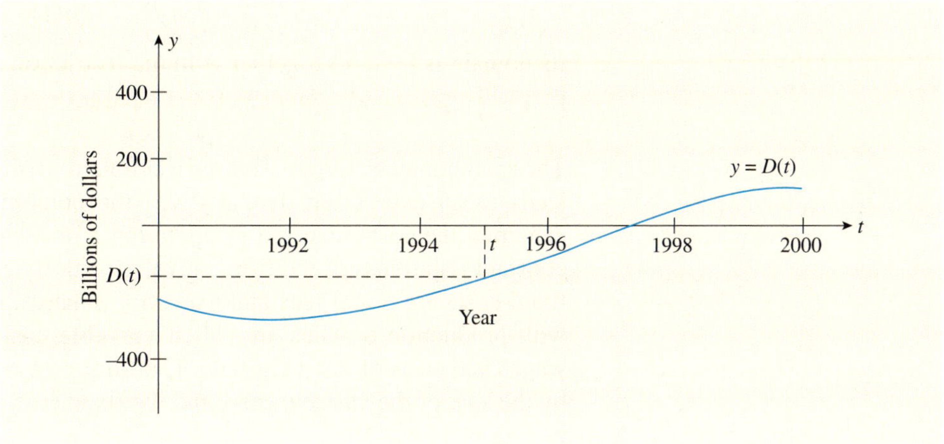

Source: Office of Management and Budget

FIGURE 12
The graph of $D(t)$

Most functions are built up from other, generally simpler functions. For example, we may view the function $f(x) = 2x + 4$ as the sum of the two functions $g(x) = 2x$

and $h(x) = 4$. The function $g(x) = 2x$ may in turn be viewed as the product of the functions $p(x) = 2$ and $q(x) = x$.

In general, given the functions f and g, we define the sum $f + g$, the difference $f - g$, the product fg, and the quotient f/g of f and g as follows.

The Sum, Difference, Product, and Quotient of Functions

Let f and g be functions with domains A and B, respectively. Then the **sum** $f + g$, **difference** $f - g$, and **product** fg of f and g are functions with domain $A \cap B$* and rule given by

$$(f + g)(x) = f(x) + g(x) \qquad \text{Sum}$$

$$(f - g)(x) = f(x) - g(x) \qquad \text{Difference}$$

$$(fg)(x) = f(x)g(x) \qquad \text{Product}$$

The **quotient** f/g of f and g has domain $A \cap B$ excluding all numbers x such that $g(x) = 0$ and rule given by

$$\left(\frac{f}{g}\right)(x) = \frac{f(x)}{g(x)} \qquad \text{Quotient}$$

*$A \cap B$ is read "A intersected with B" and denotes the set of all points common to both A and B.

EXAMPLE 1 Let $f(x) = \sqrt{x + 1}$ and $g(x) = 2x + 1$. Find the sum s, the difference d, the product p, and the quotient q of the functions f and g.

Solution Since the domain of f is $A = [-1, \infty)$ and the domain of g is $B = (-\infty, \infty)$, we see that the domain of s, d, and p is $A \cap B = [-1, \infty)$. The rules follow.

$$s(x) = (f + g)(x) = f(x) + g(x) = \sqrt{x + 1} + 2x + 1$$

$$d(x) = (f - g)(x) = f(x) - g(x) = \sqrt{x + 1} - (2x + 1) = \sqrt{x + 1} - 2x - 1$$

$$p(x) = (fg)(x) = f(x)g(x) = \sqrt{x + 1}(2x + 1) = (2x + 1)\sqrt{x + 1}$$

The quotient function q has rule

$$q(x) = \left(\frac{f}{g}\right)(x) = \frac{f(x)}{g(x)} = \frac{\sqrt{x + 1}}{2x + 1}$$

Its domain is $[-1, \infty)$ together with the restriction $x \neq -\frac{1}{2}$. We denote this by $[-1, -\frac{1}{2}) \cup (-\frac{1}{2}, \infty)$. ■

The mathematical formulation of a problem arising from a practical situation often leads to an expression that involves the combination of functions. Consider, for example, the costs incurred in operating a business. Costs that remain more or less constant regardless of the firm's level of activity are called **fixed costs.** Examples of fixed costs are rental fees and executive salaries. On the other hand, costs that vary with production or sales are called **variable costs.** Examples of variable costs are wages and costs of raw materials. The **total cost** of operating a business is thus given by the *sum* of the variable costs and the fixed costs, as illustrated in the next example.

VIDEO

APPLIED EXAMPLE 2 Cost Functions Suppose Puritron, a manufacturer of water filters, has a monthly fixed cost of \$10,000 and a variable cost of

$$-0.0001x^2 + 10x \qquad (0 \leq x \leq 40{,}000)$$

dollars, where x denotes the number of filters manufactured per month. Find a function C that gives the total monthly cost incurred by Puritron in the manufacture of x filters.

Solution Puritron's monthly fixed cost is always \$10,000, regardless of the level of production, and it is described by the constant function $F(x) = 10{,}000$. Next, the variable cost is described by the function $V(x) = -0.0001x^2 + 10x$. Since the total cost incurred by Puritron at any level of production is the sum of the variable cost and the fixed cost, we see that the required total cost function is given by

$$\begin{aligned} C(x) &= V(x) + F(x) \\ &= -0.0001x^2 + 10x + 10{,}000 \qquad (0 \le x \le 40{,}000) \end{aligned}$$

Next, the **total profit** realized by a firm in operating a business is the *difference* between the total revenue realized and the total cost incurred; that is,

$$P(x) = R(x) - C(x)$$

APPLIED EXAMPLE 3 Profit Functions Refer to Example 2. Suppose the total revenue realized by Puritron from the sale of x water filters is given by the total revenue function

$$R(x) = -0.0005x^2 + 20x \qquad (0 \le x \le 40{,}000)$$

a. Find the total profit function—that is, the function that describes the total profit Puritron realizes in manufacturing and selling x water filters per month.
b. What is the profit when the level of production is 10,000 filters per month?

Solution

a. The total profit realized by Puritron in manufacturing and selling x water filters per month is the difference between the total revenue realized and the total cost incurred. Thus, the required total profit function is given by

$$\begin{aligned} P(x) &= R(x) - C(x) \\ &= (-0.0005x^2 + 20x) - (-0.0001x^2 + 10x + 10{,}000) \\ &= -0.0004x^2 + 10x - 10{,}000 \end{aligned}$$

b. The profit realized by Puritron when the level of production is 10,000 filters per month is

$$P(10{,}000) = -0.0004(10{,}000)^2 + 10(10{,}000) - 10{,}000 = 50{,}000$$

or \$50,000 per month.

Composition of Functions

Another way to build up a function from other functions is through a process known as the *composition of functions*. Consider, for example, the function h, whose rule is given by $h(x) = \sqrt{x^2 - 1}$. Let f and g be functions defined by the rules $f(x) = x^2 - 1$ and $g(x) = \sqrt{x}$. Evaluating the function g at the point $f(x)$ [remember that for each real number x in the domain of f, $f(x)$ is simply a real number], we find that

$$g(f(x)) = \sqrt{f(x)} = \sqrt{x^2 - 1}$$

which is just the rule defining the function h!

In general, the composition of a function g with a function f is defined as follows.

The Composition of Two Functions

Let f and g be functions. Then the composition of g and f is the function $g \circ f$ defined by

$$(g \circ f)(x) = g(f(x))$$

The domain of $g \circ f$ is the set of all x in the domain of f such that $f(x)$ lies in the domain of g.

The function $g \circ f$ (read "g circle f") is also called a **composite function.** The interpretation of the function $h = g \circ f$ as a machine is illustrated in Figure 13 and its interpretation as a mapping is shown in Figure 14.

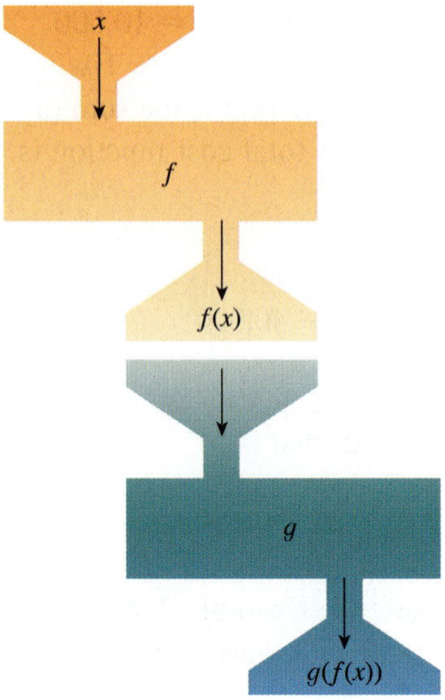

FIGURE **13**
The composite function $h = g \circ f$ viewed as a machine

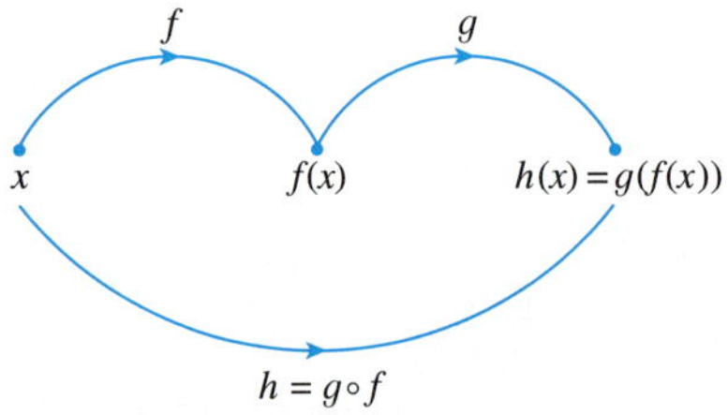

FIGURE **14**
The function $h = g \circ f$ viewed as a mapping

EXAMPLE 4 Let $f(x) = x^2 - 1$ and $g(x) = \sqrt{x} + 1$. Find:

a. The rule for the composite function $g \circ f$.
b. The rule for the composite function $f \circ g$.

Solution

a. To find the rule for the composite function $g \circ f$, evaluate the function g at $f(x)$. We obtain

$$(g \circ f)(x) = g(f(x)) = \sqrt{f(x)} + 1 = \sqrt{x^2 - 1} + 1$$

b. To find the rule for the composite function $f \circ g$, evaluate the function f at $g(x)$. Thus,

$$\begin{aligned}(f \circ g)(x) &= f(g(x)) = (g(x))^2 - 1 = (\sqrt{x} + 1)^2 - 1 \\ &= x + 2\sqrt{x} + 1 - 1 = x + 2\sqrt{x}\end{aligned}$$

Example 4 reminds us that in general $g \circ f$ is different from $f \circ g$, so care must be taken when finding the rule for a composite function.

Explore & Discuss

Let $f(x) = \sqrt{x} + 1$ for $x \geq 0$ and let $g(x) = (x - 1)^2$ for $x \geq 1$.

1. Show that $(g \circ f)(x)$ and $(f \circ g)(x) = x$. (*Note:* The function g is said to be the *inverse* of f and vice versa.)
2. Plot the graphs of f and g together with the straight line $y = x$. Describe the relationship between the graphs of f and g.

VIDEO

APPLIED EXAMPLE 5 Automobile Pollution An environmental impact study conducted for the city of Oxnard indicates that, under existing environmental protection laws, the level of carbon monoxide (CO) present in the air due to pollution from automobile exhaust will be $0.01x^{2/3}$ parts per million when the number of motor vehicles is x thousand. A separate study conducted by a state government agency estimates that t years from now the number of motor vehicles in Oxnard will be $0.2t^2 + 4t + 64$ thousand.

a. Find an expression for the concentration of CO in the air due to automobile exhaust t years from now.
b. What will be the level of concentration 5 years from now?

Solution

a. The level of CO present in the air due to pollution from automobile exhaust is described by the function $g(x) = 0.01x^{2/3}$, where x is the number (in thousands) of motor vehicles. But the number of motor vehicles x (in thousands) t years from now may be estimated by the rule $f(t) = 0.2t^2 + 4t + 64$. Therefore, the concentration of CO due to automobile exhaust t years from now is given by

$$C(t) = (g \circ f)(t) = g(f(t)) = 0.01(0.2t^2 + 4t + 64)^{2/3}$$

parts per million.

b. The level of concentration 5 years from now will be

$$C(5) = 0.01[0.2(5)^2 + 4(5) + 64]^{2/3}$$
$$= (0.01)89^{2/3} \approx 0.20$$

or approximately 0.20 parts per million. ■

2.2 Self-Check Exercises

1. Let f and g be functions defined by the rules

$$f(x) = \sqrt{x} + 1 \quad \text{and} \quad g(x) = \frac{x}{1 + x}$$

respectively. Find the rules for

a. The sum s, the difference d, the product p, and the quotient q of f and g.
b. The composite functions $f \circ g$ and $g \circ f$.

2. Health-care spending per person by the private sector includes payments by individuals, corporations, and their insurance companies and is approximated by the function

$$f(t) = 2.48t^2 + 18.47t + 509 \qquad (0 \le t \le 6)$$

where $f(t)$ is measured in dollars and t is measured in years, with $t = 0$ corresponding to the beginning of 1994. The corresponding government spending—including expenditures for Medicaid, Medicare, and other federal, state, and local government public health care—is

$$g(t) = -1.12t^2 + 29.09t + 429 \qquad (0 \le t \le 6)$$

where t has the same meaning as before.

a. Find a function that gives the difference between private and government health-care spending per person at any time t.
b. What was the difference between private and government expenditures per person at the beginning of 1995? At the beginning of 2000?

Source: Health Care Financing Administration

Solutions to Self-Check Exercises 2.2 can be found on page 74.

2.2 Concept Questions

1. a. Explain what is meant by the sum, difference, product, and quotient of the functions f and g with domains A and B, respectively.
b. If $f(2) = 3$ and $g(2) = -2$, what is $(f + g)(2)$? $(f - g)(2)$? $(fg)(2)$? $(f/g)(2)$?

2. Let f and g be functions and suppose that (x, y) is a point on the graph of h. What is the value of y for $h = f + g$? $h = f - g$? $h = fg$? $h = f/g$?

3. a. What is the composition of the functions f and g? The functions g and f?
b. If $f(2) = 3$ and $g(3) = 8$, what is $(g \circ f)(2)$? Can you conclude from the given information what $(f \circ g)(3)$ is? Explain.

4. Let f be a function with domain A and let g be a function whose domain contains the range of f. If a is any number in A, must $(g \circ f)(a)$ be defined? Explain with an example.

2.2 Exercises

In Exercises 1–8, let $f(x) = x^3 + 5$, $g(x) = x^2 - 2$, and $h(x) = 2x + 4$. Find the rule for each function.

1. $f + g$
2. $f - g$
3. fg
4. gf
5. $\frac{f}{g}$
6. $\frac{f - g}{h}$
7. $\frac{fg}{h}$
8. fgh

In Exercises 9–18, let $f(x) = x - 1$, $g(x) = \sqrt{x + 1}$, and $h(x) = 2x^3 - 1$. Find the rule for each function.

9. $f + g$
10. $g - f$
11. fg
12. gf
13. $\frac{g}{h}$
14. $\frac{h}{g}$
15. $\frac{fg}{h}$
16. $\frac{fh}{g}$
17. $\frac{f - h}{g}$
18. $\frac{gh}{g - f}$

In Exercises 19–24, find the functions $f + g$, $f - g$, fg, and f/g.

19. $f(x) = x^2 + 5;\ g(x) = \sqrt{x} - 2$
20. $f(x) = \sqrt{x - 1};\ g(x) = x^3 + 1$
21. $f(x) = \sqrt{x + 3};\ g(x) = \frac{1}{x - 1}$
22. $f(x) = \frac{1}{x^2 + 1};\ g(x) = \frac{1}{x^2 - 1}$
23. $f(x) = \frac{x + 1}{x - 1};\ g(x) = \frac{x + 2}{x - 2}$
24. $f(x) = x^2 + 1;\ g(x) = \sqrt{x + 1}$

In Exercises 25–30, find the rules for the composite functions $f \circ g$ and $g \circ f$.

25. $f(x) = x^2 + x + 1;\ g(x) = x^2$
26. $f(x) = 3x^2 + 2x + 1;\ g(x) = x + 3$
27. $f(x) = \sqrt{x} + 1;\ g(x) = x^2 - 1$
28. $f(x) = 2\sqrt{x} + 3;\ g(x) = x^2 + 1$
29. $f(x) = \frac{x}{x^2 + 1};\ g(x) = \frac{1}{x}$
30. $f(x) = \sqrt{x + 1};\ g(x) = \frac{1}{x - 1}$

In Exercises 31–34, evaluate $h(2)$, where $h = g \circ f$.

31. $f(x) = x^2 + x + 1;\ g(x) = x^2$
32. $f(x) = \sqrt[3]{x^2 - 1};\ g(x) = 3x^3 + 1$
33. $f(x) = \frac{1}{2x + 1};\ g(x) = \sqrt{x}$
34. $f(x) = \frac{1}{x - 1};\ g(x) = x^2 + 1$

In Exercises 35–42, find functions f and g such that $h = g \circ f$. (*Note:* The answer is *not* unique.)

35. $h(x) = (2x^3 + x^2 + 1)^5$
36. $h(x) = (3x^2 - 4)^{-3}$
37. $h(x) = \sqrt{x^2 - 1}$
38. $h(x) = (2x - 3)^{3/2}$
39. $h(x) = \frac{1}{x^2 - 1}$
40. $h(x) = \frac{1}{\sqrt{x^2 - 4}}$
41. $h(x) = \frac{1}{(3x^2 + 2)^{3/2}}$
42. $h(x) = \frac{1}{\sqrt{2x + 1}} + \sqrt{2x + 1}$

In Exercises 43–46, find $f(a + h) - f(a)$ for each function. Simplify your answer.

43. $f(x) = 3x + 4$
44. $f(x) = -\frac{1}{2}x + 3$
45. $f(x) = 4 - x^2$
46. $f(x) = x^2 - 2x + 1$

In Exercises 47–52, find and simplify

$$\frac{f(a + h) - f(a)}{h} \qquad (h \neq 0)$$

for each function.

47. $f(x) = x^2 + 1$
48. $f(x) = 2x^2 - x + 1$
49. $f(x) = x^3 - x$
50. $f(x) = 2x^3 - x^2 + 1$
51. $f(x) = \frac{1}{x}$
52. $f(x) = \sqrt{x}$

53. **Restaurant Revenue** Nicole owns and operates two restaurants. The revenue of the first restaurant at time t is $f(t)$ dollars, and the revenue of the second restaurant at time t is $g(t)$ dollars. What does the function $F(t) = f(t) + g(t)$ represent?

54. **Birthrate of Endangered Species** The birthrate of an endangered species of whales in year t is $f(t)$ whales/year. This species of whales is dying at the rate of $g(t)$ whales/year in year t. What does the function $F(t) = f(t) - g(t)$ represent?

55. **Value of an Investment** The number of IBM shares that Nancy owns is given by $f(t)$. The price per share of the stock of IBM at time t is $g(t)$ dollars. What does the function $f(t)g(t)$ represent?

56. **Production Costs** The total cost incurred by time t in the production of a certain commodity is $f(t)$ dollars. The number of products produced by time t is $g(t)$ units. What does the function $f(t)/g(t)$ represent?

57. **Carbon Monoxide Pollution** The number of cars running in the business district of a town at time t is given by $f(t)$. Carbon monoxide pollution coming from these cars is given by $g(x)$ parts per million, where x is the number of cars being operated in the district. What does the function $g \circ f$ represent?

58. **Effect of Advertising on Revenue** The revenue of Leisure Travel is given by $f(x)$ dollars, where x is the dollar amount spent by the company on advertising. The amount spent by Leisure at time t on advertising is given by $g(t)$ dollars. What does the function $f \circ g$ represent?

59. **Manufacturing Costs** TMI, a manufacturer of blank audiocassette tapes, has a monthly fixed cost of \$12,100 and a variable cost of \$.60/tape. Find a function C that gives the total cost incurred by TMI in the manufacture of x tapes/month.

60. **Spam Messages** The total number of email messages per day (in billions) between 2003 and 2007 is approximated by

$$f(t) = 1.54t^2 + 7.1t + 31.4 \qquad (0 \le t \le 4)$$

where t is measured in years, with $t = 0$ corresponding to 2003. Over the same period, the total number of spam messages per day (in billions) is approximated by

$$g(t) = 1.21t^2 + 6t + 14.5 \qquad (0 \le t \le 4)$$

a. Find the rule for the function $D = f - g$. Compute $D(4)$ and explain what it measures.
b. Find the rule for the function $P = f/g$. Compute $P(4)$ and explain what it means.

Source: Technology Review

61. **Global Supply of Plutonium** The global stockpile of plutonium for military applications between 1990 ($t = 0$) and 2003 ($t = 13$) stood at a constant 267 tons. On the other hand, the global stockpile of plutonium for civilian use was

$$2t^2 + 46t + 733$$

tons in year t over the same period.
a. Find the function f giving the global stockpile of plutonium for military use from 1990 through 2003 and the function g giving the global stockpile of plutonium for civilian use over the same period.
b. Find the function h giving the total global stockpile of plutonium between 1990 and 2003.
c. What was the total global stockpile of plutonium in 2003?

Source: Institute for Science and International Security

62. **Motorcycle Deaths** Suppose the fatality rate (deaths/100 million miles traveled) of motorcyclists is given by $g(x)$, where x is the percentage of motorcyclists who wear helmets. Next, suppose the percentage of motorcyclists who wear helmets at time t (t measured in years) is $f(t)$, with $t = 0$ corresponding to 2000.
a. If $f(0) = 0.64$ and $g(0.64) = 26$ find $(g \circ f)(0)$ and interpret your result.
b. If $f(6) = 0.51$ and $g(0.51) = 42$ find $(g \circ f)(6)$ and interpret your result.
c. Comment on the results of parts (a) and (b).

Source: National Highway Traffic Safety Administration

63. **Fighting Crime** Suppose the reported serious crimes (crimes that include homicide, rape, robbery, aggravated assault, burglary, and car theft) that end in arrests or in the identification of suspects is $g(x)$ percent, where x denotes the total number of detectives. Next, suppose the total number of detectives in year t is $f(t)$, with $t = 0$ corresponding to 2001.
a. If $f(1) = 406$ and $g(406) = 23$, find $(g \circ f)(1)$ and interpret your result.
b. If $f(6) = 326$ and $g(326) = 18$, find $(g \circ f)(6)$ and interpret your result.
c. Comment on the results of parts (a) and (b).

Source: Boston Police Department

64. **Cost of Producing PDAs** Apollo manufactures PDAs at a variable cost of

$$V(x) = 0.000003x^3 - 0.03x^2 + 200x$$

dollars, where x denotes the number of units manufactured per month. The monthly fixed cost attributable to the division that produces these PDAs is \$100,000. Find a function C that gives the total cost incurred by the manufacture of x PDAs. What is the total cost incurred in producing 2000 units/month?

65. **Profit from Sale of PDAs** Refer to Exercise 64. Suppose the total revenue realized by Apollo from the sale of x PDAs is given by the total revenue function

$$R(x) = -0.1x^2 + 500x \qquad (0 \le x \le 5000)$$

where $R(x)$ is measured in dollars.
a. Find the total profit function.
b. What is the profit when 1500 units are produced and sold each month?

66. **Profit from Sale of Pagers** A division of Chapman Corporation manufactures a pager. The weekly fixed cost for the division is \$20,000, and the variable cost for producing x pagers/week is

$$V(x) = 0.000001x^3 - 0.01x^2 + 50x$$

dollars. The company realizes a revenue of

$$R(x) = -0.02x^2 + 150x \qquad (0 \le x \le 7500)$$

dollars from the sale of x pagers/week.
a. Find the total cost function.
b. Find the total profit function.
c. What is the profit for the company if 2000 units are produced and sold each week?

67. **Overcrowding of Prisons** The 1980s saw a trend toward old-fashioned punitive deterrence as opposed to the more liberal penal policies and community-based corrections popular in the 1960s and early 1970s. As a result, prisons became more crowded, and the gap between the number of people in prison and the prison capacity widened. The number of prisoners (in thousands) in federal and state prisons is approximated by the function

$$N(t) = 3.5t^2 + 26.7t + 436.2 \qquad (0 \le t \le 10)$$

where t is measured in years, with $t = 0$ corresponding to 1983. The number of inmates for which prisons were designed is given by

$$C(t) = 24.3t + 365 \qquad (0 \le t \le 10)$$

where $C(t)$ is measured in thousands and t has the same meaning as before.
a. Find an expression that shows the gap between the number of prisoners and the number of inmates for which the prisons were designed at any time t.
b. Find the gap at the beginning of 1983 and at the beginning of 1986.

Source: U.S. Department of Justice

68. **Effect of Mortgage Rates on Housing Starts** A study prepared for the National Association of Realtors estimated that the number of housing starts per year over the next 5 yr will be

$$N(r) = \frac{7}{1 + 0.02r^2}$$

million units, where r (percent) is the mortgage rate. Suppose the mortgage rate over the next t mo is

$$r(t) = \frac{10t + 150}{t + 10} \qquad (0 \le t \le 24)$$

percent/year.
 a. Find an expression for the number of housing starts per year as a function of t, t mo from now.
 b. Using the result from part (a), determine the number of housing starts at present, 12 mo from now, and 18 mo from now.

69. **Hotel Occupancy Rate** The occupancy rate of the all-suite Wonderland Hotel, located near an amusement park, is given by the function

$$r(t) = \frac{10}{81}t^3 - \frac{10}{3}t^2 + \frac{200}{9}t + 55 \qquad (0 \le t \le 11)$$

where t is measured in months and $t = 0$ corresponds to the beginning of January. Management has estimated that the monthly revenue (in thousands of dollars) is approximated by the function

$$R(r) = -\frac{3}{5000}r^3 + \frac{9}{50}r^2 \qquad (0 \le r \le 100)$$

where r (percent) is the occupancy rate.
 a. What is the hotel's occupancy rate at the beginning of January? At the beginning of June?
 b. What is the hotel's monthly revenue at the beginning of January? At the beginning of June?
 Hint: Compute $R(r(0))$ and $R(r(5))$.

70. **Housing Starts and Construction Jobs** The president of a major housing construction firm reports that the number of construction jobs (in millions) created is given by

$$N(x) = 1.42x$$

where x denotes the number of housing starts. Suppose the number of housing starts in the next t mo is expected to be

$$x(t) = \frac{7(t + 10)^2}{(t + 10)^2 + 2(t + 15)^2}$$

million units/year. Find an expression for the number of jobs created per month in the next t mo. How many jobs will have been created 6 mo and 12 mo from now?

71. a. Let f, g, and h be functions. How would you define the "sum" of f, g, and h?
 b. Give a real-life example involving the sum of three functions. (*Note:* The answer is not unique.)

72. a. Let f, g, and h be functions. How would you define the "composition" of h, g, and f, in that order?
 b. Give a real-life example involving the composition of these functions. (*Note:* The answer is not unique.)

In Exercises 73–76, determine whether the statement is true or false. If it is true, explain why it is true. If it is false, give an example to show why it is false.

73. If f and g are functions with domain D, then $f + g = g + f$.

74. If $g \circ f$ is defined at $x = a$, then $f \circ g$ must also be defined at $x = a$.

75. If f and g are functions, then $f \circ g = g \circ f$.

76. If f is a function, then $f \circ f = f^2$.

2.2 Solutions to Self-Check Exercises

1. a. $s(x) = f(x) + g(x) = \sqrt{x} + 1 + \dfrac{x}{1 + x}$

$$d(x) = f(x) - g(x) = \sqrt{x} + 1 - \frac{x}{1 + x}$$

$$p(x) = f(x)g(x) = (\sqrt{x} + 1) \cdot \frac{x}{1 + x} = \frac{x(\sqrt{x} + 1)}{1 + x}$$

$$q(x) = \frac{f(x)}{g(x)} = \frac{\sqrt{x} + 1}{\dfrac{x}{1 + x}} = \frac{(\sqrt{x} + 1)(1 + x)}{x}$$

 b. $(f \circ g)(x) = f(g(x)) = \sqrt{\dfrac{x}{1 + x}} + 1$

$$(g \circ f)(x) = g(f(x)) = \frac{\sqrt{x} + 1}{1 + (\sqrt{x} + 1)} = \frac{\sqrt{x} + 1}{\sqrt{x} + 2}$$

2. a. The difference between private and government health-care spending per person at any time t is given by the function d with the rule

$$\begin{aligned} d(t) = f(t) - g(t) &= (2.48t^2 + 18.47t + 509) \\ &\quad - (-1.12t^2 + 29.09t + 429) \\ &= 3.6t^2 - 10.62t + 80 \end{aligned}$$

 b. The difference between private and government expenditures per person at the beginning of 1995 is given by

$$d(1) = 3.6(1)^2 - 10.62(1) + 80$$

or \$72.98/person.

The difference between private and government expenditures per person at the beginning of 2000 is given by

$$d(6) = 3.6(6)^2 - 10.62(6) + 80$$

or \$145.88/person.

2.3 Functions and Mathematical Models

Mathematical Models

One of the fundamental goals in this book is to show how mathematics and, in particular, calculus can be used to solve real-world problems such as those arising from the world of business and the social, life, and physical sciences. You have already seen some of these problems earlier. Here are a few more examples of real-world phenomena that we will analyze in this and ensuing chapters.

- Global warming (p. 78)
- The solvency of the U.S. Social Security trust fund (p. 79)
- The growth in the number of mobile instant messaging accounts (p. 166)
- The number of Internet users in China (p. 335)
- The size of the autistic brain (p. 368)
- The projected U.S. gasoline usage (p. 449)

Regardless of the field from which the real-world problem is drawn, the problem is analyzed using a process called **mathematical modeling.** The four steps in this process, as illustrated in Figure 15, follow.

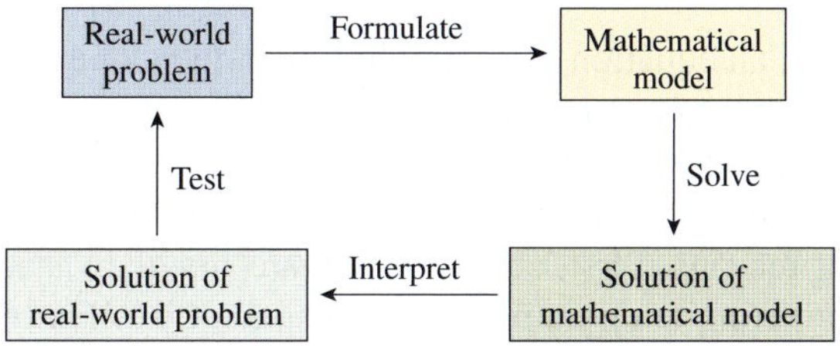

FIGURE 15

1. **Formulate** Given a real-world problem, our first task is to formulate the problem, using the language of mathematics. The many techniques used in constructing mathematical models range from theoretical consideration of the problem on the one extreme to an interpretation of data associated with the problem on the other. For example, the mathematical model giving the accumulated amount at any time when a certain sum of money is deposited in the bank can be derived theoretically (see Chapter 5). On the other hand, many of the mathematical models in this book are constructed by studying the data associated with the problem (see Using Technology, pages 93–96). In calculus, we are primarily concerned with how one (dependent) variable depends on one or more (independent) variables. Consequently, most of our mathematical models will involve functions of one or more variables or equations defining these functions (implicitly).
2. **Solve** Once a mathematical model has been constructed, we can use the appropriate mathematical techniques, which we will develop throughout the book, to solve the problem.
3. **Interpret** Bearing in mind that the solution obtained in step 2 is just the solution of the mathematical model, we need to interpret these results in the context of the original real-world problem.
4. **Test** Some mathematical models of real-world applications describe the situations with complete accuracy. For example, the model describing a deposit in a bank account gives the exact accumulated amount in the account at any time. But other mathematical models give, at best, an approximate description of the real-world problem. In this case we need to test the accuracy of the model by observing how well it describes the original real-world problem and how well it predicts past and/or future behavior. If the results are unsatisfactory, then we may have to reconsider the assumptions made in the construction of the model or, in the worst case, return to step 1.

Many real-world phenomena including those mentioned at the beginning of this section are modeled by an appropriate function.

In what follows, we will recall some familiar functions and give examples of real-world phenomena that are modeled using these functions.

Polynomial Functions

A **polynomial function** of degree n is a function of the form

$$f(x) = a_n x^n + a_{n-1}x^{n-1} + \cdots + a_2 x^2 + a_1 x + a_0 \qquad (a_n \neq 0)$$

where n is a nonnegative integer and the numbers $a_0, a_1, \ldots, a_n$ are constants called the **coefficients** of the polynomial function. For example, the functions

$$f(x) = 2x^5 - 3x^4 + \frac{1}{2}x^3 + \sqrt{2}x^2 - 6$$

$$g(x) = 0.001x^3 - 0.2x^2 + 10x + 200$$

are polynomial functions of degrees 5 and 3, respectively. Observe that a polynomial function is defined for every value of x and so its domain is $(-\infty, \infty)$.

A polynomial function of degree 1 ($n = 1$) has the form

$$y = f(x) = a_1 x + a_0 \qquad (a_1 \neq 0)$$

and is an equation of a straight line in the slope-intercept form with slope $m = a_1$ and y-intercept $b = a_0$ (see Section 2.1). For this reason, a polynomial function of degree 1 is called a **linear function.**

Linear functions are used extensively in mathematical modeling for two important reasons. First, some models are *linear* by nature. For example, the formula for converting temperature from Celsius (°C) to Fahrenheit (°F) is $F = \frac{9}{5}C + 32$, and F is a linear function of C. Second, some natural phenomena exhibit linear characteristics over a small range of values and can therefore be modeled by a linear function restricted to a small interval.

The following example uses a linear function to model the market for cholesterol-reducing drugs. In Section 8.4, we will show how this model is constructed using the *least-squares technique.* In Using Technology on pages 93–96, you will be asked to use a graphing calculator to construct other mathematical models from raw data.

VIDEO

APPLIED EXAMPLE 1 Market for Cholesterol-Reducing Drugs In a study conducted in early 2000, experts projected a rise in the market for cholesterol-reducing drugs. The U.S. market (in billions of dollars) for such drugs from 1999 through 2004 is given in the following table:

Year	1999	2000	2001	2002	2003	2004
Market	12.07	14.07	16.21	18.28	20	21.72

A mathematical model giving the approximate U.S. market over the period in question is given by

$$M(t) = 1.95t + 12.19$$

where t is measured in years, with $t = 0$ corresponding to 1999.

a. Sketch the graph of the function M and the given data on the same set of axes.

b. Assuming that the projection held and the trend continued, what was the market for cholesterol-reducing drugs in 2005 ($t = 6$)?

c. What was the rate of increase of the market for cholesterol-reducing drugs over the period in question?

Source: S. G. Cowen

Solution

a. The graph of M is shown in Figure 16.

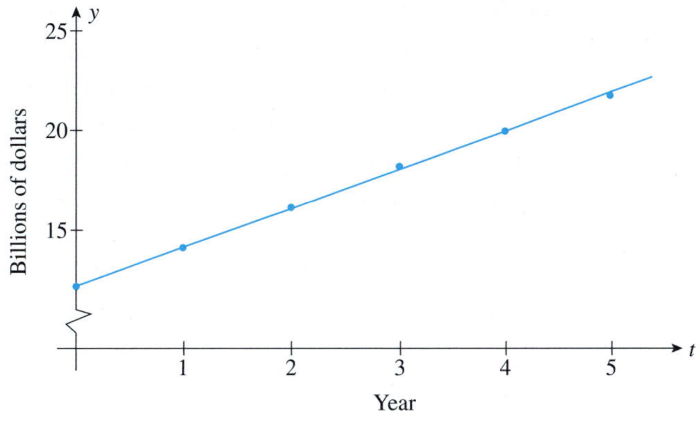

FIGURE **16**
Projected U.S. market for cholesterol-reducing drugs

b. The projected market in 2005 for cholesterol-reducing drugs was

$$M(6) = 1.95(6) + 12.19$$
$$= 23.89$$

or \$23.89 billion.

c. The function M is linear, and so we see that the rate of increase of the market for cholesterol-reducing drugs is given by the slope of the straight line represented by M, which is approximately \$1.95 billion per year.

A polynomial function of degree 2 has the form

$$y = f(x) = a_2x^2 + a_1x + a_0 \qquad (a_2 \neq 0)$$

or more simply, $y = ax^2 + bx + c$, and is called a **quadratic function.** The graph of a quadratic function is a parabola (see Figure 17).

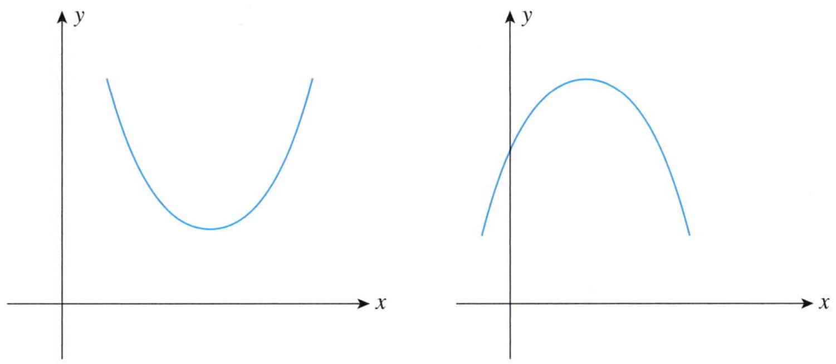

(a) If $a > 0$, the parabola opens upward.

(b) If $a < 0$, the parabola opens downward.

FIGURE **17**
The graph of a quadratic function is a parabola.

The parabola opens upward if $a > 0$ and downward if $a < 0$. To see this, we rewrite the equation for y obtaining

$$f(x) = ax^2 + bx + c = x^2\left(a + \frac{b}{x} + \frac{c}{x^2}\right) \qquad (x \neq 0)$$

Observe that if x is large in absolute value, then the expression inside the parentheses is close to a so that $f(x)$ behaves like ax^2 for large values of x. Thus, $y = f(x)$ is large and positive (the parabola opens upward) if $a > 0$ and is large in magnitude and negative if $a < 0$ (the parabola opens downward).

Quadratic functions serve as mathematical models for many phenomena, as Example 2 shows.

APPLIED EXAMPLE 2 Global Warming The increase in carbon dioxide (CO_2) in the atmosphere is a major cause of global warming. The Keeling curve, named after Charles David Keeling, a professor at Scripps Institution of Oceanography, gives the average amount of CO_2, measured in parts per million volume (ppmv), in the atmosphere from 1958 through 2007. Even though data were available for every year in this time interval, we'll construct the curve based only on the following randomly selected data points.

Year	1958	1970	1974	1978	1985	1991	1998	2003	2007
Amount	315	325	330	335	345	355	365	375	380

The **scatter plot** associated with these data is shown in Figure 18a. A mathematical model giving the approximate amount of CO_2 in the atmosphere during this period is given by

$$A(t) = 0.010716t^2 + 0.8212t + 313.4 \qquad (1 \le t \le 50)$$

where t is measured in years, with $t = 1$ corresponding to 1958. The graph of A is shown in Figure 18b.

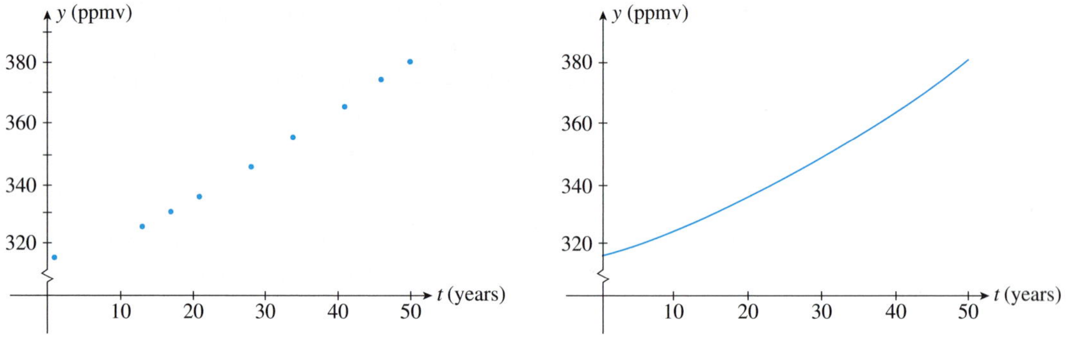

FIGURE **18** **(a)** Scatter plot **(b)** Graph of A

a. Use the model to estimate the average amount of atmospheric CO_2 in 1980 ($t = 23$).

b. Assume that the trend continued and use the model to predict the average amount of atmospheric CO_2 in 2010.

Source: Scripps Institution of Oceanography

Solution

a. The average amount of atmospheric carbon dioxide in 1980 is given by

$$A(23) = 0.010716(23)^2 + 0.8212(23) + 313.4 \approx 337.96$$

or approximately 338 ppmv.

b. Assuming that the trend continued, the average amount of atmospheric CO_2 in 2010 will be

$$A(53) = 0.010716(53)^2 + 0.8212(53) + 313.4 \approx 387.03$$

or approximately 387 ppmv. ■

The next example uses a polynomial of degree 4 to help us construct a model that describes the projected assets of the Social Security trust fund.

APPLIED EXAMPLE 3 Social Security Trust Fund Assets The projected assets of the Social Security trust fund (in trillions of dollars) from 2008 through 2040 are given in the following table:

Year	2008	2011	2014	2017	2020	2023	2026	2029	2032	2035	2038	2040
Assets	2.4	3.2	4.0	4.7	5.3	5.7	5.9	5.6	4.9	3.6	1.7	0

The scatter plot associated with these data are shown in Figure 19a, where $t = 0$ corresponds to 2008. A mathematical model giving the approximate value of the assets in the trust fund $A(t)$, (in trillions of dollars) in year t is

$$A(t) = -0.00000268t^4 - 0.000356t^3 + 0.00393t^2 + 0.2514t + 2.4094 \qquad (0 \leq t \leq 32)$$

The graph of $A(t)$ is shown in Figure 19b. (You will be asked to construct this model in Exercise 22, Using Technology Exercises 2.3.)

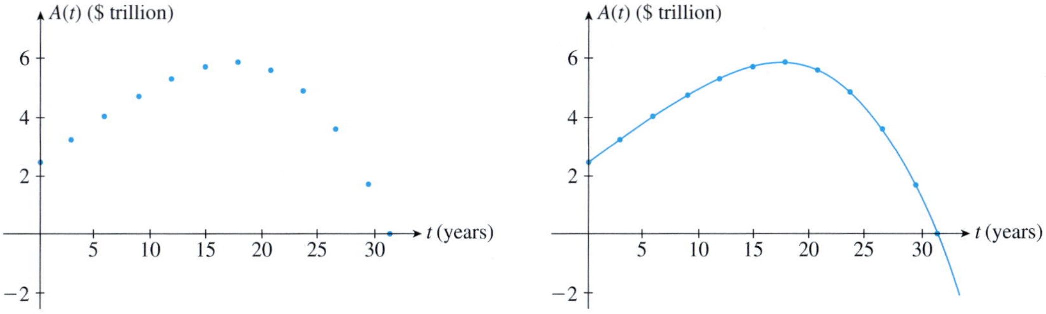

FIGURE **19** **(a)** Scatter plot **(b)** Graph of A

a. The first baby boomers will turn 65 in 2011. What will be the assets of the Social Security system trust fund at that time? The last of the baby boomers will turn 65 in 2029. What will the assets of the trust fund be at that time?

b. Unless payroll taxes are increased significantly and/or benefits are scaled back dramatically, it is a matter of time before the assets of the current system are depleted. Use the graph of the function $A(t)$ to estimate the year in which the current Social Security system is projected to go broke.

Source: Social Security Administration

Solution

a. The assets of the Social Security trust fund in 2011 ($t = 3$) will be

$$A(3) = -0.00000268(3)^4 - 0.000356(3)^3 + 0.00393(3)^2 + 0.2514(3) + 2.4094 \approx 3.19$$

or approximately \$3.19 trillion. The assets of the trust fund in 2029 ($t = 21$) will be

$$\begin{aligned} A(21) = &-0.00000268(21)^4 - 0.000356(21)^3 + 0.00393(21)^2 \\ &+ 0.2514(21) + 2.4094 \approx 5.60 \end{aligned}$$

or approximately \$5.60 trillion.

b. From Figure 19b, we see that the graph of A crosses the t-axis at approximately $t = 32$. So unless the current system is changed, it is projected to go broke in 2040. (At this time the first of the baby boomers would be 94 and the last of the baby boomers would be 76.) ■

Rational and Power Functions

Another important class of functions is rational functions. A **rational function** is simply the quotient of two polynomials. Examples of rational functions are

$$F(x) = \frac{3x^3 + x^2 - x + 1}{x - 2}$$

$$G(x) = \frac{x^2 + 1}{x^2 - 1}$$

In general, a rational function has the form

$$R(x) = \frac{f(x)}{g(x)}$$

where $f(x)$ and $g(x)$ are polynomial functions. Since division by zero is not allowed, we conclude that the domain of a rational function is the set of all real numbers except the zeros of g—that is, the roots of the equation $g(x) = 0$. Thus, the domain of the function F is the set of all numbers except $x = 2$, whereas the domain of the function G is the set of all numbers except those that satisfy $x^2 - 1 = 0$, or $x = \pm 1$.

Functions of the form

$$f(x) = x^r$$

where r is any real number, are called **power functions.** We encountered examples of power functions earlier in our work. For example, the functions

$$f(x) = \sqrt{x} = x^{1/2} \quad \text{and} \quad g(x) = \frac{1}{x^2} = x^{-2}$$

are power functions.

Many of the functions that we encounter later will involve combinations of the functions introduced here. For example, the following functions may be viewed as combinations of such functions:

$$f(x) = \sqrt{\frac{1 - x^2}{1 + x^2}}$$

$$g(x) = \sqrt{x^2 - 3x + 4}$$

$$h(x) = (1 + 2x)^{1/2} + \frac{1}{(x^2 + 2)^{3/2}}$$

As with polynomials of degree 3 or greater, analyzing the properties of these functions is facilitated by using the tools of calculus, to be developed later.

In the next example, we use a power function to construct a model that describes the driving costs of a car.

VIDEO

APPLIED EXAMPLE 4 Driving Costs A study of driving costs based on a 2008 medium-sized sedan found the following average costs (car payments, gas, insurance, upkeep, and depreciation), measured in cents per mile.

Miles/year	5000	10,000	15,000	20,000
Cost/mile, y (¢)	147.52	71.9	55.2	46.9

A mathematical model (using least-squares techniques) giving the average cost in cents per mile is

$$C(x) = \frac{1735.2}{x^{1.72}} + 38.6$$

where x (in thousands) denotes the number of miles the car is driven in 1 year. The scatter plot associated with this data and the graph of C are shown in Figure 20. Using this model, estimate the average cost of driving a 2008 medium-sized sedan 8000 miles per year and 18,000 miles per year.

Source: American Automobile Association

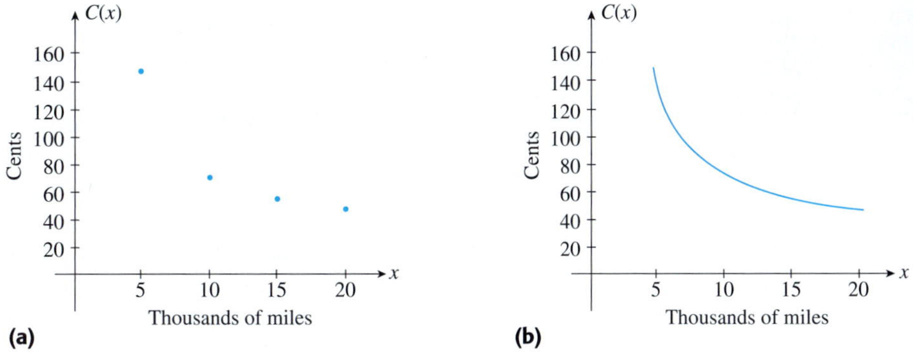

FIGURE 20
(a) The scatter plot and (b) the graph of the model for driving costs

Solution The average cost for driving a car 8000 miles per year is

$$C(8) = \frac{1735.2}{8^{1.72}} + 38.6 \approx 87.1$$

or approximately 87.1¢/mile. The average cost for driving it 18,000 miles per year is

$$C(18) = \frac{1735.2}{18^{1.72}} + 38.6 \approx 50.6$$

or approximately 50.6¢/mile.

Some Economic Models

In the remainder of this section, we look at some economic models.

In a free-market economy, consumer demand for a particular commodity depends on the commodity's unit price. A **demand equation** expresses the relationship between the unit price and the quantity demanded. The graph of the demand equation is called a **demand curve.** In general, the quantity demanded of a commodity decreases as the commodity's unit price increases, and vice versa. Accordingly, a **demand function** defined by $p = f(x)$, where p measures the unit price and x measures the number of units of the commodity in question, is generally characterized as a decreasing function of x; that is, $p = f(x)$ decreases as x increases. Since both x and p assume only nonnegative values, the demand curve is that part of the graph of $f(x)$ that lies in the first quadrant (Figure 21).

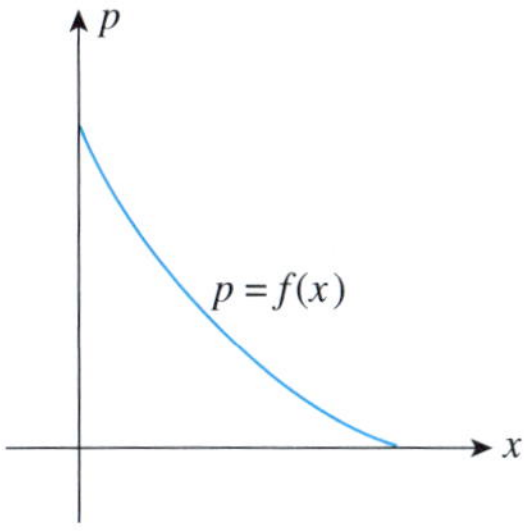

FIGURE 21
A demand curve

In a competitive market, a relationship also exists between the unit price of a commodity and the commodity's availability in the market. In general, an increase in the commodity's unit price induces the producer to increase the supply of the commodity. Conversely, a decrease in the unit price generally leads to a drop in the supply. The equation that expresses the relation between the unit price and the quantity supplied is called a **supply equation,** and its graph is called a **supply curve.** A **supply function** defined by $p = f(x)$ is generally characterized as an increasing function of x; that is, $p = f(x)$ increases as x increases. Since both x and p assume only nonnegative values, the supply curve is that part of the graph of $f(x)$ that lies in the first quadrant (Figure 22).

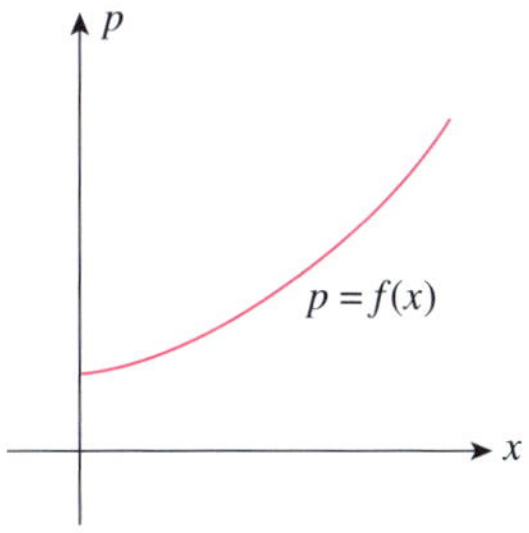

FIGURE 22
A supply curve

Under pure competition, the price of a commodity will eventually settle at a level dictated by the following condition: The supply of the commodity will be equal to the demand for it. If the price is too high, the consumer will not buy; if the price is too low, the supplier will not produce. **Market equilibrium** prevails when the quantity

PORTFOLIO Deb Farace

TITLE Sr. National Accounts Manager
INSTITUTION PepsiCo Beverages & Foods

Working for the national accounts division for PepsiCo Beverages & Foods, I need to understand applied mathematics in order to control the variables associated with making a profit, manufacturing, production, and most importantly selling our products to mass club channels. Examples of these large, "quality product at great value" outlets are Wal*Mart, Costco and Target. The types of products I handle include Gatorade, Tropicana, and Quaker foods.

Our studies show that the grocery store channels' sales are flattening or declining as a whole in lieu of large, national outlets like the above. So in order to maximize growth in this segment of our business, I meet with regional buying offices of these chains and discuss various packaging, pricing, product, promotional and shipping options so that we can successfully compete in the market.

A number of factors must be taken into consideration in order to meet my company's financial forecasts. Precision using mathematical models is key here, since so many variables can impact last-minute decision-making. Extended variables of supply-and-demand include time of year, competitive landscape, special coupon distribution and other promotions, selling cycles and holidays, size of the outlets, and yes—even the weather.

For example, it's natural to assume that when it's hot outside people will buy more thirst-quenching products like Gatorade. But since our business is so precise, we need to understand mathematically how the weather affects sales. A mathematical model developed by Gatorade analyzes longterm data that impacts sales by geographic market due to the weather. Its findings include exponentially increased sales of Gatorade for each degree above 90 degrees. I share our mathematical analysis like this study with buyers and negotiate larger orders based on up-to-the-minute weather forecasts. The result: increased sales of product based on math.

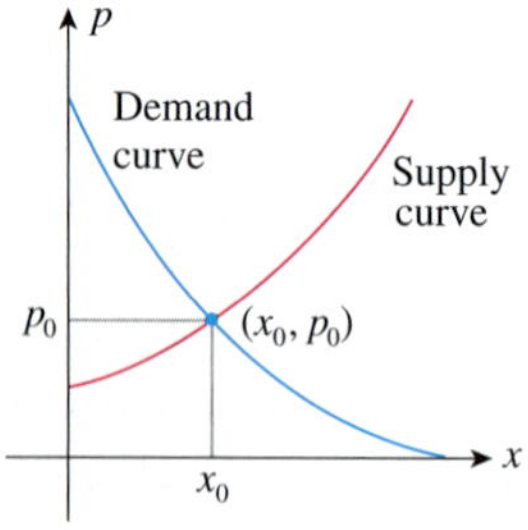

FIGURE 23
Market equilibrium corresponds to (x_0, p_0), the point at which the supply and demand curves intersect.

produced is equal to the quantity demanded. The quantity produced at market equilibrium is called the **equilibrium quantity**, and the corresponding price is called the **equilibrium price**.

Market equilibrium corresponds to the point at which the demand curve and the supply curve intersect. In Figure 23, x_0 represents the equilibrium quantity and p_0 the equilibrium price. The point (x_0, p_0) lies on the supply curve and therefore satisfies the supply equation. At the same time, it also lies on the demand curve and therefore satisfies the demand equation. Thus, to find the point (x_0, p_0), and hence the equilibrium quantity and price, we solve the demand and supply equations simultaneously for x and p. For meaningful solutions, x and p must both be positive.

APPLIED EXAMPLE 5 Supply-Demand The demand function for a certain brand of bluetooth wireless headsets is given by

$$p = d(x) = -0.025x^2 - 0.5x + 60$$

and the corresponding supply function is given by

$$p = s(x) = 0.02x^2 + 0.6x + 20$$

where p is expressed in dollars and x is measured in units of a thousand. Find the equilibrium quantity and price.

Solution We solve the following system of equations:

$$\begin{aligned} p &= -0.025x^2 - 0.5x + 60 \\ p &= 0.02x^2 + 0.6x + 20 \end{aligned}$$

Substituting the first equation into the second yields

$$-0.025x^2 - 0.5x + 60 = 0.02x^2 + 0.6x + 20$$

which is equivalent to

$$0.045x^2 + 1.1x - 40 = 0$$
$$45x^2 + 1100x - 40{,}000 = 0 \quad \text{Multiply by 1000.}$$
$$9x^2 + 220x - 8000 = 0 \quad \text{Divide by 5.}$$
$$(9x + 400)(x - 20) = 0$$

Thus, $x = -\frac{400}{9}$ or $x = 20$. Since x must be nonnegative, the root $x = -\frac{400}{9}$ is rejected. Therefore, the equilibrium quantity is 20,000 headsets. The equilibrium price is given by

$$p = 0.02(20)^2 + 0.6(20) + 20 = 40$$

or \$40 per headset (Figure 24). ■

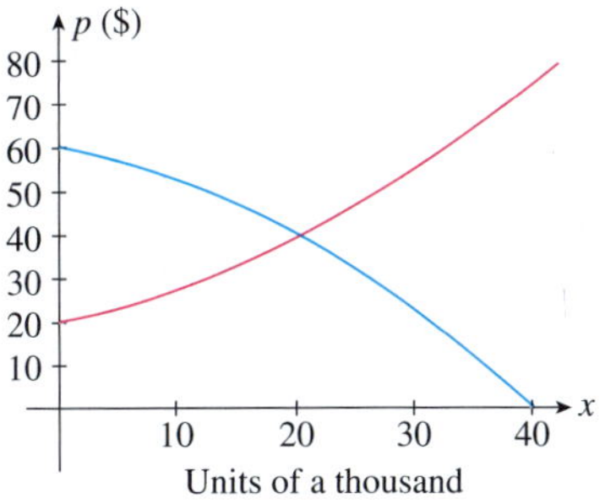

FIGURE 24
The supply curve and the demand curve intersect at the point (20, 40).

Exploring with TECHNOLOGY

1. **a.** Use a graphing utility to plot the straight lines L_1 and L_2 with equations $y = 2x - 1$ and $y = 2.1x + 3$, respectively, on the same set of axes, using the standard viewing window. Do the lines appear to intersect?
   b. Plot the straight lines L_1 and L_2, using the viewing window $[-100, 100] \times [-100, 100]$. Do the lines appear to intersect? Can you find the point of intersection using TRACE and ZOOM? Using the "intersection" function of your graphing utility?
   c. Find the point of intersection of L_1 and L_2 algebraically.
   d. Comment on the effectiveness of the methods of solutions in parts (b) and (c).
2. **a.** Use a graphing utility to plot the straight lines L_1 and L_2 with equations $y = 3x - 2$ and $y = -2x + 3$, respectively, on the same set of axes, using the standard viewing window. Then use TRACE and ZOOM to find the point of intersection of L_1 and L_2. Repeat using the "intersection" function of your graphing utility.
   b. Find the point of intersection of L_1 and L_2 algebraically.
   c. Comment on the effectiveness of the methods.

Constructing Mathematical Models

We close this section by showing how some mathematical models can be constructed using elementary geometric and algebraic arguments.

The following guidelines can be used to construct mathematical models.

Guidelines for Constructing Mathematical Models

1. Assign a letter to each variable mentioned in the problem. If appropriate, draw and label a figure.
2. Find an expression for the quantity sought.
3. Use the conditions given in the problem to write the quantity sought as a function f of one variable. Note any restrictions to be placed on the domain of f from physical considerations of the problem.

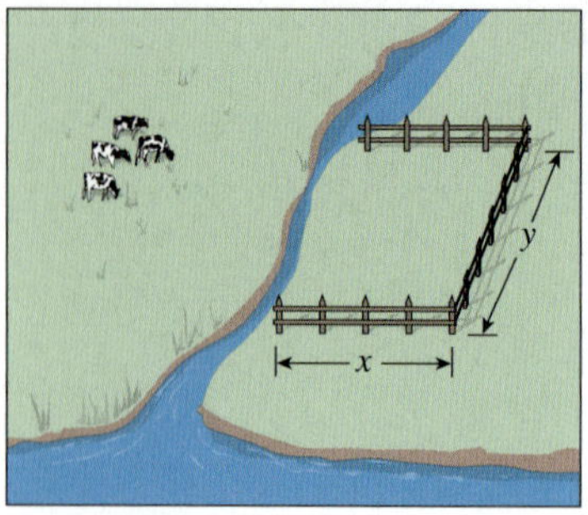

FIGURE 25
The rectangular grazing land has width x and length y.

APPLIED EXAMPLE 6 Enclosing an Area The owner of the Rancho Los Feliz has 3000 yards of fencing with which to enclose a rectangular piece of grazing land along the straight portion of a river. Fencing is not required along the river. Letting x denote the width of the rectangle, find a function f in the variable x giving the area of the grazing land if she uses all of the fencing (Figure 25).

Solution

1. This information was given.
2. The area of the rectangular grazing land is $A = xy$. Next, observe that the amount of fencing is $2x + y$ and this must be equal to 3000 since all the fencing is used; that is,

$$2x + y = 3000$$

3. From the equation we see that $y = 3000 - 2x$. Substituting this value of y into the expression for A gives

$$A = xy = x(3000 - 2x) = 3000x - 2x^2$$

Finally, observe that both x and y must be nonnegative since they represent the width and length of a rectangle, respectively. Thus, $x \geq 0$ and $y \geq 0$. But the latter is equivalent to $3000 - 2x \geq 0$, or $x \leq 1500$. So the required function is $f(x) = 3000x - 2x^2$ with domain $0 \leq x \leq 1500$. ■

Note Observe that if we view the function $f(x) = 3000x - 2x^2$ strictly as a mathematical entity, then its domain is the set of all real numbers. But physical considerations dictate that its domain should be restricted to the interval $[0, 1500]$. ■

APPLIED EXAMPLE 7 Charter-Flight Revenue If exactly 200 people sign up for a charter flight, Leisure World Travel Agency charges \$300 per person. However, if more than 200 people sign up for the flight (assume this is the case), then each fare is reduced by \$1 for each additional person. Letting x denote the number of passengers above 200, find a function giving the revenue realized by the company.

Solution

1. This information was given.
2. If there are x passengers above 200, then the number of passengers signing up for the flight is $200 + x$. Furthermore, the fare will be $(300 - x)$ dollars per passenger.
3. The revenue will be

$$R = (200 + x)(300 - x)$$
$$= -x^2 + 100x + 60{,}000$$

Number of passengers × the fare per passenger

Clearly, x must be nonnegative, and $300 - x \geq 0$, or $x \leq 300$. So the required function is $f(x) = -x^2 + 100x + 60{,}000$ with domain $[0, 300]$. ■

2.3 Self-Check Exercises

1. Thomas Young has suggested the following rule for calculating the dosage of medicine for children from ages 1 to 12 yr. If a denotes the adult dosage (in milligrams) and t is the age of the child (in years), then the child's dosage is given by

$$D(t) = \frac{at}{t + 12}$$

If the adult dose of a substance is 500 mg, how much should a 4-yr-old child receive?

2. The demand function for Mrs. Baker's cookies is given by

$$d(x) = -\frac{2}{15}x + 4$$

where $d(x)$ is the wholesale price in dollars/pound and x is the quantity demanded each week, measured in thousands of pounds. The supply function for the cookies is given by

$$s(x) = \frac{1}{75}x^2 + \frac{1}{10}x + \frac{3}{2}$$

where $s(x)$ is the wholesale price in dollars/pound and x is the quantity, in thousands of pounds, that will be made available in the market each week by the supplier.

a. Sketch the graphs of the functions d and s.
b. Find the equilibrium quantity and price.

Solutions to Self-Check Exercises 2.3 can be found on page 92.

2.3 Concept Questions

1. Describe mathematical modeling in your own words.
2. Define (a) a polynomial function and (b) a rational function. Give an example of each.
3. **a.** What is a demand function? A supply function?
   b. What is market equilibrium? Describe how you would go about finding the equilibrium quantity and equilibrium price given the demand and supply equations associated with a commodity.

2.3 Exercises

In Exercises 1–8, determine whether the equation defines *y* as a linear function of *x*. If so, write it in the form *y* = *mx* + *b*.

1. $2x + 3y = 6$
2. $-2x + 4y = 7$
3. $x = 2y - 4$
4. $2x = 3y + 8$
5. $2x - 4y + 9 = 0$
6. $3x - 6y + 7 = 0$
7. $2x^2 - 8y + 4 = 0$
8. $3\sqrt{x} + 4y = 0$

In Exercises 9–14, determine whether the given function is a polynomial function, a rational function, or some other function. State the degree of each polynomial function.

9. $f(x) = 3x^6 - 2x^2 + 1$
10. $f(x) = \dfrac{x^2 - 9}{x - 3}$
11. $G(x) = 2(x^2 - 3)^3$
12. $H(x) = 2x^{-3} + 5x^{-2} + 6$
13. $f(t) = 2t^2 + 3\sqrt{t}$
14. $f(r) = \dfrac{6r}{r^3 - 8}$

15. Find the constants m and b in the linear function $f(x) = mx + b$ so that $f(0) = 2$ and $f(3) = -1$.

16. Find the constants m and b in the linear function $f(x) = mx + b$ so that $f(2) = 4$ and the straight line represented by f has slope -1.

17. A manufacturer has a monthly fixed cost of \$40,000 and a production cost of \$8 for each unit produced. The product sells for \$12/unit.
    a. What is the cost function?
    b. What is the revenue function?
    c. What is the profit function?
    d. Compute the profit (loss) corresponding to production levels of 8000 and 12,000 units.

18. A manufacturer has a monthly fixed cost of \$100,000 and a production cost of \$14 for each unit produced. The product sells for \$20/unit.
    a. What is the cost function?
    b. What is the revenue function?
    c. What is the profit function?
    d. Compute the profit (loss) corresponding to production levels of 12,000 and 20,000 units.

19. **Disposable Income** Economists define the *disposable annual income* for an individual by the equation $D = (1 - r)T$, where T is the individual's total income and r is the net rate at which he or she is taxed. What is the disposable income for an individual whose income is \$60,000 and whose net tax rate is 28%?

20. **Drug Dosages** A method sometimes used by pediatricians to calculate the dosage of medicine for children is based on the child's surface area. If a denotes the adult dosage (in milligrams) and S is the surface area of the child (in square meters), then the child's dosage is given by

$$D(S) = \frac{Sa}{1.7}$$

If the adult dose of a substance is 500 mg, how much should a child whose surface area is 0.4 m^2 receive?

21. **Cowling's Rule** Cowling's rule is a method for calculating pediatric drug dosages. If a denotes the adult dosage (in milligrams) and t is the age of the child (in years), then the child's dosage is given by

$$D(t) = \left(\frac{t + 1}{24}\right)a$$

If the adult dose of a substance is 500 mg, how much should a 4-yr-old child receive?

22. **Worker Efficiency** An efficiency study showed that the average worker at Delphi Electronics assembled cordless telephones at the rate of

$$f(t) = -\frac{3}{2}t^2 + 6t + 10 \qquad (0 \le t \le 4)$$

phones/hour, t hr after starting work during the morning shift. At what rate does the average worker assemble telephones 2 hr after starting work?

23. **Effect of Advertising on Sales** The quarterly profit of Cunningham Realty depends on the amount of money x spent on advertising/quarter according to the rule

$$P(x) = -\frac{1}{8}x^2 + 7x + 30 \qquad (0 \le x \le 50)$$

where $P(x)$ and x are measured in thousands of dollars. What is Cunningham's profit when its quarterly advertising budget is $28,000?

24. **Instant Messaging Accounts** The number of enterprise instant messaging (IM) accounts is projected to grow according to the function

$$N(t) = 2.96t^2 + 11.37t + 59.7 \qquad (0 \le t \le 5)$$

where $N(t)$ is measured in millions and t in years, with $t = 0$ corresponding to 2006.

a. How many enterprise IM accounts were there in 2006?
b. What is the number of enterprise IM accounts expected to be in 2010?

Source: The Radical Group

25. **Solar Power** More and more businesses and homeowners are installing solar panels on their roofs to draw energy from the Sun's rays. According to the U.S. Department of Energy, the solar cell kilowatt-hour use in the United States (in millions) is projected to be

$$S(t) = 0.73t^2 + 15.8t + 2.7 \qquad (0 \le t \le 8)$$

in year t, with $t = 0$ corresponding to 2000. What was the projected solar cell kilowatt-hours used in the United States for 2006? For 2008?

Source: U.S. Department of Energy

26. **Average Single-Family Property Tax** Based on data from 298 of 351 cities and towns in Massachusetts, the average single-family tax bill from 1997 through 2007 is approximated by the function

$$T(t) = 7.26t^2 + 91.7t + 2360 \qquad (0 \le t \le 10)$$

where $T(t)$ is measured in dollars and t in years, with $t = 0$ corresponding to 1997.

a. What was the property tax on a single-family home in Massachusetts in 1997?
b. If the trend continues, what will be the property tax in 2010?

Source: Massachusetts Department of Revenue

27. **Revenue of Polo Ralph Lauren** Citing strong sales and benefits from a new arm that will design lifestyle brands for department and specialty stores, the company projects revenue (in billions of dollars) to be

$$R(t) = -0.06t^2 + 0.69t + 3.25 \qquad (0 \le t \le 3)$$

in year t, where $t = 0$ corresponds to 2005.

a. What was the revenue of the company in 2005?
b. Find $R(1)$, $R(2)$, and $R(3)$ and interpret your results.
c. Sketch the graph of R.

Source: Company reports

28. **Aging Drivers** The number of fatalities due to car crashes, based on the number of miles driven, begins to climb after the driver is past age 65. Aside from declining ability as one ages, the older driver is more fragile. The number of fatalities per 100 million vehicle miles driven is approximately

$$N(x) = 0.0336x^3 - 0.118x^2 + 0.215x + 0.7 \qquad (0 \le x \le 7)$$

where x denotes the age group of drivers, with $x = 0$ corresponding to those aged 50–54, $x = 1$ corresponding to those aged 55–59, $x = 2$ corresponding to those aged 60–64, . . . , and $x = 7$ corresponding to those aged 85–89. What is the fatality rate per 100 million vehicle miles driven for an average driver in the 50–54 age group? In the 85–89 age group?

Source: U.S. Department of Transportation

29. **Rising Water Rates** Based on records from 2001 through 2006, services paid for by households in 60 Boston-area communities that use an average of 90,000 gallons of water a year are given by

$$C(t) = 2.16t^3 + 40t + 751.5 \qquad (0 \le t \le 6)$$

Here $t = 0$ corresponds to 2001, and $C(t)$ is measured in dollars/year. What was the average amount paid by a household in 2001 for water and sewer services? If the trend continued, what was the average amount paid in 2008?

Source: Massachusetts Water Resources Authority

30. **Gift Cards** Gift cards have increased in popularity in recent years. Consumers appreciate gift cards because they get to select the present they like. The U.S. sales of gift cards (in billions of dollars) is approximated by

$$S(t) = -0.6204t^3 + 4.671t^2 + 3.354t + 47.4 \qquad (0 \le t \le 5)$$

in year t, where $t = 0$ corresponds to 2003.

a. What were the sales of gift cards for 2003?
b. What were the projected sales of gift cards for 2008?

Source: The Tower Group

31. **BlackBerry Subscribers** According to a study conducted in 2004, the number of subscribers of BlackBerry, the handheld email devices manufactured by Research in Motion Ltd., is approximated by

$$N(t) = -0.0675t^4 + 0.5083t^3 - 0.893t^2 + 0.66t + 0.32$$
$$(0 \le t \le 4)$$

where $N(t)$ is measured in millions and t in years, with $t = 0$ corresponding to the beginning of 2002.

a. How many BlackBerry subscribers were there at the beginning of 2002?

b. How many BlackBerry subscribers were there at the beginning of 2006?

Source: ThinkEquity Partners

32. Infant Mortality Rates in Massachusetts The deaths of children less than 1 year old per 1000 live births is modeled by the function

$$R(t) = 162.8t^{-3.025} \qquad (1 \le t \le 3)$$

where t is measured in 50-year intervals, with $t = 1$ corresponding to 1900.

a. Find $R(1)$, $R(2)$, and $R(3)$ and use your result to sketch the graph of the function R over the domain $[1, 3]$.

b. What was the infant mortality rate in 1900? In 1950? In 2000?

Source: Massachusetts Department of Public Health

33. Online Video Viewers As broadband Internet grows more popular, video services such as YouTube will continue to expand. The number of online video viewers (in millions) is projected to grow according to the rule

$$N(t) = 52t^{0.531} \qquad (1 \le t \le 10)$$

where $t = 1$ corresponds to 2003.

a. Sketch the graph of N.

b. How many online video viewers will there be in 2012?

Source: eMarketer.com

34. Chip Sales The worldwide flash memory chip sales (in billions of dollars) is projected to be

$$S(t) = 4.3(t + 2)^{0.94} \qquad (0 \le t \le 6)$$

where t is measured in years, with $t = 0$ corresponding to 2002. Flash chips are used in cell phones, digital cameras, and other products.

a. What were the worldwide flash memory chip sales in 2002?

b. What were the projected sales for 2008?

Source: Web-Feet Research Inc.

35. Outsourcing of Jobs According to a study conducted in 2003, the total number of U.S. jobs (in millions) that are projected to leave the country by year t, where $t = 0$ corresponds to 2000, is

$$N(t) = 0.0018425(t + 5)^{2.5} \qquad (0 \le t \le 15)$$

What was the projected number of outsourced jobs for 2005 ($t = 5$)? For 2010 ($t = 10$)?

Source: Forrester Research

36. Immigration to the United States The immigration to the United States from Europe, as a percentage of the total immigration, is approximately

$$P(t) = 0.767t^3 - 0.636t^2 - 19.17t + 52.7 \qquad (0 \le t \le 4)$$

where t is measured in decades, with $t = 0$ corresponding to the decade of the 1950s.

a. Complete the table:

t	0	1	2	3	4
$P(t)$					

b. Use the result of part (a) to sketch the graph of P.

c. Use the result of part (b) to estimate the decade when the immigration, as a percentage of the total immigration, was the greatest and the smallest.

Source: Jeffrey Williamson, Harvard University

37. Selling Price of DVD Recorders The rise of digital music and the improvement to the DVD format are part of the reasons why the average selling price of standalone DVD recorders will drop in the coming years. The function

$$A(t) = \frac{699}{(t + 1)^{0.94}} \qquad (0 \le t \le 5)$$

gives the projected average selling price (in dollars) of standalone DVD recorders in year t, where $t = 0$ corresponds to the beginning of 2002. What was the average selling price of standalone DVD recorders at the beginning of 2002? At the beginning of 2007?

Source: Consumer Electronics Association

38. Reaction of a Frog to a Drug Experiments conducted by A. J. Clark suggest that the response $R(x)$ of a frog's heart muscle to the injection of x units of acetylcholine (as a percent of the maximum possible effect of the drug) may be approximated by the rational function

$$R(x) = \frac{100x}{b + x} \qquad (x \ge 0)$$

where b is a positive constant that depends on the particular frog.

a. If a concentration of 40 units of acetylcholine produces a response of 50% for a certain frog, find the "response function" for this frog.

b. Using the model found in part (a), find the response of the frog's heart muscle when 60 units of acetylcholine are administered.

39. Digital versus Film Cameras The sales of digital cameras (in millions of units) in year t is given by the function

$$f(t) = 3.05t + 6.85 \qquad (0 \le t \le 3)$$

where $t = 0$ corresponds to 2001. Over that same period, the sales of film cameras (in millions of units) is given by

$$g(t) = -1.85t + 16.58 \qquad (0 \le t \le 3)$$

a. Show that more film cameras than digital cameras were sold in 2001.

b. When did the sales of digital cameras first exceed those of film cameras?

Source: Popular Science

40. **Walking versus Running** The oxygen consumption (in milliliter/pound/minute) for a person walking at x mph is approximated by the function

$$f(x) = \frac{5}{3}x^2 + \frac{5}{3}x + 10 \qquad (0 \le x \le 9)$$

whereas the oxygen consumption for a runner at x mph is approximated by the function

$$g(x) = 11x + 10 \qquad (4 \le x \le 9)$$

a. Sketch the graphs of f and g.
b. At what speed is the oxygen consumption the same for a walker as it is for a runner? What is the level of oxygen consumption at that speed?
c. What happens to the oxygen consumption of the walker and the runner at speeds beyond that found in part (b)?

Source: William McArdley, Frank Katch, and Victor Katch, *Exercise Physiology*

41. **Price of Automobile Parts** For years, automobile manufacturers had a monopoly on the replacement-parts market, particularly for sheet metal parts such as fenders, doors, and hoods, the parts most often damaged in a crash. Beginning in the late 1970s, however, competition appeared on the scene. In a report conducted by an insurance company to study the effects of the competition, the price of an OEM (original equipment manufacturer) fender for a particular 1983 model car was found to be

$$f(t) = \frac{110}{\frac{1}{2}t + 1} \qquad (0 \le t \le 2)$$

where $f(t)$ is measured in dollars and t is in years. Over the same period of time, the price of a non-OEM fender for the car was found to be

$$g(t) = 26\left(\frac{1}{4}t^2 - 1\right)^2 + 52 \qquad (0 \le t \le 2)$$

where $g(t)$ is also measured in dollars. Find a function $h(t)$ that gives the difference in price between an OEM fender and a non-OEM fender. Compute $h(0)$, $h(1)$, and $h(2)$. What does the result of your computation seem to say about the price gap between OEM and non-OEM fenders over the 2 yr?

42. **Cricket Chirping and Temperature** Entomologists have discovered that a linear relationship exists between the number of chirps of crickets of a certain species and the air temperature. When the temperature is 70°F, the crickets chirp at the rate of 120 times/minute, and when the temperature is 80°F, they chirp at the rate of 160 times/minute.

a. Find an equation giving the relationship between the air temperature T and the number of chirps/minute, N, of the crickets.
b. Find N as a function of T and use this formula to determine the rate at which the crickets chirp when the temperature is 102°F.

43. **Linear Depreciation** In computing income tax, businesses are allowed by law to depreciate certain assets such as buildings, machines, furniture, automobiles, and so on, over a period of time. The linear depreciation, or straight-line method, is often used for this purpose. Suppose an asset has an initial value of $\$C$ and is to be depreciated linearly over n yr with a scrap value of $\$S$. Show that the book value of the asset at any time t $(0 \le t \le n)$ is given by the linear function

$$V(t) = C - \frac{C - S}{n}t$$

Hint: Find an equation of the straight line that passes through the points $(0, C)$ and (n, S). Then rewrite the equation in the slope-intercept form.

44. **Linear Depreciation** Using the linear depreciation model of Exercise 43, find the book value of a printing machine at the end of the second year if its initial value is \$100,000 and it is depreciated linearly over 5 yr with a scrap value of \$30,000.

45. **Credit Card Debt** Following the introduction in 1950 of the nation's first credit card, the Diners Club Card, credit cards have proliferated over the years. More than 720 different cards are now used at more than 4 million locations in the United States. The average U.S. credit card debt (per household) in thousands of dollars is approximately given by

$$D(t) = \begin{cases} 4.77(1 + t)^{0.2676} & \text{if } 0 \le t \le 2 \\ 5.6423t^{0.1818} & \text{if } 2 < t \le 6 \end{cases}$$

where t is measured in years, with $t = 0$ corresponding to the beginning of 1994. What was the average U.S. credit card debt (per household) at the beginning of 1994? At the beginning of 1996? At the beginning of 1999?

Source: David Evans and Richard Schmalensee, *Paying with Plastic: The Digital Revolution in Buying and Borrowing*

46. **Obese Children in the United States** The percentage of obese children aged 12–19 in the United States is approximately

$$P(t) = \begin{cases} 0.04t + 4.6 & \text{if } 0 \le t < 10 \\ -0.01005t^2 + 0.945t - 3.4 & \text{if } 10 \le t \le 30 \end{cases}$$

where t is measured in years, with $t = 0$ corresponding to the beginning of 1970. What was the percentage of obese children aged 12–19 at the beginning of 1970? At the beginning of 1985? At the beginning of 2000?

Source: Centers for Disease Control

47. **Price of Ivory** According to the World Wildlife Fund, a group in the forefront of the fight against illegal ivory trade, the price of ivory (in dollars/kilo) compiled from a variety of legal and black market sources is approximated by the function

$$f(t) = \begin{cases} 8.37t + 7.44 & \text{if } 0 \le t \le 8 \\ 2.84t + 51.68 & \text{if } 8 < t \le 30 \end{cases}$$

where t is measured in years, with $t = 0$ corresponding to the beginning of 1970.

a. Sketch the graph of the function f.

b. What was the price of ivory at the beginning of 1970? At the beginning of 1990?

Source: World Wildlife Fund

48. WORKING-AGE POPULATION The ratio of working-age population to the elderly in the United States (including projections after 2000) is given by

$$f(t) = \begin{cases} 4.1 & \text{if } 0 \le t < 5 \\ -0.03t + 4.25 & \text{if } 5 \le t < 15 \\ -0.075t + 4.925 & \text{if } 15 \le t \le 35 \end{cases}$$

with $t = 0$ corresponding to the beginning of 1995.

a. Sketch the graph of f.

b. What was the ratio at the beginning of 2005? What will be the ratio at the beginning of 2020?

c. Over what years is the ratio constant?

d. Over what years is the decline of the ratio greatest?

Source: U.S. Census Bureau

49. SENIOR CITIZENS' HEALTH CARE According to a study, the out-of-pocket cost to senior citizens for health care, $f(t)$ (as a percentage of income), in year t where $t = 0$ corresponds to 1977, is given by

$$f(t) = \begin{cases} \frac{2}{7}t + 12 & \text{if } 0 \le t \le 7 \\ t + 7 & \text{if } 7 < t \le 10 \\ \frac{1}{3}t + \frac{41}{3} & \text{if } 10 < t \le 25 \end{cases}$$

a. Sketch the graph of f.

b. What was the out-of-pocket cost, as a percentage of income, to senior citizens for health care in 1982? In 1992?

Source: Senate Select Committee on Aging, AARP

50. SALES OF DVD PLAYERS VS. VCRS The sales of DVD players in year t (in millions of units) is given by the function

$$f(t) = 5.6(1 + t) \qquad (0 \le t \le 3)$$

where $t = 0$ corresponds to 2001. Over the same period, the sales of VCRs (in millions of units) is given by

$$g(t) = \begin{cases} -9.6t + 22.5 & \text{if } 0 \le t \le 1 \\ -0.5t + 13.4 & \text{if } 1 < t \le 2 \\ -7.8t + 28 & \text{if } 2 < t \le 3 \end{cases}$$

a. Show that more VCRs than DVD players were sold in 2001.

b. When did the sales of DVD players first exceed those of VCRs?

Source: Popular Science

For the demand equations in Exercises 51–54, where x represents the quantity demanded in units of a thousand and p is the unit price in dollars, (a) sketch the demand curve and (b) determine the quantity demanded when the unit price is set at $\$p$.

51. $p = -x^2 + 16; p = 7$ **52.** $p = -x^2 + 36; p = 11$

53. $p = \sqrt{18 - x^2}; p = 3$ **54.** $p = \sqrt{9 - x^2}; p = 2$

For the supply equations in Exercises 55–58, where x is the quantity supplied in units of a thousand and p is the unit price in dollars, (a) sketch the supply curve and (b) determine the price at which the supplier will make 2000 units of the commodity available in the market.

55. $p = x^2 + 16x + 40$ **56.** $p = 2x^2 + 18$

57. $p = x^3 + 2x + 3$ **58.** $p = x^3 + x + 10$

59. DEMAND FOR CD CLOCK RADIOS In the accompanying figure, L_1 is the demand curve for the model A CD clock radios manufactured by Ace Radio, and L_2 is the demand curve for their model B CD clock radios. Which line has the greater slope? Interpret your results.

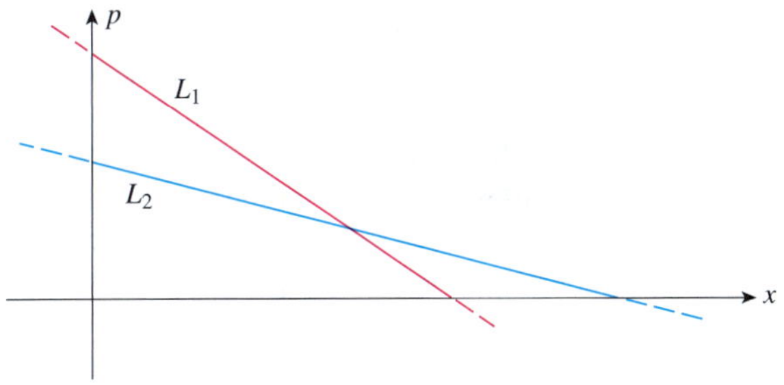

60. SUPPLY OF CD CLOCK RADIOS In the accompanying figure, L_1 is the supply curve for the model A CD clock radios manufactured by Ace Radio, and L_2 is the supply curve for their model B CD clock radios. Which line has the greater slope? Interpret your results.

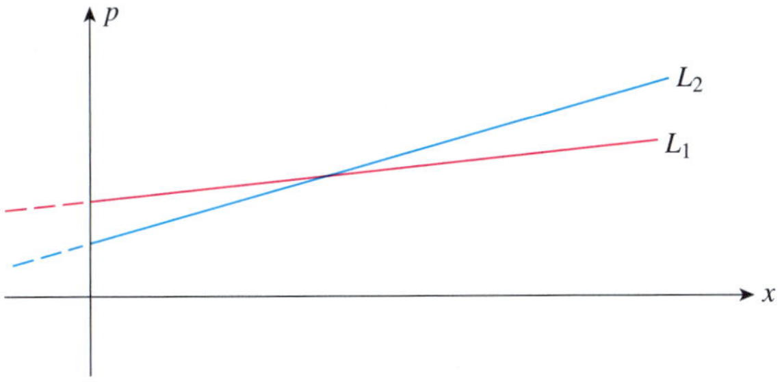

61. DEMAND FOR SMOKE ALARMS The demand function for the Sentinel smoke alarm is given by

$$p = \frac{30}{0.02x^2 + 1} \qquad (0 \le x \le 10)$$

where x (measured in units of a thousand) is the quantity demanded per week and p is the unit price in dollars.

a. Sketch the graph of the demand function.

b. What is the unit price that corresponds to a quantity demanded of 10,000 units?

62. **DEMAND FOR COMMODITIES** Assume that the demand function for a certain commodity has the form

$$p = \sqrt{-ax^2 + b} \qquad (a \geq 0, b \geq 0)$$

where x is the quantity demanded, measured in units of a thousand and p is the unit price in dollars. Suppose the quantity demanded is 6000 ($x = 6$) when the unit price is \$8.00 and 8000 ($x = 8$) when the unit price is \$6.00. Determine the demand equation. What is the quantity demanded when the unit price is set at \$7.50?

63. **SUPPLY OF DESK LAMPS** The supply function for the Luminar desk lamp is given by

$$p = 0.1x^2 + 0.5x + 15$$

where x is the quantity supplied (in thousands) and p is the unit price in dollars.

a. Sketch the graph of the supply function.
b. What unit price will induce the supplier to make 5000 lamps available in the marketplace?

64. **SUPPLY OF SATELLITE RADIOS** Suppliers of satellite radios will market 10,000 units when the unit price is \$20 and 62,500 units when the unit price is \$35. Determine the supply function if it is known to have the form

$$p = a\sqrt{x} + b \qquad (a > 0, b > 0)$$

where x is the quantity supplied and p is the unit price in dollars. Sketch the graph of the supply function. What unit price will induce the supplier to make 40,000 satellite radios available in the marketplace?

65. Suppose the demand and supply equations for a certain commodity are given by $p = ax + b$ and $p = cx + d$, respectively, where $a < 0$, $c > 0$, and $b > d > 0$ (see the accompanying figure).

a. Find the equilibrium quantity and equilibrium price in terms of a, b, c, and d.
b. Use part (a) to determine what happens to the market equilibrium if c is increased, while a, b, and d remain fixed. Interpret your answer in economic terms.
c. Use part (a) to determine what happens to the market equilibrium if b is decreased while a, c, and d remain fixed. Interpret your answer in economic terms.

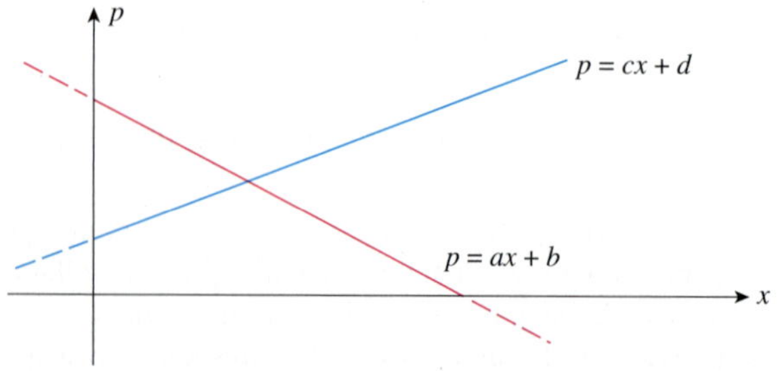

For each pair of supply and demand equations in Exercises 66–69, where x represents the quantity demanded in units of a thousand and p the unit price in dollars, find the equilibrium quantity and the equilibrium price.

66. $p = -x^2 - 2x + 100$ and $p = 8x + 25$

67. $p = -2x^2 + 80$ and $p = 15x + 30$

68. $p = 60 - 2x^2$ and $p = x^2 + 9x + 30$

69. $11p + 3x - 66 = 0$ and $2p^2 + p - x = 10$

70. **MARKET EQUILIBRIUM** The weekly demand and supply functions for Sportsman 5×7 tents are given by

$$p = -0.1x^2 - x + 40$$
$$p = 0.1x^2 + 2x + 20$$

respectively, where p is measured in dollars and x is measured in units of a hundred. Find the equilibrium quantity and price.

71. **MARKET EQUILIBRIUM** The management of Titan Tire Company has determined that the weekly demand and supply functions for their Super Titan tires are given by

$$p = 144 - x^2$$
$$p = 48 + \frac{1}{2}x^2$$

respectively, where p is measured in dollars and x is measured in units of a thousand. Find the equilibrium quantity and price.

72. **ENCLOSING AN AREA** Patricia wishes to have a rectangular-shaped garden in her backyard. She has 80 ft of fencing with which to enclose her garden. Letting x denote the width of the garden, find a function f in the variable x giving the area of the garden. What is its domain?

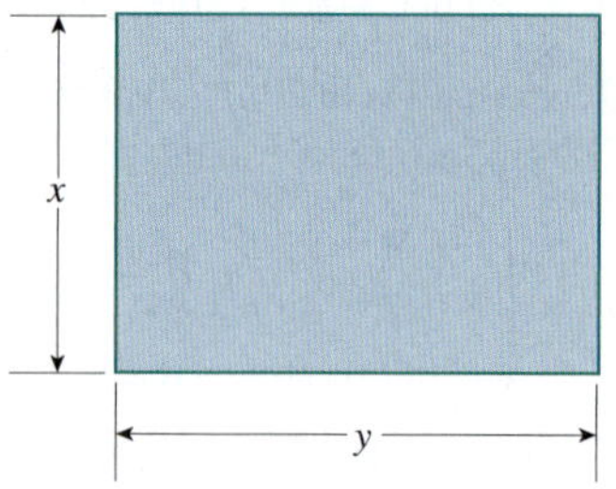

73. **ENCLOSING AN AREA** Patricia's neighbor, Juanita, also wishes to have a rectangular-shaped garden in her backyard. But Juanita wants her garden to have an area of 250 ft^2. Letting x denote the width of the garden, find a function f in the variable x giving the length of the fencing required to construct the garden. What is the domain of the function?

Hint: Refer to the figure for Exercise 72. The amount of fencing required is equal to the perimeter of the rectangle, which is twice the width plus twice the length of the rectangle.

74. Packaging By cutting away identical squares from each corner of a rectangular piece of cardboard and folding up the resulting flaps, an open box may be made. If the cardboard is 15 in. long and 8 in. wide and the square cutaways have dimensions of x in. by x in., find a function giving the volume of the resulting box.

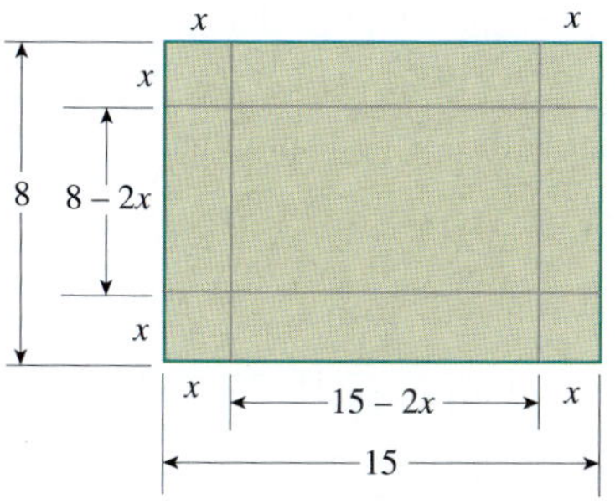

75. Construction Costs A rectangular box is to have a square base and a volume of 20 ft^3. The material for the base costs 30¢/ft^2, the material for the sides costs 10¢/ft^2, and the material for the top costs 20¢/ft^2. Letting x denote the length of one side of the base, find a function in the variable x giving the cost of constructing the box.

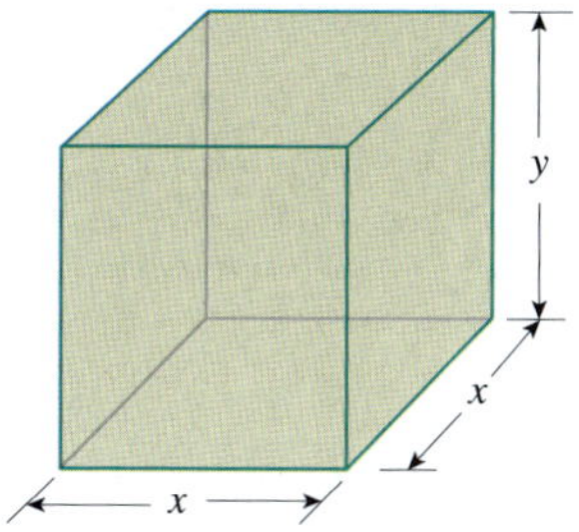

76. Area of a Norman Window A Norman window has the shape of a rectangle surmounted by a semicircle (see the accompanying figure). Suppose a Norman window is to have a perimeter of 28 ft; find a function in the variable x giving the area of the window.

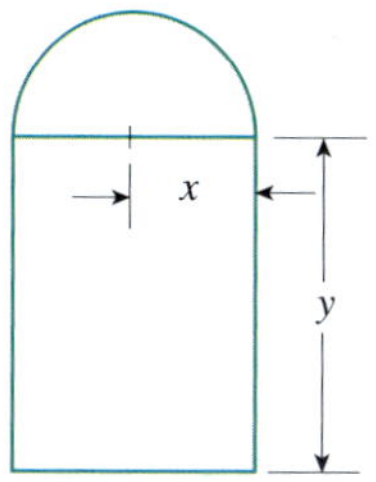

77. Yield of an Apple Orchard An apple orchard has an average yield of 36 bushels of apples/tree if tree density is 22 trees/acre. For each unit increase in tree density, the yield decreases by 2 bushels/tree. Letting x denote the number of trees beyond 22/acre, find a function in x that gives the yield of apples.

78. Book Design A book designer has decided that the pages of a book should have 1-in. margins at the top and bottom and $\frac{1}{2}$-in. margins on the sides. She further stipulated that each page should have an area of 50 in.^{2}. Find a function in the variable x, giving the area of the printed page. What is the domain of the function?

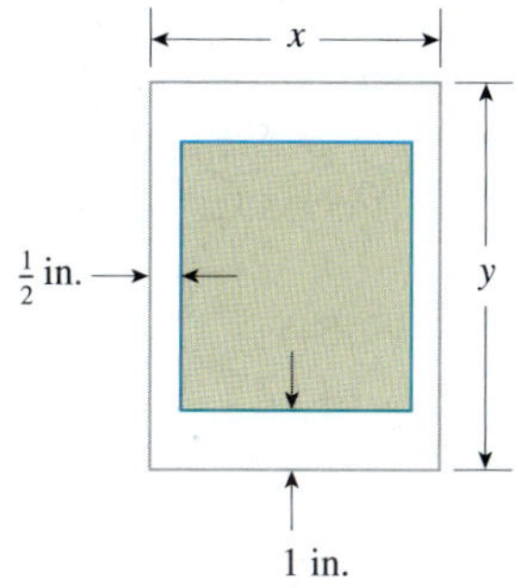

79. Profit of a Vineyard Phillip, the proprietor of a vineyard, estimates that if 10,000 bottles of wine were produced this season, then the profit would be \$5/bottle. But if more than 10,000 bottles were produced, then the profit/bottle for the entire lot would drop by \$0.0002 for each additional bottle sold. Assume at least 10,000 bottles of wine are produced and sold and let x denote the number of bottles produced and sold above 10,000.

a. Find a function P giving the profit in terms of x.

b. What is the profit Phillip can expect from the sale of 16,000 bottles of wine from his vineyard?

80. Charter Revenue The owner of a luxury motor yacht that sails among the 4000 Greek islands charges \$600/person/day if exactly 20 people sign up for the cruise. However, if more than 20 people sign up (up to the maximum capacity of 90) for the cruise, then each fare is reduced by \$4 for each additional passenger. Assume at least 20 people sign up for the cruise and let x denote the number of passengers above 20.

a. Find a function R giving the revenue/day realized from the charter.

b. What is the revenue/day if 60 people sign up for the cruise?

c. What is the revenue/day if 80 people sign up for the cruise?

In Exercises 81–84, determine whether the statement is true or false. If it is true, explain why it is true. If it is false, give an example to show why it is false.

81. A polynomial function is a sum of constant multiples of power functions.

82. A polynomial function is a rational function, but the converse is false.

83. If $r > 0$, then the power function $f(x) = x^r$ is defined for all values of x.

84. The function $f(x) = 2^x$ is a power function.

2.3 Solutions to Self-Check Exercises

1. Since the adult dose of the substance is 500 mg, $a = 500$; thus, the rule in this case is

$$D(t) = \frac{500t}{t + 12}$$

A 4-yr-old should receive

$$D(4) = \frac{500(4)}{4 + 12}$$

or 125 mg of the substance.

2. **a.** The graphs of the functions d and s are shown in the following figure:

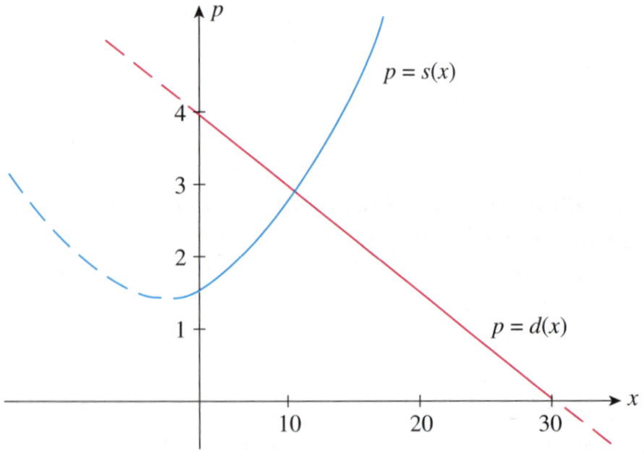

b. Solve the following system of equations:

$$p = -\frac{2}{15}x + 4$$

$$p = \frac{1}{75}x^2 + \frac{1}{10}x + \frac{3}{2}$$

Substituting the first equation into the second yields

$$\frac{1}{75}x^2 + \frac{1}{10}x + \frac{3}{2} = -\frac{2}{15}x + 4$$

$$\frac{1}{75}x^2 + \left(\frac{1}{10} + \frac{2}{15}\right)x - \frac{5}{2} = 0$$

$$\frac{1}{75}x^2 + \frac{7}{30}x - \frac{5}{2} = 0$$

Multiplying both sides of the last equation by 150, we have

$$2x^2 + 35x - 375 = 0$$

$$(2x - 15)(x + 25) = 0$$

Thus, $x = -25$ or $x = 15/2 = 7.5$. Since x must be nonnegative, we take $x = 7.5$, and the equilibrium quantity is 7500 lb. The equilibrium price is given by

$$p = -\frac{2}{15}\left(\frac{15}{2}\right) + 4$$

or \$3/lb.

USING TECHNOLOGY

Finding the Points of Intersection of Two Graphs and Modeling

A graphing utility can be used to find the point(s) of intersection of the graphs of two functions.

EXAMPLE 1 Find the points of intersection of the graphs of

$$f(x) = 0.3x^2 - 1.4x - 3 \quad \text{and} \quad g(x) = -0.4x^2 + 0.8x + 6.4$$

Solution The graphs of both f and g in the standard viewing window are shown in Figure T1a. Using TRACE and ZOOM or the function for finding the points of intersection of two graphs on your graphing utility, we find the point(s) of intersection, accurate to four decimal places, to be $(-2.4158, 2.1329)$ (Figure T1b) and $(5.5587, -1.5125)$ (Figure T1c).

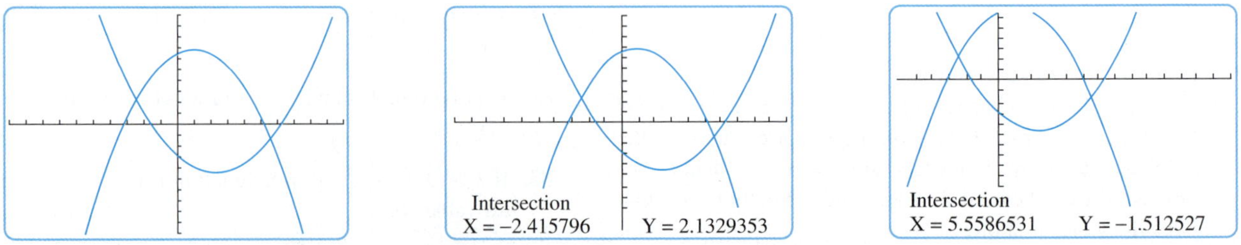

(a) (b) (c)

FIGURE **T1**
(a) The graphs of f and g in the standard viewing window; (b) and (c) the TI-83/84 intersection screens ■

EXAMPLE 2 Consider the demand and supply functions

$$p = d(x) = -0.01x^2 - 0.2x + 8 \quad \text{and} \quad p = s(x) = 0.01x^2 + 0.1x + 3$$

a. Plot the graphs of d and s in the viewing window $[0, 15] \times [0, 10]$.
b. Verify that the equilibrium point is (10, 5).

Solution

a. The graphs of d and s are shown in Figure T2a.

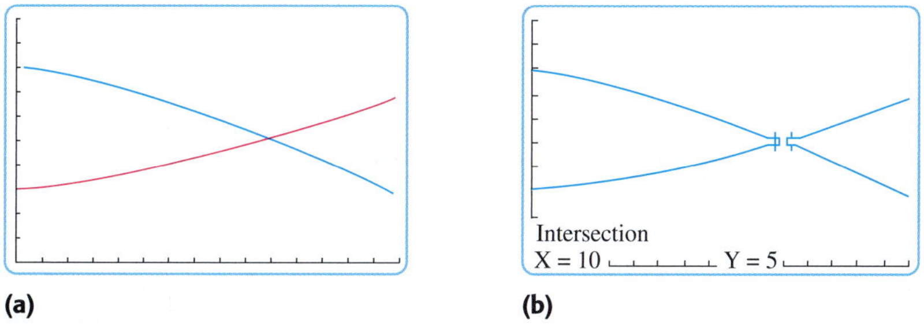

FIGURE **T2**
(a) The graphs of d and s in the window $[0, 15] \times [0, 10]$; (b) the TI-83/84 intersection screen

b. Using TRACE and ZOOM or the function for finding the point of intersection of two graphs, we see that $x = 10$ and $y = 5$ (Figure T2b), so the equilibrium point is (10, 5).

Constructing Mathematical Models from Raw Data

A graphing utility can sometimes be used to construct mathematical models from sets of data. For example, if the points corresponding to the given data are scattered about a straight line, then use **LinReg(ax+b)** (linear regression) from the statistical calculation menu of the graphing utility to obtain a function (model) that approximates the data at hand. If the points seem to be scattered along a parabola (the graph of a quadratic function), then use **QuadReg** (second-degree polynomial regression), and so on. (These are functions on the TI-83/84 calculator.)

APPLIED EXAMPLE 3 Indian Gaming Industry The following data gives the estimated gross revenues (in billions of dollars) from the Indian gaming industries from 2000 ($t = 0$) to 2005 ($t = 5$).

Year	0	1	2	3	4	5
Revenue	11.0	12.8	14.7	16.8	19.5	22.7

a. Use a graphing utility to find a polynomial function f of degree 4 that models the data.
b. Plot the graph of the function f, using the viewing window $[0, 6] \times [0, 30]$.
c. Use the function evaluation capability of the graphing utility to compute $f(0)$, $f(1), \ldots, f(5)$ and compare these values with the original data.
d. If the trend continued, what was the gross revenue for 2006 ($t = 6$)?

Source: National Indian Gaming Association

Solution

a. Choosing **QuartReg** (fourth-degree polynomial regression) from the statistical calculations menu of a graphing utility, we find

$$f(t) = -0.00417t^4 + 0.0713t^3 - 0.168t^2 + 1.920t + 11$$

(continued)

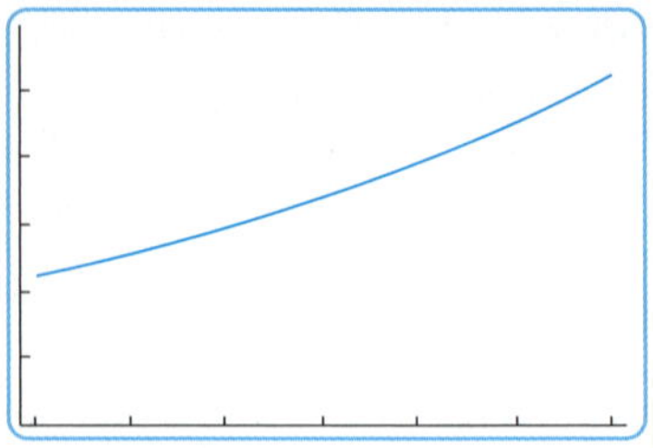

FIGURE T3
The graph of f in the viewing window $[0, 6] \times [0, 30]$

b. The graph of f is shown in Figure T3.

c. The required values, which compare favorably with the given data, follow:

t	0	1	2	3	4	5
$f(t)$	11.0	12.8	14.7	16.8	19.5	22.7

d. The gross revenue for 2006 ($t = 6$) is given by

$$f(6) = -0.00417(6)^4 + 0.0713(6)^3 - 0.168(6)^2 + 1.920(6) + 11 \approx 26.468$$

or \$26.5 billion.

TECHNOLOGY EXERCISES

In Exercises 1–6, find the points of intersection of the graphs of the functions. Express your answer accurate to four decimal places.

1. $f(x) = 1.2x + 3.8$; $g(x) = -0.4x^2 + 1.2x + 7.5$

2. $f(x) = 0.2x^2 - 1.3x - 3$; $g(x) = -1.3x + 2.8$

3. $f(x) = 0.3x^2 - 1.7x - 3.2$; $g(x) = -0.4x^2 + 0.9x + 6.7$

4. $f(x) = -0.3x^2 + 0.6x + 3.2$; $g(x) = 0.2x^2 - 1.2x - 4.8$

5. $f(x) = 0.3x^3 - 1.8x^2 + 2.1x - 2$; $g(x) = 2.1x - 4.2$

6. $f(x) = -0.2x^3 + 1.2x^2 - 1.2x + 2$; $g(x) = -0.2x^2 + 0.8x + 2.1$

7. **Market Equilibrium** The monthly demand and supply functions for a certain brand of wall clock are given by

$$p = -0.2x^2 - 1.2x + 50$$
$$p = 0.1x^2 + 3.2x + 25$$

respectively, where p is measured in dollars and x is measured in units of a hundred.
   a. Plot the graphs of both functions in an appropriate viewing window.
   b. Find the equilibrium quantity and price.

8. **Market Equilibrium** The quantity demanded x (in units of a hundred) of Mikado miniature cameras/week is related to the unit price p (in dollars) by

$$p = -0.2x^2 + 80$$

The quantity x (in units of a hundred) that the supplier is willing to make available in the market is related to the unit price p (in dollars) by

$$p = 0.1x^2 + x + 40$$

   a. Plot the graphs of both functions in an appropriate viewing window.
   b. Find the equilibrium quantity and price.

In Exercises 9–22, use the statistical calculations menu to construct a mathematical model associated with the given data.

9. **Consumption of Bottled Water** The annual per-capita consumption of bottled water (in gallons) and the scatter plot for these data follow:

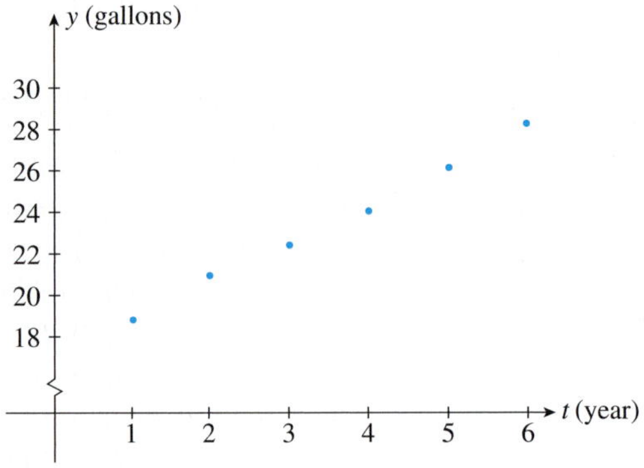

Year	2001	2002	2003	2004	2005	2006
Consumption	18.8	20.9	22.4	24	26.1	28.3

   a. Use **LinReg(ax + b)** to find a first-degree (linear) polynomial regression model for the data. Let $t = 1$ correspond to 2001.
   b. Plot the graph of the function f found in part (a), using the viewing window $[1, 6] \times [0, 30]$.
   c. Compute the values for $t = 1, 2, 3, 4, 5$, and 6. How do your figures compare with the given data?
   d. If the trend continued, what will be the annual per-capita consumption of bottled water in 2008?

   Source: Beverage Marketing Corporation

10. **Web Conferencing** Web conferencing is a big business, and it's growing rapidly. The amount (in billions of dollars) spent on Web conferencing from the beginning of 2003 through 2010, and the scatter diagram for these data follow:

Year	2003	2004	2005	2006	2007	2008	2009	2010
Amount	0.50	0.63	0.78	0.92	1.16	1.38	1.60	1.90

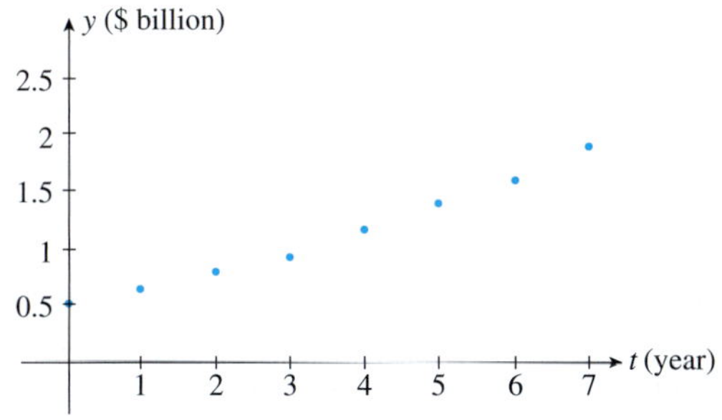

a. Let $t = 0$ correspond to the beginning of 2003 and use **QuadReg** to find a second-degree polynomial regression model based on the given data.

b. Plot the graph of the function f found in part (a) using the window $[0, 7] \times [0, 2]$.

c. Compute $f(0), f(3), f(6)$, and $f(7)$. Compare these values with the given data.

Source: Gartner Dataquest

11. Student Population The projected total number of students in elementary schools, secondary schools, and colleges (in millions) from 1995 through 2015 is given in the following table:

Year	1995	2000	2005	2010	2015
Number	64.8	68.7	72.6	74.8	78

a. Use **QuadReg** to find a second-degree polynomial regression model for the data. Let t be measured in 5-yr intervals, with $t = 0$ corresponding to the beginning of 1995.

b. Plot the graph of the function f found in part (a), using the viewing window $[0, 4] \times [0, 85]$.

c. Using the model found in part (a), what will the projected total number of students (all categories) enrolled in 2015?

Source: U.S. National Center for Education Statistics

12. Digital TV Shipments The estimated number of digital TV shipments between the year 2000 and 2006 (in millions of units) and the scatter plot for these data follow:

Year	2000	2001	2002	2003	2004	2005	2006
Units Shipped	0.63	1.43	2.57	4.1	6	8.1	10

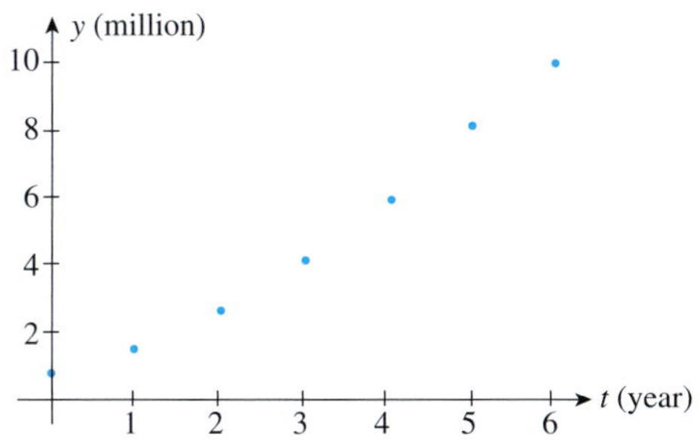

a. Use **CubicReg** to find a third-degree polynomial regression model for the data. Let $t = 0$ correspond to the beginning of 2000.

b. Plot the graph of the function f found in part (a), using the viewing window $[0, 6] \times [0, 11]$.

c. Compute the values of $f(t)$ for $t = 0, 1, 2, 3, 4, 5$, and 6.

Source: Consumer Electronics Manufacturers Association

13. Health-Care Spending Health-care spending by business (in billions of dollars) from the year 2000 through 2006 is summarized below:

Year	2000	2001	2002	2003	2004	2005	2006
Number	185	235	278	333	389	450	531

a. Plot the scatter diagram for the above data. Let $t = 0$ correspond to the beginning of 2000.

b. Use **QuadReg** to find a second-degree polynomial regression model for the data.

c. If the trend continued, what was the spending in 2007?

Source: Centers for Medicine and Medicaid Services

14. TIVO Owners The projected number of households (in millions) with digital video recorders that allow viewers to record shows onto a server and skip commercials are given in the following table:

Year	2006	2007	2008	2009	2010
Households	31.2	49.0	71.6	97.0	130.2

a. Let $t = 0$ correspond to the beginning of 2006 and use **QuadReg** to find a second-degree polynomial regression model based on the given data.

b. Obtain the scatter plot and the graph of the function f found in part (a), using the viewing window $[0, 4] \times [0, 140]$.

Source: Strategy Analytics

15. Telecommunications Industry Revenue The telecommunications industry revenue is expected to grow in the coming years, fueled by the demand for broadband and high-speed data services. The worldwide revenue for the industry (in trillions of dollars) and the scatter diagram for these data follow:

Year	2000	2002	2004	2006	2008	2010
Revenue	1.7	2.0	2.5	3.0	3.6	4.2

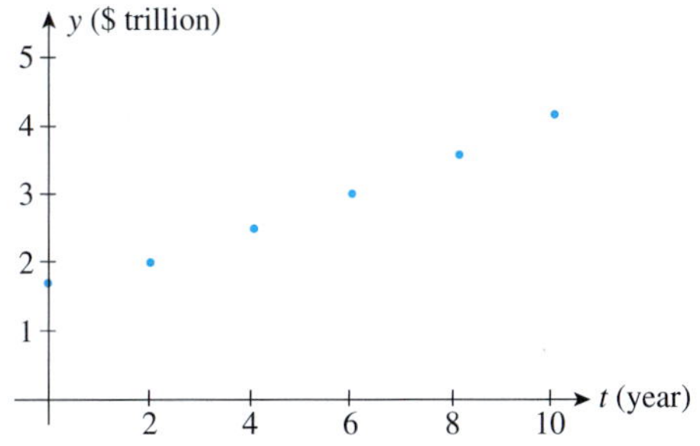

a. Let $t = 0$ correspond to the beginning of 2000 and use **CubicReg** to find a third-degree polynomial regression model based on the given data.

b. Plot the graph of the function f found in part (a), using the viewing window $[0, 10] \times [0, 5]$.

c. Find the worldwide revenue for the industry in 2001 and 2005 and find the projected revenue for 2010.

Source: Telecommunication Industry Association

(continued)

16. **POPULATION GROWTH IN CLARK COUNTY** Clark County in Nevada—dominated by greater Las Vegas—is one of the fastest-growing metropolitan areas in the United States. The population of the county from 1970 through 2000 is given in the following table:

Year	1970	1980	1990	2000
Population	273,288	463,087	741,459	1,375,765

a. Use **CubicReg** to find a third-degree polynomial regression model for the data. Let t be measured in decades, with $t = 0$ corresponding to the beginning of 1970.
b. Plot the graph of the function f found in part (a), using the viewing window $[0, 3] \times [0, 1{,}500{,}000]$.
c. Compare the values of f at $t = 0$, 1, 2, and 3, with the given data.

Source: U.S. Census Bureau

17. **HIRING LOBBYISTS** Many public entities like cities, counties, states, utilities, and Indian tribes are hiring firms to lobby Congress. One goal of such lobbying is to place earmarks—money directed at a specific project—into appropriation bills. The amount (in millions of dollars) spent by public entities on lobbying from 1998 through 2004 is shown in the following table:

Year	1998	1999	2000	2001	2002	2003	2004
Amount	43.4	51.7	62.5	76.3	92.3	101.5	107.7

a. Use **CubicReg** to find a third-degree polynomial regression model for the data, letting $t = 0$ correspond to the beginning of 1998.
b. Plot the scatter diagram and the graph of the function f found in part (a), using the viewing window $[0, 6] \times [0, 120]$.
c. Compare the values of f at $t = 0$, 3, and 6 with the given data.

Source: Center for Public Integrity

18. **MOBILE ENTERPRISE IM ACCOUNTS** The projected number of mobile enterprise instant messaging accounts (in millions) from 2006 through 2010 is given in the following table ($t = 0$ corresponds to the beginning of 2006):

Year	0	1	2	3	4
Accounts	2.3	3.6	5.8	8.7	14.9

a. Use **CubicReg** to find a third-degree polynomial regression model based on the given data.
b. Plot the graph of the function f found in part (a), using the viewing window $[0, 5] \times [0, 16]$.
c. Compute $f(0), f(1), f(2), f(3)$, and $f(4)$.

Source: The Radical Group

19. **MEASLES DEATHS** Measles is still a leading cause of vaccine-preventable death among children, but due to improvements in immunizations, measles deaths have dropped globally. The following table gives the number of measles deaths (in thousands) in sub-Saharan Africa from 1999 through 2005:

Year	1999	2001	2003	2005
Amount	506	338	250	126

a. Use **CubicReg** to find a third-degree polynomial regression model for the data, letting $t = 0$ correspond to the beginning of 1999.
b. Plot the scatter diagram and the graph of the function f found in part (a).
c. Compute the values of f for $t = 0$, 2, and 6.

Source: Centers for Disease Control and World Health Organization

20. **OFFICE VACANCY RATE** The total vacancy rate of offices in Manhattan from the year 2000 through 2006 is shown in the following table.

Year	2000	2001	2002	2003	2004	2005	2006
Vacancy Rate	3.8	8.9	12	12.5	11	8.4	6.7

a. Use **CubicReg** to find a third-degree polynomial regression model for the data, letting $t = 0$ correspond to the beginning of 2000.
b. Plot the scatter diagram and the graph of the function f found in part (a).
c. Compute the values for $t = 1$, 2, 3, 4, 5, and 6.

Source: Cushman and Wakefield

21. **NICOTINE CONTENT OF CIGARETTES** Even as measures to discourage smoking have been growing more stringent in recent years, the nicotine content of cigarettes has been rising, making it more difficult for smokers to quit. The following table gives the average amount of nicotine in cigarette smoke from 1999 through 2004:

Year	1999	2000	2001	2002	2003	2004
Yield per Cigarette (mg)	1.71	1.81	1.85	1.84	1.83	1.89

a. Use **QuartReg** to find a fourth-degree polynomial regression model for the data. Let $t = 0$ correspond to the beginning of 1999.
b. Plot the graph of the function f found in part (a), using the viewing window $[0, 5] \times [0, 2]$.
c. Compute the values of $f(t)$ for $t = 0$, 1, 2, 3, 4, and 5.
d. If the trend continued, what would have been the average amount of nicotine in cigarettes in 2005?

Source: Massachusetts Tobacco Control Program

22. **SOCIAL SECURITY TRUST FUND ASSETS** The projected assets of the Social Security trust fund (in trillions of dollars) from 2008 through 2040 are given in the following table:

Year	2008	2011	2014	2017	2020	2023	2026	2029	2032	2035	2038	2040
Assets	2.4	3.2	4.0	4.7	5.3	5.7	5.9	5.6	4.9	3.6	1.7	0

Use **QuartReg** to find a fourth-degree polynomial regression model for the data. Let $t = 0$ correspond to the beginning of 2008.

Source: Social Security Administration

2.4 Limits

Introduction to Calculus

Historically, the development of calculus by Isaac Newton (1642–1727) and Gottfried Wilhelm Leibniz (1646–1716) resulted from the investigation of the following problems:

1. Finding the tangent line to a curve at a given point on the curve (Figure 26a)
2. Finding the area of a planar region bounded by an arbitrary curve (Figure 26b)

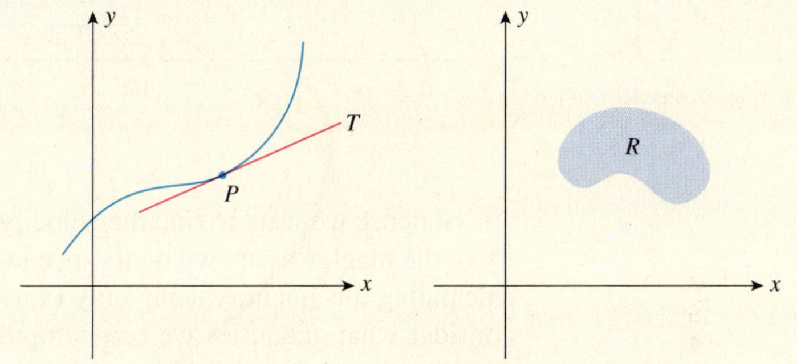

(a) What is the slope of the tangent line T at point P?

(b) What is the area of the region R?

FIGURE **26**

The tangent-line problem might appear to be unrelated to any practical applications of mathematics, but as you will see later, the problem of finding the *rate of change* of one quantity with respect to another is mathematically equivalent to the geometric problem of finding the slope of the *tangent line* to a curve at a given point on the curve. It is precisely the discovery of the relationship between these two problems that spurred the development of calculus in the 17th century and made it such an indispensable tool for solving practical problems. The following are a few examples of such problems:

- Finding the velocity of an object
- Finding the rate of change of a bacteria population with respect to time
- Finding the rate of change of a company's profit with respect to time
- Finding the rate of change of a travel agency's revenue with respect to the agency's expenditure for advertising

The study of the tangent-line problem led to the creation of *differential calculus*, which relies on the concept of the *derivative* of a function. The study of the area problem led to the creation of *integral calculus*, which relies on the concept of the *antiderivative*, or *integral*, of a function. (The derivative of a function and the integral of a function are intimately related, as you will see in Section 6.4.) Both the derivative of a function and the integral of a function are defined in terms of a more fundamental concept—the limit—our next topic.

A Real-Life Example

From data obtained in a test run conducted on a prototype of a maglev (magnetic levitation train), which moves along a straight monorail track, engineers have determined that the position of the maglev (in feet) from the origin at time t (in seconds) is given by

$$s = f(t) = 4t^2 \qquad (0 \le t \le 30) \tag{3}$$

where f is called the **position function** of the maglev. The position of the maglev at time $t = 0, 1, 2, 3, \ldots, 10$, measured from its initial position, is

$$f(0) = 0 \qquad f(1) = 4 \qquad f(2) = 16 \qquad f(3) = 36, \ldots \qquad f(10) = 400$$

feet (Figure 27).

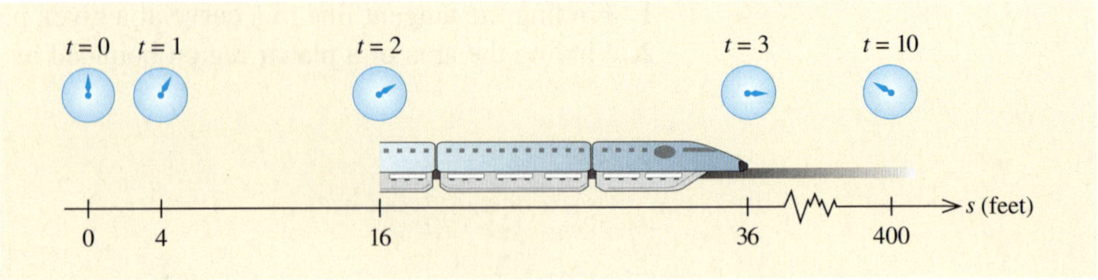

FIGURE **27**
A maglev moving along an elevated monorail track

Suppose we want to find the velocity of the maglev at $t = 2$. This is just the velocity of the maglev as shown on its speedometer at that precise instant of time. Offhand, calculating this quantity using only Equation (3) appears to be an impossible task; but consider what quantities we *can* compute using this relationship. Obviously, we can compute the position of the maglev at any time t as we did earlier for some selected values of t. Using these values, we can then compute the *average velocity* of the maglev over an interval of time. For example, the average velocity of the train over the time interval [2, 4] is given by

$$\begin{aligned}\frac{\text{Distance covered}}{\text{Time elapsed}} &= \frac{f(4) - f(2)}{4 - 2} \\ &= \frac{4(4^2) - 4(2^2)}{2} \\ &= \frac{64 - 16}{2} = 24\end{aligned}$$

or 24 feet/second.

Although this is not quite the velocity of the maglev at $t = 2$, it does provide us with an approximation of its velocity at that time.

Can we do better? Intuitively, the smaller the time interval we pick (with $t = 2$ as the left endpoint), the better the average velocity over that time interval will approximate the actual velocity of the maglev at $t = 2$.*

Now, let's describe this process in general terms. Let $t > 2$. Then, the average velocity of the maglev over the time interval $[2, t]$ is given by

$$\frac{f(t) - f(2)}{t - 2} = \frac{4t^2 - 4(2^2)}{t - 2} = \frac{4(t^2 - 4)}{t - 2} \tag{4}$$

By choosing the values of t closer and closer to 2, we obtain a sequence of numbers that give the average velocities of the maglev over smaller and smaller time intervals. As we observed earlier, this sequence of numbers should approach the *instantaneous velocity* of the train at $t = 2$.

Let's try some sample calculations. Using Equation (4) and taking the sequence $t = 2.5, 2.1, 2.01, 2.001$, and 2.0001, which approaches 2, we find

*Actually, any interval containing $t = 2$ will do.

The average velocity over [2, 2.5] is $\dfrac{4(2.5^2 - 4)}{2.5 - 2} = 18$, or 18 feet/second.

The average velocity over [2, 2.1] is $\dfrac{4(2.1^2 - 4)}{2.1 - 2} = 16.4$, or 16.4 feet/second.

and so forth. These results are summarized in Table 1.

TABLE 1

	t approaches 2 from the right.				
t	2.5	2.1	2.01	2.001	2.0001
Average Velocity over [2, t]	18	16.4	16.04	16.004	16.0004

Average velocity approaches 16 from the right.

From Table 1, we see that the average velocity of the maglev seems to approach the number 16 as it is computed over smaller and smaller time intervals. These computations suggest that the instantaneous velocity of the train at $t = 2$ is 16 feet/second.

Note Notice that we cannot obtain the instantaneous velocity for the maglev at $t = 2$ by substituting $t = 2$ into Equation (4) because this value of t is not in the domain of the average velocity function. ■

Intuitive Definition of a Limit

Consider the function g defined by

$$g(t) = \frac{4(t^2 - 4)}{t - 2}$$

which gives the average velocity of the maglev [see Equation (4)]. Suppose we are required to determine the value that $g(t)$ approaches as t approaches the (fixed) number 2. If we take the sequence of values of t approaching 2 from the right-hand side, as we did earlier, we see that $g(t)$ approaches the number 16. Similarly, if we take a sequence of values of t approaching 2 from the left, such as $t = 1.5, 1.9, 1.99, 1.999$, and 1.9999, we obtain the results shown in Table 2.

TABLE 2

	t approaches 2 from the left.				
t	1.5	1.9	1.99	1.999	1.9999
$g(t)$	14	15.6	15.96	15.996	15.9996

Average velocity approaches 16 from the left.

Observe that $g(t)$ approaches the number 16 as t approaches 2—this time from the left-hand side. In other words, as t approaches 2 from *either* side of 2, $g(t)$ approaches 16. In this situation, we say that the limit of $g(t)$ as t approaches 2 is 16, written

$$\lim_{t \to 2} g(t) = \lim_{t \to 2} \frac{4(t^2 - 4)}{t - 2} = 16$$

The graph of the function g, shown in Figure 28, confirms this observation.

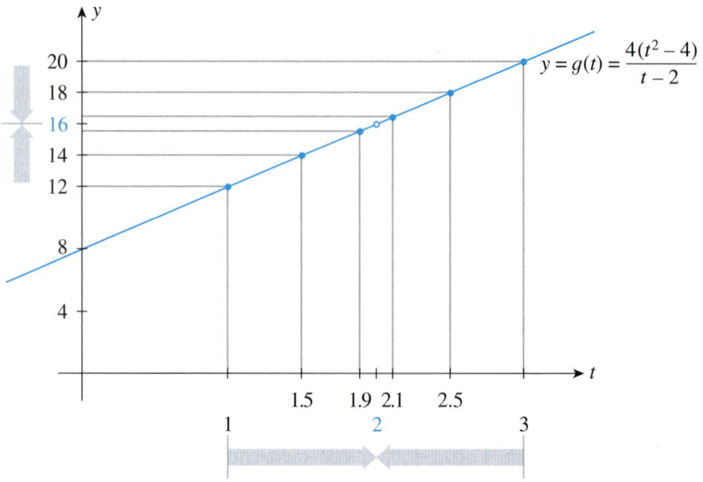

FIGURE **28**
As t approaches $t = 2$ from either direction, $g(t)$ approaches $y = 16$.

Observe that the point $t = 2$ is not in the domain of the function g [for this reason, the point (2, 16) is missing from the graph of g]. This, however, is inconsequential because the value, if any, of $g(t)$ at $t = 2$ plays no role in computing the limit.

This example leads to the following informal definition.

Limit of a Function

The function f has the **limit** L as x approaches a, written

$$\lim_{x \to a} f(x) = L$$

if the value of $f(x)$ can be made as close to the number L as we please by taking x sufficiently close to (but not equal to) a.

Exploring with TECHNOLOGY

1. Use a graphing utility to plot the graph of

$$g(x) = \frac{4(x^2 - 4)}{x - 2}$$

in the viewing window $[0, 3] \times [0, 20]$.

2. Use ZOOM and TRACE to describe what happens to the values of $g(x)$ as x approaches 2, first from the right and then from the left.
3. What happens to the y-value when you try to evaluate $g(x)$ at $x = 2$? Explain.
4. Reconcile your results with those of the preceding example.

Evaluating the Limit of a Function

Let's now consider some examples involving the computation of limits.

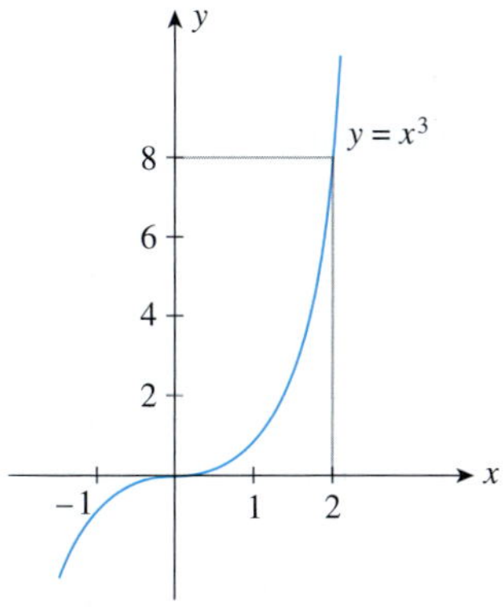

FIGURE **29**
$f(x)$ is close to 8 whenever x is close to 2.

EXAMPLE 1 Let $f(x) = x^3$ and evaluate $\lim_{x \to 2} f(x)$.

Solution The graph of f is shown in Figure 29. You can see that $f(x)$ can be made as close to the number 8 as we please by taking x sufficiently close to 2. Therefore,

$$\lim_{x \to 2} x^3 = 8$$

■

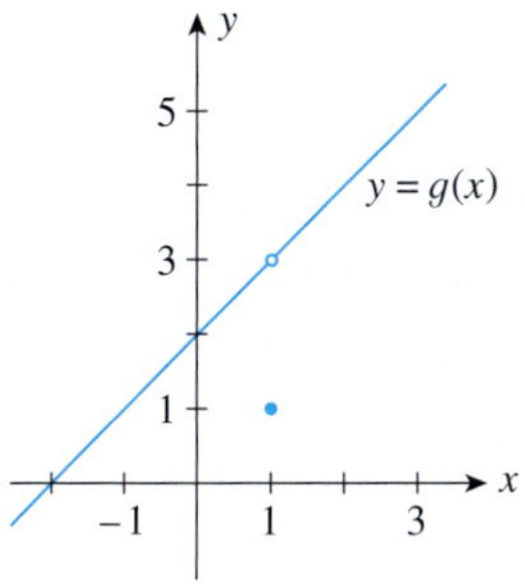

FIGURE **30**
$\lim_{x \to 1} g(x) = 3$

EXAMPLE 2 Let

$$g(x) = \begin{cases} x + 2 & \text{if } x \neq 1 \\ 1 & \text{if } x = 1 \end{cases}$$

Evaluate $\lim_{x \to 1} g(x)$.

Solution The domain of g is the set of all real numbers. From the graph of g shown in Figure 30, we see that $g(x)$ can be made as close to 3 as we please by taking x sufficiently close to 1. Therefore,

$$\lim_{x \to 1} g(x) = 3$$

Observe that $g(1) = 1$, which is not equal to the limit of the function g as x approaches 1. [Once again, the value of $g(x)$ at $x = 1$ has no bearing on the existence or value of the limit of g as x approaches 1.] ■

EXAMPLE 3 Evaluate the limit of the following functions as x approaches the indicated point.

a. $f(x) = \begin{cases} -1 & \text{if } x < 0 \\ 1 & \text{if } x \geq 0 \end{cases}$; $x = 0$ **b.** $g(x) = \dfrac{1}{x^2}$; $x = 0$

Solution The graphs of the functions f and g are shown in Figure 31.

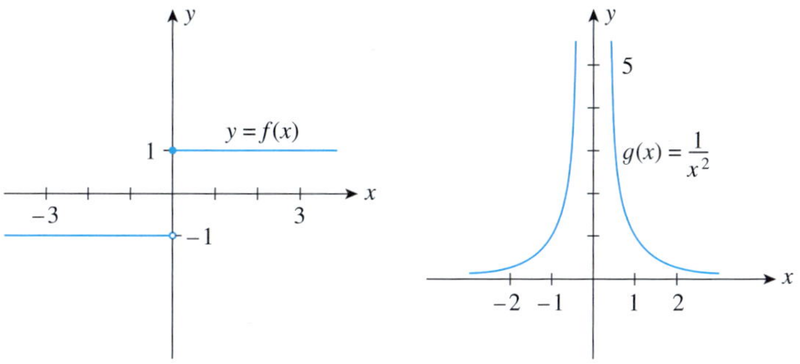

FIGURE **31**

(a) $\lim_{x \to 0} f(x)$ does not exist. **(b)** $\lim_{x \to 0} g(x)$ does not exist.

a. Referring to Figure 31a, we see that no matter how close x is to zero, $f(x)$ takes on the values 1 or -1, depending on whether x is positive or negative. Thus, there is no *single* real number L that $f(x)$ approaches as x approaches zero. We conclude that the limit of $f(x)$ does *not* exist as x approaches zero.

b. Referring to Figure 31b, we see that as x approaches zero (from either side), $g(x)$ increases without bound and thus does not approach any specific real number. We conclude, accordingly, that the limit of $g(x)$ does *not* exist as x approaches zero. ■

Explore & Discuss

Consider the graph of the function h whose graph is depicted in the following figure.

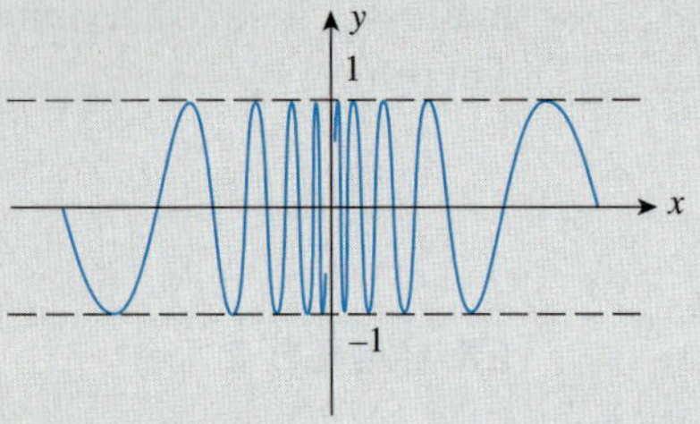

It has the property that as x approaches zero from either the right or the left, the curve oscillates more and more frequently between the lines $y = -1$ and $y = 1$.

1. Explain why $\lim_{x\to 0} h(x)$ does not exist.
2. Compare this function with those in Example 3. More specifically, discuss the different ways the functions fail to have a limit at $x = 0$.

Until now, we have relied on knowing the actual values of a function or the graph of a function near $x = a$ to help us evaluate the limit of the function $f(x)$ as x approaches a. The following properties of limits, which we list without proof, enable us to evaluate limits of functions algebraically.

THEOREM 1

Properties of Limits

Suppose

$$\lim_{x\to a} f(x) = L \quad \text{and} \quad \lim_{x\to a} g(x) = M$$

Then,

1. $\lim_{x\to a}[f(x)]^r = [\lim_{x\to a} f(x)]^r = L^r$ *r, a real number*
2. $\lim_{x\to a} cf(x) = c\lim_{x\to a} f(x) = cL$ *c, a real number*
3. $\lim_{x\to a}[f(x) \pm g(x)] = \lim_{x\to a} f(x) \pm \lim_{x\to a} g(x) = L \pm M$
4. $\lim_{x\to a}[f(x)g(x)] = [\lim_{x\to a} f(x)][\lim_{x\to a} g(x)] = LM$
5. $\lim_{x\to a} \dfrac{f(x)}{g(x)} = \dfrac{\lim_{x\to a} f(x)}{\lim_{x\to a} g(x)} = \dfrac{L}{M}$ *Provided that $M \neq 0$*

EXAMPLE 4 Use Theorem 1 to evaluate the following limits.

a. $\lim_{x\to 2} x^3$ **b.** $\lim_{x\to 4} 5x^{3/2}$ **c.** $\lim_{x\to 1}(5x^4 - 2)$

d. $\lim_{x\to 3} 2x^3\sqrt{x^2 + 7}$ **e.** $\lim_{x\to 2} \dfrac{2x^2 + 1}{x + 1}$

Solution

a. $\lim_{x\to 2} x^3 = [\lim_{x\to 2} x]^3$ *Property 1*

$= 2^3 = 8$ *$\lim_{x\to 2} x = 2$*

b. $\lim_{x\to 4} 5x^{3/2} = 5[\lim_{x\to 4} x^{3/2}]$ Property 2

$= 5(4)^{3/2} = 40$ Property 1

c. $\lim_{x\to 1}(5x^4 - 2) = \lim_{x\to 1} 5x^4 - \lim_{x\to 1} 2$ Property 3

To evaluate $\lim_{x\to 1} 2$, observe that the constant function $g(x) = 2$ has value 2 for all values of x. Therefore, $g(x)$ must approach the limit 2 as x approaches $x = 1$ (or any other point for that matter!). Therefore,

$$\lim_{x\to 1}(5x^4 - 2) = 5(1)^4 - 2 = 3$$

d. $\lim_{x\to 3} 2x^3\sqrt{x^2 + 7} = 2\lim_{x\to 3} x^3\sqrt{x^2 + 7}$ Property 2

$= 2\lim_{x\to 3} x^3 \lim_{x\to 3}\sqrt{x^2 + 7}$ Property 4

$= 2(3)^3\sqrt{3^2 + 7}$ Property 1

$= 2(27)\sqrt{16} = 216$

e. $\lim_{x\to 2}\dfrac{2x^2 + 1}{x + 1} = \dfrac{\lim_{x\to 2}(2x^2 + 1)}{\lim_{x\to 2}(x + 1)}$ Property 5

$= \dfrac{2(2)^2 + 1}{2 + 1} = \dfrac{9}{3} = 3$ ■

Indeterminate Forms

Let's emphasize once again that Property 5 of limits is valid only when the limit of the function that appears in the denominator is not equal to zero at the number in question.

If the numerator has a limit different from zero and the denominator has a limit equal to zero, then the limit of the quotient does not exist at the number in question. This is the case with the function $g(x) = 1/x^2$ in Example 3b. Here, as x approaches zero, the numerator approaches 1 but the denominator approaches zero, so the quotient becomes arbitrarily large. Thus, as observed earlier, the limit does not exist.

Next, consider

$$\lim_{x\to 2}\frac{4(x^2 - 4)}{x - 2}$$

which we evaluated earlier by looking at the values of the function for x near $x = 2$. If we attempt to evaluate this expression by applying Property 5 of limits, we see that both the numerator and denominator of the function

$$\frac{4(x^2 - 4)}{x - 2}$$

approach zero as x approaches 2; that is, we obtain an expression of the form 0/0. In this event, we say that the limit of the quotient $f(x)/g(x)$ as x approaches 2 has the **indeterminate form 0/0.**

We need to evaluate limits of this type when we discuss the derivative of a function, a fundamental concept in the study of calculus. As the name suggests, the meaningless expression 0/0 does not provide us with a solution to our problem. One strategy that can be used to solve this type of problem follows.

Strategy for Evaluating Indeterminate Forms

1. Replace the given function with an appropriate one that takes on the same values as the original function everywhere except at $x = a$.
2. Evaluate the limit of this function as x approaches a.

Examples 5 and 6 illustrate this strategy.

VIDEO

EXAMPLE 5 Evaluate:

$$\lim_{x \to 2} \frac{4(x^2 - 4)}{x - 2}$$

Solution Since both the numerator and the denominator of this expression approach zero as x approaches 2, we have the indeterminate form 0/0. We rewrite

$$\frac{4(x^2 - 4)}{x - 2} = \frac{4(x - 2)(x + 2)}{(x - 2)}$$

which, upon canceling the common factors, is equivalent to $4(x + 2)$, provided $x \neq 2$. Next, we replace $4(x^2 - 4)/(x - 2)$ with $4(x + 2)$ and find that

$$\lim_{x \to 2} \frac{4(x^2 - 4)}{x - 2} = \lim_{x \to 2} 4(x + 2) = 16$$

The graphs of the functions

$$f(x) = \frac{4(x^2 - 4)}{x - 2} \quad \text{and} \quad g(x) = 4(x + 2)$$

are shown in Figure 32. Observe that the graphs are identical except when $x = 2$. The function g is defined for all values of x and, in particular, its value at $x = 2$ is $g(2) = 4(2 + 2) = 16$. Thus, the point (2, 16) is on the graph of g. However, the function f is not defined at $x = 2$. Since $f(x) = g(x)$ for all values of x except $x = 2$, it follows that the graph of f must look exactly like the graph of g, with the exception that the point (2, 16) is missing from the graph of f. This illustrates graphically why we can evaluate the limit of f by evaluating the limit of the "equivalent" function g.

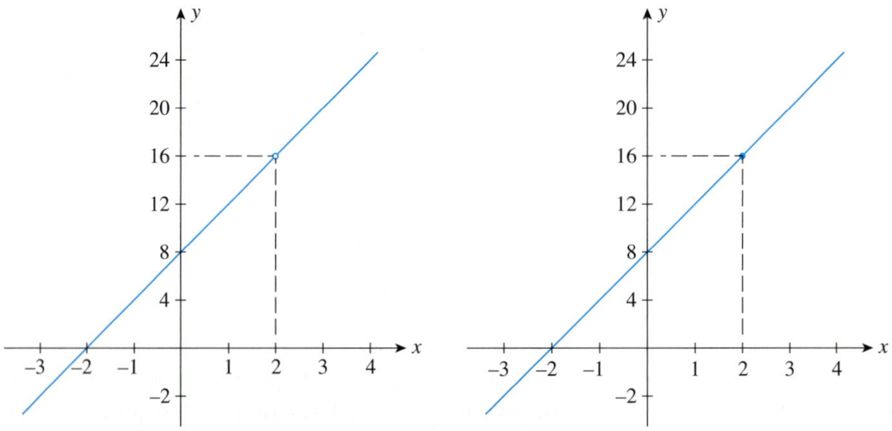

(a) $f(x) = \dfrac{4(x^2 - 4)}{x - 2}$

(b) $g(x) = 4(x + 2)$

FIGURE 32 The graphs of $f(x)$ and $g(x)$ are identical except at the point (2, 16).

Note Notice that the limit in Example 5 is the same limit that we evaluated earlier when we discussed the instantaneous velocity of a maglev at a specified time.

Exploring with TECHNOLOGY

1. Use a graphing utility to plot the graph of

$$f(x) = \frac{4(x^2 - 4)}{x - 2}$$

in the viewing window $[0, 3] \times [0, 20]$. Then use ZOOM and TRACE to find

$$\lim_{x \to 2} \frac{4(x^2 - 4)}{x - 2}$$

2. Use a graphing utility to plot the graph of $g(x) = 4(x + 2)$ in the viewing window $[0, 3] \times [0, 20]$. Then use ZOOM and TRACE to find $\lim_{x \to 2} 4(x + 2)$.

   What happens to the y-value when you try to evaluate $f(x)$ at $x = 2$? Explain.
3. Can you distinguish between the graphs of f and g?
4. Reconcile your results with those of Example 5.

EXAMPLE 6 Evaluate:

$$\lim_{h \to 0} \frac{\sqrt{1 + h} - 1}{h}$$

Solution Letting h approach zero, we obtain the indeterminate form 0/0. Next, we rationalize the numerator of the quotient by multiplying both the numerator and the denominator by the expression $(\sqrt{1 + h} + 1)$, obtaining

$$\frac{\sqrt{1 + h} - 1}{h} = \frac{(\sqrt{1 + h} - 1)(\sqrt{1 + h} + 1)}{h(\sqrt{1 + h} + 1)}$$

* See page 19.

$$= \frac{1 + h - 1}{h(\sqrt{1 + h} + 1)} \qquad (\sqrt{a} - \sqrt{b})(\sqrt{a} + \sqrt{b}) = a - b$$

$$= \frac{h}{h(\sqrt{1 + h} + 1)}$$

$$= \frac{1}{\sqrt{1 + h} + 1}$$

Therefore,

$$\lim_{h \to 0} \frac{\sqrt{1 + h} - 1}{h} = \lim_{h \to 0} \frac{1}{\sqrt{1 + h} + 1} = \frac{1}{\sqrt{1} + 1} = \frac{1}{2}$$

Exploring with TECHNOLOGY

1. Use a graphing utility to plot the graph of $g(x) = \dfrac{\sqrt{1 + x} - 1}{x}$ in the viewing window $[-1, 2] \times [0, 1]$. Then use ZOOM and TRACE to find $\lim_{x \to 0} \dfrac{\sqrt{1 + x} - 1}{x}$ by observing the values of $g(x)$ as x approaches zero from the left and from the right.
2. Use a graphing utility to plot the graph of $f(x) = \dfrac{1}{\sqrt{1 + x} + 1}$ in the viewing window $[-1, 2] \times [0, 1]$. Then use ZOOM and TRACE to find $\lim_{x \to 0} \dfrac{1}{\sqrt{1 + x} + 1}$. What happens to the y-value when x takes on the value zero? Explain.
3. Can you distinguish between the graphs of f and g?
4. Reconcile your results with those of Example 6.

*The symbol indicates that an algebraic computation or problem-solving skill used in the example is reviewed on the given page.

Limits at Infinity

Up to now we have studied the limit of a function as x approaches a (finite) number a. There are occasions, however, when we want to know whether $f(x)$ approaches a unique number as x increases without bound. Consider, for example, the function P, giving the number of fruit flies (*Drosophila*) in a container under controlled laboratory conditions, as a function of a time t. The graph of P is shown in Figure 33. You can see from the graph of P that, as t increases without bound (gets larger and larger), $P(t)$ approaches the number 400. This number, called the *carrying capacity* of the environment, is determined by the amount of living space and food available, as well as other environmental factors.

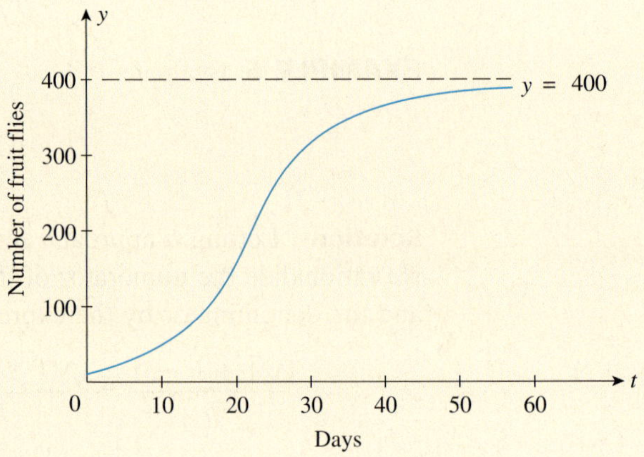

FIGURE 33
The graph of $P(t)$ gives the population of fruit flies in a laboratory experiment.

As another example, suppose we are given the function

$$f(x) = \frac{2x^2}{1 + x^2}$$

and we want to determine what happens to $f(x)$ as x gets larger and larger. Picking the sequence of numbers 1, 2, 5, 10, 100, and 1000 and computing the corresponding values of $f(x)$, we obtain the following table of values:

x	1	2	5	10	100	1000
$f(x)$	1	1.6	1.92	1.98	1.9998	1.999998

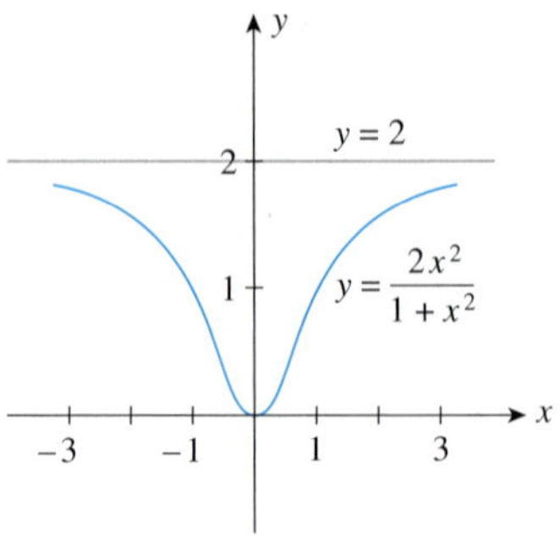

FIGURE 34
The graph of

$$y = \frac{2x^2}{1 + x^2}$$

has a horizontal asymptote at $y = 2$.

From the table, we see that as x gets larger and larger, $f(x)$ gets closer and closer to 2. The graph of the function f shown in Figure 34 confirms this observation. We call the line $y = 2$ a **horizontal asymptote.*** In this situation, we say that the limit of the function $f(x)$ as x increases without bound is 2, written

$$\lim_{x \to \infty} \frac{2x^2}{1 + x^2} = 2$$

In the general case, the following definition for a **limit of a function at infinity** is applicable.

*We will discuss asymptotes in greater detail in Section 4.3.

Limit of a Function at Infinity

The function f has the limit L as x increases without bound (or, as x approaches infinity), written

$$\lim_{x\to\infty} f(x) = L$$

if $f(x)$ can be made arbitrarily close to L by taking x large enough.

Similarly, the function f has the limit M as x decreases without bound (or as x approaches negative infinity), written

$$\lim_{x\to-\infty} f(x) = M$$

if $f(x)$ can be made arbitrarily close to M by taking x to be negative and sufficiently large in absolute value.

EXAMPLE 7 Let f and g be the functions

$$f(x) = \begin{cases} -1 & \text{if } x < 0 \\ 1 & \text{if } x \geq 0 \end{cases} \quad \text{and} \quad g(x) = \frac{1}{x^2}$$

Evaluate:

a. $\lim_{x\to\infty} f(x)$ and $\lim_{x\to-\infty} f(x)$ **b.** $\lim_{x\to\infty} g(x)$ and $\lim_{x\to-\infty} g(x)$

Solution The graphs of $f(x)$ and $g(x)$ are shown in Figure 35. Referring to the graphs of the respective functions, we see that

a. $\lim_{x\to\infty} f(x) = 1$ and $\lim_{x\to-\infty} f(x) = -1$ **b.** $\lim_{x\to\infty} \frac{1}{x^2} = 0$ and $\lim_{x\to-\infty} \frac{1}{x^2} = 0$

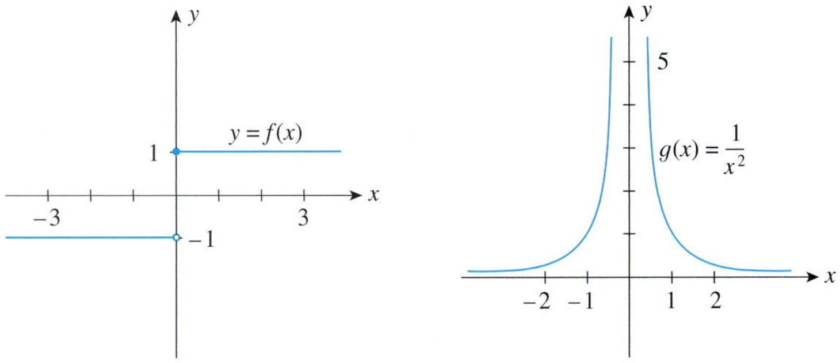

FIGURE 35

(a) $\lim_{x\to\infty} f(x) = 1$ and $\lim_{x\to-\infty} f(x) = -1$ (b) $\lim_{x\to\infty} g(x) = 0$ and $\lim_{x\to-\infty} g(x) = 0$

All the properties of limits listed in Theorem 1 are valid when a is replaced by ∞ or $-\infty$. In addition, we have the following property for the limit at infinity.

THEOREM 2

For all $n > 0$,

$$\lim_{x\to\infty} \frac{1}{x^n} = 0 \quad \text{and} \quad \lim_{x\to-\infty} \frac{1}{x^n} = 0$$

provided that $\frac{1}{x^n}$ is defined.

Exploring with TECHNOLOGY

1. Use a graphing utility to plot the graphs of

$$y_1 = \frac{1}{x^{0.5}} \qquad y_2 = \frac{1}{x} \qquad y_3 = \frac{1}{x^{1.5}}$$

in the viewing window $[0, 200] \times [0, 0.5]$. What can you say about $\lim_{x \to \infty} \frac{1}{x^n}$ if $n = 0.5$, $n = 1$, and $n = 1.5$? Are these results predicted by Theorem 2?

2. Use a graphing utility to plot the graphs of

$$y_1 = \frac{1}{x} \quad \text{and} \quad y_2 = \frac{1}{x^{5/3}}$$

in the viewing window $[-50, 0] \times [-0.5, 0]$. What can you say about $\lim_{x \to -\infty} \frac{1}{x^n}$ if $n = 1$ and $n = \frac{5}{3}$? Are these results predicted by Theorem 2?

Hint: To graph y_2, write it in the form y2 = 1/(x^(1/3))^5.

We often use the following technique to evaluate the limit at infinity of a rational function: *Divide the numerator and denominator of the expression by x^n, where n is the highest power present in the denominator of the expression.*

EXAMPLE 8 Evaluate

$$\lim_{x \to \infty} \frac{x^2 - x + 3}{2x^3 + 1}$$

Solution Since the limits of both the numerator and the denominator do not exist as x approaches infinity, the property pertaining to the limit of a quotient (Property 5) is not applicable. Let's divide the numerator and denominator of the rational expression by x^3, obtaining

$$\begin{aligned}\lim_{x \to \infty} \frac{x^2 - x + 3}{2x^3 + 1} &= \lim_{x \to \infty} \frac{\frac{1}{x} - \frac{1}{x^2} + \frac{3}{x^3}}{2 + \frac{1}{x^3}} \\ &= \frac{0 - 0 + 0}{2 + 0} = \frac{0}{2} \qquad \text{Use Theorem 2.} \\ &= 0\end{aligned}$$

■

EXAMPLE 9 Let

$$f(x) = \frac{3x^2 + 8x - 4}{2x^2 + 4x - 5}$$

Compute $\lim_{x \to \infty} f(x)$ if it exists.

Solution Again, we see that Property 5 is not applicable. Dividing the numerator and the denominator by x^2, we obtain

$$\lim_{x\to\infty} \frac{3x^2 + 8x - 4}{2x^2 + 4x - 5} = \lim_{x\to\infty} \frac{3 + \dfrac{8}{x} - \dfrac{4}{x^2}}{2 + \dfrac{4}{x} - \dfrac{5}{x^2}}$$

$$= \frac{\lim\limits_{x\to\infty} 3 + 8 \lim\limits_{x\to\infty} \dfrac{1}{x} - 4 \lim\limits_{x\to\infty} \dfrac{1}{x^2}}{\lim\limits_{x\to\infty} 2 + 4 \lim\limits_{x\to\infty} \dfrac{1}{x} - 5 \lim\limits_{x\to\infty} \dfrac{1}{x^2}}$$

$$= \frac{3 + 0 - 0}{2 + 0 - 0}$$ Use Theorem 2.

$$= \frac{3}{2}$$

EXAMPLE 10 Let $f(x) = \dfrac{2x^3 - 3x^2 + 1}{x^2 + 2x + 4}$ and evaluate:

a. $\lim\limits_{x\to\infty} f(x)$ **b.** $\lim\limits_{x\to-\infty} f(x)$

Solution

a. Dividing the numerator and the denominator of the rational expression by x^2, we obtain

$$\lim_{x\to\infty} \frac{2x^3 - 3x^2 + 1}{x^2 + 2x + 4} = \lim_{x\to\infty} \frac{2x - 3 + \dfrac{1}{x^2}}{1 + \dfrac{2}{x} + \dfrac{4}{x^2}}$$

Since the numerator becomes arbitrarily large whereas the denominator approaches 1 as x approaches infinity, we see that the quotient $f(x)$ gets larger and larger as x approaches infinity. In other words, the limit does not exist. We indicate this by writing

$$\lim_{x\to\infty} \frac{2x^3 - 3x^2 + 1}{x^2 + 2x + 4} = \infty$$

b. Once again, dividing both the numerator and the denominator by x^2, we obtain

$$\lim_{x\to-\infty} \frac{2x^3 - 3x^2 + 1}{x^2 + 2x + 4} = \lim_{x\to-\infty} \frac{2x - 3 + \dfrac{1}{x^2}}{1 + \dfrac{2}{x} + \dfrac{4}{x^2}}$$

In this case, the numerator becomes arbitrarily large in magnitude but negative in sign, whereas the denominator approaches 1 as x approaches negative infinity. Therefore, the quotient $f(x)$ decreases without bound, and the limit does not exist. We indicate this by writing

$$\lim_{x\to-\infty} \frac{2x^3 - 3x^2 + 1}{x^2 + 2x + 4} = -\infty$$

Example 11 gives an application of the concept of the limit of a function at infinity.

VIDEO

APPLIED EXAMPLE 11 Average Cost Functions Custom Office makes a line of executive desks. It is estimated that the total cost of making x Senior Executive Model desks is $C(x) = 100x + 200{,}000$ dollars per year, so the average cost of making x desks is given by

$$\overline{C}(x) = \frac{C(x)}{x} = \frac{100x + 200{,}000}{x} = 100 + \frac{200{,}000}{x}$$

dollars per desk. Evaluate $\lim_{x\to\infty} \overline{C}(x)$ and interpret your results.

Solution

$$\lim_{x\to\infty} \overline{C}(x) = \lim_{x\to\infty}\left(100 + \frac{200{,}000}{x}\right) = \lim_{x\to\infty} 100 + \lim_{x\to\infty} \frac{200{,}000}{x} = 100$$

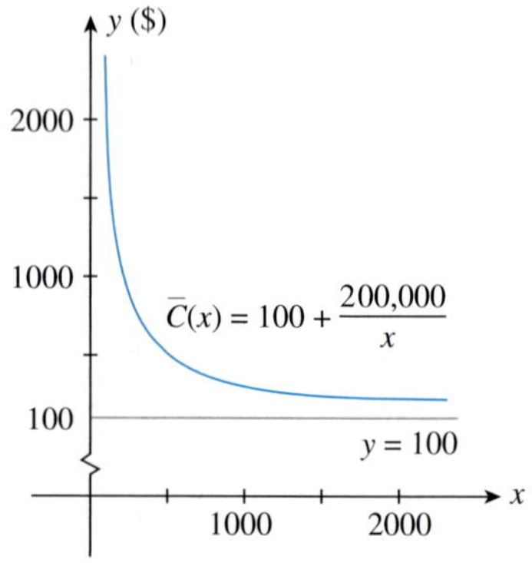

FIGURE 36
As the level of production increases, the average cost approaches $100 per desk.

A sketch of the graph of the function $\overline{C}(x)$ appears in Figure 36. The result we obtained is fully expected if we consider its economic implications. Note that as the level of production increases, the fixed cost per desk produced, represented by the term $(200{,}000/x)$, drops steadily. The average cost should approach a constant unit cost of production—$100 in this case. ■

Explore & Discuss

Consider the graph of the function f depicted in the following figure:

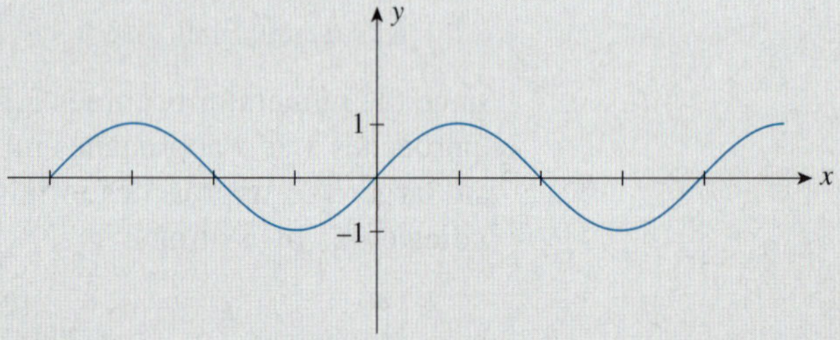

It has the property that the curve oscillates between $y = -1$ and $y = 1$ indefinitely in either direction.

1. Explain why $\lim_{x\to-\infty} f(x)$ and $\lim_{x\to\infty} f(x)$ do not exist.
2. Compare this function with those of Example 10. More specifically, discuss the different ways each function fails to have a limit at infinity or minus infinity.

2.4 Self-Check Exercises

1. Find the indicated limit if it exists.

   a. $\lim_{x\to 3} \dfrac{\sqrt{x^2+7} + \sqrt{3x-5}}{x+2}$

   b. $\lim_{x\to -1} \dfrac{x^2 - x - 2}{2x^2 - x - 3}$

2. The average cost per disc (in dollars) incurred by Herald Records in pressing x CDs is given by the average cost function

$$\overline{C}(x) = 1.8 + \frac{3000}{x}$$

Evaluate $\lim_{x\to\infty} \overline{C}(x)$ and interpret your result.

Solutions to Self-Check Exercises 2.4 can be found on page 114.

2.4 Concept Questions

1. Explain what is meant by the statement $\lim_{x\to 2} f(x) = 3$.
2. **a.** If $\lim_{x\to 3} f(x) = 5$, what can you say about $f(3)$? Explain.
   b. If $f(2) = 6$, what can you say about $\lim_{x\to 2} f(x)$? Explain.
3. Evaluate and state the property of limits that you use at each step.
   a. $\lim_{x\to 4} \sqrt{x}(2x^2 + 1)$ **b.** $\lim_{x\to 1} \left(\dfrac{2x^2 + x + 5}{x^4 + 1}\right)^{3/2}$
4. What is an indeterminate form? Illustrate with an example.
5. Explain in your own words the meaning of $\lim_{x\to\infty} f(x) = L$ and $\lim_{x\to -\infty} f(x) = M$.

2.4 Exercises

In Exercises 1–8, use the graph of the given function f to determine $\lim_{x\to a} f(x)$ at the indicated value of a, if it exists.

1.
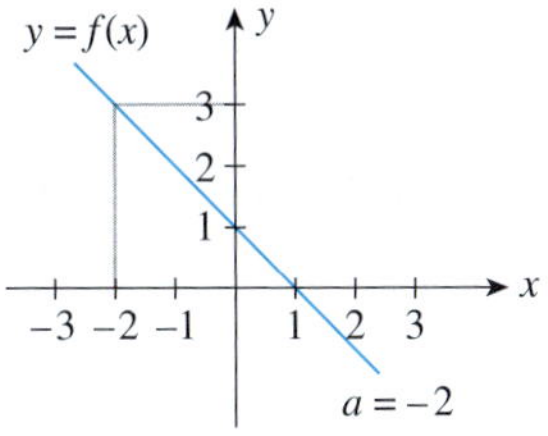

2.
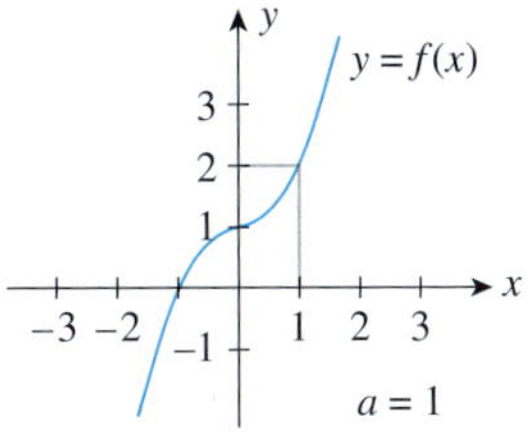

3.
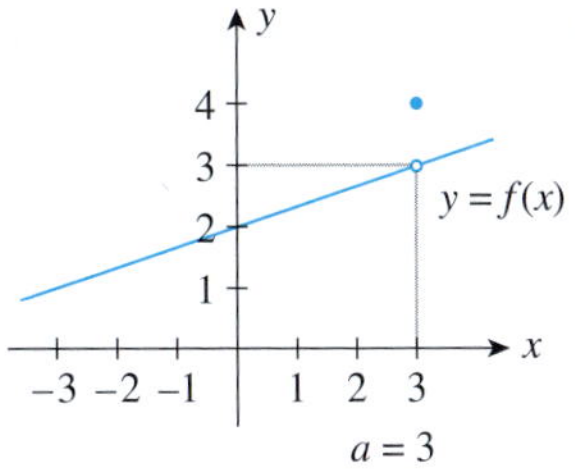

4.
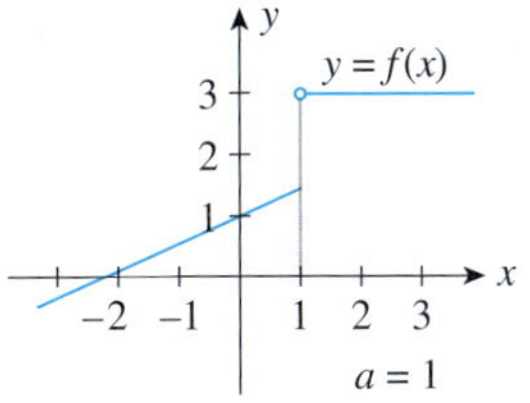

5.
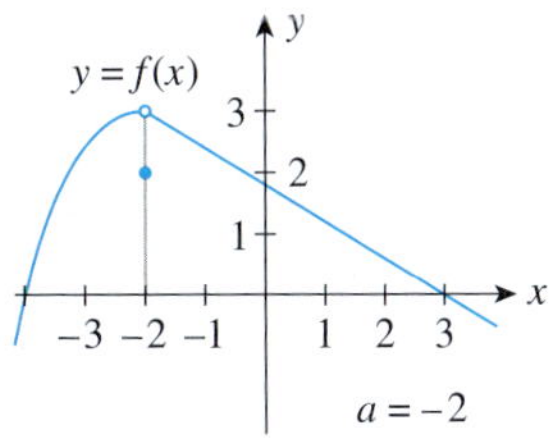

6.
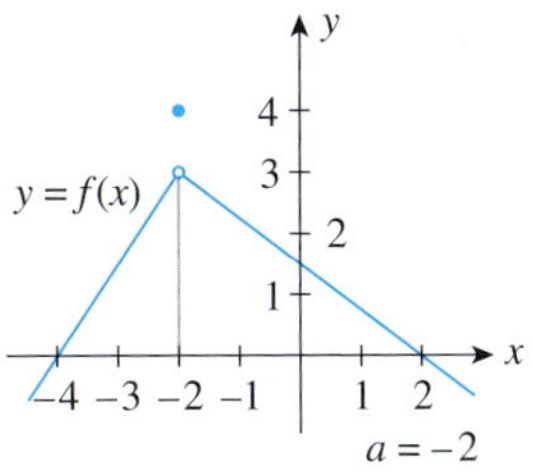

7.
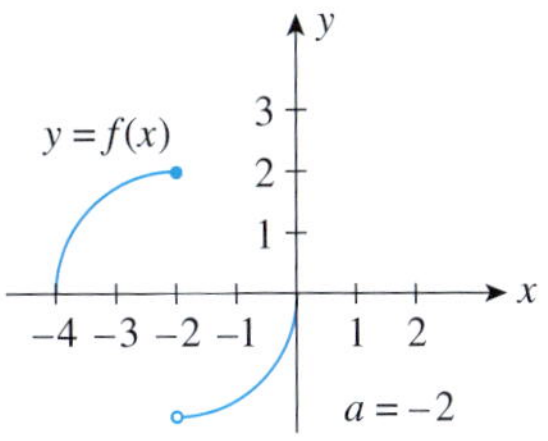

8.
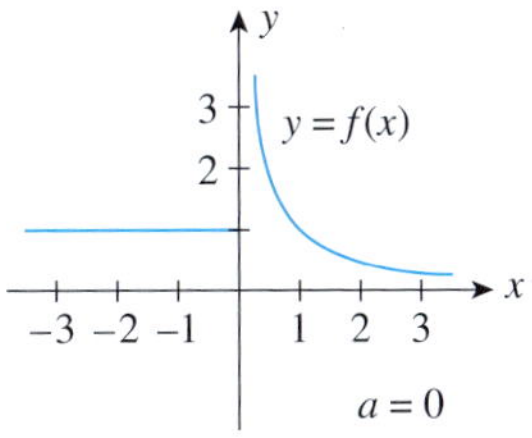

In Exercises 9–16, complete the table by computing $f(x)$ at the given values of x. Use these results to estimate the indicated limit (if it exists).

9. $f(x) = x^2 + 1;\ \lim_{x\to 2} f(x)$

x	1.9	1.99	1.999	2.001	2.01	2.1
$f(x)$						

10. $f(x) = 2x^2 - 1;\ \lim_{x\to 1} f(x)$

x	0.9	0.99	0.999	1.001	1.01	1.1
$f(x)$						

11. $f(x) = \dfrac{|x|}{x};\ \lim_{x\to 0} f(x)$

x	−0.1	−0.01	−0.001	0.001	0.01	0.1
$f(x)$						

12. $f(x) = \dfrac{|x-1|}{x-1};\ \lim_{x\to 1} f(x)$

x	0.9	0.99	0.999	1.001	1.01	1.1
$f(x)$						

13. $f(x) = \dfrac{1}{(x-1)^2};\ \lim_{x\to 1} f(x)$

x	0.9	0.99	0.999	1.001	1.01	1.1
$f(x)$						

14. $f(x) = \dfrac{1}{x-2};\ \lim_{x\to 2} f(x)$

x	1.9	1.99	1.999	2.001	2.01	2.1
$f(x)$						

15. $f(x) = \dfrac{x^2 + x - 2}{x - 1};\ \lim_{x\to 1} f(x)$

x	0.9	0.99	0.999	1.001	1.01	1.1
$f(x)$						

16. $f(x) = \dfrac{x-1}{x-1}$; $\lim_{x\to 1} f(x)$

x	0.9	0.99	0.999	1.001	1.01	1.1
$f(x)$						

In Exercises 17–22, sketch the graph of the function f and evaluate $\lim_{x\to a} f(x)$, if it exists, for the given value of a.

17. $f(x) = \begin{cases} x-1 & \text{if } x \le 0 \\ -1 & \text{if } x > 0 \end{cases}$ $(a = 0)$

18. $f(x) = \begin{cases} x-1 & \text{if } x \le 3 \\ -2x+8 & \text{if } x > 3 \end{cases}$ $(a = 3)$

19. $f(x) = \begin{cases} x & \text{if } x < 1 \\ 0 & \text{if } x = 1 \\ -x+2 & \text{if } x > 1 \end{cases}$ $(a = 1)$

20. $f(x) = \begin{cases} -2x+4 & \text{if } x < 1 \\ 4 & \text{if } x = 1 \\ x^2+1 & \text{if } x > 1 \end{cases}$ $(a = 1)$

21. $f(x) = \begin{cases} |x| & \text{if } x \ne 0 \\ 1 & \text{if } x = 0 \end{cases}$ $(a = 0)$

22. $f(x) = \begin{cases} |x-1| & \text{if } x \ne 1 \\ 0 & \text{if } x = 1 \end{cases}$ $(a = 1)$

In Exercises 23–40, find the indicated limit.

23. $\lim_{x\to 2} 3$

24. $\lim_{x\to -2} -3$

25. $\lim_{x\to 3} x$

26. $\lim_{x\to -2} -3x$

27. $\lim_{x\to 1} (1 - 2x^2)$

28. $\lim_{t\to 3} (4t^2 - 2t + 1)$

29. $\lim_{x\to 1} (2x^3 - 3x^2 + x + 2)$

30. $\lim_{x\to 0} (4x^5 - 20x^2 + 2x + 1)$

31. $\lim_{s\to 0} (2s^2 - 1)(2s + 4)$

32. $\lim_{x\to 2} (x^2 + 1)(x^2 - 4)$

33. $\lim_{x\to 2} \dfrac{2x+1}{x+2}$

34. $\lim_{x\to 1} \dfrac{x^3+1}{2x^3+2}$

35. $\lim_{x\to 2} \sqrt{x+2}$

36. $\lim_{x\to -2} \sqrt[3]{5x+2}$

37. $\lim_{x\to -3} \sqrt{2x^4 + x^2}$

38. $\lim_{x\to 2} \sqrt{\dfrac{2x^3+4}{x^2+1}}$

39. $\lim_{x\to -1} \dfrac{\sqrt{x^2+8}}{2x+4}$

40. $\lim_{x\to 3} \dfrac{x\sqrt{x^2+7}}{2x - \sqrt{2x+3}}$

In Exercises 41–48, find the indicated limit given that $\lim_{x\to a} f(x) = 3$ and $\lim_{x\to a} g(x) = 4$.

41. $\lim_{x\to a} [f(x) - g(x)]$

42. $\lim_{x\to a} 2f(x)$

43. $\lim_{x\to a} [2f(x) - 3g(x)]$

44. $\lim_{x\to a} [f(x)g(x)]$

45. $\lim_{x\to a} \sqrt{g(x)}$

46. $\lim_{x\to a} \sqrt[3]{5f(x) + 3g(x)}$

47. $\lim_{x\to a} \dfrac{2f(x) - g(x)}{f(x)g(x)}$

48. $\lim_{x\to a} \dfrac{g(x) - f(x)}{f(x) + \sqrt{g(x)}}$

In Exercises 49–62, find the indicated limit, if it exists.

49. $\lim_{x\to 1} \dfrac{x^2-1}{x-1}$

50. $\lim_{x\to -2} \dfrac{x^2-4}{x+2}$

51. $\lim_{x\to 0} \dfrac{x^2-x}{x}$

52. $\lim_{x\to 0} \dfrac{2x^2-3x}{x}$

53. $\lim_{x\to -5} \dfrac{x^2-25}{x+5}$

54. $\lim_{b\to -3} \dfrac{b+1}{b+3}$

55. $\lim_{x\to 1} \dfrac{x}{x-1}$

56. $\lim_{x\to 2} \dfrac{x+2}{x-2}$

57. $\lim_{x\to -2} \dfrac{x^2-x-6}{x^2+x-2}$

58. $\lim_{z\to 2} \dfrac{z^3-8}{z-2}$

59. $\lim_{x\to 1} \dfrac{\sqrt{x}-1}{x-1}$

Hint: Multiply by $\dfrac{\sqrt{x}+1}{\sqrt{x}+1}$.

60. $\lim_{x\to 4} \dfrac{x-4}{\sqrt{x}-2}$

Hint: See Exercise 59.

61. $\lim_{x\to 1} \dfrac{x-1}{x^3+x^2-2x}$

62. $\lim_{x\to -2} \dfrac{4-x^2}{2x^2+x^3}$

In Exercises 63–68, use the graph of the function f to determine $\lim_{x\to\infty} f(x)$ and $\lim_{x\to -\infty} f(x)$, if they exist.

63.

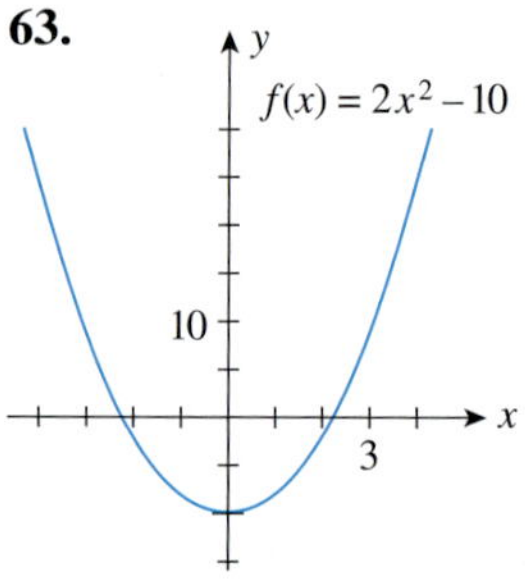

64.

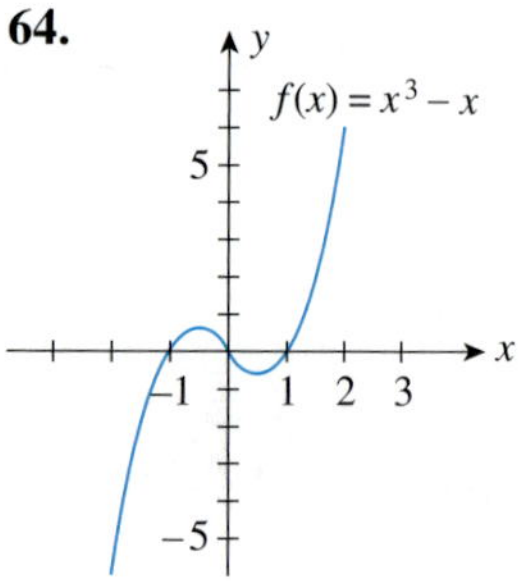

65.

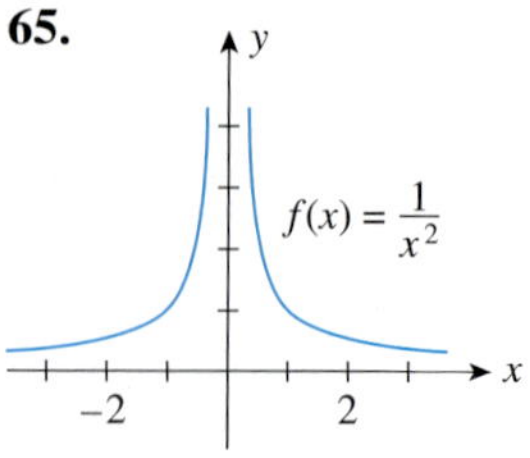

66.

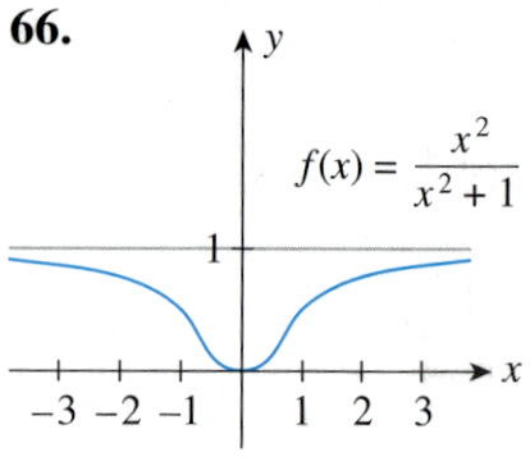

67.

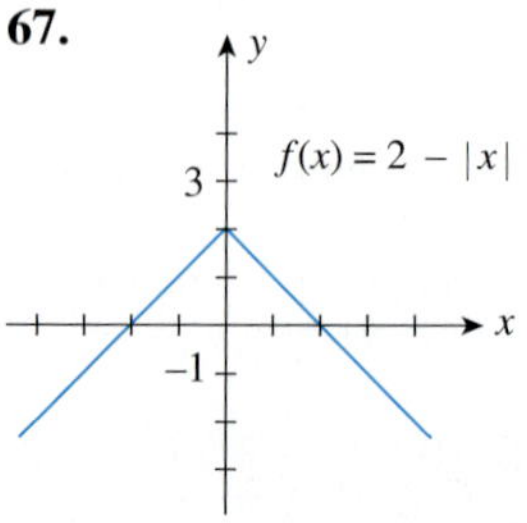

68.

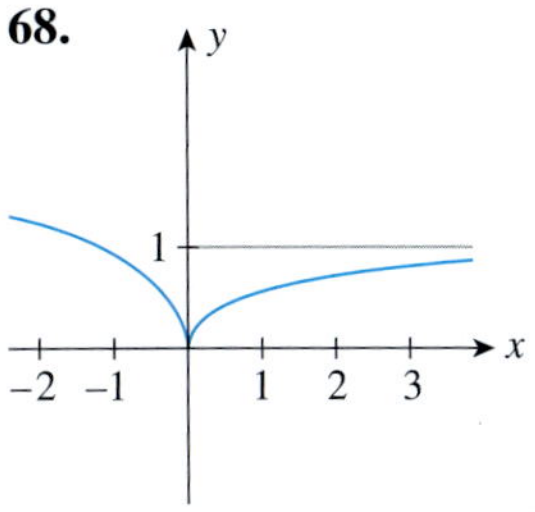

$$f(x) = \begin{cases} \sqrt{-x} & \text{if } x \le 0 \\ \dfrac{x}{x+1} & \text{if } x > 0 \end{cases}$$

In Exercises 69–72, complete the table by computing *f*(*x*) at the given values of *x*. Use the results to guess at the indicated limits, if they exist.

69. $f(x) = \dfrac{1}{x^2+1}$; $\lim_{x\to\infty} f(x)$ and $\lim_{x\to-\infty} f(x)$

x	1	10	100	1000
$f(x)$				

x	−1	−10	−100	−1000
$f(x)$				

70. $f(x) = \dfrac{2x}{x+1}$; $\lim_{x\to\infty} f(x)$ and $\lim_{x\to-\infty} f(x)$

x	1	10	100	1000
$f(x)$				

x	−5	−10	−100	−1000
$f(x)$				

71. $f(x) = 3x^3 - x^2 + 10$; $\lim_{x\to\infty} f(x)$ and $\lim_{x\to-\infty} f(x)$

x	1	5	10	100	1000
$f(x)$					

x	−1	−5	−10	−100	−1000
$f(x)$					

72. $f(x) = \dfrac{|x|}{x}$; $\lim_{x\to\infty} f(x)$ and $\lim_{x\to-\infty} f(x)$

x	1	10	100	−1	−10	−100
$f(x)$						

In Exercises 73–80, find the indicated limits, if they exist.

73. $\lim_{x\to\infty} \dfrac{3x+2}{x-5}$

74. $\lim_{x\to-\infty} \dfrac{4x^2-1}{x+2}$

75. $\lim_{x\to-\infty} \dfrac{3x^3+x^2+1}{x^3+1}$

76. $\lim_{x\to\infty} \dfrac{2x^2+3x+1}{x^4-x^2}$

77. $\lim_{x\to-\infty} \dfrac{x^4+1}{x^3-1}$

78. $\lim_{x\to\infty} \dfrac{4x^4-3x^2+1}{2x^4+x^3+x^2+x+1}$

79. $\lim_{x\to\infty} \dfrac{x^5-x^3+x-1}{x^6+2x^2+1}$

80. $\lim_{x\to\infty} \dfrac{2x^2-1}{x^3+x^2+1}$

81. Toxic Waste A city's main well was recently found to be contaminated with trichloroethylene, a cancer-causing chemical, as a result of an abandoned chemical dump leaching chemicals into the water. A proposal submitted to city council members indicates that the cost, measured in millions of dollars, of removing x% of the toxic pollutant is given by

$$C(x) = \frac{0.5x}{100-x} \qquad (0 < x < 100)$$

a. Find the cost of removing 50%, 60%, 70%, 80%, 90%, and 95% of the pollutant.

b. Evaluate

$$\lim_{x\to100} \frac{0.5x}{100-x}$$

and interpret your result.

82. A Doomsday Situation The population of a certain breed of rabbits introduced into an isolated island is given by

$$P(t) = \frac{72}{9-t} \qquad (0 \le t < 9)$$

where t is measured in months.

a. Find the number of rabbits present in the island initially (at $t = 0$).

b. Show that the population of rabbits is increasing without bound.

c. Sketch the graph of the function P.
(*Comment:* This phenomenon is referred to as a *doomsday situation.*)

83. Average Cost The average cost/disc in dollars incurred by Herald Records in pressing x DVDs is given by the average cost function

$$\overline{C}(x) = 2.2 + \frac{2500}{x}$$

Evaluate $\lim_{x\to\infty} \overline{C}(x)$ and interpret your result.

84. Concentration of a Drug in the Bloodstream The concentration of a certain drug in a patient's bloodstream t hr after injection is given by

$$C(t) = \frac{0.2t}{t^2+1}$$

mg/cm^3. Evaluate $\lim_{t\to\infty} C(t)$ and interpret your result.

85. Box-Office Receipts The total worldwide box-office receipts for a long-running blockbuster movie are approximated by the function

$$T(x) = \frac{120x^2}{x^2+4}$$

where $T(x)$ is measured in millions of dollars and x is the number of months since the movie's release.

a. What are the total box-office receipts after the first month? The second month? The third month?

b. What will the movie gross in the long run (when x is very large)?

86. **Population Growth** A major corporation is building a 4325-acre complex of homes, offices, stores, schools, and churches in the rural community of Glen Cove. As a result of this development, the planners have estimated that Glen Cove's population (in thousands) t yr from now will be given by

$$P(t) = \frac{25t^2 + 125t + 200}{t^2 + 5t + 40}$$

a. What is the current population of Glen Cove?
b. What will be the population in the long run?

87. **Driving Costs** A study of driving costs of 1992 model subcompact (four-cylinder) cars found that the average cost (car payments, gas, insurance, upkeep, and depreciation), measured in cents/mile, is approximated by the function

$$C(x) = \frac{2010}{x^{2.2}} + 17.80$$

where x denotes the number of miles (in thousands) the car is driven in a year.
a. What is the average cost of driving a subcompact car 5000 mi/yr? 10,000 mi/yr? 15,000 mi/yr? 20,000 mi/yr? 25,000 mi/yr?
b. Use part (a) to sketch the graph of the function C.
c. What happens to the average cost as the number of miles driven increases without bound?

Source: American Automobile Association

88. **Photosynthesis** The rate of production R in photosynthesis is related to the light intensity I by the function

$$R(I) = \frac{aI}{b + I^2}$$

where a and b are positive constants.
a. Taking $a = b = 1$, compute $R(I)$ for $I = 0, 1, 2, 3, 4,$ and 5.
b. Evaluate $\lim_{I \to \infty} R(I)$.
c. Use the results of parts (a) and (b) to sketch the graph of R. Interpret your results.

In Exercises 89–94, determine whether the statement is true or false. If it is true, explain why it is true. If it is false, give an example to show why it is false.

89. If $\lim_{x \to a} f(x)$ exists, then f is defined at $x = a$.

90. If $\lim_{x \to 0} f(x) = 4$ and $\lim_{x \to 0} g(x) = 0$, then $\lim_{x \to 0} f(x)g(x) = 0$.

91. If $\lim_{x \to 2} f(x) = 3$ and $\lim_{x \to 2} g(x) = 0$, then $\lim_{x \to 2} [f(x)]/[g(x)]$ does not exist.

92. If $\lim_{x \to 3} f(x) = 0$ and $\lim_{x \to 3} g(x) = 0$, then $\lim_{x \to 3} [f(x)]/[g(x)]$ does not exist.

93. $\lim_{x \to 2} \left(\frac{x}{x+1} + \frac{3}{x-1} \right) = \lim_{x \to 2} \frac{x}{x+1} + \lim_{x \to 2} \frac{3}{x-1}$

94. $\lim_{x \to 1} \left(\frac{2x}{x-1} - \frac{2}{x-1} \right) = \lim_{x \to 1} \frac{2x}{x-1} - \lim_{x \to 1} \frac{2}{x-1}$

95. **Speed of a Chemical Reaction** Certain proteins, known as enzymes, serve as catalysts for chemical reactions in living things. In 1913 Leonor Michaelis and L. M. Menten discovered the following formula giving the initial speed V (in moles/liter/second) at which the reaction begins in terms of the amount of substrate x (the substance being acted upon, measured in moles/liters) present:

$$V = \frac{ax}{x + b}$$

where a and b are positive constants. Evaluate

$$\lim_{x \to \infty} \frac{ax}{x + b}$$

and interpret your result.

96. Show by means of an example that $\lim_{x \to a} [f(x) + g(x)]$ may exist even though neither $\lim_{x \to a} f(x)$ nor $\lim_{x \to a} g(x)$ exists. Does this example contradict Theorem 1?

97. Show by means of an example that $\lim_{x \to a} [f(x)g(x)]$ may exist even though neither $\lim_{x \to a} f(x)$ nor $\lim_{x \to a} g(x)$ exists. Does this example contradict Theorem 1?

98. Show by means of an example that $\lim_{x \to a} f(x)/g(x)$ may exist even though neither $\lim_{x \to a} f(x)$ nor $\lim_{x \to a} g(x)$ exists. Does this example contradict Theorem 1?

2.4 Solutions to Self-Check Exercises

1. a.
$$\lim_{x \to 3} \frac{\sqrt{x^2 + 7} + \sqrt{3x - 5}}{x + 2} = \frac{\sqrt{9 + 7} + \sqrt{3(3) - 5}}{3 + 2} = \frac{\sqrt{16} + \sqrt{4}}{5} = \frac{6}{5}$$

b. Letting x approach -1 leads to the indeterminate form 0/0. Thus, we proceed as follows:

$$\lim_{x \to -1} \frac{x^2 - x - 2}{2x^2 - x - 3} = \lim_{x \to -1} \frac{(x + 1)(x - 2)}{(x + 1)(2x - 3)}$$

$$= \lim_{x \to -1} \frac{x - 2}{2x - 3} \quad \text{Cancel the common factors.}$$

$$= \frac{-1 - 2}{2(-1) - 3}$$

$$= \frac{3}{5}$$

2. $\lim_{x\to\infty} \overline{C}(x) = \lim_{x\to\infty}\left(1.8 + \frac{3000}{x}\right)$

$$= \lim_{x\to\infty} 1.8 + \lim_{x\to\infty} \frac{3000}{x}$$

$$= 1.8$$

Our computation reveals that, as the production of CDs increases "without bound," the average cost drops and approaches a unit cost of \$1.80/disc.

USING TECHNOLOGY

Finding the Limit of a Function

A graphing utility can be used to help us find the limit of a function, if it exists, as illustrated in the following examples.

EXAMPLE 1 Let $f(x) = \dfrac{x^3 - 1}{x - 1}$.

a. Plot the graph of f in the viewing window $[-2, 2] \times [0, 4]$.

b. Use ZOOM to find $\lim_{x\to 1} \dfrac{x^3 - 1}{x - 1}$.

c. Verify your result by evaluating the limit algebraically.

Solution

a. The graph of f in the viewing window $[-2, 2] \times [0, 4]$ is shown in Figure T1a.

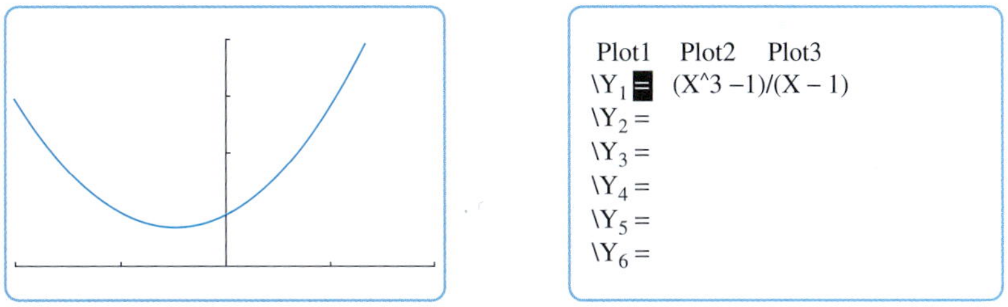

(a) (b)

FIGURE **T1**
(a) The graph of

$$f(x) = \frac{x^3 - 1}{x - 1}$$

in the viewing window $[-2, 2] \times [0, 4]$; (b) the TI-83/84 equation screen

b. Using ZOOM-IN repeatedly, we see that the y-value approaches 3 as the x-value approaches 1. We conclude, accordingly, that

$$\lim_{x\to 1} \frac{x^3 - 1}{x - 1} = 3$$

c. We compute

$$\lim_{x\to 1} \frac{x^3 - 1}{x - 1} = \lim_{x\to 1} \frac{(x - 1)(x^2 + x + 1)}{x - 1}$$

$$= \lim_{x\to 1} (x^2 + x + 1) = 3$$

Note If you attempt to find the limit in Example 1 by using the evaluation function of your graphing utility to find the value of $f(x)$ when $x = 1$, you will see that the graphing utility does not display the y-value. This happens because $x = 1$ is not in the domain of f.

(continued)

EXAMPLE 2 Use ZOOM to find $\lim_{x \to 0} (1 + x)^{1/x}$.

Solution We first plot the graph of $f(x) = (1 + x)^{1/x}$ in a suitable viewing window. Figure T2a shows a plot of f in the window $[-1, 1] \times [0, 4]$. Using ZOOM-IN repeatedly, we see that $\lim_{x \to 0} (1 + x)^{1/x} \approx 2.71828$.

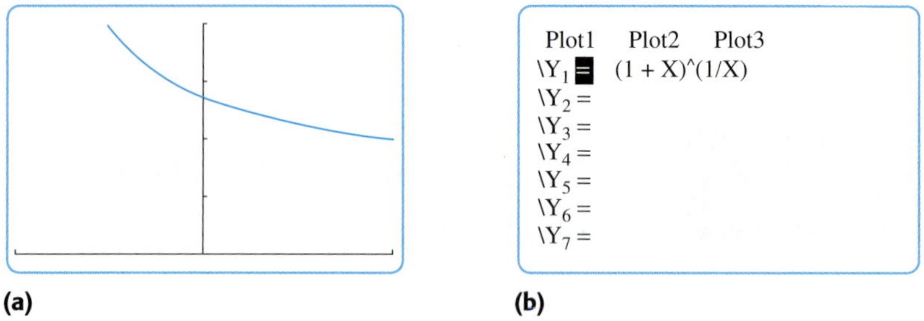

FIGURE T2
(a) The graph of $f(x) = (1 + x)^{1/x}$ in the viewing window $[-1, 1] \times [0, 4]$; (b) the T1-83/84 equation screen

The limit of $f(x) = (1 + x)^{1/x}$ as x approaches zero, denoted by the letter e, plays a very important role in the study of mathematics and its applications (see Section 5.6). Thus,

$$\lim_{x \to 0} (1 + x)^{1/x} = e$$

where, as we have just seen, $e \approx 2.71828$. ■

APPLIED EXAMPLE 3 Oxygen Content of a Pond When organic waste is dumped into a pond, the oxidation process that takes place reduces the pond's oxygen content. However, given time, nature will restore the oxygen content to its natural level. Suppose the oxygen content t days after the organic waste has been dumped into the pond is given by

$$f(t) = 100\left(\frac{t^2 + 10t + 100}{t^2 + 20t + 100}\right)$$

percent of its normal level.

a. Plot the graph of f in the viewing window $[0, 200] \times [70, 100]$.
b. What can you say about $f(t)$ when t is very large?
c. Verify your observation in part (b) by evaluating $\lim_{t \to \infty} f(t)$.

Solution

a. The graph of f is shown in Figure T3a.
b. From the graph of f, it appears that $f(t)$ approaches 100 steadily as t gets larger and larger. This observation tells us that eventually the oxygen content of the pond will be restored to its natural level.

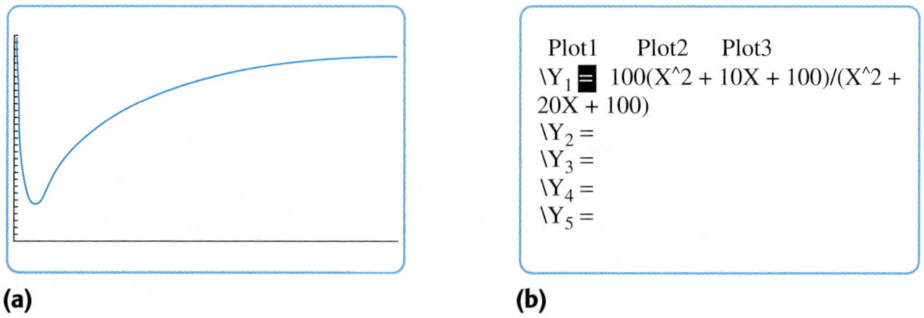

FIGURE T3
(a) The graph of f in the viewing window $[0, 200] \times [70, 100]$; (b) the TI-83/84 equation screen

c. To verify the observation made in part (b), we compute

$$\lim_{t\to\infty} f(t) = \lim_{t\to\infty} 100\left(\frac{t^2 + 10t + 100}{t^2 + 20t + 100}\right)$$

$$= 100 \lim_{t\to\infty}\left(\frac{1 + \frac{10}{t} + \frac{100}{t^2}}{1 + \frac{20}{t} + \frac{100}{t^2}}\right) = 100$$

TECHNOLOGY EXERCISES

In Exercises 1–10, find the indicated limit by first plotting the graph of the function in a suitable viewing window and then using the ZOOM-IN feature of the calculator.

1. $\lim_{x\to 1} \frac{2x^3 - 2x^2 + 3x - 3}{x - 1}$

2. $\lim_{x\to -2} \frac{2x^3 + 3x^2 - x + 2}{x + 2}$

3. $\lim_{x\to -1} \frac{x^3 + 1}{x + 1}$

4. $\lim_{x\to -1} \frac{x^4 - 1}{x - 1}$

5. $\lim_{x\to 1} \frac{x^3 - x^2 - x + 1}{x^3 - 3x + 2}$

6. $\lim_{x\to 0} \frac{\sqrt{x+1} - 1}{x}$

7. $\lim_{x\to 0} (1 + 2x)^{1/x}$

8. $\lim_{x\to 0} \frac{2^x - 1}{x}$

9. Show that $\lim_{x\to 3} \frac{2}{x - 3}$ does not exist.

10. Show that $\lim_{x\to 2} \frac{x^3 - 2x + 1}{x - 2}$ does not exist.

11. **City Planning** A major developer is building a 5000-acre complex of homes, offices, stores, schools, and churches in the rural community of Marlboro. As a result of this development, the planners have estimated that Marlboro's population (in thousands) t yr from now will be given by

$$P(t) = \frac{25t^2 + 125t + 200}{t^2 + 5t + 40}$$

a. Plot the graph of P in the viewing window $[0, 50] \times [0, 30]$.

b. What will be the population of Marlboro in the long run?
Hint: Find $\lim_{t\to\infty} P(t)$.

12. **Amount of Rainfall** The total amount of rain (in inches) after t hr during a rainfall is given by

$$T(t) = \frac{0.8t}{t + 4.1}$$

a. Plot the graph of T in the viewing window $[0, 30] \times [0, 0.8]$.

b. What is the total amount of rain during this rainfall?
Hint: Find $\lim_{t\to\infty} T(t)$.

2.5 One-Sided Limits and Continuity

One-Sided Limits

Consider the function f defined by

$$f(x) = \begin{cases} x - 1 & \text{if } x < 0 \\ x + 1 & \text{if } x \geq 0 \end{cases}$$

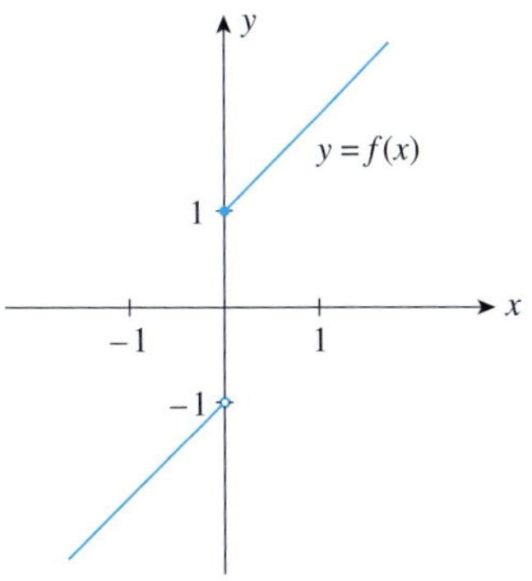

FIGURE 37
The function f does not have a limit as x approaches zero.

From the graph of f shown in Figure 37, we see that the function f does not have a limit as x approaches zero because, no matter how close x is to zero, $f(x)$ takes on values that are close to 1 if x is positive and values that are close to -1 if x is negative. Therefore, $f(x)$ cannot be close to a single number L—no matter how close x is to zero. Now, if we restrict x to be greater than zero (to the right of zero), then we see that $f(x)$ can be made as close to 1 as we please by taking x sufficiently close to zero. In this situation we say that the right-hand limit of f as x approaches zero (from the right) is 1, written

$$\lim_{x\to 0^+} f(x) = 1$$

Similarly, we see that $f(x)$ can be made as close to -1 as we please by taking x sufficiently close to, but to the left of, zero. In this situation we say that the left-hand limit of f as x approaches zero (from the left) is -1, written

$$\lim_{x\to 0^-} f(x) = -1$$

These limits are called **one-sided limits.** More generally, we have the following definitions.

One-Sided Limits

The function f has the **right-hand limit** L as x approaches a from the right, written

$$\lim_{x\to a^+} f(x) = L$$

if the values of $f(x)$ can be made as close to L as we please by taking x sufficiently close to (but not equal to) a and to the right of a.

Similarly, the function f has the **left-hand limit** M as x approaches a from the left, written

$$\lim_{x\to a^-} f(x) = M$$

if the values of $f(x)$ can be made as close to M as we please by taking x sufficiently close to (but not equal to) a and to the left of a.

The connection between one-sided limits and the two-sided limit defined earlier is given by the following theorem.

THEOREM 3

Let f be a function that is defined for all values of x close to $x = a$ with the possible exception of a itself. Then

$$\lim_{x\to a} f(x) = L \quad \text{if and only if} \quad \lim_{x\to a^+} f(x) = \lim_{x\to a^-} f(x) = L$$

Thus, the two-sided limit exists if and only if the one-sided limits exist and are equal.

EXAMPLE 1 Let

$$f(x) = \begin{cases} -x & \text{if } x \le 0 \\ \sqrt{x} & \text{if } x > 0 \end{cases} \quad \text{and} \quad g(x) = \begin{cases} -1 & \text{if } x < 0 \\ 1 & \text{if } x \ge 0 \end{cases}$$

a. Show that $\lim_{x\to 0} f(x)$ exists by studying the one-sided limits of f as x approaches $x = 0$.

b. Show that $\lim_{x\to 0} g(x)$ does not exist.

Solution

a. For $x \le 0$

$$\lim_{x\to 0^-} f(x) = \lim_{x\to 0^-} (-x) = 0$$

and for $x > 0$, we find

$$\lim_{x\to 0^+} f(x) = \lim_{x\to 0^+} \sqrt{x} = 0$$

Thus,

$$\lim_{x \to 0} f(x) = 0$$

(Figure 38a).

b. We have

$$\lim_{x \to 0^-} g(x) = -1 \quad \text{and} \quad \lim_{x \to 0^+} g(x) = 1$$

and since these one-sided limits are not equal, we conclude that $\lim_{x \to 0} g(x)$ does not exist (Figure 38b).

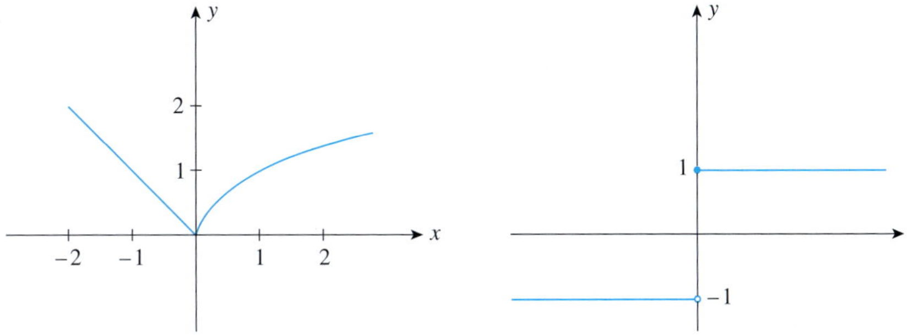

FIGURE **38**

(a) $\lim_{x \to 0} f(x)$ exists. **(b)** $\lim_{x \to 0} g(x)$ does not exist.

Continuous Functions

Continuous functions will play an important role throughout most of our study of calculus. Loosely speaking, a function is continuous at a point if the graph of the function at that point is devoid of holes, gaps, jumps, or breaks. Consider, for example, the graph of the function f depicted in Figure 39.

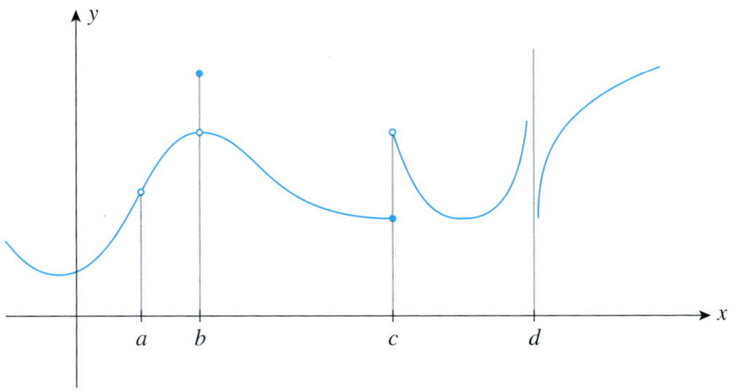

FIGURE **39**
The graph of this function is not continuous at $x = a$, $x = b$, $x = c$, and $x = d$.

Let's take a closer look at the behavior of f at or near $x = a$, $x = b$, $x = c$, and $x = d$. First, note that f is not defined at $x = a$; that is, $x = a$ is not in the domain of f, thereby resulting in a "hole" in the graph of f. Next, observe that the value of f at b, $f(b)$, is not equal to the limit of $f(x)$ as x approaches b, resulting in a "jump" in the graph of f at $x = b$. The function f does not have a limit at $x = c$ since the left-hand and right-hand limits of $f(x)$ are not equal, also resulting in a jump in the graph of f at $x = c$. Finally, the limit of f does not exist at $x = d$, resulting in a break in the graph of f. The function f is *discontinuous* at each of these numbers. It is *continuous* everywhere else.

Continuity of a Function at a Number

A function f is **continuous at a number** $x = a$ if the following conditions are satisfied.

1. $f(a)$ is defined. **2.** $\lim_{x \to a} f(x)$ exists. **3.** $\lim_{x \to a} f(x) = f(a)$

Thus, a function f is continuous at $x = a$ if the limit of f at $x = a$ exists and has the value $f(a)$. Geometrically, f is continuous at $x = a$ if the proximity of x to a implies the proximity of $f(x)$ to $f(a)$.

If f is not continuous at $x = a$, then f is said to be **discontinuous** at $x = a$. Also, f is **continuous on an interval** if f is continuous at every number in the interval.

Figure 40 depicts the graph of a continuous function on the interval (a, b). Notice that the graph of the function over the stated interval can be sketched without lifting one's pencil from the paper.

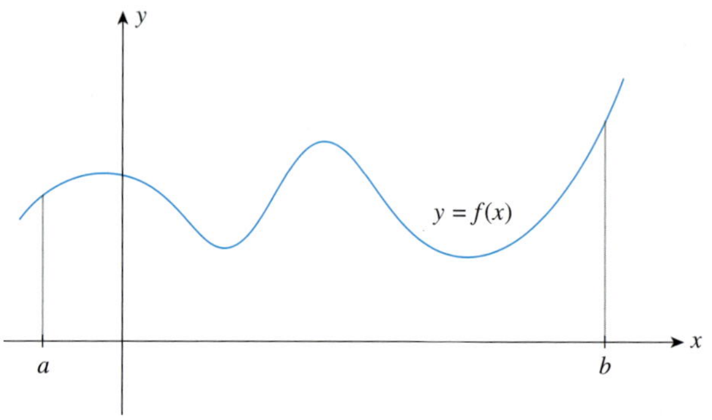

FIGURE **40**
The graph of f is continuous on the interval (a, b).

VIDEO

EXAMPLE 2 Find the values of x for which each function is continuous.

a. $f(x) = x + 2$ **b.** $g(x) = \dfrac{x^2 - 4}{x - 2}$ **c.** $h(x) = \begin{cases} x + 2 & \text{if } x \neq 2 \\ 1 & \text{if } x = 2 \end{cases}$

d. $F(x) = \begin{cases} -1 & \text{if } x < 0 \\ 1 & \text{if } x \geq 0 \end{cases}$ **e.** $G(x) = \begin{cases} \dfrac{1}{x} & \text{if } x > 0 \\ -1 & \text{if } x \leq 0 \end{cases}$

The graph of each function is shown in Figure 41.

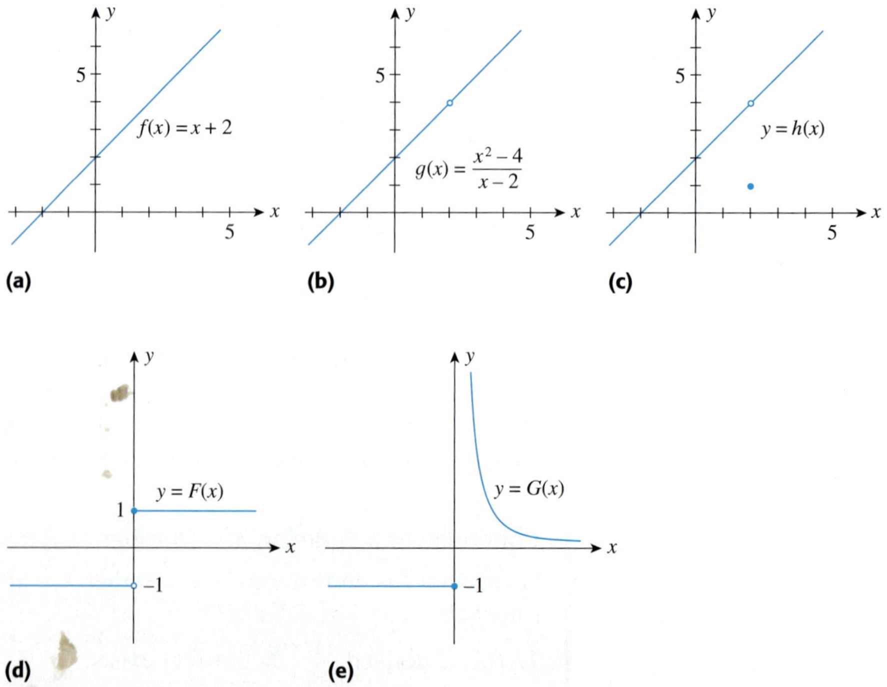

FIGURE **41**

Solution

a. The function f is continuous everywhere because the three conditions for continuity are satisfied for all values of x.

b. The function g is discontinuous at $x = 2$ because g is not defined at that number. It is continuous everywhere else.

c. The function h is discontinuous at $x = 2$ because the third condition for continuity is violated; the limit of $h(x)$ as x approaches 2 exists and has the value 4, but this limit is not equal to $h(2) = 1$. It is continuous for all other values of x.

d. The function F is continuous everywhere except at $x = 0$, where the limit of $F(x)$ fails to exist as x approaches zero (see Example 3a, Section 2.4).

e. Since the limit of $G(x)$ does not exist as x approaches zero, we conclude that G fails to be continuous at $x = 0$. The function G is continuous everywhere else. ■

Properties of Continuous Functions

The following properties of continuous functions follow directly from the definition of continuity and the corresponding properties of limits. They are stated without proof.

Properties of Continuous Functions

1. The constant function $f(x) = c$ is continuous everywhere.
2. The identity function $f(x) = x$ is continuous everywhere.

If f and g are continuous at $x = a$, then

3. $[f(x)]^n$, where n is a real number, is continuous at $x = a$ whenever it is defined at that number.
4. $f \pm g$ is continuous at $x = a$.
5. fg is continuous at $x = a$.
6. f/g is continuous at $x = a$ provided $g(a) \neq 0$.

Using these properties of continuous functions, we can prove the following results. (A proof is sketched in Exercise 100, page 130.)

Continuity of Polynomial and Rational Functions

1. A polynomial function $y = P(x)$ is continuous at every value of x.
2. A rational function $R(x) = p(x)/q(x)$ is continuous at every value of x where $q(x) \neq 0$.

EXAMPLE 3 Find the values of x for which each function is continuous.

a. $f(x) = 3x^3 + 2x^2 - x + 10$ **b.** $g(x) = \dfrac{8x^{10} - 4x + 1}{x^2 + 1}$

c. $h(x) = \dfrac{4x^3 - 3x^2 + 1}{x^2 - 3x + 2}$

Solution

a. The function f is a polynomial function of degree 3, so $f(x)$ is continuous for all values of x.

b. The function g is a rational function. Observe that the denominator of g—namely, $x^2 + 1$—is never equal to zero. Therefore, we conclude that g is continuous for all values of x.

c. The function h is a rational function. In this case, however, the denominator of h is equal to zero at $x = 1$ and $x = 2$, which can be seen by factoring it. Thus,

$$x^2 - 3x + 2 = (x - 2)(x - 1)$$

We therefore conclude that h is continuous everywhere except at $x = 1$ and $x = 2$, where it is discontinuous. ■

Up to this point, most of the applications we have discussed involved functions that are continuous everywhere. In Example 4, we consider an application from the field of educational psychology that involves a discontinuous function.

APPLIED EXAMPLE 4 Learning Curves Figure 42 depicts the learning curve associated with a certain individual. Beginning with no knowledge of the subject being taught, the individual makes steady progress toward understanding it over the time interval $0 \leq t < t_1$. In this instance, the individual's progress slows as we approach time t_1 because he fails to grasp a particularly difficult concept. All of a sudden, a breakthrough occurs at time t_1, propelling his knowledge of the subject to a higher level. The curve is discontinuous at t_1.

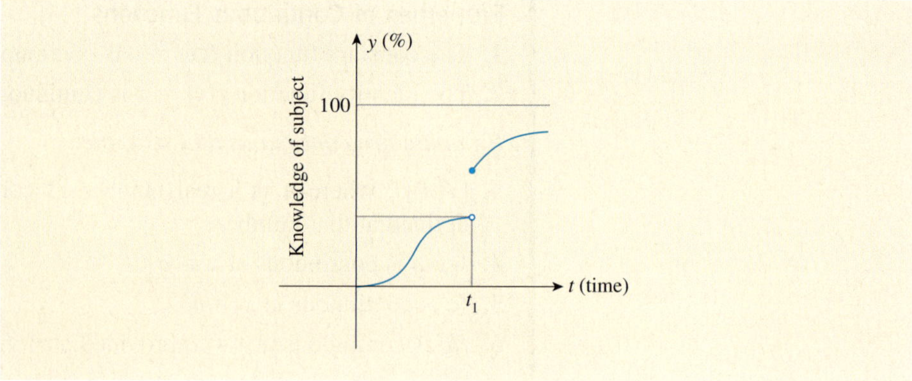

FIGURE 42
A learning curve that is discontinuous at $t = t_1$ ■

Intermediate Value Theorem

Let's look again at our model of the motion of the maglev on a straight stretch of track. We know that the train cannot vanish at any instant of time and it cannot skip portions of the track and reappear someplace else. To put it another way, the train cannot occupy the positions s_1 and s_2 without at least, at some time, occupying an intermediate position (Figure 43).

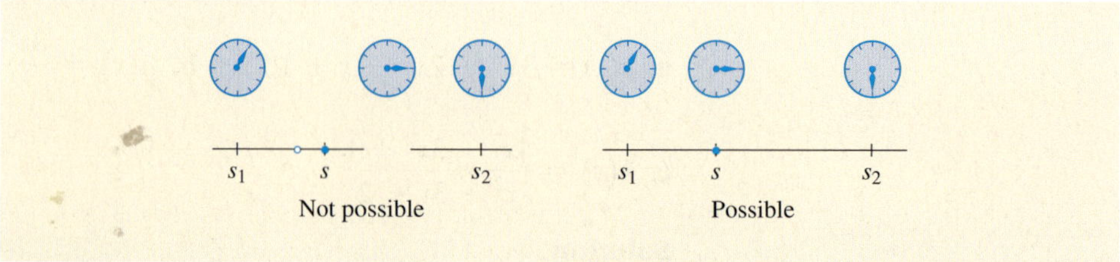

FIGURE 43
The position of the maglev

To state this fact mathematically, recall that the position of the maglev as a function of time is described by

$$f(t) = 4t^2 \qquad (0 \leq t \leq 10)$$

Suppose the position of the maglev is s_1 at some time t_1 and its position is s_2 at some time t_2 (Figure 44). Then, if s_3 is any number between s_1 and s_2 giving an intermediate position of the maglev, there must be at least one t_3 between t_1 and t_2 giving the time at which the train is at s_3—that is, $f(t_3) = s_3$.

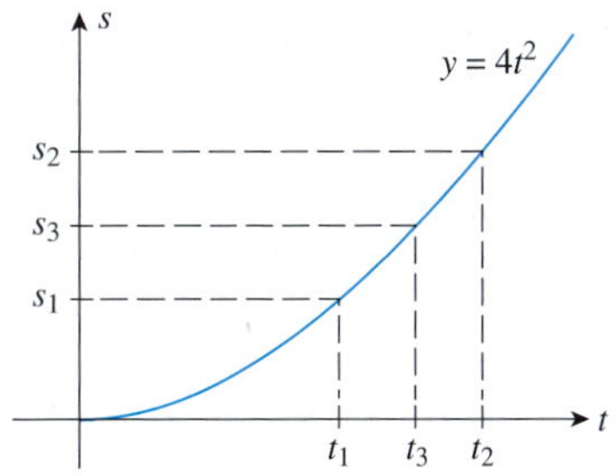

FIGURE 44
If $s_1 \le s_3 \le s_2$, then there must be at least one t_3 $(t_1 \le t_3 \le t_2)$ such that $f(t_3) = s_3$.

This discussion carries the gist of the intermediate value theorem. The proof of this theorem can be found in most advanced calculus texts.

THEOREM 4

The Intermediate Value Theorem

If f is a continuous function on a closed interval $[a, b]$ and M is any number between $f(a)$ and $f(b)$, then there is at least one number c in $[a, b]$ such that $f(c) = M$ (Figure 45).

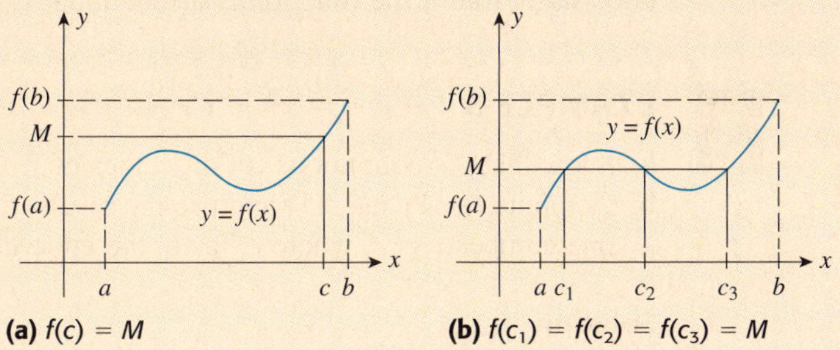

FIGURE 45

To illustrate the intermediate value theorem, let's look at the example involving the motion of the maglev again (see Figure 27, page 98). Notice that the initial position of the train is $f(0) = 0$ and the position at the end of its test run is $f(10) = 400$. Furthermore, the function f is continuous on $[0, 10]$. So, the intermediate value theorem guarantees that if we arbitrarily pick a number between 0 and 400—say, 100—giving the position of the maglev, there must be a $\bar{t}$ (read "t bar") between 0 and 10 at which time the train is at the position $s = 100$. To find the value of $\bar{t}$, we solve the equation $f(\bar{t}) = s$, or

$$4\bar{t}^2 = 100$$

giving $\bar{t} = 5$ (t must lie between 0 and 10).

It is important to remember when we use Theorem 4 that the function f must be continuous. The conclusion of the intermediate value theorem may not hold if f is not continuous (see Exercise 101, page 130).

The next theorem is an immediate consequence of the intermediate value theorem. It not only tells us when a **zero of a function** f [root of the equation $f(x) = 0$] exists but also provides the basis for a method of approximating it.

THEOREM 5

Existence of Zeros of a Continuous Function

If f is a continuous function on a closed interval $[a, b]$, and if $f(a)$ and $f(b)$ have opposite signs, then there is at least one solution of the equation $f(x) = 0$ in the interval (a, b) (Figure 46).

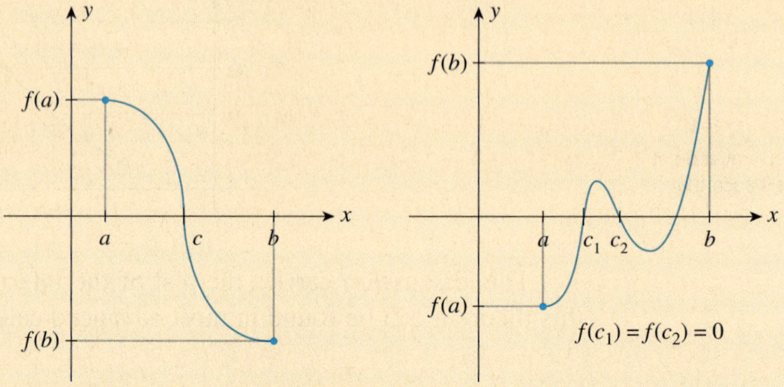

FIGURE 46
If $f(a)$ and $f(b)$ have opposite signs, there must be at least one number c ($a < c < b$) such that $f(c) = 0$.

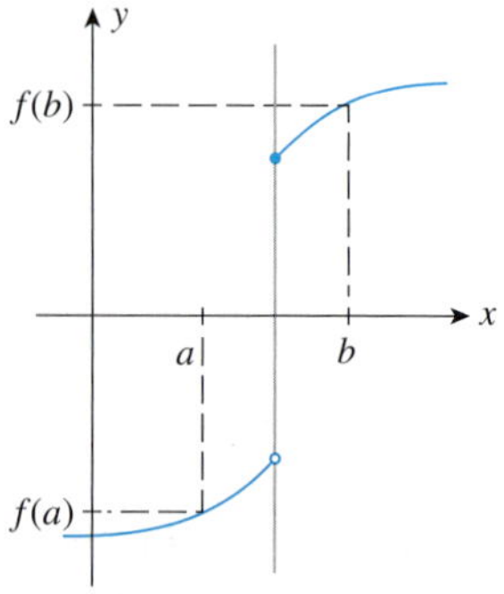

FIGURE 47
$f(a) < 0$ and $f(b) > 0$, but the graph of f does not cross the x-axis between a and b because f is discontinuous.

Geometrically, this property states that if the graph of a continuous function goes from above the x-axis to below the x-axis, or vice versa, it must *cross* the x-axis. This is not necessarily true if the function is discontinuous (Figure 47).

VIDEO

EXAMPLE 5 Let $f(x) = x^3 + x + 1$.

a. Show that f is continuous for all values of x.
b. Compute $f(-1)$ and $f(1)$ and use the results to deduce that there must be at least one number $x = c$, where c lies in the interval $(-1, 1)$ and $f(c) = 0$.

Solution

a. The function f is a polynomial function of degree 3 and is therefore continuous everywhere.
b. $f(-1) = (-1)^3 + (-1) + 1 = -1$ and $f(1) = 1^3 + 1 + 1 = 3$

Since $f(-1)$ and $f(1)$ have opposite signs, Theorem 5 tells us that there must be at least one number $x = c$ with $-1 < c < 1$ such that $f(c) = 0$.

The next example shows how the intermediate value theorem can be used to help us find the zero of a function.

EXAMPLE 6 Let $f(x) = x^3 + x - 1$. Since f is a polynomial function, it is continuous everywhere. Observe that $f(0) = -1$ and $f(1) = 1$ so that Theorem 5 guarantees the existence of at least one root of the equation $f(x) = 0$ in $(0, 1)$.*

We can locate the root more precisely by using Theorem 5 once again as follows: Evaluate $f(x)$ at the midpoint of $[0, 1]$, obtaining

$$f(0.5) = -0.375$$

*It can be shown that f has precisely one zero in $(0, 1)$ (see Exercise 105, Section 4.1).

Because $f(0.5) < 0$ and $f(1) > 0$, Theorem 5 now tells us that the root must lie in (0.5, 1).

Repeat the process: Evaluate $f(x)$ at the midpoint of [0.5, 1], which is

$$\frac{0.5 + 1}{2} = 0.75$$

Thus,

$$f(0.75) = 0.1719$$

Because $f(0.5) < 0$ and $f(0.75) > 0$, Theorem 5 tells us that the root is in (0.5, 0.75). This process can be continued. Table 3 summarizes the results of our computations through nine steps.

From Table 3 we see that the root is approximately 0.68, accurate to two decimal places. By continuing the process through a sufficient number of steps, we can obtain as accurate an approximation to the root as we please.

TABLE 3

Step	Root of $f(x) = 0$ Lies in
1	(0, 1)
2	(0.5, 1)
3	(0.5, 0.75)
4	(0.625, 0.75)
5	(0.625, 0.6875)
6	(0.65625, 0.6875)
7	(0.671875, 0.6875)
8	(0.6796875, 0.6875)
9	(0.6796875, 0.6835937)

Note The process of finding the root of $f(x) = 0$ used in Example 6 is called the **method of bisection.** It is crude but effective.

2.5 Self-Check Exercises

1. Evaluate $\lim_{x\to -1^-} f(x)$ and $\lim_{x\to -1^+} f(x)$, where

$$f(x) = \begin{cases} 1 & \text{if } x < -1 \\ 1 + \sqrt{x+1} & \text{if } x \geq -1 \end{cases}$$

Does $\lim_{x\to -1} f(x)$ exist?

2. Determine the values of x for which the given function is discontinuous. At each number where f is discontinuous, indicate which condition(s) for continuity are violated. Sketch the graph of the function.

a. $f(x) = \begin{cases} -x^2 + 1 & \text{if } x \leq 1 \\ x - 1 & \text{if } x > 1 \end{cases}$

b. $g(x) = \begin{cases} -x + 1 & \text{if } x < -1 \\ 2 & \text{if } -1 < x \leq 1 \\ -x + 3 & \text{if } x > 1 \end{cases}$

Solutions to Self-Check Exercises 2.5 can be found on page 130.

2.5 Concept Questions

1. Explain what is meant by the statement $\lim_{x\to 3^-} f(x) = 2$ and $\lim_{x\to 3^+} f(x) = 4$.

2. Suppose $\lim_{x\to 1^-} f(x) = 3$ and $\lim_{x\to 1^+} f(x) = 4$.
   a. What can you say about $\lim_{x\to 1} f(x)$? Explain.
   b. What can you say about $f(1)$? Explain.

3. Explain what it means for a function f to be continuous (a) at a number a and (b) on an interval I.

4. Determine whether each function f is continuous or discontinuous. Explain your answer.
   a. $f(t)$ gives the altitude of an airplane at time t.
   b. $f(t)$ measures the total amount of rainfall at time t at the Municipal Airport.
   c. $f(s)$ measures the fare as a function of the distance s for taking a cab from Kennedy Airport to downtown Manhattan.
   d. $f(t)$ gives the interest rate charged by a financial institution at time t.

5. Explain the intermediate value theorem in your own words.

2.5 Exercises

In Exercises 1–8, use the graph of the function f to find $\lim_{x\to a^-} f(x)$, $\lim_{x\to a^+} f(x)$, and $\lim_{x\to a} f(x)$ at the indicated value of a, if the limit exists.

1.
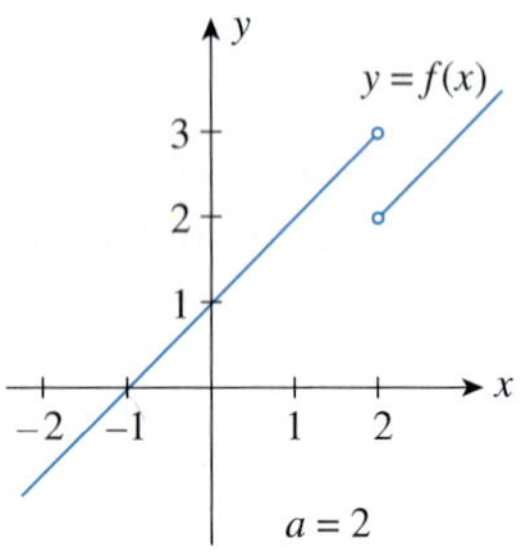

2.
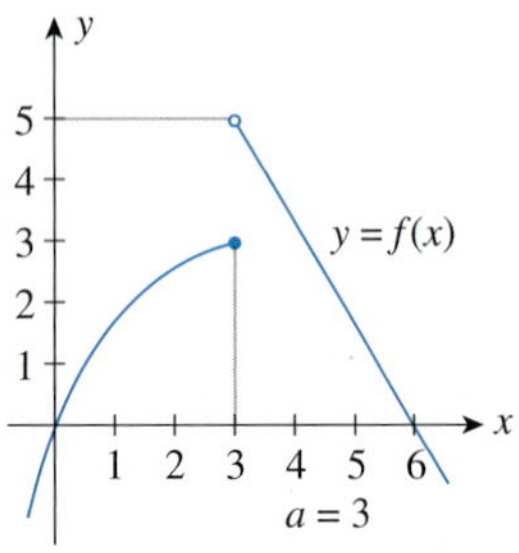

3.
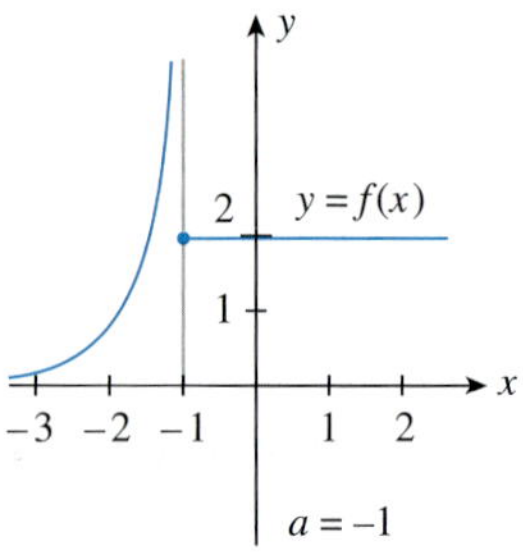

4.
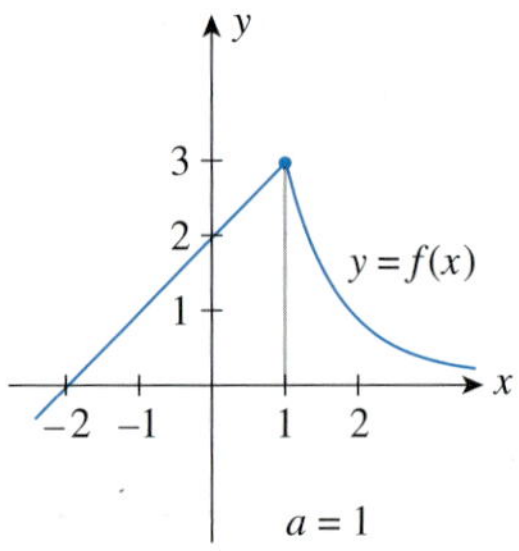

5.
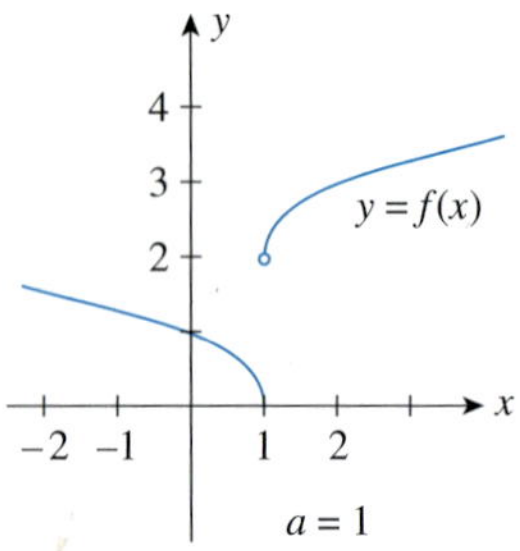

6.
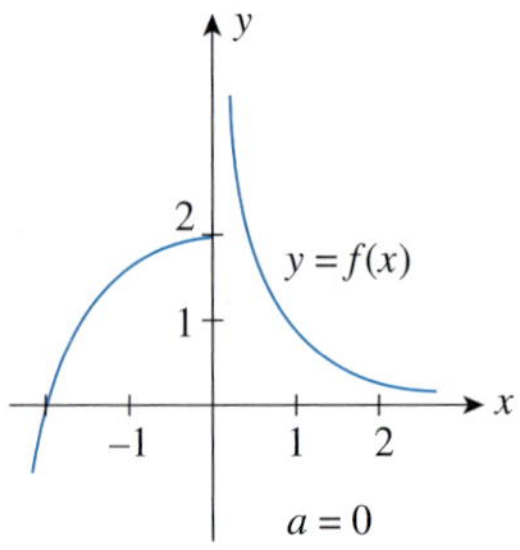

7.
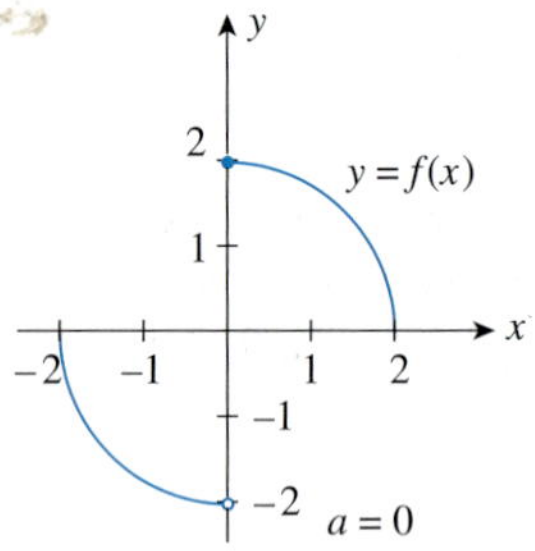

8.
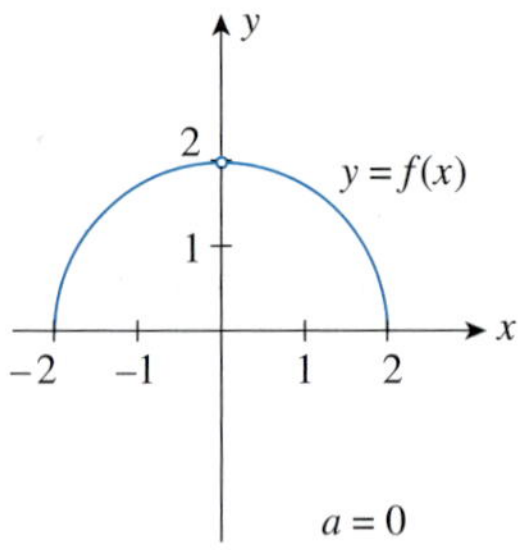

In Exercises 9–14, refer to the graph of the function f and determine whether each statement is true or false.

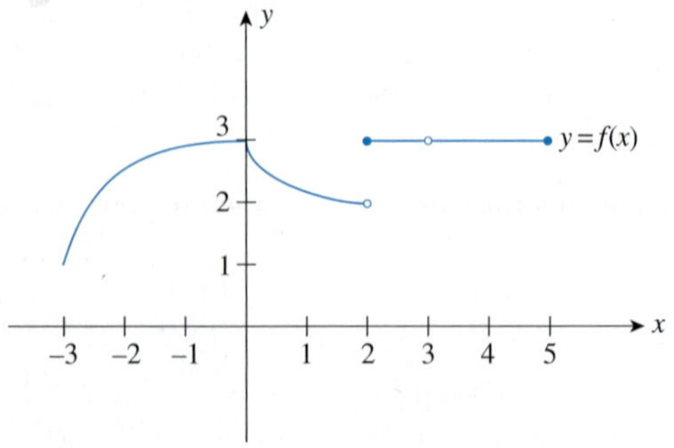

9. $\lim_{x\to -3^+} f(x) = 1$

10. $\lim_{x\to 0} f(x) = f(0)$

11. $\lim_{x\to 2^-} f(x) = 2$

12. $\lim_{x\to 2^+} f(x) = 3$

13. $\lim_{x\to 3} f(x)$ does not exist.

14. $\lim_{x\to 5^-} f(x) = 3$

In Exercises 15–20, refer to the graph of the function f and determine whether each statement is true or false.

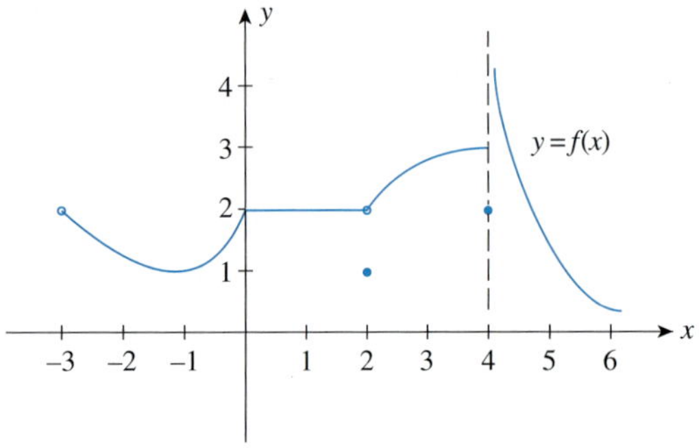

15. $\lim_{x\to -3^+} f(x) = 2$

16. $\lim_{x\to 0} f(x) = 2$

17. $\lim_{x\to 2} f(x) = 1$

18. $\lim_{x\to 4^-} f(x) = 3$

19. $\lim_{x\to 4^+} f(x)$ does not exist.

20. $\lim_{x\to 4} f(x) = 2$

In Exercises 21–38, find the indicated one-sided limit, if it exists.

21. $\lim_{x\to 1^+} (2x + 4)$

22. $\lim_{x\to 1^-} (3x - 4)$

23. $\lim_{x\to 2^-} \dfrac{x-3}{x+2}$

24. $\lim_{x\to 1^+} \dfrac{x+2}{x+1}$

25. $\lim_{x\to 0^+} \dfrac{1}{x}$

26. $\lim_{x\to 0^-} \dfrac{1}{x}$

27. $\lim_{x\to 0^+} \dfrac{x-1}{x^2+1}$

28. $\lim_{x\to 2^+} \dfrac{x+1}{x^2-2x+3}$

29. $\lim_{x\to 0^+} \sqrt{x}$

30. $\lim_{x\to 2^+} 2\sqrt{x-2}$

31. $\lim_{x\to -2^+} (2x + \sqrt{2+x})$

32. $\lim_{x\to -5^+} x(1 + \sqrt{5+x})$

33. $\lim_{x\to 1^-} \dfrac{1+x}{1-x}$

34. $\lim_{x\to 1^+} \dfrac{1+x}{1-x}$

35. $\lim_{x\to 2^-} \dfrac{x^2-4}{x-2}$

36. $\lim_{x\to -3^+} \dfrac{\sqrt{x+3}}{x^2+1}$

37. $\lim_{x\to 0^+} f(x)$ and $\lim_{x\to 0^-} f(x)$, where

$$f(x) = \begin{cases} 2x & \text{if } x < 0 \\ x^2 & \text{if } x \geq 0 \end{cases}$$

38. $\lim_{x\to 0^+} f(x)$ and $\lim_{x\to 0^-} f(x)$, where

$$f(x) = \begin{cases} -x+1 & \text{if } x \leq 0 \\ 2x+3 & \text{if } x > 0 \end{cases}$$

In Exercises 39–44, determine the values of *x*, if any, at which each function is discontinuous. At each number where *f* is discontinuous, state the condition(s) for continuity that are violated.

39. 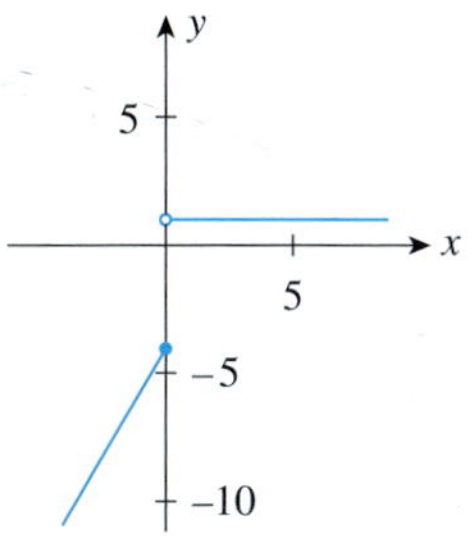

$$f(x) = \begin{cases} 2x - 4 & \text{if } x \le 0 \\ 1 & \text{if } x > 0 \end{cases}$$

40. 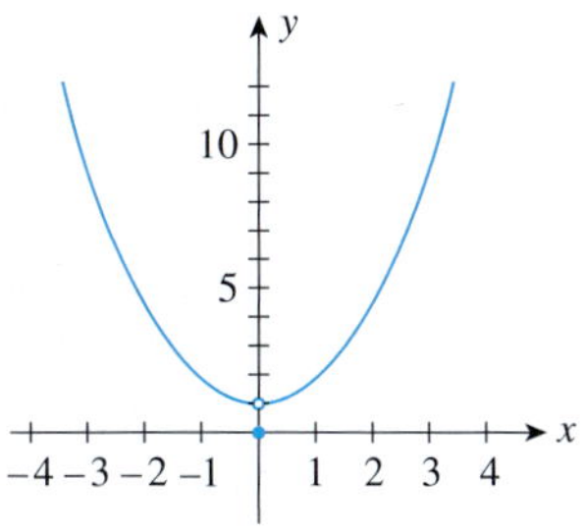

$$f(x) = \begin{cases} x^2 + 1 & \text{if } x \ne 0 \\ 0 & \text{if } x = 0 \end{cases}$$

41. 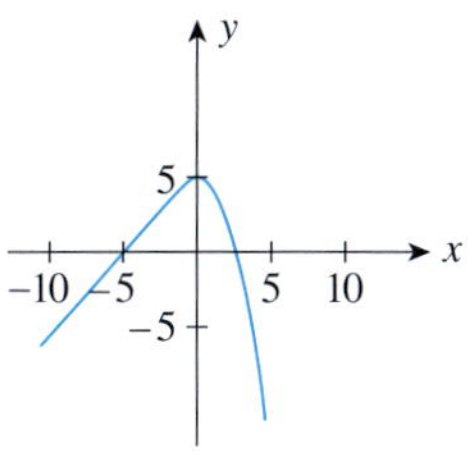

$$f(x) = \begin{cases} x + 5 & \text{if } x \le 0 \\ -x^2 + 5 & \text{if } x > 0 \end{cases}$$

42. 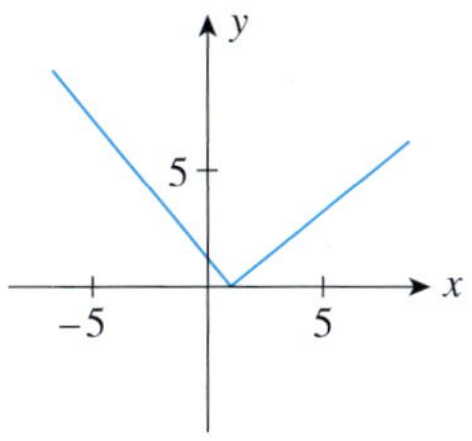

$$f(x) = |x - 1|$$

43. 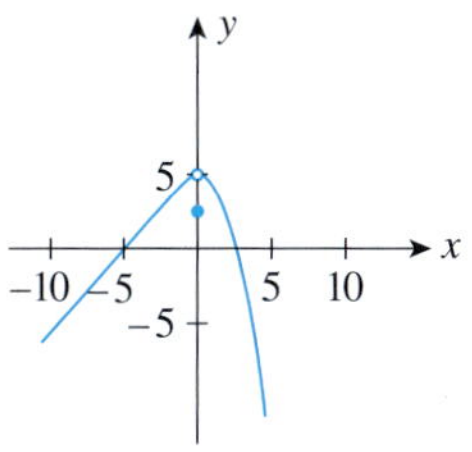

$$f(x) = \begin{cases} x + 5 & \text{if } x < 0 \\ 2 & \text{if } x = 0 \\ -x^2 + 5 & \text{if } x > 0 \end{cases}$$

44. 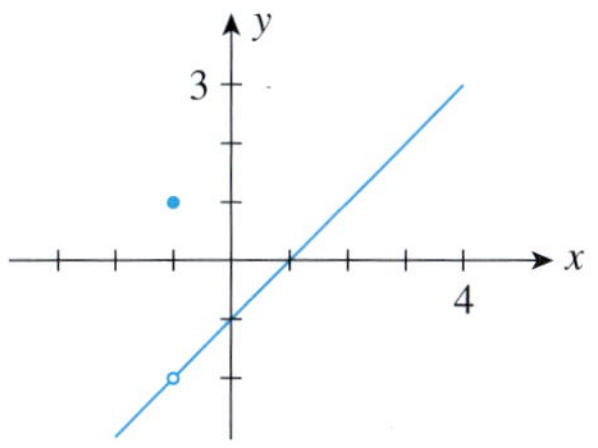

$$f(x) = \begin{cases} \dfrac{x^2 - 1}{x + 1} & \text{if } x \ne -1 \\ 1 & \text{if } x = -1 \end{cases}$$

In Exercises 45–56, find the values of *x* for which each function is continuous.

45. $f(x) = 2x^2 + x - 1$

46. $f(x) = x^3 - 2x^2 + x - 1$

47. $f(x) = \dfrac{2}{x^2 + 1}$

48. $f(x) = \dfrac{x}{2x^2 + 1}$

49. $f(x) = \dfrac{2}{2x - 1}$

50. $f(x) = \dfrac{x + 1}{x - 1}$

51. $f(x) = \dfrac{2x + 1}{x^2 + x - 2}$

52. $f(x) = \dfrac{x - 1}{x^2 + 2x - 3}$

53. $f(x) = \begin{cases} x & \text{if } x \le 1 \\ 2x - 1 & \text{if } x > 1 \end{cases}$

54. $f(x) = \begin{cases} -2x + 1 & \text{if } x < 0 \\ x^2 + 1 & \text{if } x \ge 0 \end{cases}$

55. $f(x) = |x + 1|$

56. $f(x) = \dfrac{|x - 1|}{x - 1}$

In Exercises 57–60, determine all values of *x* at which the function is discontinuous.

57. $f(x) = \dfrac{2x}{x^2 - 1}$

58. $f(x) = \dfrac{1}{(x - 1)(x - 2)}$

59. $f(x) = \dfrac{x^2 - 2x}{x^2 - 3x + 2}$

60. $f(x) = \dfrac{x^2 - 3x + 2}{x^2 - 2x}$

61. **The Postage Function** The graph of the "postage function" for 2009,

$$f(x) = \begin{cases} 122 & \text{if } 0 < x \le 1 \\ 139 & \text{if } 1 < x \le 2 \\ \vdots & \\ 326 & \text{if } 12 < x \le 13 \end{cases}$$

where x denotes the weight of a package in ounces and $f(x)$ the postage in cents, is shown in the accompanying figure. Determine the values of x for which f is discontinuous.

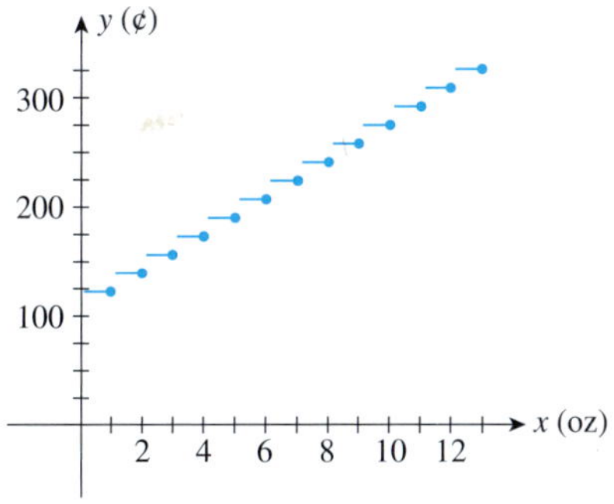

62. **Inventory Control** As part of an optimal inventory policy, the manager of an office supply company orders 500 reams of photocopy paper every 20 days. The accompanying graph shows the *actual* inventory level of paper in an office supply store during the first 60 business days of 2010. Determine the values of t for which the "inventory function" is discontinuous and give an interpretation of the graph.

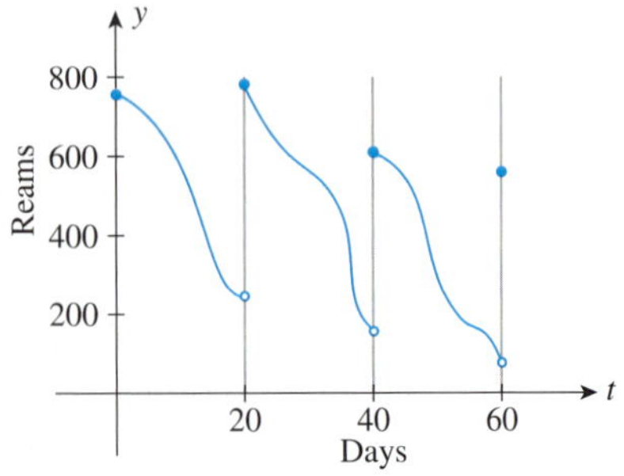

63. **Learning Curves** The following graph describes the progress Michael made in solving a problem correctly during a mathematics quiz. Here, y denotes the percent of work completed, and x is measured in minutes. Give an interpretation of the graph.

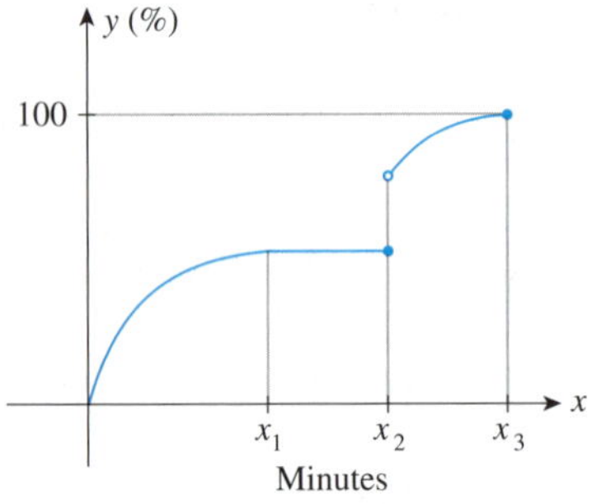

64. **Ailing Financial Institutions** Franklin Savings and Loan acquired two ailing financial institutions in 2010. One of them was acquired at time $t = T_1$, and the other was acquired at time $t = T_2$ ($t = 0$ corresponds to the beginning of 2010). The following graph shows the total amount of money on deposit with Franklin. Explain the significance of the discontinuities of the function at T_1 and T_2.

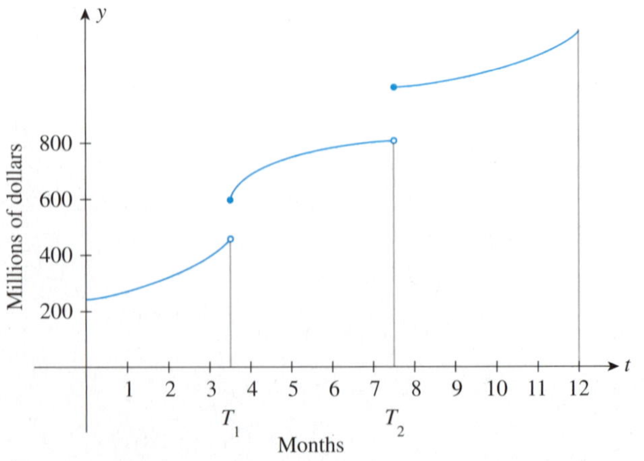

65. **Energy Consumption** The following graph shows the amount of home heating oil remaining in a 200-gal tank over a 120-day period ($t = 0$ corresponds to October 1). Explain why the function is discontinuous at $t = 40$, 70, 95, and 110.

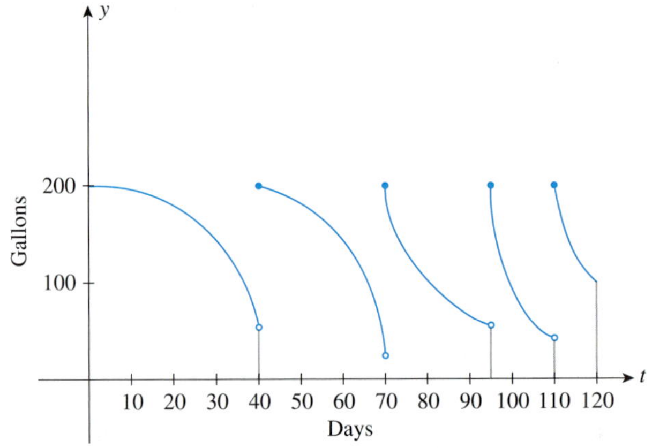

66. **Prime Interest Rate** The function P, whose graph follows, gives the prime rate (the interest rate banks charge their best corporate customers) for a certain country as a function of time for the first 32 wk in 2008. Determine the values of t for which P is discontinuous and interpret your results.

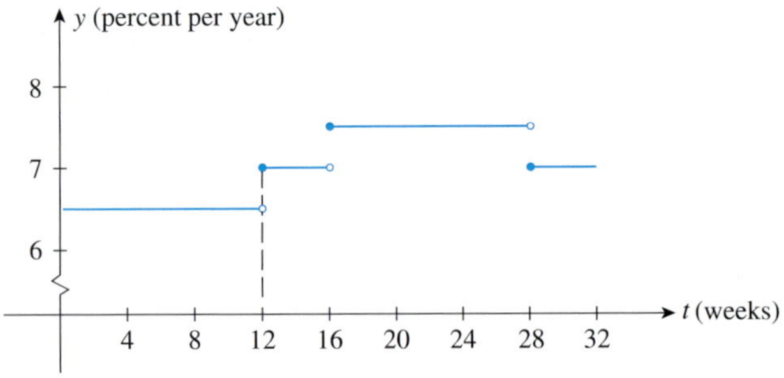

67. **Administration of an Intravenous Solution** A dextrose solution is being administered to a patient intravenously. The 1-liter (L) bottle holding the solution is removed and replaced by another as soon as the contents drop to approximately 5% of the initial (1-L) amount. The rate of discharge is constant, and it takes 6 hr to discharge 95% of the contents of a full bottle. Draw a graph showing the amount of dextrose solution in a bottle in the IV system over a 24-hr period, assuming that we started with a full bottle.

68. **Commissions** The base monthly salary of a salesman working on commission is \$22,000. For each \$50,000 of sales beyond \$100,000, he is paid a \$1000 commission. Sketch a graph showing his earnings as a function of the level of his sales x. Determine the values of x for which the function f is discontinuous.

69. **Parking Fees** The fee charged per car in a downtown parking lot is \$2.00 for the first half hour and \$1.00 for each additional half hour or part thereof, subject to a maximum of \$10.00. Derive a function f relating the parking fee to the length of time a car is left in the lot. Sketch the graph of f and determine the values of x for which the function f is discontinuous.

70. **Commodity Prices** The function that gives the cost of a certain commodity is defined by

$$C(x) = \begin{cases} 5x & \text{if } 0 < x < 10 \\ 4x & \text{if } 10 \le x < 30 \\ 3.5x & \text{if } 30 \le x < 60 \\ 3.25x & \text{if } x \ge 60 \end{cases}$$

where x is the number of pounds of a certain commodity sold and $C(x)$ is measured in dollars. Sketch the graph of the function C and determine the values of x for which the function C is discontinuous.

71. **Weiss's Law** According to Weiss's law of excitation of tissue, the strength S of an electric current is related to the time t the current takes to excite tissue by the formula

$$S(t) = \frac{a}{t} + b \qquad (t > 0)$$

where a and b are positive constants.

a. Evaluate $\lim_{t \to 0^+} S(t)$ and interpret your result.

b. Evaluate $\lim_{t \to \infty} S(t)$ and interpret your result.

(*Note:* The limit in part (b) is called the threshold strength of the current. Why?)

72. **Energy Expended by a Fish** Suppose a fish swimming a distance of L ft at a speed of v ft/sec relative to the water and against a current flowing at the rate of u ft/sec $(u < v)$ expends a total energy given by

$$E(v) = \frac{aLv^3}{v - u}$$

where E is measured in foot-pounds (ft-lb) and a is a constant.

a. Evaluate $\lim_{v \to u^+} E(v)$ and interpret your result.

b. Evaluate $\lim_{v \to \infty} E(v)$ and interpret your result.

73. Let

$$f(x) = \begin{cases} x + 2 & \text{if } x \le 1 \\ kx^2 & \text{if } x > 1 \end{cases}$$

Find the value of k that will make f continuous on $(-\infty, \infty)$.

74. Let

$$f(x) = \begin{cases} \dfrac{x^2 - 4}{x + 2} & \text{if } x \ne -2 \\ k & \text{if } x = -2 \end{cases}$$

For what value of k will f be continuous on $(-\infty, \infty)$?

75. a. Suppose f is continuous at a and g is discontinuous at a. Is the sum $f + g$ discontinuous at a? Explain.

b. Suppose f and g are both discontinuous at a. Is the sum $f + g$ necessarily discontinuous at a? Explain.

76. a. Suppose f is continuous at a and g is discontinuous at a. Is the product fg necessarily discontinuous at a? Explain.

b. Suppose f and g are both discontinuous at a. Is the product fg necessarily discontinuous at a? Explain.

In Exercises 77–80, (a) show that the function *f* is continuous for all values of *x* in the interval [*a*, *b*] and (b) prove that *f* must have at least one zero in the interval (*a*, *b*) by showing that *f*(*a*) and *f*(*b*) have opposite signs.

77. $f(x) = x^2 - 6x + 8$; $a = 1$, $b = 3$

78. $f(x) = 2x^3 - 3x^2 - 36x + 14$; $a = 0$, $b = 1$

79. $f(x) = x^3 - 2x^2 + 3x + 2$; $a = -1$, $b = 1$

80. $f(x) = 2x^{5/3} - 5x^{4/3}$; $a = 14$, $b = 16$

In Exercises 81–82, use the intermediate value theorem to find the value of *c* such that *f*(*c*) = *M*.

81. $f(x) = x^2 - 4x + 6$ on $[0, 3]$; $M = 4$

82. $f(x) = x^2 - x + 1$ on $[-1, 4]$; $M = 7$

83. Use the method of bisection (see Example 6) to find the root of the equation $x^5 + 2x - 7 = 0$ accurate to two decimal places.

84. Use the method of bisection (see Example 6) to find the root of the equation $x^3 - x + 1 = 0$ accurate to two decimal places.

85. **Falling Object** Joan is looking straight out a window of an apartment building at a height of 32 ft from the ground. A boy throws a tennis ball straight up by the side of the building where the window is located. Suppose the height of the ball (measured in feet) from the ground at time t is $h(t) = 4 + 64t - 16t^2$.

a. Show that $h(0) = 4$ and $h(2) = 68$.

b. Use the intermediate value theorem to conclude that the ball must cross Joan's line of sight at least once.

c. At what time(s) does the ball cross Joan's line of sight? Interpret your results.

86. **Oxygen Content of a Pond** The oxygen content t days after organic waste has been dumped into a pond is given by

$$f(t) = 100\left(\frac{t^2 + 10t + 100}{t^2 + 20t + 100}\right)$$

percent of its normal level.

a. Show that $f(0) = 100$ and $f(10) = 75$.

b. Use the intermediate value theorem to conclude that the oxygen content of the pond must have been at a level of 80% at some time.

c. At what time(s) is the oxygen content at the 80% level? Hint: Use the quadratic formula.

In Exercises 87–96, determine whether the statement is true or false. If it is true, explain why it is true. If it is false, give an example to show why it is false.

87. If $f(2) = 4$, then $\lim_{x \to 2} f(x) = 4$.

88. If $\lim_{x \to 0} f(x) = 3$, then $f(0) = 3$.

89. If $\lim_{x \to 2^+} f(x) = 3$ and $f(2) = 3$ then $\lim_{x \to 2^-} f(x) = 3$.

90. If $\lim_{x \to 3^-} f(x)$ and $\lim_{x \to 3^+} f(x)$ both exist, then $\lim_{x \to 3} f(x)$ exists.

91. If $f(5)$ is not defined, then $\lim_{x \to 5^-} f(x)$ does not exist.

92. Suppose the function f is defined on the interval $[a, b]$. If $f(a)$ and $f(b)$ have the same sign, then f has no zero in $[a, b]$.

93. If $\lim_{x \to a^-} f(x) = L$ and $\lim_{x \to a^+} f(x) = L$, then $f(a) = L$.

94. If $\lim_{x \to a} f(x) = L$, then $\lim_{x \to a^+} f(x) - \lim_{x \to a^-} f(x) \neq 0$.

95. If f is continuous for all $x \neq 0$ and $f(0) = 0$, then $\lim_{x \to 0} f(x) = 0$.

96. If $\lim_{x \to a} f(x) = L$ and $g(a) = M$, then $\lim_{x \to a} f(x)g(x) = LM$.

97. Suppose f is continuous on $[a, b]$ and $f(a) < f(b)$. If M is a number that lies outside the interval $[f(a), f(b)]$, then there does not exist a number $a < c < b$ such that $f(c) = M$. Does this contradict the intermediate value theorem?

98. Let $f(x) = \dfrac{x^2}{x^2 + 1}$.

a. Show that f is continuous for all values of x.
b. Show that $f(x)$ is nonnegative for all values of x.
c. Show that f has a zero at $x = 0$. Does this contradict Theorem 5?

99. Let $f(x) = x - \sqrt{1 - x^2}$.

a. Show that f is continuous for all values of x in the interval $[-1, 1]$.
b. Show that f has at least one zero in $[-1, 1]$.
c. Find the zeros of f in $[-1, 1]$ by solving the equation $f(x) = 0$.

100. a. Prove that a polynomial function $y = P(x)$ is continuous at every number x. Follow these steps:

(i) Use Properties 2 and 3 of continuous functions to establish that the function $g(x) = x^n$, where n is a positive integer, is continuous everywhere.
(ii) Use Properties 1 and 5 to show that $f(x) = cx^n$, where c is a constant and n is a positive integer, is continuous everywhere.
(iii) Use Property 4 to complete the proof of the result.

b. Prove that a rational function $R(x) = p(x)/q(x)$ is continuous at every point x, where $q(x) \neq 0$.
Hint: Use the result of part (a) and Property 6.

101. Show that the conclusion of the intermediate value theorem does not necessarily hold if f is discontinuous on $[a, b]$.

2.5 Solutions to Self-Check Exercises

1. For $x < -1$, $f(x) = 1$, and so

$$\lim_{x \to -1^-} f(x) = \lim_{x \to -1^-} 1 = 1$$

For $x \geq -1$, $f(x) = 1 + \sqrt{x + 1}$, and so

$$\lim_{x \to -1^+} f(x) = \lim_{x \to -1^+} (1 + \sqrt{x + 1}) = 1$$

Since the left-hand and right-hand limits of f exist as x approaches $x = -1$ and both are equal to 1, we conclude that

$$\lim_{x \to -1} f(x) = 1$$

2. a. The graph of f follows:

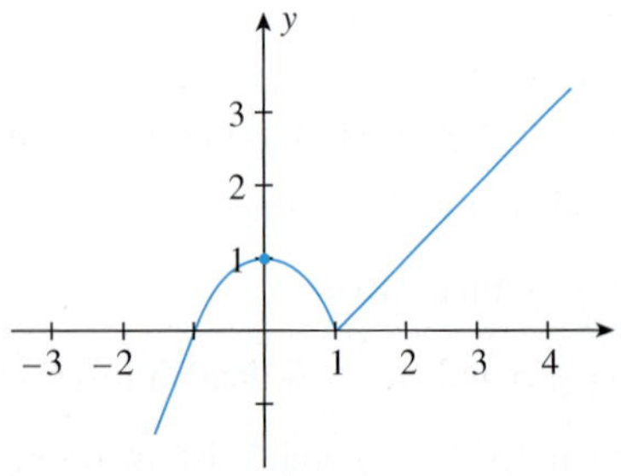

We see that f is continuous everywhere.

b. The graph of g follows:

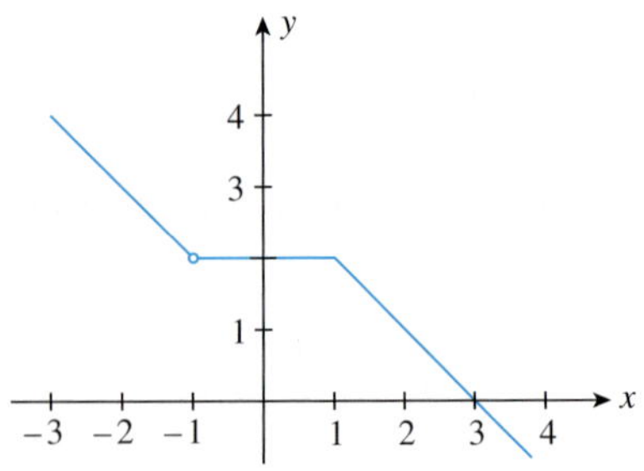

Since g is not defined at $x = -1$, it is discontinuous there. It is continuous everywhere else.

USING TECHNOLOGY

Finding the Points of Discontinuity of a Function

You can very often recognize the points of discontinuity of a function f by examining its graph. For example, Figure T1a shows the graph of $f(x) = x/(x^2 - 1)$ obtained using a graphing utility. It is evident that f is discontinuous at $x = -1$ and $x = 1$. This observation is also borne out by the fact that both these points are not in the domain of f.

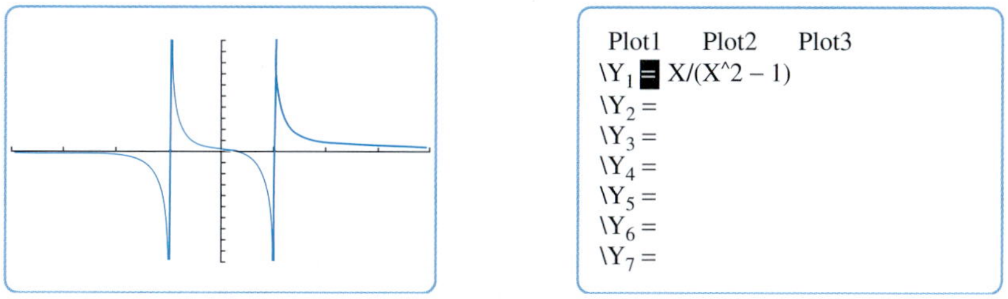

(a) (b)

FIGURE **T1**
(a) The graph of

$$f(x) = \frac{x}{x^2 - 1}$$

in the viewing window $[-4, 4] \times [-10, 10]$; (b) the TI-83/84 equation screen

Consider the function

$$g(x) = \frac{2x^3 + x^2 - 7x - 6}{x^2 - x - 2}$$

Using a graphing utility, we obtain the graph of g shown in Figure T2a.

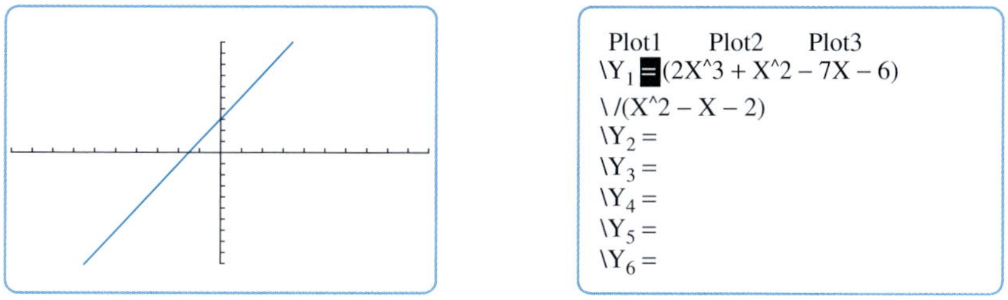

(a) (b)

FIGURE **T2**
(a) The graph of

$$g(x) = \frac{2x^3 + x^2 - 7x - 6}{x^2 - x - 2}$$

in the standard viewing window; (b) the TI-83/84 equation screen.

An examination of this graph does not reveal any points of discontinuity. However, if we factor both the numerator and the denominator of the rational expression, we see that

$$g(x) = \frac{(x + 1)(x - 2)(2x + 3)}{(x + 1)(x - 2)}$$
$$= 2x + 3$$

provided $x \neq -1$ and $x \neq 2$, so that its graph in fact looks like that shown in Figure T3.

This example shows the limitation of the graphing utility and reminds us of the importance of studying functions analytically!

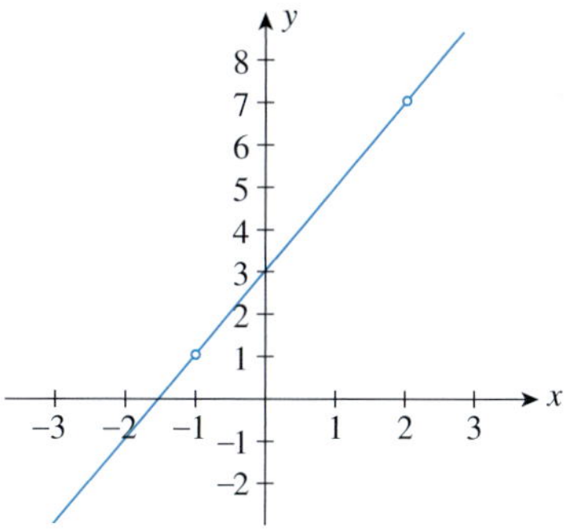

FIGURE **T3**
The graph of g has holes at $(-1, 1)$ and $(2, 7)$.

Graphing Functions Defined Piecewise

The following example illustrates how to plot the graphs of functions defined in a piecewise manner on a graphing utility.

EXAMPLE 1 Plot the graph of

$$f(x) = \begin{cases} x + 1 & \text{if } x \leq 1 \\ \dfrac{2}{x} & \text{if } x > 1 \end{cases}$$

(continued)

Solution We enter the function

$$y1 = (x + 1)(x \le 1) + (2/x)(x > 1)$$

Figure T4 shows the graph of the function in the viewing window $[-5, 5] \times [-2, 4]$.

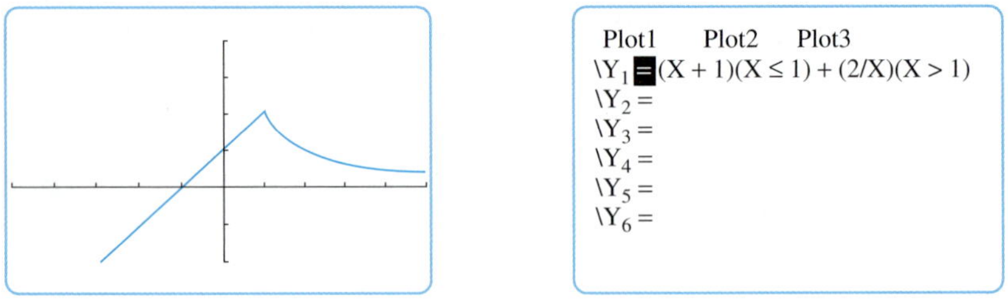

(a) (b)

FIGURE T4
(a) The graph of f in the viewing window $[-5, 5] \times [-2, 4]$; (b) the TI-83/84 equation screen.

APPLIED EXAMPLE 2 TV Viewing Patterns The percent of U.S. households, $P(t)$, watching television during weekdays between the hours of 4 p.m. and 4 a.m. is given by

$$P(t) = \begin{cases} 0.01354t^4 - 0.49375t^3 + 2.58333t^2 + 3.8t + 31.60704 & \text{if } 0 \le t \le 8 \\ 1.35t^2 - 33.05t + 208 & \text{if } 8 < t \le 12 \end{cases}$$

where t is measured in hours, with $t = 0$ corresponding to 4 p.m. Plot the graph of P in the viewing window $[0, 12] \times [0, 80]$.

Source: A. C. Nielsen Co.

Solution We enter the function

$$\begin{aligned} y1 = {} & (.01354x\hat{}4 - .49375x\hat{}3 + 2.58333x\hat{}2 + 3.8x + 31.60704)(x \ge 0)(x \le 8) \\ & + (1.35x\hat{}2 - 33.05x + 208)(x > 8)(x \le 12) \end{aligned}$$

Figure T5a shows the graph of P.

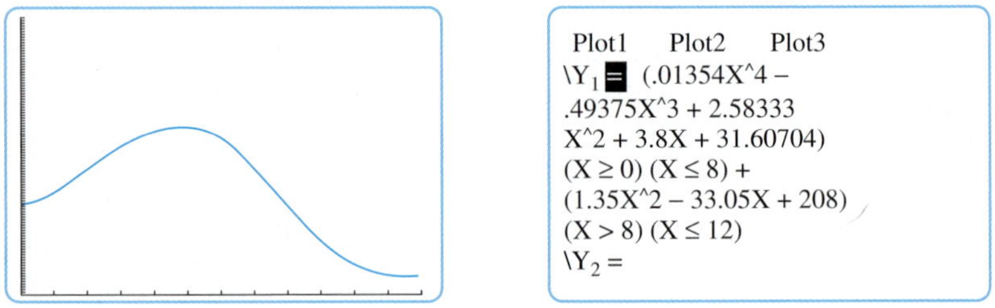

(a) (b)

FIGURE T5
(a) The graph of P in the viewing window $[0, 12] \times [0, 80]$; (b) the TI-83/84 equation screen

TECHNOLOGY EXERCISES

In Exercises 1–8, plot the graph of *f* and find the points of discontinuity of *f*. Then use analytical means to verify your observation and find all numbers where *f* is discontinuous.

1. $f(x) = \dfrac{2}{x^2 - x}$

2. $f(x) = \dfrac{3}{\sqrt{x}(x + 1)}$

3. $f(x) = \dfrac{6x^3 + x^2 - 2x}{2x^2 - x}$

4. $f(x) = \dfrac{2x^3 - x^2 - 13x - 6}{2x^2 - 5x - 3}$

5. $f(x) = \dfrac{2x^4 - 3x^3 - 2x^2}{2x^2 - 3x - 2}$

6. $f(x) = \dfrac{6x^4 - x^3 + 5x^2 - 1}{6x^2 - x - 1}$

7. $f(x) = \dfrac{x^3 + x^2 - 2x}{x^4 + 2x^3 - x - 2}$
Hint: $x^4 + 2x^3 - x - 2 = (x^3 - 1)(x + 2)$

8. $f(x) = \dfrac{x^3 - x}{x^{4/3} - x + x^{1/3} - 1}$
Hint: $x^{4/3} - x + x^{1/3} - 1 = (x^{1/3} - 1)(x + 1)$
Can you explain why part of the graph is missing?

In Exercises 9 and 10, plot the graph of f in the indicated viewing window.

9. $f(x) = \begin{cases} 2 & \text{if } x \le 0 \\ \sqrt{4 - x^2} & \text{if } x > 0; [-2, 2] \times [-4, 4] \end{cases}$

10. $f(x) = \begin{cases} -x^2 + x + 2 & \text{if } x \le 1 \\ 2x^3 - x^2 - 4 & \text{if } x > 1; [-4, 4] \times [-5, 5] \end{cases}$

11. Flight Path of a Plane The function

$$f(x) = \begin{cases} 0 & \text{if } 0 \le x < 1 \\ -0.00411523x^3 + 0.0679012x^2 \\ \quad -0.123457x + 0.0596708 & \text{if } 1 \le x < 10 \\ 1.5 & \text{if } 10 \le x \le 100 \end{cases}$$

where both x and $f(x)$ are measured in units of 1000 ft, describes the flight path of a plane taking off from the origin and climbing to an altitude of 15,000 ft. Plot the graph of f to visualize the trajectory of the plane.

12. Home Shopping Industry According to industry sources, revenue from the home shopping industry for the years since its inception may be approximated by the function

$$R(t) = \begin{cases} -0.03t^3 + 0.25t^2 - 0.12t & \text{if } 0 \le t \le 3 \\ 0.57t - 0.63 & \text{if } 3 < t \le 11 \end{cases}$$

where $R(t)$ measures the revenue in billions of dollars and t is measured in years, with $t = 0$ corresponding to the beginning of 1984. Plot the graph of R.

Source: Paul Kagan Associates

2.6 The Derivative

An Intuitive Example

We mentioned in Section 2.4 that the problem of finding the *rate of change* of one quantity with respect to another is mathematically equivalent to the problem of finding the *slope of the tangent line* to a curve at a given point on the curve. Before going on to establish this relationship, let's show its plausibility by looking at it from an intuitive point of view.

Consider the motion of the maglev discussed in Section 2.4. Recall that the position of the maglev at any time t is given by

$$s = f(t) = 4t^2 \qquad (0 \le t \le 30)$$

where s is measured in feet and t in seconds. The graph of the function f is sketched in Figure 48.

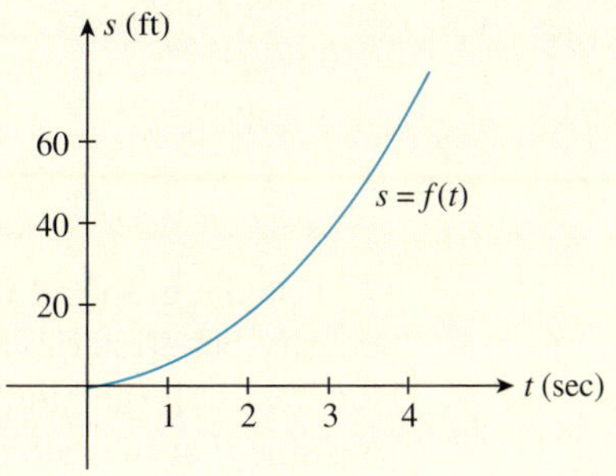

FIGURE **48**
Graph showing the position s of a maglev at time t

Observe that the graph of f rises slowly at first but more rapidly as t increases, reflecting the fact that the speed of the maglev is increasing with time. This observation suggests a relationship between the speed of the maglev at any time t and the *steepness* of the curve at the point corresponding to this value of t. Thus, it would appear that we can solve the problem of finding the speed of the maglev at any time if we can find a way to measure the steepness of the curve at any point on the curve.

To discover a yardstick that will measure the steepness of a curve, consider the graph of a function f such as the one shown in Figure 49a. Think of the curve as representing a stretch of roller coaster track (Figure 49b). When the car is at the point

P on the curve, a passenger sitting erect in the car and looking straight ahead will have a line of sight that is parallel to the line T, the tangent to the curve at P.

As Figure 49a suggests, the steepness of the curve—that is, the rate at which y is increasing or decreasing with respect to x—is given by the slope of the tangent line to the graph of f at the point $P(x, f(x))$. But for now we will show how this relationship can be used to estimate the rate of change of a function from its graph.

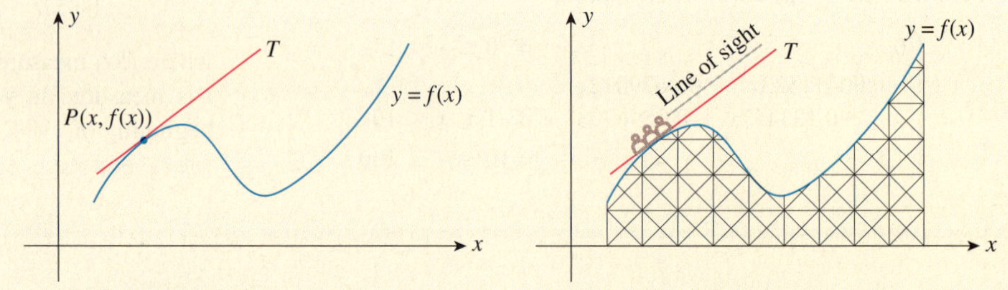

(a) T is the tangent line to the curve at P. **(b)** T is parallel to the line of sight.

FIGURE **49**

APPLIED EXAMPLE 1 Social Security Beneficiaries The graph of the function $y = N(t)$, shown in Figure 50, gives the number of Social Security beneficiaries from the beginning of 1990 ($t = 0$) through the year 2045 ($t = 55$).

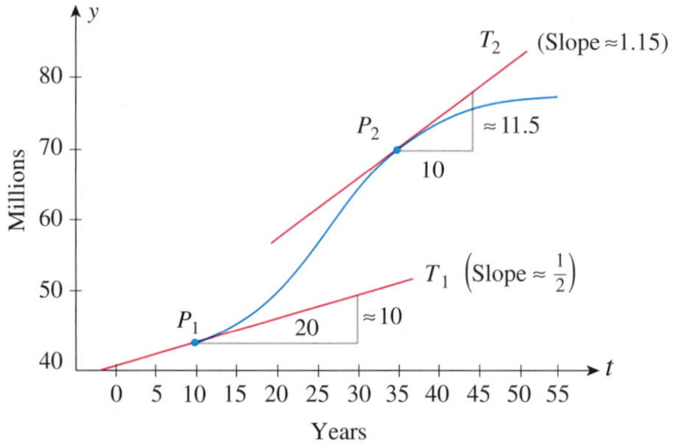

FIGURE **50**
The number of Social Security beneficiaries from 1990 through 2045. We can use the slope of the tangent line at the indicated points to estimate the rate at which the number of Social Security beneficiaries will be changing.

Use the graph of $y = N(t)$ to estimate the rate at which the number of Social Security beneficiaries was growing at the beginning of the year 2000 ($t = 10$). How fast will the number be growing at the beginning of 2025 ($t = 35$)? [Assume that the rate of change of the function N at any value of t is given by the slope of the tangent line at the point $P(t, N(t))$.]
Source: Social Security Administration

Solution From the figure, we see that the slope of the tangent line T_1 to the graph of $y = N(t)$ at $P_1(10, 44.7)$ is approximately 0.5. This tells us that the quantity y is increasing at the rate of $\frac{1}{2}$ unit per unit increase in t, when $t = 10$. In other words, at the beginning of the year 2000, the number of Social Security beneficiaries was increasing at the rate of approximately 0.5 million, or 500,000, per year.

The slope of the tangent line T_2 at $P_2(35, 71.9)$ is approximately 1.15. This tells us that at the beginning of 2025 the number of Social Security beneficiaries will be growing at the rate of approximately 1.15 million, or 1,150,000, per year. ■

Slope of a Tangent Line

In Example 1 we answered the questions raised by drawing the graph of the function N and estimating the position of the tangent lines. Ideally, however, we would like to solve a problem analytically whenever possible. To do this we need a precise definition of the slope of a tangent line to a curve.

To define the tangent line to a curve C at a point P on the curve, fix P and let Q be any point on C distinct from P (Figure 51). The straight line passing through P and Q is called a **secant line.**

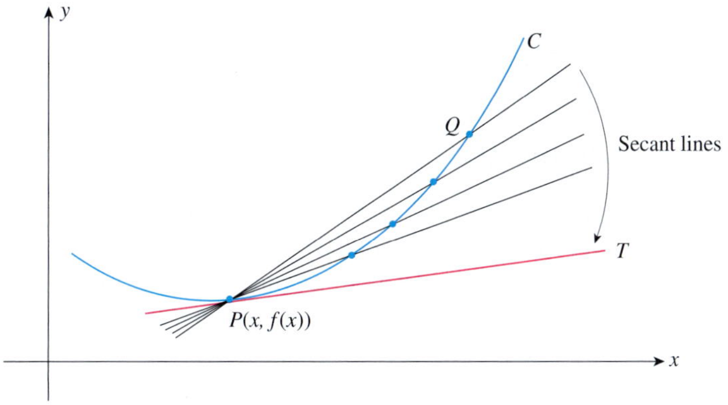

FIGURE 51
As Q approaches P along the curve C, the secant lines approach the tangent line T.

Now, as the point Q is allowed to move toward P along the curve, the secant line through P and Q rotates about the fixed point P and approaches a fixed line through P. This fixed line, which is the limiting position of the secant lines through P and Q as Q approaches P, is the **tangent line to the graph of f** at the point P.

We can describe the process more precisely as follows. Suppose the curve C is the graph of a function f defined by $y = f(x)$. Then the point P is described by $P(x, f(x))$ and the point Q by $Q(x + h, f(x + h))$, where h is some appropriate nonzero number (Figure 52a). Observe that we can make Q approach P along the curve C by letting h approach zero (Figure 52b).

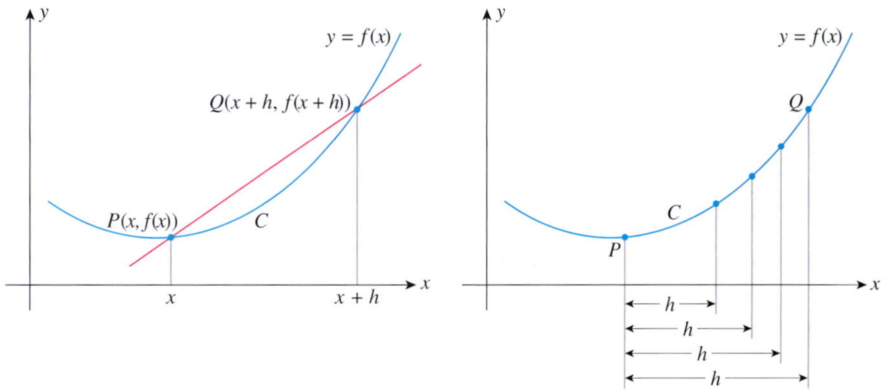

(a) The points $P(x, f(x))$ and $Q(x + h, f(x + h))$

(b) As h approaches zero, Q approaches P.

FIGURE 52

Next, using the formula for the slope of a line, we can write the slope of the secant line passing through $P(x, f(x))$ and $Q(x + h, f(x + h))$ as

$$\frac{f(x + h) - f(x)}{(x + h) - x} = \frac{f(x + h) - f(x)}{h} \tag{5}$$

As observed earlier, Q approaches P, and therefore the secant line through P and Q approaches the tangent line T as h approaches zero. Consequently, we might expect that the slope of the secant line would approach the slope of the tangent line T as h approaches zero. This leads to the following definition.

Slope of a Tangent Line

The slope of the tangent line to the graph of f at the point $P(x, f(x))$ is given by

$$\lim_{h \to 0} \frac{f(x + h) - f(x)}{h} \tag{6}$$

if it exists.

Rates of Change

We now show that the problem of finding the slope of the tangent line to the graph of a function f at the point $P(x, f(x))$ is mathematically equivalent to the problem of finding the rate of change of f at x. To see this, suppose we are given a function f that describes the relationship between the two quantities x and y—that is, $y = f(x)$. The number $f(x + h) - f(x)$ measures the change in y that corresponds to a change h in x (Figure 53).

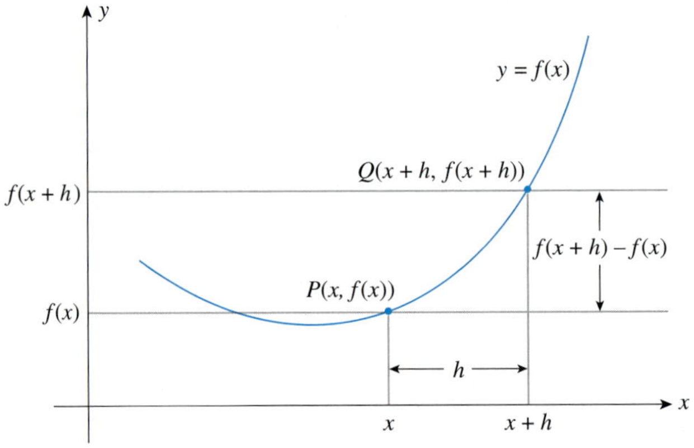

FIGURE 53
$f(x + h) - f(x)$ is the change in y that corresponds to a change h in x.

Then, the **difference quotient**

$$\frac{f(x + h) - f(x)}{h} \tag{7}$$

measures the **average rate of change of y with respect to x** over the interval $[x, x + h]$. For example, if y measures the position of a car at time x, then quotient (7) gives the average velocity of the car over the time interval $[x, x + h]$.

Observe that the difference quotient (7) is the same as (5). We conclude that the difference quotient (7) also measures the slope of the secant line that passes through the two points $P(x, f(x))$ and $Q(x + h, f(x + h))$ lying on the graph of $y = f(x)$. Next, by taking the limit of the difference quotient (7) as h goes to zero—that is, by evaluating

$$\lim_{h \to 0} \frac{f(x + h) - f(x)}{h} \tag{8}$$

we obtain the **rate of change of f at x.** For example, if y measures the position of a car at time x, then the limit (8) gives the velocity of the car at time x. For emphasis, the rate of change of a function f at x is often called the **instantaneous rate of change**

of f at x. This distinguishes it from the average rate of change of f, which is computed over an *interval* $[x, x + h]$ rather than at a *number* x.

Observe that the limit (8) is the same as (6). Therefore, the limit of the difference quotient also measures the slope of the tangent line to the graph of $y = f(x)$ at the point $(x, f(x))$. The following summarizes this discussion.

Average and Instantaneous Rates of Change

The **average rate of change** of f over the interval $[x, x + h]$ or **slope of the secant line** to the graph of f through the points $(x, f(x))$ and $(x + h, f(x + h))$ is

$$\frac{f(x+h) - f(x)}{h} \tag{9}$$

The **instantaneous rate of change** of f at x or **slope of the tangent line** to the graph of f at $(x, f(x))$ is

$$\lim_{h \to 0} \frac{f(x+h) - f(x)}{h} \tag{10}$$

Explore & Discuss

Explain the difference between the average rate of change of a function and the instantaneous rate of change of a function.

The Derivative

The limit (6) or (10), which measures both the slope of the tangent line to the graph of $y = f(x)$ at the point $P(x, f(x))$ and the (instantaneous) rate of change of f at x, is given a special name: the **derivative of f at x.**

Derivative of a Function

The derivative of a function f with respect to x is the function f' (read "f prime"),

$$f'(x) = \lim_{h \to 0} \frac{f(x+h) - f(x)}{h} \tag{11}$$

The domain of f' is the set of all x where the limit exists.

Thus, the derivative of a function f is a function f' that gives the slope of the tangent line to the graph of f at *any* point $(x, f(x))$ and also the rate of change of f at x (Figure 54).

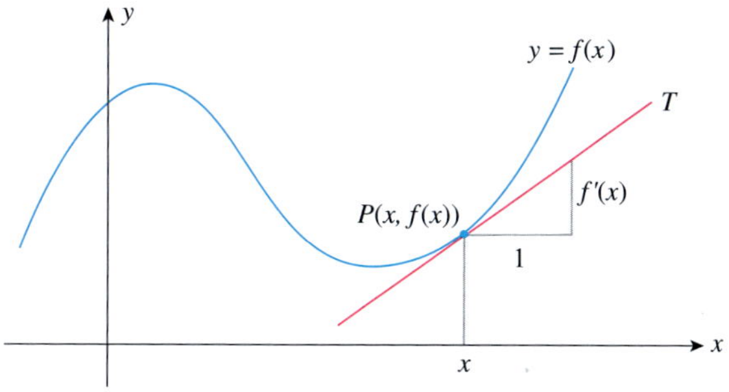

FIGURE 54
The slope of the tangent line at $P(x, f(x))$ is $f'(x)$; f changes at the rate of $f'(x)$ units per unit change in x at x.

Other notations for the derivative of f include:

$D_x f(x)$ Read "d sub x of f of x"

$\dfrac{dy}{dx}$ Read "d y d x"

y' Read "y prime"

The last two are used when the rule for f is written in the form $y = f(x)$.

The calculation of the derivative of f is facilitated using the following four-step process.

Four-Step Process for Finding $f'(x)$

1. Compute $f(x + h)$.
2. Form the difference $f(x + h) - f(x)$.
3. Form the quotient $\dfrac{f(x + h) - f(x)}{h}$.
4. Compute $f'(x) = \lim\limits_{h \to 0} \dfrac{f(x + h) - f(x)}{h}$.

EXAMPLE 2 Find the slope of the tangent line to the graph of $f(x) = 3x + 5$ at any point $(x, f(x))$.

Solution The slope of the tangent line at any point on the graph of f is given by the derivative of f at x. To find the derivative, we use the four-step process:

Step 1 $f(x + h) = 3(x + h) + 5 = 3x + 3h + 5$

Step 2 $f(x + h) - f(x) = (3x + 3h + 5) - (3x + 5) = 3h$

Step 3 $\dfrac{f(x + h) - f(x)}{h} = \dfrac{3h}{h} = 3$

Step 4 $f'(x) = \lim\limits_{h \to 0} \dfrac{f(x + h) - f(x)}{h} = \lim\limits_{h \to 0} 3 = 3$

We expect this result since the tangent line to any point on a straight line must coincide with the line itself and therefore must have the same slope as the line. In this case, the graph of f is a straight line with slope 3. ■

EXAMPLE 3 Let $f(x) = x^2$.

a. Find $f'(x)$.
b. Compute $f'(2)$ and interpret your result.

Solution

a. To find $f'(x)$, we use the four-step process:

Step 1 $f(x + h) = (x + h)^2 = x^2 + 2xh + h^2$

Step 2 $f(x + h) - f(x) = x^2 + 2xh + h^2 - x^2 = 2xh + h^2 = h(2x + h)$

Step 3 $\dfrac{f(x + h) - f(x)}{h} = \dfrac{h(2x + h)}{h} = 2x + h$

Step 4 $f'(x) = \lim\limits_{h \to 0} \dfrac{f(x + h) - f(x)}{h} = \lim\limits_{h \to 0} (2x + h) = 2x$

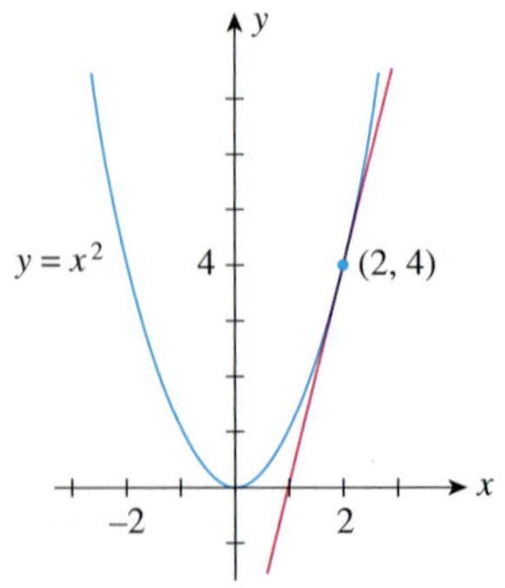

FIGURE 55
The tangent line to the graph of $f(x) = x^2$ at $(2, 4)$

b. $f'(2) = 2(2) = 4$. This result tells us that the slope of the tangent line to the graph of f at the point $(2, 4)$ is 4. It also tells us that the function f is changing at the rate of 4 units per unit change in x at $x = 2$. The graph of f and the tangent line at $(2, 4)$ are shown in Figure 55. ■

Exploring with TECHNOLOGY

1. Consider the function $f(x) = x^2$ of Example 3. Suppose we want to compute $f'(2)$, using Equation (11). Thus,

$$f'(2) = \lim_{h \to 0} \frac{f(2+h) - f(2)}{h} = \lim_{h \to 0} \frac{(2+h)^2 - 2^2}{h}$$

Use a graphing utility to plot the graph of

$$g(x) = \frac{(2+x)^2 - 4}{x}$$

in the viewing window $[-3, 3] \times [-2, 6]$.

2. Use **ZOOM** and **TRACE** to find $\lim_{x \to 0} g(x)$.
3. Explain why the limit found in part 2 is $f'(2)$.

VIDEO

EXAMPLE 4 Let $f(x) = x^2 - 4x$.

a. Find $f'(x)$.
b. Find the point on the graph of f where the tangent line to the curve is horizontal.
c. Sketch the graph of f and the tangent line to the curve at the point found in part (b).
d. What is the rate of change of f at this point?

Solution

a. To find $f'(x)$, we use the four-step process:

Step 1 $f(x+h) = (x+h)^2 - 4(x+h) = x^2 + 2xh + h^2 - 4x - 4h$

Step 2 $f(x+h) - f(x) = x^2 + 2xh + h^2 - 4x - 4h - (x^2 - 4x)$
$= 2xh + h^2 - 4h = h(2x + h - 4)$

Step 3 $\dfrac{f(x+h) - f(x)}{h} = \dfrac{h(2x + h - 4)}{h} = 2x + h - 4$

Step 4 $f'(x) = \lim_{h \to 0} \dfrac{f(x+h) - f(x)}{h} = \lim_{h \to 0} (2x + h - 4) = 2x - 4$

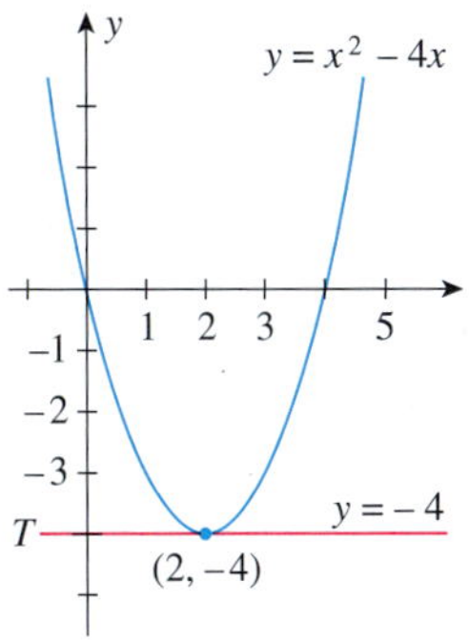

FIGURE 56
The tangent line to the graph of $y = x^2 - 4x$ at $(2, -4)$ is $y = -4$.

b. At a point on the graph of f where the tangent line to the curve is horizontal and hence has slope zero, the derivative f' of f is zero. Accordingly, to find such point(s) we set $f'(x) = 0$, which gives $2x - 4 = 0$, or $x = 2$. The corresponding value of y is given by $y = f(2) = -4$, and the required point is $(2, -4)$.
c. The graph of f and the tangent line are shown in Figure 56.
d. The rate of change of f at $x = 2$ is zero. ■

Explore & Discuss

Can the tangent line to the graph of a function intersect the graph at more than one point? Explain your answer using illustrations.

EXAMPLE 5 Let $f(x) = \dfrac{1}{x}$.

a. Find $f'(x)$.
b. Find the slope of the tangent line T to the graph of f at the point where $x = 1$.
c. Find an equation of the tangent line T in part (b).

Solution

a. To find $f'(x)$, we use the four-step process:

Step 1 $$f(x+h) = \frac{1}{x+h}$$

Step 2 $$f(x+h) - f(x) = \frac{1}{x+h} - \frac{1}{x} = \frac{x-(x+h)}{x(x+h)} = -\frac{h}{x(x+h)}$$

Step 3 $$\frac{f(x+h)-f(x)}{h} = -\frac{h}{x(x+h)} \cdot \frac{1}{h} = -\frac{1}{x(x+h)}$$

See page 17.

Step 4 $$f'(x) = \lim_{h \to 0} \frac{f(x+h)-f(x)}{h} = \lim_{h \to 0} -\frac{1}{x(x+h)} = -\frac{1}{x^2}$$

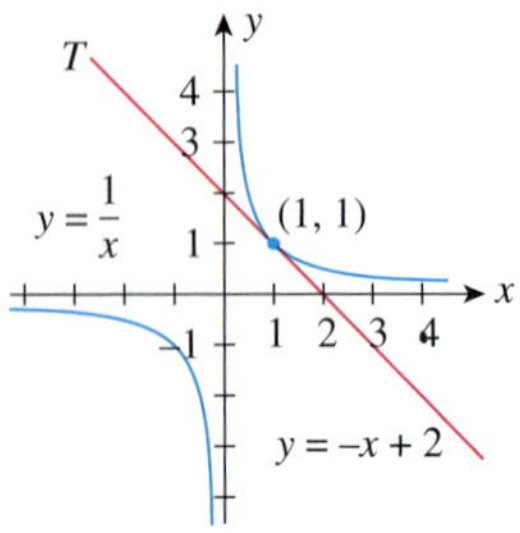

FIGURE 57
The tangent line to the graph of $f(x) = 1/x$ at $(1, 1)$

b. The slope of the tangent line T to the graph of f where $x = 1$ is given by $f'(1) = -1$.

c. When $x = 1$, $y = f(1) = 1$ and T is tangent to the graph of f at the point $(1, 1)$. From part (b), we know that the slope of T is -1. Thus, an equation of T is

$$y - 1 = -1(x - 1)$$
$$y = -x + 2$$

(Figure 57).

Exploring with TECHNOLOGY

1. Use the results of Example 5 to draw the graph of $f(x) = 1/x$ and its tangent line at the point $(1, 1)$ by plotting the graphs of $y_1 = 1/x$ and $y_2 = -x + 2$ in the viewing window $[-4, 4] \times [-4, 4]$.
2. Some graphing utilities draw the tangent line to the graph of a function at a given point automatically—you need only specify the function and give the x-coordinate of the point of tangency. If your graphing utility has this feature, verify the result of part 1 without finding an equation of the tangent line.

Explore & Discuss

Consider the following alternative approach to the definition of the derivative of a function: Let h be a positive number and suppose $P(x - h, f(x - h))$ and $Q(x + h, f(x + h))$ are two points on the graph of f.

1. Give a geometric and a physical interpretation of the quotient
$$\frac{f(x+h)-f(x-h)}{2h}$$
Make a sketch to illustrate your answer.
2. Give a geometric and a physical interpretation of the limit
$$\lim_{h \to 0} \frac{f(x+h)-f(x-h)}{2h}$$
Make a sketch to illustrate your answer.
3. Explain why it makes sense to define
$$f'(x) = \lim_{h \to 0} \frac{f(x+h)-f(x-h)}{2h}$$
4. Using the definition given in part 3, formulate a four-step process for finding $f'(x)$ similar to that given on page 138 and use it to find the derivative of $f(x) = x^2$. Compare your answer with that obtained in Example 3 on page 138.

APPLIED EXAMPLE 6 Average Velocity of a Car Suppose the distance (in feet) covered by a car moving along a straight road t seconds after starting from rest is given by the function $f(t) = 2t^2$ $(0 \le t \le 30)$.

a. Calculate the average velocity of the car over the time intervals [22, 23], [22, 22.1], and [22, 22.01].
b. Calculate the (instantaneous) velocity of the car when $t = 22$.
c. Compare the results obtained in part (a) with that obtained in part (b).

Solution

a. We first compute the average velocity (average rate of change of f) over the interval $[t, t + h]$ using Formula (9). We find

$$\frac{f(t+h) - f(t)}{h} = \frac{2(t+h)^2 - 2t^2}{h} = \frac{2t^2 + 4th + 2h^2 - 2t^2}{h} = 4t + 2h$$

Next, using $t = 22$ and $h = 1$, we find that the average velocity of the car over the time interval [22, 23] is

$$4(22) + 2(1) = 90$$

or 90 feet per second. Similarly, using $t = 22$, $h = 0.1$, and $h = 0.01$, we find that its average velocities over the time intervals [22, 22.1] and [22, 22.01] are 88.2 and 88.02 feet per second, respectively.

b. Using the limit (10), we see that the instantaneous velocity of the car at any time t is given by

$$\lim_{h \to 0} \frac{f(t+h) - f(t)}{h} = \lim_{h \to 0} (4t + 2h) = 4t$$ Use the results from part (a).

In particular, the velocity of the car 22 seconds from rest ($t = 22$) is given by

$$v = 4(22)$$

or 88 feet per second.

c. The computations in part (a) show that, as the time intervals over which the average velocity of the car are computed become smaller and smaller, the average velocities over these intervals do approach 88 feet per second, the instantaneous velocity of the car at $t = 22$. ■

APPLIED EXAMPLE 7 Demand for Tires The management of Titan Tire Company has determined that the weekly demand function of their Super Titan tires is given by

$$p = f(x) = 144 - x^2$$

where p is measured in dollars and x is measured in units of a thousand (Figure 58).

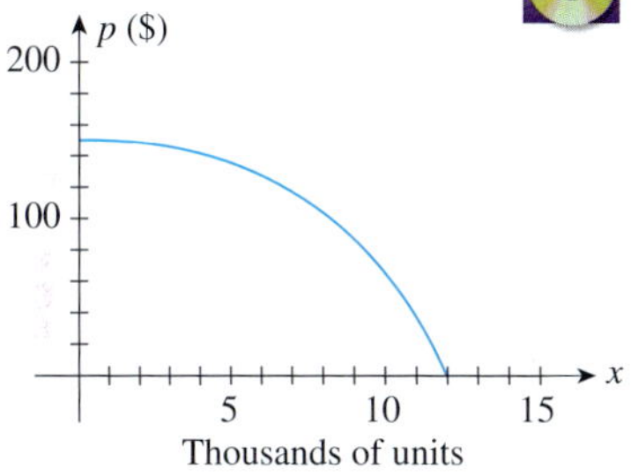

FIGURE 58
The graph of the demand function $p = 144 - x^2$

a. Find the average rate of change in the unit price of a tire if the quantity demanded is between 5000 and 6000 tires, between 5000 and 5100 tires, and between 5000 and 5010 tires.
b. What is the instantaneous rate of change of the unit price when the quantity demanded is 5000 units?

Solution

a. The average rate of change of the unit price of a tire if the quantity demanded is between x and $x + h$ is

$$\frac{f(x+h) - f(x)}{h} = \frac{[144 - (x+h)^2] - (144 - x^2)}{h}$$
$$= \frac{144 - x^2 - 2xh - h^2 - 144 + x^2}{h}$$
$$= -2x - h$$

To find the average rate of change of the unit price of a tire when the quantity demanded is between 5000 and 6000 tires (that is, over the interval [5, 6]), we take $x = 5$ and $h = 1$, obtaining

$$-2(5) - 1 = -11$$

or $-\$11$ per 1000 tires. (Remember, x is measured in units of a thousand.) Similarly, taking $h = 0.1$ and $h = 0.01$ with $x = 5$, we find that the average rates of change of the unit price when the quantities demanded are between 5000 and 5100 and between 5000 and 5010 are $-\$10.10$ and $-\$10.01$ per 1000 tires, respectively.

b. The instantaneous rate of change of the unit price of a tire when the quantity demanded is x units is given by

$$\lim_{h\to 0} \frac{f(x+h) - f(x)}{h} = \lim_{h\to 0} (-2x - h) \quad \text{Use the results from part (a).}$$
$$= -2x$$

In particular, the instantaneous rate of change of the unit price per tire when the quantity demanded is 5000 is given by $-2(5)$, or $-\$10$ per 1000 tires. ■

The derivative of a function provides us with a tool for measuring the rate of change of one quantity with respect to another. Table 4 lists several other applications involving this limit.

TABLE 4

Applications Involving Rate of Change

x Stands for	y Stands for	$\frac{f(a+h) - f(a)}{h}$ Measures	$\lim_{h\to 0} \frac{f(a+h) - f(a)}{h}$ Measures
Time	**Concentration of a drug** in the bloodstream at time x	Average rate of change in the concentration of the drug over the time interval $[a, a+h]$	Instantaneous rate of change in the concentration of the drug in the bloodstream at time $x = a$
Number of items sold	**Revenue** at a sales level of x units	Average rate of change in the revenue when the sales level is between $x = a$ and $x = a + h$	Instantaneous rate of change in the revenue when the sales level is a units
Time	**Volume of sales** at time x	Average rate of change in the volume of sales over the time interval $[a, a+h]$	Instantaneous rate of change in the volume of sales at time $x = a$
Time	**Population** of *Drosophila* (fruit flies) at time x	Average rate of growth of the fruit fly population over the time interval $[a, a+h]$	Instantaneous rate of change of the fruit fly population at time $x = a$
Temperature in a chemical reaction	**Amount of product formed in the chemical reaction** when the temperature is x degrees	Average rate of formation of chemical product over the temperature range $[a, a+h]$	Instantaneous rate of formation of chemical product when the temperature is a degrees

Differentiability and Continuity

In practical applications, one encounters continuous functions that fail to be **differentiable**—that is, do not have a derivative—at certain values in the domain of the function f. It can be shown that a continuous function f fails to be differentiable at $x = a$ when the graph of f makes an abrupt change of direction at $(a, f(a))$. We call such a point a "corner." A function also fails to be differentiable at a point where the tangent line is vertical since the slope of a vertical line is undefined. These cases are illustrated in Figure 59.

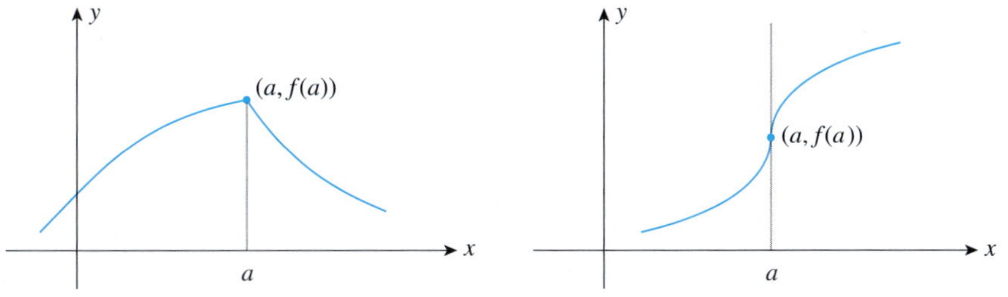

(a) The graph makes an abrupt change of direction at $x = a$.

(b) The slope at $x = a$ is undefined.

FIGURE **59**

The next example illustrates a function that is not differentiable at a point.

APPLIED EXAMPLE 8 Wages Mary works at the B&O department store, where, on a weekday, she is paid \$8 an hour for the first 8 hours and \$12 an hour for overtime. The function

$$f(x) = \begin{cases} 8x & \text{if } 0 \le x \le 8 \\ 12x - 32 & \text{if } 8 < x \end{cases}$$

gives Mary's earnings on a weekday in which she worked x hours. Sketch the graph of the function f and explain why it is not differentiable at $x = 8$.

Solution The graph of f is shown in Figure 60. Observe that the graph of f has a corner at $x = 8$ and consequently is not differentiable at $x = 8$. ■

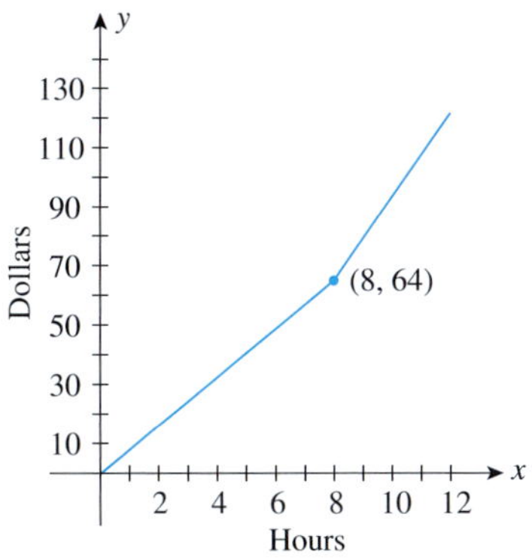

FIGURE **60**
The function f is not differentiable at (8, 64).

We close this section by mentioning the connection between the continuity and the differentiability of a function at a given value $x = a$ in the domain of f. By reexamining the function of Example 8, it becomes clear that f is continuous everywhere and, in particular, when $x = 8$. This shows that in general the continuity of a function at $x = a$ does not necessarily imply the differentiability of the function at that number. The converse, however, is true: If a function f is differentiable at $x = a$, then it is continuous there.

Differentiability and Continuity

If a function is differentiable at $x = a$, then it is continuous at $x = a$.

For a proof of this result, see Exercise 62, page 149.

Explore & Discuss

Suppose a function f is differentiable at $x = a$. Can there be two tangent lines to the graphs of f at the point $(a, f(a))$? Explain your answer.

Exploring with TECHNOLOGY

1. Use a graphing utility to plot the graph of $f(x) = x^{1/3}$ in the viewing window $[-2, 2] \times [-2, 2]$.
2. Use a graphing utility to draw the tangent line to the graph of f at the point (0, 0). Can you explain why the process breaks down?

EXAMPLE 9 Figure 61 depicts a portion of the graph of a function. Explain why the function fails to be differentiable at each of the numbers $x = a, b, c, d, e, f,$ and g.

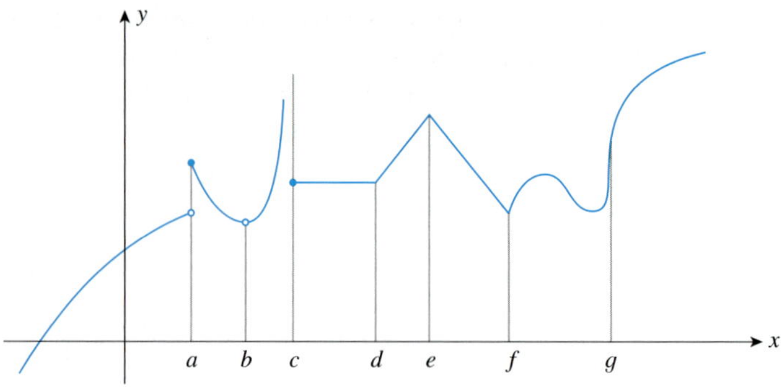

FIGURE 61
The graph of this function is not differentiable at the numbers a–g.

Solution The function fails to be differentiable at $x = a$, b, and c because it is discontinuous at each of these numbers. The derivative of the function does not exist at $x = d$, e, and f because it has a kink at each point on the graph corresponding to these numbers. Finally, the function is not differentiable at $x = g$ because the tangent line is vertical at $(g, f(g))$. ■

2.6 Self-Check Exercises

1. Let $f(x) = -x^2 - 2x + 3$.
   a. Find the derivative f' of f, using the definition of the derivative.
   b. Find the slope of the tangent line to the graph of f at the point (0, 3).
   c. Find the rate of change of f when $x = 0$.
   d. Find an equation of the tangent line to the graph of f at the point (0, 3).
   e. Sketch the graph of f and the tangent line to the curve at the point (0, 3).

2. The losses (in millions of dollars) due to bad loans extended chiefly in agriculture, real estate, shipping, and energy by the Franklin Bank are estimated to be

$$A = f(t) = -t^2 + 10t + 30 \qquad (0 \leq t \leq 10)$$

where t is the time in years ($t = 0$ corresponds to the beginning of 2003). How fast were the losses mounting at the beginning of 2006? At the beginning of 2008? At the beginning of 2010?

Solutions to Self-Check Exercises 2.6 can be found on page 149.

2.6 Concept Questions

For Questions 1 and 2, refer to the following figure.

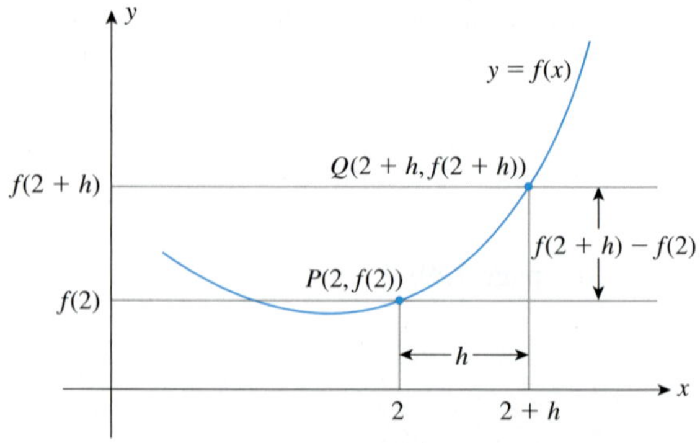

1. Let $P(2, f(2))$ and $Q(2 + h, f(2 + h))$ be points on the graph of a function f.
   a. Find an expression for the slope of the secant line passing through P and Q.
   b. Find an expression for the slope of the tangent line passing through P.

2. Refer to Question 1.
   a. Find an expression for the average rate of change of f over the interval $[2, 2 + h]$.
   b. Find an expression for the instantaneous rate of change of f at 2.
   c. Compare your answers for part (a) and (b) with those of Question 1.

3. a. Give a geometric and a physical interpretation of the expression

$$\frac{f(x + h) - f(x)}{h}$$

   b. Give a geometric and a physical interpretation of the expression

$$\lim_{h \to 0} \frac{f(x + h) - f(x)}{h}$$

4. Under what conditions does a function fail to have a derivative at a number? Illustrate your answer with sketches.

2.6 Exercises

1. **Average Weight of an Infant** The following graph shows the weight measurements of the average infant from the time of birth ($t = 0$) through age 2 ($t = 24$). By computing the slopes of the respective tangent lines, estimate the rate of change of the average infant's weight when $t = 3$ and when $t = 18$. What is the average rate of change in the average infant's weight over the first year of life?

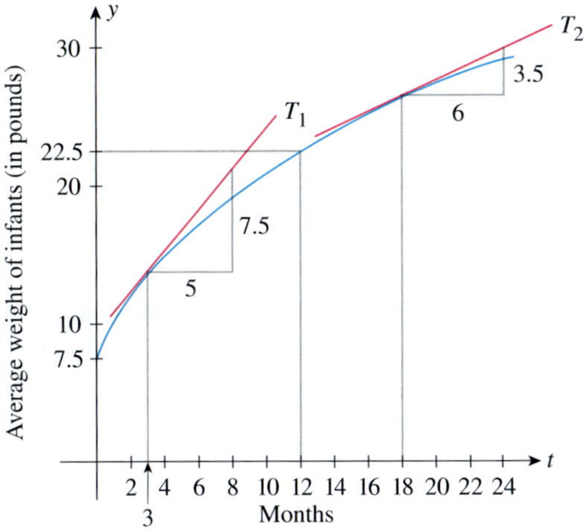

2. **Forestry** The following graph shows the volume of wood produced in a single-species forest. Here $f(t)$ is measured in cubic meters/hectare and t is measured in years. By computing the slopes of the respective tangent lines, estimate the rate at which the wood grown is changing at the beginning of year 10 and at the beginning of year 30.

Source: The Random House Encyclopedia

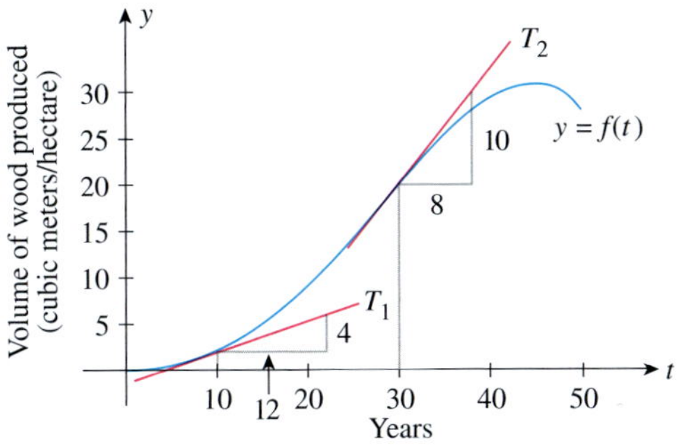

3. **TV-Viewing Patterns** The following graph shows the percent of U.S. households watching television during a 24-hr period on a weekday ($t = 0$ corresponds to 6 a.m.). By computing the slopes of the respective tangent lines, estimate the rate of change of the percent of households watching television at 4 p.m. and 11 p.m.

Source: A. C. Nielsen Company

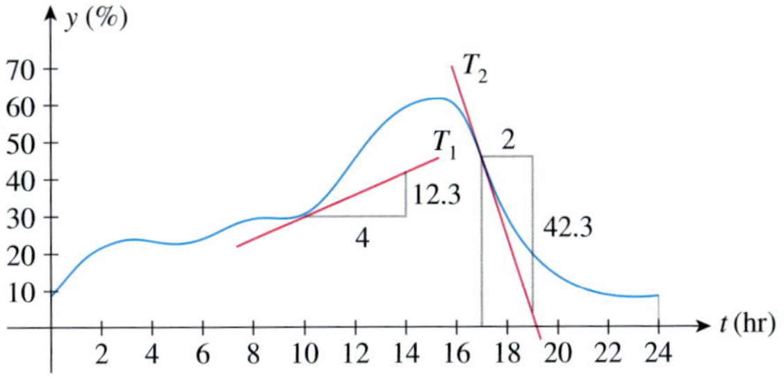

4. **Crop Yield** Productivity and yield of cultivated crops are often reduced by insect pests. The following graph shows the relationship between the yield of a certain crop, $f(x)$, as a function of the density of aphids x. (Aphids are small insects that suck plant juices.) Here, $f(x)$ is measured in kilograms/4000 square meters, and x is measured in hundreds of aphids/bean stem. By computing the slopes of the respective tangent lines, estimate the rate of change of the crop yield with respect to the density of aphids when that density is 200 aphids/bean stem and when it is 800 aphids/bean stem.

Source: The Random House Encyclopedia

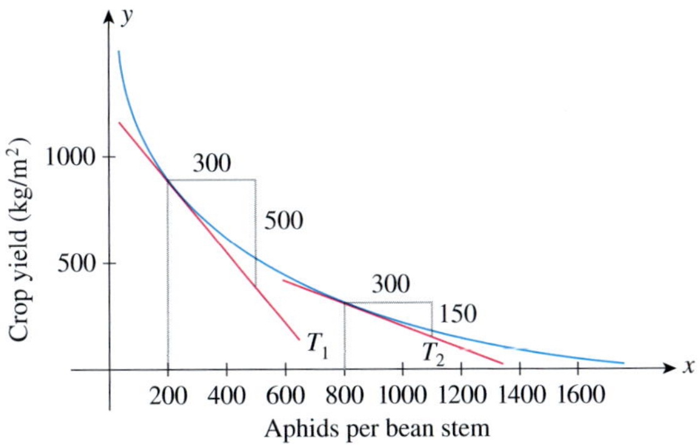

5. The position of car A and car B, starting out side by side and traveling along a straight road, is given by $s = f(t)$ and $s = g(t)$, respectively, where s is measured in feet and t is measured in seconds (see the accompanying figure).

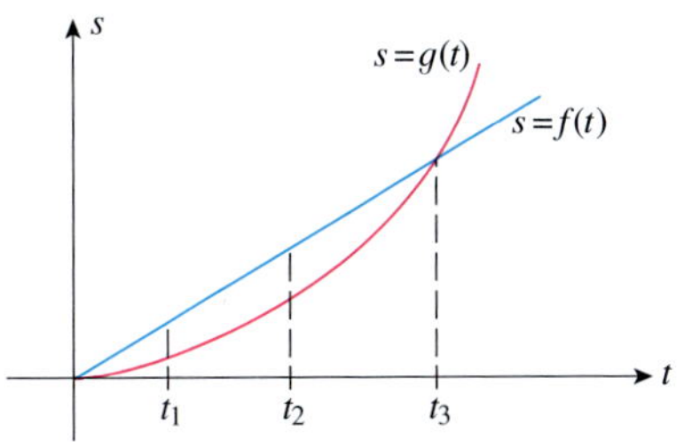

a. Which car is traveling faster at t_1?

b. What can you say about the speed of the cars at t_2?
Hint: Compare tangent lines.

c. Which car is traveling faster at t_3?

d. What can you say about the positions of the cars at t_3?

6. The velocity of car A and car B, starting out side by side and traveling along a straight road, is given by $v = f(t)$ and $v = g(t)$, respectively, where v is measured in feet/second and t is measured in seconds (see the accompanying figure).

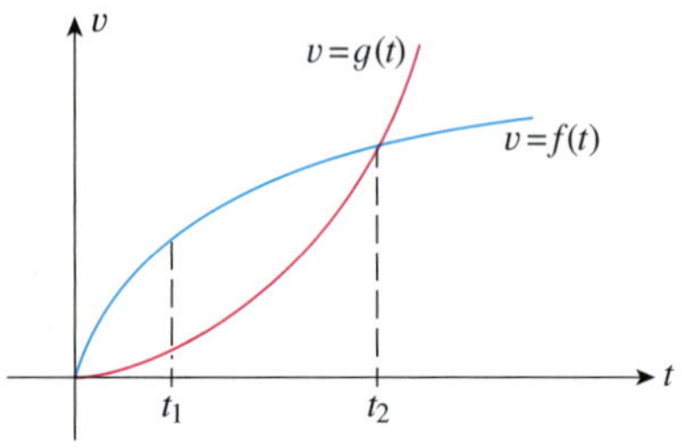

a. What can you say about the velocity and acceleration of the two cars at t_1? (Acceleration is the rate of change of velocity.)

b. What can you say about the velocity and acceleration of the two cars at t_2?

7. Effect of a Bactericide on Bacteria In the following figure, $f(t)$ gives the population P_1 of a certain bacteria culture at time t after a portion of bactericide A was introduced into the population at $t = 0$. The graph of g gives the population P_2 of a similar bacteria culture at time t after a portion of bactericide B was introduced into the population at $t = 0$.

a. Which population is decreasing faster at t_1?

b. Which population is decreasing faster at t_2?

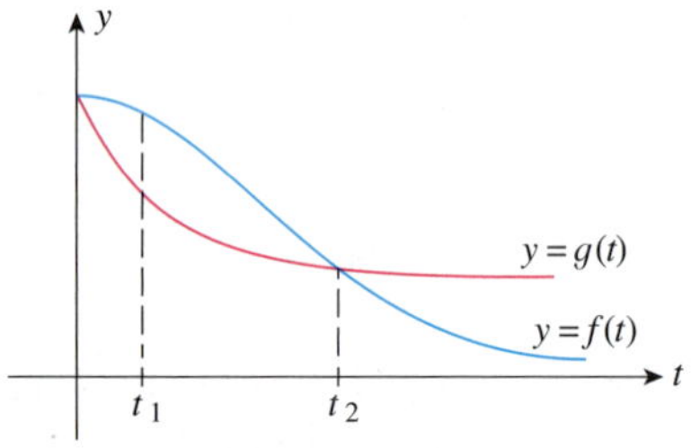

c. Which bactericide is more effective in reducing the population of bacteria in the short run? In the long run?

8. Market Share The following figure shows the devastating effect the opening of a new discount department store had on an established department store in a small town. The revenue of the discount store at time t (in months) is given by $f(t)$ million dollars, whereas the revenue of the established department store at time t is given by $g(t)$ million dollars. Answer the following questions by giving the value of t at which the specified event took place.

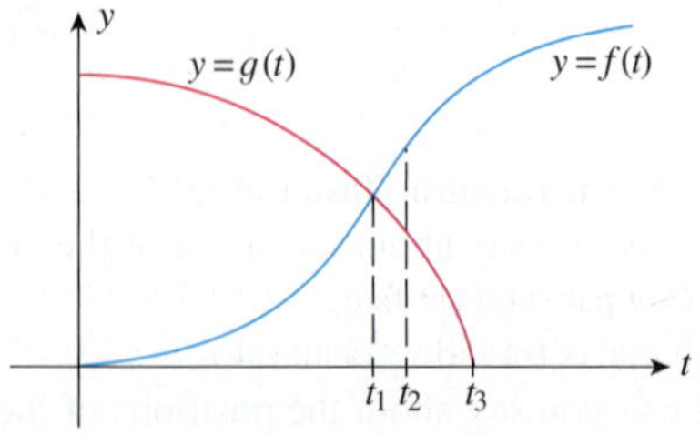

a. The revenue of the established department store is decreasing at the slowest rate.

b. The revenue of the established department store is decreasing at the fastest rate.

c. The revenue of the discount store first overtakes that of the established store.

d. The revenue of the discount store is increasing at the fastest rate.

In Exercises 9–16, use the four-step process to find the slope of the tangent line to the graph of the given function at any point.

9. $f(x) = 13$

10. $f(x) = -6$

11. $f(x) = 2x + 7$

12. $f(x) = 8 - 4x$

13. $f(x) = 3x^2$

14. $f(x) = -\frac{1}{2}x^2$

15. $f(x) = -x^2 + 3x$

16. $f(x) = 2x^2 + 5x$

In Exercises 17–22, find the slope of the tangent line to the graph of each function at the given point and determine an equation of the tangent line.

17. $f(x) = 2x + 7$ at $(2, 11)$

18. $f(x) = -3x + 4$ at $(-1, 7)$

19. $f(x) = 3x^2$ at $(1, 3)$

20. $f(x) = 3x - x^2$ at $(-2, -10)$

21. $f(x) = -\dfrac{1}{x}$ at $\left(3, -\dfrac{1}{3}\right)$

22. $f(x) = \dfrac{3}{2x}$ at $\left(1, \dfrac{3}{2}\right)$

23. Let $f(x) = 2x^2 + 1$.

a. Find the derivative f' of f.

b. Find an equation of the tangent line to the curve at the point $(1, 3)$.

c. Sketch the graph of f.

24. Let $f(x) = x^2 + 6x$.

a. Find the derivative f' of f.

b. Find the point on the graph of f where the tangent line to the curve is horizontal.
Hint: Find the value of x for which $f'(x) = 0$.

c. Sketch the graph of f and the tangent line to the curve at the point found in part (b).

25. Let $f(x) = x^2 - 2x + 1$.

a. Find the derivative f' of f.

b. Find the point on the graph of f where the tangent line to the curve is horizontal.

c. Sketch the graph of f and the tangent line to the curve at the point found in part (b).

d. What is the rate of change of f at this point?

26. Let $f(x) = \dfrac{1}{x - 1}$.

a. Find the derivative f' of f.

b. Find an equation of the tangent line to the curve at the point $(-1, -\frac{1}{2})$.

c. Sketch the graph of f and the tangent line to the curve at $(-1, -\frac{1}{2})$.

27. Let $y = f(x) = x^2 + x$.

a. Find the average rate of change of y with respect to x in the interval from $x = 2$ to $x = 3$, from $x = 2$ to $x = 2.5$, and from $x = 2$ to $x = 2.1$.

b. Find the (instantaneous) rate of change of y at $x = 2$.

c. Compare the results obtained in part (a) with that of part (b).

28. Let $y = f(x) = x^2 - 4x$.

a. Find the average rate of change of y with respect to x in the interval from $x = 3$ to $x = 4$, from $x = 3$ to $x = 3.5$, and from $x = 3$ to $x = 3.1$.

b. Find the (instantaneous) rate of change of y at $x = 3$.

c. Compare the results obtained in part (a) with that of part (b).

29. Velocity of a Car Suppose the distance s (in feet) covered by a car moving along a straight road after t sec is given by the function $s = f(t) = 2t^2 + 48t$.

a. Calculate the average velocity of the car over the time intervals [20, 21], [20, 20.1], and [20, 20.01].

b. Calculate the (instantaneous) velocity of the car when $t = 20$.

c. Compare the results of part (a) with that of part (b).

30. Velocity of a Ball Thrown into the Air A ball is thrown straight up with an initial velocity of 128 ft/sec, so that its height (in feet) after t sec is given by $s(t) = 128t - 16t^2$.

a. What is the average velocity of the ball over the time intervals [2, 3], [2, 2.5], and [2, 2.1]?

b. What is the instantaneous velocity at time $t = 2$?

c. What is the instantaneous velocity at time $t = 5$? Is the ball rising or falling at this time?

d. When will the ball hit the ground?

31. During the construction of a high-rise building, a worker accidentally dropped his portable electric screwdriver from a height of 400 ft. After t sec, the screwdriver had fallen a distance of $s = 16t^2$ ft.

a. How long did it take the screwdriver to reach the ground?

b. What was the average velocity of the screwdriver between the time it was dropped and the time it hit the ground?

c. What was the velocity of the screwdriver at the time it hit the ground?

32. A hot-air balloon rises vertically from the ground so that its height after t sec is $h = \frac{1}{2}t^2 + \frac{1}{2}t$ ft $(0 \le t \le 60)$.

a. What is the height of the balloon at the end of 40 sec?

b. What is the average velocity of the balloon between $t = 0$ and $t = 40$?

c. What is the velocity of the balloon at the end of 40 sec?

33. At a temperature of 20°C, the volume V (in liters) of 1.33 g of O_2 is related to its pressure p (in atmospheres) by the formula $V = 1/p$.

a. What is the average rate of change of V with respect to p as p increases from $p = 2$ to $p = 3$?

b. What is the rate of change of V with respect to p when $p = 2$?

34. Cost of Producing Surfboards The total cost $C(x)$ (in dollars) incurred by Aloha Company in manufacturing x surfboards a day is given by

$$C(x) = -10x^2 + 300x + 130 \qquad (0 \le x \le 15)$$

a. Find $C'(x)$.

b. What is the rate of change of the total cost when the level of production is ten surfboards a day?

35. Effect of Advertising on Profit The quarterly profit (in thousands of dollars) of Cunningham Realty is given by

$$P(x) = -\frac{1}{3}x^2 + 7x + 30 \qquad (0 \le x \le 50)$$

where x (in thousands of dollars) is the amount of money Cunningham spends on advertising per quarter.

a. Find $P'(x)$.

b. What is the rate of change of Cunningham's quarterly profit if the amount it spends on advertising is \$10,000/quarter ($x = 10$) and \$30,000/quarter ($x = 30$)?

36. Demand for Tents The demand function for Sportsman 5×7 tents is given by

$$p = f(x) = -0.1x^2 - x + 40$$

where p is measured in dollars and x is measured in units of a thousand.

a. Find the average rate of change in the unit price of a tent if the quantity demanded is between 5000 and 5050 tents; between 5000 and 5010 tents.

b. What is the rate of change of the unit price if the quantity demanded is 5000?

37. A Country's GDP The gross domestic product (GDP) of a certain country is projected to be

$$N(t) = t^2 + 2t + 50 \qquad (0 \le t \le 5)$$

billion dollars t yr from now. What will be the rate of change of the country's GDP 2 yr and 4 yr from now?

38. Growth of Bacteria Under a set of controlled laboratory conditions, the size of the population of a certain bacteria culture at time t (in minutes) is described by the function

$$P = f(t) = 3t^2 + 2t + 1$$

Find the rate of population growth at $t = 10$ min.

39. Air Temperature The air temperature at a height of h ft from the surface of the earth is $T = f(h)$ degrees Fahrenheit.

a. Give a physical interpretation of $f'(h)$. Give units.

b. Generally speaking, what do you expect the sign of $f'(h)$ to be?

c. If you know that $f'(1000) = -0.05$, estimate the change in the air temperature if the altitude changes from 1000 ft to 1001 ft.

40. **Revenue of a Travel Agency** Suppose that the total revenue realized by the Odyssey Travel Agency is $R = f(x)$ thousand dollars if x thousand dollars are spent on advertising.

a. What does

$$\frac{f(b) - f(a)}{b - a} \quad (0 < a < b)$$

measure? What are the units?

b. What does $f'(x)$ measure? Give units.

c. Given that $f'(20) = 3$, what is the approximate change in the revenue if Odyssey increases its advertising budget from \$20,000 to \$21,000?

In Exercises 41–46, let x and $f(x)$ represent the given quantities. Fix $x = a$ and let h be a small positive number. Give an interpretation of the quantities

$$\frac{f(a+h) - f(a)}{h} \quad \text{and} \quad \lim_{h \to 0} \frac{f(a+h) - f(a)}{h}$$

41. x denotes time and $f(x)$ denotes the population of seals at time x.

42. x denotes time and $f(x)$ denotes the prime interest rate at time x.

43. x denotes time and $f(x)$ denotes a country's industrial production.

44. x denotes the level of production of a certain commodity, and $f(x)$ denotes the total cost incurred in producing x units of the commodity.

45. x denotes altitude and $f(x)$ denotes atmospheric pressure.

46. x denotes the speed of a car (in mph), and $f(x)$ denotes the fuel economy of the car measured in miles per gallon (mpg).

In each of Exercises 47–52, the graph of a function is shown. For each function, state whether or not (a) $f(x)$ has a limit at $x = c$, (b) $f(x)$ is continuous at $x = a$, and (c) $f(x)$ is differentiable at $x = a$. Justify your answers.

47.

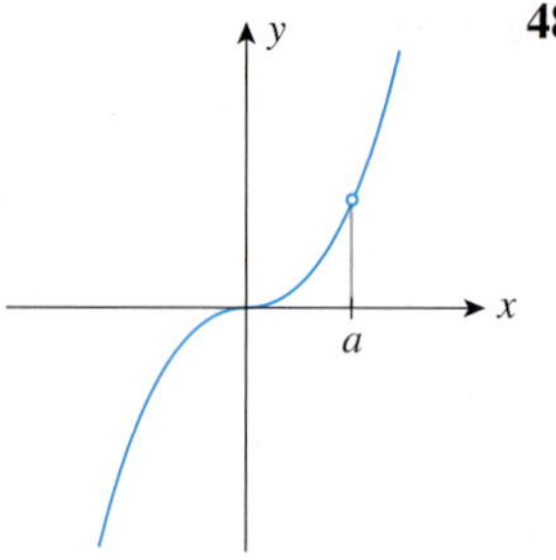

48.

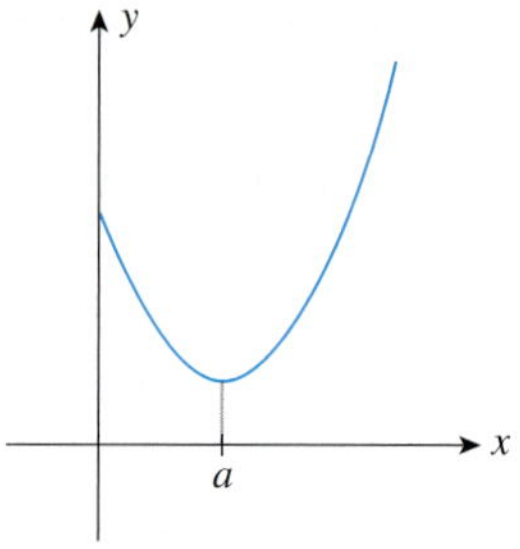

49.

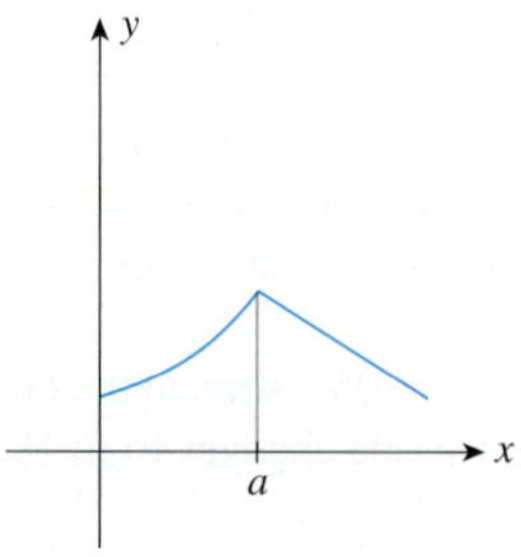

50.

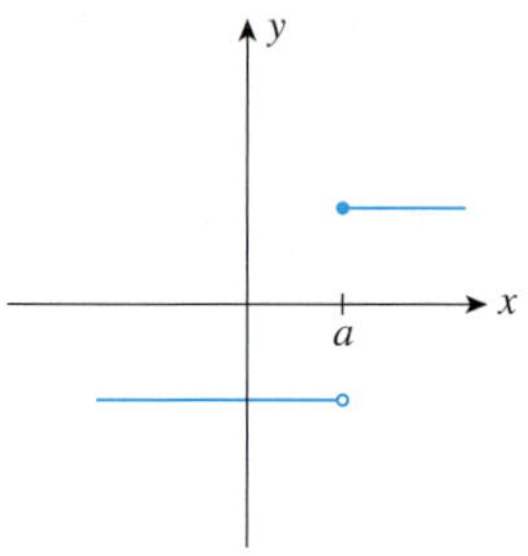

51.

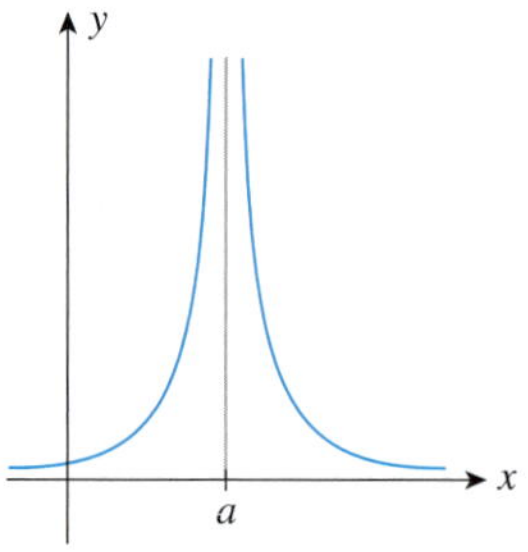

52.

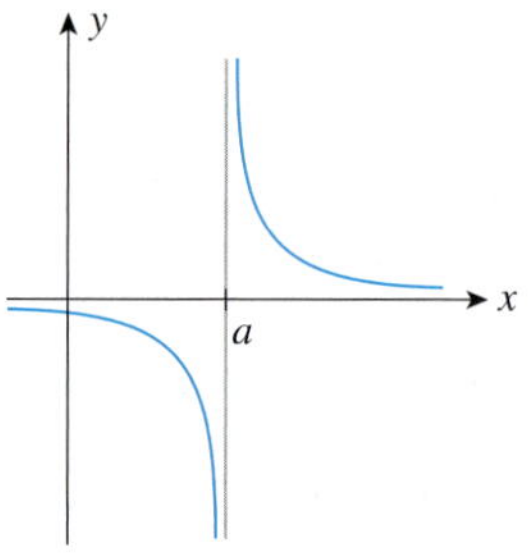

53. The distance s (in feet) covered by a motorcycle traveling in a straight line and starting from rest in t sec is given by the function

$$s(t) = -0.1t^3 + 2t^2 + 24t$$

Calculate the motorcycle's average velocity over the time interval $[2, 2 + h]$ for $h = 1, 0.1, 0.01, 0.001, 0.0001$, and 0.00001 and use your results to guess at the motorcycle's instantaneous velocity at $t = 2$.

54. The daily total cost $C(x)$ incurred by Trappee and Sons for producing x cases of TexaPep hot sauce is given by

$$C(x) = 0.000002x^3 + 5x + 400$$

Calculate

$$\frac{C(100 + h) - C(100)}{h}$$

for $h = 1, 0.1, 0.01, 0.001$, and 0.0001 and use your results to estimate the rate of change of the total cost function when the level of production is 100 cases/day.

In Exercises 55 and 56, determine whether the statement is true or false. If it is true, explain why it is true. If it is false, give an example to show why it is false.

55. If f is continuous at $x = a$, then f is differentiable at $x = a$.

56. If f is continuous at $x = a$ and g is differentiable at $x = a$, then $\lim_{x \to a} f(x)g(x) = f(a)g(a)$.

57. Sketch the graph of the function $f(x) = |x + 1|$ and show that the function does not have a derivative at $x = -1$.

58. Sketch the graph of the function $f(x) = 1/(x - 1)$ and show that the function does not have a derivative at $x = 1$.

59. Let

$$f(x) = \begin{cases} x^2 & \text{if } x \leq 1 \\ ax + b & \text{if } x > 1 \end{cases}$$

Find the values of a and b so that f is continuous and has a derivative at $x = 1$. Sketch the graph of f.

60. Sketch the graph of the function $f(x) = x^{2/3}$. Is the function continuous at $x = 0$? Does $f'(0)$ exist? Why or why not?

61. Prove that the derivative of the function $f(x) = |x|$ for $x \neq 0$ is given by

$$f'(x) = \begin{cases} 1 & \text{if } x > 0 \\ -1 & \text{if } x < 0 \end{cases}$$

Hint: Recall the definition of the absolute value of a number.

62. Show that if a function f is differentiable at $x = a$, then f must be continuous at that number.
Hint: Write

$$f(x) - f(a) = \left[\frac{f(x) - f(a)}{x - a}\right](x - a)$$

Use the product rule for limits and the definition of the derivative to show that

$$\lim_{x \to a} [f(x) - f(a)] = 0$$

2.6 Solutions to Self-Check Exercises

1. a. $f'(x)$

$$\begin{aligned}
&= \lim_{h \to 0} \frac{f(x + h) - f(x)}{h} \\
&= \lim_{h \to 0} \frac{[-(x + h)^2 - 2(x + h) + 3] - (-x^2 - 2x + 3)}{h} \\
&= \lim_{h \to 0} \frac{-x^2 - 2xh - h^2 - 2x - 2h + 3 + x^2 + 2x - 3}{h} \\
&= \lim_{h \to 0} \frac{h(-2x - h - 2)}{h} \\
&= \lim_{h \to 0} (-2x - h - 2) = -2x - 2
\end{aligned}$$

b. From the result of part (a), we see that the slope of the tangent line to the graph of f at any point $(x, f(x))$ is given by

$$f'(x) = -2x - 2$$

In particular, the slope of the tangent line to the graph of f at $(0, 3)$ is

$$f'(0) = -2$$

c. The rate of change of f when $x = 0$ is given by $f'(0) = -2$, or -2 units/unit change in x.

d. Using the result from part (b), we see that an equation of the required tangent line is

$$\begin{aligned} y - 3 &= -2(x - 0) \\ y &= -2x + 3 \end{aligned}$$

e.

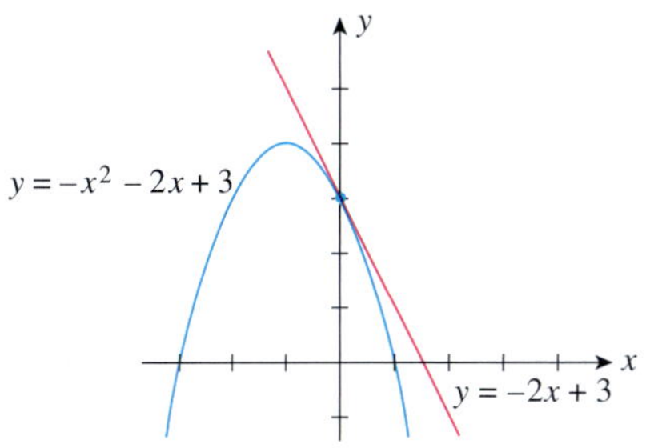

2. The rate of change of the losses at any time t is given by $f'(t)$

$$\begin{aligned}
&= \lim_{h \to 0} \frac{f(t + h) - f(t)}{h} \\
&= \lim_{h \to 0} \frac{[-(t + h)^2 + 10(t + h) + 30] - (-t^2 + 10t + 30)}{h} \\
&= \lim_{h \to 0} \frac{-t^2 - 2th - h^2 + 10t + 10h + 30 + t^2 - 10t - 30}{h} \\
&= \lim_{h \to 0} \frac{h(-2t - h + 10)}{h} \\
&= \lim_{h \to 0} (-2t - h + 10) \\
&= -2t + 10
\end{aligned}$$

Therefore, the rate of change of the losses suffered by the bank at the beginning of 2006 ($t = 3$) was

$$f'(3) = -2(3) + 10 = 4$$

In other words, the losses were increasing at the rate of \$4 million/year. At the beginning of 2008 ($t = 5$),

$$f'(5) = -2(5) + 10 = 0$$

and we see that the growth in losses due to bad loans was zero at this point. At the beginning of 2010 ($t = 7$),

$$f'(7) = -2(7) + 10 = -4$$

and we conclude that the losses were decreasing at the rate of \$4 million/year.

USING TECHNOLOGY

Graphing a Function and Its Tangent Line

We can use a graphing utility to plot the graph of a function f and the tangent line at any point on the graph.

EXAMPLE 1 Let $f(x) = x^2 - 4x$.

a. Find an equation of the tangent line to the graph of f at the point $(3, -3)$.
b. Plot both the graph of f and the tangent line found in part (a) on the same set of axes.

Solution

a. The slope of the tangent line at any point on the graph of f is given by $f'(x)$. But from Example 4 (page 139) we find $f'(x) = 2x - 4$. Using this result, we see that the slope of the required tangent line is

$$f'(3) = 2(3) - 4 = 2$$

Finally, using the point-slope form of the equation of a line, we find that an equation of the tangent line is

$$\begin{aligned} y - (-3) &= 2(x - 3) \\ y + 3 &= 2x - 6 \\ y &= 2x - 9 \end{aligned}$$

b. Figure T1a shows the graph of f in the standard viewing window and the tangent line of interest.

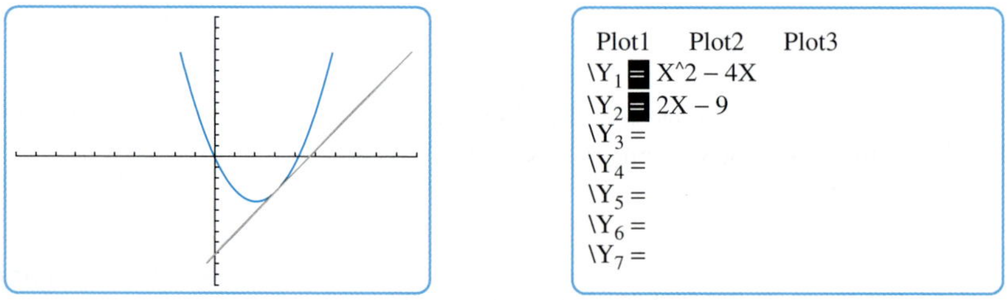

(a) (b)

FIGURE **T1**
(a) The graph of $f(x) = x^2 - 4x$ and the tangent line $y = 2x - 9$ in the standard viewing window; (b) the TI-83/84 equation screen.

Note Some graphing utilities will draw both the graph of a function f and the tangent line to the graph of f at a specified point when the function and the specified value of x are entered.

Finding the Derivative of a Function at a Given Point

The numerical derivative operation of a graphing utility can be used to give an approximate value of the derivative of a function for a given value of x.

EXAMPLE 2 Let $f(x) = \sqrt{x}$.

a. Use the numerical derivative operation of a graphing utility to find the derivative of f at $(4, 2)$.
b. Find an equation of the tangent line to the graph of f at $(4, 2)$.
c. Plot the graph of f and the tangent line on the same set of axes.

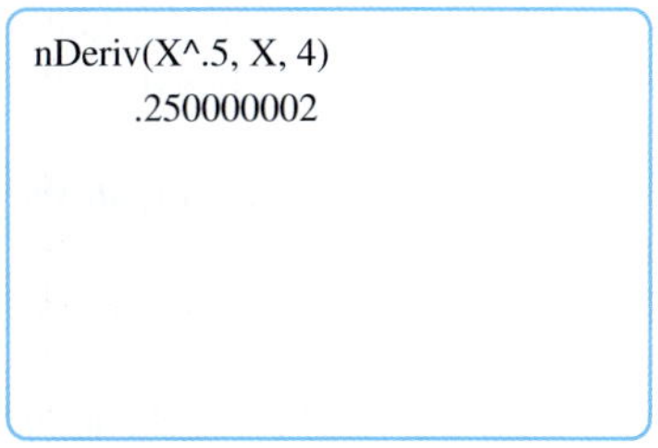

FIGURE T2
The TI-83/84 numerical derivative screen.

Solution

a. Using the numerical derivative operation of a graphing utility, we find that

$$f'(4) = \frac{1}{4}$$

(Figure T2).

b. An equation of the required tangent line is

$$y - 2 = \frac{1}{4}(x - 4)$$

$$y = \frac{1}{4}x + 1$$

c. Figure T3a shows the graph of f and the tangent line in the viewing window $[0, 15] \times [0, 4]$.

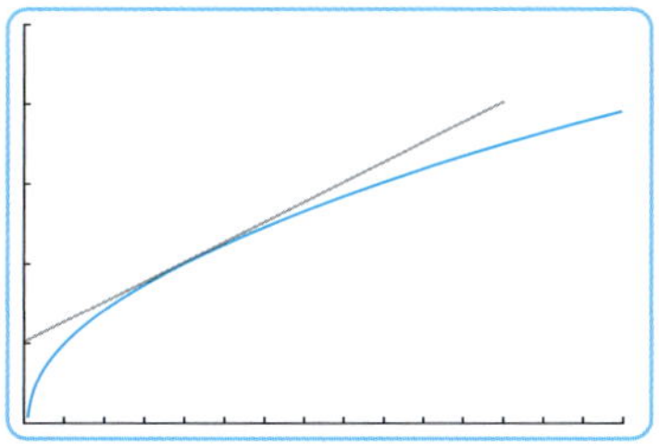

(a)

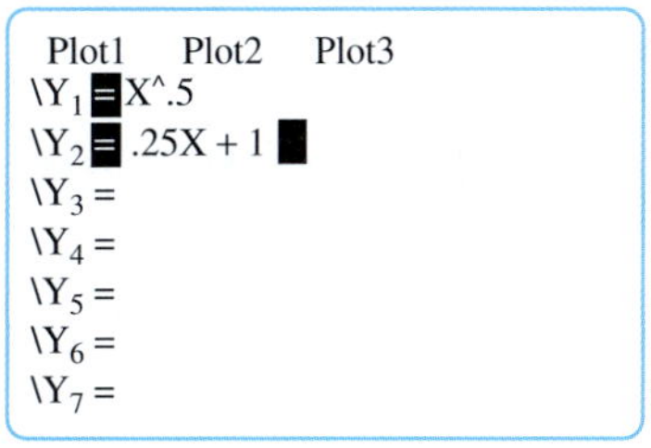

(b)

FIGURE T3
(a) The graph of $f(x) = \sqrt{x}$ and the tangent line $y = \frac{1}{4}x + 1$ in the viewing window $[0, 15] \times [0, 4]$; (b) the TI-83/84 equation screen

TECHNOLOGY EXERCISES

In Exercises 1–4, (a) find an equation of the tangent line to the graph of *f* at the indicated point and (b) plot the graph of *f* and the tangent line on the same set of axes. Use a suitable viewing window.

1. $f(x) = 2x^2 + x - 3$; $(2, 7)$

2. $f(x) = x + \frac{1}{x}$; $(1, 2)$

3. $f(x) = \sqrt{x}$; $(4, 2)$

4. $f(x) = \frac{1}{\sqrt{x}}$; $\left(4, \frac{1}{2}\right)$

In Exercises 5–8, (a) use the numerical derivative operation to find the derivative of *f* for the given value of *x* (to two desired places of accuracy), (b) find an equation of the tangent line to the graph of *f* at the indicated point, and (c) plot the graph of *f* and the tangent line on the same set of axes. Use a suitable viewing window.

5. $f(x) = x^3 + x + 1$; $x = 1$; $(1, 3)$

6. $f(x) = \frac{1}{x + 1}$; $x = 1$; $\left(1, \frac{1}{2}\right)$

7. $f(x) = x\sqrt{x^2 + 1}$; $x = 2$; $(2, 2\sqrt{5})$

8. $f(x) = \frac{x}{\sqrt{x^2 + 1}}$; $x = 1$; $\left(1, \frac{\sqrt{2}}{2}\right)$

9. Driving Costs The average cost of owning and operating a car in the United States from 1991 through 2001 is approximated by the function

$$C(t) = 0.06t^2 + 0.74t + 37.3 \qquad (0 \le t \le 11)$$

where $C(t)$ is measured in cents/mile and t is measured in years, with $t = 0$ corresponding to the beginning of 1991.

a. Plot the graph of C in the viewing window $[0, 10] \times [35, 52]$.

b. What was the average cost of driving a car at the beginning of 1995?

c. How fast was the average cost of driving a car changing at the beginning of 1995?

Source: Automobile Association of America

(continued)

10. **Modeling with Data** Annual retail sales in the United States from 1990 through the year 2000 (in billions of dollars) are given in the following table:

Year	1990	1991	1992	1993	1994	1995
Sales	471.6	485.4	519.2	553.4	595	625.5

Year	1996	1997	1998	1999	2000
Sales	656.6	685.6	727.2	781.7	877.7

a. Let $t = 0$ correspond to 1990 and use **QuadReg** to find a second-degree polynomial regression model based on the given data.

b. Plot the graph of the function found in part (a) in the viewing window $[0, 10] \times [0, 1000]$.

c. What were the annual retail sales in the United States in 1999 ($t = 9$)?

d. Approximately, how fast were the retail sales changing in 1999 ($t = 9$)?

Source: National Retail Federation

CHAPTER 2 Summary of Principal Formulas and Terms

FORMULAS

1. Average rate of change of f over $[x, x + h]$ or Slope of the secant line to the graph of f through $(x, f(x))$ and $(x + h, f(x + h))$ or Difference quotient	$\dfrac{f(x + h) - f(x)}{h}$
2. Instantaneous rate of change of f at $(x, f(x))$ or Slope of tangent line to the graph of f at $(x, f(x))$ at x or Derivative of f	$\lim_{h \to 0} \dfrac{f(x + h) - f(x)}{h}$

TERMS

function (50)
domain (50)
range (50)
independent variable (51)
dependent variable (51)
ordered pairs (53)
function (alternative definition) (53)
graph of a function (53)
graph of an equation (56)
vertical-line test (56)
composite function (70)
polynomial function (76)
linear function (76)
quadratic function (77)
rational function (80)
power function (80)
demand function (81)
supply function (81)
market equilibrium (81)
equilibrium quantity (82)
equilibrium price (82)
limit of a function (100)
indeterminate form (103)
limit of a function at infinity (107)
right-hand limit of a function (118)
left-hand limit of a function (118)
continuity of a function at a number (119)
zero of a function (123)
secant line (135)
tangent line to the graph of f (135)
differentiable function (143)

CHAPTER 2 Concept Review Questions

Fill in the blanks.

1. If f is a function from the set A to the set B, then A is called the _____ of f, and the set of all values of $f(x)$ as x takes on all possible values in A is called the _____ of f. The range of f is contained in the set _____.

2. The graph of a function is the set of all points (x, y) in the xy-plane such that x is in the _____ of f and $y =$ _____. The vertical-line test states that a curve in the xy-plane is the graph of a function $y = f(x)$ if and only if each _____ line intersects it in at most one _____.

3. If f and g are functions with domains A and B, respectively, then (a) $(f \pm g)(x) =$ _____, (b) $(fg)(x) =$ _____, and (c) $\left(\frac{f}{g}\right)(x) =$ _____. The domain of $f + g$ is _____. The domain of $\frac{f}{g}$ is _____ with the additional condition that $g(x)$ is never _____.

4. The composition of g and f is the function with rule $(g \circ f)(x) =$ _____. Its domain is the set of all x in the domain of _____ such that _____ lies in the domain of _____.

5. **a.** A polynomial function of degree n is a function of the form _____.
   b. A polynomial function of degree 1 is called a/an _____ function; one of degree 2 is called a/an _____ function; one of degree 3 is called a/an _____ function.
   c. A rational function is a/an _____ of two _____.
   d. A power function has the form $f(x) =$ _____.

6. The statement $\lim_{x \to a} f(x) = L$ means that there is a number _____ such that the values of _____ can be made as close to _____ as we please by taking x sufficiently close to _____.

7. If $\lim_{x \to a} f(x) = L$ and $\lim_{x \to a} g(x) = M$, then
   a. $\lim_{x \to a} [f(x)]^r =$ _____, where r is a real number.
   b. $\lim_{x \to a} [f(x) \pm g(x)] =$ _____.
   c. $\lim_{x \to a} [f(x)g(x)] =$ _____.
   d. $\lim_{x \to a} \dfrac{f(x)}{g(x)} =$ _____ provided that _____.

8. **a.** The statement $\lim_{x \to \infty} f(x) = L$ means that $f(x)$ can be made arbitrarily close to _____ by taking _____ large enough.
   b. The statement $\lim_{x \to -\infty} f(x) = M$ means that $f(x)$ can be made arbitrarily close to _____ by taking x to be _____ and sufficiently large in _____ value.

9. **a.** The statement $\lim_{x \to a^+} f(x) = L$ is similar to the statement $\lim_{x \to a} f(x) = L$, but here x is required to lie to the _____ of a.
   b. The statement $\lim_{x \to a^-} f(x) = L$ is similar to the statement $\lim_{x \to a} f(x) = L$, but here x is required to lie to the _____ of a.
   c. $\lim_{x \to a} f(x) = L$ if and only if both $\lim_{x \to a^-} f(x) =$ _____ and $\lim_{x \to a^+} f(x) =$ _____.

10. **a.** If $f(a)$ is defined, $\lim_{x \to a} f(x)$ exists and $\lim_{x \to a} f(x) = f(a)$, then f is _____ at a.
    b. If f is not continuous at a, then it is _____ at a.
    c. f is continuous on an interval I if f is continuous at _____ number in the interval.

11. **a.** If f and g are continuous at a, then $f \pm g$ and fg are continuous at _____. Also, $\frac{f}{g}$ is continuous at _____, provided _____ $\neq 0$.
    b. A polynomial function is continuous _____.
    c. A rational function $R = \frac{P}{Q}$ is continuous everywhere except at values of x where _____ $= 0$.

12. **a.** Suppose f is continuous on $[a, b]$ and $f(a) < M < f(b)$. Then the intermediate value theorem guarantees the existence of at least one number c in _____ such that _____.
    b. If f is continuous on $[a, b]$ and $f(a)f(b) < 0$, then there must be at least one solution of the equation _____ in the interval _____.

13. **a.** The tangent line at $P(a, f(a))$ to the graph of f is the line passing through P and having slope _____.
    b. If the slope of the tangent line at $P(a, f(a))$ is m, then an equation of the tangent line at P is _____.

14. **a.** The slope of the secant line passing through $P(a, f(a))$ and $Q(a + h, f(a + h))$ and the average rate of change of f over the interval $[a, a + h]$ are both given by _____.
    b. The slope of the tangent line at $P(a, f(a))$ and the instantaneous rate of change of f at a are both given by _____.

CHAPTER 2 Review Exercises

1. Find the domain of each function:
   a. $f(x) = \sqrt{9 - x}$ **b.** $f(x) = \dfrac{x + 3}{2x^2 - x - 3}$

2. Find the domain of the function.
   a. $f(x) = \dfrac{\sqrt{2 - x}}{x + 3}$ **b.** $f(x) = \dfrac{x^2 + 3x + 4}{\sqrt{x^2 + 1}}$

3. Let $f(x) = 3x^2 + 5x - 2$. Find:
   a. $f(-2)$ **b.** $f(a + 2)$
   c. $f(2a)$ **d.** $f(a + h)$

4. Let $f(x) = 2x^2 - x + 1$. Find:
   a. $f(x - 1) + f(x + 1)$
   b. $f(x + 2h)$

5. Let $y^2 = 2x + 1$.
   a. Sketch the graph of this equation.
   b. Is y a function of x? Why?
   c. Is x a function of y? Why?

6. Sketch the graph of the function defined by

$$f(x) = \begin{cases} x + 1 & \text{if } x < 1 \\ -x^2 + 4x - 1 & \text{if } x \geq 1 \end{cases}$$

7. Let $f(x) = 1/x$ and $g(x) = 2x + 3$. Find:
   a. $f(x)g(x)$ **b.** $f(x)/g(x)$
   c. $f(g(x))$ **d.** $g(f(x))$

8. Find the rules for the composite functions $f \circ g$ and $g \circ f$.

a. $f(x) = 2x - 1$; $g(x) = x^2 + 4$

b. $f(x) = 1 - x$; $g(x) = \dfrac{1}{3x + 4}$

c. $f(x) = x - 3$; $g(x) = \dfrac{1}{\sqrt{x + 1}}$

9. Find functions f and g such that $h = g \circ f$. (*Note:* The answer is *not* unique.)

a. $h(x) = \dfrac{1}{(2x^2 + x + 1)^3}$ **b.** $h(x) = \sqrt{x^2 + x + 4}$

10. Find the value of c such that the point (4, 2) lies on the graph of $f(x) = cx^2 + 3x - 4$.

In Exercises 11–24, find the indicated limits, if they exist.

11. $\lim_{x\to 0}(5x - 3)$

12. $\lim_{x\to 1}(x^2 + 1)$

13. $\lim_{x\to -1}(3x^2 + 4)(2x - 1)$

14. $\lim_{x\to 3}\dfrac{x - 3}{x + 4}$

15. $\lim_{x\to 2}\dfrac{x + 3}{x^2 - 9}$

16. $\lim_{x\to -2}\dfrac{x^2 - 2x - 3}{x^2 + 5x + 6}$

17. $\lim_{x\to 3}\sqrt{2x^3 - 5}$

18. $\lim_{x\to 3}\dfrac{4x - 3}{\sqrt{x + 1}}$

19. $\lim_{x\to 1^+}\dfrac{x - 1}{x(x - 1)}$

20. $\lim_{x\to 1^-}\dfrac{\sqrt{x} - 1}{x - 1}$

21. $\lim_{x\to \infty}\dfrac{x^2}{x^2 - 1}$

22. $\lim_{x\to -\infty}\dfrac{x + 1}{x}$

23. $\lim_{x\to \infty}\dfrac{3x^2 + 2x + 4}{2x^2 - 3x + 1}$

24. $\lim_{x\to -\infty}\dfrac{x^2}{x + 1}$

25. Sketch the graph of the function

$$f(x) = \begin{cases} 2x - 3 & \text{if } x \le 2 \\ -x + 3 & \text{if } x > 2 \end{cases}$$

and evaluate $\lim_{x\to a^+} f(x)$, $\lim_{x\to a^-} f(x)$, and $\lim_{x\to a} f(x)$ at the point $a = 2$, if the limits exist.

26. Sketch the graph of the function

$$f(x) = \begin{cases} 4 - x & \text{if } x \le 2 \\ x + 2 & \text{if } x > 2 \end{cases}$$

and evaluate $\lim_{x\to a^+} f(x)$, $\lim_{x\to a^-} f(x)$, and $\lim_{x\to a} f(x)$ at the point $a = 2$, if the limits exist.

In Exercises 27–30, determine all values of *x* for which each function is discontinuous.

27. $g(x) = \begin{cases} x + 3 & \text{if } x \ne 2 \\ 0 & \text{if } x = 2 \end{cases}$

28. $f(x) = \dfrac{3x + 4}{4x^2 - 2x - 2}$

29. $f(x) = \begin{cases} \dfrac{1}{(x + 1)^2} & \text{if } x \ne -1 \\ 2 & \text{if } x = -1 \end{cases}$

30. $f(x) = \dfrac{|2x|}{x}$

31. Let $y = x^2 + 2$.

a. Find the average rate of change of y with respect to x over the intervals [1, 2], [1, 1.5], and [1, 1.1].

b. Find the (instantaneous) rate of change of y at $x = 1$.

32. Use the definition of the derivative to find the slope of the tangent line to the graph of the function $f(x) = 4x + 5$ at any point $P(x, f(x))$ on the graph.

33. Use the definition of the derivative to find the slope of the tangent line to the graph of the function $f(x) = -1/x$ at any point $P(x, f(x))$ on the graph.

34. Use the definition of the derivative to find the slope of the tangent line to the graph of the function $f(x) = \frac{3}{2}x + 5$ at the point $(-2, 2)$ and determine an equation of the tangent line.

35. Use the definition of the derivative to find the slope of the tangent line to the graph of the function $f(x) = -x^2$ at the point $(2, -4)$ and determine an equation of the tangent line.

36. The graph of the function f is shown in the accompanying figure.

a. Is f continuous at $x = a$? Why?

b. Is f differentiable at $x = a$? Justify your answers.

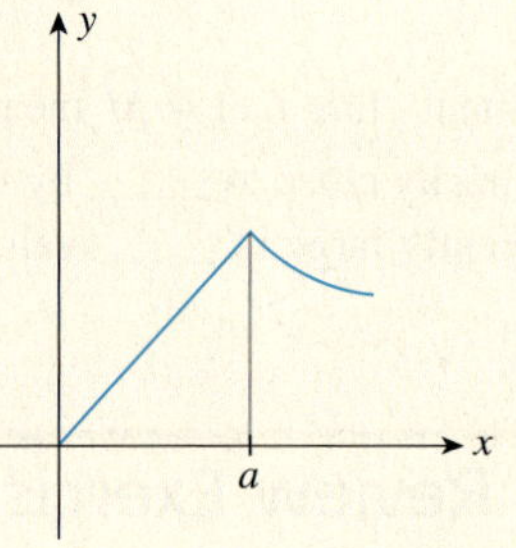

37. Sales of Clock Radios Sales of a certain stereo clock radio are approximated by the relationship $S(x) = 6000x + 30{,}000$ $(0 \le x \le 5)$, where $S(x)$ denotes the number of clock radios sold in year x ($x = 0$ corresponds to the year 2002). Find the number of clock radios expected to be sold in 2006.

38. Sales of a Company A company's total sales (in millions of dollars) are approximately linear as a function of time (in years). Sales in 2005 were \$2.4 million, whereas sales in 2010 amounted to \$7.4 million.

a. Find an equation that gives the company's sales as a function of time.

b. What were the sales in 2008?

39. Profit Functions A company has a fixed cost of \$30,000 and a production cost of \$6 for each unit it manufactures. A unit sells for \$10.

a. What is the cost function?
b. What is the revenue function?
c. What is the profit function?
d. Compute the profit (loss) corresponding to production levels of 6000, 8000, and 12,000 units, respectively.

40. Find the point of intersection of the two straight lines having the equations $y = \frac{3}{4}x + 6$ and $3x - 2y + 3 = 0$.

41. The cost and revenue functions for a certain firm are given by $C(x) = 12x + 20{,}000$ and $R(x) = 20x$, respectively. Find the company's profit function.

42. Market Equilibrium Given the demand equation $3x + p - 40 = 0$ and the supply equation $2x - p + 10 = 0$, where p is the unit price in dollars and x represents the quantity in units of a thousand, determine the equilibrium quantity and the equilibrium price.

43. Clark's Rule Clark's rule is a method for calculating pediatric drug dosages based on a child's weight. If a denotes the adult dosage (in milligrams) and if w is the weight of the child (in pounds), then the child's dosage is given by

$$D(w) = \frac{aw}{150}$$

If the adult dose of a substance is 500 mg, how much should a child who weighs 35 lb receive?

44. Revenue Functions The revenue (in dollars) realized by Apollo from the sale of its ink-jet printers is given by

$$R(x) = -0.1x^2 + 500x$$

where x denotes the number of units manufactured each month. What is Apollo's revenue when 1000 units are produced?

45. Revenue Functions The monthly revenue R (in hundreds of dollars) realized in the sale of Royal electric shavers is related to the unit price p (in dollars) by the equation

$$R(p) = -\frac{1}{2}p^2 + 30p$$

Find the revenue when an electric shaver is priced at \$30.

46. Health Club Membership The membership of the newly opened Venus Health Club is approximated by the function

$$N(x) = 200(4 + x)^{1/2} \qquad (1 \le x \le 24)$$

where $N(x)$ denotes the number of members x mo after the club's grand opening. Find $N(0)$ and $N(12)$ and interpret your results.

47. Population Growth A study prepared for a Sunbelt town's Chamber of Commerce projected that the population of the town in the next 3 yr will grow according to the rule

$$P(x) = 50{,}000 + 30x^{3/2} + 20x$$

where $P(x)$ denotes the population x mo from now. By how much will the population increase during the next 9 mo? During the next 16 mo?

48. Thurstone Learning Curve Psychologist L. L. Thurstone discovered the following model for the relationship between the learning time T and the length of a list n:

$$T = f(n) = An\sqrt{n - b}$$

where A and b are constants that depend on the person and the task. Suppose that, for a certain person and a certain task, $A = 4$ and $b = 4$. Compute $f(4), f(5), \ldots, f(12)$ and use this information to sketch the graph of the function f. Interpret your results.

49. Forecasting Sales The annual sales of Crimson Drug Store are expected to be given by

$$S_1(t) = 2.3 + 0.4t$$

million dollars t yr from now, whereas the annual sales of Cambridge Drug Store are expected to be given by

$$S_2(t) = 1.2 + 0.6t$$

million dollars t yr from now. When will the annual sales of Cambridge first surpass the annual sales of Crimson?

50. Market Equilibrium The monthly demand and supply functions for the Luminar desk lamp are given by

$$p = d(x) = -1.1x^2 + 1.5x + 40$$
$$p = s(x) = 0.1x^2 + 0.5x + 15$$

respectively, where p is measured in dollars and x in units of a thousand. Find the equilibrium quantity and price.

51. Sales of Digital TVs The number of homes with digital TVs is expected to grow according to the function

$$f(t) = 0.1714t^2 + 0.6657t + 0.7143 \qquad (0 \le t \le 6)$$

where t is measured in years, with $t = 0$ corresponding to the beginning of 2000, and $f(t)$ is measured in millions of homes. How many homes had digital TVs
a. At the beginning of 2000?
b. At the beginning of 2005?

Source: Consumer Electronics Manufacturers Association

52. Testosterone Use Fueled by the promotion of testosterone as an antiaging elixir, use of the hormone by middle-age and older men grew dramatically. The total number of prescriptions for testosterone from 1999 through 2002 is given by

$$N(t) = -35.8t^3 + 202t^2 + 87.8t + 648 \qquad (0 \le t \le 3)$$

where $N(t)$ is measured in thousands and t is measured in years, with $t = 0$ corresponding to the beginning of 1999. Find the total number of prescriptions for testosterone in 1999, 2000, 2001, and 2002.

Source: IMS Health

53. U.S. Nutritional Supplements Market The size of the U.S. nutritional supplements market from 1999 through 2003 is approximated by the function

$$A(t) = 16.4(t + 1)^{0.1} \qquad (0 \le t \le 4)$$

where $A(t)$ is measured in billions of dollars and t is measured in years, with $t = 0$ corresponding to the beginning of 1999.
a. Compute $A(0)$, $A(1)$, $A(2)$, $A(3)$, and $A(4)$. Interpret your results.
b. Use the results of part (a) to sketch the graph of A.

Source: Nutrition Business Journal

54. Hotel Occupancy Rate A forecast released by PricewaterhouseCoopers in June of 2004 predicted the occupancy rate of U.S. hotels between 2001 ($t = 0$) and 2005 ($t = 4$) to be

$$P(t) = \begin{cases} -0.9t + 59.8 & \text{if } 0 \le t < 1 \\ 0.3t + 58.6 & \text{if } 1 \le t < 2 \\ 56.79t^{0.06} & \text{if } 2 \le t \le 4 \end{cases}$$

percent.

a. Compute $P(0)$, $P(1)$, $P(2)$, $P(3)$, and $P(4)$.
b. Sketch the graph of P.
c. What was the predicted occupancy rate of hotels for 2004?

Source: PricewaterhouseCoopers LLP Hospitality & Leisure Research

55. Oil Spills The oil spilling from the ruptured hull of a grounded tanker spreads in all directions in calm waters. Suppose the area polluted is a circle of radius r and the radius is increasing at the rate of 2 ft/sec.

a. Find a function f giving the area polluted in terms of r.
b. Find a function g giving the radius of the polluted area in terms of t.
c. Find a function h giving the area polluted in terms of t.
d. What is the size of the polluted area 30 sec after the hull was ruptured?

56. Packaging By cutting away identical squares from each corner of a 20-in. × 20-in. piece of cardboard and folding up the resulting flaps, an open box may be made. Denoting the length of a side of a cutaway by x, find a function of x giving the volume of the resulting box.

57. Construction Costs The length of a rectangular box is to be twice that of its width and its volume is to be 30 ft³. The material for the base costs 30¢/ft², the material for the sides costs 15¢/ft², and the material for the top costs 20¢/ft². Letting x denote the width of the box, find a function in the variable x giving the cost of constructing the box.

58. Film Conversion Prices PhotoMart transfers movie films to DVDs. The fees charged for this service are shown in the following table. Find a function C relating the cost $C(x)$ to the number of feet x of film transferred. Sketch the graph of the function C and discuss its continuity.

Length of Film in Feet, x	Price ($) for Conversion
$1 \le x \le 100$	5.00
$100 < x \le 200$	9.00
$200 < x \le 300$	12.50
$300 < x \le 400$	15.00
$x > 400$	$7 + 0.02x$

59. Average Price of a Commodity The average cost (in dollars) of producing x units of a certain commodity is given by

$$\overline{C}(x) = 20 + \frac{400}{x}$$

Evaluate $\lim_{x \to \infty} \overline{C}(x)$ and interpret your results.

60. Manufacturing Costs Suppose that the total cost in manufacturing x units of a certain product is $C(x)$ dollars.

a. What does $C'(x)$ measure? Give units.
b. What can you say about the sign of C'?
c. Given that $C'(1000) = 20$, estimate the additional cost to be incurred by the company in producing the 1001st unit of the product.

CHAPTER 2 Before Moving On . . .

1. Let

$$f(x) = \begin{cases} -2x + 1 & \text{if } -1 \le x < 0 \\ x^2 + 2 & \text{if } 0 \le x \le 2 \end{cases}$$

Find (a) $f(-1)$, (b) $f(0)$, and (c) $f(\frac{3}{2})$.

2. Let $f(x) = \frac{1}{x+1}$ and $g(x) = x^2 + 1$. Find the rules for (a) $f + g$, (b) fg, (c) $f \circ g$, and (d) $g \circ f$.

3. Postal regulations specify that a parcel sent by priority mail may have a combined length and girth of no more than 108 in. Suppose a rectangular package that has a square cross section of x in. × x in. is to have a combined length and girth of exactly 108 in. Find a function in terms of x giving the volume of the package.

Hint: The length plus the girth is $4x + h$ (see the accompanying figure).

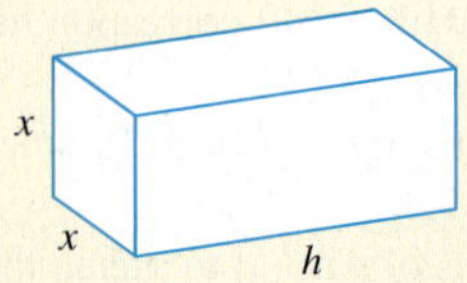

4. Find $\lim_{x \to -1} \frac{x^2 + 4x + 3}{x^2 + 3x + 2}$.

5. Let

$$f(x) = \begin{cases} x^2 - 1 & \text{if } -2 \le x < 1 \\ x^3 & \text{if } \ \ 1 \le x \le 2 \end{cases}$$

Find (a) $\lim_{x \to 1^-} f(x)$ and (b) $\lim_{x \to 1^+} f(x)$. Is f continuous at $x = 1$? Explain.

6. Find the slope of the tangent line to the graph of $f(x) = x^2 - 3x + 1$ at the point $(1, -1)$. What is an equation of the tangent line?

DIFFERENTIATION 3

THIS CHAPTER GIVES several rules that will greatly simplify the task of finding the derivative of a function, thus enabling us to study how fast one quantity is changing with respect to another in many real-world situations. For example, we will be able to find how fast the population of an endangered species of whales grows after certain conservation measures have been implemented, how fast an economy's consumer price index (CPI) is changing at any time, and how fast the time taken to learn the items on a list changes with respect to the length of a list. We also see how these rules of differentiation facilitate the study of the rate of change of economic quantities—that is, the study of marginal analysis. Finally, we introduce the notion of the differential of a function. Differentials are used to approximate the change in one quantity due to a small change in a related quantity.

What happens to the sales of a DVD recording of a certain hit movie over a 10-year period after it is first released into the market? In Example 6, page 174, you will see how to find the rate of change of sales for the DVD over the first 10 years after its release.

3.1 Basic Rules of Differentiation

Four Basic Rules

The method used in Chapter 2 for computing the derivative of a function is based on a faithful interpretation of the definition of the derivative as the limit of a quotient. To find the rule for the derivative f' of a function f, we first computed the difference quotient

$$\frac{f(x+h)-f(x)}{h}$$

and then evaluated its limit as h approached zero. As you have probably observed, this method is tedious even for relatively simple functions.

The main purpose of this chapter is to derive certain rules that will simplify the process of finding the derivative of a function. We will use the notation

$$\frac{d}{dx}[f(x)] \qquad \text{Read "d, d x of f of x"}$$

to mean "the derivative of f with respect to x at x." In stating the rules of differentiation, we assume that the functions f and g are differentiable.

Rule 1: Derivative of a Constant

$$\frac{d}{dx}(c) = 0 \qquad (c\text{, a constant})$$

The derivative of a constant function is equal to zero.

FIGURE 1
The slope of the tangent line to the graph of $f(x) = c$, where c is a constant, is zero.

We can see this from a geometric viewpoint by recalling that the graph of a constant function is a straight line parallel to the x-axis (Figure 1). Since the tangent line to a straight line at any point on the line coincides with the straight line itself, its slope [as given by the derivative of $f(x) = c$] must be zero. We can also use the definition of the derivative to prove this result by computing

$$\begin{aligned} f'(x) &= \lim_{h\to 0}\frac{f(x+h)-f(x)}{h} \\ &= \lim_{h\to 0}\frac{c-c}{h} \\ &= \lim_{h\to 0} 0 = 0 \end{aligned}$$

EXAMPLE 1

a. If $f(x) = 28$, then

$$f'(x) = \frac{d}{dx}(28) = 0$$

b. If $f(x) = -2$, then

$$f'(x) = \frac{d}{dx}(-2) = 0$$

■

Rule 2: The Power Rule

If n is any real number, then $\frac{d}{dx}(x^n) = nx^{n-1}$.

Let's verify the power rule for the special case $n = 2$. If $f(x) = x^2$, then

$$\begin{aligned} f'(x) = \frac{d}{dx}(x^2) &= \lim_{h \to 0} \frac{f(x+h) - f(x)}{h} \\ &= \lim_{h \to 0} \frac{(x+h)^2 - x^2}{h} \\ &= \lim_{h \to 0} \frac{x^2 + 2xh + h^2 - x^2}{h} \\ &= \lim_{h \to 0} \frac{2xh + h^2}{h} = \lim_{h \to 0} \frac{h(2x+h)}{h} \\ &= \lim_{h \to 0} (2x + h) = 2x \end{aligned}$$

as we set out to show.

The proof of the power rule for the general case is not easy to prove and will be omitted. However, you will be asked to prove the rule for the special case $n = 3$ in Exercise 79, page 168.

EXAMPLE 2

a. If $f(x) = x$, then

$$f'(x) = \frac{d}{dx}(x) = 1 \cdot x^{1-1} = x^0 = 1$$

b. If $f(x) = x^8$, then

$$f'(x) = \frac{d}{dx}(x^8) = 8x^7$$

c. If $f(x) = x^{5/2}$, then

$$f'(x) = \frac{d}{dx}(x^{5/2}) = \frac{5}{2}x^{3/2}$$

To differentiate a function whose rule involves a radical, we first rewrite the rule using fractional powers. The resulting expression can then be differentiated using the power rule.

EXAMPLE 3 Find the derivative of the following functions:

a. $f(x) = \sqrt{x}$ **b.** $g(x) = \dfrac{1}{\sqrt[3]{x}}$

Solution

a. Rewriting $\sqrt{x}$ in the form $x^{1/2}$, we obtain

See page 6.

$$\begin{aligned} f'(x) &= \frac{d}{dx}(x^{1/2}) \\ &= \frac{1}{2}x^{-1/2} = \frac{1}{2x^{1/2}} = \frac{1}{2\sqrt{x}} \end{aligned}$$

b. Rewriting $\dfrac{1}{\sqrt[3]{x}}$ in the form $x^{-1/3}$, we obtain

$$g'(x) = \frac{d}{dx}(x^{-1/3})$$
$$= -\frac{1}{3}x^{-4/3} = -\frac{1}{3x^{4/3}}$$

Rule 3: Derivative of a Constant Multiple of a Function

$$\frac{d}{dx}[cf(x)] = c\frac{d}{dx}[f(x)] \quad (c\text{, a constant})$$

The derivative of a constant times a differentiable function is equal to the constant times the derivative of the function.

This result follows from the following computations.
If $g(x) = cf(x)$, then

$$g'(x) = \lim_{h\to 0}\frac{g(x+h) - g(x)}{h} = \lim_{h\to 0}\frac{cf(x+h) - cf(x)}{h}$$
$$= c\lim_{h\to 0}\frac{f(x+h) - f(x)}{h}$$
$$= cf'(x)$$

EXAMPLE 4

a. If $f(x) = 5x^3$, then

$$f'(x) = \frac{d}{dx}(5x^3) = 5\frac{d}{dx}(x^3)$$
$$= 5(3x^2) = 15x^2$$

b. If $f(x) = \dfrac{3}{\sqrt{x}}$, then

$$f'(x) = \frac{d}{dx}(3x^{-1/2})$$
$$= 3\left(-\frac{1}{2}x^{-3/2}\right) = -\frac{3}{2x^{3/2}}$$

Rule 4: The Sum Rule

$$\frac{d}{dx}[f(x) \pm g(x)] = \frac{d}{dx}[f(x)] \pm \frac{d}{dx}[g(x)]$$

The derivative of the sum (difference) of two differentiable functions is equal to the sum (difference) of their derivatives.

This result may be extended to the sum and difference of any finite number of differentiable functions. Let's verify the rule for a sum of two functions.

If $s(x) = f(x) + g(x)$, then

$$\begin{aligned}
s'(x) &= \lim_{h\to 0} \frac{s(x+h) - s(x)}{h} \\
&= \lim_{h\to 0} \frac{[f(x+h) + g(x+h)] - [f(x) + g(x)]}{h} \\
&= \lim_{h\to 0} \frac{[f(x+h) - f(x)] + [g(x+h) - g(x)]}{h} \\
&= \lim_{h\to 0} \frac{f(x+h) - f(x)}{h} + \lim_{h\to 0} \frac{g(x+h) - g(x)}{h} \\
&= f'(x) + g'(x)
\end{aligned}$$

VIDEO

EXAMPLE 5 Find the derivatives of the following functions:

a. $f(x) = 4x^5 + 3x^4 - 8x^2 + x + 3$ **b.** $g(t) = \frac{t^2}{5} + \frac{5}{t^3}$

Solution

a.
$$\begin{aligned}
f'(x) &= \frac{d}{dx}(4x^5 + 3x^4 - 8x^2 + x + 3) \\
&= \frac{d}{dx}(4x^5) + \frac{d}{dx}(3x^4) - \frac{d}{dx}(8x^2) + \frac{d}{dx}(x) + \frac{d}{dx}(3) \\
&= 20x^4 + 12x^3 - 16x + 1
\end{aligned}$$

b. Here, the independent variable is t instead of x, so we differentiate with respect to t. Thus,

$$\begin{aligned}
g'(t) &= \frac{d}{dt}\left(\frac{1}{5}t^2 + 5t^{-3}\right) && \text{Rewrite } \frac{1}{t^3} \text{ as } t^{-3}. \\
&= \frac{2}{5}t - 15t^{-4} = \frac{2}{5}t - \frac{15}{t^4} && \text{Rewrite } t^{-4} \text{ as } \frac{1}{t^4}. \\
&= \frac{2t^5 - 75}{5t^4} && \text{Simplify.}
\end{aligned}$$

EXAMPLE 6 Find the slope and an equation of the tangent line to the graph of $f(x) = 2x + 1/\sqrt{x}$ at the point $(1, 3)$.

Solution The slope of the tangent line at any point on the graph of f is given by

$$\begin{aligned}
f'(x) &= \frac{d}{dx}\left(2x + \frac{1}{\sqrt{x}}\right) \\
&= \frac{d}{dx}(2x + x^{-1/2}) && \text{Rewrite } \frac{1}{\sqrt{x}} \text{ as } \frac{1}{x^{1/2}} = x^{-1/2}. \\
&= 2 - \frac{1}{2}x^{-3/2} && \text{Use the sum rule.} \\
&= 2 - \frac{1}{2x^{3/2}} && \text{Rewrite } \frac{1}{2}x^{-3/2} \text{ as } \frac{1}{2x^{3/2}}.
\end{aligned}$$

In particular, the slope of the tangent line to the graph of f at $(1, 3)$ (where $x = 1$) is

$$f'(1) = 2 - \frac{1}{2(1^{3/2})} = 2 - \frac{1}{2} = \frac{3}{2}$$

Using the point-slope form of the equation of a line with slope $\frac{3}{2}$ and the point (1, 3), we see that an equation of the tangent line is

$$y - 3 = \frac{3}{2}(x - 1) \qquad y - y_1 = m(x - x_1)$$

See page 36.

or, upon simplification,

$$y = \frac{3}{2}x + \frac{3}{2}$$

(see Figure 2).

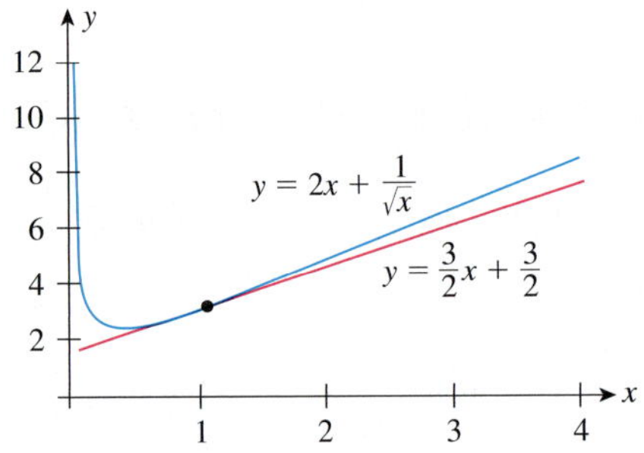

FIGURE 2
The tangent line to the graph of $f(x) = 2x + 1/\sqrt{x}$ at (1, 3).

APPLIED EXAMPLE 7 Conservation of a Species A group of marine biologists at the Neptune Institute of Oceanography recommended that a series of conservation measures be carried out over the next decade to save a certain species of whale from extinction. After implementing the conservation measures, the population of this species is expected to be

$$N(t) = 3t^3 + 2t^2 - 10t + 600 \qquad (0 \le t \le 10)$$

where $N(t)$ denotes the population at the end of year t. Find the rate of growth of the whale population when $t = 2$ and $t = 6$. How large will the whale population be 8 years after implementing the conservation measures?

Solution The rate of growth of the whale population at any time t is given by

$$N'(t) = 9t^2 + 4t - 10$$

In particular, when $t = 2$ and $t = 6$, we have

$$\begin{aligned} N'(2) &= 9(2)^2 + 4(2) - 10 \\ &= 34 \\ N'(6) &= 9(6)^2 + 4(6) - 10 \\ &= 338 \end{aligned}$$

Thus, the whale population's rate of growth will be 34 whales per year after 2 years and 338 per year after 6 years.

The whale population at the end of the eighth year will be

$$\begin{aligned} N(8) &= 3(8)^3 + 2(8)^2 - 10(8) + 600 \\ &= 2184 \end{aligned}$$

The graph of the function N appears in Figure 3. Note the rapid growth of the population in the later years, as the conservation measures begin to pay off, compared with the growth in the early years.

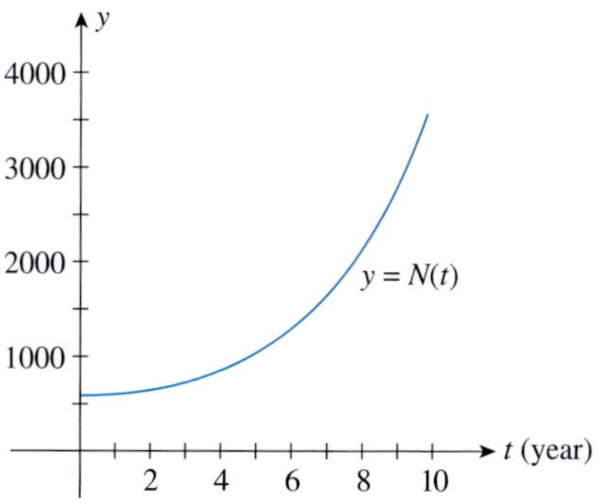

FIGURE 3
The whale population after year t is given by $N(t)$.

■

VIDEO

APPLIED EXAMPLE 8 Altitude of a Rocket The altitude of a rocket (in feet) t seconds into flight is given by

$$s = f(t) = -t^3 + 96t^2 + 195t + 5 \qquad (t \geq 0)$$

a. Find an expression v for the rocket's velocity at any time t.
b. Compute the rocket's velocity when $t = 0$, 30, 50, 65, and 70. Interpret your results.
c. Using the results from the solution to part (b) and the observation that at the highest point in its trajectory the rocket's velocity is zero, find the maximum altitude attained by the rocket.

Solution

a. The rocket's velocity at any time t is given by

$$v = f'(t) = -3t^2 + 192t + 195$$

b. The rocket's velocity when $t = 0$, 30, 50, 65, and 70 is given by

$$\begin{aligned} f'(0) &= -3(0)^2 + 192(0) + 195 = 195 \\ f'(30) &= -3(30)^2 + 192(30) + 195 = 3255 \\ f'(50) &= -3(50)^2 + 192(50) + 195 = 2295 \\ f'(65) &= -3(65)^2 + 192(65) + 195 = 0 \\ f'(70) &= -3(70)^2 + 192(70) + 195 = -1065 \end{aligned}$$

or 195, 3255, 2295, 0, and -1065 feet per second (ft/sec).

Thus, the rocket has an initial velocity of 195 ft/sec at $t = 0$ and accelerates to a velocity of 3255 ft/sec at $t = 30$. Fifty seconds into the flight, the rocket's velocity is 2295 ft/sec, which is less than the velocity at $t = 30$. This means that the rocket begins to decelerate after an initial period of acceleration. (Later on we will learn how to determine the rocket's maximum velocity.)

The deceleration continues: The velocity is 0 ft/sec at $t = 65$ and -1065 ft/sec when $t = 70$. This result tells us that 70 seconds into flight the rocket is heading back to Earth with a speed of 1065 ft/sec.

c. The results of part (b) show that the rocket's velocity is zero when $t = 65$. At this instant, the rocket's maximum altitude is

$$\begin{aligned} s = f(65) &= -(65)^3 + 96(65)^2 + 195(65) + 5 \\ &= 143{,}655 \end{aligned}$$

or 143,655 feet. A sketch of the graph of f appears in Figure 4. ■

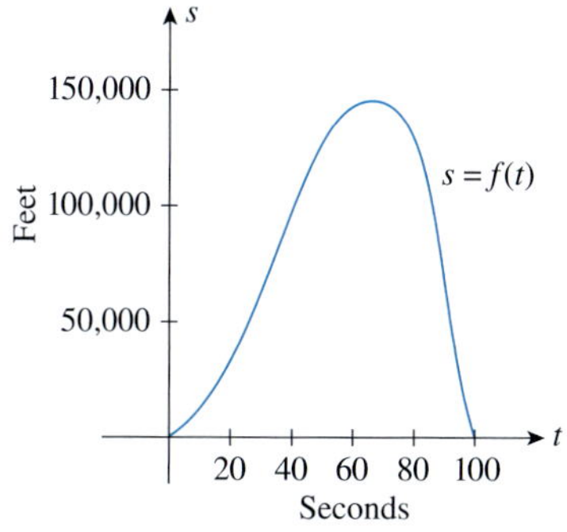

FIGURE 4
The rocket's altitude t seconds into flight is given by $f(t)$.

Exploring with TECHNOLOGY

Refer to Example 8.

1. Use a graphing utility to plot the graph of the velocity function

$$v = f'(t) = -3t^2 + 192t + 195$$

using the viewing window $[0, 120] \times [-5000, 5000]$. Then, using **ZOOM** and **TRACE** or the root-finding capability of your graphing utility, verify that $f'(65) = 0$.

2. Plot the graph of the position function of the rocket

$$s = f(t) = -t^3 + 96t^2 + 195t + 5$$

using the viewing window $[0, 120] \times [0, 150{,}000]$. Then, using **ZOOM** and **TRACE** repeatedly, verify that the maximum altitude of the rocket is 143,655 feet.

3. Use **ZOOM** and **TRACE** or the root-finding capability of your graphing utility to find when the rocket returns to Earth.

3.1 Self-Check Exercises

1. Find the derivative of each function using the rules of differentiation.
   a. $f(x) = 1.5x^2 + 2x^{1.5}$
   b. $g(x) = 2\sqrt{x} + \dfrac{3}{\sqrt{x}}$
2. Let $f(x) = 2x^3 - 3x^2 + 2x - 1$.
   a. Compute $f'(x)$.
   b. What is the slope of the tangent line to the graph of f when $x = 2$?
   c. What is the rate of change of the function f at $x = 2$?
3. A certain country's gross domestic product (GDP) (in millions of dollars) is described by the function

$$G(t) = -2t^3 + 45t^2 + 20t + 6000 \qquad (0 \leq t \leq 11)$$

where $t = 0$ corresponds to the beginning of 2000.
   a. At what rate was the GDP changing at the beginning of 2005? At the beginning of 2007? At the beginning of 2010?
   b. What was the average rate of growth of the GDP over the period 2005–2010?

Solutions to Self-Check Exercises 3.1 can be found on page 168.

3.1 Concept Questions

1. State the following rules of differentiation in your own words.
   a. The rule for differentiating a constant function
   b. The power rule
   c. The constant multiple rule
   d. The sum rule
2. If $f'(2) = 3$ and $g'(2) = -2$, find
   a. $h'(2)$ if $h(x) = 2f(x)$
   b. $F'(2)$ if $F(x) = 3f(x) - 4g(x)$
3. Suppose f and g are differentiable functions and a and b are nonzero numbers. Find $F'(x)$ if
   a. $F(x) = af(x) + bg(x)$
   b. $F(x) = \dfrac{f(x)}{a}$

3.1 Exercises

In Exercises 1–34, find the derivative of the function *f* by using the rules of differentiation.

1. $f(x) = -3$
2. $f(x) = 365$
3. $f(x) = x^5$
4. $f(x) = x^7$
5. $f(x) = x^{2.1}$
6. $f(x) = x^{0.8}$
7. $f(x) = 3x^2$
8. $f(x) = -2x^3$
9. $f(r) = \pi r^2$
10. $f(r) = \dfrac{4}{3}\pi r^3$
11. $f(x) = 9x^{1/3}$
12. $f(x) = \dfrac{5}{4}x^{4/5}$

13. $f(x) = 3\sqrt{x}$

14. $f(u) = \dfrac{2}{\sqrt{u}}$

15. $f(x) = 7x^{-12}$

16. $f(x) = 0.3x^{-1.2}$

17. $f(x) = 5x^2 - 3x + 7$

18. $f(x) = x^3 - 3x^2 + 1$

19. $f(x) = -x^3 + 2x^2 - 6$

20. $f(x) = x^4 - 2x^2 + 5$

21. $f(x) = 0.03x^2 - 0.4x + 10$

22. $f(x) = 0.002x^3 - 0.05x^2 + 0.1x - 20$

23. $f(x) = \dfrac{x^3 - 4x^2 + 3}{x}$

24. $f(x) = \dfrac{x^3 + 2x^2 + x - 1}{x}$

25. $f(x) = 4x^4 - 3x^{5/2} + 2$

26. $f(x) = 5x^{4/3} - \dfrac{2}{3}x^{3/2} + x^2 - 3x + 1$

27. $f(x) = 3x^{-1} + 4x^{-2}$

28. $f(x) = -\dfrac{1}{3}(x^{-3} - x^6)$

29. $f(t) = \dfrac{4}{t^4} - \dfrac{3}{t^3} + \dfrac{2}{t}$

30. $f(x) = \dfrac{5}{x^3} - \dfrac{2}{x^2} - \dfrac{1}{x} + 200$

31. $f(x) = 2x - 5\sqrt{x}$

32. $f(t) = 2t^2 + \sqrt{t^3}$

33. $f(x) = \dfrac{2}{x^2} - \dfrac{3}{x^{1/3}}$

34. $f(x) = \dfrac{3}{x^3} + \dfrac{4}{\sqrt{x}} + 1$

35. Let $f(x) = 2x^3 - 4x$. Find:
a. $f'(-2)$ **b.** $f'(0)$ **c.** $f'(2)$

36. Let $f(x) = 4x^{5/4} + 2x^{3/2} + x$. Find:
a. $f'(0)$ **b.** $f'(16)$

In Exercises 37–40, find each limit by evaluating the derivative of a suitable function at an appropriate point.
Hint: Look at the definition of the derivative.

37. $\displaystyle\lim_{h\to 0} \frac{(1 + h)^3 - 1}{h}$

38. $\displaystyle\lim_{x\to 1} \frac{x^5 - 1}{x - 1}$
Hint: Let $h = x - 1$.

39. $\displaystyle\lim_{h\to 0} \frac{3(2 + h)^2 - (2 + h) - 10}{h}$

40. $\displaystyle\lim_{t\to 0} \frac{1 - (1 + t)^2}{t(1 + t)^2}$

In Exercises 41–44, find the slope and an equation of the tangent line to the graph of the function *f* at the specified point.

41. $f(x) = 2x^2 - 3x + 4$; $(2, 6)$

42. $f(x) = -\dfrac{5}{3}x^2 + 2x + 2$; $\left(-1, -\dfrac{5}{3}\right)$

43. $f(x) = x^4 - 3x^3 + 2x^2 - x + 1$; $(1, 0)$

44. $f(x) = \sqrt{x} + \dfrac{1}{\sqrt{x}}$; $\left(4, \dfrac{5}{2}\right)$

45. Let $f(x) = x^3$.
a. Find the point on the graph of f where the tangent line is horizontal.
b. Sketch the graph of f and draw the horizontal tangent line.

46. Let $f(x) = x^3 - 4x^2$. Find the point(s) on the graph of f where the tangent line is horizontal.

47. Let $f(x) = x^3 + 1$.
a. Find the point(s) on the graph of f where the slope of the tangent line is equal to 12.
b. Find the equation(s) of the tangent line(s) of part (a).
c. Sketch the graph of f showing the tangent line(s).

48. Let $f(x) = \frac{2}{3}x^3 + x^2 - 12x + 6$. Find the values of x for which:
a. $f'(x) = -12$ **b.** $f'(x) = 0$
c. $f'(x) = 12$

49. Let $f(x) = \frac{1}{4}x^4 - \frac{1}{3}x^3 - x^2$. Find the point(s) on the graph of f where the slope of the tangent line is equal to:
a. $-2x$ **b.** 0 **c.** $10x$

50. A straight line perpendicular to and passing through the point of tangency of the tangent line is called the *normal* to the curve. Find an equation of the tangent line and the normal to the curve $y = x^3 - 3x + 1$ at the point $(2, 3)$.

51. **Growth of a Cancerous Tumor** The volume of a spherical cancerous tumor is given by the function

$$V(r) = \frac{4}{3}\pi r^3$$

where r is the radius of the tumor in centimeters. Find the rate of change in the volume of the tumor when

a. $r = \dfrac{2}{3}$ cm **b.** $r = \dfrac{5}{4}$ cm

52. **Velocity of Blood in an Artery** The velocity (in centimeters/second) of blood r cm from the central axis of an artery is given by

$$v(r) = k(R^2 - r^2)$$

where k is a constant and R is the radius of the artery (see the accompanying figure). Suppose $k = 1000$ and $R = 0.2$ cm. Find $v(0.1)$ and $v'(0.1)$ and interpret your results.

Blood vessel

53. Sales of Digital Cameras According to projections made in 2004, the worldwide shipments of digital point-and-shoot cameras are expected to grow in accordance with the rule

$$N(t) = 16.3t^{0.8766} \qquad (1 \le t \le 6)$$

where $N(t)$ is measured in millions and t is measured in years, with $t = 1$ corresponding to 2001.

a. How many digital cameras were sold in 2001 ($t = 1$)?

b. How fast were sales increasing in 2001?

c. What were the projected sales in 2005?

d. How fast were the sales projected to grow in 2005?

Source: International Data Corp.

54. Online Buyers As use of the Internet grows, so does the number of consumers who shop online. The number of online buyers, as a percent of net users, is expected to be

$$P(t) = 53t^{0.12} \qquad (1 \le t \le 7)$$

where t is measured in years, with $t = 1$ corresponding to the beginning of 2002.

a. How many online buyers, as a percent of net users, were there at the beginning of 2007?

b. How fast was the number of online buyers, as a percentage of net users, changing at the beginning of 2007?

Source: Strategy Analytics

55. Married Households with Children The percentage of families that were married households with children between 1970 and 2000 is approximately

$$P(t) = \frac{49.6}{t^{0.27}} \qquad (1 \le t \le 4)$$

where t is measured in decades, with $t = 1$ corresponding to 1970.

a. What percentage of families were married households with children in 1970? In 1980? In 1990? In 2000?

b. How fast was the percentage of families that were married households with children changing in 1980? In 1990?

Source: U.S. Census Bureau

56. Effect of Stopping on Average Speed According to data from a study, the average speed of your trip A (in mph) is related to the number of stops/mile you make on the trip x by the equation

$$A = \frac{26.5}{x^{0.45}}$$

Compute dA/dx for $x = 0.25$ and $x = 2$. How is the rate of change of the average speed of your trip affected by the number of stops/mile?

Source: General Motors

57. Online Video Viewers As broadband Internet grows more popular, video services such as YouTube will continue to expand. The number of online video viewers (in millions) is projected to grow according to the rule

$$N(t) = 52t^{0.531} \qquad (1 \le t \le 10)$$

where $t = 1$ corresponds to 2003.

a. What will be the projected number of online video viewers in 2010?

b. How fast will the projected number of online video viewers be changing in 2010?

Source: eMarketer.com

58. Demand Functions The demand function for the Luminar desk lamp is given by

$$p = f(x) = -0.1x^2 - 0.4x + 35$$

where x is the quantity demanded in thousands and p is the unit price in dollars.

a. Find $f'(x)$.

b. What is the rate of change of the unit price when the quantity demanded is 10,000 units ($x = 10$)? What is the unit price at that level of demand?

59. Stopping Distance of a Racing Car During a test by the editors of an auto magazine, the stopping distance s (in feet) of the MacPherson X-2 racing car conformed to the rule

$$s = f(t) = 120t - 15t^2 \qquad (t \ge 0)$$

where t was the time (in seconds) after the brakes were applied.

a. Find an expression for the car's velocity v at any time t.

b. What was the car's velocity when the brakes were first applied?

c. What was the car's stopping distance for that particular test?

Hint: The stopping time is found by setting $v = 0$.

60. Instant Messaging Accounts Mobile instant messaging (IM) is a small portion of total IM usage, but it is expected to grow sharply. The function

$$P(t) = 0.257t^2 + 0.57t + 3.9 \qquad (0 \le t \le 4)$$

gives the projected mobile IM accounts as a percentage of total enterprise IM accounts from 2006 ($t = 0$) through 2010 ($t = 4$).

a. What percentage of total enterprise IM accounts are the mobile accounts expected to be in 2008?

b. How fast is this percentage expected to change in 2008?

Source: The Radical Group

61. Child Obesity The percentage of obese children, ages 12–19, in the United States has grown dramatically in recent years. The percentage of obese children from 1980 through the year 2000 is approximated by the function

$$P(t) = -0.0105t^2 + 0.735t + 5 \qquad (0 \le t \le 20)$$

where t is measured in years, with $t = 0$ corresponding to the beginning of 1980.

a. What percentage of children were obese at the beginning of 1980? At the beginning of 1990? At the beginning of the year 2000?

b. How fast was the percentage of obese children changing at the beginning of 1985? At the beginning of 1990?

Source: Centers for Disease Control and Prevention

62. **Spending on Medicare** Based on the current eligibility requirement, a study conducted in 2004 showed that federal spending on entitlement programs, particularly Medicare, would grow enormously in the future. The study predicted that spending on Medicare, as a percentage of the gross domestic product (GDP), will be

$$P(t) = 0.27t^2 + 1.4t + 2.2 \qquad (0 \leq t \leq 5)$$

percent in year t, where t is measured in decades, with $t = 0$ corresponding to 2000.

a. How fast will the spending on Medicare, as a percentage of the GDP, be growing in 2010? In 2020?

b. What will the predicted spending on Medicare be in 2010? In 2020?

Source: Congressional Budget Office

63. **Fisheries** The total groundfish population on Georges Bank in New England between 1989 and 1999 is approximated by the function

$$f(t) = 5.303t^2 - 53.977t + 253.8 \qquad (0 \leq t \leq 10)$$

where $f(t)$ is measured in thousands of metric tons and t in years, with $t = 0$ corresponding to the beginning of 1989.

a. What was the rate of change of the groundfish population at the beginning of 1994? At the beginning of 1996?

b. Fishing restrictions were imposed on Dec. 7, 1994. Were the conservation measures effective?

Source: New England Fishery Management Council

64. **Worker Efficiency** An efficiency study conducted for Elektra Electronics showed that the number of Space Commander walkie-talkies assembled by the average worker t hr after starting work at 8 a.m. is given by

$$N(t) = -t^3 + 6t^2 + 15t$$

a. Find the rate at which the average worker will be assembling walkie-talkies t hr after starting work.

b. At what rate will the average worker be assembling walkie-talkies at 10 a.m.? At 11 a.m.?

c. How many walkie-talkies will the average worker assemble between 10 a.m. and 11 a.m.?

65. **Consumer Price Index** An economy's consumer price index (CPI) is described by the function

$$I(t) = -0.2t^3 + 3t^2 + 100 \qquad (0 \leq t \leq 10)$$

where $t = 0$ corresponds to 2000.

a. At what rate was the CPI changing in 2005? In 2007? In 2010?

b. What was the average rate of increase in the CPI over the period from 2005 to 2010?

66. **Effect of Advertising on Sales** The relationship between the amount of money x that Cannon Precision Instruments spends on advertising and the company's total sales $S(x)$ is given by the function

$$S(x) = -0.002x^3 + 0.6x^2 + x + 500 \qquad (0 \leq x \leq 200)$$

where x is measured in thousands of dollars. Find the rate of change of the sales with respect to the amount of money spent on advertising. Are Cannon's total sales increasing at a faster rate when the amount of money spent on advertising is (a) \$100,000 or (b) \$150,000?

67. **Supply Functions** The supply function for a certain make of satellite radio is given by

$$p = f(x) = 0.0001x^{5/4} + 10$$

where x is the quantity supplied and p is the unit price in dollars.

a. Find $f'(x)$.

b. What is the rate of change of the unit price if the quantity supplied is 10,000 satellite radios?

68. **Population Growth** A study prepared for a Sunbelt town's chamber of commerce projected that the town's population in the next 3 yr will grow according to the rule

$$P(t) = 50{,}000 + 30t^{3/2} + 20t$$

where $P(t)$ denotes the population t mo from now. How fast will the population be increasing 9 mo and 16 mo from now?

69. **Average Speed of a Vehicle on a Highway** The average speed of a vehicle on a stretch of Route 134 between 6 a.m. and 10 a.m. on a typical weekday is approximated by the function

$$f(t) = 20t - 40\sqrt{t} + 50 \qquad (0 \leq t \leq 4)$$

where $f(t)$ is measured in mph and t is measured in hours, with $t = 0$ corresponding to 6 a.m.

a. Compute $f'(t)$.

b. What is the average speed of a vehicle on that stretch of Route 134 at 6 a.m.? At 7 a.m.? At 8 a.m.?

c. How fast is the average speed of a vehicle on that stretch of Route 134 changing at 6:30 a.m.? At 7 a.m.? At 8 a.m.?

70. **Curbing Population Growth** Five years ago, the government of a Pacific Island state launched an extensive propaganda campaign toward curbing the country's population growth. According to the Census Department, the population (measured in thousands of people) for the following 4 yr was

$$P(t) = -\frac{1}{3}t^3 + 64t + 3000$$

where t is measured in years and $t = 0$ corresponds to the start of the campaign. Find the rate of change of the population at the end of years 1, 2, 3, and 4. Was the plan working?

71. **Conservation of Species** A certain species of turtle faces extinction because dealers collect truckloads of turtle eggs to be sold as aphrodisiacs. After severe conservation measures are implemented, it is hoped that the turtle population will grow according to the rule

$$N(t) = 2t^3 + 3t^2 - 4t + 1000 \qquad (0 \leq t \leq 10)$$

where $N(t)$ denotes the population at the end of year t. Find the rate of growth of the turtle population when $t = 2$ and $t = 8$. What will be the population 10 yr after the conservation measures are implemented?

72. FLIGHT OF A ROCKET The altitude (in feet) of a rocket t sec into flight is given by

$$s = f(t) = -2t^3 + 114t^2 + 480t + 1 \qquad (t \geq 0)$$

a. Find an expression v for the rocket's velocity at any time t.

b. Compute the rocket's velocity when $t = 0$, 20, 40, and 60. Interpret your results.

c. Using the results from the solution to part (b), find the maximum altitude attained by the rocket.

Hint: At its highest point, the velocity of the rocket is zero.

73. OBESITY IN AMERICA The body mass index (BMI) measures body weight in relation to height. A BMI of 25 to 29.9 is considered overweight, a BMI of 30 or more is considered obese, and a BMI of 40 or more is morbidly obese. The percentage of the U.S. population that is obese is approximated by the function

$$P(t) = 0.0004t^3 + 0.0036t^2 + 0.8t + 12 \qquad (0 \leq t \leq 13)$$

where t is measured in years, with $t = 0$ corresponding to the beginning of 1991.

a. What percentage of the U.S. population was deemed obese at the beginning of 1991? At the beginning of 2004?

b. How fast was the percentage of the U.S. population that is deemed obese changing at the beginning of 1991? At the beginning of 2004?

(*Note:* A formula for calculating the BMI of a person is given in Exercise 29, page 538.)

Source: Centers for Disease Control and Prevention

74. HEALTH-CARE SPENDING Despite efforts at cost containment, the cost of the Medicare program is increasing. Two major reasons for this increase are an aging population and extensive use by physicians of new technologies. Based on data from the Health Care Financing Administration and the U.S. Census Bureau, health-care spending through the year 2000 may be approximated by the function

$$S(t) = 0.02836t^3 - 0.05167t^2 + 9.60881t + 41.9 \qquad (0 \leq t \leq 35)$$

where $S(t)$ is the spending in billions of dollars and t is measured in years, with $t = 0$ corresponding to the beginning of 1965.

a. Find an expression for the rate of change of health-care spending at any time t.

b. How fast was health-care spending changing at the beginning of 1980? At the beginning of 2000?

c. What was the amount of health-care spending at the beginning of 1980? At the beginning of 2000?

Source: Health Care Financing Administration and U.S. Census Bureau

75. AGING POPULATION The population (in millions) of developed countries from 2005 through 2034 is projected to be

$$f(t) = 3.567t + 175.2 \qquad (5 \leq t \leq 35)$$

where t is measured in years. On the other hand, the population of underdeveloped/emerging countries over the same period is projected to be

$$g(t) = 0.46t^2 + 0.16t + 287.8 \qquad (5 \leq t \leq 35)$$

a. What does the function $D = g + f$ represent?

b. Find D' and $D'(10)$ and interpret your results.

Source: U.S. Census Bureau, United Nations

76. SHORTAGE OF NURSES The projected number of nurses (in millions) from the year 2000 through 2015 is given by

$$N(t) = \begin{cases} 1.9 & \text{if } 0 \leq t < 5 \\ -0.0004t^2 + 0.038t + 1.72 & \text{if } 5 \leq t \leq 15 \end{cases}$$

where $t = 0$ corresponds to 2000. The projected number of nursing jobs (in millions) over the same period is

$$J(t) = \begin{cases} -0.0002t^2 + 0.032t + 2 & \text{if } 0 \leq t < 10 \\ -0.0016t^2 + 0.12t + 1.26 & \text{if } 10 \leq t \leq 15 \end{cases}$$

a. Find the rule for the function $G = J - N$ giving the gap between the supply and the demand of nurses from 2000 through 2015.

b. How fast was the gap between the supply and the demand of nurses changing in 2008? In 2012?

Source: Department of Health and Human Services

In Exercises 77 and 78, determine whether the statement is true or false. If it is true, explain why it is true. If it is false, give an example to show why it is false.

77. If f and g are differentiable, then

$$\frac{d}{dx}[2f(x) - 5g(x)] = 2f'(x) - 5g'(x)$$

78. If $f(x) = \pi^x$, then $f'(x) = x\pi^{x-1}$.

79. Prove the power rule (Rule 2) for the special case $n = 3$.

Hint: Compute $\lim_{h \to 0}\left[\frac{(x+h)^3 - x^3}{h}\right]$.

3.1 Solutions to Self-Check Exercises

1. a.
$$\begin{aligned} f'(x) &= \frac{d}{dx}(1.5x^2) + \frac{d}{dx}(2x^{1.5}) \\ &= (1.5)(2x) + (2)(1.5x^{0.5}) \\ &= 3x + 3\sqrt{x} = 3(x + \sqrt{x}) \end{aligned}$$

b.
$$\begin{aligned} g'(x) &= \frac{d}{dx}(2x^{1/2}) + \frac{d}{dx}(3x^{-1/2}) \\ &= (2)\left(\frac{1}{2}x^{-1/2}\right) + (3)\left(-\frac{1}{2}x^{-3/2}\right) \\ &= x^{-1/2} - \frac{3}{2}x^{-3/2} = \frac{1}{2}x^{-3/2}(2x - 3) = \frac{2x - 3}{2x^{3/2}} \end{aligned}$$

2. a. $f'(x) = \frac{d}{dx}(2x^3) - \frac{d}{dx}(3x^2) + \frac{d}{dx}(2x) - \frac{d}{dx}(1)$

$$= (2)(3x^2) - (3)(2x) + 2$$
$$= 6x^2 - 6x + 2$$

b. The slope of the tangent line to the graph of f when $x = 2$ is given by

$$f'(2) = 6(2)^2 - 6(2) + 2 = 14$$

c. The rate of change of f at $x = 2$ is given by $f'(2)$. Using the results of part (b), we see that the required rate of change is 14 units/unit change in x.

3. a. The rate at which the GDP was changing at any time t $(0 < t < 11)$ is given by

$$G'(t) = -6t^2 + 90t + 20$$

In particular, the rates of change of the GDP at the beginning of the years 2005 $(t = 5)$, 2007 $(t = 7)$, and 2010 $(t = 10)$ are given by

$$G'(5) = 320 \qquad G'(7) = 356 \qquad G'(10) = 320$$

respectively—that is, by \$320 million/year, \$356 million/year, and \$320 million/year, respectively.

b. The average rate of growth of the GDP over the period from the beginning of 2005 $(t = 5)$ to the beginning of 2010 $(t = 10)$ is given by

$$\frac{G(10) - G(5)}{10 - 5} = \frac{[-2(10)^3 + 45(10)^2 + 20(10) + 6000]}{5} - \frac{[-2(5)^3 + 45(5)^2 + 20(5) + 6000]}{5}$$
$$= \frac{8700 - 6975}{5}$$

or \$345 million/year.

USING TECHNOLOGY

Finding the Rate of Change of a Function

We can use the numerical derivative operation of a graphing utility to obtain the value of the derivative at a given value of x. Since the derivative of a function $f(x)$ measures the rate of change of the function with respect to x, the numerical derivative operation can be used to answer questions pertaining to the rate of change of one quantity y with respect to another quantity x, where $y = f(x)$, for a specific value of x.

```
nDeriv((3X^3+2X^
.5), X, 1)
        10.00000313
```

FIGURE **T1**
The TI-83/84 numerical derivative screen for computing $f'(1)$

EXAMPLE 1 Let $y = 3t^3 + 2\sqrt{t}$.

a. Use the numerical derivative operation of a graphing utility to find how fast y is changing with respect to t when $t = 1$.

b. Verify the result of part (a), using the rules of differentiation of this section.

Solution

a. Write $f(t) = 3t^3 + 2\sqrt{t}$. Using the numerical derivative operation of a graphing utility, we find that the rate of change of y with respect to t when $t = 1$ is given by $f'(1) = 10$ (Figure T1).

b. Here, $f(t) = 3t^3 + 2t^{1/2}$ and

$$f'(t) = 9t^2 + 2\left(\frac{1}{2}t^{-1/2}\right) = 9t^2 + \frac{1}{\sqrt{t}}$$

Using this result, we see that when $t = 1$, y is changing at the rate of

$$f'(1) = 9(1^2) + \frac{1}{\sqrt{1}} = 10$$

units per unit change in t, as obtained earlier. ■

(continued)

APPLIED EXAMPLE 2 Fuel Economy of Cars According to data obtained from the U.S. Department of Energy and the Shell Development Company, a typical car's fuel economy depends on the speed it is driven and is approximated by the function

$$f(x) = 0.00000310315x^4 - 0.000455174x^3 + 0.00287869x^2 + 1.25986x \qquad (0 \le x \le 75)$$

where x is measured in mph and $f(x)$ is measured in miles per gallon (mpg).

a. Use a graphing utility to graph the function f on the interval [0, 75].

b. Find the rate of change of f when $x = 20$ and when $x = 50$.

c. Interpret your results.

Source: U.S. Department of Energy and the Shell Development Company

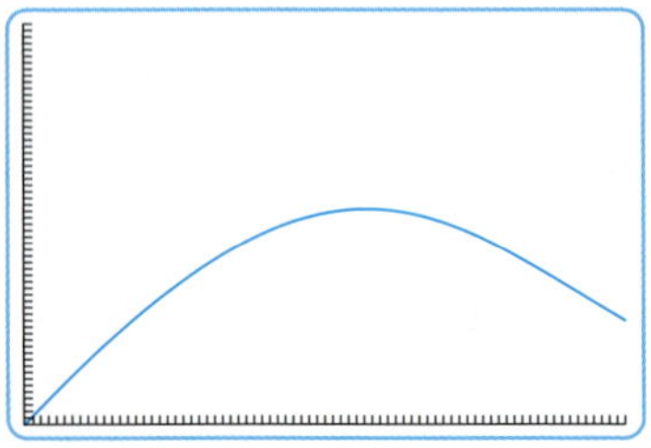

FIGURE **T2**
The graph of the function f on the interval [0, 75]

Solution

a. The graph is shown in Figure T2.

b. Using the numerical derivative operation of a graphing utility, we see that $f'(20) = .9280996$. The rate of change of f when $x = 50$ is given by $f'(50) = -.3145009995$. (See Figure T3a and T3b.)

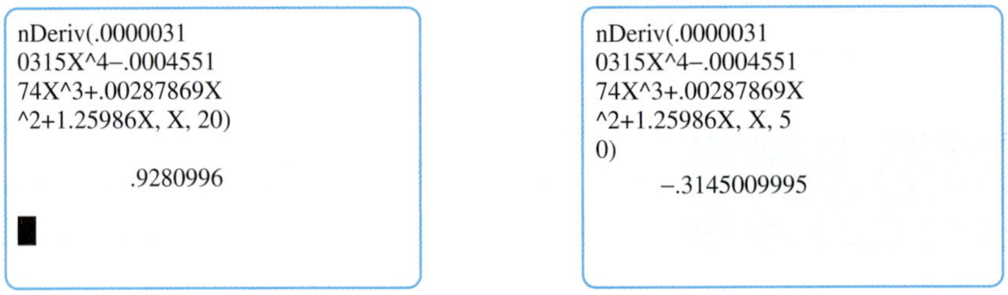

(a) **(b)**

FIGURE **T3**
The TI-83/84 numerical derivative screen for computing (a) $f'(20)$ and (b) $f'(50)$

c. The results of part (b) tell us that when a typical car is being driven at 20 mph, its fuel economy increases at the rate of approximately 0.9 mpg per 1 mph increase in its speed. At a speed of 50 mph, its fuel economy decreases at the rate of approximately 0.3 mpg per 1 mph increase in its speed. ■

TECHNOLOGY EXERCISES

In Exercises 1–6, use the numerical derivative operation to find the rate of change of $f(x)$ at the given value of x. Give your answer accurate to four decimal places.

1. $f(x) = 4x^5 - 3x^3 + 2x^2 + 1;\ x = 0.5$

2. $f(x) = -x^5 + 4x^2 + 3;\ x = 0.4$

3. $f(x) = x - 2\sqrt{x};\ x = 3$

4. $f(x) = \dfrac{\sqrt{x} - 1}{x};\ x = 2$

5. $f(x) = x^{1/2} - x^{1/3};\ x = 1.2$

6. $f(x) = 2x^{5/4} + x;\ x = 2$

7. **Carbon Monoxide in the Atmosphere** The projected average global atmospheric concentration of carbon monoxide is approximated by the function

$$f(t) = 0.881443t^4 - 1.45533t^3 + 0.695876t^2 + 2.87801t + 293 \qquad (0 \le t \le 4)$$

where t is measured in 40-yr intervals with $t = 0$ corresponding to the beginning of 1860, and $f(t)$ is measured in parts per million by volume.

a. Plot the graph of f in the viewing window $[0, 4] \times [280, 400]$.

b. Use a graphing utility to estimate how fast the projected average global atmospheric concentration of carbon monoxide was changing at the beginning of 1900 ($t = 1$) and at the beginning of 2000 ($t = 3.5$).

Source: Meadows et al. "Beyond the Limits"

8. **Spread of HIV** The estimated number of children newly infected with HIV through mother-to-child contact worldwide is given by

$$f(t) = -0.2083t^3 + 3.0357t^2 + 44.0476t + 200.2857 \qquad (0 \le t \le 12)$$

where $f(t)$ is measured in thousands and t is measured in years, with $t = 0$ corresponding to the beginning of 1990.

a. Plot the graph of f in the viewing window $[0, 12] \times [0, 800]$.

b. How fast was the estimated number of children newly infected with HIV through mother-to-child contact worldwide increasing at the beginning of the year 2000?

Source: United Nations

9. **Modeling with Data** A hedge fund is a lightly regulated pool of professionally managed money. The assets (in billions of dollars) of hedge funds from the beginning of 1999 ($t = 0$) through the beginning of 2004 are given in the following table:

Year	1999	2000	2001	2002	2003	2004
Assets ($ billions)	472	517	594	650	817	950

a. Use **CubicReg** to find a third-degree polynomial function for the data, letting $t = 0$ correspond to the beginning of 1999.

b. Plot the graph of the function found in part (a).

c. Use the numerical derivative capability of your graphing utility to find the rate at which the assets of hedge funds were increasing at the beginning of 2000 and the beginning of 2003.

Sources: Hennessee Group; Institutional Investor

10. **Modeling with Data** The number of people (in millions) enrolled in HMOs from 1994 through 2002 is given in the following table:

Year	1994	1995	1996	1997	1998	1999	2000	2001	2002
People	45.4	50.6	58.7	67.0	76.4	81.3	80.9	80.0	74.2

a. Use **QuartReg** to find a fourth-degree polynomial regression model for this data. Let $t = 0$ correspond to 1994.

b. Use the model to estimate the number of people enrolled in HMOs in 2000. How does this number compare with the actual number?

c. How fast was the number of people receiving their care in an HMO changing at the beginning of 2001?

Source: Group Health Association of America

3.2 The Product and Quotient Rules

In this section we study two more rules of differentiation: the **product rule** and the **quotient rule.**

The Product Rule

The derivative of the product of two differentiable functions is given by the following rule:

Rule 5: The Product Rule

$$\frac{d}{dx}[f(x)g(x)] = f(x)g'(x) + g(x)f'(x)$$

The derivative of the product of two functions is the first function times the derivative of the second plus the second function times the derivative of the first.

The product rule may be extended to the case involving the product of any finite number of functions (see Exercise 65, p. 179). We prove the product rule at the end of this section.

The derivative of the product of two functions is *not* given by the product of the derivatives of the functions; that is, in general

$$\frac{d}{dx}[f(x)g(x)] \neq f'(x)g'(x)$$

EXAMPLE 1 Find the derivative of the function

$$f(x) = (2x^2 - 1)(x^3 + 3)$$

Solution By the product rule,

$$\begin{aligned} f'(x) &= (2x^2 - 1)\frac{d}{dx}(x^3 + 3) + (x^3 + 3)\frac{d}{dx}(2x^2 - 1) \\ &= (2x^2 - 1)(3x^2) + (x^3 + 3)(4x) \\ &= 6x^4 - 3x^2 + 4x^4 + 12x \\ &= 10x^4 - 3x^2 + 12x \quad \text{Combine like terms.} \\ &= x(10x^3 - 3x + 12) \quad \text{Factor out } x. \end{aligned}$$

See page 8.

EXAMPLE 2 Differentiate (that is, find the derivative of) the function

$$f(x) = x^3(\sqrt{x} + 1)$$

Solution First, we express the function in exponential form, obtaining

$$f(x) = x^3(x^{1/2} + 1)$$

By the product rule,

$$\begin{aligned} f'(x) &= x^3\frac{d}{dx}(x^{1/2} + 1) + (x^{1/2} + 1)\frac{d}{dx}x^3 \\ &= x^3\left(\frac{1}{2}x^{-1/2}\right) + (x^{1/2} + 1)(3x^2) \\ &= \frac{1}{2}x^{5/2} + 3x^{5/2} + 3x^2 \\ &= \frac{7}{2}x^{5/2} + 3x^2 \end{aligned}$$

Note We can also solve the problem by first expanding the product before differentiating f. Examples for which this is not possible will be considered in Section 3.3, where the true value of the product rule will be appreciated.

The Quotient Rule

The derivative of the quotient of two differentiable functions is given by the following rule:

Rule 6: The Quotient Rule

$$\frac{d}{dx}\left[\frac{f(x)}{g(x)}\right] = \frac{g(x)f'(x) - f(x)g'(x)}{[g(x)]^2} \qquad [g(x) \neq 0]$$

As an aid to remembering this expression, observe that it has the following form:

$$\frac{d}{dx}\left[\frac{f(x)}{g(x)}\right] = \frac{(\text{Denominator})\begin{pmatrix}\text{Derivative of}\\ \text{numerator}\end{pmatrix} - (\text{Numerator})\begin{pmatrix}\text{Derivative of}\\ \text{denominator}\end{pmatrix}}{(\text{Square of denominator})}$$

For a proof of the quotient rule, see Exercise 66, page 179.

The derivative of a quotient is *not* equal to the quotient of the derivatives; that is,

$$\frac{d}{dx}\left[\frac{f(x)}{g(x)}\right] \neq \frac{f'(x)}{g'(x)}$$

For example, if $f(x) = x^3$ and $g(x) = x^2$, then

$$\frac{d}{dx}\left[\frac{f(x)}{g(x)}\right] = \frac{d}{dx}\left(\frac{x^3}{x^2}\right) = \frac{d}{dx}(x) = 1$$

which is *not* equal to

$$\frac{f'(x)}{g'(x)} = \frac{\frac{d}{dx}(x^3)}{\frac{d}{dx}(x^2)} = \frac{3x^2}{2x} = \frac{3}{2}x$$

EXAMPLE 3 Find $f'(x)$ if $f(x) = \dfrac{x}{2x - 4}$.

Solution Using the quotient rule, we obtain

$$\begin{aligned} f'(x) &= \frac{(2x - 4)\frac{d}{dx}(x) - x\frac{d}{dx}(2x - 4)}{(2x - 4)^2} \\ &= \frac{(2x - 4)(1) - x(2)}{(2x - 4)^2} \\ &= \frac{2x - 4 - 2x}{(2x - 4)^2} = -\frac{4}{(2x - 4)^2} \end{aligned}$$

■

EXAMPLE 4 Find $f'(x)$ if $f(x) = \dfrac{x^2 + 1}{x^2 - 1}$.

Solution By the quotient rule,

$$\begin{aligned} f'(x) &= \frac{(x^2 - 1)\frac{d}{dx}(x^2 + 1) - (x^2 + 1)\frac{d}{dx}(x^2 - 1)}{(x^2 - 1)^2} \\ &= \frac{(x^2 - 1)(2x) - (x^2 + 1)(2x)}{(x^2 - 1)^2} \\ &= \frac{2x^3 - 2x - 2x^3 - 2x}{(x^2 - 1)^2} \\ &= -\frac{4x}{(x^2 - 1)^2} \end{aligned}$$

■

EXAMPLE 5 Find $h'(x)$ if $h(x) = \dfrac{\sqrt{x}}{x^2 + 1}$.

Solution Rewrite $h(x)$ in the form $h(x) = \dfrac{x^{1/2}}{x^2 + 1}$. By the quotient rule, we find

$$\begin{aligned} h'(x) &= \frac{(x^2 + 1)\frac{d}{dx}(x^{1/2}) - x^{1/2}\frac{d}{dx}(x^2 + 1)}{(x^2 + 1)^2} \\ &= \frac{(x^2 + 1)(\frac{1}{2}x^{-1/2}) - x^{1/2}(2x)}{(x^2 + 1)^2} \\ &= \frac{\frac{1}{2}x^{-1/2}(x^2 + 1 - 4x^2)}{(x^2 + 1)^2} \\ &= \frac{1 - 3x^2}{2\sqrt{x}(x^2 + 1)^2} \end{aligned}$$

Factor out $\frac{1}{2}x^{-1/2}$ from the numerator.

See page 9.

■

APPLIED EXAMPLE 6 Rate of Change of DVD Sales The sales (in millions of dollars) of a DVD recording of a hit movie t years from the date of release is given by

$$S(t) = \frac{5t}{t^2 + 1}$$

a. Find the rate at which the sales are changing at time t.
b. How fast are the sales changing at the time the DVDs are released ($t = 0$)? Two years from the date of release?

Solution

a. The rate at which the sales are changing at time t is given by $S'(t)$. Using the quotient rule, we obtain

$$\begin{aligned} S'(t) &= \frac{d}{dt}\left[\frac{5t}{t^2 + 1}\right] = 5\frac{d}{dt}\left[\frac{t}{t^2 + 1}\right] \\ &= 5\left[\frac{(t^2 + 1)(1) - t(2t)}{(t^2 + 1)^2}\right] \\ &= 5\left[\frac{t^2 + 1 - 2t^2}{(t^2 + 1)^2}\right] = \frac{5(1 - t^2)}{(t^2 + 1)^2} \end{aligned}$$

See page 16.

b. The rate at which the sales are changing at the time the DVDs are released is given by

$$S'(0) = \frac{5(1 - 0)}{(0 + 1)^2} = 5$$

That is, they are increasing at the rate of \$5 million per year.

Two years from the date of release, the sales are changing at the rate of

$$S'(2) = \frac{5(1 - 4)}{(4 + 1)^2} = -\frac{3}{5} = -0.6$$

That is, they are decreasing at the rate of \$600,000 per year.

The graph of the function S is shown in Figure 5.

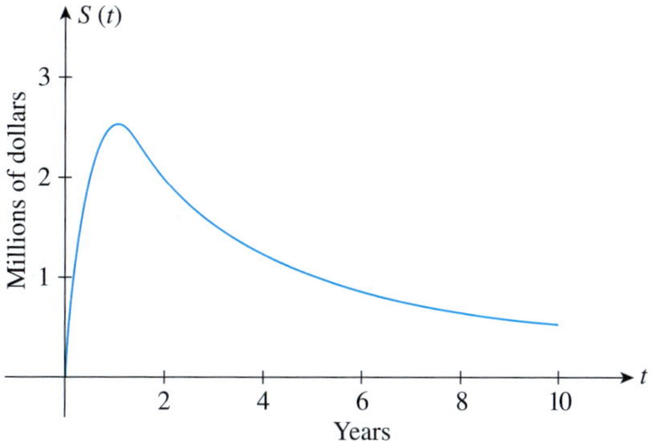

FIGURE 5
After a spectacular rise, the sales begin to taper off.

Exploring with TECHNOLOGY

Refer to Example 6.

1. Use a graphing utility to plot the graph of the function S, using the viewing window $[0, 10] \times [0, 3]$.
2. Use TRACE and ZOOM to determine the coordinates of the highest point on the graph of S in the interval $[0, 10]$. Interpret your results.

Explore & Discuss

Suppose the revenue of a company is given by $R(x) = xp(x)$, where x is the number of units of the product sold at a unit price of $p(x)$ dollars.

1. Compute $R'(x)$ and explain, in words, the relationship between $R'(x)$ and $p(x)$ and/or its derivative.
2. What can you say about $R'(x)$ if $p(x)$ is constant? Is this expected?

APPLIED EXAMPLE 7 Oxygen-Restoration Rate in a Pond When organic waste is dumped into a pond, the oxidation process that takes place reduces the pond's oxygen content. However, given time, nature will restore the oxygen content to its natural level. Suppose the oxygen content t days after organic waste has been dumped into the pond is given by

$$f(t) = 100\left[\frac{t^2 + 10t + 100}{t^2 + 20t + 100}\right] \qquad (0 < t < \infty)$$

percent of its normal level.

a. Derive a general expression that gives the rate of change of the pond's oxygen level at any time t.
b. How fast is the pond's oxygen content changing 1 day, 10 days, and 20 days after the organic waste has been dumped?

Solution

a. The rate of change of the pond's oxygen level at any time t is given by the derivative of the function f. Thus, the required expression is

$$f'(t) = 100\frac{d}{dt}\left[\frac{t^2 + 10t + 100}{t^2 + 20t + 100}\right]$$

$$= 100\left[\frac{(t^2 + 20t + 100)\frac{d}{dt}(t^2 + 10t + 100) - (t^2 + 10t + 100)\frac{d}{dt}(t^2 + 20t + 100)}{(t^2 + 20t + 100)^2}\right]$$

$$= 100\left[\frac{(t^2 + 20t + 100)(2t + 10) - (t^2 + 10t + 100)(2t + 20)}{(t^2 + 20t + 100)^2}\right]$$

See page 16.

$$= 100\left[\frac{2t^3 + 10t^2 + 40t^2 + 200t + 200t + 1000 - 2t^3 - 20t^2 - 20t^2 - 200t - 200t - 2000}{(t^2 + 20t + 100)^2}\right]$$

$$= 100\left[\frac{10t^2 - 1000}{(t^2 + 20t + 100)^2}\right]$$

Combine like terms in the numerator.

b. The rate at which the pond's oxygen content is changing 1 day after the organic waste has been dumped is given by

$$f'(1) = 100\left[\frac{10 - 1000}{(1 + 20 + 100)^2}\right] \approx -6.76$$

That is, it is dropping at the rate of 6.8% per day. After 10 days, the rate is

$$f'(10) = 100\left[\frac{10(10)^2 - 1000}{(10^2 + 20(10) + 100)^2}\right] = 0$$

That is, it is neither increasing nor decreasing. After 20 days, the rate is

$$f'(20) = 100\left[\frac{10(20)^2 - 1000}{(20^2 + 20(20) + 100)^2}\right] \approx 0.37$$

That is, the oxygen content is increasing at the rate of 0.37% per day, and the restoration process has indeed begun.

Verification of the Product Rule

We will now verify the product rule. If $p(x) = f(x)g(x)$, then

$$p'(x) = \lim_{h\to 0}\frac{p(x + h) - p(x)}{h} = \lim_{h\to 0}\frac{f(x + h)g(x + h) - f(x)g(x)}{h}$$

By adding $-f(x + h)g(x) + f(x + h)g(x)$ (which is zero!) to the numerator and factoring, we have

$$p'(x) = \lim_{h\to 0}\frac{f(x + h)[g(x + h) - g(x)] + g(x)[f(x + h) - f(x)]}{h}$$

$$= \lim_{h\to 0}\left\{f(x + h)\left[\frac{g(x + h) - g(x)}{h}\right] + g(x)\left[\frac{f(x + h) - f(x)}{h}\right]\right\}$$

$$= \lim_{h\to 0} f(x + h)\left[\frac{g(x + h) - g(x)}{h}\right] + \lim_{h\to 0} g(x)\left[\frac{f(x + h) - f(x)}{h}\right]$$

By Property 3 of limits

$$= \lim_{h\to 0} f(x + h)\cdot\lim_{h\to 0}\frac{g(x + h) - g(x)}{h}$$

$$+ \lim_{h\to 0} g(x)\cdot\lim_{h\to 0}\frac{f(x + h) - f(x)}{h}$$

By Property 4 of limits

$$= f(x)g'(x) + g(x)f'(x)$$

Observe that in the second from the last link in the chain of equalities, we have used the fact that $\lim_{h\to 0} f(x+h) = f(x)$ because f is continuous at x.

3.2 Self-Check Exercises

1. Find the derivative of $f(x) = \dfrac{2x+1}{x^2-1}$.

2. What is the slope of the tangent line to the graph of
$$f(x) = (x^2+1)(2x^3-3x^2+1)$$
at the point (2, 25)? How fast is the function f changing when $x = 2$?

3. The total sales of Security Products in its first 2 yr of operation are given by
$$S = f(t) = \frac{0.3t^3}{1+0.4t^2} \qquad (0 \le t \le 2)$$
where S is measured in millions of dollars and $t = 0$ corresponds to the date Security Products began operations. How fast were the sales increasing at the beginning of the company's second year of operation?

Solutions to Self-Check Exercises 3.2 can be found on page 180.

3.2 Concept Questions

1. State the rule of differentiation in your own words.
   a. Product rule **b.** Quotient rule

2. If $f(1) = 3$, $g(1) = 2$, $f'(1) = -1$, and $g'(1) = 4$, find
   a. $h'(1)$ if $h(x) = f(x)g(x)$ **b.** $F'(1)$ if $F(x) = \dfrac{f(x)}{g(x)}$

3.2 Exercises

In Exercises 1–30, find the derivative of each function.

1. $f(x) = 2x(x^2+1)$
2. $f(x) = 3x^2(x-1)$
3. $f(t) = (t-1)(2t+1)$
4. $f(x) = (2x+3)(3x-4)$
5. $f(x) = (3x+1)(x^2-2)$
6. $f(x) = (x+1)(2x^2-3x+1)$
7. $f(x) = (x^3-1)(x+1)$
8. $f(x) = (x^3-12x)(3x^2+2x)$
9. $f(w) = (w^3-w^2+w-1)(w^2+2)$
10. $f(x) = \frac{1}{5}x^5 + (x^2+1)(x^2-x-1) + 28$
11. $f(x) = (5x^2+1)(2\sqrt{x}-1)$
12. $f(t) = (1+\sqrt{t})(2t^2-3)$
13. $f(x) = (x^2-5x+2)\left(x - \dfrac{2}{x}\right)$
14. $f(x) = (x^3+2x+1)\left(2 + \dfrac{1}{x^2}\right)$
15. $f(x) = \dfrac{1}{x-2}$
16. $g(x) = \dfrac{3}{2x+4}$
17. $f(x) = \dfrac{x-1}{2x+1}$
18. $f(t) = \dfrac{1-2t}{1+3t}$
19. $f(x) = \dfrac{1}{x^2+1}$
20. $f(u) = \dfrac{u}{u^2+1}$
21. $f(s) = \dfrac{s^2-4}{s+1}$
22. $f(x) = \dfrac{x^3-2}{x^2+1}$
23. $f(x) = \dfrac{\sqrt{x}+1}{x^2+1}$
24. $f(x) = \dfrac{x^2+1}{\sqrt{x}}$
25. $f(x) = \dfrac{x^2+2}{x^2+x+1}$
26. $f(x) = \dfrac{x+1}{2x^2+2x+3}$
27. $f(x) = \dfrac{(x+1)(x^2+1)}{x-2}$
28. $f(x) = (3x^2-1)\left(x^2 - \dfrac{1}{x}\right)$
29. $f(x) = \dfrac{x}{x^2-4} - \dfrac{x-1}{x^2+4}$
30. $f(x) = \dfrac{x+\sqrt{3x}}{3x-1}$

In Exercises 31–34, suppose *f* and *g* are functions that are differentiable at *x* = 1 and that $f(1) = 2$, $f'(1) = -1$, $g(1) = -2$, and $g'(1) = 3$. Find the value of $h'(1)$.

31. $h(x) = f(x)g(x)$

32. $h(x) = (x^2 + 1)g(x)$

33. $h(x) = \dfrac{xf(x)}{x + g(x)}$

34. $h(x) = \dfrac{f(x)g(x)}{f(x) - g(x)}$

In Exercises 35–38, find the derivative of each function and evaluate $f'(x)$ at the given value of *x*.

35. $f(x) = (2x - 1)(x^2 + 3)$; $x = 1$

36. $f(x) = \dfrac{2x + 1}{2x - 1}$; $x = 2$

37. $f(x) = \dfrac{x}{x^4 - 2x^2 - 1}$; $x = -1$

38. $f(x) = (\sqrt{x} + 2x)(x^{3/2} - x)$; $x = 4$

In Exercises 39–42, find the slope and an equation of the tangent line to the graph of the function *f* at the specified point.

39. $f(x) = (x^3 + 1)(x^2 - 2)$; $(2, 18)$

40. $f(x) = \dfrac{x^2}{x + 1}$; $\left(2, \dfrac{4}{3}\right)$

41. $f(x) = \dfrac{x + 1}{x^2 + 1}$; $(1, 1)$

42. $f(x) = \dfrac{1 + 2x^{1/2}}{1 + x^{3/2}}$; $\left(4, \dfrac{5}{9}\right)$

43. Suppose $g(x) = x^2 f(x)$ and it is known that $f(2) = 3$ and $f'(2) = -1$. Evaluate $g'(2)$.

44. Suppose $g(x) = (x^2 + 1)f(x)$ and it is known that $f(2) = 3$ and $f'(2) = -1$. Evaluate $g'(2)$.

45. Find an equation of the tangent line to the graph of the function $f(x) = (x^3 + 1)(3x^2 - 4x + 2)$ at the point $(1, 2)$.

46. Find an equation of the tangent line to the graph of the function $f(x) = \dfrac{3x}{x^2 - 2}$ at the point $(2, 3)$.

47. Let $f(x) = (x^2 + 1)(2 - x)$. Find the point(s) on the graph of *f* where the tangent line is horizontal.

48. Let $f(x) = \dfrac{x}{x^2 + 1}$. Find the point(s) on the graph of *f* where the tangent line is horizontal.

49. Find the point(s) on the graph of the function $f(x) = (x^2 + 6)(x - 5)$ where the slope of the tangent line is equal to -2.

50. Find the point(s) on the graph of the function $f(x) = \dfrac{x + 1}{x - 1}$ where the slope of the tangent line is equal to $-\frac{1}{2}$.

51. A straight line perpendicular to and passing through the point of tangency of the tangent line is called the *normal* to the curve. Find the equation of the tangent line and the normal to the curve

$$y = \frac{1}{1 + x^2}$$

at the point $(1, \frac{1}{2})$.

52. **Concentration of a Drug in the Bloodstream** The concentration of a certain drug in a patient's bloodstream *t* hr after injection is given by

$$C(t) = \frac{0.2t}{t^2 + 1}$$

a. Find the rate at which the concentration of the drug is changing with respect to time.
b. How fast is the concentration changing $\frac{1}{2}$ hr, 1 hr, and 2 hr after the injection?

53. **Cost of Removing Toxic Waste** A city's main well was recently found to be contaminated with trichloroethylene, a cancer-causing chemical, as a result of an abandoned chemical dump leaching chemicals into the water. A proposal submitted to the city's council members indicates that the cost, measured in millions of dollars, of removing *x*% of the toxic pollutant is given by

$$C(x) = \frac{0.5x}{100 - x}$$

Find $C'(80)$, $C'(90)$, $C'(95)$, and $C'(99)$. What does your result tell you about the cost of removing *all* of the pollutant?

54. **Drug Dosages** Thomas Young has suggested the following rule for calculating the dosage of medicine for children 1 to 12 yr old. If *a* denotes the adult dosage (in milligrams) and if *t* is the child's age (in years), then the child's dosage is given by

$$D(t) = \frac{at}{t + 12}$$

Suppose the adult dosage of a substance is 500 mg. Find an expression that gives the rate of change of a child's dosage with respect to the child's age. What is the rate of change of a child's dosage with respect to his or her age for a 6-yr-old child? A 10-yr-old child?

55. **Effect of Bactericide** The number of bacteria $N(t)$ in a certain culture *t* min after an experimental bactericide is introduced obeys the rule

$$N(t) = \frac{10{,}000}{1 + t^2} + 2000$$

Find the rate of change of the number of bacteria in the culture 1 min and 2 min after the bactericide is introduced. What is the population of the bacteria in the culture 1 min and 2 min after the bactericide is introduced?

56. **DEMAND FUNCTIONS** The demand function for the Sicard wristwatch is given by

$$d(x) = \frac{50}{0.01x^2 + 1} \qquad (0 \le x \le 20)$$

where x (measured in units of a thousand) is the quantity demanded per week and $d(x)$ is the unit price in dollars.

a. Find $d'(x)$.

b. Find $d'(5)$, $d'(10)$, and $d'(15)$ and interpret your results.

57. **LEARNING CURVES** From experience, Emory Secretarial School knows that the average student taking Advanced Typing will progress according to the rule

$$N(t) = \frac{60t + 180}{t + 6} \qquad (t \ge 0)$$

where $N(t)$ measures the number of words/minute the student can type after t wk in the course.

a. Find an expression for $N'(t)$.

b. Compute $N'(t)$ for $t = 1$, 3, 4, and 7 and interpret your results.

c. Sketch the graph of the function N. Does it confirm the results obtained in part (b)?

d. What will be the average student's typing speed at the end of the 12-wk course?

58. **BOX-OFFICE RECEIPTS** The total worldwide box-office receipts for a long-running movie are approximated by the function

$$T(x) = \frac{120x^2}{x^2 + 4}$$

where $T(x)$ is measured in millions of dollars and x is the number of years since the movie's release. How fast are the total receipts changing 1 yr, 3 yr, and 5 yr after its release?

59. **FORMALDEHYDE LEVELS** A study on formaldehyde levels in 900 homes indicates that emissions of various chemicals can decrease over time. The formaldehyde level (parts per million) in an average home in the study is given by

$$f(t) = \frac{0.055t + 0.26}{t + 2} \qquad (0 \le t \le 12)$$

where t is the age of the house in years. How fast is the formaldehyde level of the average house dropping when it is new? At the beginning of its fourth year?

Source: Bonneville Power Administration

60. **POPULATION GROWTH** A major corporation is building a 4325-acre complex of homes, offices, stores, schools, and churches in the rural community of Glen Cove. As a result of this development, the planners have estimated that Glen Cove's population (in thousands) t yr from now will be given by

$$P(t) = \frac{25t^2 + 125t + 200}{t^2 + 5t + 40}$$

a. Find the rate at which Glen Cove's population is changing with respect to time.

b. What will be the population after 10 yr? At what rate will the population be increasing when $t = 10$?

In Exercises 61–64, determine whether the statement is true or false. If it is true, explain why it is true. If it is false, give an example to show why it is false.

61. If f and g are differentiable, then

$$\frac{d}{dx}[f(x)g(x)] = f'(x)g'(x)$$

62. If f is differentiable, then

$$\frac{d}{dx}[xf(x)] = f(x) + xf'(x)$$

63. If f is differentiable, then

$$\frac{d}{dx}\left[\frac{f(x)}{x^2}\right] = \frac{f'(x)}{2x}$$

64. If f, g, and h are differentiable, then

$$\frac{d}{dx}\left[\frac{f(x)g(x)}{h(x)}\right] = \frac{f'(x)g(x)h(x) + f(x)g'(x)h(x) - f(x)g(x)h'(x)}{[h(x)]^2}$$

65. Extend the product rule for differentiation to the following case involving the product of three differentiable functions: Let $h(x) = u(x)v(x)w(x)$ and show that $h'(x) = u(x)v(x)w'(x) + u(x)v'(x)w(x) + u'(x)v(x)w(x)$.

Hint: Let $f(x) = u(x)v(x)$, $g(x) = w(x)$, and $h(x) = f(x)g(x)$ and apply the product rule to the function h.

66. Prove the quotient rule for differentiation (Rule 6).

Hint: Let $k(x) = f(x)/g(x)$ and verify the following steps:

a. $$\frac{k(x+h) - k(x)}{h} = \frac{f(x+h)g(x) - f(x)g(x+h)}{hg(x+h)g(x)}$$

b. By adding $[-f(x)g(x) + f(x)g(x)]$ to the numerator and simplifying, show that

$$\frac{k(x+h) - k(x)}{h} = \frac{1}{g(x+h)g(x)} \times \left\{\left[\frac{f(x+h) - f(x)}{h}\right]\cdot g(x) - \left[\frac{g(x+h) - g(x)}{h}\right]\cdot f(x)\right\}$$

c. $$k'(x) = \lim_{h \to 0} \frac{k(x+h) - k(x)}{h} = \frac{g(x)f'(x) - f(x)g'(x)}{[g(x)]^2}$$

3.2 Solutions to Self-Check Exercises

1. We use the quotient rule to obtain

$$f'(x) = \frac{(x^2-1)\frac{d}{dx}(2x+1) - (2x+1)\frac{d}{dx}(x^2-1)}{(x^2-1)^2}$$
$$= \frac{(x^2-1)(2) - (2x+1)(2x)}{(x^2-1)^2}$$
$$= \frac{2x^2 - 2 - 4x^2 - 2x}{(x^2-1)^2}$$
$$= \frac{-2x^2 - 2x - 2}{(x^2-1)^2}$$
$$= \frac{-2(x^2 + x + 1)}{(x^2-1)^2}$$

2. The slope of the tangent line to the graph of f at any point is given by

$$f'(x) = (x^2+1)\frac{d}{dx}(2x^3 - 3x^2 + 1) + (2x^3 - 3x^2 + 1)\frac{d}{dx}(x^2+1)$$
$$= (x^2+1)(6x^2 - 6x) + (2x^3 - 3x^2 + 1)(2x)$$

In particular, the slope of the tangent line to the graph of f when $x = 2$ is

$$f'(2) = (2^2+1)[6(2^2) - 6(2)] + [2(2^3) - 3(2^2) + 1][2(2)]$$
$$= 60 + 20 = 80$$

Note that it is not necessary to simplify the expression for $f'(x)$ since we are required only to evaluate the expression at $x = 2$. We also conclude, from this result, that the function f is changing at the rate of 80 units/unit change in x when $x = 2$.

3. The rate at which the company's total sales are changing at any time t is given by

$$S'(t) = \frac{(1 + 0.4t^2)\frac{d}{dt}(0.3t^3) - (0.3t^3)\frac{d}{dt}(1 + 0.4t^2)}{(1 + 0.4t^2)^2}$$
$$= \frac{(1 + 0.4t^2)(0.9t^2) - (0.3t^3)(0.8t)}{(1 + 0.4t^2)^2}$$

Therefore, at the beginning of the second year of operation, Security Products' sales were increasing at the rate of

$$S'(1) = \frac{(1 + 0.4)(0.9) - (0.3)(0.8)}{(1 + 0.4)^2} = 0.520408$$

or \$520,408/year.

USING TECHNOLOGY

The Product and Quotient Rules

EXAMPLE 1 Let $f(x) = (2\sqrt{x} + 0.5x)(0.3x^3 + 2x - \frac{0.3}{x})$. Find $f'(0.2)$.

Solution Using the numerical derivative operation of a graphing utility, we find

$$f'(0.2) = 6.4797499802$$

See Figure T1.

```
nDeriv((2X^.5+.5
X)(.3X^3+2X–.3/X),
X,.2)
        6.4797499802
█
```

FIGURE **T1**
The TI-83/84 numerical derivative screen for computing $f'(0.2)$

APPLIED EXAMPLE 2 Importance of Time in Treating Heart Attacks According to the American Heart Association, the treatment benefit for heart attacks depends on the time until treatment and is described by the function

$$f(t) = \frac{0.44t^4 + 700}{0.1t^4 + 7} \qquad (0 \le t \le 24)$$

where t is measured in hours and $f(t)$ is expressed as a percent.

a. Use a graphing utility to graph the function f using the viewing window $[0, 24] \times [0, 100]$.
b. Use a graphing utility to find the derivative of f when $t = 0$ and $t = 2$.
c. Interpret the results obtained in part (b).

Source: American Heart Association

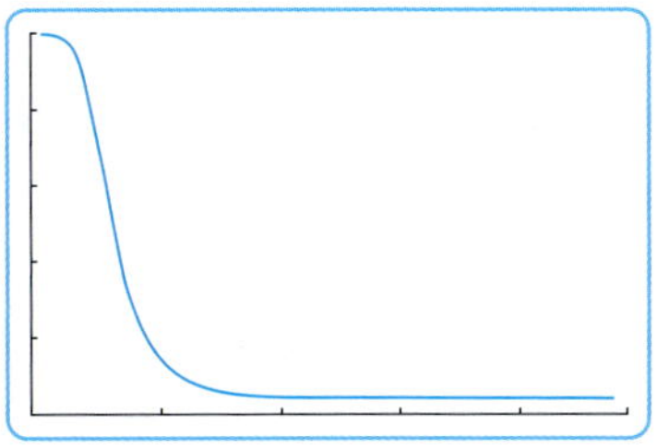

FIGURE **T2**

Solution

a. The graph of f is shown in Figure T2.
b. Using the numerical derivative operation of a graphing utility, we find

$$f'(0) \approx 0$$
$$f'(2) \approx -28.95402429$$

(see Figure T3).

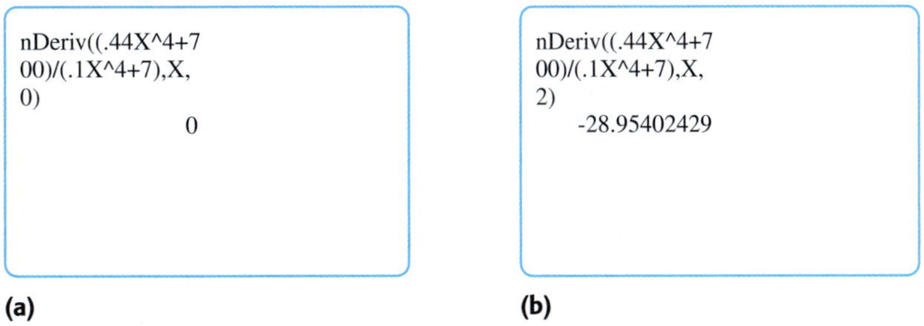

FIGURE **T3**
TI-83/84 numerical derivative screens (a) for computing $f'(0)$ and (b) for computing $f'(2)$

c. The results of part (b) show that there is no drop in the treatment benefit when the heart attack is treated immediately. But the treatment benefit drops off at the rate of approximately 29% per hour when the time to treatment is 2 hours. Thus, it is extremely urgent that a patient suffering a heart attack receive medical attention as soon as possible. ■

TECHNOLOGY EXERCISES

In Exercises 1–6, use the numerical derivative operation to find the rate of change of $f(x)$ at the given value of x. Give your answer accurate to four decimal places.

1. $f(x) = (2x^2 + 1)(x^3 + 3x + 4);\ x = -0.5$
2. $f(x) = (\sqrt{x} + 1)(2x^2 + x - 3);\ x = 1.5$
3. $f(x) = \dfrac{\sqrt{x} - 1}{\sqrt{x} + 1};\ x = 3$
4. $f(x) = \dfrac{\sqrt{x}(x^2 + 4)}{x^3 + 1};\ x = 4$
5. $f(x) = \dfrac{\sqrt{x}(1 + x^{-1})}{x + 1};\ x = 1$
6. $f(x) = \dfrac{x^2(2 + \sqrt{x})}{1 + \sqrt{x}};\ x = 1$
7. **New Construction Jobs** The president of a major housing construction company claims that the number of construction jobs created in the next t mo is given by

$$f(t) = 1.42\left(\frac{7t^2 + 140t + 700}{3t^2 + 80t + 550}\right)$$

where $f(t)$ is measured in millions of jobs/year. At what rate will construction jobs be created 1 yr from now, assuming her projection is correct?

8. **Population Growth** A major corporation is building a 4325-acre complex of homes, offices, stores, schools, and churches in the rural community of Glen Cove. As a result of this development, the planners have estimated that Glen Cove's population (in thousands) t yr from now will be given by

$$P(t) = \frac{25t^2 + 125t + 200}{t^2 + 5t + 40}$$

a. What will be the population 10 yr from now?
b. At what rate will the population be increasing 10 yr from now?

3.3 The Chain Rule

The population of Americans age 55 years and older as a percentage of the total population is approximated by the function

$$f(t) = 10.72(0.9t + 10)^{0.3} \qquad (0 \leq t \leq 20)$$

where t is measured in years with $t = 0$ corresponding to the year 2000 (Figure 6).

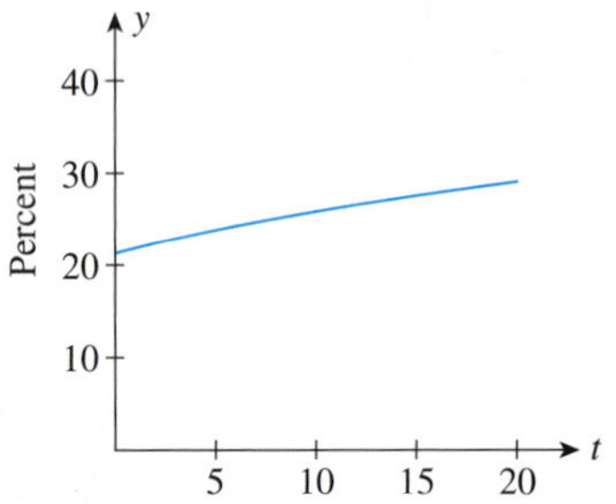

FIGURE 6
Population of Americans age 55 years and older
Source: U.S. Census Bureau

How fast will the population age 55 years and older be increasing at the beginning of 2012? To answer this question, we have to evaluate $f'(12)$, where f' is the derivative of f. But the rules of differentiation that we have developed up to now will not help us find the derivative of f.

In this section, we will introduce another rule of differentiation called the **chain rule.** When used in conjunction with the rules of differentiation developed in the last two sections, the chain rule enables us to greatly enlarge the class of functions that we are able to differentiate. (In Exercise 70, page 190, we will use the chain rule to answer the question posed in the introductory example.)

The Chain Rule

Consider the function $h(x) = (x^2 + x + 1)^2$. If we were to compute $h'(x)$ using only the rules of differentiation from the previous sections, then our approach might be to expand $h(x)$. Thus,

$$\begin{aligned} h(x) &= (x^2 + x + 1)^2 = (x^2 + x + 1)(x^2 + x + 1) \\ &= x^4 + 2x^3 + 3x^2 + 2x + 1 \end{aligned}$$

from which we find

$$h'(x) = 4x^3 + 6x^2 + 6x + 2$$

But what about the function $H(x) = (x^2 + x + 1)^{100}$? The same technique may be used to find the derivative of the function H, but the amount of work involved in this case would be prodigious! Consider, also, the function $G(x) = \sqrt{x^2 + 1}$. For each of the two functions H and G, the rules of differentiation of the previous sections cannot be applied directly to compute the derivatives H' and G'.

Observe that both H and G are **composite functions;** that is, each is composed of, or built up from, simpler functions. For example, the function H is composed of the two simpler functions $f(x) = x^2 + x + 1$ and $g(x) = x^{100}$ as follows:

$$\begin{aligned} H(x) &= g[f(x)] = [f(x)]^{100} \\ &= (x^2 + x + 1)^{100} \end{aligned}$$

In a similar manner, we see that the function G is composed of the two simpler functions $f(x) = x^2 + 1$ and $g(x) = \sqrt{x}$. Thus,

$$G(x) = g[f(x)] = \sqrt{f(x)}$$
$$= \sqrt{x^2 + 1}$$

As a first step toward finding the derivative h' of a composite function $h = g \circ f$ defined by $h(x) = g[f(x)]$, we write

$$u = f(x) \quad \text{and} \quad y = g[f(x)] = g(u)$$

The dependency of h on g and f is illustrated in Figure 7. Since u is a function of x, we may compute the derivative of u with respect to x, if f is a differentiable function, obtaining $du/dx = f'(x)$. Next, if g is a differentiable function of u, we may compute the derivative of g with respect to u, obtaining $dy/du = g'(u)$. Now, since the function h is composed of the function g and the function f, we might suspect that the rule $h'(x)$ for the derivative h' of h will be given by an expression that involves the rules for the derivatives of f and g. But how do we combine these derivatives to yield h'?

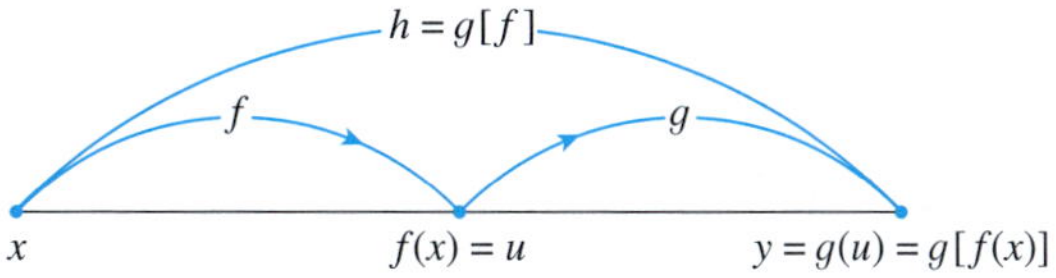

FIGURE 7
The composite function $h(x) = g[f(x)]$

This question can be answered by interpreting the derivative of each function as the rate of change of that function. For example, suppose $u = f(x)$ changes three times as fast as x—that is,

$$f'(x) = \frac{du}{dx} = 3$$

And suppose $y = g(u)$ changes twice as fast as u—that is,

$$g'(u) = \frac{dy}{du} = 2$$

Then, we would expect $y = h(x)$ to change six times as fast as x—that is,

$$h'(x) = g'(u)f'(x) = (2)(3) = 6$$

or equivalently,

$$\frac{dy}{dx} = \frac{dy}{du} \cdot \frac{du}{dx} = (2)(3) = 6$$

This observation suggests the following result, which we state without proof.

Rule 7: The Chain Rule

If $h(x) = g[f(x)]$, then

$$h'(x) = \frac{d}{dx} g(f(x)) = g'(f(x))f'(x) \tag{1}$$

Equivalently, if we write $y = h(x) = g(u)$, where $u = f(x)$, then

$$\frac{dy}{dx} = \frac{dy}{du} \cdot \frac{du}{dx} \tag{2}$$

Notes

1. If we label the composite function h in the following manner

$$h(x) = \underset{\substack{\uparrow \\ \text{Outside function}}}{g}[\overset{\substack{\text{Inside function} \\ \downarrow}}{f(x)}]$$

then $h'(x)$ is just the *derivative* of the "outside function" *evaluated at* the "inside function" times the *derivative* of the "inside function."

2. Equation (2) can be remembered by observing that if we "cancel" the du's, then

$$\frac{dy}{dx} = \frac{dy}{\cancel{du}} \cdot \frac{\cancel{du}}{dx} = \frac{dy}{dx}$$

The Chain Rule for Powers of Functions

Many composite functions have the special form $h(x) = g(f(x))$, where g is defined by the rule $g(x) = x^n$ (n, a real number)—that is,

$$h(x) = [f(x)]^n$$

In other words, the function h is given by the power of a function f. The functions

$$h(x) = (x^2 + x + 1)^2 \qquad H(x) = (x^2 + x + 1)^{100} \qquad G(x) = \sqrt{x^2 + 1}$$

discussed earlier are examples of this type of composite function. By using the following corollary of the chain rule, the general power rule, we can find the derivative of this type of function much more easily than by using the chain rule directly.

The General Power Rule

If the function f is differentiable and $h(x) = [f(x)]^n$ (n, a real number), then

$$h'(x) = \frac{d}{dx}[f(x)]^n = n[f(x)]^{n-1}f'(x) \qquad \textbf{(3)}$$

To see this, we observe that $h(x) = g(f(x))$, where $g(x) = x^n$, so that, by virtue of the chain rule, we have

$$\begin{aligned} h'(x) &= g'(f(x))f'(x) \\ &= n[f(x)]^{n-1}f'(x) \end{aligned}$$

since $g'(x) = nx^{n-1}$.

EXAMPLE 1 Let $F(x) = (3x + 1)^2$.

a. Find $F'(x)$, using the general power rule.
b. Verify your result without the benefit of the general power rule.

Solution

a. Using the general power rule, we obtain

$$\begin{aligned} F'(x) &= 2(3x + 1)^1 \frac{d}{dx}(3x + 1) \\ &= 2(3x + 1)(3) \\ &= 6(3x + 1) \end{aligned}$$

b. We first expand $F(x)$. Thus,

$$F(x) = (3x + 1)^2 = 9x^2 + 6x + 1$$

Next, differentiating, we have

$$\begin{aligned} F'(x) &= \frac{d}{dx}(9x^2 + 6x + 1) \\ &= 18x + 6 \\ &= 6(3x + 1) \end{aligned}$$

as before. ■

EXAMPLE 2 Differentiate the function $G(x) = \sqrt{x^2 + 1}$.

Solution We rewrite the function $G(x)$ as

$$G(x) = (x^2 + 1)^{1/2}$$

and apply the general power rule, obtaining

$$\begin{aligned} G'(x) &= \frac{1}{2}(x^2 + 1)^{-1/2}\frac{d}{dx}(x^2 + 1) \\ &= \frac{1}{2}(x^2 + 1)^{-1/2} \cdot 2x = \frac{x}{\sqrt{x^2 + 1}} \end{aligned}$$ ■

VIDEO

EXAMPLE 3 Differentiate the function $f(x) = x^2(2x + 3)^5$.

Solution Applying the product rule followed by the general power rule, we obtain

$$\begin{aligned} f'(x) &= x^2\frac{d}{dx}(2x + 3)^5 + (2x + 3)^5\frac{d}{dx}(x^2) \\ &= (x^2)5(2x + 3)^4 \cdot \frac{d}{dx}(2x + 3) + (2x + 3)^5(2x) \\ &= 5x^2(2x + 3)^4(2) + 2x(2x + 3)^5 \\ &= 2x(2x + 3)^4(5x + 2x + 3) = 2x(7x + 3)(2x + 3)^4 \end{aligned}$$ ■

EXAMPLE 4 Find $f'(x)$ if $f(x) = (2x^2 + 3)^4(3x - 1)^5$.

Solution Applying the product rule, we have

$$f'(x) = (2x^2 + 3)^4\frac{d}{dx}(3x - 1)^5 + (3x - 1)^5\frac{d}{dx}(2x^2 + 3)^4$$

Next, we apply the general power rule to each term, obtaining

$$\begin{aligned} f'(x) &= (2x^2 + 3)^4 \cdot 5(3x - 1)^4\frac{d}{dx}(3x - 1) + (3x - 1)^5 \cdot 4(2x^2 + 3)^3\frac{d}{dx}(2x^2 + 3) \\ &= 5(2x^2 + 3)^4(3x - 1)^4 \cdot 3 + 4(3x - 1)^5(2x^2 + 3)^3(4x) \end{aligned}$$

Finally, observing that $(2x^2 + 3)^3(3x - 1)^4$ is common to both terms, we can factor and simplify as follows:

$$\begin{aligned} f'(x) &= (2x^2 + 3)^3(3x - 1)^4[15(2x^2 + 3) + 16x(3x - 1)] \\ &= (2x^2 + 3)^3(3x - 1)^4(30x^2 + 45 + 48x^2 - 16x) \\ &= (2x^2 + 3)^3(3x - 1)^4(78x^2 - 16x + 45) \end{aligned}$$ ■

EXAMPLE 5 Find $f'(x)$ if $f(x) = \dfrac{1}{(4x^2 - 7)^2}$.

Solution Rewriting $f(x)$ and then applying the general power rule, we obtain

$$
\begin{aligned}
f'(x) &= \frac{d}{dx}\left[\frac{1}{(4x^2 - 7)^2}\right] = \frac{d}{dx}(4x^2 - 7)^{-2} \\
&= -2(4x^2 - 7)^{-3}\frac{d}{dx}(4x^2 - 7) \\
&= -2(4x^2 - 7)^{-3}(8x) = -\frac{16x}{(4x^2 - 7)^3}
\end{aligned}
$$

EXAMPLE 6 Find the slope of the tangent line to the graph of the function

$$f(x) = \left(\frac{2x + 1}{3x + 2}\right)^3$$

at the point $(0, \frac{1}{8})$.

Solution The slope of the tangent line to the graph of f at any point is given by $f'(x)$. To compute $f'(x)$, we use the general power rule followed by the quotient rule, obtaining

$$
\begin{aligned}
f'(x) &= 3\left(\frac{2x + 1}{3x + 2}\right)^2 \frac{d}{dx}\left(\frac{2x + 1}{3x + 2}\right) \\
&= 3\left(\frac{2x + 1}{3x + 2}\right)^2 \left[\frac{(3x + 2)(2) - (2x + 1)(3)}{(3x + 2)^2}\right] \\
&= 3\left(\frac{2x + 1}{3x + 2}\right)^2 \left[\frac{6x + 4 - 6x - 3}{(3x + 2)^2}\right] \\
&= \frac{3(2x + 1)^2}{(3x + 2)^4}
\end{aligned}
$$

See page 16.

Combine like terms and simplify.

In particular, the slope of the tangent line to the graph of f at $(0, \frac{1}{8})$ is given by

$$f'(0) = \frac{3(0 + 1)^2}{(0 + 2)^4} = \frac{3}{16}$$

Exploring with TECHNOLOGY

Refer to Example 6.

1. Use a graphing utility to plot the graph of the function f, using the viewing window $[-2, 1] \times [-1, 2]$. Then draw the tangent line to the graph of f at the point $(0, \frac{1}{8})$.
2. For a better picture, repeat part 1 using the viewing window $[-1, 1] \times [-0.1, 0.3]$.
3. Use the numerical differentiation capability of the graphing utility to verify that the slope of the tangent line at $(0, \frac{1}{8})$ is $\frac{3}{16}$.

VIDEO

APPLIED EXAMPLE 7 Growth in a Health Club Membership The membership of The Fitness Center, which opened a few years ago, is approximated by the function

$$N(t) = 100(64 + 4t)^{2/3} \qquad (0 \le t \le 52)$$

where $N(t)$ gives the number of members at the beginning of week t.

a. Find $N'(t)$.
b. How fast was the center's membership increasing initially ($t = 0$)?
c. How fast was the membership increasing at the beginning of the 40th week?
d. What was the membership when the center first opened? At the beginning of the 40th week?

Solution

a. Using the general power rule, we obtain

$$\begin{aligned} N'(t) &= \frac{d}{dt}[100(64 + 4t)^{2/3}] \\ &= 100\frac{d}{dt}(64 + 4t)^{2/3} \\ &= 100\left(\frac{2}{3}\right)(64 + 4t)^{-1/3}\frac{d}{dt}(64 + 4t) \\ &= \frac{200}{3}(64 + 4t)^{-1/3}(4) \\ &= \frac{800}{3(64 + 4t)^{1/3}} \end{aligned}$$

b. The rate at which the membership was increasing when the center first opened is given by

$$N'(0) = \frac{800}{3(64)^{1/3}} \approx 66.7$$

or approximately 67 people per week.

c. The rate at which the membership was increasing at the beginning of the 40th week is given by

$$N'(40) = \frac{800}{3(64 + 160)^{1/3}} \approx 43.9$$

or approximately 44 people per week.

d. The membership when the center first opened is given by

$$N(0) = 100(64)^{2/3} = 100(16)$$

or approximately 1600 people. The membership at the beginning of the 40th week is given by

$$N(40) = 100(64 + 160)^{2/3} \approx 3688.3$$

or approximately 3688 people. ■

Explore & Discuss

The profit P of a one-product software manufacturer depends on the number of units of its products sold. The manufacturer estimates that it will sell x units of its product per week. Suppose $P = g(x)$ and $x = f(t)$, where g and f are differentiable functions.

1. Write an expression giving the rate of change of the profit with respect to the number of units sold.

2. Write an expression giving the rate of change of the number of units sold per week.

3. Write an expression giving the rate of change of the profit per week.

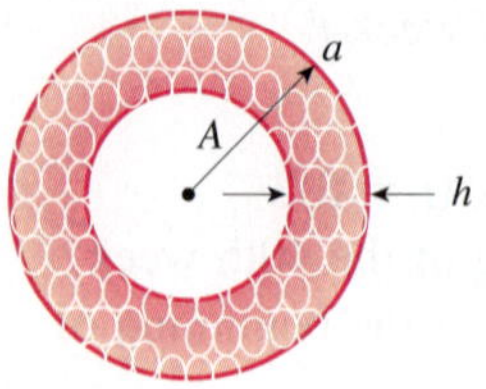

FIGURE 8
Cross section of the aorta

APPLIED EXAMPLE 8 Arteriosclerosis Arteriosclerosis begins during childhood when plaque (soft masses of fatty material) forms in the arterial walls, blocking the flow of blood through the arteries and leading to heart attacks, strokes, and gangrene. Suppose the idealized cross section of the aorta is circular with radius a cm and by year t the thickness of the plaque (assume it is uniform) is $h = g(t)$ cm (Figure 8). Then the area of the opening is given by $A = \pi(a - h)^2$ square centimeters (cm^2).

Suppose the radius of an individual's artery is 1 cm ($a = 1$) and the thickness of the plaque in year t is given by

$$h = g(t) = 1 - 0.01(10{,}000 - t^2)^{1/2} \text{ cm}$$

Since the area of the arterial opening is given by

$$A = f(h) = \pi(1 - h)^2$$

the rate at which A is changing with respect to time is given by

$$\frac{dA}{dt} = \frac{dA}{dh} \cdot \frac{dh}{dt} = f'(h) \cdot g'(t) \qquad \text{By the chain rule}$$

$$= 2\pi(1 - h)(-1)\left[-0.01\left(\frac{1}{2}\right)(10{,}000 - t^2)^{-1/2}(-2t)\right] \qquad \text{Use the chain rule twice.}$$

$$= -2\pi(1 - h)\left[\frac{0.01t}{(10{,}000 - t^2)^{1/2}}\right]$$

$$= -\frac{0.02\pi(1 - h)t}{\sqrt{10{,}000 - t^2}}$$

For example, when $t = 50$,

$$h = g(50) = 1 - 0.01(10{,}000 - 2500)^{1/2} \approx 0.134$$

so that

$$\frac{dA}{dt} = -\frac{0.02\pi(1 - 0.134)50}{\sqrt{10{,}000 - 2500}} \approx -0.03$$

That is, the area of the arterial opening is decreasing at the rate of 0.03 cm^2 per year. ■

Explore & Discuss

Suppose the population P of a certain bacteria culture is given by $P = f(T)$, where T is the temperature of the medium. Further, suppose the temperature T is a function of time t in seconds—that is, $T = g(t)$. Give an interpretation of each of the following quantities:

1. $\dfrac{dP}{dT}$ **2.** $\dfrac{dT}{dt}$ **3.** $\dfrac{dP}{dt}$ **4.** $(f \circ g)(t)$ **5.** $f'(g(t))g'(t)$

3.3 Self-Check Exercises

1. Find the derivative of

$$f(x) = -\frac{1}{\sqrt{2x^2 - 1}}$$

2. Suppose the life expectancy at birth (in years) of a female in a certain country is described by the function

$$g(t) = 50.02(1 + 1.09t)^{0.1} \qquad (0 \le t \le 150)$$

where t is measured in years, with $t = 0$ corresponding to the beginning of 1900.

a. What is the life expectancy at birth of a female born at the beginning of 1980? At the beginning of 2000?

b. How fast is the life expectancy at birth of a female born at any time t changing?

Solutions to Self-Check Exercises 3.3 can be found on page 192.

3.3 Concept Questions

1. In your own words, state the chain rule for differentiating the composite function $h(x) = g[f(x)]$.
2. In your own words, state the general power rule for differentiating the function $h(x) = [f(x)]^n$, where n is a real number.
3. If $f(t)$ gives the number of units of a certain product sold by a company after t days, and $g(x)$ gives the revenue (in dollars) realized from the sale of x units of the company's products, what does $(g \circ f)'(t)$ describe?
4. Suppose $f(x)$ gives the air temperature in the gondola of a hot-air balloon when it is at an altitude of x ft from the ground and $g(t)$ gives the altitude of the balloon t min after lifting off from the ground. Find a function giving the rate of change of the air temperature in the gondola at time t.

3.3 Exercises

In Exercises 1–48, find the derivative of each function.

1. $f(x) = (2x - 1)^4$
2. $f(x) = (1 - x)^3$
3. $f(x) = (x^2 + 2)^5$
4. $f(t) = 2(t^3 - 1)^5$
5. $f(x) = (2x - x^2)^3$
6. $f(x) = 3(x^3 - x)^4$
7. $f(x) = (2x + 1)^{-2}$
8. $f(t) = \frac{1}{2}(2t^2 + t)^{-3}$
9. $f(x) = (x^2 - 4)^{3/2}$
10. $f(t) = (3t^2 - 2t + 1)^{3/2}$
11. $f(x) = \sqrt{3x - 2}$
12. $f(t) = \sqrt{3t^2 - t}$
13. $f(x) = \sqrt[3]{1 - x^2}$
14. $f(x) = \sqrt{2x^2 - 2x + 3}$
15. $f(x) = \dfrac{1}{(2x + 3)^3}$
16. $f(x) = \dfrac{2}{(x^2 - 1)^4}$
17. $f(t) = \dfrac{1}{\sqrt{2t - 3}}$
18. $f(x) = \dfrac{1}{\sqrt{2x^2 - 1}}$
19. $y = \dfrac{1}{(4x^4 + x)^{3/2}}$
20. $f(t) = \dfrac{4}{\sqrt[3]{2t^2 + t}}$
21. $f(x) = (3x^2 + 2x + 1)^{-2}$
22. $f(t) = (5t^3 + 2t^2 - t + 4)^{-3}$
23. $f(x) = (x^2 + 1)^3 - (x^3 + 1)^2$
24. $f(t) = (2t - 1)^4 + (2t + 1)^4$
25. $f(t) = (t^{-1} - t^{-2})^3$
26. $f(v) = (v^{-3} + 4v^{-2})^3$
27. $f(x) = \sqrt{x + 1} + \sqrt{x - 1}$
28. $f(u) = (2u + 1)^{3/2} + (u^2 - 1)^{-3/2}$
29. $f(x) = 2x^2(3 - 4x)^4$
30. $h(t) = t^2(3t + 4)^3$
31. $f(x) = (x - 1)^2(2x + 1)^4$
32. $g(u) = (1 + u^2)^5(1 - 2u^2)^8$
33. $f(x) = \left(\dfrac{x + 3}{x - 2}\right)^3$
34. $f(x) = \left(\dfrac{x + 1}{x - 1}\right)^5$
35. $s(t) = \left(\dfrac{t}{2t + 1}\right)^{3/2}$
36. $g(s) = \left(s^2 + \dfrac{1}{s}\right)^{3/2}$
37. $g(u) = \sqrt{\dfrac{u + 1}{3u + 2}}$
38. $g(x) = \sqrt{\dfrac{2x + 1}{2x - 1}}$
39. $f(x) = \dfrac{x^2}{(x^2 - 1)^4}$
40. $g(u) = \dfrac{2u^2}{(u^2 + u)^3}$
41. $h(x) = \dfrac{(3x^2 + 1)^3}{(x^2 - 1)^4}$
42. $g(t) = \dfrac{(2t - 1)^2}{(3t + 2)^4}$
43. $f(x) = \dfrac{\sqrt{2x + 1}}{x^2 - 1}$
44. $f(t) = \dfrac{4t^2}{\sqrt{2t^2 + 2t - 1}}$
45. $g(t) = \dfrac{\sqrt{t + 1}}{\sqrt{t^2 + 1}}$
46. $f(x) = \dfrac{\sqrt{x^2 + 1}}{\sqrt{x^2 - 1}}$
47. $f(x) = (3x + 1)^4(x^2 - x + 1)^3$
48. $g(t) = (2t + 3)^2(3t^2 - 1)^{-3}$

In Exercises 49–54, find $\dfrac{dy}{du}$, $\dfrac{du}{dx}$, and $\dfrac{dy}{dx}$.

49. $y = u^{4/3}$ and $u = 3x^2 - 1$
50. $y = \sqrt{u}$ and $u = 7x - 2x^2$
51. $y = u^{-2/3}$ and $u = 2x^3 - x + 1$
52. $y = 2u^2 + 1$ and $u = x^2 + 1$
53. $y = \sqrt{u} + \dfrac{1}{\sqrt{u}}$ and $u = x^3 - x$
54. $y = \dfrac{1}{u}$ and $u = \sqrt{x} + 1$
55. Suppose $F(x) = g(f(x))$ and $f(2) = 3$, $f'(2) = -3$, $g(3) = 5$, and $g'(3) = 4$. Find $F'(2)$.
56. Suppose $h = f \circ g$. Find $h'(0)$ given that $f(0) = 6$, $f'(5) = -2$, $g(0) = 5$, and $g'(0) = 3$.
57. Suppose $F(x) = f(x^2 + 1)$. Find $F'(1)$ if $f'(2) = 3$.
58. Let $F(x) = f(f(x))$. Does it follow that $F'(x) = [f'(x)]^2$?
Hint: Let $f(x) = x^2$.
59. Suppose $h = g \circ f$. Does it follow that $h' = g' \circ f'$?
Hint: Let $f(x) = x$ and $g(x) = x^2$.
60. Suppose $h = f \circ g$. Show that $h' = (f' \circ g)g'$.

In Exercises 61–64, find an equation of the tangent line to the graph of the function at the given point.

61. $f(x) = (1 - x)(x^2 - 1)^2$; $(2, -9)$

62. $f(x) = \left(\frac{x + 1}{x - 1}\right)^2$; $(3, 4)$

63. $f(x) = x\sqrt{2x^2 + 7}$; $(3, 15)$

64. $f(x) = \frac{8}{\sqrt{x^2 + 6x}}$; $(2, 2)$

65. **Television Viewing** The number of viewers of a television series introduced several years ago is approximated by the function

$$N(t) = (60 + 2t)^{2/3} \qquad (1 \le t \le 26)$$

where $N(t)$ (measured in millions) denotes the number of weekly viewers of the series in the tth week. Find the rate of increase of the weekly audience at the end of week 2 and at the end of week 12. How many viewers were there in week 2? In week 24?

66. **Outsourcing of Jobs** According to a study conducted in 2003, the total number of U.S. jobs that are projected to leave the country by year t, where $t = 0$ corresponds to the beginning of 2000, is

$$N(t) = 0.0018425(t + 5)^{2.5} \qquad (0 \le t \le 15)$$

where $N(t)$ is measured in millions. How fast was the number of U.S. jobs that were outsourced changing in 2005? In 2010 ($t = 10$)?

Source: Forrester Research

67. **Working Mothers** The percentage of mothers who work outside the home and have children younger than age 6 yr is approximated by the function

$$P(t) = 33.55(t + 5)^{0.205} \qquad (0 \le t \le 21)$$

where t is measured in years, with $t = 0$ corresponding to the beginning of 1980. Compute $P'(t)$. At what rate was the percentage of these mothers changing at the beginning of 2000? What was the percentage of these mothers at the beginning of 2000?

Source: U.S. Bureau of Labor Statistics

68. **Selling Price of DVD Recorders** The rise of digital music and the improvement to the DVD format are some of the reasons why the average selling price of standalone DVD recorders will drop in the coming years. The function

$$A(t) = \frac{699}{(t + 1)^{0.94}} \qquad (0 \le t \le 5)$$

gives the projected average selling price (in dollars) of stand-alone DVD recorders in year t, where $t = 0$ corresponds to the beginning of 2002. How fast was the average selling price of standalone DVD recorders falling at the beginning of 2002? How fast was it falling at the beginning of 2006?

Source: Consumer Electronics Association

69. **Socially Responsible Funds** Since its inception in 1971, socially responsible investments, or SRIs, have yielded returns to investors on par with investments in general. The assets of socially responsible funds (in billions of dollars) from 1991 through 2001 is given by

$$f(t) = 23.7(0.2t + 1)^{1.32} \qquad (0 \le t \le 11)$$

where $t = 0$ corresponds to the beginning of 1991.

a. Find the rate at which the assets of SRIs were changing at the beginning of 2000.
b. What were the assets of SRIs at the beginning of 2000?

Source: Thomson Financial Wiesenberger

70. **Aging Population** The population of Americans age 55 yr and older as a percentage of the total population is approximated by the function

$$f(t) = 10.72(0.9t + 10)^{0.3} \qquad (0 \le t \le 20)$$

where t is measured in years, with $t = 0$ corresponding to the year 2000. At what rate was the percentage of Americans age 55 and over changing at the beginning of 2000? At what rate will the percentage of Americans age 55 yr and older be changing in 2010? What will be the percentage of the population of Americans age 55 yr and older in 2010?

Source: U.S. Census Bureau

71. **Concentration of Carbon Monoxide (CO) in the Air** According to a joint study conducted by Oxnard's Environmental Management Department and a state government agency, the concentration of CO in the air due to automobile exhaust t yr from now is given by

$$C(t) = 0.01(0.2t^2 + 4t + 64)^{2/3}$$

parts per million.

a. Find the rate at which the level of CO is changing with respect to time.
b. Find the rate at which the level of CO will be changing 5 yr from now.

72. **Continuing Education Enrollment** The registrar of Kellogg University estimates that the total student enrollment in the Continuing Education division will be given by

$$N(t) = -\frac{20{,}000}{\sqrt{1 + 0.2t}} + 21{,}000$$

where $N(t)$ denotes the number of students enrolled in the division t yr from now. Find an expression for $N'(t)$. How fast is the student enrollment increasing currently? How fast will it be increasing 5 yr from now?

73. **Air Pollution** According to the South Coast Air Quality Management District, the level of nitrogen dioxide, a brown gas that impairs breathing, present in the atmosphere on a certain May day in downtown Los Angeles is approximated by

$$A(t) = 0.03t^3(t - 7)^4 + 60.2 \qquad (0 \le t \le 7)$$

where $A(t)$ is measured in pollutant standard index and t is measured in hours, with $t = 0$ corresponding to 7 a.m.

a. Find $A'(t)$.
b. Find $A'(1)$, $A'(3)$, and $A'(4)$ and interpret your results.
Source: Los Angeles Times

74. Effect of Luxury Tax on Consumption Government economists of a developing country determined that the purchase of imported perfume is related to a proposed "luxury tax" by the formula

$$N(x) = \sqrt{10{,}000 - 40x - 0.02x^2} \qquad (0 \le x \le 200)$$

where $N(x)$ measures the percentage of normal consumption of perfume when a "luxury tax" of $x\%$ is imposed on it. Find the rate of change of $N(x)$ for taxes of 10%, 100%, and 150%.

75. Pulse Rate of an Athlete The pulse rate (the number of heartbeats/minute) of a long-distance runner t sec after leaving the starting line is given by

$$P(t) = \frac{300\sqrt{\frac{1}{2}t^2 + 2t + 25}}{t + 25} \qquad (t \ge 0)$$

Compute $P'(t)$. How fast is the athlete's pulse rate increasing 10 sec, 60 sec, and 2 min into the run? What is her pulse rate 2 min into the run?

76. Thurstone Learning Model Psychologist L. L. Thurstone suggested the following relationship between learning time T and the length of a list n:

$$T = f(n) = An\sqrt{n - b}$$

where A and b are constants that depend on the person and the task.
a. Compute dT/dn and interpret your result.
b. For a certain person and a certain task, suppose $A = 4$ and $b = 4$. Compute $f'(13)$ and $f'(29)$ and interpret your results.

77. Oil Spills In calm waters, the oil spilling from the ruptured hull of a grounded tanker spreads in all directions. Assuming that the area polluted is a circle and that its radius is increasing at a rate of 2 ft/sec, determine how fast the area is increasing when the radius of the circle is 40 ft.

78. Arteriosclerosis Refer to Example 8, page 188. Suppose the radius of an individual's artery is 1 cm and the thickness of the plaque (in centimeters) t yr from now is given by

$$h = g(t) = \frac{0.5t^2}{t^2 + 10} \qquad (0 \le t \le 10)$$

How fast will the arterial opening be decreasing 5 yr from now?

79. Traffic Flow Opened in the late 1950s, the Central Artery in downtown Boston was designed to move 75,000 vehicles a day. The number of vehicles moved per day is approximated by the function

$$x = f(t) = 6.25t^2 + 19.75t + 74.75 \qquad (0 \le t \le 5)$$

where x is measured in thousands and t in decades, with $t = 0$ corresponding to the beginning of 1959. Suppose the average speed of traffic flow in mph is given by

$$S = g(x) = -0.00075x^2 + 67.5 \qquad (75 \le x \le 350)$$

where x has the same meaning as before. What was the rate of change of the average speed of traffic flow at the beginning of 1999? What was the average speed of traffic flow at that time?
Hint: $S = g[f(t)]$.

80. Hotel Occupancy Rates The occupancy rate of the all-suite Wonderland Hotel, located near an amusement park, is given by the function

$$r(t) = \frac{10}{81}t^3 - \frac{10}{3}t^2 + \frac{200}{9}t + 60 \qquad (0 \le t \le 12)$$

where t is measured in months, with $t = 0$ corresponding to the beginning of January. Management has estimated that the monthly revenue (in thousands of dollars/month) is approximated by the function

$$R(r) = -\frac{3}{5000}r^3 + \frac{9}{50}r^2 \qquad (0 \le r \le 100)$$

where r is the occupancy rate.
a. Find an expression that gives the rate of change of Wonderland's occupancy rate with respect to time.
b. Find an expression that gives the rate of change of Wonderland's monthly revenue with respect to the occupancy rate.
c. What is the rate of change of Wonderland's monthly revenue with respect to time at the beginning of January? At the beginning of July?
Hint: Use the chain rule to find $R'(r(0))r'(0)$ and $R'(r(6))r'(6)$.

81. Effect of Housing Starts on Jobs The president of a major housing construction firm claims that the number of construction jobs created is given by

$$N(x) = 1.42x$$

where x denotes the number of housing starts. Suppose the number of housing starts in the next t mo is expected to be

$$x(t) = \frac{7t^2 + 140t + 700}{3t^2 + 80t + 550}$$

million units/year. Find an expression that gives the rate at which the number of construction jobs will be created t mo from now. At what rate will construction jobs be created 1 yr from now?

82. Demand for PCs The quantity demanded per month, x, of a certain make of personal computer (PC) is related to the average unit price, p (in dollars), of PCs by the equation

$$x = f(p) = \frac{100}{9}\sqrt{810{,}000 - p^2}$$

It is estimated that t mo from now, the average price of a PC will be given by

$$p(t) = \frac{400}{1 + \frac{1}{8}\sqrt{t}} + 200 \qquad (0 \le t \le 60)$$

dollars. Find the rate at which the quantity demanded per month of the PCs will be changing 16 mo from now.

83. **Demand for Watches** The demand equation for the Sicard wristwatch is given by

$$x = f(p) = 10\sqrt{\frac{50 - p}{p}} \qquad (0 < p \leq 50)$$

where x (measured in units of a thousand) is the quantity demanded each week and p is the unit price in dollars. Find the rate of change of the quantity demanded of the wristwatches with respect to the unit price when the unit price is \$25.

84. **Cruise Ship Bookings** The management of Cruise World, operators of Caribbean luxury cruises, expects that the percentage of young adults booking passage on their cruises in the years ahead will rise dramatically. They have constructed the following model, which gives the percentage of young adult passengers in year t:

$$p = f(t) = 50\left(\frac{t^2 + 2t + 4}{t^2 + 4t + 8}\right) \qquad (0 \leq t \leq 5)$$

Young adults normally pick shorter cruises and generally spend less on their passage. The following model gives an approximation of the average amount of money R (in dollars) spent per passenger on a cruise when the percentage of young adults is p:

$$R(p) = 1000\left(\frac{p + 4}{p + 2}\right)$$

Find the rate at which the price of the average passage will be changing 2 yr from now.

In Exercises 85–88, determine whether the statement is true or false. If it is true, explain why it is true. If it is false, give an example to show why it is false.

85. If f and g are differentiable and $h = f \circ g$, then $h'(x) = f'[g(x)]g'(x)$.

86. If f is differentiable and c is a constant, then

$$\frac{d}{dx}[f(cx)] = cf'(cx)$$

87. If f is differentiable, then

$$\frac{d}{dx}\sqrt{f(x)} = \frac{f'(x)}{2\sqrt{f(x)}}$$

88. If f is differentiable, then

$$\frac{d}{dx}\left[f\left(\frac{1}{x}\right)\right] = f'\left(\frac{1}{x}\right)$$

89. In Section 3.1, we proved that

$$\frac{d}{dx}(x^n) = nx^{n-1}$$

for the special case when $n = 2$. Use the chain rule to show that

$$\frac{d}{dx}(x^{1/n}) = \frac{1}{n}x^{1/n-1}$$

for any nonzero integer n, assuming that $f(x) = x^{1/n}$ is differentiable.

Hint: Let $f(x) = x^{1/n}$ so that $[f(x)]^n = x$. Differentiate both sides with respect to x.

90. With the aid of Exercise 89, prove that

$$\frac{d}{dx}(x^r) = rx^{r-1}$$

for any rational number r.

Hint: Let $r = m/n$, where m and n are integers, with $n \neq 0$, and write $x^r = (x^m)^{1/n}$.

3.3 Solutions to Self-Check Exercises

1. Rewriting, we have

$$f(x) = -(2x^2 - 1)^{-1/2}$$

Using the general power rule, we find

$$\begin{aligned} f'(x) &= -\frac{d}{dx}(2x^2 - 1)^{-1/2} \\ &= -\left(-\frac{1}{2}\right)(2x^2 - 1)^{-3/2}\frac{d}{dx}(2x^2 - 1) \\ &= \frac{1}{2}(2x^2 - 1)^{-3/2}(4x) \\ &= \frac{2x}{(2x^2 - 1)^{3/2}} \end{aligned}$$

2. a. The life expectancy at birth of a female born at the beginning of 1980 is given by

$$g(80) = 50.02[1 + 1.09(80)]^{0.1} \approx 78.29$$

or approximately 78 yr. Similarly, the life expectancy at birth of a female born at the beginning of the year 2000 is given by

$$g(100) = 50.02[1 + 1.09(100)]^{0.1} \approx 80.04$$

or approximately 80 yr.

b. The rate of change of the life expectancy at birth of a female born at any time t is given by $g'(t)$. Using the general power rule, we have

$$\begin{aligned} g'(t) &= 50.02\frac{d}{dt}(1 + 1.09t)^{0.1} \\ &= (50.02)(0.1)(1 + 1.09t)^{-0.9}\frac{d}{dt}(1 + 1.09t) \\ &= (50.02)(0.1)(1.09)(1 + 1.09t)^{-0.9} \\ &= 5.45218(1 + 1.09t)^{-0.9} \\ &= \frac{5.45218}{(1 + 1.09t)^{0.9}} \end{aligned}$$

USING TECHNOLOGY

Finding the Derivative of a Composite Function

EXAMPLE 1 Find the rate of change of $f(x) = \sqrt{x}(1 + 0.02x^2)^{3/2}$ when $x = 2.1$.

Solution Using the numerical derivative operation of a graphing utility, we find

$$f'(2.1) = 0.5821463392$$

or approximately 0.58 unit per unit change in x. (See Figure T1.)

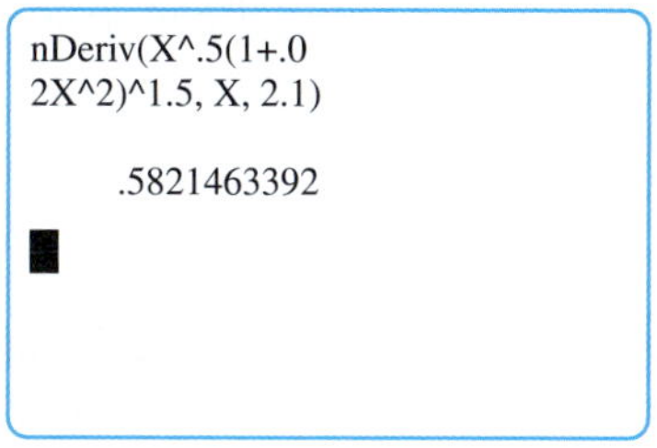

FIGURE **T1**
The TI-83/84 numerical derivative screen for computing $f'(2.1)$

APPLIED EXAMPLE 2 Amusement Park Attendance The management of AstroWorld ("The Amusement Park of the Future") estimates that the total number of visitors (in thousands) to the amusement park t hours after opening time at 9 a.m. is given by

$$N(t) = \frac{30t}{\sqrt{2 + t^2}}$$

What is the rate at which visitors are admitted to the amusement park at 10:30 a.m.?

Solution Using the numerical derivative operation of a graphing utility, we find

$$N'(1.5) \approx 6.8481$$

or approximately 6848 visitors per hour. (See Figure T2.)

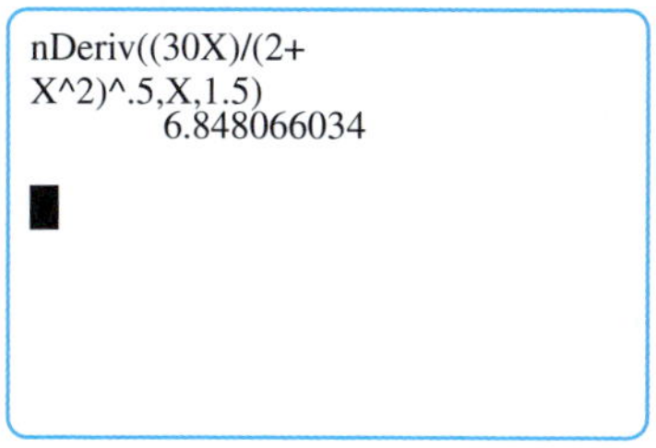

FIGURE **T2**
The TI-83/84 numerical derivative screen for computing $N'(1.5)$

TECHNOLOGY EXERCISES

In Exercises 1–6, use the numerical derivative operation to find the rate of change of $f(x)$ at the given value of x. Give your answer accurate to four decimal places.

1. $f(x) = \sqrt{x^2 - x^4}$; $x = 0.5$
2. $f(x) = x - \sqrt{1 - x^2}$; $x = 0.4$
3. $f(x) = x\sqrt{1 - x^2}$; $x = 0.2$
4. $f(x) = (x + \sqrt{x^2 + 4})^{3/2}$; $x = 1$
5. $f(x) = \dfrac{\sqrt{1 + x^2}}{x^3 + 2}$; $x = -1$
6. $f(x) = \dfrac{x^3}{1 + (1 + x^2)^{3/2}}$; $x = 3$
7. **Worldwide Production of Vehicles** The worldwide production of vehicles between 1960 and 1990 is given by the function

$$f(t) = 16.5\sqrt{1 + 2.2t} \qquad (0 \le t \le 3)$$

where $f(t)$ is measured in units of a million and t is measured in decades, with $t = 0$ corresponding to the beginning of 1960. What was the rate of change of the worldwide production of vehicles at the beginning of 1970? At the beginning of 1980?
Source: Automotive News

8. **Accumulation Years** People from their mid-40s to their mid-50s are in the prime investing years. Demographic studies of this type are of particular importance to financial institutions. The function

$$N(t) = 34.4(1 + 0.32125t)^{0.15} \qquad (0 \le t \le 12)$$

gives the projected number of people in this age group in the United States (in millions) in year t, where $t = 0$ corresponds to the beginning of 1996.
  a. How large was this segment of the population projected to be at the beginning of 2005?
  b. How fast was this segment of the population growing at the beginning of 2005?
Source: U.S. Census Bureau

3.4 Marginal Functions in Economics

Marginal analysis is the study of the rate of change of economic quantities. For example, an economist is not merely concerned with the value of an economy's gross domestic product (GDP) at a given time but is equally concerned with the rate at which it is growing or declining. In the same vein, a manufacturer is not only interested in the total cost corresponding to a certain level of production of a commodity but also is interested in the rate of change of the total cost with respect to the level of production, and so on. Let's begin with an example to explain the meaning of the adjective *marginal*, as used by economists.

Cost Functions

APPLIED EXAMPLE 1 Rate of Change of Cost Functions Suppose the total cost in dollars incurred each week by Polaraire for manufacturing x refrigerators is given by the total cost function

$$C(x) = 8000 + 200x - 0.2x^2 \qquad (0 \leq x \leq 400)$$

a. What is the actual cost incurred for manufacturing the 251st refrigerator?
b. Find the rate of change of the total cost function with respect to x when $x = 250$.
c. Compare the results obtained in parts (a) and (b).

Solution

a. The actual cost incurred in producing the 251st refrigerator is the difference between the total cost incurred in producing the first 251 refrigerators and the total cost of producing the first 250 refrigerators:

$$\begin{aligned} C(251) - C(250) &= [8000 + 200(251) - 0.2(251)^2] \\ &\quad - [8000 + 200(250) - 0.2(250)^2] \\ &= 45{,}599.8 - 45{,}500 \\ &= 99.80 \end{aligned}$$

or \$99.80.

b. The rate of change of the total cost function C with respect to x is given by the derivative of C—that is, $C'(x) = 200 - 0.4x$. Thus, when the level of production is 250 refrigerators, the rate of change of the total cost with respect to x is given by

$$\begin{aligned} C'(250) &= 200 - 0.4(250) \\ &= 100 \end{aligned}$$

or \$100.

c. From the solution to part (a), we know that the actual cost for producing the 251st refrigerator is \$99.80. This answer is very closely approximated by the answer to part (b), \$100. To see why this is so, observe that the difference $C(251) - C(250)$ may be written in the form

$$\frac{C(251) - C(250)}{1} = \frac{C(250 + 1) - C(250)}{1} = \frac{C(250 + h) - C(250)}{h}$$

where $h = 1$. In other words, the difference $C(251) - C(250)$ is precisely the average rate of change of the total cost function C over the interval [250, 251], or, equivalently, the slope of the secant line through the points (250, 45,500) and (251, 45,599.8). However, the number $C'(250) = 100$ is the instantaneous rate of change of the total cost function C at $x = 250$, or, equivalently, the slope of the tangent line to the graph of C at $x = 250$.

Now when h is small, the average rate of change of the function C is a good approximation to the instantaneous rate of change of the function C, or, equivalently, the slope of the secant line through the points in question is a good approximation to the slope of the tangent line through the point in question. Thus, we may expect

$$C(251) - C(250) = \frac{C(251) - C(250)}{1} \approx \frac{C(250 + h) - C(250)}{h} \quad (h \text{ small})$$

$$\approx \lim_{h \to 0} \frac{C(250 + h) - C(250)}{h} = C'(250)$$

which is precisely the case in this example.

The actual cost incurred in producing an additional unit of a certain commodity given that a plant is already at a certain level of operation is called the **marginal cost.** Knowing this cost is very important to management. As we saw in Example 1, the marginal cost is approximated by the rate of change of the total cost function evaluated at the appropriate point. For this reason, economists have defined the **marginal cost function** to be the derivative of the corresponding total cost function. In other words, if C is a total cost function, then the marginal cost function is defined to be its derivative C'. Thus, the adjective *marginal* is synonymous with *derivative of.*

APPLIED EXAMPLE 2 Marginal Cost Functions A subsidiary of Elektra Electronics manufactures a portable DVD player. Management determined that the daily total cost of producing these DVD players (in dollars) is given by

$$C(x) = 0.0001x^3 - 0.08x^2 + 40x + 5000$$

where x stands for the number of DVD players produced.

a. Find the marginal cost function.
b. What is the marginal cost when $x = 200, 300, 400$, and 600?
c. Interpret your results.

Solution

a. The marginal cost function C' is given by the derivative of the total cost function C. Thus,

$$C'(x) = 0.0003x^2 - 0.16x + 40$$

b. The marginal cost when $x = 200, 300, 400$, and 600 is given by

$$C'(200) = 0.0003(200)^2 - 0.16(200) + 40 = 20$$
$$C'(300) = 0.0003(300)^2 - 0.16(300) + 40 = 19$$
$$C'(400) = 0.0003(400)^2 - 0.16(400) + 40 = 24$$
$$C'(600) = 0.0003(600)^2 - 0.16(600) + 40 = 52$$

or \$20/unit, \$19/unit, \$24/unit, and \$52/unit, respectively.

c. From the results of part (b), we see that Elektra's actual cost for producing the 201st DVD player is approximately \$20. The actual cost incurred for producing one additional DVD player when the level of production is already 300

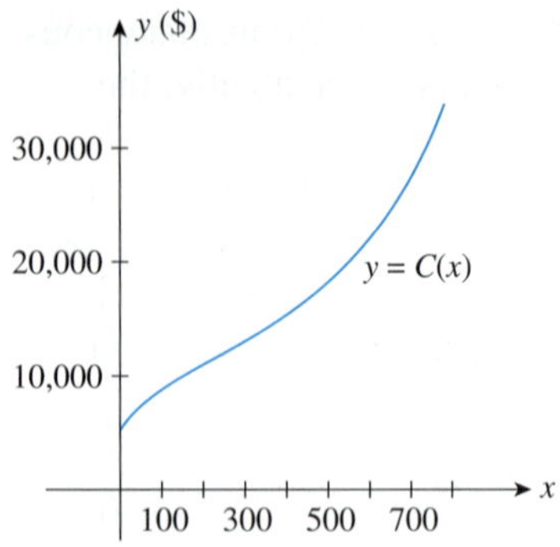

FIGURE 9
The cost of producing x DVD players is given by $C(x)$.

players is approximately \$19, and so on. Observe that when the level of production is already 600 units, the actual cost of producing one additional unit is approximately \$52. The higher cost for producing this additional unit when the level of production is 600 units may be the result of several factors, among them excessive costs incurred because of overtime or higher maintenance, production breakdown caused by greater stress and strain on the equipment, and so on. The graph of the total cost function appears in Figure 9. ■

Average Cost Functions

Let's now introduce another marginal concept closely related to the marginal cost. Let $C(x)$ denote the total cost incurred in producing x units of a certain commodity. Then the **average cost** of producing x units of the commodity is obtained by dividing the total production cost by the number of units produced. This leads to the following definition:

Average Cost Function

Suppose $C(x)$ is a total cost function. Then the **average cost function,** denoted by $\overline{C}(x)$ (read "C bar of x"), is

$$\frac{C(x)}{x} \tag{4}$$

The derivative $\overline{C}'(x)$ of the average cost function, called the **marginal average cost function,** measures the rate of change of the average cost function with respect to the number of units produced.

VIDEO

APPLIED EXAMPLE 3 Marginal Average Cost Functions The total cost of producing x units of a certain commodity is given by

$$C(x) = 400 + 20x$$

dollars.

a. Find the average cost function $\overline{C}$.
b. Find the marginal average cost function $\overline{C}'$.
c. What are the economic implications of your results?

Solution

a. The average cost function is given by

$$\overline{C}(x) = \frac{C(x)}{x} = \frac{400 + 20x}{x}$$
$$= 20 + \frac{400}{x}$$

b. The marginal average cost function is

$$\overline{C}'(x) = -\frac{400}{x^2}$$

c. Since the marginal average cost function is negative for all admissible values of x, the rate of change of the average cost function is negative for all $x > 0$;

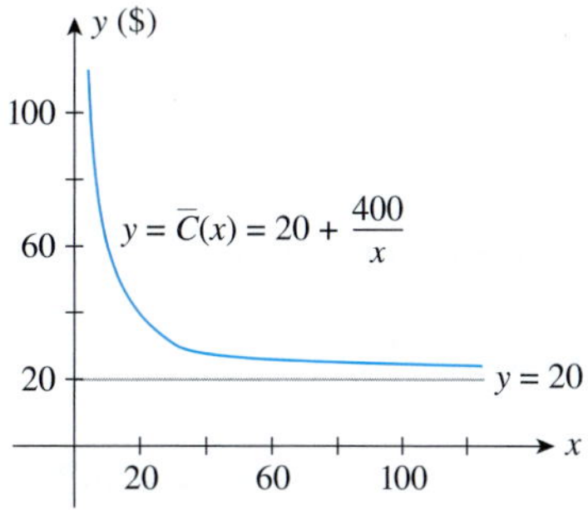

FIGURE 10
As the level of production increases, the average cost approaches \$20.

that is, $\overline{C}(x)$ decreases as x increases. However, the graph of $\overline{C}$ always lies above the horizontal line $y = 20$, but it approaches the line since

$$\lim_{x \to \infty} \overline{C}(x) = \lim_{x \to \infty} \left(20 + \frac{400}{x}\right) = 20$$

A sketch of the graph of the function $\overline{C}(x)$ appears in Figure 10. This result is fully expected if we consider the economic implications. Note that as the level of production increases, the fixed cost per unit of production, represented by the term $(400/x)$, drops steadily. The average cost approaches the constant unit cost of production, which is \$20 in this case. ■

APPLIED EXAMPLE 4 Marginal Average Cost Functions Once again consider the subsidiary of Elektra Electronics. The daily total cost for producing its portable DVD players is given by

$$C(x) = 0.0001x^3 - 0.08x^2 + 40x + 5000$$

dollars, where x stands for the number of DVD players produced (see Example 2).

a. Find the average cost function $\overline{C}$.
b. Find the marginal average cost function $\overline{C}'$. Compute $\overline{C}'(500)$.
c. Sketch the graph of the function $\overline{C}$ and interpret the results obtained in parts (a) and (b).

Solution

a. The average cost function is given by

$$\overline{C}(x) = \frac{C(x)}{x} = 0.0001x^2 - 0.08x + 40 + \frac{5000}{x}$$

b. The marginal average cost function is given by

$$\overline{C}'(x) = 0.0002x - 0.08 - \frac{5000}{x^2}$$

Also,

$$\overline{C}'(500) = 0.0002(500) - 0.08 - \frac{5000}{(500)^2} = 0$$

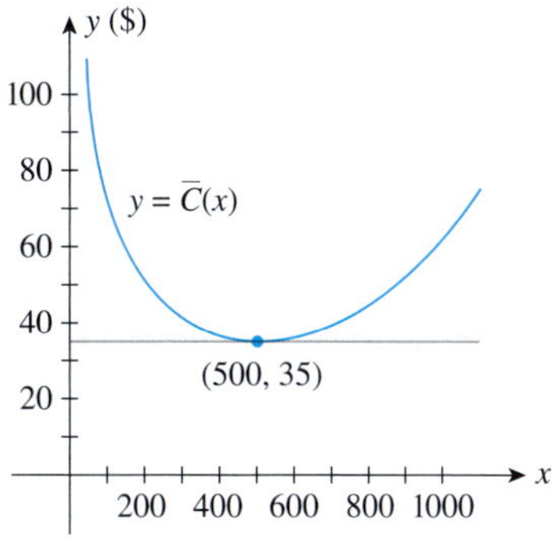

FIGURE 11
The average cost reaches a minimum of \$35 when 500 DVD players are produced.

c. To sketch the graph of the function $\overline{C}$, observe that if x is a small positive number, then $\overline{C}(x) > 0$. Furthermore, $\overline{C}(x)$ becomes arbitrarily large as x approaches zero from the right, since the term $(5000/x)$ becomes arbitrarily large as x approaches zero. Next, the result $\overline{C}'(500) = 0$ obtained in part (b) tells us that the tangent line to the graph of the function $\overline{C}$ is horizontal at the point (500, 35) on the graph. Finally, plotting the points on the graph corresponding to, say, $x = 100, 200, 300, \ldots, 900$, we obtain the sketch in Figure 11. As expected, the average cost drops as the level of production increases. But in this case, as opposed to the case in Example 3, the average cost reaches a minimum value of \$35, corresponding to a production level of 500, and *increases* thereafter.

This phenomenon is typical in situations where the marginal cost increases from some point on as production increases, as in Example 2. This situation is in contrast to that of Example 3, in which the marginal cost remains constant at any level of production. ■

Exploring with TECHNOLOGY

Refer to Example 4.

1. Use a graphing utility to plot the graph of the average cost function

$$\overline{C}(x) = 0.0001x^2 - 0.08x + 40 + \frac{5000}{x}$$

using the viewing window $[0, 1000] \times [0, 100]$. Then, using ZOOM and TRACE, show that the lowest point on the graph of $\overline{C}$ is (500, 35).

2. Draw the tangent line to the graph of $\overline{C}$ (500, 35). What is its slope? Is this expected?

3. Plot the graph of the marginal average cost function

$$\overline{C}'(x) = 0.0002x - 0.08 - \frac{5000}{x^2}$$

using the viewing window $[0, 2000] \times [-1, 1]$. Then use ZOOM and TRACE to show that the zero of the function $\overline{C}'$ occurs at $x = 500$. Verify this result using the root-finding capability of your graphing utility. Is this result compatible with that obtained in part 2? Explain your answer.

Revenue Functions

Recall that a revenue function $R(x)$ gives the revenue realized by a company from the sale of x units of a certain commodity. If the company charges p dollars per unit, then

$$R(x) = px \tag{5}$$

However, the price that a company can command for the product depends on the market in which it operates. If the company is one of many—none of which is able to dictate the price of the commodity—then in this competitive market environment the price is determined by market equilibrium (see Section 2.3). On the other hand, if the company is the sole supplier of the product, then under this monopolistic situation it can manipulate the price of the commodity by controlling the supply. The unit selling price p of the commodity is related to the quantity x of the commodity demanded. This relationship between p and x is called a *demand equation* (see Section 2.3). Solving the demand equation for p in terms of x, we obtain the unit price function f. Thus,

$$p = f(x)$$

and the revenue function R is given by

$$R(x) = px = xf(x)$$

The **marginal revenue** gives the actual revenue realized from the sale of an additional unit of the commodity given that sales are already at a certain level. Following an argument parallel to that applied to the cost function in Example 1, you can convince yourself that the marginal revenue is approximated by $R'(x)$. Thus, we define the **marginal revenue function** to be $R'(x)$, where R is the revenue function. The derivative R' of the function R measures the rate of change of the revenue function.

VIDEO

APPLIED EXAMPLE 5 Marginal Revenue Functions Suppose the relationship between the unit price p in dollars and the quantity demanded x of the Acrosonic model F loudspeaker system is given by the equation

$$p = -0.02x + 400 \qquad (0 \le x \le 20{,}000)$$

a. Find the revenue function R.
b. Find the marginal revenue function R'.
c. Compute $R'(2000)$ and interpret your result.

Solution

a. The revenue function R is given by

$$\begin{aligned} R(x) &= px \\ &= x(-0.02x + 400) \\ &= -0.02x^2 + 400x \qquad (0 \le x \le 20{,}000) \end{aligned}$$

b. The marginal revenue function R' is given by

$$R'(x) = -0.04x + 400$$

c. $$R'(2000) = -0.04(2000) + 400 = 320$$

Thus, the actual revenue to be realized from the sale of the 2001st loudspeaker system is approximately \$320.

Profit Functions

Our final example of a marginal function involves the profit function. The profit function P is given by

$$P(x) = R(x) - C(x) \tag{6}$$

where R and C are the revenue and cost functions and x is the number of units of a commodity produced and sold. The **marginal profit function** $P'(x)$ measures the rate of change of the profit function P and provides us with a good approximation of the actual profit or loss realized from the sale of the $(x + 1)$st unit of the commodity (assuming the xth unit has been sold).

APPLIED EXAMPLE 6 Marginal Profit Functions Refer to Example 5. Suppose the cost of producing x units of the Acrosonic model F loudspeaker is

$$C(x) = 100x + 200{,}000$$

dollars.
a. Find the profit function P.
b. Find the marginal profit function P'.
c. Compute $P'(2000)$ and interpret your result.
d. Sketch the graph of the profit function P.

Solution

a. From the solution to Example 5a, we have

$$R(x) = -0.02x^2 + 400x$$

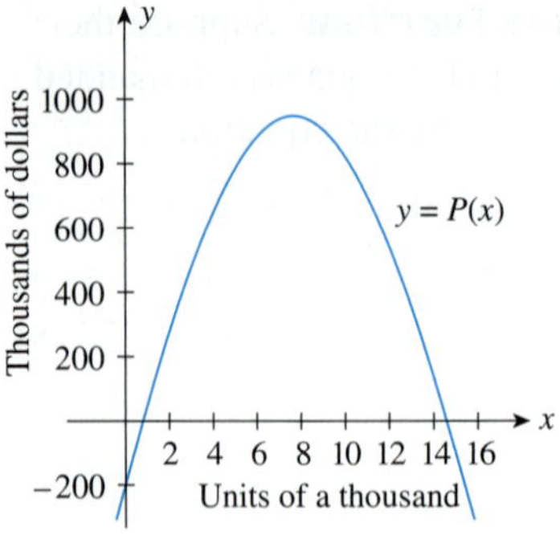

FIGURE 12
The total profit made when x loudspeakers are produced is given by $P(x)$.

Thus, the required profit function P is given by

$$\begin{aligned} P(x) &= R(x) - C(x) \\ &= (-0.02x^2 + 400x) - (100x + 200{,}000) \\ &= -0.02x^2 + 300x - 200{,}000 \end{aligned}$$

b. The marginal profit function P' is given by

$$P'(x) = -0.04x + 300$$

c. $$P'(2000) = -0.04(2000) + 300 = 220$$

Thus, the actual profit realized from the sale of the 2001st loudspeaker system is approximately \$220.

d. The graph of the profit function P appears in Figure 12.

Elasticity of Demand

Finally, let's use the marginal concepts introduced in this section to derive an important criterion used by economists to analyze a demand function: elasticity of demand.

In what follows, it will be convenient to write the demand function f in the form $x = f(p)$; that is, we will think of the quantity demanded of a certain commodity as a function of its unit price. Since the quantity demanded of a commodity usually decreases as its unit price increases, the function f is typically a decreasing function of p (Figure 13a).

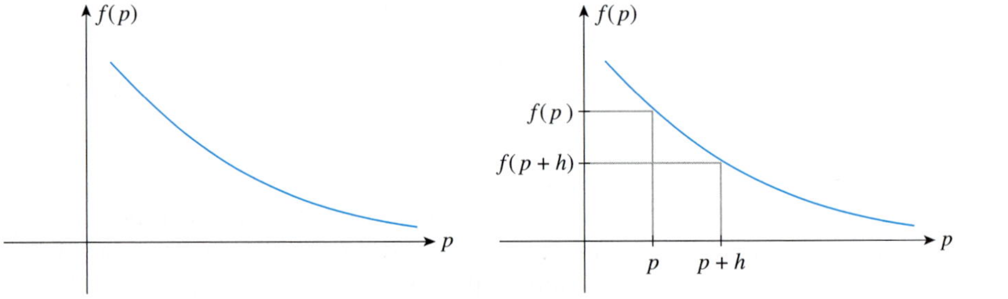

FIGURE 13

(a) A demand function

(b) $f(p + h)$ is the quantity demanded when the unit price increases from p to $p + h$ dollars.

Suppose the unit price of a commodity is increased by h dollars from p dollars to $(p + h)$ dollars (Figure 13b). Then the quantity demanded drops from $f(p)$ units to $f(p + h)$ units, a change of $[f(p + h) - f(p)]$ units. The percentage change in the unit price is

$$\frac{h}{p}(100)$$ $\dfrac{\text{Change in unit price}}{\text{Price } p}(100)$

and the corresponding percentage change in the quantity demanded is

$$100\left[\frac{f(p + h) - f(p)}{f(p)}\right]$$ $\dfrac{\text{Change in quantity demanded}}{\text{Quantity demanded at price } p}(100)$

Now, one good way to measure the effect that a percentage change in price has on the percentage change in the quantity demanded is to look at the ratio of the latter to the former. We find

$$\frac{\text{Percentage change in the quantity demanded}}{\text{Percentage change in the unit price}} = \frac{100\left[\dfrac{f(p+h)-f(p)}{f(p)}\right]}{100\left(\dfrac{h}{p}\right)}$$

$$= \frac{\dfrac{f(p+h)-f(p)}{h}}{\dfrac{f(p)}{p}}$$

If f is differentiable at p, then

$$\frac{f(p+h)-f(p)}{h} \approx f'(p)$$

when h is small. Therefore, if h is small, then the ratio is approximately equal to

$$\frac{f'(p)}{\dfrac{f(p)}{p}} = \frac{pf'(p)}{f(p)}$$

Economists call the negative of this quantity the elasticity of demand.

Elasticity of Demand

If f is a differentiable demand function defined by $x = f(p)$, then the **elasticity of demand** at price p is given by

$$E(p) = -\frac{pf'(p)}{f(p)} \tag{7}$$

Note It will be shown later (Section 4.1) that if f is decreasing on an interval, then $f'(p) < 0$ for p in that interval. In light of this, we see that since both p and $f(p)$ are positive, the quantity $\dfrac{pf'(p)}{f(p)}$ is negative. Because economists would rather work with a positive value, the elasticity of demand $E(p)$ is defined to be the negative of this quantity. ■

APPLIED EXAMPLE 7 Elasticity of Demand Consider the demand equation

$$p = -0.02x + 400 \qquad (0 \le x \le 20{,}000)$$

which describes the relationship between the unit price in dollars and the quantity demanded x of the Acrosonic model F loudspeaker systems.

a. Find the elasticity of demand $E(p)$.
b. Compute $E(100)$ and interpret your result.
c. Compute $E(300)$ and interpret your result.

Solution

a. Solving the given demand equation for x in terms of p, we find

$$x = f(p) = -50p + 20{,}000$$

from which we see that

$$f'(p) = -50$$

Therefore,

$$E(p) = -\frac{pf'(p)}{f(p)} = -\frac{p(-50)}{-50p + 20{,}000} = \frac{p}{400 - p}$$

b.

$$E(100) = \frac{100}{400 - 100} = \frac{1}{3}$$

which is the elasticity of demand when $p = 100$. To interpret this result, recall that $E(100)$ is the negative of the ratio of the percentage change in the quantity demanded to the percentage change in the unit price when $p = 100$. Therefore, our result tells us that when the unit price p is set at \$100 per speaker, an increase of 1% in the unit price will cause a decrease of approximately 0.33% in the quantity demanded.

c.

$$E(300) = \frac{300}{400 - 300} = 3$$

which is the elasticity of demand when $p = 300$. It tells us that when the unit price is set at \$300 per speaker, an increase of 1% in the unit price will cause a decrease of approximately 3% in the quantity demanded. ■

Economists often use the following terminology to describe demand in terms of elasticity.

Elasticity of Demand

The demand is said to be **elastic** if $E(p) > 1$.

The demand is said to be **unitary** if $E(p) = 1$.

The demand is said to be **inelastic** if $E(p) < 1$.

As an illustration, our computations in Example 7 revealed that demand for Acrosonic loudspeakers is elastic when $p = 300$ but inelastic when $p = 100$. These computations confirm that when demand is elastic, a small percentage change in the unit price will result in a greater percentage change in the quantity demanded; and when demand is inelastic, a small percentage change in the unit price will cause a smaller percentage change in the quantity demanded. Finally, when demand is unitary, a small percentage change in the unit price will result in the same percentage change in the quantity demanded.

We can describe the way revenue responds to changes in the unit price using the notion of elasticity. If the quantity demanded of a certain commodity is related to its unit price by the equation $x = f(p)$, then the revenue realized through the sale of x units of the commodity at a price of p dollars each is

$$R(p) = px = pf(p)$$

The rate of change of the revenue with respect to the unit price p is given by

$$\begin{aligned} R'(p) &= f(p) + pf'(p) \\ &= f(p)\left[1 + \frac{pf'(p)}{f(p)}\right] \\ &= f(p)[1 - E(p)] \end{aligned}$$

Now, suppose demand is elastic when the unit price is set at a dollars. Then $E(a) > 1$, and so $1 - E(a) < 0$. Since $f(p)$ is positive for all values of p, we see that

$$R'(a) = f(a)[1 - E(a)] < 0$$

and so $R(p)$ is decreasing at $p = a$. This implies that a small increase in the unit price when $p = a$ results in a decrease in the revenue, whereas a small decrease in the unit price will result in an increase in the revenue. Similarly, you can show that if the demand is inelastic when the unit price is set at a dollars, then a small increase in the unit price will cause the revenue to increase, and a small decrease in the unit price will cause the revenue to decrease. Finally, if the demand is unitary when the unit price is set at a dollars, then $E(a) = 1$ and $R'(a) = 0$. This implies that a small increase or decrease in the unit price will not result in a change in the revenue. The following statements summarize this discussion.

1. If the demand is elastic at p $[E(p) > 1]$, then an increase in the unit price will cause the revenue to decrease, whereas a decrease in the unit price will cause the revenue to increase.
2. If the demand is inelastic at p $[E(p) < 1]$, then an increase in the unit price will cause the revenue to increase, and a decrease in the unit price will cause the revenue to decrease.
3. If the demand is unitary at p $[E(p) = 1]$, then an increase in the unit price will cause the revenue to stay about the same.

These results are illustrated in Figure 14.

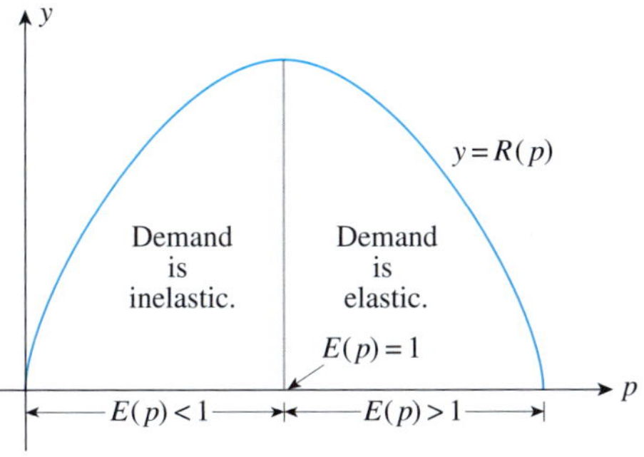

FIGURE 14 The revenue is increasing on an interval where the demand is inelastic, decreasing on an interval where the demand is elastic, and stationary at the point where the demand is unitary.

Note As an aid to remembering this, note the following:

1. If demand is elastic, then the change in revenue and the change in the unit price move in opposite directions.
2. If demand is inelastic, then they move in the same direction.

APPLIED EXAMPLE 8 Elasticity of Demand Refer to Example 7.

a. Is demand elastic, unitary, or inelastic when $p = 100$? When $p = 300$?
b. If the price is \$100, will raising the unit price slightly cause the revenue to increase or decrease?

Solution

a. From the results of Example 7, we see that $E(100) = \frac{1}{3} < 1$ and $E(300) = 3 > 1$. We conclude accordingly that demand is inelastic when $p = 100$ and elastic when $p = 300$.
b. Since demand is inelastic when $p = 100$, raising the unit price slightly will cause the revenue to increase.

3.4 Self-Check Exercises

1. The weekly demand for Pulsar DVD recorders is given by the demand equation

$$p = -0.02x + 300 \qquad (0 \leq x \leq 15{,}000)$$

where p denotes the wholesale unit price in dollars and x denotes the quantity demanded. The weekly total cost function associated with manufacturing these recorders is

$$C(x) = 0.000003x^3 - 0.04x^2 + 200x + 70{,}000$$

dollars.

a. Find the revenue function R and the profit function P.

b. Find the marginal cost function C', the marginal revenue function R', and the marginal profit function P'.

c. Find the marginal average cost function $\overline{C}'$.

d. Compute $C'(3000)$, $R'(3000)$, and $P'(3000)$ and interpret your results.

2. Refer to the preceding exercise. Determine whether the demand is elastic, unitary, or inelastic when $p = 100$ and when $p = 200$.

Solutions to Self-Check Exercises 3.4 can be found on page 207.

3.4 Concept Questions

1. Explain each term in your own words:
a. Marginal cost function
b. Average cost function
c. Marginal average cost function
d. Marginal revenue function
e. Marginal profit function

2. **a.** Define the elasticity of demand.
b. When is the elasticity of demand elastic? Unitary? Inelastic? Explain the meaning of each term.

3.4 Exercises

1. **Production Costs** The graph of a typical total cost function $C(x)$ associated with the manufacture of x units of a certain commodity is shown in the following figure.
a. Explain why the function C is always increasing.
b. As the level of production x increases, the cost/unit drops so that $C(x)$ increases but at a slower pace. However, a level of production is soon reached at which the cost/unit begins to increase dramatically (due to a shortage of raw material, overtime, breakdown of machinery due to excessive stress and strain) so that $C(x)$ continues to increase at a faster pace. Use the graph of C to find the approximate level of production x_0 where this occurs.

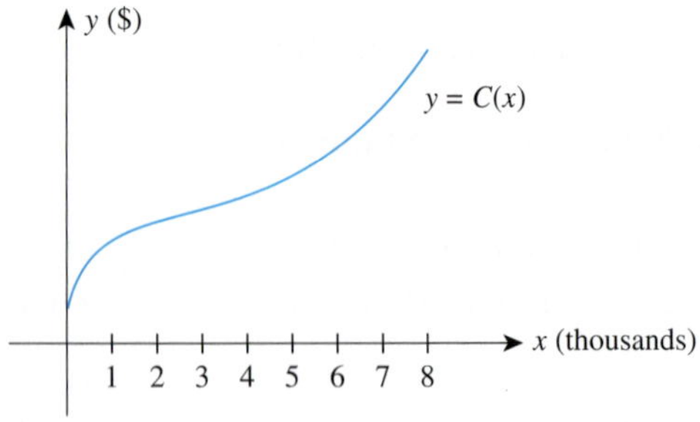

2. **Production Costs** The graph of a typical average cost function $A(x) = C(x)/x$, where $C(x)$ is a total cost function associated with the manufacture of x units of a certain commodity is shown in the following figure.
a. Explain in economic terms why $A(x)$ is large if x is small and why $A(x)$ is large if x is large.
b. What is the significance of the numbers x_0 and y_0, the x- and y-coordinates of the lowest point on the graph of the function A?

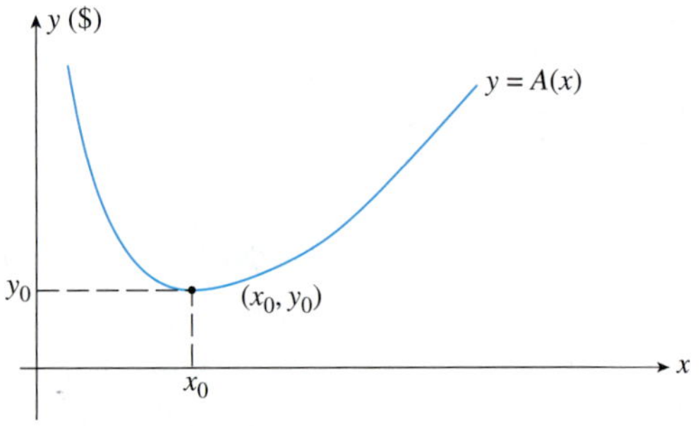

3. **Marginal Cost** The total weekly cost (in dollars) incurred by Lincoln Records in pressing x compact discs is

$$C(x) = 2000 + 2x - 0.0001x^2 \qquad (0 \leq x \leq 6000)$$

a. What is the actual cost incurred in producing the 1001st and the 2001st disc?
b. What is the marginal cost when $x = 1000$ and 2000?

4. **Marginal Cost** A division of Ditton Industries manufactures the Futura model microwave oven. The daily cost (in dollars) of producing these microwave ovens is

$$C(x) = 0.0002x^3 - 0.06x^2 + 120x + 5000$$

where x stands for the number of units produced.

a. What is the actual cost incurred in manufacturing the 101st oven? The 201st oven? The 301st oven?
b. What is the marginal cost when $x = 100$, 200, and 300?

5. Marginal Average Cost Custom Office makes a line of executive desks. It is estimated that the total cost for making x units of their Senior Executive model is

$$C(x) = 100x + 200{,}000$$

dollars/year.
a. Find the average cost function $\overline{C}$.
b. Find the marginal average cost function $\overline{C}'$.
c. What happens to $\overline{C}(x)$ when x is very large? Interpret your results.

6. Marginal Average Cost The management of ThermoMaster Company, whose Mexican subsidiary manufactures an indoor–outdoor thermometer, has estimated that the total weekly cost (in dollars) for producing x thermometers is

$$C(x) = 5000 + 2x$$

a. Find the average cost function $\overline{C}$.
b. Find the marginal average cost function $\overline{C}'$.
c. Interpret your results.

7. Find the average cost function $\overline{C}$ and the marginal average cost function $\overline{C}'$ associated with the total cost function C of Exercise 3.

8. Find the average cost function $\overline{C}$ and the marginal average cost function $\overline{C}'$ associated with the total cost function C of Exercise 4.

9. Marginal Revenue Williams Commuter Air Service realizes a monthly revenue of

$$R(x) = 8000x - 100x^2$$

dollars when the price charged per passenger is x dollars.
a. Find the marginal revenue R'.
b. Compute $R'(39)$, $R'(40)$, and $R'(41)$.
c. Based on the results of part (b), what price should the airline charge in order to maximize their revenue?

10. Marginal Revenue The management of Acrosonic plans to market the ElectroStat, an electrostatic speaker system. The marketing department has determined that the demand for these speakers is

$$p = -0.04x + 800 \qquad (0 \le x \le 20{,}000)$$

where p denotes the speaker's unit price (in dollars) and x denotes the quantity demanded.
a. Find the revenue function R.
b. Find the marginal revenue function R'.
c. Compute $R'(5000)$ and interpret your results.

11. Marginal Profit Refer to Exercise 10. Acrosonic's production department estimates that the total cost (in dollars) incurred in manufacturing x ElectroStat speaker systems in the first year of production will be

$$C(x) = 200x + 300{,}000$$

a. Find the profit function P.
b. Find the marginal profit function P'.
c. Compute $P'(5000)$ and $P'(8000)$.
d. Sketch the graph of the profit function and interpret your results.

12. Marginal Profit Lynbrook West, an apartment complex, has 100 two-bedroom units. The monthly profit (in dollars) realized from renting x apartments is

$$P(x) = -10x^2 + 1760x - 50{,}000$$

a. What is the actual profit realized from renting the 51st unit, assuming that 50 units have already been rented?
b. Compute the marginal profit when $x = 50$ and compare your results with that obtained in part (a).

13. Marginal Cost, Revenue, and Profit The weekly demand for the Pulsar 25 color LED television is

$$p = 600 - 0.05x \qquad (0 \le x \le 12{,}000)$$

where p denotes the wholesale unit price in dollars and x denotes the quantity demanded. The weekly total cost function associated with manufacturing the Pulsar 25 is given by

$$C(x) = 0.000002x^3 - 0.03x^2 + 400x + 80{,}000$$

where $C(x)$ denotes the total cost incurred in producing x sets.
a. Find the revenue function R and the profit function P.
b. Find the marginal cost function C', the marginal revenue function R', and the marginal profit function P'.
c. Compute $C'(2000)$, $R'(2000)$, and $P'(2000)$ and interpret your results.
d. Sketch the graphs of the functions C, R, and P and interpret parts (b) and (c), using the graphs obtained.

14. Marginal Cost, Revenue, and Profit Pulsar also manufactures a series of 20-in. flat-tube digital televisions. The quantity x of these sets demanded each week is related to the wholesale unit price p by the equation

$$p = -0.006x + 180$$

The weekly total cost incurred by Pulsar for producing x sets is

$$C(x) = 0.000002x^3 - 0.02x^2 + 120x + 60{,}000$$

dollars. Answer the questions in Exercise 13 for these data.

15. Marginal Average Cost Find the average cost function $\overline{C}$ associated with the total cost function C of Exercise 13.
a. What is the marginal average cost function $\overline{C}'$?
b. Compute $\overline{C}'(5000)$ and $\overline{C}'(10{,}000)$ and interpret your results.
c. Sketch the graph of $\overline{C}$.

16. Marginal Average Cost Find the average cost function $\overline{C}$ associated with the total cost function C of Exercise 14.
a. What is the marginal average cost function $\overline{C}'$?
b. Compute $\overline{C}'(5000)$ and $\overline{C}'(10{,}000)$ and interpret your results.

17. **Marginal Revenue** The quantity of Sicard wristwatches demanded each month is related to the unit price by the equation

$$p = \frac{50}{0.01x^2 + 1} \qquad (0 \le x \le 20)$$

where p is measured in dollars and x in units of a thousand.
a. Find the revenue function R.
b. Find the marginal revenue function R'.
c. Compute $R'(2)$ and interpret your result.

18. **Marginal Propensity to Consume** The consumption function of the U.S. economy from 1929 to 1941 is

$$C(x) = 0.712x + 95.05$$

where $C(x)$ is the personal consumption expenditure and x is the personal income, both measured in billions of dollars. Find the rate of change of consumption with respect to income, dC/dx. This quantity is called the *marginal propensity to consume.*

19. **Marginal Propensity to Consume** Refer to Exercise 18. Suppose a certain economy's consumption function is

$$C(x) = 0.873x^{1.1} + 20.34$$

where $C(x)$ and x are measured in billions of dollars. Find the marginal propensity to consume when $x = 10$.

20. **Marginal Propensity to Save** Suppose $C(x)$ measures an economy's personal consumption expenditure and x the personal income, both in billions of dollars. Then,

$$S(x) = x - C(x) \qquad \text{Income minus consumption}$$

measures the economy's savings corresponding to an income of x billion dollars. Show that

$$\frac{dS}{dx} = 1 - \frac{dC}{dx}$$

The quantity dS/dx is called the *marginal propensity to save.*

21. Refer to Exercise 20. For the consumption function of Exercise 18, find the marginal propensity to save.

22. Refer to Exercise 20. For the consumption function of Exercise 19, find the marginal propensity to save when $x = 10$.

For each demand equation in Exercises 23–28, compute the elasticity of demand and determine whether the demand is elastic, unitary, or inelastic at the indicated price.

23. $x = -\frac{5}{4}p + 20; p = 10$

24. $x = -\frac{3}{2}p + 9; p = 2$

25. $x + \frac{1}{3}p - 20 = 0; p = 30$

26. $0.4x + p - 20 = 0; p = 10$

27. $p = 169 - x^2; p = 29$

28. $p = 144 - x^2; p = 96$

29. **Elasticity of Demand** The demand equation for the Roland portable hair dryer is given by

$$x = \frac{1}{5}(225 - p^2) \qquad (0 \le p \le 15)$$

where x (measured in units of a hundred) is the quantity demanded per week and p is the unit price in dollars.
a. Is the demand elastic or inelastic when $p = 8$ and when $p = 10$?
b. When is the demand unitary?
Hint: Solve $E(p) = 1$ for p.
c. If the unit price is lowered slightly from \$10, will the revenue increase or decrease?
d. If the unit price is increased slightly from \$8, will the revenue increase or decrease?

30. **Elasticity of Demand** The management of Titan Tire Company has determined that the quantity demanded x of their Super Titan tires per week is related to the unit price p by the equation

$$x = \sqrt{144 - p} \qquad (0 \le p \le 144)$$

where p is measured in dollars and x in units of a thousand.
a. Compute the elasticity of demand when $p = 63$, 96, and 108.
b. Interpret the results obtained in part (a).
c. Is the demand elastic, unitary, or inelastic when $p = 63$, 96, and 108?

31. **Elasticity of Demand** The proprietor of the Showplace, a video store, has estimated that the rental price p (in dollars) of prerecorded DVDs is related to the quantity x (in thousands) rented/day by the demand equation

$$x = \frac{2}{3}\sqrt{36 - p^2} \qquad (0 \le p \le 6)$$

Currently, the rental price is \$2/disc.
a. Is the demand elastic or inelastic at this rental price?
b. If the rental price is increased, will the revenue increase or decrease?

32. **Elasticity of Demand** The quantity demanded each week x (in units of a hundred) of the Mikado digital camera is related to the unit price p (in dollars) by the demand equation

$$x = \sqrt{400 - 5p} \qquad (0 \le p \le 80)$$

a. Is the demand elastic or inelastic when $p = 40$? When $p = 60$?
b. When is the demand unitary?
c. If the unit price is lowered slightly from \$60, will the revenue increase or decrease?
d. If the unit price is increased slightly from \$40, will the revenue increase or decrease?

33. **Elasticity of Demand** The demand function for a certain make of exercise bicycle sold exclusively through cable television is

$$p = \sqrt{9 - 0.02x} \qquad (0 \leq x \leq 450)$$

where p is the unit price in hundreds of dollars and x is the quantity demanded/week. Compute the elasticity of demand and determine the range of prices corresponding to inelastic, unitary, and elastic demand.
Hint: Solve the equation $E(p) = 1$.

34. **Elasticity of Demand** The demand equation for the Sicard wristwatch is given by

$$x = 10\sqrt{\frac{50 - p}{p}} \qquad (0 < p \leq 50)$$

where x (measured in units of a thousand) is the quantity demanded/week and p is the unit price in dollars. Compute the elasticity of demand and determine the range of prices corresponding to inelastic, unitary, and elastic demand.

In Exercises 35 and 36, determine whether the statement is true or false. If it is true, explain why it is true. If it is false, give an example to show why it is false.

35. If C is a differentiable total cost function, then the marginal average cost function is

$$\overline{C}'(x) = \frac{xC'(x) - C(x)}{x^2}$$

36. If the marginal profit function is positive at $x = a$, then it makes sense to decrease the level of production.

3.4 Solutions to Self-Check Exercises

1. **a.**

$$\begin{aligned} R(x) &= px \\ &= x(-0.02x + 300) \\ &= -0.02x^2 + 300x \qquad (0 \leq x \leq 15{,}000) \\ P(x) &= R(x) - C(x) \\ &= -0.02x^2 + 300x \\ &\quad - (0.000003x^3 - 0.04x^2 + 200x + 70{,}000) \\ &= -0.000003x^3 + 0.02x^2 + 100x - 70{,}000 \end{aligned}$$

b.

$$\begin{aligned} C'(x) &= 0.000009x^2 - 0.08x + 200 \\ R'(x) &= -0.04x + 300 \\ P'(x) &= -0.000009x^2 + 0.04x + 100 \end{aligned}$$

c. The average cost function is

$$\begin{aligned} \overline{C}(x) &= \frac{C(x)}{x} \\ &= \frac{0.000003x^3 - 0.04x^2 + 200x + 70{,}000}{x} \\ &= 0.000003x^2 - 0.04x + 200 + \frac{70{,}000}{x} \end{aligned}$$

Therefore, the marginal average cost function is

$$\overline{C}'(x) = 0.000006x - 0.04 - \frac{70{,}000}{x^2}$$

d. Using the results from part (b), we find

$$\begin{aligned} C'(3000) &= 0.000009(3000)^2 - 0.08(3000) + 200 \\ &= 41 \end{aligned}$$

That is, when the level of production is already 3000 recorders, the actual cost of producing one additional recorder is approximately \$41. Next,

$$R'(3000) = -0.04(3000) + 300 = 180$$

That is, the actual revenue to be realized from selling the 3001st recorder is approximately \$180. Finally,

$$\begin{aligned} P'(3000) &= -0.000009(3000)^2 + 0.04(3000) + 100 \\ &= 139 \end{aligned}$$

That is, the actual profit realized from selling the 3001st DVD recorder is approximately \$139.

2. We first solve the given demand equation for x in terms of p, obtaining

$$\begin{aligned} x = f(p) &= -50p + 15{,}000 \\ f'(p) &= -50 \end{aligned}$$

Therefore,

$$\begin{aligned} E(p) &= -\frac{pf'(p)}{f(p)} = -\frac{p}{-50p + 15{,}000}(-50) \\ &= \frac{p}{300 - p} \qquad (0 \leq p < 300) \end{aligned}$$

Next, we compute

$$E(100) = \frac{100}{300 - 100} = \frac{1}{2} < 1$$

and we conclude that demand is inelastic when $p = 100$. Also,

$$E(200) = \frac{200}{300 - 200} = 2 > 1$$

and we see that demand is elastic when $p = 200$.

3.5 Higher-Order Derivatives

Higher-Order Derivatives

The derivative f' of a function f is also a function. As such, the differentiability of f' may be considered. Thus, the function f' has a derivative f'' at a point x in the domain of f' if the limit of the quotient

$$\frac{f'(x+h)-f'(x)}{h}$$

exists as h approaches zero. In other words, it is the derivative of the first derivative.

The function f'' obtained in this manner is called the **second derivative of the function *f***, just as the derivative f' of f is often called the first derivative of f. Continuing in this fashion, we are led to considering the third, fourth, and higher-order derivatives of f whenever they exist. Notations for the first, second, third, and, in general, nth derivatives of a function f at a point x are

$$f'(x), f''(x), f'''(x), \ldots, f^{(n)}(x)$$

or

$$D^1f(x), D^2f(x), D^3f(x), \ldots, D^nf(x)$$

If f is written in the form $y = f(x)$, then the notations for its derivatives are

$$y', y'', y''', \ldots, y^{(n)}$$
$$\frac{dy}{dx}, \frac{d^2y}{dx^2}, \frac{d^3y}{dx^3}, \ldots, \frac{d^ny}{dx^n}$$

or

$$D^1y, D^2y, D^3y, \ldots, D^ny$$

respectively.

EXAMPLE 1 Find the derivatives of all orders of the polynomial function

$$f(x) = x^5 - 3x^4 + 4x^3 - 2x^2 + x - 8$$

Solution We have

$$f'(x) = 5x^4 - 12x^3 + 12x^2 - 4x + 1$$
$$f''(x) = \frac{d}{dx}f'(x) = 20x^3 - 36x^2 + 24x - 4$$
$$f'''(x) = \frac{d}{dx}f''(x) = 60x^2 - 72x + 24$$
$$f^{(4)}(x) = \frac{d}{dx}f'''(x) = 120x - 72$$
$$f^{(5)}(x) = \frac{d}{dx}f^{(4)}(x) = 120$$

and, in general,

$$f^{(n)}(x) = 0 \qquad (\text{for } n > 5)$$

■

EXAMPLE 2 Find the third derivative of the function f defined by $y = x^{2/3}$. What is its domain?

Solution We have

$$y' = \frac{2}{3}x^{-1/3}$$

$$y'' = \left(\frac{2}{3}\right)\left(-\frac{1}{3}\right)x^{-4/3} = -\frac{2}{9}x^{-4/3}$$

so the required derivative is

$$y''' = \left(-\frac{2}{9}\right)\left(-\frac{4}{3}\right)x^{-7/3} = \frac{8}{27}x^{-7/3} = \frac{8}{27x^{7/3}}$$

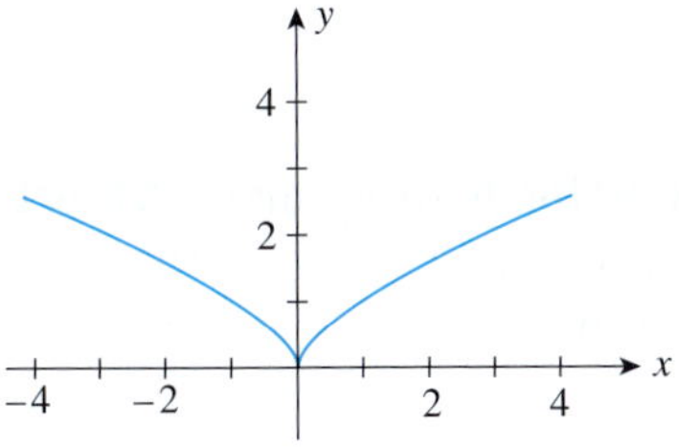

FIGURE **15**
The graph of the function $y = x^{2/3}$

The common domain of the functions f', f'', and f''' is the set of all real numbers except $x = 0$. The domain of $y = x^{2/3}$ is the set of all real numbers. The graph of the function $y = x^{2/3}$ appears in Figure 15. ■

Note Always simplify an expression before differentiating it to obtain the next order derivative. ■

VIDEO

EXAMPLE 3 Find the second derivative of the function $y = (2x^2 + 3)^{3/2}$.

Solution We have, using the general power rule,

$$y' = \frac{3}{2}(2x^2 + 3)^{1/2}(4x) = 6x(2x^2 + 3)^{1/2}$$

Next, using the product rule and then the chain rule, we find

$$\begin{aligned} y'' &= (6x) \cdot \frac{d}{dx}(2x^2 + 3)^{1/2} + \left[\frac{d}{dx}(6x)\right](2x^2 + 3)^{1/2} \\ &= (6x)\left(\frac{1}{2}\right)(2x^2 + 3)^{-1/2}(4x) + 6(2x^2 + 3)^{1/2} \\ &= 12x^2(2x^2 + 3)^{-1/2} + 6(2x^2 + 3)^{1/2} \\ &= 6(2x^2 + 3)^{-1/2}[2x^2 + (2x^2 + 3)] \\ &= \frac{6(4x^2 + 3)}{\sqrt{2x^2 + 3}} \end{aligned}$$

See page 9.

Factor out $(2x^2 + 3)^{-1/2}$.

■

Just as the derivative of a function f at a point x measures the rate of change of the function f at that point, the second derivative of f (the derivative of f') measures the rate of change of the derivative f' of the function f. The third derivative of the function f, f''', measures the rate of change of f'', and so on.

In Chapter 4, we will discuss applications involving the geometric interpretation of the second derivative of a function. The following example gives an interpretation of the second derivative in a familiar role.

APPLIED EXAMPLE 4 Acceleration of a Maglev Refer to the example on pages 97–98. The distance s (in feet) covered by a maglev moving along a straight track t seconds after starting from rest is given by the function $s = 4t^2$ $(0 \le t \le 10)$. What is the maglev's acceleration at any time t?

Solution The velocity of the maglev t seconds from rest is given by

$$v = \frac{ds}{dt} = \frac{d}{dt}(4t^2) = 8t$$

The acceleration of the maglev t seconds from rest is given by the rate of change of the velocity of t—that is,

$$a = \frac{d}{dt}v = \frac{d}{dt}\left(\frac{ds}{dt}\right) = \frac{d^2s}{dt^2} = \frac{d}{dt}(8t) = 8$$

or 8 feet per second per second, normally abbreviated 8 ft/sec^2. ■

VIDEO

APPLIED EXAMPLE 5 Acceleration and Velocity of a Falling Object A ball is thrown straight up into the air from the roof of a building. The height of the ball as measured from the ground is given by

$$s = -16t^2 + 24t + 120$$

where s is measured in feet and t in seconds. Find the velocity and acceleration of the ball 3 seconds after it is thrown into the air.

Solution The velocity v and acceleration a of the ball at any time t are given by

$$v = \frac{ds}{dt} = \frac{d}{dt}(-16t^2 + 24t + 120) = -32t + 24$$

and

$$a = \frac{d^2s}{dt^2} = \frac{d}{dt}\left(\frac{ds}{dt}\right) = \frac{d}{dt}(-32t + 24) = -32$$

Therefore, the velocity of the ball 3 seconds after it is thrown into the air is

$$v = -32(3) + 24 = -72$$

That is, the ball is falling downward at a speed of 72 ft/sec. The acceleration of the ball is 32 ft/sec^2 downward at any time during the motion. ■

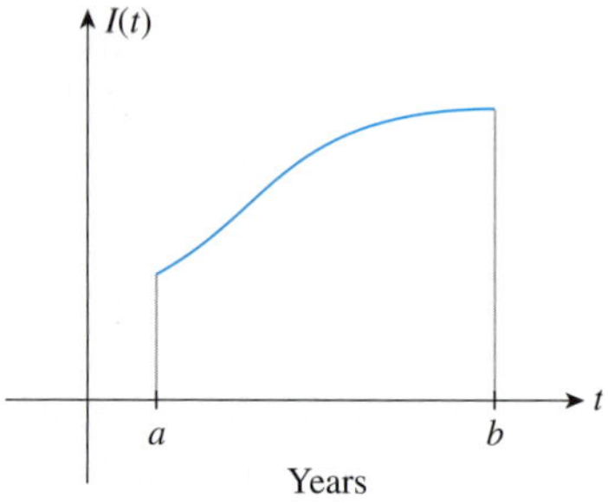

FIGURE 16
The CPI of a certain economy from year a to year b is given by $I(t)$.

Another interpretation of the second derivative of a function—this time from the field of economics—follows. Suppose the consumer price index (CPI) of an economy between the years a and b is described by the function $I(t)$ $(a \le t \le b)$ (Figure 16). Then the first derivative of I at $t = c$, $I'(c)$, where $a < c < b$, gives the rate of change of I at c. The quantity

$$\frac{I'(c)}{I(c)}$$

called the *relative rate of change of* $I(t)$ with respect to t at $t = c$, measures the *inflation rate* of the economy at $t = c$. The second derivative of I at $t = c$, $I''(c)$, gives the rate of change of I' at $t = c$. Now, it is possible for $I'(t)$ to be positive and $I''(t)$ to be negative at $t = c$ (see Example 6). This tells us that at $t = c$ the economy is experiencing inflation (the CPI is increasing) but the rate at which inflation is growing is in fact decreasing. This is precisely the situation described by an economist or a politician when she claims that "inflation is slowing." One may not jump to the conclusion from the aforementioned quote that prices of goods and services are about to drop!

APPLIED EXAMPLE 6 Inflation Rate of an Economy The function

$$I(t) = -0.2t^3 + 3t^2 + 100 \qquad (0 \le t \le 9)$$

gives the CPI of an economy, where $t = 0$ corresponds to the beginning of 2002.

a. Find the inflation rate at the beginning of 2008 ($t = 6$).
b. Show that inflation was moderating at that time.

Solution

a. We find $I'(t) = -0.6t^2 + 6t$. Next, we compute

$$I'(6) = -0.6(6)^2 + 6(6) = 14.4 \quad \text{and} \quad I(6) = -0.2(6)^3 + 3(6)^2 + 100 = 164.8$$

from which we see that the inflation rate is

$$\frac{I'(6)}{I(6)} = \frac{14.4}{164.8} \approx 0.0874$$

or approximately 8.7%.

b. We find

$$I''(t) = \frac{d}{dt}(-0.6t^2 + 6t) = -1.2t + 6$$

Since

$$I''(6) = -1.2(6) + 6 = -1.2$$

we see that I' is indeed decreasing at $t = 6$ and conclude that inflation was moderating at that time (Figure 17).

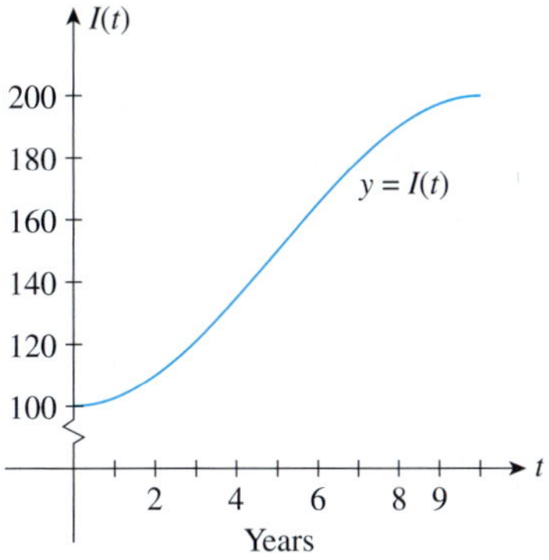

FIGURE 17
The CPI of an economy is given by $I(t)$.

3.5 Self-Check Exercises

1. Find the third derivative of

$$f(x) = 2x^5 - 3x^3 + x^2 - 6x + 10$$

2. Let

$$f(x) = \frac{1}{1 + x}$$

Find $f'(x)$, $f''(x)$, and $f'''(x)$.

3. **Conservation of Species** A certain species of turtles faces extinction because dealers collect truckloads of turtle eggs to be sold as aphrodisiacs. After severe conservation measures are implemented, it is hoped that the turtle population will grow according to the rule

$$N(t) = 2t^3 + 3t^2 - 4t + 1000 \qquad (0 \le t \le 10)$$

where $N(t)$ denotes the population at the end of year t. Compute $N''(2)$ and $N''(8)$. What do your results tell you about the effectiveness of the program?

Solutions to Self-Check Exercises 3.5 can be found on page 213.

3.5 Concept Questions

1. **a.** What is the second derivative of a function f?
   b. How do you find the second derivative of a function f, assuming that it exists?

2. If $s = f(t)$ gives the position of an object moving on the coordinate line, what do $f'(t)$ and $f''(t)$ measure?

3. Suppose $f(t)$ measures the population of a country at time t. What can you say about the signs of $f'(t)$ and $f''(t)$ over the time interval (a, b)
   a. If the population is increasing at an increasing rate over (a, b)?
   b. If the population is increasing at a decreasing rate over (a, b)?
   c. If the population is decreasing at an increasing rate over (a, b)?
   d. If the population is decreasing at a decreasing rate over (a, b)?

4. Suppose $f(t)$ measures the population of a country at time t. What can you say about the signs of $f'(t)$ and $f''(t)$ over the time interval
   a. If the population is increasing at a constant rate over (a, b)?
   b. If the population is decreasing at a constant rate over (a, b)?
   c. If the population is constant over (a, b)?

3.5 Exercises

In Exercises 1–20, find the first and second derivatives of the function.

1. $f(x) = 4x^2 - 2x + 1$
2. $f(x) = -0.2x^2 + 0.3x + 4$
3. $f(x) = 2x^3 - 3x^2 + 1$
4. $g(x) = -3x^3 + 24x^2 + 6x - 64$
5. $h(t) = t^4 - 2t^3 + 6t^2 - 3t + 10$
6. $f(x) = x^5 - x^4 + x^3 - x^2 + x - 1$
7. $f(x) = (x^2 + 2)^5$
8. $g(t) = t^2(3t + 1)^4$
9. $g(t) = (2t^2 - 1)^2(3t^2)$
10. $h(x) = (x^2 + 1)^2(x - 1)$
11. $f(x) = (2x^2 + 2)^{7/2}$
12. $h(w) = (w^2 + 2w + 4)^{5/2}$
13. $f(x) = x(x^2 + 1)^2$
14. $g(u) = u(2u - 1)^3$
15. $f(x) = \dfrac{x}{2x + 1}$
16. $g(t) = \dfrac{t^2}{t - 1}$
17. $f(s) = \dfrac{s - 1}{s + 1}$
18. $f(u) = \dfrac{u}{u^2 + 1}$
19. $f(u) = \sqrt{4 - 3u}$
20. $f(x) = \sqrt{2x - 1}$

In Exercises 21–28, find the third derivative of the given function.

21. $f(x) = 3x^4 - 4x^3$
22. $f(x) = 3x^5 - 6x^4 + 2x^2 - 8x + 12$
23. $f(x) = \dfrac{1}{x}$
24. $f(x) = \dfrac{2}{x^2}$
25. $g(s) = \sqrt{3s - 2}$
26. $g(t) = \sqrt{2t + 3}$
27. $f(x) = (2x - 3)^4$
28. $g(t) = \left(\dfrac{1}{2}t^2 - 1\right)^5$

29. **Acceleration of a Falling Object** During the construction of an office building, a hammer is accidentally dropped from a height of 256 ft. The distance (in feet) the hammer falls in t sec is $s = 16t^2$. What is the hammer's velocity when it strikes the ground? What is its acceleration?

30. **Acceleration of a Car** The distance s (in feet) covered by a car after t sec is given by

$$s = -t^3 + 8t^2 + 20t \qquad (0 \leq t \leq 6)$$

Find a general expression for the car's acceleration at any time t $(0 \leq t \leq 6)$. Show that the car is decelerating after $2\frac{2}{3}$ sec.

31. **Crime Rates** The number of major crimes committed in Bronxville between 1998 and 2005 is approximated by the function

$$N(t) = -0.1t^3 + 1.5t^2 + 100 \qquad (0 \leq t \leq 7)$$

where $N(t)$ denotes the number of crimes committed in year t, with $t = 0$ corresponding to 1998. Enraged by the dramatic increase in the crime rate, Bronxville's citizens, with the help of the local police, organized "Neighborhood Crime Watch" groups in early 2002 to combat this menace.

a. Verify that the crime rate was increasing from 1998 through 2005.
Hint: Compute $N'(0), N'(1), \ldots, N'(7)$.

b. Show that the Neighborhood Crime Watch program was working by computing $N''(4)$, $N''(5)$, $N''(6)$, and $N''(7)$.

32. **GDP of a Developing Country** A developing country's gross domestic product (GDP) from 2000 to 2008 is approximated by the function

$$G(t) = -0.2t^3 + 2.4t^2 + 60 \qquad (0 \leq t \leq 8)$$

where $G(t)$ is measured in billions of dollars, with $t = 0$ corresponding to 2000.

a. Compute $G'(0), G'(1), \ldots, G'(8)$.

b. Compute $G''(0), G''(1), \ldots, G''(8)$.

c. Using the results obtained in parts (a) and (b), show that after a spectacular growth rate in the early years, the growth of the GDP cooled off.

33. **Disability Benefits** The number of persons age 18–64 yr receiving disability benefits through Social Security, the Supplemental Security income, or both, from 1990 through 2000 is approximated by the function

$$N(t) = 0.00037t^3 - 0.0242t^2 + 0.52t + 5.3 \qquad (0 \leq t \leq 10)$$

where $N(t)$ is measured in units of a million and t is measured in years, with $t = 0$ corresponding to the beginning of 1990. Compute $N(8)$, $N'(8)$, and $N''(8)$ and interpret your results.

Source: Social Security Administration

34. **Obesity in America** The body mass index (BMI) measures body weight in relation to height. A BMI of 25 to 29.9 is considered overweight, a BMI of 30 or more is considered obese, and a BMI of 40 or more is morbidly obese. The percent of the U.S. population that is obese is approximated by the function

$$P(t) = 0.0004t^3 + 0.0036t^2 + 0.8t + 12 \qquad (0 \leq t \leq 13)$$

where t is measured in years, with $t = 0$ corresponding to the beginning of 1991. Show that the rate of the rate of change of the percent of the U.S. population that is deemed obese was positive from 1991 to 2004. What does this mean?

Source: Centers for Disease Control and Prevention

35. **Test Flight of a VTOL** In a test flight of the McCord Terrier, McCord Aviation's experimental VTOL (vertical takeoff and landing) aircraft, it was determined that t sec after liftoff, when the craft was operated in the vertical takeoff mode, its altitude (in feet) was

$$h(t) = \frac{1}{16}t^4 - t^3 + 4t^2 \qquad (0 \le t \le 8)$$

a. Find an expression for the craft's velocity at time t.
b. Find the craft's velocity when $t = 0$ (the initial velocity), $t = 4$, and $t = 8$.
c. Find an expression for the craft's acceleration at time t.
d. Find the craft's acceleration when $t = 0$, 4, and 8.
e. Find the craft's height when $t = 0$, 4, and 8.

36. **Air Purification** During testing of a certain brand of air purifier, the amount of smoke remaining t min after the start of the test was

$$A(t) = -0.00006t^5 + 0.00468t^4 - 0.1316t^3 + 1.915t^2 - 17.63t + 100$$

percent of the original amount. Compute $A'(10)$ and $A''(10)$ and interpret your results.
Source: Consumer Reports

37. **Aging Population** The population of Americans age 55 yr and older as a percentage of the total population is approximated by the function

$$f(t) = 10.72(0.9t + 10)^{0.3} \qquad (0 \le t \le 20)$$

where t is measured in years, with $t = 0$ corresponding to 2000. Compute $f''(10)$ and interpret your result.
Source: U.S. Census Bureau

38. **Working Mothers** The percentage of mothers who work outside the home and have children younger than age 6 yr is approximated by the function

$$P(t) = 33.55(t + 5)^{0.205} \qquad (0 \le t \le 21)$$

where t is measured in years, with $t = 0$ corresponding to the beginning of 1980. Compute $P''(20)$ and interpret your result.
Source: U.S. Bureau of Labor Statistics

In Exercises 39–43, determine whether the statement is true or false. If it is true, explain why it is true. If it is false, give an example to show why it is false.

39. If the second derivative of f exists at $x = a$, then $f''(a) = [f'(a)]^2$.

40. If $h = fg$ where f and g have second-order derivatives, then

$$h''(x) = f''(x)g(x) + 2f'(x)g'(x) + f(x)g''(x)$$

41. If $f(x)$ is a polynomial function of degree n, then $f^{(n+1)}(x) = 0$.

42. Suppose $P(t)$ represents the population of bacteria at time t and suppose $P'(t) > 0$ and $P''(t) < 0$; then the population is increasing at time t but at a decreasing rate.

43. If $h(x) = f(2x)$, then $h''(x) = 4f''(2x)$.

44. Let f be the function defined by the rule $f(x) = x^{7/3}$. Show that f has first- and second-order derivatives at all points x, and in particular at $x = 0$. Also show that the third derivative of f does *not* exist at $x = 0$.

45. Construct a function f that has derivatives of order up through and including n at a point a but fails to have the $(n + 1)$st derivative there.
Hint: See Exercise 44.

46. Show that a polynomial function has derivatives of all orders.
Hint: Let $P(x) = a_nx^n + a_{n-1}x^{n-1} + a_{n-2}x^{n-2} + \cdots + a_0$ be a polynomial of degree n, where n is a positive integer and $a_0, a_1, \ldots, a_n$ are constants with $a_n \neq 0$. Compute $P'(x), P''(x), \ldots$.

3.5 Solutions to Self-Check Exercises

1. $f'(x) = 10x^4 - 9x^2 + 2x - 6$
$f''(x) = 40x^3 - 18x + 2$
$f'''(x) = 120x^2 - 18$

2. We write $f(x) = (1 + x)^{-1}$ and use the general power rule, obtaining

$$f'(x) = (-1)(1 + x)^{-2}\frac{d}{dx}(1 + x) = -(1 + x)^{-2}(1)$$
$$= -(1 + x)^{-2} = -\frac{1}{(1 + x)^2}$$

Continuing, we find

$$f''(x) = -(-2)(1 + x)^{-3}$$
$$= 2(1 + x)^{-3} = \frac{2}{(1 + x)^3}$$

$$f'''(x) = 2(-3)(1 + x)^{-4}$$
$$= -6(1 + x)^{-4}$$
$$= -\frac{6}{(1 + x)^4}$$

3. $N'(t) = 6t^2 + 6t - 4$ and $N''(t) = 12t + 6 = 6(2t + 1)$

Therefore, $N''(2) = 30$ and $N''(8) = 102$. The results of our computations reveal that at the end of year 2, the *rate* of growth of the turtle population is increasing at the rate of 30 turtles/year/year. At the end of year 8, the rate is increasing at the rate of 102 turtles/year/year. Clearly, the conservation measures are paying off handsomely.

USING TECHNOLOGY

Finding the Second Derivative of a Function at a Given Point

Some graphing utilities have the capability of numerically computing the second derivative of a function at a point. If your graphing utility has this capability, use it to work through the examples and exercises of this section.

EXAMPLE 1 Use the (second) numerical derivative operation of a graphing utility to find the second derivative of $f(x) = \sqrt{x}$ when $x = 4$.

Solution Using the (second) numerical derivative operation, we find

$$f''(4) = \text{der2}(x\text{\^{}}.5, x, 4) = -.03125$$

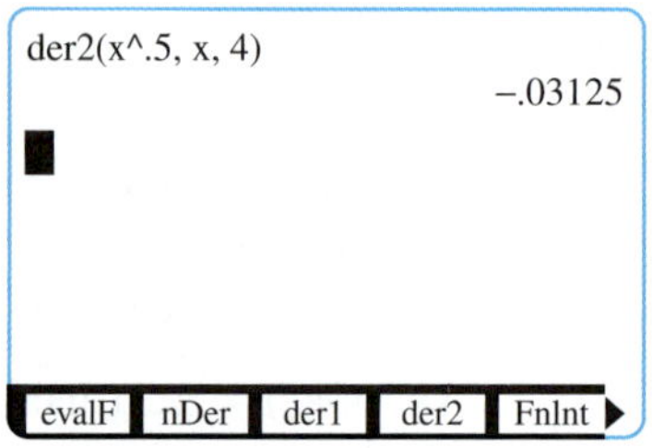

FIGURE **T1**
The TI-86 second derivative screen for computing $f''(4)$

(Figure T1). ■

APPLIED EXAMPLE 2 Prevalence of Alzheimer's Patients The number of Alzheimer's patients in the United States is given by

$$f(t) = -0.02765t^4 + 0.3346t^3 - 1.1261t^2 + 1.7575t + 3.7745 \qquad (0 \le t \le 5)$$

where $f(t)$ is measured in millions and t is measured in decades, with $t = 0$ corresponding to the beginning of 1990.

a. How fast is the number of Alzheimer's patients in the United States anticipated to be changing at the beginning of 2030?
b. How fast is the rate of change of the number of Alzheimer's patients in the United States anticipated to be changing at the beginning of 2030?
c. Plot the graph of f in the viewing window $[0, 5] \times [0, 12]$.

Source: Alzheimer's Association

Solution

a. Using the numerical derivative operation of a graphing utility, we find that the number of Alzheimer's patients at the beginning of 2030 can be anticipated to be changing at the rate of

$$f'(4) = 1.7311$$

That is, the number is increasing at the rate of approximately 1.7 million patients per decade.

b. Using the (second) numerical derivative operation of a graphing utility, we find that

$$f''(4) = 0.4694$$

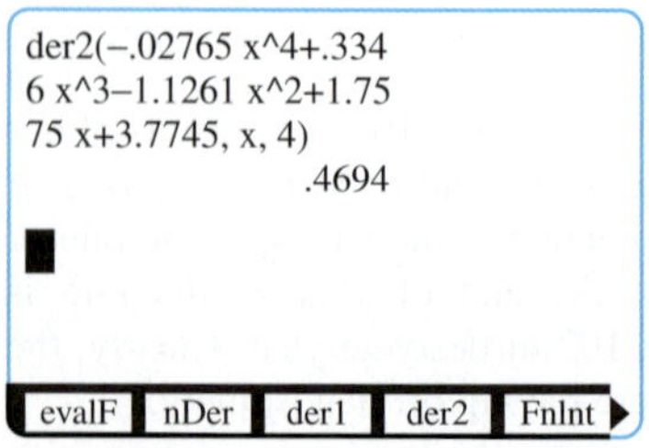

FIGURE **T2**
The TI-86 second derivative screen for computing $f''(4)$

(Figure T2); that is, the rate of change of the number of Alzheimer's patients is increasing at the rate of approximately 0.5 million patients per decade per decade.

c. Figure T3 shows the graph.

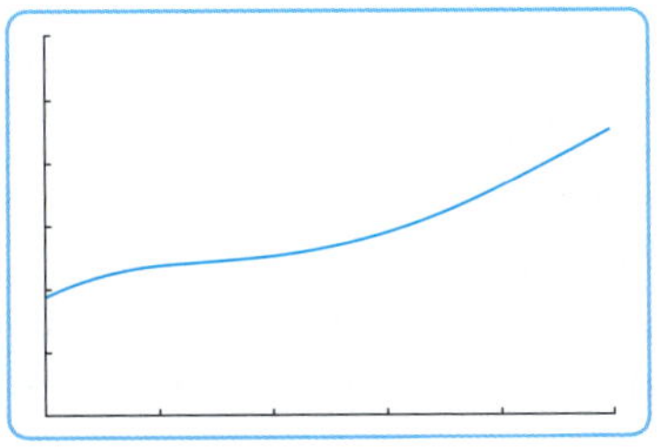

FIGURE **T3**
The graph of f in the viewing window $[0, 5] \times [0, 12]$

TECHNOLOGY EXERCISES

In Exercises 1–8, find the value of the second derivative of f at the given value of x. Express your answer correct to four decimal places.

1. $f(x) = 2x^3 - 3x^2 + 1;\ x = -1$
2. $f(x) = 2.5x^5 - 3x^3 + 1.5x + 4;\ x = 2.1$
3. $f(x) = 2.1x^{3.1} - 4.2x^{1.7} + 4.2;\ x = 1.4$
4. $f(x) = 1.7x^{4.2} - 3.2x^{1.3} + 4.2x - 3.2;\ x = 2.2$
5. $f(x) = \dfrac{x^2 + 2x - 5}{x^3 + 1};\ x = 2.1$
6. $f(x) = \dfrac{x^3 + x + 2}{2x^2 - 5x + 4};\ x = 1.2$
7. $f(x) = \dfrac{x^{1/2} + 2x^{3/2} + 1}{2x^{1/2} + 3};\ x = 0.5$
8. $f(x) = \dfrac{\sqrt{x} - 1}{2x + \sqrt{x} + 4};\ x = 2.3$
9. **Rate of Bank Failures** The rate at which banks were failing between 1982 and 1994 is given by

$$f(t) = -0.063447t^4 - 1.953283t^3 + 14.632576t^2 - 6.684704t + 47.458874 \quad (0 \le t \le 12)$$

where $f(t)$ is measured in the number of banks/year and t is measured in years, with $t = 0$ corresponding to the beginning of 1982. Compute $f''(6)$ and interpret your results.
Source: Federal Deposit Insurance Corporation

10. **Modeling with Data** The revenues (in billions of dollars) from cable advertisement for the years 1995 through 2000 follow:

Year	1995	1996	1997	1998	1999	2000
Revenue	5.1	6.6	8.1	9.4	11.1	13.7

a. Use **CubicReg** to find a third-degree polynomial regression model for the data. Let $t = 0$ correspond to 1995.
b. Plot the graph of the function f found in part (a), using the viewing window $[0, 6] \times [0, 14]$.
c. Compute $f''(5)$ and interpret your results.
Source: National Cable Television Association

3.6 Implicit Differentiation and Related Rates

Differentiating Implicitly

Up to now we have dealt with functions expressed in the form $y = f(x)$; that is, functions in which the dependent variable y is expressed *explicitly* in terms of the independent variable x. However, not all functions are expressed in this form. Consider, for example, the equation

$$x^2y + y - x^2 + 1 = 0 \tag{8}$$

This equation does express y *implicitly* as a function of x. In fact, solving (8) for y in terms of x, we obtain

$$(x^2 + 1)y = x^2 - 1 \qquad \text{Implicit equation}$$

$$y = f(x) = \frac{x^2 - 1}{x^2 + 1} \qquad \text{Explicit equation}$$

which gives an explicit representation of f.

Next, consider the equation

$$y^4 - y^3 - y + 2x^3 - x = 8$$

If certain restrictions are placed on x and y, this equation defines y as a function of x. But in this instance, we would be hard pressed to find y explicitly in terms of x. The following question arises naturally: How do we compute dy/dx in this case?

As it turns out, thanks to the chain rule, a method *does* exist for computing the derivative of a function directly from the implicit equation defining the function. This method is called **implicit differentiation** and is demonstrated in the next several examples.

EXAMPLE 1 Given the equation $y^2 = x$, find $\dfrac{dy}{dx}$.

Solution Differentiating both sides of the equation with respect to x, we obtain

$$\frac{d}{dx}(y^2) = \frac{d}{dx}(x)$$

To carry out the differentiation of the term $\dfrac{d}{dx}(y^2)$, we note that y (with suitable restrictions) is a function of x. Writing $y = f(x)$ to remind us of this fact, we find that

$$\begin{aligned}\frac{d}{dx}(y^2) &= \frac{d}{dx}[f(x)]^2 && \text{Write } y = f(x).\\ &= 2f(x)f'(x) && \text{Use the chain rule.}\\ &= 2y\frac{dy}{dx} && \text{Replace } f(x) \text{ with } y.\end{aligned}$$

Therefore, the equation

$$\frac{d}{dx}(y^2) = \frac{d}{dx}(x)$$

is equivalent to

$$2y\frac{dy}{dx} = 1$$

Solving for $\dfrac{dy}{dx}$ yields

$$\frac{dy}{dx} = \frac{1}{2y}$$

Before considering other examples, let's summarize the important steps involved in implicit differentiation. (Here we assume that dy/dx exists.)

Finding $\dfrac{dy}{dx}$ by Implicit Differentiation

1. Differentiate both sides of the equation *with respect to x.* (Make sure that the derivative of any term involving y includes the factor dy/dx.)
2. Solve the resulting equation for dy/dx in terms of x and y.

VIDEO

EXAMPLE 2 Find $\dfrac{dy}{dx}$ given the equation

$$y^3 - y + 2x^3 - x = 8$$

Solution Differentiating both sides of the given equation with respect to x, we obtain

$$\frac{d}{dx}(y^3 - y + 2x^3 - x) = \frac{d}{dx}(8)$$

$$\frac{d}{dx}(y^3) - \frac{d}{dx}(y) + \frac{d}{dx}(2x^3) - \frac{d}{dx}(x) = 0$$

Now, recalling that y is a function of x, we apply the chain rule to the first two terms on the left. Thus,

$$3y^2\frac{dy}{dx} - \frac{dy}{dx} + 6x^2 - 1 = 0$$

$$(3y^2 - 1)\frac{dy}{dx} = 1 - 6x^2$$

$$\frac{dy}{dx} = \frac{1 - 6x^2}{3y^2 - 1}$$

Explore & Discuss

Refer to Example 2. Suppose we think of the equation $y^3 - y + 2x^3 - x = 8$ as defining x implicitly as a function of y. Find dx/dy and justify your method of solution.

EXAMPLE 3 Consider the equation $x^2 + y^2 = 4$.

a. Find dy/dx by implicit differentiation.
b. Find the slope of the tangent line to the graph of the function $y = f(x)$ at the point $(1, \sqrt{3})$.
c. Find an equation of the tangent line of part (b).

Solution

a. Differentiating both sides of the equation with respect to x, we obtain

$$\frac{d}{dx}(x^2 + y^2) = \frac{d}{dx}(4)$$

$$\frac{d}{dx}(x^2) + \frac{d}{dx}(y^2) = 0$$

$$2x + 2y\frac{dy}{dx} = 0$$

$$2y\frac{dy}{dx} = -2x$$

$$\frac{dy}{dx} = -\frac{x}{y} \qquad (y \neq 0)$$

b. The slope of the tangent line to the graph of the function at the point $(1, \sqrt{3})$ is given by

$$\left.\frac{dy}{dx}\right|_{(1, \sqrt{3})} = \left.-\frac{x}{y}\right|_{(1, \sqrt{3})} = -\frac{1}{\sqrt{3}}$$

(*Note:* This notation is read "dy/dx evaluated at the point $(1, \sqrt{3})$.")

c. An equation of the tangent line in question is found by using the point-slope form of the equation of a line with the slope $m = -1/\sqrt{3}$ and the point $(1, \sqrt{3})$. Thus,

$$y - \sqrt{3} = -\frac{1}{\sqrt{3}}(x - 1)$$

See page 36.

$$\sqrt{3}y - 3 = -x + 1$$

$$x + \sqrt{3}y - 4 = 0$$

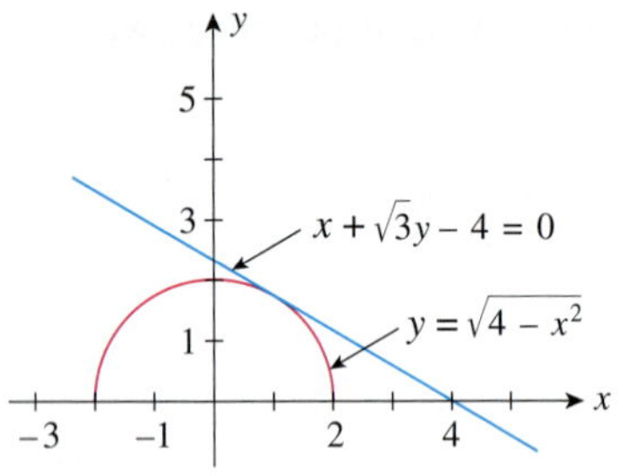

FIGURE 18
The line $x + \sqrt{3}y - 4 = 0$ is tangent to the graph of the function $y = f(x)$.

A sketch of this tangent line is shown in Figure 18.

We can also solve the equation $x^2 + y^2 = 4$ explicitly for y in terms of x. If we do this, we obtain

$$y = \pm\sqrt{4 - x^2}$$

From this, we see that the equation $x^2 + y^2 = 4$ defines the two functions

$$y = f(x) = \sqrt{4 - x^2}$$
$$y = g(x) = -\sqrt{4 - x^2}$$

Since the point $(1, \sqrt{3})$ does not lie on the graph of $y = g(x)$, we conclude that

$$y = f(x) = \sqrt{4 - x^2}$$

is the required function. The graph of f is the upper semicircle shown in Figure 18. ■

Note The notation

$$\left.\frac{dy}{dx}\right|_{(a,\, b)}$$

is used to denote the value of dy/dx at the point (a, b). ■

Explore & Discuss

Refer to Example 3. Yet another function defined implicitly by the equation $x^2 + y^2 = 4$ is the function

$$y = h(x) = \begin{cases} \sqrt{4 - x^2} & \text{if } -2 \le x < 0 \\ -\sqrt{4 - x^2} & \text{if } 0 \le x \le 2 \end{cases}$$

1. Sketch the graph of h.
2. Show that $h'(x) = -x/y$.
3. Find an equation of the tangent line to the graph of h at the point $(1, -\sqrt{3})$.

To find dy/dx at a *specific point* (a, b), differentiate the given equation implicitly with respect to x and then replace x and y by a and b, respectively, *before* solving the equation for dy/dx. This often simplifies the amount of algebra involved.

EXAMPLE 4 Find $\dfrac{dy}{dx}$ given that x and y are related by the equation

$$x^2y^3 + 6x^2 = y + 12$$

and that $y = 2$ when $x = 1$.

Solution Differentiating both sides of the given equation with respect to x, we obtain

$$\frac{d}{dx}(x^2y^3) + \frac{d}{dx}(6x^2) = \frac{d}{dx}(y) + \frac{d}{dx}(12)$$

$$x^2 \cdot \frac{d}{dx}(y^3) + y^3 \cdot \frac{d}{dx}(x^2) + 12x = \frac{dy}{dx}$$ Use the product rule on $\frac{d}{dx}(x^2y^3)$.

$$3x^2y^2\frac{dy}{dx} + 2xy^3 + 12x = \frac{dy}{dx}$$

Substituting $x = 1$ and $y = 2$ into this equation gives

$$3(1)^2(2)^2\frac{dy}{dx} + 2(1)(2)^3 + 12(1) = \frac{dy}{dx}$$

$$12\frac{dy}{dx} + 16 + 12 = \frac{dy}{dx}$$

and, solving for $\frac{dy}{dx}$,

$$\frac{dy}{dx} = -\frac{28}{11}$$

Note that it is not necessary to find an explicit expression for dy/dx.

Note In Examples 3 and 4, you can verify that the points at which we evaluated dy/dx actually lie on the curve in question by showing that the coordinates of the points satisfy the given equations.

EXAMPLE 5 Find $\frac{dy}{dx}$ given that x and y are related by the equation

$$\sqrt{x^2 + y^2} - x^2 = 5$$

Solution Differentiating both sides of the given equation with respect to x, we obtain

$$\frac{d}{dx}(x^2 + y^2)^{1/2} - \frac{d}{dx}(x^2) = \frac{d}{dx}(5)$$

Write $\sqrt{x^2 + y^2} = (x^2 + y^2)^{1/2}$.

$$\frac{1}{2}(x^2 + y^2)^{-1/2}\frac{d}{dx}(x^2 + y^2) - 2x = 0$$

Use the general power rule on the first term.

$$\frac{1}{2}(x^2 + y^2)^{-1/2}\left(2x + 2y\frac{dy}{dx}\right) - 2x = 0$$

$$\frac{1}{2}(x^2 + y^2)^{-1/2}\left(2x + 2y\frac{dy}{dx}\right) = 2x$$

Transpose $2x$.

$$2x + 2y\frac{dy}{dx} = 4x(x^2 + y^2)^{1/2}$$

Multiply both sides by $2(x^2 + y^2)^{1/2}$.

$$2y\frac{dy}{dx} = 4x(x^2 + y^2)^{1/2} - 2x$$

$$\frac{dy}{dx} = \frac{2x\sqrt{x^2 + y^2} - x}{y}$$

Related Rates

Implicit differentiation is a useful technique for solving a class of problems known as **related-rates** problems. The following is a typical related-rates problem: Suppose x and y are two quantities that depend on a third quantity t and we know the relationship between x and y in the form of an equation. Can we find a relationship between dx/dt and dy/dt? In particular, if we know one of the rates of change at a specific value of t—say, dx/dt—can we find the other rate, dy/dt, at that value of t?

VIDEO

APPLIED EXAMPLE 6 Rate of Change of Housing Starts A study prepared for the National Association of Realtors estimates that the number of housing starts in the southwest, $N(t)$ (in units of a million), over the next 5 years is related to the mortgage rate $r(t)$ (percent per year) by the equation

$$9N^2 + r = 36$$

What is the rate of change of the number of housing starts with respect to time when the mortgage rate is 11% per year and is increasing at the rate of 1.5% per year?

Solution We are given that

$$r = 11 \quad \text{and} \quad \frac{dr}{dt} = 1.5$$

at a certain instant of time, and we are required to find dN/dt. First, by substituting $r = 11$ into the given equation, we find

$$9N^2 + 11 = 36$$

$$N^2 = \frac{25}{9}$$

or $N = 5/3$ (we reject the negative root). Next, differentiating the given equation implicitly on both sides with respect to t, we obtain

$$\frac{d}{dt}(9N^2) + \frac{d}{dt}(r) = \frac{d}{dt}(36)$$

$$18N\frac{dN}{dt} + \frac{dr}{dt} = 0$$

Use the chain rule on the first term.

Then, substituting $N = 5/3$ and $dr/dt = 1.5$ into this equation gives

$$18\left(\frac{5}{3}\right)\frac{dN}{dt} + 1.5 = 0$$

Solving this equation for dN/dt then gives

$$\frac{dN}{dt} = -\frac{1.5}{30} = -0.05$$

Thus, at the instant of time under consideration, the number of housing starts is decreasing at the rate of 50,000 units per year. ■

APPLIED EXAMPLE 7 Supply–Demand Texar Inc., a manufacturer of disk drives is willing to make x thousand IGB USB flash drives available in the marketplace each week when the wholesale price is $\$p$ per drive. It is known that the relationship between x and p is governed by the supply equation

$$x^2 - 3xp + p^2 = 5$$

How fast is the supply of drives changing when the price per drive is \$11, the quantity supplied is 4000 drives, and the wholesale price per drive is increasing at the rate of \$.10 per drive each week?

Solution We are given that

$$p = 11 \qquad x = 4 \qquad \frac{dp}{dt} = 0.1$$

at a certain instant of time, and we are required to find dx/dt. Differentiating the given equation on both sides with respect to t, we obtain

$$\frac{d}{dt}(x^2) - \frac{d}{dt}(3xp) + \frac{d}{dt}(p^2) = \frac{d}{dt}(5)$$

$$2x\frac{dx}{dt} - 3\left(p\frac{dx}{dt} + x\frac{dp}{dt}\right) + 2p\frac{dp}{dt} = 0$$

Use the product rule on the second term.

Substituting the given values of p, x, and dp/dt into the last equation, we have

$$2(4)\frac{dx}{dt} - 3\left[(11)\frac{dx}{dt} + 4(0.1)\right] + 2(11)(0.1) = 0$$

$$8\frac{dx}{dt} - 33\frac{dx}{dt} - 1.2 + 2.2 = 0$$

$$25\frac{dx}{dt} = 1$$

$$\frac{dx}{dt} = 0.04$$

Thus, at the instant of time under consideration the supply of drives is increasing at the rate of (0.04)(1000), or 40, drives per week. ■

In certain related-rates problems, we need to formulate the problem mathematically before analyzing it. The following guidelines can be used to help solve problems of this type.

Solving Related-Rates Problems

1. Assign a variable to each quantity. Draw a diagram if needed.
2. Write the *given* values of the variables and their rates of change with respect to t.
3. Find an equation giving the relationship between the variables.
4. Differentiate both sides of this equation implicitly with respect to t.
5. Replace the variables and their derivatives by the numerical data found in step 2 and solve the equation for the required rate of change.

APPLIED EXAMPLE 8 Watching a Rocket Launch At a distance of 4000 feet from the launch site, a spectator is observing a rocket being launched. If the rocket lifts off vertically and is rising at a speed of 600 feet/second when it is at an altitude of 3000 feet, how fast is the distance between the rocket and the spectator changing at that instant?

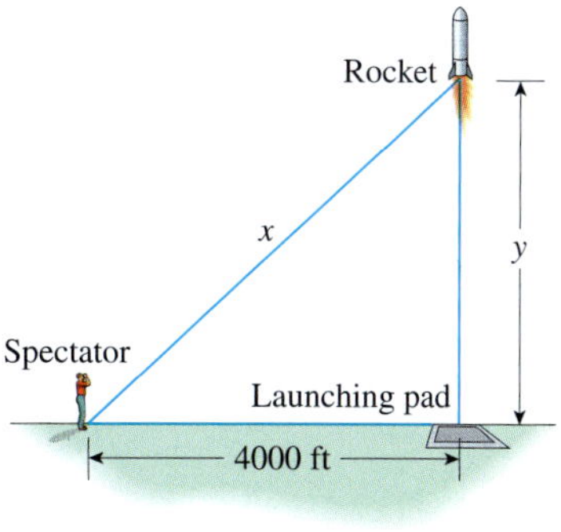

FIGURE 19
The rate at which x is changing with respect to time is related to the rate of change of y with respect to time.

Solution

Step 1 Let

y = altitude of the rocket

x = distance between the rocket and the spectator

at any time t (Figure 19).

Step 2 We are given that at a certain instant of time

$$y = 3000 \quad \text{and} \quad \frac{dy}{dt} = 600$$

and are asked to find dx/dt at that instant.

Step 3 Applying the Pythagorean theorem to the right triangle in Figure 19, we find that

$$x^2 = y^2 + 4000^2$$

Therefore, when $y = 3000$,

$$x = \sqrt{3000^2 + 4000^2} = 5000$$

Step 4 Next, we differentiate the equation $x^2 = y^2 + 4000^2$ with respect to t, obtaining

$$2x \frac{dx}{dt} = 2y \frac{dy}{dt}$$

(Remember, both x and y are functions of t.)

Step 5 Substituting $x = 5000$, $y = 3000$, and $dy/dt = 600$, we find

$$2(5000)\frac{dx}{dt} = 2(3000)(600)$$

$$\frac{dx}{dt} = 360$$

Therefore, the distance between the rocket and the spectator is changing at a rate of 360 feet/second.

Be sure that you do *not* replace the variables in the equation found in Step 3 by their numerical values before differentiating the equation.

EXAMPLE 9 A passenger ship and an oil tanker left port sometime in the morning; the former headed north, and the latter headed east. At noon, the passenger ship was 40 miles from port and sailing at 30 mph, while the oil tanker was 30 miles from port and sailing at 20 mph. How fast was the distance between the two ships changing at that time?

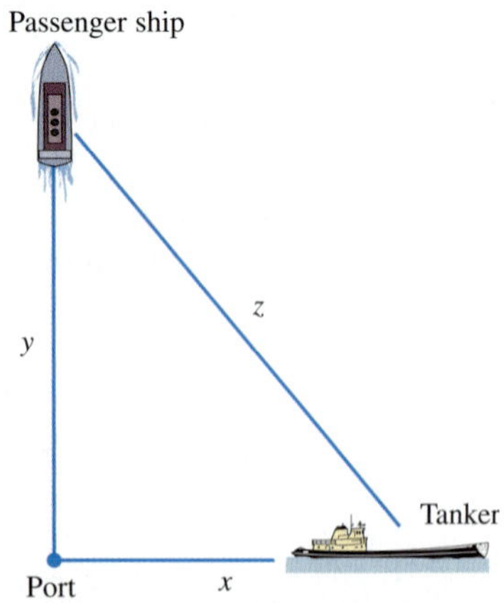

FIGURE 20
We want to find dz/dt, the rate at which the distance between the two ships is changing at a certain instant of time.

Solution

Step 1 Let

x = distance of the oil tanker from port
y = distance of the passenger ship from port
z = distance between the two ships

See Figure 20.

Step 2 We are given that at noon

$$x = 30 \qquad y = 40 \qquad \frac{dx}{dt} = 20 \qquad \frac{dy}{dt} = 30$$

and we are required to find $\frac{dz}{dt}$ at that time.

Step 3 Applying the Pythagorean theorem to the right triangle in Figure 20, we find that

$$z^2 = x^2 + y^2 \tag{9}$$

In particular, when $x = 30$ and $y = 40$, we have

$$z^2 = 30^2 + 40^2 = 2500 \quad \text{or} \quad z = 50.$$

Step 4 Differentiating (9) implicitly with respect to t, we obtain

$$2z\frac{dz}{dt} = 2x\frac{dx}{dt} + 2y\frac{dy}{dt}$$

$$z\frac{dz}{dt} = x\frac{dx}{dt} + y\frac{dy}{dt}$$

Step 4 Finally, substituting $x = 30$, $y = 40$, $z = 50$, $dx/dt = 20$, and $dy/dt = 30$ into the last equation, we find

$$50\frac{dz}{dt} = (30)(20) + (40)(30) \quad \text{and} \quad \frac{dz}{dt} = 36$$

Therefore, at noon on the day in question, the ships are moving apart at the rate of 36 mph.

3.6 Self-Check Exercises

1. Given the equation $x^3 + 3xy + y^3 = 4$, find dy/dx by implicit differentiation.

2. Find an equation of the tangent line to the graph of

$$16x^2 + 9y^2 = 144$$

at the point

$$\left(2, -\frac{4\sqrt{5}}{3}\right)$$

Solutions to Self-Check Exercises 3.6 can be found on page 226.

3.6 Concept Questions

1. a. Suppose the equation $F(x, y) = 0$ defines y as a function of x. Explain how implicit differentiation can be used to find dy/dx.
   b. What is the role of the chain rule in implicit differentiation?
2. Suppose the equation $xg(y) + yf(x) = 0$, where f and g are differentiable functions, defines y as a function of x. Find an expression for dy/dx.
3. In your own words, describe what a related-rates problem is.
4. Give the steps that you would use to solve a related-rates problem.

3.6 Exercises

In Exercises 1–8, find the derivative *dy/dx* (a) by solving each of the implicit equations for *y* explicitly in terms of *x* and (b) by differentiating each of the equations implicitly. Show that, in each case, the results are equivalent.

1. $x + 2y = 5$
2. $3x + 4y = 6$
3. $xy = 1$
4. $xy - y - 1 = 0$
5. $x^3 - x^2 - xy = 4$
6. $x^2y - x^2 + y - 1 = 0$
7. $\dfrac{x}{y} - x^2 = 1$
8. $\dfrac{y}{x} - 2x^3 = 4$

In Exercises 9–30, find *dy/dx* by implicit differentiation.

9. $x^2 + y^2 = 16$
10. $2x^2 + y^2 = 16$
11. $x^2 - 2y^2 = 16$
12. $x^3 + y^3 + y - 4 = 0$
13. $x^2 - 2xy = 6$
14. $x^2 + 5xy + y^2 = 10$
15. $x^2y^2 - xy = 8$
16. $x^2y^3 - 2xy^2 = 5$
17. $x^{1/2} + y^{1/2} = 1$
18. $x^{1/3} + y^{1/3} = 1$
19. $\sqrt{x + y} = x$
20. $(2x + 3y)^{1/3} = x^2$
21. $\dfrac{1}{x^2} + \dfrac{1}{y^2} = 1$
22. $\dfrac{1}{x^3} + \dfrac{1}{y^3} = 5$
23. $\sqrt{xy} = x + y$
24. $\sqrt{xy} = 2x + y^2$
25. $\dfrac{x + y}{x - y} = 3x$
26. $\dfrac{x - y}{2x + 3y} = 2x$
27. $xy^{3/2} = x^2 + y^2$
28. $x^2y^{1/2} = x + 2y^3$
29. $(x + y)^3 + x^3 + y^3 = 0$
30. $(x + y^2)^{10} = x^2 + 25$

In Exercises 31–34, find an equation of the tangent line to the graph of the function *f* defined by the equation at the indicated point.

31. $4x^2 + 9y^2 = 36$; $(0, 2)$

32. $y^2 - x^2 = 16$; $(2, 2\sqrt{5})$

33. $x^2y^3 - y^2 + xy - 1 = 0$; $(1, 1)$

34. $(x - y - 1)^3 = x$; $(1, -1)$

In Exercises 35–38, find the second derivative d^2y/dx^2 of each of the functions defined implicitly by the equation.

35. $xy = 1$

36. $x^3 + y^3 = 28$

37. $y^2 - xy = 8$

38. $x^{1/3} + y^{1/3} = 1$

39. The volume of a right-circular cylinder of radius r and height h is $V = \pi r^2 h$. Suppose the radius and height of the cylinder are changing with respect to time t.

a. Find a relationship between dV/dt, dr/dt, and dh/dt.

b. At a certain instant of time, the radius and height of the cylinder are 2 and 6 in. and are increasing at the rate of 0.1 and 0.3 in./sec, respectively. How fast is the volume of the cylinder increasing?

40. A car leaves an intersection traveling west. Its position 4 sec later is 20 ft from the intersection. At the same time, another car leaves the same intersection heading north so that its position 4 sec later is 28 ft from the intersection. If the speed of the cars at that instant of time is 9 ft/sec and 11 ft/sec, respectively, find the rate at which the distance between the two cars is changing.

41. PRICE-DEMAND Suppose the quantity demanded weekly of the Super Titan radial tires is related to its unit price by the equation

$$p + x^2 = 144$$

where p is measured in dollars and x is measured in units of a thousand. How fast is the quantity demanded changing when $x = 9$, $p = 63$, and the price/tire is increasing at the rate of \$2/week?

42. PRICE-SUPPLY Suppose the quantity x of Super Titan radial tires made available each week in the marketplace is related to the unit-selling price by the equation

$$p - \frac{1}{2}x^2 = 48$$

where x is measured in units of a thousand and p is in dollars. How fast is the weekly supply of Super Titan radial tires being introduced into the marketplace when $x = 6$, $p = 66$, and the price/tire is decreasing at the rate of \$3/week?

43. PRICE-DEMAND The demand equation for a certain brand of two-way headphones is

$$100x^2 + 9p^2 = 3600$$

where x represents the number (in thousands) of headphones demanded each week when the unit price is \$$p$. How fast is the quantity demanded increasing when the unit price/headphone is \$14 and the selling price is dropping at the rate of \$0.15/headphone/week?

Hint: To find the value of x when $p = 14$, solve the equation $100x^2 + 9p^2 = 3600$ for x when $p = 14$.

44. EFFECT OF PRICE ON SUPPLY Suppose the wholesale price of a certain brand of medium-sized eggs p (in dollars/carton) is related to the weekly supply x (in thousands of cartons) by the equation

$$625p^2 - x^2 = 100$$

If 25,000 cartons of eggs are available at the beginning of a certain week and the price is falling at the rate of 2¢/carton/week, at what rate is the supply falling?

Hint: To find the value of p when $x = 25$, solve the supply equation for p when $x = 25$.

45. SUPPLY-DEMAND Refer to Exercise 44. If 25,000 cartons of eggs are available at the beginning of a certain week and the supply is falling at the rate of 1000 cartons/week, at what rate is the wholesale price changing?

46. ELASTICITY OF DEMAND The demand function for a certain make of ink-jet cartridge is

$$p = -0.01x^2 - 0.1x + 6$$

where p is the unit price in dollars and x is the quantity demanded each week, measured in units of a thousand. Compute the elasticity of demand and determine whether the demand is inelastic, unitary, or elastic when $x = 10$.

47. ELASTICITY OF DEMAND The demand function for a certain brand of compact disc is

$$p = -0.01x^2 - 0.2x + 8$$

where p is the wholesale unit price in dollars and x is the quantity demanded each week, measured in units of a thousand. Compute the elasticity of demand and determine whether the demand is inelastic, unitary, or elastic when $x = 15$.

48. The volume V of a cube with sides of length x in. is changing with respect to time. At a certain instant of time, the sides of the cube are 5 in. long and increasing at the rate of 0.1 in./sec. How fast is the volume of the cube changing at that instant of time?

49. OIL SPILLS In calm waters, oil spilling from the ruptured hull of a grounded tanker spreads in all directions. If the area polluted is a circle and its radius is increasing at a rate of 2 ft/sec, determine how fast the area is increasing when the radius of the circle is 40 ft.

50. Two ships leave the same port at noon. Ship A sails north at 15 mph, and ship B sails east at 12 mph. How fast is the distance between them changing at 1 p.m.?

51. **Oil Spills** In calm waters, the oil spilling from the ruptured hull of a grounded tanker spreads in all directions. Assuming that the area polluted is circular, determine how fast the area is increasing when the radius of the circle is 60 ft and is increasing at the rate of $\frac{1}{2}$ ft/sec?

52. A car leaves an intersection traveling east. Its position t sec later is given by $x = t^2 + t$ ft. At the same time, another car leaves the same intersection heading north, traveling $y = t^2 + 3t$ ft in t sec. Find the rate at which the distance between the two cars will be changing 5 sec later.

53. A car leaves an intersection traveling west. Its position 4 sec later is 20 ft from the intersection. At the same time, another car leaves the same intersection heading north so that its position t sec later is $t^2 + 2t$ ft from the intersection. If the speed of the first car 4 sec after leaving the intersection is 9 ft/sec, find the rate at which the distance between the two cars is changing at that instant of time.

54. At a distance of 50 ft from the pad, a man observes a helicopter taking off from a heliport. If the helicopter lifts off vertically and is rising at a speed of 44 ft/sec when it is at an altitude of 120 ft, how fast is the distance between the helicopter and the man changing at that instant?

55. A spectator watches a rowing race from the edge of a river bank. The lead boat is moving in a straight line that is 120 ft from the river bank. If the boat is moving at a constant speed of 20 ft/sec, how fast is the boat moving away from the spectator when it is 50 ft past her?

56. A boat is pulled toward a dock by means of a rope wound on a drum that is located 4 ft above the bow of the boat. If the rope is being pulled in at the rate of 3 ft/sec, how fast is the boat approaching the dock when it is 25 ft from the dock?

57. Assume that a snowball is in the shape of a sphere. If the snowball melts at a rate that is proportional to its surface area, show that its radius decreases at a constant rate.
Hint: Its volume is $V = (4/3)\pi r^3$, and its surface area is $S = 4\pi r^2$.

58. **Blowing Soap Bubbles** Carlos is blowing air into a soap bubble at the rate of 8 cm^3/sec. Assuming that the bubble is spherical, how fast is its radius changing at the instant of time when the radius is 10 cm? How fast is the surface area of the bubble changing at that instant of time?

59. **Coast Guard Patrol Search Mission** The pilot of a Coast Guard patrol aircraft on a search mission had just spotted a disabled fishing trawler and decided to go in for a closer look. Flying in a straight line at a constant altitude of 1000 ft and at a steady speed of 264 ft/sec, the aircraft passed directly over the trawler. How fast was the aircraft receding from the trawler when it was 1500 ft from it?

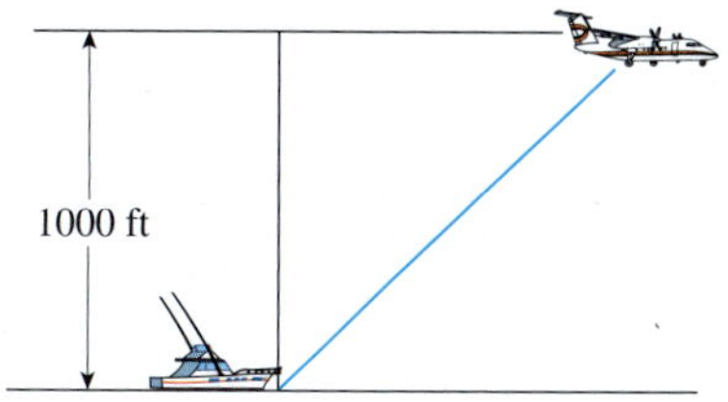

60. A coffee pot in the form of a circular cylinder of radius 4 in. is being filled with water flowing at a constant rate. If the water level is rising at the rate of 0.4 in./sec, what is the rate at which water is flowing into the coffee pot?

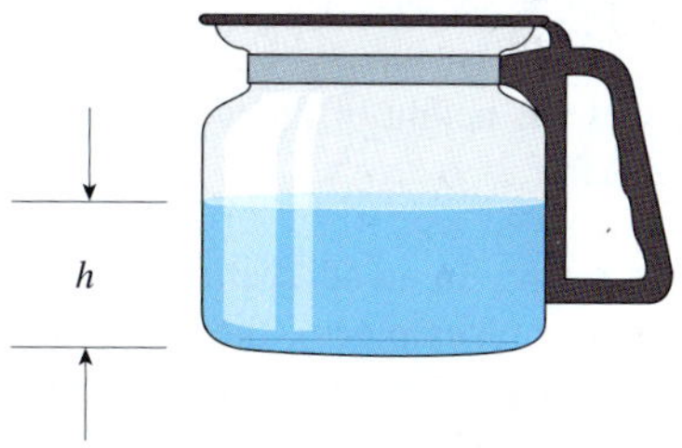

61. A 6-ft tall man is walking away from a street light 18 ft high at a speed of 6 ft/sec. How fast is the tip of his shadow moving along the ground?

62. A 20-ft ladder leaning against a wall begins to slide. How fast is the top of the ladder sliding down the wall at the instant of time when the bottom of the ladder is 12 ft from the wall and sliding away from the wall at the rate of 5 ft/sec?
Hint: Refer to the accompanying figure. By the Pythagorean theorem, $x^2 + y^2 = 400$. Find dy/dt when $x = 12$ and $dx/dt = 5$.

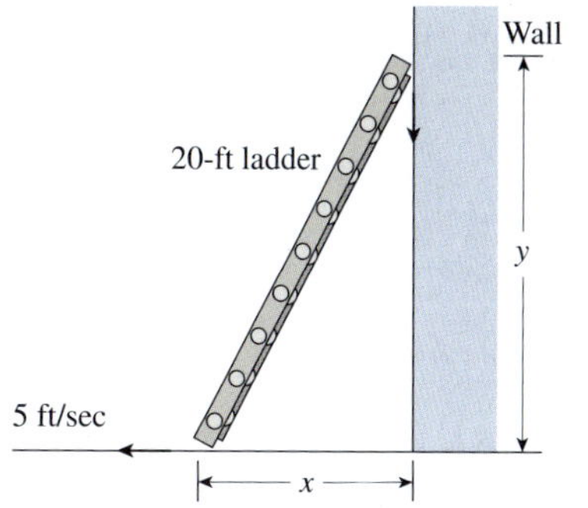

63. The base of a 13-ft ladder leaning against a wall begins to slide away from the wall. At the instant of time when the base is 12 ft from the wall, the base is moving at the rate of 8 ft/sec. How fast is the top of the ladder sliding down the wall at that instant of time?
Hint: Refer to the hint in Problem 62.

64. Water flows from a tank of constant cross-sectional area 50 ft^2 through an orifice of constant cross-sectional area 1.4 ft^2 located at the bottom of the tank (see the figure).

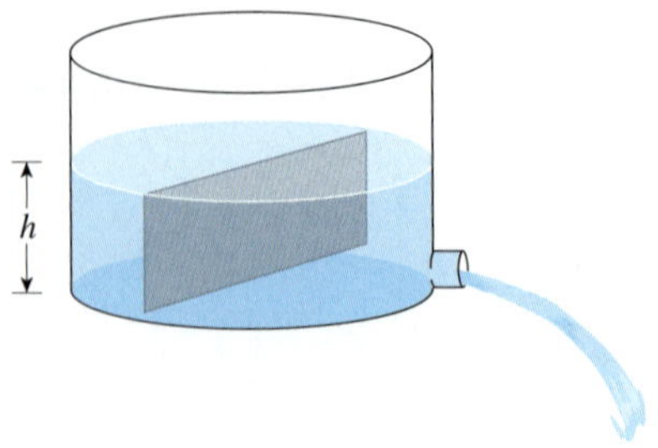

Initially, the height of the water in the tank was 20 ft and its height t sec later is given by the equation

$$2\sqrt{h} + \frac{1}{25}t - 2\sqrt{20} = 0 \qquad (0 \le t \le 50\sqrt{20})$$

How fast was the height of the water decreasing when its height was 8 ft?

65. Volume of a Gas In an adiabatic process (one in which no heat transfer takes place), the pressure P and volume V of an ideal gas such as oxygen satisfy the equation $P^5V^7 = C$, where C is a constant. Suppose that at a certain instant of time, the volume of the gas is 4L, the pressure is 100 kPa, and the pressure is decreasing at the rate of 5 kPa/sec. Find the rate at which the volume is changing.

66. Mass of a Moving Particle The mass m of a particle moving at a velocity v is related to its rest mass m_0 by the equation

$$m = \frac{m_0}{\sqrt{1 - \dfrac{v^2}{c^2}}}$$

where c (2.98×10^8 m/sec) is the speed of light. Suppose an electron of mass 9.11×10^{-31} kg is being accelerated in a particle accelerator. When its velocity is 2.92×10^8 m/sec and its acceleration is 2.42×10^5 m/sec^2, how fast is the mass of the electron changing?

In Exercises 67–70, determine whether the statement is true or false. If it is true, explain why it is true. If it is false, give an example to show why it is false.

67. The equation $x^2 + y^2 + 1 = 0$ defines y as a function of x.

68. The function

$$f(x) = \begin{cases} \sqrt{1 - x^2} & \text{if } -1 \le x < 0 \\ -\sqrt{1 - x^2} & \text{if } \ 0 \le x \le 1 \end{cases}$$

may be defined implicitly by the equation $x^2 + y^2 = 1$.

69. If f and g are differentiable and $f(x)g(y) = 0$, then

$$\frac{dy}{dx} = -\frac{f'(x)g(y)}{f(x)g'(y)} \qquad [f(x) \ne 0 \text{ and } g'(y) \ne 0]$$

70. If f and g are differentiable and $f(x) + g(y) = 0$, then

$$\frac{dy}{dx} = -\frac{f'(x)}{g'(y)}$$

3.6 Solutions to Self-Check Exercises

1. Differentiating both sides of the equation with respect to x, we have

$$3x^2 + 3y + 3xy' + 3y^2y' = 0$$
$$(x^2 + y) + (x + y^2)y' = 0$$
$$(x + y^2)y' = -(x^2 + y)$$
$$y' = -\frac{x^2 + y}{x + y^2}$$

2. To find the slope of the tangent line to the graph of the function at any point, we differentiate the equation implicitly with respect to x, obtaining

$$32x + 18yy' = 0$$
$$y' = -\frac{16x}{9y}$$

In particular, the slope of the tangent line at $\left(2, -\frac{4\sqrt{5}}{3}\right)$ is

$$m = -\frac{16(2)}{9\left(-\frac{4\sqrt{5}}{3}\right)} = \frac{8}{3\sqrt{5}}$$

Using the point-slope form of the equation of a line, we find

$$y - \left(-\frac{4\sqrt{5}}{3}\right) = \frac{8}{3\sqrt{5}}(x - 2)$$
$$y = \frac{8\sqrt{5}}{15}x - \frac{36\sqrt{5}}{15} = \frac{8\sqrt{5}}{15}x - \frac{12\sqrt{5}}{5}$$

3.7 Differentials

The Millers are planning to buy a house in the near future and estimate that they will need a 30-year fixed-rate mortgage of \$240,000. If the interest rate increases from the present rate of 7% per year to 7.4% per year between now and the time the Millers decide to secure the loan, approximately how much more per month will their mortgage be? (You will be asked to answer this question in Exercise 44, page 235.)

Questions such as this, in which one wishes to *estimate* the change in the dependent variable (monthly mortgage payment) corresponding to a small change in the independent variable (interest rate per year), occur in many real-life applications. For example:

- An economist would like to know how a small increase in a country's capital expenditure will affect the country's gross domestic output.
- A sociologist would like to know how a small increase in the amount of capital investment in a housing project will affect the crime rate.
- A businesswoman would like to know how raising a product's unit price by a small amount will affect her profit.
- A bacteriologist would like to know how a small increase in the amount of a bactericide will affect a population of bacteria.

To calculate these changes and estimate their effects, we use the *differential* of a function, a concept that will be introduced shortly.

Increments

Let x denote a variable quantity and suppose x changes from x_1 to x_2. This change in x is called the **increment in x** and is denoted by the symbol Δx (read "delta x"). Thus,

$$\Delta x = x_2 - x_1 \quad \text{Final value} - \text{initial value} \qquad (10)$$

EXAMPLE 1 Find the increment in x as x changes (a) from 3 to 3.2 and (b) from 3 to 2.7.

Solution

a. Here, $x_1 = 3$ and $x_2 = 3.2$, so

$$\Delta x = x_2 - x_1 = 3.2 - 3 = 0.2$$

b. Here, $x_1 = 3$ and $x_2 = 2.7$. Therefore,

$$\Delta x = x_2 - x_1 = 2.7 - 3 = -0.3$$

Observe that Δx plays the same role that h played in Section 2.4.

Now, suppose two quantities, x and y, are related by an equation $y = f(x)$, where f is a function. If x changes from x to $x + \Delta x$, then the corresponding change in y is called the **increment in y.** It is denoted by Δy and is defined by

$$\Delta y = f(x + \Delta x) - f(x) \qquad (11)$$

(see Figure 21).

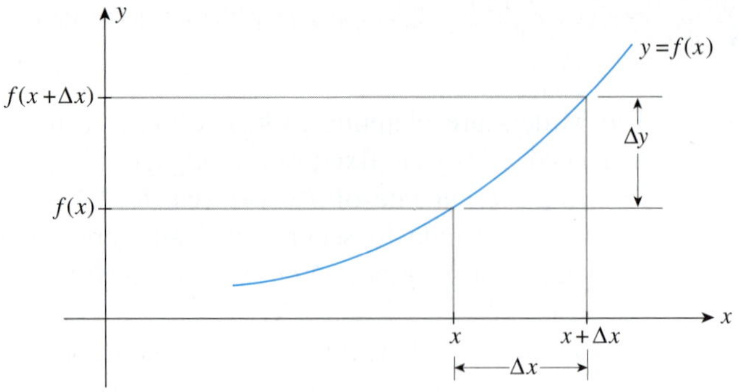

FIGURE 21
An increment of Δx in x induces an increment of $\Delta y = f(x + \Delta x) - f(x)$ in y.

EXAMPLE 2 Let $y = x^3$. Find Δx and Δy when x changes (a) from 2 to 2.01 and (b) from 2 to 1.98.

Solution Let $f(x) = x^3$.

a. Here, $\Delta x = 2.01 - 2 = 0.01$. Next,

$$\begin{aligned}\Delta y &= f(x + \Delta x) - f(x) = f(2.01) - f(2)\\ &= (2.01)^3 - 2^3 = 8.120601 - 8 = 0.120601\end{aligned}$$

b. Here, $\Delta x = 1.98 - 2 = -0.02$. Next,

$$\begin{aligned}\Delta y &= f(x + \Delta x) - f(x) = f(1.98) - f(2)\\ &= (1.98)^3 - 2^3 = 7.762392 - 8 = -0.237608\end{aligned}$$

Differentials

We can obtain a relatively quick and simple way of approximating Δy, the change in y due to a small change Δx, by examining the graph of the function f shown in Figure 22.

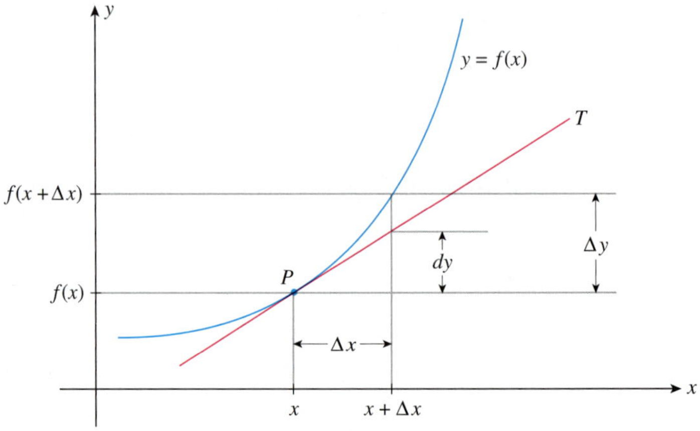

FIGURE 22
If Δx is small, dy is a good approximation of Δy.

Observe that near the point of tangency P, the tangent line T is close to the graph of f. Therefore, if Δx is small, then dy is a good approximation of Δy. We can find an expression for dy as follows: Notice that the slope of T is given by

$$\frac{dy}{\Delta x} \qquad \text{Rise} \div \text{run}$$

However, the slope of T is given by $f'(x)$. Therefore, we have

$$\frac{dy}{\Delta x} = f'(x)$$

or $dy = f'(x)\Delta x$. Thus, we have the approximation

$$\Delta y \approx dy = f'(x)\Delta x$$

in terms of the derivative of f at x. The quantity dy is called the *differential of y*.

The Differential

Let $y = f(x)$ define a differentiable function of x. Then,

1. The **differential dx** of the independent variable x is $dx = \Delta x$.
2. The **differential dy** of the dependent variable y is

$$dy = f'(x)\Delta x = f'(x)\,dx \tag{12}$$

Notes

1. For the independent variable x: There is no difference between Δx and dx—both measure the change in x from x to $x + \Delta x$.
2. For the dependent variable y: Δy measures the *actual* change in y as x changes from x to $x + \Delta x$, whereas dy measures the *approximate* change in y corresponding to the same change in x.
3. The differential dy depends on both x and dx, but for fixed x, dy is a linear function of dx. ■

EXAMPLE 3 Let $y = x^3$.

a. Find the differential dy of y.
b. Use dy to approximate Δy when x changes from 2 to 2.01.
c. Use dy to approximate Δy when x changes from 2 to 1.98.
d. Compare the results of part (b) with those of Example 2.

Solution

a. Let $f(x) = x^3$. Then,

$$dy = f'(x)\,dx = 3x^2\,dx$$

b. Here, $x = 2$ and $dx = 2.01 - 2 = 0.01$. Therefore,

$$dy = 3x^2\,dx = 3(2)^2(0.01) = 0.12$$

c. Here, $x = 2$ and $dx = 1.98 - 2 = -0.02$. Therefore,

$$dy = 3x^2\,dx = 3(2)^2(-0.02) = -0.24$$

d. As you can see, both approximations 0.12 and -0.24 are quite close to the actual changes Δy obtained in Example 2: 0.120601 and -0.237608. ■

Observe how much easier it is to find an approximation to the exact change in a function with the help of the differential, rather than calculating the exact change in the function itself. In the following examples, we take advantage of this fact.

EXAMPLE 4 Approximate the value of $\sqrt{26.5}$ using differentials. Verify your result using the $\boxed{\sqrt{\ }}$ key on your calculator.

Solution Since we want to compute the square root of a number, let's consider the function $y = f(x) = \sqrt{x}$. Since 25 is the number nearest 26.5 whose square root is readily recognized, let's take $x = 25$. We want to know the change in y, Δy, as x

changes from $x = 25$ to $x = 26.5$, an increase of $\Delta x = 1.5$ units. Using Equation (12), we find

$$\begin{aligned}\Delta y \approx dy &= f'(x)\Delta x \\ &= \left(\frac{1}{2\sqrt{x}}\bigg|_{x=25}\right)\cdot(1.5) = \left(\frac{1}{10}\right)(1.5) = 0.15\end{aligned}$$

Therefore,

$$\begin{aligned}\sqrt{26.5} - \sqrt{25} &= \Delta y \approx 0.15 \\ \sqrt{26.5} &\approx \sqrt{25} + 0.15 = 5.15\end{aligned}$$

The exact value of $\sqrt{26.5}$, rounded off to five decimal places, is 5.14782. Thus, the error incurred in the approximation is 0.00218. ■

APPLIED EXAMPLE 5 The Effect of Speed on Vehicular Operating Cost The total cost incurred in operating a certain type of truck on a 500-mile trip, traveling at an average speed of v mph, is estimated to be

$$C(v) = 125 + v + \frac{4500}{v}$$

dollars. Find the approximate change in the total operating cost when the average speed is increased from 55 mph to 58 mph.

Solution With $v = 55$ and $\Delta v = dv = 3$, we find

$$\begin{aligned}\Delta C \approx dC = C'(v)dv &= \left(1 - \frac{4500}{v^2}\right)\bigg|_{v=55}(3) \\ &= \left(1 - \frac{4500}{3025}\right)(3) \approx -1.46\end{aligned}$$

so the total operating cost is found to decrease by \$1.46. This might explain why so many independent truckers often exceed the 55 mph speed limit. ■

VIDEO

APPLIED EXAMPLE 6 The Effect of Advertising on Sales The relationship between the amount of money x spent by Cannon Precision Instruments on advertising and Cannon's total sales $S(x)$ is given by the function

$$S(x) = -0.002x^3 + 0.6x^2 + x + 500 \qquad (0 \le x \le 200)$$

where x is measured in thousands of dollars. Use differentials to estimate the change in Cannon's total sales if advertising expenditures are increased from \$100,000 ($x = 100$) to \$105,000 ($x = 105$).

Solution The required change in sales is given by

$$\begin{aligned}\Delta S \approx dS &= S'(100)dx \\ &= -0.006x^2 + 1.2x + 1\bigg|_{x=100}\cdot(5) \qquad dx = 105 - 100 = 5 \\ &= (-60 + 120 + 1)(5) = 305\end{aligned}$$

—that is, an increase of \$305,000. ■

APPLIED EXAMPLE 7 The Rings of Neptune

a. A ring has an inner radius of r units and an outer radius of R units, where $(R - r)$ is small in comparison to r (Figure 23a). Use differentials to estimate the area of the ring.

b. Recent observations, including those of *Voyager I* and *II*, showed that Neptune's ring system is considerably more complex than had been believed. For one thing, it is made up of a large number of distinguishable rings rather than one continuous great ring as previously thought (Figure 23b). The outermost ring, 1989N1R, has an inner radius of approximately 62,900 kilometers (measured from the center of the planet), and a radial width of approximately 50 kilometers. Using these data, estimate the area of the ring.

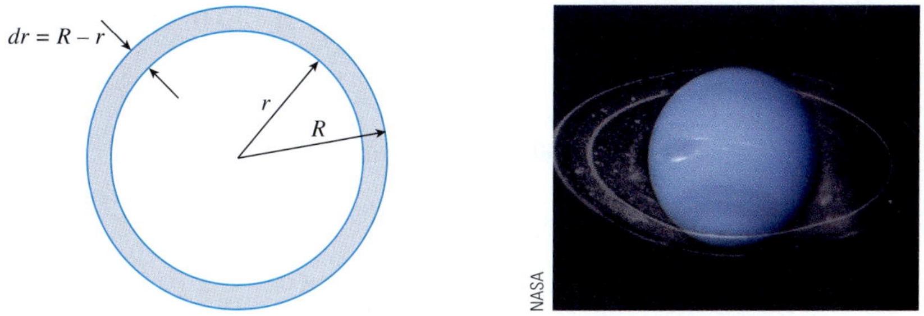

NASA

FIGURE **23**

(a) The area of the ring is the circumference of the inner circle times the thickness.

(b) Neptune and its rings

Solution

a. Since the area of a circle of radius x is $A = f(x) = \pi x^2$, we find

$$\begin{aligned}\pi R^2 - \pi r^2 &= f(R) - f(r) \\ &= \Delta A \\ &\approx dA \\ &= f'(r)dr\end{aligned}$$

Remember, ΔA = change in f when x changes from $x = r$ to $x = R$.

where $dr = R - r$. So, we see that the area of the ring is approximately $2\pi r(R - r)$ square units. In words, the area of the ring is approximately equal to

Circumference of the inner circle $\times$ Thickness of the ring

b. Applying the results of part (a) with $r = 62{,}900$ and $dr = 50$, we find that the area of the ring is approximately $2\pi(62{,}900)(50)$, or 19,760,000 square kilometers, which is roughly 4% of Earth's surface. ■

Before looking at the next example, we need to familiarize ourselves with some terminology. If a quantity with exact value q is measured or calculated with an error of Δq, then the quantity $\Delta q/q$ is called the *relative error* in the measurement or calculation of q. If the quantity $\Delta q/q$ is expressed as a percentage, it is then called the *percentage error*. Because Δq is approximated by dq, we normally approximate the relative error $\Delta q/q$ by dq/q.

APPLIED EXAMPLE 8 Estimating Errors in Measurement Suppose the radius of a ball-bearing is measured to be 0.5 inch, with a maximum error of ± 0.0002 inch. Then, the relative error in r is

$$\frac{dr}{r} = \frac{\pm 0.0002}{0.5} = \pm 0.0004$$

and the percentage error is $\pm 0.04\%$. ■

APPLIED EXAMPLE 9 Estimating Errors in Measurement Suppose the side of a cube is measured with a maximum percentage error of 2%. Use differentials to estimate the maximum percentage error in the calculated volume of the cube.

Solution Suppose the side of the cube is x, so its volume is

$$V = x^3$$

We are given that $\left|\frac{dx}{x}\right| \leq 0.02$. Now,

$$dV = 3x^2dx$$

and so

$$\frac{dV}{V} = \frac{3x^2dx}{x^3} = 3\,\frac{dx}{x}$$

Therefore,

$$\left|\frac{dV}{V}\right| = 3\left|\frac{dx}{x}\right| \leq 3(0.02) = 0.06$$

and we see that the maximum percentage error in the measurement of the volume of the cube is 6%. ■

Finally, if at some point in reading this section you have a sense of déjà vu, do not be surprised, because the notion of the differential was first used in Section 3.4 (see Example 1). There we took $\Delta x = 1$ since we were interested in finding the marginal cost when the level of production was increased from $x = 250$ to $x = 251$. If we had used differentials, we would have found

$$C(251) - C(250) \approx C'(250)\,dx$$

so that taking $dx = \Delta x = 1$, we have $C(251) - C(250) \approx C'(250)$, which agrees with the result obtained in Example 1. Thus, in Section 3.4, we touched upon the notion of the differential, albeit in the special case in which $dx = 1$.

3.7 Self-Check Exercises

1. Find the differential of $f(x) = \sqrt{x} + 1$.

2. A certain country's government economists have determined that the demand equation for corn in that country is given by

$$p = f(x) = \frac{125}{x^2 + 1}$$

where p is expressed in dollars/bushel and x, the quantity demanded each year, is measured in billions of bushels. The economists are forecasting a harvest of 6 billion bushels for the year. If the actual production of corn were 6.2 billion bushels for the year instead, what would be the approximate drop in the predicted price of corn/bushel?

Solutions to Self-Check Exercises 3.7 can be found on page 235.

3.7 Concept Questions

1. If $y = f(x)$, what is the differential of x? Write an expression for the differential dy.
2. Refer to the following figure.

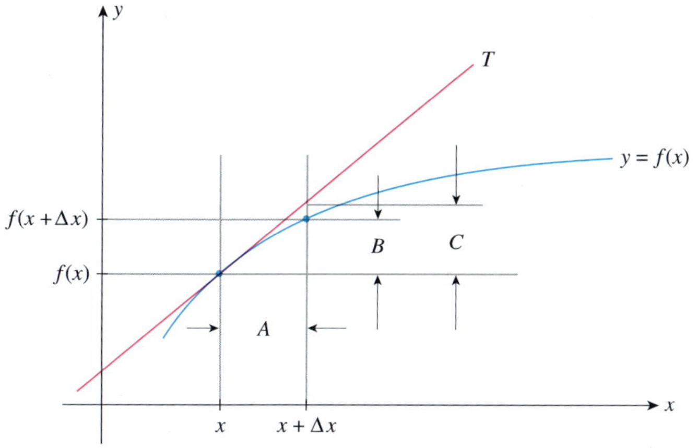

a. Write A, B, and C in terms of Δx, Δy, and dy.
b. Write $f'(x)$ (the slope of the tangent line T at x) in terms of Δx and dy.
c. Observing that $B \approx C$, explain why $\Delta y \approx f'(x)\Delta x = f'(x)dx$.

3.7 Exercises

In Exercises 1–14, find the differential of the function.

1. $f(x) = 2x^2$
2. $f(x) = 3x^2 + 1$
3. $f(x) = x^3 - x$
4. $f(x) = 2x^3 + x$
5. $f(x) = \sqrt{x + 1}$
6. $f(x) = \dfrac{3}{\sqrt{x}}$
7. $f(x) = 2x^{3/2} + x^{1/2}$
8. $f(x) = 3x^{5/6} + 7x^{2/3}$
9. $f(x) = x + \dfrac{2}{x}$
10. $f(x) = \dfrac{3}{x - 1}$
11. $f(x) = \dfrac{x - 1}{x^2 + 1}$
12. $f(x) = \dfrac{2x^2 + 1}{x + 1}$
13. $f(x) = \sqrt{3x^2 - x}$
14. $f(x) = (2x^2 + 3)^{1/3}$

15. Let f be the function defined by

$$y = f(x) = x^2 - 1$$

a. Find the differential of f.
b. Use your result from part (a) to find the approximate change in y if x changes from 1 to 1.02.
c. Find the actual change in y if x changes from 1 to 1.02 and compare your result with that obtained in part (b).

16. Let f be the function defined by

$$y = f(x) = 3x^2 - 2x + 6$$

a. Find the differential of f.
b. Use your result from part (a) to find the approximate change in y if x changes from 2 to 1.97.
c. Find the actual change in y if x changes from 2 to 1.97 and compare your result with that obtained in part (b).

17. Let f be the function defined by

$$y = f(x) = \frac{1}{x}$$

a. Find the differential of f.
b. Use your result from part (a) to find the approximate change in y if x changes from -1 to -0.95.
c. Find the actual change in y if x changes from -1 to -0.95 and compare your result with that obtained in part (b).

18. Let f be the function defined by

$$y = f(x) = \sqrt{2x + 1}$$

a. Find the differential of f.
b. Use your result from part (a) to find the approximate change in y if x changes from 4 to 4.1.
c. Find the actual change in y if x changes from 4 to 4.1 and compare your result with that obtained in part (b).

In Exercises 19–26, use differentials to approximate the quantity.

19. $\sqrt{10}$
20. $\sqrt{17}$
21. $\sqrt{49.5}$
22. $\sqrt{99.7}$
23. $\sqrt[3]{7.8}$
24. $\sqrt[4]{81.6}$
25. $\sqrt{0.089}$
26. $\sqrt[3]{0.00096}$

27. Use a differential to approximate $\sqrt{4.02} + \dfrac{1}{\sqrt{4.02}}$.

Hint: Let $f(x) = \sqrt{x} + \dfrac{1}{\sqrt{x}}$ and compute dy with $x = 4$ and $dx = 0.02$.

28. Use a differential to approximate $\dfrac{2(4.98)}{(4.98)^2 + 1}$.

Hint: Study the hint for Exercise 27.

29. **ERROR ESTIMATION** The length of each edge of a cube is 12 cm, with a possible error in measurement of 0.02 cm. Use differentials to estimate the error that might occur when the volume of the cube is calculated.

30. **ESTIMATING THE AMOUNT OF PAINT REQUIRED** A coat of paint of thickness 0.05 cm is to be applied uniformly to the faces of a cube of edge 30 cm. Use differentials to find the approximate amount of paint required for the job.

31. **ERROR ESTIMATION** A hemisphere-shaped dome of radius 60 ft is to be coated with a layer of rust-proofer before painting. Use differentials to estimate the amount of rust-proofer needed if the coat is to be 0.01 in. thick.
Hint: The volume of a hemisphere of radius r is $V = \frac{2}{3}\pi r^3$.

32. **GROWTH OF A CANCEROUS TUMOR** The volume of a spherical cancerous tumor is given by

$$V(r) = \frac{4}{3}\pi r^3$$

If the radius of a tumor is estimated at 1.1 cm, with a maximum error in measurement of 0.005 cm, determine the error that might occur when the volume of the tumor is calculated.

33. **UNCLOGGING ARTERIES** Research done in the 1930s by the French physiologist Jean Poiseuille showed that the resistance R of a blood vessel of length l and radius r is $R = kl/r^4$, where k is a constant. Suppose a dose of the drug TPA increases r by 10%. How will this affect the resistance R? Assume that l is constant.

34. **GROSS DOMESTIC PRODUCT** An economist has determined that a certain country's gross domestic product (GDP) is approximated by the function $f(x) = 640x^{1/5}$, where $f(x)$ is measured in billions of dollars and x is the capital outlay in billions of dollars. Use differentials to estimate the change in the country's GDP if the country's capital expenditure changes from \$243 billion to \$248 billion.

35. **LEARNING CURVES** The length of time (in seconds) a certain individual takes to learn a list of n items is approximated by

$$f(n) = 4n\sqrt{n - 4}$$

Use differentials to approximate the additional time it takes the individual to learn the items on a list when n is increased from 85 to 90 items.

36. **EFFECT OF ADVERTISING ON PROFITS** The relationship between Cunningham Realty's quarterly profits, $P(x)$, and the amount of money x spent on advertising per quarter is described by the function

$$P(x) = -\frac{1}{8}x^2 + 7x + 30 \qquad (0 \le x \le 50)$$

where both $P(x)$ and x are measured in thousands of dollars. Use differentials to estimate the increase in profits when advertising expenditure each quarter is increased from \$24,000 to \$26,000.

37. **EFFECT OF MORTGAGE RATES ON HOUSING STARTS** A study prepared for the National Association of Realtors estimates that the number of housing starts per year over the next 5 yr will be

$$N(r) = \frac{7}{1 + 0.02r^2}$$

million units, where r (percent) is the mortgage rate. Use differentials to estimate the decrease in the number of housing starts when the mortgage rate is increased from 6% to 6.5%.

38. **SUPPLY-PRICE** The supply equation for a certain brand of radio is given by

$$p = s(x) = 0.3\sqrt{x} + 10$$

where x is the quantity supplied and p is the unit price in dollars. Use differentials to approximate the change in price when the quantity supplied is increased from 10,000 units to 10,500 units.

39. **DEMAND-PRICE** The demand function for the Sentinel smoke alarm is given by

$$p = d(x) = \frac{30}{0.02x^2 + 1}$$

where x is the quantity demanded (in units of a thousand) and p is the unit price in dollars. Use differentials to estimate the change in the price p when the quantity demanded changes from 5000 to 5500 units/week.

40. **SURFACE AREA OF AN ANIMAL** Animal physiologists use the formula

$$S = kW^{2/3}$$

to calculate an animal's surface area (in square meters) from its weight W (in kilograms), where k is a constant that depends on the animal under consideration. Suppose a physiologist calculates the surface area of a horse ($k = 0.1$). If the horse's weight is estimated at 300 kg, with a maximum error in measurement of 0.6 kg, determine the percentage error in the calculation of the horse's surface area.

41. **FORECASTING PROFITS** The management of Trappee and Sons forecast that they will sell 200,000 cases of their TexaPep hot sauce next year. Their annual profit is described by

$$P(x) = -0.000032x^3 + 6x - 100$$

thousand dollars, where x is measured in thousands of cases. If the maximum error in the forecast is 15%, determine the corresponding error in Trappee's profits.

42. **FORECASTING COMMODITY PRICES** A certain country's government economists have determined that the demand equation for soybeans in that country is given by

$$p = f(x) = \frac{55}{2x^2 + 1}$$

where p is expressed in dollars/bushel and x, the quantity demanded each year, is measured in billions of bushels. The economists are forecasting a harvest of 1.8 billion bushels for the year, with a maximum error of 15% in their forecast. Determine the corresponding maximum error in the predicted price per bushel of soybeans.

43. **CRIME STUDIES** A sociologist has found that the number of serious crimes in a certain city each year is described by the function

$$N(x) = \frac{500(400 + 20x)^{1/2}}{(5 + 0.2x)^2}$$

where x (in cents/dollar deposited) is the level of reinvestment in the area in conventional mortgages by the city's ten largest banks. Use differentials to estimate the change in the number of crimes if the level of reinvestment changes from 20¢/dollar deposited to 22¢/dollar deposited.

44. **FINANCING A HOME** The Millers are planning to buy a home in the near future and estimate that they will need a 30-yr fixed-rate mortgage for $240,000. Their monthly payment P (in dollars) can be computed using the formula

$$P = \frac{20{,}000r}{1 - \left(1 + \frac{r}{12}\right)^{-360}}$$

where r is the interest rate per year.

a. Find the differential of P.

b. If the interest rate increases from the present rate of 7%/year to 7.2%/year between now and the time the Millers decide to secure the loan, approximately how much more will their monthly mortgage payment be? How much more will it be if the interest rate increases to 7.3%/year? To 7.4%/year? To 7.5%/year?

45. **INVESTMENTS** Lupé deposits a sum of $10,000 into an account that pays interest at the rate of r/year compounded monthly. Her investment at the end of 10 yr is given by

$$A = 10{,}000\left(1 + \frac{r}{12}\right)^{120}$$

a. Find the differential of A.

b. Approximately how much more would Lupé's account be worth at the end of the term if her account paid 8.1%/year instead of 8%/year? 8.2%/year instead of 8%/year? 8.3%/year instead of 8%/year?

46. **KEOGH ACCOUNTS** Ian, who is self-employed, contributes $2000 a month into a Keogh account earning interest at the rate of r/year compounded monthly. At the end of 25 yr, his account will be worth

$$S = \frac{24{,}000\left[\left(1 + \frac{r}{12}\right)^{300} - 1\right]}{r}$$

dollars.

a. Find the differential of S.

b. Approximately how much more would Ian's account be worth at the end of 25 yr if his account earned 6.1%/year instead of 6%/year? 6.2%/year instead of 6%/year? 6.3%/year instead of 6%/year?

In Exercises 47 and 48, determine whether the statement is true or false. If it is true, explain why it is true. If it is false, give an example to show why it is false.

47. If $y = ax + b$ where a and b are constants, then $\Delta y = dy$.

48. If $A = f(x)$, then the percentage change in A is

$$\frac{100f'(x)}{f(x)}\,dx$$

3.7 Solutions to Self-Check Exercises

1. We find

$$f'(x) = \frac{1}{2}x^{-1/2} = \frac{1}{2\sqrt{x}}$$

Therefore, the required differential of f is

$$dy = \frac{1}{2\sqrt{x}}\,dx$$

2. We first compute the differential

$$dp = -\frac{250x}{(x^2 + 1)^2}\,dx$$

Next, using Equation (12) with $x = 6$ and $dx = 0.2$, we find

$$\Delta p \approx dp = -\frac{250(6)}{(36 + 1)^2}(0.2) = -0.22$$

or a drop in price of 22¢/bushel.

USING TECHNOLOGY

Finding the Differential of a Function

The calculation of the differential of f at a given value of x involves the evaluation of the derivative of f at that point and can be facilitated through the use of the numerical derivative function.

EXAMPLE 1 Use dy to approximate Δy if $y = x^2(2x^2 + x + 1)^{2/3}$ and x changes from 2 to 1.98.

Solution Let $f(x) = x^2(2x^2 + x + 1)^{2/3}$. Since $dx = 1.98 - 2 = -0.02$, we find the required approximation to be

$$dy = f'(2)(-0.02)$$

But using the numerical derivative operation, we find

$$f'(2) = 30.57581679$$

(see Figure T1). Thus,

$$dy = (-0.02)(30.57581679) = -0.6115163358$$ ■

```
nDeriv(X^2(2X^2+
X+1)^(2/3), X, 2)
        30.57581679
```

FIGURE **T1**
The TI-83/84 numerical derivative screen for computing $f'(2)$

APPLIED EXAMPLE 2 Financing a Home The Meyers are considering the purchase of a house in the near future and estimate that they will need a loan of \$240,000. Based on a 30-year conventional mortgage with an annual interest rate of r, their monthly repayment will be

$$P = \frac{20{,}000r}{1 - \left(1 + \frac{r}{12}\right)^{-360}}$$

dollars. If the interest rate increases from 7% per year to 7.2% per year between now and the time the Meyers decide to secure the loan, approximately how much more will their monthly mortgage payment be?

Solution Let's write

$$P = f(r) = \frac{20{,}000r}{1 - \left(1 + \frac{r}{12}\right)^{-360}}$$

Then the increase in the mortgage payment will be approximately

$$dP = f'(0.07)\,dr = f'(0.07)(0.002) \quad \text{Since } dr = 0.072 - 0.07$$
$$\approx 32.2364 \quad \text{Use the numerical derivative operation.}$$

or approximately \$32.24 per month. (See Figure T2.)

```
nDeriv((20000X)/
(1-(1+X/12)^-360
),X,.07)
        16118.19243
```

FIGURE T2
The TI-83/84 numerical derivative screen for computing $f'(0.07)$

TECHNOLOGY EXERCISES

In Exercises 1–6, use dy to approximate Δy for the function $y = f(x)$ when x changes from $x = a$ to $x = b$.

1. $f(x) = 0.21x^7 - 3.22x^4 + 5.43x^2 + 1.42x + 12.42$; $a = 3$, $b = 3.01$
2. $f(x) = \dfrac{0.2x^2 + 3.1}{1.2x + 1.3}$; $a = 2$, $b = 1.96$
3. $f(x) = \sqrt{2.2x^2 + 1.3x + 4}$; $a = 1$, $b = 1.03$
4. $f(x) = x\sqrt{2x^3 - x + 4}$; $a = 2$, $b = 1.98$
5. $f(x) = \dfrac{\sqrt{x^2 + 4}}{x - 1}$; $a = 4$, $b = 4.1$
6. $f(x) = 2.1x^2 + \dfrac{3}{\sqrt{x}} + 5$; $a = 3$, $b = 2.95$
7. **Calculating Mortgage Payments** Refer to Example 2. How much more will the Meyers' mortgage payment be each month if the interest rate increases from 7% to 7.3%/year? To 7.4%/year? To 7.5%/year?
8. **Estimating the Area of a Ring of Neptune** The ring 1989N2R of the planet Neptune has an inner radius of approximately 53,200 km (measured from the center of the planet) and a radial width of 15 km. Use differentials to estimate the area of the ring.
9. **Effect of Price Increase on Quantity Demanded** The quantity demanded each week of the Alpha Sports Watch, x (in thousands), is related to its unit price of p dollars by the equation
$$x = f(p) = 10\sqrt{\frac{50 - p}{p}} \qquad (0 \leq p \leq 50)$$
Use differentials to find the decrease in the quantity of the watches demanded each week if the unit price is increased from \$40 to \$42.
10. **Period of a Communications Satellite** According to Kepler's third law, the period T (in days) of a satellite moving in a circular orbit d mi above the surface of Earth is given by
$$T = 0.0588\left(1 + \frac{d}{3959}\right)^{3/2}$$
Suppose a communications satellite that was moving in a circular orbit 22,000 mi above Earth's surface at one time has, because of friction, dropped down to a new orbit that is 21,500 mi above Earth's surface. Estimate the decrease in the period of the satellite to the nearest $\frac{1}{100}$th hr.

CHAPTER 3 Summary of Principal Formulas and Terms

FORMULAS

1. Derivative of a constant	$\dfrac{d}{dx}(c) = 0 \qquad (c, \text{ a constant})$
2. Power rule	$\dfrac{d}{dx}(x^n) = nx^{n-1}$
3. Constant multiple rule	$\dfrac{d}{dx}[cf(x)] = cf'(x)$
4. Sum rule	$\dfrac{d}{dx}[f(x) \pm g(x)] = f'(x) \pm g'(x)$

5. Product rule	$\frac{d}{dx}[f(x)g(x)] = f(x)g'(x) + g(x)f'(x)$
6. Quotient rule	$\frac{d}{dx}\left[\frac{f(x)}{g(x)}\right] = \frac{g(x)f'(x) - f(x)g'(x)}{[g(x)]^2}$
7. Chain rule	$\frac{d}{dx}g(f(x)) = g'(f(x))f'(x)$
8. General power rule	$\frac{d}{dx}[f(x)]^n = n[f(x)]^{n-1}f'(x)$
9. Average cost function	$\overline{C}(x) = \frac{C(x)}{x}$
10. Revenue function	$R(x) = px$
11. Profit function	$P(x) = R(x) - C(x)$
12. Elasticity of demand	$E(p) = -\frac{pf'(p)}{f(p)}$
13. Differential of y	$dy = f'(x)dx$

TERMS

marginal cost (195)
marginal cost function (195)
average cost (196)
marginal average cost function (196)
marginal revenue (198)
marginal revenue function (198)
marginal profit function (199)
elasticity of demand (201)
elastic demand (202)
unitary demand (202)
inelastic demand (202)
second derivative of f (208)
implicit differentiation (216)
related rates (219)

CHAPTER 3 Concept Review Questions

Fill in the blanks.

1. a. If c is a constant, then $\frac{d}{dx}(c)$ = _____.

b. The power rule states that if n is any real number, then $\frac{d}{dx}(x^n)$ = _____.

c. The constant multiple rule states that if c is a constant, then $\frac{d}{dx}[cf(x)]$ = _____.

d. The sum rule states that $\frac{d}{dx}[f(x) \pm g(x)]$ = _____.

2. a. The product rule states that $\frac{d}{dx}[f(x)g(x)]$ = _____.

b. The quotient rule states that $\frac{d}{dx}[f(x)/g(x)]$ = _____.

3. a. The chain rule states that if $h(x) = g[f(x)]$, then $h'(x)$ = _____.

b. The general power rule states that if $h(x) = [f(x)]^n$, then $h'(x)$ = _____.

4. If C, R, P, and $\overline{C}$ denote the total cost function, the total revenue function, the profit function, and the average cost function, respectively, then C' denotes the _____ _____ function, R' denotes the _____ _____ function, P' denotes the _____ _____ function, and $\overline{C}'$ denotes the _____ _____ _____ function.

5. a. If f is a differentiable demand function defined by $x = f(p)$, then the elasticity of demand at price p is given by $E(p)$ = _____.

b. The demand is _____ if $E(p) > 1$; it is _____ if $E(p) = 1$; it is _____ if $E(p) < 1$.

6. Suppose a function $y = f(x)$ is defined implicitly by an equation in x and y. To find $\frac{dy}{dx}$, we differentiate _____ _____ of the equation with respect to x and then solve the resulting equation for $\frac{dy}{dx}$. The derivative of a term involving y includes _____ as a factor.

7. In a related-rates problem, we are given a relationship between x and _____ that depends on a third variable t. Knowing the values of x, y, and $\frac{dx}{dt}$ at a, we want to find _____ at _____.

8. Let $y = f(t)$ and $x = g(t)$. If $x^2 + y^2 = 4$, then $\frac{dx}{dt}$ = _____. If $xy = 1$, then $\frac{dy}{dt}$ = _____.

9. a. If a variable quantity x changes from x_1 to x_2, then the increment in x is Δx = _____.

b. If $y = f(x)$ and x changes from x to $x + \Delta x$, then the increment in y is Δy = _____.

10. If $y = f(x)$, where f is a differentiable function, then the differential dx of x is dx = _____, where _____ is an increment in _____, and the differential dy of y is dy = _____.

CHAPTER 3 Review Exercises

In Exercises 1–30, find the derivative of the function.

1. $f(x) = 3x^5 - 2x^4 + 3x^2 - 2x + 1$
2. $f(x) = 4x^6 + 2x^4 + 3x^2 - 2$
3. $g(x) = -2x^{-3} + 3x^{-1} + 2$
4. $f(t) = 2t^2 - 3t^3 - t^{-1/2}$
5. $g(t) = 2t^{-1/2} + 4t^{-3/2} + 2$
6. $h(x) = x^2 + \dfrac{2}{x}$
7. $f(t) = t + \dfrac{2}{t} + \dfrac{3}{t^2}$
8. $g(s) = 2s^2 - \dfrac{4}{s} + \dfrac{2}{\sqrt{s}}$
9. $h(x) = x^2 - \dfrac{2}{x^{3/2}}$
10. $f(x) = \dfrac{x + 1}{2x - 1}$
11. $g(t) = \dfrac{t^2}{2t^2 + 1}$
12. $h(t) = \dfrac{\sqrt{t}}{\sqrt{t} + 1}$
13. $f(x) = \dfrac{\sqrt{x} - 1}{\sqrt{x} + 1}$
14. $f(t) = \dfrac{t}{2t^2 + 1}$
15. $f(x) = \dfrac{x^2(x^2 + 1)}{x^2 - 1}$
16. $f(x) = (2x^2 + x)^3$
17. $f(x) = (3x^3 - 2)^8$
18. $h(x) = (\sqrt{x} + 2)^5$
19. $f(t) = \sqrt{2t^2 + 1}$
20. $g(t) = \sqrt[3]{1 - 2t^3}$
21. $s(t) = (3t^2 - 2t + 5)^{-2}$
22. $f(x) = (2x^3 - 3x^2 + 1)^{-3/2}$
23. $h(x) = \left(x + \dfrac{1}{x}\right)^2$
24. $h(x) = \dfrac{1 + x}{(2x^2 + 1)^2}$
25. $h(t) = (t^2 + t)^4(2t^2)$
26. $f(x) = (2x + 1)^3(x^2 + x)^2$
27. $g(x) = \sqrt{x}(x^2 - 1)^3$
28. $f(x) = \dfrac{x}{\sqrt{x^3 + 2}}$
29. $h(x) = \dfrac{\sqrt{3x + 2}}{4x - 3}$
30. $f(t) = \dfrac{\sqrt{2t + 1}}{(t + 1)^3}$

In Exercises 31–36, find the second derivative of the function.

31. $f(x) = 2x^4 - 3x^3 + 2x^2 + x + 4$
32. $g(x) = \sqrt{x} + \dfrac{1}{\sqrt{x}}$
33. $h(t) = \dfrac{t}{t^2 + 4}$
34. $f(x) = (x^3 + x + 1)^2$
35. $f(x) = \sqrt{2x^2 + 1}$
36. $f(t) = t(t^2 + 1)^3$

In Exercises 37–42, find *dy/dx* by implicit differentiation.

37. $6x^2 - 3y^2 = 9$
38. $2x^3 - 3xy = 4$
39. $y^3 + 3x^2 = 3y$
40. $x^2 + 2x^2y^2 + y^2 = 10$
41. $x^2 - 4xy - y^2 = 12$
42. $3x^2y - 4xy + x - 2y = 6$

43. Find the differential of $f(x) = x^2 + \dfrac{1}{x^2}$.

44. Find the differential of $f(x) = \dfrac{1}{\sqrt{x^3 + 1}}$.

45. Let f be the function defined by $f(x) = \sqrt{2x^2 + 4}$.
   a. Find the differential of f.
   b. Use your result from part (a) to find the approximate change in $y = f(x)$ if x changes from 4 to 4.1.
   c. Find the actual change in y if x changes from 4 to 4.1 and compare your result with that obtained in part (b).

46. Use a differential to approximate $\sqrt[3]{26.8}$.

47. Let $f(x) = 2x^3 - 3x^2 - 16x + 3$.
   a. Find the points on the graph of f at which the slope of the tangent line is equal to -4.
   b. Find the equation(s) of the tangent line(s) of part (a).

48. Let $f(x) = \frac{1}{3}x^3 + \frac{1}{2}x^2 - 4x + 1$.
   a. Find the points on the graph of f at which the slope of the tangent line is equal to -2.
   b. Find the equation(s) of the tangent line(s) of part (a).

49. Find an equation of the tangent line to the graph of $y = \sqrt{4 - x^2}$ at the point $(1, \sqrt{3})$.

50. Find an equation of the tangent line to the graph of $y = x(x + 1)^5$ at the point $(1, 32)$.

51. Find the third derivative of the function

$$f(x) = \frac{1}{2x - 1}$$

What is its domain?

52. The demand equation for a certain product is $2x + 5p - 60 = 0$, where p is the unit price and x is the quantity demanded of the product. Find the elasticity of demand and determine whether the demand is elastic or inelastic, at the indicated prices.
   a. $p = 3$ b. $p = 6$ c. $p = 9$

53. The demand equation for a certain product is

$$x = \frac{25}{\sqrt{p}} - 1$$

where p is the unit price and x is the quantity demanded for the product. Compute the elasticity of demand and determine the range of prices corresponding to inelastic, unitary, and elastic demand.

54. The demand equation for a certain product is $x = 100 - 0.01p^2$.
a. Is the demand elastic, unitary, or inelastic when $p = 40$?
b. If the price is \$40, will raising the price slightly cause the revenue to increase or decrease?

55. The demand equation for a certain product is

$$p = 9\sqrt[3]{1000 - x}$$

a. Is the demand elastic, unitary, or inelastic when $p = 60$?
b. If the price is \$60, will raising the price slightly cause the revenue to increase or decrease?

56. GDP of a Country The gross domestic product (GDP) of a certain country was

$$f(t) = 0.1t^3 + 0.5t^2 + 2t + 20 \qquad (0 \leq t \leq 4)$$

billion dollars in year t, where t is measured in years with $t = 0$ corresponding to 2007.
a. What was the GDP of the country in 2010?
b. How fast was the GDP of the country changing in 2010?

57. Cell Phones The percent of the U.S. population with cell phones is projected to be

$$P(t) = 24.4t^{0.34} \qquad (1 \leq t \leq 10)$$

where t is measured in years, with $t = 1$ corresponding to the beginning of 1998.
a. What percentage of the U.S. population is expected to have cell phones by the beginning of 2006?
b. How fast was the percentage of the U.S. population with cell phones expected to be changing at the beginning of 2006?
Source: BancAmerica Robertson Stephens

58. Sales of Cameras The shipments of the Lica digital single-lens reflex cameras, or SLRs, are projected to be

$$N(t) = 6t^2 + 200t + 4\sqrt{t} + 20{,}000 \qquad (0 \leq t \leq 4)$$

units t years from now.
a. How many Lica SLRs will be shipped after 2 yr?
b. At what rate will the number of Lica SLRs shipped be changing after 2 yr?

59. Sales of DSPs The sales of digital signal processors (DSPs) in billions of dollars is projected to be

$$S(t) = 0.14t^2 + 0.68t + 3.1 \qquad (0 \leq t \leq 6)$$

where t is measured in years, with $t = 0$ corresponding to the beginning of 1997.
a. What were the sales of DSPs at the beginning of 1997? What were the sales at the beginning of 2002?
b. How fast was the level of sales increasing at the beginning of 1997? How fast were sales increasing at the beginning of 2002?
Source: World Semiconductor Trade Statistics

60. Adult Obesity In the United States, the percent of adults (age 20–74) classified as obese held steady through the 1960s and 1970s at around 14% but began to rise rapidly during the 1980s and 1990s. This rise in adult obesity coincided with the period when an increasing number of Americans began eating more sugar and fats. The function

$$P(t) = 0.01484t^2 + 0.446t + 15 \qquad (0 \leq t \leq 22)$$

gives the percentage of obese adults from 1978 ($t = 0$) through the year 2000 ($t = 22$).
a. What percentage of adults were obese in 1978? In 2000?
b. How fast was the percentage of obese adults increasing in 1980 ($t = 2$)? In 1998 ($t = 20$)?
Source: Journal of the American Medical Association

61. Population Growth The population of a certain suburb is expected to be

$$P(t) = 30 - \frac{20}{2t + 3} \qquad (0 \leq t \leq 5)$$

thousand t years from now.
a. By how much will the population have grown after 3 yr?
b. How fast is the population changing after 3 yr?

62. Best-Selling Novel The number of copies of a best-selling novel sold t wk after it was introduced is given by

$$N(t) = (4 + 5t)^{5/3} \qquad (1 \leq t \leq 30)$$

where $N(t)$ is measured in thousands.
a. How many copies of the novel were sold after 12 wk?
b. How fast were the sales of the novel changing after 12 wk?

63. Cable TV Subscribers The number of subscribers to CNC Cable Television in the town of Randolph is approximated by the function

$$N(x) = 1000(1 + 2x)^{1/2} \qquad (1 \leq x \leq 30)$$

where $N(x)$ denotes the number of subscribers to the service in the xth week. Find the rate of increase in the number of subscribers at the end of the 12th week.

64. Cost of Wireless Phone Calls As cellular phone usage continues to soar, the airtime costs have dropped. The average price per minute of use (in cents) is projected to be

$$f(t) = 31.88(1 + t)^{-0.45} \qquad (0 \le t \le 6)$$

where t is measured in years and $t = 0$ corresponds to the beginning of 1998. Compute $f'(t)$. How fast was the average price/minute of use changing at the beginning of 2000? What was the average price/minute of use at the beginning of 2000?

Source: Cellular Telecommunications Industry Association

65. **Male Life Expectancy** Suppose the life expectancy of a male at birth in a certain country is described by the function

$$f(t) = 46.9(1 + 1.09t)^{0.1} \qquad (0 \le t \le 150)$$

where t is measured in years, with $t = 0$ corresponding to the beginning of 1900. How long can a male born at the beginning of 2000 in that country expect to live? What is the rate of change of the life expectancy of a male born in that country at the beginning of 2000?

66. **Cost of Producing DVDs** The total weekly cost in dollars incurred by Herald Media Corp. in producing x DVDs is given by the total cost function

$$C(x) = 2500 + 2.2x \qquad (0 \le x \le 8000)$$

a. What is the marginal cost when $x = 1000$ and 2000?
b. Find the average cost function $\overline{C}$ and the marginal average cost function $\overline{C}'$.
c. Using the results from part (b), show that the average cost incurred by Herald in pressing a DVD approaches \$2.20/disc when the level of production is high enough.

67. **Supply Function** The supply function for a certain brand of satellite radios is given by

$$p = \frac{1}{10}x^{3/2} + 10 \qquad (0 \le x \le 50)$$

where x is the quantity demanded (in thousands) if the unit price is \$$p$. Find $p'(40)$ and interpret your result.

68. **Demand Function** The demand for a certain brand of electric shavers is given by

$$p = 20\sqrt{-x^2 + 100} \qquad (0 \le x \le 10)$$

where x (in thousands) is the quantity demanded if the unit price is \$$p$. Find $p'(6)$ and interpret your result.

69. **Marginal Cost** The total daily cost (in dollars) incurred by Delta Electronics in producing x MP3 players is

$$C(x) = 0.0001x^3 - 0.02x^2 + 24x + 2000 \qquad (0 \le x \le 500)$$

where x stands for the number of units produced.
a. What is the actual cost incurred in the manufacturing the 301st MP3 player, assuming that the 300th player was manufactured?
b. What is the marginal cost when $x = 300$?

70. **Demand for Cordless Phones** The marketing department of Telecon has determined that the demand for their cordless phones obeys the relationship

$$p = -0.02x + 600 \qquad (0 \le x \le 30{,}000)$$

where p denotes the phone's unit price (in dollars) and x denotes the quantity demanded.
a. Find the revenue function R.
b. Find the marginal revenue function R'.
c. Compute $R'(10{,}000)$ and interpret your result.

71. **Demand for Photocopying Machines** The weekly demand for the LectroCopy photocopying machine is given by the demand equation

$$p = 2000 - 0.04x \qquad (0 \le x \le 50{,}000)$$

where p denotes the wholesale unit price in dollars and x denotes the quantity demanded. The weekly total cost function for manufacturing these copiers is given by

$$C(x) = 0.000002x^3 - 0.02x^2 + 1000x + 120{,}000$$

where $C(x)$ denotes the total cost incurred in producing x units.
a. Find the revenue function R, the profit function P, and the average cost function $\overline{C}$.
b. Find the marginal cost function C', the marginal revenue function R', the marginal profit function P', and the marginal average cost function $\overline{C}'$.
c. Compute $C'(3000)$, $R'(3000)$, and $P'(3000)$.
d. Compute $\overline{C}'(5000)$ and $\overline{C}'(8000)$ and interpret your results.

72. **Marginal Average Cost** The Custom Office makes a line of executive desks. It is estimated that the total cost for making x units of the Junior Executive model is

$$C(x) = 80x + 150{,}000 \qquad (0 \le x \le 20{,}000)$$

dollars/year.
a. Find the average cost function $\overline{C}$.
b. Find the marginal average cost function $\overline{C}'$.
c. What happens to $\overline{C}(x)$ when x is very large? Interpret your result.

73. **GDP of a Country** The GDP of a country from the years 2000 to 2007 is approximated by the function

$$G(t) = -0.3t^3 + 1.2t^2 + 500 \qquad (0 \le t \le 7)$$

where $G(t)$ is measured in billions of dollars and $t = 0$ corresponds to 2000. Find $G'(2)$ and $G''(2)$ and interpret your results.

74. **Motion of an Object** The position of an object moving along a straight line is given by

$$s = t\sqrt{2t^2 + 1} \qquad (0 \le t \le 5)$$

where s is measured in feet and t in seconds. Find the velocity and acceleration of the object after 2 sec.

CHAPTER 3 Before Moving On . . .

1. Find the derivative of $f(x) = 2x^3 - 3x^{1/3} + 5x^{-2/3}$.

2. Differentiate $g(x) = x\sqrt{2x^2 - 1}$.

3. Find $\frac{dy}{dx}$ if $y = \frac{2x + 1}{x^2 + x + 1}$.

4. Find the first three derivatives of $f(x) = \frac{1}{\sqrt{x + 1}}$.

5. Find $\frac{dy}{dx}$ given that $xy^2 - x^2y + x^3 = 4$.

6. Let $y = x\sqrt{x^2 + 5}$.
   a. Find the differential of y.
   b. If x changes from $x = 2$ to $x = 2.01$, what is the approximate change in y?

APPLICATIONS OF THE DERIVATIVE

4

THIS CHAPTER FURTHER EXPLORES the power of the derivative as a tool to help analyze the properties of functions. The information obtained can then be used to accurately sketch graphs of functions. We also see how the derivative is used in solving a large class of optimization problems, including finding what level of production will yield a maximum profit for a company, finding what level of production will result in minimal cost to a company, finding the maximum height attained by a rocket, finding the maximum velocity at which air is expelled when a person coughs, and a host of other problems.

How many loudspeaker systems should the Aerosonic company produce to maximize its profit? In Example 4, page 301, you will see how the techniques of calculus can be used to help answer this question.

4.1 Applications of the First Derivative

Determining the Intervals Where a Function Is Increasing or Decreasing

According to a study by the U.S. Department of Energy and the Shell Development Company, a typical car's fuel economy as a function of its speed is described by the graph shown in Figure 1. Observe that the fuel economy $f(x)$ in miles per gallon (mpg) improves as x, the vehicle's speed in miles per hour (mph), increases from 0 to 42, and then drops as the speed increases beyond 42 mph. We use the terms *increasing* and *decreasing* to describe the behavior of a function as we move from left to right along its graph.

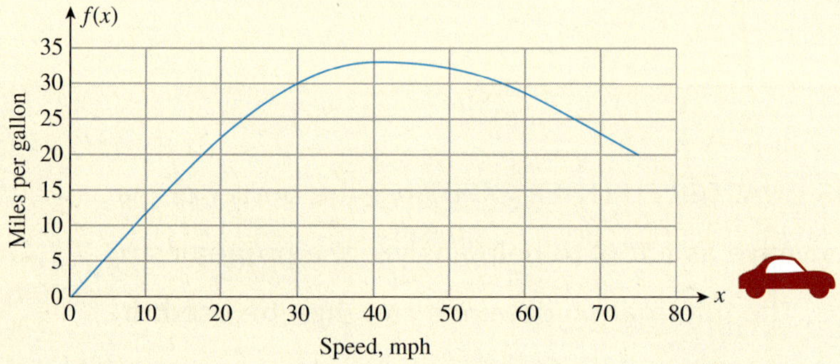

FIGURE 1 A typical car's fuel economy improves as the speed at which it is driven increases from 0 mph to 42 mph and drops at speeds greater than 42 mph.

Source: U.S. Department of Energy and Shell Development Co.

More precisely, we have the following definitions.

Increasing and Decreasing Functions

A function f is **increasing** on an interval (a, b) if for any two numbers x_1 and x_2 in (a, b), $f(x_1) < f(x_2)$ whenever $x_1 < x_2$ (Figure 2a).

A function f is **decreasing** on an interval (a, b) if for any two numbers x_1 and x_2 in (a, b), $f(x_1) > f(x_2)$ whenever $x_1 < x_2$ (Figure 2b).

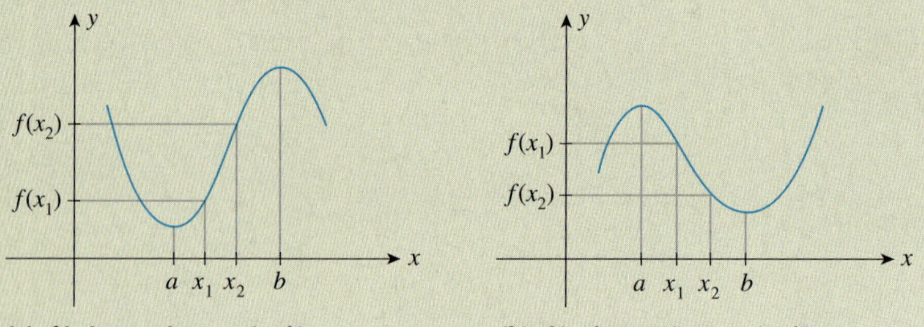

FIGURE 2

(a) f is increasing on (a, b). **(b)** f is decreasing on (a, b).

We say that f is *increasing at a number c* if there exists an interval (a, b) containing c such that f is increasing on (a, b). Similarly, we say that f is *decreasing at a number c* if there exists an interval (a, b) containing c such that f is decreasing on (a, b).

Since the rate of change of a function at $x = c$ is given by the derivative of the function at that number, the derivative lends itself naturally to being a tool for determining the intervals where a differentiable function is increasing or decreasing. Indeed, as we saw in Chapter 2, the derivative of a function at a number measures both the slope of the tangent line to the graph of the function at the point on the graph of f corresponding to that number and the rate of change of the function at that number. In fact, at a number where the derivative is positive, the slope of the tangent line to the graph is positive, and the function is increasing. At a number where the derivative is negative, the slope of the tangent line to the graph is negative, and the function is decreasing (Figure 3).

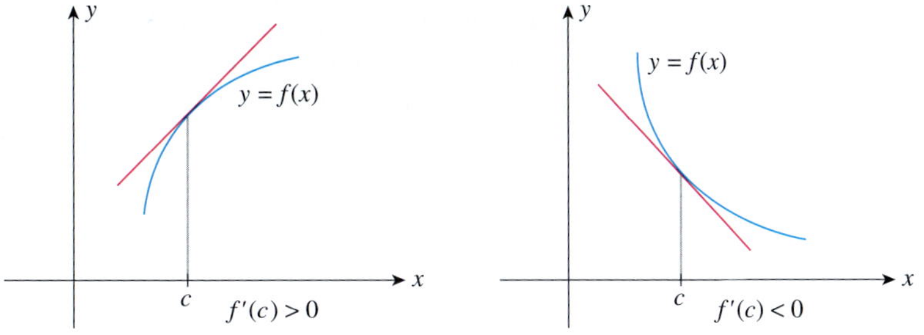

FIGURE **3** **(a)** f is increasing at $x = c$. **(b)** f is decreasing at $x = c$.

These observations lead to the following important theorem, which we state without proof.

THEOREM 1

a. If $f'(x) > 0$ for each value of x in an interval (a, b), then f is increasing on (a, b).

b. If $f'(x) < 0$ for each value of x in an interval (a, b), then f is decreasing on (a, b).

c. If $f'(x) = 0$ for each value of x in an interval (a, b), then f is constant on (a, b).

EXAMPLE 1 Find the interval where the function $f(x) = x^2$ is increasing and the interval where it is decreasing.

Solution The derivative of $f(x) = x^2$ is $f'(x) = 2x$. Since

$$f'(x) = 2x > 0 \quad \text{if } x > 0 \qquad \text{and} \qquad f'(x) = 2x < 0 \quad \text{if } x < 0$$

f is increasing on the interval $(0, \infty)$ and decreasing on the interval $(-\infty, 0)$ (Figure 4). ■

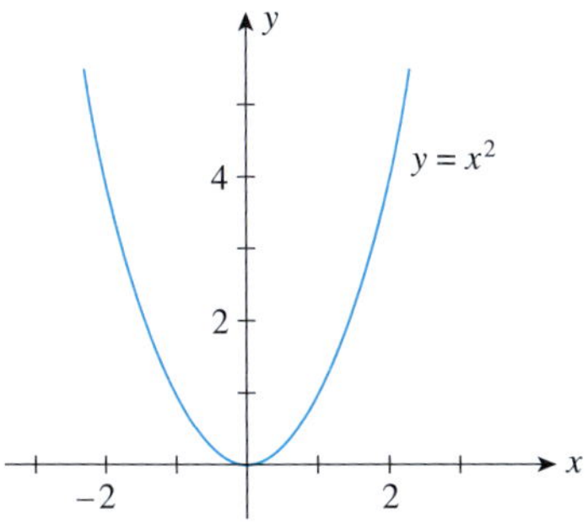

FIGURE **4**
The graph of f falls on $(-\infty, 0)$ where $f'(x) < 0$ and rises on $(0, \infty)$ where $f'(x) > 0$.

Recall that the graph of a continuous function cannot have any breaks. As a consequence, a continuous function cannot change sign unless it equals zero for some value of x. (See Theorem 5, page 124.) This observation suggests the following procedure for determining the sign of the derivative f' of a function f, and hence the intervals where the function f is increasing and where it is decreasing.

Determining the Intervals Where a Function Is Increasing or Decreasing

1. Find all values of x for which $f'(x) = 0$ or f' is discontinuous and identify the open intervals determined by these numbers.
2. Select a test number c in each interval found in step 1 and determine the sign of $f'(c)$ in that interval.
   a. If $f'(c) > 0$, f is increasing on that interval.
   b. If $f'(c) < 0$, f is decreasing on that interval.

Explore & Discuss

True or false? If f is continuous at c and f is increasing at c, then $f'(c) \neq 0$. Explain your answer. Hint: Consider $f(x) = x^3$ and $c = 0$.

EXAMPLE 2 Determine the intervals where the function $f(x) = x^3 - 3x^2 - 24x + 32$ is increasing and where it is decreasing.

Solution

1. The derivative of f is

$$f'(x) = 3x^2 - 6x - 24 = 3(x + 2)(x - 4)$$

See page 11.

and it is continuous everywhere. The zeros of $f'(x)$ are $x = -2$ and $x = 4$, and these numbers divide the real line into the intervals $(-\infty, -2)$, $(-2, 4)$, and $(4, \infty)$.

2. To determine the sign of $f'(x)$ in the intervals $(-\infty, -2)$, $(-2, 4)$, and $(4, \infty)$, compute $f'(x)$ at a convenient test point in each interval. The results are shown in the following table.

Interval	Test Point c	$f'(c)$	Sign of $f'(x)$
$(-\infty, -2)$	-3	21	+
$(-2, 4)$	0	-24	−
$(4, \infty)$	5	21	+

See page 20.

Using these results, we obtain the sign diagram shown in Figure 5. We conclude that f is increasing on the intervals $(-\infty, -2)$ and $(4, \infty)$ and is decreasing on the interval $(-2, 4)$. Figure 6 shows the graph of f.

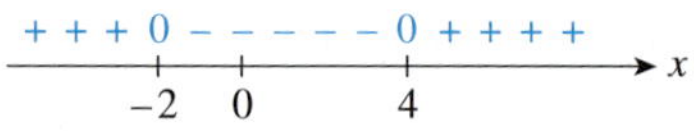

FIGURE 5
Sign diagram for f'

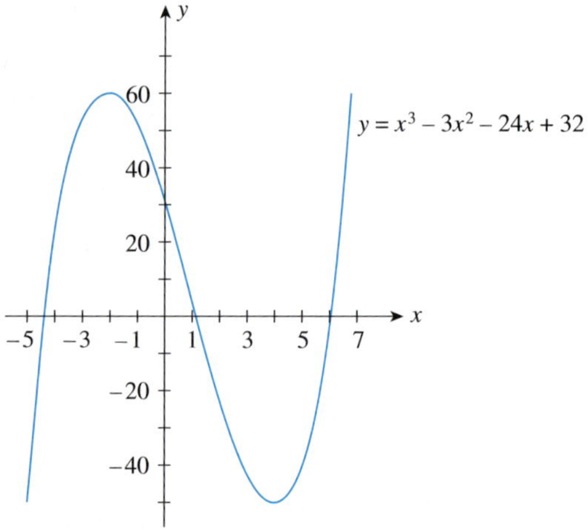

FIGURE 6
The graph of f rises on $(-\infty, -2)$, falls on $(-2, 4)$, and rises again on $(4, \infty)$.

Note We will learn how to sketch these graphs later. However, if you are familiar with the use of a graphing utility, you may go ahead and verify each graph.

Exploring with TECHNOLOGY

Refer to Example 2.

1. Use a graphing utility to plot the graphs of $f(x) = x^3 - 3x^2 - 24x + 32$ and its derivative function $f'(x) = 3x^2 - 6x - 24$ using the viewing window $[-10, 10] \times [-50, 70]$.

2. By looking at the graph of f', determine the intervals where $f'(x) > 0$ and the intervals where $f'(x) < 0$. Next, look at the graph of f and determine the intervals where it is increasing and the intervals where it is decreasing. Describe the relationship. Is it what you expected?

EXAMPLE 3 Find the interval where the function $f(x) = x^{2/3}$ is increasing and the interval where it is decreasing.

Solution

1. The derivative of f is

$$f'(x) = \frac{2}{3}x^{-1/3} = \frac{2}{3x^{1/3}}$$

The function f' is not defined at $x = 0$, so f' is discontinuous there. It is continuous everywhere else. Furthermore, f' is not equal to zero anywhere. The number 0 divides the real line (the domain of f) into the intervals $(-\infty, 0)$ and $(0, \infty)$.

2. Pick a test point (say, $x = -1$) in the interval $(-\infty, 0)$ and compute

$$f'(-1) = -\frac{2}{3}$$

Since $f'(-1) < 0$, we see that $f'(x) < 0$ on $(-\infty, 0)$. Next, we pick a test point (say, $x = 1$) in the interval $(0, \infty)$ and compute

$$f'(1) = \frac{2}{3}$$

Since $f'(1) > 0$, we see that $f'(x) > 0$ on $(0, \infty)$. Figure 7 shows these results in the form of a sign diagram.

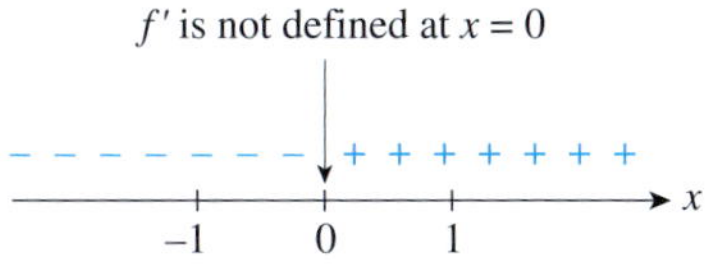

FIGURE 7
Sign diagram for f'

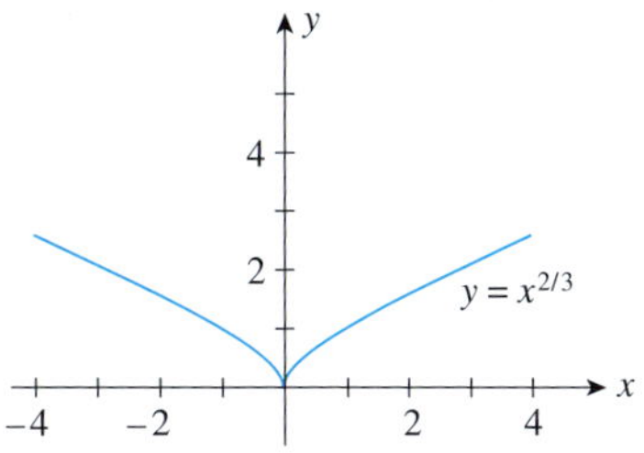

FIGURE 8
f decreases on $(-\infty, 0)$ and increases on $(0, \infty)$.

We conclude that f is decreasing on the interval $(-\infty, 0)$ and increasing on the interval $(0, \infty)$. The graph of f, shown in Figure 8, confirms these results.

EXAMPLE 4 Find the intervals where the function $f(x) = x + \dfrac{1}{x}$ is increasing and where it is decreasing.

Solution

1. The derivative of f is

$$f'(x) = 1 - \frac{1}{x^2} = \frac{x^2 - 1}{x^2}$$

See page 17.

Since f' is not defined at $x = 0$, it is discontinuous there. Furthermore, $f'(x)$ is equal to zero when $x^2 - 1 = 0$ or $x = \pm 1$. These values of x partition the domain of f' into the open intervals $(-\infty, -1)$, $(-1, 0)$, $(0, 1)$, and $(1, \infty)$, where the sign of f' is different from zero.

2. To determine the sign of f' in each of these intervals, we compute $f'(x)$ at the test points $x = -2, -\frac{1}{2}, \frac{1}{2}$, and 2, respectively, obtaining $f'(-2) = \frac{3}{4}$, $f'(-\frac{1}{2}) = -3$,

$f'(\frac{1}{2}) = -3$, and $f'(2) = \frac{3}{4}$. From the sign diagram for f' (Figure 9), we conclude that f is increasing on $(-\infty, -1)$ and $(1, \infty)$ and decreasing on $(-1, 0)$ and $(0, 1)$.

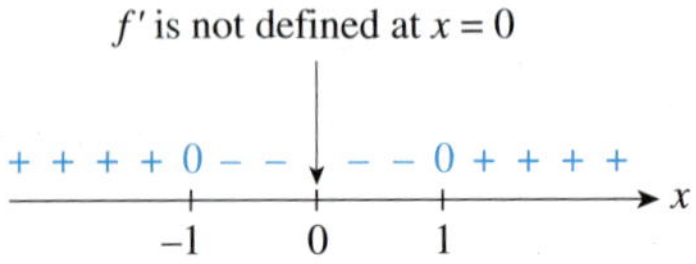

FIGURE 9
f' does not change sign as we move across $x = 0$.

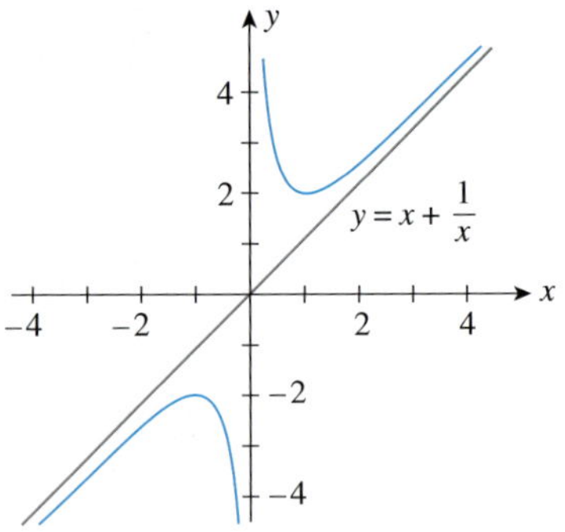

FIGURE 10
The graph of f rises on $(-\infty, -1)$, falls on $(-1, 0)$ and $(0, 1)$, and rises again on $(1, \infty)$.

The graph of f appears in Figure 10. Note that f' does not change sign as we move across $x = 0$. (Compare this with Example 3.)

Example 4 reminds us that we must *not* automatically conclude that the derivative f' must change sign when we move across a number where f' is discontinous or a zero of f'.

Explore & Discuss

Consider the profit function P associated with a certain commodity defined by

$$P(x) = R(x) - C(x) \qquad (x \geq 0)$$

where R is the revenue function, C is the total cost function, and x is the number of units of the product produced and sold.

1. Find an expression for $P'(x)$.
2. Find relationships in terms of the derivatives of R and C so that
   a. P is increasing at $x = a$.
   b. P is decreasing at $x = a$.
   c. P is neither increasing nor decreasing at $x = a$.
   Hint: Recall that the derivative of a function at $x = a$ measures the rate of change of the function at that number.
3. Explain the results of part 2 in economic terms.

Exploring with TECHNOLOGY

1. Use a graphing utility to sketch the graphs of $f(x) = x^3 - ax$ for $a = -2, -1, 0, 1$, and 2, using the viewing window $[-2, 2] \times [-2, 2]$.
2. Use the results of part 1 to guess at the values of a so that f is increasing on $(-\infty, \infty)$.
3. Prove your conjecture analytically.

Relative Extrema

Besides helping us determine where the graph of a function is increasing and decreasing, the first derivative may be used to help us locate certain "high points" and "low points" on the graph of f. Knowing these points is invaluable in sketching the graphs of functions and solving optimization problems. These "high points" and "low points" correspond to the *relative (local) maxima* and *relative minima* of a function. They are so called because they are the highest or the lowest points when compared with points nearby.

The graph shown in Figure 11 gives the U.S. budget surplus (deficit) from 1996 ($t = 0$) to 2007. The relative maxima and the relative minima of the function f are indicated on the graph.

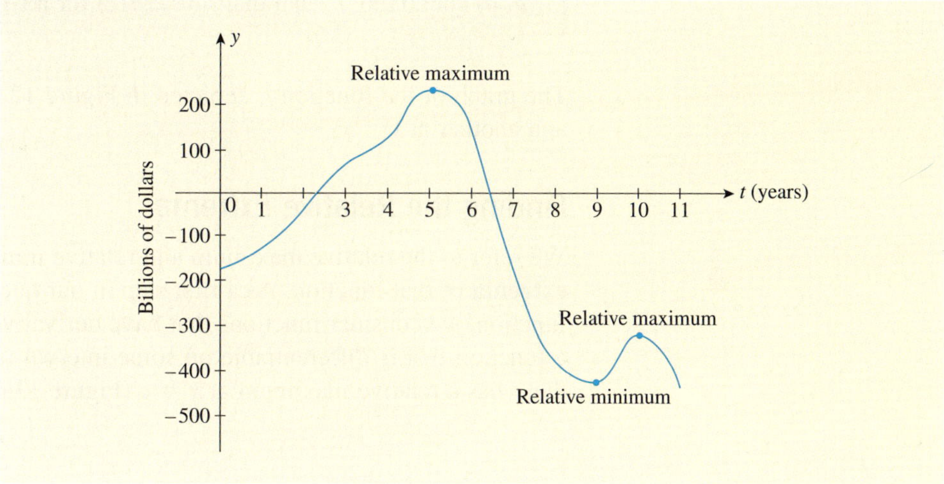

Source: Office of Management and Budget

FIGURE 11
U.S. budget surplus (deficit) from 1996 to 2007

More generally, we have the following definition:

> **Relative Maximum**
>
> A function f has a **relative maximum** at $x = c$ if there exists an open interval (a, b) containing c such that $f(x) \leq f(c)$ for all x in (a, b).

Geometrically, this means that there is *some* interval containing $x = c$ such that no point on the graph of f with its x-coordinate in that interval can lie above the point $(c, f(c))$; that is, $f(c)$ is the largest value of $f(x)$ in some interval around $x = c$. Figure 12 depicts the graph of a function f that has a relative maximum at $x = x_1$ and another at $x = x_3$.

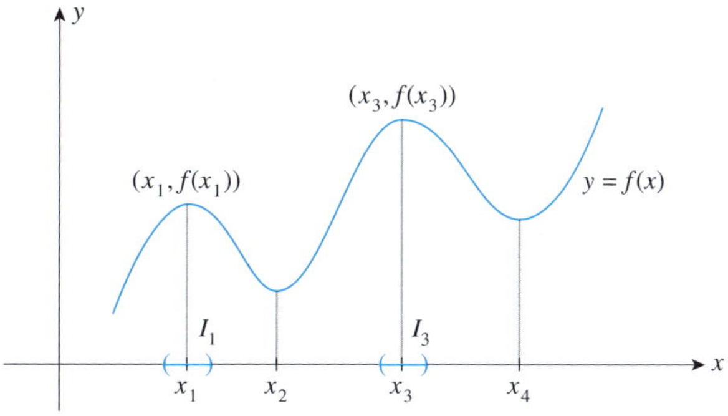

FIGURE 12
f has a relative maximum at $x = x_1$ and at $x = x_3$.

Observe that all the points on the graph of f with x-coordinates in the interval I_1 containing x_1 (shown in blue) lie on or below the point $(x_1, f(x_1))$. This is also true for the point $(x_3, f(x_3))$ and the interval I_3. Thus, even though there are points on the graph of f that are "higher" than the points $(x_1, f(x_1))$ and $(x_3, f(x_3))$, the latter points are "highest" relative to points in their respective neighborhoods (intervals). Points on the graph of a function f that are "highest" and "lowest" with respect to *all* points in the domain of f will be studied in Section 4.4.

The definition of the relative minimum of a function parallels that of the relative maximum of a function.

Relative Minimum

A function f has a **relative minimum** at $x = c$ if there exists an open interval (a, b) containing c such that $f(x) \geq f(c)$ for all x in (a, b).

The graph of the function f, depicted in Figure 12, has a relative minimum at $x = x_2$ and another at $x = x_4$.

Finding the Relative Extrema

We refer to the relative maximum and relative minimum of a function as the **relative extrema** of that function. As a first step in our quest to find the relative extrema of a function, we consider functions that have derivatives at such points. Suppose that f is a function that is differentiable on some interval (a, b) that contains a number c and that f has a relative maximum at $x = c$ (Figure 13a).

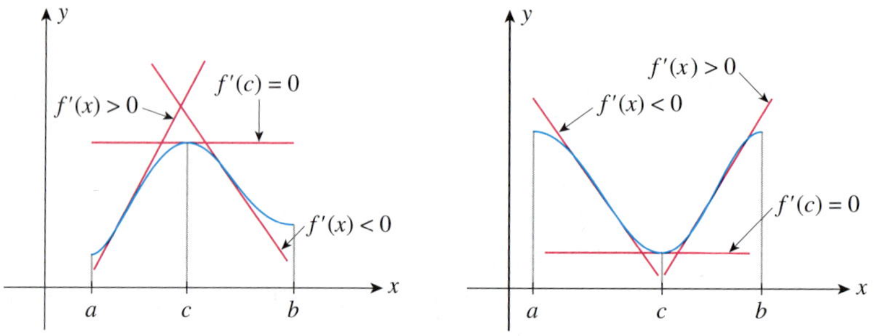

FIGURE 13

(a) f has a relative maximum at $x = c$. **(b)** f has a relative minimum at $x = c$.

Observe that the slope of the tangent line to the graph of f must change from positive to negative as we move across $x = c$ from left to right. Therefore, the tangent line to the graph of f at the point $(c, f(c))$ must be horizontal; that is, $f'(c) = 0$ (Figure 13a).

Using a similar argument, it may be shown that the derivative f' of a differentiable function f must also be equal to zero at $x = c$, where f has a relative minimum (Figure 13b).

This analysis reveals an important characteristic of the relative extrema of a differentiable function f: *At any number c where f has a relative extremum, $f'(c) = 0$.*

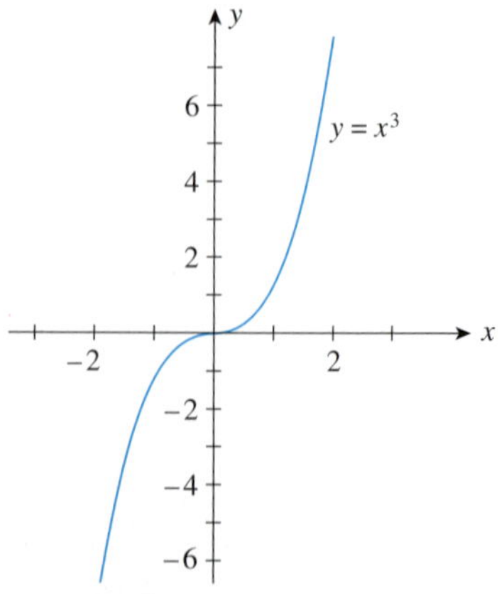

FIGURE 14
$f'(0) = 0$, but f does not have a relative extremum at $(0, 0)$.

Before we develop a procedure for finding such numbers, a few words of caution are in order. First, this result tells us that if a differentiable function f has a relative extremum at a number $x = c$, then $f'(c) = 0$. The converse of this statement—if $f'(c) = 0$ at $x = c$, then f must have a relative extremum at that number—is *not* true. Consider, for example, the function $f(x) = x^3$. Here, $f'(x) = 3x^2$, so $f'(0) = 0$. Yet, f has neither a relative maximum nor a relative minimum at $x = 0$ (Figure 14).

Second, our result assumes that the function is differentiable and thus has a derivative at a number that gives rise to a relative extremum. The functions $f(x) = |x|$ and $g(x) = x^{2/3}$ demonstrate that a relative extremum of a function may exist at a number at which the derivative does not exist. Both these functions fail to be differentiable at $x = 0$, but each has a relative minimum there. Figure 15 shows the graphs of these functions. Note that the slopes of the tangent lines change from negative to positive as we move across $x = 0$, just as in the case of a function that is differentiable at a value of x that gives rise to a relative minimum.

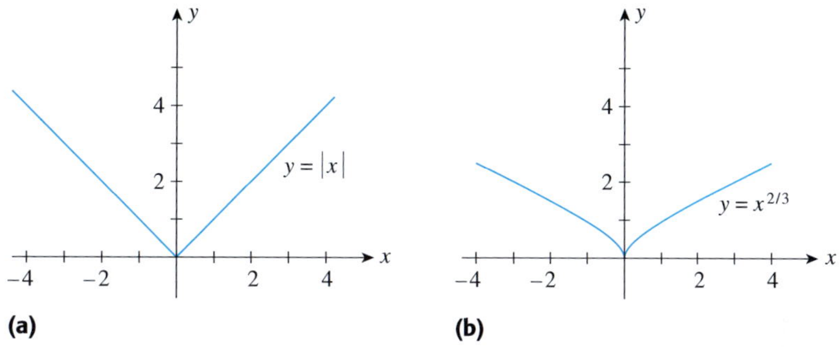

FIGURE 15 Each of these functions has a relative extremum at (0, 0), but the derivative does not exist there.

We refer to a number in the domain of f that *may* give rise to a relative extremum as a critical number.

Critical Number of f

A **critical number** of a function f is any number x in the domain of f such that $f'(x) = 0$ or $f'(x)$ does not exist.

Figure 16 depicts the graph of a function that has critical numbers at $x = a, b, c, d$, and e. Observe that $f'(x) = 0$ at $x = a, b$, and c. Next, since there is a corner at $x = d$, $f'(x)$ does not exist there. Finally, $f'(x)$ does not exist at $x = e$ because the tangent line there is vertical. Also, observe that the critical numbers $x = a, b$, and d give rise to relative extrema of f, whereas the critical numbers $x = c$ and $x = e$ do not.

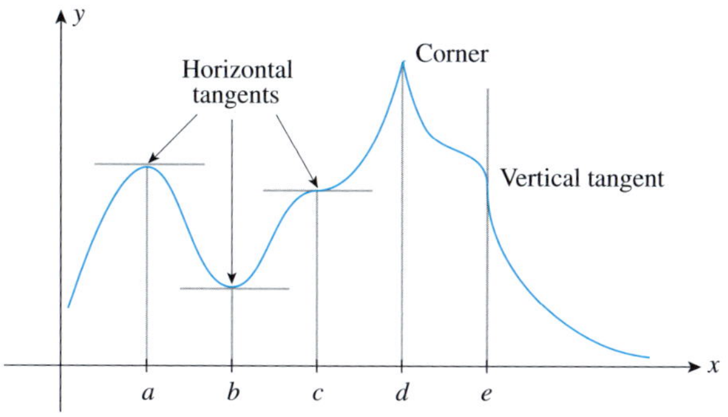

FIGURE 16 Critical numbers of f

Having defined what a critical number is, we can now state a formal procedure for finding the relative extrema of a continuous function that is differentiable everywhere except at isolated values of x. Incorporated into the procedure is the so-called **first derivative test,** which helps us determine whether a number gives rise to a relative maximum or a relative minimum of the function f.

The First Derivative Test

Procedure for Finding Relative Extrema of a Continuous Function f

1. Determine the critical numbers of f.
2. Determine the sign of $f'(x)$ to the left and right of each critical number.
   a. If $f'(x)$ changes sign from *positive* to *negative* as we move across a critical number c, then $f(c)$ is a relative maximum.
   b. If $f'(x)$ changes sign from *negative* to *positive* as we move across a critical number c, then $f(c)$ is a relative minimum.
   c. If $f'(x)$ does not change sign as we move across a critical number c, then $f(c)$ is not a relative extremum.

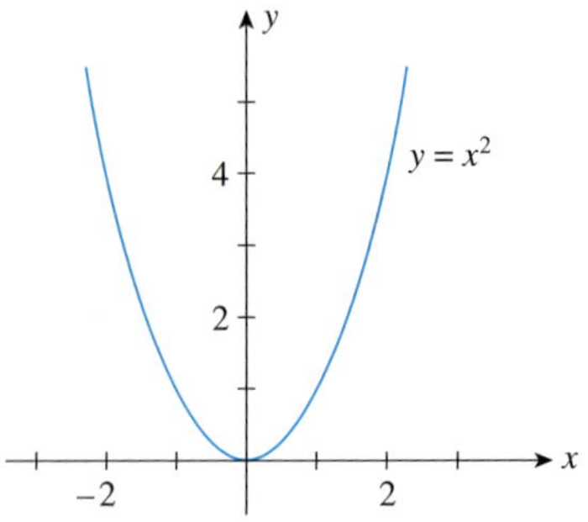

FIGURE 17
f has a relative minimum at $x = 0$.

EXAMPLE 5 Find the relative maxima and relative minima of the function $f(x) = x^2$.

Solution The derivative of $f(x) = x^2$ is given by $f'(x) = 2x$. Setting $f'(x) = 0$ yields $x = 0$ as the only critical number of f. Since

$$f'(x) < 0 \quad \text{if } x < 0 \qquad \text{and} \qquad f'(x) > 0 \quad \text{if } x > 0$$

we see that $f'(x)$ changes sign from negative to positive as we move across the critical number 0. Thus, we conclude that $f(0) = 0$ is a relative minimum of f (Figure 17). ■

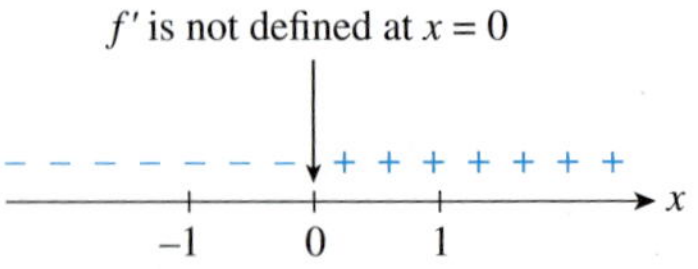

FIGURE 18
Sign diagram for f'

EXAMPLE 6 Find the relative maxima and relative minima of the function $f(x) = x^{2/3}$ (see Example 3).

Solution The derivative of f is $f'(x) = \frac{2}{3}x^{-1/3}$. As noted in Example 3, f' is not defined at $x = 0$, is continuous everywhere else, and is not equal to zero in its domain. Thus, $x = 0$ is the only critical number of the function f.

The sign diagram obtained in Example 3 is reproduced in Figure 18. We can see that the sign of $f'(x)$ changes from negative to positive as we move across $x = 0$ from left to right. Thus, an application of the first derivative test tells us that $f(0) = 0$ is a relative minimum of f (Figure 19). ■

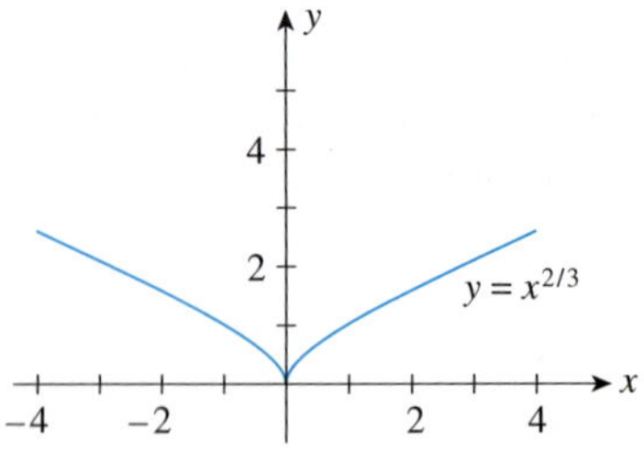

FIGURE 19
f has a relative minimum at $x = 0$.

Explore & Discuss

Recall that the average cost function $\overline{C}$ is defined by

$$\overline{C} = \frac{C(x)}{x}$$

where $C(x)$ is the total cost function and x is the number of units of a commodity manufactured (see Section 3.4).

1. Show that

$$\overline{C}'(x) = \frac{C'(x) - \overline{C}(x)}{x} \qquad (x > 0)$$

2. Use the result of part 1 to conclude that $\overline{C}$ is decreasing for values of x at which $C'(x) < \overline{C}(x)$. Find similar conditions for which $\overline{C}$ is increasing and for which $\overline{C}$ is constant.
3. Explain the results of part 2 in economic terms.

EXAMPLE 7 Find the relative maxima and relative minima of the function

$$f(x) = x^3 - 3x^2 - 24x + 32$$

Solution The derivative of f is

$$f'(x) = 3x^2 - 6x - 24 = 3(x + 2)(x - 4)$$

See page 11.

and it is continuous everywhere. The zeros of $f'(x)$, $x = -2$ and $x = 4$, are the only critical numbers of the function f. The sign diagram for f' is shown in Figure 20. Examine the two critical numbers $x = -2$ and $x = 4$ for a relative extremum using the first derivative test and the sign diagram for f':

1. *The critical number* -2: Since the function $f'(x)$ changes sign from positive to negative as we move across $x = -2$ from left to right, we conclude that a relative maximum of f occurs at $x = -2$. The value of $f(x)$ when $x = -2$ is

$$f(-2) = (-2)^3 - 3(-2)^2 - 24(-2) + 32 = 60$$

2. *The critical number* 4: $f'(x)$ changes sign from negative to positive as we move across $x = 4$ from left to right, so $f(4) = -48$ is a relative minimum of f. The graph of f appears in Figure 21.

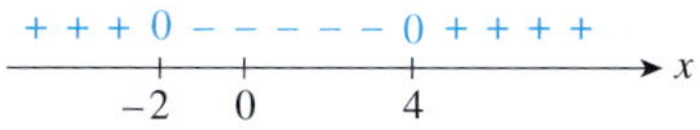

FIGURE 20
Sign diagram for f'

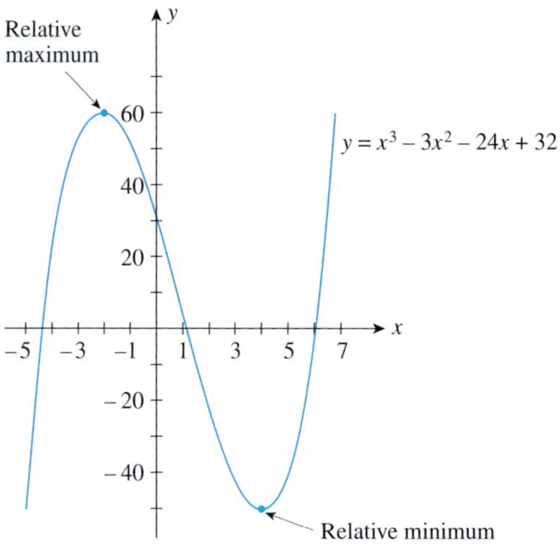

FIGURE 21
f has a relative maximum at $x = -2$ and a relative minimum at $x = 4$.

EXAMPLE 8 Find the relative maxima and the relative minima of the function

$$f(x) = x + \frac{1}{x}$$

Solution The derivative of f is

$$f'(x) = 1 - \frac{1}{x^2} = \frac{x^2 - 1}{x^2} = \frac{(x + 1)(x - 1)}{x^2}$$

Since f' is equal to zero at $x = -1$ and $x = 1$, these are critical numbers for the function f. Next, observe that f' is discontinuous at $x = 0$. However, because *f is not defined at that number,* $x = 0$ does not qualify as a critical number of f. Figure 22 shows the sign diagram for f'.

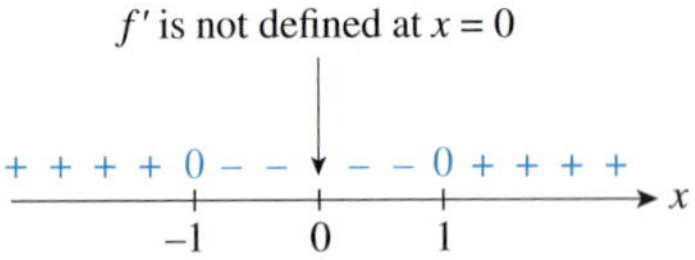

FIGURE 22
$x = 0$ is not a critical number because f is not defined at $x = 0$.

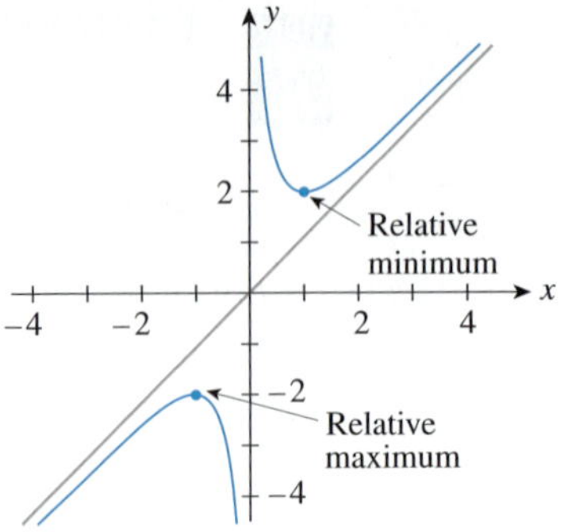

FIGURE 23
$f(x) = x + \dfrac{1}{x}$

Since $f'(x)$ changes sign from positive to negative as we move across $x = -1$ from left to right, the first derivative test implies that $f(-1) = -2$ is a relative maximum of the function f. Next, $f'(x)$ changes sign from negative to positive as we move across $x = 1$ from left to right, so $f(1) = 2$ is a relative minimum of the function f. The graph of f appears in Figure 23. Note that this function has a relative maximum that lies below its relative minimum. ■

Exploring with TECHNOLOGY

Refer to Example 8.

1. Use a graphing utility to plot the graphs of $f(x) = x + 1/x$ and its derivative function $f'(x) = 1 - 1/x^2$, using the viewing window $[-4, 4] \times [-8, 8]$.
2. By studying the graph of f', determine the critical numbers of f. Next, note the sign of $f'(x)$ immediately to the left and to the right of each critical number. What can you conclude about each critical number? Are your conclusions borne out by the graph of f?

APPLIED EXAMPLE 9 Profit Functions The profit function of Acrosonic Company is given by

$$P(x) = -0.02x^2 + 300x - 200{,}000$$

dollars, where x is the number of Acrosonic model F loudspeaker systems produced. Find where the function P is increasing and where it is decreasing.

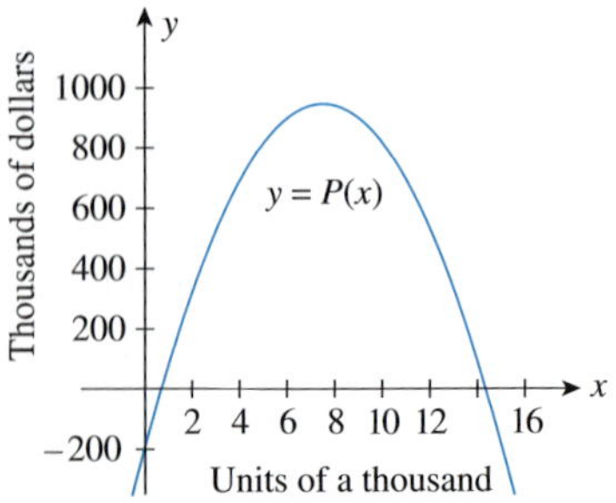

FIGURE 24
The profit function is increasing on (0, 7500) and decreasing on (7500, ∞).

Solution The derivative P' of the function P is

$$P'(x) = -0.04x + 300 = -0.04(x - 7500)$$

Thus, $P'(x) = 0$ when $x = 7500$. Furthermore, $P'(x) > 0$ for x in the interval $(0, 7500)$, and $P'(x) < 0$ for x in the interval $(7500, \infty)$. This means that the profit function P is increasing on $(0, 7500)$ and decreasing on $(7500, \infty)$ (Figure 24). ■

VIDEO

APPLIED EXAMPLE 10 Crime Rates The number of major crimes committed in the city of Bronxville from 2001 to 2008 is approximated by the function

$$N(t) = -0.1t^3 + 1.5t^2 + 100 \qquad (0 \le t \le 7)$$

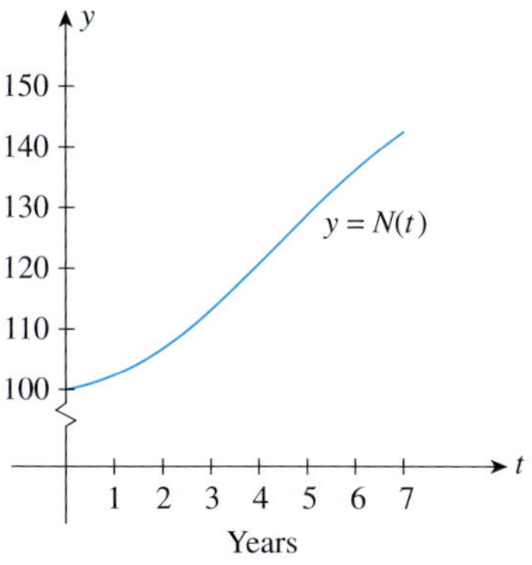

FIGURE 25
The number of crimes, $N(t)$, is increasing over the 7-year interval.

where $N(t)$ denotes the number of crimes committed in year t, with $t = 0$ corresponding to the beginning of 2001. Find where the function N is increasing and where it is decreasing.

Solution The derivative N' of the function N is

$$N'(t) = -0.3t^2 + 3t = -0.3t(t - 10)$$

Since $N'(t) > 0$ for t in the interval (0, 7), the function N is increasing throughout that interval (Figure 25). ■

4.1 Self-Check Exercises

1. Find the intervals where the function $f(x) = \frac{2}{3}x^3 - x^2 - 12x + 3$ is increasing and the intervals where it is decreasing.

2. Find the relative extrema of $f(x) = \dfrac{x^2}{1 - x^2}$.

Solutions to Self-Check Exercises 4.1 can be found on page 261.

4.1 Concept Questions

1. Explain each of the following:
   a. f is increasing on an interval I.
   b. f is decreasing on an interval I.

2. Describe a procedure for determining where a function is increasing and where it is decreasing.

3. Explain each term: **(a)** relative maximum and **(b)** relative minimum.

4. **a.** What is a critical number of a function f?
   b. Explain the role of a critical number in determining the relative extrema of a function.

5. Describe the first derivative test and describe a procedure for finding the relative extrema of a function.

4.1 Exercises

In Exercises 1–8, you are given the graph of a function f. Determine the intervals where f is increasing, constant, or decreasing.

1.
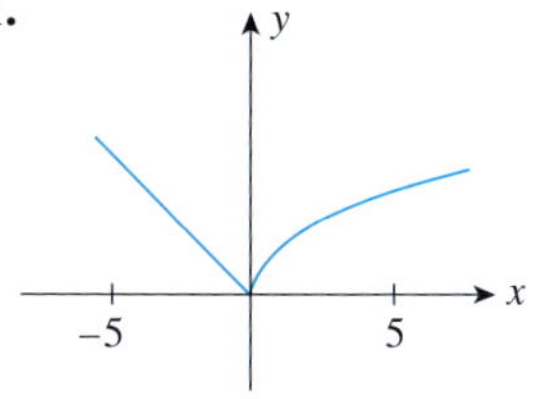

2.
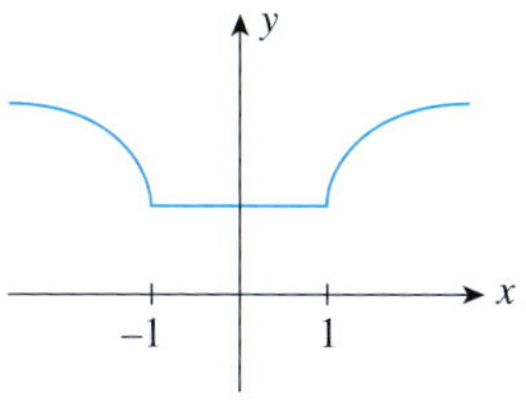

3.
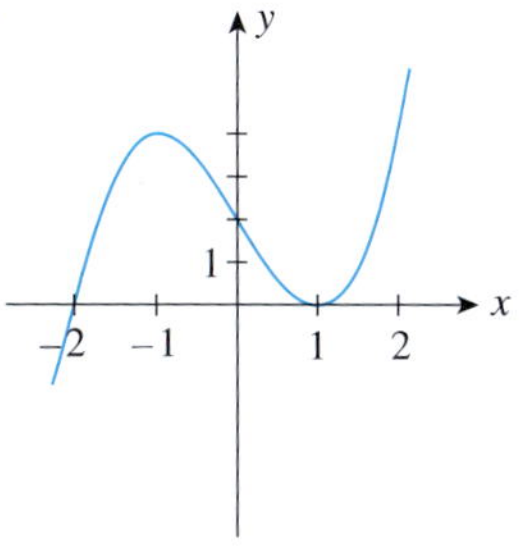

4.
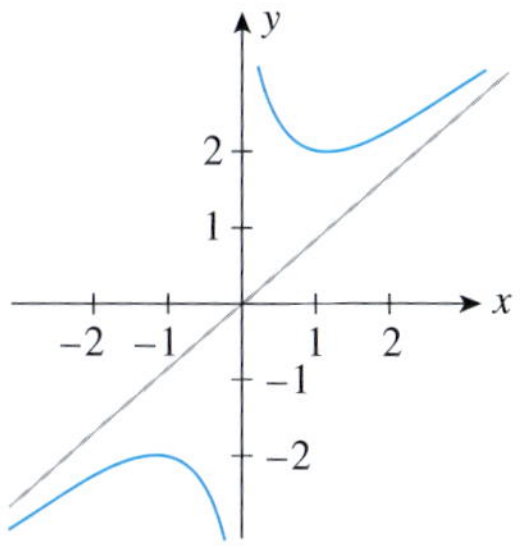

5.

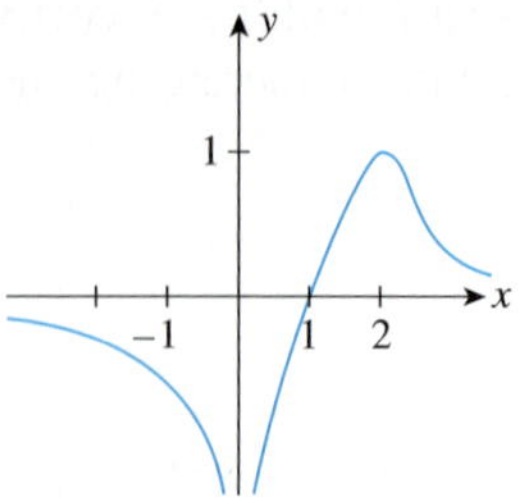

6.

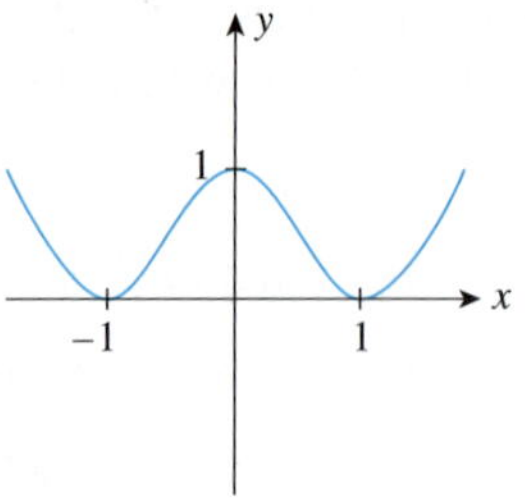

7.

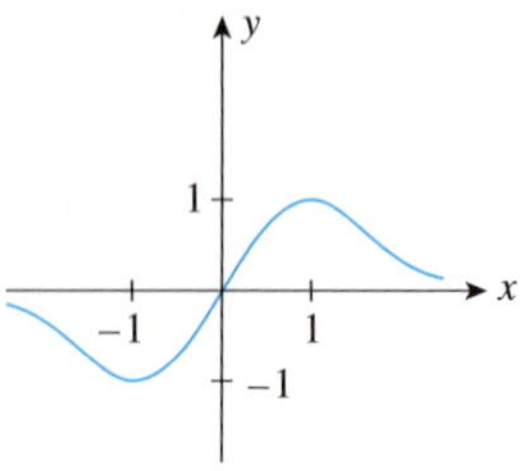

8.

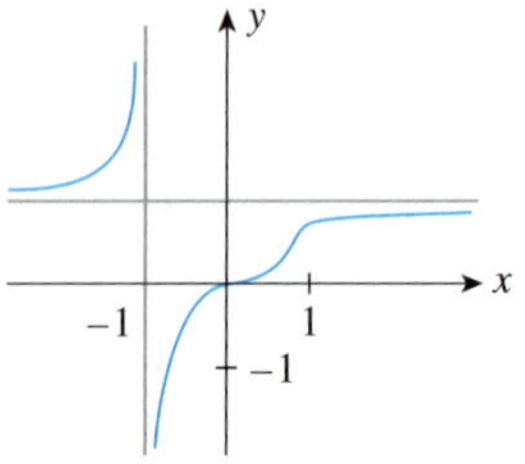

9. **THE BOSTON MARATHON** The graph of the function f shown in the accompanying figure gives the elevation of that part of the Boston Marathon course that includes the notorious Heartbreak Hill. Determine the intervals (stretches of the course) where the function f is increasing (the runner is laboring), where it is constant (the runner is taking a breather), and where it is decreasing (the runner is coasting).

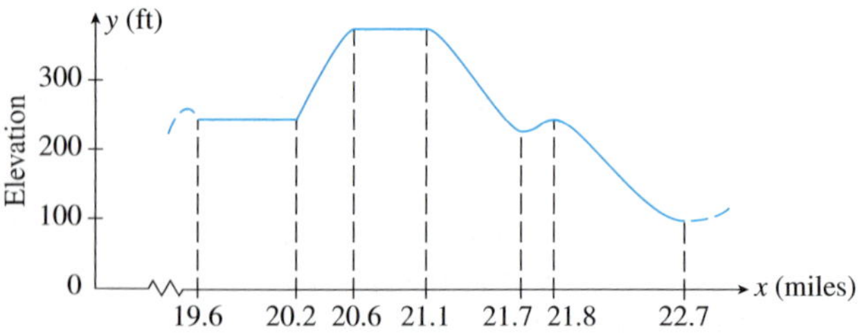

10. **AIRCRAFT STRUCTURAL INTEGRITY** Among the important factors in determining the structural integrity of an aircraft is its age. Advancing age makes planes more likely to crack. The graph of the function f, shown in the accompanying figure, is referred to as a "bathtub curve" in the airline industry. It gives the fleet damage rate (damage due to corrosion, accident, and metal fatigue) of a typical fleet of commercial aircraft as a function of the number of years of service.

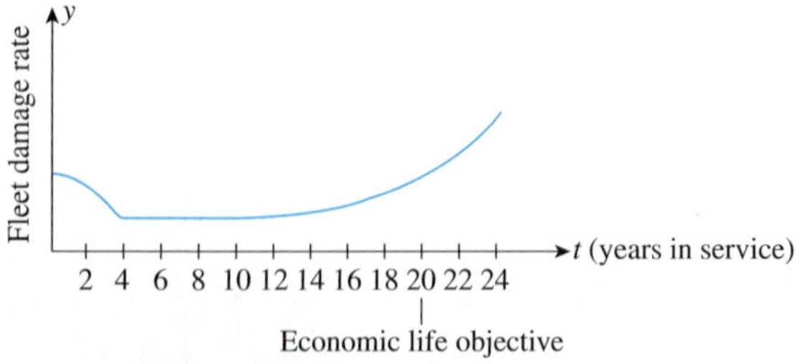

a. Determine the interval where f is decreasing. This corresponds to the time period when the fleet damage rate is dropping as problems are found and corrected during the initial "shakedown" period.

b. Determine the interval where f is constant. After the initial shakedown period, planes have few structural problems, and this is reflected by the fact that the function is constant on this interval.

c. Determine the interval where f is increasing. Beyond the time period mentioned in part (b), the function is increasing—reflecting an increase in structural defects due mainly to metal fatigue.

11. Refer to the following figure:

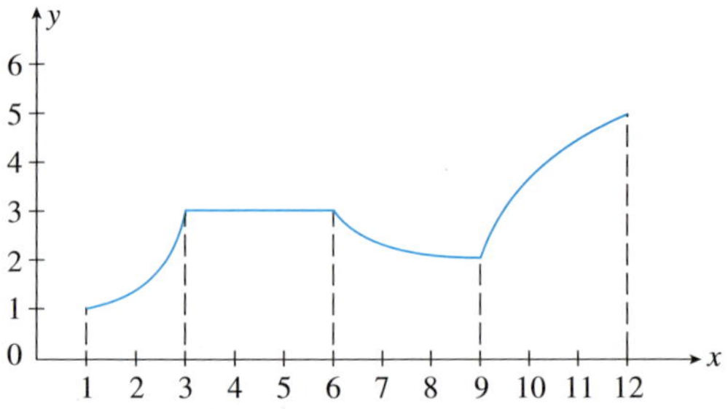

What is the sign of the following?

a. $f'(2)$

b. $f'(x)$ in the interval $(1, 3)$

c. $f'(4)$

d. $f'(x)$ in the interval $(3, 6)$

e. $f'(7)$

f. $f'(x)$ in the interval $(6, 9)$

g. $f'(x)$ in the interval $(9, 12)$

12. Refer to the following figure:

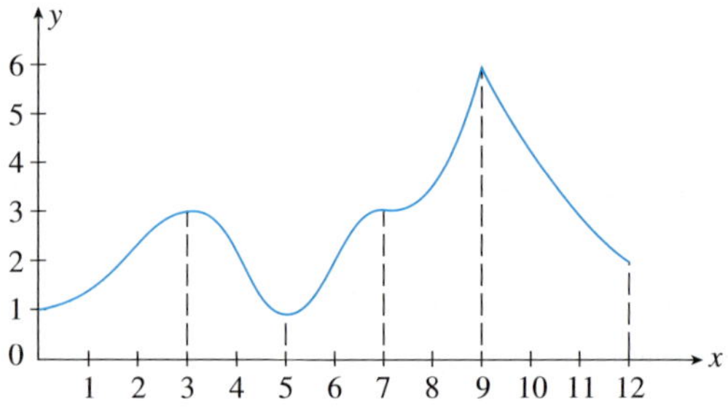

a. What are the critical numbers of f. Give reasons for your answers.

b. Draw the sign diagram for f'.

c. Find the relative extrema of f.

In Exercises 13–36, find the interval(s) where the function is increasing and the interval(s) where it is decreasing.

13. $f(x) = 3x + 5$

14. $f(x) = 4 - 5x$

15. $f(x) = x^2 - 3x$

16. $f(x) = 2x^2 + x + 1$

17. $g(x) = x - x^3$

18. $f(x) = x^3 - 3x^2$

19. $g(x) = x^3 + 3x^2 + 1$

20. $f(x) = x^3 - 3x + 4$

21. $f(x) = \frac{1}{3}x^3 - 3x^2 + 9x + 20$

22. $f(x) = \frac{2}{3}x^3 - 2x^2 - 6x - 2$

23. $h(x) = x^4 - 4x^3 + 10$

24. $g(x) = x^4 - 2x^2 + 4$

25. $f(x) = \frac{1}{x - 2}$

26. $h(x) = \frac{1}{2x + 3}$

27. $h(t) = \frac{t}{t - 1}$

28. $g(t) = \frac{2t}{t^2 + 1}$

29. $f(x) = x^{3/5}$

30. $f(x) = x^{2/3} + 5$

31. $f(x) = \sqrt{x + 1}$

32. $f(x) = (x - 5)^{2/3}$

33. $f(x) = \sqrt{16 - x^2}$

34. $g(x) = x\sqrt{x + 1}$

35. $f(x) = \frac{x^2 - 1}{x}$

36. $h(x) = \frac{x^2}{x - 1}$

In Exercises 37–44, you are given the graph of a function *f*. Determine the relative maxima and relative minima, if any.

37.

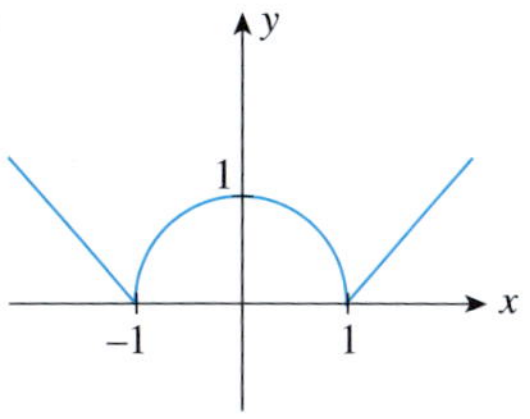

38.

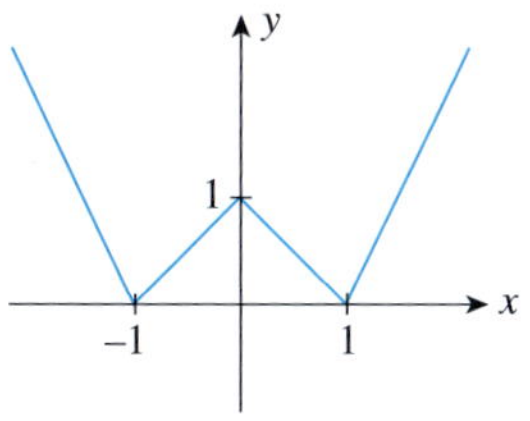

39.

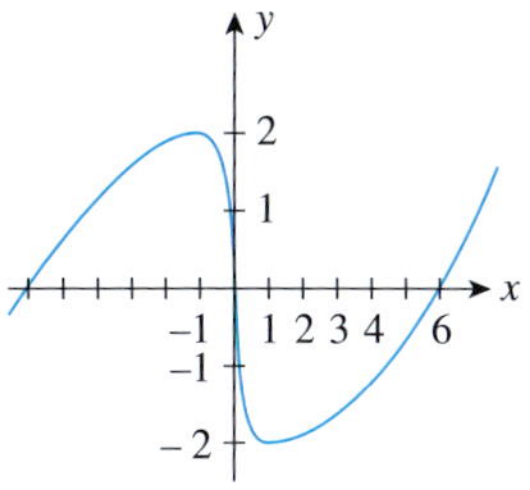

40.

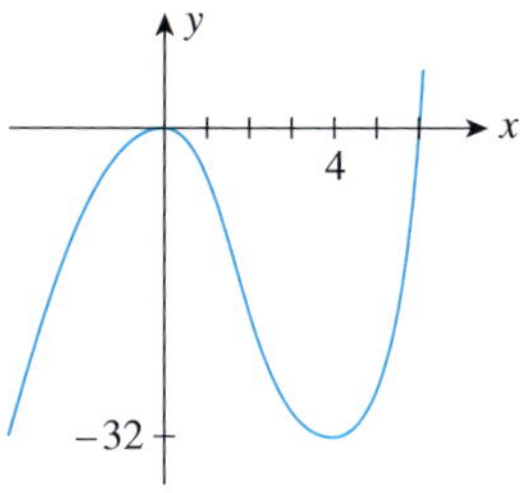

41.

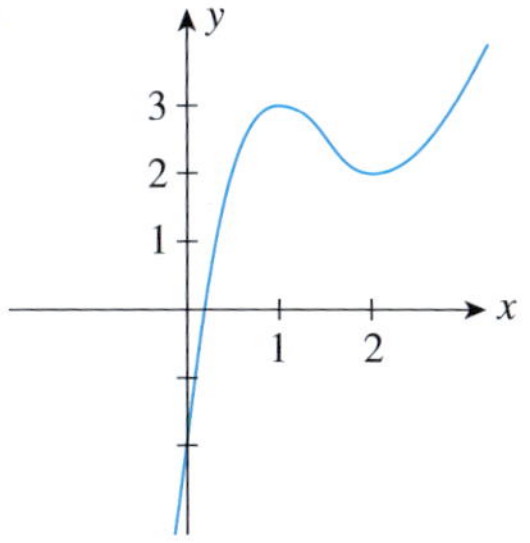

42.

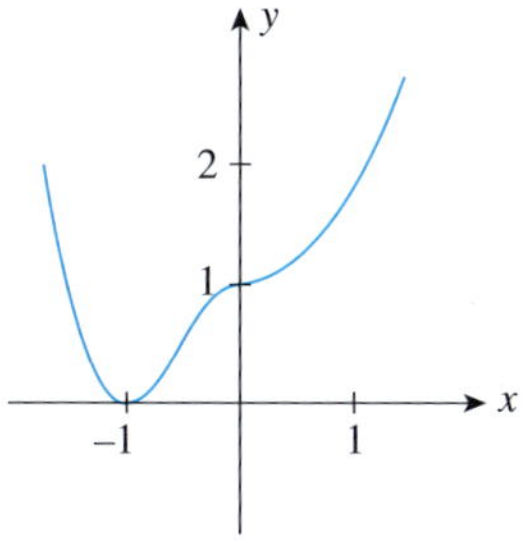

43.

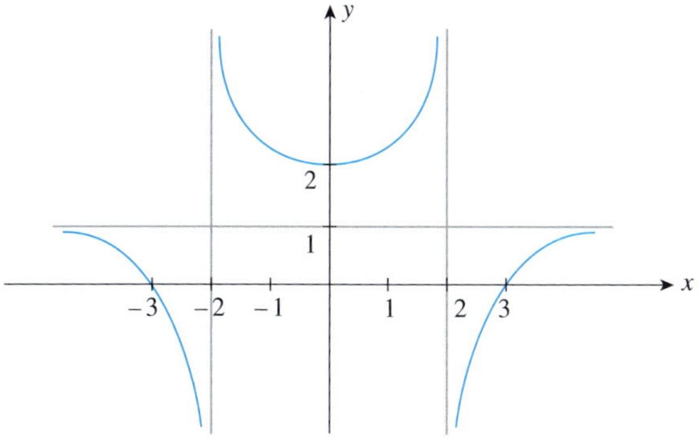

44.

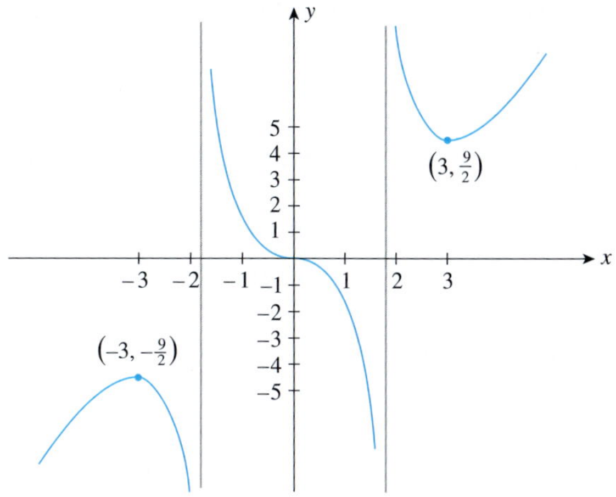

In Exercises 45–48, match the graph of the function with the graph of its derivative in (a)–(d).

45.

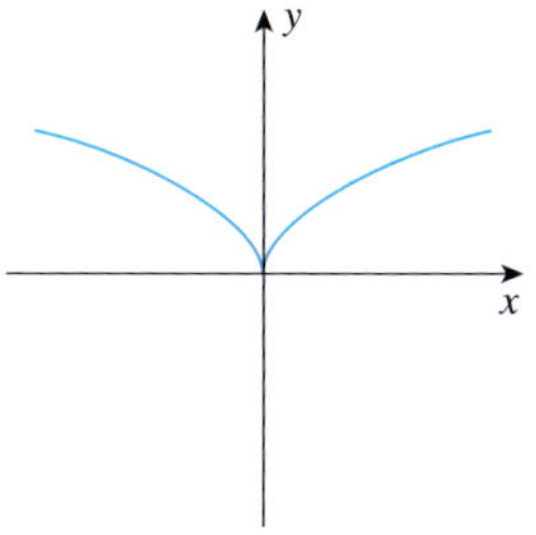

46.

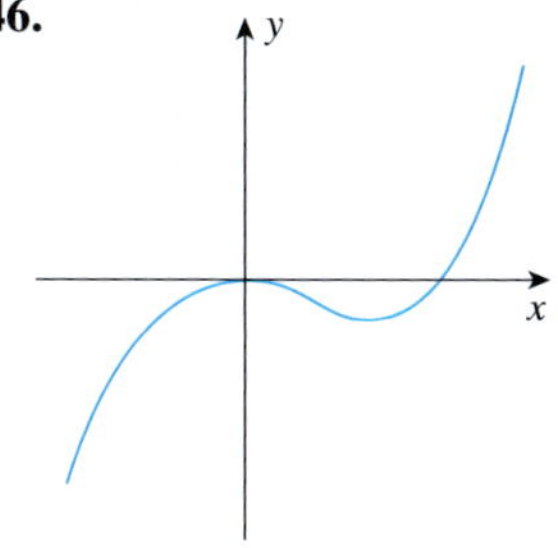

47.

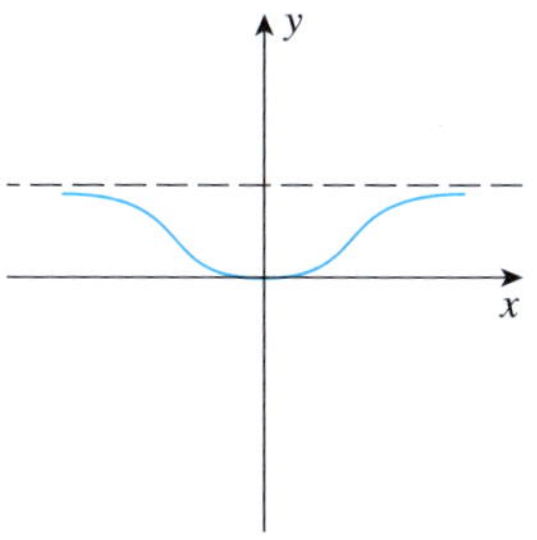

48.

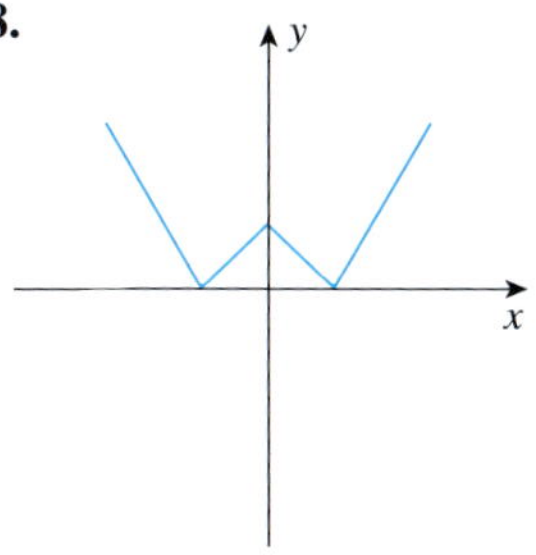

(a)

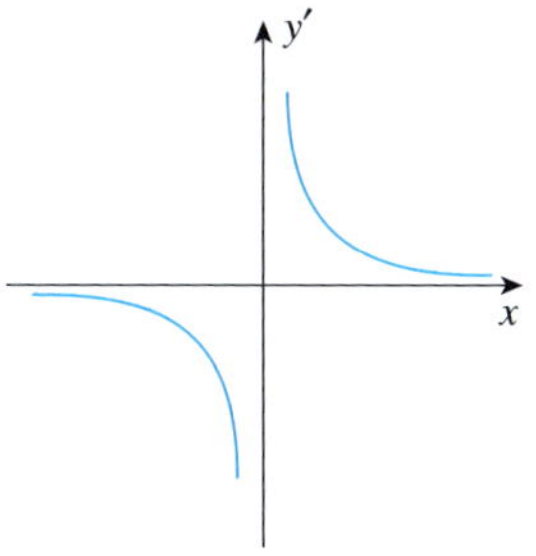

(b)

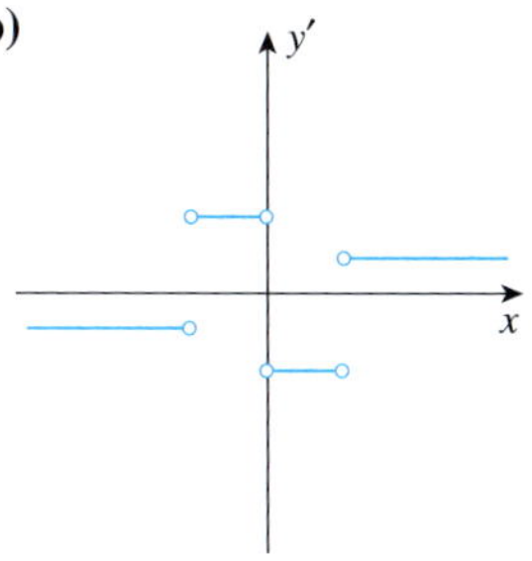

(c)

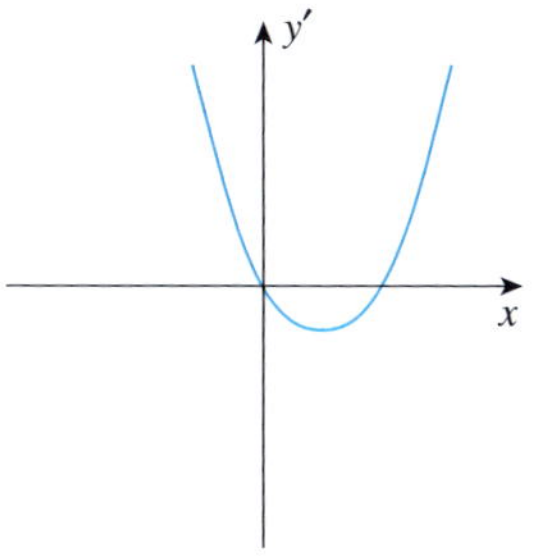

(d)

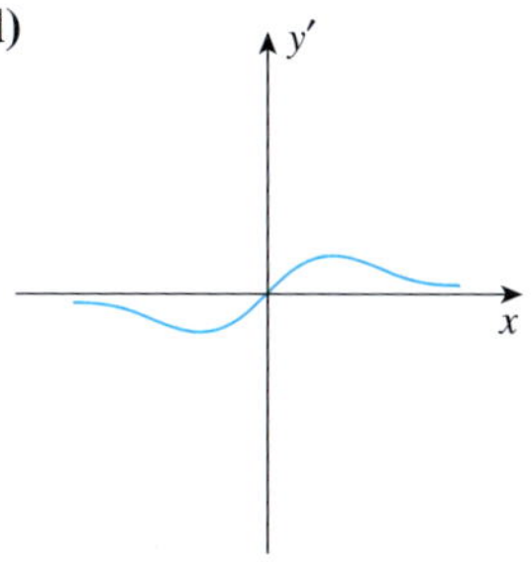

In Exercises 49–70, find the relative maxima and relative minima, if any, of each function.

49. $f(x) = x^2 - 4x$

50. $g(x) = x^2 + 3x + 8$

51. $h(t) = -t^2 + 6t + 6$

52. $f(x) = \frac{1}{2}x^2 - 2x + 4$

53. $f(x) = x^{5/3}$

54. $f(x) = x^{2/3} + 2$

55. $g(x) = x^3 - 3x^2 + 4$

56. $f(x) = x^3 - 3x + 6$

57. $f(x) = \frac{1}{2}x^4 - x^2$

58. $h(x) = \frac{1}{2}x^4 - 3x^2 + 4x - 8$

59. $F(x) = \frac{1}{3}x^3 - x^2 - 3x + 4$

60. $F(t) = 3t^5 - 20t^3 + 20$

61. $g(x) = x^4 - 4x^3 + 8$

62. $f(x) = 3x^4 - 2x^3 + 4$

63. $g(x) = \dfrac{x+1}{x}$

64. $h(x) = \dfrac{x}{x+1}$

65. $f(x) = x + \dfrac{9}{x} + 2$

66. $g(x) = 2x^2 + \dfrac{4000}{x} + 10$

67. $f(x) = \dfrac{x}{1+x^2}$

68. $g(x) = \dfrac{x}{x^2-1}$

69. $f(x) = (x-1)^{2/3}$

70. $g(x) = x\sqrt{x-4}$

71. A stone is thrown straight up from the roof of an 80-ft building. The distance (in feet) of the stone from the ground at any time t (in seconds) is given by

$$h(t) = -16t^2 + 64t + 80$$

When is the stone rising, and when is it falling? If the stone were to miss the building, when would it hit the ground? Sketch the graph of h.
Hint: The stone is on the ground when $h(t) = 0$.

72. **Profit Functions** The Mexican subsidiary of ThermoMaster manufactures an indoor–outdoor thermometer. Management estimates that the profit (in dollars) realizable by the company for the manufacture and sale of x units of thermometers each week is

$$P(x) = -0.001x^2 + 8x - 5000$$

Find the intervals where the profit function P is increasing and the intervals where P is decreasing.

73. **Prevalence of Alzheimer's Patients** Based on a study conducted in 1997, the percent of the U.S. population by age afflicted with Alzheimer's disease is given by the function

$$P(x) = 0.0726x^2 + 0.7902x + 4.9623 \qquad (0 \le x \le 25)$$

where x is measured in years, with $x = 0$ corresponding to age 65 yr. Show that P is an increasing function of x on the interval (0, 25). What does your result tell you about the relationship between Alzheimer's disease and age for the population that is age 65 yr and older?
Source: Alzheimer's Association

74. **Growth of Managed Services** Almost half of companies let other firms manage some of their Web operations—a practice called Web hosting. Managed services—monitoring a customer's technology services—is the fastest growing part of Web hosting. Managed services sales are expected to grow in accordance with the function

$$f(t) = 0.469t^2 + 0.758t + 0.44 \qquad (0 \le t \le 6)$$

where $f(t)$ is measured in billions of dollars and t is measured in years, with $t = 0$ corresponding to 1999.
a. Find the interval where f is increasing and the interval where f is decreasing.
b. What does your result tell you about sales in managed services from 1999 through 2005?
Source: International Data Corp.

75. **Flight of a Rocket** The height (in feet) attained by a rocket t sec into flight is given by the function

$$h(t) = -\frac{1}{3}t^3 + 16t^2 + 33t + 10 \qquad (t \ge 0)$$

When is the rocket rising, and when is it descending?

76. **Environment of Forests** Following the lead of the National Wildlife Federation, the Department of the Interior of a South American country began to record an index of environmental quality that measured progress and decline in the environmental quality of its forests. The index for the years 1998 through 2008 is approximated by the function

$$I(t) = \frac{1}{3}t^3 - \frac{5}{2}t^2 + 80 \qquad (0 \le t \le 10)$$

where $t = 0$ corresponds to 1998. Find the intervals where the function I is increasing and the intervals where it is decreasing. Interpret your results.

77. **Average Speed of a Highway Vehicle** The average speed of a vehicle on a stretch of Route 134 between 6 a.m. and 10 a.m. on a typical weekday is approximated by the function

$$f(t) = 20t - 40\sqrt{t} + 50 \qquad (0 \le t \le 4)$$

where $f(t)$ is measured in miles per hour and t is measured in hours, with $t = 0$ corresponding to 6 a.m. Find the interval where f is increasing and the interval where f is decreasing and interpret your results.

78. **Average Cost** The average cost (in dollars) incurred by Lincoln Records each week in pressing x compact discs is given by

$$\overline{C}(x) = -0.0001x + 2 + \frac{2000}{x} \qquad (0 < x \le 6000)$$

Show that $\overline{C}(x)$ is always decreasing over the interval (0, 6000).

79. **WEB HOSTING** Refer to Exercise 74. Sales in the Web-hosting industry are projected to grow in accordance with the function

$$f(t) = -0.05t^3 + 0.56t^2 + 5.47t + 7.5 \qquad (0 \le t \le 6)$$

where $f(t)$ is measured in billions of dollars and t is measured in years, with $t = 0$ corresponding to 1999.

a. Find the interval where f is increasing and the interval where f is decreasing.
Hint: Use the quadratic formula.

b. What does your result tell you about sales in the Web-hosting industry from 1999 through 2005?

Source: International Data Corp.

80. **MEDICAL SCHOOL APPLICANTS** According to a study from the American Medical Association, the number of medical school applicants from academic year 1997–1998 ($t = 0$) through the academic year 2002–2003 is approximated by the function

$$N(t) = -0.0333t^3 + 0.47t^2 - 3.8t + 47 \qquad (0 \le t \le 5)$$

where $N(t)$ measured in thousands.

a. Show that the number of medical school applicants had been declining over the period in question.
Hint: Use the quadratic formula.

b. What was the largest number of medical school applicants in any one academic year for the period in question? In what academic year did that occur?

Source: Journal of the American Medical Association

81. **SALES OF FUNCTIONAL FOOD PRODUCTS** The sales of functional food products—those that promise benefits beyond basic nutrition—have risen sharply in recent years. The sales (in billions of dollars) of foods and beverages with herbal and other additives is approximated by the function

$$S(t) = 0.46t^3 - 2.22t^2 + 6.21t + 17.25 \qquad (0 \le t \le 4)$$

where t is measured in years, with $t = 0$ corresponding to the beginning of 1997. Show that S is increasing on the interval $[0, 4]$.
Hint: Use the quadratic formula.

Source: Frost & Sullivan

82. **PROJECTED RETIREMENT FUNDS** Based on data from the Central Provident Fund of a certain country (a government agency similar to the Social Security Administration), the estimated cash in the fund in 2003 is given by

$$A(t) = -96.6t^4 + 403.6t^3 + 660.9t^2 + 250 \qquad (0 \le t \le 5)$$

where $A(t)$ is measured in billions of dollars and t is measured in decades, with $t = 0$ corresponding to 2003. Find the interval where A is increasing and the interval where A is decreasing and interpret your results.
Hint: Use the quadratic formula.

83. **SPENDING ON FIBER-OPTIC LINKS** U.S. telephone company spending on fiber-optic links to homes and businesses from 2001 to 2006 is projected to be

$$S(t) = -2.315t^3 + 34.325t^2 + 1.32t + 23 \qquad (0 \le t \le 5)$$

billion dollars in year t, where t is measured in years with $t = 0$ corresponding to 2001. Show that $S'(t) > 0$ for all t in the interval $[0, 5]$. What conclusion can you draw from this result?
Hint: Use the quadratic formula.

Source: RHK Inc.

84. **AIR POLLUTION** According to the South Coast Air Quality Management District, the level of nitrogen dioxide, a brown gas that impairs breathing, present in the atmosphere on a certain May day in downtown Los Angeles is approximated by

$$A(t) = 0.03t^3(t - 7)^4 + 60.2 \qquad (0 \le t \le 7)$$

where $A(t)$ is measured in pollutant standard index (PSI) and t is measured in hours, with $t = 0$ corresponding to 7 a.m. At what time of day is the air pollution increasing, and at what time is it decreasing?

85. **DRUG CONCENTRATION IN THE BLOOD** The concentration (in milligrams/cubic centimeter) of a certain drug in a patient's body t hr after injection is given by

$$C(t) = \frac{t^2}{2t^3 + 1} \qquad (0 \le t \le 4)$$

When is the concentration of the drug increasing, and when is it decreasing?

86. **AGE OF DRIVERS IN CRASH FATALITIES** The number of crash fatalities per 100,000 vehicle miles of travel (based on 1994 data) is approximated by the model

$$f(x) = \frac{15}{0.08333x^2 + 1.91667x + 1} \qquad (0 \le x \le 11)$$

where x is the age of the driver in years, with $x = 0$ corresponding to age 16. Show that f is decreasing on $(0, 11)$ and interpret your result.

Source: National Highway Traffic Safety Administration

87. **AIR POLLUTION** The amount of nitrogen dioxide, a brown gas that impairs breathing, present in the atmosphere on a certain May day in the city of Long Beach is approximated by

$$A(t) = \frac{136}{1 + 0.25(t - 4.5)^2} + 28 \qquad (0 \le t \le 11)$$

where $A(t)$ is measured in pollutant standard index (PSI) and t is measured in hours, with $t = 0$ corresponding to 7 a.m. Find the intervals where A is increasing and where A is decreasing and interpret your results.

Source: Los Angeles Times

88. PRISON OVERCROWDING The 1980s saw a trend toward old-fashioned punitive deterrence as opposed to the more liberal penal policies and community-based corrections popular in the 1960s and early 1970s. As a result, prisons became more crowded, and the gap between the number of people in prison and the prison capacity widened. The number of prisoners (in thousands) in federal and state prisons is approximated by the function

$$N(t) = 3.5t^2 + 26.7t + 436.2 \qquad (0 \le t \le 10)$$

where t is measured in years, with $t = 0$ corresponding to 1984. The number of inmates for which prisons were designed is given by

$$C(t) = 24.3t + 365 \qquad (0 \le t \le 10)$$

where $C(t)$ is measured in thousands and t has the same meaning as before. Show that the gap between the number of prisoners and the number for which the prisons were designed has been widening at any time t.

Hint: First, write a function G that gives the gap between the number of prisoners and the number for which the prisons were designed at any time t. Then show that $G'(t) > 0$ for all values of t in the interval (0, 10).

Source: U.S. Department of Justice

89. U.S. NURSING SHORTAGE The demand for nurses between 2000 and 2015 is estimated to be

$$D(t) = 0.0007t^2 + 0.0265t + 2 \qquad (0 \le t \le 15)$$

where $D(t)$ is measured in millions and $t = 0$ corresponds to the year 2000. The supply of nurses over the same time period is estimated to be

$$S(t) = -0.0014t^2 + 0.0326t + 1.9 \qquad (0 \le t \le 15)$$

where $S(t)$ is also measured in millions.

a. Find an expression $G(t)$ giving the gap between the demand and supply of nurses over the period in question.
b. Find the interval where G is decreasing and where it is increasing. Interpret your result.
c. Find the relative extrema of G. Interpret your result.

Source: U.S. Department of Health and Human Services

In Exercises 90–95, determine whether the statement is true or false. If it is true, explain why it is true. If it is false, give an example to show why it is false.

90. If f is decreasing on (a, b), then $f'(x) < 0$ for each x in (a, b).

91. If f and g are both increasing on (a, b), then $f + g$ is increasing on (a, b).

92. If f and g are both decreasing on (a, b), then $f - g$ is decreasing on (a, b).

93. If $f(x)$ and $g(x)$ are positive on (a, b) and both f and g are increasing on (a, b), then fg is increasing on (a, b).

94. If $f'(c) = 0$, then f has a relative maximum or a relative minimum at $x = c$.

95. If f has a relative minimum at $x = c$, then $f'(c) = 0$.

96. Using Theorem 1, verify that the linear function $f(x) = mx + b$ is (a) increasing everywhere if $m > 0$, (b) decreasing everywhere if $m < 0$, and (c) constant if $m = 0$.

97. Show that the function $f(x) = x^3 + x + 1$ has no relative extrema on $(-\infty, \infty)$.

98. Let $f(x) = -x^2 + ax + b$. Determine the constants a and b so that f has a relative maximum at $x = 2$ and the relative maximum value is 7.

99. Let $f(x) = ax^3 + 6x^2 + bx + 4$. Determine the constants a and b so that f has a relative minimum at $x = -1$ and a relative maximum at $x = 2$.

100. Let

$$f(x) = \begin{cases} -3x & \text{if } x < 0 \\ 2x + 4 & \text{if } x \ge 0 \end{cases}$$

a. Compute $f'(x)$ and show that it changes sign from negative to positive as we move across $x = 0$.
b. Show that f does not have a relative minimum at $x = 0$. Does this contradict the first derivative test? Explain your answer.

101. Let

$$f(x) = \begin{cases} -x^2 + 3 & \text{if } x \ne 0 \\ 2 & \text{if } x = 0 \end{cases}$$

a. Compute $f'(x)$ and show that it changes sign from positive to negative as we move across $x = 0$.
b. Show that f does not have a relative maximum at $x = 0$. Does this contradict the first derivative test? Explain your answer.

102. Let

$$f(x) = \begin{cases} \dfrac{1}{x^2} & \text{if } x > 0 \\ x^2 & \text{if } x \le 0 \end{cases}$$

a. Compute $f'(x)$ and show that it does not change sign as we move across $x = 0$.
b. Show that f has a relative minimum at $x = 0$. Does this contradict the first derivative test? Explain your answer.

103. Show that the quadratic function

$$f(x) = ax^2 + bx + c \qquad (a \ne 0)$$

has a relative extremum when $x = -b/2a$. Also, show that the relative extremum is a relative maximum if $a < 0$ and a relative minimum if $a > 0$.

104. Show that the cubic function

$$f(x) = ax^3 + bx^2 + cx + d \qquad (a \neq 0)$$

has no relative extremum if and only if $b^2 - 3ac \leq 0$.

105. Refer to Example 6, page 124.

a. Show that f is increasing on the interval $(0, 1)$.

b. Show that $f(0) = -1$ and $f(1) = 1$ and use the result of part (a) together with the intermediate value theorem to conclude that there is exactly one root of $f(x) = 0$ in $(0, 1)$.

106. Show that the function

$$f(x) = \frac{ax + b}{cx + d}$$

does not have a relative extremum if $ad - bc \neq 0$. What can you say about f if $ad - bc = 0$?

4.1 Solutions to Self-Check Exercises

1. The derivative of f is

$$f'(x) = 2x^2 - 2x - 12 = 2(x + 2)(x - 3)$$

and it is continuous everywhere. The zeros of $f'(x)$ are $x = -2$ and $x = 3$. The sign diagram of f' is shown in the accompanying figure. We conclude that f is increasing on the intervals $(-\infty, -2)$ and $(3, \infty)$ and decreasing on the interval $(-2, 3)$.

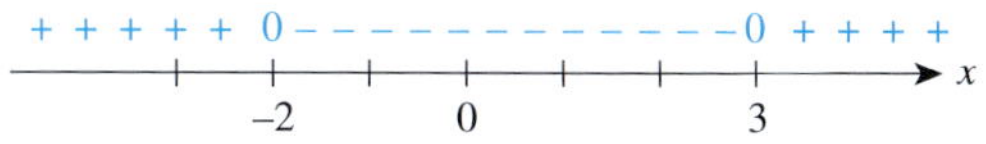

2. The derivative of f is

$$f'(x) = \frac{(1 - x^2)\dfrac{d}{dx}(x^2) - x^2\dfrac{d}{dx}(1 - x^2)}{(1 - x^2)^2}$$

$$= \frac{(1 - x^2)(2x) - x^2(-2x)}{(1 - x^2)^2} = \frac{2x}{(1 - x^2)^2}$$

and it is continuous everywhere except at $x = \pm 1$. Since $f'(x)$ is equal to zero at $x = 0$, $x = 0$ is a critical number of f. Next, observe that $f'(x)$ is discontinuous at $x = \pm 1$, but since these numbers are not in the domain of f, they do not qualify as critical numbers of f. Finally, from the sign diagram of f' shown in the accompanying figure, we conclude that $f(0) = 0$ is a relative minimum of f.

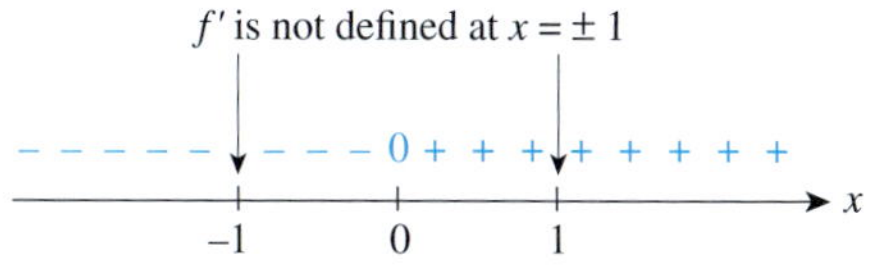

USING TECHNOLOGY

Using the First Derivative to Analyze a Function

A graphing utility is an effective tool for analyzing the properties of functions. This is especially true when we also bring into play the power of calculus, as the following examples show.

EXAMPLE 1 Let $f(x) = 2.4x^4 - 8.2x^3 + 2.7x^2 + 4x + 1$.

a. Use a graphing utility to plot the graph of f.

b. Find the intervals where f is increasing and the intervals where f is decreasing.

c. Find the relative extrema of f.

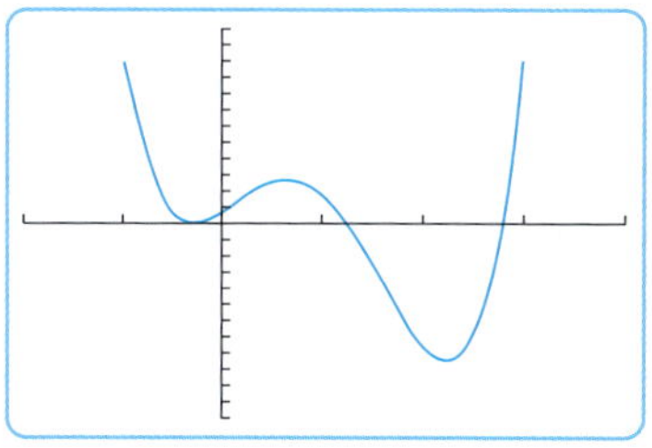

FIGURE T1
The graph of f in the viewing window $[-2, 4] \times [-10, 10]$

Solution

a. The graph of f in the viewing window $[-2, 4] \times [-10, 10]$ is shown in Figure T1.

b. We compute

$$f'(x) = 9.6x^3 - 24.6x^2 + 5.4x + 4$$

and observe that f' is continuous everywhere, so the critical numbers of f occur at values of x where $f'(x) = 0$. To solve this last equation, observe that $f'(x)$ is

(continued)

a *polynomial function* of degree 3. The easiest way to solve the polynomial equation

$$9.6x^3 - 24.6x^2 + 5.4x + 4 = 0$$

is to use the function on a graphing utility for solving polynomial equations. (Not all graphing utilities have this function.) You can also use **TRACE** and **ZOOM**, but this will not give the same accuracy without a much greater effort.

We find

$$x_1 \approx 2.22564943249 \qquad x_2 \approx 0.63272944121 \qquad x_3 \approx -0.295878873696$$

Referring to Figure T1, we conclude that f is decreasing on $(-\infty, -0.2959)$ and $(0.6327, 2.2256)$ (correct to four decimal places) and f is increasing on $(-0.2959, 0.6327)$ and $(2.2256, \infty)$.

c. Using the evaluation function of a graphing utility, we find the value of f at each of the critical numbers found in part (b). Upon referring to Figure T1 once again, we see that $f(x_3) \approx 0.2836$ and $f(x_1) \approx -8.2366$ are relative minimum values of f and $f(x_2) \approx 2.9194$ is a relative maximum value of f.

Note The equation $f'(x) = 0$ in Example 1 is a polynomial equation, and so it is easily solved using the function for solving polynomial equations. We could also solve the equation using the function for finding the roots of equations, but that would require much more work. For equations that are *not* polynomial equations, however, our only choice is to use the function for finding the roots of equations.

If the derivative of a function is difficult to compute or simplify and we do not require great precision in the solution, we can find the relative extrema of the function using a combination of **ZOOM** and **TRACE.** This technique, which does not require the use of the derivative of f, is illustrated in the following example.

EXAMPLE 2 Let $f(x) = x^{1/3}(x^2 + 1)^{-3/2}3^{-x}$.

a. Use a graphing utility to plot the graph of f.*

b. Find the relative extrema of f.

Solution

a. The graph of f in the viewing window $[-4, 2] \times [-2, 1]$ is shown in Figure T2.

b. From the graph of f in Figure T2, we see that f has relative maxima when $x \approx -2$ and $x \approx 0.25$ and a relative minimum when $x \approx -0.75$. To obtain a better approximation of the first relative maximum, we zoom-in with the cursor at approximately the point on the graph corresponding to $x \approx -2$. Then, using **TRACE,** we see that a relative maximum occurs when $x \approx -1.76$ with value $y \approx -1.01$. Similarly, we find the other relative maximum where $x \approx 0.20$ with value $y \approx 0.44$.

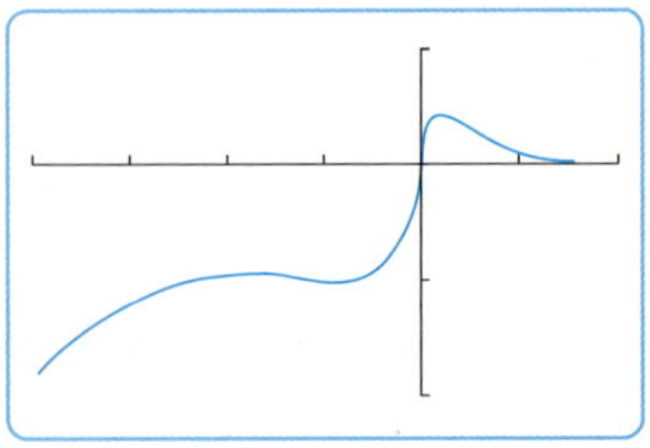

FIGURE **T2**
The graph of f in the viewing window $[-4, 2] \times [-2, 1]$

*Functions of the form $f(x) = 3^{-x}$ are called *exponential functions*, and we will study them in greater detail in Chapter 5.

Repeating the procedure, we find the relative minimum at $x \approx -0.86$ and $y \approx -1.07$. ■

You can also use the "minimum" and "maximum" functions of a graphing utility to find the relative extrema of the function. See the Web site for the procedure.

Finally, we comment that if you have access to a computer and software such as Derive, Maple, or Mathematica, then symbolic differentiation will yield the derivative $f'(x)$ of any differentiable function. This software will also solve the equation $f'(x) = 0$ with ease. Thus, the use of a computer will simplify even more greatly the analysis of functions.

TECHNOLOGY EXERCISES

In Exercises 1–4, find (a) the intervals where *f* is increasing and the intervals where *f* is decreasing and (b) the relative extrema of *f*. Express your answers accurate to four decimal places.

1. $f(x) = 3.4x^4 - 6.2x^3 + 1.8x^2 + 3x - 2$
2. $f(x) = 1.8x^4 - 9.1x^3 + 5x - 4$
3. $f(x) = 2x^5 - 5x^3 + 8x^2 - 3x + 2$
4. $f(x) = 3x^5 - 4x^2 + 3x - 1$

In Exercises 5–8, use the ZOOM and TRACE features to find (a) the intervals where *f* is increasing and the intervals where *f* is decreasing and (b) the relative extrema of *f*. Express your answers accurate to two decimal places.

5. $f(x) = (2x + 1)^{1/3}(x^2 + 1)^{-2/3}$
6. $f(x) = [x^2(x^3 - 1)]^{1/3} + \dfrac{1}{x}$
7. $f(x) = x - \sqrt{1 - x^2}$
8. $f(x) = \dfrac{\sqrt{x}(x^2 - 1)^2}{x - 2}$
9. **MANUFACTURING CAPACITY** Data show that the annual increase in manufacturing capacity between 1994 and 2000 is given by

$$f(t) = 0.009417t^3 - 0.426571t^2 + 2.74894t + 5.54 \quad (0 \le t \le 6)$$

percent where t is measured in years, with $t = 0$ corresponding to the beginning of 1994.

a. Plot the graph of f in the viewing window $[0, 6] \times [0, 11]$.

b. Determine the interval where f is increasing and the interval where f is decreasing and interpret your result.

Source: Federal Reserve

10. **SURGERIES IN PHYSICIANS' OFFICES** Driven by technological advances and financial pressures, the number of surgeries performed in physicians' offices nationwide has been increasing over the years. The function

$$f(t) = -0.00447t^3 + 0.09864t^2 + 0.05192t + 0.8 \quad (0 \le t \le 15)$$

gives the number of surgeries (in millions) performed in physicians' offices in year t, with $t = 0$ corresponding to the beginning of 1986.

a. Plot the graph of f in the viewing window $[0, 15] \times [0, 10]$.

b. Prove that f is increasing on the interval $[0, 15]$.
Hint: Show that f' is positive on the interval.

Source: SMG Marketing Group

11. **AIR POLLUTION** The amount of nitrogen dioxide, a brown gas that impairs breathing, present in the atmosphere on a certain May day in the city of Long Beach, is approximated by

$$A(t) = \frac{136}{1 + 0.25(t - 4.5)^2} + 28 \quad (0 \le t \le 11)$$

where $A(t)$ is measured in pollutant standard index (PSI) and t is measured in hours, with $t = 0$ corresponding to 7 a.m. When is the PSI increasing and when is it decreasing? At what time is the PSI highest, and what is its value at that time?

12. **MODELING WITH DATA** The following data gives the volume of cargo (in millions of tons) moved in the port of New York/New Jersey from 1991 through 2002.

Year	1991	1992	1993	1994	1995	1996	1997	1998	1999	2000	2001	2002
Volume	13.5	14.1	14.6	15.6	15.1	14.9	15.7	16.9	18.8	20.8	21.9	24

a. Use **CubicReg** to find a third-degree polynomial regression model for the data. Let $t = 0$ correspond to 1991.

b. Plot the graph of V in the viewing window $[0, 11] \times [0, 25]$.

c. Where is V increasing? What does this tell us?

d. Verify the result of part (c) analytically.

Source: Port Authority of New York/New Jersey

4.2 Applications of the Second Derivative

Determining the Intervals of Concavity

Consider the graphs shown in Figure 26, which give the estimated population of the world and of the United States through the year 2000. Both graphs are rising, indicating that both the U.S. population and the world population continued to increase through the year 2000. But observe that the graph in Figure 26a opens upward, whereas the graph in Figure 26b opens downward. What is the significance of this? To answer this question, let's look at the slopes of the tangent lines to various points on each graph (Figure 27).

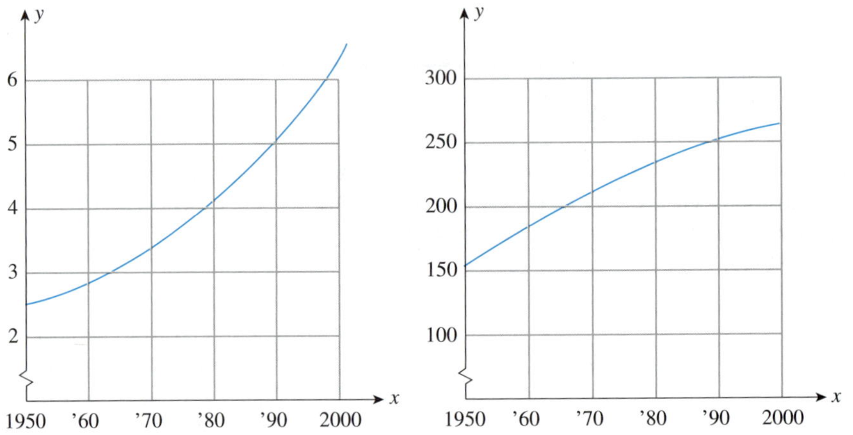

(a) World population in billions

(b) U.S. population in millions

Source: U.S. Department of Commerce and Worldwatch Institute

FIGURE **26**

In Figure 27a, we see that the slopes of the tangent lines to the graph are increasing as we move from left to right. Since the slope of the tangent line to the graph at a point on the graph measures the rate of change of the function at that point, we conclude that the world population was not only increasing through the year 2000 but was also increasing at an *increasing* pace. A similar analysis of Figure 27b reveals that the U.S. population was increasing, but at a *decreasing* pace.

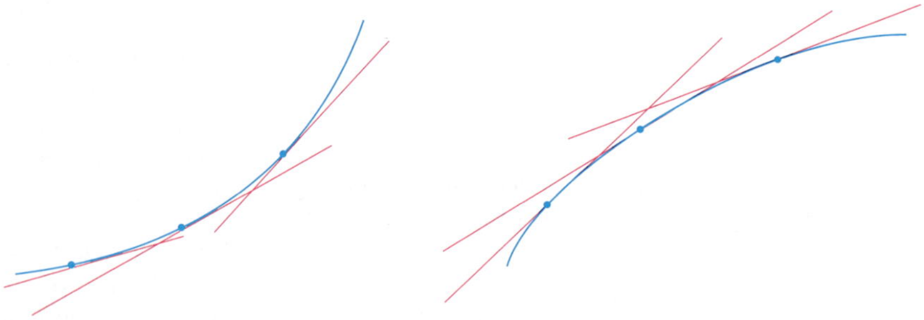

(a) Slopes of tangent lines are increasing.

(b) Slopes of tangent lines are decreasing.

FIGURE **27**

The shape of a curve can be described using the notion of concavity.

> **Concavity of a Function f**
>
> Let the function f be differentiable on an interval (a, b). Then,
>
> **1.** f is **concave upward** on (a, b) if f' is increasing on (a, b).
>
> **2.** f is **concave downward** on (a, b) if f' is decreasing on (a, b).

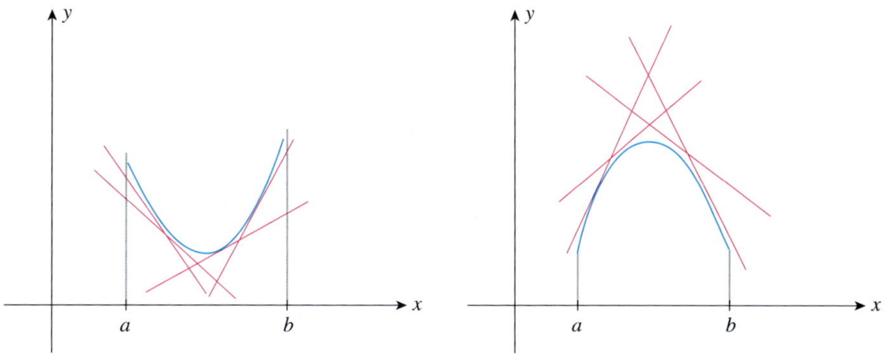

FIGURE 28

(a) f is concave upward on (a, b).

(b) f is concave downward on (a, b).

Geometrically, a curve is concave upward if it lies above its tangent lines (Figure 28a). Similarly, a curve is concave downward if it lies below its tangent lines (Figure 28b).

We also say that f is *concave upward at a number c* if there exists an interval (a, b) containing c in which f is concave upward. Similarly, we say that f is *concave downward at a number c* if there exists an interval (a, b) containing c in which f is concave downward.

If a function f has a second derivative f'', we can use f'' to determine the intervals of concavity of the function. Recall that $f''(x)$ measures the rate of change of the slope $f'(x)$ of the tangent line to the graph of f at the point $(x, f(x))$. Thus, if $f''(x) > 0$ on an interval (a, b), then the slopes of the tangent lines to the graph of f are increasing on (a, b), and so f is concave upward on (a, b). Similarly, if $f''(x) < 0$ on (a, b), then f is concave downward on (a, b). These observations suggest the following theorem.

THEOREM 2

a. If $f''(x) > 0$ for each value of x in (a, b), then f is concave upward on (a, b).

b. If $f''(x) < 0$ for each value of x in (a, b), then f is concave downward on (a, b).

The following procedure, based on the conclusions of Theorem 2, may be used to determine the intervals of concavity of a function.

Determining the Intervals of Concavity of f

1. Determine the values of x for which f'' is zero or where f'' is not defined, and identify the open intervals determined by these numbers.

2. Determine the sign of f'' in each interval found in step 1. To do this, compute $f''(c)$, where c is any conveniently chosen test number in the interval.

   a. If $f''(c) > 0$, f is concave upward on that interval.

   b. If $f''(c) < 0$, f is concave downward on that interval.

EXAMPLE 1 Determine where the function $f(x) = x^3 - 3x^2 - 24x + 32$ is concave upward and where it is concave downward.

Solution Here,

$$f'(x) = 3x^2 - 6x - 24$$
$$f''(x) = 6x - 6 = 6(x - 1)$$

and f'' is defined everywhere. Setting $f''(x) = 0$ gives $x = 1$. The sign diagram of f'' appears in Figure 29. We conclude that f is concave downward on the interval $(-\infty, 1)$ and is concave upward on the interval $(1, \infty)$. Figure 30 shows the graph of f.

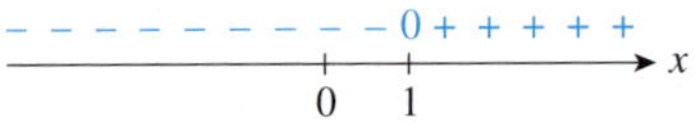

FIGURE 29
Sign diagram for f''

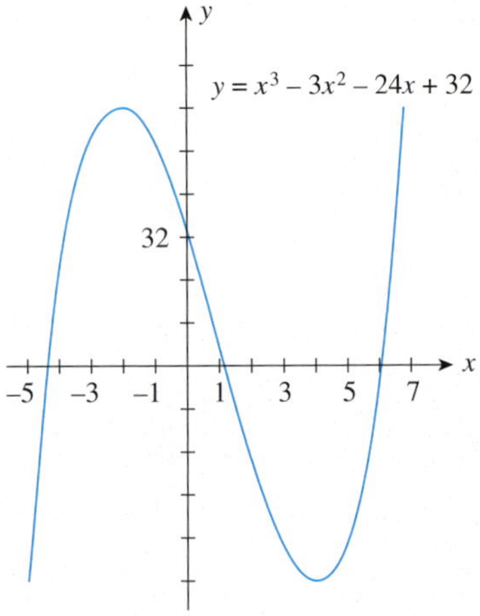

FIGURE 30
f is concave downward on $(-\infty, 1)$ and concave upward on $(1, \infty)$.

Exploring with TECHNOLOGY

Refer to Example 1.

1. Use a graphing utility to plot the graph of $f(x) = x^3 - 3x^2 - 24x + 32$ and its second derivative $f''(x) = 6x - 6$ using the viewing window $[-10, 10] \times [-80, 90]$.
2. By studying the graph of f'', determine the intervals where $f''(x) > 0$ and the intervals where $f''(x) < 0$. Next, look at the graph of f and determine the intervals where the graph of f is concave upward and the intervals where the graph of f is concave downward. Are these observations what you might have expected?

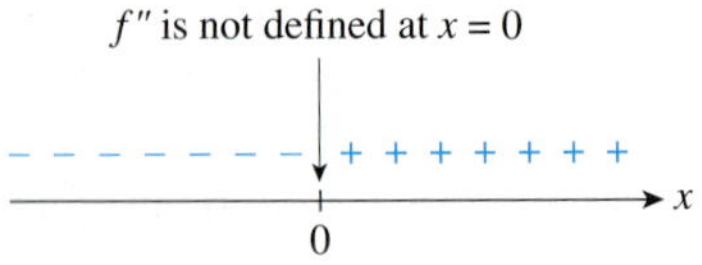

FIGURE 31
The sign diagram for f''

EXAMPLE 2 Determine the intervals where the function $f(x) = x + \dfrac{1}{x}$ is concave upward and where it is concave downward.

Solution We have

$$f'(x) = 1 - \frac{1}{x^2}$$
$$f''(x) = \frac{2}{x^3}$$

We deduce from the sign diagram for f'' (Figure 31) that the function f is concave downward on the interval $(-\infty, 0)$ and concave upward on the interval $(0, \infty)$. The graph of f is sketched in Figure 32.

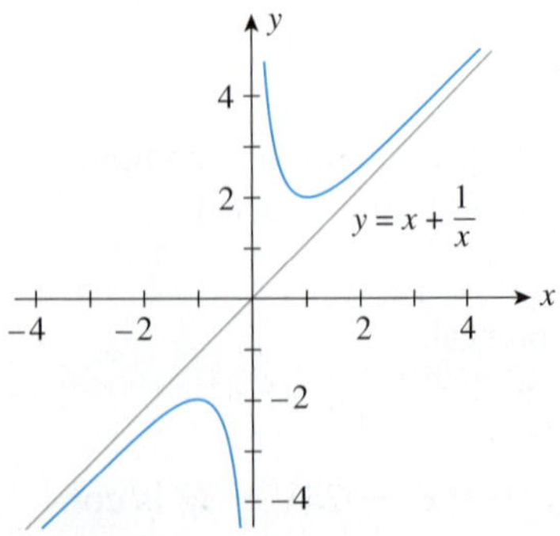

FIGURE 32
f is concave downward on $(-\infty, 0)$ and concave upward on $(0, \infty)$.

Inflection Points

Figure 33 shows the total sales S of a manufacturer of automobile air conditioners versus the amount of money x that the company spends on advertising its product.

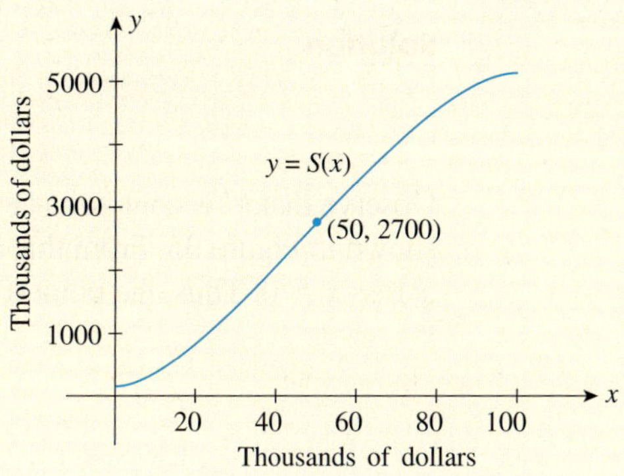

FIGURE 33
The graph of S has a point of inflection at (50, 2700).

Notice that the graph of the continuous function $y = S(x)$ changes concavity—from upward to downward—at the point (50, 2700). This point is called an inflection point of S. To understand the significance of this inflection point, observe that the total sales increase rather slowly at first, but as more money is spent on advertising, the total sales increase rapidly. This rapid increase reflects the effectiveness of the company's ads. However, a point is soon reached after which any additional advertising expenditure results in increased sales but at a slower rate of increase. This point, commonly known as the *point of diminishing returns,* is the point of inflection of the function S. We will return to this example later.

Let's now state formally the definition of an inflection point.

Inflection Point

A point on the graph of a continuous function f where the tangent line exists and where the concavity changes is called an **inflection point.**

Observe that the graph of a function crosses its tangent line at a point of inflection (Figure 34).

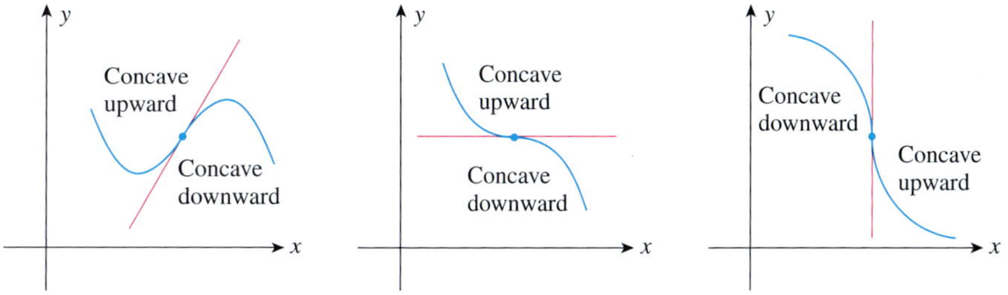

FIGURE 34
At each point of inflection, the graph of a function crosses its tangent line.

The following procedure may be used to find inflection points.

Finding Inflection Points

1. Compute $f''(x)$.
2. Determine the numbers in the domain of f for which $f''(x) = 0$ or $f''(x)$ does not exist.
3. Determine the sign of $f''(x)$ to the left and right of each number c found in step 2. If there is a change in the sign of $f''(x)$ as we move across $x = c$, then $(c, f(c))$ is an inflection point of f.

The numbers determined in step 2 are only *candidates* for the inflection points of f. For example, you can easily verify that $f''(0) = 0$ if $f(x) = x^4$, but a sketch of the graph of f will show that (0, 0) is *not* an inflection point of f.

EXAMPLE 3 Find the point of inflection of the function $f(x) = x^3$.

Solution

$$f'(x) = 3x^2$$
$$f''(x) = 6x$$

Observe that f'' is continuous everywhere and is zero if $x = 0$. The sign diagram of f'' is shown in Figure 35. From this diagram, we see that $f''(x)$ changes sign as we move across $x = 0$. Thus, the point (0, 0) is an inflection point of the function f (Figure 36).

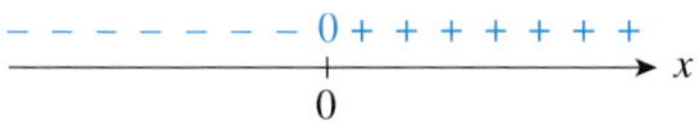

FIGURE 35
Sign diagram for f''

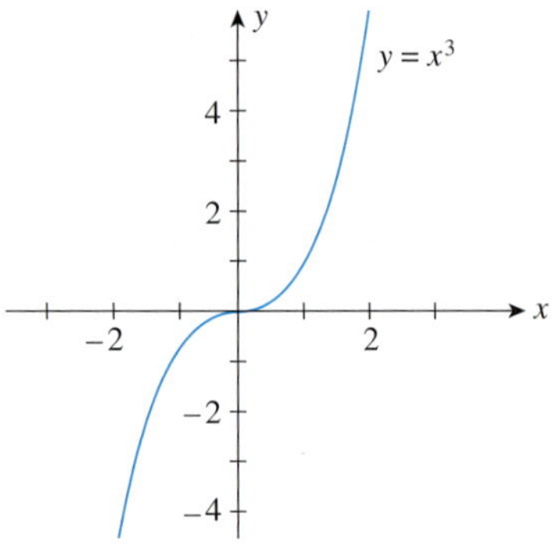

FIGURE 36
f has an inflection point at (0, 0).

EXAMPLE 4 Determine the intervals where the function $f(x) = (x - 1)^{5/3}$ is concave upward and where it is concave downward and find the inflection points of f.

Solution The first derivative of f is

$$f'(x) = \frac{5}{3}(x - 1)^{2/3}$$

and the second derivative of f is

$$f''(x) = \frac{10}{9}(x - 1)^{-1/3} = \frac{10}{9(x - 1)^{1/3}}$$

We see that f'' is not defined at $x = 1$. Furthermore, $f''(x)$ is not equal to zero anywhere. The sign diagram of f'' is shown in Figure 37. From the sign diagram, we see that f is concave downward on $(-\infty, 1)$ and concave upward on $(1, \infty)$. Next, since $x = 1$ does lie in the domain of f, our computations also reveal that the point (1, 0) is an inflection point of f (Figure 38).

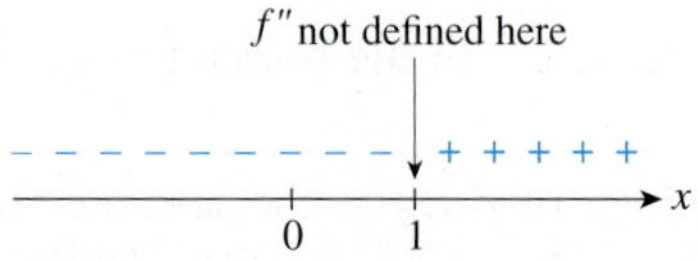

FIGURE 37
Sign diagram for f''

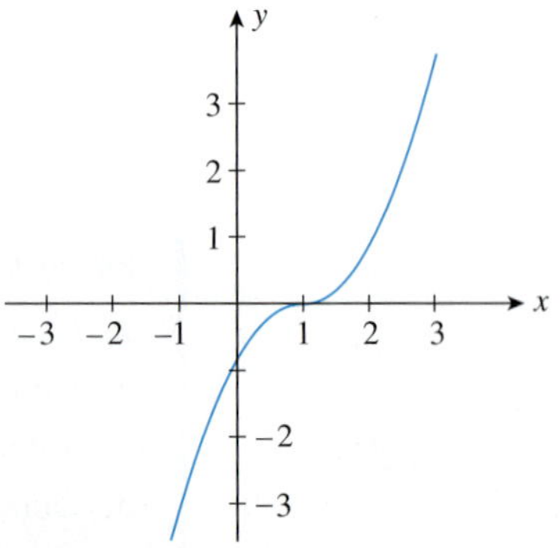

FIGURE 38
f has an inflection point at (1, 0).

EXAMPLE 5 Determine the intervals where the function

$$f(x) = \frac{1}{x^2 + 1}$$

is concave upward and where it is concave downward and find the inflection points of f.

Solution The first derivative of f is

$$f'(x) = \frac{d}{dx}(x^2 + 1)^{-1} = -2x(x^2 + 1)^{-2}$$ Rewrite the original function and use the general power rule.

$$= -\frac{2x}{(x^2 + 1)^2}$$

Next, using the quotient rule, we find

$$f''(x) = \frac{(x^2 + 1)^2(-2) + (2x)2(x^2 + 1)(2x)}{(x^2 + 1)^4}$$ See page 16.

$$= \frac{(x^2 + 1)[-2(x^2 + 1) + 8x^2]}{(x^2 + 1)^4} = \frac{(x^2 + 1)(6x^2 - 2)}{(x^2 + 1)^4}$$

$$= \frac{2(3x^2 - 1)}{(x^2 + 1)^3}$$ Cancel the common factors.

Observe that f'' is continuous everywhere and is zero if

$$3x^2 - 1 = 0$$

$$x^2 = \frac{1}{3}$$

or $x = \pm\sqrt{3}/3$. The sign diagram for f'' is shown in Figure 39. From the sign diagram for f'', we see that f is concave upward on $(-\infty, -\sqrt{3}/3) \cup (\sqrt{3}/3, \infty)$ and concave downward on $(-\sqrt{3}/3, \sqrt{3}/3)$. Also, observe that $f''(x)$ changes sign as we move across the numbers $x = -\sqrt{3}/3$ and $x = \sqrt{3}/3$. Since

$$f\left(-\frac{\sqrt{3}}{3}\right) = \frac{1}{\frac{1}{3} + 1} = \frac{3}{4} \quad \text{and} \quad f\left(\frac{\sqrt{3}}{3}\right) = \frac{3}{4}$$

we see that the points $(-\sqrt{3}/3, 3/4)$ and $(\sqrt{3}/3, 3/4)$ are inflection points of f. The graph of f is shown in Figure 40.

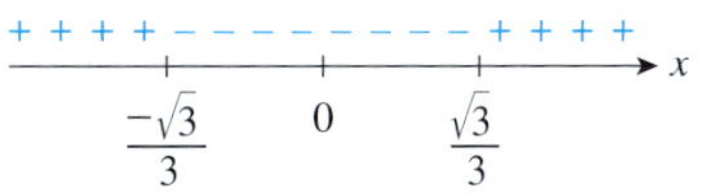

FIGURE 39
Sign diagram for f''

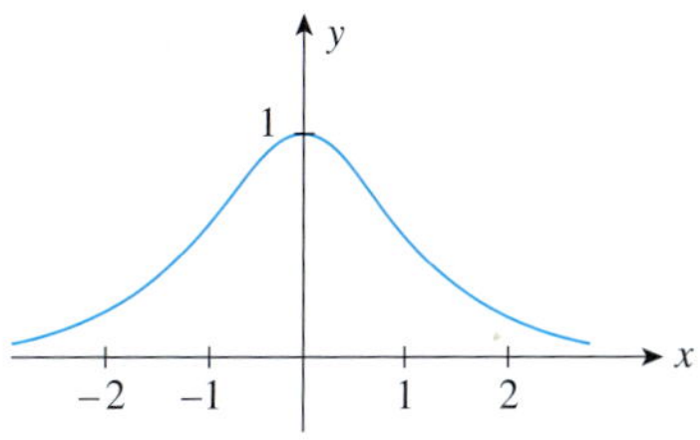

FIGURE 40
The graph of $f(x) = \frac{1}{x^2 + 1}$ is concave upward on $(-\infty, -\sqrt{3}/3) \cup (\sqrt{3}/3, \infty)$ and concave downward on $(-\sqrt{3}/3, \sqrt{3}/3)$.

Explore & Discuss

1. Suppose $(c, f(c))$ is an inflection point of f. Can you conclude that f has no relative extremum at $x = c$? Explain your answer.
2. True or false: A polynomial function of degree 3 has exactly one inflection point.
   Hint: Study the function $f(x) = ax^3 + bx^2 + cx + d$ $(a \neq 0)$.

The next example uses an interpretation of the first and second derivatives to help us sketch a graph of a function.

EXAMPLE 6 Sketch the graph of a function having the following properties:

$$\begin{aligned} f(-1) &= 4 \\ f(0) &= 2 \\ f(1) &= 0 \\ f'(-1) &= 0 \\ f'(1) &= 0 \\ f'(x) &> 0 \quad \text{on } (-\infty, -1) \cup (1, \infty) \\ f'(x) &< 0 \quad \text{on } (-1, 1) \\ f''(x) &< 0 \quad \text{on } (-\infty, 0) \\ f''(x) &> 0 \quad \text{on } (0, \infty) \end{aligned}$$

Solution First, we plot the points $(-1, 4)$, $(0, 2)$, and $(1, 0)$ that lie on the graph of f. Since $f'(-1) = 0$ and $f'(1) = 0$, the tangent lines at the points $(-1, 4)$ and $(1, 0)$ are horizontal. Since $f'(x) > 0$ on $(-\infty, -1)$ and $f'(x) < 0$ on $(-1, 1)$, we see that f has a relative maximum at the point $(-1, 4)$. Also, $f'(x) < 0$ on $(-1, 1)$ and $f'(x) > 0$ on $(1, \infty)$ implies that f has a relative minimum at the point $(1, 0)$ (Figure 41a).

Since $f''(x) < 0$ on $(-\infty, 0)$ and $f''(x) > 0$ on $(0, \infty)$, we see that the point $(0, 2)$ is an inflection point. Finally, we complete the graph making use of the fact that f is increasing on $(-\infty, -1) \cup (1, \infty)$, where it is given that $f'(x) > 0$, and f is decreasing on $(-1, 1)$, where $f'(x) < 0$. Also, make sure that f is concave downward on $(-\infty, 0)$ and concave upward on $(0, \infty)$ (Figure 41b).

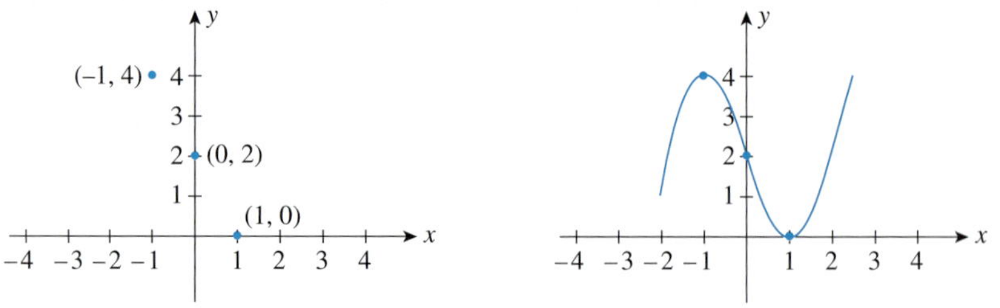

FIGURE 41 (a) (b)

Examples 7 and 8 illustrate familiar interpretations of the significance of the inflection point of a function.

APPLIED EXAMPLE 7 Effect of Advertising on Sales The total sales S (in thousands of dollars) of Arctic Air Corporation, a manufacturer of automobile air conditioners, is related to the amount of money x (in thousands of dollars) the company spends on advertising its products by the formula

$$S(x) = -0.01x^3 + 1.5x^2 + 200 \qquad (0 \leq x \leq 100)$$

Find the inflection point of the function S.

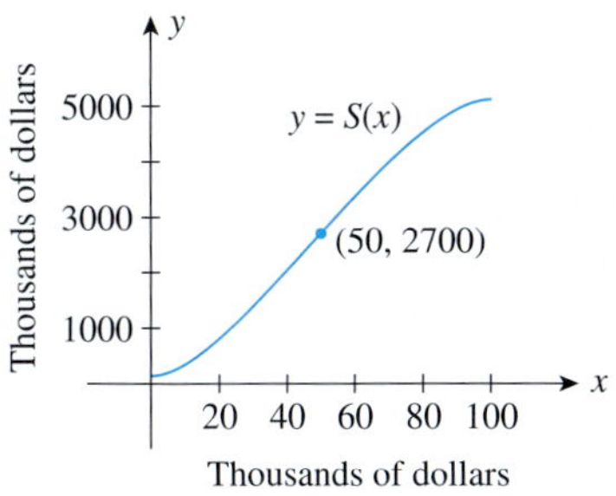

FIGURE **42**
The graph of $S(x)$ has a point of inflection at (50, 2700).

Solution The first two derivatives of S are given by

$$S'(x) = -0.03x^2 + 3x$$
$$S''(x) = -0.06x + 3$$

Setting $S''(x) = 0$ gives $x = 50$. So $(50, S(50))$ is the only candidate for an inflection point of S. Moreover, since

$$S''(x) > 0 \quad \text{for} \quad x < 50$$

and

$$S''(x) < 0 \quad \text{for} \quad x > 50$$

the point (50, 2700) is an inflection point of the function S. The graph of S appears in Figure 42. Notice that this is the graph of the function we discussed earlier. ■

VIDEO

APPLIED EXAMPLE 8 Consumer Price Index An economy's consumer price index (CPI) is described by the function

$$I(t) = -0.2t^3 + 3t^2 + 100 \qquad (0 \le t \le 10)$$

where $t = 0$ corresponds to the beginning of 2000. Find the point of inflection of the function I and discuss its significance.

Solution The first two derivatives of I are given by

$$I'(t) = -0.6t^2 + 6t$$
$$I''(t) = -1.2t + 6 = -1.2(t - 5)$$

Setting $I''(t) = 0$ gives $t = 5$. So $(5, I(5))$ is the only candidate for an inflection point of I. Next, we observe that

$$I''(x) > 0 \quad \text{for} \quad t < 5$$
$$I''(x) < 0 \quad \text{for} \quad t > 5$$

so the point (5, 150) is an inflection point of I. The graph of I is sketched in Figure 43.

Since the second derivative of I measures the *rate of change* of the rate of change of the CPI, our computations reveal that the rate of inflation had in fact peaked at $t = 5$. Thus, relief actually began at the beginning of 2005.

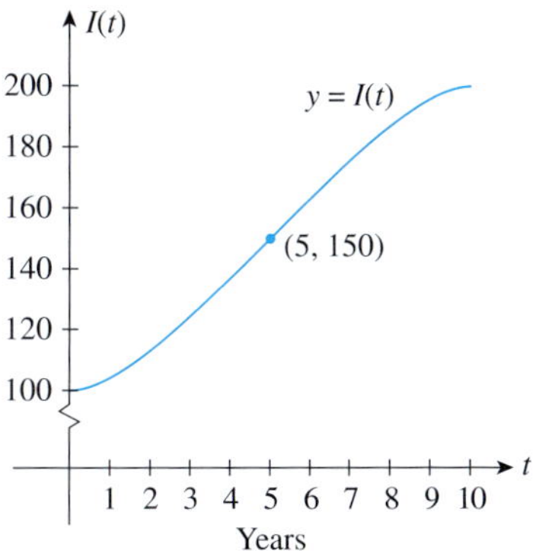

FIGURE **43**
The graph of $I(t)$ has a point of inflection at (5, 150).

■

The Second Derivative Test

We now show how the second derivative f'' of a function f can be used to help us determine whether a critical number of f gives rise to a relative extremum of f. Figure 44a shows the graph of a function that has a relative maximum at $x = c$.

Observe that f is concave downward at that number. Similarly, Figure 44b shows that at a relative minimum of f the graph is concave upward. But from our previous work, we know that f is concave downward at $x = c$ if $f''(c) < 0$ and f is concave upward at $x = c$ if $f''(c) > 0$. These observations suggest the following alternative procedure for determining whether a critical number of f gives rise to a relative extremum of f. This result is called the **second derivative test** and is applicable when f'' exists.

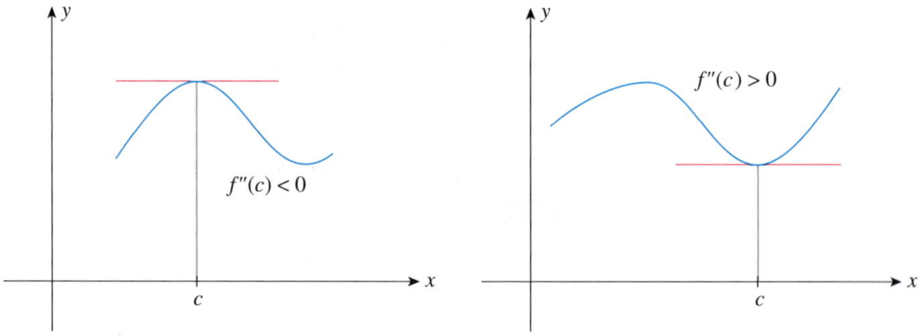

FIGURE **44** **(a)** f has a relative maximum at $x = c$. **(b)** f has a relative minimum at $x = c$.

The Second Derivative Test

1. Compute $f'(x)$ and $f''(x)$.
2. Find all the critical numbers of f at which $f'(x) = 0$.
3. Compute $f''(c)$ for each such critical number c.
   - **a.** If $f''(c) < 0$, then f has a relative maximum at c.
   - **b.** If $f''(c) > 0$, then f has a relative minimum at c.
   - **c.** If $f''(c) = 0$, the test fails; that is, it is inconclusive.

Note The second derivative test does not yield a conclusion if $f''(c) = 0$ or if $f''(c)$ does not exist. In other words, $x = c$ may give rise to a relative extremum or an inflection point (see Exercise 108, page 281). In such cases, you should revert to the first derivative test. ■

VIDEO

EXAMPLE 9 Determine the relative extrema of the function

$$f(x) = x^3 - 3x^2 - 24x + 32$$

using the second derivative test. (See Example 7, Section 4.1.)

Solution We have

$$f'(x) = 3x^2 - 6x - 24 = 3(x + 2)(x - 4)$$

so $f'(x) = 0$ gives $x = -2$ and $x = 4$, the critical numbers of f, as in Example 7. Next, we compute

$$f''(x) = 6x - 6 = 6(x - 1)$$

Since

$$f''(-2) = 6(-2 - 1) = -18 < 0$$

the second derivative test implies that $f(-2) = 60$ is a relative maximum of f. Also,

$$f''(4) = 6(4 - 1) = 18 > 0$$

and the second derivative test implies that $f(4) = -48$ is a relative minimum of f, which confirms the results obtained earlier. ■

Explore & Discuss

Suppose a function f has the following properties:

1. $f''(x) > 0$ for all x in an interval (a, b).
2. There is a number c between a and b such that $f'(c) = 0$.

What special property can you ascribe to the point $(c, f(c))$? Answer the question if Property 1 is replaced by the property that $f''(x) < 0$ for all x in (a, b).

Comparing the First and Second Derivative Tests

Notice that both the first derivative test and the second derivative test are used to classify the critical numbers of f. What are the pros and cons of the two tests? Since the second derivative test is applicable only when f'' exists, it is less versatile than the first derivative test. For example, it cannot be used to locate the relative minimum $f(0) = 0$ of the function $f(x) = x^{2/3}$.

Furthermore, the second derivative test is inconclusive when f'' is equal to zero at a critical number of f, whereas the first derivative test always yields positive conclusions. The second derivative test is also inconvenient to use when f'' is difficult to compute. On the plus side, if f'' is computed easily, then we use the second derivative test since it involves just the evaluation of f'' at the critical number(s) of f. Also, the conclusions of the second derivative test are important in theoretical work.

We close this section by summarizing the different roles played by the first derivative f' and the second derivative f'' of a function f in determining the properties of the graph of f. The first derivative f' tells us where f is increasing and where f is decreasing, whereas the second derivative f'' tells us where f is concave upward and where f is concave downward. These different properties of f are reflected by the signs of f' and f'' in the interval of interest. The following table shows the general characteristics of the function f for various possible combinations of the signs of f' and f'' in the interval (a, b).

Signs of f' and f''	**Properties of the Graph of f**	**General Shape of the Graph of f**
$f'(x) > 0$ $f''(x) > 0$	f increasing f concave upward	
$f'(x) > 0$ $f''(x) < 0$	f increasing f concave downward	
$f'(x) < 0$ $f''(x) > 0$	f decreasing f concave upward	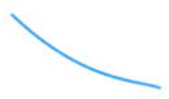
$f'(x) < 0$ $f''(x) < 0$	f decreasing f concave downward	

4.2 Self-Check Exercises

1. Determine where the function $f(x) = 4x^3 - 3x^2 + 6$ is concave upward and where it is concave downward.

2. Using the second derivative test, if applicable, find the relative extrema of the function $f(x) = 2x^3 - \frac{1}{2}x^2 - 12x - 10$.

3. A certain country's gross domestic product (GDP) (in millions of dollars) in year t is described by the function

$$G(t) = -2t^3 + 45t^2 + 20t + 6000 \qquad (0 \le t \le 11)$$

where $t = 0$ corresponds to the beginning of 1995. Find the inflection point of the function G and discuss its significance.

Solutions to Self-Check Exercises 4.2 can be found on page 281.

4.2 Concept Questions

1. Explain what it means for a function f to be (a) concave upward and (b) concave downward on an open interval I. Given that f has a second derivative on I (except at isolated numbers), how do you determine where the graph of f is concave upward and where it is concave downward?

2. What is an inflection point of the graph of a function f? How do you find the inflection point(s) of the graph of a function f whose rule is given?

3. State the second derivative test. What are the pros and cons of using the first derivative test and the second derivative test?

4.2 Exercises

In Exercises 1–8, you are given the graph of a function *f*. Determine the intervals where *f* is concave upward and where it is concave downward. Also, find all inflection points of *f*, if any.

1.
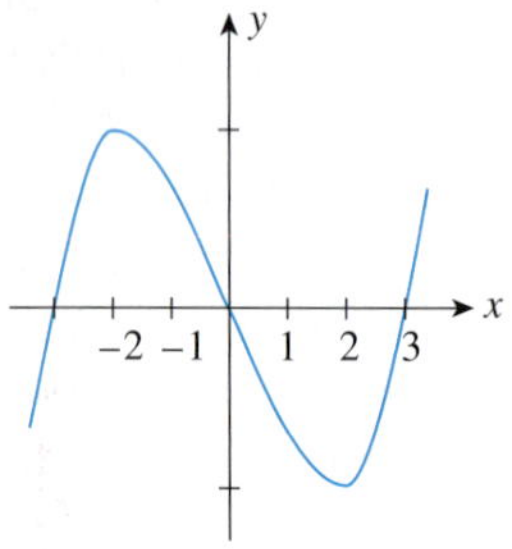

2.
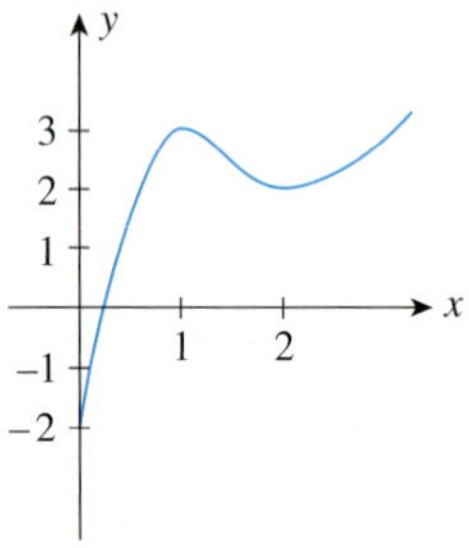

3.
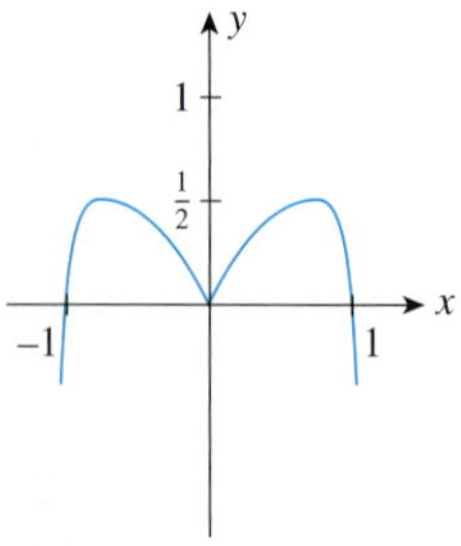

4.
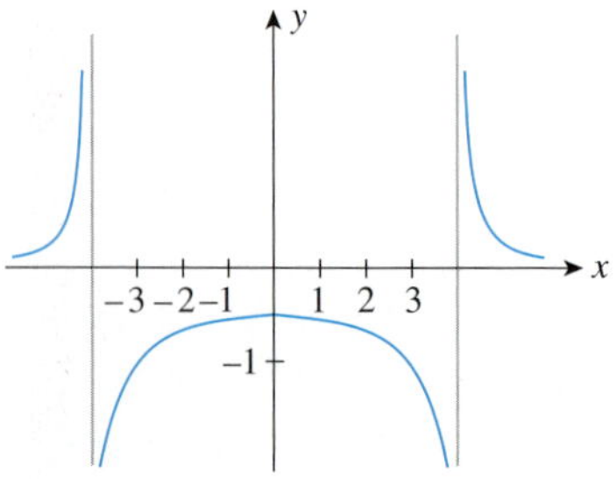

5.
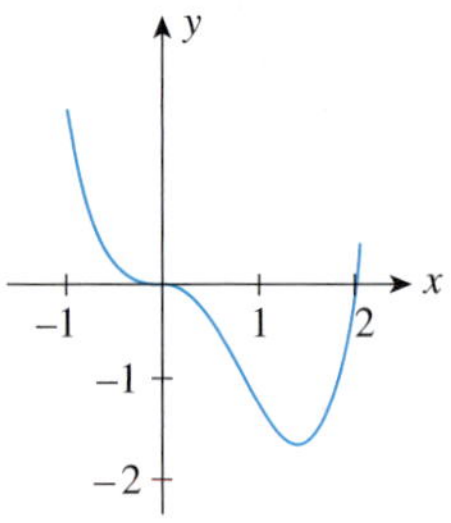

6.
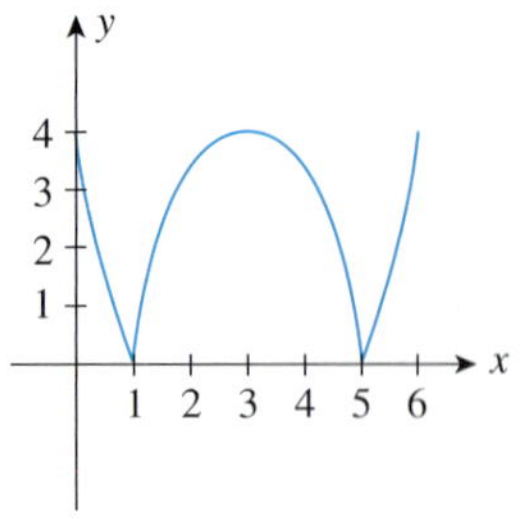

7.
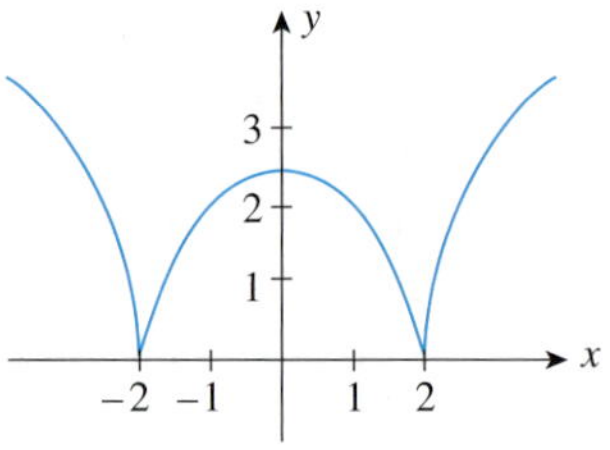

8.
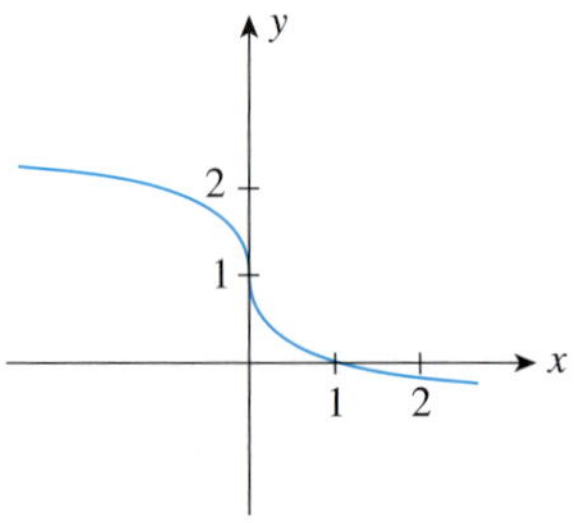

9. Refer to the graph of f shown in the following figure:

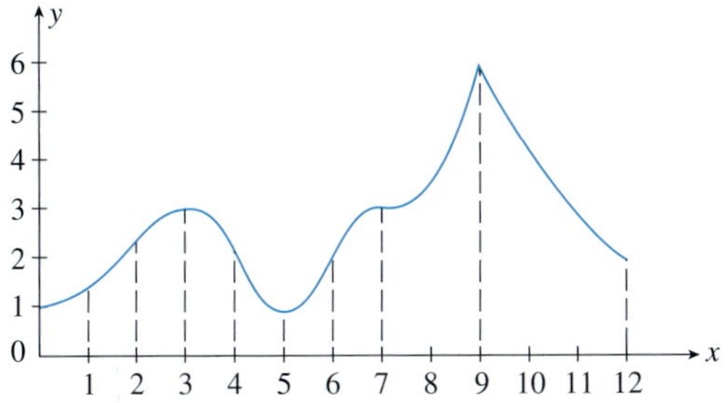

a. Find the intervals where f is concave upward and the intervals where f is concave downward.
b. Find the inflection points of f.

10. Refer to the figure for Exercise 9.
a. Explain how the second derivative test can be used to show that the critical number 3 gives rise to a relative maximum of f and the critical number 5 gives rise to a relative minimum of f.
b. Explain why the second derivative test cannot be used to show that the critical number 7 does not give rise to a relative extremum of f nor can it be used to show that the critical number 9 gives rise to a relative maximum of f.

In Exercises 11–14, determine which graph—(a), (b), or (c)—is the graph of the function f with the specified properties.

11. $f(2) = 1, f'(2) > 0$, and $f''(2) < 0$

(a)

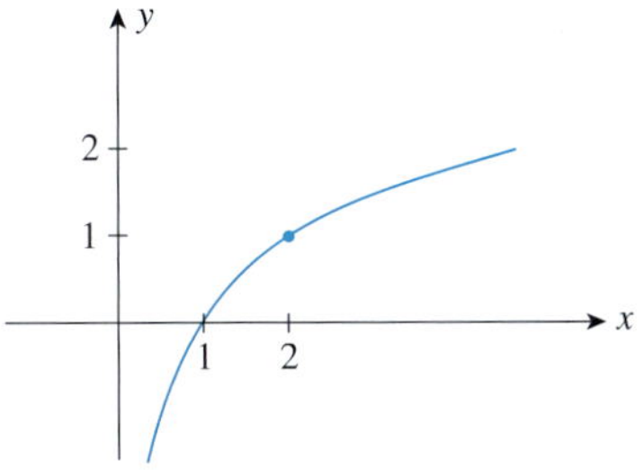

(b)

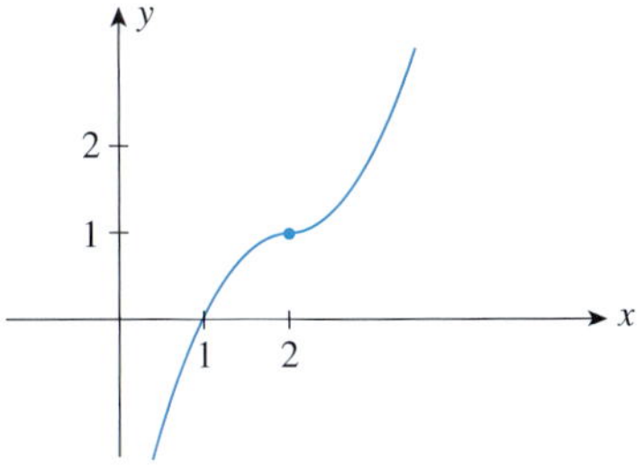

(c)

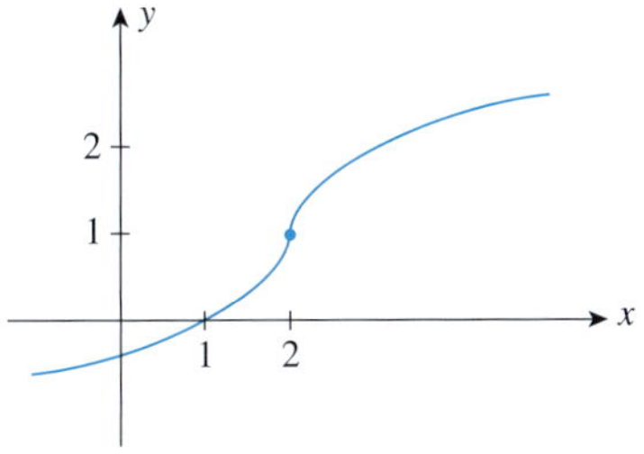

12. $f(1) = 2, f'(x) > 0$ on $(-\infty, 1) \cup (1, \infty)$, and $f''(1) = 0$

(a)

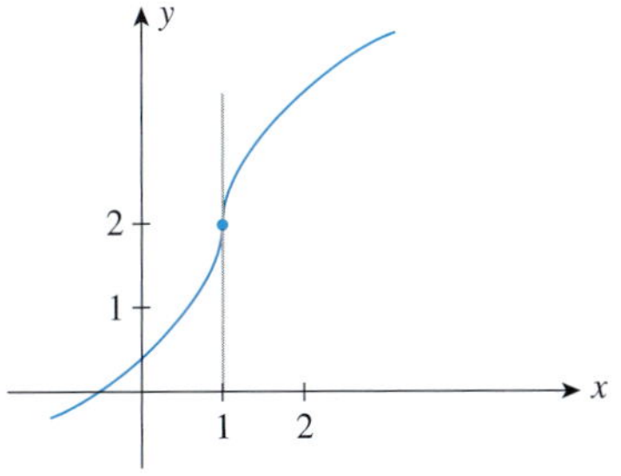

(b)

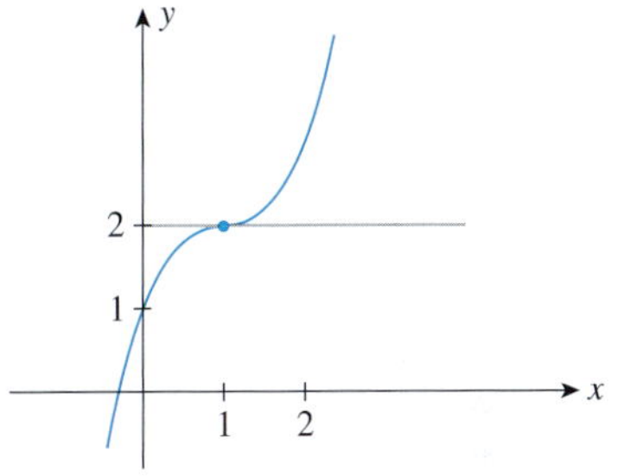

(c)

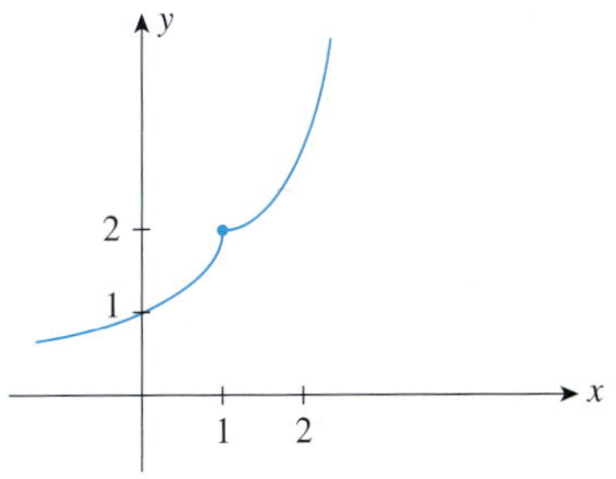

13. $f'(0)$ is undefined, f is decreasing on $(-\infty, 0)$, f is concave downward on $(0, 3)$, and f has an inflection point at $x = 3$.

(a)

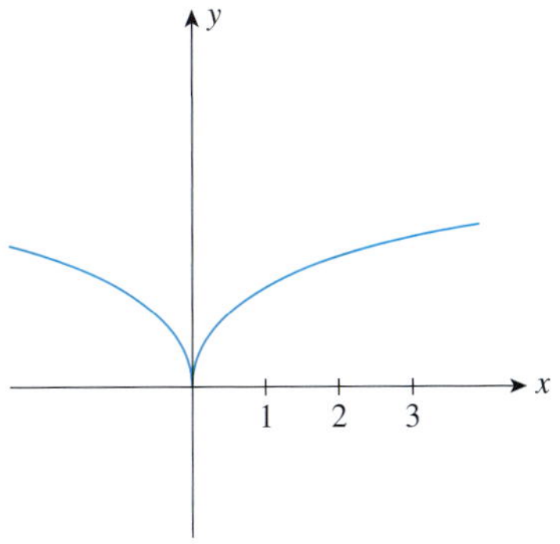

(b)

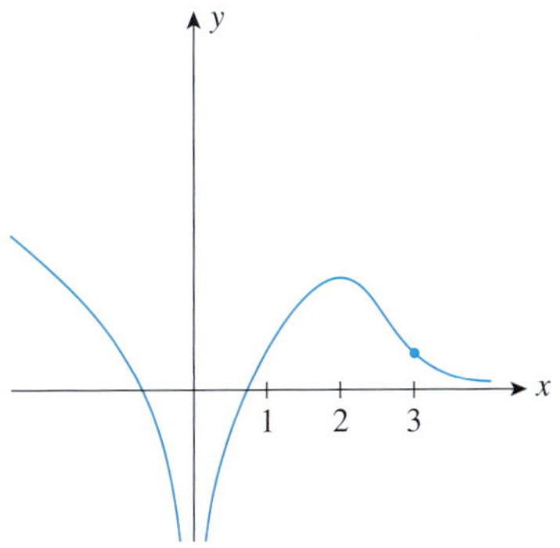

(c)

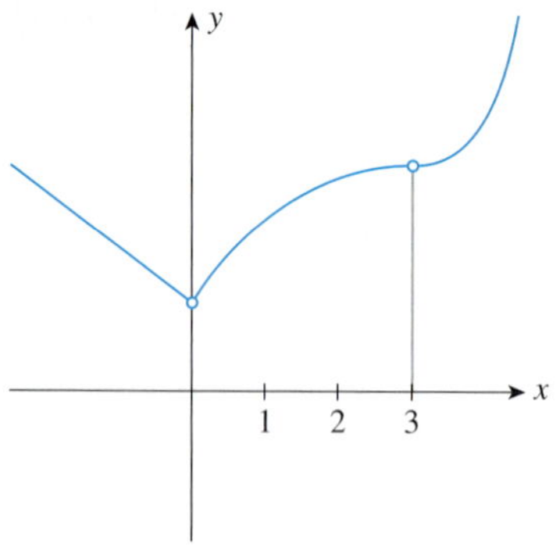

14. f is decreasing on $(-\infty, 2)$ and increasing on $(2, \infty)$, f is concave upward on $(1, \infty)$, and f has inflection points at $x = 0$ and $x = 1$.

(a)

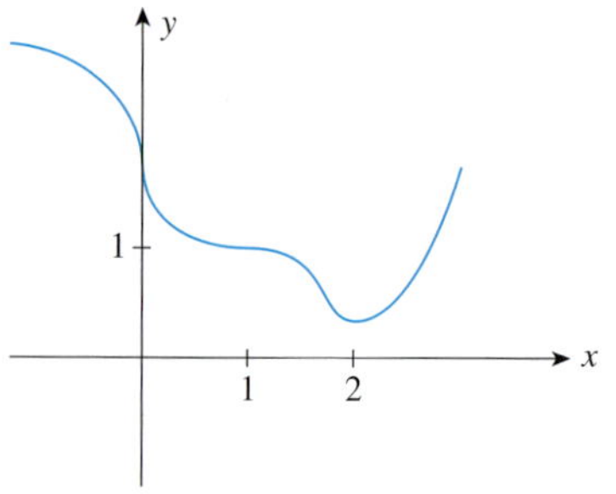

(b)

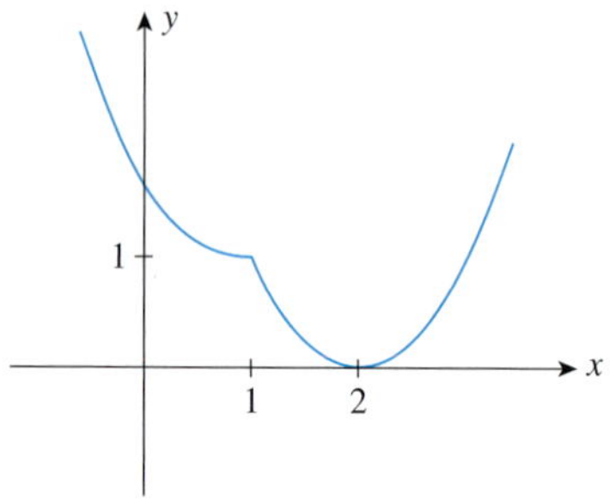

(c)

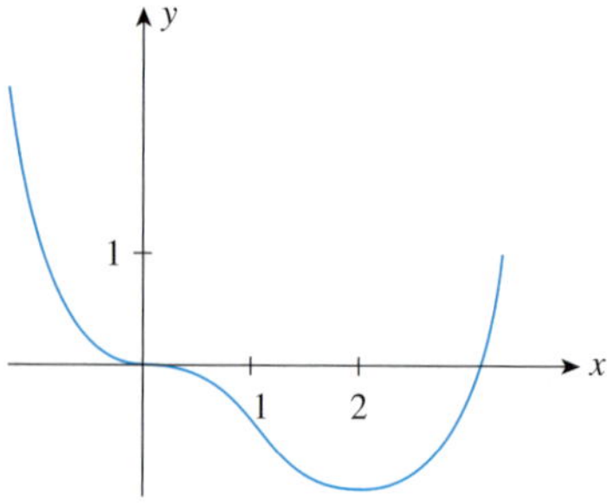

15. EFFECT OF ADVERTISING ON BANK DEPOSITS The following graphs were used by the CEO of the Madison Savings Bank to illustrate what effect a projected promotional campaign would have on its deposits over the next year. The functions D_1 and D_2 give the projected amount of money on deposit with the bank over the next 12 mo with and without the proposed promotional campaign, respectively.

a. Determine the signs of $D_1'(t)$, $D_2'(t)$, $D_1''(t)$, and $D_2''(t)$ on the interval $(0, 12)$.

b. What can you conclude about the rate of change of the growth rate of the money on deposit with the bank with and without the proposed promotional campaign?

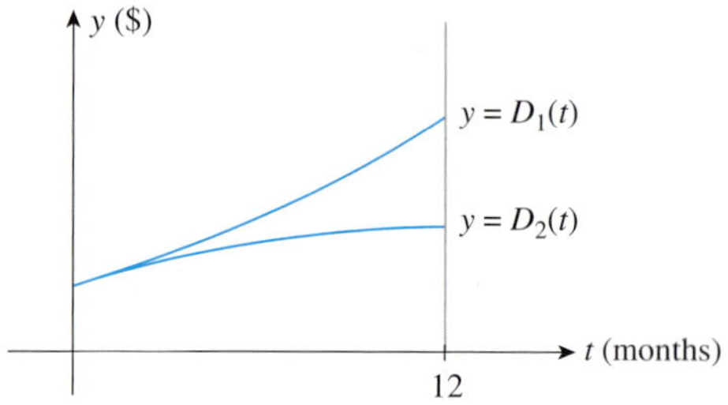

16. ASSEMBLY TIME OF A WORKER In the following graph, $N(t)$ gives the number of personal radios assembled by the average worker by the tth hr, where $t = 0$ corresponds to 8 a.m. and $0 \le t \le 4$. The point P is an inflection point of N.

a. What can you say about the rate of change of the rate of the number of personal radios assembled by the average worker between 8 a.m. and 10 a.m.? Between 10 a.m. and 12 p.m.?

b. At what time is the rate at which the personal radios are being assembled by the average worker greatest?

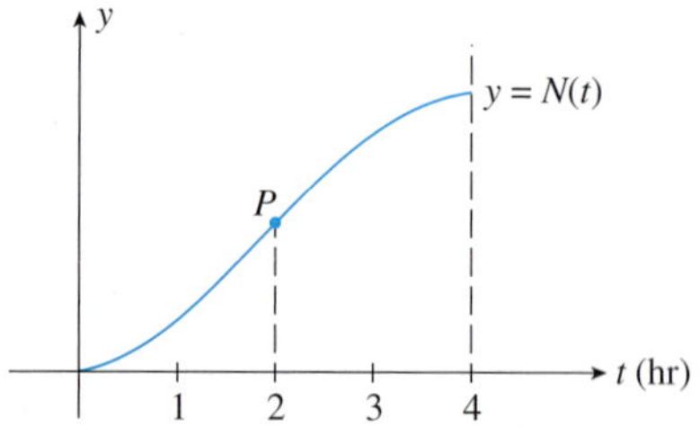

17. WATER POLLUTION When organic waste is dumped into a pond, the oxidation process that takes place reduces the pond's oxygen content. However, given time, nature will restore the oxygen content to its natural level. In the following graph, $P(t)$ gives the oxygen content (as a percent of its normal level) t days after organic waste has been dumped into the pond. Explain the significance of the inflection point Q.

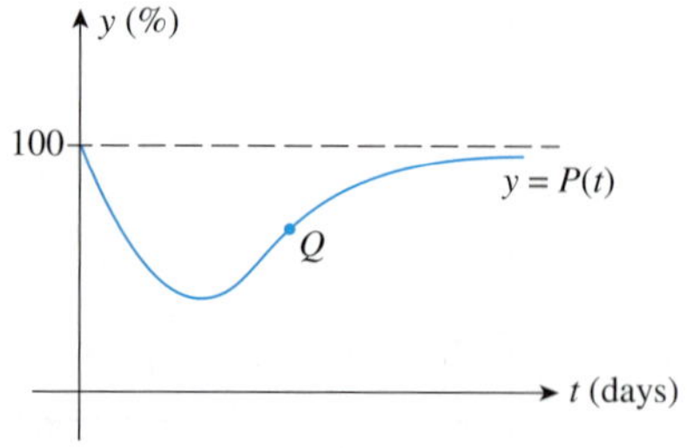

18. **Spread of a Rumor** Initially, a handful of students heard a rumor on campus. The rumor spread and, after t hr, the number had grown to $N(t)$. The graph of the function N is shown in the following figure:

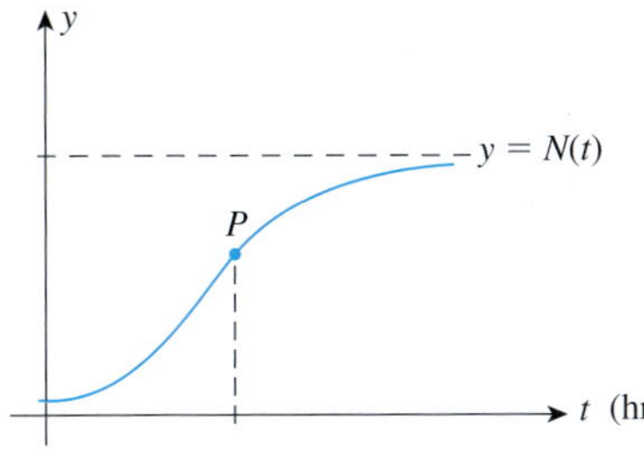

Describe the spread of the rumor in terms of the speed it was spread. In particular, explain the significance of the inflection point P of the graph of N.

In Exercises 19–22, show that the function is concave upward wherever it is defined.

19. $f(x) = 4x^2 - 12x + 7$

20. $g(x) = x^4 + \frac{1}{2}x^2 + 6x + 10$

21. $f(x) = \frac{1}{x^4}$

22. $g(x) = -\sqrt{4 - x^2}$

In Exercises 23–42, determine where the function is concave upward and where it is concave downward.

23. $f(x) = 2x^2 - 3x + 4$

24. $g(x) = -x^2 + 3x + 4$

25. $f(x) = x^3 - 1$

26. $g(x) = x^3 - x$

27. $f(x) = x^4 - 6x^3 + 2x + 8$

28. $f(x) = 3x^4 - 6x^3 + x - 8$

29. $f(x) = x^{4/7}$

30. $f(x) = \sqrt[3]{x}$

31. $f(x) = \sqrt{4 - x}$

32. $g(x) = \sqrt{x - 2}$

33. $f(x) = \frac{1}{x - 2}$

34. $g(x) = \frac{x}{x + 1}$

35. $f(x) = \frac{1}{2 + x^2}$

36. $g(x) = \frac{x}{1 + x^2}$

37. $h(t) = \frac{t^2}{t - 1}$

38. $f(x) = \frac{x + 1}{x - 1}$

39. $g(x) = x + \frac{1}{x^2}$

40. $h(r) = -\frac{1}{(r - 2)^2}$

41. $g(t) = (2t - 4)^{1/3}$

42. $f(x) = (x - 2)^{2/3}$

In Exercises 43–54, find the inflection point(s), if any, of each function.

43. $f(x) = x^3 - 2$

44. $g(x) = x^3 - 6x$

45. $f(x) = 6x^3 - 18x^2 + 12x - 15$

46. $g(x) = 2x^3 - 3x^2 + 18x - 8$

47. $f(x) = 3x^4 - 4x^3 + 1$

48. $f(x) = x^4 - 2x^3 + 6$

49. $g(t) = \sqrt[3]{t}$

50. $f(x) = \sqrt[5]{x}$

51. $f(x) = (x - 1)^3 + 2$

52. $f(x) = (x - 2)^{4/3}$

53. $f(x) = \frac{2}{1 + x^2}$

54. $f(x) = 2 + \frac{3}{x}$

In Exercises 55–70, find the relative extrema, if any, of each function. Use the second derivative test, if applicable.

55. $f(x) = -x^2 + 2x + 4$

56. $g(x) = 2x^2 + 3x + 7$

57. $f(x) = 2x^3 + 1$

58. $g(x) = x^3 - 6x$

59. $f(x) = \frac{1}{3}x^3 - 2x^2 - 5x - 10$

60. $f(x) = 2x^3 + 3x^2 - 12x - 4$

61. $g(t) = t + \frac{9}{t}$

62. $f(t) = 2t + \frac{3}{t}$

63. $f(x) = \frac{x}{1 - x}$

64. $f(x) = \frac{2x}{x^2 + 1}$

65. $f(t) = t^2 - \frac{16}{t}$

66. $g(x) = x^2 + \frac{2}{x}$

67. $g(s) = \frac{s}{1 + s^2}$

68. $g(x) = \frac{1}{1 + x^2}$

69. $f(x) = \frac{x^4}{x - 1}$

70. $f(x) = \frac{x^2}{x^2 + 1}$

In Exercises 71–76, sketch the graph of a function having the given properties.

71. $f(2) = 4, f'(2) = 0, f''(x) < 0$ on $(-\infty, \infty)$

72. $f(2) = 2, f'(2) = 0, f'(x) > 0$ on $(-\infty, 2), f'(x) > 0$ on $(2, \infty), f''(x) < 0$ on $(-\infty, 2), f''(x) > 0$ on $(2, \infty)$

73. $f(-2) = 4, f(3) = -2, f'(-2) = 0, f'(3) = 0, f'(x) > 0$ on $(-\infty, -2) \cup (3, \infty), f'(x) < 0$ on $(-2, 3)$, inflection point at $(1, 1)$

74. $f(0) = 0, f'(0)$ does not exist, $f''(x) < 0$ if $x \neq 0$

75. $f(0) = 1, f'(0) = 0, f(x) > 0$ on $(-\infty, \infty), f''(x) < 0$ on $(-\sqrt{2}/2, \sqrt{2}/2), f''(x) > 0$ on $(-\infty, -\sqrt{2}/2) \cup (\sqrt{2}/2, \infty)$

76. f has domain $[-1, 1], f(-1) = -1, f(-\frac{1}{2}) = -2, f'(-\frac{1}{2}) = 0, f''(x) > 0$ on $(-1, 1)$

77. **Demand for RNs** The following graph gives the total number of help-wanted ads for RNs (registered nurses) in 22 cities over the last 12 mo as a function of time t (t measured in months).

a. Explain why $N'(t)$ is positive on the interval $(0, 12)$.

b. Determine the signs of $N''(t)$ on the interval $(0, 6)$ and the interval $(6, 12)$.

c. Interpret the results of part (b).

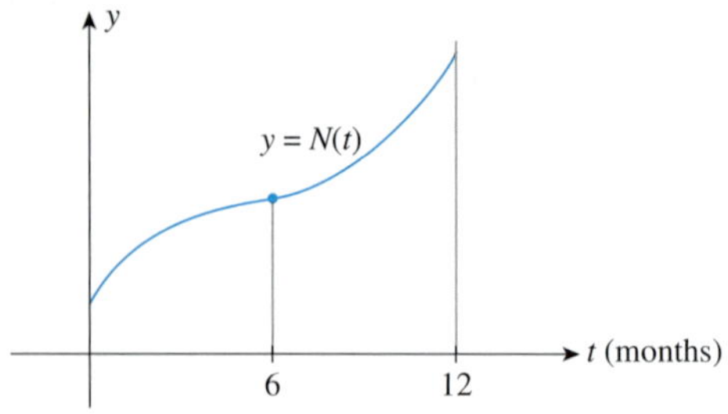

78. Effect of Budget Cuts on Drug-Related Crimes The graphs below were used by a police commissioner to illustrate what effect a budget cut would have on crime in the city. The number $N_1(t)$ gives the projected number of drug-related crimes in the next 12 mo. The number $N_2(t)$ gives the projected number of drug-related crimes in the same time frame if next year's budget is cut.

a. Explain why $N_1'(t)$ and $N_2'(t)$ are both positive on the interval (0, 12).

b. What are the signs of $N_1''(t)$ and $N_2''(t)$ on the interval (0, 12)?

c. Interpret the results of part (b).

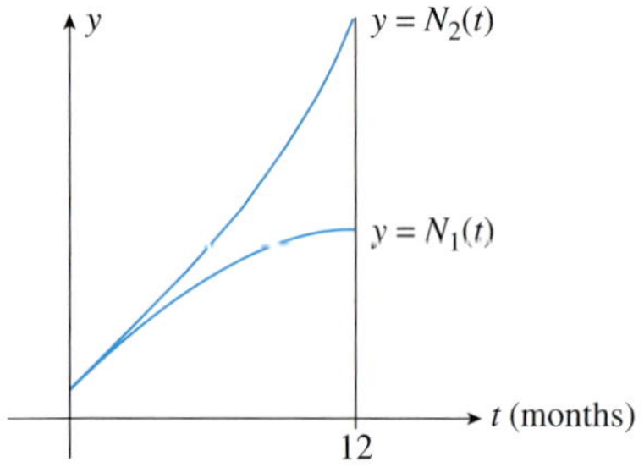

79. In the following figure, water is poured into the vase at a constant rate (in appropriate units), and the water level rises to a height of $f(t)$ units at time t as measured from the base of the vase. The graph of f follows. Explain the shape of the curve in terms of its concavity. What is the significance of the inflection point?

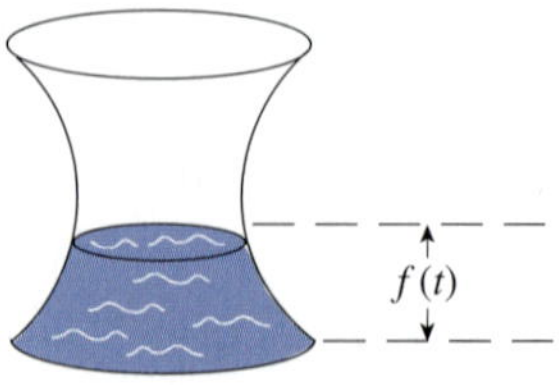

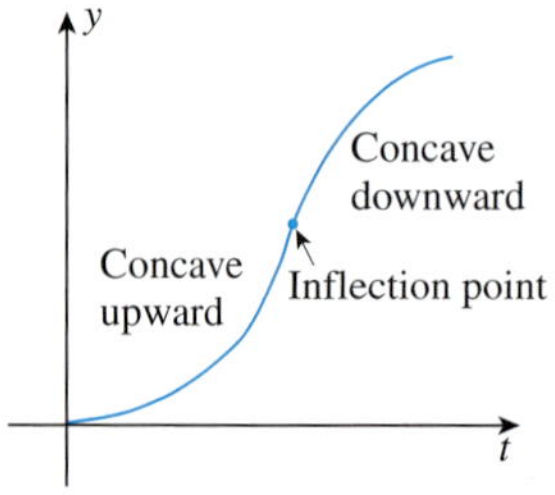

80. In the following figure, water is poured into an urn at a constant rate (in appropriate units), and the water level rises to a height of $f(t)$ units at time t as measured from the base of the urn. Sketch the graph of f and explain its shape, indicating where it is concave upward and concave downward. Indicate the inflection point on the graph and explain its significance.
Hint: Study Exercise 79.

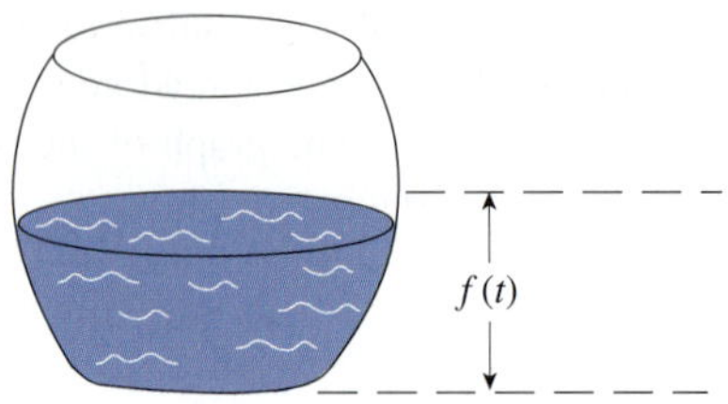

81. State Cigarette Taxes The average state cigarette tax per pack (in dollars) from 2001 through 2007 is approximated by the function

$$T(t) = 0.43t^{0.43} \qquad (1 \le t \le 7)$$

where t is measured in years, with $t = 1$ corresponding to 2001.

a. Show that the average state cigarette tax per pack was increasing throughout the period in question.

b. What can you say about the rate at which the average state cigarette tax per pack was increasing over the period in question?

Source: Campaign for Tobacco-Free Kids

82. Global Warming The increase in carbon dioxide (CO_2) in the atmosphere is a major cause of global warming. Using data obtained by Charles David Keeling, professor at Scripps Institution of Oceanography, the average amount of CO_2 in the atmosphere from 1958 through 2007 is approximated by

$$A(t) = 0.010716t^2 + 0.8212t + 313.4 \qquad (1 \le t \le 50)$$

where $A(t)$ is measured in parts per million volume (ppmv) and t in years, with $t = 1$ corresponding to 1958.

a. What can you say about the rate of change of the average amount of atmospheric CO_2 from 1958 through 2007?

b. What can you say about the rate of the rate of change of the average amount of atmospheric CO_2 from 1958 through 2007?

Source: Scripps Institution of Oceanography

83. Effect of Smoking Bans The sales (in billions of dollars) in restaurants and bars in California from 1993 ($t = 0$) through 2000 ($t = 7$) are approximated by the function

$$S(t) = 0.195t^2 + 0.32t + 23.7 \qquad (0 \le t \le 7)$$

a. Show that the sales in restaurants and bars continued to rise after smoking bans were implemented in restaurants in 1995 and in bars in 1998.
Hint: Show that S is increasing on the interval (2, 7).

b. What can you say about the rate at which the sales were rising after smoking bans were implemented?

Source: California Board of Equalization

84. Digital Television Sales Since their introduction into the market in the late 1990s, the sales of digital televisions, including high-definition television sets, have slowly gathered momentum. The model

$$S(t) = 0.164t^2 + 0.85t + 0.3 \qquad (0 \le t \le 4)$$

describes the sales of digital television sets (in billions of dollars) between 1999 ($t = 0$) and 2003 ($t = 4$).

a. Find $S'(t)$ and $S''(t)$.

b. Use the results of part (a) to conclude that the sales of digital TVs were increasing between 1999 and 2003 and that the sales were increasing at an increasing rate over that time interval.

Source: Consumer Electronics Association

85. Worker Efficiency An efficiency study conducted for Elektra Electronics showed that the number of Space Commander walkie-talkies assembled by the average worker t hr after starting work at 8 a.m. is given by

$$N(t) = -t^3 + 6t^2 + 15t \qquad (0 \le t \le 4)$$

At what time during the morning shift is the average worker performing at peak efficiency?

86. Flight of a Rocket The altitude (in feet) of a rocket t sec into flight is given by

$$s = f(t) = -t^3 + 54t^2 + 480t + 6 \qquad (t \ge 0)$$

Find the point of inflection of the function f and interpret your result. What is the maximum velocity attained by the rocket?

87. Business Spending on Technology In a study conducted in 2003, business spending on technology (in billions of dollars) from 2000 through 2005 was projected to be

$$S(t) = -1.88t^3 + 30.33t^2 - 76.14t + 474 \qquad (0 \le t \le 5)$$

where t is measured in years, with $t = 0$ corresponding to 2000. Show that the graph of S is concave upward on the interval (0, 5). What does this result tell you about the rate of business spending on technology over the period in question?

Source: Quantit Economic Group

88. Alternative Minimum Tax Congress created the alternative minimum tax (AMT) in the late 1970s to ensure that wealthy people paid their fair share of taxes. But because of quirks in the law, even middle-income taxpayers have started to get hit with the tax. The AMT (in billions of dollars) projected to be collected by the IRS from 2001 through 2010 is

$$f(t) = 0.0117t^3 + 0.0037t^2 + 0.7563t + 4.1 \qquad (0 \le t \le 9)$$

where t is measured in years, with $t = 0$ corresponding to 2001.

a. Show that f is increasing on the interval (0, 9). What does this result tell you about the projected amount of AMT paid over the years in question?

b. Show that f' is increasing on the interval (0, 9). What conclusion can you draw from this result concerning the rate of growth at which the AMT is paid over the years in question?

Source: U.S. Congress Joint Economic Committee

89. Effect of Advertising on Hotel Revenue The total annual revenue R of the Miramar Resorts Hotel is related to the amount of money x the hotel spends on advertising its services by the function

$$R(x) = -0.003x^3 + 1.35x^2 + 2x + 8000 \qquad (0 \le x \le 400)$$

where both R and x are measured in thousands of dollars.

a. Find the interval where the graph of R is concave upward and the interval where the graph of R is concave downward. What is the inflection point of the graph of R?

b. Would it be more beneficial for the hotel to increase its advertising budget slightly when the budget is \$140,000 or when it is \$160,000?

90. Forecasting Profits As a result of increasing energy costs, the growth rate of the profit of the 4-yr old Venice Glassblowing Company has begun to decline. Venice's management, after consulting with energy experts, decides to implement certain energy-conservation measures aimed at cutting energy bills. The general manager reports that, according to his calculations, the growth rate of Venice's profit should be on the increase again within 4 yr. If Venice's profit (in hundreds of dollars) t yr from now is given by the function

$$P(t) = t^3 - 9t^2 + 40t + 50 \qquad (0 \le t \le 8)$$

determine whether the general manager's forecast will be accurate.

Hint: Find the inflection point of the function P and study the concavity of P.

91. Outsourcing The amount (in billions of dollars) spent by the top 15 U.S. financial institutions on IT (information technology) offshore outsourcing is projected to be

$$A(t) = 0.92(t + 1)^{0.61} \qquad (0 \le t \le 4)$$

where t is measured in years, with $t = 0$ corresponding to 2004.

a. Show that A is increasing on (0, 4) and interpret your result.

b. Show that A is concave downward on (0, 4). Interpret your result.

Source: Tower Group

92. Sales of Mobile Processors The rising popularity of notebook computers is fueling the sales of mobile PC processors. In a study conducted in 2003, the sales of these chips (in billions of dollars) was projected to be

$$S(t) = 6.8(t + 1.03)^{0.49} \qquad (0 \le t \le 4)$$

where t is measured in years, with $t = 0$ corresponding to 2003.

a. Show that S is increasing on the interval (0, 4) and interpret your result.

b. Show that the graph of S is concave downward on the interval (0, 4). Interpret your result.

Source: International Data Corp.

93. **Drug Spending** Medicaid spending on drugs in Massachusetts started slowing down in part after the state demanded that patients use more generic drugs and limited the range of drugs available to the program. The annual pharmacy spending (in millions of dollars) from 1999 through 2004 is given by

$$S(t) = -1.806t^3 + 10.238t^2 + 93.35t + 583 \qquad (0 \le t \le 5)$$

where t is measured in years with $t = 0$ corresponding to 1999. Find the inflection point of S and interpret your result.

Source: MassHealth

94. **Surveillance Cameras** Research reports indicate that surveillance cameras at major intersections dramatically reduce the number of drivers who barrel through red lights. The cameras automatically photograph vehicles that drive into intersections after the light turns red. Vehicle owners are then mailed citations instructing them to pay a fine or sign an affidavit that they weren't driving at the time. The function

$$N(t) = 6.08t^3 - 26.79t^2 + 53.06t + 69.5 \qquad (0 \le t \le 4)$$

gives the number, $N(t)$, of U.S. communities using surveillance cameras at intersections in year t, with $t = 0$ corresponding to 2003.

a. Show that N is increasing on $(0, 4)$.

b. When was the number of communities using surveillance cameras at intersections increasing least rapidly? What is the rate of increase?

Source: Insurance Institute for Highway Safety

95. **Google's Revenue** The revenue for Google from 1999 ($t = 0$) through 2003 ($t = 4$) is approximated by the function

$$R(t) = 24.975t^3 - 49.81t^2 + 41.25t + 0.2 \qquad (0 \le t \le 4)$$

where $R(t)$ is measured in millions of dollars.

a. Find $R'(t)$ and $R''(t)$.

b. Show that $R'(t) > 0$ for all t in the interval $(0, 4)$ and interpret your result.

Hint: Use the quadratic formula.

c. Find the inflection point of R and interpret your result.

Source: Company Report

96. **Population Growth in Clark County** Clark County in Nevada—dominated by greater Las Vegas—is one of the fastest-growing metropolitan areas in the United States. The population of the county from 1970 through 2000 is approximated by the function

$$P(t) = 44560t^3 - 89394t^2 + 234633t + 273288 \qquad (0 \le t \le 4)$$

where t is measured in decades, with $t = 0$ corresponding to the beginning of 1970.

a. Show that the population of Clark County was always increasing over the time period in question.

Hint: Show that $P'(t) > 0$ for all t in the interval $(0, 4)$.

b. Show that the population of Clark County was increasing at the slowest pace some time toward the middle of August 1976.

Hint: Find the inflection point of P in the interval $(0, 4)$.

Source: U.S. Census Bureau

97. **Measles Deaths** Measles is still a leading cause of vaccine-preventable death among children, but due to improvements in immunizations, measles deaths have dropped globally. The function

$$N(t) = -2.42t^3 + 24.5t^2 - 123.3t + 506 \qquad (0 \le t \le 6)$$

gives the number of measles deaths (in thousands) in sub-Saharan Africa in year t, with $t = 0$ corresponding to 1999.

a. How many measles deaths were there in 1999? In 2005?

b. Show that $N'(t) < 0$ on $(0, 6)$. What does this say about the number of measles deaths from 1999 through 2005?

c. When was the number of measles deaths decreasing most rapidly? What was the rate of measles death at that instant of time?

Source: Centers for Disease Control and World Health Organization

98. **Hiring Lobbyists** Many public entities like cities, counties, states, utilities, and Indian tribes are hiring firms to lobby Congress. One goal of such lobbying is to place earmarks—money directed at a specific project—into appropriation bills. The amount (in millions of dollars) spent by public entities on lobbying from 1998 through 2004 is given by

$$f(t) = -0.425t^3 + 3.6571t^2 + 4.018t + 43.7 \qquad (0 \le t \le 6)$$

where t is measured in years, with $t = 0$ corresponding to 1998.

a. Show that f is increasing on $(0, 6)$. What does this say about the spending by public entities on lobbying over the years in question?

b. Find the inflection point of f. What does your result tell you about the growth of spending by the public entities on lobbying?

Source: Center for Public Integrity

99. **Air Pollution** The level of ozone, an invisible gas that irritates and impairs breathing, present in the atmosphere on a certain May day in the city of Riverside was approximated by

$$A(t) = 1.0974t^3 - 0.0915t^4 \qquad (0 \le t \le 11)$$

where $A(t)$ is measured in pollutant standard index (PSI) and t is measured in hours, with $t = 0$ corresponding to 7 a.m. Use the second derivative test to show that the function A has a relative maximum at approximately $t = 9$. Interpret your results.

100. **Cash Reserves at Blue Cross and Blue Shield** Based on company financial reports, the cash reserves of Blue Cross and Blue Shield as of the beginning of year t is approximated by the function

$$R(t) = -1.5t^4 + 14t^3 - 25.4t^2 + 64t + 290 \qquad (0 \le t \le 6)$$

where $R(t)$ is measured in millions of dollars and t is measured in years, with $t = 0$ corresponding to the beginning of 1998.

a. Find the inflection points of R.
Hint: Use the quadratic formula.

b. Use the result of part (a) to show that the cash reserves of the company was growing at the greatest rate at the beginning of 2002.

Source: Blue Cross and Blue Shield

101. Women's Soccer Starting with the youth movement that took hold in the 1970s and buoyed by the success of the U.S. national women's team in international competition in recent years, girls and women have taken to soccer in ever-growing numbers. The function

$$N(t) = -0.9307t^3 + 74.04t^2 + 46.8667t + 3967 \qquad (0 \le t \le 16)$$

gives the number of participants in women's soccer in year t, with $t = 0$ corresponding to the beginning of 1985.

a. Verify that the number of participants in women's soccer had been increasing from 1985 through 2000.
Hint: Use the quadratic formula.

b. Show that the number of participants in women's soccer had been increasing at an increasing rate from 1985 through 2000.
Hint: Show that the sign of N'' is positive on the interval in question.

Source: NCCA News

102. Dependency Ratio The share of the world population that is over 60 years of age compared to the rest of the working population in the world is of concern to economists. An increasing dependency ratio means that there will be fewer workers to support an aging population. The dependency ratio over the next century is forecast to be

$$R(t) = 0.00731t^4 - 0.174t^3 + 1.528t^2 + 0.48t + 19.3 \qquad (0 \le t \le 10)$$

in year t, where t is measured in decades with $t = 0$ corresponding to 2000.

a. Show that the dependency ratio will be increasing at the fastest pace around 2052.
Hint: Use the quadratic formula.

b. What will the dependency ratio be at that time?

Source: International Institute for Applied Systems Analysis

In Exercises 103–106, determine whether the statement is true or false. If it is true, explain why it is true. If it is false, give an example to show why it is false.

103. If the graph of f is concave upward on (a, b), then the graph of $-f$ is concave downward on (a, b).

104. If the graph of f is concave upward on (a, c) and concave downward on (c, b), where $a < c < b$, then f has an inflection point at $(c, f(c))$.

105. If c is a critical number of f where $a < c < b$ and $f''(x) < 0$ on (a, b), then f has a relative maximum at $x = c$.

106. A polynomial function of degree n $(n \ge 3)$ can have at most $(n - 2)$ inflection points.

107. Show that the quadratic function

$$f(x) = ax^2 + bx + c \qquad (a \ne 0)$$

is concave upward if $a > 0$ and concave downward if $a < 0$. Thus, by examining the sign of the coefficient of x^2, one can tell immediately whether the parabola opens upward or downward.

108. Consider the functions $f(x) = x^3$, $g(x) = x^4$, and $h(x) = -x^4$.

a. Show that $x = 0$ is a critical number of each of the functions f, g, and h.

b. Show that the second derivative of each of the functions f, g, and h equals zero at $x = 0$.

c. Show that f has neither a relative maximum nor a relative minimum at $x = 0$, that g has a relative minimum at $x = 0$, and that h has a relative maximum at $x = 0$.

4.2 Solutions to Self-Check Exercises

1. We first compute

$$f'(x) = 12x^2 - 6x$$
$$f''(x) = 24x - 6 = 6(4x - 1)$$

Observe that f'' is continuous everywhere and has a zero at $x = \frac{1}{4}$. The sign diagram of f'' is shown in the accompanying figure.

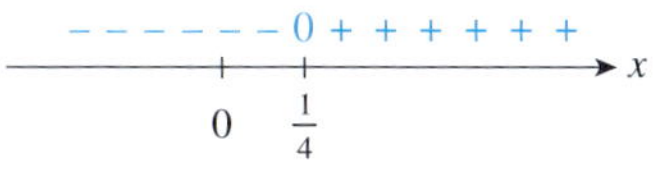

From the sign diagram for f'', we see that f is concave upward on $(\frac{1}{4}, \infty)$ and concave downward on $(-\infty, \frac{1}{4})$.

2. First, we find the critical numbers of f by solving the equation

$$f'(x) = 6x^2 - x - 12 = 0$$

That is,

$$(3x + 4)(2x - 3) = 0$$

giving $x = -\frac{4}{3}$ and $x = \frac{3}{2}$. Next, we compute

$$f''(x) = 12x - 1$$

Since

$$f''\left(-\frac{4}{3}\right) = 12\left(-\frac{4}{3}\right) - 1 = -17 < 0$$

the second derivative test implies that $f(-\frac{4}{3}) = \frac{10}{27}$ is a relative maximum of f. Also,

$$f''\left(\frac{3}{2}\right) = 12\left(\frac{3}{2}\right) - 1 = 17 > 0$$

and we see that $f(\frac{3}{2}) = -\frac{179}{8}$ is a relative minimum.

3. We compute the second derivative of G. Thus,

$$G'(t) = -6t^2 + 90t + 20$$
$$G''(t) = -12t + 90$$

Now, G'' is continuous everywhere, and $G''(t) = 0$, when $t = \frac{15}{2}$, giving $t = \frac{15}{2}$ as the only candidate for an inflection point of G. Since $G''(t) > 0$ for $t < \frac{15}{2}$ and $G''(t) < 0$ for $t > \frac{15}{2}$, we see that $(\frac{15}{2}, \frac{15,675}{2})$ is an inflection point of G. The results of our computations tell us that the country's GDP was increasing most rapidly at the beginning of July 2002.

USING TECHNOLOGY

Finding the Inflection Points of a Function

A graphing utility can be used to find the inflection points of a function and hence the intervals where the graph of the function is concave upward and the intervals where it is concave downward. Some graphing utilities have an operation for finding inflection points directly. For example, both the TI-85 and TI-86 graphing calculators have this capability. If your graphing utility has this capability, use it to work through the example and exercises in this section.

EXAMPLE 1 Let $f(x) = 2.5x^5 - 12.4x^3 + 4.2x^2 - 5.2x + 4$.

a. Use a graphing utility to plot the graph of f.
b. Find the inflection points of f.
c. Find the intervals where f is concave upward and where it is concave downward.

Solution

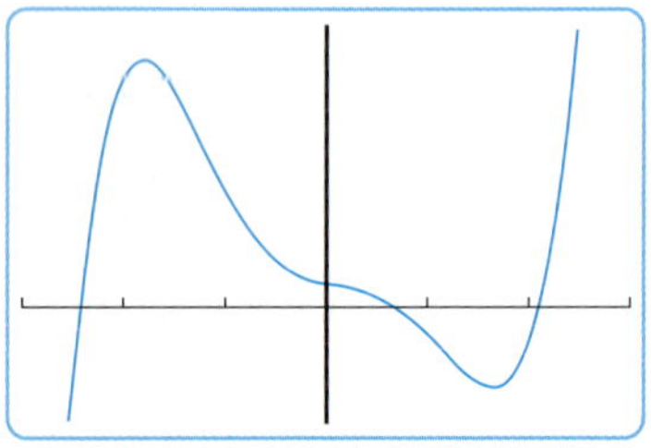

FIGURE T1
The graph of f in the viewing window $[-3, 3] \times [-25, 60]$

a. The graph of f, using the viewing window $[-3, 3] \times [-25, 60]$, is shown in Figure T1.

b. We describe here the procedure using the TI-85. See the Web site for instructions for using the TI-86. From Figure T1 we see that f has three inflection points—one occurring at the point where the x-coordinate is approximately -1, another at the point where $x \approx 0$, and the third at the point where $x \approx 1$. To find the first inflection point, we use the inflection operation, moving the cursor to the point on the graph of f where $x \approx -1$. We obtain the point $(-1.2728, 34.6395)$ (accurate to four decimal places). Next, setting the cursor near $x = 0$ yields the inflection point $(0.1139, 3.4440)$. Finally, with the cursor set at $x = 1$, we obtain the third inflection point $(1.1589, -10.4594)$. (See Figure T2a–c.)

c. From the results of part (b), we see that f is concave upward on the intervals $(-1.2728, 0.1139)$ and $(1.1589, \infty)$ and concave downward on $(-\infty, -1.2728)$ and $(0.1139, 1.1589)$.

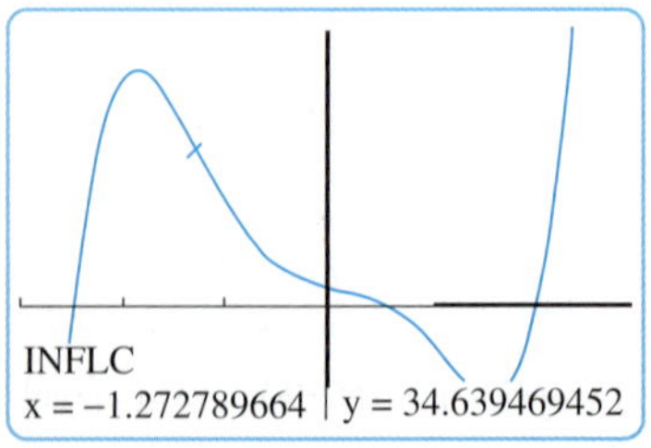

(a)

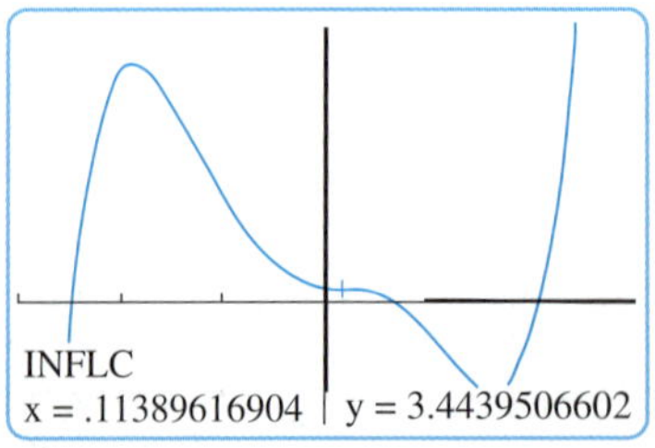

(b)

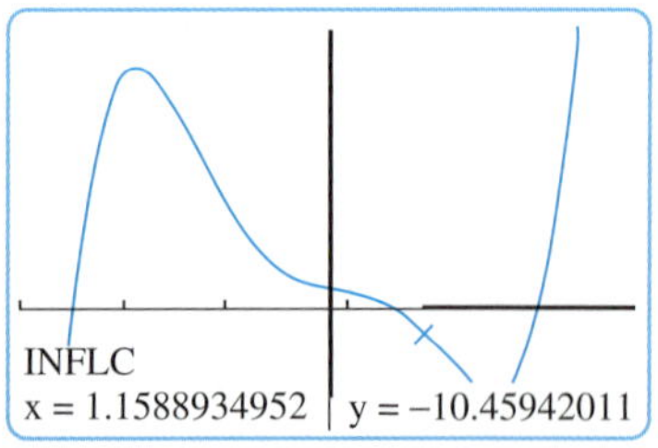

(c)

FIGURE T2
The TI-85 inflection point screens showing the points (a) $(-1.2728, 34.6395)$, (b) $(0.1139, 3.4440)$, and (c) $(1.1589, -10.4594)$

TECHNOLOGY EXERCISES

In Exercises 1–8, find (a) the intervals where f is concave upward and the intervals where f is concave downward and (b) the inflection points of f. Express your answers accurate to four decimal places.

1. $f(x) = 1.8x^4 - 4.2x^3 + 2.1x + 2$
2. $f(x) = -2.1x^4 + 3.1x^3 + 2x^2 - x + 1.2$
3. $f(x) = 1.2x^5 - 2x^4 + 3.2x^3 - 4x + 2$
4. $f(x) = -2.1x^5 + 3.2x^3 - 2.2x^2 + 4.2x - 4$
5. $f(x) = x^3(x^2 + 1)^{-1/3}$
6. $f(x) = x^2(x^3 - 1)^3$
7. $f(x) = \dfrac{x^2 - 1}{x^3}$
8. $f(x) = \dfrac{x + 1}{\sqrt{x}}$
9. **Time on the Market** The average number of days a single-family home remains for sale from listing to accepted offer (in the greater Boston area) is approximated by the function

$$f(t) = 0.0171911t^4 - 0.662121t^3 + 6.18083t^2 - 8.97086t + 53.3357 \qquad (0 \le t \le 10)$$

where t is measured in years, with $t = 0$ corresponding to the beginning of 1984.
   a. Plot the graph of f in the viewing window $[0, 12] \times [0, 120]$.
   b. Find the points of inflection and interpret your result.

   Source: Greater Boston Real Estate Board—Multiple Listing Service

10. **Multimedia Sales** Sales in the multimedia market (hardware and software) are approximated by the function

$$S(t) = -0.0094t^4 + 0.1204t^3 - 0.0868t^2 + 0.0195t + 3.3325 \qquad (0 \le t \le 10)$$

where $S(t)$ is measured in billions of dollars and t is measured in years, with $t = 0$ corresponding to 1990.
   a. Plot the graph of S in the viewing window $[0, 12] \times [0, 25]$.
   b. Find the inflection point of S and interpret your result.

   Source: Electronic Industries Association

11. **Surgeries in Physicians' Offices** Driven by technological advances and financial pressures, the number of surgeries performed in physicians' offices nationwide has been increasing over the years. The function

$$f(t) = -0.00447t^3 + 0.09864t^2 + 0.05192t + 0.8 \qquad (0 \le t \le 15)$$

gives the number of surgeries (in millions) performed in physicians' offices in year t, with $t = 0$ corresponding to the beginning of 1986.
   a. Plot the graph of f in the viewing window $[0, 15] \times [0, 10]$.
   b. At what time in the period under consideration is the number of surgeries performed in physicians' offices increasing at the fastest rate?

   Source: SMG Marketing Group

12. **Modeling with Data** The following data gives the number of computer-security incidents (in thousands), including computer viruses and intrusions, in which the same tool is used by an intruder, from 1999 through 2003.

Year	1999	2000	2001	2002	2003
Number of incidents	10	21	53	83	137

   a. Use **QuartReg** to find a fourth-degree polynomial regression model for the data. Let $t = 0$ correspond to 1999.
   b. Show that the number $N(t)$ of computer-security incidents was always increasing between 2000 and 2003.
   Hint: Plot the graph of N' and show that it always lies above the t-axis for $1 \le t \le 4$.
   c. Show that between 2000 and 2001, the number of computer-security incidents was increasing at the fastest rate in the middle of 2000 and that between 2001 and 2003 the number of incidents was increasing at the slowest rate in the middle of 2001.
   Hint: Study the nature of the inflection points of N.

   Source: CERT Coordination Center

4.3 Curve Sketching

A Real-Life Example

As we have seen on numerous occasions, the graph of a function is a useful aid for visualizing the function's properties. From a practical point of view, the graph of a function also gives, at one glance, a complete summary of all the information captured by the function.

Consider, for example, the graph of the function giving the Dow-Jones Industrial Average (DJIA) on Black Monday, October 19, 1987 (Figure 45). Here, $t = 0$ corresponds to 9:30 a.m., when the market was open for business, and $t = 6.5$ corresponds to 4 p.m., the closing time. The following information may be gleaned from studying the graph.

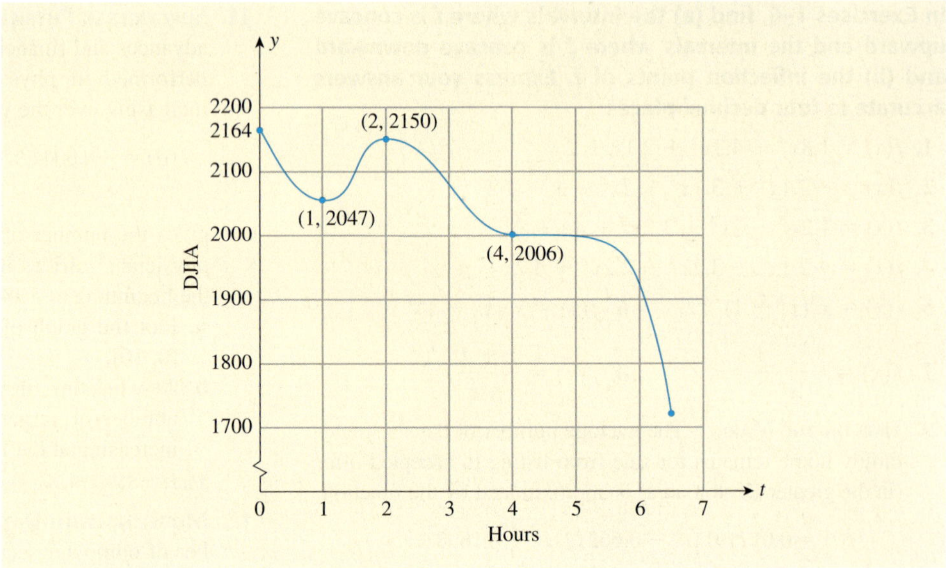

FIGURE 45
The Dow-Jones Industrial Average on Black Monday.

Source: Wall Street Journal

The graph is *decreasing* rapidly from $t = 0$ to $t = 1$, reflecting the sharp drop in the index in the first hour of trading. The point (1, 2047) is a *relative minimum* point of the function, and this turning point coincides with the start of an aborted recovery. The short-lived rally, represented by the portion of the graph that is *increasing* on the interval (1, 2), quickly fizzled out at $t = 2$ (11:30 a.m.). The *relative maximum* point (2, 2150) marks the highest point of the recovery. The function is decreasing in the rest of the interval. The point (4, 2006) is an *inflection point* of the function; it shows that there was a temporary respite at $t = 4$ (1:30 p.m.). However, selling pressure continued unabated, and the DJIA continued to fall until the closing bell. Finally, the graph also shows that the index opened at the high of the day [$f(0) = 2164$ is the *absolute maximum* of the function] and closed at the low of the day [$f(\frac{13}{2}) = 1739$ is the *absolute minimum* of the function], a drop of 508 points from the previous close!*

Before we turn our attention to the actual task of sketching the graph of a function, let's look at some properties of graphs that will be helpful in this connection.

Vertical Asymptotes

Before going on, you might want to review the material on one-sided limits and the limit at infinity of a function (Sections 2.4 and 2.5).

Consider the graph of the function

$$f(x) = \frac{x + 1}{x - 1}$$

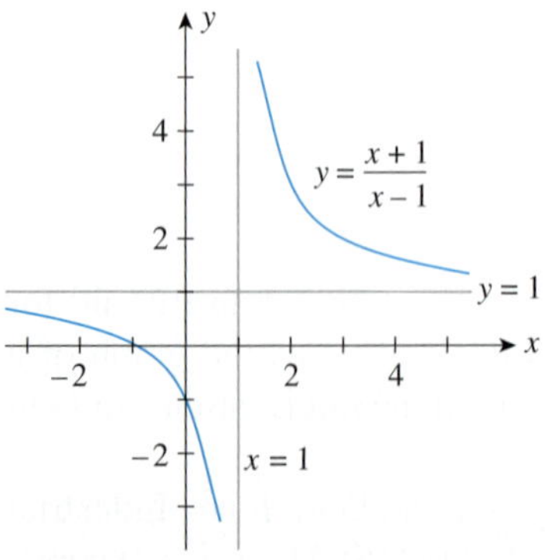

FIGURE 46
The graph of f has a vertical asymptote at $x = 1$.

shown in Figure 46. Observe that $f(x)$ increases without bound (tends to infinity) as x approaches $x = 1$ from the right; that is,

$$\lim_{x \to 1^+} \frac{x + 1}{x - 1} = \infty$$

You can verify this by taking a sequence of values of x approaching $x = 1$ from the right and looking at the corresponding values of $f(x)$.

*Absolute maxima and absolute minima of functions are covered in Section 4.4.

Here is another way of looking at the situation: Observe that if x is a number that is a little larger than 1, then both $(x + 1)$ and $(x - 1)$ are positive, so $(x + 1)/(x - 1)$ is also positive. As x approaches $x = 1$, the numerator $(x + 1)$ approaches the number 2, but the denominator $(x - 1)$ approaches zero, so the quotient $(x + 1)/(x - 1)$ approaches infinity, as observed earlier. The line $x = 1$ is called a vertical asymptote of the graph of f.

For the function $f(x) = (x + 1)/(x - 1)$, you can show that

$$\lim_{x\to 1^-} \frac{x+1}{x-1} = -\infty$$

and this tells us how $f(x)$ approaches the asymptote $x = 1$ from the left.

More generally, we have the following definition:

Vertical Asymptote

The line $x = a$ is a **vertical asymptote** of the graph of a function f if either

$$\lim_{x\to a^+} f(x) = \infty \quad \text{or} \quad -\infty$$

or

$$\lim_{x\to a^-} f(x) = \infty \quad \text{or} \quad -\infty$$

Note Although a vertical asymptote of a graph is not part of the graph, it serves as a useful aid for sketching the graph. ■

For rational functions

$$f(x) = \frac{P(x)}{Q(x)}$$

there is a simple criterion for determining whether the graph of f has any vertical asymptotes.

Finding Vertical Asymptotes of Rational Functions

Suppose f is a rational function

$$f(x) = \frac{P(x)}{Q(x)}$$

where P and Q are polynomial functions. Then, the line $x = a$ is a vertical asymptote of the graph of f if $Q(a) = 0$ but $P(a) \neq 0$.

For the function

$$f(x) = \frac{x+1}{x-1}$$

considered earlier, $P(x) = x + 1$ and $Q(x) = x - 1$. Observe that $Q(1) = 0$ but $P(1) = 2 \neq 0$, so $x = 1$ is a vertical asymptote of the graph of f.

EXAMPLE 1 Find the vertical asymptotes of the graph of the function

$$f(x) = \frac{x^2}{4 - x^2}$$

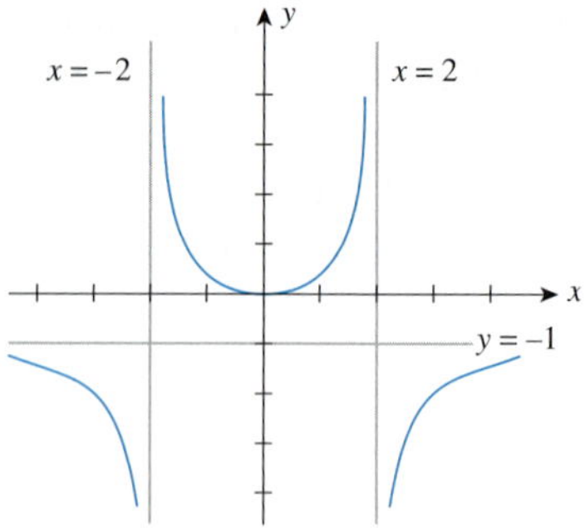

FIGURE 47
$x = -2$ and $x = 2$ are vertical asymptotes of the graph of f.

Solution The function f is a rational function with $P(x) = x^2$ and $Q(x) = 4 - x^2$. The zeros of Q are found by solving

$$4 - x^2 = 0$$

—that is,

$$(2 - x)(2 + x) = 0$$

giving $x = -2$ and $x = 2$. These are candidates for the vertical asymptotes of the graph of f. Examining $x = -2$, we compute $P(-2) = (-2)^2 = 4 \neq 0$, and we see that $x = -2$ is indeed a vertical asymptote of the graph of f. Similarly, we find $P(2) = 2^2 = 4 \neq 0$, and so $x = 2$ is also a vertical asymptote of the graph of f. The graph of f sketched in Figure 47 confirms these results. ■

Recall that in order for the line $x = a$ to be a vertical asymptote of the graph of a rational function f, *only* the denominator of $f(x)$ must be equal to zero at $x = a$. If *both* $P(a)$ and $Q(a)$ are equal to zero, then $x = a$ need *not* be a vertical asymptote. For example, look at the function

$$f(x) = \frac{4(x^2 - 4)}{x - 2}$$

whose graph appears in Figure 32a, page 104.

Horizontal Asymptotes

Let's return to the function f defined by

$$f(x) = \frac{x + 1}{x - 1}$$

(Figure 48).

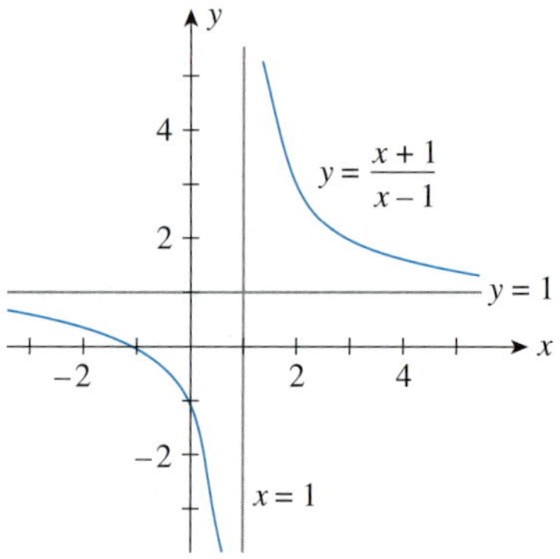

FIGURE 48
The graph of f has a horizontal asymptote at $y = 1$.

Observe that $f(x)$ approaches the horizontal line $y = 1$ as x approaches infinity, and, in this case, $f(x)$ approaches $y = 1$ as x approaches minus infinity as well. The line $y = 1$ is called a horizontal asymptote of the graph of f. More generally, we have the following definition:

Horizontal Asymptote

The line $y = b$ is a **horizontal asymptote** of the graph of a function f if either

$$\lim_{x \to \infty} f(x) = b \quad \text{or} \quad \lim_{x \to -\infty} f(x) = b$$

For the function

$$f(x) = \frac{x+1}{x-1}$$

we see that

$$\lim_{x\to\infty} \frac{x+1}{x-1} = \lim_{x\to\infty} \frac{1+\frac{1}{x}}{1-\frac{1}{x}} \quad \text{Divide numerator and denominator by } x.$$

$$= 1$$

Also,

$$\lim_{x\to-\infty} \frac{x+1}{x-1} = \lim_{x\to-\infty} \frac{1+\frac{1}{x}}{1-\frac{1}{x}}$$

$$= 1$$

In either case, we conclude that $y = 1$ is a horizontal asymptote of the graph of f, as observed earlier.

EXAMPLE 2 Find the horizontal asymptotes of the graph of the function

$$f(x) = \frac{x^2}{4 - x^2}$$

Solution We compute

$$\lim_{x\to\infty} \frac{x^2}{4-x^2} = \lim_{x\to\infty} \frac{1}{\frac{4}{x^2} - 1} \quad \text{Divide numerator and denominator by } x^2.$$

$$= -1$$

and so $y = -1$ is a horizontal asymptote, as before. (Similarly, $\lim_{x\to-\infty} f(x) = -1$, as well.) The graph of f sketched in Figure 49 confirms this result.

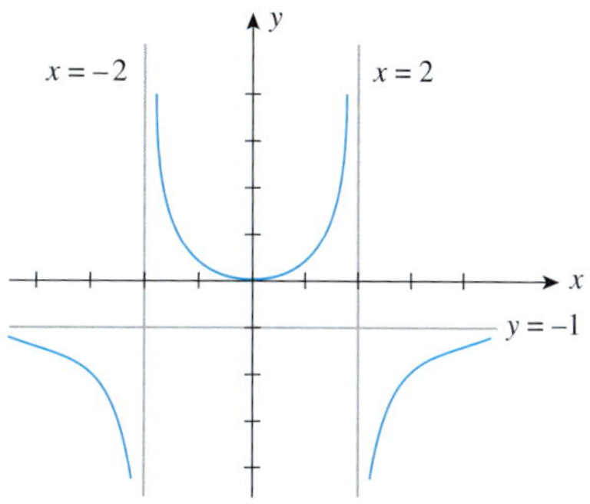

FIGURE 49
The graph of f has a horizontal asymptote at $y = -1$.

We next state an important property of polynomial functions.

A polynomial function has no vertical or horizontal asymptotes.

To see this, note that a polynomial function $P(x)$ can be written as a rational function with denominator equal to 1. Thus,

$$P(x) = \frac{P(x)}{1}$$

Since the denominator is never equal to zero, P has no vertical asymptotes. Next, if P is a polynomial of degree greater than or equal to 1, then

$$\lim_{x\to\infty} P(x) \quad \text{and} \quad \lim_{x\to-\infty} P(x)$$

are either infinity or minus infinity; that is, they do not exist. Therefore, P has no horizontal asymptotes.

In the last two sections, we saw how the first and second derivatives of a function are used to reveal various properties of the graph of a function f. We now show how this information can be used to help us sketch the graph of f. We begin by giving a general procedure for curve sketching.

A Guide to Curve Sketching

1. Determine the domain of f.
2. Find the x- and y-intercepts of f.*
3. Determine the behavior of f for large absolute values of x.
4. Find all horizontal and vertical asymptotes of f.
5. Determine the intervals where f is increasing and where f is decreasing.
6. Find the relative extrema of f.
7. Determine the concavity of f.
8. Find the inflection points of f.
9. Plot a few additional points to help further identify the shape of the graph of f and sketch the graph.

*The equation $f(x) = 0$ may be difficult to solve, in which case one may decide against finding the x-intercepts or to use technology, if available, for assistance.

We now illustrate the techniques of curve sketching in the next two examples.

Two Step-by-Step Examples

VIDEO

EXAMPLE 3 Sketch the graph of the function

$$y = f(x) = x^3 - 6x^2 + 9x + 2$$

Solution Obtain the following information on the graph of f.

1. The domain of f is the interval $(-\infty, \infty)$.
2. By setting $x = 0$, we find that the y-intercept is 2. The x-intercept is found by setting $y = 0$, which in this case leads to a cubic equation. Since the solution is not readily found, we will not use this information.
3. Since

$$\lim_{x\to-\infty} f(x) = \lim_{x\to-\infty} (x^3 - 6x^2 + 9x + 2) = -\infty$$

$$\lim_{x\to\infty} f(x) = \lim_{x\to\infty} (x^3 - 6x^2 + 9x + 2) = \infty$$

we see that f decreases without bound as x decreases without bound and that f increases without bound as x increases without bound.

4. Since f is a polynomial function, there are no asymptotes.
5.
$$f'(x) = 3x^2 - 12x + 9 = 3(x^2 - 4x + 3) = 3(x - 3)(x - 1)$$

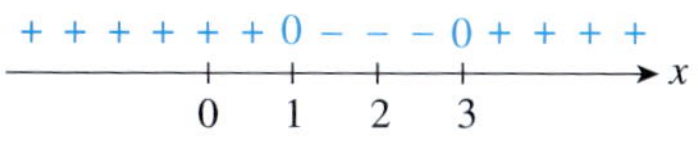

FIGURE 50
Sign diagram for f'

Setting $f'(x) = 0$ gives $x = 1$ or $x = 3$. The sign diagram for f' shows that f is increasing on the intervals $(-\infty, 1)$ and $(3, \infty)$ and decreasing on the interval $(1, 3)$ (Figure 50).

6. From the results of step 5, we see that $x = 1$ and $x = 3$ are critical numbers of f. Furthermore, f' changes sign from positive to negative as we move across $x = 1$, so a relative maximum of f occurs at $x = 1$. Similarly, we see that a relative minimum of f occurs at $x = 3$. Now,

$$f(1) = 1 - 6 + 9 + 2 = 6$$
$$f(3) = 3^3 - 6(3)^2 + 9(3) + 2 = 2$$

so $f(1) = 6$ is a relative maximum of f and $f(3) = 2$ is a relative minimum of f.

7. $$f''(x) = 6x - 12 = 6(x - 2)$$

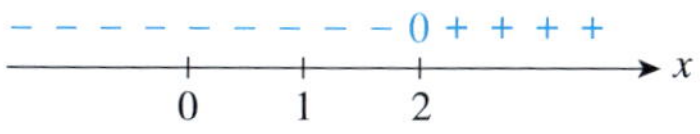

FIGURE 51
Sign diagram for f''

which is equal to zero when $x = 2$. The sign diagram of f'' shows that f is concave downward on the interval $(-\infty, 2)$ and concave upward on the interval $(2, \infty)$ (Figure 51).

8. From the results of step 7, we see that f'' changes sign as we move across $x = 2$. Next,

$$f(2) = 2^3 - 6(2)^2 + 9(2) + 2 = 4$$

and so the required inflection point of f is $(2, 4)$.

Summarizing, we have the following:

Domain: $(-\infty, \infty)$
Intercept: $(0, 2)$
$\lim_{x \to -\infty} f(x)$; $\lim_{x \to \infty} f(x)$: $-\infty$; ∞
Asymptotes: None
Intervals where f is ↗ or ↘: ↗ on $(-\infty, 1) \cup (3, \infty)$; ↘ on $(1, 3)$
Relative extrema: Relative maximum at $(1, 6)$; relative minimum at $(3, 2)$
Concavity: Downward on $(-\infty, 2)$; upward on $(2, \infty)$
Point of inflection: $(2, 4)$

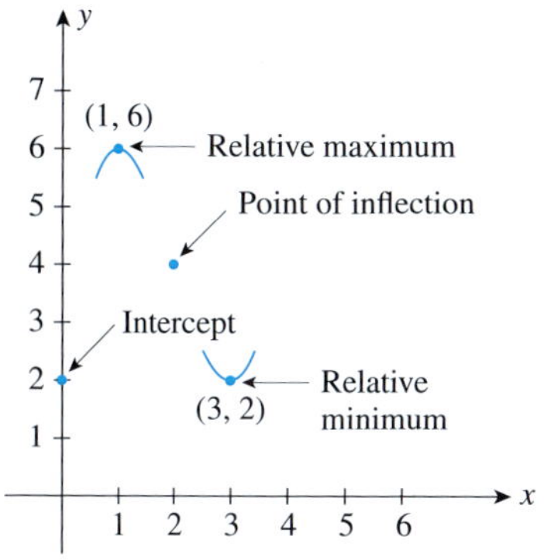

FIGURE 52
We first plot the intercept, the relative extrema, and the inflection point.

In general, it is a good idea to start graphing by plotting the intercept(s), relative extrema, and inflection point(s) (Figure 52). Then, using the rest of the information, we complete the graph of f, as sketched in Figure 53. ■

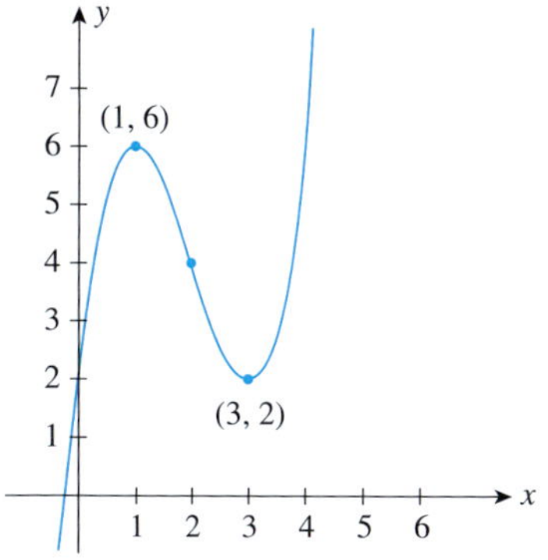

FIGURE 53
The graph of $y = x^3 - 6x^2 + 9x + 2$

Explore & Discuss

The average price of gasoline at the pump over a 3-month period, during which there was a temporary shortage of oil, is described by the function f defined on the interval $[0, 3]$. During the first month, the price was increasing at an increasing rate. Starting with the second month, the good news was that the rate of increase was slowing down, although the price of gas was still increasing. This pattern continued until the end of the second month. The price of gas peaked at the end of $t = 2$ and began to fall at an increasing rate until $t = 3$.

1. Describe the signs of $f'(t)$ and $f''(t)$ over each of the intervals $(0, 1)$, $(1, 2)$, and $(2, 3)$.
2. Make a sketch showing a plausible graph of f over $[0, 3]$.

EXAMPLE 4 Sketch the graph of the function

$$y = f(x) = \frac{x + 1}{x - 1}$$

Solution Obtain the following information:

1. f is undefined when $x = 1$, so the domain of f is the set of all real numbers other than $x = 1$.
2. Setting $y = 0$ gives -1, the x-intercept of f. Next, setting $x = 0$ gives -1 as the y-intercept of f.
3. Earlier we found that

$$\lim_{x \to \infty} \frac{x + 1}{x - 1} = 1 \quad \text{and} \quad \lim_{x \to -\infty} \frac{x + 1}{x - 1} = 1$$

(see pp. 286–287). Consequently, we see that $f(x)$ approaches the line $y = 1$ as $|x|$ becomes arbitrarily large. For $x > 1, f(x) > 1$ and $f(x)$ approaches the line $y = 1$ from above. For $x < 1, f(x) < 1$, so $f(x)$ approaches the line $y = 1$ from below.

4. The straight line $x = 1$ is a vertical asymptote of the graph of f. Also, from the results of step 3, we conclude that $y = 1$ is a horizontal asymptote of the graph of f.
5. $$f'(x) = \frac{(x - 1)(1) - (x + 1)(1)}{(x - 1)^2} = -\frac{2}{(x - 1)^2}$$

and is discontinuous at $x = 1$. The sign diagram of f' shows that $f'(x) < 0$ whenever it is defined. Thus, f is decreasing on the intervals $(-\infty, 1)$ and $(1, \infty)$ (Figure 54).

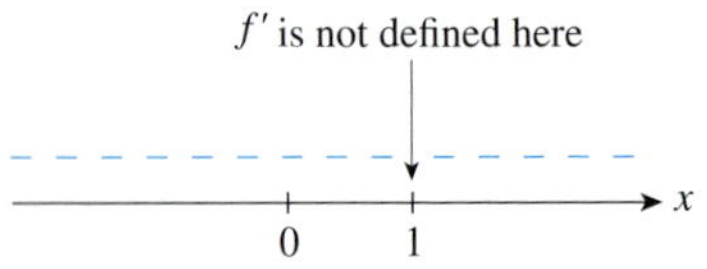

FIGURE 54
The sign diagram for f'

6. From the results of step 5, we see that there are no critical numbers of f since $f'(x)$ is never equal to zero for any value of x in the domain of f.
7. $$f''(x) = \frac{d}{dx}[-2(x - 1)^{-2}] = 4(x - 1)^{-3} = \frac{4}{(x - 1)^3}$$

The sign diagram of f'' shows immediately that f is concave downward on the interval $(-\infty, 1)$ and concave upward on the interval $(1, \infty)$ (Figure 55).

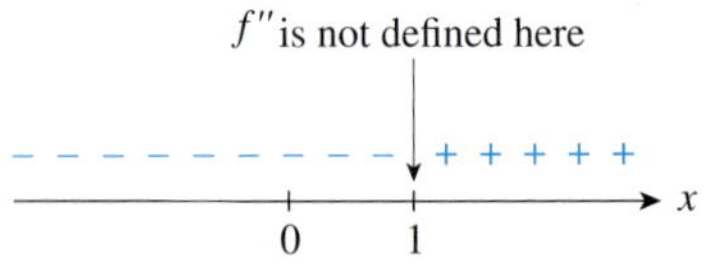

FIGURE 55
The sign diagram for f''

8. From the results of step 7, we see that there are no candidates for inflection points of f since $f''(x)$ is never equal to zero for any value of x in the domain of f. Hence, f has no inflection points.

Summarizing, we have the following:

Domain: $(-\infty, 1) \cup (1, \infty)$
Intercepts: $(0, -1)$; $(-1, 0)$
$\lim_{x \to -\infty} f(x)$; $\lim_{x \to \infty} f(x)$: 1; 1
Asymptotes: $x = 1$ is a vertical asymptote
$y = 1$ is a horizontal asymptote
Intervals where f is ↗ or ↘: ↘ on $(-\infty, 1) \cup (1, \infty)$
Relative extrema: None
Concavity: Downward on $(-\infty, 1)$; upward on $(1, \infty)$
Points of inflection: None

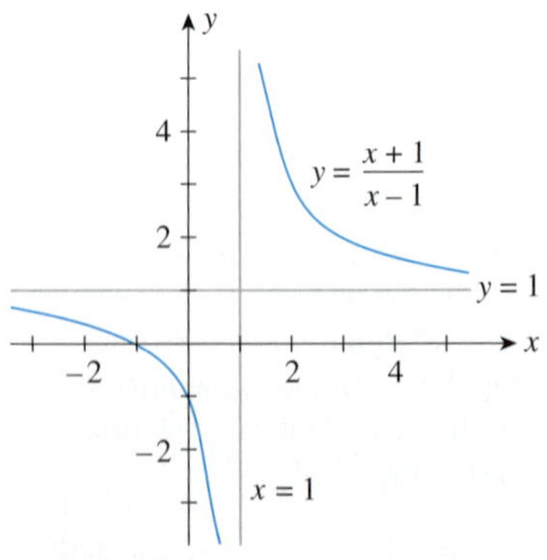

FIGURE 56
The graph of f has a horizontal asymptote at $y = 1$ and a vertical asymptote at $x = 1$.

The graph of f is sketched in Figure 56. ■

4.3 Self-Check Exercises

1. Find the horizontal and vertical asymptotes of the graph of the function

$$f(x) = \frac{2x^2}{x^2 - 1}$$

2. Sketch the graph of the function

$$f(x) = \frac{2}{3}x^3 - 2x^2 - 6x + 4$$

Solutions to Self-Check Exercises 4.3 can be found on page 294.

4.3 Concept Questions

1. Explain the following terms in your own words:
   a. Vertical asymptote **b.** Horizontal asymptote
2. **a.** How many vertical asymptotes can the graph of a function f have? Explain using graphs.
   b. How many horizontal asymptotes can the graph of a function f have? Explain using graphs.
3. How do you find the vertical asymptotes of a rational function?
4. Give a procedure for sketching the graph of a function.

4.3 Exercises

In Exercises 1–10, find the horizontal and vertical asymptotes of the graph.

1.

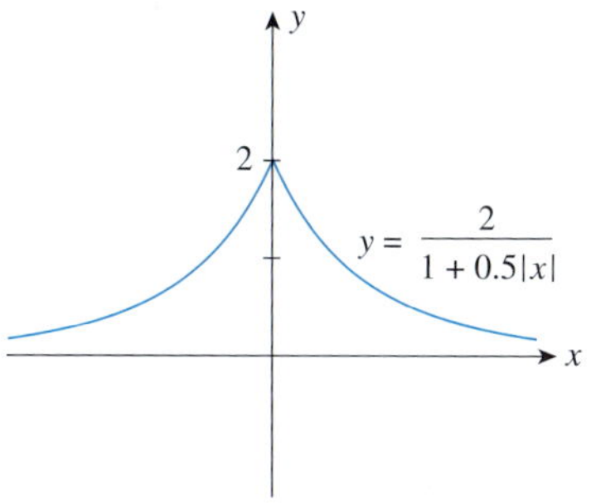

2.

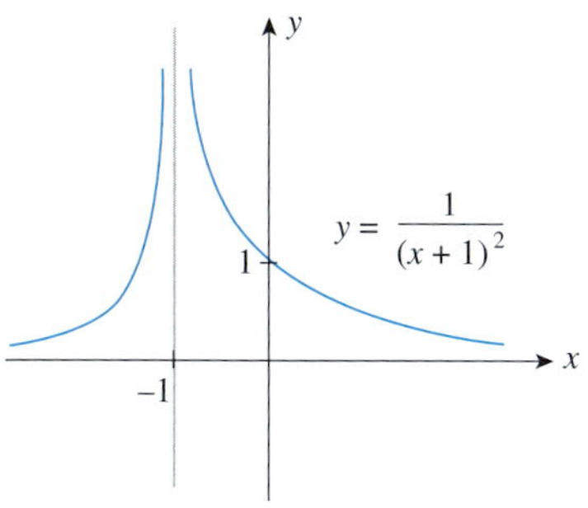

3.

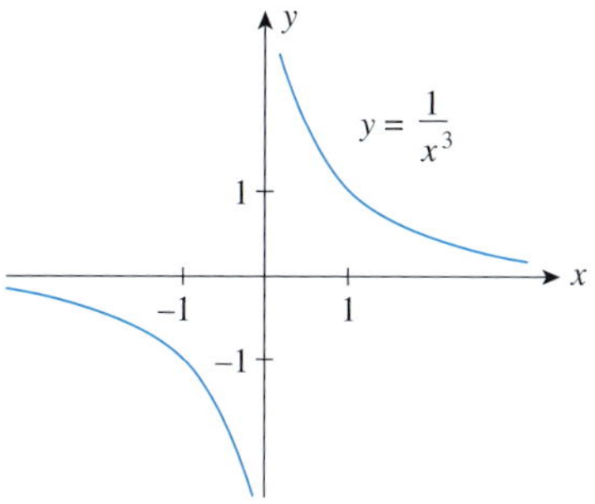

4.

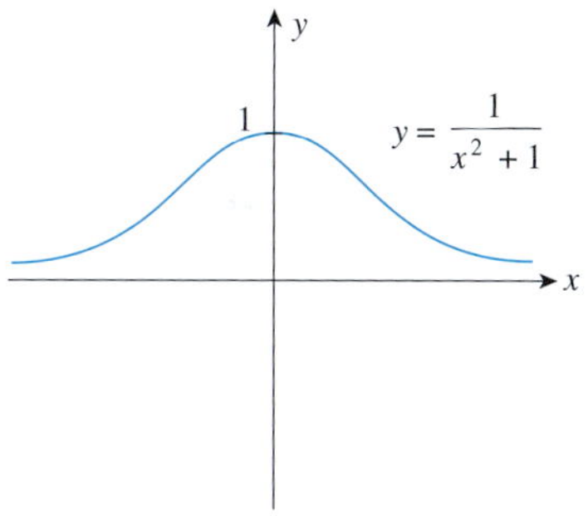

5.

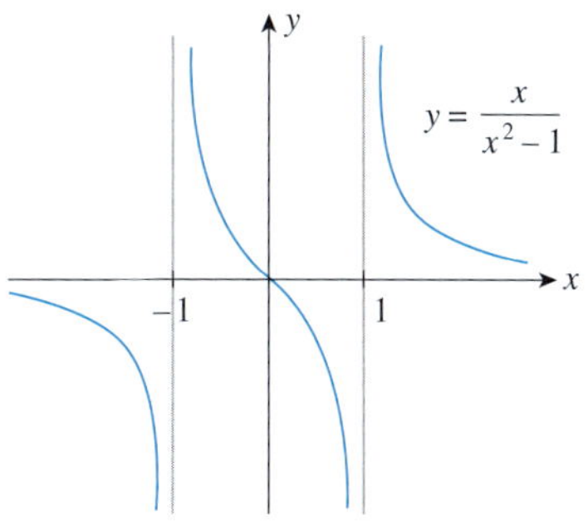

6.

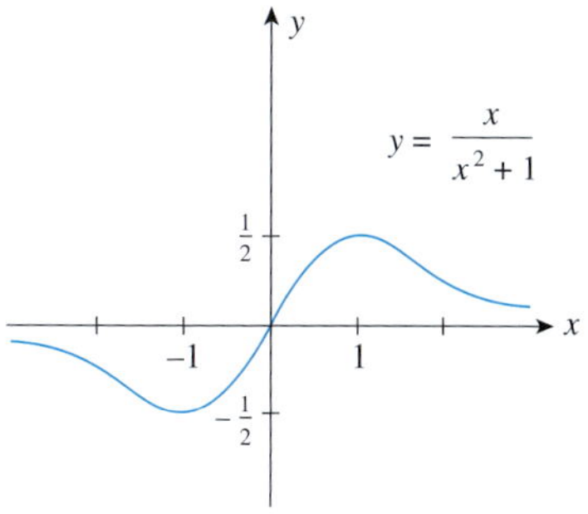

7.

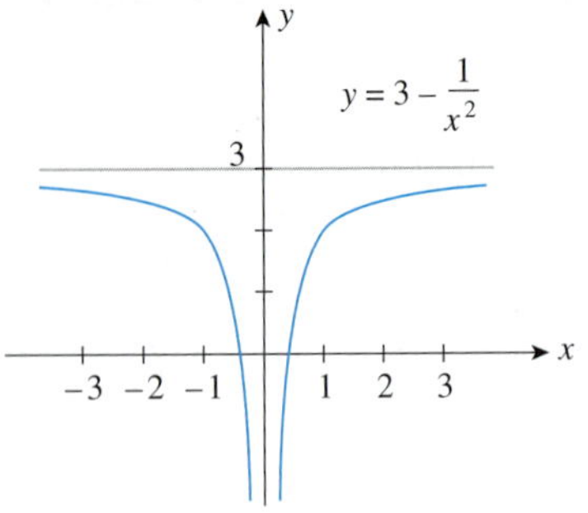

8.

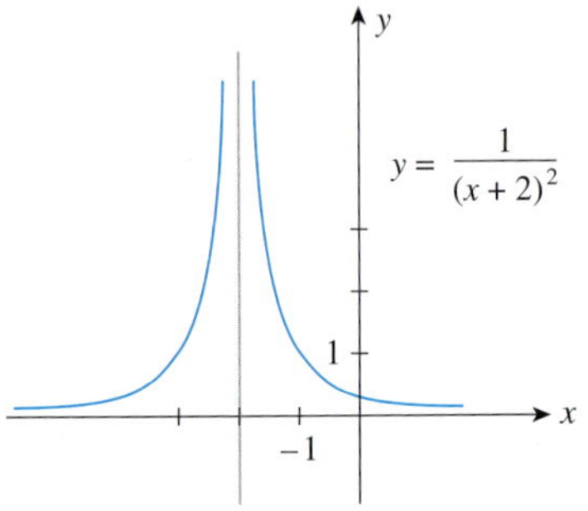

9.

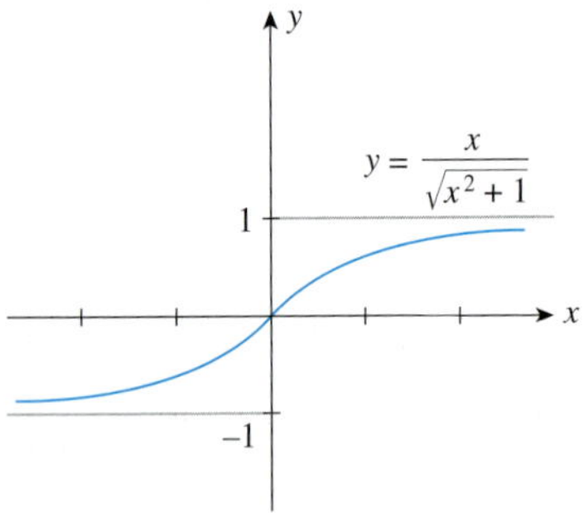

10.

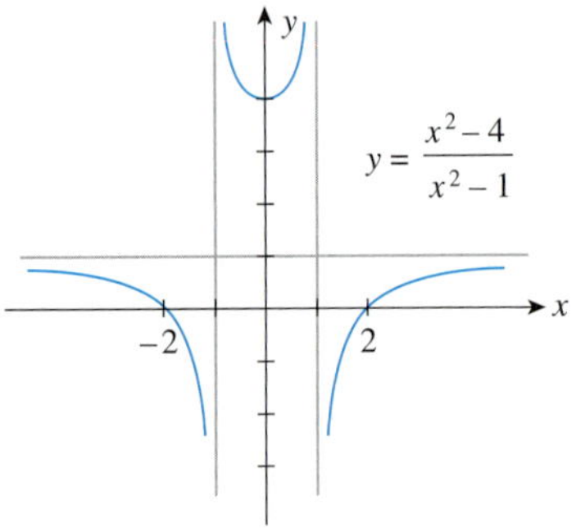

In Exercises 11–28, find the horizontal and vertical asymptotes of the graph of the function. (You need not sketch the graph.)

11. $f(x) = \frac{1}{x}$

12. $f(x) = \frac{1}{x+2}$

13. $f(x) = -\frac{2}{x^2}$

14. $g(x) = \frac{1}{1 + 2x^2}$

15. $f(x) = \frac{x-1}{x+1}$

16. $g(t) = \frac{t+1}{2t-1}$

17. $h(x) = x^3 - 3x^2 + x + 1$

18. $g(x) = 2x^3 + x^2 + 1$

19. $f(t) = \frac{t^2}{t^2 - 9}$

20. $g(x) = \frac{x^3}{x^2 - 4}$

21. $f(x) = \frac{3x}{x^2 - x - 6}$

22. $g(x) = \frac{2x}{x^2 + x - 2}$

23. $g(t) = 2 + \frac{5}{(t-2)^2}$

24. $f(x) = 1 + \frac{2}{x-3}$

25. $f(x) = \frac{x^2 - 2}{x^2 - 4}$

26. $h(x) = \frac{2 - x^2}{x^2 + x}$

27. $g(x) = \frac{x^3 - x}{x(x+1)}$

28. $f(x) = \frac{x^4 - x^2}{x(x-1)(x+2)}$

In Exercises 29 and 30, you are given the graphs of two functions *f* and *g*. One function is the derivative function of the other. Identify each of them.

29.

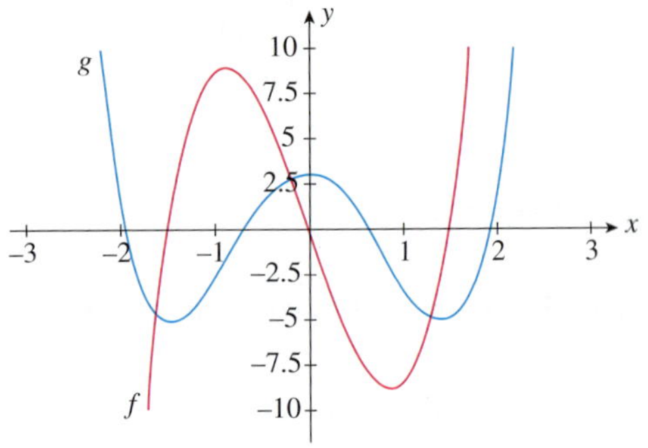

30.

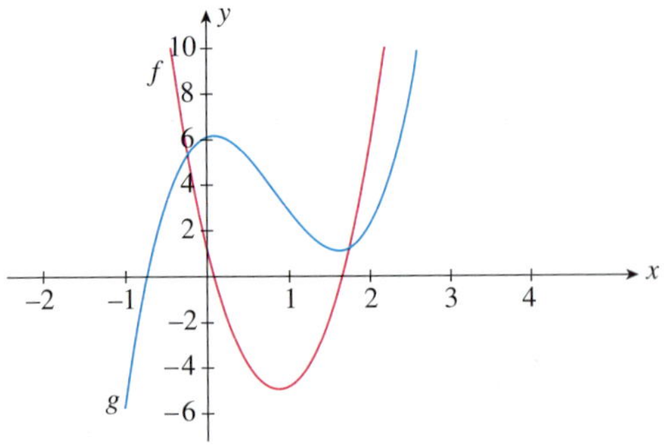

31. **Terminal Velocity** A skydiver leaps from the gondola of a hot-air balloon. As she free-falls, air resistance, which is proportional to her velocity, builds up to a point where it balances the force due to gravity. The resulting motion may be described in terms of her velocity as follows: Starting at rest (zero velocity), her velocity increases and approaches a constant velocity, called the *terminal velocity*. Sketch a graph of her velocity v versus time t.

32. **Spread of a Flu Epidemic** Initially, 10 students at a junior high school contracted influenza. The flu spread over time, and the total number of students who eventually contracted the flu approached but never exceeded 200. Let $P(t)$ denote the number of students who had contracted the flu after t days, where P is an appropriate function.

a. Make a sketch of the graph of P. (Your answer will *not* be unique.)

b. Where is the function increasing?
c. Does P have a horizontal asymptote? If so, what is it?
d. Discuss the concavity of P. Explain its significance.
e. Is there an inflection point on the graph of P? If so, explain its significance.

In Exercises 33–36, use the information summarized in the table to sketch the graph of *f*.

33. $f(x) = x^3 - 3x^2 + 1$

Domain: $(-\infty, \infty)$
Intercept: y-intercept: 1
Asymptotes: None
Intervals where f is ↗ and ↘: ↗ on $(-\infty, 0) \cup (2, \infty)$; ↘ on $(0, 2)$
Relative extrema: Rel. max. at $(0, 1)$; rel. min. at $(2, -3)$
Concavity: Downward on $(-\infty, 1)$; upward on $(1, \infty)$
Point of inflection: $(1, -1)$

34. $f(x) = \frac{1}{9}(x^4 - 4x^3)$

Domain: $(-\infty, \infty)$
Intercepts: x-intercepts: 0, 4; y-intercept: 0
Asymptotes: None
Intervals where f is ↗ and ↘: ↗ on $(3, \infty)$; ↘ on $(-\infty, 0) \cup (0, 3)$
Relative extrema: Rel. min. at $(3, -3)$
Concavity: Downward on $(0, 2)$; upward on $(-\infty, 0) \cup (2, \infty)$
Points of inflection: $(0, 0)$ and $(2, -\frac{16}{9})$

35. $f(x) = \dfrac{4x - 4}{x^2}$

Domain: $(-\infty, 0) \cup (0, \infty)$
Intercept: x-intercept: 1
Asymptotes: x-axis and y-axis
Intervals where f is ↗ and ↘: ↗ on $(0, 2)$; ↘ on $(-\infty, 0) \cup (2, \infty)$
Relative extrema: Rel. max. at $(2, 1)$
Concavity: Downward on $(-\infty, 0) \cup (0, 3)$; upward on $(3, \infty)$
Point of inflection: $(3, \frac{8}{9})$

36. $f(x) = x - 3x^{1/3}$

Domain: $(-\infty, \infty)$
Intercepts: x-intercepts: $\pm 3\sqrt{3}, 0$
Asymptotes: None
Intervals where f is ↗ and ↘: ↗ on $(-\infty, -1) \cup (1, \infty)$; ↘ on $(-1, 1)$
Relative extrema: Rel. max. at $(-1, 2)$; rel. min. at $(1, -2)$
Concavity: Downward on $(-\infty, 0)$; upward on $(0, \infty)$
Point of inflection: $(0, 0)$

In Exercises 37–60, sketch the graph of the function, using the curve-sketching guide of this section.

37. $g(x) = 4 - 3x - 2x^3$
38. $f(x) = x^2 - 2x + 3$
39. $h(x) = x^3 - 3x + 1$
40. $f(x) = 2x^3 + 1$
41. $f(x) = -2x^3 + 3x^2 + 12x + 2$
42. $f(t) = 2t^3 - 15t^2 + 36t - 20$
43. $h(x) = \frac{3}{2}x^4 - 2x^3 - 6x^2 + 8$
44. $f(t) = 3t^4 + 4t^3$
45. $f(t) = \sqrt{t^2 - 4}$
46. $f(x) = \sqrt{x^2 + 5}$
47. $g(x) = \frac{1}{2}x - \sqrt{x}$
48. $f(x) = \sqrt[3]{x^2}$
49. $g(x) = \dfrac{2}{x - 1}$
50. $f(x) = \dfrac{1}{x + 1}$
51. $h(x) = \dfrac{x + 2}{x - 2}$
52. $g(x) = \dfrac{x}{x - 1}$
53. $f(t) = \dfrac{t^2}{1 + t^2}$
54. $g(x) = \dfrac{x}{x^2 - 4}$
55. $g(t) = -\dfrac{t^2 - 2}{t - 1}$
56. $f(x) = \dfrac{x^2 - 9}{x^2 - 4}$
57. $g(t) = \dfrac{t^2}{t^2 - 1}$
58. $h(x) = \dfrac{1}{x^2 - x - 2}$
59. $h(x) = (x - 1)^{2/3} + 1$
60. $g(x) = (x + 2)^{3/2} + 1$

61. Cost of Removing Toxic Pollutants A city's main well was recently found to be contaminated with trichloroethylene (a cancer-causing chemical) as a result of an abandoned chemical dump that leached chemicals into the water. A proposal submitted to the city council indicated that the cost, measured in millions of dollars, of removing x% of the toxic pollutants is given by

$$C(x) = \frac{0.5x}{100 - x}$$

a. Find the vertical asymptote of $C(x)$.
b. Is it possible to remove 100% of the toxic pollutant from the water?

62. Average Cost of Producing DVDs The average cost per disc (in dollars) incurred by Herald Media Corporation in pressing x DVDs is given by the average cost function

$$\overline{C}(x) = 2.2 + \frac{2500}{x}$$

a. Find the horizontal asymptote of $\overline{C}(x)$.
b. What is the limiting value of the average cost?

63. Concentration of a Drug in the Bloodstream The concentration (in milligrams/cubic centimeter) of a certain drug in a patient's bloodstream t hr after injection is given by

$$C(t) = \frac{0.2t}{t^2 + 1}$$

a. Find the horizontal asymptote of $C(t)$.
b. Interpret your result.

64. **Effect of Enzymes on Chemical Reactions** Certain proteins, known as enzymes, serve as catalysts for chemical reactions in living things. In 1913 Leonor Michaelis and L. M. Menten discovered the following formula giving the initial speed V (in moles/liter/second) at which the reaction begins in terms of the amount of substrate x (the substance that is being acted upon, measured in moles/liter):

$$V = \frac{ax}{x + b}$$

where a and b are positive constants.

a. Find the horizontal asymptote of V.

b. What does the result of part (a) tell you about the initial speed at which the reaction begins, if the amount of substrate is very large?

65. **GDP of a Developing Country** A developing country's gross domestic product (GDP) from 2000 to 2008 is approximated by the function

$$G(t) = -0.2t^3 + 2.4t^2 + 60 \qquad (0 \leq t \leq 8)$$

where $G(t)$ is measured in billions of dollars, with $t = 0$ corresponding to 2000. Sketch the graph of the function G and interpret your results.

66. **Crime Rate** The number of major crimes per 100,000 committed in a city between 2000 and 2007 is approximated by the function

$$N(t) = -0.1t^3 + 1.5t^2 + 80 \qquad (0 \leq t \leq 7)$$

where $N(t)$ denotes the number of crimes per 100,000 committed in year t, with $t = 0$ corresponding to 2000. Enraged by the dramatic increase in the crime rate, the citizens, with the help of the local police, organized Neighborhood Crime Watch groups in early 2004 to combat this menace. Sketch the graph of the function N' and interpret your results. Is the Neighborhood Crime Watch program working?

67. **Worker Efficiency** An efficiency study showed that the total number of cordless telephones assembled by an average worker at Delphi Electronics t hr after starting work at 8 a.m. is given by

$$N(t) = -\frac{1}{2}t^3 + 3t^2 + 10t \qquad (0 \leq t \leq 4)$$

Sketch the graph of the function N and interpret your results.

68. **Concentration of a Drug in the Bloodstream** The concentration (in millimeters/cubic centimeter) of a certain drug in a patient's bloodstream t hr after injection is given by

$$C(t) = \frac{0.2t}{t^2 + 1}$$

Sketch the graph of the function C and interpret your results.

69. **Box-Office Receipts** The total worldwide box-office receipts for a long-running movie are approximated by the function

$$T(x) = \frac{120x^2}{x^2 + 4}$$

where $T(x)$ is measured in millions of dollars and x is the number of years since the movie's release. Sketch the graph of the function T and interpret your results.

70. **Oxygen Content of a Pond** When organic waste is dumped into a pond, the oxidation process that takes place reduces the pond's oxygen content. However, given time, nature will restore the oxygen content to its natural level. Suppose the oxygen content t days after organic waste has been dumped into the pond is given by

$$f(t) = 100\left(\frac{t^2 - 4t + 4}{t^2 + 4}\right) \qquad (0 \leq t < \infty)$$

percent of its normal level. Sketch the graph of the function f and interpret your results.

71. **Cost of Removing Toxic Pollutants** Refer to Exercise 61. The cost, measured in millions of dollars, of removing x% of a toxic pollutant is given by

$$C(x) = \frac{0.5x}{100 - x}$$

Sketch the graph of the function C and interpret your results.

72. **Traffic Flow Analysis** The speed of traffic flow in miles per hour on a stretch of Route 123 between 6 a.m. and 10 a.m. on a typical workday is approximated by the function

$$f(t) = 20t - 40\sqrt{t} + 52 \qquad (0 \leq t \leq 4)$$

where t is measured in hours, with $t = 0$ corresponding to 6 a.m. Sketch the graph of f and interpret your results.

4.3 Solutions to Self-Check Exercises

1. Since

$$\lim_{x\to\infty} \frac{2x^2}{x^2 - 1} = \lim_{x\to\infty} \frac{2}{1 - \dfrac{1}{x^2}} = 2$$

Divide the numerator and denominator by x^2.

we see that $y = 2$ is a horizontal asymptote. Next, since

$$x^2 - 1 = (x + 1)(x - 1) = 0$$

implies $x = -1$ or $x = 1$, these are candidates for the vertical asymptotes of f. Since the numerator of f is not equal to zero for $x = -1$ or $x = 1$, we conclude that $x = -1$ and $x = 1$ are vertical asymptotes of the graph of f.

2. We obtain the following information on the graph of f.

(1) The domain of f is the interval $(-\infty, \infty)$.

(2) By setting $x = 0$, we find the y-intercept is 4.

(3) Since

$$\lim_{x\to-\infty} f(x) = \lim_{x\to-\infty} \left(\frac{2}{3}x^3 - 2x^2 - 6x + 4\right) = -\infty$$

$$\lim_{x\to\infty} f(x) = \lim_{x\to\infty} \left(\frac{2}{3}x^3 - 2x^2 - 6x + 4\right) = \infty$$

we see that $f(x)$ decreases without bound as x decreases without bound and that $f(x)$ increases without bound as x increases without bound.

(4) Since f is a polynomial function, there are no asymptotes.

(5) $$\begin{aligned} f'(x) &= 2x^2 - 4x - 6 = 2(x^2 - 2x - 3) \\ &= 2(x + 1)(x - 3) \end{aligned}$$

Setting $f'(x) = 0$ gives $x = -1$ or $x = 3$. The accompanying sign diagram for f' shows that f is increasing on the intervals $(-\infty, -1)$ and $(3, \infty)$ and decreasing on $(-1, 3)$.

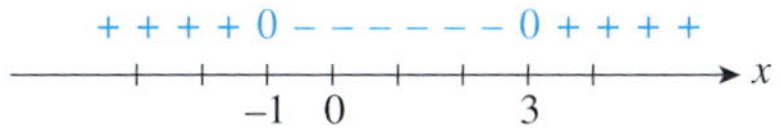

Sign diagram for f'

(6) From the results of step 5, we see that $x = -1$ and $x = 3$ are critical numbers of f. Furthermore, the sign diagram of f' tells us that $x = -1$ gives rise to a relative maximum of f and $x = 3$ gives rise to a relative minimum of f. Now,

$$f(-1) = \frac{2}{3}(-1)^3 - 2(-1)^2 - 6(-1) + 4 = \frac{22}{3}$$

$$f(3) = \frac{2}{3}(3)^3 - 2(3)^2 - 6(3) + 4 = -14$$

so $f(-1) = \frac{22}{3}$ is a relative maximum of f and $f(3) = -14$ is a relative minimum of f.

(7) $$f''(x) = 4x - 4 = 4(x - 1)$$

which is equal to zero when $x = 1$. The accompanying sign diagram of f'' shows that f is concave downward on the interval $(-\infty, 1)$ and concave upward on the interval $(1, \infty)$.

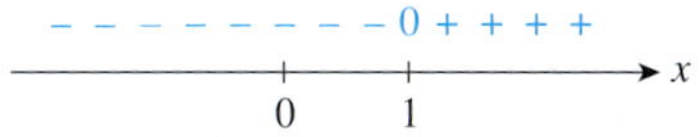

Sign diagram for f''

(8) From the results of step 7, we see that $x = 1$ is the only candidate for an inflection point of f. Since $f''(x)$ changes sign as we move across the point $x = 1$ and

$$f(1) = \frac{2}{3}(1)^3 - 2(1)^2 - 6(1) + 4 = -\frac{10}{3}$$

we see that the required inflection point is $(1, -\frac{10}{3})$.

(9) Summarizing this information, we have the following:

Domain: $(-\infty, \infty)$
Intercept: $(0, 4)$
$\lim_{x\to-\infty} f(x)$; $\lim_{x\to\infty} f(x)$: $-\infty$; ∞
Asymptotes: None
Intervals where f is ↗ or ↘: ↗ on $(-\infty, -1) \cup (3, \infty)$; ↘ on $(-1, 3)$
Relative extrema: Rel. max. at $(-1, \frac{22}{3})$; rel. min. at $(3, -14)$
Concavity: Downward on $(-\infty, 1)$; upward on $(1, \infty)$
Point of inflection: $(1, -\frac{10}{3})$

The graph of f is sketched in the accompanying figure.

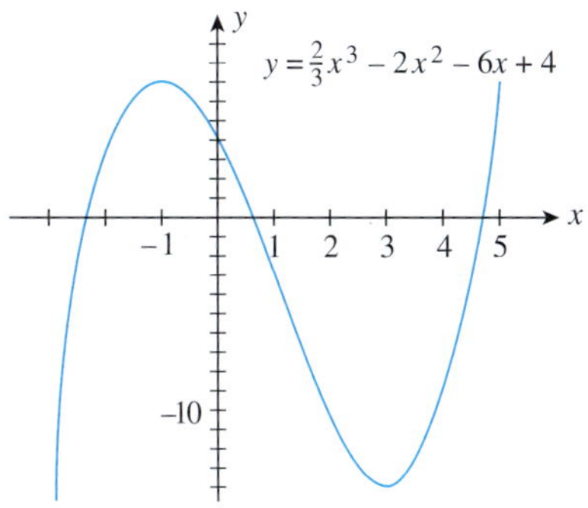

USING TECHNOLOGY

Analyzing the Properties of a Function

One of the main purposes of studying Section 4.3 is to see how the many concepts of calculus come together to paint a picture of a function. The techniques of graphing also play a very practical role. For example, using the techniques of graphing developed in Section 4.3, you can tell if the graph of a function generated by a graphing utility is reasonably complete. Furthermore, these techniques can often reveal details that are missing from a graph.

EXAMPLE 1 Consider the function $f(x) = 2x^3 - 3.5x^2 + x - 10$. A plot of the graph of f in the standard viewing window is shown in Figure T1. Since the domain of f is the interval $(-\infty, \infty)$, we see that Figure T1 does not reveal the part of the graph to

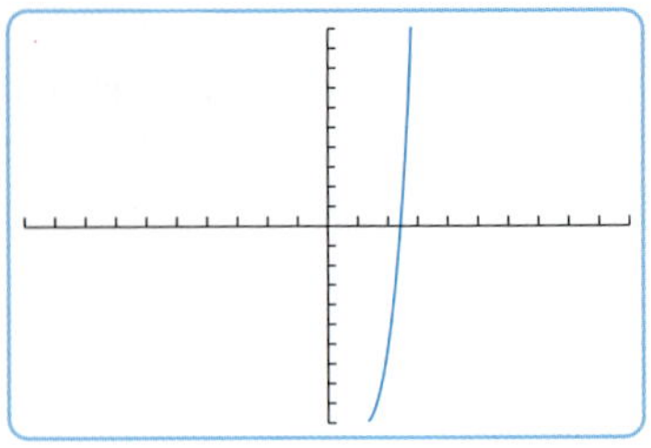

FIGURE T1
The graph of f in the standard viewing window

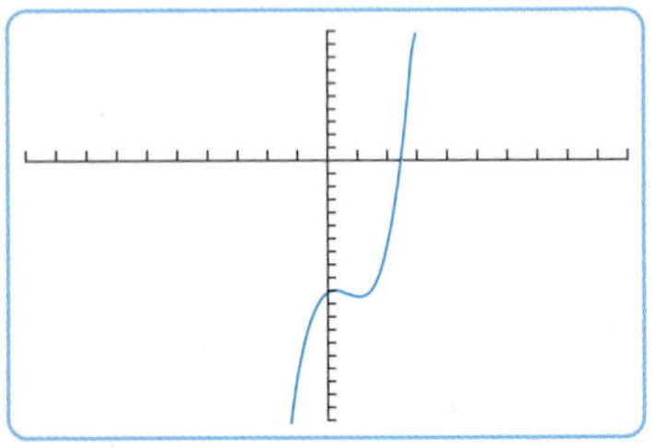

FIGURE T2
The graph of f in the viewing window $[-10, 10] \times [-20, 10]$

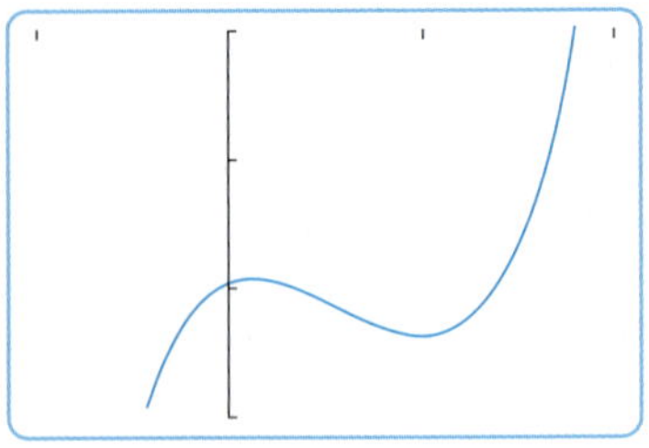

FIGURE T3
The graph of f in the viewing window $[-1, 2] \times [-11, -9]$

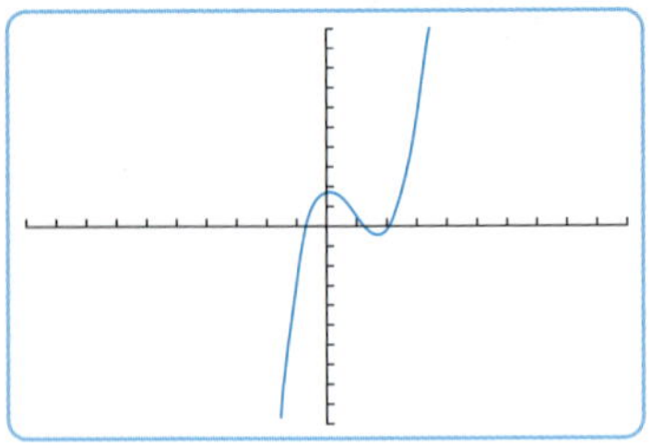

FIGURE T4
The graph of $f(x) = x^3 - 3x^2 + x + 1.5$

the left of the y-axis. This suggests that we enlarge the viewing window accordingly. Figure T2 shows the graph of f in the viewing window $[-10, 10] \times [-20, 10]$.

The behavior of f for large values of x

$$\lim_{x \to -\infty} f(x) = -\infty \quad \text{and} \quad \lim_{x \to \infty} f(x) = \infty$$

suggests that this viewing window has captured a sufficiently complete picture of f. Next, an analysis of the first derivative of x,

$$f'(x) = 6x^2 - 7x + 1 = (6x - 1)(x - 1)$$

reveals that f has critical values at $x = \frac{1}{6}$ and $x = 1$. In fact, a sign diagram of f' shows that f has a relative maximum at $x = \frac{1}{6}$ and a relative minimum at $x = 1$, details that are not revealed in the graph of f shown in Figure T2. To examine this portion of the graph of f, we use, say, the viewing window $[-1, 2] \times [-11, -9]$. The resulting graph of f is shown in Figure T3, which certainly reveals the hitherto missing details! Thus, through an interaction of calculus and a graphing utility, we are able to obtain a good picture of the properties of f. ■

Finding x-Intercepts

As noted in Section 4.3, it is not always easy to find the x-intercepts of the graph of a function. But this information is very important in applications. By using the function for solving polynomial equations or the function for finding the roots of an equation, we can solve the equation $f(x) = 0$ quite easily and hence yield the x-intercepts of the graph of a function.

EXAMPLE 2 Let $f(x) = x^3 - 3x^2 + x + 1.5$.

a. Use the function for solving polynomial equations on a graphing utility to find the x-intercepts of the graph of f.

b. Use the function for finding the roots of an equation on a graphing utility to find the x-intercepts of the graph of f.

Solution

a. Observe that f is a polynomial function of degree 3, and so we may use the function for solving polynomial equations to solve the equation $x^3 - 3x^2 + x + 1.5 = 0$ $[f(x) = 0]$. We find that the solutions (x-intercepts) are

$$x_1 \approx -0.525687120865 \qquad x_2 \approx 1.2586520225 \qquad x_3 \approx 2.26703509836$$

b. Using the graph of f (Figure T4), we see that $x_1 \approx -0.5$, $x_2 \approx 1$, and $x_3 \approx 2$. Using the function for finding the roots of an equation on a graphing utility, and these values of x as initial guesses, we find

$$x_1 \approx -0.5256871209 \qquad x_2 \approx 1.2586520225 \qquad x_3 \approx 2.2670350984$$ ■

Note The function for solving polynomial equations on a graphing utility will solve a polynomial equation $f(x) = 0$, where f is a polynomial function. The function for finding the roots of a polynomial, however, will solve equations $f(x) = 0$ even if f is not a polynomial. ■

APPLIED EXAMPLE 3 TV Mobile Phones The number of people watching TV on mobile phones (in millions) is expected to be

$$N(t) = 11.9\sqrt{1 + 0.91t} \qquad (0 \le t \le 4)$$

where t is measured in years, with $t = 0$ corresponding to the beginning of 2007.

a. Use a graphing calculator to plot the graph of N.

b. Based on this model, when will the number of people watching TV on mobile phones first exceed 20 million?

Solution

a. The graph of N in the window $[0, 4] \times [0, 30]$ is shown in Figure T5a.

b. Using the function for finding the intersection of the graphs of $y_1 = N(t)$ and $y_2 = 20$, we find $t \approx 2.005$ (see Figure T5b). So the number of people watching TV on mobile phones will first exceed 20 million at the beginning of January 2009.

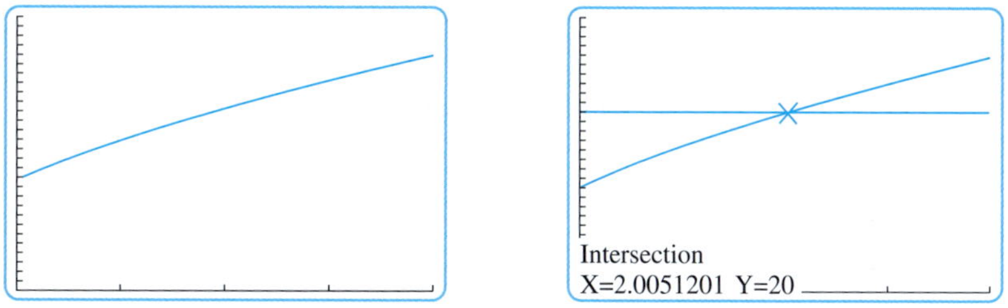

FIGURE **T5**

(a) The graph of N in the viewing window $[0, 4] \times [0, 30]$

(b) The graph showing the intersection of $y_1 = N(t)$ and $y_2 = 20$ on the TI 83/84.

TECHNOLOGY EXERCISES

In Exercises 1–4, use the method of Example 1 to analyze the function. (*Note:* Your answers will *not* be unique.)

1. $f(x) = 4x^3 - 4x^2 + x + 10$

2. $f(x) = x^3 + 2x^2 + x - 12$

3. $f(x) = \frac{1}{2}x^4 + x^3 + \frac{1}{2}x^2 - 10$

4. $f(x) = 2.25x^4 - 4x^3 + 2x^2 + 2$

In Exercises 5–8, find the *x*-intercepts of the graph of *f*. Give your answers accurate to four decimal places.

5. $f(x) = 0.2x^3 - 1.2x^2 + 0.8x + 2.1$

6. $f(x) = -0.2x^4 + 0.8x^3 - 2.1x + 1.2$

7. $f(x) = 2x^2 - \sqrt{x + 1} - 3$

8. $f(x) = x - \sqrt{1 - x^2}$

9. AIR POLLUTION The level of ozone, an invisible gas that irritates and impairs breathing, present in the atmosphere on a certain day in June in the city of Riverside is approximated by

$$S(t) = 1.0974t^3 - 0.0915t^4 \qquad (0 \le t \le 11)$$

where $S(t)$ is measured in pollutant standard index (PSI) and t is measured in hours, with $t = 0$ corresponding to 7 a.m. Sketch the graph of S and interpret your results.

Source: Los Angeles Times

10. FLIGHT PATH OF A PLANE The function

$$f(x) = \begin{cases} 0 & \text{if } 0 \le x < 1 \\ -0.0411523x^3 + 0.679012x^2 \\ \quad -1.23457x + 0.596708 & \text{if } 1 \le x < 10 \\ 15 & \text{if } 10 \le x \le 11 \end{cases}$$

where both x and $f(x)$ are measured in units of 1000 ft, describes the flight path of a plane taking off from the origin and climbing to an altitude of 15,000 ft. Sketch the graph of f to visualize the trajectory of the plane.

4.4 Optimization I

Absolute Extrema

The graph of the function f in Figure 57 shows the average age of cars in use in the United States from the beginning of 1946 ($t = 0$) to the beginning of 2002 ($t = 56$). Observe that the highest average age of cars in use during this period is 9 years, whereas the lowest average age of cars in use during the same period is $5\frac{1}{2}$ years. The number 9, the largest value of $f(t)$ for all values of t in the interval $[0, 56]$ (the domain of f), is called the *absolute maximum value of f* on that interval. The number $5\frac{1}{2}$, the smallest value of $f(t)$ for all values of t in $[0, 56]$, is called the *absolute minimum value of f* on that interval. Notice, too, that the absolute maximum value of f is attained at the endpoint $t = 0$ of the interval, whereas the absolute minimum value of f is attained at the points $t = 12$ (corresponding to 1958) and $t = 23$ (corresponding to 1969) that lie within the interval $(0, 56)$.

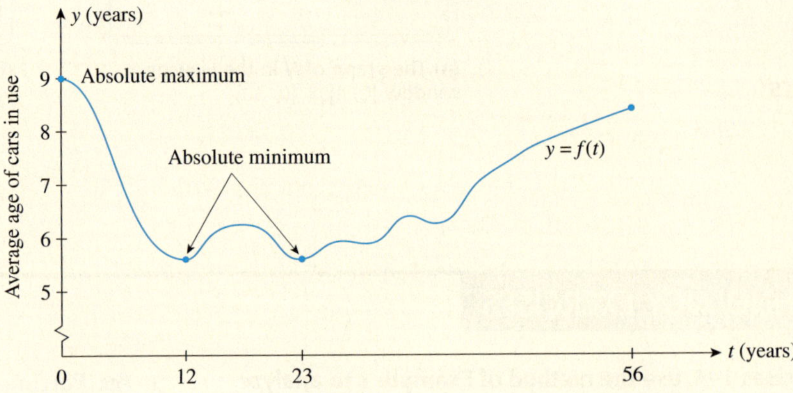

Source: American Automobile Association

FIGURE 57
$f(t)$ gives the average age of cars in use in year t, t in $[0, 56]$.

(Incidentally, it is interesting to note that 1946 marked the first year of peace following World War II, and the two years, 1958 and 1969, marked the end of two periods of prosperity in recent U.S. history!)

A precise definition of the **absolute extrema** (absolute maximum or absolute minimum) of a function follows.

The Absolute Extrema of a Function f

If $f(x) \leq f(c)$ for all x in the domain of f, then $f(c)$ is called the **absolute maximum value** of f.

If $f(x) \geq f(c)$ for all x in the domain of f, then $f(c)$ is called the **absolute minimum value** of f.

Figure 58 shows the graphs of several functions and gives the absolute maximum and absolute minimum of each function, if they exist.

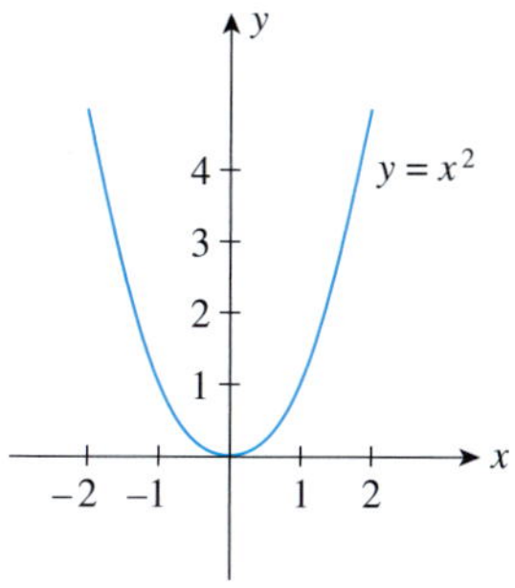

(a) $f(0) = 0$ is the absolute minimum of f; f has no absolute maximum.

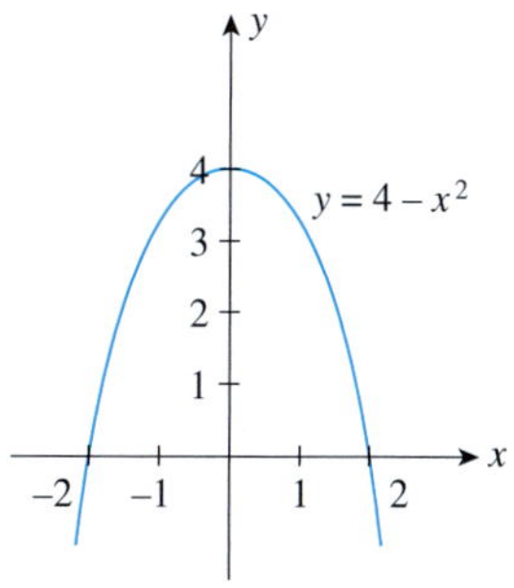

(b) $f(0) = 4$ is the absolute maximum of f; f has no absolute minimum.

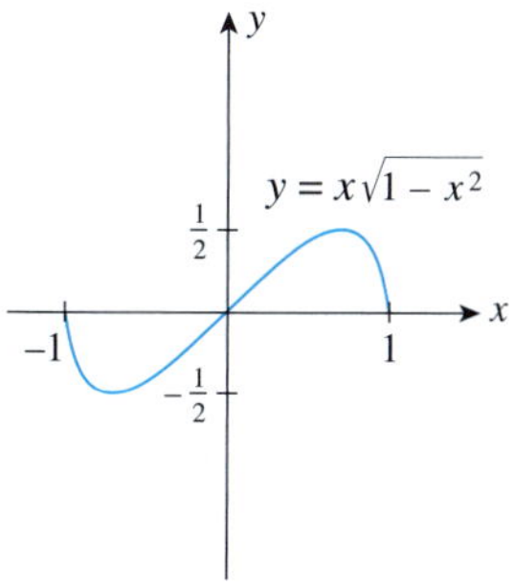

(c) $f(\sqrt{2}/2) = 1/2$ is the absolute maximum of f; $f(-\sqrt{2}/2) = -1/2$ is the absolute minimum of f.

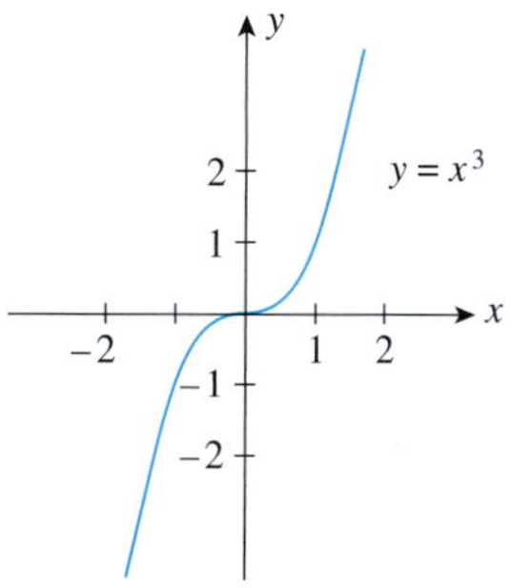

(d) f has no absolute extrema.

FIGURE **58**

Absolute Extrema on a Closed Interval

As the preceding examples show, a continuous function defined on an arbitrary interval does not always have an absolute maximum or an absolute minimum. But an important case arises often in practical applications in which both the absolute maximum and the absolute minimum of a function are guaranteed to exist. This occurs when a continuous function is defined on a *closed* interval. Let's state this important result in the form of a theorem, whose proof we will omit.

THEOREM 3

If a function f is continuous on a closed interval $[a, b]$, then f has both an absolute maximum value and an absolute minimum value on $[a, b]$.

Observe that if an absolute extremum of a continuous function f occurs at a point in an open interval (a, b), then it must be a relative extremum of f and hence its x-coordinate must be a critical number of f. Otherwise, the absolute extremum of f must occur at one or both of the endpoints of the interval $[a, b]$. A typical situation is illustrated in Figure 59.

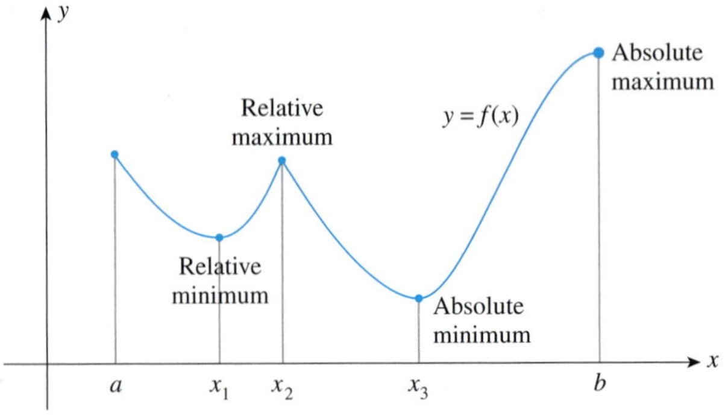

FIGURE 59
The relative minimum of f at x_3 is the absolute minimum of f. The right endpoint b of the interval $[a, b]$ gives rise to the absolute maximum value $f(b)$ of f.

Here x_1, x_2, and x_3 are critical numbers of f. The absolute minimum of f occurs at x_3, which lies in the open interval (a, b) and is a critical number of f. The absolute maximum of f occurs at b, an endpoint. This observation suggests the following procedure for finding the absolute extrema of a continuous function on a closed interval.

Finding the Absolute Extrema of f on a Closed Interval

1. Find the critical numbers of f that lie in (a, b).
2. Compute the value of f at each critical number found in step 1 and compute $f(a)$ and $f(b)$.
3. The absolute maximum value and absolute minimum value of f will correspond to the largest and smallest numbers, respectively, found in step 2.

EXAMPLE 1 Find the absolute extrema of the function $F(x) = x^2$ defined on the interval $[-1, 2]$.

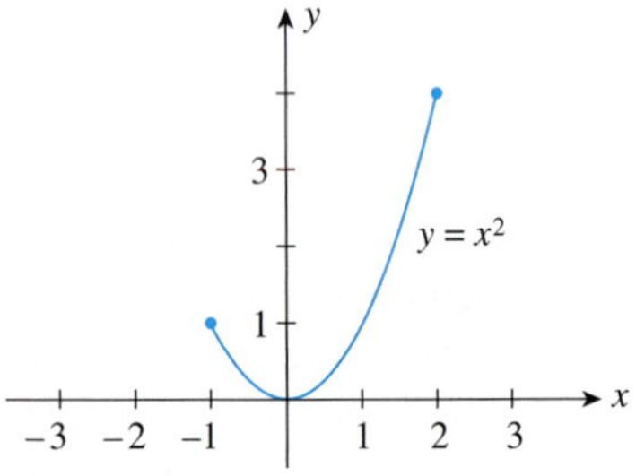

FIGURE 60
F has an absolute minimum value of 0 and an absolute maximum value of 4.

Solution The function F is continuous on the closed interval $[-1, 2]$ and differentiable on the open interval $(-1, 2)$. The derivative of F is

$$F'(x) = 2x$$

so 0 is the only critical number of F. Next, evaluate $F(x)$ at $x = -1$, $x = 0$, and $x = 2$. Thus,

$$F(-1) = 1 \qquad F(0) = 0 \qquad F(2) = 4$$

It follows that 0 is the absolute minimum value of F and 4 is the absolute maximum value of F. The graph of F, in Figure 60, confirms our results. ■

EXAMPLE 2 Find the absolute extrema of the function

$$f(x) = x^3 - 2x^2 - 4x + 4$$

defined on the interval $[0, 3]$.

Solution The function f is continuous on the closed interval $[0, 3]$ and differentiable on the open interval $(0, 3)$. The derivative of f is

$$f'(x) = 3x^2 - 4x - 4 = (3x + 2)(x - 2)$$

and it is equal to zero when $x = -\frac{2}{3}$ and $x = 2$. Since $x = -\frac{2}{3}$ lies outside the interval $[0, 3]$, it is dropped from further consideration, and $x = 2$ is seen to be the sole critical number of f. Next, we evaluate $f(x)$ at the critical number of f as well as the endpoints of f, obtaining

$$f(0) = 4 \qquad f(2) = -4 \qquad f(3) = 1$$

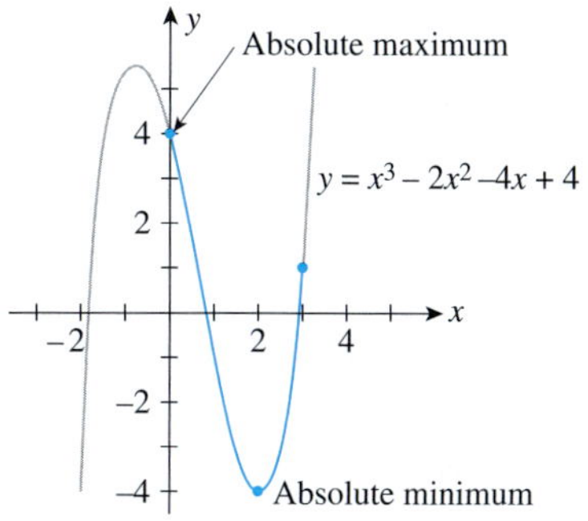

FIGURE 61
f has an absolute maximum value of 4 and an absolute minimum value of −4.

From these results, we conclude that -4 is the absolute minimum value of f and 4 is the absolute maximum value of f. The graph of f, which appears in Figure 61, confirms our results. Observe that the absolute maximum of f occurs at the endpoint $x = 0$ of the interval $[0, 3]$, while the absolute minimum of f occurs at $x = 2$, which lies in the interval $(0, 3)$. ■

Exploring with TECHNOLOGY

Let $f(x) = x^3 - 2x^2 - 4x + 4$. (This is the function of Example 2.)

1. Use a graphing utility to plot the graph of f, using the viewing window $[0, 3] \times [-5, 5]$. Use **TRACE** to find the absolute extrema of f on the interval $[0, 3]$ and thus verify the results obtained analytically in Example 2.
2. Plot the graph of f, using the viewing window $[-2, 1] \times [-5, 6]$. Use **ZOOM** and **TRACE** to find the absolute extrema of f on the interval $[-2, 1]$. Verify your results analytically.

VIDEO

EXAMPLE 3 Find the absolute maximum and absolute minimum values of the function $f(x) = x^{2/3}$ on the interval $[-1, 8]$.

Solution The derivative of f is

$$f'(x) = \frac{2}{3}x^{-1/3} = \frac{2}{3x^{1/3}}$$

Note that f' is not defined at $x = 0$, is continuous everywhere else, and does not equal zero for all x. Therefore, 0 is the only critical number of f. Evaluating $f(x)$ at $x = -1$, 0, and 8, we obtain

$$f(-1) = 1 \qquad f(0) = 0 \qquad f(8) = 4$$

We conclude that the absolute minimum value of f is 0, attained at $x = 0$, and the absolute maximum value of f is 4, attained at $x = 8$ (Figure 62). ■

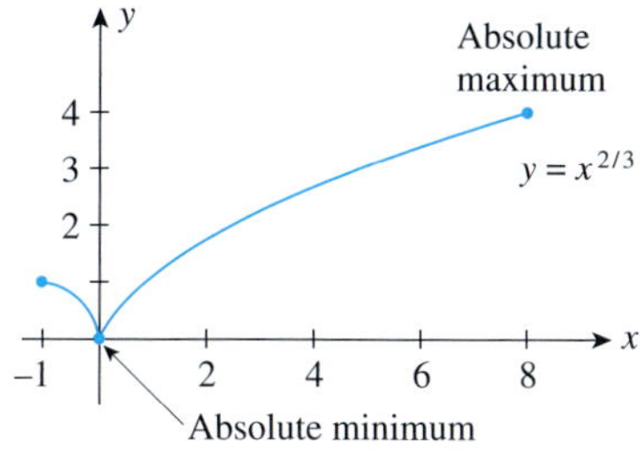

FIGURE 62
f has an absolute minimum value of $f(0) = 0$ and an absolute maximum value of $f(8) = 4$.

Many real-world applications call for finding the absolute maximum value or the absolute minimum value of a given function. For example, management is interested in finding what level of production will yield the maximum profit for a company; a farmer is interested in finding the right amount of fertilizer to maximize crop yield; a doctor is interested in finding the maximum concentration of a drug in a patient's body and the time at which it occurs; and an engineer is interested in finding the dimension of a container with a specified shape and volume that can be constructed at a minimum cost.

VIDEO

APPLIED EXAMPLE 4 Maximizing Profits Acrosonic's total profit (in dollars) from manufacturing and selling x units of their model F loudspeaker systems is given by

$$P(x) = -0.02x^2 + 300x - 200{,}000 \qquad (0 \le x \le 20{,}000)$$

How many units of the loudspeaker system must Acrosonic produce to maximize its profits?

Solution To find the absolute maximum of P on $[0, 20{,}000]$, first find the critical points of P on the interval $(0, 20{,}000)$. To do this, compute

$$P'(x) = -0.04x + 300$$

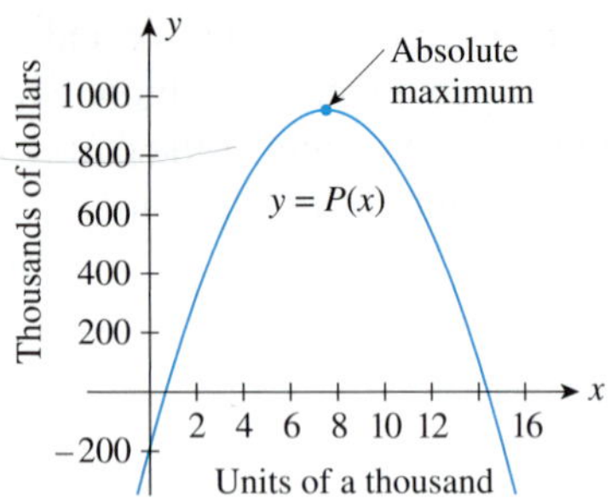

FIGURE 63
P has an absolute maximum at (7500, 925,000).

Solving the equation $P'(x) = 0$ gives $x = 7500$. Next, evaluate $P(x)$ at $x = 7500$ as well as the endpoints $x = 0$ and $x = 20{,}000$ of the interval [0, 20,000], obtaining

$$\begin{aligned} P(0) &= -200{,}000 \\ P(7500) &= 925{,}000 \\ P(20{,}000) &= -2{,}200{,}000 \end{aligned}$$

From these computations we see that the absolute maximum value of the function P is 925,000. Thus, by producing 7500 units, Acrosonic will realize a maximum profit of \$925,000. The graph of P is sketched in Figure 63. ■

Explore & Discuss

Recall that the total profit function P is defined as $P(x) = R(x) - C(x)$, where R is the total revenue function, C is the total cost function, and x is the number of units of a product produced and sold. (Assume all derivatives exist.)

1. Show that at the level of production x_0 that yields the maximum profit for the company, the following two conditions are satisfied:

$$R'(x_0) = C'(x_0) \quad \text{and} \quad R''(x_0) < C''(x_0)$$

2. Interpret the two conditions in part 1 in economic terms and explain why they make sense.

APPLIED EXAMPLE 5 Trachea Contraction during a Cough When a person coughs, the trachea (windpipe) contracts, allowing air to be expelled at a maximum velocity. It can be shown that during a cough the velocity v of airflow is given by the function

$$v = f(r) = kr^2(R - r)$$

where r is the trachea's radius (in centimeters) during a cough, R is the trachea's normal radius (in centimeters), and k is a positive constant that depends on the length of the trachea. Find the radius r for which the velocity of airflow is greatest.

Solution To find the absolute maximum of f on $[0, R]$, first find the critical numbers of f on the interval $(0, R)$. We compute

$$\begin{aligned} f'(r) &= 2kr(R - r) - kr^2 \qquad \text{Use the product rule.} \\ &= -3kr^2 + 2kRr = kr(-3r + 2R) \end{aligned}$$

Setting $f'(r) = 0$ gives $r = 0$ or $r = \frac{2}{3}R$, and so $\frac{2}{3}R$ is the sole critical number of f ($r = 0$ is an endpoint). Evaluating $f(r)$ at $r = \frac{2}{3}R$, as well as at the endpoints $r = 0$ and $r = R$, we obtain

$$\begin{aligned} f(0) &= 0 \\ f\left(\frac{2}{3}R\right) &= \frac{4k}{27}R^3 \\ f(R) &= 0 \end{aligned}$$

from which we deduce that the velocity of airflow is greatest when the radius of the contracted trachea is $\frac{2}{3}R$—that is, when the radius is contracted by approximately 33%. The graph of the function f is shown in Figure 64. ■

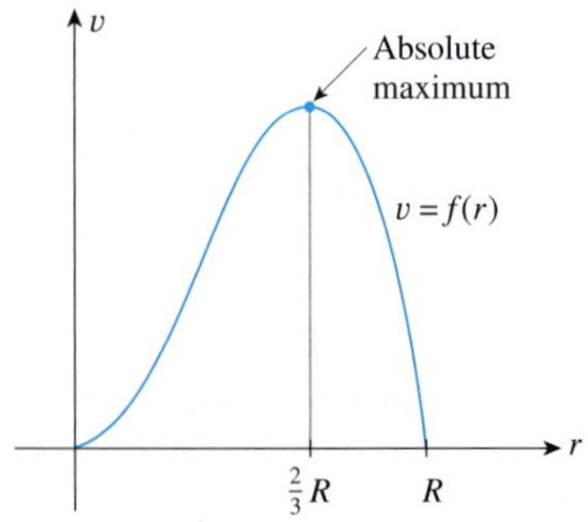

FIGURE 64
The velocity of airflow is greatest when the radius of the contracted trachea is $\frac{2}{3}R$.

Explore & Discuss

Prove that if a cost function $C(x)$ is concave upward $[C''(x) > 0]$, then the level of production that will result in the smallest average production cost occurs when

$$\overline{C}(x) = C'(x)$$

—that is, when the average cost $\overline{C}(x)$ is equal to the marginal cost $C'(x)$.

Hints:

1. Show that

$$\overline{C}'(x) = \frac{xC'(x) - C(x)}{x^2}$$

so that the critical number of the function $\overline{C}$ occurs when

$$xC'(x) - C(x) = 0$$

2. Show that at a critical number of $\overline{C}$

$$\overline{C}''(x) = \frac{C''(x)}{x}$$

Use the second derivative test to reach the desired conclusion.

APPLIED EXAMPLE 6 Minimizing Average Cost The daily average cost function (in dollars per unit) of Elektra Electronics is given by

$$\overline{C}(x) = 0.0001x^2 - 0.08x + 40 + \frac{5000}{x} \qquad (x > 0)$$

where x stands for the number of graphing calculators that Elektra produces. Show that a production level of 500 units per day results in a minimum average cost for the company.

Solution The domain of the function $\overline{C}$ is the interval $(0, \infty)$, which is not closed. To solve the problem, we resort to the graphical method. Using the techniques of graphing from the last section, we sketch the graph of $\overline{C}$ (Figure 65).

Now,

$$\overline{C}'(x) = 0.0002x - 0.08 - \frac{5000}{x^2}$$

Substituting the given value of x, 500, into $\overline{C}'(x)$ gives $\overline{C}'(500) = 0$, so 500 is a critical number of $\overline{C}$. Next,

$$\overline{C}''(x) = 0.0002 + \frac{10{,}000}{x^3}$$

Thus,

$$\overline{C}''(500) = 0.0002 + \frac{10{,}000}{(500)^3} > 0$$

and by the second derivative test, a relative minimum of the function $\overline{C}$ occurs at 500. Furthermore, $\overline{C}''(x) > 0$ for $x > 0$, which implies that the graph of $\overline{C}$ is

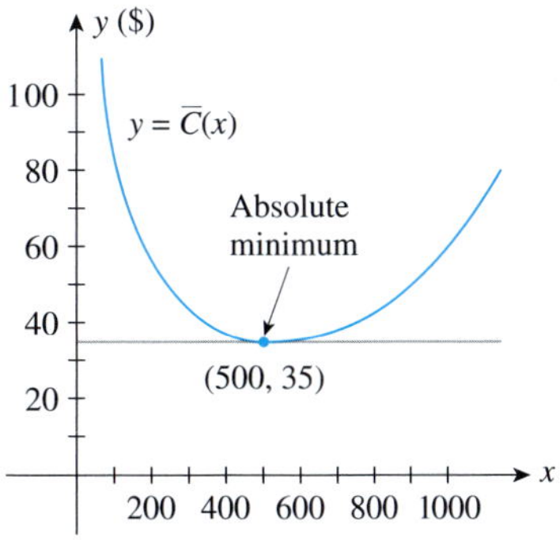

FIGURE 65
The minimum average cost is $35 per unit.

concave upward everywhere, so the relative minimum of $\overline{C}$ must be the absolute minimum of $\overline{C}$. The minimum average cost is given by

$$\begin{aligned}\overline{C}(500) &= 0.0001(500)^2 - 0.08(500) + 40 + \frac{5000}{500} \\ &= 35\end{aligned}$$

or \$35 per unit.

Exploring with TECHNOLOGY

Refer to the preceding Explore & Discuss and Example 6.

1. Using a graphing utility, plot the graphs of

$$\overline{C}(x) = 0.0001x^2 - 0.08x + 40 + \frac{5000}{x}$$
$$C'(x) = 0.0003x^2 - 0.16x + 40$$

using the viewing window $[0, 1000] \times [0, 150]$.

Note: $C(x) = 0.0001x^3 - 0.08x^2 + 40x + 5000$ (Why?)

2. Find the point of intersection of the graphs of $\overline{C}$ and C' and thus verify the assertion in the Explore & Discuss for the special case studied in Example 6.

APPLIED EXAMPLE 7 Flight of a Rocket The altitude (in feet) of a rocket t seconds into flight is given by

$$s = f(t) = -t^3 + 96t^2 + 195t + 5 \qquad (t \geq 0)$$

a. Find the maximum altitude attained by the rocket.
b. Find the maximum velocity attained by the rocket.

Solution

a. The maximum altitude attained by the rocket is given by the largest value of the function f in the closed interval $[0, T]$, where T denotes the time the rocket impacts Earth. We know that such a number exists because the dominant term in the expression for the continuous function f is $-t^3$. So for t large enough, the value of $f(t)$ must change from positive to negative and, in particular, it must attain the value 0 for some T.

To find the absolute maximum of f, compute

$$\begin{aligned}f'(t) &= -3t^2 + 192t + 195 \\ &= -3(t - 65)(t + 1)\end{aligned}$$

See page 11.

and solve the equation $f'(t) = 0$, obtaining $t = -1$ and $t = 65$. Ignore $t = -1$ since it lies outside the interval $[0, T]$. This leaves the critical number 65 of f. Continuing, we compute

$$f(0) = 5 \qquad f(65) = 143{,}655 \qquad f(T) = 0$$

and conclude, accordingly, that the absolute maximum value of f is 143,655. Thus, the maximum altitude of the rocket is 143,655 feet, attained 65 seconds into flight. The graph of f is sketched in Figure 66.

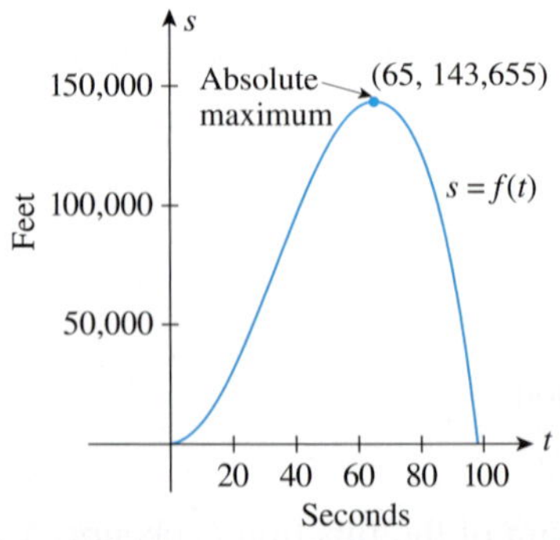

FIGURE 66
The maximum altitude of the rocket is 143,655 feet.

b. To find the maximum velocity attained by the rocket, find the largest value of the function that describes the rocket's velocity at any time t—namely,

$$v = f'(t) = -3t^2 + 192t + 195 \qquad (t \geq 0)$$

We find the critical number of v by setting $v' = 0$. But

$$v' = -6t + 192$$

and the critical number of v is 32. Since

$$v'' = -6 < 0$$

the second derivative test implies that a relative maximum of v occurs at $t = 32$. Our computation has in fact clarified the property of the "velocity curve." Since $v'' < 0$ everywhere, the velocity curve is concave downward everywhere. With this observation, we assert that the relative maximum must in fact be the absolute maximum of v. The maximum velocity of the rocket is given by evaluating v at $t = 32$,

$$f'(32) = -3(32)^2 + 192(32) + 195$$

or 3267 feet per second. The graph of the velocity function v is sketched in Figure 67. ■

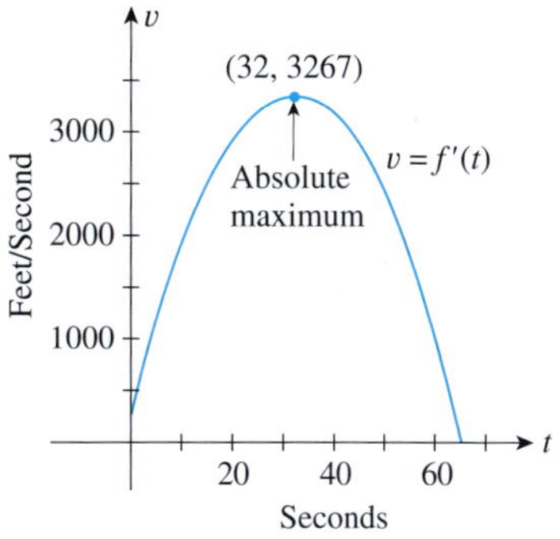

FIGURE 67
The maximum velocity of the rocket is 3267 feet per second.

4.4 Self-Check Exercises

1. Let $f(x) = x - 2\sqrt{x}$
   a. Find the absolute extrema of f on the interval [0, 9].
   b. Find the absolute extrema of f.
2. Find the absolute extrema of $f(x) = 3x^4 + 4x^3 + 1$ on $[-2, 1]$.
3. The operating rate (expressed as a percent) of factories, mines, and utilities in a certain region of the country on the tth day of 2006 is given by the function

$$f(t) = 80 + \frac{1200t}{t^2 + 40,000} \qquad (0 \leq t \leq 250)$$

On which of the first 250 days of 2006 was the manufacturing capacity operating rate highest?

Solutions to Self-Check Exercises 4.4 can be found on page 310.

4.4 Concept Questions

1. Explain the following terms: (a) absolute maximum and (b) absolute minimum.
2. Describe the procedure for finding the absolute extrema of a continuous function on a closed interval.

4.4 Exercises

In Exercises 1–8, you are given the graph of a function *f* defined on the indicated interval. Find the absolute maximum and the absolute minimum of *f*, if they exist.

1.

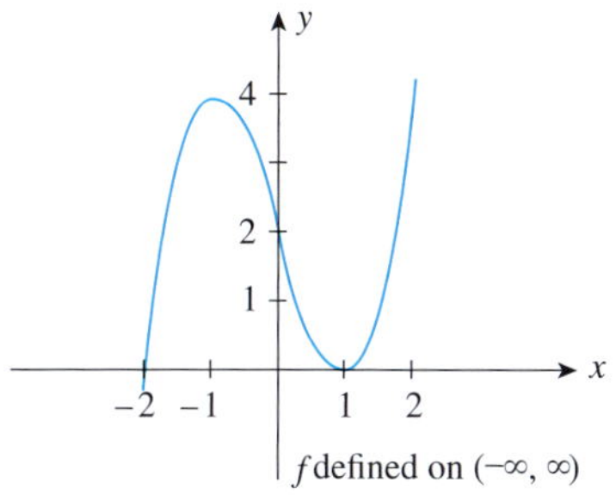

2.

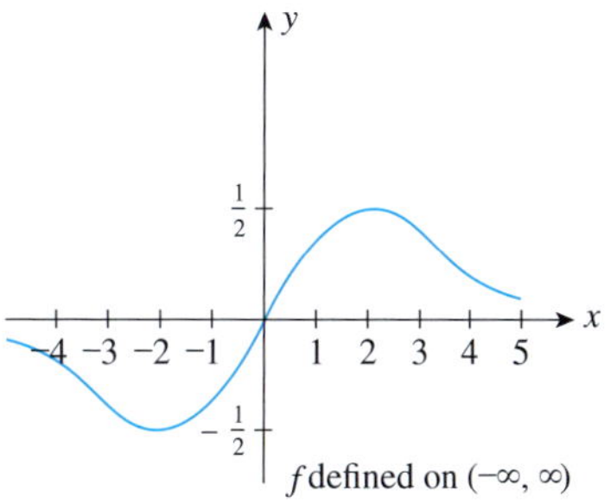

3.

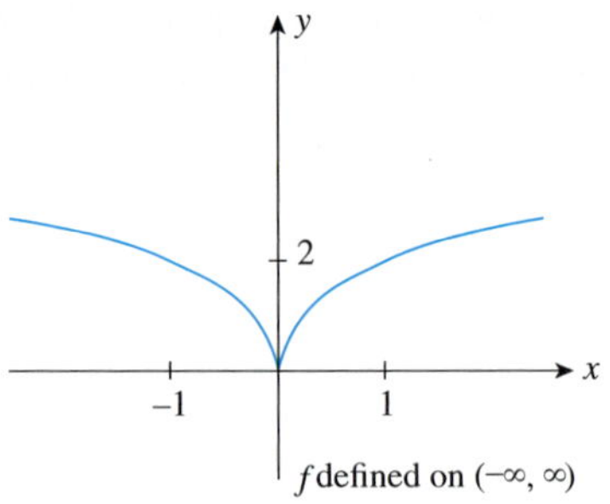

4.

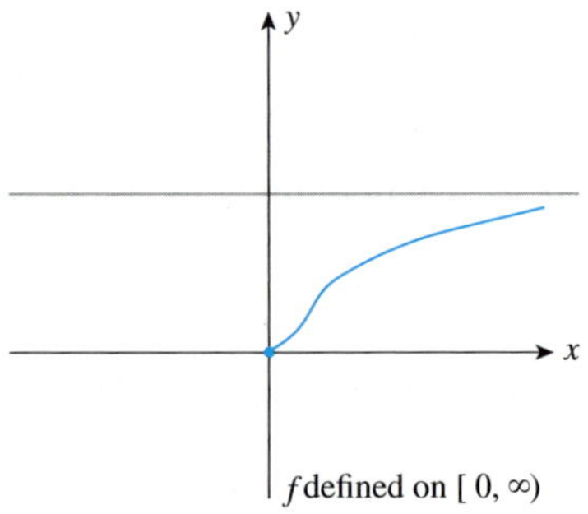

5.

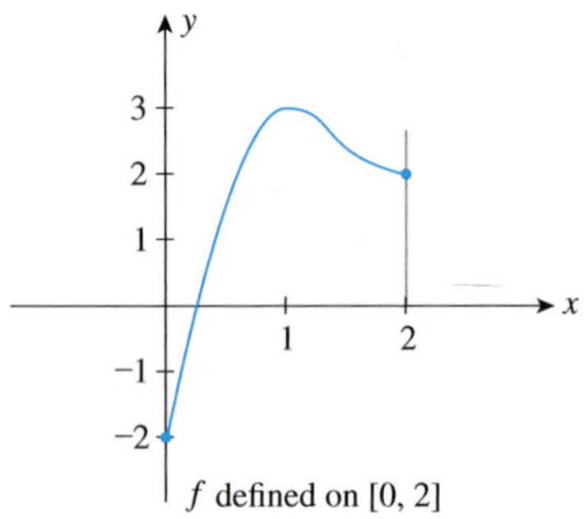

6.

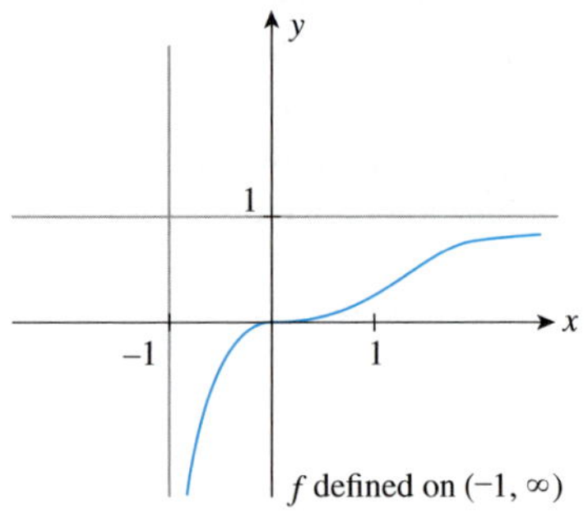

7.

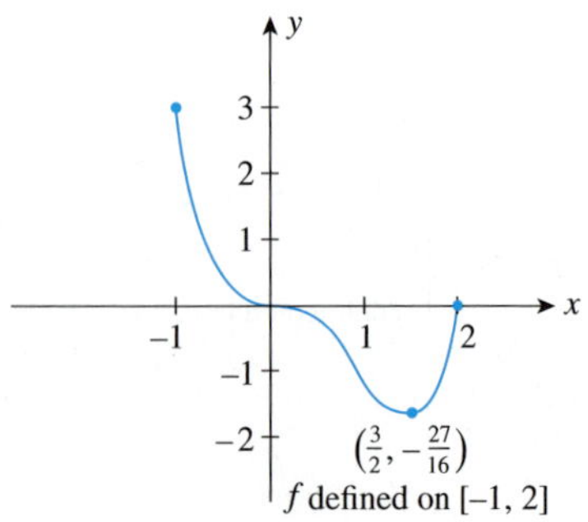

8.

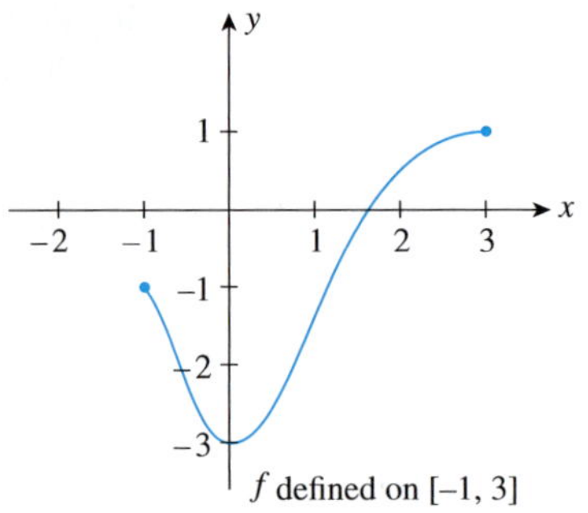

In Exercises 9–38, find the absolute maximum value and the absolute minimum value, if any, of each function.

9. $f(x) = 2x^2 + 3x - 4$

10. $g(x) = -x^2 + 4x + 3$

11. $h(x) = x^{1/3}$

12. $f(x) = x^{2/3}$

13. $f(x) = \dfrac{1}{1 + x^2}$

14. $f(x) = \dfrac{x}{1 + x^2}$

15. $f(x) = x^2 - 2x - 3$ on $[-2, 3]$

16. $g(x) = x^2 - 2x - 3$ on $[0, 4]$

17. $f(x) = -x^2 + 4x + 6$ on $[0, 5]$

18. $f(x) = -x^2 + 4x + 6$ on $[3, 6]$

19. $f(x) = x^3 + 3x^2 - 1$ on $[-3, 2]$

20. $g(x) = x^3 + 3x^2 - 1$ on $[-3, 1]$

21. $g(x) = 3x^4 + 4x^3$ on $[-2, 1]$

22. $f(x) = \dfrac{1}{2}x^4 - \dfrac{2}{3}x^3 - 2x^2 + 3$ on $[-2, 3]$

23. $f(x) = \dfrac{x + 1}{x - 1}$ on $[2, 4]$

24. $g(t) = \dfrac{t}{t - 1}$ on $[2, 4]$

25. $f(x) = 4x + \dfrac{1}{x}$ on $[1, 3]$

26. $f(x) = 9x - \dfrac{1}{x}$ on $[1, 3]$

27. $f(x) = \dfrac{1}{2}x^2 - 2\sqrt{x}$ on $[0, 3]$

28. $g(x) = \dfrac{1}{8}x^2 - 4\sqrt{x}$ on $[0, 9]$

29. $f(x) = \dfrac{1}{x}$ on $(0, \infty)$

30. $g(x) = \dfrac{1}{x + 1}$ on $(0, \infty)$

31. $f(x) = 3x^{2/3} - 2x$ on $[0, 3]$

32. $g(x) = x^2 + 2x^{2/3}$ on $[-2, 2]$

33. $f(x) = x^{2/3}(x^2 - 4)$ on $[-1, 2]$

34. $f(x) = x^{2/3}(x^2 - 4)$ on $[-1, 3]$

35. $f(x) = \dfrac{x}{x^2 + 2}$ on $[-1, 2]$

36. $f(x) = \dfrac{1}{x^2 + 2x + 5}$ on $[-2, 1]$

37. $f(x) = \dfrac{x}{\sqrt{x^2 + 1}}$ on $[-1, 1]$

38. $g(x) = x\sqrt{4 - x^2}$ on $[0, 2]$

39. A stone is thrown straight up from the roof of an 80-ft building. The height (in feet) of the stone at any time t (in seconds), measured from the ground, is given by

$$h(t) = -16t^2 + 64t + 80$$

What is the maximum height the stone reaches?

40. Maximizing Profits Lynbrook West, an apartment complex, has 100 two-bedroom units. The monthly profit (in dollars) realized from renting out x apartments is given by

$$P(x) = -10x^2 + 1760x - 50{,}000$$

To maximize the monthly rental profit, how many units should be rented out? What is the maximum monthly profit realizable?

41. Seniors in the Workforce The percentage of men, age 65 yr and older, in the workforce from 1950 ($t = 0$) through 2000 ($t = 50$) is approximately

$$P(t) = 0.0135t^2 - 1.126t + 41.2 \qquad (0 \leq t \leq 50)$$

Show that the percentage of men, age 65 yr and older, in the workforce in the period of time under consideration was smallest around mid-September 1991. What is that percent?

Source: U.S. Census Bureau

42. Flight of a Rocket The altitude (in feet) attained by a model rocket t sec into flight is given by the function

$$h(t) = -\frac{1}{3}t^3 + 4t^2 + 20t + 2 \qquad (t \geq 0)$$

Find the maximum altitude attained by the rocket.

43. Female Self-Employed Workforce Data show that the number of nonfarm, full-time, self-employed women can be approximated by

$$N(t) = 0.81t - 1.14\sqrt{t} + 1.53 \qquad (0 \leq t \leq 6)$$

where $N(t)$ is measured in millions and t is measured in 5-yr intervals, with $t = 0$ corresponding to the beginning of 1963. Determine the absolute extrema of the function N on the interval $[0, 6]$. Interpret your results.

Source: U.S. Department of Labor

44. Average Speed of a Vehicle The average speed of a vehicle on a stretch of Route 134 between 6 a.m. and 10 a.m. on a typical weekday is approximated by the function

$$f(t) = 20t - 40\sqrt{t} + 50 \qquad (0 \leq t \leq 4)$$

where $f(t)$ is measured in miles per hour and t is measured in hours, with $t = 0$ corresponding to 6 a.m. At what time of the morning commute is the traffic moving at the slowest rate? What is the average speed of a vehicle at that time?

45. Maximizing Profits The management of Trappee and Sons, producers of the famous TexaPep hot sauce, estimate that their profit (in dollars) from the daily production and sale of x cases (each case consisting of 24 bottles) of the hot sauce is given by

$$P(x) = -0.000002x^3 + 6x - 400$$

What is the largest possible profit Trappee can make in 1 day?

46. Maximizing Profits The quantity demanded each month of the Walter Serkin recording of Beethoven's *Moonlight Sonata*, manufactured by Phonola Record Industries, is related to the price/compact disc. The equation

$$p = -0.00042x + 6 \qquad (0 \leq x \leq 12{,}000)$$

where p denotes the unit price in dollars and x is the number of discs demanded, relates the demand to the price. The total monthly cost (in dollars) for pressing and packaging x copies of this classical recording is given by

$$C(x) = 600 + 2x - 0.00002x^2 \qquad (0 \leq x \leq 20{,}000)$$

To maximize its profits, how many copies should Phonola produce each month?

Hint: The revenue is $R(x) = px$, and the profit is $P(x) = R(x) - C(x)$.

47. Maximizing Profit A manufacturer of tennis rackets finds that the total cost $C(x)$ (in dollars) of manufacturing x rackets/day is given by $C(x) = 400 + 4x + 0.0001x^2$. Each racket can be sold at a price of p dollars, where p is related to x by the demand equation $p = 10 - 0.0004x$. If all rackets that are manufactured can be sold, find the daily level of production that will yield a maximum profit for the manufacturer.

48. Maximizing Profit The weekly demand for the Pulsar 25-in. color console television is given by the demand equation

$$p = -0.05x + 600 \qquad (0 \leq x \leq 12{,}000)$$

where p denotes the wholesale unit price in dollars and x denotes the quantity demanded. The weekly total cost function associated with manufacturing these sets is given by

$$C(x) = 0.000002x^3 - 0.03x^2 + 400x + 80{,}000$$

where $C(x)$ denotes the total cost incurred in producing x sets. Find the level of production that will yield a maximum profit for the manufacturer.

Hint: Use the quadratic formula.

49. Maximizing Profit A division of Chapman Corporation manufactures a pager. The weekly fixed cost for the division is \$20,000, and the variable cost for producing x pagers/week is

$$V(x) = 0.000001x^3 - 0.01x^2 + 50x$$

dollars. The company realizes a revenue of

$$R(x) = -0.02x^2 + 150x \qquad (0 \leq x \leq 7500)$$

dollars from the sale of x pagers/week. Find the level of production that will yield a maximum profit for the manufacturer.

Hint: Use the quadratic formula.

50. **Minimizing Average Cost** Suppose the total cost function for manufacturing a certain product is $C(x) = 0.2(0.01x^2 + 120)$ dollars, where x represents the number of units produced. Find the level of production that will minimize the average cost.

51. **Minimizing Production Costs** The total monthly cost (in dollars) incurred by Cannon Precision Instruments for manufacturing x units of the model M1 camera is given by the function

$$C(x) = 0.0025x^2 + 80x + 10{,}000$$

a. Find the average cost function $\overline{C}$.
b. Find the level of production that results in the smallest average production cost.
c. Find the level of production for which the average cost is equal to the marginal cost.
d. Compare the result of part (c) with that of part (b).

52. **Minimizing Production Costs** The daily total cost (in dollars) incurred by Trappee and Sons for producing x cases of TexaPep hot sauce is given by the function

$$C(x) = 0.000002x^3 + 5x + 400$$

Using this function, answer the questions posed in Exercise 51.

53. **Maximizing Revenue** Suppose the quantity demanded per week of a certain dress is related to the unit price p by the demand equation $p = \sqrt{800 - x}$, where p is in dollars and x is the number of dresses made. To maximize the revenue, how many dresses should be made and sold each week?
Hint: $R(x) = px$.

54. **Maximizing Revenue** The quantity demanded each month of the Sicard wristwatch is related to the unit price by the equation

$$p = \frac{50}{0.01x^2 + 1} \qquad (0 \le x \le 20)$$

where p is measured in dollars and x is measured in units of a thousand. To yield a maximum revenue, how many watches must be sold?

55. **Oxygen Content of a Pond** When organic waste is dumped into a pond, the oxidation process that takes place reduces the pond's oxygen content. However, given time, nature will restore the oxygen content to its natural level. Suppose the oxygen content t days after organic waste has been dumped into the pond is given by

$$f(t) = 100\left[\frac{t^2 - 4t + 4}{t^2 + 4}\right] \qquad (0 \le t < \infty)$$

percent of its normal level.
a. When is the level of oxygen content lowest?
b. When is the rate of oxygen regeneration greatest?

56. **Air Pollution** The amount of nitrogen dioxide, a brown gas that impairs breathing, present in the atmosphere on a certain May day in the city of Long Beach is approximated by

$$A(t) = \frac{136}{1 + 0.25(t - 4.5)^2} + 28 \qquad (0 \le t \le 11)$$

where $A(t)$ is measured in pollutant standard index (PSI) and t is measured in hours, with $t = 0$ corresponding to 7 a.m. Determine the time of day when the pollution is at its highest level.

57. **Maximizing Revenue** The average revenue is defined as the function

$$\overline{R}(x) = \frac{R(x)}{x} \qquad (x > 0)$$

Prove that if a revenue function $R(x)$ is concave downward $[R''(x) < 0]$, then the level of sales that will result in the largest average revenue occurs when $\overline{R}(x) = R'(x)$.

58. **Velocity of Blood** According to a law discovered by the 19th-century physician Jean Louis Marie Poiseuille, the velocity (in centimeters/second) of blood r cm from the central axis of an artery is given by

$$v(r) = k(R^2 - r^2)$$

where k is a constant and R is the radius of the artery. Show that the velocity of blood is greatest along the central axis.

59. **GDP of a Developing Country** A developing country's gross domestic product (GDP) from 2000 to 2008 is approximated by the function

$$G(t) = -0.2t^3 + 2.4t^2 + 60 \qquad (0 \le t \le 8)$$

where $G(t)$ is measured in billions of dollars and $t = 0$ corresponds to 2000. Show that the growth rate of the country's GDP was maximal in 2004.

60. **Crime Rates** The number of major crimes committed in the city of Bronxville between 2000 and 2007 is approximated by the function

$$N(t) = -0.1t^3 + 1.5t^2 + 100 \qquad (0 \le t \le 7)$$

where $N(t)$ denotes the number of crimes committed in year t ($t = 0$ corresponds to 2000). Enraged by the dramatic increase in the crime rate, the citizens of Bronxville, with the help of the local police, organized "Neighborhood Crime Watch" groups in early 2004 to combat this menace. Show that the growth in the crime rate was maximal in 2005, giving credence to the claim that the Neighborhood Crime Watch program was working.

61. **Foreign-Born Residents** The percentage of foreign-born residents in the United States from 1910 through 2000 is approximated by the function

$$P(t) = 0.04363t^3 - 0.267t^2 - 1.59t + 14.7 \qquad (0 \le t \le 9)$$

where t is measured in decades, with $t = 0$ corresponding to 1910. Show that the percentage of foreign-born residents was lowest in early 1970.
Hint: Use the quadratic formula.
Source: Journal of American Medical Association

62. **Brain Growth and IQs** In a study conducted at the National Institute of Mental Health, researchers followed the development of the cortex, the thinking part of the brain, in 307 children. Using repeated magnetic resonance imaging scans from childhood to the latter teens, they measured the thickness (in millimeters) of the cortex of children of age t yr with the highest IQs—121 to 149. These data lead to the model

$$S(t) = 0.000989t^3 - 0.0486t^2 + 0.7116t + 1.46 \qquad (5 \le t \le 19)$$

Show that the cortex of children with superior intelligence reaches maximum thickness around age 11.
Hint: Use the quadratic formula.
Source: Nature

63. **Brain Growth and IQs** Refer to Exercise 62. The researchers at the Institute also measured the thickness (also in millimeters) of the cortex of children of age t yr who were of average intelligence. These data lead to the model

$$A(t) = -0.00005t^3 - 0.000826t^2 + 0.0153t + 4.55 \qquad (5 \le t \le 19)$$

Show that the cortex of children with average intelligence reaches maximum thickness at age 6 yr.
Source: Nature

64. **Average Prices of Homes** The average annual price of single-family homes in Massachusetts between 1990 and 2002 is approximated by the function

$$P(t) = -0.183t^3 + 4.65t^2 - 17.3t + 200 \qquad (0 \le t \le 12)$$

where $P(t)$ is measured in thousands of dollars and t is measured in years, with $t = 0$ corresponding to 1990. In what year was the average annual price of single-family homes in Massachusetts lowest? What was the approximate lowest average annual price?
Hint: Use the quadratic formula.
Source: Massachusetts Association of Realtors

65. **Office Rents** After the economy softened, the sky-high office space rents of the late 1990s started to come down to earth. The function R gives the approximate price per square foot in dollars, $R(t)$, of prime space in Boston's Back Bay and Financial District from 1997 ($t = 0$) through 2002, where

$$R(t) = -0.711t^3 + 3.76t^2 + 0.2t + 36.5 \qquad (0 \le t \le 5)$$

Show that the office space rents peaked at about the middle of 2000. What was the highest office space rent during the period in question?
Hint: Use the quadratic formula.
Source: Meredith & Grew Inc./Oncor

66. **World Population** The total world population is forecast to be

$$P(t) = 0.00074t^3 - 0.0704t^2 + 0.89t + 6.04 \qquad (0 \le t \le 10)$$

in year t, where t is measured in decades with $t = 0$ corresponding to 2000 and $P(t)$ is measured in billions.

a. Show that the world population is forecast to peak around 2071.
Hint: Use the quadratic formula.
b. What will the population peak at?

Source: International Institute for Applied Systems Analysis

67. **Venture-Capital Investment** Venture-capital investment increased dramatically in the late 1990s but came to a screeching halt after the dot-com bust. The venture-capital investment (in billions of dollars) from 1995 ($t = 0$) through 2003 ($t = 8$) is approximated by the function

$$C(t) = \begin{cases} 0.6t^2 + 2.4t + 7.6 & \text{if } 0 \le t < 3 \\ 3t^2 + 18.8t - 63.2 & \text{if } 3 \le t < 5 \\ -3.3167t^3 + 80.1t^2 - 642.583t + 1730.8025 & \text{if } 5 \le t < 8 \end{cases}$$

a. In what year did venture-capital investment peak over the period under consideration? What was the amount of that investment?
b. In what year was the venture-capital investment lowest over this period? What was the amount of that investment?
Hint: Find the absolute extrema of C on each of the closed intervals [0, 3], [3, 5], and [5, 8].

Sources: Venture One; Ernst & Young

68. **Energy Expended by a Fish** It has been conjectured that a fish swimming a distance of L ft at a speed of v ft/sec relative to the water and against a current flowing at the rate of u ft/sec ($u < v$) expends a total energy given by

$$E(v) = \frac{aLv^3}{v - u}$$

where E is measured in foot-pounds (ft-lb) and a is a constant. Find the speed v at which the fish must swim in order to minimize the total energy expended. (*Note:* This result has been verified by biologists.)

69. **Reaction to a Drug** The strength of a human body's reaction R to a dosage D of a certain drug is given by

$$R = D^2\left(\frac{k}{2} - \frac{D}{3}\right)$$

where k is a positive constant. Show that the maximum reaction is achieved if the dosage is k units.

70. Refer to Exercise 69. Show that the rate of change in the reaction R with respect to the dosage D is maximal if $D = k/2$.

71. **Maximum Power Output** Suppose the source of current in an electric circuit is a battery. Then the power output P (in watts) obtained if the circuit has a resistance of R ohms is given by

$$P = \frac{E^2R}{(R + r)^2}$$

where E is the electromotive force in volts and r is the internal resistance of the battery in ohms. If E and r are constant, find the value of R that will result in the greatest power output. What is the maximum power output?

72. **Velocity of a Wave** In deep water, a wave of length L travels with a velocity

$$v = k\sqrt{\frac{L}{C} + \frac{C}{L}}$$

where k and C are positive constants. Find the length of the wave that has a minimum velocity.

73. **Chemical Reaction** In an autocatalytic chemical reaction, the product formed acts as a catalyst for the reaction. If Q is the amount of the original substrate present initially and x is the amount of catalyst formed, then the rate of change of the chemical reaction with respect to the amount of catalyst present in the reaction is

$$R(x) = kx(Q - x) \qquad (0 \le x \le Q)$$

where k is a constant. Show that the rate of the chemical reaction is greatest at the point when exactly half of the original substrate has been transformed.

74. **A Mixture Problem** A tank initially contains 10 gal of brine with 2 lb of salt. Brine with 1.5 lb of salt per gallon enters the tank at the rate of 3 gal/min, and the well-stirred mixture leaves the tank at the rate of 4 gal/min. It can be shown that the amount of salt in the tank after t minutes is x lb where

$$x = f(t) = 1.5(10 - t) - 0.0013(10 - t)^4 \qquad (0 \le t \le 10)$$

What is the maximum amount of salt present in the tank at any time?

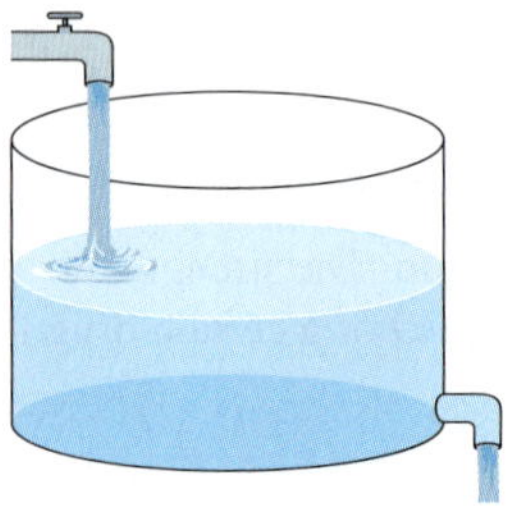

In Exercises 75–78, determine whether the statement is true or false. If it is true, explain why it is true. If it is false, give an example to show why it is false.

75. If f is defined on a closed interval $[a, b]$, then f has an absolute maximum value.

76. If f is continuous on an open interval (a, b), then f does not have an absolute minimum value.

77. If f is not continuous on the closed interval $[a, b]$, then f cannot have an absolute maximum value.

78. If $f''(x) < 0$ on (a, b) and $f'(c) = 0$ where $a < c < b$, then $f(c)$ is the absolute maximum value of f on $[a, b]$.

79. Let f be a constant function—that is, let $f(x) = c$, where c is some real number. Show that every number a gives rise to an absolute maximum and, at the same time, an absolute minimum of f.

80. Show that a polynomial function defined on the interval $(-\infty, \infty)$ cannot have both an absolute maximum and an absolute minimum unless it is a constant function.

81. One condition that must be satisfied before Theorem 3 (p. 299) is applicable is that the function f must be continuous on the closed interval $[a, b]$. Define a function f on the closed interval $[-1, 1]$ by

$$f(x) = \begin{cases} \dfrac{1}{x} & \text{if } x \in [-1, 1] \quad (x \neq 0) \\ 0 & \text{if } x = 0 \end{cases}$$

a. Show that f is not continuous at $x = 0$.
b. Show that $f(x)$ does not attain an absolute maximum or an absolute minimum on the interval $[-1, 1]$.
c. Confirm your results by sketching the function f.

82. One condition that must be satisfied before Theorem 3 (page 299) is applicable is that the interval on which f is defined must be a closed interval $[a, b]$. Define a function f on the *open* interval $(-1, 1)$ by $f(x) = x$. Show that f does not attain an absolute maximum or an absolute minimum on the interval $(-1, 1)$.
Hint: What happens to $f(x)$ if x is close to but not equal to $x = -1$? If x is close to but not equal to $x = 1$?

4.4 Solutions to Self-Check Exercises

1. a. The function f is continuous in its domain and differentiable on the interval $(0, 9)$. The derivative of f is

$$f'(x) = 1 - x^{-1/2} = \frac{x^{1/2} - 1}{x^{1/2}}$$

and it is equal to zero when $x = 1$. Evaluating $f(x)$ at the endpoints $x = 0$ and $x = 9$ and at the critical number 1 of f, we have

$$f(0) = 0 \qquad f(1) = -1 \qquad f(9) = 3$$

From these results, we see that -1 is the absolute minimum value of f and 3 is the absolute maximum value of f.

b. In this case, the domain of f is the interval $[0, \infty)$, which is not closed. Therefore, we resort to the graphical method. Using the techniques of graphing, we sketch the graph of f in the accompanying figure.

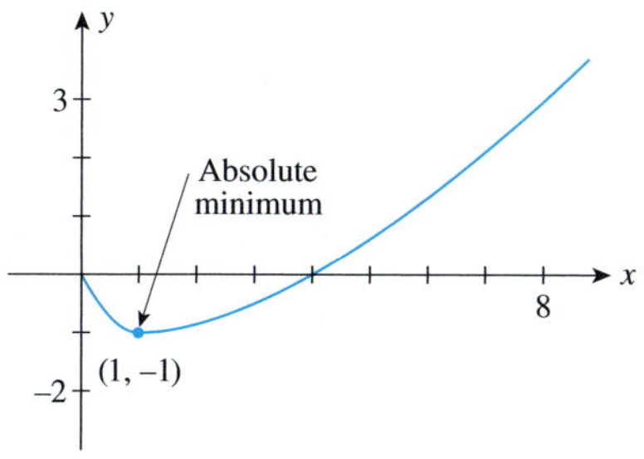

The graph of f shows that -1 is the absolute minimum value of f, but f has no absolute maximum since $f(x)$ increases without bound as x increases without bound.

2. The function f is continuous on the interval $[-2, 1]$. It is also differentiable on the open interval $(-2, 1)$. The derivative of f is

$$f'(x) = 12x^3 + 12x^2 = 12x^2(x + 1)$$

and it is continuous on $(-2, 1)$. Setting $f'(x) = 0$ gives -1 and 0 as critical numbers of f. Evaluating $f(x)$ at these critical numbers of f as well as at the endpoints of the interval $[-2, 1]$, we obtain

$$f(-2) = 17 \qquad f(-1) = 0 \qquad f(0) = 1 \qquad f(1) = 8$$

From these results, we see that 0 is the absolute minimum value of f and 17 is the absolute maximum value of f.

3. The problem is solved by finding the absolute maximum of the function f on $[0, 250]$. Differentiating $f(t)$, we obtain

$$f'(t) = \frac{(t^2 + 40{,}000)(1200) - 1200t(2t)}{(t^2 + 40{,}000)^2} = \frac{-1200(t^2 - 40{,}000)}{(t^2 + 40{,}000)^2}$$

Upon setting $f'(t) = 0$ and solving the resulting equation, we obtain $t = -200$ or 200. Since -200 lies outside the interval $[0, 250]$, we are interested only in the critical number 200 of f. Evaluating $f(t)$ at $t = 0$, $t = 200$, and $t = 250$, we find

$$f(0) = 80 \qquad f(200) = 83 \qquad f(250) = 82.93$$

We conclude that the manufacturing capacity operating rate was the highest on the 200th day of 2006—that is, a little past the middle of July 2006.

USING TECHNOLOGY

Finding the Absolute Extrema of a Function

Some graphing utilities have a function for finding the absolute maximum and the absolute minimum values of a continuous function on a closed interval. If your graphing utility has this capability, use it to work through the example and exercises of this section.

EXAMPLE 1 Let $f(x) = \dfrac{2x + 4}{(x^2 + 1)^{3/2}}$.

a. Use a graphing utility to plot the graph of f in the viewing window $[-3, 3] \times [-1, 5]$.

b. Find the absolute maximum and absolute minimum values of f on the interval $[-3, 3]$. Express your answers accurate to four decimal places.

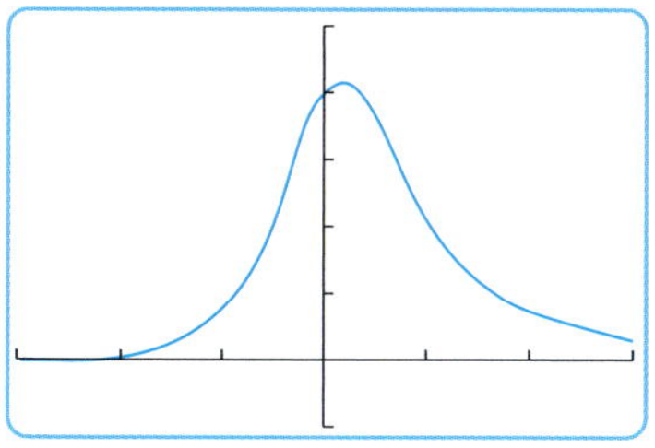

FIGURE **T1**
The graph of f in the viewing window $[-3, 3] \times [-1, 5]$

Solution

a. The graph of f is shown in Figure T1.

b. Using the function on a graphing utility for finding the absolute minimum value of a continuous function on a closed interval, we find the absolute minimum value of f to be -0.0632. Similarly, using the function for finding the absolute maximum value, we find the absolute maximum value to be 4.1593. ■

Note Some graphing utilities will enable you to find the absolute minimum and absolute maximum values of a continuous function on a closed interval without having to graph the function. For example, using **fMax** on the TI-83/84 will yield the x-coordinate of the absolute maximum of f. The absolute maximum value can then be found by evaluating f at that value of x. Figure T2 shows the work involved in finding the absolute maximum of the function of Example 1. ■

```
fMax((2X+4)/(X^2
+1)^1.5,X,-3,3)
        .1583117413
(2*.1583+4)/(.15
83^2+1)^1.5
        4.15928406
```

FIGURE **T2**
The TI-83/84 screen for Example 1

(continued)

TECHNOLOGY EXERCISES

In Exercises 1–6, find the absolute maximum and the absolute minimum values of *f* in the given interval using the method of Example 1. Express your answers accurate to four decimal places.

1. $f(x) = 3x^4 - 4.2x^3 + 6.1x - 2$; $[-2, 3]$
2. $f(x) = 2.1x^4 - 3.2x^3 + 4.1x^2 + 3x - 4$; $[-1, 2]$
3. $f(x) = \dfrac{2x^3 - 3x^2 + 1}{x^2 + 2x - 8}$; $[-3, 1]$
4. $f(x) = \sqrt{x}(x^3 - 4)^2$; $[0.5, 1]$
5. $f(x) = \dfrac{x^3 - 1}{x^2}$; $[1, 3]$
6. $f(x) = \dfrac{x^3 - x^2 + 1}{x - 2}$; $[1, 3]$
7. **Use of Diesel Engines** Diesel engines are popular in cars in Europe, where fuel prices are high. The percentage of new vehicles in Western Europe equipped with diesel engines is approximated by the function

$$f(t) = 0.3t^4 - 2.58t^3 + 8.11t^2 - 7.71t + 23.75 \quad (0 \le t \le 4)$$

where t is measured in years, with $t = 0$ corresponding to the beginning of 1996.

a. Use a graphing utility to sketch the graph of f on $[0, 4] \times [0, 40]$.

b. What was the lowest percentage of new vehicles equipped with diesel engines for the period in question?

Source: German Automobile Industry Association

8. **Demand for Electricity** The demand for electricity from 1 a.m. through 7 p.m. on August 1, 2006, in Boston is described by the function

$$D(t) = -11.3975t^3 + 285.991t^2 - 1467.73t + 23{,}755 \quad (0 \le t \le 18)$$

where $D(t)$ is measured in megawatts (MW), with $t = 0$ corresponding to 1 a.m. Driven overwhelmingly by air-conditioning and refrigeration systems, the demand for electricity reached a new record high that day. Show that the demand for electricity did not exceed the system capacity of 31,000 MW, thus negating the necessity for imposing rolling blackouts if electricity demand were to exceed supply.

Source: ISO New England

9. **Sickouts** In a sickout by pilots of American Airlines in February 1999, the number of canceled flights from February 6 ($t = 0$) through February 14 ($t = 8$) is approximated by the function

$$N(t) = 1.2576t^4 - 26.357t^3 + 127.98t^2 + 82.3t + 43 \quad (0 \le t \le 8)$$

where t is measured in days. The sickout ended after the union was threatened with millions of dollars of fines.

a. Show that the number of canceled flights was increasing at the fastest rate on February 8.

b. Estimate the maximum number of canceled flights in a day during the sickout.

Source: Associated Press

10. **Modeling with Data** The following data give the average account balance (in thousands of dollars) of a 401(k) investor from 1996 through 2002.

Year	1996	1997	1998	1999	2000	2001	2002
Account Balance	37.5	40.8	47.3	55.5	49.4	43	40

a. Use **QuartReg** to find a fourth-degree polynomial regression model for the data. Let $t = 0$ correspond to 1996.

b. Plot the graph of A, using the viewing window $[0, 6] \times [0, 60]$.

c. When was the average account balance lowest in the period under consideration? When was it highest?

d. What were the lowest average account balance and the highest average account balance during the period under consideration?

Source: Investment Company Institute

4.5 Optimization II

Section 4.4 outlined how to find the solution to certain optimization problems in which the objective function is given. In this section, we consider problems in which we are required to first find the appropriate function to be optimized. The following guidelines will be useful for solving these problems.

Guidelines for Solving Optimization Problems

1. Assign a letter to each variable mentioned in the problem. If appropriate, draw and label a figure.
2. Find an expression for the quantity to be optimized.

3. Use the conditions given in the problem to write the quantity to be optimized as a function f of *one* variable. Note any restrictions to be placed on the domain of f from physical considerations of the problem.
4. Optimize the function f over its domain using the methods of Section 4.4.

Note In carrying out step 4, remember that if the function f to be optimized is continuous on a closed interval, then the absolute maximum and absolute minimum of f are, respectively, the largest and smallest values of $f(x)$ on the set composed of the critical numbers of f and the endpoints of the interval. If the domain of f is not a closed interval, then we resort to the graphical method. ■

Maximization Problems

APPLIED EXAMPLE 1 Fencing a Garden A man wishes to have a rectangular-shaped garden in his backyard. He has 50 feet of fencing with which to enclose his garden. Find the dimensions for the largest garden he can have if he uses all of the fencing.

Solution

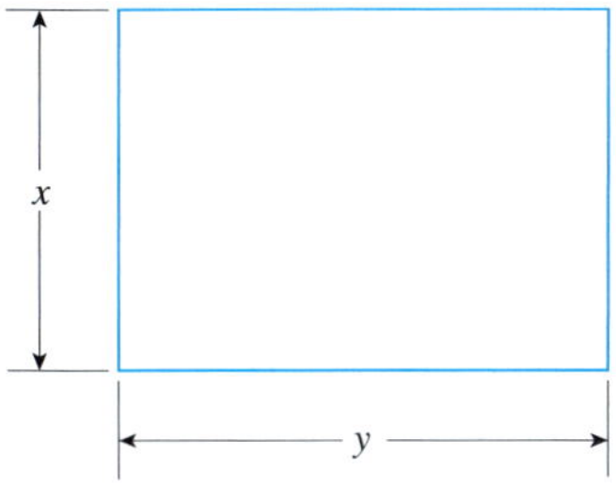

FIGURE 68
What is the maximum rectangular area that can be enclosed with 50 feet of fencing?

Step 1 Let x and y denote the dimensions (in feet) of two adjacent sides of the garden (Figure 68) and let A denote its area. See pages 83–84.

Step 2 The area of the garden

$$A = xy \tag{1}$$

is the quantity to be maximized.

Step 3 The perimeter of the rectangle, $(2x + 2y)$ feet, must equal 50 feet. Therefore, we have the equation

$$2x + 2y = 50$$

Next, solving this equation for y in terms of x yields

$$y = 25 - x \tag{2}$$

which, when substituted into Equation (1), gives

$$\begin{aligned} A &= x(25 - x) \\ &= -x^2 + 25x \end{aligned}$$

(Remember, the function to be optimized must involve just one variable.) Since the sides of the rectangle must be nonnegative, we must have $x \geq 0$ and $y = 25 - x \geq 0$; that is, we must have $0 \leq x \leq 25$. Thus, the problem is reduced to that of finding the absolute maximum of $A = f(x) = -x^2 + 25x$ on the closed interval $[0, 25]$.

Step 4 Observe that f is continuous on $[0, 25]$, so the absolute maximum value of f must occur at the endpoint(s) of the interval or at the critical number(s) of f. The derivative of the function A is given by

$$A' = f'(x) = -2x + 25$$

Setting $A' = 0$ gives

$$-2x + 25 = 0$$

or 12.5, as the critical number of A. Next, we evaluate the function $A = f(x)$ at $x = 12.5$ and at the endpoints $x = 0$ and $x = 25$ of the interval $[0, 25]$, obtaining

$$f(0) = 0 \qquad f(12.5) = 156.25 \qquad f(25) = 0$$

We see that the absolute maximum value of the function f is 156.25. From Equation (2) we see that $y = 12.5$ when $x = 12.5$. Thus, the garden of maximum area (156.25 square feet) is a square with sides of length 12.5 feet. ■

APPLIED EXAMPLE 2 Packaging By cutting away identical squares from each corner of a rectangular piece of cardboard and folding up the resulting flaps, the cardboard may be turned into an open box. If the cardboard is 16 inches long and 10 inches wide, find the dimensions of the box that will yield the maximum volume.

Solution

Step 1 Let x denote the length (in inches) of one side of each of the identical squares to be cut out of the cardboard (Figure 69) and let V denote the volume of the resulting box.

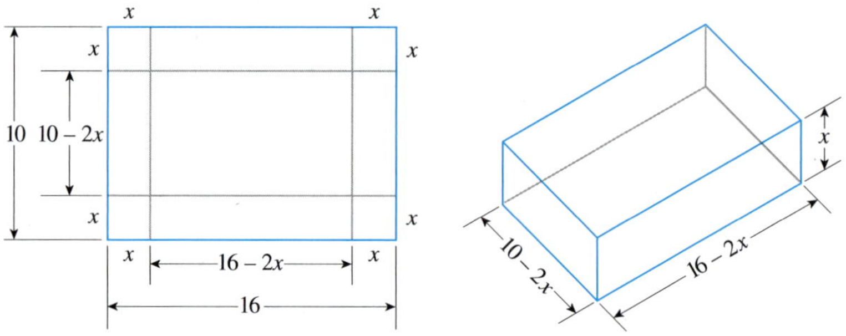

FIGURE 69
The dimensions of the open box are $(16 - 2x)$ inches by $(10 - 2x)$ inches by x inches.

Step 2 The dimensions of the box are $(16 - 2x)$ inches by $(10 - 2x)$ inches by x inches. Therefore, its volume (in cubic inches),

$$\begin{aligned} V &= (16 - 2x)(10 - 2x)x \\ &= 4(x^3 - 13x^2 + 40x) \qquad \text{Expand the expression.} \end{aligned}$$

is the quantity to be maximized.

Step 3 Since each side of the box must be nonnegative, x must satisfy the inequalities $x \geq 0$, $16 - 2x \geq 0$, and $10 - 2x \geq 0$. This set of inequalities is satisfied if $0 \leq x \leq 5$. Thus, the problem at hand is equivalent to that of finding the absolute maximum of

$$V = f(x) = 4(x^3 - 13x^2 + 40x)$$

on the closed interval [0, 5].

Step 4 Observe that f is continuous on [0, 5], so the absolute maximum value of f must be attained at the endpoint(s) or at the critical number(s) of f.

Differentiating $f(x)$, we obtain

$$\begin{aligned} f'(x) &= 4(3x^2 - 26x + 40) \\ &= 4(3x - 20)(x - 2) \end{aligned}$$

Upon setting $f'(x) = 0$ and solving the resulting equation for x, we obtain $x = \frac{20}{3}$ or $x = 2$. Since $\frac{20}{3}$ lies outside the interval [0, 5], it is no longer considered, and we are interested only in the critical number 2 of f. Next, evaluating $f(x)$ at $x = 0$, $x = 5$ (the endpoints of the interval [0, 5]), and $x = 2$, we obtain

$$f(0) = 0 \qquad f(2) = 144 \qquad f(5) = 0$$

Thus, the volume of the box is maximized by taking $x = 2$. The dimensions of the box are $12'' \times 6'' \times 2''$, and the volume is 144 cubic inches. ■

Exploring with TECHNOLOGY

Refer to Example 2.

1. Use a graphing utility to plot the graph of

$$f(x) = 4(x^3 - 13x^2 + 40x)$$

using the viewing window $[0, 5] \times [0, 150]$. Explain what happens to $f(x)$ as x increases from $x = 0$ to $x = 5$ and give a physical interpretation.

2. Using ZOOM and TRACE, find the absolute maximum of f on the interval $[0, 5]$ and thus verify the solution for Example 2 obtained analytically.

APPLIED EXAMPLE 3 Optimal Subway Fare A city's Metropolitan Transit Authority (MTA) operates a subway line for commuters from a certain suburb to the downtown metropolitan area. Currently, an average of 6000 passengers a day take the trains, paying a fare of \$3.00 per ride. The board of the MTA, contemplating raising the fare to \$3.50 per ride in order to generate a larger revenue, engages the services of a consulting firm. The firm's study reveals that for each \$.50 increase in fare, the ridership will be reduced by an average of 1000 passengers a day. Thus, the consulting firm recommends that MTA stick to the current fare of \$3.00 per ride, which already yields a maximum revenue. Show that the consultants are correct.

Solution

Step 1 Let x denote the number of passengers per day, p denote the fare per ride, and R be MTA's revenue. See pages 83–84.

Step 2 To find a relationship between x and p, observe that the given data imply that when $x = 6000$, $p = 3$, and when $x = 5000$, $p = 3.50$. Therefore, the points (6000, 3) and (5000, 3.50) lie on a straight line. (Why?) To find the linear relationship between p and x, use the point-slope form of the equation of a straight line. Now, the slope of the line is

$$m = \frac{3.50 - 3}{5000 - 6000} = -0.0005$$

Therefore, the required equation is

$$\begin{aligned} p - 3 &= -0.0005(x - 6000) \\ &= -0.0005x + 3 \\ p &= -0.0005x + 6 \end{aligned}$$

Therefore, the revenue

$$R = f(x) = xp = -0.0005x^2 + 6x \quad \text{Number of riders} \times \text{unit fare}$$

is the quantity to be maximized.

Step 3 Since both p and x must be nonnegative, we see that $0 \le x \le 12{,}000$, and the problem is that of finding the absolute maximum of the function f on the closed interval [0, 12,000].

Step 4 Observe that f is continuous on [0, 12,000]. To find the critical number of R, we compute

$$f'(x) = -0.001x + 6$$

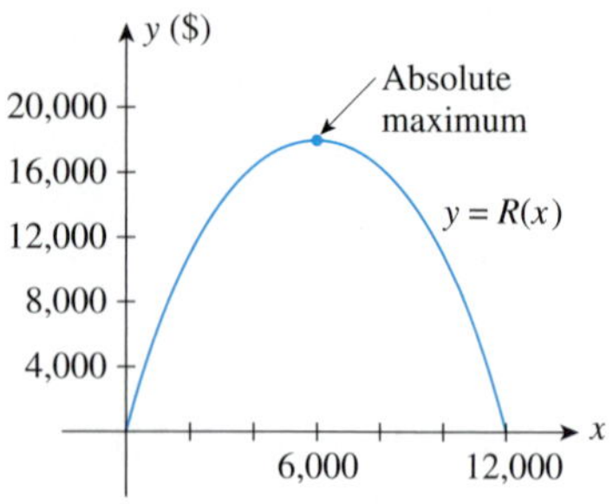

FIGURE 70
f has an absolute maximum of 18,000 when $x = 6000$.

and set it equal to zero, giving $x = 6000$. Evaluating the function f at $x = 6000$, as well as at the endpoints $x = 0$ and $x = 12{,}000$, yields

$$f(0) = 0$$
$$f(6000) = 18{,}000$$
$$f(12{,}000) = 0$$

We conclude that a maximum revenue of \$18,000 per day is realized when the ridership is 6000 per day. The optimum price of the fare per ride is therefore \$3.00, as recommended by the consultants. The graph of the revenue function R is shown in Figure 70. ■

Minimization Problems

APPLIED EXAMPLE 4 Packaging Betty Moore Company requires that its corned beef hash containers have a capacity of 54 cubic inches, have the shape of right circular cylinders, and be made of aluminum. Determine the radius and height of the container that requires the least amount of metal.

Solution

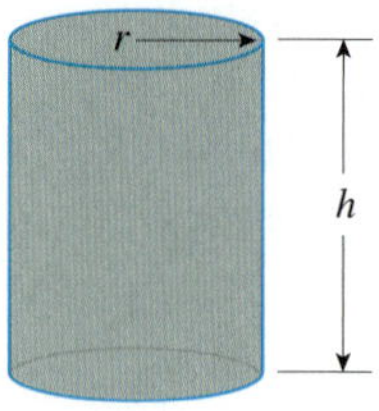

FIGURE 71
We want to minimize the amount of material used to construct the container.

Step 1 Let the radius and height of the container be r and h inches, respectively, and let S denote the surface area of the container (Figure 71).

Step 2 The amount of aluminum used to construct the container is given by the total surface area of the cylinder. Now, the area of the base and the top of the cylinder are each πr^2 square inches and the area of the side is $2\pi rh$ square inches. Therefore,

$$S = 2\pi r^2 + 2\pi rh \tag{3}$$

is the quantity to be minimized.

Step 3 The requirement that the volume of a container be 54 cubic inches implies that

$$\pi r^2 h = 54 \tag{4}$$

Solving Equation (4) for h, we obtain

$$h = \frac{54}{\pi r^2} \tag{5}$$

which, when substituted into (3), yields

$$S = 2\pi r^2 + 2\pi r\left(\frac{54}{\pi r^2}\right)$$
$$= 2\pi r^2 + \frac{108}{r}$$

Clearly, the radius r of the container must satisfy the inequality $r > 0$. The problem now is reduced to finding the absolute minimum of the function $S = f(r)$ on the interval $(0, \infty)$.

Step 4 Using the curve-sketching techniques of Section 4.3, we obtain the graph of f in Figure 72.

To find the critical number of f, we compute

$$S' = 4\pi r - \frac{108}{r^2}$$

and solve the equation $S' = 0$ for r:

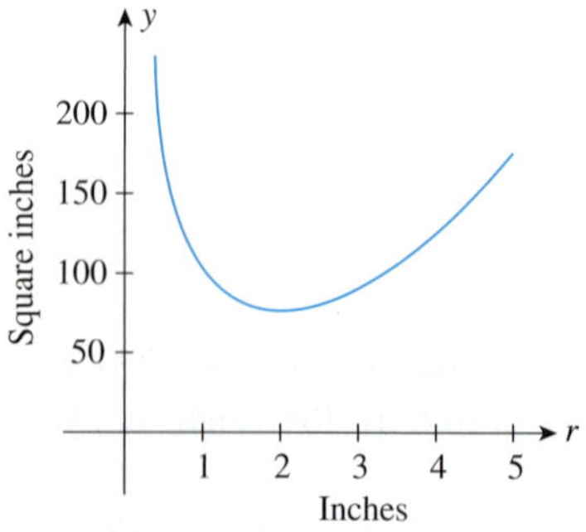

FIGURE 72
The total surface area of the right cylindrical container is graphed as a function of r.

$$
\begin{aligned}
4\pi r - \frac{108}{r^2} &= 0 \\
4\pi r^3 - 108 &= 0 \\
r^3 &= \frac{27}{\pi} \\
r &= \frac{3}{\sqrt[3]{\pi}} \approx 2
\end{aligned} \tag{6}
$$

Next, let's show that this value of r gives rise to the absolute minimum of f. To show this, we first compute

$$S'' = 4\pi + \frac{216}{r^3}$$

Since $S'' > 0$ for $r = 3/\sqrt[3]{\pi}$, the second derivative test implies that the value of r in Equation (6) gives rise to a relative minimum of f. Finally, this relative minimum of f is also the absolute minimum of f since f is always concave upward ($S'' > 0$ for all $r > 0$). To find the height of the given container, we substitute the value of r given in (6) into (5). Thus,

$$
\begin{aligned}
h = \frac{54}{\pi r^2} &= \frac{54}{\pi\left(\dfrac{3}{\pi^{1/3}}\right)^2} \\
&= \frac{54\pi^{2/3}}{(\pi)9} \\
&= \frac{6}{\pi^{1/3}} = \frac{6}{\sqrt[3]{\pi}} \\
&= 2r
\end{aligned}
$$

We conclude that the required container has a radius of approximately 2 inches and a height of approximately 4 inches, or twice the size of the radius. ■

An Inventory Problem

One problem faced by many companies is that of controlling the inventory of goods carried. Ideally, the manager must ensure that the company has sufficient stock to meet customer demand at all times. At the same time, she must make sure that this is accomplished without overstocking (incurring unnecessary storage costs) and also without having to place orders too frequently (incurring reordering costs).

APPLIED EXAMPLE 5 Inventory Control and Planning Dixie Import-Export is the sole agent for the Excalibur 250-cc motorcycle. Management estimates that the demand for these motorcycles is 10,000 per year and that they will sell at a uniform rate throughout the year. The cost incurred in ordering each shipment of motorcycles is \$10,000, and the cost per year of storing each motorcycle is \$200.

Dixie's management faces the following problem: Ordering too many motorcycles at one time ties up valuable storage space and increases the storage cost. On the other hand, placing orders too frequently increases the ordering costs. How large should each order be, and how often should orders be placed, to minimize ordering and storage costs?

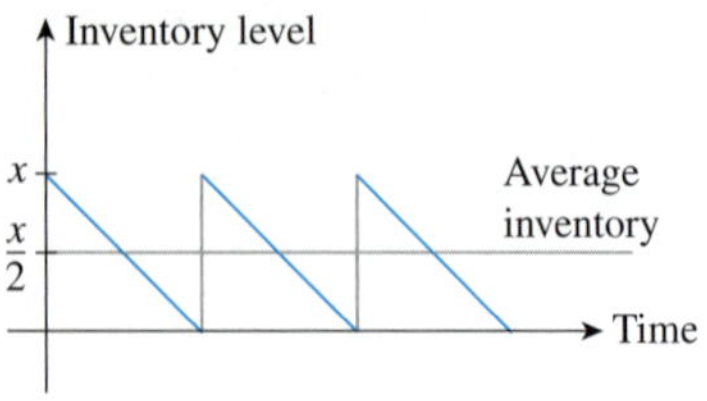

FIGURE 73
As each lot is depleted, the new lot arrives. The average inventory level is $x/2$ if x is the lot size.

Solution Let x denote the number of motorcycles in each order (the lot size). Then, assuming that each shipment arrives just as the previous shipment has been sold, the average number of motorcycles in storage during the year is $x/2$. You can see that this is the case by examining Figure 73. Thus, Dixie's storage cost for the year is given by $200(x/2)$, or $100x$ dollars.

Next, since the company requires 10,000 motorcycles for the year and since each order is for x motorcycles, the number of orders required is

$$\frac{10{,}000}{x}$$

This gives an ordering cost of

$$10{,}000\left(\frac{10{,}000}{x}\right) = \frac{100{,}000{,}000}{x}$$

dollars for the year. Thus, the total yearly cost incurred by Dixie, which includes the ordering and storage costs attributed to the sale of these motorcycles, is given by

$$C(x) = 100x + \frac{100{,}000{,}000}{x}$$

The problem is reduced to finding the absolute minimum of the function C on the interval $(0, 10{,}000]$. To accomplish this, we compute

$$C'(x) = 100 - \frac{100{,}000{,}000}{x^2}$$

Setting $C'(x) = 0$ and solving the resulting equation, we obtain $x = \pm 1000$. Since the number -1000 is outside the domain of the function C, it is rejected, leaving 1000 as the only critical number of C. Next, we find

$$C''(x) = \frac{200{,}000{,}000}{x^3}$$

Since $C''(1000) > 0$, the second derivative test implies that the critical number 1000 is a relative minimum of the function C (Figure 74). Also, since $C''(x) > 0$ for all x in $(0, 10{,}000)$, the function C is concave upward everywhere so that $x = 1000$ also gives the absolute minimum of C. Thus, to minimize the ordering and storage costs, Dixie should place 10,000/1000, or 10, orders a year, each for a shipment of 1000 motorcycles. ■

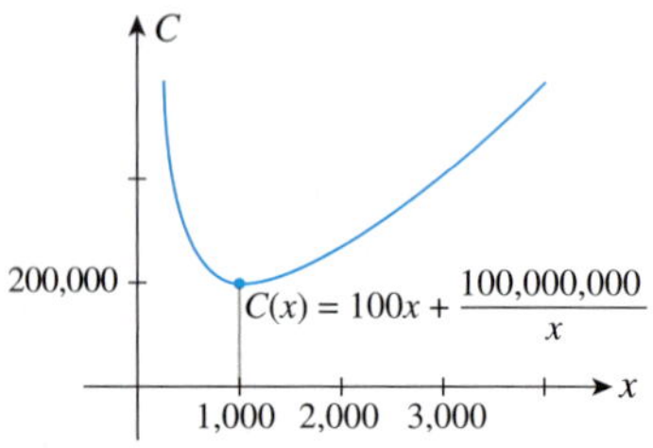

FIGURE 74
C has an absolute minimum at (1000, 200,000).

4.5 Self-Check Exercises

1. A man wishes to have an enclosed vegetable garden in his backyard. If the garden is to be a rectangular area of 300 ft^2, find the dimensions of the garden that will minimize the amount of fencing needed.

2. The demand for the Super Titan tires is 1,000,000/year. The setup cost for each production run is \$4000, and the manufacturing cost is \$20/tire. The cost of storing each tire over the year is \$2. Assuming uniformity of demand throughout the year and instantaneous production, determine how many tires should be manufactured per production run in order to keep the production cost to a minimum.

Solutions to Self-Check Exercises 4.5 can be found on page 323.

4.5 Concept Questions

1. If the domain of a function f is not a closed interval, how would you find the absolute extrema of f, if they exist?

2. Refer to Example 4 (page 316). In the solution given in the example, we solved for h in terms of r, resulting in a function of r, which we then optimized with respect to r. Write S in terms of h and re-solve the problem. Which choice is better?

4.5 Exercises

1. Find the dimensions of a rectangle with a perimeter of 100 ft that has the largest possible area.

2. Find the dimensions of a rectangle of area 144 sq ft that has the smallest possible perimeter.

3. **Enclosing the Largest Area** The owner of the Rancho Los Feliz has 3000 yd of fencing with which to enclose a rectangular piece of grazing land along the straight portion of a river. If fencing is not required along the river, what are the dimensions of the largest area that he can enclose? What is this area?

4. **Enclosing the Largest Area** Refer to Exercise 3. As an alternative plan, the owner of the Rancho Los Feliz might use the 3000 yd of fencing to enclose the rectangular piece of grazing land along the straight portion of the river and then subdivide it by means of a fence running parallel to the sides. Again, no fencing is required along the river. What are the dimensions of the largest area that can be enclosed? What is this area? (See the accompanying figure.)

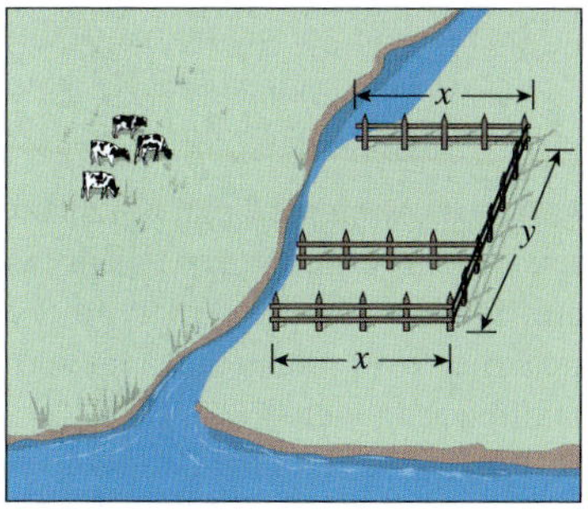

5. **Minimizing Construction Costs** The management of the UNICO department store has decided to enclose an 800-ft^2 area outside the building for displaying potted plants and flowers. One side will be formed by the external wall of the store, two sides will be constructed of pine boards, and the fourth side will be made of galvanized steel fencing. If the pine board fencing costs \$6/running foot and the steel fencing costs \$3/running foot, determine the dimensions of the enclosure that can be erected at minimum cost.

6. **Packaging** By cutting away identical squares from each corner of a rectangular piece of cardboard and folding up the resulting flaps, an open box may be made. If the cardboard is 15 in. long and 8 in. wide, find the dimensions of the box that will yield the maximum volume.

7. **Metal Fabrication** If an open box is made from a tin sheet 8 in. square by cutting out identical squares from each corner and bending up the resulting flaps, determine the dimensions of the largest box that can be made.

8. **Minimizing Packaging Costs** If an open box has a square base and a volume of 108 in.^{3} and is constructed from a tin sheet, find the dimensions of the box, assuming a minimum amount of material is used in its construction.

9. **Minimizing Packaging Costs** What are the dimensions of a closed rectangular box that has a square cross section, a capacity of 128 in.^{3}, and is constructed using the least amount of material?

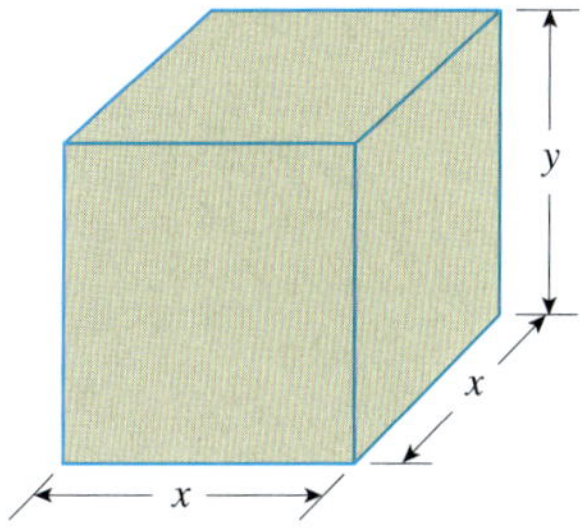

10. **Minimizing Packaging Costs** A rectangular box is to have a square base and a volume of 20 ft^3. If the material for the base costs 30¢/square foot, the material for the sides costs 10¢/square foot, and the material for the top costs 20¢/square foot, determine the dimensions of the box that can be constructed at minimum cost. (Refer to the figure for Exercise 9.)

11. **Parcel Post Regulations** Postal regulations specify that a parcel sent by priority mail may have a combined length and girth of no more than 108 in. Find the dimensions of a rectangular package that has a square cross section and the largest volume that may be sent via priority mail. What is the volume of such a package?
Hint: The length plus the girth is $4x + h$ (see the accompanying figure).

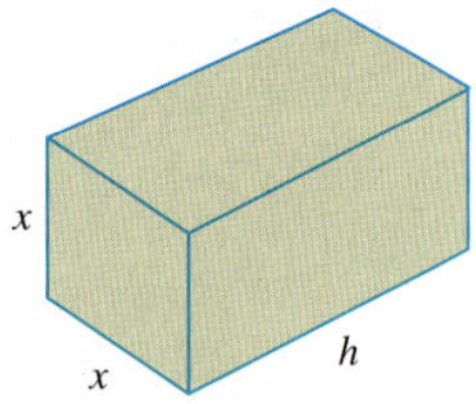

12. **Book Design** A book designer has decided that the pages of a book should have 1-in. margins at the top and bottom and $\frac{1}{2}$-in. margins on the sides. She further stipulated that each page should have an area of 50 $in.^2$ (see the accompanying figure). Determine the page dimensions that will result in the maximum printed area on the page.

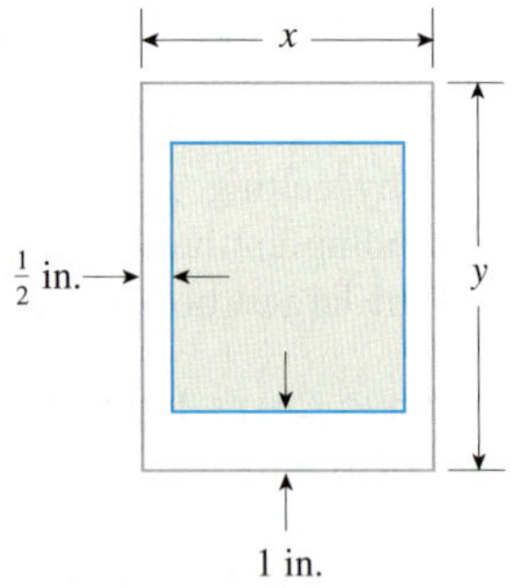

13. **Parcel Post Regulations** Postal regulations specify that a parcel sent by priority mail may have a combined length and girth of no more than 108 in. Find the dimensions of the cylindrical package of greatest volume that may be sent via priority mail. What is the volume of such a package? Compare with Exercise 11.
Hint: The length plus the girth is $2\pi r + l$.

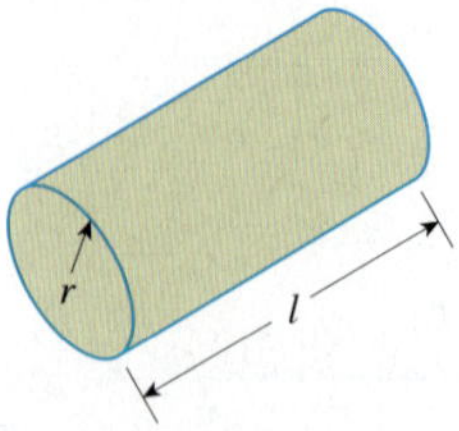

14. **Minimizing Costs** For its beef stew, Betty Moore Company uses aluminum containers that have the form of right circular cylinders. Find the radius and height of a container if it has a capacity of 36 $in.^3$ and is constructed using the least amount of metal.

15. **Product Design** The cabinet that will enclose the Acrosonic model D loudspeaker system will be rectangular and will have an internal volume of 2.4 ft^3. For aesthetic reasons, it has been decided that the height of the cabinet is to be 1.5 times its width. If the top, bottom, and sides of the cabinet are constructed of veneer costing 40¢/square foot and the front (ignore the cutouts in the baffle) and rear are constructed of particle board costing 20¢/square foot, what are the dimensions of the enclosure that can be constructed at a minimum cost?

16. **Designing a Norman Window** A Norman window has the shape of a rectangle surmounted by a semicircle (see the accompanying figure). If a Norman window is to have a perimeter of 28 ft, what should its dimensions be in order to allow the maximum amount of light through the window?

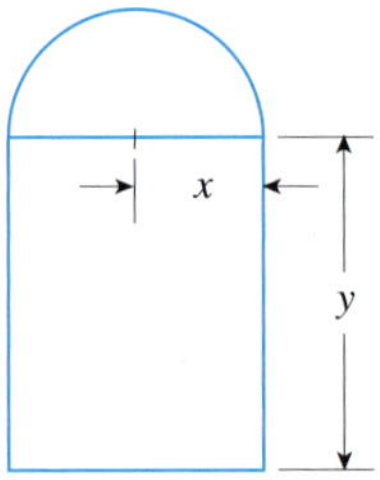

17. **Optimal Charter-Flight Fare** If exactly 200 people sign up for a charter flight, Leisure World Travel Agency charges \$300/person. However, if more than 200 people sign up for the flight (assume this is the case), then each fare is reduced by \$1 for each additional person. Determine how many passengers will result in a maximum revenue for the travel agency. What is the maximum revenue? What would be the fare per passenger in this case?
Hint: Let x denote the number of passengers above 200. Show that the revenue function R is given by $R(x) = (200 + x)(300 - x)$.

18. **Maximizing Yield** An apple orchard has an average yield of 36 bushels of apples/tree if tree density is 22 trees/acre. For each unit increase in tree density, the yield decreases by 2 bushels/tree. How many trees should be planted in order to maximize the yield?

19. **Charter Revenue** The owner of a luxury motor yacht that sails among the 4000 Greek islands charges \$600/person/day if exactly 20 people sign up for the cruise. However, if more than 20 people sign up (up to the maximum capacity of 90) for the cruise, then each fare is reduced by \$4 for each additional passenger. Assuming at least 20 people sign up for the cruise, determine how many passengers will result in the maximum revenue for the owner of the yacht. What is the maximum revenue? What would be the fare/passenger in this case?

20. **Profit of a Vineyard** Phillip, the proprietor of a vineyard, estimates that the first 10,000 bottles of wine produced this season will fetch a profit of \$5/bottle. But if more than

10,000 bottles were produced, then the profit/bottle for the entire lot would drop by \$0.0002 for each additional bottle sold. Assuming at least 10,000 bottles of wine are produced and sold, what is the maximum profit?

21. **Optimal Speed of a Truck** A truck gets $600/x$ mpg when driven at a constant speed of x mph (between 50 and 70 mph). If the price of fuel is \$3/gallon and the driver is paid \$18/hour, at what speed between 50 and 70 mph is it most economical to drive?

22. **Minimizing Costs** Suppose the cost incurred in operating a cruise ship for one hour is $a + bv^3$ dollars, where a and b are positive constants and v is the ship's speed in miles per hour. At what speed should the ship be operated between two ports, to minimize the cost?

23. **Strength of a Beam** A wooden beam has a rectangular cross section of height h in. and width w in. (see the accompanying figure). The strength S of the beam is directly proportional to its width and the square of its height. What are the dimensions of the cross section of the strongest beam that can be cut from a round log of diameter 24 in.?
Hint: $S = kh^2w$, where k is a constant of proportionality.

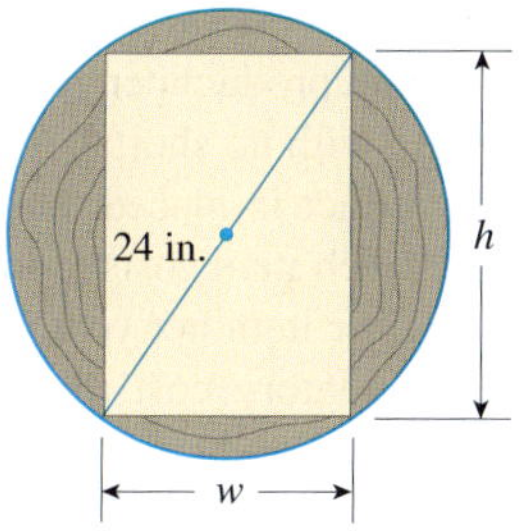

24. **Designing a Grain Silo** A grain silo has the shape of a right circular cylinder surmounted by a hemisphere (see the accompanying figure). If the silo is to have a capacity of 504π ft^3, find the radius and height of the silo that requires the least amount of material to construct.
Hint: The volume of the silo is $\pi r^2h + \frac{2}{3}\pi r^3$, and the surface area (including the floor) is $\pi(3r^2 + 2rh)$.

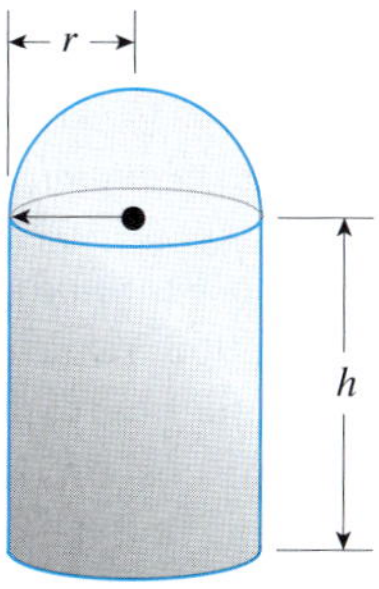

25. **Minimizing Cost of Laying Cable** In the following diagram, S represents the position of a power relay station located on a straight coast, and E shows the location of a marine biology experimental station on an island. A cable is to be laid connecting the relay station with the experimental station. If the cost of running the cable on land is \$1.50/running foot and the cost of running the cable under water is \$2.50/running foot, locate the point P that will result in a minimum cost (solve for x).

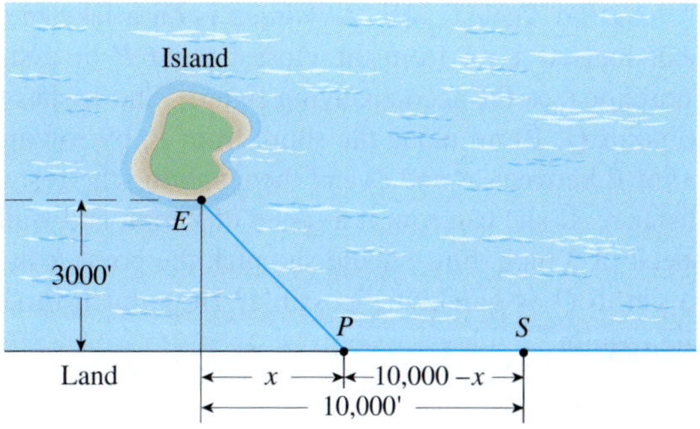

26. **Storing Radioactive Waste** A cylindrical container for storing radioactive waste is to be constructed from lead and have a thickness of 6 in. (see the accompanying figure). If the volume of the outside cylinder is to be 16π ft^3, find the radius and the height of the inside cylinder that will result in a container of maximum storage capacity.

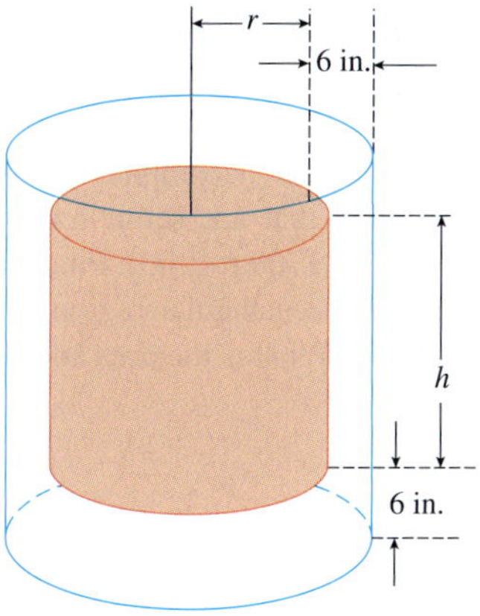

Hint: Show that the storage capacity (inside volume) is given by

$$V(r) = \pi r^2\left[\frac{16}{(r + \frac{1}{2})^2} - 1\right] \qquad (0 \le r \le \tfrac{7}{2})$$

27. **Flights of Birds** During daylight hours, some birds fly more slowly over water than over land because some of their energy is expended in overcoming the downdrafts of air over open bodies of water. Suppose a bird that flies at a constant speed of 4 mph over water and 6 mph over land starts its journey at the point E on an island and ends at its nest N on the shore of the mainland, as shown in the accompanying figure. Find the location of the point P that allows the bird to complete its journey in the minimum time (solve for x).

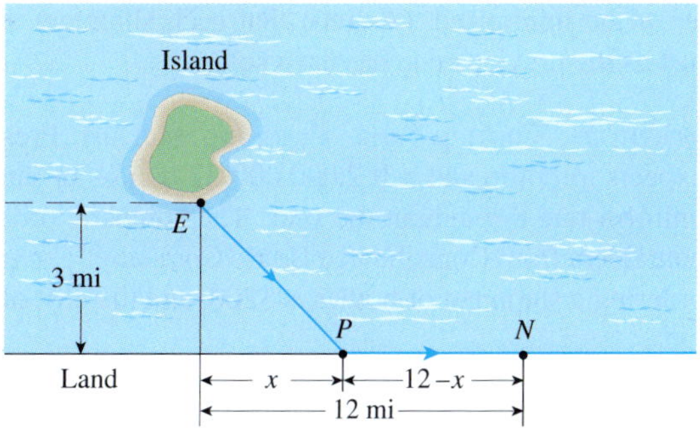

28. **Minimizing Travel Time** A woman is on a lake in a row boat located 1 mi from the closest point P of a straight shoreline (see the accompanying figure). She wishes to get to point Q, 10 mi along the shore from P, by rowing to a point R between P and Q and then walking the rest of the distance. If she can row at a speed of 2 mph and walk at a speed of 3 mph, how should she pick the point R in order to get to Q as quickly as possible? How much time does she require?

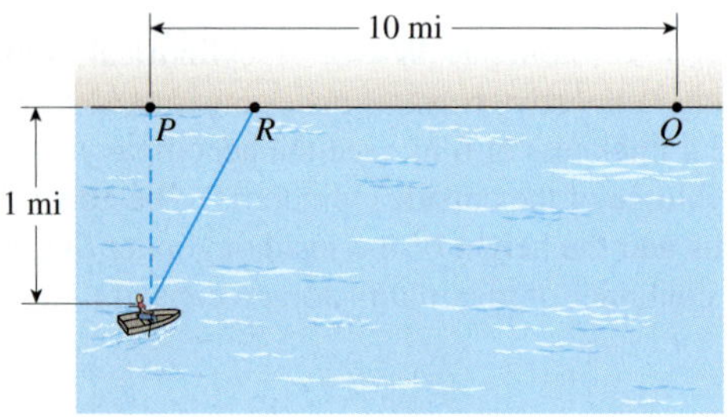

29. **Racetrack Design** The accompanying figure depicts a racetrack with ends that are semicircular in shape. The length of the track is 1760 ft ($\frac{1}{3}$ mi). Find l and r so that the area enclosed by the rectangular region of the racetrack is as large as possible. What is the area enclosed by the track in this case?

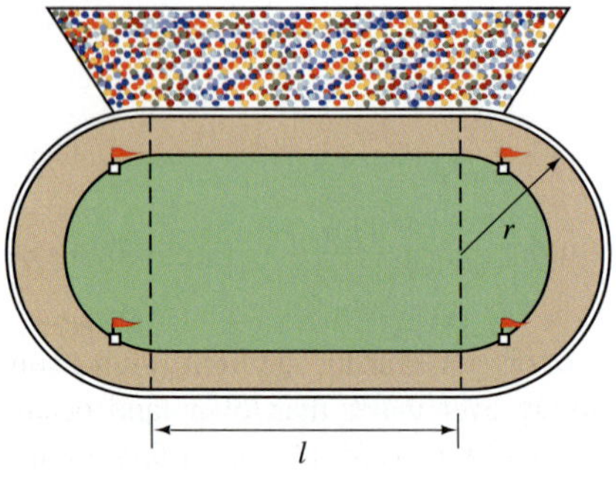

30. **Inventory Control and Planning** The demand for motorcycle tires imported by Dixie Import-Export is 40,000/year and may be assumed to be uniform throughout the year. The cost of ordering a shipment of tires is \$400, and the cost of storing each tire for a year is \$2. Determine how many tires should be in each shipment if the ordering and storage costs are to be minimized. (Assume that each shipment arrives just as the previous one has been sold.)

31. **Inventory Control and Planning** McDuff Preserves expects to bottle and sell 2,000,000 32-oz jars of jam at a uniform rate throughout the year. The company orders its containers from Consolidated Bottle Company. The cost of ordering a shipment of bottles is \$200, and the cost of storing each empty bottle for a year is \$.40. How many orders should McDuff place per year and how many bottles should be in each shipment if the ordering and storage costs are to be minimized? (Assume that each shipment of bottles is used up before the next shipment arrives.)

32. **Inventory Control and Planning** Neilsen Cookie Company sells its assorted butter cookies in containers that have a net content of 1 lb. The estimated demand for the cookies is 1,000,000 1-lb containers. The setup cost for each production run is \$500, and the manufacturing cost is \$.50 for each container of cookies. The cost of storing each container of cookies over the year is \$.40. Assuming uniformity of demand throughout the year and instantaneous production, how many containers of cookies should Neilsen produce per production run in order to minimize the production cost?

Hint: Following the method of Example 5, show that the total production cost is given by the function

$$C(x) = \frac{500{,}000{,}000}{x} + 0.2x + 500{,}000$$

Then minimize the function C on the interval (0, 1,000,000).

33. **Inventory Control and Planning** A company expects to sell D units of a certain product per year. Sales are assumed to be at a steady rate with no shortages allowed. Each time an order for the product is placed, an ordering cost of K dollars is incurred. Each item costs p dollars, and the holding cost is h dollars per item per year.

a. Show that the inventory cost (the combined ordering cost, purchasing cost, and holding cost) is

$$C(x) = \frac{KD}{x} + pD + \frac{hx}{2} \qquad (x > 0)$$

where x is the order quantity (the number of items in each order).

b. Use the result of part (a) to show that the inventory cost is minimized if

$$x = \sqrt{\frac{2KD}{h}}$$

This quantity is called the *economic order quantity* (EOQ).

34. **Inventory Control and Planning** Refer to Exercise 33. The Camera Store sells 960 Yamaha A35 digital cameras per year. Each time an order for cameras is placed with the manufacturer, an ordering cost of \$10 is incurred. The store pays \$80 for each camera, and the cost for holding a camera (mainly due to the opportunity cost incurred in tying up capital in inventory) is \$12/year. Assume that the cameras sell at a uniform rate and no shortages are allowed.

a. What is the EOQ?

b. How many orders will be placed each year?

c. What is the interval between orders?

4.5 Solutions to Self-Check Exercises

1. Let x and y (measured in feet) denote the length and width of the rectangular garden.

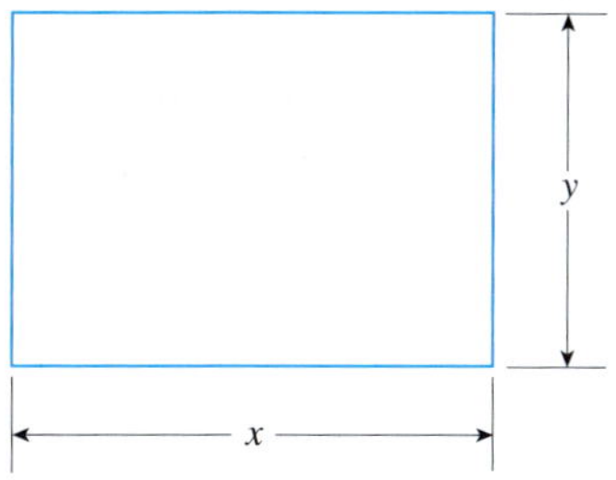

Since the area is to be 300 ft^2, we have

$$xy = 300$$

Next, the amount of fencing to be used is given by the perimeter, and this quantity is to be minimized. Thus, we want to minimize

$$2x + 2y$$

or, since $y = 300/x$ (obtained by solving for y in the first equation), we see that the expression to be minimized is

$$\begin{aligned} f(x) &= 2x + 2\left(\frac{300}{x}\right) \\ &= 2x + \frac{600}{x} \end{aligned}$$

for positive values of x. Now

$$f'(x) = 2 - \frac{600}{x^2}$$

Setting $f'(x) = 0$ yields $x = -\sqrt{300}$ or $x = \sqrt{300}$. We consider only the critical number $\sqrt{300}$ since $-\sqrt{300}$ lies outside the interval $(0, \infty)$. We then compute

$$f''(x) = \frac{1200}{x^3}$$

Since

$$f''(300) > 0$$

the second derivative test implies that a relative minimum of f occurs at $x = \sqrt{300}$. In fact, since $f''(x) > 0$ for all x in $(0, \infty)$, we conclude that $x = \sqrt{300}$ gives rise to the absolute minimum of f. The corresponding value of y, obtained by substituting this value of x into the equation $xy = 300$, is $y = \sqrt{300}$. Therefore, the required dimensions of the vegetable garden are approximately 17.3 ft × 17.3 ft.

2. Let x denote the number of tires in each production run. Then, the average number of tires in storage is $x/2$, so the storage cost incurred by the company is $2(x/2)$, or x dollars. Next, since the company needs to manufacture 1,000,000 tires for the year in order to meet the demand, the number of production runs is $1{,}000{,}000/x$. This gives setup costs amounting to

$$4000\left(\frac{1{,}000{,}000}{x}\right) = \frac{4{,}000{,}000{,}000}{x}$$

dollars for the year. The total manufacturing cost is \$20,000,000. Thus, the total yearly cost incurred by the company is given by

$$C(x) = x + \frac{4{,}000{,}000{,}000}{x} + 20{,}000{,}000$$

Differentiating $C(x)$, we find

$$C'(x) = 1 - \frac{4{,}000{,}000{,}000}{x^2}$$

Setting $C'(x) = 0$ gives 63,246 as the critical number in the interval (0, 1,000,000). Next, we find

$$C''(x) = \frac{8{,}000{,}000{,}000}{x^3}$$

Since $C''(x) > 0$ for all $x > 0$, we see that C is concave upward for all $x > 0$. Furthermore, $C''(63{,}246) > 0$ implies that $x = 63{,}246$ gives rise to a relative minimum of C (by the second derivative test). Since C is always concave upward for $x > 0$, $x = 63{,}246$ gives the absolute minimum of C. Therefore, the company should manufacture 63,246 tires in each production run.

CHAPTER 4 Summary of Principal Terms

TERMS

increasing function (244)
decreasing function (244)
relative maximum (249)
relative minimum (250)
relative extrema (250)
critical number (251)
first derivative test (252)
concave upward (264)
concave downward (264)
inflection point (267)
second derivative test (272)
vertical asymptote (285)
horizontal asymptote (286)
absolute extrema (298)
absolute maximum value (298)
absolute minimum value (298)

CHAPTER 4 Concept Review Questions

Fill in the blanks.

1. a. A function f is increasing on an interval I, if for any two numbers x_1 and x_2 in I, $x_1 < x_2$ implies that _____.
   b. A function f is decreasing on an interval I, if for any two numbers x_1 and x_2 in I, $x_1 < x_2$ implies that _____.
2. a. If f is differentiable on an open interval (a, b) and $f'(x) > 0$ on (a, b), then f is _____ on (a, b).
   b. If f is differentiable on an open interval (a, b) and _____ on (a, b), then f is decreasing on (a, b).
   c. If $f'(x) = 0$ for each value of x in the interval (a, b), then f is _____ on (a, b).
3. a. A function f has a relative maximum at c if there exists an open interval (a, b) containing c such that _____ for all x in (a, b).
   b. A function f has a relative minimum at c if there exists an open interval (a, b) containing c such that _____ for all x in (a, b).
4. a. A critical number of a function f is any number in the _____ of f at which $f'(c)$ _____ or $f'(c)$ does *not* _____.
   b. If f has a relative extremum at c, then c must be a/an _____ _____ of f.
   c. If c is a critical number of f, then f may or may not have a/an _____ _____ at c.
5. a. A differentiable function f is concave upward on an interval I if _____ is increasing on I.
   b. If f has a second derivative on an open interval I and $f''(x)$ _____ on I, then the graph of f is concave upward on I.
   c. If the graph of a continuous function f has a tangent line at $P(c, f(c))$ and the graph of f changes _____ at P, then P is called an inflection point of the graph of f.
   d. Suppose f has a continuous second derivative on an interval (a, b), containing a critical number c of f. If $f''(c) < 0$, then f has a/an _____ _____ at c. If $f''(c) = 0$, then f may or may not have a/an _____ _____ at c.
6. The line $x = a$ is a vertical asymptote of the graph f if at least one of the following is true: $\lim_{x \to a^+} f(x) =$ _____ or $\lim_{x \to a^-} f(x) =$ _____.
7. For a rational function $f(x) = \dfrac{P(x)}{Q(x)}$, the line $x = a$ is a vertical asymptote of the graph of f if $Q(a) =$ _____ but $P(a) \neq$ _____.
8. The line $y = b$ is a horizontal asymptote of the graph of a function f if either $\lim_{x \to \infty} f(x) =$ _____ or $\lim_{x \to -\infty} f(x) =$ _____.
9. a. A function f has an absolute maximum at c if _____ for all x in the domain D of f. The number $f(c)$ is called the _____ _____ _____ of f on D.
   b. A function f has a relative minimum at c if _____ for all values of x in some _____ _____ containing c.
10. The extreme value theorem states that if f is _____ on the closed interval $[a, b]$, then f has both a/an _____ maximum value and a/an _____ minimum value on $[a, b]$.

CHAPTER 4 Review Exercises

In Exercises 1–10, (a) find the intervals where the function *f* is increasing and where it is decreasing, (b) find the relative extrema of *f*, (c) find the intervals where *f* is concave upward and where it is concave downward, and (d) find the inflection points, if any, of *f*.

1. $f(x) = \frac{1}{3}x^3 - x^2 + x - 6$
2. $f(x) = (x - 2)^3$
3. $f(x) = x^4 - 2x^2$
4. $f(x) = x + \dfrac{4}{x}$
5. $f(x) = \dfrac{x^2}{x - 1}$
6. $f(x) = \sqrt{x - 1}$
7. $f(x) = (1 - x)^{1/3}$
8. $f(x) = x\sqrt{x - 1}$
9. $f(x) = \dfrac{2x}{x + 1}$
10. $f(x) = \dfrac{-1}{1 + x^2}$

In Exercises 11–18, use the curve-sketching guide on page 288 to sketch the graph of the function.

11. $f(x) = x^2 - 5x + 5$
12. $f(x) = -2x^2 - x + 1$
13. $g(x) = 2x^3 - 6x^2 + 6x + 1$
14. $g(x) = \frac{1}{3}x^3 - x^2 + x - 3$
15. $h(x) = x\sqrt{x - 2}$
16. $h(x) = \dfrac{2x}{1 + x^2}$
17. $f(x) = \dfrac{x - 2}{x + 2}$
18. $f(x) = x - \dfrac{1}{x}$

In Exercises 19–22, find the horizontal and vertical asymptotes of the graph of each function. Do not sketch the graph.

19. $f(x) = \dfrac{1}{2x + 3}$
20. $f(x) = \dfrac{2x}{x + 1}$
21. $f(x) = \dfrac{5x}{x^2 - 2x - 8}$
22. $f(x) = \dfrac{x^2 + x}{x(x - 1)}$

In Exercises 23–32, find the absolute maximum value and the absolute minimum value, if any, of the function.

23. $f(x) = 2x^2 + 3x - 2$

24. $g(x) = x^{2/3}$

25. $g(t) = \sqrt{25 - t^2}$

26. $f(x) = \frac{1}{3}x^3 - x^2 + x + 1$ on $[0, 2]$

27. $h(t) = t^3 - 6t^2$ on $[2, 5]$

28. $g(x) = \frac{x}{x^2 + 1}$ on $[0, 5]$

29. $f(x) = x - \frac{1}{x}$ on $[1, 3]$

30. $h(t) = 8t - \frac{1}{t^2}$ on $[1, 3]$

31. $f(s) = s\sqrt{1 - s^2}$ on $[-1, 1]$

32. $f(x) = \frac{x^2}{x - 1}$ on $[-1, 3]$

33. **Maximizing Profits** Odyssey Travel Agency's monthly profit (in thousands of dollars) depends on the amount of money x (in thousands of dollars) spent on advertising each month according to the rule

$$P(x) = -x^2 + 8x + 20$$

To maximize its monthly profits, what should be Odyssey's monthly advertising budget?

34. **Online Hotel Reservations** The online lodging industry is expected to grow dramatically. In a study conducted in 1999, analysts projected the U.S. online travel spending for lodging to be approximately

$$f(t) = 0.157t^2 + 1.175t + 2.03 \qquad (0 \leq t \leq 6)$$

billion dollars, where t is measured in years, with $t = 0$ corresponding to 1999.

a. Show that f is increasing on the interval (0, 6).
b. Show that the graph of f is concave upward on (0, 6).
c. What do your results from parts (a) and (b) tell you about the growth of online travel spending over the years in question?

Source: International Data Corp.

35. **Cell Phone Revenue** According to a study conducted in 1997, the revenue (in millions of dollars) in the U.S. cellular phone market in the next 6 yr is approximated by the function

$$R(t) = 0.03056t^3 - 0.45357t^2 + 4.81111t + 31.7 \qquad (0 \leq t \leq 6)$$

where t is measured in years, with $t = 0$ corresponding to 1997.

a. Find the interval where R is increasing and the interval where R is decreasing.
b. What does your result tell you about the revenue in the cell phone market in the years under consideration?
Hint: Use the quadratic formula.

Source: Paul Kagan Associates, Inc.

36. **Elderly Workforce** The percentage of men 65 yr and older in the workforce from 1970 through the year 2000 is approximated by the function

$$P(t) = 0.00093t^3 - 0.018t^2 - 0.51t + 25 \qquad (0 \leq t \leq 30)$$

where t is measured in years, with $t = 0$ corresponding to the beginning of 1970.

a. Find the interval where P is decreasing and the interval where P is increasing.
b. Interpret the results of part (a).

Source: U.S. Census Bureau

37. **PC Shipments** In a study conducted in 2003, it was projected that worldwide PC shipments (in millions) through 2005 will be given by

$$N(t) = 0.75t^3 - 1.5t^2 + 8.25t + 133 \qquad (0 \leq t \leq 4)$$

where t is measured in years, with $t = 0$ corresponding to 2001.

a. Determine the intervals where N is concave upward and where it is concave downward.
b. Find the inflection point of N and interpret your result.

Source: International Data Corporation

38. **Sales of Camera Phones** Camera phones, virtually nonexistent ten years ago, are quickly gaining in popularity. The function

$$N(t) = 8.125t^2 + 24.625t + 18.375 \qquad (0 \leq t \leq 3)$$

gives the projected worldwide shipments of camera phones (in millions of units) in year t, with $t = 0$ corresponding to 2002.

a. Find $N'(t)$. What does this say about the sales of camera phones between 2002 and 2005?
b. Find $N''(t)$. What does this say about the rate of the rate of sales of camera phones between 2002 and 2005?

Source: In-Stat/MDR

39. **Effect of Advertising on Sales** The total sales S of Cannon Precision Instruments is related to the amount of money x that Cannon spends on advertising its products by the function

$$S(x) = -0.002x^3 + 0.6x^2 + x + 500 \qquad (0 \leq x \leq 200)$$

where S and x are measured in thousands of dollars. Find the inflection point of the function S and discuss its significance.

40. **Cost of Producing Calculators** A subsidiary of Elektra Electronics manufactures graphing calculators. Management determines that the daily cost $C(x)$ (in dollars) of producing these calculators is

$$C(x) = 0.0001x^3 - 0.08x^2 + 40x + 5000$$

where x is the number of calculators produced. Find the inflection point of the function C and interpret your result.

41. **Index of Environmental Quality** The Department of the Interior of an African country began to record an index of environmental quality to measure progress or decline in the environmental quality of its wildlife. The index for the years 1998 through 2008 is approximated by the function

$$I(t) = \frac{50t^2 + 600}{t^2 + 10} \qquad (0 \le t \le 10)$$

a. Compute $I'(t)$ and show that $I(t)$ is decreasing on the interval $(0, 10)$.
b. Compute $I''(t)$. Study the concavity of the graph of I.
c. Sketch the graph of I.
d. Interpret your results.

42. **Maximizing Profits** The weekly demand for DVDs manufactured by Herald Media Corporation is given by

$$p = -0.0005x^2 + 60$$

where p denotes the unit price in dollars and x denotes the quantity demanded. The weekly total cost function associated with producing these discs is given by

$$C(x) = -0.001x^2 + 18x + 4000$$

where $C(x)$ denotes the total cost (in dollars) incurred in pressing x discs. Find the production level that will yield a maximum profit for the manufacturer.
Hint: Use the quadratic formula.

43. **Maximizing Profits** The estimated monthly profit (in dollars) realizable by Cannon Precision Instruments for manufacturing and selling x units of its model M1 digital camera is

$$P(x) = -0.04x^2 + 240x - 10{,}000$$

To maximize its profits, how many cameras should Cannon produce each month?

44. **Minimizing Average Cost** The total monthly cost (in dollars) incurred by Carlota Music in manufacturing x units of its Professional Series guitars is given by the function

$$C(x) = 0.001x^2 + 100x + 4000$$

a. Find the average cost function $\overline{C}$.
b. Determine the production level that will result in the smallest average production cost.

45. **Worker Efficiency** The average worker at Wakefield Avionics can assemble

$$N(t) = -2t^3 + 12t^2 + 2t \qquad (0 \le t \le 4)$$

ready-to-fly radio-controlled model airplanes t hr into the 8 a.m. to 12 noon morning shift. At what time during this shift is the average worker performing at peak efficiency?

46. **Senior Workforce** The percentage of women 65 yr and older in the workforce from 1970 through the year 2000 is approximated by the function

$$P(t) = -0.0002t^3 + 0.018t^2 - 0.36t + 10 \qquad (0 \le t \le 30)$$

where t is measured in years, with $t = 0$ corresponding to the beginning of 1970.
a. Find the interval where P is decreasing and the interval where P is increasing.
b. Find the absolute minimum of P.
c. Interpret the results of parts (a) and (b).
Source: U.S. Census Bureau

47. **Spread of a Contagious Disease** The incidence (number of new cases/day) of a contagious disease spreading in a population of M people is given by

$$R(x) = kx(M - x)$$

where k is a positive constant and x denotes the number of people already infected. Show that the incidence R is greatest when half the population is infected.

48. **Maximizing the Volume of a Box** A box with an open top is to be constructed from a square piece of cardboard, 10 in. wide, by cutting out a square from each of the four corners and bending up the sides. What is the maximum volume of such a box?

49. **Minimizing Construction Costs** A man wishes to construct a cylindrical barrel with a capacity of 32π ft^3. The cost/square foot of the material for the side of the barrel is half that of the cost/square foot for the top and bottom. Help him find the dimensions of the barrel that can be constructed at a minimum cost in terms of material used.

50. **Packaging** You wish to construct a closed rectangular box that has a volume of 4 ft^3. The length of the base of the box will be twice as long as its width. The material for the top and bottom of the box costs 30¢/square foot. The material for the sides of the box costs 20¢/square foot. Find the dimensions of the least expensive box that can be constructed.

51. **Inventory Control and Planning** Lehen Vinters imports a certain brand of beer. The demand, which may be assumed to be uniform, is 800,000 cases/year. The cost of ordering a shipment of beer is \$500, and the cost of storing each case of beer for a year is \$2. Determine how many cases of beer should be in each shipment if the ordering and storage costs are to be kept at a minimum. (Assume that each shipment of beer arrives just as the previous one has been sold.)

52. In what interval is the quadratic function

$$f(x) = ax^2 + bx + c \qquad (a \ne 0)$$

increasing? In what interval is f decreasing?

53. Let $f(x) = x^2 + ax + b$. Determine the constants a and b so that f has a relative minimum at $x = 2$ and the relative minimum value is 7.

54. Find the values of c so that the graph of

$$f(x) = x^4 + 2x^3 + cx^2 + 2x + 2$$

is concave upward everywhere.

55. Suppose that the point $(a, f(a))$ is an inflection point of the graph of $y = f(x)$. Show that the number a gives rise to a relative extremum of the function f'.

56. Let

$$f(x) = \begin{cases} x^3 + 1 & \text{if } x \neq 0 \\ 2 & \text{if } x = 0 \end{cases}$$

a. Compute $f'(x)$ and show that it does not change sign as we move across $x = 0$.

b. Show that f has a relative maximum at $x = 0$. Does this contradict the first derivative test? Explain your answer.

CHAPTER 4 Before Moving On . . .

1. Find the interval(s) where $f(x) = \dfrac{x^2}{1 - x}$ is increasing and where it is decreasing.

2. Find the relative maxima and relative minima, if any, of $f(x) = 2x^2 - 12x^{1/3}$.

3. Find the intervals where $f(x) = \frac{1}{3}x^3 - \frac{1}{4}x^2 - \frac{1}{2}x + 1$ is concave upward, the intervals where f is concave downward, and the inflection point(s) of f.

4. Sketch the graph of $f(x) = 2x^3 - 9x^2 + 12x - 1$.

5. Find the absolute maximum and absolute minimum values of $f(x) = 2x^3 + 3x^2 - 1$ on the interval $[-2, 3]$.

6. An open bucket in the form of a right circular cylinder is to be constructed with a capacity of 1 ft^3. Find the radius and height of the cylinder if the amount of material used is minimal.

EXPONENTIAL AND LOGARITHMIC FUNCTIONS

5

THE EXPONENTIAL FUNCTION is, without doubt, the most important function in mathematics and its applications. After a brief introduction to the exponential function and its *inverse,* the logarithmic function, we learn how to differentiate such functions. This lays the foundation for exploring the many applications involving exponential functions. For example, we look at the role played by exponential functions in computing earned interest on a bank account and in studying the growth of a bacteria population in the laboratory, the rate at which radioactive matter decays, the rate at which a factory worker learns a certain process, and the rate at which a communicable disease is spread over time.

How many cameras can a new employee at Eastman Optical assemble after completing the basic training program, and how many cameras can he assemble after being on the job for 6 months? In Example 5, page 384, you will see how to answer these questions.

5.1 Exponential Functions

Exponential Functions and Their Graphs

Suppose you deposit a sum of \$1000 in an account earning interest at the rate of 10% per year *compounded continuously* (the way most financial institutions compute interest). Then, the accumulated amount at the end of t years ($0 \leq t \leq 20$) is described by the function f, whose graph appears in Figure 1.* This function is called an *exponential function*. Observe that the graph of f rises rather slowly at first but very rapidly as time goes by. For purposes of comparison, we have also shown the graph of the function $y = g(t) = 1000(1 + 0.10t)$, giving the accumulated amount for the same principal (\$1000) but earning *simple* interest at the rate of 10% per year. The moral of the story: It is never too early to save.

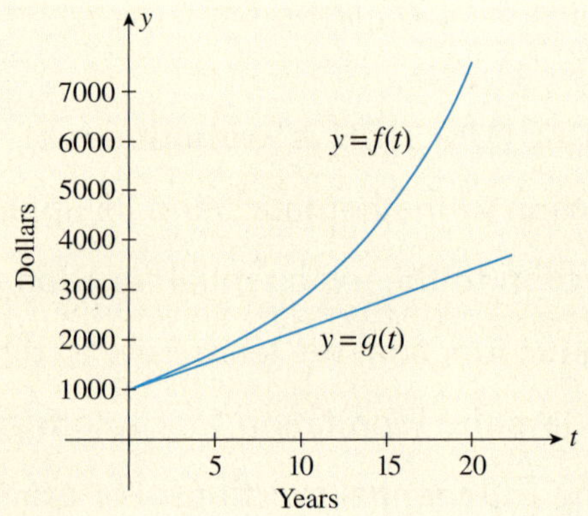

FIGURE 1
Under continuous compounding, a sum of money grows exponentially.

Exponential functions play an important role in many real-world applications, as you will see throughout this chapter.

Recall that whenever b is a positive number and n is any real number, the expression b^n is a real number. This enables us to define an exponential function as follows:

> **Exponential Function**
>
> The function defined by
>
> $$f(x) = b^x \qquad (b > 0, b \neq 1)$$
>
> is called an **exponential function with base b and exponent x.** The domain of f is the set of all real numbers.

For example, the exponential function with base 2 is the function

$$f(x) = 2^x$$

with domain $(-\infty, \infty)$. The values of $f(x)$ for selected values of x follow:

$$f(3) = 2^3 = 8 \qquad f\left(\frac{3}{2}\right) = 2^{3/2} = 2 \cdot 2^{1/2} = 2\sqrt{2} \qquad f(0) = 2^0 = 1$$

$$f(-1) = 2^{-1} = \frac{1}{2} \qquad f\left(-\frac{2}{3}\right) = 2^{-2/3} = \frac{1}{2^{2/3}} = \frac{1}{\sqrt[3]{4}}$$

*We will derive the rule for f in Section 5.3.

Computations involving exponentials are facilitated by the laws of exponents. These laws were stated in Section 1.1, and you might want to review the material there. For convenience, however, we will restate these laws.

Laws of Exponents

Let a and b be positive numbers and let x and y be real numbers. Then,

1. $b^x \cdot b^y = b^{x+y}$

2. $\dfrac{b^x}{b^y} = b^{x-y}$

3. $(b^x)^y = b^{xy}$

4. $(ab)^x = a^x b^x$

5. $\left(\dfrac{a}{b}\right)^x = \dfrac{a^x}{b^x}$

The use of the laws of exponents is illustrated in the next two examples.

EXAMPLE 1

a. $16^{7/4} \cdot 16^{-1/2} = 16^{7/4-1/2} = 16^{5/4} = 2^5 = 32$ Law 1

b. $\dfrac{8^{5/3}}{8^{-1/3}} = 8^{5/3-(-1/3)} = 8^2 = 64$ Law 2

c. $(64^{4/3})^{-1/2} = 64^{(4/3)(-1/2)} = 64^{-2/3}$

$= \dfrac{1}{64^{2/3}} = \dfrac{1}{(64^{1/3})^2} = \dfrac{1}{4^2} = \dfrac{1}{16}$ Law 3

d. $(16 \cdot 81)^{-1/4} = 16^{-1/4} \cdot 81^{-1/4} = \dfrac{1}{16^{1/4}} \cdot \dfrac{1}{81^{1/4}} = \dfrac{1}{2} \cdot \dfrac{1}{3} = \dfrac{1}{6}$ Law 4

e. $\left(\dfrac{3^{1/2}}{2^{1/3}}\right)^4 = \dfrac{3^{4/2}}{2^{4/3}} = \dfrac{9}{2^{4/3}}$ Law 5

EXAMPLE 2 Let $f(x) = 2^{2x-1}$. Find the value of x for which $f(x) = 16$.

Solution We want to solve the equation

$$2^{2x-1} = 16 = 2^4$$

But this equation holds if and only if

$$2x - 1 = 4 \qquad b^m = b^n \Rightarrow m = n$$

giving $x = \frac{5}{2}$.

Exponential functions play an important role in mathematical analysis. Because of their special characteristics, they are some of the most useful functions and are found in virtually every field where mathematics is applied. To mention a few examples: Under ideal conditions, the number of bacteria present at any time t in a culture may be described by an exponential function of t; radioactive substances decay over time in accordance with an "exponential" law of decay; money left on fixed deposit and earning compound interest grows exponentially; and some of the most important distribution functions encountered in statistics are exponential.

Let's begin our investigation into the properties of exponential functions by studying their graphs.

EXAMPLE 3 Sketch the graph of the exponential function $y = 2^x$.

Solution First, as discussed earlier, the domain of the exponential function $y = f(x) = 2^x$ is the set of real numbers. Next, putting $x = 0$ gives $y = 2^0 = 1$, the

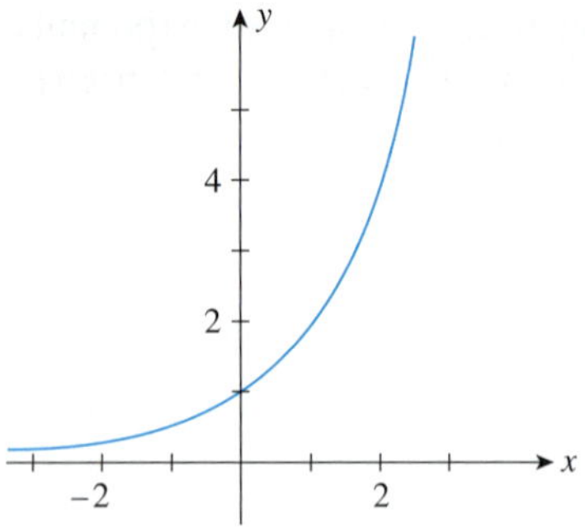

FIGURE 2
The graph of $y = 2^x$

y-intercept of f. There is no x-intercept since there is no value of x for which $y = 0$. To find the range of f, consider the following table of values:

x	−5	−4	−3	−2	−1	0	1	2	3	4	5
y	$\frac{1}{32}$	$\frac{1}{16}$	$\frac{1}{8}$	$\frac{1}{4}$	$\frac{1}{2}$	1	2	4	8	16	32

We see from these computations that 2^x decreases and approaches zero as x decreases without bound and that 2^x increases without bound as x increases without bound. Thus, the range of f is the interval $(0, \infty)$—that is, the set of positive real numbers. Finally, we sketch the graph of $y = f(x) = 2^x$ in Figure 2. ■

VIDEO

EXAMPLE 4 Sketch the graph of the exponential function $y = (1/2)^x$.

Solution The domain of the exponential function $y = (1/2)^x$ is the set of all real numbers. The y-intercept is $(1/2)^0 = 1$; there is no x-intercept since there is no value of x for which $y = 0$. From the following table of values

x	−5	−4	−3	−2	−1	0	1	2	3	4	5
y	32	16	8	4	2	1	$\frac{1}{2}$	$\frac{1}{4}$	$\frac{1}{8}$	$\frac{1}{16}$	$\frac{1}{32}$

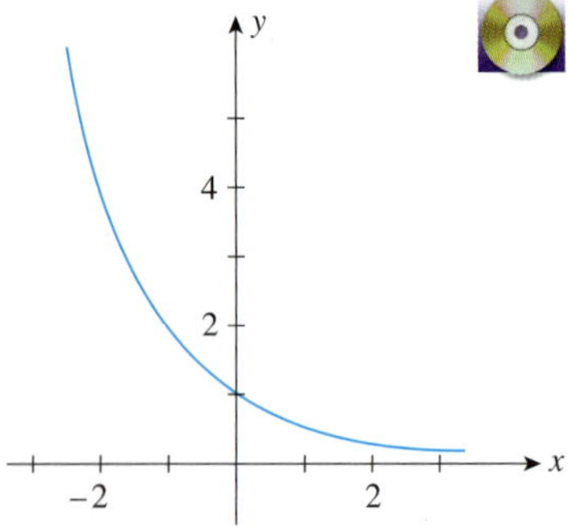

FIGURE 3
The graph of $y = \left(\frac{1}{2}\right)^x$

we deduce that $(1/2)^x = 1/2^x$ increases without bound as x decreases without bound and that $(1/2)^x$ decreases and approaches zero as x increases without bound. Thus, the range of f is the interval $(0, \infty)$. The graph of $y = f(x) = (1/2)^x$ is sketched in Figure 3. ■

The functions $y = 2^x$ and $y = (1/2)^x$, whose graphs you studied in Examples 3 and 4, are special cases of the exponential function $y = f(x) = b^x$, obtained by setting $b = 2$ and $b = 1/2$, respectively. In general, the exponential function $y = b^x$ with $b > 1$ has a graph similar to $y = 2^x$, whereas the graph of $y = b^x$ for $0 < b < 1$ is similar to that of $y = (1/2)^x$ (Exercises 27 and 28 on page 334). When $b = 1$, the function $y = b^x$ reduces to the constant function $y = 1$. For comparison, the graphs of all three functions are sketched in Figure 4.

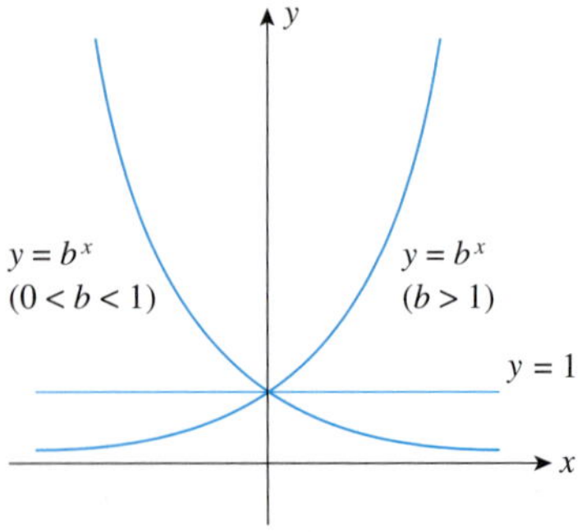

FIGURE 4
$y = b^x$ is an increasing function of x if $b > 1$, a constant function if $b = 1$, and a decreasing function if $0 < b < 1$.

> **Properties of the Exponential Function**
>
> The exponential function $y = b^x$ ($b > 0$, $b \neq 1$) has the following properties:
>
> 1. Its domain is $(-\infty, \infty)$.
> 2. Its range is $(0, \infty)$.
> 3. Its graph passes through the point $(0, 1)$.
> 4. It is continuous on $(-\infty, \infty)$.
> 5. It is increasing on $(-\infty, \infty)$ if $b > 1$ and decreasing on $(-\infty, \infty)$ if $b < 1$.

The Base e

Exponential functions to the base e, where e is an irrational number whose value is 2.7182818. . . , play an important role in both theoretical and applied problems. It can be shown, although we will not do so here, that

$$e = \lim_{m \to \infty} \left(1 + \frac{1}{m}\right)^m \tag{1}$$

TABLE 1

m	$\left(1+\frac{1}{m}\right)^m$
10	2.59374
100	2.70481
1000	2.71692
10,000	2.71815
100,000	2.71827
1,000,000	2.71828

However, you may convince yourself of the plausibility of this definition of the number e by examining Table 1, which may be constructed with the help of a calculator.

Exploring with TECHNOLOGY

To obtain a visual confirmation of the fact that the expression $(1 + 1/m)^m$ approaches the number $e = 2.71828\ldots$ as m increases without bound, plot the graph of $f(x) = (1 + 1/x)^x$ in a suitable viewing window and observe that $f(x)$ approaches $2.71828\ldots$ as x increases without bound. Use ZOOM and TRACE to find the value of $f(x)$ for large values of x.

EXAMPLE 5 Sketch the graph of the function $y = e^x$.

Solution Since $e > 1$, it follows from our previous discussion that the graph of $y = e^x$ is similar to the graph of $y = 2^x$ (see Figure 2). With the aid of a calculator, we obtain the following table:

x	−3	−2	−1	0	1	2	3
y	0.05	0.14	0.37	1	2.72	7.39	20.09

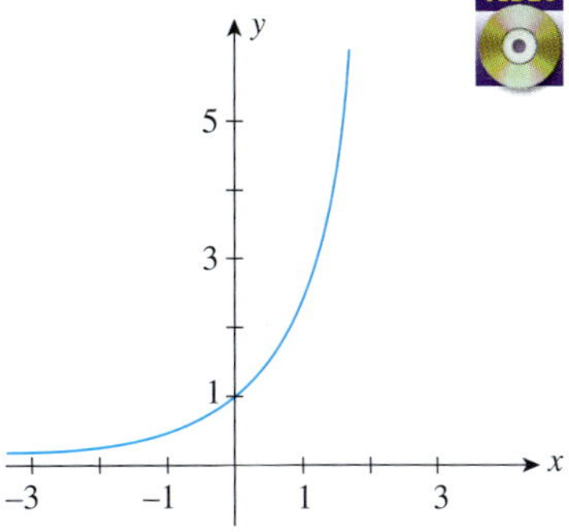

FIGURE 5
The graph of $y = e^x$

The graph of $y = e^x$ is sketched in Figure 5. ■

Next, we consider another exponential function to the base e that is closely related to the previous function and is particularly useful in constructing models that describe "exponential decay."

EXAMPLE 6 Sketch the graph of the function $y = e^{-x}$.

Solution Since $e > 1$, it follows that $0 < 1/e < 1$, so $f(x) = e^{-x} = 1/e^x = (1/e)^x$ is an exponential function with base less than 1. Therefore, it has a graph similar to that of the exponential function $y = (1/2)^x$. As before, we construct the following table of values of $y = e^{-x}$ for selected values of x:

x	−3	−2	−1	0	1	2	3
y	20.09	7.39	2.72	1	0.37	0.14	0.05

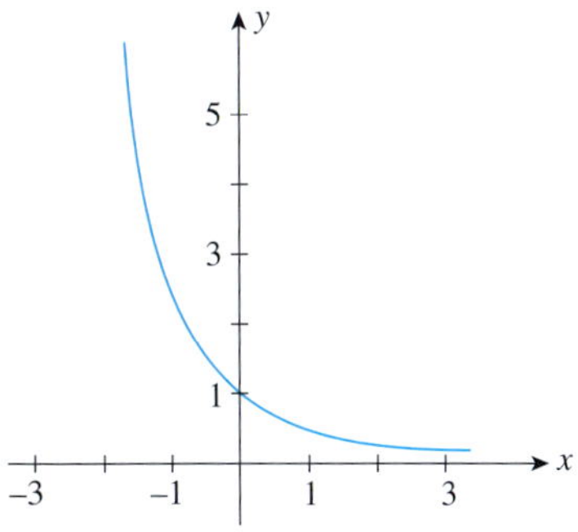

FIGURE 6
The graph of $y = e^{-x}$

Using this table, we sketch the graph of $y = e^{-x}$ in Figure 6. ■

5.1 Self-Check Exercises

1. Solve the equation $2^{2x+1} \cdot 2^{-3} = 2^{x-1}$.
2. Sketch the graph of $y = e^{0.4x}$.

Solutions to Self-Check Exercises 5.1 can be found on page 336.

5.1 Concept Questions

1. Define the exponential function f with base b and exponent x. What restrictions, if any, are placed on b?

2. For the exponential function $y = b^x$ $(b > 0, b \neq 1)$, state (a) its domain and range, (b) its y-intercept, (c) where it is continuous, and (d) where it is increasing and where it is decreasing for the case $b > 1$ and the case $b < 1$.

5.1 Exercises

In Exercises 1–8, evaluate the expression.

1. a. $4^{-3} \cdot 4^5$ b. $3^{-3} \cdot 3^6$

2. a. $(2^{-1})^3$ b. $(3^{-2})^3$

3. a. $9(9)^{-1/2}$ b. $5(5)^{-1/2}$

4. a. $\left[\left(-\frac{1}{2}\right)^3\right]^{-2}$ b. $\left[\left(-\frac{1}{3}\right)^2\right]^{-3}$

5. a. $\dfrac{(-3)^4(-3)^5}{(-3)^8}$ b. $\dfrac{(2^{-4})(2^6)}{2^{-1}}$

6. a. $3^{1/4} \cdot 9^{-5/8}$ b. $2^{3/4} \cdot 4^{-3/2}$

7. a. $\dfrac{5^{3.3} \cdot 5^{-1.6}}{5^{-0.3}}$ b. $\dfrac{4^{2.7} \cdot 4^{-1.3}}{4^{-0.4}}$

8. a. $\left(\frac{1}{16}\right)^{-1/4}\left(\frac{27}{64}\right)^{-1/3}$ b. $\left(\frac{8}{27}\right)^{-1/3}\left(\frac{81}{256}\right)^{-1/4}$

In Exercises 9–16, simplify the expression.

9. a. $(64x^9)^{1/3}$ b. $(25x^3y^4)^{1/2}$

10. a. $(2x^3)(-4x^{-2})$ b. $(4x^{-2})(-3x^5)$

11. a. $\dfrac{6a^{-5}}{3a^{-3}}$ b. $\dfrac{4b^{-4}}{12b^{-6}}$

12. a. $y^{-3/2}y^{5/3}$ b. $x^{-3/5}x^{8/3}$

13. a. $(2x^3y^2)^3$ b. $(4x^2y^2z^3)^2$

14. a. $(x^{r/s})^{s/r}$ b. $(x^{-b/a})^{-a/b}$

15. a. $\dfrac{5^0}{(2^{-3}x^{-3}y^2)^2}$ b. $\dfrac{(x+y)(x-y)}{(x-y)^0}$

16. a. $\dfrac{(a^m \cdot a^{-n})^{-2}}{(a^{m+n})^2}$ b. $\left(\dfrac{x^{2n-2}y^{2n}}{x^{5n+1}y^{-n}}\right)^{1/3}$

In Exercises 17–26, solve the equation for x.

17. $6^{2x} = 6^4$

18. $5^{-x} = 5^3$

19. $3^{3x-4} = 3^5$

20. $10^{2x-1} = 10^{x+3}$

21. $(2.1)^{x+2} = (2.1)^5$

22. $(-1.3)^{x-2} = (-1.3)^{2x+1}$

23. $8^x = \left(\frac{1}{32}\right)^{x-2}$

24. $3^{x-x^2} = \dfrac{1}{9^x}$

25. $3^{2x} - 12 \cdot 3^x + 27 = 0$

26. $2^{2x} - 4 \cdot 2^x + 4 = 0$

In Exercises 27–36, sketch the graphs of the given functions on the same axes.

27. $y = 2^x$, $y = 3^x$, and $y = 4^x$

28. $y = \left(\frac{1}{2}\right)^x$, $y = \left(\frac{1}{3}\right)^x$, and $y = \left(\frac{1}{4}\right)^x$

29. $y = 2^{-x}$, $y = 3^{-x}$, and $y = 4^{-x}$

30. $y = 4^{0.5x}$ and $y = 4^{-0.5x}$

31. $y = 4^{0.5x}$, $y = 4^x$, and $y = 4^{2x}$

32. $y = e^x$, $y = 2e^x$, and $y = 3e^x$

33. $y = e^{0.5x}$, $y = e^x$, and $y = e^{1.5x}$

34. $y = e^{-0.5x}$, $y = e^{-x}$, and $y = e^{-1.5x}$

35. $y = 0.5e^{-x}$, $y = e^{-x}$, and $y = 2e^{-x}$

36. $y = 1 - e^{-x}$ and $y = 1 - e^{-0.5x}$

37. A function f has the form $f(x) = Ae^{kx}$. Find f if it is known that $f(0) = 100$ and $f(1) = 120$.
Hint: $e^{kx} = (e^k)^x$.

38. If $f(x) = Axe^{-kx}$, find $f(3)$ if $f(1) = 5$ and $f(2) = 7$.
Hint: $e^{kx} = (e^k)^x$.

39. If

$$f(t) = \frac{1000}{1 + Be^{-kt}}$$

find $f(5)$ given that $f(0) = 20$ and $f(2) = 30$.
Hint: $e^{kx} = (e^k)^x$.

40. **Tracking with GPS** Employers are increasingly turning to GPS (global positioning system) technology to keep track of their fleet vehicles. The estimated number of automatic vehicle trackers installed on fleet vehicles in the United States is approximated by

$$N(t) = 0.6e^{0.17t} \qquad (0 \le t \le 5)$$

where $N(t)$ is measured in millions and t is measured in years, with $t = 0$ corresponding to 2000.
a. What was the number of automatic vehicle trackers installed in the year 2000? How many were projected to be installed in 2005?
b. Sketch the graph of N.
Source: C. J. Driscoll Associates

41. **Disability Rates** Because of medical technology advances, the disability rates for people over 65 yr old have been dropping rather dramatically. The function

$$R(t) = 26.3e^{-0.016t} \qquad (0 \le t \le 18)$$

gives the disability rate $R(t)$, in percent, for people over age 65 from 1982 ($t = 0$) through 2000, where t is measured in years.

a. What was the disability rate in 1982? In 1986? In 1994? In 2000?

b. Sketch the graph of R.

Source: Frost and Sullivan

42. **Married Households** The percentage of families that were married households between 1970 and 2000 is approximately

$$P(t) = 86.9e^{-0.05t} \qquad (0 \le t \le 3)$$

where t is measured in decades, with $t = 0$ corresponding to 1970.

a. What percentage of families were married households in 1970? In 1980? In 1990? In 2000?

b. Sketch the graph of P.

Source: U.S. Census Bureau

43. **Growth of Web Sites** According to a study conducted in 2000, the projected number of Web addresses (in billions) is approximated by the function

$$N(t) = 0.45e^{0.5696t} \qquad (0 \le t \le 5)$$

where t is measured in years, with $t = 0$ corresponding to 1997.

a. Complete the following table by finding the number of Web addresses in each year:

Year	0	1	2	3	4	5
Number of Web Addresses (billions)						

b. Sketch the graph of N.

44. **Internet Users in China** The number of Internet users in China is projected to be

$$N(t) = 94.5e^{0.2t} \qquad (1 \le t \le 6)$$

where $N(t)$ is measured in millions and t is measured in years, with $t = 1$ corresponding to 2005.

a. How many Internet users were there in 2005? In 2006?

b. How many Internet users are there expected to be in 2010?

c. Sketch the graph of N.

Source: C. E. Unterberg

45. **Alternative Minimum Tax** The alternative minimum tax was created in 1969 to prevent the very wealthy from using creative deductions and shelters to avoid having to pay anything to the Internal Revenue Service. But it has increasingly hit the middle class. The number of taxpayers subjected to an alternative minimum tax is projected to be

$$N(t) = \frac{35.5}{1 + 6.89e^{-0.8674t}} \qquad (0 \le t \le 6)$$

where $N(t)$ is measured in millions and t is measured in years, with $t = 0$ corresponding to 2004. What is the projected number of taxpayers subjected to an alternative minimum tax in 2010?

Source: Brookings Institution

46. **Absorption of Drugs** The concentration of a drug in an organ at any time t (in seconds) is given by

$$C(t) = \begin{cases} 0.3t - 18(1 - e^{-t/60}) & \text{if } 0 \le t \le 20 \\ 18e^{-t/60} - 12e^{-(t-20)/60} & \text{if } t > 20 \end{cases}$$

where $C(t)$ is measured in grams/cubic centimeter (g/cm^3).

a. What is the initial concentration of the drug in the organ?

b. What is the concentration of the drug in the organ after 10 sec?

c. What is the concentration of the drug in the organ after 30 sec?

d. What will be the concentration of the drug in the long run?

Hint: Evaluate $\lim_{t \to \infty} C(t)$.

47. **Absorption of Drugs** The concentration of a drug in an organ at any time t (in seconds) is given by

$$x(t) = 0.08 + 0.12(1 - e^{-0.02t})$$

where $x(t)$ is measured in grams/cubic centimeter (g/cm^3).

a. What is the initial concentration of the drug in the organ?

b. What is the concentration of the drug in the organ after 20 sec?

c. What will be the concentration of the drug in the organ in the long run?

Hint: Evaluate $\lim_{t \to \infty} x(t)$.

d. Sketch the graph of x.

48. **Absorption of Drugs** Jane took 100 mg of a drug in the morning and another 100 mg of the same drug at the same time the following morning. The amount of the drug in her body t days after the first dosage was taken is given by

$$A(t) = \begin{cases} 100e^{-1.4t} & \text{if } 0 \le t < 1 \\ 100(1 + e^{1.4})e^{-1.4t} & \text{if } t \ge 1 \end{cases}$$

a. What was the amount of drug in Jane's body immediately after taking the second dose? After 2 days? In the long run?

b. Sketch the graph of A.

In Exercises 49–52, determine whether the statement is true or false. If it is true, explain why it is true. If it is false, give an example to show why it is false.

49. $(x^2 + 1)^3 = x^6 + 1$

50. $e^{xy} = e^x e^y$

51. If $x < y$, then $e^x < e^y$.

52. If $0 < b < 1$ and $x < y$, then $b^x > b^y$.

5.1 Solutions to Self-Check Exercises

1. $2^{2x+1} \cdot 2^{-3} = 2^{x-1}$

$$\frac{2^{2x+1}}{2^{x-1}} \cdot 2^{-3} = 1 \quad \text{Divide both sides by } 2^{x-1}.$$

$$2^{(2x+1)-(x-1)-3} = 1$$

$$2^{x-1} = 1$$

This is true if and only if $x - 1 = 0$ or $x = 1$.

2. We first construct the following table of values:

x	−3	−2	−1	0	1	2	3	4
$y = e^{0.4x}$	0.3	0.4	0.7	1	1.5	2.2	3.3	5

Next, we plot these points and join them by a smooth curve to obtain the graph of f shown in the accompanying figure.

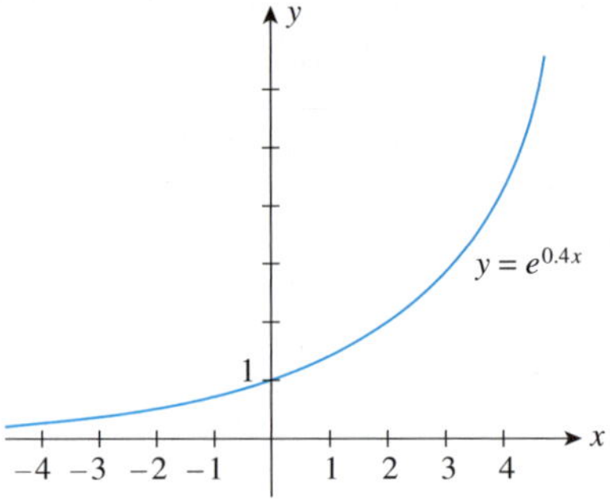

USING TECHNOLOGY

Although the proof is outside the scope of this book, it can be proved that an exponential function of the form $f(x) = b^x$, where $b > 1$, will ultimately grow faster than the power function $g(x) = x^n$ for *any* positive real number n. To give a visual demonstration of this result for the special case of the exponential function $f(x) = e^x$, we can use a graphing utility to plot the graphs of both f and g (for selected values of n) on the same set of axes in an appropriate viewing window and observe that the graph of f ultimately lies above that of g.

EXAMPLE 1 Use a graphing utility to plot the graphs of (a) $f(x) = e^x$ and $g(x) = x^3$ on the same set of axes in the viewing window $[0, 6] \times [0, 250]$ and (b) $f(x) = e^x$ and $g(x) = x^5$ in the viewing window $[0, 20] \times [0, 1{,}000{,}000]$.

Solution

a. The graphs of $f(x) = e^x$ and $g(x) = x^3$ in the viewing window $[0, 6] \times [0, 250]$ are shown in Figure T1a.

b. The graphs of $f(x) = e^x$ and $g(x) = x^5$ in the viewing window $[0, 20] \times [0, 1{,}000{,}000]$ are shown in Figure T1b.

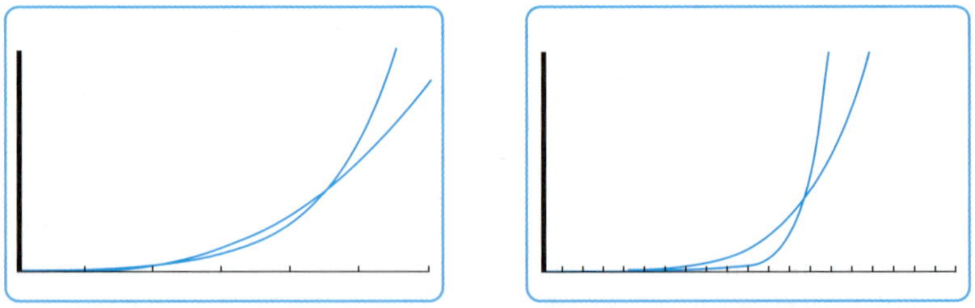

FIGURE **T1**

(a) The graphs of $f(x) = e^x$ and $g(x) = x^3$ in the viewing window $[0, 6] \times [0, 250]$

(b) The graphs of $f(x) = e^x$ and $g(x) = x^5$ in the viewing window $[0, 20] \times [0, 1{,}000{,}000]$

In the exercises that follow, you are asked to use a graphing utility to reveal the properties of exponential functions.

TECHNOLOGY EXERCISES

In Exercises 1 and 2, plot the graphs of the functions f and g on the same set of axes in the specified viewing window.

1. $f(x) = e^x$ and $g(x) = x^2$; $[0, 4] \times [0, 30]$

2. $f(x) = e^x$ and $g(x) = x^4$; $[0, 15] \times [0, 20{,}000]$

In Exercises 3 and 4, plot the graphs of the functions f and g on the same set of axes in an appropriate viewing window to demonstrate that f ultimately grows faster than g. (*Note:* Your answer will *not* be unique.)

3. $f(x) = 2^x$ and $g(x) = x^{2.5}$

4. $f(x) = 3^x$ and $g(x) = x^3$

5. Plot the graphs of $f(x) = 2^x$, $g(x) = 3^x$, and $h(x) = 4^x$ on the same set of axes in the viewing window $[0, 5] \times [0, 100]$. Comment on the relationship between the base b and the growth of the function $f(x) = b^x$.

6. Plot the graphs of $f(x) = (1/2)^x$, $g(x) = (1/3)^x$, and $h(x) = (1/4)^x$ on the same set of axes in the viewing window $[0, 4] \times [0, 1]$. Comment on the relationship between the base b and the growth of the function $f(x) = b^x$.

7. Plot the graphs of $f(x) = e^x$, $g(x) = 2e^x$, and $h(x) = 3e^x$ on the same set of axes in the viewing window $[-3, 3] \times [0, 10]$. Comment on the role played by the constant k in the graph of $f(x) = ke^x$.

8. Plot the graphs of $f(x) = -e^x$, $g(x) = -2e^x$, and $h(x) = -3e^x$ on the same set of axes in the viewing window $[-3, 3] \times [-10, 0]$. Comment on the role played by the constant k in the graph of $f(x) = ke^x$.

9. Plot the graphs of $f(x) = e^{0.5x}$, $g(x) = e^x$, and $h(x) = e^{1.5x}$ on the same set of axes in the viewing window $[-2, 2] \times [0, 4]$. Comment on the role played by the constant k in the graph of $f(x) = e^{kx}$.

10. Plot the graphs of $f(x) = e^{-0.5x}$, $g(x) = e^{-x}$, and $h(x) = e^{-1.5x}$ on the same set of axes in the viewing window $[-2, 2] \times [0, 4]$. Comment on the role played by the constant k in the graph of $f(x) = e^{kx}$.

11. **Absorption of Drugs** The concentration of a drug in an organ at any time t (in seconds) is given by

$$x(t) = 0.08 + 0.12(1 - e^{-0.02t})$$

where $x(t)$ is measured in grams/cubic centimeter (g/cm^3).

a. Plot the graph of the function x in the viewing window $[0, 200] \times [0, 0.2]$.

b. What is the initial concentration of the drug in the organ?

c. What is the concentration of the drug in the organ after 20 sec?

d. What will be the concentration of the drug in the organ in the long run?
Hint: Evaluate $\lim_{t \to \infty} x(t)$.

12. **Absorption of Drugs** Jane took 100 mg of a drug in the morning and another 100 mg of the same drug at the same time the following morning. The amount of the drug in her body t days after the first dosage was taken is given by

$$A(t) = \begin{cases} 100e^{-1.4t} & \text{if } 0 \le t < 1 \\ 100(1 + e^{1.4})e^{-1.4t} & \text{if } t \ge 1 \end{cases}$$

a. Plot the graph of the function A in the viewing window $[0, 5] \times [0, 140]$.

b. Verify the results of Exercise 48, page 335.

13. **Absorption of Drugs** The concentration of a drug in an organ at any time t (in seconds) is given by

$$C(t) = \begin{cases} 0.3t - 18(1 - e^{-t/60}) & \text{if } 0 \le t \le 20 \\ 18e^{-t/60} - 12e^{-(t-20)/60} & \text{if } t > 20 \end{cases}$$

where $C(t)$ is measured in grams/cubic centimeter (g/cm^3).

a. Plot the graph of the function C in the viewing window $[0, 120] \times [0, 1]$.

b. How long after the drug is first introduced will it take for the concentration of the drug to reach a peak?

c. How long after the concentration of the drug has peaked will it take for the concentration of the drug to fall back to 0.5 g/cm^3?
Hint: Plot the graphs of $y_1 = C(x)$ and $y_2 = 0.5$ and use the ISECT function of your graphing utility.

14. **Modeling with Data** The number of Internet users in China (in millions) from 2005 through 2010 are shown in the following table:

Year	2005	2006	2007	2008	2009	2010
Number	116.1	141.9	169.0	209.0	258.1	314.8

(*Note:* The number of users for the years 2007 through 2010 are projections.)

a. Use **ExpReg** to find an exponential regression model for the data. Let $t = 1$ correspond to 2005.
Hint: $a^x = e^{x \ln a}$.

b. Plot the scatter diagram and the graph of the function f found in part (a).

5.2 Logarithmic Functions

Logarithms

You are already familiar with exponential equations of the form

$$b^y = x \qquad (b > 0, b \neq 1)$$

where the variable x is expressed in terms of a real number b and a variable y. But what about solving this same equation for y? You may recall from your study of algebra that the number y is called the **logarithm of x to the base b** and is denoted by **$\log_b x$.** It is the power to which the base b must be raised in order to obtain the number x.

> **Logarithm of x to the Base b**
>
> $$y = \log_b x \quad \text{if and only if} \quad x = b^y \qquad (x > 0)$$

Observe that the logarithm $\log_b x$ is defined only for positive values of x.

EXAMPLE 1

a. $\log_{10} 100 = 2$ since $100 = 10^2$
b. $\log_5 125 = 3$ since $125 = 5^3$
c. $\log_3 \dfrac{1}{27} = -3$ since $\dfrac{1}{27} = \dfrac{1}{3^3} = 3^{-3}$
d. $\log_{20} 20 = 1$ since $20 = 20^1$

EXAMPLE 2 Solve each of the following equations for x.

a. $\log_3 x = 4$ **b.** $\log_{16} 4 = x$ **c.** $\log_x 8 = 3$

Solution

a. By definition, $\log_3 x = 4$ implies $x = 3^4 = 81$.
b. $\log_{16} 4 = x$ is equivalent to $4 = 16^x = (4^2)^x = 4^{2x}$, or $4^1 = 4^{2x}$, from which we deduce that

$$2x = 1 \qquad b^m = b^n \Rightarrow m = n$$

$$x = \frac{1}{2}$$

c. Referring once again to the definition, we see that the equation $\log_x 8 = 3$ is equivalent to

$$8 = 2^3 = x^3$$

$$x = 2 \qquad a^m = b^m \Rightarrow a = b$$

The two most widely used systems of logarithms are the system of **common logarithms,** which uses the number 10 as its base, and the system of **natural logarithms,** which uses the irrational number $e = 2.71828\ldots$ as its base. Also, it is standard practice to write **log** for $\log_{10}$ and **ln** for $\log_e$.

> **Logarithmic Notation**
>
> $\log x = \log_{10} x$ — Common logarithm
>
> $\ln x = \log_e x$ — Natural logarithm

The system of natural logarithms is widely used in theoretical work. Using natural logarithms rather than logarithms to other bases often leads to simpler expressions.

Laws of Logarithms

Computations involving logarithms are facilitated by the following **laws of logarithms.**

Laws of Logarithms

If m and n are positive numbers, then

1. $\log_b mn = \log_b m + \log_b n$
2. $\log_b \dfrac{m}{n} = \log_b m - \log_b n$
3. $\log_b m^n = n \log_b m$
4. $\log_b 1 = 0$
5. $\log_b b = 1$

Do not confuse the expression $\log m/n$ (Law 2) with the expression $\log m/\log n$. For example,

$$\log \frac{100}{10} = \log 100 - \log 10 = 2 - 1 = 1 \neq \frac{\log 100}{\log 10} = \frac{2}{1} = 2$$

You will be asked to prove these laws in Exercises 62–64 on page 345. Their derivations are based on the definition of a logarithm and the corresponding laws of exponents. The following examples illustrate the properties of logarithms.

EXAMPLE 3

a. $\log(2 \cdot 3) = \log 2 + \log 3$ **b.** $\ln \dfrac{5}{3} = \ln 5 - \ln 3$

c. $\log\sqrt{7} = \log 7^{1/2} = \dfrac{1}{2}\log 7$ **d.** $\log_5 1 = 0$

e. $\log_{45} 45 = 1$ ■

EXAMPLE 4 Given that $\log 2 \approx 0.3010$, $\log 3 \approx 0.4771$, and $\log 5 \approx 0.6990$, use the laws of logarithms to find

a. $\log 15$ **b.** $\log 7.5$ **c.** $\log 81$ **d.** $\log 50$

Solution

a. Note that $15 = 3 \cdot 5$, so by Law 1 for logarithms,

$$\begin{aligned} \log 15 &= \log 3 \cdot 5 \\ &= \log 3 + \log 5 \\ &\approx 0.4771 + 0.6990 \\ &= 1.1761 \end{aligned}$$

b. Observing that $7.5 = 15/2 = (3 \cdot 5)/2$, we apply Laws 1 and 2, obtaining

$$\begin{aligned}\log 7.5 &= \log \frac{(3)(5)}{2}\\ &= \log 3 + \log 5 - \log 2\\ &\approx 0.4771 + 0.6990 - 0.3010\\ &= 0.8751\end{aligned}$$

c. Since $81 = 3^4$, we apply Law 3 to obtain

$$\begin{aligned}\log 81 &= \log 3^4\\ &= 4 \log 3\\ &\approx 4(0.4771)\\ &= 1.9084\end{aligned}$$

d. We write $50 = 5 \cdot 10$ and find

$$\begin{aligned}\log 50 &= \log(5)(10)\\ &= \log 5 + \log 10\\ \log 50 &\approx 0.6990 + 1 && \text{Use Law 5}\\ &= 1.6990\end{aligned}$$

EXAMPLE 5 Expand and simplify the following expressions:

a. $\log_3 x^2y^3$ **b.** $\log_2 \dfrac{x^2 + 1}{2^x}$ **c.** $\ln \dfrac{x^2\sqrt{x^2 - 1}}{e^x}$

Solution

a.
$$\begin{aligned}\log_3 x^2y^3 &= \log_3 x^2 + \log_3 y^3 && \text{Law 1}\\ &= 2 \log_3 x + 3 \log_3 y && \text{Law 3}\end{aligned}$$

b.
$$\begin{aligned}\log_2 \frac{x^2 + 1}{2^x} &= \log_2(x^2 + 1) - \log_2 2^x && \text{Law 2}\\ &= \log_2(x^2 + 1) - x \log_2 2 && \text{Law 3}\\ &= \log_2(x^2 + 1) - x && \text{Law 5}\end{aligned}$$

c.
$$\begin{aligned}\ln \frac{x^2\sqrt{x^2 - 1}}{e^x} &= \ln \frac{x^2(x^2 - 1)^{1/2}}{e^x} && \text{Rewrite}\\ &= \ln x^2 + \ln(x^2 - 1)^{1/2} - \ln e^x && \text{Laws 1 and 2}\\ &= 2 \ln x + \frac{1}{2} \ln(x^2 - 1) - x \ln e && \text{Law 3}\\ &= 2 \ln x + \frac{1}{2} \ln(x^2 - 1) - x && \text{Law 5}\end{aligned}$$

Logarithmic Functions and Their Graphs

The definition of a logarithm implies that if b and n are positive numbers and b is different from 1, then the expression $\log_b n$ is a real number. This enables us to define a logarithmic function as follows.

Logarithmic Function

The function defined by

$$f(x) = \log_b x \qquad (b > 0, b \neq 1)$$

is called the **logarithmic function with base *b*.** The domain of f is the set of all positive numbers.

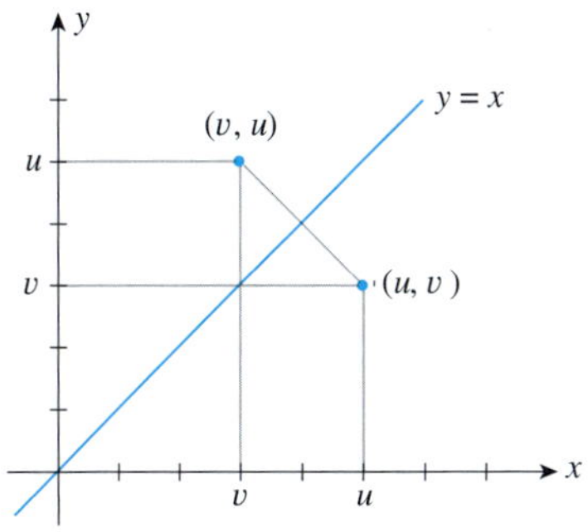

FIGURE 7
The points (u, v) and (v, u) are mirror reflections of each other.

One easy way to obtain the graph of the logarithmic function $y = \log_b x$ is to construct a table of values of the logarithm (base b). However, another method—and a more instructive one—is based on exploiting the intimate relationship between logarithmic and exponential functions.

If a point (u, v) lies on the graph of $y = \log_b x$, then

$$v = \log_b u$$

But we can also write this equation in exponential form as

$$u = b^v$$

So the point (v, u) also lies on the graph of the function $y = b^x$. Let's look at the relationship between the points (u, v) and (v, u) and the line $y = x$ (Figure 7). If we think of the line $y = x$ as a mirror, then the point (v, u) is the mirror reflection of the point (u, v). Similarly, the point (u, v) is a mirror reflection of the point (v, u). We can take advantage of this relationship to help us draw the graph of logarithmic functions. For example, if we wish to draw the graph of $y = \log_b x$, where $b > 1$, then we need only draw the mirror reflection of the graph of $y = b^x$ with respect to the line $y = x$ (Figure 8).

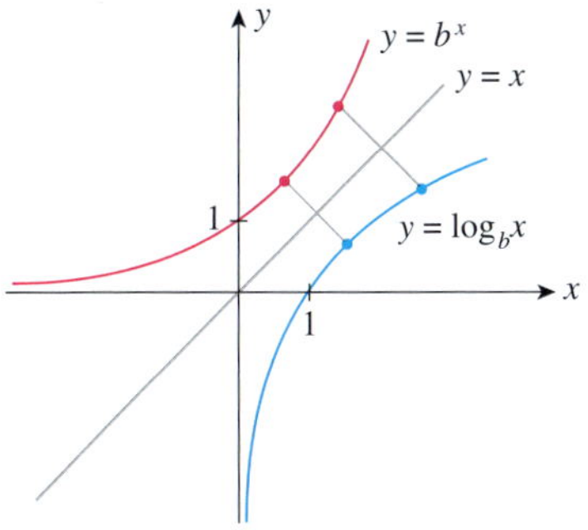

FIGURE 8
The graphs of $y = b^x$ and $y = \log_b x$ are mirror reflections of each other.

You may discover the following properties of the logarithmic function by taking the reflection of the graph of an appropriate exponential function (Exercises 33 and 34 on page 344).

Properties of the Logarithmic Function

The logarithmic function $y = \log_b x$ $(b > 0, b \neq 1)$ has the following properties:

1. Its domain is $(0, \infty)$.
2. Its range is $(-\infty, \infty)$.
3. Its graph passes through the point $(1, 0)$.
4. It is continuous on $(0, \infty)$.
5. It is increasing on $(0, \infty)$ if $b > 1$ and decreasing on $(0, \infty)$ if $b < 1$.

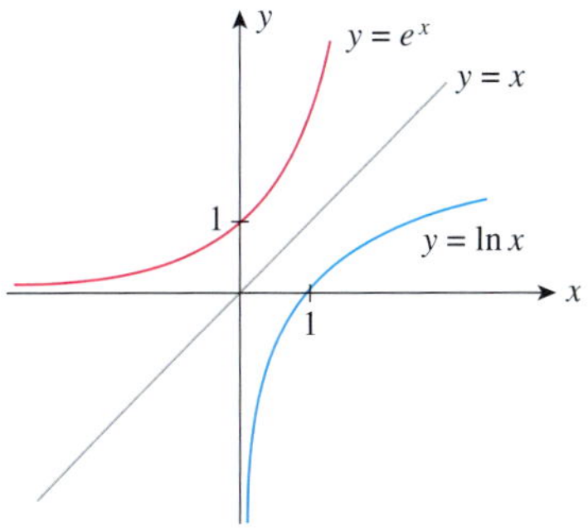

FIGURE 9
The graph of $y = \ln x$ is the mirror reflection of the graph of $y = e^x$.

EXAMPLE 6 Sketch the graph of the function $y = \ln x$.

Solution We first sketch the graph of $y = e^x$. Then, the required graph is obtained by tracing the mirror reflection of the graph of $y = e^x$ with respect to the line $y = x$ (Figure 9). ■

Properties Relating the Exponential and Logarithmic Functions

We made use of the relationship that exists between the exponential function $f(x) = e^x$ and the logarithmic function $g(x) = \ln x$ when we sketched the graph of g in Example 6. This relationship is further described by the following properties, which are an immediate consequence of the definition of the logarithm of a number.

Properties Relating e^x and $\ln x$

$$e^{\ln x} = x \qquad (x > 0) \qquad \textbf{(2)}$$

$$\ln e^x = x \qquad \text{(for any real number } x\text{)} \qquad \textbf{(3)}$$

(Try to verify these properties.)

From Properties 2 and 3, we conclude that the composite function

$$\begin{aligned}(f \circ g)(x) &= f[g(x)] \\ &= e^{\ln x} = x \\ (g \circ f)(x) &= g[f(x)] \\ &= \ln e^x = x\end{aligned}$$

Thus,

$$\begin{aligned}f[g(x)] &= g[f(x)] \\ &= x\end{aligned}$$

Any two functions f and g that satisfy this relationship are said to be **inverses** of each other. Note that the function f undoes what the function g does, and vice versa, so the composition of the two functions in any order results in the identity function $F(x) = x$.*

The relationships expressed in Equations (2) and (3) are useful in solving equations that involve exponentials and logarithms.

Exploring with TECHNOLOGY

You can demonstrate the validity of Properties 2 and 3, which state that the exponential function $f(x) = e^x$ and the logarithmic function $g(x) = \ln x$ are inverses of each other, as follows:

1. Sketch the graph of $(f \circ g)(x) = e^{\ln x}$, using the viewing window $[0, 10] \times [0, 10]$. Interpret the result.
2. Sketch the graph of $(g \circ f)(x) = \ln e^x$, using the standard viewing window. Interpret the result.

VIDEO

EXAMPLE 7 Solve the equation $2e^{x+2} = 5$.

Solution We first divide both sides of the equation by 2 to obtain

$$e^{x+2} = \frac{5}{2} = 2.5$$

Next, taking the natural logarithm of each side of the equation and using Equation (3), we have

$$\begin{aligned}\ln e^{x+2} &= \ln 2.5 \\ x + 2 &= \ln 2.5 \\ x &= -2 + \ln 2.5 \\ &\approx -1.08\end{aligned}$$

■

Explore & Discuss

Consider the equation $y = y_0 b^{kx}$, where y_0 and k are positive constants and $b > 0$, $b \neq 1$. Suppose we want to express y in the form $y = y_0 e^{px}$. Use the laws of logarithms to show that $p = k \ln b$ and hence that $y = y_0 e^{(k \ln b)x}$ is an alternative form of $y = y_0 b^{kx}$ using the base e.

*For a more extensive treatment of inverse functions, see the appendix.

EXAMPLE 8 Solve the equation $5 \ln x + 3 = 0$.

Solution Adding -3 to both sides of the equation leads to

$$5 \ln x = -3$$
$$\ln x = -\frac{3}{5} = -0.6$$

and so

$$e^{\ln x} = e^{-0.6}$$

Using Equation (2), we conclude that

$$x = e^{-0.6}$$
$$\approx 0.55$$

5.2 Self-Check Exercises

1. Sketch the graph of $y = 3^x$ and $y = \log_3 x$ on the same set of axes.
2. Solve the equation $3e^{x+1} - 2 = 4$.

Solutions to Self-Check Exercises 5.2 can be found on page 345.

5.2 Concept Questions

1. **a.** Define $y = \log_b x$.
   b. Define the logarithmic function f with base b. What restrictions, if any, are placed on b?
2. For the logarithmic function $y = \log_b x$ $(b > 0, b \neq 1)$, state (a) its domain and range, (b) its x-intercept, (c) where it is continuous, and (d) where it is increasing and where it is decreasing for the case $b > 1$ and the case $b < 1$.
3. **a.** If $x > 0$, what is $e^{\ln x}$?
   b. If x is any real number, what is $\ln e^x$?
4. Let $f(x) = \ln x^2$ and $g(x) = 2 \ln x$. Are f and g identical? Hint: Look at their domains.

5.2 Exercises

In Exercises 1–10, express each equation in logarithmic form.

1. $2^6 = 64$
2. $3^5 = 243$
3. $3^{-2} = \frac{1}{9}$
4. $5^{-3} = \frac{1}{125}$
5. $\left(\frac{1}{3}\right)^1 = \frac{1}{3}$
6. $\left(\frac{1}{2}\right)^{-4} = 16$
7. $32^{3/5} = 8$
8. $81^{3/4} = 27$
9. $10^{-3} = 0.001$
10. $16^{-1/4} = 0.5$

In Exercises 11–16, given that $\log 3 \approx 0.4771$ and $\log 4 \approx 0.6021$, find the value of each logarithm.

11. $\log 12$
12. $\log \frac{3}{4}$
13. $\log 16$
14. $\log \sqrt{3}$
15. $\log 48$
16. $\log \frac{1}{300}$

In Exercises 17–20, write the expression as the logarithm of a single quantity.

17. $2 \ln a + 3 \ln b$
18. $\frac{1}{2} \ln x + 2 \ln y - 3 \ln z$
19. $\ln 3 + \frac{1}{2} \ln x + \ln y - \frac{1}{3} \ln z$
20. $\ln 2 + \frac{1}{2} \ln(x + 1) - 2 \ln(1 + \sqrt{x})$

In Exercises 21–28, use the laws of logarithms to expand and simplify the expression.

21. $\log x(x + 1)^4$

22. $\log x(x^2 + 1)^{-1/2}$

23. $\log \dfrac{\sqrt{x + 1}}{x^2 + 1}$

24. $\ln \dfrac{e^x}{1 + e^x}$

25. $\ln xe^{-x^2}$

26. $\ln x(x + 1)(x + 2)$

27. $\ln \dfrac{x^{1/2}}{x^2\sqrt{1 + x^2}}$

28. $\ln \dfrac{x^2}{\sqrt{x}(1 + x)^2}$

In Exercises 29–32, sketch the graph of the equation.

29. $y = \log_3 x$

30. $y = \log_{1/3} x$

31. $y = \ln 2x$

32. $y = \ln \dfrac{1}{2}x$

In Exercises 33 and 34, sketch the graphs of the equations on the same coordinate axes.

33. $y = 2^x$ and $y = \log_2 x$

34. $y = e^{3x}$ and $y = \dfrac{1}{3}\ln x$

In Exercises 35–44, use logarithms to solve the equation for *t*.

35. $e^{0.4t} = 8$

36. $\dfrac{1}{3}e^{-3t} = 0.9$

37. $5e^{-2t} = 6$

38. $4e^{t-1} = 4$

39. $2e^{-0.2t} - 4 = 6$

40. $12 - e^{0.4t} = 3$

41. $\dfrac{50}{1 + 4e^{0.2t}} = 20$

42. $\dfrac{200}{1 + 3e^{-0.3t}} = 100$

43. $A = Be^{-t/2}$

44. $\dfrac{A}{1 + Be^{t/2}} = C$

45. A function f has the form $f(x) = a + b \ln x$. Find f if it is known that $f(1) = 2$ and $f(2) = 4$.

46. **Average Life Span** One reason for the increase in the life span over the years has been the advances in medical technology. The average life span for American women from 1907 through 2007 is given by

$$W(t) = 49.9 + 17.1 \ln t \qquad (1 \le t \le 6)$$

where $W(t)$ is measured in years and t is measured in 20-yr intervals, with $t = 1$ corresponding to 1907.

a. What was the average life expectancy for women in 1907?

b. If the trend continues, what will be the average life expectancy for women in 2027?

Source: American Association of Retired People

47. **Blood Pressure** A normal child's systolic blood pressure may be approximated by the function

$$p(x) = m(\ln x) + b$$

where $p(x)$ is measured in millimeters of mercury, x is measured in pounds, and m and b are constants. Given that $m = 19.4$ and $b = 18$, determine the systolic blood pressure of a child who weighs 92 lb.

48. **Magnitude of Earthquakes** On the Richter scale, the magnitude R of an earthquake is given by the formula

$$R = \log \frac{I}{I_0}$$

where I is the intensity of the earthquake being measured and I_0 is the standard reference intensity.

a. Express the intensity I of an earthquake of magnitude $R = 5$ in terms of the standard intensity I_0.

b. Express the intensity I of an earthquake of magnitude $R = 8$ in terms of the standard intensity I_0. How many times greater is the intensity of an earthquake of magnitude 8 than one of magnitude 5?

c. In modern times, the greatest loss of life attributable to an earthquake occurred in eastern China in 1976. Known as the Tangshan earthquake, it registered 8.2 on the Richter scale. How does the intensity of this earthquake compare with the intensity of an earthquake of magnitude $R = 5$?

49. **Sound Intensity** The relative loudness of a sound D of intensity I is measured in decibels (db), where

$$D = 10 \log \frac{I}{I_0}$$

and I_0 is the standard threshold of audibility.

a. Express the intensity I of a 30-db sound (the sound level of normal conversation) in terms of I_0.

b. Determine how many times greater the intensity of an 80-db sound (rock music) is than that of a 30-db sound.

c. Prolonged noise above 150 db causes permanent deafness. How does the intensity of a 150-db sound compare with the intensity of an 80-db sound?

50. **Barometric Pressure** Halley's law states that the barometric pressure (in inches of mercury) at an altitude of x mi above sea level is approximated by the equation

$$p(x) = 29.92e^{-0.2x} \qquad (x \ge 0)$$

If the barometric pressure as measured by a hot-air balloonist is 20 in. of mercury, what is the balloonist's altitude?

51. **Height of Trees** The height (in feet) of a certain kind of tree is approximated by

$$h(t) = \frac{160}{1 + 240e^{-0.2t}}$$

where t is the age of the tree in years. Estimate the age of an 80-ft tree.

52. **Newton's Law of Cooling** The temperature of a cup of coffee t min after it is poured is given by

$$T = 70 + 100e^{-0.0446t}$$

where T is measured in degrees Fahrenheit.

a. What was the temperature of the coffee when it was poured?

b. When will the coffee be cool enough to drink (say, 120°F)?

53. **Lengths of Fish** The length (in centimeters) of a typical Pacific halibut t yr old is approximately

$$f(t) = 200(1 - 0.956e^{-0.18t})$$

Suppose a Pacific halibut caught by Mike measures 140 cm. What is its approximate age?

54. **Absorption of Drugs** The concentration of a drug in an organ at any time t (in seconds) is given by

$$x(t) = 0.08(1 - e^{-0.02t})$$

where $x(t)$ is measured in grams/cubic centimeter (g/cm^3).
a. How long would it take for the concentration of the drug in the organ to reach 0.02 g/cm^3?
b. How long would it take for the concentration of the drug in the organ to reach 0.04 g/cm^3?

55. **Absorption of Drugs** The concentration of a drug in an organ at any time t (in seconds) is given by

$$x(t) = 0.08 + 0.12e^{-0.02t}$$

where $x(t)$ is measured in grams/cubic centimeter (g/cm^3).
a. How long would it take for the concentration of the drug in the organ to reach 0.18 g/cm^3?
b. How long would it take for the concentration of the drug in the organ to reach 0.16 g/cm^3?

56. **Forensic Science** Forensic scientists use the following law to determine the time of death of accident or murder victims. If T denotes the temperature of a body t hr after death, then

$$T = T_0 + (T_1 - T_0)(0.97)^t$$

where T_0 is the air temperature and T_1 is the body temperature at the time of death. John Doe was found murdered at midnight in his house, when the room temperature was 70°F and his body temperature was 80°F. When was he killed? Assume that the normal body temperature is 98.6°F.

In Exercises 57–60, determine whether the statement is true or false. If it is true, explain why it is true. If it is false, give an example to show why it is false.

57. $(\ln x)^3 = 3 \ln x$ for all x in $(0, \infty)$.

58. $\ln a - \ln b = \ln(a - b)$ for all positive real numbers a and b.

59. The function $f(x) = \dfrac{1}{\ln x}$ is continuous on $(1, \infty)$.

60. The function $f(x) = \ln|x|$ is continuous for all $x \neq 0$.

61. **a.** Given that $2^x = e^{kx}$, find k.
b. Show that, in general, if b is a nonnegative real number, then any equation of the form $y = b^x$ may be written in the form $y = e^{kx}$, for some real number k.

62. Use the definition of a logarithm to prove
a. $\log_b mn = \log_b m + \log_b n$
b. $\log_b \dfrac{m}{n} = \log_b m - \log_b n$
Hint: Let $\log_b m = p$ and $\log_b n = q$. Then, $b^p = m$ and $b^q = n$.

63. Use the definition of a logarithm to prove

$$\log_b m^n = n \log_b m$$

64. Use the definition of a logarithm to prove
a. $\log_b 1 = 0$
b. $\log_b b = 1$

5.2 Solutions to Self-Check Exercises

1. First, sketch the graph of $y = 3^x$ with the help of the following table of values:

x	-3	-2	-1	0	1	2	3
$y = 3^x$	1/27	1/9	1/3	0	3	9	27

Next, take the mirror reflection of this graph with respect to the line $y = x$ to obtain the graph of $y = \log_3 x$.

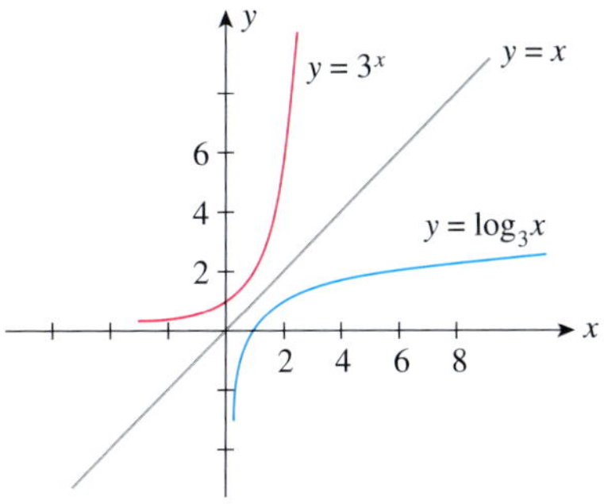

2. $3e^{x+1} - 2 = 4$

$$3e^{x+1} = 6$$

$$e^{x+1} = 2$$

$$\ln e^{x+1} = \ln 2 \quad \text{Take the logarithm of both sides.}$$

$$(x + 1)\ln e = \ln 2 \quad \text{Law 3}$$

$$x + 1 = \ln 2 \quad \text{Law 5}$$

$$x = \ln 2 - 1$$

$$\approx -0.3069$$

5.3 Compound Interest

Compound Interest

Compound interest is a natural application of the exponential function to the business world. We begin by recalling that simple interest is interest that is computed only on the original principal. Thus, if I denotes the interest on a principal P (in dollars) at an interest rate of r per year for t years, then we have

$$I = Prt$$

The **accumulated amount** A, the sum of the principal and interest after t years, is given by

$$\begin{aligned} A &= P + I = P + Prt \\ &= P(1 + rt) \quad \text{Simple interest formula} \end{aligned} \tag{4}$$

Frequently, interest earned is periodically added to the principal and thereafter earns interest itself at the same rate. This is called **compound interest.** To find a formula for the accumulated amount, let's consider a numerical example. Suppose \$1000 (the principal) is deposited in a bank for a **term** of 3 years, earning interest at the rate of 8% per year (called the **nominal,** or **stated, rate**) compounded annually. Then, using Formula (4) with $P = 1000$, $r = 0.08$, and $t = 1$, we see that the accumulated amount at the end of the first year is

$$\begin{aligned} A_1 &= P(1 + rt) \\ &= 1000[1 + 0.08(1)] = 1000(1.08) = 1080 \end{aligned}$$

or \$1080.

To find the accumulated amount A_2 at the end of the second year, we use (4) once again, this time with $P = A_1$. (Remember, the principal *and* interest now earn interest over the second year.) We obtain

$$\begin{aligned} A_2 &= P(1 + rt) = A_1(1 + rt) \\ &= 1000[1 + 0.08(1)][1 + 0.08(1)] \\ &= 1000(1 + 0.08)^2 = 1000(1.08)^2 \approx 1166.40 \end{aligned}$$

or approximately \$1166.40.

Finally, the accumulated amount A_3 at the end of the third year is found using (4) with $P = A_2$, giving

$$\begin{aligned} A_3 &= P(1 + rt) = A_2(1 + rt) \\ &= 1000[1 + 0.08(1)]^2[1 + 0.08(1)] \\ &= 1000(1 + 0.08)^3 = 1000(1.08)^3 \approx 1259.71 \end{aligned}$$

or approximately \$1259.71.

If you reexamine our calculations in this example, you will see that the accumulated amounts at the end of each year have the following form:

$$\begin{array}{llll} \text{First year:} & A_1 = 1000(1 + 0.08) & \text{or} & A_1 = P(1 + r) \\ \text{Second year:} & A_2 = 1000(1 + 0.08)^2 & \text{or} & A_2 = P(1 + r)^2 \\ \text{Third year:} & A_3 = 1000(1 + 0.08)^3 & \text{or} & A_3 = P(1 + r)^3 \end{array}$$

These observations suggest the following general result: If P dollars are invested over a term of t years earning interest at the rate of r per year compounded annually, then the accumulated amount is

$$A = P(1 + r)^t \tag{5}$$

Formula (5) was derived under the assumption that interest was compounded *annually*. In practice, however, interest is usually compounded more than once a year. The interval of time between successive interest calculations is called the **conversion period.**

If interest at a nominal rate of r per year is compounded m times a year on a principal of P dollars, then the simple interest rate per conversion period is

$$i = \frac{r}{m} \qquad \frac{\text{Annual interest rate}}{\text{Periods per year}}$$

For example, if the nominal interest rate is 8% per year ($r = 0.08$) and interest is compounded quarterly ($m = 4$), then

$$i = \frac{r}{m} = \frac{0.08}{4} = 0.02$$

or 2% per period.

To find a general formula for the accumulated amount when a principal of P dollars is deposited in a bank for a term of t years and earns interest at the (nominal) rate of r per year compounded m times per year, we proceed as before using Formula (5) repeatedly with the interest rate $i = r/m$. We see that the accumulated amount at the end of each period is as follows:

$$\begin{aligned}
&\text{First period:} && A_1 = P(1 + i) \\
&\text{Second period:} && A_2 = A_1(1 + i) = [P(1 + i)](1 + i) = P(1 + i)^2 \\
&\text{Third period:} && A_3 = A_2(1 + i) = [P(1 + i)^2](1 + i) = P(1 + i)^3 \\
&\quad\vdots && \qquad\vdots \\
&n\text{th period:} && A_n = A_{n-1}(1 + i) = [P(1 + i)^{n-1}](1 + i) = P(1 + i)^n
\end{aligned}$$

But there are $n = mt$ periods in t years (number of conversion periods times the term). Therefore, the accumulated amount at the end of t years is given by

Compound Interest Formula

$$A = P\left(1 + \frac{r}{m}\right)^{mt} \tag{6}$$

where

A = Accumulated amount at the end of t years
P = Principal
r = Nominal interest rate per year
m = Number of conversion periods per year
t = Term (number of years)

EXAMPLE 1 Find the accumulated amount after 3 years if \$1000 is invested at 8% per year compounded (a) annually, (b) semiannually, (c) quarterly, (d) monthly, and (e) daily.

Solution

a. Here, $P = 1000$, $r = 0.08$, $m = 1$, and $t = 3$, so Formula (6) gives

$$\begin{aligned} A &= 1000(1 + 0.08)^3 \\ &= 1259.71 \end{aligned}$$

or \$1259.71.

b. Here, $P = 1000$, $r = 0.08$, $m = 2$, and $t = 3$, so (6) gives

$$A = 1000\left(1 + \frac{0.08}{2}\right)^{(2)(3)}$$
$$= 1265.32$$

or \$1265.32.

c. In this case, $P = 1000$, $r = 0.08$, $m = 4$, and $t = 3$, so (6) gives

$$A = 1000\left(1 + \frac{0.08}{4}\right)^{(4)(3)}$$
$$= 1268.24$$

or \$1268.24

d. Here, $P = 1000$, $r = 0.08$, $m = 12$, and $t = 3$, so (6) gives

$$A = 1000\left(1 + \frac{0.08}{12}\right)^{(12)(3)}$$
$$= 1270.24$$

or \$1270.24.

e. Here, $P = 1000$, $r = 0.08$, $m = 365$, and $t = 3$, so (6) gives

$$A = 1000\left(1 + \frac{0.08}{365}\right)^{(365)(3)}$$
$$= 1271.22$$

or \$1271.22. These results are summarized in Table 2.

TABLE 2

Nominal Rate, r	Conversion Period	Term in Years	Initial Investment	Accumulated Amount
8%	Annual ($m = 1$)	3	\$1000	\$1259.71
8	Semiannual ($m = 2$)	3	1000	1265.32
8	Quarterly ($m = 4$)	3	1000	1268.24
8	Monthly ($m = 12$)	3	1000	1270.24
8	Daily ($m = 365$)	3	1000	1271.22

Effective Rate of Interest

In the last example, we saw that the interest actually earned on an investment depends on the frequency with which the interest is compounded. Thus, the stated, or nominal, rate of 8% per year does not reflect the actual rate at which interest is earned. This suggests that we need to find a common basis for comparing interest rates. One way of comparing interest rates is provided by using the effective rate. The **effective rate** is the *simple* interest rate that would produce the same accumulated amount in 1 year as the nominal rate compounded m times a year. The effective rate is also called the **true rate.**

To derive a relation between the nominal interest rate, r per year compounded m times, and its corresponding effective rate, r_{eff} per year, let's assume an initial investment of P dollars. Then, the accumulated amount after 1 year at a simple interest rate of r_{eff} per year is

$$A = P(1 + r_{\text{eff}})$$

PORTFOLIO Richard Mizak

TITLE Director
INSTITUTION Kroll Zolfo Cooper

Kroll Zolfo Cooper is one of the world's leading restructuring firms and works with companies experiencing critical financial or operational problems. We are retained as interim management or advisors to assist in stabilizing, rehabilitating and reorganizing the business, whether through performance improvement changes or Chapter 11 (bankruptcy) reorganization. We then make recommendations based on the financial analysis to the company's senior management team and use the analysis to develop concrete action plans to help get the company back on track. As a Director at Kroll Zolfo Cooper, I use mathematics when I perform financial analysis on companies in financial distress and/or bankruptcy.

At a large merchant power company we built a highly complex mathematical model to analyze the financial capabilities of the corporation and to project the company's financial performance into the future. The model was used by the company to forecast performance and by the lenders, investors and other key stakeholders to determine whether or not to lend money, invest and/or partner with the company.

At a major jewelry manufacturer, we determined that the company was losing money on most of its smaller orders. We used our mathematical analysis to work with the company to eliminate a series of costly activities that were unimportant to the smaller customers, determined the true cost of handling smaller orders, revised their computer pricing models, improved the company's pricing, and created incentives for customers to place larger orders. Almost immediately our financial analysis resulted in higher profitability for the department and higher overall sales results.

While these are just a few examples, the bulk of the work performed in the financial world has its basis in the same mathematical principals taught in your college coursework.

Also, the accumulated amount after 1 year at an interest rate of r per year compounded m times a year is

$$A = P\left(1 + \frac{r}{m}\right)^m \qquad \text{Since } t = 1$$

Equating the two expressions gives

$$P(1 + r_{\text{eff}}) = P\left(1 + \frac{r}{m}\right)^m$$

$$1 + r_{\text{eff}} = \left(1 + \frac{r}{m}\right)^m \qquad \text{Divide both sides by } P.$$

or, upon solving for r_{eff}, we obtain the formula for computing the effective rate of interest:

Effective Rate of Interest Formula

$$r_{\text{eff}} = \left(1 + \frac{r}{m}\right)^m - 1 \tag{7}$$

where

r_{eff} = Effective rate of interest
r = Nominal interest rate per year
m = Number of conversion periods per year

VIDEO

EXAMPLE 2 Find the effective rate of interest corresponding to a nominal rate of 8% per year compounded (a) annually, (b) semiannually, (c) quarterly, (d) monthly, and (e) daily.

Solution

a. The effective rate of interest corresponding to a nominal rate of 8% per year compounded annually is of course given by 8% per year. This result is also confirmed by using Formula (7) with $r = 0.08$ and $m = 1$. Thus,

$$r_{\text{eff}} = (1 + 0.08) - 1 = 0.08$$

b. Let $r = 0.08$ and $m = 2$. Then, (7) yields

$$\begin{aligned} r_{\text{eff}} &= \left(1 + \frac{0.08}{2}\right)^2 - 1 \\ &= 0.0816 \end{aligned}$$

so the required effective rate is 8.16% per year.

c. Let $r = 0.08$ and $m = 4$. Then, (7) yields

$$\begin{aligned} r_{\text{eff}} &= \left(1 + \frac{0.08}{4}\right)^4 - 1 \\ &= 0.08243 \end{aligned}$$

so the corresponding effective rate in this case is 8.243% per year.

d. Let $r = 0.08$ and $m = 12$. Then, (7) yields

$$\begin{aligned} r_{\text{eff}} &= \left(1 + \frac{0.08}{12}\right)^{12} - 1 \\ &= 0.08300 \end{aligned}$$

so the corresponding effective rate in this case is 8.3% per year.

e. Let $r = 0.08$ and $m = 365$. Then, (7) yields

$$\begin{aligned} r_{\text{eff}} &= \left(1 + \frac{0.08}{365}\right)^{365} - 1 \\ &= 0.08328 \end{aligned}$$

so the corresponding effective rate in this case is 8.328% per year. ■

Now, if the effective rate of interest r_{eff} is known, then the accumulated amount after t years on an investment of P dollars may be more readily computed by using the formula

$$A = P(1 + r_{\text{eff}})^t$$

The 1968 Truth in Lending Act passed by Congress requires that the effective rate of interest be disclosed in all contracts involving interest charges. The passage of this act has benefited consumers because they now have a common basis for comparing the various nominal rates quoted by different financial institutions. Furthermore, knowing the effective rate enables consumers to compute the actual charges involved in a transaction. Thus, if the effective rates of interest found in Example 2 were known, the accumulated values of Example 1, shown in Table 3, could have been readily found.

TABLE 3

Nominal Rate, r	Frequency of Interest Payment	Effective Rate	Initial Investment	Accumulated Amount after 3 Years
8%	Annually	8%	\$1000	$1000(1 + 0.08)^3 =$ \$1259.71
8	Semiannually	8.16	1000	$1000(1 + 0.0816)^3 =$ 1265.32
8	Quarterly	8.243	1000	$1000(1 + 0.08243)^3 =$ 1268.23
8	Monthly	8.300	1000	$1000(1 + 0.08300)^3 =$ 1270.24
8	Daily	8.328	1000	$1000(1 + 0.08328)^3 =$ 1271.22

Present Value

Let's return to the compound interest Formula (6), which expresses the accumulated amount at the end of t years when interest at the rate of r is compounded m times a year. The principal P in (6) is often referred to as the **present value,** and the accumulated value A is called the **future value** since it is realized at a future date. In certain instances an investor may wish to determine how much money he should invest now, at a fixed rate of interest, so that he will realize a certain sum at some future date. This problem may be solved by expressing P in terms of A. Thus, from (6) we find

Present Value Formula for Compound Interest

$$P = A\left(1 + \frac{r}{m}\right)^{-mt} \tag{8}$$

EXAMPLE 3 How much money should be deposited in a bank paying interest at the rate of 6% per year compounded monthly so that at the end of 3 years the accumulated amount will be \$20,000?

Solution Here, $A = 20{,}000$, $r = 0.06$, $m = 12$, and $t = 3$. Using Formula (8), we obtain

$$\begin{aligned} P &= 20{,}000\left(1 + \frac{0.06}{12}\right)^{-(12)(3)} \\ &\approx 16{,}713 \end{aligned}$$

or \$16,713.

EXAMPLE 4 Find the present value of \$49,158.60 due in 5 years at an interest rate of 10% per year compounded quarterly.

Solution Using Formula (8) with $A = 49{,}158.60$, $r = 0.1$, $m = 4$, and $t = 5$, we obtain

$$P = (49{,}158.6)\left(1 + \frac{0.1}{4}\right)^{-(4)(5)} \approx 30{,}000$$

or \$30,000.

Continuous Compounding of Interest

One question that arises naturally in the study of compound interest is: What happens to the accumulated amount over a fixed period of time if the interest is computed more and more frequently?

Intuition suggests that the more often interest is compounded, the larger the accumulated amount will be. This is confirmed by the results of Example 1, where we found that the accumulated amounts did in fact increase when we increased the number of conversion periods per year.

This leads us to another question: Does the accumulated amount approach a limit when the interest is computed more and more frequently over a fixed period of time?

To answer this question, let's look again at the compound interest formula:

$$A = P\left(1 + \frac{r}{m}\right)^{mt} \tag{9}$$

Recall that m is the number of conversion periods per year. So to find an answer to our problem, we should let m get larger and larger (approach infinity) in (9). But first we will rewrite this equation in the form

$$A = P\left[\left(1 + \frac{r}{m}\right)^{m}\right]^{t}$$ Since $b^{xy} = (b^x)^y$

Now, letting $m \to \infty$, we find that

$$\lim_{m\to\infty}\left[P\left(1 + \frac{r}{m}\right)^{m}\right]^{t} = P\left[\lim_{m\to\infty}\left(1 + \frac{r}{m}\right)^{m}\right]^{t}$$ Why?

Next, upon making the substitution $u = m/r$ and observing that $u \to \infty$ as $m \to \infty$, the foregoing expression reduces to

$$P\left[\lim_{u\to\infty}\left(1 + \frac{1}{u}\right)^{ur}\right]^{t} = P\left[\lim_{u\to\infty}\left(1 + \frac{1}{u}\right)^{u}\right]^{rt}$$

But

$$\lim_{u\to\infty}\left(1 + \frac{1}{u}\right)^{u} = e$$ Use (1).

so

$$\lim_{m\to\infty} P\left[\left(1 + \frac{r}{m}\right)^{m}\right]^{t} = Pe^{rt}$$

Our computations tell us that as the frequency with which interest is compounded increases without bound, the accumulated amount approaches Pe^{rt}. In this situation, we say that interest is *compounded continuously*. Let's summarize this important result.

Continuous Compound Interest Formula

$$A = Pe^{rt} \tag{10}$$

where

P = Principal

r = Annual interest rate compounded continuously

t = Time in years

A = Accumulated amount at the end of t years

EXAMPLE 5 Find the accumulated amount after 3 years if \$1000 is invested at 8% per year compounded (a) daily (assume a 365-day year) and (b) continuously.

Solution

a. Using Formula (6) with $P = 1000$, $r = 0.08$, $m = 365$, and $t = 3$, we find

$$A = 1000\left(1 + \frac{0.08}{365}\right)^{(365)(3)} \approx 1271.22$$

or \$1271.22.

b. Here we use Formula (10) with $P = 1000$, $r = 0.08$, and $t = 3$, obtaining

$$A = 1000e^{(0.08)(3)}$$
$$\approx 1271.25$$

or \$1271.25. ■

Observe that the accumulated amounts corresponding to interest compounded daily and interest compounded continuously differ by very little. The continuous compound interest formula is a very important tool in theoretical work in financial analysis.

Exploring with TECHNOLOGY

In the opening paragraph of Section 5.1, we pointed out that the accumulated amount of an account earning interest *compounded continuously* will eventually outgrow by far the accumulated amount of an account earning interest at the same nominal rate but earning simple interest. Illustrate this fact using the following example.

Suppose you deposit \$1000 in account I, earning interest at the rate of 10% per year compounded continuously so that the accumulated amount at the end of t years is $A_1(t) = 1000e^{0.1t}$. Suppose you also deposit \$1000 in account II, earning simple interest at the rate of 10% per year so that the accumulated amount at the end of t years is $A_2(t) = 1000(1 + 0.1t)$. Use a graphing utility to sketch the graphs of the functions A_1 and A_2 in the viewing window $[0, 20] \times [0, 10{,}000]$ to see the accumulated amounts $A_1(t)$ and $A_2(t)$ over a 20-year period.

If we solve Formula (10) for P, we obtain

$$P = Ae^{-rt} \tag{11}$$

which gives the present value in terms of the future (accumulated) value for the case of continuous compounding.

VIDEO

APPLIED EXAMPLE 6 Real Estate Investment Blakely Investment Company owns an office building located in the commercial district of a city. As a result of the continued success of an urban renewal program, local business is enjoying a miniboom. The market value of Blakely's property is

$$V(t) = 300{,}000e^{\sqrt{t}/2}$$

where $V(t)$ is measured in dollars and t is the time in years from the present. If the expected rate of appreciation is 9% compounded continuously for the next 10 years, find an expression for the present value $P(t)$ of the market price of the property valid for the next 10 years. Compute $P(7)$, $P(8)$, and $P(9)$, and interpret your results.

Solution Using Formula (11) with $A = V(t)$ and $r = 0.09$, we find that the present value of the market price of the property t years from now is

$$\begin{aligned} P(t) &= V(t)e^{-0.09t} \\ &= 300{,}000e^{-0.09t+\sqrt{t}/2} \qquad (0 \leq t \leq 10) \end{aligned}$$

Letting $t = 7$, 8, and 9, respectively, we find that

$$\begin{aligned} P(7) &= 300{,}000e^{-0.09(7)+\sqrt{7}/2} \approx 599{,}837 \quad \text{or} \quad \$599{,}837 \\ P(8) &= 300{,}000e^{-0.09(8)+\sqrt{8}/2} \approx 600{,}640 \quad \text{or} \quad \$600{,}640 \\ P(9) &= 300{,}000e^{-0.09(9)+\sqrt{9}/2} \approx 598{,}115 \quad \text{or} \quad \$598{,}115 \end{aligned}$$

From the results of these computations, we see that the present value of the property's market price seems to decrease after a certain period of growth. This suggests that there is an optimal time for the owners to sell. Later we will show that the highest present value of the property's market price is \$600,779, which occurs at time $t = 7.72$ years.

Exploring with TECHNOLOGY

The effective rate of interest is given by

$$r_{\text{eff}} = \left(1 + \frac{r}{m}\right)^m - 1$$

where the number of conversion periods per year is m. In Exercise 47 you will be asked to show that the effective rate of interest r_{eff} corresponding to a nominal interest rate r per year compounded continuously is given by

$$\hat{r}_{\text{eff}} = e^r - 1$$

To obtain a visual confirmation of this result, consider the special case where $r = 0.1$ (10% per year).

1. Use a graphing utility to plot the graph of both

$$y_1 = \left(1 + \frac{0.1}{x}\right)^x - 1 \quad \text{and} \quad y_2 = e^{0.1} - 1$$

in the viewing window $[0, 3] \times [0, 0.12]$.

2. Does your result seem to imply that

$$\left(1 + \frac{r}{m}\right)^m - 1$$

approaches

$$\hat{r}_{\text{eff}} = e^r - 1$$

as m increases without bound for the special case $r = 0.1$?

The next two examples show how logarithms can be used to solve problems involving compound interest.

EXAMPLE 7 How long will it take \$10,000 to grow to \$15,000 if the investment earns an interest rate of 12% per year compounded quarterly?

Solution Using Formula (6) with $A = 15{,}000$, $P = 10{,}000$, $r = 0.12$, and $m = 4$, we obtain

$$15{,}000 = 10{,}000\left(1 + \frac{0.12}{4}\right)^{4t}$$

$$(1.03)^{4t} = \frac{15{,}000}{10{,}000} = 1.5$$

Taking the logarithm on each side of the equation gives

$$\ln(1.03)^{4t} = \ln 1.5$$

$$4t \ln 1.03 = \ln 1.5 \qquad \log_b m^n = n \log_b m$$

$$4t = \frac{\ln 1.5}{\ln 1.03}$$

$$t = \frac{\ln 1.5}{4 \ln 1.03} \approx 3.43$$

So it will take approximately 3.4 years for the investment to grow from \$10,000 to \$15,000.

EXAMPLE 8 Find the interest rate needed for an investment of \$10,000 to grow to an amount of \$18,000 in 5 years if the interest is compounded monthly.

Solution Using Formula (6) with $A = 18{,}000$, $P = 10{,}000$, $m = 12$, and $t = 5$, we obtain

$$18{,}000 = 10{,}000\left(1 + \frac{r}{12}\right)^{12(5)}$$

Dividing both sides of the equation by 10,000 gives

$$\frac{18{,}000}{10{,}000} = \left(1 + \frac{r}{12}\right)^{60}$$

or, upon simplification,

$$\left(1 + \frac{r}{12}\right)^{60} = 1.8$$

Now, we take the logarithm on each side of the equation, obtaining

$$\ln\left(1 + \frac{r}{12}\right)^{60} = \ln 1.8$$

$$60 \ln\left(1 + \frac{r}{12}\right) = \ln 1.8$$

$$\ln\left(1 + \frac{r}{12}\right) = \frac{\ln 1.8}{60} \approx 0.009796$$

$$\left(1 + \frac{r}{12}\right) \approx e^{0.009796} \qquad \text{By Property 2}$$

$$\approx 1.009844$$

and

$$\frac{r}{12} \approx 1.009844 - 1$$

$$r \approx 0.1181$$

or 11.81% per year.

5.3 Self-Check Exercises

1. Find the present value of \$20,000 due in 3 yr at an interest rate of 12%/year compounded monthly.
2. Glen is a retiree living on Social Security and the income from his investment. Currently, his \$100,000 investment in a 1-yr CD is yielding 6.6% interest compounded daily. If he reinvests the principal (\$100,000) on the due date of the CD in another 1-yr CD paying 5.2% interest compounded daily, find the net decrease in his yearly income from his investment.
3. **a.** What is the accumulated amount after 5 yr if \$10,000 is invested at 8%/year compounded continuously?
   b. Find the present value of \$10,000 due in 5 yr at an interest rate of 8%/year compounded continuously.

Solutions to Self-Check Exercises 5.3 can be found on page 358.

5.3 Concept Questions

1. **a.** What is the difference between simple interest and compound interest?
   b. State the simple interest formula and the compound interest formula.
2. **a.** What is the effective rate of interest?
   b. State the formula for computing the effective rate of interest.
3. What is the present value formula for compound interest?
4. State the continuous compound interest formula.

5.3 Exercises

In Exercises 1–4, find the accumulated amount *A* if the principal *P* is invested at an interest rate of *r* per year for *t* years.

1. $P = \$2500$, $r = 7\%$, $t = 10$, compounded semiannually
2. $P = \$12{,}000$, $r = 8\%$, $t = 10$, compounded quarterly
3. $P = \$150{,}000$, $r = 10\%$, $t = 4$, compounded monthly
4. $P = \$150{,}000$, $r = 9\%$, $t = 3$, compounded daily

In Exercises 5 and 6, find the effective rate corresponding to the given nominal rate.

5. **a.** 10%/year compounded semiannually
   b. 9%/year compounded quarterly
6. **a.** 8%/year compounded monthly
   b. 8%/year compounded daily

In Exercises 7 and 8, find the present value of \$40,000 due in 4 yr at the given rate of interest.

7. **a.** 8%/year compounded semiannually
   b. 8%/year compounded quarterly
8. **a.** 7%/year compounded monthly
   b. 9%/year compounded daily
9. Find the accumulated amount after 4 yr if \$5000 is invested at 8%/year compounded continuously.
10. An amount of \$25,000 is deposited in a bank that pays interest at the rate of 7%/year, compounded annually. What is the total amount on deposit at the end of 6 yr, assuming there are no deposits or withdrawals during those 6 yr? What is the interest earned in that period of time?
11. How much money should be deposited in a bank paying interest at the rate of 7%/year compounded daily (assume a 365-day year) so that at the end of 2 yr the accumulated amount will be \$10,000?
12. Jada deposited an amount of money in a bank 3 yr ago. If the bank had been paying interest at the rate of 5%/year compounded daily (assume a 365-day year) and she has \$15,000 on deposit today, what was her initial deposit?
13. How much money should Jack deposit in a bank paying interest at the rate of 6%/year compounded continuously so that at the end of 3 yr the accumulated amount will be \$20,000?
14. Diego deposited a certain sum of money in a bank 2 yr ago. If the bank had been paying interest at the rate of 6% compounded continuously and he has \$12,000 on deposit today, what was his initial deposit?
15. Find the interest rate needed for an investment of \$5000 to grow to an amount of \$7500 in 3 yr if interest is compounded monthly.
16. Find the interest rate needed for an investment of \$5000 to grow to an amount of \$7500 in 3 yr if interest is compounded quarterly.
17. Find the interest rate needed for an investment of \$5000 to grow to an amount of \$8000 in 4 yr if interest is compounded semiannually.
18. Find the interest rate needed for an investment of \$5000 to grow to an amount of \$5500 in 6 mo if interest is compounded monthly.

19. Find the interest rate needed for an investment of \$2000 to double in 5 yr if interest is compounded annually.

20. Find the interest rate needed for an investment of \$2000 to triple in 5 yr if interest is compounded monthly.

21. How long will it take \$5000 to grow to \$6500 if the investment earns interest at the rate of 12%/year compounded monthly?

22. How long will it take \$12,000 to grow to \$15,000 if the investment earns interest at the rate of 8%/year compounded monthly?

23. How long will it take an investment of \$2000 to double if the investment earns interest at the rate of 9%/year compounded monthly?

24. How long will it take an investment of \$5000 to triple if the investment earns interest at the rate of 8%/year compounded daily?

25. Find the interest rate needed for an investment of \$5000 to grow to an amount of \$6000 in 3 yr if interest is compounded continuously.

26. Find the interest rate needed for an investment of \$4000 to double in 5 yr if interest is compounded continuously.

27. How long will it take an investment of \$6000 to grow to \$7000 if the investment earns interest at the rate of $7\frac{1}{2}\%$ compounded continuously?

28. How long will it take an investment of \$8000 to double if the investment earns interest at the rate of 8% compounded continuously?

29. **Housing Prices** The Estradas are planning to buy a house 4 yr from now. Housing experts in their area have estimated that the cost of a home will increase at a rate of 9%/year during that 4-yr period. If this economic prediction holds true, how much can they expect to pay for a house that currently costs \$180,000?

30. **Energy Consumption** A metropolitan utility company in a western city of the United States expects the consumption of electricity to increase by 8%/year during the next decade, due mainly to the expected population increase. If consumption does increase at this rate, find the amount by which the utility company will have to increase its generating capacity in order to meet the area's needs at the end of the decade.

31. **Pension Funds** The managers of a pension fund have invested \$1.5 million in U.S. government certificates of deposit (CDs) that pay interest at the rate of 6.5%/year compounded semiannually over a period of 10 yr. At the end of this period, how much will the investment be worth?

32. **Savings Accounts** Bernie invested a sum of money 5 yr ago in a savings account, which has since paid interest at the rate of 8%/yr compounded quarterly. His investment is now worth \$22,289.22. How much did he originally invest?

33. **Loan Consolidation** The proprietors of the Coachmen Inn secured two loans from the Union Bank: one for \$8000 due in 3 yr and one for \$15,000 due in 6 yr, both at an interest rate of 10%/year compounded semiannually. The bank agreed to allow the two loans to be consolidated into one loan payable in 5 yr at the same interest rate. How much will the proprietors have to pay the bank at the end of 5 yr?

34. **Tax-Deferred Annuities** Kate is in the 28% tax bracket and has \$25,000 available for investment during her current tax year. Assume that she remains in the same tax bracket over the next 10 yr and determine the accumulated amount of her investment if she puts the \$25,000 into a
    a. Tax-deferred annuity that pays 12%/year, tax deferred for 10 yr.
    b. Taxable instrument that pays 12%/year for 10 yr.
    Hint: In this case the yield after taxes is 8.64%/year.

35. **Revenue Growth of a Home Theater Business** Maxwell started a home theater business in 2005. The revenue of his company for that year was \$240,000. The revenue grew by 20% in 2006 and 30% in 2007. Maxwell projected that the revenue growth for his company in the next 3 yr will be at least 25%/year. How much does Maxwell expect his minimum revenue to be for 2010?

36. **Online Retail Sales** Online retail sales stood at \$23.5 billion for 2000. For the next 2 yr, they grew by 33.2% and 27.8% per year, respectively. For the next 6 yr, online retail sales are projected to grow at 30.5%, 19.9%, 24.3%, 14.0%, 17.6%, and 10.5% per year, respectively. What are the projected online sales for 2008?
    Source: Jupiter Research

37. **Purchasing Power** The inflation rates in the U.S. economy in 2000 through 2003 were 3.4%, 2.8%, 1.6%, and 2.3%, respectively. What was the purchasing power of a dollar at the beginning of 2004 compared to that at the beginning of 2000?
    Source: U.S. Census Bureau

38. **Investment Returns** Zoe purchased a house in 1999 for \$160,000. In 2005 she sold the house and made a net profit of \$56,000. Find the effective annual rate of return on her investment over the 6-yr period.

39. **Investment Returns** Julio purchased 1000 shares of a certain stock for \$25,250 (including commissions). He sold the shares 2 yr later and received \$32,100 after deducting commissions. Find the effective annual rate of return on his investment over the 2-yr period.

40. **Investment Options** Investment A offers a 10% return compounded semiannually, and investment B offers a 9.75% return compounded continuously. Which investment has a higher rate of return over a 4-yr period?

41. **Present Value** Find the present value of \$59,673 due in 5 yr at an interest rate of 8%/year compounded continuously.

42. **Real Estate Investments** A condominium complex was purchased by a group of private investors for \$1.4 million and sold 6 yr later for \$3.6 million. Find the annual rate of return (compounded continuously) on their investment.

43. **SAVING FOR COLLEGE** Having received a large inheritance, a child's parents wish to establish a trust for the child's college education. If 7 yr from now they need an estimated \$70,000, how much should they set aside in trust now, if they invest the money at 10.5% compounded (a) quarterly? (b) Continuously?

44. **EFFECT OF INFLATION ON SALARIES** Omar's current annual salary is \$35,000. How much will he need to earn 10 yr from now in order to retain his present purchasing power if the rate of inflation over that period is 6%/year? Assume that inflation is continuously compounded.

45. **PENSIONS** Eleni, who is now 50 years old, is employed by a firm that guarantees her a pension of \$40,000/year at age 65. What is the present value of her first year's pension if inflation over the next 15 yr is (a) 6%? (b) 8%? (c) 12%? Assume that inflation is continuously compounded.

46. **REAL ESTATE INVESTMENTS** An investor purchased a piece of waterfront property. Because of the development of a marina in the vicinity, the market value of the property is expected to increase according to the rule

$$V(t) = 80{,}000e^{\sqrt{t}/2}$$

where $V(t)$ is measured in dollars and t is the time in years from the present. If the rate of appreciation is expected to be 9% compounded continuously for the next 8 yr, find an expression for the present value $P(t)$ of the property's market price valid for the next 8 yr. What is $P(t)$ expected to be in 4 yr?

47. Show that the effective interest rate $\hat{r}_{\text{eff}}$ that corresponds to a nominal interest rate r per year compounded continuously is given by

$$\hat{r}_{\text{eff}} = e^r - 1$$

Hint: From Formula (7) we see that the effective rate $\hat{r}_{\text{eff}}$ corresponding to a nominal interest rate r per year compounded m times a year is given by

$$\hat{r}_{\text{eff}} = \left(1 + \frac{r}{m}\right)^m - 1$$

Let m tend to infinity in this expression.

48. Refer to Exercise 47. Find the effective interest rate that corresponds to a nominal rate of 10%/year compounded (a) quarterly, (b) monthly, and (c) continuously.

49. **INVESTMENT ANALYSIS** Refer to Exercise 47. Bank A pays interest on deposits at a 7% annual rate compounded quarterly, and Bank B pays interest on deposits at a $7\frac{1}{8}$% annual rate compounded continuously. Which bank has the higher effective rate of interest?

50. **INVESTMENT ANALYSIS** Find the nominal interest rate that, when compounded monthly, yields an effective interest rate of 10%/year.
Hint: Use Equation (7).

51. **INVESTMENT ANALYSIS** Find the nominal interest rate that, when compounded continuously, yields an effective interest rate of 10%/year.
Hint: See Exercise 47.

52. **ANNUITIES** An annuity is a sequence of payments made at regular time intervals. The future value of an annuity of n payments of R dollars each paid at the end of each investment period into an account that earns an interest rate of i/period is

$$S = R\left[\frac{(1+i)^n - 1}{i}\right]$$

Determine

$$\lim_{i \to 0} R\left[\frac{(1+i)^n - 1}{i}\right]$$

and interpret your result.
Hint: Use the definition of the derivative.

5.3 Solutions to Self-Check Exercises

1. Using Formula (8) with $A = 20{,}000$, $r = 0.12$, $m = 12$, and $t = 3$, we find the required present value to be

$$P = 20{,}000\left(1 + \frac{0.12}{12}\right)^{-(12)(3)}$$
$$= 13{,}978.50$$

or \$13,978.50.

2. The accumulated amount of Glen's current investment is found by using Formula (6) with $P = 100{,}000$, $r = 0.066$, and $m = 365$. Thus, the required accumulated amount is

$$A = 100{,}000\left(1 + \frac{0.066}{365}\right)^{365}$$
$$= 106{,}822.03$$

or \$106,822.03. Next, we compute the accumulated amount of Glen's reinvestment. Once again, using (6) with $P = 100{,}000$, $r = 0.052$, and $m = 365$, we find the required accumulated amount in this case to be

$$\bar{A} = 100{,}000\left(1 + \frac{0.052}{365}\right)^{365}$$

or \$105,337.18. Therefore, Glen can expect to experience a net decrease in yearly income of

$$106{,}822.03 - 105{,}337.18$$

or \$1,484.85.

3. a. Using Formula (10) with $P = 10{,}000$, $r = 0.08$, and $t = 5$, we find that the required accumulated amount is given by

$$\begin{aligned} A &= 10{,}000e^{(0.08)(5)} \\ &= 14{,}918.25 \end{aligned}$$

or \$14,918.25.

b. Using Formula (11) with $A = 10{,}000$, $r = 0.08$, and $t = 5$, we see that the required present value is given by

$$\begin{aligned} P &= 10{,}000e^{-(0.08)(5)} \\ &= 6703.20 \end{aligned}$$

or \$6703.20.

USING TECHNOLOGY

Finding the Accumulated Amount of an Investment, the Effective Rate of Interest, and the Present Value of an Investment

Graphing Utility

Some graphing utilities have built-in routines for solving problems involving the mathematics of finance. For example, the TI-83/84 Solver function incorporates several functions that can be used to solve the problems that are encountered in this section. To access the TVM SOLVER on the TI-83 press **2ND**, press **FINANCE**, and then select **1: TVM Solver**. To access the TVM SOLVER on the TI-83 plus and the TI-84, press **APPS**, press **1: Finance**, and then select **1: TVM Solver**. Step-by-step procedures for using these functions can be found on our Companion Web site.

N = 120
I% = 10
PV = −5000
PMT = 0
■ FV = 13535.20745
P/Y = 12
C/Y = 12
PMT : END BEGIN

FIGURE T1
The TI-83/84 screen showing the future value (FV) of an investment

EXAMPLE 1 Finding the Accumulated Amount of an Investment Find the accumulated amount after 10 years if \$5000 is invested at a rate of 10% per year compounded monthly.

Solution Using the TI-83/84 TVM SOLVER with the following inputs,

N = 120 (10)(12)
I% = 10
PV = −5000 Recall that an investment is an outflow.
PMT = 0
FV = 0
P/Y = 12 The number of payments each year
C/Y = 12 The number of conversion periods each year
PMT:END BEGIN

we obtain the display shown in Figure T1. We conclude that the required accumulated amount is \$13,535.21. ■

▶ Eff (10, 4)
10.38128906

FIGURE T2
The TI-83/84 screen showing the effective rate of interest (Eff)

EXAMPLE 2 Finding the Effective Rate of Interest Find the effective rate of interest corresponding to a nominal rate of 10% per year compounded quarterly.

Solution Here we use the **Eff** function of the TI-83/84 calculator to obtain the result shown in Figure T2. The required effective rate is approximately 10.38% per year. ■

(continued)

EXAMPLE 3 Finding the Present Value of an Investment Find the present value of \$20,000 due in 5 years if the interest rate is 7.5% per year compounded daily.

Solution Using the TI-83/84 TVM SOLVER with the following inputs,

$$
\begin{aligned}
N &= 1825 && (5)(365)\\
I\% &= 7.5\\
PV &= 0\\
PMT &= 0\\
FV &= 20000\\
P/Y &= 365 && \text{The number of payments each year}\\
C/Y &= 365 && \text{The number of conversions each year}
\end{aligned}
$$

PMT:END BEGIN

we obtain the display shown in Figure T3. We see that the required present value is approximately \$13,746.32. Note that PV is negative because an investment is an outflow (money is paid out).

```
  N    = 1825
  I%   = 7.5
■ PV   = −13746.3151
  PMT  = 0
  FV   = 20000
  P/Y  = 365
  C/Y  = 365
  PMT : END     BEGIN
```

FIGURE **T3**
The TI-83/84 screen showing the present value (PV) of an investment

TECHNOLOGY EXERCISES

1. Find the accumulated amount A if \$5000 is invested at the interest rate of $5\frac{3}{8}\%$/year compounded monthly for 3 yr.
2. Find the accumulated amount A if \$327.35 is invested at the interest rate of $5\frac{1}{3}\%$/year compounded daily for 7 yr.
3. Find the effective rate corresponding to $8\frac{2}{3}\%$/year compounded quarterly.
4. Find the effective rate corresponding to $10\frac{5}{8}\%$/year compounded monthly.
5. Find the present value of \$38,000 due in 3 yr at $8\frac{1}{4}\%$/year compounded quarterly.
6. Find the present value of \$150,000 due in 5 yr at $9\frac{3}{8}\%$/year compounded monthly.

5.4 Differentiation of Exponential Functions

The Derivative of the Exponential Function

To study the effects of budget deficit-reduction plans at different income levels, it is important to know the income distribution of American families. Based on data from the House Budget Committee, the House Ways and Means Committee, and the U.S. Census Bureau, the graph of f shown in Figure 10 gives the number of American families y (in millions) as a function of their annual income x (in thousands of dollars) in 1990.

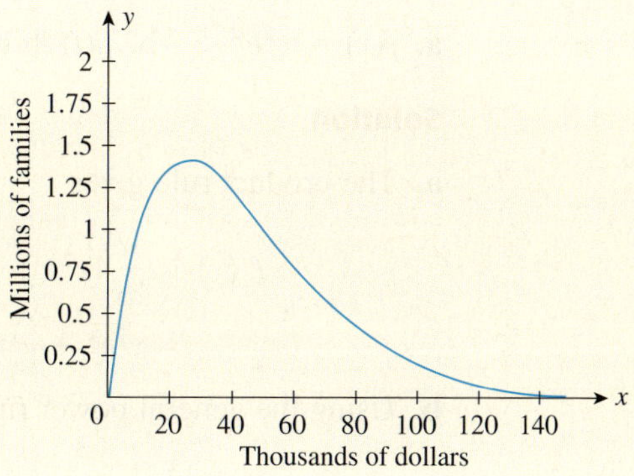

FIGURE 10
The graph of f shows the number of families versus their annual income.
Source: House Budget Committee, House Ways and Means Committee, and U.S. Census Bureau

Observe that the graph of f rises very quickly and then tapers off. From the graph of f, you can see that the bulk of American families earned less than \$100,000 per year. In fact, 95% of U.S. families earned less than \$102,358 per year in 1990. (We will refer to this model again in Using Technology at the end of this section.)

To analyze mathematical models involving exponential and logarithmic functions in greater detail, we need to develop rules for computing the derivative of these functions. We begin by looking at the rule for computing the derivative of the exponential function.

Rule 1: Derivative of the Exponential Function

$$\frac{d}{dx}(e^x) = e^x$$

Thus, the derivative of the exponential function with base e is equal to the function itself. To demonstrate the validity of this rule, we compute

$$\begin{aligned} f'(x) &= \lim_{h\to 0}\frac{f(x+h)-f(x)}{h} \\ &= \lim_{h\to 0}\frac{e^{x+h}-e^x}{h} \\ &= \lim_{h\to 0}\frac{e^x(e^h-1)}{h} && \text{Write } e^{x+h} = e^x e^h \text{ and factor.} \\ &= e^x \lim_{h\to 0}\frac{e^h-1}{h} && \text{Why?} \end{aligned}$$

To evaluate

$$\lim_{h\to 0}\frac{e^h-1}{h}$$

let's refer to Table 4, which is constructed with the aid of a calculator. From the table, we see that

TABLE 4

h	$\dfrac{e^h-1}{h}$
0.1	1.0517
0.01	1.0050
0.001	1.0005
−0.1	0.9516
−0.01	0.9950
−0.001	0.9995

$$\lim_{h\to 0}\frac{e^h-1}{h} = 1$$

(Although a rigorous proof of this fact is possible, it is beyond the scope of this book. Also see Example 1, Using Technology, page 370.) Using this result, we conclude that

$$f'(x) = e^x \cdot 1 = e^x$$

as we set out to show.

EXAMPLE 1 Find the derivative of each of the following functions:

a. $f(x) = x^2e^x$ **b.** $g(t) = (e^t + 2)^{3/2}$

Solution

a. The product rule gives

$$f'(x) = \frac{d}{dx}(x^2e^x) = x^2\frac{d}{dx}(e^x) + e^x\frac{d}{dx}(x^2)$$
$$= x^2e^x + e^x(2x) = xe^x(x + 2)$$

See page 10.

b. Using the general power rule, we find

$$g'(t) = \frac{3}{2}(e^t + 2)^{1/2}\frac{d}{dt}(e^t + 2) = \frac{3}{2}(e^t + 2)^{1/2}e^t = \frac{3}{2}e^t(e^t + 2)^{1/2}$$

Exploring with TECHNOLOGY

Consider the exponential function $f(x) = b^x$ $(b > 0, b \neq 1)$.

1. Use the definition of the derivative of a function to show that

$$f'(x) = b^x \cdot \lim_{h\to 0}\frac{b^h - 1}{h}$$

2. Use the result of part 1 to show that

$$\frac{d}{dx}(2^x) = 2^x \cdot \lim_{h\to 0}\frac{2^h - 1}{h}$$

$$\frac{d}{dx}(3^x) = 3^x \cdot \lim_{h\to 0}\frac{3^h - 1}{h}$$

3. Use the technique in Using Technology, page 370, to show that (to two decimal places)

$$\lim_{h\to 0}\frac{2^h - 1}{h} = 0.69 \quad \text{and} \quad \lim_{h\to 0}\frac{3^h - 1}{h} = 1.10$$

4. Conclude from the results of parts 2 and 3 that

$$\frac{d}{dx}(2^x) \approx (0.69)2^x \quad \text{and} \quad \frac{d}{dx}(3^x) \approx (1.10)3^x$$

Thus,

$$\frac{d}{dx}(b^x) = k \cdot b^x$$

where k is an appropriate constant.

5. The results of part 4 suggest that, for convenience, we pick the base b, where $2 < b < 3$, so that $k = 1$. This value of b is $e \approx 2.718281828\ldots$. Thus,

$$\frac{d}{dx}(e^x) = e^x$$

This is why we prefer to work with the exponential function $f(x) = e^x$.

Applying the Chain Rule to Exponential Functions

To enlarge the class of exponential functions to be differentiated, we appeal to the chain rule to obtain the following rule for differentiating composite functions of the form $h(x) = e^{f(x)}$. An example of such a function is $h(x) = e^{x^2-2x}$. Here, $f(x) = x^2 - 2x$.

Rule 2: The Chain Rule for Exponential Functions

If $f(x)$ is a differentiable function, then

$$\frac{d}{dx}(e^{f(x)}) = e^{f(x)}f'(x)$$

To see this, observe that if $h(x) = g[f(x)]$, where $g(x) = e^x$, then by virtue of the chain rule,

$$h'(x) = g'(f(x))f'(x) = e^{f(x)}f'(x)$$

since $g'(x) = e^x$.

As an aid to remembering the chain rule for exponential functions, observe that it has the following form:

$$\frac{d}{dx}(e^{f(x)}) = e^{f(x)} \cdot \text{derivative of exponent}$$

Same (the two exponents $f(x)$ are the same)

EXAMPLE 2 Find the derivative of each of the following functions:

a. $f(x) = e^{2x}$ **b.** $y = e^{-3x}$ **c.** $g(t) = e^{2t^2+t}$

Solution

a. $f'(x) = e^{2x}\frac{d}{dx}(2x) = e^{2x} \cdot 2 = 2e^{2x}$

b. $\frac{dy}{dx} = e^{-3x}\frac{d}{dx}(-3x) = -3e^{-3x}$

c. $g'(t) = e^{2t^2+t} \cdot \frac{d}{dt}(2t^2 + t) = (4t + 1)e^{2t^2+t}$ ■

EXAMPLE 3 Differentiate the function $y = xe^{-2x}$.

Solution Using the product rule, followed by the chain rule, we find

$$\begin{aligned}\frac{dy}{dx} &= x\frac{d}{dx}e^{-2x} + e^{-2x}\frac{d}{dx}(x)\\ &= xe^{-2x}\frac{d}{dx}(-2x) + e^{-2x} \qquad \text{Use the chain rule on the first term.}\\ &= -2xe^{-2x} + e^{-2x}\\ &= e^{-2x}(1 - 2x)\end{aligned}$$

■

EXAMPLE 4 Differentiate the function $g(t) = \dfrac{e^t}{e^t + e^{-t}}$.

Solution Using the quotient rule, followed by the chain rule, we find

$$
\begin{aligned}
g'(t) &= \frac{(e^t + e^{-t})\dfrac{d}{dt}(e^t) - e^t\dfrac{d}{dt}(e^t + e^{-t})}{(e^t + e^{-t})^2} \\
&= \frac{(e^t + e^{-t})e^t - e^t(e^t - e^{-t})}{(e^t + e^{-t})^2} \\
&= \frac{e^{2t} + 1 - e^{2t} + 1}{(e^t + e^{-t})^2} \qquad e^0 = 1 \\
&= \frac{2}{(e^t + e^{-t})^2}
\end{aligned}
$$

See page 9.

EXAMPLE 5 In Section 5.6, we will discuss some practical applications of the exponential function

$$Q(t) = Q_0e^{kt}$$

where Q_0 and k are positive constants and $t \in [0, \infty)$. A quantity $Q(t)$ growing according to this law experiences exponential growth. Show that for a quantity $Q(t)$ experiencing exponential growth, the rate of growth of the quantity $Q'(t)$ at any time t is directly proportional to the amount of the quantity present.

Solution Using the chain rule for exponential functions, we compute the derivative Q' of the function Q. Thus,

$$
\begin{aligned}
Q'(t) &= Q_0e^{kt}\frac{d}{dt}(kt) \\
&= Q_0e^{kt}(k) \\
&= kQ_0e^{kt} \\
&= kQ(t) \qquad Q(t) = Q_0e^{kt}
\end{aligned}
$$

which is the desired conclusion.

EXAMPLE 6 Find the inflection points of the function $f(x) = e^{-x^2}$.

Solution The first derivative of f is

$$f'(x) = -2xe^{-x^2}$$

Differentiating $f'(x)$ with respect to x yields

$$
\begin{aligned}
f''(x) &= (-2x)(-2xe^{-x^2}) - 2e^{-x^2} \\
&= 2e^{-x^2}(2x^2 - 1)
\end{aligned}
$$

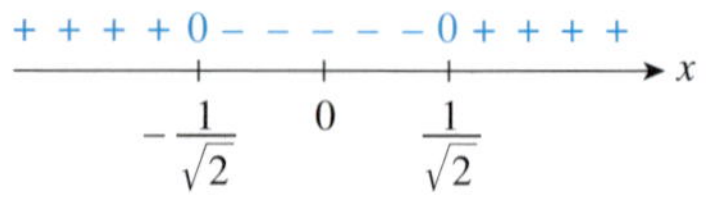

FIGURE 11
Sign diagram for f''

Setting $f''(x) = 0$ gives

$$2e^{-x^2}(2x^2 - 1) = 0$$

Since e^{-x^2} never equals zero for any real value of x, we see that $x = \pm 1/\sqrt{2}$ are the only candidates for inflection points of f. The sign diagram of f'', shown in Figure 11, tells us that both $x = -1/\sqrt{2}$ and $x = 1/\sqrt{2}$ give rise to inflection points of f.

Next,

$$f\left(-\frac{1}{\sqrt{2}}\right) = f\left(\frac{1}{\sqrt{2}}\right) = e^{-1/2}$$

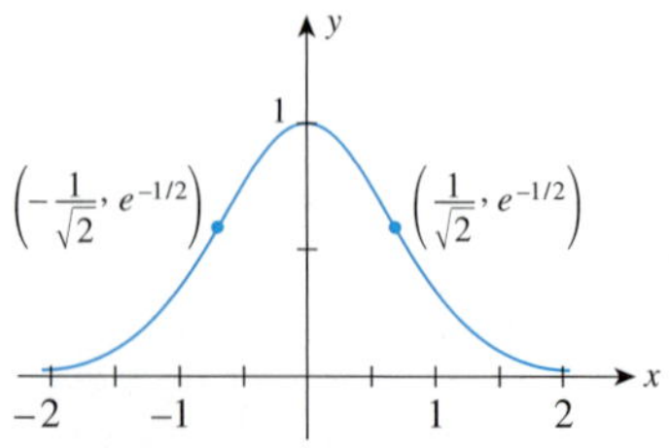

FIGURE 12
The graph of $y = e^{-x^2}$ has two inflection points.

and the inflection points of f are $(-1/\sqrt{2}, e^{-1/2})$ and $(1/\sqrt{2}, e^{-1/2})$. The graph of f appears in Figure 12.

Our final example involves finding the absolute maximum of an exponential function.

VIDEO

APPLIED EXAMPLE 7 Optimal Market Price Refer to Example 6, Section 5.3. The present value of the market price of the Blakely Office Building is given by

$$P(t) = 300{,}000e^{-0.09t+\sqrt{t}/2} \qquad (0 \le t \le 10)$$

Find the optimal present value of the building's market price.

Solution To find the maximum value of P over [0, 10], we compute

$$\begin{aligned} P'(t) &= 300{,}000e^{-0.09t+\sqrt{t}/2}\frac{d}{dt}\left(-0.09t + \frac{1}{2}t^{1/2}\right) \\ &= 300{,}000e^{-0.09t+\sqrt{t}/2}\left(-0.09 + \frac{1}{4}t^{-1/2}\right) \end{aligned}$$

Setting $P'(t) = 0$ gives

$$-0.09 + \frac{1}{4t^{1/2}} = 0$$

since $e^{-0.09t+\sqrt{t}/2}$ is never zero for any value of t. Solving this equation, we find

$$\begin{aligned} \frac{1}{4t^{1/2}} &= 0.09 \\ t^{1/2} &= \frac{1}{4(0.09)} \\ &= \frac{1}{0.36} \\ t &= \left(\frac{1}{0.36}\right)^2 \approx 7.72 \end{aligned}$$

the sole critical number of the function P. Finally, evaluating $P(t)$ at the critical number as well as at the endpoints of [0, 10], we have

t	0	7.72	10
$P(t)$	300,000	600,779	592,838

We conclude, accordingly, that the optimal present value of the property's market price is \$600,779 and that this will occur 7.72 years from now. ■

5.4 Self-Check Exercises

1. Let $f(x) = xe^{-x}$.
   a. Find the first and second derivatives of f.
   b. Find the relative extrema of f.
   c. Find the inflection points of f.

2. An industrial asset is being depreciated at a rate so that its book value t yr from now will be

$$V(t) = 50{,}000e^{-0.4t}$$

dollars. How fast will the book value of the asset be changing 3 yr from now?

Solutions to Self-Check Exercises 5.4 can be found on page 370.

5.4 Concept Questions

1. State the rule for differentiating (a) $f(x) = e^x$ and (b) $g(x) = e^{f(x)}$, where f is a differentiable function.

2. Let $f(x) = e^{kx}$.
   a. Compute $f'(x)$.
   b. Use the result to deduce the behavior of f for the case $k > 0$ and the case $k < 0$.

5.4 Exercises

In Exercises 1–28, find the derivative of the function.

1. $f(x) = e^{3x}$
2. $f(x) = 3e^x$
3. $g(t) = e^{-t}$
4. $f(x) = e^{-2x}$
5. $f(x) = e^x + x$
6. $f(x) = 2e^x - x^2$
7. $f(x) = x^3e^x$
8. $f(u) = u^2e^{-u}$
9. $f(x) = \dfrac{2e^x}{x}$
10. $f(x) = \dfrac{x}{e^x}$
11. $f(x) = 3(e^x + e^{-x})$
12. $f(x) = \dfrac{e^x + e^{-x}}{2}$
13. $f(w) = \dfrac{e^w + 1}{e^w}$
14. $f(x) = \dfrac{e^x}{e^x + 1}$
15. $f(x) = 2e^{3x-1}$
16. $f(t) = 4e^{3t+2}$
17. $h(x) = e^{-x^2}$
18. $f(x) = e^{x^2-1}$
19. $f(x) = 3e^{-1/x}$
20. $f(x) = e^{1/(2x)}$
21. $f(x) = (e^x + 1)^{25}$
22. $f(x) = (4 - e^{-3x})^3$
23. $f(x) = e^{\sqrt{x}}$
24. $f(t) = -e^{-\sqrt{2t}}$
25. $f(x) = (x - 1)e^{3x+2}$
26. $f(s) = (s^2 + 1)e^{-s^2}$
27. $f(x) = \dfrac{e^x - 1}{e^x + 1}$
28. $g(t) = \dfrac{e^{-t}}{1 + t^2}$

In Exercises 29–32, find the second derivative of the function.

29. $f(x) = e^{-4x} + 2e^{3x}$
30. $f(t) = 3e^{-2t} - 5e^{-t}$
31. $f(x) = 2xe^{3x}$
32. $f(t) = t^2e^{-2t}$

33. Find an equation of the tangent line to the graph of $y = e^{2x-3}$ at the point $(\frac{3}{2}, 1)$.

34. Find an equation of the tangent line to the graph of $y = e^{-x^2}$ at the point $(1, 1/e)$.

35. Determine the intervals where the function $f(x) = e^{-x^2/2}$ is increasing and where it is decreasing.

36. Determine the intervals where the function $f(x) = x^2e^{-x}$ is increasing and where it is decreasing.

37. Determine the intervals of concavity for the function $f(x) = \dfrac{e^x - e^{-x}}{2}$.

38. Determine the intervals of concavity for the function $f(x) = xe^x$.

39. Find the inflection point of the function $f(x) = xe^{-2x}$.

40. Find the inflection point(s) of the function $f(x) = 2e^{-x^2}$.

41. Find the equations of the tangent lines to the graph of $f(x) = e^{-x^2}$ at its inflection points.

42. Find an equation of the tangent line to the graph of $f(x) = xe^{-x}$ at its inflection point.

In Exercises 43–46, find the absolute extrema of the function.

43. $f(x) = e^{-x^2}$ on $[-1, 1]$
44. $h(x) = e^{x^2-4}$ on $[-2, 2]$
45. $g(x) = (2x - 1)e^{-x}$ on $[0, \infty)$
46. $f(x) = xe^{-x^2}$ on $[0, 2]$

In Exercises 47–50, use the curve-sketching guidelines of Chapter 4, page 288, to sketch the graph of the function.

47. $f(t) = e^t - t$
48. $h(x) = \dfrac{e^x + e^{-x}}{2}$
49. $f(x) = 2 - e^{-x}$
50. $f(x) = \dfrac{3}{1 + e^{-x}}$

51. **Percentage of Population Relocating** Based on data obtained from the Census Bureau, the manager of Plymouth Van Lines estimates that the percent of the total population relocating in year t ($t = 0$ corresponds to the year 1960) may be approximated by the formula

$$P(t) = 20.6e^{-0.009t} \qquad (0 \le t \le 35)$$

Compute $P'(10)$, $P'(20)$, and $P'(30)$ and interpret your results.

52. **Online Banking** In a study prepared in 2000, the percentage of households using online banking was projected to be

$$f(t) = 1.5e^{0.78t} \qquad (0 \le t \le 4)$$

where t is measured in years, with $t = 0$ corresponding to the beginning of 2000.

a. What was the projected percentage of households using online banking at the beginning of 2003?

b. How fast was the projected percentage of households using online banking changing at the beginning of 2003?

c. How fast was the rate of the projected percentage of households using online banking changing at the beginning of 2003?

Hint: We want $f''(3)$. Why?

Source: Online Banking Report

53. Over-100 Population Based on data obtained from the Census Bureau, the number of Americans over age 100 yr is expected to be

$$P(t) = 0.07e^{0.54t} \qquad (0 \le t \le 4)$$

where $P(t)$ is measured in millions and t is measured in decades, with $t = 0$ corresponding to the beginning of 2000.

a. What was the population of Americans over age 100 at the beginning of 2000? What will it be at the beginning of 2030?

b. How fast was the population of Americans over age 100 yr changing at the beginning of 2000? How fast will it be changing at the beginning of 2030?

Source: U.S. Census Bureau

54. World Population Growth After its fastest rate of growth ever during the 1980s and 1990s, the rate of growth of world population is expected to slow dramatically in the 21st century. The function

$$G(t) = 1.58e^{-0.213t}$$

gives the projected annual average percent population growth/decade in the tth decade, with $t = 1$ corresponding to 2000.

a. What will the projected annual average population growth rate be in 2020 ($t = 3$)?

b. How fast will the projected annual average population growth rate be changing in 2020?

Source: U.S. Census Bureau

55. Loans at Japanese Banks The total loans outstanding at all Japanese banks have been declining in recent years. The function

$$L(t) = 4.6e^{-0.04t} \qquad (0 \le t \le 6)$$

gives the approximate total loans outstanding from 1998 ($t = 0$) through 2004 in trillions of dollars.

a. What were the total loans outstanding in 1998? In 2004?

b. How fast were the total loans outstanding declining in 1998? In 2004?

c. Show that the total loans outstanding were declining but at a slower rate over the years between 1998 and 2004.

Hint: Look at L'' on the interval $(0, 6)$.

Source: Bank of Japan

56. Energy Consumption of Appliances The average energy consumption of the typical refrigerator/freezer manufactured by York Industries is approximately

$$C(t) = 1486e^{-0.073t} + 500 \qquad (0 \le t \le 20)$$

kilowatt-hours (kWh) per year, where t is measured in years, with $t = 0$ corresponding to 1972.

a. What was the average energy consumption of the York refrigerator/freezer at the beginning of 1972?

b. Prove that the average energy consumption of the York refrigerator/freezer is decreasing over the years in question.

c. All refrigerator/freezers manufactured as of January 1, 1990, must meet the 950-kWh/year maximum energy-consumption standard set by the National Appliance Conservation Act. Show that the York refrigerator/freezer satisfies this requirement.

57. Sales Promotion The Lady Bug, a women's clothing chain store, found that t days after the end of a sales promotion the volume of sales was given by

$$S(t) = 20{,}000(1 + e^{-0.5t}) \qquad (0 \le t \le 5)$$

dollars.

a. Find the rate of change of The Lady Bug's sales volume when $t = 1$, $t = 2$, $t = 3$, and $t = 4$.

b. After how many days will the sales volume drop below \$27,400?

58. Blood Alcohol Level The percentage of alcohol in a person's bloodstream t hr after drinking 8 fluid oz of whiskey is given by

$$A(t) = 0.23te^{-0.4t} \qquad (0 \le t \le 12)$$

a. What is the percentage of alcohol in a person's bloodstream after $\frac{1}{2}$ hr? After 8 hr?

b. How fast is the percentage of alcohol in a person's bloodstream changing after $\frac{1}{2}$ hr? After 8 hr?

Source: Encyclopedia Britannica

59. Polio Immunization Polio, a once-feared killer, declined markedly in the United States in the 1950s after Jonas Salk developed the inactivated polio vaccine and mass immunization of children took place. The number of polio cases in the United States from the beginning of 1959 to the beginning of 1963 is approximated by the function

$$N(t) = 5.3e^{0.095t^2 - 0.85t} \qquad (0 \le t \le 4)$$

where $N(t)$ gives the number of polio cases (in thousands) and t is measured in years with $t = 0$ corresponding to the beginning of 1959.

a. Show that the function N is decreasing over the time interval under consideration.

b. How fast was the number of polio cases decreasing at the beginning of 1959? At the beginning of 1962? (*Comment:* Following the introduction of the oral vaccine developed by Dr. Albert B. Sabin in 1963, polio in the United States has, for all practical purposes, been eliminated.)

60. **Autistic Brain** At birth, the autistic brain is similar in size to a healthy child's brain. Between birth and 2 yr, it grows to be abnormally large, reaching its maximum size between 3 and 6 yr of age. The percentage difference in size between the autistic brain and the normal brain to age 40 is approximated by

$$D(t) = 6.9te^{-0.24t} \qquad (0 \le t \le 40)$$

where t is measured in years.

a. At what age is the difference in the size between the autistic brain and the normal brain increasing? Decreasing?
b. At what age is the difference in the size between the autistic brain and the normal brain the greatest? What is the maximum difference?
c. At what age is the difference in the size between the autistic brain and the normal brain decreasing at the fastest rate?
d. Sketch the graph of D on the interval [0, 40].

Source: Newsweek

61. **Death Due to Strokes** Before 1950 little was known about strokes. By 1960, however, risk factors such as hypertension were identified. In recent years, CAT scans used as a diagnostic tool have helped prevent strokes. As a result, death due to strokes have fallen dramatically. The function

$$N(t) = 130.7e^{-0.1155t^2} + 50 \qquad (0 \le t \le 6)$$

gives the number of deaths per 100,000 people from 1950 through 2010, where t is measured in decades, with $t = 0$ corresponding to 1950.

a. How many deaths due to strokes per 100,000 people were there in 1950?
b. How fast was the number of deaths due to strokes per 100,000 people changing in 1950? In 1960? In 1970? In 1980?
c. When was the decline in the number of deaths due to strokes per 100,000 people greatest?
d. If the trend continues, how many deaths due to strokes per 100,000 people will there be in 2010?

Source: American Heart Association, Centers for Disease Control, and National Institutes of Health.

62. **Marginal Revenue** The unit selling price p (in dollars) and the quantity demanded x (in pairs) of a certain brand of women's gloves is given by the demand equation

$$p = 100e^{-0.0001x} \qquad (0 \le x \le 20{,}000)$$

a. Find the revenue function R.
Hint: $R(x) = px$.
b. Find the marginal revenue function R'.
c. What is the marginal revenue when $x = 10{,}000$?

63. **Maximizing Revenue** Refer to Exercise 62. How many pairs of the gloves must be sold to yield a maximum revenue? What will be the maximum revenue?

64. **Price of Perfume** The monthly demand for a certain brand of perfume is given by the demand equation

$$p = 100e^{-0.0002x} + 150$$

where p denotes the retail unit price (in dollars) and x denotes the quantity (in 1-oz bottles) demanded.

a. Find the rate of change of the price per bottle when $x = 1000$ and when $x = 2000$.
b. What is the price per bottle when $x = 1000$? When $x = 2000$?

65. **Price of Wine** The monthly demand for a certain brand of table wine is given by the demand equation

$$p = 240\left(1 - \frac{3}{3 + e^{-0.0005x}}\right)$$

where p denotes the wholesale price per case (in dollars) and x denotes the number of cases demanded.

a. Find the rate of change of the price per case when $x = 1000$.
b. What is the price per case when $x = 1000$?

66. **Spread of an Epidemic** During a flu epidemic, the total number of students on a state university campus who had contracted influenza by the xth day was given by

$$N(x) = \frac{3000}{1 + 99e^{-x}} \qquad (x \ge 0)$$

a. How many students had influenza initially?
b. Derive an expression for the rate at which the disease was being spread and prove that the function N is increasing on the interval $(0, \infty)$.
c. Sketch the graph of N. What was the total number of students who contracted influenza during that particular epidemic?

67. **Weights of Children** The Ehrenberg equation

$$W = 2.4e^{1.84h}$$

gives the relationship between the height h (in meters) and the average weight W (in kilograms) for children between 5 and 13 yr of age.

a. What is the average weight of a 10-yr-old child who stands at 1.6 m tall?
b. Use differentials to estimate the change in the average weight of a 10-yr-old child whose height increases from 1.6 m to 1.65 m.

68. **Population Distribution** The number of people living x mi from the center of town is given by

$$P(x) = 50{,}000(1 - e^{-0.01x^2}) \qquad (0 < x < 25)$$

Use differentials to estimate the number of people living between 10 and 10.1 mi from the center of town.

69. **Optimal Selling Time** Refer to Exercise 46, page 358. The present value of a piece of waterfront property purchased by an investor is given by the function

$$P(t) = 80{,}000e^{\sqrt{t/2} - 0.09t} \qquad (0 \le t \le 8)$$

Determine the optimal time (based on present value) for the investor to sell the property. What is the property's optimal present value?

70. **Maximum Oil Production** It has been estimated that the total production of oil from a certain oil well is given by

$$T(t) = -1000(t + 10)e^{-0.1t} + 10,000$$

thousand barrels t yr after production has begun. Determine the year when the oil well will be producing at maximum capacity.

71. **Blood Alcohol Level** Refer to Exercise 58, p. 367. At what time after drinking the alcohol is the percentage of alcohol in the person's bloodstream at its highest level? What is that level?

72. **Price of a Commodity** The price of a certain commodity in dollars per unit at time t (measured in weeks) is given by $p = 8 + 4e^{-2t} + te^{-2t}$.
   a. What is the price of the commodity at $t = 0$?
   b. How fast is the price of the commodity changing at $t = 0$?
   c. Find the equilibrium price of the commodity.
   Hint: It's given by $\lim_{t\to\infty} p$. Also, use the fact that $\lim_{t\to\infty} te^{-2t} = 0$.

73. **Thermometer Readiness** A thermometer is moved from inside a house out to the deck. Its temperature t min after it has been moved is given by

$$T(t) = 30 + 40e^{-0.98t}$$

   a. What is the temperature inside the house?
   b. How fast is the reading on the thermometer changing 1 min after it has been taken out of the house?
   c. What is the outdoor temperature?
   Hint: Evaluate $\lim_{t\to\infty} T(t)$.

74. **Chemical Reaction** Two chemicals, A and B, interact to form a chemical C. Suppose the amount (in grams) of chemical C formed t min after the interaction first takes place is

$$A(t) = \frac{150(1 - e^{0.022662t})}{1 - 2.5e^{0.022662t}}$$

   a. How fast is chemical C being formed 1 min after the interaction first began?
   b. How much chemical C will there be eventually?
   Hint: Evaluate $\lim_{t\to\infty} A(t)$.

75. **Concentration of a Drug in the Bloodstream** The concentration of a drug in the bloodstream t sec after injection into a muscle is given by

$$y = c(e^{-bt} - e^{-at}) \qquad (\text{for } t \geq 0)$$

where a, b, and c are positive constants, with $a > b$.
   a. Find the time at which the concentration is maximal.
   b. Find the time at which the concentration of the drug in the bloodstream is decreasing most rapidly.

76. **Absorption of Drugs** A liquid carries a drug into an organ of volume V cm^3 at the rate of a cm^3/sec and leaves at the same rate. The concentration of the drug in the entering liquid is c g/cm^3. Letting $x(t)$ denote the concentration of the drug in the organ at any time t, we have $x(t) = c(1 - e^{-at/V})$.
   a. Show that x is an increasing function on $(0, \infty)$.
   b. Sketch the graph of x.

77. **Absorption of Drugs** Refer to Exercise 76. Suppose the maximum concentration of the drug in the organ must *not* exceed m g/cm^3, where $m < c$. Show that the liquid must not be allowed to enter the organ for a time longer than

$$T = \left(\frac{V}{a}\right) \ln \left(\frac{c}{c - m}\right)$$

minutes.

78. **Absorption of Drugs** Jane took 100 mg of a drug in the morning and another 100 mg of the same drug at the same time the following morning. The amount of the drug in her body t days after the first dosage was taken is given by

$$A(t) = \begin{cases} 100e^{-1.4t} & \text{if } 0 \leq t < 1 \\ 100(1 + e^{1.4})e^{-1.4t} & \text{if } t \geq 1 \end{cases}$$

   a. How fast was the amount of drug in Jane's body changing after 12 hr ($t = \frac{1}{2}$)? After 2 days?
   b. When was the amount of drug in Jane's body a maximum?
   c. What was the maximum amount of drug in Jane's body?

79. **Absorption of Drugs** The concentration of a drug in an organ at any time t (in seconds) is given by

$$C(t) = \begin{cases} 0.3t - 18(1 - e^{-t/60}) & \text{if } 0 \leq t \leq 20 \\ 18e^{-t/60} - 12e^{-(t-20)/60} & \text{if } t > 20 \end{cases}$$

where $C(t)$ is measured in grams/cubic centimeter (g/cm^3).
   a. How fast is the concentration of the drug in the organ changing after 10 sec?
   b. How fast is the concentration of the drug in the organ changing after 30 sec?
   c. When will the concentration of the drug in the organ reach a maximum?
   d. What is the maximum drug concentration in the organ?

In Exercises 80–83, determine whether the statement is true or false. If it is true, explain why it is true. If it is false, give an example to show why it is false.

80. If $f(x) = 3^x$, then $f'(x) = x \cdot 3^{x-1}$.

81. If $f(x) = e^{\pi}$, then $f'(x) = e^{\pi}$.

82. If $f(x) = \pi^x$, then $f'(x) = \pi^x$.

83. If $x^2 + e^y = 10$, then $y' = \dfrac{-2x}{e^y}$.

5.4 Solutions to Self-Check Exercises

1. a. Using the product rule, we obtain

$$f'(x) = x\frac{d}{dx}e^{-x} + e^{-x}\frac{d}{dx}x$$
$$= -xe^{-x} + e^{-x} = (1 - x)e^{-x}$$

Using the product rule once again, we obtain

$$f''(x) = (1 - x)\frac{d}{dx}e^{-x} + e^{-x}\frac{d}{dx}(1 - x)$$
$$= (1 - x)(-e^{-x}) + e^{-x}(-1)$$
$$= -e^{-x} + xe^{-x} - e^{-x} = (x - 2)e^{-x}$$

b. Setting $f'(x) = 0$ gives

$$(1 - x)e^{-x} = 0$$

Since $e^{-x} \neq 0$, we see that $1 - x = 0$, and this gives 1 as the only critical number of f. The sign diagram of f' shown in the accompanying figure tells us that the point $(1, e^{-1})$ is a relative maximum of f.

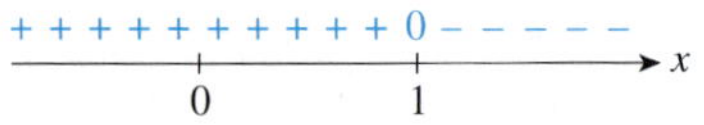

c. Setting $f''(x) = 0$ gives $x - 2 = 0$, so $x = 2$ is a candidate for an inflection point of f. The sign diagram of f'' (see the accompanying figure) shows that $(2, 2e^{-2})$ is an inflection point of f.

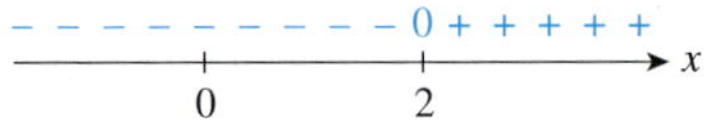

2. The rate change of the book value of the asset t yr from now is

$$V'(t) = 50{,}000\frac{d}{dt}e^{-0.4t}$$
$$= 50{,}000(-0.4)e^{-0.4t} = -20{,}000e^{-0.4t}$$

Therefore, 3 yr from now the book value of the asset will be changing at the rate of

$$V'(3) = -20{,}000e^{-0.4(3)} = -20{,}000e^{-1.2} \approx -6023.88$$

—that is, decreasing at the rate of approximately \$6024/year.

USING TECHNOLOGY

EXAMPLE 1 At the beginning of Section 5.4, we demonstrated via a table of values of $(e^h - 1)/h$ for selected values of h the plausibility of the result

$$\lim_{h\to 0}\frac{e^h - 1}{h} = 1$$

To obtain a visual confirmation of this result, we plot the graph of

$$f(x) = \frac{e^x - 1}{x}$$

in the viewing window $[-1, 1] \times [0, 2]$ (Figure T1). From the graph of f, we see that $f(x)$ appears to approach 1 as x approaches 0. ■

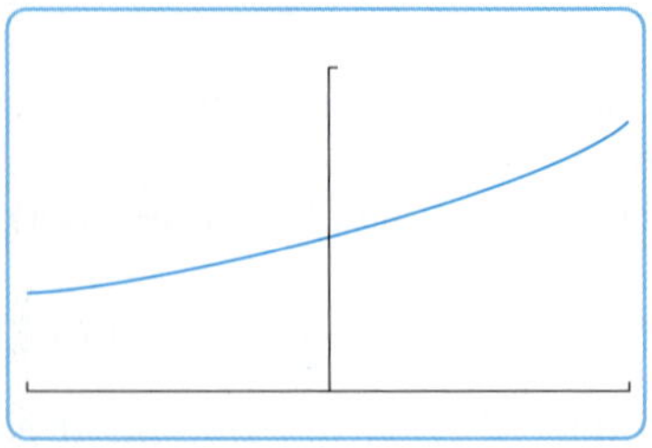

FIGURE **T1**
The graph of f in the viewing window $[-1, 1] \times [0, 2]$

The numerical derivative function of a graphing utility will yield the derivative of an exponential or logarithmic function for any value of x, just as it did for algebraic functions.*

*The rules for differentiating logarithmic functions will be covered in Section 5.5. However, the exercises given here can be done without using these rules.

TECHNOLOGY EXERCISES

In Exercises 1–6, use the numerical derivative operation to find the rate of change of $f(x)$ at the given value of x. Give your answer accurate to four decimal places.

1. $f(x) = x^3e^{-1/x}$; $x = -1$
2. $f(x) = (\sqrt{x} + 1)^{3/2}e^{-x}$; $x = 0.5$
3. $f(x) = x^3\sqrt{\ln x}$; $x = 2$
4. $f(x) = \dfrac{\sqrt{x}\ln x}{x + 1}$; $x = 3.2$
5. $f(x) = e^{-x}\ln(2x + 1)$; $x = 0.5$

6. $f(x) = \dfrac{e^{-\sqrt{x}}}{\ln(x^2 + 1)}; x = 1$

7. **An Extinction Situation** The number of saltwater crocodiles in a certain area of northern Australia is given by

$$P(t) = \frac{300e^{-0.024t}}{5e^{-0.024t} + 1}$$

a. How many crocodiles were in the population initially?
b. Show that $\lim_{t\to\infty} P(t) = 0$.
c. Plot the graph of P in the viewing window $[0, 200] \times [0, 70]$.

(*Comment:* This phenomenon is referred to as an *extinction situation.*)

8. **Income of American Families** Based on data, it is estimated that the number of American families y (in millions) who earned x thousand dollars in 1990 is related by the equation

$$y = 0.1584xe^{-0.0000016x^3+0.00011x^2-0.04491x} \qquad (x > 0)$$

a. Plot the graph of the equation in the viewing window $[0, 150] \times [0, 2]$.
b. How fast is y changing with respect to x when $x = 10$? When $x = 50$? Interpret your results.

Source: House Budget Committee, House Ways and Means Committee, and U.S. Census Bureau

9. **World Population Growth** Based on data obtained in a study, the world population (in billions) is approximated by the function

$$f(t) = \frac{12}{1 + 3.74914e^{-1.42804t}} \qquad (0 \le t \le 4)$$

where t is measured in half centuries, with $t = 0$ corresponding to the beginning of 1950.

a. Plot the graph of f in the viewing window $[0, 5] \times [0, 14]$.
b. How fast was the world population expected to increase at the beginning of 2000?

Source: United Nations Population Division

10. **Loan Amortization** The Sotos plan to secure a loan of \$160,000 to purchase a house. They are considering a conventional 30-yr home mortgage at 9%/year on the unpaid balance. It can be shown that the Sotos will have an outstanding principal of

$$B(x) = \frac{160{,}000(1.0075^{360} - 1.0075^{x})}{1.0075^{360} - 1}$$

dollars after making x monthly payments of \$1287.40.

a. Plot the graph of $B(x)$, using the viewing window $[0, 360] \times [0, 160{,}000]$.
b. Compute $B(0)$ and $B'(0)$ and interpret your results; compute $B(180)$ and $B'(180)$ and interpret your results.

11. **Increase in Juvenile Offenders** The number of youths aged 15 to 19 increased by 21% between 1994 and 2005, pushing up the crime rate. According to the National Council on Crime and Delinquency, the number of violent crime arrests of juveniles under age 18 in year t is given by

$$f(t) = -0.438t^2 + 9.002t + 107 \qquad (0 \le t \le 13)$$

where $f(t)$ is measured in thousands and t in years, with $t = 0$ corresponding to 1989. According to the same source, if trends like inner-city drug use and wider availability of guns continues, then the number of violent crime arrests of juveniles under age 18 in year t is given by

$$g(t) = \begin{cases} -0.438t^2 + 9.002t + 107 & \text{if } 0 \le t < 4 \\ 99.456e^{0.07824t} & \text{if } 4 \le t \le 13 \end{cases}$$

where $g(t)$ is measured in thousands and $t = 0$ corresponds to 1989.

a. Compute $f(11)$ and $g(11)$ and interpret your results.
b. Compute $f'(11)$ and $g'(11)$ and interpret your results.

Source: National Council on Crime and Delinquency

12. **Increasing Crop Yields** If left untreated on bean stems, aphids (small insects that suck plant juices) will multiply at an increasing rate during the summer months and reduce productivity and crop yield of cultivated crops. But if the aphids are treated in mid-June, the numbers decrease sharply to less than 100/bean stem, allowing for steep rises in crop yield. The function

$$F(t) = \begin{cases} 62e^{1.152t} & \text{if } 0 \le t < 1.5 \\ 349e^{-1.324(t-1.5)} & \text{if } 1.5 \le t \le 3 \end{cases}$$

gives the number of aphids on a typical bean stem at time t, where t is measured in months, with $t = 0$ corresponding to the beginning of May.

a. How many aphids are there on a typical bean stem at the beginning of June ($t = 1$)? At the beginning of July ($t = 2$)?
b. How fast is the population of aphids changing at the beginning of June? At the beginning of July?

Source: The Random House Encyclopedia

13. **Women in the Labor Force** Based on data from the U.S. Census Bureau, the chief economist of Manpower, Inc., constructed the following formula giving the percentage of the total female population in the civilian labor force, $P(t)$, at the beginning of the tth decade ($t = 0$ corresponds to the year 1900):

$$P(t) = \frac{74}{1 + 2.6e^{-0.166t+0.04536t^2-0.0066t^3}} \qquad (0 \le t \le 11)$$

Assume that this trend continued for the rest of the 20th century.

a. What percentage of the total female population was in the civilian labor force at the beginning of 2000?
b. What was the growth rate of the percentage of the total female population in the civilian labor force at the beginning of 2000?

Source: U.S. Census Bureau

5.5 Differentiation of Logarithmic Functions

The Derivative of ln *x*

Let's now turn our attention to the differentiation of logarithmic functions.

Rule 3: Derivative of ln *x*

$$\frac{d}{dx}\ln|x| = \frac{1}{x} \qquad (x \neq 0)$$

To derive Rule 3, suppose $x > 0$ and write $f(x) = \ln x$ in the equivalent form

$$x = e^{f(x)}$$

Differentiating both sides of the equation with respect to x, we find, using the chain rule,

$$1 = e^{f(x)} \cdot f'(x)$$

from which we see that

$$f'(x) = \frac{1}{e^{f(x)}}$$

or, since $e^{f(x)} = x$,

$$f'(x) = \frac{1}{x}$$

as we set out to show. You are asked to prove the rule for the case $x < 0$ in Exercise 88, page 379.

EXAMPLE 1 Find the derivative of each function:

a. $f(x) = x \ln x$ **b.** $g(x) = \dfrac{\ln x}{x}$

Solution

a. Using the product rule, we obtain

$$\begin{aligned} f'(x) &= \frac{d}{dx}(x \ln x) = x\frac{d}{dx}(\ln x) + (\ln x)\frac{d}{dx}(x) \\ &= x\left(\frac{1}{x}\right) + \ln x = 1 + \ln x \end{aligned}$$

b. Using the quotient rule, we obtain

$$g'(x) = \frac{x\dfrac{d}{dx}(\ln x) - (\ln x)\dfrac{d}{dx}(x)}{x^2} = \frac{x\left(\dfrac{1}{x}\right) - \ln x}{x^2} = \frac{1 - \ln x}{x^2}$$

Explore & Discuss

You can derive the formula for the derivative of $f(x) = \ln x$ directly from the definition of the derivative, as follows.

1. Show that

$$f'(x) = \lim_{h \to 0} \frac{f(x+h) - f(x)}{h} = \lim_{h \to 0} \ln\left(1 + \frac{h}{x}\right)^{1/h}$$

2. Put $m = x/h$ and note that $m \to \infty$ as $h \to 0$. Then, $f'(x)$ can be written in the form

$$f'(x) = \lim_{m \to \infty} \ln \left(1 + \frac{1}{m}\right)^{m/x}$$

3. Finally, use both the fact that the natural logarithmic function is continuous and the definition of the number e to show that

$$f'(x) = \frac{1}{x} \ln \left[\lim_{m \to \infty} \left(1 + \frac{1}{m}\right)^{m}\right] = \frac{1}{x}$$

The Chain Rule for Logarithmic Functions

To enlarge the class of logarithmic functions to be differentiated, we appeal once again to the chain rule to obtain the following rule for differentiating composite functions of the form $h(x) = \ln f(x)$, where $f(x)$ is assumed to be a positive differentiable function.

Rule 4: Chain Rule for Logarithmic Functions

If $f(x)$ is a differentiable function, then

$$\frac{d}{dx}[\ln f(x)] = \frac{f'(x)}{f(x)} \qquad [f(x) > 0]$$

To see this, observe that $h(x) = g[f(x)]$, where $g(x) = \ln x$ $(x > 0)$. Since $g'(x) = 1/x$, we have, using the chain rule,

$$\begin{aligned} h'(x) &= g'(f(x))f'(x) \\ &= \frac{1}{f(x)} f'(x) = \frac{f'(x)}{f(x)} \end{aligned}$$

Observe that in the special case $f(x) = x$, $h(x) = \ln x$, so the derivative of h is, by Rule 3, given by $h'(x) = 1/x$.

VIDEO

EXAMPLE 2 Find the derivative of the function $f(x) = \ln(x^2 + 1)$.

Solution Using Rule 4, we see immediately that

$$f'(x) = \frac{\frac{d}{dx}(x^2 + 1)}{x^2 + 1} = \frac{2x}{x^2 + 1}$$

When differentiating functions involving logarithms, the rules of logarithms may be used to advantage, as shown in Examples 3 and 4.

EXAMPLE 3 Differentiate the function $y = \ln[(x^2 + 1)(x^3 + 2)^6]$.

Solution We first rewrite the given function using the properties of logarithms:

$$\begin{aligned} y &= \ln[(x^2 + 1)(x^3 + 2)^6] \\ &= \ln(x^2 + 1) + \ln(x^3 + 2)^6 && \ln mn = \ln m + \ln n \\ &= \ln(x^2 + 1) + 6 \ln(x^3 + 2) && \ln m^n = n \ln m \end{aligned}$$

Differentiating and using Rule 4, we obtain

$$y' = \frac{\frac{d}{dx}(x^2+1)}{x^2+1} + \frac{6\frac{d}{dx}(x^3+2)}{x^3+2}$$

$$= \frac{2x}{x^2+1} + \frac{6(3x^2)}{x^3+2} = \frac{2x}{x^2+1} + \frac{18x^2}{x^3+2}$$

Exploring with TECHNOLOGY

Use a graphing utility to plot the graphs of $f(x) = \ln x$; its first derivative function, $f'(x) = 1/x$; and its second derivative function $f''(x) = -1/x^2$, using the same viewing window $[0, 4] \times [-3, 3]$.

1. Describe the properties of the graph of f revealed by studying the graph of $f'(x)$. What can you say about the rate of increase of f for large values of x?
2. Describe the properties of the graph of f revealed by studying the graph of $f''(x)$. What can you say about the concavity of f for large values of x?

EXAMPLE 4 Find the derivative of the function $g(t) = \ln(t^2e^{-t^2})$.

Solution Here again, to save a lot of work, we first simplify the given expression using the properties of logarithms. We have

$$\begin{aligned} g(t) &= \ln(t^2e^{-t^2}) \\ &= \ln t^2 + \ln e^{-t^2} && \ln mn = \ln m + \ln n \\ &= 2\ln t - t^2 && \ln m^n = n\ln m \quad \text{and} \quad \ln e = 1 \end{aligned}$$

Therefore,

$$g'(t) = \frac{2}{t} - 2t = \frac{2(1-t^2)}{t}$$

Logarithmic Differentiation

As we saw in the last two examples, the task of finding the derivative of a given function can sometimes be made easier by first applying the laws of logarithms to simplify the function. We now illustrate a process called **logarithmic differentiation**, which not only simplifies the calculation of the derivatives of certain functions but also enables us to compute the derivatives of functions we could not otherwise differentiate using the techniques developed thus far.

EXAMPLE 5 Differentiate $y = x(x+1)(x^2+1)$, using logarithmic differentiation.

Solution First, we take the natural logarithm on both sides of the given equation, obtaining

$$\ln y = \ln x(x+1)(x^2+1)$$

Next, we use the properties of logarithms to rewrite the right-hand side of this equation, obtaining

$$\ln y = \ln x + \ln(x+1) + \ln(x^2+1)$$

If we differentiate both sides of this equation, we have

$$\begin{aligned}\frac{d}{dx}\ln y &= \frac{d}{dx}[\ln x + \ln(x+1) + \ln(x^2+1)]\\ &= \frac{1}{x} + \frac{1}{x+1} + \frac{2x}{x^2+1} \qquad \text{Use Rule 4.}\end{aligned}$$

To evaluate the expression on the left-hand side, note that y is a function of x. Therefore, writing $y = f(x)$ to remind us of this fact, we have

$$\begin{aligned}\frac{d}{dx}\ln y &= \frac{d}{dx}\ln[f(x)] \qquad \text{Write } y = f(x).\\ &= \frac{f'(x)}{f(x)} \qquad \text{Use Rule 4.}\\ &= \frac{y'}{y} \qquad \text{Return to using } y \text{ instead of } f(x).\end{aligned}$$

Therefore, we have

$$\frac{y'}{y} = \frac{1}{x} + \frac{1}{x+1} + \frac{2x}{x^2+1}$$

Finally, solving for y', we have

$$\begin{aligned}y' &= y\left(\frac{1}{x} + \frac{1}{x+1} + \frac{2x}{x^2+1}\right)\\ &= x(x+1)(x^2+1)\left(\frac{1}{x} + \frac{1}{x+1} + \frac{2x}{x^2+1}\right)\end{aligned}$$

Before considering other examples, let's summarize the important steps involved in logarithmic differentiation.

Finding $\frac{dy}{dx}$ by Logarithmic Differentiation

1. Take the natural logarithm on both sides of the equation and use the properties of logarithms to write any "complicated expression" as a sum of simpler terms.
2. Differentiate both sides of the equation with respect to x.
3. Solve the resulting equation for $\frac{dy}{dx}$.

VIDEO

EXAMPLE 6 Differentiate $y = x^2(x-1)(x^2+4)^3$.

Solution Taking the natural logarithm on both sides of the given equation and using the laws of logarithms, we obtain

$$\begin{aligned}\ln y &= \ln x^2(x-1)(x^2+4)^3\\ &= \ln x^2 + \ln(x-1) + \ln(x^2+4)^3\\ &= 2\ln x + \ln(x-1) + 3\ln(x^2+4)\end{aligned}$$

Differentiating both sides of the equation with respect to x, we have

$$\frac{d}{dx}\ln y = \frac{y'}{y} = \frac{2}{x} + \frac{1}{x-1} + 3\cdot\frac{2x}{x^2+4}$$

Finally, solving for y', we have

$$\begin{aligned} y' &= y\left(\frac{2}{x} + \frac{1}{x-1} + \frac{6x}{x^2+4}\right) \\ &= x^2(x-1)(x^2+4)^3\left(\frac{2}{x} + \frac{1}{x-1} + \frac{6x}{x^2+4}\right) \end{aligned}$$

■

EXAMPLE 7 Find the derivative of $f(x) = x^x$ $(x > 0)$.

Solution A word of caution! This function is neither a power function nor an exponential function. Taking the natural logarithm on both sides of the equation gives

$$\ln f(x) = \ln x^x = x \ln x$$

Differentiating both sides of the equation with respect to x, we obtain

$$\begin{aligned} \frac{f'(x)}{f(x)} &= x\frac{d}{dx}\ln x + (\ln x)\frac{d}{dx}x \\ &= x\left(\frac{1}{x}\right) + \ln x \\ &= 1 + \ln x \end{aligned}$$

Therefore,

$$f'(x) = f(x)(1 + \ln x) = x^x(1 + \ln x)$$

■

Exploring with TECHNOLOGY

Refer to Example 7.

1. Use a graphing utility to plot the graph of $f(x) = x^x$, using the viewing window $[0, 2] \times [0, 2]$. Then use ZOOM and TRACE to show that

$$\lim_{x \to 0^+} f(x) = 1$$

2. Use the results of part 1 and Example 7 to show that $\lim_{x \to 0^+} f'(x) = -\infty$. Justify your answer.

5.5 Self-Check Exercises

1. Find an equation of the tangent line to the graph of $f(x) = x \ln(2x + 3)$ at the point $(-1, 0)$.

2. Use logarithmic differentiation to compute y', given $y = (2x + 1)^3(3x + 4)^5$.

Solutions to Self-Check Exercises 5.5 can be found on page 379.

5.5 Concept Questions

1. State the rule for differentiating (a) $f(x) = \ln|x|$ $(x \neq 0)$, and (b) $g(x) = \ln f(x)$ $[f(x) > 0]$, where f is a differentiable function.

2. Explain the technique of logarithmic differentiation.

5.5 Exercises

In Exercises 1–34, find the derivative of the function.

1. $f(x) = 5 \ln x$
2. $f(x) = \ln 5x$
3. $f(x) = \ln(x + 1)$
4. $g(x) = \ln(2x + 1)$
5. $f(x) = \ln x^8$
6. $h(t) = 2 \ln t^5$
7. $f(x) = \ln \sqrt{x}$
8. $f(x) = \ln(\sqrt{x} + 1)$
9. $f(x) = \ln \dfrac{1}{x^2}$
10. $f(x) = \ln \dfrac{1}{2x^3}$
11. $f(x) = \ln(4x^2 - 6x + 3)$
12. $f(x) = \ln(3x^2 - 2x + 1)$
13. $f(x) = \ln \dfrac{2x}{x + 1}$
14. $f(x) = \ln \dfrac{x + 1}{x - 1}$
15. $f(x) = x^2 \ln x$
16. $f(x) = 3x^2 \ln 2x$
17. $f(x) = \dfrac{2 \ln x}{x}$
18. $f(x) = \dfrac{3 \ln x}{x^2}$
19. $f(u) = \ln(u - 2)^3$
20. $f(x) = \ln(x^3 - 3)^4$
21. $f(x) = \sqrt{\ln x}$
22. $f(x) = \sqrt{\ln x + x}$
23. $f(x) = (\ln x)^3$
24. $f(x) = 2(\ln x)^{3/2}$
25. $f(x) = \ln(x^3 + 1)$
26. $f(x) = \ln \sqrt{x^2 - 4}$
27. $f(x) = e^x \ln x$
28. $f(x) = e^x \ln \sqrt{x + 3}$
29. $f(t) = e^{2t} \ln(t + 1)$
30. $g(t) = t^2 \ln(e^{2t} + 1)$
31. $f(x) = \dfrac{\ln x}{x^2}$
32. $g(t) = \dfrac{t}{\ln t}$
33. $f(x) = \ln(\ln x)$
34. $g(x) = \ln(e^x + \ln x)$

In Exercises 35–40, find the second derivative of the function.

35. $f(x) = \ln 2x$
36. $f(x) = \ln(x + 5)$
37. $f(x) = \ln(x^2 + 2)$
38. $f(x) = (\ln x)^2$
39. $f(x) = x^2 \ln x$
40. $g(x) = e^{2x} \ln x$

In Exercises 41–50, use logarithmic differentiation to find the derivative of the function.

41. $y = (x + 1)^2(x + 2)^3$
42. $y = (3x + 2)^4(5x - 1)^2$
43. $y = (x - 1)^2(x + 1)^3(x + 3)^4$
44. $y = \sqrt{3x + 5}(2x - 3)^4$
45. $y = \dfrac{(2x^2 - 1)^5}{\sqrt{x + 1}}$
46. $y = \dfrac{\sqrt{4 + 3x^2}}{\sqrt[3]{x^2 + 1}}$
47. $y = 3^x$
48. $y = x^{x+2}$
49. $y = (x^2 + 1)^x$
50. $y = x^{\ln x}$

In Exercises 51 and 52, use implicit differentiation to find *dy/dx*.

51. $\ln y - x \ln x = -1$
52. $\ln xy - y^2 = 5$

53. Find an equation of the tangent line to the graph of $y = x \ln x$ at the point $(1, 0)$.

54. Find an equation of the tangent line to the graph of $y = \ln x^2$ at the point $(2, \ln 4)$.

55. Determine the intervals where the function $f(x) = \ln x^2$ is increasing and where it is decreasing.

56. Determine the intervals where the function $f(x) = \dfrac{\ln x}{x}$ is increasing and where it is decreasing.

57. Determine the intervals of concavity for the function $f(x) = x^2 + \ln x^2$.

58. Determine the intervals of concavity for the function $f(x) = \dfrac{\ln x}{x}$.

59. Find the inflection points of the function $f(x) = \ln(x^2 + 1)$.

60. Find the inflection points of the function $f(x) = x^2 \ln x$.

61. Find an equation of the tangent line to the graph of $f(x) = x^2 + 2 \ln x$ at its inflection point.

62. Find an equation of the tangent line to the graph of $f(x) = e^{x/2} \ln x$ at its inflection point.
Hint: Show that $(1, 0)$ is the only inflection point of f.

63. Find the absolute extrema of the function $f(x) = x - \ln x$ on $[\frac{1}{2}, 3]$.

64. Find the absolute extrema of the function $g(x) = \dfrac{x}{\ln x}$ on $[2, 5]$.

65. **Strain on Vertebrae** The strain (percentage of compression) on the lumbar vertebral disks in an adult human as a function of the load x (in kilograms) is given by

$$f(x) = 7.2956 \ln(0.0645012x^{0.95} + 1)$$

What is the rate of change of the strain with respect to the load when the load is 100 kg? When the load is 500 kg?
Source: Benedek and Villars, *Physics with Illustrative Examples from Medicine and Biology*

66. **Heights of Children** For children between the ages of 5 and 13 yr, the Ehrenberg equation

$$\ln W = \ln 2.4 + 1.84h$$

gives the relationship between the weight W (in kilograms) and the height h (in meters) of a child. Use differentials to estimate the change in the weight of a child who grows from 1 m to 1.1 m.

67. **Yahoo! in Europe** Yahoo! is putting more emphasis on Western Europe, where the number of online households is expected to grow steadily. In a study conducted in 2004, the number of online households (in millions) in Western Europe was projected to be

$$N(t) = 34.68 + 23.88 \ln(1.05t + 5.3) \qquad (0 \le t \le 2)$$

where $t = 0$ corresponds to the beginning of 2004.

a. What was the projected number of online households in Western Europe at the beginning of 2005?

b. How fast was the projected number of online households in Western Europe increasing at the beginning of 2005?

Source: Jupiter Research

68. **Depreciation of Equipment** For assets such as machines, whose market values drop rapidly in the early years of usage, businesses often use the double declining–balance method. In practice, a business firm normally employs the double declining–balance method for depreciating such assets for a certain number of years and then switches over to the linear method (see Exercise 43, page 88). The double declining–balance formula is

$$V(n) = C\left(1 - \frac{2}{N}\right)^n$$

where $V(n)$ denotes the book value of the assets at the end of n years and N is the number of years over which the asset is depreciated.

a. Find $V'(n)$.
Hint: Use logarithmic differentiation.

b. What is the relative rate of change of $V(n)$?
Hint: Find $[V'(n)]/[V(n)]$. See Section 3.5.

69. **Depreciation of Equipment** Refer to Exercise 68. A tractor purchased at a cost of \$60,000 is to be depreciated by the double declining–balance method over 10 yrs.

a. What is the book value of the tractor at the end of 2 yr?

b. What is the relative rate of change of the book value of the tractor at the end of 2 yr?

70. **Online Buyers** The number of online buyers in Western Europe is expected to grow steadily in the coming years. The function

$$P(t) = 28.5 + 14.42 \ln t \qquad (1 \le t \le 7)$$

gives the estimated online buyers as a percentage of the total population, where t is measured in years, with $t = 1$ corresponding to 2001.

a. What was the percentage of online buyers in 2001 ($t = 1$)? How fast was it changing in 2001?

b. What was the percentage of online buyers expected to be in 2006 ($t = 6$)? How fast was it expected to be changing in 2006?

Source: Jupiter Research

71. **Average Life Span** One reason for the increase in the life span over the years has been the advances in medical technology. The average life span for American women from 1907 through 2007 is given by

$$W(t) = 49.9 + 17.1 \ln t \qquad (1 \le t \le 6)$$

where $W(t)$ is measured in years and t is measured in 20-yr intervals, with $t = 1$ corresponding to the beginning of 1907.

a. Show that W is increasing on $(0, 6)$.

b. What can you say about the concavity of the graph of W on the interval $(0, 6)$?

72. **Lambert's Law of Absorption** Lambert's law of absorption states that the light intensity $I(x)$ (in calories/square centimeter/second) at a depth of x m as measured from the surface of a material is given by $I = I_0 a^x$, where I_0 and a are positive constants.

a. Find the rate of change of the light intensity with respect to x at a depth of x m from the surface of the material.

b. Using the result of part (a), conclude that the rate of change $I'(x)$ at a depth of x m is proportional to $I(x)$. What is the constant of proportion?

73. **Absorption of Light** When light passes through a window glass, some of it is absorbed. It can be shown that if $r\%$ of the light is absorbed by a glass of thickness w, then the percentage of light absorbed by a piece of glass of thickness nw is

$$A(n) = 100\left[1 - \left(1 - \frac{r}{100}\right)^n\right] \qquad (0 \le r \le 100)$$

a. Show that A is an increasing function of n on $(0, \infty)$ if $0 < r < 100$.
Hint: Use logarithmic differentiation.

b. Sketch the graph of A for the special case where $r = 10$.

c. Evaluate $\lim_{n \to \infty} A(n)$ and interpret your result.

74. **Magnitude of Earthquakes** On the Richter scale, the magnitude R of an earthquake is given by the formula

$$R = \log \frac{I}{I_0}$$

where I is the intensity of the earthquake being measured and I_0 is the standard reference intensity.

a. What is the magnitude of an earthquake that has intensity 1 million times that of I_0?

b. Suppose an earthquake is measured with a magnitude of 6 on the Richter scale with an error of at most 2%. Use differentials to find the error in the intensity of the earthquake.
Hint: Observe that $I = I_0 10^R$ and use logarithmic differentiation.

75. **Weber–Fechner Law** The Weber–Fechner law

$$R = k \ln \frac{S}{S_0}$$

where k is a positive constant, describes the relationship between a stimulus S and the resulting response R. Here, S_0, a positive constant, is the threshold level.

a. Show that $R = 0$ if the stimulus is at the threshold level S_0.

b. The derivative dR/dS is the *sensitivity* corresponding to the stimulus level S and measures the capability to detect small changes in the stimulus level. Show that dR/dS is inversely proportional to S and interpret your result.

76. **Predator–Prey Model** The relationship between the number of rabbits $y(t)$ and the number of foxes $x(t)$ at any time t is given by

$$-C \ln y + Dy = A \ln x - Bx + E$$

where A, B, C, D, and E are constants. This relationship is based on a model by Lotka (1880–1949) and Volterra (1860–1940) for analyzing the ecological balance between two species of animals, one of which is a prey and the other a predator. Use implicit differentiation to find the relationship between the rate of change of the rabbit population in terms of the rate of change of the fox population.

77. **Rate of a Catalytic Chemical Reaction** A catalyst is a substance that either accelerates a chemical reaction or is necessary for the reaction to occur. Suppose an enzyme E (a catalyst) combines with a substrate S (a reacting chemical) to form an intermediate product X that then produces a product P and releases the enzyme. If initially there are x_0 moles/liter of S and there is no P, then based on the theory of Michaelis and Menten, the concentration of P, $p(t)$, after t hr is given by the equation

$$Vt = p - k \ln\left(1 - \frac{p}{x_0}\right)$$

where the constant V is the maximum possible speed of the reaction and the constant k is called the **Michaelis constant** for the reaction. Find the rate of change of the formation of the product P in this reaction.

In Exercises 78 and 79, use the guidelines on page 288 to sketch the graph of the given function.

78. $f(x) = \ln(x - 1)$

79. $f(x) = 2x - \ln x$

80. **Derivative of b^x**
   a. Let $f(x) = b^x$ $(b > 0, b \neq 1)$. Show that $f'(x) = (\ln b)b^x$.
   b. Use the result of part (a) to find the derivative of $f(x) = 3^x$.

81. **Derivative of $\log_b x$**
   a. Let $f(x) = \log_b x$ $(b > 0, b \neq 1)$. Use the result of Exercise 80 to show that $f'(x) = \dfrac{1}{\ln b} \cdot \dfrac{1}{x}$.
   b. Use the result of part (a) to find the derivative of $f(x) = \log_{10} x$.

In Exercises 82–85, use the results of Exercises 80 and 81 to find the derivative of the given function.

82. $f(x) = x^3 2^x$

83. $g(x) = \dfrac{10^x}{x + 1}$

84. $h(x) = x^2 \log_{10} x$

85. $f(x) = 3^{x^2} + \log_2(x^2 + 1)$

In Exercises 86 and 87, determine whether the statement is true or false. If it is true, explain why it is true. If it is false, give an example to show why it is false.

86. If $f(x) = \ln 5$, then $f'(x) = \frac{1}{5}$.

87. If $f(x) = \ln a^x$, then $f'(x) = \ln a$.

88. Prove that $\dfrac{d}{dx} \ln|x| = \dfrac{1}{x}$ $(x \neq 0)$ for the case $x < 0$.

89. Use the definition of the derivative to show that

$$\lim_{x \to 0} \frac{\ln(x + 1)}{x} = 1$$

5.5 Solutions to Self-Check Exercises

1. The slope of the tangent line to the graph of f at any point $(x, f(x))$ lying on the graph of f is given by $f'(x)$. Using the product rule, we find

$$\begin{aligned} f'(x) &= \frac{d}{dx}[x \ln(2x + 3)] \\ &= x \frac{d}{dx} \ln(2x + 3) + \ln(2x + 3) \cdot \frac{d}{dx}(x) \\ &= x\left(\frac{2}{2x + 3}\right) + \ln(2x + 3) \cdot 1 \\ &= \frac{2x}{2x + 3} + \ln(2x + 3) \end{aligned}$$

In particular, the slope of the tangent line to the graph of f at the point $(-1, 0)$ is

$$f'(-1) = \frac{-2}{-2 + 3} + \ln 1 = -2$$

Therefore, using the point-slope form of the equation of a line, we see that a required equation is

$$\begin{aligned} y - 0 &= -2(x + 1) \\ y &= -2x - 2 \end{aligned}$$

2. Taking the logarithm on both sides of the equation gives

$$\begin{aligned} \ln y &= \ln(2x + 1)^3(3x + 4)^5 \\ &= \ln(2x + 1)^3 + \ln(3x + 4)^5 \\ &= 3 \ln(2x + 1) + 5 \ln(3x + 4) \end{aligned}$$

Differentiating both sides of the equation with respect to x, keeping in mind that y is a function of x, we obtain

$$\begin{aligned} \frac{d}{dx}(\ln y) = \frac{y'}{y} &= 3 \cdot \frac{2}{2x + 1} + 5 \cdot \frac{3}{3x + 4} \\ &= 3\left(\frac{2}{2x + 1} + \frac{5}{3x + 4}\right) \end{aligned}$$

and

$$y' = 3(2x + 1)^3(3x + 4)^5\left(\frac{2}{2x + 1} + \frac{5}{3x + 4}\right)$$

5.6 Exponential Functions as Mathematical Models

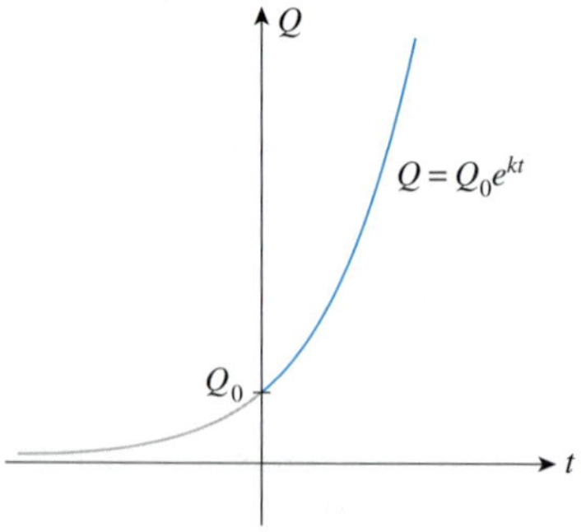

FIGURE **13**
Exponential growth

Exponential Growth

Many problems arising from practical situations can be described mathematically in terms of exponential functions or functions closely related to the exponential function. In this section, we look at some applications involving exponential functions from the fields of the life and social sciences.

In Section 5.1, we saw that the exponential function $f(x) = b^x$ is an increasing function when $b > 1$. In particular, the function $f(x) = e^x$ shares this property. From this result, one may deduce that the function $Q(t) = Q_0e^{kt}$, where Q_0 and k are positive constants, has the following properties:

1. $Q(0) = Q_0$
2. $Q(t)$ increases "rapidly" without bound as t increases without bound (Figure 13).

Property 1 follows from the computation

$$Q(0) = Q_0e^0 = Q_0$$

Next, to study the rate of change of the function $Q(t)$, we differentiate it with respect to t, obtaining

$$\begin{aligned} Q'(t) &= \frac{d}{dt}(Q_0e^{kt}) \\ &= Q_0\frac{d}{dt}(e^{kt}) \\ &= kQ_0e^{kt} \\ &= kQ(t) \end{aligned} \tag{12}$$

Since $Q(t) > 0$ (because Q_0 is assumed to be positive) and $k > 0$, we see that $Q'(t) > 0$ and so $Q(t)$ is an increasing function of t. Our computation has in fact shed more light on an important property of the function $Q(t)$. Equation (12) says that the rate of increase of the function $Q(t)$ is proportional to the amount $Q(t)$ of the quantity present at time t. The implication is that as $Q(t)$ increases, so does the *rate of increase* of $Q(t)$, resulting in a very rapid increase in $Q(t)$ as t increases without bound.

Thus, the exponential function

$$Q(t) = Q_0e^{kt} \qquad (0 \le t < \infty) \tag{13}$$

provides us with a mathematical model of a quantity $Q(t)$ that is initially present in the amount of $Q(0) = Q_0$ and whose rate of growth at any time t is directly proportional to the amount of the quantity present at time t. Such a quantity is said to exhibit **exponential growth**, and the constant k of proportionality is called the **growth constant.** Interest earned on a fixed deposit when compounded continuously exhibits exponential growth. Other examples of unrestricted exponential growth follow.

APPLIED EXAMPLE 1 Growth of Bacteria Under ideal laboratory conditions, the number of bacteria in a culture grows in accordance with the law $Q(t) = Q_0e^{kt}$, where Q_0 denotes the number of bacteria initially present in the culture, k is a constant determined by the strain of bacteria under consideration, and t is the elapsed time measured in hours. Suppose 10,000 bacteria are present initially in the culture and 60,000 present 2 hours later.

a. How many bacteria will there be in the culture at the end of 4 hours?
b. What is the rate of growth of the population after 4 hours?

Solution

a. We are given that $Q(0) = Q_0 = 10{,}000$, so $Q(t) = 10{,}000e^{kt}$. Next, the fact that 60,000 bacteria are present 2 hours later translates into $Q(2) = 60{,}000$. Thus,

$$60{,}000 = 10{,}000e^{2k}$$
$$e^{2k} = 6$$

Taking the natural logarithm on both sides of the equation, we obtain

$$\ln e^{2k} = \ln 6$$
$$2k = \ln 6 \qquad \text{Since } \ln e = 1$$
$$k \approx 0.8959$$

Thus, the number of bacteria present at any time t is given by

$$Q(t) = 10{,}000e^{0.8959t}$$

In particular, the number of bacteria present in the culture at the end of 4 hours is given by

$$Q(4) = 10{,}000e^{0.8959(4)}$$
$$\approx 360{,}029$$

b. The rate of growth of the bacteria population at any time t is given by

$$Q'(t) = kQ(t)$$

Thus, using the result from part (a), we find that the rate at which the population is growing at the end of 4 hours is

$$Q'(4) = kQ(4)$$
$$\approx (0.8959)(360{,}029)$$
$$\approx 322{,}550$$

or approximately 322,550 bacteria per hour.

Exponential Decay

In contrast to exponential growth, a quantity exhibits **exponential decay** if it decreases at a rate that is directly proportional to its size. Such a quantity may be described by the exponential function

$$Q(t) = Q_0e^{-kt} \qquad (0 \le t < \infty) \qquad \textbf{(14)}$$

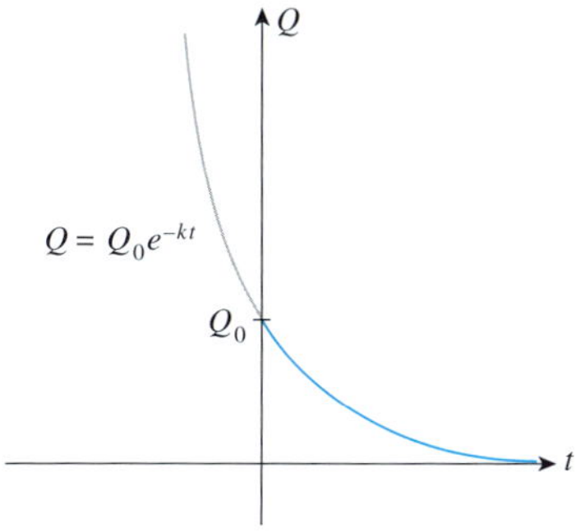

FIGURE 14
Exponential decay

where the positive constant Q_0 measures the amount present initially ($t = 0$) and k is some suitable positive number, called the **decay constant**. The choice of this number is determined by the nature of the substance under consideration. The graph of this function is sketched in Figure 14.

To verify the properties ascribed to the function $Q(t)$, we simply compute

$$Q(0) = Q_0e^0 = Q_0$$
$$Q'(t) = \frac{d}{dt}(Q_0e^{-kt})$$
$$= Q_0\frac{d}{dt}(e^{-kt})$$
$$= -kQ_0e^{-kt} = -kQ(t)$$

APPLIED EXAMPLE 2 Radioactive Decay Radioactive substances decay exponentially. For example, the amount of radium present at any time t obeys the law $Q(t) = Q_0e^{-kt}$, where Q_0 is the initial amount present and k is a suitable positive constant. The **half-life of a radioactive substance** is the time required for a given amount to be reduced by one-half. Now, it is known that the half-life of radium is approximately 1600 years. Suppose initially there are 200 milligrams of pure radium. Find the amount left after t years. What is the amount left after 800 years?

Solution The initial amount of radium present is 200 milligrams, so $Q(0) = Q_0 = 200$. Thus, $Q(t) = 200e^{-kt}$. Next, the datum concerning the half-life of radium implies that $Q(1600) = 100$, and this gives

$$100 = 200e^{-1600k}$$

$$e^{-1600k} = \frac{1}{2}$$

Taking the natural logarithm on both sides of this equation yields

$$-1600k \ln e = \ln \frac{1}{2}$$

$$-1600k = \ln \frac{1}{2} \qquad \ln e = 1$$

$$k = -\frac{1}{1600} \ln \left(\frac{1}{2}\right) = 0.0004332$$

Therefore, the amount of radium left after t years is

$$Q(t) = 200e^{-0.0004332t}$$

In particular, the amount of radium left after 800 years is

$$Q(800) = 200e^{-0.0004332(800)} \approx 141.42$$

or approximately 141 milligrams. ■

VIDEO

APPLIED EXAMPLE 3 Carbon-14 Decay Carbon 14, a radioactive isotope of carbon, has a half-life of 5770 years. What is its decay constant?

Solution We have $Q(t) = Q_0e^{-kt}$. Since the half-life of the element is 5770 years, half of the substance is left at the end of that period; that is,

$$Q(5770) = Q_0e^{-5770k} = \frac{1}{2}Q_0$$

$$e^{-5770k} = \frac{1}{2}$$

Taking the natural logarithm on both sides of this equation, we have

$$\ln e^{-5770k} = \ln \frac{1}{2}$$

$$-5770k = -0.693147$$

$$k \approx 0.00012$$

■

Carbon-14 dating is a well-known method used by anthropologists to establish the age of animal and plant fossils. This method assumes that the proportion of carbon 14 (C-14) present in the atmosphere has remained constant over the past 50,000 years.

Professor Willard Libby, recipient of the Nobel Prize in chemistry in 1960, proposed this theory.

The amount of C-14 in the tissues of a living plant or animal is constant. However, when an organism dies, it stops absorbing new quantities of C-14, and the amount of C-14 in the remains diminishes because of the natural decay of the radioactive substance. Thus, the approximate age of a plant or animal fossil can be determined by measuring the amount of C-14 present in the remains.

APPLIED EXAMPLE 4 Carbon-14 Dating A skull from an archeological site has one-tenth the amount of C-14 that it originally contained. Determine the approximate age of the skull.

Solution Here,

$$\begin{aligned} Q(t) &= Q_0 e^{-kt} \\ &= Q_0 e^{-0.00012t} \end{aligned}$$

where Q_0 is the amount of C-14 present originally and k, the decay constant, is equal to 0.00012 (see Example 3). Since $Q(t) = (1/10)Q_0$, we have

$$\begin{aligned} \frac{1}{10} Q_0 &= Q_0 e^{-0.00012t} \\ \ln \frac{1}{10} &= -0.00012t \qquad \text{Take the natural logarithm on both sides.} \\ t &= \frac{\ln \frac{1}{10}}{-0.00012} \\ &\approx 19{,}200 \end{aligned}$$

or approximately 19,200 years. ■

Learning Curves

The next example shows how the exponential function may be applied to describe certain types of learning processes. Consider the function

$$Q(t) = C - Ae^{-kt}$$

where C, A, and k are positive constants. To sketch the graph of the function Q, observe that its y-intercept is given by $Q(0) = C - A$. Next, we compute

$$Q'(t) = kAe^{-kt}$$

Since both k and A are positive, we see that $Q'(t) > 0$ for all values of t. Thus, $Q(t)$ is an increasing function of t. Also,

$$\begin{aligned} \lim_{t\to\infty} Q(t) &= \lim_{t\to\infty} (C - Ae^{-kt}) \\ &= \lim_{t\to\infty} C - \lim_{t\to\infty} Ae^{-kt} \\ &= C \end{aligned}$$

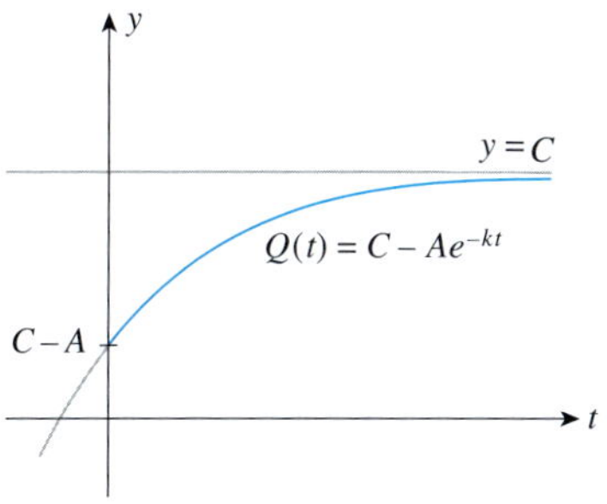

FIGURE 15
A learning curve

so $y = C$ is a horizontal asymptote of Q. Thus, $Q(t)$ increases and approaches the number C as t increases without bound. The graph of the function Q is shown in Figure 15, where that part of the graph corresponding to the negative values of t is drawn with a gray line since, in practice, one normally restricts the domain of the function to the interval $[0, \infty)$.

Observe that $Q(t)$ $(t > 0)$ increases rather rapidly initially but that the rate of increase slows down considerably after a while. To see this, we compute

$$\lim_{t \to \infty} Q'(t) = \lim_{t \to \infty} kAe^{-kt} = 0$$

This behavior of the graph of the function Q closely resembles the learning pattern experienced by workers engaged in highly repetitive work. For example, the productivity of an assembly-line worker increases very rapidly in the early stages of the training period. This productivity increase is a direct result of the worker's training and accumulated experience. But the rate of increase of productivity slows as time goes by, and the worker's productivity level approaches some fixed level due to the limitations of the worker and the machine. Because of this characteristic, the graph of the function $Q(t) = C - Ae^{-kt}$ is often called a **learning curve.**

APPLIED EXAMPLE 5 Assembly Time The Camera Division of Eastman Optical produces a 35-mm single-lens reflex camera. Eastman's training department determines that after completing the basic training program, a new, previously inexperienced employee will be able to assemble

$$Q(t) = 50 - 30e^{-0.5t}$$

model F cameras per day, t months after the employee starts work on the assembly line.

a. How many model F cameras can a new employee assemble per day after basic training?
b. How many model F cameras can an employee with 1 month of experience assemble per day? An employee with 2 months of experience? An employee with 6 months of experience?
c. How many model F cameras can the average experienced employee assemble per day?

Solution

a. The number of model F cameras a new employee can assemble is given by

$$Q(0) = 50 - 30 = 20$$

b. The number of model F cameras that an employee with 1 month of experience, 2 months of experience, and 6 months of experience can assemble per day is given by

$$Q(1) = 50 - 30e^{-0.5} \approx 31.80$$
$$Q(2) = 50 - 30e^{-1} \approx 38.96$$
$$Q(6) = 50 - 30e^{-3} \approx 48.51$$

or approximately 32, 39, and 49, respectively.

c. As t increases without bound, $Q(t)$ approaches 50. Hence, the average experienced employee can ultimately be expected to assemble 50 model F cameras per day. ■

Other applications of the learning curve are found in models that describe the dissemination of information about a product or the velocity of an object dropped into a viscous medium.

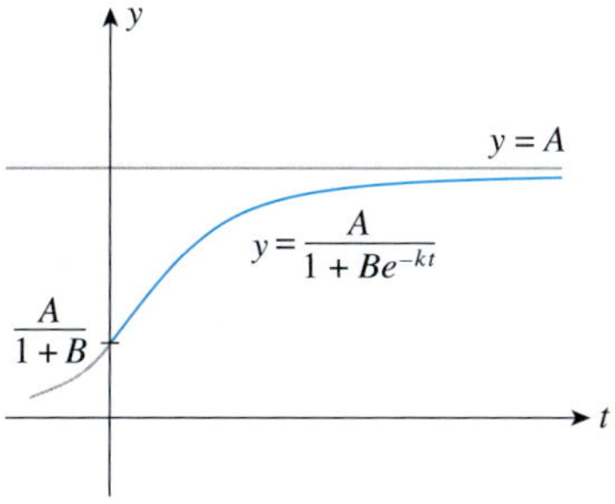

FIGURE 16
A logistic curve

Logistic Growth Functions

Our last example of an application of exponential functions to the description of natural phenomena involves the **logistic** (also called the **S-shaped,** or **sigmoidal**) **curve,** which is the graph of the function

$$Q(t) = \frac{A}{1 + Be^{-kt}}$$

where A, B, and k are positive constants. The function Q is called a **logistic growth function,** and the graph of the function Q is sketched in Figure 16.

Observe that $Q(t)$ increases rather rapidly for small values of t. In fact, for small values of t, the logistic curve resembles an exponential growth curve. However, the *rate of growth* of $Q(t)$ decreases quite rapidly as t increases and $Q(t)$ approaches the number A as t increases without bound.

Thus, the logistic curve exhibits both the property of rapid growth of the exponential growth curve as well as the "saturation" property of the learning curve. Because of these characteristics, the logistic curve serves as a suitable mathematical model for describing many natural phenomena. For example, if a small number of rabbits were introduced to a tiny island in the South Pacific, the rabbit population might be expected to grow very rapidly at first, but the growth rate would decrease quickly as overcrowding, scarcity of food, and other environmental factors affected it. The population would eventually stabilize at a level compatible with the life-support capacity of the environment. This level, given by A, is called the *carrying capacity* of the environment. Models describing the spread of rumors and epidemics are other examples of the application of the logistic curve.

VIDEO

APPLIED EXAMPLE 6 Spread of Flu The number of soldiers at Fort MacArthur who contracted influenza after t days during a flu epidemic is approximated by the exponential model

$$Q(t) = \frac{5000}{1 + 1249e^{-kt}}$$

If 40 soldiers contracted the flu by day 7, find how many soldiers contracted the flu by day 15.

Solution The given information implies that

$$Q(7) = \frac{5000}{1 + 1249e^{-7k}} = 40$$

Thus,

$$40(1 + 1249e^{-7k}) = 5000$$

$$1 + 1249e^{-7k} = \frac{5000}{40} = 125$$

$$e^{-7k} = \frac{124}{1249}$$

$$-7k = \ln \frac{124}{1249}$$

$$k = -\frac{\ln \frac{124}{1249}}{7} \approx 0.33$$

Therefore, the number of soldiers who contracted the flu after t days is given by

$$Q(t) = \frac{5000}{1 + 1249e^{-0.33t}}$$

In particular, the number of soldiers who contracted the flu by day 15 is given by

$$Q(15) = \frac{5000}{1 + 1249e^{-15(0.33)}} \approx 508$$

or approximately 508 soldiers.

Exploring with TECHNOLOGY

Refer to Example 6.

1. Use a graphing utility to plot the graph of the function Q, using the viewing window $[0, 40] \times [0, 5000]$.
2. Find how long it takes for the first 1000 soldiers to contract the flu.
   Hint: Plot the graphs of $y_1 = Q(t)$ and $y_2 = 1000$ and find the point of intersection of the two graphs.

5.6 Self-Check Exercise

Suppose the population (in millions) of a country at any time t grows in accordance with the rule

$$P = \left(P_0 + \frac{I}{k}\right)e^{kt} - \frac{I}{k}$$

where P denotes the population at any time t, k is a constant reflecting the natural growth rate of the population, I is a constant giving the (constant) rate of immigration into the country, and P_0 is the total population of the country at time $t = 0$. The population of the United States in 1980 ($t = 0$) was 226.5 million. If the natural growth rate is 0.8% annually ($k = 0.008$) and net immigration is allowed at the rate of half a million people per year ($I = 0.5$), what is the expected population of the United States in 2010?

The solution to Self-Check Exercise 5.6 can be found on page 390.

5.6 Concept Questions

1. Give the model for unrestricted exponential growth and the model for exponential decay. What effect does the magnitude of the growth (decay) constant have on the growth (decay) of a quantity?
2. What is the half-life of a radioactive substance?
3. What is the logistic growth function? What are its characteristics?

5.6 Exercises

1. **Exponential Growth** Given that a quantity $Q(t)$ is described by the exponential growth function

$$Q(t) = 400e^{0.05t}$$

where t is measured in minutes, answer the following questions:

a. What is the growth constant?
b. What quantity is present initially?
c. Complete the following table of values:

t	0	10	20	100	1000
Q					

2. **Exponential Decay** Given that a quantity $Q(t)$ exhibiting exponential decay is described by the function

$$Q(t) = 2000e^{-0.06t}$$

where t is measured in years, answer the following questions:
   a. What is the decay constant?
   b. What quantity is present initially?
   c. Complete the following table of values:

t	0	5	10	20	100
Q					

3. **Growth of Bacteria** The growth rate of the bacterium *Escherichia coli*, a common bacterium found in the human intestine, is proportional to its size. Under ideal laboratory conditions, when this bacterium is grown in a nutrient broth medium, the number of cells in a culture doubles approximately every 20 min.
   a. If the initial cell population is 100, determine the function $Q(t)$ that expresses the exponential growth of the number of cells of this bacterium as a function of time t (in minutes).
   b. How long will it take for a colony of 100 cells to increase to a population of 1 million?
   c. If the initial cell population were 1000, how would this alter our model?

4. **World Population** The world population at the beginning of 1990 was 5.3 billion. Assume that the population continues to grow at the rate of approximately 2%/year and find the function $Q(t)$ that expresses the world population (in billions) as a function of time t (in years), with $t = 0$ corresponding to the beginning of 1990.
   a. Using this function, complete the following table of values and sketch the graph of the function Q.

Year	1990	1995	2000	2005
World Population				

Year	2010	2015	2020	2025
World Population				

   b. Find the estimated rate of growth in 2010.

5. **World Population** Refer to Exercise 4.
   a. If the world population continues to grow at the rate of approximately 2%/year, find the length of time t_0 required for the world population to triple in size.
   b. Using the time t_0 found in part (a), what would be the world population if the growth rate were reduced to 1.8%/year?

6. **Resale Value** A certain piece of machinery was purchased 3 yr ago by Garland Mills for \$500,000. Its present resale value is \$320,000. Assuming that the machine's resale value decreases exponentially, what will it be 4 yr from now?

7. **Atmospheric Pressure** If the temperature is constant, then the atmospheric pressure P (in pounds/square inch) varies with the altitude above sea level h in accordance with the law

$$P = p_0 e^{-kh}$$

where p_0 is the atmospheric pressure at sea level and k is a constant. If the atmospheric pressure is 15 lb/in.^{2} at sea level and 12.5 lb/in.^{2} at 4000 ft, find the atmospheric pressure at an altitude of 12,000 ft. How fast is the atmospheric pressure changing with respect to altitude at an altitude of 12,000 ft?

8. **Radioactive Decay** The radioactive element polonium decays according to the law

$$Q(t) = Q_0 \cdot 2^{-(t/140)}$$

where Q_0 is the initial amount and the time t is measured in days. If the amount of polonium left after 280 days is 20 mg, what was the initial amount present?

9. **Radioactive decay** Phosphorus 32 (P-32) has a half-life of 14.2 days. If 100 g of this substance are present initially, find the amount present after t days. What amount will be left after 7.1 days? How fast is P-32 decaying when $t = 7.1$?

10. **Nuclear Fallout** Strontium 90 (Sr-90), a radioactive isotope of strontium, is present in the fallout resulting from nuclear explosions. It is especially hazardous to animal life, including humans, because, upon ingestion of contaminated food, it is absorbed into the bone structure. Its half-life is 27 yr. If the amount of Sr-90 in a certain area is found to be four times the "safe" level, find how much time must elapse before an "acceptable level" is reached.

11. **Carbon-14 Dating** Wood deposits recovered from an archeological site contain 20% of the C-14 they originally contained. How long ago did the tree from which the wood was obtained die?

12. **Carbon-14 Dating** Skeletal remains of the so-called Pittsburgh Man, unearthed in Pennsylvania, had lost 82% of the C-14 they originally contained. Determine the approximate age of the bones.

13. **Learning Curves** The American Court Reporting Institute finds that the average student taking Advanced Machine Shorthand, an intensive 20-wk course, progresses according to the function

$$Q(t) = 120(1 - e^{-0.05t}) + 60 \qquad (0 \le t \le 20)$$

where $Q(t)$ measures the number of words (per minute) of dictation that the student can take in machine shorthand after t wk in the course. Sketch the graph of the function Q and answer the following questions:
   a. What is the beginning shorthand speed for the average student in this course?
   b. What shorthand speed does the average student attain halfway through the course?
   c. How many words per minute can the average student take after completing this course?

14. **People Living with HIV** Based on data compiled by WHO, the number of people living with HIV (human immunodeficiency virus) worldwide from 1985 through 2006 is estimated to be

$$N(t) = \frac{39.88}{1 + 18.94e^{-0.2957t}} \qquad (0 \le t \le 21)$$

where $N(t)$ is measured in millions and t in years, with $t = 0$ corresponding to the beginning of 1985.
   a. How many people were living with HIV worldwide at the beginning of 1985? At the beginning of 2005?
   b. Assuming that the trend continued, how many people were living with HIV worldwide at the beginning of 2008?
   Source: World Health Organization

15. **Federal Debt** According to data obtained from the CBO, the total federal debt (in trillions of dollars) from 2001 through 2006 is given by

$$f(t) = 5.37e^{0.078t} \qquad (1 \le t \le 6)$$

where t is measured in years, with $t = 1$ corresponding to 2001.
   a. What was the total federal debt in 2001? In 2006?
   b. How fast was the total federal debt increasing in 2001? In 2006?
   Source: Congressional Budget Office

16. **Effect of Advertising on Sales** Metro Department Store found that t wk after the end of a sales promotion the volume of sales was given by

$$S(t) = B + Ae^{-kt} \qquad (0 \le t \le 4)$$

where $B = 50{,}000$ and is equal to the average weekly volume of sales before the promotion. The sales volumes at the end of the first and third weeks were \$83,515 and \$65,055, respectively. Assume that the sales volume is decreasing exponentially.
   a. Find the decay constant k.
   b. Find the sales volume at the end of the fourth week.
   c. How fast is the sales volume dropping at the end of the fourth week?

17. **Demand for Computers** Universal Instruments found that the monthly demand for its new line of Galaxy Home Computers t mo after placing the line on the market was given by

$$D(t) = 2000 - 1500e^{-0.05t} \qquad (t > 0)$$

Graph this function and answer the following questions:
   a. What is the demand after 1 mo? After 1 yr? After 2 yr? After 5 yr?
   b. At what level is the demand expected to stabilize?
   c. Find the rate of growth of the demand after the tenth month.

18. **Reliability of Computer Chips** The percentage of a certain brand of computer chips that will fail after t yr of use is estimated to be

$$P(t) = 100(1 - e^{-0.1t})$$

   a. What percentage of this brand of computer chips are expected to be usable after 3 yr?
   b. Evaluate $\lim_{t\to\infty} P(t)$. Did you expect this result?

19. **Lengths of Fish** The length (in centimeters) of a typical Pacific halibut t yr old is approximately

$$f(t) = 200(1 - 0.956e^{-0.18t})$$

   a. What is the length of a typical 5-yr-old Pacific halibut?
   b. How fast is the length of a typical 5-yr-old Pacific halibut increasing?
   c. What is the maximum length a typical Pacific halibut can attain?

20. **Spread of an Epidemic** During a flu epidemic, the number of children in the Woodbridge Community School System who contracted influenza after t days was given by

$$Q(t) = \frac{1000}{1 + 199e^{-0.8t}}$$

   a. How many children were stricken by the flu after the first day?
   b. How many children had the flu after 10 days?
   c. How many children eventually contracted the disease?

21. **Lay Teachers at Roman Catholic Schools** The change from religious to lay teachers at Roman Catholic schools has been partly attributed to the decline in the number of women and men entering religious orders. The percentage of teachers who are lay teachers is given by

$$f(t) = \frac{98}{1 + 2.77e^{-t}} \qquad (0 \le t \le 4)$$

where t is measured in decades, with $t = 0$ corresponding to the beginning of 1960.
   a. What percentage of teachers were lay teachers at the beginning of 1990?
   b. How fast was the percentage of lay teachers changing at the beginning of 1990?
   c. Find the year when the percentage of lay teachers was increasing most rapidly.
   Sources: National Catholic Education Association and the Department of Education

22. **Growth of a Fruit Fly Population** On the basis of data collected during an experiment, a biologist found that the growth of a fruit fly (*Drosophila*) with a limited food supply could be approximated by the exponential model

$$N(t) = \frac{400}{1 + 39e^{-0.16t}}$$

where t denotes the number of days since the beginning of the experiment.
   a. What was the initial fruit fly population in the experiment?
   b. What was the maximum fruit fly population that could be expected under this laboratory condition?
   c. What was the population of the fruit fly colony on the 20th day?
   d. How fast was the population changing on the 20th day?

23. **Demographics** The number of citizens aged 45–64 yr is projected to be

$$P(t) = \frac{197.9}{1 + 3.274e^{-0.0361t}} \qquad (0 \le t \le 20)$$

where $P(t)$ is measured in millions and t is measured in years, with $t = 0$ corresponding to the beginning of 1990. People belonging to this age group are the targets of insurance companies that want to sell them annuities. What is the projected population of citizens aged 45–64 yr in 2010?

Source: K. G. Securities

24. **Population Growth in the 21st Century** The U.S. population is approximated by the function

$$P(t) = \frac{616.5}{1 + 4.02e^{-0.5t}}$$

where $P(t)$ is measured in millions of people and t is measured in 30-yr intervals, with $t = 0$ corresponding to 1930. What is the expected population of the United States in 2020 ($t = 3$)?

25. **Dissemination of Information** Three hundred students attended the dedication ceremony of a new building on a college campus. The president of the traditionally female college announced a new expansion program, which included plans to make the college coeducational. The number of students who learned of the new program t hr later is given by the function

$$f(t) = \frac{3000}{1 + Be^{-kt}}$$

If 600 students on campus had heard about the new program 2 hr after the ceremony, how many students had heard about the policy after 4 hr? How fast was the news spreading 4 hr after the ceremony?

26. **Price of a Commodity** The unit price of a certain commodity is given by

$$p = f(t) = 6 + 4e^{-2t}$$

where p is measured in dollars and t is measured in months.

a. Show that f is decreasing on $(0, \infty)$.
b. Show that the graph of f is concave upward on $(0, \infty)$.
c. Evaluate $\lim_{t \to \infty} f(t)$. (*Note:* This value is called the *equilibrium price* of the commodity, and in this case, we have *price stability*.)
d. Sketch the graph of f.

27. **Chemical Mixtures** Two chemicals react to form another chemical. Suppose the amount of the chemical formed in time t (in hours) is given by

$$x(t) = \frac{15\left[1 - \left(\frac{2}{3}\right)^{3t}\right]}{1 - \frac{1}{4}\left(\frac{2}{3}\right)^{3t}}$$

where $x(t)$ is measured in pounds. How many pounds of the chemical are formed eventually?

Hint: You need to evaluate $\lim_{t \to \infty} x(t)$.

28. **Von Bertalanffy Growth Function** The length (in centimeters) of a common commercial fish is approximated by the von Bertalanffy growth function

$$f(t) = a(1 - be^{-kt})$$

where a, b, and k are positive constants.

a. Show that f is increasing on the interval $(0, \infty)$.
b. Show that the graph of f is concave downward on $(0, \infty)$.
c. Show that $\lim_{t \to \infty} f(t) = a$.
d. Use the results of parts (a)–(c) to sketch the graph of f.

29. **Absorption of Drugs** The concentration of a drug in grams/cubic centimeter (g/cm^3) t min after it has been injected into the bloodstream is given by

$$C(t) = \frac{k}{b - a}(e^{-at} - e^{-bt})$$

where a, b, and k are positive constants, with $b > a$.

a. At what time is the concentration of the drug the greatest?
b. What will be the concentration of the drug in the long run?

30. **Concentration of Glucose in the Bloodstream** A glucose solution is administered intravenously into the bloodstream at a constant rate of r mg/hr. As the glucose is being administered, it is converted into other substances and removed from the bloodstream. Suppose the concentration of the glucose solution at time t is given by

$$C(t) = \frac{r}{k} - \left[\left(\frac{r}{k}\right) - C_0\right]e^{-kt}$$

where C_0 is the concentration at time $t = 0$ and k is a positive constant. Assuming that $C_0 < r/k$, evaluate $\lim_{t \to \infty} C(t)$.

a. What does your result say about the concentration of the glucose solution in the long run?
b. Show that the function C is increasing on $(0, \infty)$.
c. Show that the graph of C is concave downward on $(0, \infty)$.
d. Sketch the graph of the function C.

31. **Radioactive Decay** A radioactive substance decays according to the formula

$$Q(t) = Q_0e^{-kt}$$

where $Q(t)$ denotes the amount of the substance present at time t (measured in years), Q_0 denotes the amount of the substance present initially, and k (a positive constant) is the decay constant.

a. Show that half-life of the substance is $\bar{t} = \ln 2/k$.
b. Suppose a radioactive substance decays according to the formula

$$Q(t) = 20e^{-0.0001238\bar{t}}$$

How long will it take for the substance to decay to half the original amount?

32. **Logistic Growth Function** Consider the logistic growth function

$$Q(t) = \frac{A}{1 + Be^{-kt}}$$

where A, B, and k are positive constants.

a. Show that Q satisfies the equation

$$Q'(t) = kQ\left(1 - \frac{Q}{A}\right)$$

b. Show that $Q(t)$ is increasing on $[0, \infty)$.

33. **a.** Use the results of Exercise 32 to show that the graph of Q has an inflection point when $Q = \frac{A}{2}$ and that this occurs when $t = \frac{\ln B}{k}$.

b. Interpret your results.

34. **Logistic Growth Function** Consider the logistic growth function

$$Q(t) = \frac{A}{1 + Be^{-kt}}$$

Suppose the population is Q_1 when $t = t_1$ and Q_2 when $t = t_2$. Show that the value of k is

$$k = \frac{1}{t_2 - t_1} \ln \left[\frac{Q_2(A - Q_1)}{Q_1(A - Q_2)}\right]$$

35. **Logistic Growth Function** The carrying capacity of a colony of fruit flies (*Drosophila*) is 600. The population of fruit flies after 14 days is 76, and the population after 21 days is 167. What is the value of the growth constant k?
Hint: Use the result of Exercise 34.

36. **Gompertz Growth Curve** Consider the function

$$Q(t) = Ce^{-Ae^{-kt}}$$

where $Q(t)$ is the size of a quantity at time t and A, C, and k are positive constants. The graph of this function, called the *Gompertz growth curve*, is used by biologists to describe restricted population growth.

a. Show that the function Q is always increasing.

b. Find the time t at which the growth rate $Q'(t)$ is increasing most rapidly.
Hint: Find the inflection point of Q.

c. Show that $\lim_{t \to \infty} Q(t) = C$ and interpret your result.

5.6 Solution to Self-Check Exercise

We are given that $P_0 = 226.5$, $k = 0.008$, and $I = 0.5$. So

$$P = \left(226.5 + \frac{0.5}{0.008}\right)e^{0.008t} - \frac{0.5}{0.008}$$
$$= 289e^{0.008t} - 62.5$$

Therefore, the expected population in 2010 is given by

$$P(30) = 289e^{0.24} - 62.5$$
$$\approx 304.9$$

or approximately 304.9 million.

USING TECHNOLOGY

Analyzing Mathematical Models

We can use a graphing utility to analyze the mathematical models encountered in this section.

APPLIED EXAMPLE 1 Internet Gaming Sales The estimated growth in global Internet-gaming revenue (in billions of dollars), as predicted by industry analysts, is given in the following table:

Year	2001	2002	2003	2004	2005	2006	2007	2008	2009	2010
Revenue	3.1	3.9	5.6	8.0	11.8	15.2	18.2	20.4	22.7	24.5

a. Use **Logistic** to find a regression model for the data. Let $t = 0$ correspond to 2001.

b. Plot the scatter diagram and the graph of the function f found in part (a) using the viewing window $[0, 9] \times [0, 30]$.

c. How fast was the revenue from global gaming on the Internet changing in 2001? In 2007?

d. At what time was the revenue from global gaming on the Internet increasing most rapidly?

Source: Christiansen Capital/Advisors

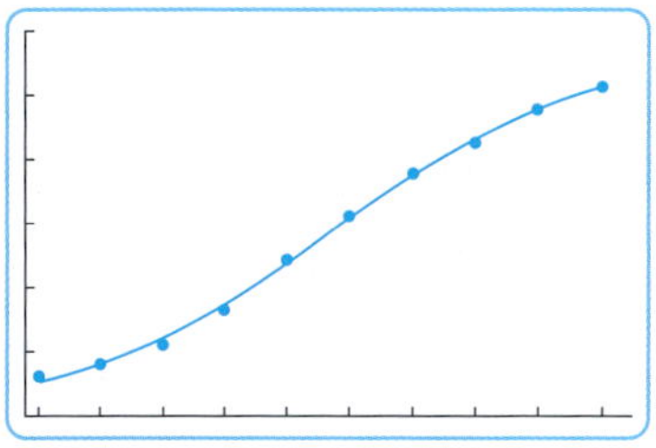

FIGURE T1
The graph of f in the viewing window $[0, 9] \times [0, 30]$

Solution

a. Using **Logistic** we find

$$f(t) = \frac{27.11}{1 + 9.64e^{-0.49t}} \qquad (0 \le t \le 9)$$

b. The scatter plot for the data, and the graph of f in the viewing window $[0, 9] \times [0, 30]$ are shown in Figure T1.

c. Using the numerical derivative operation, we find $f'(0) \approx 1.131$ and $f'(6) \approx 2.971$. We conclude that the revenue was increasing at the rate of \$1.13 billion/yr in 2001, and at the rate of \$2.97 billion/yr in 2007.

d. Using INFLC (TI-85), or otherwise, we see that the inflection point of f occurs at (4.62, 13.56). Thus the largest increase in the revenue occurred at the time when $t \approx 4.6$, or a little past July of 2005. ■

TECHNOLOGY EXERCISES

1. **Online Banking** In a study prepared in 2000, the percentage of households using online banking was projected to be

$$f(t) = 1.5e^{0.78t} \qquad (0 \le t \le 4)$$

where t is measured in years, with $t = 0$ corresponding to the beginning of 2000.

a. Plot the graph of f, using the viewing window $[0, 4] \times [0, 40]$.

b. How fast was the projected percentage of households using online banking changing at the beginning of 2003?

c. How fast was the rate of the projected percentage of households using online banking changing at the beginning of 2003?
Hint: We want $f''(3)$. Why?

Source: Online Banking Report

2. **Newton's Law of Cooling** The temperature of a cup of coffee t min after it is poured is given by

$$T = 70 + 100e^{-0.0446t}$$

where T is measured in degrees Fahrenheit.

a. Plot the graph of T, using the viewing window $[0, 30] \times [0, 200]$.

b. When will the coffee be cool enough to drink (say, 120°)?
Hint: Use the ISECT function.

3. **Air Travel** Air travel has been rising dramatically in the past 30 yr. In a study conducted in 2000, the FAA projected further exponential growth for air travel through 2010. The function

$$f(t) = 666e^{0.0413t} \qquad (0 \le t \le 10)$$

gives the number of passengers (in millions) in year t, with $t = 0$ corresponding to 2000.

a. Plot the graph of f, using the viewing window $[0, 10] \times [0, 1000]$.

b. How many air passengers were there in 2000? What was the projected number of air passengers for 2008?

c. What was the rate of change of the number of air passengers in 2008?

Source: Federal Aviation Administration

4. **Computer Game Sales** The total number of Starr Communication's newest game, Laser Beams, sold t mo after its release is given by

$$N(t) = -20(t + 20)e^{-0.05t} + 400$$

thousand units.

a. Plot the graph of N, using the viewing window $[0, 500] \times [0, 500]$.

b. Use the result of part (a) to find $\lim_{t \to \infty} N(t)$ and interpret this result.

5. **Population Growth in the 21st Century** The U.S. population is approximated by the function

$$P(t) = \frac{616.5}{1 + 4.02e^{-0.5t}}$$

where $P(t)$ is measured in millions of people and t is measured in 30-yr intervals, with $t = 0$ corresponding to 1930.

a. Plot the graph of f, using the viewing window $[0, 4] \times [0, 650]$.

b. What is the expected population of the United States in 2020 ($t = 3$)?

c. What is the expected rate of growth of the U.S. population in 2020?

6. **Time Rate of Growth of a Tumor** The rate at which a tumor grows, with respect to time, is given by

$$R = Ax \ln \frac{B}{x} \qquad (\text{for } 0 < x < B)$$

where A and B are positive constants and x is the radius of the tumor.

a. Plot the graph of R for the case $A = B = 10$.

b. Find the radius of the tumor when the tumor is growing most rapidly with respect to time.

7. **Absorption of Drugs** The concentration of a drug in an organ at any time t (in seconds) is given by

$$C(t) = \begin{cases} 0.3t - 18(1 - e^{-t/60}) & \text{if } 0 \le t \le 20 \\ 18e^{-t/60} - 12e^{-(t-20)/60} & \text{if } t > 20 \end{cases}$$

where $C(t)$ is measured in grams/cubic centimeter (g/cm^3).

a. Plot the graph of C, using the viewing window $[0, 120] \times [0, 1]$.

b. What is the initial concentration of the drug in the organ?

c. What is the concentration of the drug in the organ after 10 sec?

d. What is the concentration of the drug in the organ after 30 sec?

e. What will be the concentration of the drug in the long run?

8. **Annuities** At the time of retirement, Christine expects to have a sum of \$500,000 in her retirement account. Assuming that the account pays interest at the rate of 5%/year compounded continuously, her accountant pointed out to her that if she made withdrawals amounting to x dollars per year ($x > 25{,}000$), then the time required to deplete her savings would be T years, where

$$T = f(x) = 20 \ln\left(\frac{x}{x - 25{,}000}\right) \qquad (x > 25{,}000)$$

a. Plot the graph of f, using the viewing window $[25{,}000, 50{,}000] \times [0, 100]$.

b. How much should Christine plan to withdraw from her retirement account each year if she wants it to last for 25 yr?

c. Evaluate $\lim_{x \to 25{,}000^+} f(x)$. Is the result expected? Explain.

d. Evaluate $\lim_{x \to \infty} f(x)$. Is the result expected? Explain.

9. **Households with Microwaves** The number of households with microwave ovens increased greatly in the 1980s and 1990s. The percentage of households with microwave ovens from 1981 through 1999 is given by

$$f(t) = \frac{87}{1 + 4.209e^{-0.3727t}} \qquad (0 \le t \le 18)$$

where t is measured in years, with $t = 0$ corresponding to the beginning of 1981.

a. Use a graphing utility to plot the graph of f on the interval $[0, 18]$.

b. What percentage of households owned microwave ovens at the beginning of 1984? At the beginning of 1994?

c. At what rate was the ownership of microwave ovens increasing at the beginning of 1984? At the beginning of 1994?

d. At what time was the increase in ownership of microwave ovens greatest?

Source: Energy Information Agency

10. **Modeling with Data** The snowfall accumulation at Logan Airport (in inches), t hr after a 33-hr snowstorm in Boston in 1995, follows:

Hour	0	3	6	9	12	15	18	21	24	27	30	33
Inches	0.1	0.4	3.6	6.5	9.1	14.4	19.5	22	23.6	24.8	26.6	27

Here, $t = 0$ corresponds to noon of February 6.

a. Use **Logistic** to find a regression model for the data.

b. Plot the scatter diagram and the graph of the function f found in part (a), using the viewing window $[0, 33] \times [0, 30]$.

c. How fast was the snowfall accumulating at midnight on February 6? At noon on February 7?

d. At what time during the storm was the snowfall accumulating at the greatest rate? What was the rate of accumulation?

Source: Boston Globe

11. **Worldwide PC Shipments** According to an IDC forecast made in 2007, worldwide PC shipments (in millions of units) from 2005 through 2009 is given in the following table:

Year	2005	2006	2007	2008	2009
PCs	207.1	226.2	252.9	283.3	302.4

a. Use **Logistic** to find a regression model for the data. Let $t = 0$ correspond to 2005.

b. Plot the graph of the function f found in part (a), using the viewing window $[0, 4] \times [200, 300]$.

c. How fast were the worldwide PC shipments increasing in 2006? In 2008?

Source: International Data Corporation

12. **Federal Debt** According to data obtained from the CBO, the total federal debt (in trillions of dollars) from 2001 through 2006 is given in the following table:

Year	2001	2002	2003	2004	2005	2006
Debt	5.81	6.23	6.78	7.40	7.93	8.51

a. Use **ExpReg** to find a regression model for the data. Let $t = 1$ correspond to 2001.

b. Plot the graph of the function f found in part (a), using the viewing window $[1, 6] \times [4, 10]$.

Source: Congressional Budget Office

CHAPTER 5 Summary of Principal Formulas and Terms

FORMULAS

1. Exponential function with base b	$y = b^x$
2. The number e	$e = \lim_{m \to \infty} \left(1 + \frac{1}{m}\right)^m = 2.71828\ldots$
3. Exponential function with base e	$y = e^x$
4. Logarithmic function with base b	$y = \log_b x$
5. Logarithmic function with base e	$y = \ln x$
6. Inverse properties of $\ln x$ and e^x	$\ln e^x = x$ and $e^{\ln x} = x$
7. Compound interest (accumulated amount)	$A = P\left(1 + \frac{r}{m}\right)^{mt}$
8. Effective rate of interest	$r_{\text{eff}} = \left(1 + \frac{r}{m}\right)^m - 1$
9. Compound interest (present value)	$P = A\left(1 + \frac{r}{m}\right)^{-mt}$
10. Continuous compound interest	$A = Pe^{rt}$
11. Derivative of the exponential function	$\frac{d}{dx}(e^x) = e^x$
12. Chain rule for exponential functions	$\frac{d}{dx}(e^u) = e^u \frac{du}{dx}$
13. Derivative of the logarithmic function	$\frac{d}{dx} \ln\lvert x\rvert = \frac{1}{x}$
14. Chain rule for logarithmic functions	$\frac{d}{dx} \ln\lvert u\rvert = \frac{1}{u}\frac{du}{dx}$

TERMS

common logarithm (338)
natural logarithm (338)
compound interest (346)
logarithmic differentiation (374)
exponential growth (380)
growth constant (380)
exponential decay (381)
decay constant (381)
half-life of a radioactive substance (382)
logistic growth function (385)

CHAPTER 5 Concept Review Questions

Fill in the blanks.

1. The function $f(x) = x^b$ (b, a real number) is called a/an _____ function, whereas the function $g(x) = b^x$, where $b >$ _____ and $b \neq$ _____, is called a/an _____ function.

2. **a.** The domain of the function $y = 3^x$ is _____, and its range is _____.
   b. The graph of the function $y = 0.3^x$ passes through the point _____ and is decreasing on _____.

3. **a.** If $b > 0$ and $b \neq 1$, then the logarithmic function $y = \log_b x$ has domain _____ and range _____; its graph passes through the point _____.
   b. The graph of $y = \log_b x$ is decreasing if b _____ and increasing if b _____.

4. **a.** If $x > 0$, then $e^{\ln x} =$ _____.
   b. If x is any real number, then $\ln e^x =$ _____.

5. In the compound interest formula $A = P(1 + \frac{r}{m})^{mt}$, A stands for the _____ _____, P stands for the _____, r stands for the _____ _____ _____ per year, m stands for the _____ _____ _____ _____ per year, and t stands for the _____.

6. The effective rate r_{eff} is related to the nominal interest rate r per year and the number of conversion periods per year by $r_{\text{eff}} =$ _____.

7. If interest earned at the rate of r per year is compounded continuously over t years, then a principal of P dollars will have an accumulated value of $A =$ _____ dollars.

8. a. If $g(x) = e^{f(x)}$, where f is a differentiable function, then $g'(x) =$ _____.

b. If $g(x) = \ln f(x)$, where $f(x) > 0$ is differentiable, then $g'(x) =$ _____.

9. a. In the unrestricted exponential growth model $Q = Q_0e^{kt}$, Q_0 represents the quantity present _____, and k is called the _____ constant.

b. In the exponential decay model $Q = Q_0e^{-kt}$, k is called the _____ constant.

c. The half-life of a radioactive substance is the _____ required for a substance to decay to _____ _____ of its original amount.

10. a. For the model $Q(t) = C - Ae^{-kt}$ describing a learning curve, $y = C$ is a/an _____ _____ of the graph of Q. The value of $Q(t)$ never exceeds _____.

b. For the logistic growth model $Q(t) = \dfrac{A}{1 + Be^{-kt}}$, $y = A$ is a/an _____ _____ of the graph of Q. If the quantity $Q(t)$ is initially smaller than A, then $Q(t)$ will eventually approach _____ as t increases; the number A represents the life-support capacity of the environment and is called the _____ _____ of the environment.

CHAPTER 5 Review Exercises

1. Sketch on the same set of coordinate axes the graphs of the exponential functions defined by the equations.

a. $y = 2^{-x}$ **b.** $y = \left(\dfrac{1}{2}\right)^x$

In Exercises 2 and 3, express each equation in logarithmic form.

2. $\left(\dfrac{2}{3}\right)^{-3} = \dfrac{27}{8}$ **3.** $16^{-3/4} = 0.125$

In Exercises 4 and 5, solve each equation for *x*.

4. $\log_4(2x + 1) = 2$

5. $\ln(x - 1) + \ln 4 = \ln(2x + 4) - \ln 2$

In Exercises 6–8, given that ln 2 = *x*, ln 3 = *y*, and ln 5 = *z*, express each of the given logarithmic values in terms of *x*, *y*, and *z*.

6. $\ln 30$ **7.** $\ln 3.6$ **8.** $\ln 75$

9. Sketch the graph of the function $y = \log_2(x + 3)$.

10. Sketch the graph of the function $y = \log_3(x + 1)$.

11. A sum of \$10,000 is deposited in a bank. Find the amount on deposit after 2 yr if the bank pays interest at the rate of 6%/year compounded (a) daily (assume a 365-day year) and (b) continuously.

12. Find the interest rate needed for an investment of \$10,000 to grow to an amount of \$12,000 in 3 yr if interest is compounded quarterly.

13. How long will it take for an investment of \$10,000 to grow to \$15,000 if the investment earns interest at the rate of 6%/year compounded quarterly?

14. Find the nominal interest rate that yields an effective interest rate of 8%/year compounded quarterly.

In Exercises 15–32, find the derivative of the function.

15. $f(x) = xe^{2x}$ **16.** $f(t) = \sqrt{t}e^t + t$

17. $g(t) = \sqrt{t}e^{-2t}$ **18.** $g(x) = e^x\sqrt{1 + x^2}$

19. $y = \dfrac{e^{2x}}{1 + e^{-2x}}$ **20.** $f(x) = e^{2x^2-1}$

21. $f(x) = xe^{-x^2}$ **22.** $g(x) = (1 + e^{2x})^{3/2}$

23. $f(x) = x^2e^x + e^x$ **24.** $g(t) = t \ln t$

25. $f(x) = \ln(e^{x^2} + 1)$ **26.** $f(x) = \dfrac{x}{\ln x}$

27. $f(x) = \dfrac{\ln x}{x + 1}$ **28.** $y = (x + 1)e^x$

29. $y = \ln(e^{4x} + 3)$ **30.** $f(r) = \dfrac{re^r}{1 + r^2}$

31. $f(x) = \dfrac{\ln x}{1 + e^x}$ **32.** $g(x) = \dfrac{e^{x^2}}{1 + \ln x}$

33. Find the second derivative of the function $y = \ln(3x + 1)$.

34. Find the second derivative of the function $y = x \ln x$.

35. Find $h'(0)$ if $h(x) = g(f(x))$, $g(x) = x + \dfrac{1}{x}$, and $f(x) = e^x$.

36. Find $h'(1)$ if $h(x) = g(f(x))$, $g(x) = \dfrac{x + 1}{x - 1}$, and $f(x) = \ln x$.

37. Use logarithmic differentiation to find the derivative of $f(x) = (2x^3 + 1)(x^2 + 2)^3$.

38. Use logarithmic differentiation to find the derivative of

$$f(x) = \frac{x(x^2 - 2)^2}{(x - 1)}$$

39. Find an equation of the tangent line to the graph of $y = e^{-2x}$ at the point $(1, e^{-2})$.

40. Find an equation of the tangent line to the graph of $y = xe^{-x}$ at the point $(1, e^{-1})$.

41. Sketch the graph of the function $f(x) = xe^{-2x}$.

42. Sketch the graph of the function $f(x) = x^2 - \ln x$.

43. Find the absolute extrema of the function $f(t) = te^{-t}$.

44. Find the absolute extrema of the function

$$g(t) = \frac{\ln t}{t}$$

on [1, 2].

45. Investment Return A hotel was purchased by a conglomerate for \$4.5 million and sold 5 yr later for \$8.2 million. Find the annual rate of return (compounded continuously).

46. Find the present value of \$30,000 due in 5 yr at an interest rate of 8%/year compounded monthly.

47. Find the present value of \$119,346 due in 4 yr at an interest rate of 10%/year compounded continuously.

48. Consumer Price Index At an annual inflation rate of 7.5%, how long will it take the Consumer Price Index (CPI) to double?

49. Growth of Bacteria A culture of bacteria that initially contained 2000 bacteria has a count of 18,000 bacteria after 2 hr.

a. Determine the function $Q(t)$ that expresses the exponential growth of the number of cells of this bacterium as a function of time t (in minutes).

b. Find the number of bacteria present after 4 hr.

50. Radioactive Decay The radioactive element radium has a half-life of 1600 yr. What is its decay constant?

51. Demand for DVD Players VCA Television found that the monthly demand for its new line of DVD players t mo after placing the players on the market is given by

$$D(t) = 4000 - 3000e^{-0.06t} \qquad (t \geq 0)$$

Graph this function and answer the following questions:

a. What was the demand after 1 mo? After 1 yr? After 2 yr?

b. At what level is the demand expected to stabilize?

52. Radioactivity The mass of a radioactive isotope at time t (in years) is $M(t) = 200e^{-0.14t}$ g. What is the mass of the isotope initially? How fast is the mass of the isotope changing 2 yr later?

53. Oil Used to Fuel Productivity A study on worldwide oil use was prepared for a major oil company. The study predicted that the amount of oil used to fuel productivity in a certain country is given by

$$f(t) = 1.5 + 1.8te^{-1.2t} \qquad (0 \leq t \leq 4)$$

where $f(t)$ denotes the number of barrels per \$1000 of economic output and t is measured in decades ($t = 0$ corresponds to 1965). Compute $f'(0), f'(1), f'(2)$, and $f'(3)$ and interpret your results.

54. Price of a Commodity The price of a certain commodity in dollars per unit at time t (measured in weeks) is given by $p = 18 - 3e^{-2t} - 6e^{-t/3}$.

a. What is the price of the commodity at $t = 0$?

b. How fast is the price of the commodity changing at $t = 0$?

c. Find the equilibrium price of the commodity.
Hint: It is given by $\lim_{t \to \infty} p$.

55. Flu Epidemic During a flu epidemic, the number of students at a certain university who contracted influenza after t days could be approximated by the exponential model

$$Q(t) = \frac{3000}{1 + 499e^{-kt}}$$

If 90 students contracted the flu by day 10, how many students contracted the flu by day 20?

56. U.S. Infant Mortality Rate The U.S. infant mortality rate (per 1000 live births) is approximated by the function

$$N(t) = 12.5e^{-0.0294t} \qquad (0 \leq t \leq 21)$$

where t is measured in years, with $t = 0$ corresponding to 1980.

a. What was the mortality rate in 1980? In 1990? In 2000?

b. Sketch the graph of N.

Source: U.S. Department of Health and Human Services

57. Maximizing Revenue The unit selling price p (in dollars) and the quantity demanded x (in pairs) of a certain brand of men's socks is given by the demand equation

$$p = 20e^{-0.0002x} \qquad (0 \leq x \leq 10{,}000)$$

How many pairs of socks must be sold to yield a maximum revenue? What will the maximum revenue be?

58. Absorption of Drugs The concentration of a drug in an organ at any time t (in seconds) is given by

$$x(t) = 0.08(1 - e^{-0.02t})$$

where $x(t)$ is measured in grams/cubic centimeter (g/cm^3).

a. What is the initial concentration of the drug in the organ?

b. What is the concentration of the drug in the organ after 30 sec?

c. What will be the concentration of the drug in the organ in the long run?

d. Sketch the graph of x.

CHAPTER 5 Before Moving On . . .

1. Solve the equation $\dfrac{100}{1 + 2e^{0.3t}} = 40$ for t.

2. Find the accumulated amount after 4 yr if \$3000 is invested at 8%/year compounded weekly.

3. Find the slope of the tangent line to the graph of $f(x) = e^{\sqrt{x}}$.

4. Find the rate at which $y = x \ln(x^2 + 1)$ is changing at $x = 1$.

5. Find the second derivative of $y = e^{2x} \ln 3x$.

6. The temperature of a cup of coffee at time t (in minutes) is

$$T(t) = 70 + ce^{-kt}$$

Initially, the temperature of the coffee was 200°F. Three minutes later, it was 180°. When will the temperature of the coffee be 150°F?

INTEGRATION

6

DIFFERENTIAL CALCULUS IS concerned with the problem of finding the rate of change of one quantity with respect to another. In this chapter, we begin the study of the other branch of calculus, known as integral calculus. Here we are interested in precisely the opposite problem: If we know the rate of change of one quantity with respect to another, can we find the relationship between the two quantities? The principal tool used in the study of integral calculus is the *antiderivative* of a function, and we develop rules for antidifferentiation, or *integration*, as the process of finding the antiderivative is called. We also show that a link is established between differential and integral calculus—via the fundamental theorem of calculus.

How much electricity should be produced over the next 3 years to meet the projected demand? In Example 9, page 436, you will see how the current rate of consumption can be used to answer this question.

6.1 Antiderivatives and the Rules of Integration

Antiderivatives

Let's return, once again, to the example involving the motion of the maglev (Figure 1).

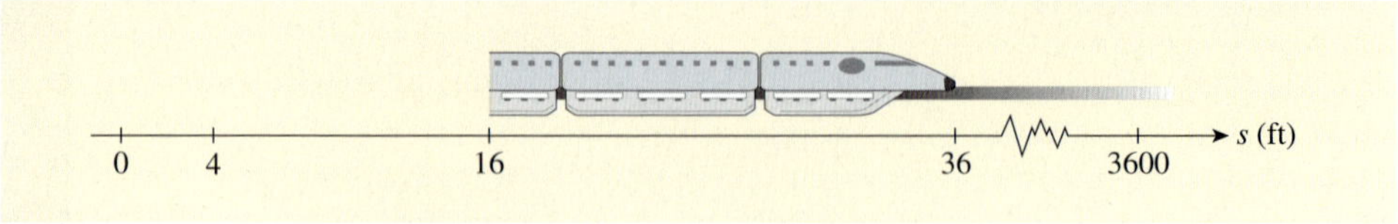

FIGURE 1
A maglev moving along an elevated monorail track

In Chapter 2, we discussed the following problem:

If we know the position of the maglev at any time t, can we find its velocity at time t?

As it turns out, if the position of the maglev is described by the position function f, then its velocity at any time t is given by $f'(t)$. Here f'—the velocity function of the maglev—is just the derivative of f.

Now, in Chapters 6 and 7, we will consider precisely the opposite problem:

If we know the velocity of the maglev at any time t, can we find its position at time t?

Stated another way, if we know the velocity function f' of the maglev, can we find its position function f?

To solve this problem, we need the concept of an antiderivative of a function.

Antiderivative

A function F is an **antiderivative** of f on an interval I if $F'(x) = f(x)$ for all x in I.

Thus, an antiderivative of a function f is a function F whose derivative is f. For example, $F(x) = x^2$ is an antiderivative of $f(x) = 2x$ because

$$F'(x) = \frac{d}{dx}(x^2) = 2x = f(x)$$

and $F(x) = x^3 + 2x + 1$ is an antiderivative of $f(x) = 3x^2 + 2$ because

$$F'(x) = \frac{d}{dx}(x^3 + 2x + 1) = 3x^2 + 2 = f(x)$$

EXAMPLE 1 Let $F(x) = \frac{1}{3}x^3 - 2x^2 + x - 1$. Show that F is an antiderivative of $f(x) = x^2 - 4x + 1$.

Solution Differentiating the function F, we obtain

$$F'(x) = x^2 - 4x + 1 = f(x)$$

and the desired result follows. ■

EXAMPLE 2 Let $F(x) = x$, $G(x) = x + 2$, and $H(x) = x + C$, where C is a constant. Show that F, G, and H are all antiderivatives of the function f defined by $f(x) = 1$.

Solution Since

$$F'(x) = \frac{d}{dx}(x) = 1 = f(x)$$
$$G'(x) = \frac{d}{dx}(x + 2) = 1 = f(x)$$
$$H'(x) = \frac{d}{dx}(x + C) = 1 = f(x)$$

we see that F, G, and H are indeed antiderivatives of f. ■

Example 2 shows that once an antiderivative G of a function f is known, then another antiderivative of f may be found by adding an arbitrary constant to the function G. The following theorem states that no function other than one obtained in this manner can be an antiderivative of f. (We omit the proof.)

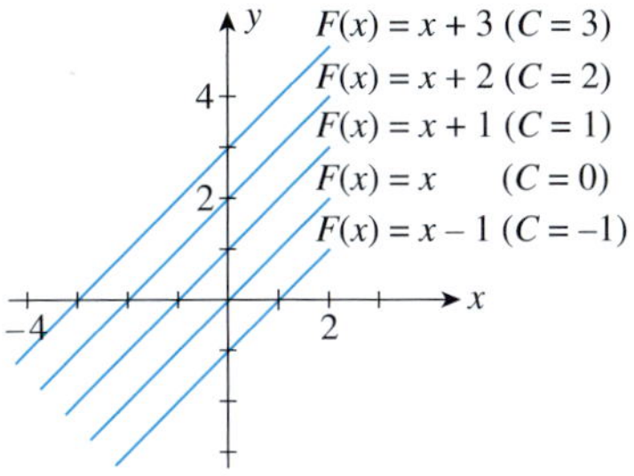

FIGURE 2
The graphs of some antiderivatives of $f(x) = 1$

THEOREM 1

Let G be an antiderivative of a function f. Then, every antiderivative F of f must be of the form $F(x) = G(x) + C$, where C is a constant.

Returning to Example 2, we see that there are infinitely many antiderivatives of the function $f(x) = 1$. We obtain each one by specifying the constant C in the function $F(x) = x + C$. Figure 2 shows the graphs of some of these antiderivatives for selected values of C. These graphs constitute part of a family of infinitely many parallel straight lines, each having a slope equal to 1. This result is expected since there are infinitely many curves (straight lines) with a given slope equal to 1. The antiderivatives $F(x) = x + C$ (C, a constant) are precisely the functions representing this family of straight lines.

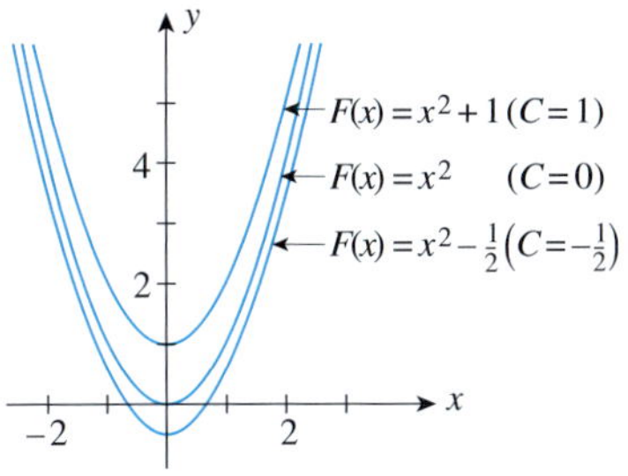

FIGURE 3
The graphs of some antiderivatives of $f(x) = 2x$

EXAMPLE 3 Prove that the function $G(x) = x^2$ is an antiderivative of the function $f(x) = 2x$. Write a general expression for the antiderivatives of f.

Solution Since $G'(x) = 2x = f(x)$, we have shown that $G(x) = x^2$ is an antiderivative of $f(x) = 2x$. By Theorem 1, every antiderivative of the function $f(x) = 2x$ has the form $F(x) = x^2 + C$, where C is some constant. The graphs of a few of the antiderivatives of f are shown in Figure 3. ■

Exploring with TECHNOLOGY

Let $f(x) = x^2 - 1$.

1. Show that $F(x) = \frac{1}{3}x^3 - x + C$, where C is an arbitrary constant, is an antiderivative of f.
2. Use a graphing utility to plot the graphs of the antiderivatives of f corresponding to $C = -2$, $C = -1$, $C = 0$, $C = 1$, and $C = 2$ on the same set of axes, using the viewing window $[-4, 4] \times [-4, 4]$.
3. If your graphing utility has the capability, draw the tangent line to each of the graphs in part 2 at the point whose x-coordinate is 2. What can you say about this family of tangent lines?
4. What is the slope of a tangent line in this family? Explain how you obtained your answer.

The Indefinite Integral

The process of finding all antiderivatives of a function is called **antidifferentiation,** or **integration.** We use the symbol $\int$, called an **integral sign,** to indicate that the operation of integration is to be performed on some function f. Thus,

$$\int f(x)\,dx = F(x) + C$$

[read "the indefinite integral of $f(x)$ with respect to x equals $F(x)$ plus C"] tells us that the **indefinite integral** of f is the family of functions given by $F(x) + C$, where $F'(x) = f(x)$. The function f to be integrated is called the **integrand,** and the constant C is called a **constant of integration.** The expression dx following the integrand $f(x)$ reminds us that the operation is performed with respect to x. If the independent variable is t, we write $\int f(t)\,dt$ instead. In this sense both t and x are "dummy variables."

Using this notation, we can write the results of Examples 2 and 3 as

$$\int 1\,dx = x + C \quad \text{and} \quad \int 2x\,dx = x^2 + K$$

where C and K are arbitrary constants.

Basic Integration Rules

Our next task is to develop some rules for finding the indefinite integral of a given function f. Because integration and differentiation are reverse operations, we discover many of the rules of integration by first making an "educated guess" at the antiderivative F of the function f to be integrated. Then this result is verified by demonstrating that $F' = f$.

Rule 1: The Indefinite Integral of a Constant

$$\int k\,dx = kx + C \qquad (k, \text{ a constant})$$

To prove this result, observe that

$$F'(x) = \frac{d}{dx}(kx + C) = k$$

EXAMPLE 4 Find each of the following indefinite integrals:

a. $\int 2\,dx$ **b.** $\int \pi^2\,dx$

Solution Each of the integrands has the form $f(x) = k$, where k is a constant. Applying Rule 1 in each case yields

a. $\int 2\,dx = 2x + C$

b. $\int \pi^2\,dx = \pi^2 x + C$ ■

Next, from the rule of differentiation,

$$\frac{d}{dx}x^n = nx^{n-1}$$

we obtain the following rule of integration.

Rule 2: The Power Rule

$$\int x^n\,dx = \frac{1}{n+1}x^{n+1} + C \qquad (n \neq -1)$$

An antiderivative of a power function is another power function obtained from the integrand by increasing its power by 1 and dividing the resulting expression by the new power.

To prove this result, observe that

$$\begin{aligned} F'(x) &= \frac{d}{dx}\left(\frac{1}{n+1}x^{n+1} + C\right) \\ &= \frac{n+1}{n+1}x^n \\ &= x^n \\ &= f(x) \end{aligned}$$

EXAMPLE 5 Find each of the following indefinite integrals:

a. $\int x^3\,dx$ **b.** $\int x^{3/2}\,dx$ **c.** $\int \frac{1}{x^{3/2}}\,dx$

Solution Each integrand is a power function with exponent $n \neq -1$. Applying Rule 2 in each case yields the following results:

a. $\int x^3\,dx = \frac{1}{4}x^4 + C$

b. $\int x^{3/2}\,dx = \frac{1}{\frac{5}{2}}x^{5/2} + C = \frac{2}{5}x^{5/2} + C$

c. $\int \frac{1}{x^{3/2}}\,dx = \int x^{-3/2}\,dx = \frac{1}{-\frac{1}{2}}x^{-1/2} + C = -2x^{-1/2} + C = -\frac{2}{x^{1/2}} + C$

These results may be verified by differentiating each of the antiderivatives and showing that the result is equal to the corresponding integrand. ■

The next rule tells us that a constant factor may be moved through an integral sign.

Rule 3: The Indefinite Integral of a Constant Multiple of a Function

$$\int cf(x)\,dx = c\int f(x)\,dx \qquad (c\text{, a constant})$$

The indefinite integral of a constant multiple of a function is equal to the constant multiple of the indefinite integral of the function.

This result follows from the corresponding rule of differentiation (see Rule 3, Section 3.1).

⚠ Only a constant can be "moved out" of an integral sign. For example, it would be incorrect to write

$$\int x^2\,dx = x^2\int 1\,dx$$

In fact, $\int x^2\,dx = \frac{1}{3}x^3 + C$, whereas $x^2\int 1\,dx = x^2(x + C) = x^3 + Cx^2$.

EXAMPLE 6 Find each of the following indefinite integrals:

a. $\int 2t^3\,dt$ **b.** $\int -3x^{-2}\,dx$

Solution Each integrand has the form $cf(x)$, where c is a constant. Applying Rule 3, we obtain:

a. $\int 2t^3\,dt = 2\int t^3\,dt = 2\left(\frac{1}{4}t^4 + K\right) = \frac{1}{2}t^4 + 2K = \frac{1}{2}t^4 + C$

where $C = 2K$. From now on, we will write the constant of integration as C, since any nonzero multiple of an arbitrary constant is an arbitrary constant.

b. $\int -3x^{-2}\,dx = -3\int x^{-2}\,dx = (-3)(-1)x^{-1} + C = \frac{3}{x} + C$ ■

Rule 4: The Sum Rule

$$\int [f(x) + g(x)]\,dx = \int f(x)\,dx + \int g(x)\,dx$$
$$\int [f(x) - g(x)]\,dx = \int f(x)\,dx - \int g(x)\,dx$$

The indefinite integral of a sum (difference) of two integrable functions is equal to the sum (difference) of their indefinite integrals.

This result is easily extended to the case involving the sum and difference of any finite number of functions. As in Rule 3, the proof of Rule 4 follows from the corresponding rule of differentiation (see Rule 4, Section 3.1).

VIDEO

EXAMPLE 7 Find the indefinite integral

$$\int (3x^5 + 4x^{3/2} - 2x^{-1/2})\,dx$$

Solution Applying the extended version of Rule 4, we find that

$$\begin{aligned}
&\int (3x^5 + 4x^{3/2} - 2x^{-1/2})\,dx \\
&\quad = \int 3x^5\,dx + \int 4x^{3/2}\,dx - \int 2x^{-1/2}\,dx \\
&\quad = 3\int x^5\,dx + 4\int x^{3/2}\,dx - 2\int x^{-1/2}\,dx && \text{Rule 3} \\
&\quad = (3)\left(\frac{1}{6}\right)x^6 + (4)\left(\frac{2}{5}\right)x^{5/2} - (2)(2)x^{1/2} + C && \text{Rule 2} \\
&\quad = \frac{1}{2}x^6 + \frac{8}{5}x^{5/2} - 4x^{1/2} + C
\end{aligned}$$
■

Observe that we have combined the three constants of integration, which arise from evaluating the three indefinite integrals, to obtain one constant C. After all, the sum of three arbitrary constants is also an arbitrary constant.

Rule 5: The Indefinite Integral of the Exponential Function

$$\int e^x\,dx = e^x + C$$

The indefinite integral of the exponential function with base e is equal to the function itself (except, of course, for the constant of integration).

EXAMPLE 8 Find the indefinite integral

$$\int (2e^x - x^3)\,dx$$

Solution We have

$$\begin{aligned}\int (2e^x - x^3)\,dx &= \int 2e^x\,dx - \int x^3\,dx\\ &= 2\int e^x\,dx - \int x^3\,dx\\ &= 2e^x - \frac{1}{4}x^4 + C\end{aligned}$$

The last rule of integration in this section covers the integration of the function $f(x) = x^{-1}$. Remember that this function constituted the only exceptional case in the integration of the power function $f(x) = x^n$ (see Rule 2).

Rule 6: The Indefinite Integral of the Function $f(x) = x^{-1}$

$$\int x^{-1}\,dx = \int \frac{1}{x}\,dx = \ln|x| + C \qquad (x \neq 0)$$

To prove Rule 6, observe that

$$\frac{d}{dx}\ln|x| = \frac{1}{x}$$

See Rule 3, Section 5.5.

EXAMPLE 9 Find the indefinite integral

$$\int \left(2x + \frac{3}{x} + \frac{4}{x^2}\right) dx$$

Solution

$$\begin{aligned}\int \left(2x + \frac{3}{x} + \frac{4}{x^2}\right) dx &= \int 2x\,dx + \int \frac{3}{x}\,dx + \int \frac{4}{x^2}\,dx\\ &= 2\int x\,dx + 3\int \frac{1}{x}\,dx + 4\int x^{-2}\,dx\\ &= 2\left(\frac{1}{2}\right)x^2 + 3\ln|x| + 4(-1)x^{-1} + C\\ &= x^2 + 3\ln|x| - \frac{4}{x} + C\end{aligned}$$

Differential Equations

Let's return to the problem posed at the beginning of the section: *Given the derivative of a function, f', can we find the function f?* As an example, suppose we are given the function

$$f'(x) = 2x - 1 \tag{1}$$

and we wish to find $f(x)$. From what we now know, we can find f by integrating Equation (1). Thus,

$$f(x) = \int f'(x)\,dx = \int (2x - 1)\,dx = x^2 - x + C \tag{2}$$

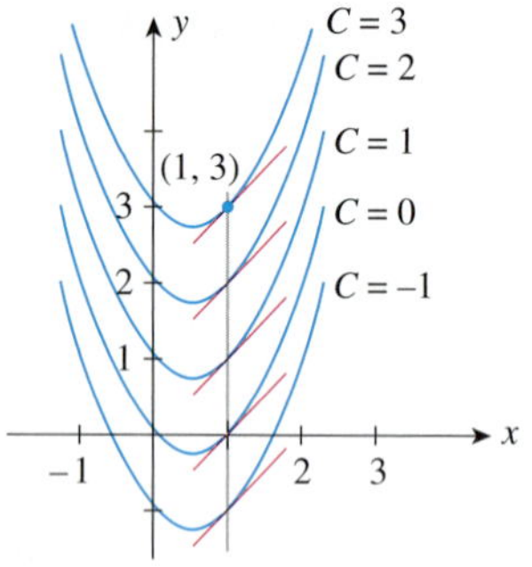

FIGURE 4
The graphs of some of the functions having the derivative $f'(x) = 2x - 1$. Observe that the slopes of the tangent lines to the graphs are the same for a fixed value of x.

where C is an arbitrary constant. Thus, infinitely many functions have the derivative f', each differing from the other by a constant.

Equation (1) is called a differential equation. In general, a **differential equation** is an equation that involves the derivative or differential of an unknown function. [In the case of Equation (1), the unknown function is f.] A **solution** of a differential equation is any function that satisfies the differential equation. Thus, Equation (2) gives *all* the solutions of the differential Equation (1), and it is, accordingly, called the **general solution** of the differential equation $f'(x) = 2x - 1$.

The graphs of $f(x) = x^2 - x + C$ for selected values of C are shown in Figure 4. These graphs have one property in common: For any fixed value of x, the tangent lines to these graphs have the same slope. This follows because any member of the family $f(x) = x^2 - x + C$ must have the same slope at x—namely, $2x - 1$!

Although there are infinitely many solutions to the differential equation $f'(x) = 2x - 1$, we can obtain a **particular solution** by specifying the value the function must assume at a certain value of x. For example, suppose we stipulate that the function f under consideration must satisfy the condition $f(1) = 3$ or, equivalently, the graph of f must pass through the point (1, 3). Then, using the condition on the general solution $f(x) = x^2 - x + C$, we find that

$$f(1) = 1 - 1 + C = 3$$

and $C = 3$. Thus, the particular solution is $f(x) = x^2 - x + 3$ (see Figure 4).

The condition $f(1) = 3$ is an example of an initial condition. More generally, an **initial condition** is a condition imposed on the value of f at $x = a$.

Initial Value Problems

An **initial value problem** is one in which we are required to find a function satisfying (1) a differential equation and (2) one or more initial conditions. The following are examples of initial value problems.

EXAMPLE 10 Find the function f if it is known that

$$f'(x) = 3x^2 - 4x + 8 \quad \text{and} \quad f(1) = 9$$

Solution We are required to solve the initial value problem

$$\left.\begin{aligned} f'(x) &= 3x^2 - 4x + 8 \\ f(1) &= 9 \end{aligned}\right\}$$

Integrating the function f', we find

$$\begin{aligned} f(x) &= \int f'(x)\,dx \\ &= \int (3x^2 - 4x + 8)\,dx \\ &= x^3 - 2x^2 + 8x + C \end{aligned}$$

Using the condition $f(1) = 9$, we have

$$9 = f(1) = 1^3 - 2(1)^2 + 8(1) + C = 7 + C \quad \text{or} \quad C = 2$$

Therefore, the required function f is given by $f(x) = x^3 - 2x^2 + 8x + 2$. ■

APPLIED EXAMPLE 11 Velocity of a Maglev In a test run of a maglev along a straight elevated monorail track, data obtained from reading its speedometer indicate that the velocity of the maglev at time t can be described by the velocity function

$$v(t) = 8t \qquad (0 \le t \le 30)$$

Find the position function of the maglev. Assume that initially the maglev is located at the origin of a coordinate line.

Solution Let $s(t)$ denote the position of the maglev at any time t $(0 \leq t \leq 30)$. Then, $s'(t) = v(t)$. So, we have the initial value problem

$$\left.\begin{aligned} s'(t) &= 8t \\ s(0) &= 0 \end{aligned}\right\}$$

Integrating both sides of the differential equation $s'(t) = 8t$, we obtain

$$s(t) = \int s'(t)\,dt = \int 8t\,dt = 4t^2 + C$$

where C is an arbitrary constant. To evaluate C, we use the initial condition $s(0) = 0$ to write

$$s(0) = 4(0) + C = 0 \quad \text{or} \quad C = 0$$

Therefore, the required position function is $s(t) = 4t^2$ $(0 \leq t \leq 30)$. ■

VIDEO

APPLIED EXAMPLE 12 Magazine Circulation The current circulation of the *Investor's Digest* is 3000 copies per week. The managing editor of the weekly projects a growth rate of

$$4 + 5t^{2/3}$$

copies per week, t weeks from now, for the next 3 years. Based on her projection, what will be the circulation of the digest 125 weeks from now?

Solution Let $S(t)$ denote the circulation of the digest t weeks from now. Then $S'(t)$ is the rate of change in the circulation in the tth week and is given by

$$S'(t) = 4 + 5t^{2/3}$$

Furthermore, the current circulation of 3000 copies per week translates into the initial condition $S(0) = 3000$. Integrating the differential equation with respect to t gives

$$\begin{aligned} S(t) &= \int S'(t)\,dt = \int (4 + 5t^{2/3})\,dt \\ &= 4t + 5\left(\frac{t^{5/3}}{\frac{5}{3}}\right) + C = 4t + 3t^{5/3} + C \end{aligned}$$

To determine the value of C, we use the condition $S(0) = 3000$ to write

$$S(0) = 4(0) + 3(0) + C = 3000$$

which gives $C = 3000$. Therefore, the circulation of the digest t weeks from now will be

$$S(t) = 4t + 3t^{5/3} + 3000$$

In particular, the circulation 125 weeks from now will be

$$S(125) = 4(125) + 3(125)^{5/3} + 3000 = 12{,}875$$

copies per week. ■

6.1 Self-Check Exercises

1. Evaluate $\int \left(\frac{1}{\sqrt{x}} - \frac{2}{x} + 3e^x \right) dx$.

2. Find the rule for the function f given that (1) the slope of the tangent line to the graph of f at any point $P(x, f(x))$ is given by the expression $3x^2 - 6x + 3$ and (2) the graph of f passes through the point (2, 9).

3. Suppose United Motors' share of the new cars sold in a certain country is changing at the rate of

$$f(t) = -0.01875t^2 + 0.15t - 1.2 \qquad (0 \le t \le 12)$$

percent/year at year t ($t = 0$ corresponds to the beginning of 1996). The company's market share at the beginning of 1996 was 48.4%. What was United Motors' market share at the beginning of 2008?

Solutions to Self-Check Exercises 6.1 can be found on page 411.

6.1 Concept Questions

1. What is an antiderivative? Give an example.

2. If $f'(x) = g'(x)$ for all x in an interval I, what is the relationship between f and g?

3. What is the difference between an antiderivative of f and the indefinite integral of f?

4. Can the power rule be used to integrate $\int \frac{1}{x} dx$? Explain your answer.

6.1 Exercises

In Exercises 1–4, verify directly that *F* is an antiderivative of *f*.

1. $F(x) = \frac{1}{3}x^3 + 2x^2 - x + 2; f(x) = x^2 + 4x - 1$

2. $F(x) = xe^x + \pi; f(x) = e^x(1 + x)$

3. $F(x) = \sqrt{2x^2 - 1}; f(x) = \frac{2x}{\sqrt{2x^2 - 1}}$

4. $F(x) = x \ln x - x; f(x) = \ln x$

In Exercises 5–8, (a) verify that *G* is an antiderivative of *f*, (b) find all antiderivatives of *f*, and (c) sketch the graphs of a few of the family of antiderivatives found in part (b).

5. $G(x) = 2x; f(x) = 2$

6. $G(x) = 2x^2; f(x) = 4x$

7. $G(x) = \frac{1}{3}x^3; f(x) = x^2$

8. $G(x) = e^x; f(x) = e^x$

In Exercises 9–50, find the indefinite integral.

9. $\int 6\, dx$

10. $\int \sqrt{2}\, dx$

11. $\int x^3\, dx$

12. $\int 2x^5\, dx$

13. $\int x^{-4}\, dx$

14. $\int 3t^{-7}\, dt$

15. $\int x^{2/3}\, dx$

16. $\int 2u^{3/4}\, du$

17. $\int x^{-5/4}\, dx$

18. $\int 3x^{-2/3}\, dx$

19. $\int \frac{2}{x^2}\, dx$

20. $\int \frac{1}{3x^5}\, dx$

21. $\int \pi\sqrt{t}\, dt$

22. $\int \frac{3}{\sqrt{t}}\, dt$

23. $\int (3 - 2x)\, dx$

24. $\int (1 + u + u^2)\, du$

25. $\int (x^2 + x + x^{-3})\, dx$

26. $\int (0.3t^2 + 0.02t + 2)\, dt$

27. $\int 4e^x\, dx$

28. $\int (1 + e^x)\, dx$

29. $\int (1 + x + e^x)\, dx$

30. $\int (2 + x + 2x^2 + e^x)\, dx$

31. $\int \left(4x^3 - \frac{2}{x^2} - 1\right) dx$

32. $\int \left(6x^3 + \frac{3}{x^2} - x\right) dx$

33. $\int (x^{5/2} + 2x^{3/2} - x)\, dx$

34. $\int (t^{3/2} + 2t^{1/2} - 4t^{-1/2})\, dt$

35. $\int \left(\sqrt{x} + \frac{3}{\sqrt{x}}\right) dx$

36. $\int \left(\sqrt[3]{x^2} - \frac{1}{x^2}\right) dx$

37. $\int \left(\frac{u^3 + 2u^2 - u}{3u}\right) du$

Hint: $\frac{u^3 + 2u^2 - u}{3u} = \frac{1}{3}u^2 + \frac{2}{3}u - \frac{1}{3}$

38. $\int \frac{x^4 - 1}{x^2}\, dx$

Hint: $\frac{x^4 - 1}{x^2} = x^2 - x^{-2}$

39. $\int (2t + 1)(t - 2)\, dt$

40. $\int u^{-2}(1 - u^2 + u^4)\, du$

41. $\int \frac{1}{x^2}(x^4 - 2x^2 + 1)\,dx$ **42.** $\int \sqrt{t}\,(t^2 + t - 1)\,dt$

43. $\int \frac{ds}{(s+1)^{-2}}$ **44.** $\int \left(\sqrt{x} + \frac{3}{x} - 2e^x\right) dx$

45. $\int (e^t + t^e)\,dt$ **46.** $\int \left(\frac{1}{x^2} - \frac{1}{\sqrt[3]{x^2}} + \frac{1}{\sqrt{x}}\right) dx$

47. $\int \left(\frac{x^3 + x^2 - x + 1}{x^2}\right) dx$

Hint: Simplify the integrand first.

48. $\int \frac{t^3 + \sqrt[3]{t}}{t^2}\,dt$

Hint: Simplify the integrand first.

49. $\int \frac{(\sqrt{x} - 1)^2}{x^2}\,dx$

Hint: Simplify the integrand first.

50. $\int (x+1)^2\left(1 - \frac{1}{x}\right) dx$

Hint: Simplify the integrand first.

In Exercises 51–58, find *f*(*x*) by solving the initial value problem.

51. $f'(x) = 2x + 1; f(1) = 3$

52. $f'(x) = 3x^2 - 6x; f(2) = 4$

53. $f'(x) = 3x^2 + 4x - 1; f(2) = 9$

54. $f'(x) = \frac{1}{\sqrt{x}}; f(4) = 2$

55. $f'(x) = 1 + \frac{1}{x^2}; f(1) = 2$

56. $f'(x) = e^x - 2x; f(0) = 2$

57. $f'(x) = \frac{x+1}{x}; f(1) = 1$

58. $f'(x) = 1 + e^x + \frac{1}{x}; f(1) = 3 + e$

In Exercises 59–62, find the function *f* given that the slope of the tangent line to the graph of *f* at any point (*x*, *f*(*x*)) is *f*′(*x*) and that the graph of *f* passes through the given point.

59. $f'(x) = \frac{1}{2}x^{-1/2}; (2, \sqrt{2})$

60. $f'(t) = t^2 - 2t + 3; (1, 2)$

61. $f'(x) = e^x + x; (0, 3)$ **62.** $f'(x) = \frac{2}{x} + 1; (1, 2)$

63. Bank Deposits Madison Finance opened two branches on September 1 ($t = 0$). Branch A is located in an established industrial park, and branch B is located in a fast-growing new development. The net rate at which money was deposited into branch A and branch B in the first 180 business days is given by the graphs of f and g, respectively (see the figure). Which branch has a larger amount on deposit at the end of 180 business days? Justify your answer.

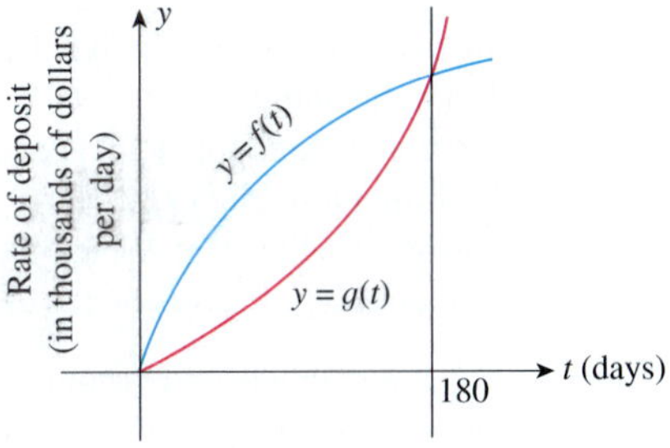

64. Velocity of a Car Two cars, side by side, start from rest and travel along a straight road. The velocity of car A is given by $v = f(t)$, and the velocity of car B is given by $v = g(t)$. The graphs of f and g are shown in the figure below. Are the cars still side by side after T sec? If not, which car is ahead of the other? Justify your answer.

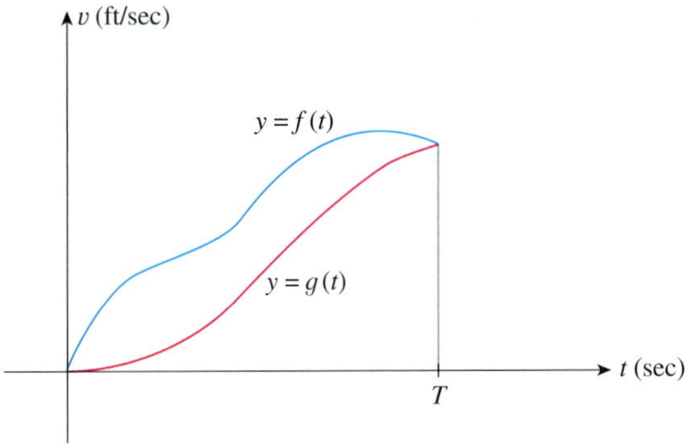

65. Velocity of a Car The velocity of a car (in feet/second) t sec after starting from rest is given by the function

$$f(t) = 2\sqrt{t} \qquad (0 \le t \le 30)$$

Find the car's position, $s(t)$, at any time t. Assume $s(0) = 0$.

66. Velocity of a Maglev The velocity (in feet/second) of a maglev is

$$v(t) = 0.2t + 3 \qquad (0 \le t \le 120)$$

At $t = 0$, it is at the station. Find the function giving the position of the maglev at time t, assuming that the motion takes place along a straight stretch of track.

67. Cost of Producing Clocks Lorimar Watch Company manufactures travel clocks. The daily marginal cost function associated with producing these clocks is

$$C'(x) = 0.000009x^2 - 0.009x + 8$$

where $C'(x)$ is measured in dollars/unit and x denotes the number of units produced. Management has determined that the daily fixed cost incurred in producing these clocks is \$120. Find the total cost incurred by Lorimar in producing the first 500 travel clocks/day.

68. **REVENUE FUNCTIONS** The management of Lorimar Watch Company has determined that the daily marginal revenue function associated with producing and selling their travel clocks is given by

$$R'(x) = -0.009x + 12$$

where x denotes the number of units produced and sold and $R'(x)$ is measured in dollars/unit.

a. Determine the revenue function $R(x)$ associated with producing and selling these clocks.

b. What is the demand equation that relates the wholesale unit price with the quantity of travel clocks demanded?

69. **PROFIT FUNCTIONS** Cannon Precision Instruments makes an automatic electronic flash with Thyrister circuitry. The estimated marginal profit associated with producing and selling these electronic flashes is

$$P'(x) = -0.004x + 20$$

dollars/unit/month when the production level is x units per month. Cannon's fixed cost for producing and selling these electronic flashes is \$16,000/month. At what level of production does Cannon realize a maximum profit? What is the maximum monthly profit?

70. **COST OF PRODUCING GUITARS** Carlota Music Company estimates that the marginal cost of manufacturing its Professional Series guitars is

$$C'(x) = 0.002x + 100$$

dollars/month when the level of production is x guitars/month. The fixed costs incurred by Carlota are \$4000/month. Find the total monthly cost incurred by Carlota in manufacturing x guitars/month.

71. **HEALTH COSTS** The national health expenditures are projected to grow at the rate of

$$r(t) = 0.0058t + 0.159 \qquad (0 \le t \le 13)$$

trillion dollars/year from 2002 through 2015. Here, $t = 0$ corresponds to 2002. The expenditure in 2002 was \$1.60 trillion.

a. Find a function f giving the projected national health expenditures in year t.

b. What does your model project the national health expenditure to be in 2015?

Source: National Health Expenditures

72. **QUALITY CONTROL** As part of a quality-control program, the chess sets manufactured by Jones Brothers are subjected to a final inspection before packing. The rate of increase in the number of sets checked per hour by an inspector t hr into the 8 a.m. to 12 noon morning shift is approximately

$$N'(t) = -3t^2 + 12t + 45 \qquad (0 \le t \le 4)$$

a. Find an expression $N(t)$ that approximates the number of sets inspected at the end of t hours.
Hint: $N(0) = 0$.

b. How many sets does the average inspector check during a morning shift?

73. **SATELLITE RADIO SUBSCRIPTIONS** Based on data obtained by polling automobile buyers, the number of subscribers of satellite radios is expected to grow at the rate of

$$r(t) = -0.375t^2 + 2.1t + 2.45 \qquad (0 \le t \le 5)$$

million subscribers/year between 2003 ($t = 0$) and 2008 ($t = 5$). The number of satellite radio subscribers at the beginning of 2003 was 1.5 million.

a. Find an expression giving the number of satellite radio subscribers in year t $(0 \le t \le 5)$.

b. Based on this model, what was the number of satellite radio subscribers in 2008?

Source: Carmel Group

74. **RISK OF DOWN SYNDROME** The rate at which the risk of Down syndrome is changing is approximated by the function

$$r(x) = 0.004641x^2 - 0.3012x + 4.9 \qquad (20 \le x \le 45)$$

where $r(x)$ is measured in percentage of all births/year and x is the maternal age at delivery.

a. Find a function f giving the risk as a percentage of all births when the maternal age at delivery is x years, given that the risk of down syndrome at age 30 is 0.14% of all births.

b. Based on this model, what is the risk of Down syndrome when the maternal age at delivery is 40 years? 45 years?

Source: New England Journal of Medicine

75. **CREDIT CARD DEBT** The average credit card debt per U.S. household between 1990 ($t = 0$) and 2003 ($t = 13$) was growing at the rate of approximately

$$D(t) = -4.479t^2 + 69.8t + 279.5 \qquad (0 \le t \le 13)$$

dollars/year. The average credit card debt per U.S. household stood at \$2917 in 1990.

a. Find an expression giving the approximate average credit card debt per U.S. household in year t $(0 \le t \le 13)$.

b. Use the result of part (a) to estimate the average credit card debt per U.S. household in 2003.

Source: Encore Capital Group

76. **GENETICALLY MODIFIED CROPS** The total number of acres of genetically modified crops grown worldwide from 1997 through 2003 was changing at the rate of

$$R(t) = 2.718t^2 - 19.86t + 50.18 \qquad (0 \le t \le 6)$$

million acres/year. The total number of acres of such crops grown in 1997 ($t = 0$) was 27.2 million acres. How many acres of genetically modified crops were grown worldwide in 2003?

Source: International Services for the Acquisition of Agri-biotech Applications

77. **Gastric Bypass Surgeries** One method of weight loss gaining in popularity is stomach-reducing surgery. It is generally reserved for people at least 100 lb overweight because the procedure carries a serious risk of death or complications. According to the American Society of Bariatric Surgery, the number of morbidly obese patients undergoing the procedure was increasing at the rate of

$$R(t) = 9.399t^2 - 13.4t + 14.07 \qquad (0 \leq t \leq 3)$$

thousands/year, with $t = 0$ corresponding to 2000. The number of gastric bypass surgeries performed in 2000 was 36.7 thousand.

a. Find an expression giving the number of gastric bypass surgeries performed in year t $(0 \leq t \leq 3)$.

b. Use the result of part (a) to find the number of gastric bypass surgeries performed in 2003.

Source: American Society for Bariatric Surgery

78. **Online Ad Sales** According to a study conducted in 2004, the share of online advertisement, worldwide, as a percentage of the total ad market, is expected to grow at the rate of

$$R(t) = -0.033t^2 + 0.3428t + 0.07 \qquad (0 \leq t \leq 6)$$

percent/year at time t (in years), with $t = 0$ corresponding to the beginning of 2000. The online ad market at the beginning of 2000 was 2.9% of the total ad market.

a. What is the projected online ad market share at any time t?

b. What was the projected online ad market share at the beginning of 2005?

Source: Jupiter Media Metrix, Inc.

79. **Health-Care Costs** The average out-of-pocket costs for beneficiaries in traditional Medicare (including premiums, cost sharing, and prescription drugs not covered by Medicare) is projected to grow at the rate of

$$C'(t) = 12.288t^2 - 150.5594t + 695.23$$

dollars/year, where t is measured in 5-yr intervals, with $t = 0$ corresponding to 2000. The out-of-pocket costs for beneficiaries in 2000 were \$3142.

a. Find an expression giving the average out-of-pocket costs for beneficiaries in year t.

b. What is the projected average out-of-pocket costs for beneficiaries in 2010?

Source: The Urban Institute

80. **Ballast Dropped from a Balloon** A ballast is dropped from a stationary hot-air balloon that is hovering at an altitude of 400 ft. Its velocity after t sec is $-32t$ ft/sec.

a. Find the height $h(t)$ of the ballast from the ground at time t.
Hint: $h'(t) = -32t$ and $h(0) = 400$.

b. When will the ballast strike the ground?

c. Find the velocity of the ballast when it hits the ground.

81. **Cable TV Subscribers** A study conducted by TeleCable estimates that the number of cable TV subscribers will grow at the rate of

$$100 + 210t^{3/4}$$

new subscribers/month, t mo from the start date of the service. If 5000 subscribers signed up for the service before the starting date, how many subscribers will there be 16 mo from that date?

82. **Ozone Pollution** The rate of change of the level of ozone, an invisible gas that is an irritant and impairs breathing, present in the atmosphere on a certain May day in the city of Riverside is given by

$$R(t) = 3.2922t^2 - 0.366t^3 \qquad (0 < t < 11)$$

(measured in pollutant standard index/hour). Here, t is measured in hours, with $t = 0$ corresponding to 7 a.m. Find the ozone level $A(t)$ at any time t, assuming that at 7 a.m. it is zero.
Hint: $A'(t) = R(t)$ and $A(0) = 0$.

Source: Los Angeles Times

83. **Flight of a Rocket** The velocity, in feet/second, of a rocket t sec into vertical flight is given by

$$v(t) = -3t^2 + 192t + 120$$

Find an expression $h(t)$ that gives the rocket's altitude, in feet, t sec after liftoff. What is the altitude of the rocket 30 sec after liftoff?
Hint: $h'(t) = v(t)$; $h(0) = 0$.

84. **Population Growth** The development of AstroWorld ("The Amusement Park of the Future") on the outskirts of a city will increase the city's population at the rate of

$$4500\sqrt{t} + 1000$$

people/year, t yr from the start of construction. The population before construction is 30,000. Determine the projected population 9 yr after construction of the park has begun.

85. **U.S. Sales of Organic Milk** The sales of organic milk from 1999 through 2004 grew at the rate of approximately

$$R(t) = 3t^3 - 17.9445t^2 + 28.7222t + 26.632 \quad (0 \le t \le 5)$$

million dollars/year, where t is measured in years, with $t = 0$ corresponding to 1999. Sales of organic milk in 1999 totaled \$108 million.

a. Find an expression giving the total sales of organic milk by year t $(0 \le t \le 5)$.

b. According to this model, what were the total sales of organic milk in 2004?

Source: Resource, Inc.

86. **Surface Area of a Human** Empirical data suggest that the surface area of a 180-cm-tall human body changes at the rate of

$$S'(W) = 0.131773W^{-0.575}$$

square meters/kilogram, where W is the weight of the body in kilograms. If the surface area of a 180-cm-tall human body weighing 70 kg is 1.886277 m^2, what is the surface area of a human body of the same height weighing 75 kg?

87. **Outpatient Service Companies** The number of Medicare-certified home-health-care agencies (70% are freestanding, and 30% are owned by a hospital or other large facility) has been declining at the rate of

$$0.186e^{-0.02t} \quad (0 \le t \le 14)$$

thousand agencies/year between 1988 $(t = 0)$ and 2002 $(t = 14)$. The number of such agencies stood at 9.3 thousand units in 1988.

a. Find an expression giving the number of health-care agencies in year t.

b. What was the number of health-care agencies in 2002?

c. If this model held true through 2005, how many care agencies were there in 2005?

Source: Centers for Medicare and Medicaid Services

88. **Heights of Children** According to the Jenss model for predicting the height of preschool children, the rate of growth of a typical preschool child is

$$R(t) = 25.8931e^{-0.993t} + 6.39 \quad (\tfrac{1}{4} \le t \le 6)$$

centimeters/year, where t is measured in years. The height of a typical 3-mo-old preschool child is 60.2952 cm.

a. Find a model for predicting the height of a typical preschool child at age t.

b. Use the result of part (a) to estimate the height of a typical 1-yr-old child.

89. **Blood Flow in an Artery** Nineteenth-century physician Jean Louis Marie Poiseuille discovered that the rate of change of the velocity of blood r cm from the central axis of an artery (in centimeters/second/centimeter) is given by

$$a(r) = -kr$$

where k is a constant. If the radius of an artery is R cm, find an expression for the velocity of blood as a function of r (see the accompanying figure).

Hint: $v'(r) = a(r)$ and $v(R) = 0$. (Why?)

Blood vessel

90. **Acceleration of a Car** A car traveling along a straight road at 66 ft/sec accelerated to a speed of 88 ft/sec over a distance of 440 ft. What was the acceleration of the car, assuming it was constant?

91. **Deceleration of a Car** What constant deceleration would a car moving along a straight road have to be subjected to if it were brought to rest from a speed of 88 ft/sec in 9 sec? What would be the stopping distance?

92. **Carrier Landing** A pilot lands a fighter aircraft on an aircraft carrier. At the moment of touchdown, the speed of the aircraft is 160 mph. If the aircraft is brought to a complete stop in 1 sec and the deceleration is assumed to be constant, find the number of g's the pilot is subjected to during landing (1 g = 32 ft/sec^2).

93. **Crossing the Finish Line** After rounding the final turn in the bell lap, two runners emerged ahead of the pack. When runner A is 200 ft from the finish line, his speed is 22 ft/sec, a speed that he maintains until he crosses the line. At that instant of time, runner B, who is 20 ft behind runner A and running at a speed of 20 ft/sec, begins to sprint. Assuming that runner B sprints with a constant acceleration, what minimum acceleration will enable him to cross the finish line ahead of runner A?

94. **Draining a Tank** A tank has a constant cross-sectional area of 50 ft^2 and an orifice of constant cross-sectional area of $\frac{1}{2}$ ft^2 located at the bottom of the tank (see the accompanying figure).

If the tank is filled with water to a height of h ft and allowed to drain, then the height of the water decreases at a rate that is described by the equation

$$\frac{dh}{dt} = -\frac{1}{25}\left(\sqrt{20} - \frac{t}{50}\right) \quad (0 \le t \le 50\sqrt{20})$$

Find an expression for the height of the water at any time t if its height initially is 20 ft.

95. **Amount of Rainfall** During a thunderstorm, rain was falling at the rate of

$$\frac{8}{(t+4)^2} \qquad (0 \leq t \leq 2)$$

inches/hour.

a. Find an expression giving the total amount of rainfall after t hr.
Hint: The total amount of rainfall at $t = 0$ is zero.

b. How much rain had fallen after 1 hr? After 2 hr?

96. **Launching a Fighter Aircraft** A fighter aircraft is launched from the deck of a Nimitz-class aircraft carrier with the help of a steam catapult. If the aircraft is to attain a takeoff speed of at least 240 ft/sec after traveling 800 ft along the flight deck, find the minimum acceleration it must be subjected to, assuming it is constant.

In Exercises 97–100, determine whether the statement is true or false. If it is true, explain why it is true. If it is false, give an example to show why it is false.

97. If F and G are antiderivatives of f on an interval I, then $F(x) = G(x) + C$ on I.

98. If F is an antiderivative of f on an interval I, then $\int f(x)\,dx = F(x)$.

99. If f and g are integrable, then $\int [2f(x) - 3g(x)]\,dx = 2\int f(x)\,dx - 3\int g(x)\,dx$.

100. If f and g are integrable, then $\int f(x)g(x)\,dx = [\int f(x)\,dx][\int g(x)\,dx]$.

6.1 Solutions to Self-Check Exercises

1. $$\int\left(\frac{1}{\sqrt{x}} - \frac{2}{x} + 3e^x\right)dx = \int\left(x^{-1/2} - \frac{2}{x} + 3e^x\right)dx$$
$$= \int x^{-1/2}\,dx - 2\int\frac{1}{x}\,dx + 3\int e^x\,dx$$
$$= 2x^{1/2} - 2\ln|x| + 3e^x + C$$
$$= 2\sqrt{x} - 2\ln|x| + 3e^x + C$$

2. The slope of the tangent line to the graph of the function f at any point $P(x, f(x))$ is given by the derivative f' of f. Thus, the first condition implies that

$$f'(x) = 3x^2 - 6x + 3$$

which, upon integration, yields

$$f(x) = \int (3x^2 - 6x + 3)\,dx$$
$$= x^3 - 3x^2 + 3x + k$$

where k is the constant of integration.

To evaluate k, we use the initial condition (2), which implies that $f(2) = 9$, or

$$9 = f(2) = 2^3 - 3(2)^2 + 3(2) + k$$

or $k = 7$. Hence, the required rule of definition of the function f is

$$f(x) = x^3 - 3x^2 + 3x + 7$$

3. Let $M(t)$ denote United Motors' market share at year t. Then,

$$M(t) = \int f(t)\,dt$$
$$= \int (-0.01875t^2 + 0.15t - 1.2)\,dt$$
$$= -0.00625t^3 + 0.075t^2 - 1.2t + C$$

To determine the value of C, we use the initial condition $M(0) = 48.4$, obtaining $C = 48.4$. Therefore,

$$M(t) = -0.00625t^3 + 0.075t^2 - 1.2t + 48.4$$

In particular, United Motors' market share of new cars at the beginning of 2008 is given by

$$M(12) = -0.00625(12)^3 + 0.075(12)^2 - 1.2(12) + 48.4 = 34$$

or 34%.

6.2 Integration by Substitution

In Section 6.1, we developed certain rules of integration that are closely related to the corresponding rules of differentiation in Chapters 3 and 5. In this section, we introduce a method of integration called the **method of substitution,** which is related to the chain rule for differentiating functions. When used in conjunction with the rules of integration developed earlier, the method of substitution is a powerful tool for integrating a large class of functions.

How the Method of Substitution Works

Consider the indefinite integral

$$\int 2(2x + 4)^5\,dx \tag{3}$$

One way of evaluating this integral is to expand the expression $(2x + 4)^5$ and then integrate the resulting integrand term by term. As an alternative approach, let's see if we can simplify the integral by making a change of variable. Write

$$u = 2x + 4$$

with differential

$$du = 2\,dx$$

If we formally substitute these quantities into Equation (3), we obtain

$$\int 2(2x + 4)^5\,dx = \underset{\uparrow \text{ Rewrite}}{\int (2x + 4)^5(2\,dx)} = \int u^5\,du \qquad \begin{cases} u = 2x + 4 \\ du = 2\,dx \end{cases}$$

Now, the last integral involves a power function and is easily evaluated using Rule 2 of Section 6.1. Thus,

$$\int u^5\,du = \frac{1}{6}u^6 + C$$

Therefore, using this result and replacing u by $u = 2x + 4$, we obtain

$$\int 2(2x + 4)^5\,dx = \frac{1}{6}(2x + 4)^6 + C$$

We can verify that the foregoing result is indeed correct by computing

$$\begin{aligned} \frac{d}{dx}\left[\frac{1}{6}(2x + 4)^6 + C\right] &= \frac{1}{6}\cdot 6(2x + 4)^5(2) \qquad \text{Use the chain rule.} \\ &= 2(2x + 4)^5 \end{aligned}$$

and observing that the last expression is just the integrand of (3).

The Method of Integration by Substitution

To see why the approach used in evaluating the integral in (3) is successful, write

$$f(x) = x^5 \quad \text{and} \quad g(x) = 2x + 4$$

Then, $g'(x) = 2$. Furthermore, the integrand of (3) is just the composition of f and g. Thus,

$$\begin{aligned} (f \circ g)(x) &= f(g(x)) \\ &= [g(x)]^5 = (2x + 4)^5 \end{aligned}$$

Therefore, (3) can be written as

$$\int f(g(x))g'(x)\,dx \tag{4}$$

Next, let's show that an integral having the form (4) can always be written as

$$\int f(u)\,du \tag{5}$$

Suppose F is an antiderivative of f. By the chain rule, we have

$$\frac{d}{dx}[F(g(x))] = F'(g(x))g'(x)$$

Therefore,

$$\int F'(g(x))g'(x)\,dx = F(g(x)) + C$$

Letting $F' = f$ and making the substitution $u = g(x)$, we have

$$\int f(g(x))g'(x)\,dx = F(u) + C = \int F'(u)\,du = \int f(u)\,du$$

as we wished to show. Thus, if the transformed integral is readily evaluated, as is the case with the integral (3), then the method of substitution will prove successful.

Before we look at more examples, let's summarize the steps involved in integration by substitution.

Integration by Substitution

Step 1 Let $u = g(x)$, where $g(x)$ is part of the integrand, usually the "inside function" of the composite function $f(g(x))$.

Step 2 Find $du = g'(x)\,dx$.

Step 3 Use the substitution $u = g(x)$ and $du = g'(x)\,dx$ to convert the *entire* integral into one involving *only* u.

Step 4 Evaluate the resulting integral.

Step 5 Replace u by $g(x)$ to obtain the final solution as a function of x.

Note Sometimes we need to consider different choices of g for the substitution $u = g(x)$ in order to carry out Step 3 and/or Step 4. ■

EXAMPLE 1 Find $\int 2x(x^2 + 3)^4\,dx$.

Solution

Step 1 Observe that the integrand involves the composite function $(x^2 + 3)^4$ with "inside function" $g(x) = x^2 + 3$. So, we choose $u = x^2 + 3$.

Step 2 Find $du = 2x\,dx$.

Step 3 Making the substitution $u = x^2 + 3$ and $du = 2x\,dx$, we obtain

$$\int 2x(x^2 + 3)^4\,dx = \underset{\substack{\uparrow \\ \text{Rewrite}}}{\int} (x^2 + 3)^4(2x\,dx) = \int u^4\,du$$

an integral involving only the variable u.

Step 4 Evaluate

$$\int u^4\,du = \frac{1}{5}u^5 + C$$

Step 5 Replacing u by $x^2 + 3$, we obtain

$$\int 2x(x^2 + 3)^4\,dx = \frac{1}{5}(x^2 + 3)^5 + C$$ ■

EXAMPLE 2 Find $\int 3\sqrt{3x + 1}\,dx$.

Solution

Step 1 The integrand involves the composite function $\sqrt{3x + 1}$ with "inside function" $g(x) = 3x + 1$. So, let $u = 3x + 1$.

Step 2 Find $du = 3\,dx$.

Step 3 Making the substitution $u = 3x + 1$ and $du = 3\,dx$, we obtain

$$\int 3\sqrt{3x + 1}\,dx = \int \sqrt{3x + 1}(3\,dx) = \int \sqrt{u}\,du$$

an integral involving only the variable u.

Step 4 Evaluate

$$\int \sqrt{u}\,du = \int u^{1/2}\,du = \frac{2}{3}u^{3/2} + C$$

Step 5 Replacing u by $3x + 1$, we obtain

$$\int 3\sqrt{3x + 1}\,dx = \frac{2}{3}(3x + 1)^{3/2} + C$$

■

EXAMPLE 3 Find $\int x^2(x^3 + 1)^{3/2}\,dx$.

Solution

Step 1 The integrand contains the composite function $(x^3 + 1)^{3/2}$ with "inside function" $g(x) = x^3 + 1$. So, let $u = x^3 + 1$.

Step 2 Find $du = 3x^2\,dx$.

Step 3 Making the substitution $u = x^3 + 1$ and $du = 3x^2\,dx$, or $x^2\,dx = \frac{1}{3}\,du$, we obtain

$$\int x^2(x^3 + 1)^{3/2}\,dx = \int (x^3 + 1)^{3/2}(x^2\,dx)$$
$$= \int u^{3/2}\left(\frac{1}{3}\,du\right) = \frac{1}{3}\int u^{3/2}\,du$$

an integral involving only the variable u.

Step 4 We evaluate

$$\frac{1}{3}\int u^{3/2}\,du = \frac{1}{3}\cdot\frac{2}{5}u^{5/2} + C = \frac{2}{15}u^{5/2} + C$$

Step 5 Replacing u by $x^3 + 1$, we obtain

$$\int x^2(x^3 + 1)^{3/2}\,dx = \frac{2}{15}(x^3 + 1)^{5/2} + C$$

■

Explore & Discuss

Let $f(x) = x^2(x^3 + 1)^{3/2}$. Using the result of Example 3, we see that an antiderivative of f is $F(x) = \frac{2}{15}(x^3 + 1)^{5/2}$. However, in terms of u (where $u = x^3 + 1$), an antiderivative of f is $G(u) = \frac{2}{15}u^{5/2}$. Compute $F(2)$. Next, suppose we want to compute $F(2)$ using the function G instead. At what value of u should you evaluate $G(u)$ in order to obtain the desired result? Explain your answer.

In the remaining examples, we drop the practice of labeling the steps involved in evaluating each integral.

VIDEO

EXAMPLE 4 Find $\int e^{-3x}\,dx$.

Solution Let $u = -3x$ so that $du = -3\,dx$, or $dx = -\frac{1}{3}\,du$. Then,

$$\int e^{-3x}\,dx = \int e^u\left(-\frac{1}{3}\,du\right) = -\frac{1}{3}\int e^u\,du$$
$$= -\frac{1}{3}e^u + C = -\frac{1}{3}e^{-3x} + C$$

■

EXAMPLE 5 Find $\displaystyle\int \frac{x}{3x^2+1}\,dx$.

Solution Let $u = 3x^2 + 1$. Then, $du = 6x\,dx$, or $x\,dx = \frac{1}{6}\,du$. Making the appropriate substitutions, we have

$$\begin{aligned}
\int \frac{x}{3x^2+1}\,dx &= \int \frac{\frac{1}{6}}{u}\,du \\
&= \frac{1}{6}\int \frac{1}{u}\,du \\
&= \frac{1}{6}\ln|u| + C \\
&= \frac{1}{6}\ln(3x^2+1) + C \qquad \text{Since } 3x^2 + 1 > 0
\end{aligned}$$

EXAMPLE 6 Find $\displaystyle\int \frac{(\ln x)^2}{2x}\,dx$.

Solution Let $u = \ln x$. Then,

$$du = \frac{d}{dx}(\ln x)\,dx = \frac{1}{x}\,dx$$

$$\begin{aligned}
\int \frac{(\ln x)^2}{2x}\,dx &= \frac{1}{2}\int \frac{(\ln x)^2}{x}\,dx \\
&= \frac{1}{2}\int u^2\,du \\
&= \frac{1}{6}u^3 + C \\
&= \frac{1}{6}(\ln x)^3 + C
\end{aligned}$$

Explore & Discuss

Suppose $\int f(u)\,du = F(u) + C$.

1. Show that $\int f(ax+b)\,dx = \frac{1}{a}F(ax+b) + C$.
2. How can you use this result to facilitate the evaluation of integrals such as $\int (2x+3)^5\,dx$ and $\int e^{3x-2}\,dx$? Explain your answer.

Examples 7 and 8 show how the method of substitution can be used in practical situations.

VIDEO

APPLIED EXAMPLE 7 Cost of Producing Solar Cell Panels In 1990 the head of the research and development department of Soloron Corporation claimed that the cost of producing solar cell panels would drop at the rate of

$$\frac{58}{(3t+2)^2} \qquad (0 \le t \le 10)$$

dollars per peak watt for the next t years, with $t = 0$ corresponding to the beginning of 1990. (A peak watt is the power produced at noon on a sunny day.) In 1990 the panels, which are used for photovoltaic power systems, cost \$10 per peak watt. Find an expression giving the cost per peak watt of producing solar cell panels at the beginning of year t. What was the cost at the beginning of 2000?

Solution Let $C(t)$ denote the cost per peak watt for producing solar cell panels at the beginning of year t. Then,

$$C'(t) = -\frac{58}{(3t+2)^2}$$

Integrating, we find that

$$\begin{aligned} C(t) &= \int \frac{-58}{(3t+2)^2}\,dt \\ &= -58\int (3t+2)^{-2}\,dt \end{aligned}$$

Let $u = 3t + 2$ so that

$$du = 3\,dt \quad \text{or} \quad dt = \frac{1}{3}\,du$$

Then,

$$\begin{aligned} C(t) &= -58\left(\frac{1}{3}\right)\int u^{-2}\,du \\ &= -\frac{58}{3}(-1)u^{-1} + k \\ &= \frac{58}{3(3t+2)} + k \end{aligned}$$

where k is an arbitrary constant. To determine the value of k, note that the cost per peak watt of producing solar cell panels at the beginning of 1990 ($t = 0$) was 10, or $C(0) = 10$. This gives

$$C(0) = \frac{58}{3(2)} + k = 10$$

or $k = \frac{1}{3}$. Therefore, the required expression is given by

$$\begin{aligned} C(t) &= \frac{58}{3(3t+2)} + \frac{1}{3} \\ &= \frac{58 + (3t+2)}{3(3t+2)} = \frac{3t+60}{3(3t+2)} \\ &= \frac{t+20}{3t+2} \end{aligned}$$

The cost per peak watt for producing solar cell panels at the beginning of 2000 is given by

$$C(10) = \frac{10+20}{3(10)+2} \approx 0.94$$

or approximately \$0.94 per peak watt. ■

Exploring with TECHNOLOGY

Refer to Example 7.

1. Use a graphing utility to plot the graph of

$$C(t) = \frac{t + 20}{3t + 2}$$

using the viewing window $[0, 10] \times [0, 5]$. Then, use the numerical differentiation capability of the graphing utility to compute $C'(10)$.

2. Plot the graph of

$$C'(t) = -\frac{58}{(3t + 2)^2}$$

using the viewing window $[0, 10] \times [-10, 0]$. Then, use the evaluation capability of the graphing utility to find $C'(10)$. Is this value of $C'(10)$ the same as that obtained in part 1? Explain your answer.

APPLIED EXAMPLE 8 Computer Sales Projections A study prepared by the marketing department of Universal Instruments forecasts that, after its new line of Galaxy Home Computers is introduced into the market, sales will grow at the rate of

$$2000 - 1500e^{-0.05t} \qquad (0 \leq t \leq 60)$$

units per month. Find an expression that gives the total number of computers that will sell t months after they become available on the market. How many computers will Universal sell in the first year they are on the market?

Solution Let $N(t)$ denote the total number of computers that may be expected to be sold t months after their introduction in the market. Then, the rate of growth of sales is given by $N'(t)$ units per month. Thus,

$$N'(t) = 2000 - 1500e^{-0.05t}$$

so that

$$\begin{aligned} N(t) &= \int (2000 - 1500e^{-0.05t})\, dt \\ &= \int 2000\, dt - 1500 \int e^{-0.05t}\, dt \end{aligned}$$

Upon integrating the second integral by the method of substitution, we obtain

$$\begin{aligned} N(t) &= 2000t + \frac{1500}{0.05} e^{-0.05t} + C \\ &= 2000t + 30{,}000e^{-0.05t} + C \end{aligned}$$

Let $u = -0.05t$; then $du = -0.05\, dt$.

To determine the value of C, note that the number of computers sold at the end of month 0 is nil, so $N(0) = 0$. This gives

$$N(0) = 30{,}000 + C = 0$$

Since $e^0 = 1$

or $C = -30{,}000$. Therefore, the required expression is given by

$$\begin{aligned} N(t) &= 2000t + 30{,}000e^{-0.05t} - 30{,}000 \\ &= 2000t + 30{,}000(e^{-0.05t} - 1) \end{aligned}$$

The number of computers that Universal can expect to sell in the first year is given by

$$N(12) = 2000(12) + 30{,}000(e^{-0.05(12)} - 1) \approx 10{,}464$$

6.2 Self-Check Exercises

1. Evaluate $\int \sqrt{2x + 5}\, dx$.

2. Evaluate $\int \frac{x^2}{(2x^3 + 1)^{3/2}}\, dx$.

3. Evaluate $\int xe^{2x^2-1}\, dx$.

4. According to a joint study conducted by Oxnard's Environmental Management Department and a state government agency, the concentration of carbon monoxide (CO) in the air due to automobile exhaust is increasing at the rate given by

$$f(t) = \frac{8(0.1t + 1)}{300(0.2t^2 + 4t + 64)^{1/3}}$$

parts per million (ppm) per year t. Currently, the CO concentration due to automobile exhaust is 0.16 ppm. Find an expression giving the CO concentration t yr from now.

Solutions to Self-Check Exercises 6.2 can be found on page 420.

6.2 Concept Questions

1. Explain how the method of substitution works by showing the steps used to find $\int f(g(x))g'(x)\, dx$.

2. Explain why the method of substitution works for the integral $\int xe^{-x^2}\, dx$, but not for the integral $\int e^{-x^2}\, dx$.

6.2 Exercises

In Exercises 1–50, find the indefinite integral.

1. $\int 4(4x + 3)^4\, dx$
2. $\int 4x(2x^2 + 1)^7\, dx$
3. $\int (x^3 - 2x)^2(3x^2 - 2)\, dx$
4. $\int (3x^2 - 2x + 1)(x^3 - x^2 + x)^4\, dx$
5. $\int \frac{4x}{(2x^2 + 3)^3}\, dx$
6. $\int \frac{3x^2 + 2}{(x^3 + 2x)^2}\, dx$
7. $\int 3t^2\sqrt{t^3 + 2}\, dt$
8. $\int 3t^2(t^3 + 2)^{3/2}\, dt$
9. $\int (x^2 - 1)^9 x\, dx$
10. $\int x^2(2x^3 + 3)^4\, dx$
11. $\int \frac{x^4}{1 - x^5}\, dx$
12. $\int \frac{x^2}{\sqrt{x^3 - 1}}\, dx$
13. $\int \frac{2}{x - 2}\, dx$
14. $\int \frac{x^2}{x^3 - 3}\, dx$
15. $\int \frac{0.3x - 0.2}{0.3x^2 - 0.4x + 2}\, dx$
16. $\int \frac{2x^2 + 1}{0.2x^3 + 0.3x}\, dx$
17. $\int \frac{x}{3x^2 - 1}\, dx$
18. $\int \frac{x^2 - 1}{x^3 - 3x + 1}\, dx$
19. $\int e^{-2x}\, dx$
20. $\int e^{-0.02x}\, dx$
21. $\int e^{2-x}\, dx$
22. $\int e^{2t+3}\, dt$
23. $\int xe^{-x^2}\, dx$
24. $\int x^2e^{x^3-1}\, dx$
25. $\int (e^x - e^{-x})\, dx$
26. $\int (e^{2x} + e^{-3x})\, dx$
27. $\int \frac{e^x}{1 + e^x}\, dx$
28. $\int \frac{e^{2x}}{1 + e^{2x}}\, dx$
29. $\int \frac{e^{\sqrt{x}}}{\sqrt{x}}\, dx$
30. $\int \frac{e^{-1/x}}{x^2}\, dx$
31. $\int \frac{e^{3x} + x^2}{(e^{3x} + x^3)^3}\, dx$
32. $\int \frac{e^x - e^{-x}}{(e^x + e^{-x})^{3/2}}\, dx$
33. $\int e^{2x}(e^{2x} + 1)^3\, dx$
34. $\int e^{-x}(1 + e^{-x})\, dx$
35. $\int \frac{\ln 5x}{x}\, dx$
36. $\int \frac{(\ln u)^3}{u}\, du$
37. $\int \frac{1}{x \ln x}\, dx$
38. $\int \frac{1}{x(\ln x)^2}\, dx$
39. $\int \frac{\sqrt{\ln x}}{x}\, dx$
40. $\int \frac{(\ln x)^{7/2}}{x}\, dx$
41. $\int \left(xe^{x^2} - \frac{x}{x^2 + 2}\right) dx$
42. $\int \left(xe^{-x^2} + \frac{e^x}{e^x + 3}\right) dx$

43. $\int \frac{x+1}{\sqrt{x}-1}\,dx$

Hint: Let $u = \sqrt{x} - 1$.

44. $\int \frac{e^{-u}-1}{e^{-u}+u}\,du$

Hint: Let $v = e^{-u} + u$.

45. $\int x(x-1)^5\,dx$

Hint: $u = x - 1$ implies $x = u + 1$.

46. $\int \frac{t}{t+1}\,dt$

Hint: $\frac{t}{t+1} = 1 - \frac{1}{t+1}$.

47. $\int \frac{1-\sqrt{x}}{1+\sqrt{x}}\,dx$

Hint: Let $u = 1 + \sqrt{x}$.

48. $\int \frac{1+\sqrt{x}}{1-\sqrt{x}}\,dx$

Hint: Let $u = 1 - \sqrt{x}$.

49. $\int v^2(1-v)^6\,dv$

Hint: Let $u = 1 - v$.

50. $\int x^3(x^2+1)^{3/2}\,dx$

Hint: Let $u = x^2 + 1$.

In Exercises 51–54, find the function f given that the slope of the tangent line to the graph of f at any point $(x, f(x))$ is $f'(x)$ and that the graph of f passes through the given point.

51. $f'(x) = 5(2x-1)^4$; $(1, 3)$

52. $f'(x) = \frac{3x^2}{2\sqrt{x^3-1}}$; $(1, 1)$

53. $f'(x) = -2xe^{-x^2+1}$; $(1, 0)$

54. $f'(x) = 1 - \frac{2x}{x^2+1}$; $(0, 2)$

55. Cable Telephone Subscribers The number of cable telephone subscribers stood at 3.2 million at the beginning of 2004 ($t = 0$). For the next 5 yr, the number was projected to grow at the rate of

$$R(t) = 3.36(t+1)^{0.05} \qquad (0 \le t \le 5)$$

million subscribers/year. If the projection held true, how many cable telephone subscribers were there at the beginning of 2008 ($t = 4$)?

Source: Sanford C. Bernstein

56. TV Viewers: Newsmagazine Shows The number of viewers of a weekly TV newsmagazine show, introduced in the 2003 season, has been increasing at the rate of

$$3\left(2 + \frac{1}{2}t\right)^{-1/3} \qquad (1 \le t \le 6)$$

million viewers/year in its tth year on the air. The number of viewers of the program during its first year on the air is given by $9(5/2)^{2/3}$ million. Find how many viewers were expected in the 2008 season.

57. Student Enrollment The registrar of Kellogg University estimates that the total student enrollment in the Continuing Education division will grow at the rate of

$$N'(t) = 2000(1 + 0.2t)^{-3/2}$$

students/year, t yr from now. If the current student enrollment is 1000, find an expression giving the total student enrollment t yr from now. What will be the student enrollment 5 yr from now?

58. TV on Mobile Phones The number of people watching TV on mobile phones is expected to grow at the rate of

$$N'(t) = \frac{5.4145}{\sqrt{1 + 0.91t}} \qquad (0 \le t \le 4)$$

million/year. The number of people watching TV on mobile phones at the beginning of 2007 ($t = 0$) was 11.9 million.

a. Find an expression giving the number of people watching TV on mobile phones in year t.

b. According to this projection, how many people will be watching TV on mobile phones at the beginning of 2011?

Source: International Data Corporation, U.S. forecast

59. Demand: Women's Boots The rate of change of the unit price p (in dollars) of Apex women's boots is given by

$$p'(x) = \frac{-250x}{(16 + x^2)^{3/2}}$$

where x is the quantity demanded daily in units of a hundred. Find the demand function for these boots if the quantity demanded daily is 300 pairs ($x = 3$) when the unit price is \$50/pair.

60. Population Growth The population of a certain city is projected to grow at the rate of

$$r(t) = 400\left(1 + \frac{2t}{24 + t^2}\right) \qquad (0 \le t \le 5)$$

people/year, t years from now. The current population is 60,000. What will be the population 5 yr from now?

61. Oil Spill In calm waters, the oil spilling from the ruptured hull of a grounded tanker forms an oil slick that is circular in shape. If the radius r of the circle is increasing at the rate of

$$r'(t) = \frac{30}{\sqrt{2t + 4}}$$

feet/minute t min after the rupture occurs, find an expression for the radius at any time t. How large is the polluted area 16 min after the rupture occurred?

Hint: $r(0) = 0$.

62. Life Expectancy of a Female Suppose in a certain country the life expectancy at birth of a female is changing at the rate of

$$g'(t) = \frac{5.45218}{(1 + 1.09t)^{0.9}}$$

years/year. Here, t is measured in years, with $t = 0$ corresponding to the beginning of 1900. Find an expression $g(t)$ giving the life expectancy at birth (in years) of a female in that country if the life expectancy at the beginning of 1900 is 50.02 yr. What is the life expectancy at birth of a female born in 2000 in that country?

63. **Average Birth Height of Boys** Using data collected at Kaiser Hospital, pediatricians estimate that the average height of male children changes at the rate of

$$h'(t) = \frac{52.8706e^{-0.3277t}}{(1 + 2.449e^{-0.3277t})^2}$$

inches/year, where the child's height $h(t)$ is measured in inches and t, the child's age, is measured in years, with $t = 0$ corresponding to the age at birth. Find an expression $h(t)$ for the average height of a boy at age t if the height at birth of an average child is 19.4 in. What is the height of an average 8-yr-old boy?

64. **Learning Curves** The average student enrolled in the 20-wk Court Reporting I course at the American Institute of Court Reporting progresses according to the rule

$$N'(t) = 6e^{-0.05t} \qquad (0 \le t \le 20)$$

where $N'(t)$ measures the rate of change in the number of words/minute of dictation the student takes in machine shorthand after t wk in the course. Assuming that the average student enrolled in this course begins with a dictation speed of 60 words/minute, find an expression $N(t)$ that gives the dictation speed of the student after t wk in the course.

65. **Amount of Glucose in the Bloodstream** Suppose a patient is given a continuous intravenous infusion of glucose at a constant rate of r mg/min. Then, the rate at which the amount of glucose in the bloodstream is changing at time t due to this infusion is given by

$$A'(t) = re^{-at}$$

mg/min, where a is a positive constant associated with the rate at which excess glucose is eliminated from the bloodstream and is dependent on the patient's metabolism rate. Derive an expression for the amount of glucose in the bloodstream at time t.
Hint: $A(0) = 0$.

66. **Concentration of a Drug in an Organ** A drug is carried into an organ of volume V cm^3 by a liquid that enters the organ at the rate of a cm^3/sec and leaves it at the rate of b cm^3/sec. The concentration of the drug in the liquid entering the organ is c g/cm^3. If the concentration of the drug in the organ at time t is increasing at the rate of

$$x'(t) = \frac{1}{V}(ac - bx_0)e^{-bt/V}$$

g/cm^3/sec, and the concentration of the drug in the organ initially is x_0 g/cm^3, show that the concentration of the drug in the organ at time t is given by

$$x(t) = \frac{ac}{b} + \left(x_0 - \frac{ac}{b}\right)e^{-bt/V}$$

In Exercises 67 and 68, determine whether the statement is true or false. If it is true, explain why it is true. If it is false, give an example to show why it is false.

67. If f is continuous, then $\int xf(x^2)\,dx = \frac{1}{2}\int f(x)\,dx$.

68. If f is continuous, then $\int f(ax + b)\,dx = a\int f(x)\,dx$.

6.2 Solutions to Self-Check Exercises

1. Let $u = 2x + 5$. Then, $du = 2\,dx$, or $dx = \frac{1}{2}\,du$. Making the appropriate substitutions, we have

$$\begin{aligned}\int \sqrt{2x+5}\,dx &= \int \sqrt{u}\left(\frac{1}{2}\,du\right) = \frac{1}{2}\int u^{1/2}\,du \\ &= \frac{1}{2}\left(\frac{2}{3}\right)u^{3/2} + C \\ &= \frac{1}{3}(2x+5)^{3/2} + C\end{aligned}$$

2. Let $u = 2x^3 + 1$, so that $du = 6x^2\,dx$, or $x^2\,dx = \frac{1}{6}\,du$. Making the appropriate substitutions, we have

$$\begin{aligned}\int \frac{x^2}{(2x^3+1)^{3/2}}\,dx &= \int \frac{\left(\frac{1}{6}\right)du}{u^{3/2}} = \frac{1}{6}\int u^{-3/2}\,du \\ &= \left(\frac{1}{6}\right)(-2)u^{-1/2} + C \\ &= -\frac{1}{3}(2x^3+1)^{-1/2} + C \\ &= -\frac{1}{3\sqrt{2x^3+1}} + C\end{aligned}$$

3. Let $u = 2x^2 - 1$, so that $du = 4x\,dx$, or $x\,dx = \frac{1}{4}\,du$. Then,

$$\begin{aligned}\int xe^{2x^2-1}\,dx &= \frac{1}{4}\int e^u\,du \\ &= \frac{1}{4}e^u + C \\ &= \frac{1}{4}e^{2x^2-1} + C\end{aligned}$$

4. Let $C(t)$ denote the CO concentration in the air due to automobile exhaust t yr from now. Then,

$$\begin{aligned}C'(t) = f(t) &= \frac{8(0.1t+1)}{300(0.2t^2 + 4t + 64)^{1/3}} \\ &= \frac{8}{300}(0.1t+1)(0.2t^2 + 4t + 64)^{-1/3}\end{aligned}$$

Integrating, we find

$$\begin{aligned}C(t) &= \int \frac{8}{300}(0.1t+1)(0.2t^2 + 4t + 64)^{-1/3}\,dt \\ &= \frac{8}{300}\int (0.1t+1)(0.2t^2 + 4t + 64)^{-1/3}\,dt\end{aligned}$$

Let $u = 0.2t^2 + 4t + 64$, so that $du = (0.4t + 4)\,dt = 4(0.1t + 1)\,dt$, or

$$(0.1t + 1)\,dt = \frac{1}{4}\,du$$

Then,

$$\begin{aligned} C(t) &= \frac{8}{300}\left(\frac{1}{4}\right)\int u^{-1/3}\,du \\ &= \frac{1}{150}\left(\frac{3}{2}u^{2/3}\right) + k \\ &= 0.01(0.2t^2 + 4t + 64)^{2/3} + k \end{aligned}$$

where k is an arbitrary constant. To determine the value of k, we use the condition $C(0) = 0.16$, obtaining

$$\begin{aligned} C(0) = 0.16 &= 0.01(64)^{2/3} + k \\ 0.16 &= 0.16 + k \\ k &= 0 \end{aligned}$$

Therefore,

$$C(t) = 0.01(0.2t^2 + 4t + 64)^{2/3}$$

6.3 Area and the Definite Integral

An Intuitive Look

Suppose a certain state's annual rate of petroleum consumption over a 4-year period is constant and is given by the function

$$f(t) = 1.2 \qquad (0 \leq t \leq 4)$$

where t is measured in years and $f(t)$ in millions of barrels per year. Then, the state's total petroleum consumption over the period of time in question is

$$(1.2)(4 - 0) \qquad \text{Rate of consumption} \times \text{Time elapsed}$$

or 4.8 million barrels. If you examine the graph of f shown in Figure 5, you will see that this total is just the area of the rectangular region bounded above by the graph of f, below by the t-axis, and to the left and right by the vertical lines $t = 0$ (the y-axis) and $t = 4$, respectively.

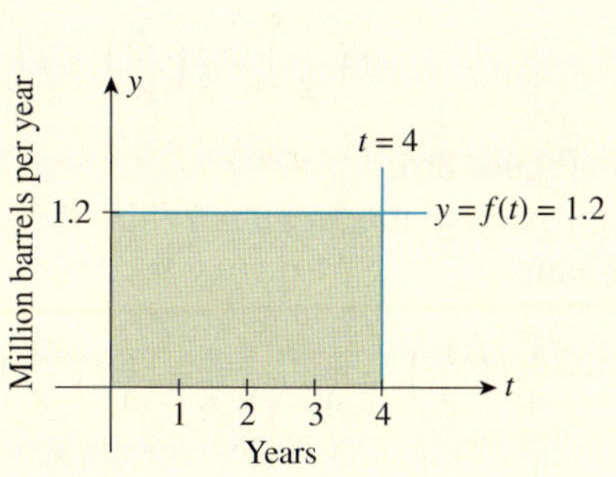

FIGURE 5
The total petroleum consumption is given by the area of the rectangular region.

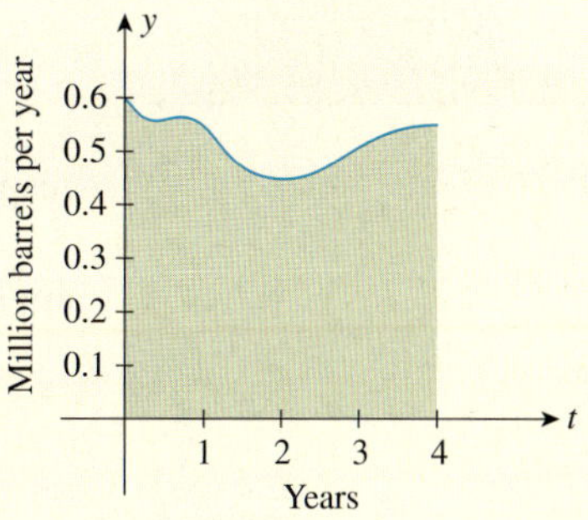

FIGURE 6
The daily petroleum consumption is given by the "area" of the shaded region.

Figure 6 shows the actual petroleum consumption of a certain New England state over a 4-year period from 1990 ($t = 0$) to 1994 ($t = 4$). Observe that the rate of consumption is not constant; that is, the function f is not a constant function. What is the state's total petroleum consumption over this 4-year period? It seems reasonable to conjecture that it is given by the "area" of the region bounded above by the graph of f, below by the t-axis, and to the left and right by the vertical lines $t = 0$ and $t = 4$, respectively.

This example raises two questions:

1. What is the "area" of the region shown in Figure 6?
2. How do we compute this area?

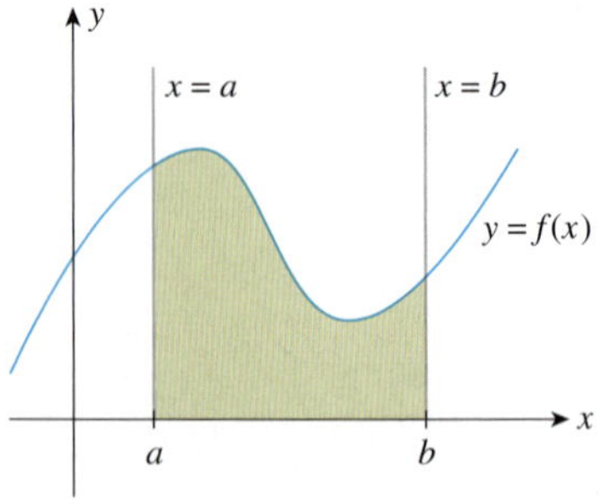

FIGURE 7
The area under the graph of f on $[a, b]$

The Area Problem

The preceding example touches on the second fundamental problem in calculus: Calculate the area of the region bounded by the graph of a nonnegative function f, the x-axis, and the vertical lines $x = a$ and $x = b$ (Figure 7). This area is called the **area under the graph of *f*** on the interval $[a, b]$, or from a to b.

Defining Area—Two Examples

Just as we used the slopes of secant lines (quantities that we could compute) to help us define the slope of the tangent line to a point on the graph of a function, we now adopt a parallel approach and use the areas of rectangles (quantities that we can compute) to help us define the area under the graph of a function. We begin by looking at a specific example.

VIDEO

EXAMPLE 1 Let $f(x) = x^2$ and consider the region R under the graph of f on the interval $[0, 1]$ (Figure 8a). To obtain an approximation of the area of R, let's construct four nonoverlapping rectangles as follows: Divide the interval $[0, 1]$ into four subintervals

$$\left[0, \frac{1}{4}\right], \quad \left[\frac{1}{4}, \frac{1}{2}\right], \quad \left[\frac{1}{2}, \frac{3}{4}\right], \quad \left[\frac{3}{4}, 1\right]$$

of equal length $\frac{1}{4}$. Next, construct four rectangles with these subintervals as bases and with heights given by the values of the function at the midpoints

$$\frac{1}{8}, \quad \frac{3}{8}, \quad \frac{5}{8}, \quad \frac{7}{8}$$

of each subinterval. Then, each of these rectangles has width $\frac{1}{4}$ and height

$$f\left(\frac{1}{8}\right), \quad f\left(\frac{3}{8}\right), \quad f\left(\frac{5}{8}\right), \quad f\left(\frac{7}{8}\right)$$

respectively (Figure 8b).

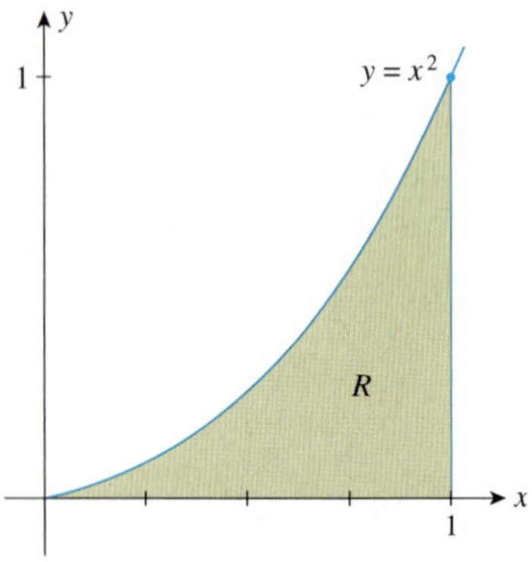

(a)

If we approximate the area A of R by the sum of the areas of the four rectangles, we obtain

$$\begin{aligned}
A &\approx \frac{1}{4}f\left(\frac{1}{8}\right) + \frac{1}{4}f\left(\frac{3}{8}\right) + \frac{1}{4}f\left(\frac{5}{8}\right) + \frac{1}{4}f\left(\frac{7}{8}\right) \\
&= \frac{1}{4}\left[f\left(\frac{1}{8}\right) + f\left(\frac{3}{8}\right) + f\left(\frac{5}{8}\right) + f\left(\frac{7}{8}\right)\right] \\
&= \frac{1}{4}\left[\left(\frac{1}{8}\right)^2 + \left(\frac{3}{8}\right)^2 + \left(\frac{5}{8}\right)^2 + \left(\frac{7}{8}\right)^2\right] \qquad \text{Recall that } f(x) = x^2. \\
&= \frac{1}{4}\left(\frac{1}{64} + \frac{9}{64} + \frac{25}{64} + \frac{49}{64}\right) = \frac{21}{64}
\end{aligned}$$

or approximately 0.328125 square unit. ■

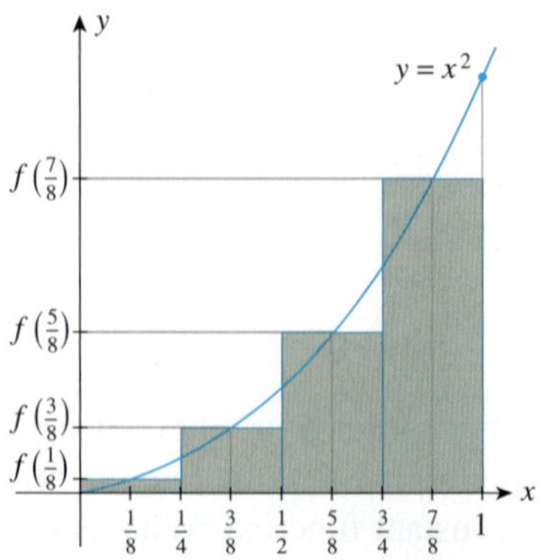

(b)

FIGURE 8
The area of the region under the graph of f on $[0, 1]$ in (a) is approximated by the sum of the areas of the four rectangles in (b).

Following the procedure of Example 1, we can obtain approximations of the area of the region R using any number n of rectangles ($n = 4$ in Example 1). Figure 9a shows the approximation of the area A of R using 8 rectangles ($n = 8$), and Figure 9b shows the approximation of the area A of R using 16 rectangles.

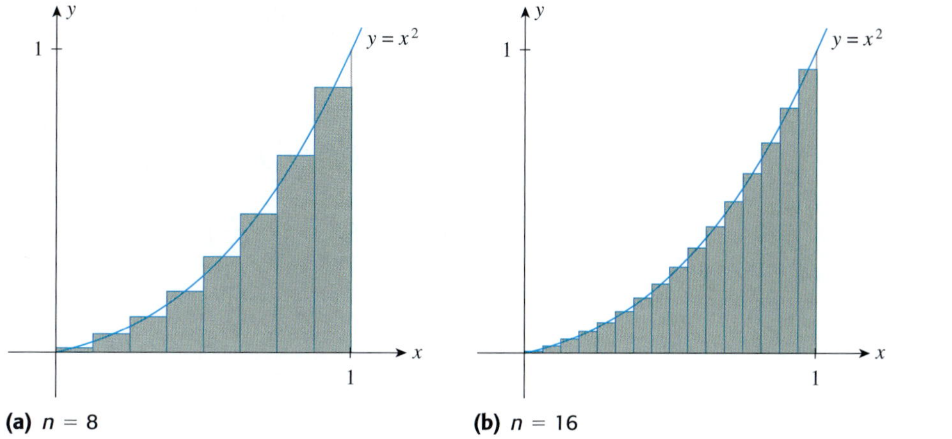

FIGURE 9
As n increases, the number of rectangles increases, and the approximation improves.

These figures suggest that the approximations seem to get better as n increases. This is borne out by the results given in Table 1, which were obtained using a computer.

TABLE 1

Number of Rectangles, n	4	8	16	32	64	100	200
Approximation of A	0.328125	0.332031	0.333008	0.333252	0.333313	0.333325	0.333331

Our computations seem to suggest that the approximations approach the number $\frac{1}{3}$ as n gets larger and larger. This result suggests that we *define* the area of the region under the graph of $f(x) = x^2$ on the interval $[0, 1]$ to be $\frac{1}{3}$ square unit.

In Example 1, we chose the *midpoint* of each subinterval as the point at which to evaluate $f(x)$ to obtain the height of the approximating rectangle. Let's consider another example, this time choosing the *left endpoint* of each subinterval.

VIDEO

EXAMPLE 2 Let R be the region under the graph of $f(x) = 16 - x^2$ on the interval $[1, 3]$. Find an approximation of the area A of R using four subintervals of $[1, 3]$ of equal length and picking the left endpoint of each subinterval to evaluate $f(x)$ to obtain the height of the approximating rectangle.

Solution The graph of f is sketched in Figure 10a. Since the length of $[1, 3]$ is 2, we see that the length of each subinterval is $\frac{2}{4}$, or $\frac{1}{2}$. Therefore, the four subintervals are

$$\left[1, \frac{3}{2}\right], \quad \left[\frac{3}{2}, 2\right], \quad \left[2, \frac{5}{2}\right], \quad \left[\frac{5}{2}, 3\right]$$

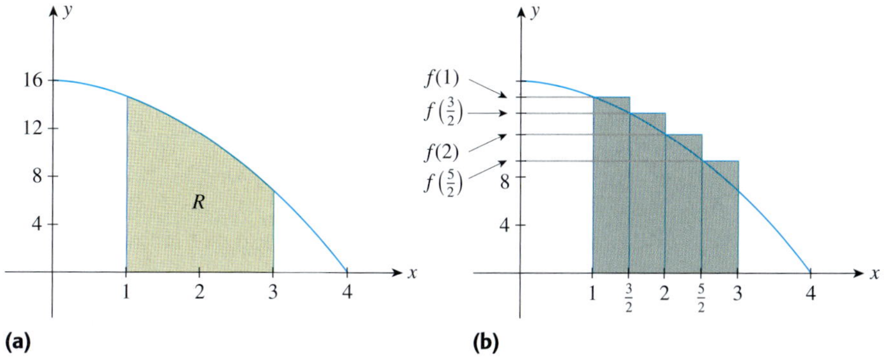

FIGURE 10
The area of R in (a) is approximated by the sum of the areas of the four rectangles in (b).

The left endpoints of these subintervals are 1, $\frac{3}{2}$, 2, and $\frac{5}{2}$, respectively, so the heights of the approximating rectangles are $f(1)$, $f(\frac{3}{2})$, $f(2)$, and $f(\frac{5}{2})$, respectively (Figure 10b). Therefore, the required approximation is

$$\begin{aligned} A &\approx \frac{1}{2}f(1) + \frac{1}{2}f\left(\frac{3}{2}\right) + \frac{1}{2}f(2) + \frac{1}{2}f\left(\frac{5}{2}\right) \\ &= \frac{1}{2}\left[f(1) + f\left(\frac{3}{2}\right) + f(2) + f\left(\frac{5}{2}\right)\right] \\ &= \frac{1}{2}\left\{[16 - (1)^2] + \left[16 - \left(\frac{3}{2}\right)^2\right]\right. \\ &\quad \left. + [16 - (2)^2] + \left[16 - \left(\frac{5}{2}\right)^2\right]\right\} \\ &= \frac{1}{2}\left(15 + \frac{55}{4} + 12 + \frac{39}{4}\right) = \frac{101}{4} \end{aligned}$$

Recall that $f(x) = 16 - x^2$.

or approximately 25.25 square units.

Table 2 shows the approximations of the area A of the region R of Example 2 when n rectangles are used for the approximation and the heights of the approximating rectangles are found by evaluating $f(x)$ at the left endpoints.

TABLE 2

Number of Rectangles, n	4	10	100	1,000	10,000	50,000	100,000
Approximation of A	25.2500	24.1200	23.4132	23.3413	23.3341	23.3335	23.3334

Once again, we see that the approximations seem to approach a unique number as n gets larger and larger—this time the number is $23\frac{1}{3}$. This result suggests that we *define* the area of the region under the graph of $f(x) = 16 - x^2$ on the interval [1, 3] to be $23\frac{1}{3}$ square units.

Defining Area—The General Case

Examples 1 and 2 point the way to defining the area A of the region R under the graph of an arbitrary but continuous, nonnegative function f on an interval $[a, b]$ (Figure 11a).

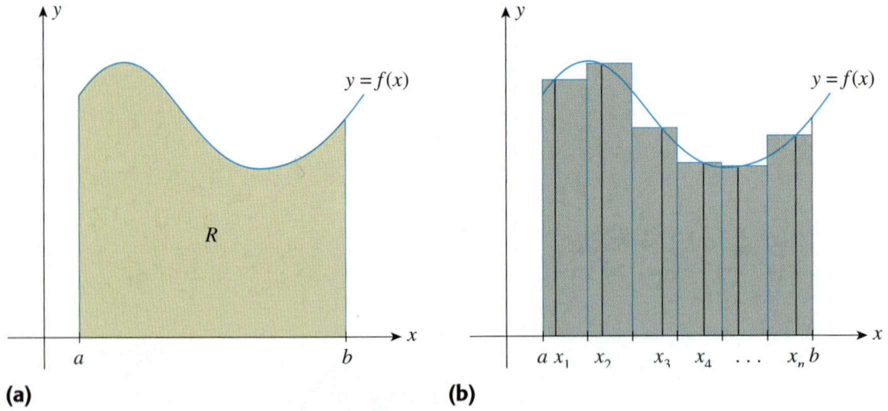

FIGURE 11
The area of the region under the graph of f on $[a, b]$ in (a) is approximated by the sum of the areas of the n rectangles shown in (b).

Divide the interval $[a, b]$ into n subintervals of equal length $\Delta x = (b - a)/n$. Next, pick n arbitrary points $x_1, x_2, \ldots, x_n$, called *representative points*, from the first, second, . . . , and nth subintervals, respectively (Figure 11b). Then, approximating the area A of the region R by the n rectangles of width Δx and heights $f(x_1), f(x_2), \ldots, f(x_n)$, so that the areas of the rectangles are $f(x_1)\Delta x, f(x_2)\Delta x, \ldots, f(x_n)\Delta x$, we have

$$A \approx f(x_1)\Delta x + f(x_2)\Delta x + \cdots + f(x_n)\Delta x$$

The sum on the right-hand side of this expression is called a **Riemann sum** in honor of the German mathematician Bernhard Riemann (1826–1866). Now, as the earlier examples seem to suggest, the Riemann sum will approach a unique number as n becomes arbitrarily large.* We define this number to be the area A of the region R.

The Area under the Graph of a Function

Let f be a nonnegative continuous function on $[a, b]$. Then, the area of the region under the graph of f is

$$A = \lim_{n \to \infty} [f(x_1) + f(x_2) + \cdots + f(x_n)]\Delta x \quad \textbf{(6)}$$

where $x_1, x_2, \ldots, x_n$ are arbitrary points in the n subintervals of $[a, b]$ of equal width $\Delta x = (b - a)/n$.

The Definite Integral

As we have just seen, the area under the graph of a continuous *nonnegative* function f on an interval $[a, b]$ is defined by the limit of the Riemann sum

$$\lim_{n \to \infty} [f(x_1)\Delta x + f(x_2)\Delta x + \cdots + f(x_n)\Delta x]$$

We now turn our attention to the study of limits of Riemann sums involving functions that are not necessarily nonnegative. Such limits arise in many applications of calculus.

For example, the calculation of the distance covered by a body traveling along a straight line involves evaluating a limit of this form. The computation of the total revenue realized by a company over a certain time period, the calculation of the total amount of electricity consumed in a typical home over a 24-hour period, the average concentration of a drug in a body over a certain interval of time, and the volume of a solid—all involve limits of this type.

We begin with the following definition.

The Definite Integral

Let f be a continuous function defined on $[a, b]$. If

$$\lim_{n \to \infty} [f(x_1)\Delta x + f(x_2)\Delta x + \cdots + f(x_n)\Delta x]$$

exists for all choices of representative points $x_1, x_2, \ldots, x_n$ in the n subintervals of $[a, b]$ of equal width $\Delta x = (b - a)/n$, then this limit is called the **definite integral** of f from a to b and is denoted by $\int_a^b f(x)\,dx$. Thus,

$$\int_a^b f(x)\,dx = \lim_{n \to \infty} [f(x_1)\Delta x + f(x_2)\Delta x + \cdots + f(x_n)\Delta x] \quad \textbf{(7)}$$

The number a is the **lower limit of integration**, and the number b is the **upper limit of integration.**

---

*Even though we chose the representative points to be the midpoints of the subintervals in Example 1 and the left endpoints in Example 2, it can be shown that each of the respective sums will always approach a unique number as n approaches infinity.

Notes

1. If f is nonnegative, then the limit in (7) is the same as the limit in (6); therefore, the definite integral gives the area under the graph of f on $[a, b]$.
2. The limit in (7) is denoted by the integral sign $\int$ because, as we will see later, the definite integral and the antiderivative of a function f are related.
3. It is important to realize that the definite integral $\int_a^b f(x)\,dx$ is a *number*, whereas the indefinite integral $\int f(x)\,dx$ represents a *family of functions* (the antiderivatives of f).
4. If the limit in (7) exists, we say that f is **integrable** on the interval $[a, b]$. ■

When Is a Function Integrable?

The following theorem, which we state without proof, guarantees that a continuous function is integrable.

Integrability of a Function

Let f be continuous on $[a, b]$. Then, f is integrable on $[a, b]$; that is, the definite integral $\int_a^b f(x)\,dx$ exists.

Geometric Interpretation of the Definite Integral

If f is nonnegative and integrable on $[a, b]$, then we have the following geometric interpretation of the definite integral $\int_a^b f(x)\,dx$.

Geometric Interpretation of $\int_a^b f(x)\,dx$ for $f(x) \geq 0$ on $[a, b]$

If f is nonnegative and continuous on $[a, b]$, then

$$\int_a^b f(x)\,dx \tag{8}$$

is equal to the area of the region under the graph of f on $[a, b]$ (Figure 12).

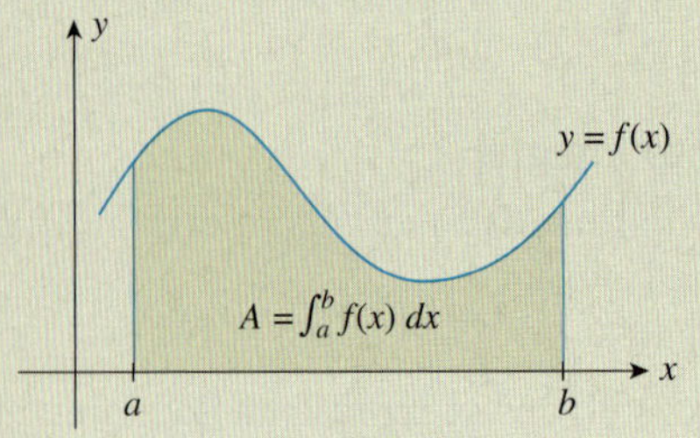

FIGURE 12
If $f(x) \geq 0$ on $[a, b]$, then $\int_a^b f(x)\,dx$ = area of the region under the graph of f on $[a, b]$.

Explore & Discuss

Suppose f is nonpositive [that is, $f(x) \leq 0$] and continuous on $[a, b]$. Explain why the area of the region below the x-axis and above the graph of f is given by $-\int_a^b f(x)\,dx$.

Next, let's extend our geometric interpretation of the definite integral to include the case where f assumes both positive as well as negative values on $[a, b]$. Consider a typical Riemann sum of the function f,

$$f(x_1)\Delta x + f(x_2)\Delta x + \cdots + f(x_n)\Delta x$$

corresponding to a partition of $[a, b]$ into n subintervals of equal width $(b - a)/n$, where $x_1, x_2, \ldots, x_n$ are representative points in the subintervals. The sum consists of n terms in which a positive term corresponds to the area of a rectangle of height $f(x_k)$ (for some positive integer k) lying above the x-axis and a negative term corresponds to the area of a rectangle of height $-f(x_k)$ lying below the x-axis. (See Figure 13, which depicts a situation with $n = 6$.)

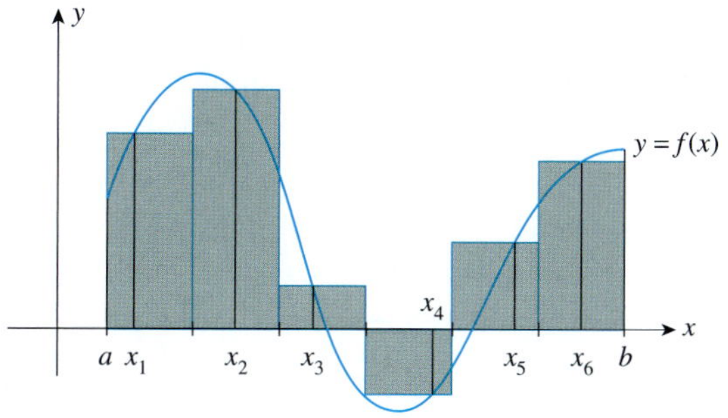

FIGURE 13
The positive terms in the Riemann sum are associated with the areas of the rectangles that lie above the x-axis, and the negative terms are associated with the areas of those that lie below the x-axis.

As n gets larger and larger, the sums of the areas of the rectangles lying above the x-axis seem to give a better and better approximation of the area of the region lying above the x-axis (Figure 14). Similarly, the sums of the areas of those rectangles lying below the x-axis seem to give a better and better approximation of the area of the region lying below the x-axis.

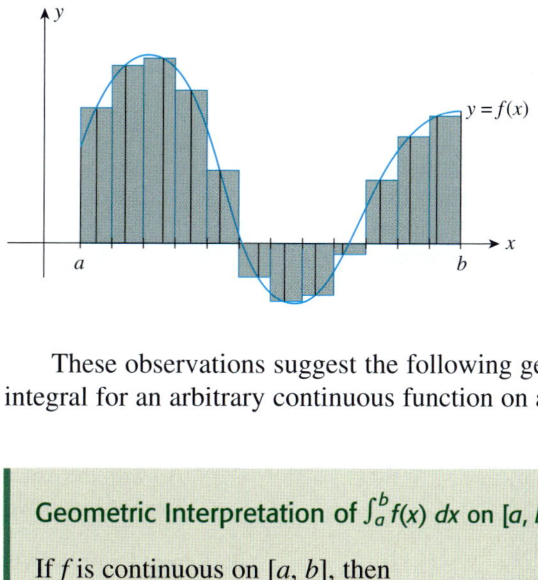

FIGURE 14
As n gets larger, the approximations get better. Here, $n = 12$ and we are approximating with twice as many rectangles as in Figure 13.

These observations suggest the following geometric interpretation of the definite integral for an arbitrary continuous function on an interval $[a, b]$.

Geometric Interpretation of $\int_a^b f(x)\,dx$ on $[a, b]$

If f is continuous on $[a, b]$, then

$$\int_a^b f(x)\,dx$$

is equal to the area of the region above $[a, b]$ minus the area of the region below $[a, b]$ (Figure 15).

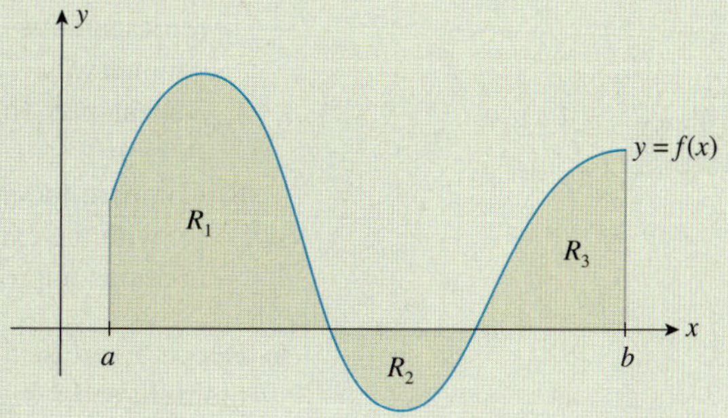

FIGURE 15
$\int_a^b f(x)\,dx$ = Area of R_1
− Area of R_2
+ Area of R_3

6.3 Self-Check Exercise

Find an approximation of the area of the region R under the graph of $f(x) = 2x^2 + 1$ on the interval [0, 3], using four subintervals of [0, 3] of equal length and picking the midpoint of each subinterval as a representative point.

The solution to Self-Check Exercise 6.3 can be found on page 430.

6.3 Concept Questions

1. Explain how you would define the area of the region under the graph of a nonnegative continuous function f on the interval $[a, b]$.

2. Define the definite integral of a continuous function on the interval $[a, b]$. Give a geometric interpretation of $\int_a^b f(x)\,dx$ for the case where (a) f is nonnegative on $[a, b]$ and (b) f assumes both positive as well as negative values on $[a, b]$. Illustrate your answers graphically.

6.3 Exercises

In Exercises 1 and 2, find an approximation of the area of the region *R* under the graph of *f* by computing the Riemann sum of *f* corresponding to the partition of the interval into the subintervals shown in the accompanying figures. In each case, use the midpoints of the subintervals as the representative points.

1.

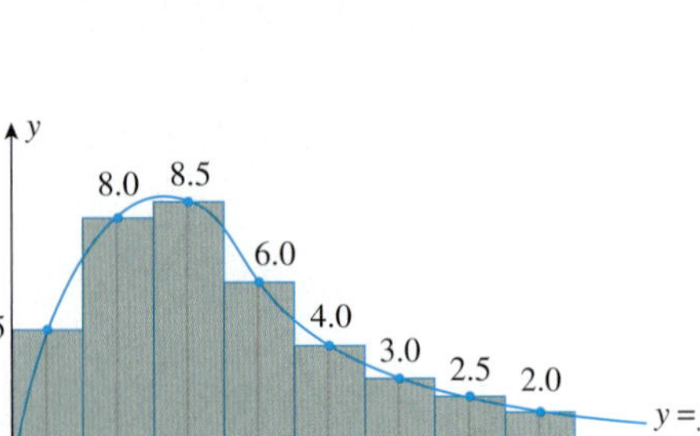

2.

3. Let $f(x) = 3x$.
   a. Sketch the region R under the graph of f on the interval [0, 2] and find its exact area using geometry.
   b. Use a Riemann sum with four subintervals of equal length ($n = 4$) to approximate the area of R. Choose the representative points to be the left endpoints of the subintervals.
   c. Repeat part (b) with eight subintervals of equal length ($n = 8$).
   d. Compare the approximations obtained in parts (b) and (c) with the exact area found in part (a). Do the approximations improve with larger n?

4. Repeat Exercise 3, choosing the representative points to be the right endpoints of the subintervals.

5. Let $f(x) = 4 - 2x$.
   a. Sketch the region R under the graph of f on the interval [0, 2] and find its exact area using geometry.
   b. Use a Riemann sum with five subintervals of equal length ($n = 5$) to approximate the area of R. Choose the representative points to be the left endpoints of the subintervals.
   c. Repeat part (b) with ten subintervals of equal length ($n = 10$).
   d. Compare the approximations obtained in parts (b) and (c) with the exact area found in part (a). Do the approximations improve with larger n?

6. Repeat Exercise 5, choosing the representative points to be the right endpoints of the subintervals.

7. Let $f(x) = x^2$ and compute the Riemann sum of f over the interval [2, 4], using
   a. Two subintervals of equal length ($n = 2$).
   b. Five subintervals of equal length ($n = 5$).
   c. Ten subintervals of equal length ($n = 10$).
   In each case, choose the representative points to be the midpoints of the subintervals.
   d. Can you guess at the area of the region under the graph of f on the interval [2, 4]?

8. Repeat Exercise 7, choosing the representative points to be the left endpoints of the subintervals.

9. Repeat Exercise 7, choosing the representative points to be the right endpoints of the subintervals.

10. Let $f(x) = x^3$ and compute the Riemann sum of f over the interval [0, 1], using
   a. Two subintervals of equal length ($n = 2$).
   b. Five subintervals of equal length ($n = 5$).
   c. Ten subintervals of equal length ($n = 10$).
   In each case, choose the representative points to be the midpoints of the subintervals.
   d. Can you guess at the area of the region under the graph of f on the interval [0, 1]?

11. Repeat Exercise 10, choosing the representative points to be the left endpoints of the subintervals.

12. Repeat Exercise 10, choosing the representative points to be the right endpoints of the subintervals.

In Exercises 13–16, find an approximation of the area of the region *R* under the graph of the function *f* on the interval [*a*, *b*]. In each case, use *n* subintervals and choose the representative points as indicated.

13. $f(x) = x^2 + 1$; [0, 2]; $n = 5$; midpoints

14. $f(x) = 4 - x^2$; [−1, 2]; $n = 6$; left endpoints

15. $f(x) = \dfrac{1}{x}$; [1, 3]; $n = 4$; right endpoints

16. $f(x) = e^x$; [0, 3]; $n = 5$; midpoints

17. **Real Estate** Figure (a) shows a vacant lot with a 100-ft frontage in a development. To estimate its area, we introduce a coordinate system so that the x-axis coincides with the edge of the straight road forming the lower boundary of the property, as shown in Figure (b). Then, thinking of the upper boundary of the property as the graph of a continuous function f over the interval [0, 100], we see that the problem is mathematically equivalent to that of finding the area under the graph of f on [0, 100]. To estimate the area of the lot using a Riemann sum, we divide the interval [0, 100] into five equal subintervals of length 20 ft. Then, using surveyor's equipment, we measure the distance from the midpoint of each of these subintervals to the upper boundary of the property. These measurements give the values of $f(x)$ at $x = 10, 30, 50, 70$, and 90. What is the approximate area of the lot?

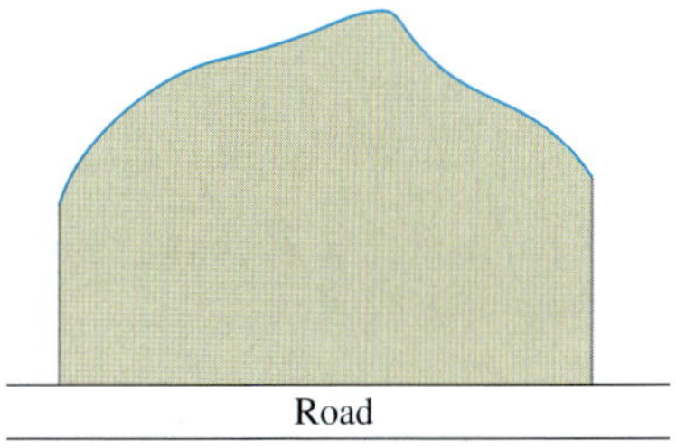

(a)

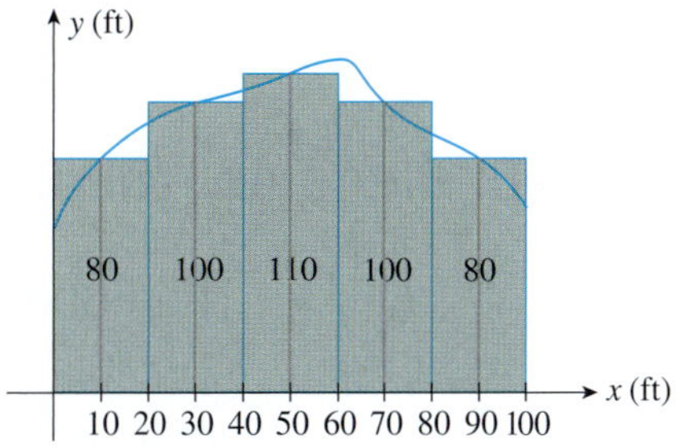

(b)

18. **Real Estate** Use the technique of Exercise 17 to obtain an estimate of the area of the vacant lot shown in the accompanying figures.

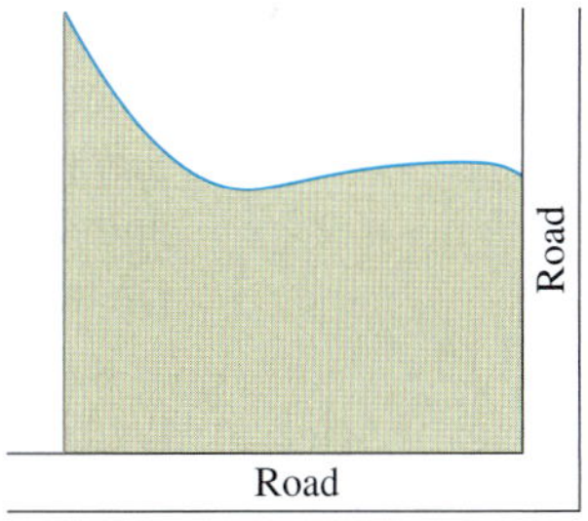

(a)

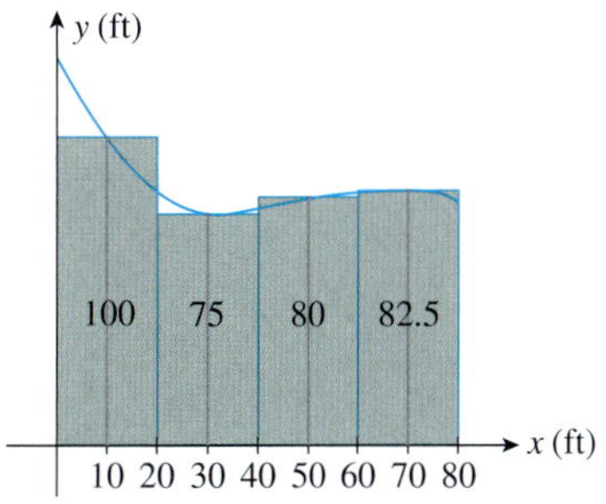

(b)

6.3 Solutions to Self-Check Exercises

The length of each subinterval is $\frac{3}{4}$. Therefore, the four subintervals are

$$\left[0, \frac{3}{4}\right], \quad \left[\frac{3}{4}, \frac{3}{2}\right], \quad \left[\frac{3}{2}, \frac{9}{4}\right], \quad \left[\frac{9}{4}, 3\right]$$

The representative points are $\frac{3}{8}$, $\frac{9}{8}$, $\frac{15}{8}$, and $\frac{21}{8}$, respectively. Therefore, the required approximation is

$$\begin{aligned} A &= \frac{3}{4}f\left(\frac{3}{8}\right) + \frac{3}{4}f\left(\frac{9}{8}\right) + \frac{3}{4}f\left(\frac{15}{8}\right) + \frac{3}{4}f\left(\frac{21}{8}\right) \\ &= \frac{3}{4}\left[f\left(\frac{3}{8}\right) + f\left(\frac{9}{8}\right) + f\left(\frac{15}{8}\right) + f\left(\frac{21}{8}\right)\right] \\ &= \frac{3}{4}\left\{\left[2\left(\frac{3}{8}\right)^2 + 1\right] + \left[2\left(\frac{9}{8}\right)^2 + 1\right] + \left[2\left(\frac{15}{8}\right)^2 + 1\right] \right. \\ &\qquad \left. + \left[2\left(\frac{21}{8}\right)^2 + 1\right]\right\} \\ &= \frac{3}{4}\left(\frac{41}{32} + \frac{113}{32} + \frac{257}{32} + \frac{473}{32}\right) = \frac{663}{32} \end{aligned}$$

or approximately 20.72 square units.

6.4 The Fundamental Theorem of Calculus

The Fundamental Theorem of Calculus

In Section 6.3, we defined the definite integral of an arbitrary continuous function on an interval $[a, b]$ as a limit of Riemann sums. Calculating the value of a definite integral by actually taking the limit of such sums is tedious and in most cases impractical. It is important to realize that the numerical results we obtained in Examples 1 and 2 of Section 6.3 were *approximations* of the respective areas of the regions in question, even though these results enabled us to *conjecture* what the actual areas might be. Fortunately, there is a much better way of finding the exact value of a definite integral.

The following theorem shows how to evaluate the definite integral of a continuous function provided we can find an antiderivative of that function. Because of its importance in establishing the relationship between differentiation and integration, this theorem—discovered independently by Sir Isaac Newton (1642–1727) in England and Gottfied Wilhelm Leibniz (1646–1716) in Germany—is called the **fundamental theorem of calculus.**

THEOREM 2

The Fundamental Theorem of Calculus

Let f be continuous on $[a, b]$. Then,

$$\int_a^b f(x)\,dx = F(b) - F(a) \tag{9}$$

where F is any antiderivative of f; that is, $F'(x) = f(x)$.

We will explain why this theorem is true at the end of this section.

When applying the fundamental theorem of calculus, it is convenient to use the notation

$$F(x)\Big|_a^b = F(b) - F(a)$$

For example, using this notation, Equation (9) is written

$$\int_a^b f(x)\,dx = F(x)\Big|_a^b = F(b) - F(a)$$

VIDEO

EXAMPLE 1 Let R be the region under the graph of $f(x) = x$ on the interval $[1, 3]$. Use the fundamental theorem of calculus to find the area A of R and verify your result by elementary means.

Solution The region R is shown in Figure 16a. Since f is nonnegative on $[1, 3]$, the area of R is given by the definite integral of f from 1 to 3; that is,

$$A = \int_1^3 x\,dx$$

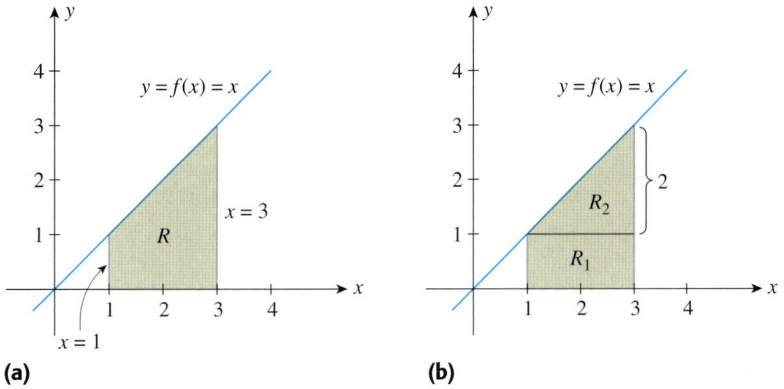

FIGURE **16**
The area of R can be computed in two different ways.

To evaluate the definite integral, observe that an antiderivative of $f(x) = x$ is $F(x) = \frac{1}{2}x^2 + C$, where C is an arbitrary constant. Therefore, by the fundamental theorem of calculus, we have

$$\begin{aligned} A = \int_1^3 x\,dx &= \frac{1}{2}x^2 + C\Big|_1^3 \\ &= \left(\frac{9}{2} + C\right) - \left(\frac{1}{2} + C\right) = 4 \text{ square units} \end{aligned}$$

To verify this result by elementary means, observe that the area A is the area of the rectangle R_1 (width $\times$ height) plus the area of the triangle R_2 ($\frac{1}{2}$ base $\times$ height) (see Figure 16b); that is,

$$2(1) + \frac{1}{2}(2)(2) = 2 + 2 = 4$$

which agrees with the result obtained earlier.

Observe that in evaluating the definite integral in Example 1, the constant of integration "dropped out." This is true in general, for if $F(x) + C$ denotes an antiderivative of some function f, then

$$\begin{aligned} F(x) + C\Big|_a^b &= [F(b) + C] - [F(a) + C] \\ &= F(b) + C - F(a) - C \\ &= F(b) - F(a) \end{aligned}$$

With this fact in mind, we may, in all future computations involving the evaluation of a definite integral, drop the constant of integration from our calculations.

Finding the Area under a Curve

Having seen how effective the fundamental theorem of calculus is in helping us find the area of simple regions, we now use it to find the area of more complicated regions.

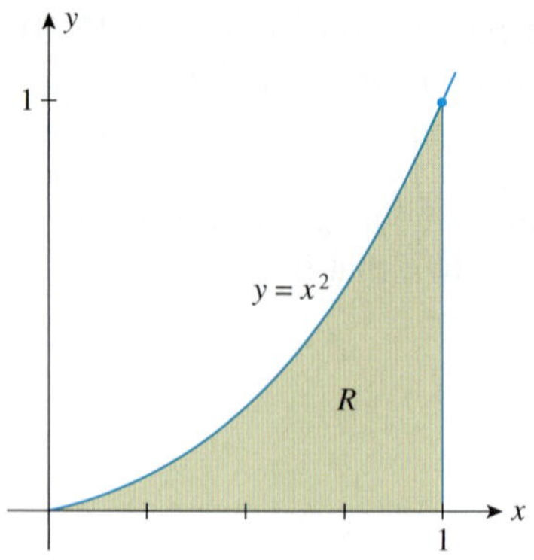

FIGURE 17
The area of R is $\int_0^1 x^2\,dx = \frac{1}{3}$.

EXAMPLE 2 In Section 6.3, we conjectured that the area of the region R under the graph of $f(x) = x^2$ on the interval [0, 1] was $\frac{1}{3}$ square unit. Use the fundamental theorem of calculus to verify this conjecture.

Solution The region R is reproduced in Figure 17. Observe that f is nonnegative on [0, 1], so the area of R is given by $A = \int_0^1 x^2\,dx$. Since an antiderivative of $f(x) = x^2$ is $F(x) = \frac{1}{3}x^3$, we see, using the fundamental theorem of calculus, that

$$A = \int_0^1 x^2\,dx = \frac{1}{3}x^3\Big|_0^1 = \frac{1}{3}(1) - \frac{1}{3}(0) = \frac{1}{3} \text{ square unit}$$

as we wished to show. ■

Note It is important to realize that the value, $\frac{1}{3}$, is by definition the exact value of the area of R. ■

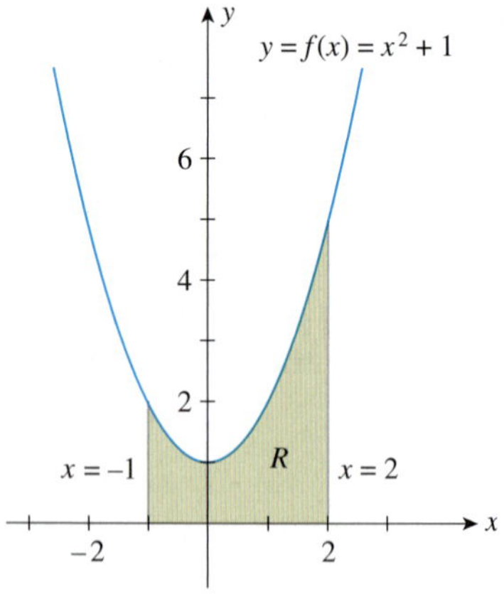

FIGURE 18
The area of R is $\int_{-1}^{2} (x^2 + 1)\,dx$.

EXAMPLE 3 Find the area of the region R under the graph of $y = x^2 + 1$ from $x = -1$ to $x = 2$.

Solution The region R under consideration is shown in Figure 18. Using the fundamental theorem of calculus, we find that the required area is

$$\int_{-1}^{2} (x^2 + 1)\,dx = \left(\frac{1}{3}x^3 + x\right)\Big|_{-1}^{2}$$
$$= \left[\frac{1}{3}(8) + 2\right] - \left[\frac{1}{3}(-1)^3 + (-1)\right] = 6$$

or 6 square units. ■

Evaluating Definite Integrals

In Examples 4 and 5, we use the rules of integration of Section 6.1 to help us evaluate the definite integrals.

EXAMPLE 4 Evaluate $\int_1^3 (3x^2 + e^x)\,dx$.

Solution

$$\int_1^3 (3x^2 + e^x)\,dx = x^3 + e^x\Big|_1^3$$
$$= (27 + e^3) - (1 + e) = 26 + e^3 - e$$ ■

EXAMPLE 5 Evaluate $\int_1^2 \left(\frac{1}{x} - \frac{1}{x^2}\right) dx$.

Solution

$$\int_1^2 \left(\frac{1}{x} - \frac{1}{x^2}\right) dx = \int_1^2 \left(\frac{1}{x} - x^{-2}\right) dx$$
$$= \ln|x| + \frac{1}{x}\Big|_1^2$$
$$= \left(\ln 2 + \frac{1}{2}\right) - (\ln 1 + 1)$$
$$= \ln 2 - \frac{1}{2}$$ Recall, $\ln 1 = 0$. ■

Explore & Discuss

Consider the definite integral $\int_{-1}^{1} \frac{1}{x^2}\, dx$.

1. Show that a formal application of Equation (9) leads to

$$\int_{-1}^{1} \frac{1}{x^2}\, dx = -\frac{1}{x}\bigg|_{-1}^{1} = -1 - 1 = -2$$

2. Observe that $f(x) = 1/x^2$ is positive at each value of x in $[-1, 1]$ where it is defined. So, one might expect that the definite integral with integrand f has a positive value, if it exists.
3. Explain this apparent contradiction in the result (1) and the observation (2).

The Definite Integral as a Measure of Net Change

In real-world applications, we are often interested in the net change of a quantity over a period of time. For example, suppose P is a function giving the population, $P(t)$, of a city at time t. Then the *net change* in the population over the period from $t = a$ to $t = b$ is given by

$$P(b) - P(a) \qquad \text{Population at } t = b \text{ minus population at } t = a$$

If P has a continuous derivative P' in $[a, b]$, then we can invoke the fundamental theorem of calculus to write

$$P(b) - P(a) = \int_a^b P'(t)\, dt \qquad P \text{ is an antiderivative of } P'.$$

Thus, if we know the *rate of change* of the population at any time t, then we can calculate the net change in the population from $t = a$ to $t = b$ by evaluating an appropriate definite integral.

APPLIED EXAMPLE 6 Population Growth in Clark County Clark County in Nevada—dominated by Las Vegas—is one of the fastest-growing metropolitan areas in the United States. From 1970 through 2000, the population was growing at the rate of

$$R(t) = 133{,}680t^2 - 178{,}788t + 234{,}633 \qquad (0 \le t \le 3)$$

people per decade, where $t = 0$ corresponds to the beginning of 1970. What was the net change in the population over the decade from 1980 to 1990?

Source: U.S. Census Bureau

Solution The net change in the population over the decade from 1980 ($t = 1$) to 1990 ($t = 2$) is given by $P(2) - P(1)$, where P denotes the population in the county at time t. But $P' = R$, and so

$$\begin{aligned}
P(2) - P(1) &= \int_1^2 P'(t)\, dt = \int_1^2 R(t)\, dt \\
&= \int_1^2 (133{,}680t^2 - 178{,}788t + 234{,}633)\, dt \\
&= 44{,}560t^3 - 89{,}394t^2 + 234{,}633t\bigg|_1^2 \\
&= [44{,}560(2)^3 - 89{,}394(2)^2 + 234{,}633(2)] \\
&\quad - [44{,}560 - 89{,}394 + 234{,}633] \\
&= 278{,}371
\end{aligned}$$

and so the net change is 278,371. ■

More generally, we have the following result. We assume that f has a continuous derivative, even though the integrability of f' is sufficient.

Net Change Formula

The net change in a function f over an interval $[a, b]$ is given by

$$f(b) - f(a) = \int_a^b f'(x)\,dx \qquad \textbf{(10)}$$

provided f' is continuous on $[a, b]$.

As another example of the net change of a function, let's consider the following example.

APPLIED EXAMPLE 7 Production Costs The management of Staedtler Office Equipment has determined that the daily marginal cost function associated with producing battery-operated pencil sharpeners is given by

$$C'(x) = 0.000006x^2 - 0.006x + 4$$

where $C'(x)$ is measured in dollars per unit and x denotes the number of units produced. Management has also determined that the daily fixed cost incurred in producing these pencil sharpeners is \$100. Find Staedtler's daily total cost for producing (a) the first 500 units and (b) the 201st through 400th units.

Solution

a. Since $C'(x)$ is the marginal cost function, its antiderivative $C(x)$ is the total cost function. The daily fixed cost incurred in producing the pencil sharpeners is $C(0)$ dollars. Since the daily fixed cost is given as \$100, we have $C(0) = 100$. We are required to find $C(500)$. Let's compute $C(500) - C(0)$, the net change in the total cost function $C(x)$ over the interval $[0, 500]$. Using the fundamental theorem of calculus, we find

$$\begin{aligned} C(500) - C(0) &= \int_0^{500} C'(x)\,dx \\ &= \int_0^{500} (0.000006x^2 - 0.006x + 4)\,dx \\ &= 0.000002x^3 - 0.003x^2 + 4x \Big|_0^{500} \\ &= [0.000002(500)^3 - 0.003(500)^2 + 4(500)] \\ &\quad - [0.000002(0)^3 - 0.003(0)^2 + 4(0)] \\ &= 1500 \end{aligned}$$

Therefore, $C(500) = 1500 + C(0) = 1500 + 100 = 1600$, so the total cost incurred daily by Staedtler in producing 500 pencil sharpeners is \$1600.

b. The daily total cost incurred by Staedtler in producing the 201st through 400th units of battery-operated pencil sharpeners is given by

$$
\begin{aligned}
C(400) - C(200) &= \int_{200}^{400} C'(x)\,dx \\
&= \int_{200}^{400} (0.000006x^2 - 0.006x + 4)\,dx \\
&= 0.000002x^3 - 0.003x^2 + 4x \Big|_{200}^{400} \\
&= [0.000002(400)^3 - 0.003(400)^2 + 4(400)] \\
&\quad - [0.000002(200)^3 - 0.003(200)^2 + 4(200)] \\
&= 552
\end{aligned}
$$

or \$552. ■

Since $C'(x)$ is nonnegative for x in the interval $(0, \infty)$, we have the following geometric interpretation of the two definite integrals in Example 7: $\int_0^{500} C'(x)\,dx$ is the area of the region under the graph of the function C' from $x = 0$ to $x = 500$, shown in Figure 19a, and $\int_{200}^{400} C'(x)\,dx$ is the area of the region from $x = 200$ to $x = 400$, shown in Figure 19b.

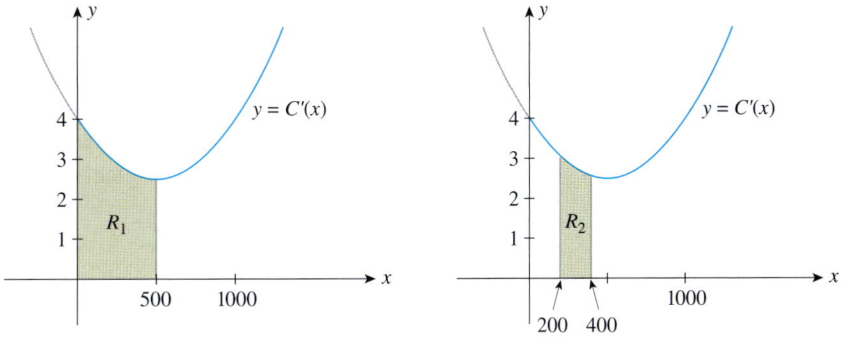

FIGURE **19**

(a) Area of $R_1 = \int_0^{500} C'(x)\,dx$ **(b)** Area of $R_2 = \int_{200}^{400} C'(x)\,dx$

VIDEO

APPLIED EXAMPLE 8 Assembly Time of Workers An efficiency study conducted for Elektra Electronics showed that the rate at which Space Commander walkie-talkies are assembled by the average worker t hours after starting work at 8 a.m. is given by the function

$$f(t) = -3t^2 + 12t + 15 \qquad (0 \le t \le 4)$$

Determine how many walkie-talkies can be assembled by the average worker in the first hour of the morning shift.

Solution Let $N(t)$ denote the number of walkie-talkies assembled by the average worker t hours after starting work in the morning shift. Then, we have

$$N'(t) = f(t) = -3t^2 + 12t + 15$$

Therefore, the number of units assembled by the average worker in the first hour of the morning shift is

$$
\begin{aligned}
N(1) - N(0) &= \int_0^1 N'(t)\,dt = \int_0^1 (-3t^2 + 12t + 15)\,dt \\
&= -t^3 + 6t^2 + 15t \Big|_0^1 = -1 + 6 + 15 \\
&= 20
\end{aligned}
$$

or 20 units. ■

Exploring with TECHNOLOGY

You can demonstrate graphically that $\int_0^x t\,dt = \frac{1}{2}x^2$ as follows:

1. Plot the graphs of $y1 = \text{fnInt}(t, t, 0, x) = \int_0^x t\,dt$ and $y2 = \frac{1}{2}x^2$ on the same set of axes, using the viewing window $[-5, 5] \times [0, 10]$.
2. Compare the graphs of $y1$ and $y2$ and draw the desired conclusion.

Explore & Discuss

The definite integral $\int_{-3}^{3} \sqrt{9 - x^2}\,dx$ cannot be evaluated here using the fundamental theorem of calculus because the method of this section does not enable us to find an antiderivative of the integrand. But the integral can be evaluated by interpreting it as the area of a certain plane region. What is the region? And what is the value of the integral?

APPLIED EXAMPLE 9 Projected Demand for Electricity A certain city's rate of electricity consumption is expected to grow exponentially with a growth constant of $k = 0.04$. If the present rate of consumption is 40 million kilowatt-hours (kWh) per year, what should be the total production of electricity over the next 3 years in order to meet the projected demand?

Solution If $R(t)$ denotes the expected rate of consumption of electricity t years from now, then

$$R(t) = 40e^{0.04t}$$

million kWh per year. Next, if $C(t)$ denotes the expected total consumption of electricity over a period of t years, then

$$C'(t) = R(t)$$

Therefore, the total consumption of electricity expected over the next 3 years is given by

$$\begin{aligned}\int_0^3 C'(t)\,dt &= \int_0^3 40e^{0.04t}\,dt \\ &= \frac{40}{0.04}\,e^{0.04t}\Big|_0^3 \\ &= 1000(e^{0.12} - 1) \\ &= 127.5\end{aligned}$$

or 127.5 million kWh, the amount that must be produced over the next 3 years in order to meet the demand. ■

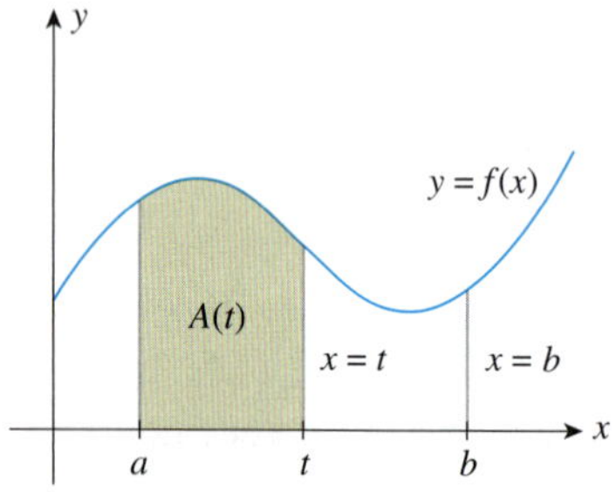

FIGURE 20
$A(t)$ = area under the graph of f from $x = a$ to $x = t$

Validity of the Fundamental Theorem of Calculus

To demonstrate the plausibility of the fundamental theorem of calculus for the case where f is nonnegative on an interval $[a, b]$, let's define an "area function" A as follows. Let $A(t)$ denote the area of the region R under the graph of $y = f(x)$ from $x = a$ to $x = t$, where $a \leq t \leq b$ (Figure 20).

If h is a small positive number, then $A(t + h)$ is the area of the region under the graph of $y = f(x)$ from $x = a$ to $x = t + h$. Therefore, the difference

$$A(t + h) - A(t)$$

is the area under the graph of $y = f(x)$ from $x = t$ to $x = t + h$ (Figure 21).

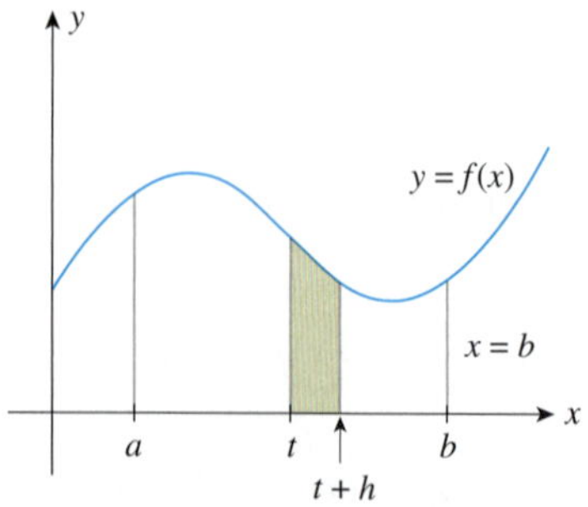

FIGURE 21
$A(t + h) - A(t)$ = area under the graph of f from $x = t$ to $x = t + h$

Now, the area of this last region can be approximated by the area of the rectangle of width h and height $f(t)$—that is, by the expression $h \cdot f(t)$ (Figure 22). Thus,

$$A(t + h) - A(t) \approx h \cdot f(t)$$

where the approximations improve as h is taken to be smaller and smaller.

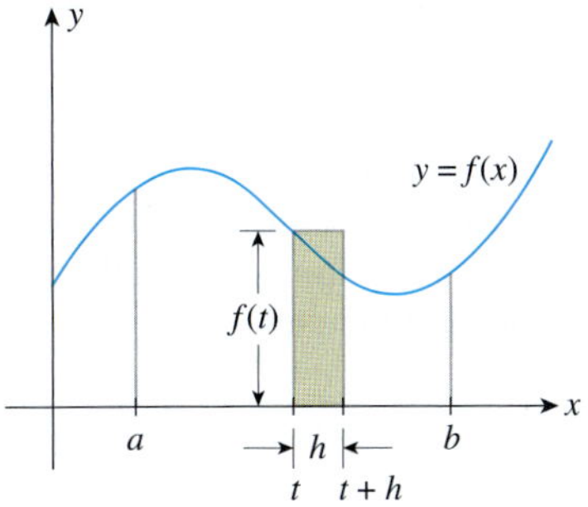

FIGURE 22
The area of the rectangle is $h \cdot f(t)$.

Dividing both sides of the foregoing relationship by h, we obtain

$$\frac{A(t+h)-A(t)}{h} \approx f(t)$$

Taking the limit as h approaches zero, we find, by the definition of the derivative, that the left-hand side is

$$\lim_{h \to 0} \frac{A(t+h)-A(t)}{h} = A'(t)$$

The right-hand side, which is independent of h, remains constant throughout the limiting process. Because the approximation becomes exact as h approaches zero, we find that

$$A'(t) = f(t)$$

Since the foregoing equation holds for all values of t in the interval $[a, b]$, we have shown that the *area function* A is an antiderivative of the function $f(x)$. By Theorem 1 of Section 6.1, we conclude that $A(x)$ must have the form

$$A(x) = F(x) + C$$

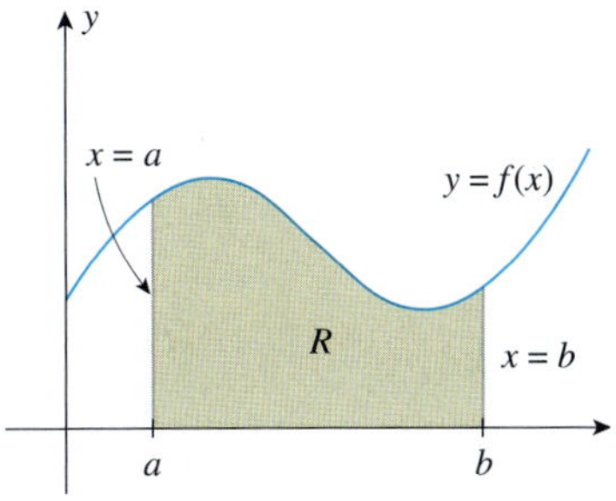

FIGURE 23
The area of R is given by $A(b)$.

where F is any antiderivative of f and C is a constant. To determine the value of C, observe that $A(a) = 0$. This condition implies that

$$A(a) = F(a) + C = 0$$

or $C = -F(a)$. Next, since the area of the region R is $A(b)$ (Figure 23), we see that the required area is

$$\begin{aligned} A(b) &= F(b) + C \\ &= F(b) - F(a) \end{aligned}$$

Since the area of the region R is

$$\int_a^b f(x)\,dx$$

we have

$$\int_a^b f(x)\,dx = F(b) - F(a)$$

as we set out to show.

6.4 Self-Check Exercises

1. Evaluate $\int_0^2 (x + e^x)\,dx$.
2. The daily marginal profit function associated with producing and selling TexaPep hot sauce is

$$P'(x) = -0.000006x^2 + 6$$

where x denotes the number of cases (each case contains 24 bottles) produced and sold daily and $P'(x)$ is measured in dollars/unit. The fixed cost is \$400.

a. What is the total profit realizable from producing and selling 1000 cases of TexaPep per day?

b. What is the additional profit realizable if the production and sale of TexaPep is increased from 1000 to 1200 cases/day?

Solutions to Self-Check Exercises 6.4 can be found on page 440.

6.4 Concept Questions

1. State the fundamental theorem of calculus.
2. State the net change formula and use it to answer the following questions:
   a. If a company generates income at the rate of R dollars/day, explain what $\int_a^b R(t)\,dt$ measures, where a and b are measured in days with $a < b$.
   b. If a private jet airplane consumes fuel at the rate of R gal/min, write an integral giving the net fuel consumption by the airplane between times $t = a$ and $t = b$ $(a < b)$, where t is measured in minutes.

6.4 Exercises

In Exercises 1–4, find the area of the region under the graph of the function *f* on the interval [*a*, *b*], using the fundamental theorem of calculus. Then verify your result using geometry.

1. $f(x) = 2;\ [1, 4]$
2. $f(x) = 4;\ [-1, 2]$
3. $f(x) = 2x;\ [1, 3]$
4. $f(x) = -\frac{1}{4}x + 1;\ [1, 4]$

In Exercises 5–16, find the area of the region under the graph of the function *f* on the interval [*a*, *b*].

5. $f(x) = 2x + 3;\ [-1, 2]$
6. $f(x) = 4x - 1;\ [2, 4]$
7. $f(x) = -x^2 + 4;\ [-1, 2]$
8. $f(x) = 4x - x^2;\ [0, 4]$
9. $f(x) = \frac{1}{x};\ [1, 2]$
10. $f(x) = \frac{1}{x^2};\ [2, 4]$
11. $f(x) = \sqrt{x};\ [1, 9]$
12. $f(x) = x^3;\ [1, 3]$
13. $f(x) = 1 - \sqrt[3]{x};\ [-8, -1]$
14. $f(x) = \frac{1}{\sqrt{x}};\ [1, 9]$
15. $f(x) = e^x;\ [0, 2]$
16. $f(x) = e^x - x;\ [1, 2]$

In Exercises 17–40, evaluate the definite integral.

17. $\int_2^4 3\,dx$
18. $\int_{-1}^2 -2\,dx$
19. $\int_1^3 (2x + 3)\,dx$
20. $\int_{-1}^0 (4 - x)\,dx$
21. $\int_{-1}^3 2x^2\,dx$
22. $\int_0^2 8x^3\,dx$
23. $\int_{-2}^2 (x^2 - 1)\,dx$
24. $\int_1^4 \sqrt{u}\,du$
25. $\int_1^8 4x^{1/3}\,dx$
26. $\int_1^4 2x^{-3/2}\,dx$
27. $\int_0^1 (x^3 - 2x^2 + 1)\,dx$
28. $\int_1^2 (t^5 - t^3 + 1)\,dt$
29. $\int_2^4 \frac{1}{x}\,dx$
30. $\int_1^3 \frac{2}{x}\,dx$
31. $\int_0^4 x(x^2 - 1)\,dx$
32. $\int_0^2 (x - 4)(x - 1)\,dx$
33. $\int_1^3 (t^2 - t)^2\,dt$
34. $\int_{-1}^1 (x^2 - 1)^2\,dx$
35. $\int_{-3}^{-1} \frac{1}{x^2}\,dx$
36. $\int_1^2 \frac{2}{x^3}\,dx$
37. $\int_1^4 \left(\sqrt{x} - \frac{1}{\sqrt{x}}\right) dx$
38. $\int_0^1 \sqrt{2x}(\sqrt{x} + \sqrt{2})\,dx$
39. $\int_1^4 \frac{3x^3 - 2x^2 + 4}{x^2}\,dx$
40. $\int_1^2 \left(1 + \frac{1}{u} + \frac{1}{u^2}\right) du$

41. **Marginal Cost** A division of Ditton Industries manufactures a deluxe toaster oven. Management has determined that the daily marginal cost function associated with producing these toaster ovens is given by

$$C'(x) = 0.0003x^2 - 0.12x + 20$$

where $C'(x)$ is measured in dollars/unit and x denotes the number of units produced. Management has also determined that the daily fixed cost incurred in the production is \$800.
   a. Find the total cost incurred by Ditton in producing the first 300 units of these toaster ovens per day.
   b. What is the total cost incurred by Ditton in producing the 201st through 300th units/day?

42. **Marginal Revenue** The management of Ditton Industries has determined that the daily marginal revenue function associated with selling x units of their deluxe toaster ovens is given by

$$R'(x) = -0.1x + 40$$

where $R'(x)$ is measured in dollars/unit.
   a. Find the daily total revenue realized from the sale of 200 units of the toaster oven.
   b. Find the additional revenue realized when the production (and sales) level is increased from 200 to 300 units.

43. **Marginal Profit** Refer to Exercise 41. The daily marginal profit function associated with the production and sales of the deluxe toaster ovens is known to be

$$P'(x) = -0.0003x^2 + 0.02x + 20$$

where x denotes the number of units manufactured and sold daily and $P'(x)$ is measured in dollars/unit.

a. Find the total profit realizable from the manufacture and sale of 200 units of the toaster ovens per day.
Hint: $P(200) - P(0) = \int_0^{200} P'(x)\,dx$, $P(0) = -800$.

b. What is the additional daily profit realizable if the production and sale of the toaster ovens are increased from 200 to 220 units/day?

44. Internet Advertising U.S. Internet advertising revenue grew at the rate of

$$R(t) = 0.82t + 1.14 \qquad (0 \le t \le 4)$$

billion dollars/year between 2002 ($t = 0$) and 2006 ($t = 4$). The advertising revenue in 2002 was \$5.9 billion.

a. Find an expression giving the advertising revenue in year t.

b. If the trend continued, what was the Internet advertising revenue in 2007?

Source: Interactive Advertising Bureau

45. Mobile-Phone Ad Spending Mobile-phone ad spending is expected to grow at the rate of

$$R(t) = 0.8256t^{-0.04} \qquad (1 \le t \le 5)$$

billion dollars/year between 2007 ($t = 1$) and 2011 ($t = 5$). The mobile-phone ad spending in 2007 was \$0.9 billion.

a. Find an expression giving the mobile-phone ad spending in year t.

b. If the trend continued, what will be the mobile-phone ad spending in 2012?

Source: Interactive Advertising Bureau

46. Efficiency Studies Tempco Electronics, a division of Tempco Toys, manufactures an electronic football game. An efficiency study showed that the rate at which the games are assembled by the average worker t hr after starting work at 8 a.m. is

$$-\frac{3}{2}t^2 + 6t + 20 \qquad (0 \le t \le 4)$$

units/hour.

a. Find the total number of games the average worker can be expected to assemble in the 4-hr morning shift.

b. How many units can the average worker be expected to assemble in the first hour of the morning shift? In the second hour of the morning shift?

47. Speedboat Racing In a recent pretrial run for the world water speed record, the velocity of the *Sea Falcon II* t sec after firing the booster rocket was given by

$$v(t) = -t^2 + 20t + 440 \qquad (0 \le t \le 20)$$

feet/second. Find the distance covered by the boat over the 20-sec period after the booster rocket was activated.
Hint: The distance is given by $\int_0^{20} v(t)\,dt$.

48. Pocket Computers Annual sales (in millions of units) of pocket computers are expected to grow in accordance with the function

$$f(t) = 0.18t^2 + 0.16t + 2.64 \qquad (0 \le t \le 6)$$

where t is measured in years, with $t = 0$ corresponding to 1997. How many pocket computers were sold over the 6-yr period between the beginning of 1997 and the end of 2002?

Source: Dataquest, Inc.

49. Single Female-Headed Households with Children The percentage of families with children that are headed by single females grew at the rate of

$$R(t) = 0.8499t^2 - 3.872t + 5 \qquad (0 \le t \le 3)$$

households/decade between 1970 ($t = 0$) and 2000 ($t = 3$). The number of such households stood at 5.6% of all families in 1970.

a. Find an expression giving the percentage of these households in the tth decade.

b. If the trend continued, estimate the percentage of these households in 2010.

c. What was the net increase in the percentage of these households from 1970 to 2000?

Source: U.S. Census Bureau

50. Air Purification To test air purifiers, engineers ran a purifier in a smoke-filled 10-ft × 20-ft room. While conducting a test for a certain brand of air purifier, it was determined that the amount of smoke in the room was decreasing at the rate of

$$R(t) = 0.00032t^4 - 0.01872t^3 + 0.3948t^2 - 3.83t + 17.63 \qquad (0 \le t \le 20)$$

percent of the (original) amount of the smoke per minute, t min after the start of the test. How much smoke was left in the room 5 min after the start of the test? Ten min after the start of the test?

Source: Consumer Reports

51. TV Set-Top Boxes The number of television set-top boxes shipped worldwide from the beginning of 2003 until the beginning of 2009 was projected to be

$$f(t) = -0.05556t^3 + 0.262t^2 + 17.46t + 63.4 \qquad (0 \le t \le 6)$$

million units/year, where t is measured in years, with $t = 0$ corresponding to 2003. If the projection held true, how many set-top boxes were shipped from the beginning of 2003 until the beginning of 2009?

Source: In-Stat.

52. Canadian Oil-Sands Production The production of oil (in millions of barrels per day) extracted from oil sands in Canada is projected to grow according to the function

$$P(t) = \frac{4.76}{1 + 4.11e^{-0.22t}} \qquad (0 \le t \le 15)$$

where t is measured in years, with $t = 0$ corresponding to 2005. What is the expected total production of oil from oil sands over the years from 2005 until 2020 ($t = 15$)?

Hint: Multiply the integrand by $\frac{e^{0.22t}}{e^{0.22t}}$.

Source: Canadian Association of Petroleum Producers.

53. **Senior Citizens** The population aged 65 yr old and older (in millions) from 2000 to 2050 is projected to be

$$f(t) = \frac{85}{1 + 1.859e^{-0.66t}} \qquad (0 \le t \le 5)$$

where t is measured in decades, with $t = 0$ corresponding to 2000. By how much will the population aged 65 yr and older increase from the beginning of 2000 until the beginning of 2030?
Hint: Multiply the integrand by $e^{0.66t}/e^{0.66t}$.
Source: U.S. Census Bureau

54. **Blood Flow** Consider an artery of length L cm and radius R cm. Using Poiseuille's law (page 61), it can be shown that the rate at which blood flows through the artery (measured in cubic centimeters/second) is given by

$$V = \int_0^R \frac{k}{L} x(R^2 - x^2)\, dx$$

where k is a constant. Find an expression for V that does *not* involve an integral.
Hint: Use the substitution $u = R^2 - x^2$.

55. Find the area of the region bounded by the graph of the function $f(x) = x^4 - 2x^2 + 2$, the x-axis, and the lines $x = a$ and $x = b$, where $a < b$ and a and b are the x-coordinates of the relative maximum point and a relative minimum point of f, respectively.

56. Find the area of the region bounded by the graph of the function $f(x) = (x + 1)/\sqrt{x}$, the x-axis, and the lines $x = a$ and $x = b$ where a and b are, respectively, the x-coordinates of the relative minimum point and the inflection point of f.

In Exercises 57–60, determine whether the statement is true or false. Give a reason for your answer.

57. $\displaystyle\int_{-1}^{1} \frac{1}{x^3}\, dx = -\frac{1}{2x^2}\Big|_{-1}^{1} = -\frac{1}{2} - \left(-\frac{1}{2}\right) = 0$

58. $\displaystyle\int_{-1}^{1} \frac{1}{x}\, dx = \ln|x|\Big|_{-1}^{1} = \ln|1| - \ln|-1| = \ln 1 - \ln 1 = 0$

59. $\int_0^2 (1 - x)\, dx$ gives the area of the region under the graph of $f(x) = 1 - x$ on the interval $[0, 2]$.

60. The total revenue realized in selling the first 500 units of a product is given by

$$\int_0^{500} R'(x)\, dx = R(500) - R(0)$$

where $R(x)$ is the total revenue.

6.4 Solutions to Self-Check Exercises

1. $$\int_0^2 (x + e^x)\, dx = \frac{1}{2}x^2 + e^x \Big|_0^2$$
$$= \left[\frac{1}{2}(2)^2 + e^2\right] - \left[\frac{1}{2}(0) + e^0\right]$$
$$= 2 + e^2 - 1$$
$$= e^2 + 1$$

2. **a.** We want $P(1000)$, but

$$P(1000) - P(0) = \int_0^{1000} P'(x)\, dx = \int_0^{1000} (-0.000006x^2 + 6)\, dx$$
$$= -0.000002x^3 + 6x \Big|_0^{1000}$$
$$= -0.000002(1000)^3 + 6(1000)$$
$$= 4000$$

So, $P(1000) = 4000 + P(0) = 4000 - 400$, or \$3600/day $[P(0) = -C(0)]$.

b. The additional profit realizable is given by

$$\int_{1000}^{1200} P'(x)\, dx = -0.000002x^3 + 6x \Big|_{1000}^{1200}$$
$$= [-0.000002(1200)^3 + 6(1200)] - [-0.000002(1000)^3 + 6(1000)]$$
$$= 3744 - 4000$$
$$= -256$$

That is, the company sustains a loss of \$256/day if production is increased from 1000 to 1200 cases/day.

USING TECHNOLOGY

Evaluating Definite Integrals

Some graphing utilities have an operation for finding the definite integral of a function. If your graphing utility has this capability, use it to work through the example and exercises of this section.

EXAMPLE 1 Use the numerical integral operation of a graphing utility to evaluate

$$\int_{-1}^{2} \frac{2x + 4}{(x^2 + 1)^{3/2}}\, dx$$

Solution Using the numerical integral operation of a graphing utility, we find

$$\int_{-1}^{2} \frac{2x + 4}{(x^2 + 1)^{3/2}}\, dx = \text{fnInt}((2x + 4)/(x^2 + 1)\text{^}1.5, x, -1, 2) \approx 6.92592226$$

TECHNOLOGY EXERCISES

In Exercises 1–4, find the area of the region under the graph of *f* on the interval [*a*, *b*]. Express your answer to four decimal places.

1. $f(x) = 0.002x^5 + 0.032x^4 - 0.2x^2 + 2$; $[-1.1, 2.2]$
2. $f(x) = x\sqrt{x^3 + 1}$; $[1, 2]$
3. $f(x) = \sqrt{x}e^{-x}$; $[0, 3]$
4. $f(x) = \dfrac{\ln x}{\sqrt{1 + x^2}}$; $[1, 2]$

In Exercises 5–10, evaluate the definite integral.

5. $\int_{-1.2}^{2.3} (0.2x^4 - 0.32x^3 + 1.2x - 1)\, dx$
6. $\int_{1}^{3} x(x^4 - 1)^{3.2}\, dx$
7. $\int_{0}^{2} \dfrac{3x^3 + 2x^2 + 1}{2x^2 + 3}\, dx$
8. $\int_{1}^{2} \dfrac{\sqrt{x} + 1}{2x^2 + 1}\, dx$
9. $\int_{0}^{2} \dfrac{e^x}{\sqrt{x^2 + 1}}\, dx$
10. $\int_{1}^{3} e^{-x} \ln(x^2 + 1)\, dx$
11. Rework Exercise 50, Exercises 6.4.
12. Rework Exercise 53, Exercises 6.4.
13. **The Global Epidemic** The number of AIDS-related deaths/year in the United States is given by the function

$$f(t) = -53.254t^4 + 673.7t^3 - 2801.07t^2 + 8833.379t + 20{,}000 \qquad (0 \le t \le 9)$$

with $t = 0$ corresponding to the beginning of 1988. Find the total number of AIDS-related deaths in the United States between the beginning of 1988 and the end of 1996.
Source: Centers for Disease Control

14. **Marijuana Arrests** The number of arrests for marijuana sales and possession in New York City grew at the rate of approximately

$$f(t) = 0.0125t^4 - 0.01389t^3 + 0.55417t^2 + 0.53294t + 4.95238 \qquad (0 \le t \le 5)$$

thousand/year, where t is measured in years, with $t = 0$ corresponding to the beginning of 1992. Find the approximate number of marijuana arrests in the city from the beginning of 1992 to the end of 1997.
Source: State Division of Criminal Justice Services

15. **Population Growth** The population of a certain city is projected to grow at the rate of $18\sqrt{t + 1} \ln \sqrt{t + 1}$ thousand people/year t yr from now. If the current population is 800,000, what will be the population 45 yr from now?

6.5 Evaluating Definite Integrals

This section continues our discussion of the applications of the fundamental theorem of calculus.

Properties of the Definite Integral

Before going on, we list the following useful properties of the definite integral, some of which parallel the rules of integration of Section 6.1.

Properties of the Definite Integral

Let f and g be integrable functions; then,

1. $\displaystyle\int_a^a f(x)\,dx = 0$

2. $\displaystyle\int_a^b f(x)\,dx = -\int_b^a f(x)\,dx$

3. $\displaystyle\int_a^b cf(x)\,dx = c\int_a^b f(x)\,dx \qquad (c\text{, a constant})$

4. $\displaystyle\int_a^b [f(x) \pm g(x)]\,dx = \int_a^b f(x)\,dx \pm \int_a^b g(x)\,dx$

5. $\displaystyle\int_a^b f(x)\,dx = \int_a^c f(x)\,dx + \int_c^b f(x)\,dx \qquad (a < c < b)$

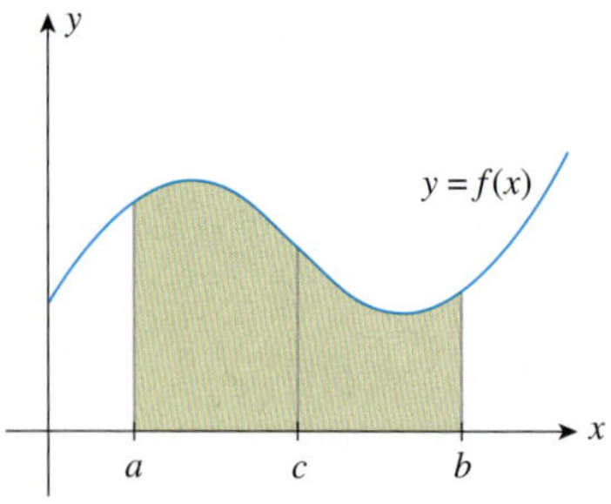

FIGURE 24
$\int_a^b f(x)\,dx = \int_a^c f(x)\,dx + \int_c^b f(x)\,dx$

Property 5 states that if c is a number lying between a and b so that the interval $[a, b]$ is divided into the intervals $[a, c]$ and $[c, b]$, then the integral of f over the interval $[a, b]$ may be expressed as the sum of the integral of f over the interval $[a, c]$ and the integral of f over the interval $[c, b]$.

Property 5 has the following geometric interpretation when f is nonnegative. By definition

$$\int_a^b f(x)\,dx$$

is the area of the region under the graph of $y = f(x)$ from $x = a$ to $x = b$ (Figure 24). Similarly, we interpret the definite integrals

$$\int_a^c f(x)\,dx \quad \text{and} \quad \int_c^b f(x)\,dx$$

as the areas of the regions under the graph of $y = f(x)$ from $x = a$ to $x = c$ and from $x = c$ to $x = b$, respectively. Since the two regions do not overlap, we see that

$$\int_a^b f(x)\,dx = \int_a^c f(x)\,dx + \int_c^b f(x)\,dx$$

The Method of Substitution for Definite Integrals

Our first example shows two approaches generally used when evaluating a definite integral using the method of substitution.

EXAMPLE 1 Evaluate $\displaystyle\int_0^4 x\sqrt{9 + x^2}\,dx$.

Solution

Method 1 We first find the corresponding indefinite integral:

$$I = \int x\sqrt{9 + x^2}\,dx$$

Make the substitution $u = 9 + x^2$ so that

$$du = \frac{d}{dx}(9 + x^2)\,dx$$
$$= 2x\,dx$$
$$x\,dx = \frac{1}{2}\,du \quad \text{Divide both sides by 2.}$$

Then,

$$I = \int \frac{1}{2}\sqrt{u}\,du = \frac{1}{2}\int u^{1/2}\,du$$
$$= \frac{1}{3}u^{3/2} + C = \frac{1}{3}(9 + x^2)^{3/2} + C \quad \text{Substitute } 9 + x^2 \text{ for } u.$$

Using this result, we now evaluate the given definite integral:

$$\int_0^4 x\sqrt{9 + x^2}\,dx = \frac{1}{3}(9 + x^2)^{3/2}\Big|_0^4$$
$$= \frac{1}{3}[(9 + 16)^{3/2} - 9^{3/2}]$$
$$= \frac{1}{3}(125 - 27) = \frac{98}{3} = 32\tfrac{2}{3}$$

Method 2 *Changing the Limits of Integration.* As before, we make the substitution

$$u = 9 + x^2 \tag{11}$$

so that

$$du = 2x\,dx$$
$$x\,dx = \frac{1}{2}\,du$$

Next, observe that the given definite integral is evaluated *with respect to x* with the range of integration given by the interval [0, 4]. If we perform the integration *with respect to u* via the substitution (11), then we must adjust the range of integration to reflect the fact that the integration is being performed with respect to the new variable u. To determine the proper range of integration, note that when $x = 0$, Equation (11) implies that

$$u = 9 + 0^2 = 9$$

which gives the required lower limit of integration with respect to u. Similarly, when $x = 4$,

$$u = 9 + 16 = 25$$

is the required upper limit of integration with respect to u. Thus, the range of integration when the integration is performed with respect to u is given by the interval [9, 25]. Therefore, we have

$$\int_0^4 x\sqrt{9 + x^2}\,dx = \int_9^{25} \frac{1}{2}\sqrt{u}\,du = \frac{1}{2}\int_9^{25} u^{1/2}\,du$$
$$= \frac{1}{3}u^{3/2}\Big|_9^{25} = \frac{1}{3}(25^{3/2} - 9^{3/2})$$
$$= \frac{1}{3}(125 - 27) = \frac{98}{3} = 32\tfrac{2}{3}$$

which agrees with the result obtained using Method 1. ■

When you use the method of substitution, make sure you adjust the limits of integration to reflect integrating with respect to the new variable u.

Exploring with TECHNOLOGY

Refer to Example 1. You can confirm the results obtained there by using a graphing utility as follows:

1. Use the numerical integration operation of the graphing utility to evaluate

$$\int_0^4 x\sqrt{9 + x^2}\,dx$$

2. Evaluate $\frac{1}{2}\int_9^{25} \sqrt{u}\,du$.

3. Conclude that $\int_0^4 x\sqrt{9 + x^2}\,dx = \frac{1}{2}\int_9^{25} \sqrt{u}\,du$.

EXAMPLE 2 Evaluate $\int_0^2 xe^{2x^2}\,dx$.

Solution Let $u = 2x^2$ so that $du = 4x\,dx$, or $x\,dx = \frac{1}{4}\,du$. When $x = 0$, $u = 0$, and when $x = 2$, $u = 8$. This gives the lower and upper limits of integration with respect to u. Making the indicated substitutions, we find

$$\int_0^2 xe^{2x^2}\,dx = \int_0^8 \frac{1}{4}e^u\,du = \frac{1}{4}e^u\Big|_0^8 = \frac{1}{4}(e^8 - 1)$$

EXAMPLE 3 Evaluate $\displaystyle\int_0^1 \frac{x^2}{x^3 + 1}\,dx$.

Solution Let $u = x^3 + 1$ so that $du = 3x^2\,dx$, or $x^2\,dx = \frac{1}{3}\,du$. When $x = 0$, $u = 1$, and when $x = 1$, $u = 2$. This gives the lower and upper limits of integration with respect to u. Making the indicated substitutions, we find

$$\int_0^1 \frac{x^2}{x^3 + 1}\,dx = \frac{1}{3}\int_1^2 \frac{du}{u} = \frac{1}{3}\ln|u|\Big|_1^2$$
$$= \frac{1}{3}(\ln 2 - \ln 1) = \frac{1}{3}\ln 2$$

Finding the Area under a Curve

VIDEO

EXAMPLE 4 Find the area of the region R under the graph of $f(x) = e^{(1/2)x}$ from $x = -1$ to $x = 1$.

Solution The region R is shown in Figure 25. Its area is given by

$$A = \int_{-1}^1 e^{(1/2)x}\,dx$$

To evaluate this integral, we make the substitution

$$u = \frac{1}{2}x$$

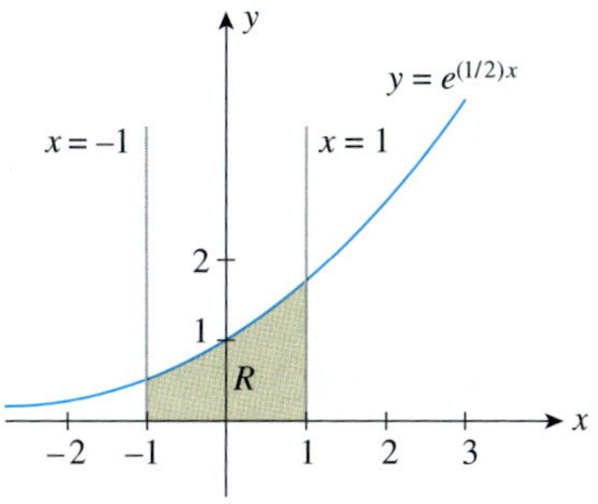

FIGURE 25
Area of $R = \int_{-1}^{1} e^{(1/2)x}\,dx$

so that

$$du = \frac{1}{2}\,dx$$
$$dx = 2\,du$$

When $x = -1$, $u = -\frac{1}{2}$, and when $x = 1$, $u = \frac{1}{2}$. Making the indicated substitutions, we obtain

$$A = \int_{-1}^{1} e^{(1/2)x}\,dx = 2\int_{-1/2}^{1/2} e^{u}\,du$$
$$= 2e^{u}\Big|_{-1/2}^{1/2} = 2(e^{1/2} - e^{-1/2})$$

or approximately 2.08 square units.

Explore & Discuss

Let f be a function defined piecewise by the rule

$$f(x) = \begin{cases} \sqrt{x} & \text{if } 0 \le x \le 1 \\ \dfrac{1}{x} & \text{if } 1 < x \le 2 \end{cases}$$

How would you use Property 5 of definite integrals to find the area of the region under the graph of f on [0, 2]? What is the area?

Average Value of a Function

The *average value* of a function over an interval provides us with an application of the definite integral. Recall that the average value of a set of n numbers is the number

$$\frac{y_1 + y_2 + \cdots + y_n}{n}$$

Now, suppose f is a continuous function defined on $[a, b]$. Let's divide the interval $[a, b]$ into n subintervals of equal length $(b - a)/n$. Choose points $x_1, x_2, \ldots, x_n$ in the first, second, . . . , and nth subintervals, respectively. Then, the average value of the numbers $f(x_1), f(x_2), \ldots, f(x_n)$, given by

$$\frac{f(x_1) + f(x_2) + \cdots + f(x_n)}{n}$$

is an approximation of the average of all the values of $f(x)$ on the interval $[a, b]$. This expression can be written in the form

$$\frac{(b-a)}{(b-a)}\left[f(x_1)\cdot\frac{1}{n} + f(x_2)\cdot\frac{1}{n} + \cdots + f(x_n)\cdot\frac{1}{n}\right]$$
$$= \frac{1}{b-a}\left[f(x_1)\cdot\frac{b-a}{n} + f(x_2)\cdot\frac{b-a}{n} + \cdots + f(x_n)\cdot\frac{b-a}{n}\right]$$
$$= \frac{1}{b-a}\left[f(x_1)\Delta x + f(x_2)\Delta x + \cdots + f(x_n)\Delta x\right] \qquad \textbf{(12)}$$

As n gets larger and larger, the expression (12) approximates the average value of $f(x)$ over $[a, b]$ with increasing accuracy. But the sum inside the brackets in (12) is a Riemann sum of the function f over $[a, b]$. In view of this, we have

$$\lim_{n\to\infty}\left[\frac{f(x_1) + f(x_2) + \cdots + f(x_n)}{n}\right]$$
$$= \frac{1}{b-a}\lim_{n\to\infty}\left[f(x_1)\Delta x + f(x_2)\Delta x + \cdots + f(x_n)\Delta x\right]$$
$$= \frac{1}{b-a}\int_a^b f(x)\,dx$$

This discussion motivates the following definition.

The Average Value of a Function

Suppose f is integrable on $[a, b]$. Then the **average value** of f over $[a, b]$ is

$$\frac{1}{b-a}\int_a^b f(x)\,dx$$

EXAMPLE 5 Find the average value of the function $f(x) = \sqrt{x}$ over the interval $[0, 4]$.

Solution The required average value is given by

$$\frac{1}{4-0}\int_0^4 \sqrt{x}\,dx = \frac{1}{4}\int_0^4 x^{1/2}\,dx$$
$$= \frac{1}{6}x^{3/2}\Big|_0^4 = \frac{1}{6}(4^{3/2})$$
$$= \frac{4}{3}$$

APPLIED EXAMPLE 6 Automobile Financing The interest rates charged by Madison Finance on auto loans for used cars over a certain 6-month period in 2010 are approximated by the function

$$r(t) = -\frac{1}{12}t^3 + \frac{7}{8}t^2 - 3t + 12 \qquad (0 \le t \le 6)$$

where t is measured in months and $r(t)$ is the annual percentage rate. What is the average rate on auto loans extended by Madison over the 6-month period?

Solution The average rate over the 6-month period in question is given by

$$\frac{1}{6-0}\int_0^6\left(-\frac{1}{12}t^3 + \frac{7}{8}t^2 - 3t + 12\right)dt$$
$$= \frac{1}{6}\left(-\frac{1}{48}t^4 + \frac{7}{24}t^3 - \frac{3}{2}t^2 + 12t\right)\Big|_0^6$$
$$= \frac{1}{6}\left[-\frac{1}{48}(6^4) + \frac{7}{24}(6^3) - \frac{3}{2}(6^2) + 12(6)\right]$$
$$= 9$$

or 9% per year.

APPLIED EXAMPLE 7 Drug Concentration in a Body The amount of a certain drug in a patient's body t days after it has been administered is

$$C(t) = 5e^{-0.2t}$$

units. Determine the average amount of the drug present in the patient's body for the first 4 days after the drug has been administered.

Solution The average amount of the drug present in the patient's body for the first 4 days after it has been administered is given by

$$\begin{aligned}\frac{1}{4-0}\int_0^4 5e^{-0.2t}\,dt &= \frac{5}{4}\int_0^4 e^{-0.2t}\,dt\\ &= \frac{5}{4}\left[\left(-\frac{1}{0.2}\right)e^{-0.2t}\Big|_0^4\right]\\ &= \frac{5}{4}(-5e^{-0.8}+5)\\ &\approx 3.44\end{aligned}$$

or approximately 3.44 units. ■

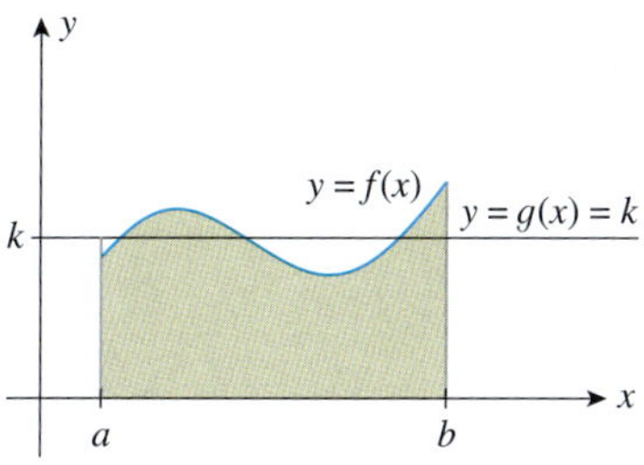

FIGURE 26
The average value of f over $[a, b]$ is k.

We now give a geometric interpretation of the average value of a function f over an interval $[a, b]$. Suppose $f(x)$ is nonnegative so that the definite integral

$$\int_a^b f(x)\,dx$$

gives the area under the graph of f from $x = a$ to $x = b$ (Figure 26). Observe that, in general, the "height" $f(x)$ varies from point to point. Can we replace $f(x)$ by a constant function $g(x) = k$ (which has constant height) such that the areas under each of the two functions f and g are the same? If so, since the area under the graph of g from $x = a$ to $x = b$ is $k(b - a)$, we have

$$\begin{aligned}k(b-a) &= \int_a^b f(x)\,dx\\ k &= \frac{1}{b-a}\int_a^b f(x)\,dx\end{aligned}$$

so that k is the average value of f over $[a, b]$. Thus, the average value of a function f over an interval $[a, b]$ is the height of a rectangle with base of length $(b - a)$ that has the same area as that of the region under the graph of f from $x = a$ to $x = b$.

6.5 Self-Check Exercises

1. Evaluate $\int_0^2 \sqrt{2x+5}\,dx$.

2. Find the average value of the function $f(x) = 1 - x^2$ over the interval $[-1, 2]$.

3. The median price of a house in a southwestern state between January 1, 2003, and January 1, 2008, is approximated by the function

$$f(t) = t^3 - 7t^2 + 17t + 280 \qquad (0 \le t \le 5)$$

where $f(t)$ is measured in thousands of dollars and t is expressed in years, with $t = 0$ corresponding to the beginning of 2003. Determine the average median price of a house over that time interval.

Solutions to Self-Check Exercises 6.5 can be found on page 451.

6.5 Concept Questions

1. Describe two approaches used to evaluate a definite integral using the method of substitution. Illustrate with the integral $\int_0^1 x^2(x^3+1)^2\,dx$.

2. Define the average value of a function f over an interval $[a, b]$. Give a geometric interpretation.

6.5 Exercises

In Exercises 1–28, evaluate the definite integral.

1. $\int_0^2 x(x^2-1)^3\,dx$
2. $\int_0^1 x^2(2x^3-1)^4\,dx$
3. $\int_0^1 x\sqrt{5x^2+4}\,dx$
4. $\int_1^3 x\sqrt{3x^2-2}\,dx$
5. $\int_0^2 x^2(x^3+1)^{3/2}\,dx$
6. $\int_1^5 (2x-1)^{5/2}\,dx$
7. $\int_0^1 \frac{1}{\sqrt{2x+1}}\,dx$
8. $\int_0^2 \frac{x}{\sqrt{x^2+5}}\,dx$
9. $\int_1^2 (2x-1)^4\,dx$
10. $\int_1^2 (2x+4)(x^2+4x-8)^3\,dx$
11. $\int_{-1}^1 x^2(x^3+1)^4\,dx$
12. $\int_1^2 \left(x^3+\frac{3}{4}\right)(x^4+3x)^{-2}\,dx$
13. $\int_1^5 x\sqrt{x-1}\,dx$
14. $\int_1^4 x\sqrt{x+1}\,dx$
    Hint: Let $u = x + 1$.
15. $\int_0^2 xe^{x^2}\,dx$
16. $\int_0^1 e^{-x}\,dx$
17. $\int_0^1 (e^{2x}+x^2+1)\,dx$
18. $\int_0^2 (e^t-e^{-t})\,dt$
19. $\int_{-1}^1 xe^{x^2+1}\,dx$
20. $\int_1^4 \frac{e^{\sqrt{x}}}{\sqrt{x}}\,dx$
21. $\int_3^6 \frac{2}{x-2}\,dx$
22. $\int_0^1 \frac{x}{1+2x^2}\,dx$
23. $\int_1^2 \frac{x^2+2x}{x^3+3x^2-1}\,dx$
24. $\int_0^1 \frac{e^x}{1+e^x}\,dx$
25. $\int_1^2 \left(4e^{2u}-\frac{1}{u}\right)du$
26. $\int_1^2 \left(1+\frac{1}{x}+e^x\right)dx$
27. $\int_1^2 \left(2e^{-4x}-\frac{1}{x^2}\right)dx$
28. $\int_1^2 \frac{\ln x}{x}\,dx$

In Exercises 29–34, find the area of the region under the graph of f on $[a, b]$.

29. $f(x) = x^2 - 2x + 2;\ [-1, 2]$
30. $f(x) = x^3 + x;\ [0, 1]$
31. $f(x) = \frac{1}{x^2};\ [1, 2]$
32. $f(x) = 2 + \sqrt{x+1};\ [0, 3]$
33. $f(x) = e^{-x/2};\ [-1, 2]$
34. $f(x) = \frac{\ln x}{4x};\ [1, 2]$

In Exercises 35–44, find the average value of the function f over the indicated interval $[a, b]$.

35. $f(x) = 2x + 3;\ [0, 2]$
36. $f(x) = 8 - x;\ [1, 4]$
37. $f(x) = 2x^2 - 3;\ [1, 3]$
38. $f(x) = 4 - x^2;\ [-2, 3]$
39. $f(x) = x^2 + 2x - 3;\ [-1, 2]$
40. $f(x) = x^3;\ [-1, 1]$
41. $f(x) = \sqrt{2x+1};\ [0, 4]$
42. $f(x) = e^{-x};\ [0, 4]$
43. $f(x) = xe^{x^2};\ [0, 2]$
44. $f(x) = \frac{1}{x+1};\ [0, 2]$

45. **World Production of Coal** A study proposed in 1980 by researchers from the major producers and consumers of the world's coal concluded that coal could and must play an important role in fueling global economic growth over the next 20 yr. The world production of coal in 1980 was 3.5 billion metric tons. If output increased at the rate of $3.5e^{0.05t}$ billion metric tons/year in year t ($t = 0$ corresponding to 1980), determine how much coal was produced worldwide between 1980 and the end of the 20th century.

46. **Newton's Law of Cooling** A bottle of white wine at room temperature (68°F) is placed in a refrigerator at 4 p.m. Its temperature after t hr is changing at the rate of

$$-18e^{-0.6t}$$

°F/hour. By how many degrees will the temperature of the wine have dropped by 7 p.m.? What will the temperature of the wine be at 7 p.m.?

47. **Net Investment Flow** The net investment flow (rate of capital formation) of the giant conglomerate LTF incorporated is projected to be

$$t\sqrt{\frac{1}{2}t^2+1}$$

million dollars/year in year t. Find the accruement on the company's capital stock in the second year.
Hint: The amount is given by

$$\int_1^2 t\sqrt{\frac{1}{2}t^2 + 1}\,dt$$

48. **Oil Production** Based on a preliminary report by a geological survey team, it is estimated that a newly discovered oil field can be expected to produce oil at the rate of

$$R(t) = \frac{600t^2}{t^3 + 32} + 5 \qquad (0 \le t \le 20)$$

thousand barrels/year, t yr after production begins. Find the amount of oil that the field can be expected to yield during the first 5 yr of production, assuming that the projection holds true.

49. **Depreciation: Double Declining-Balance Method** Suppose a tractor purchased at a price of \$60,000 is to be depreciated by the *double declining-balance method* over a 10-yr period. It can be shown that the rate at which the book value will be decreasing is given by

$$R(t) = 13388.61e^{-0.22314t} \qquad (0 \le t \le 10)$$

dollars/year at year t. Find the amount by which the book value of the tractor will depreciate over the first 5 yr of its life.

50. **Velocity of a Car** A car moves along a straight road in such a way that its velocity (in feet/second) at any time t (in seconds) is given by

$$v(t) = 3t\sqrt{16 - t^2} \qquad (0 \le t \le 4)$$

Find the distance traveled by the car in the 4 sec from $t = 0$ to $t = 4$.

51. **Mobile-Phone Ad Spending** Mobile-phone ad spending between 2005 ($t = 1$) and 2011 ($t = 7$) is projected to be

$$S(t) = 0.86t^{0.96} \qquad (1 \le t \le 7)$$

where $S(t)$ is measured in billions of dollars and t is measured in years. What is the projected average spending per year on mobile-phone spending between 2005 and 2011?
Source: Interactive Advertising Bureau

52. **Global Warming** The increase in carbon dioxide (CO_2) in the atmosphere is a major cause of global warming. Using data obtained by Charles David Keeling, professor at Scripps Institution of Oceanography, the average amount of CO_2 in the atmosphere from 1958 through 2007 is approximated by

$$A(t) = 0.010716t^2 + 0.8212t + 313.4 \qquad (1 \le t \le 50)$$

where $A(t)$ is measured in parts per million volume (ppmv) and t in years, with $t = 1$ corresponding to 1958. Find the average rate of increase of the average amount of CO_2 in the atmosphere from 1958 through 2007.
Source: Scripps Institution of Oceanography

53. **Projected U.S. Gasoline Usage** The White House wants to cut the gasoline usage from 140 billion gallons per year in 2007 to 128 billion gallons per year in 2017. But estimates by the Department of Energy's Energy Information Agency suggest that won't happen. In fact, the agency's projection of gasoline usage from the beginning of 2007 until the beginning of 2017 is given by

$$A(t) = 0.014t^2 + 1.93t + 140 \qquad (0 \le t \le 10)$$

where $A(t)$ is measured in billions of gallons/year and t is in years, with $t = 0$ corresponding to 2007.
a. According to the agency's projection, what will be gasoline consumption at the beginning of 2017?
b. What will be the average consumption/year over the period from the beginning of 2007 until the beginning of 2017?
Source: U.S. Department of Energy, Energy Information Agency

54. **U.S. Citizens 65 Years and Older** The number of U.S. citizens aged 65 yr and older from 1900 through 2050 is estimated to be growing at the rate of

$$R(t) = 0.063t^2 - 0.48t + 3.87 \qquad (0 \le t \le 15)$$

million people/decade, where t is measured in decades, with $t = 0$ corresponding to 1900. Show that the average rate of growth of U.S. citizens aged 65 yr and older between 2000 and 2050 will be growing at twice the rate of that between 1950 and 2000.
Source: American Heart Association

55. **Office Vacancy Rate** The total vacancy rate (in percent) of offices in Manhattan from 2000 through 2006 is approximated by the function

$$R(t) = 0.032t^4 - 0.26t^3 - 0.478t^2 + 5.82t + 3.8 \qquad (0 \le t \le 6)$$

where t is measured in years, with $t = 0$ corresponding to 2000. What was the average vacancy rate of offices in Manhattan over the period from 2000 through 2006?
Source: Cushman and Wakefield

56. **Average Yearly Sales** The sales of Universal Instruments in the first t yr of its operation are approximated by the function

$$S(t) = t\sqrt{0.2t^2 + 4}$$

where $S(t)$ is measured in millions of dollars. What were Universal's average yearly sales over its first 5 yr of operation?

57. **Cable TV Subscribers** The manager of TeleStar Cable Service estimates that the total number of subscribers to the service in a certain city t yr from now will be

$$N(t) = -\frac{40{,}000}{\sqrt{1 + 0.2t}} + 50{,}000$$

Find the average number of cable television subscribers over the next 5 yr if this prediction holds true.

58. Concentration of a Drug in the Bloodstream The concentration of a certain drug in a patient's bloodstream t hr after injection is

$$C(t) = \frac{0.2t}{t^2 + 1}$$

mg/cm^3. Determine the average concentration of the drug in the patient's bloodstream over the first 4 hr after the drug is injected.

59. Average Price of a Commodity The price of a certain commodity in dollars/unit at time t (measured in weeks) is given by

$$p = 18 - 3e^{-2t} - 6e^{-t/3}$$

What is the average price of the commodity over the 5-wk period from $t = 0$ to $t = 5$?

60. Flow of Blood in an Artery According to a law discovered by 19th-century physician Jean Louis Marie Poiseuille, the velocity of blood (in centimeters/second) r cm from the central axis of an artery is given by

$$v(r) = k(R^2 - r^2)$$

where k is a constant and R is the radius of the artery. Find the average velocity of blood along a radius of the artery (see the accompanying figure).

Hint: Evaluate $\frac{1}{R}\int_0^R v(r)\,dr$.

Blood vessel

61. Waste Disposal When organic waste is dumped into a pond, the oxidization process that takes place reduces the pond's oxygen content. However, in time, nature will restore the oxygen content to its natural level. Suppose that the oxygen content t days after organic waste has been dumped into a pond is given by

$$f(t) = 100\left(\frac{t^2 + 10t + 100}{t^2 + 20t + 100}\right)$$

percent of its normal level. Find the average content of oxygen in the pond over the first 10 days after organic waste has been dumped into it.
Hint: Show that

$$\frac{t^2 + 10t + 100}{t^2 + 20t + 100} = 1 - \frac{10}{t + 10} + \frac{100}{(t + 10)^2}$$

62. Velocity of a Falling Hammer During the construction of a high-rise apartment building, a construction worker accidently drops a hammer that falls vertically a distance of h ft. The velocity of the hammer after falling a distance of x ft is $v = \sqrt{2gx}$ ft/sec $(0 \le x \le h)$. Show that the average velocity of the hammer over this path is $\bar{v} = \frac{2}{3}\sqrt{2gh}$ ft/sec.

63. Prove Property 1 of the definite integral.
Hint: Let F be an antiderivative of f and use the definition of the definite integral.

64. Prove Property 2 of the definite integral.
Hint: See Exercise 63.

65. Verify by direct computation that

$$\int_1^3 x^2\,dx = -\int_3^1 x^2\,dx$$

66. Prove Property 3 of the definite integral.
Hint: See Exercise 63.

67. Verify by direct computation that

$$\int_1^9 2\sqrt{x}\,dx = 2\int_1^9 \sqrt{x}\,dx$$

68. Verify by direct computation that

$$\int_0^1 (1 + x - e^x)\,dx = \int_0^1 dx + \int_0^1 x\,dx - \int_0^1 e^x\,dx$$

What property of the definite integral is demonstrated in this exercise?

69. Verify by direct computation that

$$\int_0^3 (1 + x^3)\,dx = \int_0^1 (1 + x^3)\,dx + \int_1^3 (1 + x^3)\,dx$$

What property of the definite integral is demonstrated here?

70. Verify by direct computation that

$$\int_0^3 (1 + x^3)\,dx$$
$$= \int_0^1 (1 + x^3)\,dx + \int_1^2 (1 + x^3)\,dx + \int_2^3 (1 + x^3)\,dx$$

hence showing that Property 5 may be extended.

71. Evaluate $\int_3^3 (1 + \sqrt{x})e^{-x}\,dx$.

72. Evaluate $\int_3^0 f(x)\,dx$, given that $\int_0^3 f(x)\,dx = 4$.

73. Given that $\int_{-1}^2 f(x)\,dx = -2$ and $\int_{-1}^2 g(x)\,dx = 3$, evaluate

a. $\int_{-1}^2 [2f(x) + g(x)]\,dx$
b. $\int_{-1}^2 [g(x) - f(x)]\,dx$
c. $\int_{-1}^2 [2f(x) - 3g(x)]\,dx$

74. Given that $\int_{-1}^2 f(x)\,dx = 2$ and $\int_0^2 f(x)\,dx = 3$, evaluate

a. $\int_{-1}^0 f(x)\,dx$ b. $\int_0^2 f(x)\,dx - \int_{-1}^0 f(x)\,dx$

In Exercises 75–80, determine whether the statement is true or false. If it is true, explain why it is true. If it is false, explain why or give an example to show why it is false.

75. $\displaystyle\int_2^2 \frac{e^x}{\sqrt{1+x}}\,dx = 0$

76. $\displaystyle\int_1^3 \frac{dx}{x-2} = -\int_3^1 \frac{dx}{x-2}$

77. $\displaystyle\int_0^1 x\sqrt{x+1}\,dx = \sqrt{x+1}\int_0^1 x\,dx = \frac{1}{2}x^2\sqrt{x+1}\Big|_0^1 = \frac{\sqrt{2}}{2}$

78. If f' is continuous on $[0, 2]$, then $\int_0^2 f'(x)\,dx = f(2) - f(0)$.

79. If f and g are continuous on $[a, b]$ and k is a constant, then

$$\int_a^b [kf(x) + g(x)]\,dx = k\int_a^b f(x)\,dx + \int_a^b g(x)\,dx$$

80. If f is continuous on $[a, b]$ and $a < c < b$, then

$$\int_b^c f(x)\,dx = \int_a^c f(x)\,dx - \int_a^b f(x)\,dx$$

6.5 Solutions to Self-Check Exercises

1. Let $u = 2x + 5$. Then, $du = 2\,dx$, or $dx = \frac{1}{2}\,du$. Also, when $x = 0$, $u = 5$, and when $x = 2$, $u = 9$. Therefore,

$$\begin{aligned}\int_0^2 \sqrt{2x+5}\,dx &= \int_0^2 (2x+5)^{1/2}\,dx\\ &= \frac{1}{2}\int_5^9 u^{1/2}\,du\\ &= \left(\frac{1}{2}\right)\left(\frac{2}{3}u^{3/2}\right)\Big|_5^9\\ &= \frac{1}{3}[9^{3/2} - 5^{3/2}]\\ &= \frac{1}{3}(27 - 5\sqrt{5})\end{aligned}$$

2. The required average value is given by

$$\begin{aligned}\frac{1}{2-(-1)}\int_{-1}^2 (1-x^2)\,dx &= \frac{1}{3}\int_{-1}^2 (1-x^2)\,dx\\ &= \frac{1}{3}\left(x - \frac{1}{3}x^3\right)\Big|_{-1}^2\\ &= \frac{1}{3}\left[\left(2-\frac{8}{3}\right) - \left(-1+\frac{1}{3}\right)\right] = 0\end{aligned}$$

3. The average median price of a house over the stated time interval is given by

$$\begin{aligned}&\frac{1}{5-0}\int_0^5 (t^3 - 7t^2 + 17t + 280)\,dt\\ &\quad= \frac{1}{5}\left(\frac{1}{4}t^4 - \frac{7}{3}t^3 + \frac{17}{2}t^2 + 280t\right)\Big|_0^5\\ &\quad= \frac{1}{5}\left[\frac{1}{4}(5)^4 - \frac{7}{3}(5)^3 + \frac{17}{2}(5)^2 + 280(5)\right]\\ &\quad\approx 295.417\end{aligned}$$

or approximately \$295,417.

USING TECHNOLOGY

Evaluating Definite Integrals for Piecewise-Defined Functions

We continue using graphing utilities to find the definite integral of a function. But here we will make use of Property 5 of the properties of the definite integral (p. 442).

EXAMPLE 1 Use the numerical integral operation of a graphing utility to evaluate

$$\int_{-1}^2 f(x)\,dx$$

where

$$f(x) = \begin{cases} -x^2 & \text{if } x < 0 \\ \sqrt{x} & \text{if } x \geq 0 \end{cases}$$

Solution Using Property 5 of the definite integral, we can write

$$\int_{-1}^{2} f(x)\,dx = \int_{-1}^{0} -x^2\,dx + \int_{0}^{2} x^{1/2}\,dx$$

Using a graphing utility, we find

$$\begin{aligned} \int_{-1}^{2} f(x)\,dx &= \text{fnInt}(-x\text{^}2, x, -1, 0) + \text{fnInt}(x\text{^}0.5, x, 0, 2) \\ &\approx -0.333333 + 1.885618 \\ &= 1.552285 \end{aligned}$$

TECHNOLOGY EXERCISES

In Exercises 1–4, use Property 5 of the properties of the definite integral (page 442) to evaluate the definite integral accurate to six decimal places.

1. $\int_{-1}^{2} f(x)\,dx$, where

$$f(x) = \begin{cases} 2.3x^3 - 3.1x^2 + 2.7x + 3 & \text{if } x < 1 \\ -1.7x^2 + 2.3x + 4.3 & \text{if } x \geq 1 \end{cases}$$

2. $\int_{0}^{3} f(x)\,dx$, where $f(x) = \begin{cases} \dfrac{\sqrt{x}}{1 + x^2} & \text{if } 0 \leq x < 1 \\ 0.5e^{-0.1x^2} & \text{if } x \geq 1 \end{cases}$

3. $\int_{-2}^{2} f(x)\,dx$, where $f(x) = \begin{cases} x^4 - 2x^2 + 4 & \text{if } x < 0 \\ 2\ln(x + e^2) & \text{if } x \geq 0 \end{cases}$

4. $\int_{-2}^{6} f(x)\,dx$, where

$$f(x) = \begin{cases} 2x^3 - 3x^2 + x + 2 & \text{if } x < -1 \\ \sqrt{3x + 4} - 5 & \text{if } -1 \leq x \leq 4 \\ x^2 - 3x - 5 & \text{if } x > 4 \end{cases}$$

5. **AIDS in Massachusetts** The rate of growth (and decline) of the number of AIDS cases diagnosed in Massachusetts from the beginning of 1989 ($t = 0$) through the end of 1997 ($t = 8$) is approximated by the function

$$f(t) = \begin{cases} 69.83333t^2 + 30.16667t + 1000 & \text{if } 0 \leq t < 3 \\ 1719 & \text{if } 3 \leq t < 4 \\ -28.79167t^3 + 491.37500t^2 - 2985.083333t + 7640 & \text{if } 4 \leq t \leq 8 \end{cases}$$

where $f(t)$ is measured in the number of cases/year. Estimate the total number of AIDS cases diagnosed in Massachusetts from the beginning of 1989 through the end of 1997.

Source: Massachusetts Department of Health

6. **Crop Yield** If left untreated on bean stems, aphids (small insects that suck plant juices) will multiply at an increasing rate during the summer months and reduce productivity and crop yield of cultivated crops. But if the aphids are treated in mid-June, the numbers decrease sharply to less than 100/bean stem, allowing for steep rises in crop yield. The function

$$F(t) = \begin{cases} 62e^{1.152t} & \text{if } 0 \leq t < 1.5 \\ 349e^{-1.324(t-1.5)} & \text{if } 1.5 \leq t \leq 3 \end{cases}$$

gives the number of aphids on a typical bean stem at time t, where t is measured in months, $t = 0$ corresponding to the beginning of May. Find the average number of aphids on a typical bean stem over the period from the beginning of May to the beginning of August.

7. **Absorption of Drugs** Jane took 100 mg of a drug in the morning and another 100 mg of the same drug at the same time the following morning. The amount of the drug in her body t days after the first dosage was taken is given by

$$A(t) = \begin{cases} 100e^{-1.4t} & \text{if } 0 \leq t < 1 \\ 100(1 + e^{1.4})e^{-1.4t} & \text{if } t \geq 1 \end{cases}$$

Find the average amount of the drug in Jane's body over the first 2 days.

8. **Absorption of Drugs** The concentration of a drug in an organ at any time t (in seconds) is given by

$$C(t) = \begin{cases} 0.3t - 18(1 - e^{-t/60}) & \text{if } 0 \leq t \leq 20 \\ 18e^{-t/60} - 12e^{-(t-20)/60} & \text{if } t > 20 \end{cases}$$

where $C(t)$ is measured in grams/cubic centimeter (g/cm^3). Find the average concentration of the drug in the organ over the first 30 sec after it is administered.

6.6 Area between Two Curves

Suppose a certain country's petroleum consumption is expected to grow at the rate of $f(t)$ million barrels per year, t years from now, for the next 5 years. Then, the country's total petroleum consumption over the period of time in question is given by the area under the graph of f on the interval [0, 5] (Figure 27).

Next, suppose that because of the implementation of certain energy-conservation measures, the rate of growth of petroleum consumption is expected to be $g(t)$ million barrels per year instead. Then, the country's projected total petroleum consumption over the 5-year period is given by the area under the graph of g on the interval [0, 5] (Figure 28).

Therefore, the area of the shaded region S lying between the graphs of f and g on the interval [0, 5] (Figure 29) gives the amount of petroleum that would be saved over the 5-year period because of the conservation measures.

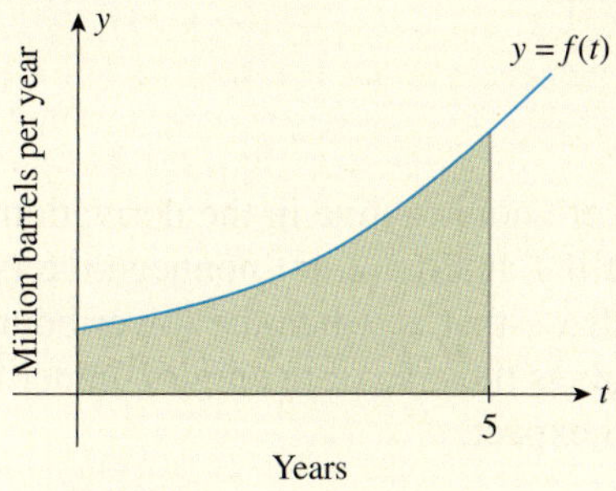

FIGURE 27
At a rate of consumption of $f(t)$ million barrels per year, the total petroleum consumption is given by the area of the region under the graph of f.

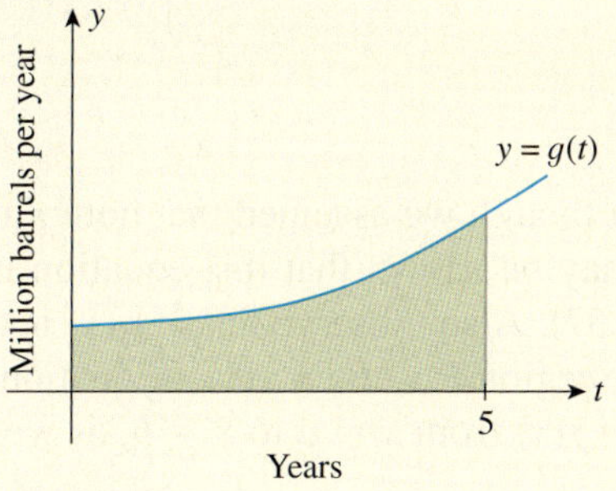

FIGURE 28
At a rate of consumption of $g(t)$ million barrels per year, the total petroleum consumption is given by the area of the region under the graph of g.

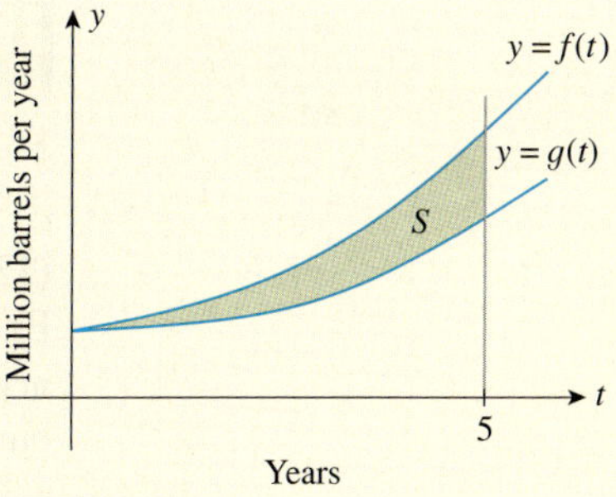

FIGURE 29
The area of S gives the amount of petroleum that would be saved over the 5-year period.

But the area of S is given by

$$\text{Area under the graph of } f \text{ on } [a, b] - \text{Area under the graph of } g \text{ on } [a, b]$$

$$= \int_0^5 f(t)\,dt - \int_0^5 g(t)\,dt$$

$$= \int_0^5 [f(t) - g(t)]\,dt \qquad \text{By Property 4, Section 6.5}$$

This example shows that some practical problems can be solved by finding the area of a region between two curves, which in turn can be found by evaluating an appropriate definite integral.

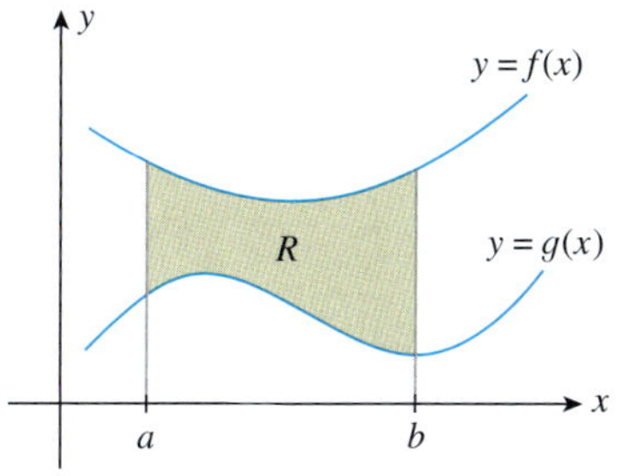

FIGURE 30
Area of $R = \int_a^b [f(x) - g(x)]\,dx$

Finding the Area between Two Curves

We now turn our attention to the general problem of finding the area of a plane region bounded both above and below by the graphs of functions. First, consider the situation in which the graph of one function lies above that of another. More specifically, let R be the region in the xy-plane (Figure 30) that is bounded above by the graph of a continuous function f, below by a continuous function g where

$f(x) \geq g(x)$ on $[a, b]$, and to the left and right by the vertical lines $x = a$ and $x = b$, respectively. From the figure, we see that

$$\begin{aligned}\text{Area of } R &= \text{Area under } f(x) - \text{Area under } g(x)\\ &= \int_a^b f(x)\,dx - \int_a^b g(x)\,dx\\ &= \int_a^b [f(x) - g(x)]\,dx\end{aligned}$$

upon using Property 4 of the definite integral.

The Area between Two Curves

Let f and g be continuous functions such that $f(x) \geq g(x)$ on the interval $[a, b]$. Then, the area of the region bounded above by $y = f(x)$ and below by $y = g(x)$ on $[a, b]$ is given by

$$\int_a^b [f(x) - g(x)]\,dx \tag{13}$$

Even though we assumed that both f and g were nonnegative in the derivation of (13), it may be shown that this equation is valid if f and g are not nonnegative (see Exercise 57). Also, observe that if $g(x)$ is 0 for all x—that is, when the lower boundary of the region R is the x-axis—Equation (13) gives the area of the region under the curve $y = f(x)$ from $x = a$ to $x = b$, as we would expect.

EXAMPLE 1 Find the area of the region bounded by the x-axis, the graph of $y = -x^2 + 4x - 8$, and the lines $x = -1$ and $x = 4$.

Solution The region R under consideration is shown in Figure 31. We can view R as the region bounded above by the graph of $f(x) = 0$ (the x-axis) and below by the graph of $g(x) = -x^2 + 4x - 8$ on $[-1, 4]$. Therefore, the area of R is given by

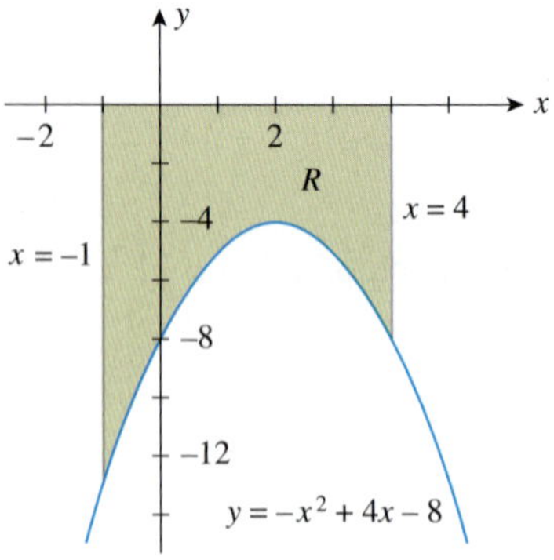

FIGURE 31
Area of $R = -\int_{-1}^{4} g(x)\,dx$

$$\begin{aligned}\int_a^b [f(x) - g(x)]\,dx &= \int_{-1}^{4} [0 - (-x^2 + 4x - 8)]\,dx\\ &= \int_{-1}^{4} (x^2 - 4x + 8)\,dx\\ &= \frac{1}{3}x^3 - 2x^2 + 8x\Big|_{-1}^{4}\\ &= \left[\frac{1}{3}(64) - 2(16) + 8(4)\right] - \left[\frac{1}{3}(-1) - 2(1) + 8(-1)\right]\\ &= 31\tfrac{2}{3}\end{aligned}$$

or $31\frac{2}{3}$ square units. ■

EXAMPLE 2 Find the area of the region R bounded by the graphs of

$$f(x) = 2x - 1 \quad \text{and} \quad g(x) = x^2 - 4$$

and the vertical lines $x = 1$ and $x = 2$.

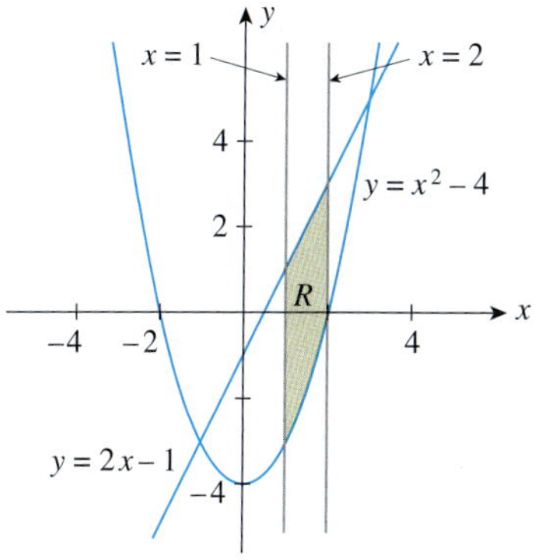

FIGURE 32
Area of $R = \int_1^2 [f(x) - g(x)]\,dx$

Solution We first sketch the graphs of the functions $f(x) = 2x - 1$ and $g(x) = x^2 - 4$ and the vertical lines $x = 1$ and $x = 2$, and then we identify the region R whose area is to be calculated (Figure 32).

Since the graph of f always lies above that of g for x in the interval $[1, 2]$, we see by Equation (13) that the required area is given by

$$\begin{aligned}\int_1^2 [f(x) - g(x)]\,dx &= \int_1^2 [(2x - 1) - (x^2 - 4)]\,dx \\ &= \int_1^2 (-x^2 + 2x + 3)\,dx \\ &= -\frac{1}{3}x^3 + x^2 + 3x\Big|_1^2 \\ &= \left(-\frac{8}{3} + 4 + 6\right) - \left(-\frac{1}{3} + 1 + 3\right) = \frac{11}{3}\end{aligned}$$

or $\frac{11}{3}$ square units. ■

EXAMPLE 3 Find the area of the region R that is completely enclosed by the graphs of the functions

$$f(x) = 2x - 1 \quad \text{and} \quad g(x) = x^2 - 4$$

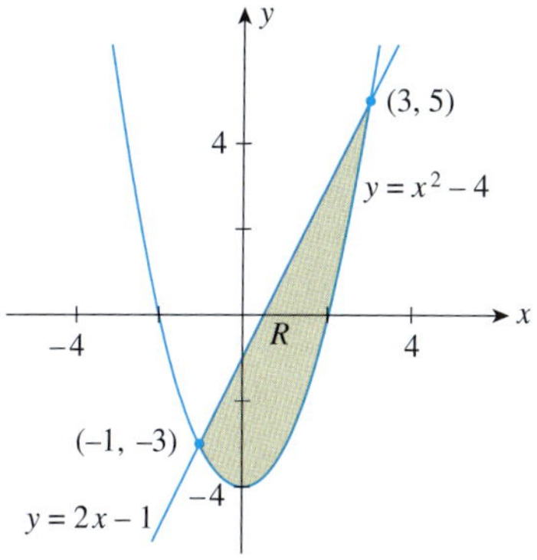

FIGURE 33
Area of $R = \int_{-1}^{3} [f(x) - g(x)]\,dx$

Solution The region R is shown in Figure 33. First, we find the points of intersection of the two curves. To do this, we solve the system that consists of the two equations $y = 2x - 1$ and $y = x^2 - 4$. Equating the two values of y gives

$$\begin{aligned}x^2 - 4 &= 2x - 1 \\ x^2 - 2x - 3 &= 0 \\ (x + 1)(x - 3) &= 0\end{aligned}$$

so $x = -1$ or $x = 3$. That is, the two curves intersect when $x = -1$ and $x = 3$.

Observe that we could also view the region R as the region bounded above by the graph of the function $f(x) = 2x - 1$, below by the graph of the function $g(x) = x^2 - 4$, and to the left and right by the vertical lines $x = -1$ and $x = 3$, respectively.

Next, since the graph of the function f always lies above that of the function g on $[-1, 3]$, we can use (13) to compute the desired area:

$$\begin{aligned}\int_a^b [f(x) - g(x)]\,dx &= \int_{-1}^3 [(2x - 1) - (x^2 - 4)]\,dx \\ &= \int_{-1}^3 (-x^2 + 2x + 3)\,dx \\ &= -\frac{1}{3}x^3 + x^2 + 3x\Big|_{-1}^3 \\ &= (-9 + 9 + 9) - \left(\frac{1}{3} + 1 - 3\right) = \frac{32}{3} \\ &= 10\tfrac{2}{3}\end{aligned}$$

or $10\frac{2}{3}$ square units. ■

EXAMPLE 4 Find the area of the region R bounded by the graphs of the functions

$$f(x) = x^2 - 2x - 1 \quad \text{and} \quad g(x) = -e^x - 1$$

and the vertical lines $x = -1$ and $x = 1$.

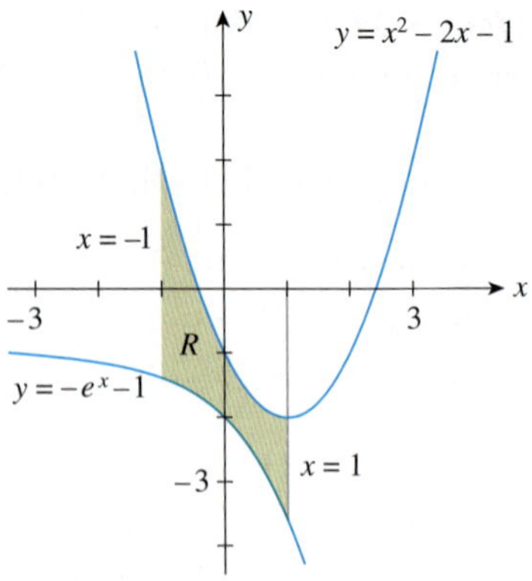

FIGURE 34
Area of $R = \int_{-1}^{1} [f(x) - g(x)]\,dx$

Solution The region R is shown in Figure 34. Since the graph of the function f always lies above that of the function g, the area of the region R is given by

$$\begin{aligned}\int_a^b [f(x) - g(x)]\,dx &= \int_{-1}^{1} [(x^2 - 2x - 1) - (-e^x - 1)]\,dx \\ &= \int_{-1}^{1} (x^2 - 2x + e^x)\,dx \\ &= \frac{1}{3}x^3 - x^2 + e^x\Big|_{-1}^{1} \\ &= \left(\frac{1}{3} - 1 + e\right) - \left(-\frac{1}{3} - 1 + e^{-1}\right) \\ &= \frac{2}{3} + e - \frac{1}{e}, \text{ or approximately 3.02 square units}\end{aligned}$$

Equation (13), which gives the area of the region between the curves $y = f(x)$ and $y = g(x)$ for $a \le x \le b$, is valid when the graph of the function f lies above that of the function g over the interval $[a, b]$. Example 5 shows how to use (13) to find the area of a region when the latter condition does not hold.

EXAMPLE 5 Find the area of the region bounded by the graph of the function $f(x) = x^3$, the x-axis, and the lines $x = -1$ and $x = 1$.

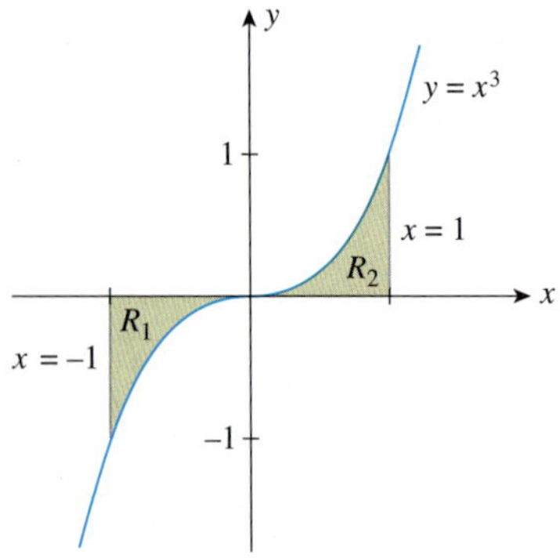

FIGURE 35
Area of R_1 = Area of R_2

Solution The region R under consideration can be thought of as composed of the two subregions R_1 and R_2, as shown in Figure 35.

Recall that the x-axis is represented by the function $g(x) = 0$. Since $g(x) \ge f(x)$ on $[-1, 0]$, we see that the area of R_1 is given by

$$\begin{aligned}\int_a^b [g(x) - f(x)]\,dx &= \int_{-1}^{0} (0 - x^3)\,dx = -\int_{-1}^{0} x^3\,dx \\ &= -\frac{1}{4}x^4\Big|_{-1}^{0} = 0 - \left(-\frac{1}{4}\right) = \frac{1}{4}\end{aligned}$$

To find the area of R_2, we observe that $f(x) \ge g(x)$ on $[0, 1]$, so it is given by

$$\begin{aligned}\int_a^b [f(x) - g(x)]\,dx &= \int_0^1 (x^3 - 0)\,dx = \int_0^1 x^3\,dx \\ &= \frac{1}{4}x^4\Big|_0^1 = \left(\frac{1}{4}\right) - 0 = \frac{1}{4}\end{aligned}$$

Therefore, the area of R is $\frac{1}{4} + \frac{1}{4}$, or $\frac{1}{2}$, square units.

By making use of symmetry, we could have obtained the same result by computing

$$-2\int_{-1}^{0} x^3\,dx \quad \text{or} \quad 2\int_0^1 x^3\,dx$$

as you may verify.

Explore & Discuss

A function is *even* if it satisfies the condition $f(-x) = f(x)$, and it is *odd* if it satisfies the condition $f(-x) = -f(x)$. Show that the graph of an even function is symmetric with respect to the y-axis while the graph of an odd function is symmetric with respect to the origin. Explain why

$$\int_{-a}^{a} f(x)\,dx = 2\int_0^a f(x)\,dx \quad \text{if } f \text{ is even}$$

and

$$\int_{-a}^{a} f(x)\,dx = 0 \quad \text{if } f \text{ is odd}$$

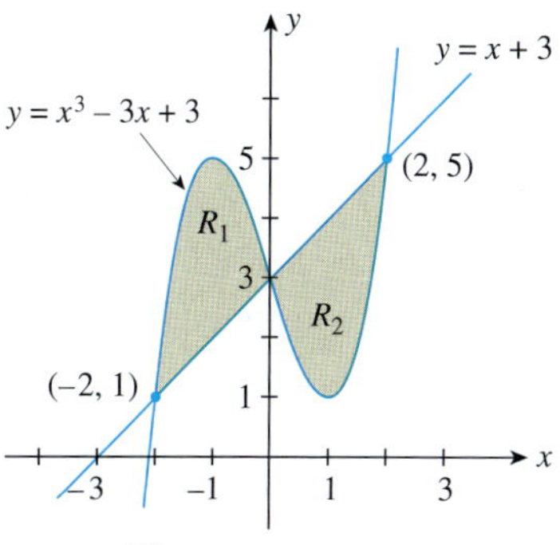

FIGURE 36
Area of R_1 + Area of R_2
$= \int_{-2}^{0}[f(x) - g(x)]\,dx + \int_0^2 [g(x) - f(x)]\,dx$

EXAMPLE 6 Find the area of the region completely enclosed by the graphs of the functions

$$f(x) = x^3 - 3x + 3 \quad \text{and} \quad g(x) = x + 3$$

Solution First, sketch the graphs of $y = x^3 - 3x + 3$ and $y = x + 3$ and then identify the required region R. We can view the region R as being composed of the two subregions R_1 and R_2, as shown in Figure 36. By solving the equations $y = x + 3$ and $y = x^3 - 3x + 3$ simultaneously, we find the points of intersection of the two curves. Equating the two values of y, we have

$$\begin{aligned} x^3 - 3x + 3 &= x + 3 \\ x^3 - 4x &= 0 \\ x(x^2 - 4) &= 0 \\ x(x + 2)(x - 2) &= 0 \\ x &= 0, -2, 2 \end{aligned}$$

Hence, the points of intersection of the two curves are $(-2, 1)$, $(0, 3)$, and $(2, 5)$.

For $-2 \leq x \leq 0$, we see that the graph of the function f lies above that of the function g, so the area of the region R_1 is, by virtue of (13),

$$\begin{aligned} \int_{-2}^{0} [(x^3 - 3x + 3) - (x + 3)]\,dx &= \int_{-2}^{0} (x^3 - 4x)\,dx \\ &= \frac{1}{4}x^4 - 2x^2 \Big|_{-2}^{0} \\ &= -(4 - 8) \\ &= 4 \end{aligned}$$

or 4 square units. For $0 \leq x \leq 2$, the graph of the function g lies above that of the function f, and the area of R_2 is given by

$$\begin{aligned} \int_{0}^{2} [(x + 3) - (x^3 - 3x + 3)]\,dx &= \int_{0}^{2} (-x^3 + 4x)\,dx \\ &= -\frac{1}{4}x^4 + 2x^2 \Big|_{0}^{2} \\ &= -4 + 8 \\ &= 4 \end{aligned}$$

or 4 square units. Therefore, the required area is the sum of the area of the two regions $R_1 + R_2$—that is, $4 + 4$, or 8 square units. ■

VIDEO

APPLIED EXAMPLE 7 Conservation of Oil In a 2004 study for a developing country's Economic Development Board, government economists and energy experts concluded that if the Energy Conservation Bill were implemented in 2005, the country's oil consumption for the next 5 years would be expected to grow in accordance with the model

$$R(t) = 20e^{0.05t}$$

where t is measured in years ($t = 0$ corresponding to the year 2005) and $R(t)$ in millions of barrels per year. Without the government-imposed conservation measures, however, the expected rate of growth of oil consumption would be given by

$$R_1(t) = 20e^{0.08t}$$

millions of barrels per year. Using these models, determine how much oil would have been saved from 2005 through 2010 if the bill had been implemented.

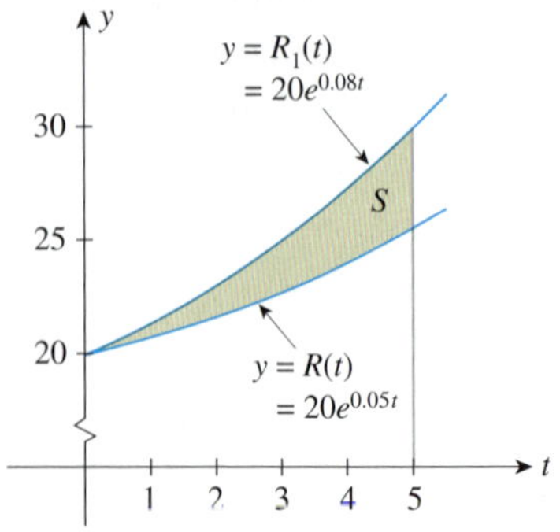

FIGURE 37
Area of $S = \int_0^5 [R_1(t) - R(t)]\,dt$

Solution Under the Energy Conservation Bill, the total amount of oil that would have been consumed between 2005 and 2010 is given by

$$\int_0^5 R(t)\,dt = \int_0^5 20e^{0.05t}\,dt \tag{14}$$

Without the bill, the total amount of oil that would have been consumed between 2005 and 2010 is given by

$$\int_0^5 R_1(t)\,dt = \int_0^5 20e^{0.08t}\,dt \tag{15}$$

Equation (14) may be interpreted as the area of the region under the curve $y = R(t)$ from $t = 0$ to $t = 5$. Similarly, we interpret (15) as the area of the region under the curve $y = R_1(t)$ from $t = 0$ to $t = 5$. Furthermore, note that the graph of $y = R_1(t) = 20e^{0.08t}$ always lies on or above the graph of $y = R(t) = 20e^{0.05t}$ $(t \geq 0)$. Thus, the area of the shaded region S in Figure 37 shows the amount of oil that would have been saved from 2005 to 2010 if the Energy Conservation Bill had been implemented. But the area of the region S is given by

$$\begin{aligned}\int_0^5 [R_1(t) - R(t)]\,dt &= \int_0^5 [20e^{0.08t} - 20e^{0.05t}]\,dt \\ &= 20\int_0^5 (e^{0.08t} - e^{0.05t})\,dt \\ &= 20\left(\frac{e^{0.08t}}{0.08} - \frac{e^{0.05t}}{0.05}\right)\Bigg|_0^5 \\ &= 20\left[\left(\frac{e^{0.4}}{0.08} - \frac{e^{0.25}}{0.05}\right) - \left(\frac{1}{0.08} - \frac{1}{0.05}\right)\right] \\ &\approx 9.3\end{aligned}$$

or approximately 9.3 square units. Thus, the amount of oil that would have been saved is 9.3 million barrels.

Exploring with TECHNOLOGY

Refer to Example 7. Suppose we want to construct a mathematical model giving the amount of oil saved from 2005 through the year $(2005 + x)$ where $x \geq 0$. In Example 7, for instance, we take $x = 5$.

1. Show that this model is given by

$$\begin{aligned}F(x) &= \int_0^x [R_1(t) - R(t)]\,dt \\ &= 250e^{0.08x} - 400e^{0.05x} + 150\end{aligned}$$

   Hint: You may find it helpful to use some of the results of Example 7.
2. Use a graphing utility to plot the graph of F, using the viewing window $[0, 10] \times [0, 50]$.
3. Find $F(5)$ and thus confirm the result of Example 7.
4. What is the main advantage of this model?

6.6 Self-Check Exercises

1. Find the area of the region bounded by the graphs of $f(x) = x^2 + 2$ and $g(x) = 1 - x$ and the vertical lines $x = 0$ and $x = 1$.

2. Find the area of the region completely enclosed by the graphs of $f(x) = -x^2 + 6x + 5$ and $g(x) = x^2 + 5$.

3. The management of Kane Corporation, which operates a chain of hotels, expects its profits to grow at the rate of $1 + t^{2/3}$ million dollars/year t yr from now. However, with renovations and improvements of existing hotels and proposed acquisitions of new hotels, Kane's profits are expected to grow at the rate of $t - 2\sqrt{t} + 4$ million dollars/year in the next decade. What additional profits are expected over the next 10 yr if the group implements the proposed plans?

Solutions to Self-Check Exercises 6.6 can be found on page 463.

6.6 Concept Questions

1. Suppose f and g are continuous functions such that $f(x) \geq g(x)$ on the interval $[a, b]$. Write an integral giving the area of the region bounded above by the graph of f, below by the graph of g, and on the left and right by the lines $x = a$ and $x = b$.

2. Write an expression in terms of definite integrals giving the area of the shaded region in the following figure:

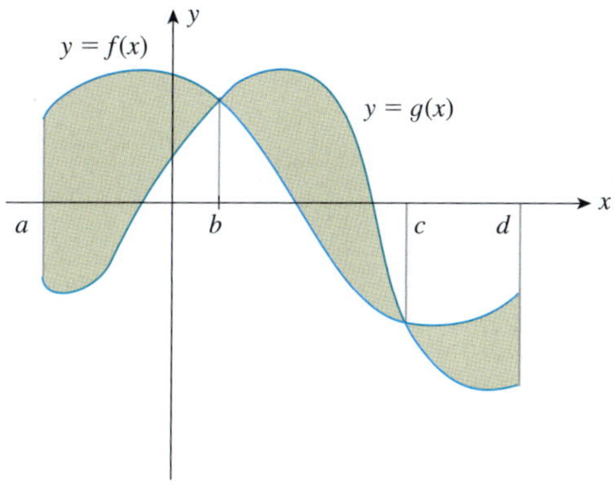

6.6 Exercises

In Exercises 1–8, find the area of the shaded region.

1.

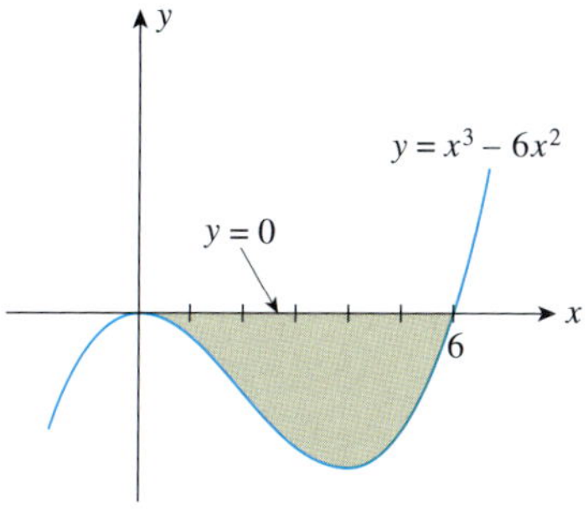

2.

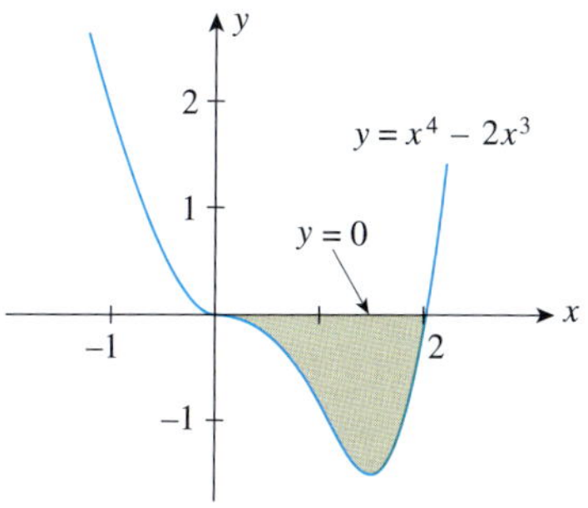

3.

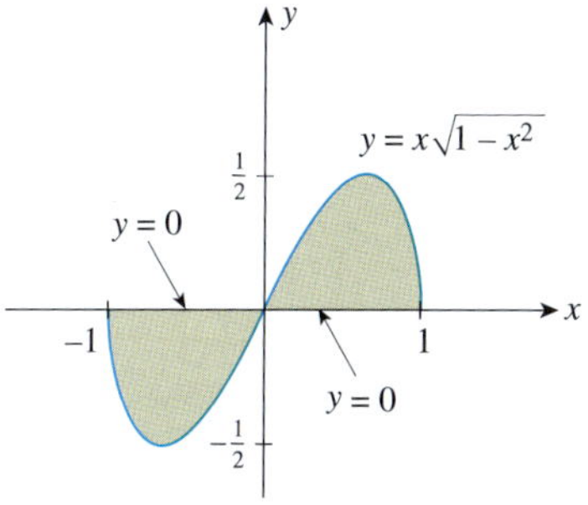

4.

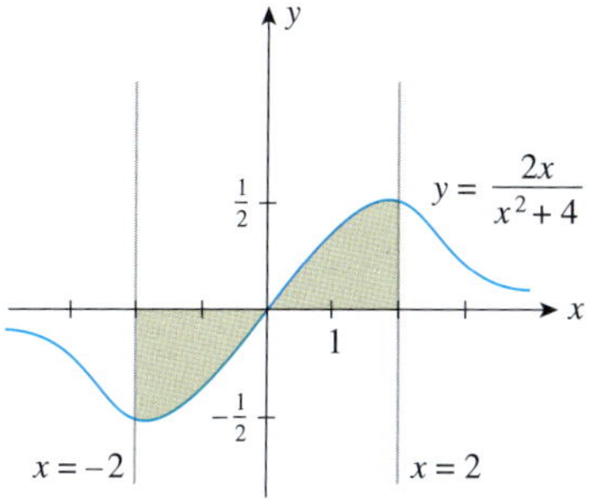

5.

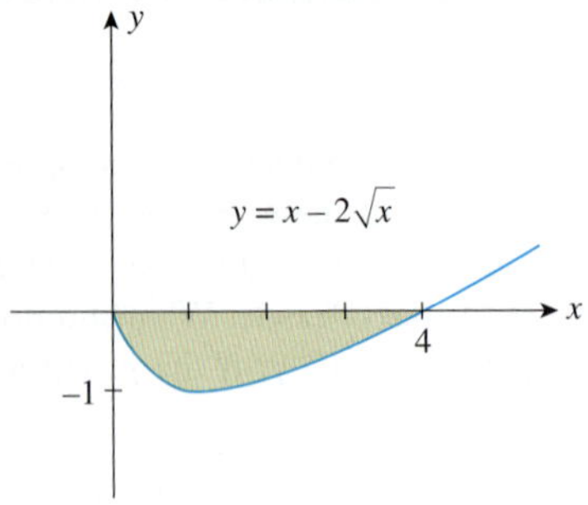

6.

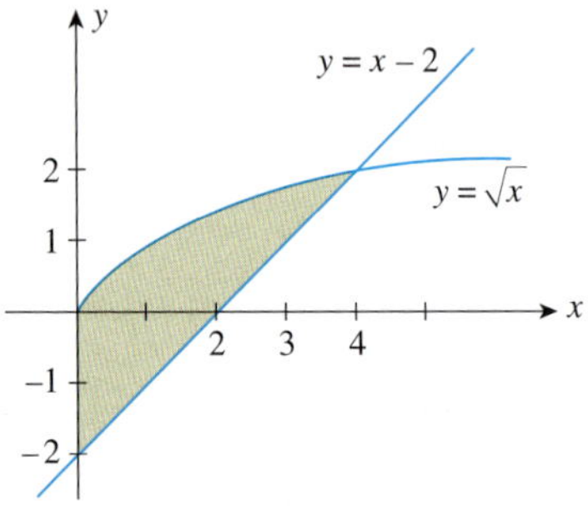

7.

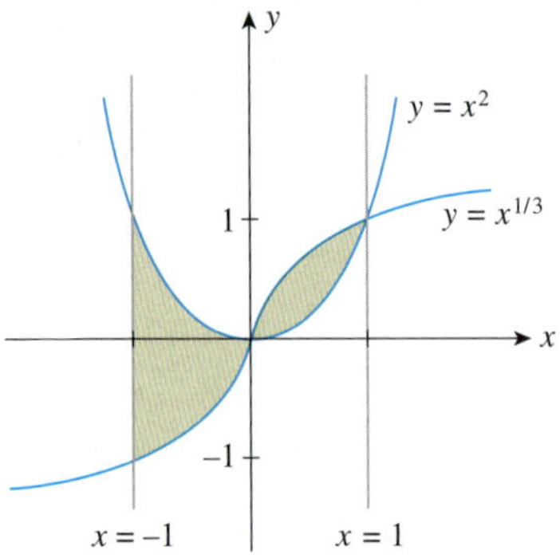

8.

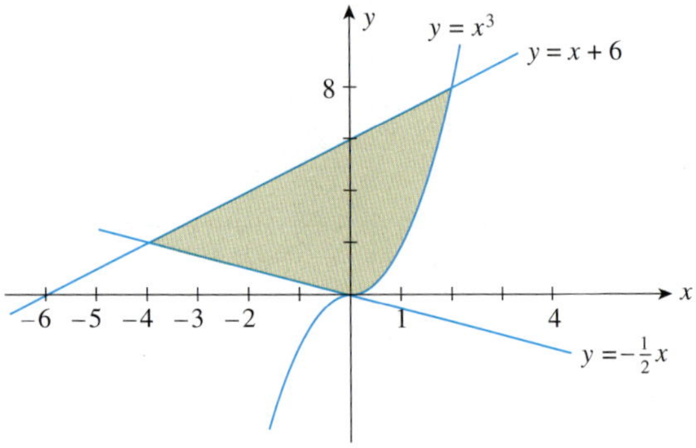

In Exercises 9–16, sketch the graph and find the area of the region bounded below by the graph of each function and above by the x-axis from $x = a$ to $x = b$.

9. $f(x) = -x^2$; $a = -1$, $b = 2$

10. $f(x) = x^2 - 4$; $a = -2$, $b = 2$

11. $f(x) = x^2 - 5x + 4$; $a = 1$, $b = 3$

12. $f(x) = x^3$; $a = -1$, $b = 0$

13. $f(x) = -1 - \sqrt{x}$; $a = 0$, $b = 9$

14. $f(x) = \frac{1}{2}x - \sqrt{x}$; $a = 0$, $b = 4$

15. $f(x) = -e^{(1/2)x}$; $a = -2$, $b = 4$

16. $f(x) = -xe^{-x^2}$; $a = 0$, $b = 1$

In Exercises 17–26, sketch the graphs of the functions f and g and find the area of the region enclosed by these graphs and the vertical lines $x = a$ and $x = b$.

17. $f(x) = x^2 + 3$, $g(x) = 1$; $a = 1$, $b = 3$

18. $f(x) = x + 2$, $g(x) = x^2 - 4$; $a = -1$, $b = 2$

19. $f(x) = -x^2 + 2x + 3$, $g(x) = -x + 3$; $a = 0$, $b = 2$

20. $f(x) = 9 - x^2$, $g(x) = 2x + 3$; $a = -1$, $b = 1$

21. $f(x) = x^2 + 1$, $g(x) = \frac{1}{3}x^3$; $a = -1$, $b = 2$

22. $f(x) = \sqrt{x}$, $g(x) = -\frac{1}{2}x - 1$; $a = 1$, $b = 4$

23. $f(x) = \frac{1}{x}$, $g(x) = 2x - 1$; $a = 1$, $b = 4$

24. $f(x) = x^2$, $g(x) = \frac{1}{x^2}$; $a = 1$, $b = 3$

25. $f(x) = e^x$, $g(x) = \frac{1}{x}$; $a = 1$, $b = 2$

26. $f(x) = x$, $g(x) = e^{2x}$; $a = 1$, $b = 3$

In Exercises 27–34, sketch the graph and find the area of the region bounded by the graph of the function f and the lines $y = 0$, $x = a$, and $x = b$.

27. $f(x) = x$; $a = -1$, $b = 2$

28. $f(x) = x^2 - 2x$; $a = -1$, $b = 1$

29. $f(x) = -x^2 + 4x - 3$; $a = -1$, $b = 2$

30. $f(x) = x^3 - x^2$; $a = -1$, $b = 1$

31. $f(x) = x^3 - 4x^2 + 3x$; $a = 0$, $b = 2$

32. $f(x) = 4x^{1/3} + x^{4/3}$; $a = -1$, $b = 8$

33. $f(x) = e^x - 1$; $a = -1$, $b = 3$

34. $f(x) = xe^{x^2}$; $a = 0$, $b = 2$

In Exercises 35–42, sketch the graph and find the area of the region completely enclosed by the graphs of the given functions f and g.

35. $f(x) = x + 2$ and $g(x) = x^2 - 4$

36. $f(x) = -x^2 + 4x$ and $g(x) = 2x - 3$

37. $f(x) = x^2$ and $g(x) = x^3$

38. $f(x) = x^3 + 2x^2 - 3x$ and $g(x) = 0$

39. $f(x) = x^3 - 6x^2 + 9x$ and $g(x) = x^2 - 3x$

40. $f(x) = \sqrt{x}$ and $g(x) = x^2$

41. $f(x) = x\sqrt{9 - x^2}$ and $g(x) = 0$

42. $f(x) = 2x$ and $g(x) = x\sqrt{x + 1}$

43. **Effect of Advertising on Revenue** In the accompanying figure, the function f gives the rate of change of Odyssey Travel's revenue with respect to the amount x it spends on advertising with their current advertising agency. By engaging the services of a different advertising agency, it is expected that Odyssey's revenue will grow at the rate given by the function g. Give an interpretation of the area A of the region S and find an expression for A in terms of a definite integral involving f and g.

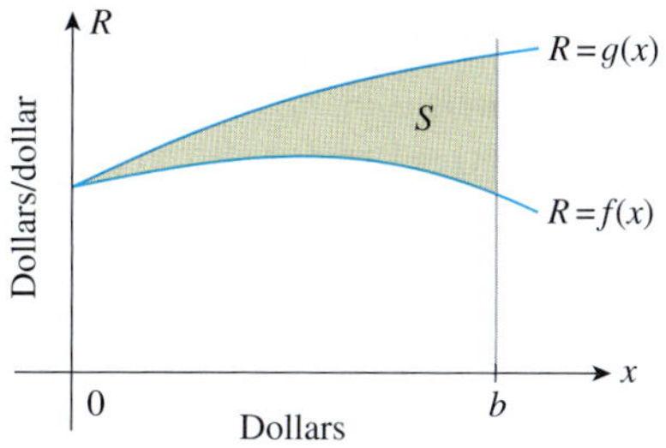

44. **Pulse Rate During Exercise** In the accompanying figure, the function f gives the rate of increase of an individual's pulse rate when he walked a prescribed course on a treadmill 6 mo ago. The function g gives the rate of increase of his pulse rate when he recently walked the same prescribed course. Give an interpretation of the area A of the region S and find an expression for A in terms of a definite integral involving f and g.

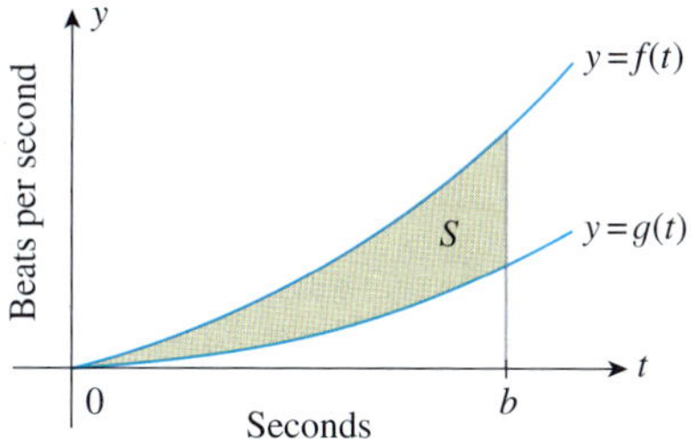

45. **Oil Production Shortfall** Energy experts disagree about when global oil production will begin to decline. In the following figure, the function f gives the annual world oil production in billions of barrels from 1980 to 2050, according to the Department of Energy projection. The function g gives the world oil production in billions of barrels per year over the same period, according to longtime petroleum geologist Colin Campbell. Find an expression in terms of the definite integrals involving f and g, giving the shortfall in the total oil production over the period in question heeding Campbell's dire warnings.

Source: U.S. Department of Energy; Colin Campbell

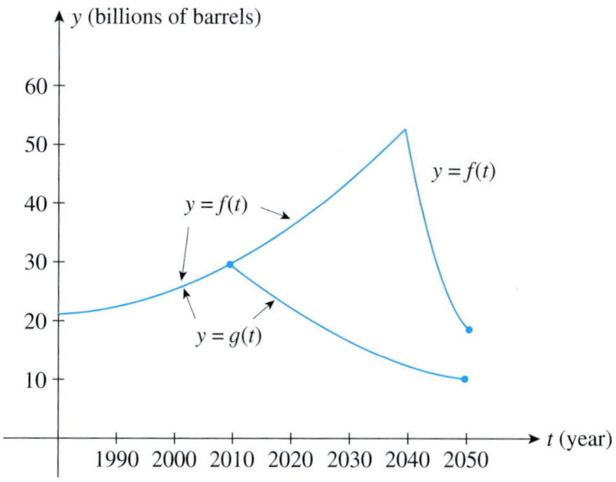

46. **Air Purification** To study the effectiveness of air purifiers in removing smoke, engineers ran each purifier in a smoke-filled 10-ft × 20-ft room. In the accompanying figure, the function f gives the rate of change of the smoke level/minute, t min after the start of the test, when a brand A purifier is used. The function g gives the rate of change of the smoke level/minute when a brand B purifier is used.

a. Give an interpretation of the area of the region S.

b. Find an expression for the area of S in terms of a definite integral involving f and g.

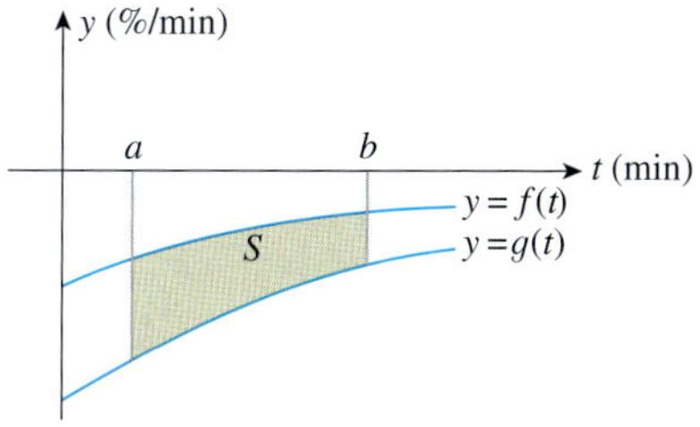

47. Two cars start out side by side and travel along a straight road. The velocity of car 1 is $f(t)$ ft/sec, the velocity of car 2 is $g(t)$ ft/sec over the interval $[0, T]$, and $0 < T_1 < T$. Furthermore, suppose the graphs of f and g are as depicted in the accompanying figure. Let A_1 and A_2 denote the areas of the regions (shown shaded).

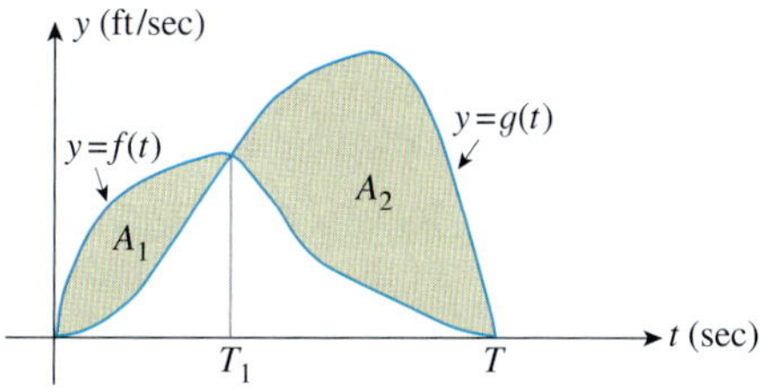

a. Write the number

$$\int_{T_1}^{T} [g(t) - f(t)]\, dt - \int_{0}^{T_1} [f(t) - g(t)]\, dt$$

in terms of A_1 and A_2.

b. What does the number obtained in part (a) represent?

48. The rate of change of the revenue of company A over the (time) interval $[0, T]$ is $f(t)$ dollars/week, whereas the rate of change of the revenue of company B over the same period is $g(t)$ dollars/week. The graphs of f and g are depicted in the accompanying figure. Find an expression in terms of definite integrals involving f and g giving the additional revenue that company B will have over company A in the period $[0, T]$.

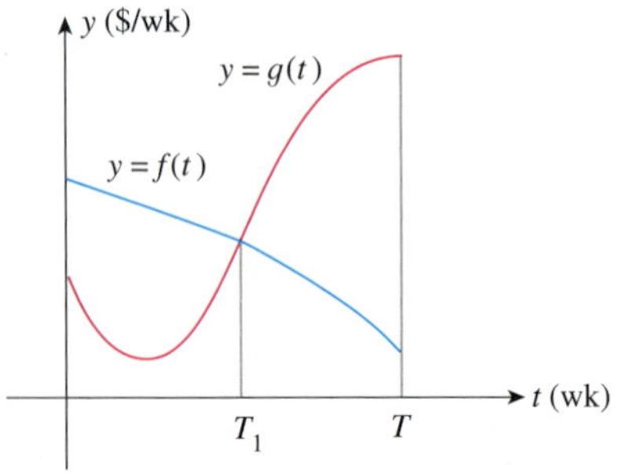

49. **Turbo-Charged Engine vs. Standard Engine** In tests conducted by *Auto Test Magazine* on two identical models of the Phoenix Elite—one equipped with a standard engine and the other with a turbo-charger—it was found that the acceleration of the former is given by

$$a = f(t) = 4 + 0.8t \qquad (0 \leq t \leq 12)$$

ft/sec/sec, t sec after starting from rest at full throttle, whereas the acceleration of the latter is given by

$$a = g(t) = 4 + 1.2t + 0.03t^2 \qquad (0 \leq t \leq 12)$$

ft/sec/sec. How much faster is the turbo-charged model moving than the model with the standard engine at the end of a 10-sec test run at full throttle?

50. **Alternative Energy Sources** Because of the increasingly important role played by coal as a viable alternative energy source, the production of coal has been growing at the rate of

$$3.5e^{0.05t}$$

billion metric tons/year, t yr from 1980 (which corresponds to $t = 0$). Had it not been for the energy crisis, the rate of production of coal since 1980 might have been only

$$3.5e^{0.01t}$$

billion metric tons/year, t yr from 1980. Determine how much additional coal was produced between 1980 and the end of the century as an alternate energy source.

51. **Effect of TV Advertising on Car Sales** Carl Williams, the proprietor of Carl Williams Auto Sales, estimates that with extensive television advertising, car sales over the next several years could be increasing at the rate of

$$5e^{0.3t}$$

thousand cars/year, t yr from now, instead of at the current rate of

$$(5 + 0.5t^{3/2})$$

thousand cars/year, t yr from now. Find how many more cars Carl expects to sell over the next 5 yr by implementing his advertising plans.

52. **Population Growth** In an endeavor to curb population growth in a Southeast Asian island state, the government has decided to launch an extensive propaganda campaign. Without curbs, the government expects the rate of population growth to have been

$$60e^{0.02t}$$

thousand people/year, t yr from now, over the next 5 yr. However, successful implementation of the proposed campaign is expected to result in a population growth rate of

$$-t^2 + 60$$

thousand people/year, t yr from now, over the next 5 yr. Assuming that the campaign is mounted, how many fewer people will there be in that country 5 yr from now than there would have been if no curbs had been imposed?

In Exercises 53–56, determine whether the statement is true or false. If it is true, explain why it is true. If it is false, give an example to show why it is false.

53. If f and g are continuous on $[a, b]$ and either $f(x) \geq g(x)$ for all x in $[a, b]$ or $f(x) \leq g(x)$ for all x in $[a, b]$, then the area of the region bounded by the graphs of f and g and the vertical lines $x = a$ and $x = b$ is given by $\int_a^b |f(x) - g(x)|\,dx$.

54. The area of the region bounded by the graphs of $f(x) = 2 - x$ and $g(x) = 4 - x^2$ and the vertical lines $x = 0$ and $x = 2$ is given by $\int_0^2 [f(x) - g(x)]\,dx$.

55. If A denotes the area bounded by the graphs of f and g on $[a, b]$, then

$$A^2 = \int_a^b [f(x) - g(x)]^2\,dx.$$

56. If f and g are continuous on $[a, b]$ and $\int_a^b [f(t) - g(t)]\,dt > 0$, then $f(t) \geq g(t)$ for all t in $[a, b]$.

57. Show that the area of a region R bounded above by the graph of a function f and below by the graph of a function g from $x = a$ to $x = b$ is given by

$$\int_a^b [f(x) - g(x)]\,dx$$

Hint: The validity of the formula was verified earlier for the case when both f and g were nonnegative. Now, let f and g be two functions such that $f(x) \geq g(x)$ for $a \leq x \leq b$. Then, there exists some nonnegative constant c such that the curves $y = f(x) + c$ and $y = g(x) + c$ are translated in the y-direction in such a way that the region R' has the same area as the region R (see the accompanying figures). Show that the area of R' is given by

$$\int_a^b \{[f(x) + c] - [g(x) + c]\}\,dx = \int_a^b [f(x) - g(x)]\,dx$$

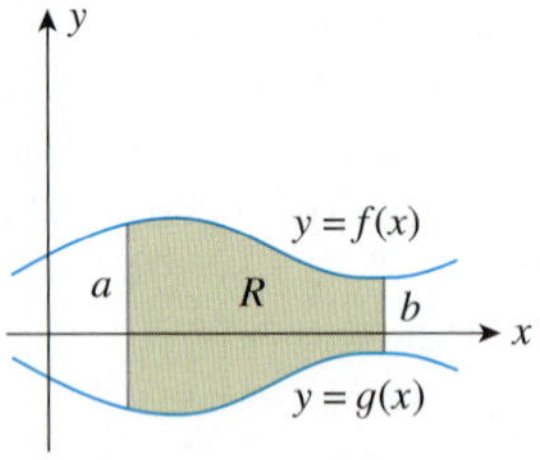

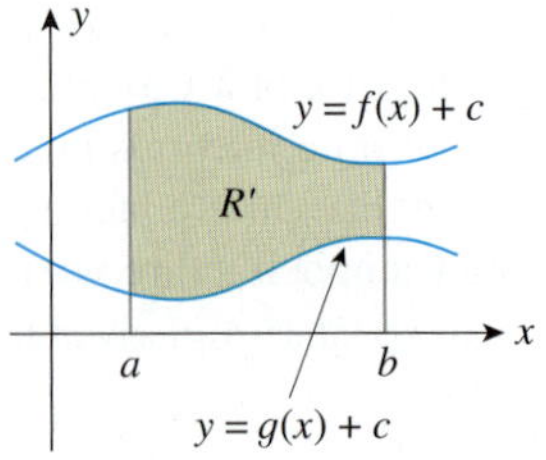

6.6 Solutions to Self-Check Exercises

1. The region in question is shown in the accompanying figure. Since the graph of the function f lies above that of the function g for $0 \leq x \leq 1$, we see that the required area is given by

$$\int_0^1 [(x^2 + 2) - (1 - x)]\,dx = \int_0^1 (x^2 + x + 1)\,dx$$
$$= \frac{1}{3}x^3 + \frac{1}{2}x^2 + x\Big|_0^1$$
$$= \frac{1}{3} + \frac{1}{2} + 1$$
$$= \frac{11}{6}$$

or $\frac{11}{6}$ square units.

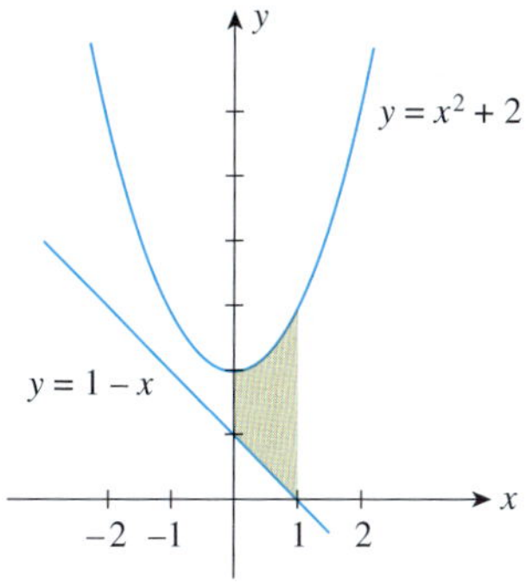

2. The region in question is shown in the accompanying figure. To find the points of intersection of the two curves, we solve the equations

$$-x^2 + 6x + 5 = x^2 + 5$$
$$2x^2 - 6x = 0$$
$$2x(x - 3) = 0$$

giving $x = 0$ or $x = 3$. Therefore, the points of intersection are (0, 5) and (3, 14).

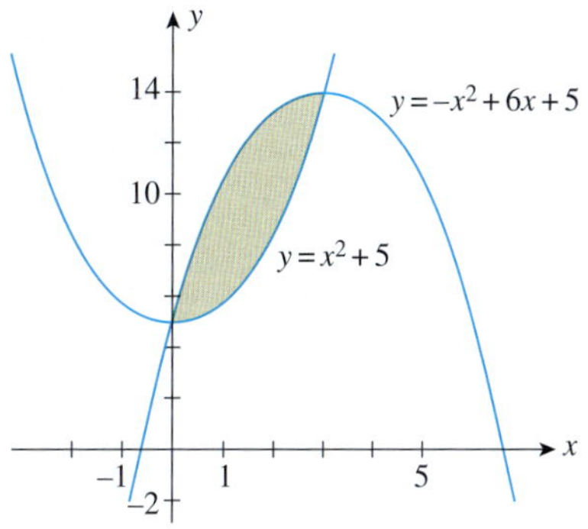

Since the graph of f always lies above that of g for $0 \leq x \leq 3$, we see that the required area is given by

$$\int_0^3 [(-x^2 + 6x + 5) - (x^2 + 5)]\,dx = \int_0^3 (-2x^2 + 6x)\,dx$$
$$= -\frac{2}{3}x^3 + 3x^2\Big|_0^3$$
$$= -18 + 27$$
$$= 9$$

or 9 square units.

3. The additional profits realizable over the next 10 yr are given by

$$\int_0^{10} [(t - 2\sqrt{t} + 4) - (1 + t^{2/3})]\,dt$$
$$= \int_0^{10} (t - 2t^{1/2} + 3 - t^{2/3})\,dt$$
$$= \frac{1}{2}t^2 - \frac{4}{3}t^{3/2} + 3t - \frac{3}{5}t^{5/3}\Big|_0^{10}$$
$$= \frac{1}{2}(10)^2 - \frac{4}{3}(10)^{3/2} + 3(10) - \frac{3}{5}(10)^{5/3}$$
$$\approx 9.99$$

or approximately \$10 million.

USING TECHNOLOGY

Finding the Area between Two Curves

The numerical integral operation can also be used to find the area between two curves. We do this by using the numerical integral operation to evaluate an appropriate definite integral or the sum (difference) of appropriate definite integrals. In the following example, the intersection operation is also used to advantage to help us find the limits of integration.

EXAMPLE 1 Use a graphing utility to find the area of the smaller region R that is completely enclosed by the graphs of the functions

$$f(x) = 2x^3 - 8x^2 + 4x - 3 \quad \text{and} \quad g(x) = 3x^2 + 10x - 11$$

(continued)

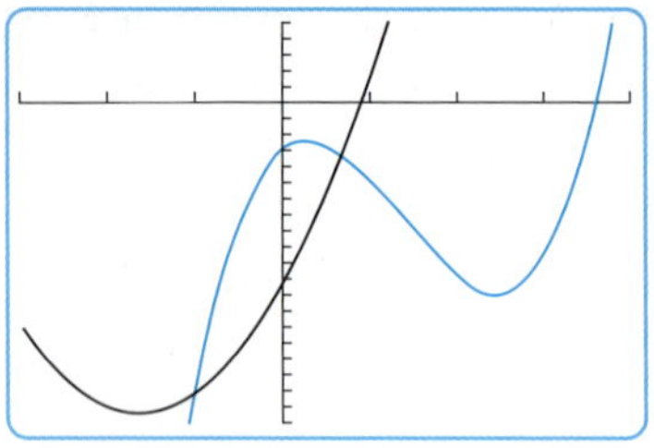

FIGURE T1
The region R is completely enclosed by the graphs of f and g.

Solution The graphs of f and g in the viewing window $[-3, 4] \times [-20, 5]$ are shown in Figure T1.

Using the intersection operation of a graphing utility, we find the x-coordinates of the points of intersection of the two graphs to be approximately -1.04 and 0.65, respectively. Since the graph of f lies above that of g on the interval $[-1.04, 0.65]$, we see that the area of R is given by

$$A = \int_{-1.04}^{0.65} [(2x^3 - 8x^2 + 4x - 3) - (3x^2 + 10x - 11)]\,dx$$
$$= \int_{-1.04}^{0.65} (2x^3 - 11x^2 - 6x + 8)\,dx$$

Using the numerical integral function of a graphing utility, we find $A \approx 9.87$, and so the area of R is approximately 9.87 square units. ■

TECHNOLOGY EXERCISES

In Exercises 1–6, (a) plot the graphs of the functions f and g and (b) find the area of the region enclosed by these graphs and the vertical lines $x = a$ and $x = b$. Express your answers accurate to four decimal places.

1. $f(x) = x^3(x - 5)^4$, $g(x) = 0$; $a = 1$, $b = 3$
2. $f(x) = x - \sqrt{1 - x^2}$, $g(x) = 0$; $a = -\frac{1}{2}$, $b = \frac{1}{2}$
3. $f(x) = x^{1/3}(x + 1)^{1/2}$, $g(x) = x^{-1}$; $a = 1.2$, $b = 2$
4. $f(x) = 2$, $g(x) = \ln(1 + x^2)$; $a = -1$, $b = 1$
5. $f(x) = \sqrt{x}$, $g(x) = \dfrac{x^2 - 3}{x^2 + 1}$; $a = 0$, $b = 3$
6. $f(x) = \dfrac{4}{x^2 + 1}$, $g(x) = x^4$; $a = -1$, $b = 1$

In Exercises 7–12, (a) plot the graphs of the functions f and g and (b) find the area of the region totally enclosed by the graphs of these functions.

7. $f(x) = 2x^3 - 8x^2 + 4x - 3$ and $g(x) = -3x^2 + 10x - 10$
8. $f(x) = x^4 - 2x^2 + 2$ and $g(x) = 4 - 2x^2$
9. $f(x) = 2x^3 - 3x^2 + x + 5$ and $g(x) = e^{2x} - 3$
10. $f(x) = \frac{1}{2}x^2 - 3$ and $g(x) = \ln x$
11. $f(x) = xe^{-x}$ and $g(x) = x - 2\sqrt{x}$
12. $f(x) = e^{-x^2}$ and $g(x) = x^4$
13. Refer to Example 1. Find the area of the larger region that is completely enclosed by the graphs of the functions f and g.

6.7 Applications of the Definite Integral to Business and Economics

In this section, we consider several applications of the definite integral in the fields of business and economics.

Consumers' and Producers' Surplus

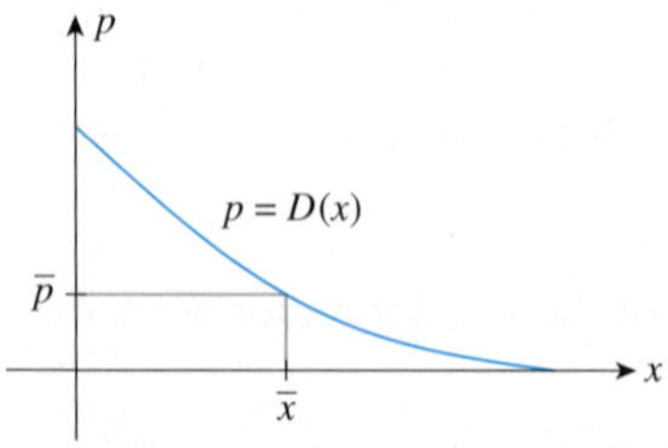

FIGURE 38
$D(x)$ is a demand function.

We begin by deriving a formula for computing the consumers' surplus. Suppose $p = D(x)$ is the demand function that relates the unit price p of a commodity to the quantity x demanded of it. Furthermore, suppose a fixed unit market price $\bar{p}$ has been established for the commodity and corresponding to this unit price the quantity demanded is $\bar{x}$ units (Figure 38). Then, those consumers who would be willing to pay a unit price higher than $\bar{p}$ for the commodity would in effect experience a savings. This difference between what the consumers *would* be willing to pay for $\bar{x}$ units of the commodity and what they *actually* pay for them is called the **consumers' surplus.**

To derive a formula for computing the consumers' surplus, divide the interval $[0, \bar{x}]$ into n subintervals, each of length $\Delta x = \bar{x}/n$, and denote the right endpoints of these subintervals by $x_1, x_2, \ldots, x_n = \bar{x}$ (Figure 39).

We observe in Figure 39 that there are consumers who would pay a unit price of at least $D(x_1)$ dollars for the first Δx units of the commodity instead of the market price of $\bar{p}$ dollars per unit. The savings to these consumers is approximated by

$$D(x_1)\Delta x - \bar{p}\Delta x = [D(x_1) - \bar{p}]\Delta x$$

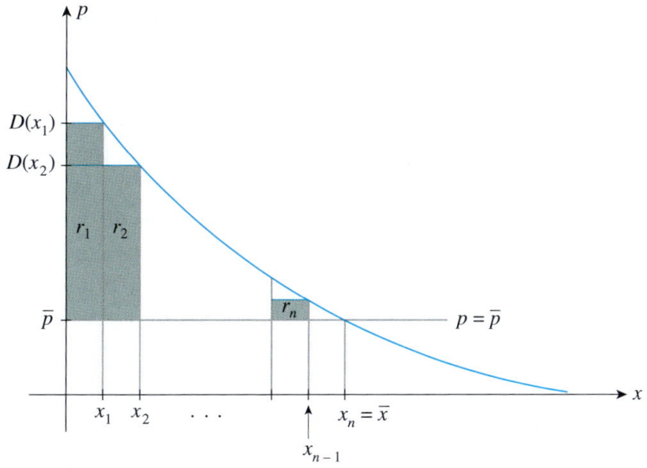

FIGURE 39
Approximating consumers' surplus by the sum of the areas of the rectangles r_1, $r_2, \ldots, r_n$

which is the area of the rectangle r_1. Pursuing the same line of reasoning, we find that the savings to the consumers who would be willing to pay a unit price of at least $D(x_2)$ dollars for the next Δx units (from x_1 through x_2) of the commodity, instead of the market price of $\bar{p}$ dollars per unit, is approximated by

$$D(x_2)\Delta x - \bar{p}\Delta x = [D(x_2) - \bar{p}]\Delta x$$

Continuing, we approximate the total savings to the consumers in purchasing $\bar{x}$ units of the commodity by the sum

$$\begin{aligned}
&[D(x_1) - \bar{p}]\Delta x + [D(x_2) - \bar{p}]\Delta x + \cdots + [D(x_n) - \bar{p}]\Delta x \\
&\quad = [D(x_1) + D(x_2) + \cdots + D(x_n)]\Delta x - \underbrace{[\bar{p}\Delta x + \bar{p}\Delta x + \cdots + \bar{p}\Delta x]}_{n \text{ terms}} \\
&\quad = [D(x_1) + D(x_2) + \cdots + D(x_n)]\Delta x - n\bar{p}\Delta x \\
&\quad = [D(x_1) + D(x_2) + \cdots + D(x_n)]\Delta x - \bar{p}\,\bar{x}
\end{aligned}$$

Now, the first term in the last expression is the Riemann sum of the demand function $p = D(x)$ over the interval $[0, \bar{x}]$ with representative points $x_1, x_2, \ldots, x_n$. Letting n approach infinity, we obtain the following formula for the consumers' surplus CS.

Consumers' Surplus

The consumers' surplus is given by

$$CS = \int_0^{\bar{x}} D(x)\,dx - \bar{p}\,\bar{x} \tag{16}$$

where D is the demand function, $\bar{p}$ is the unit market price, and $\bar{x}$ is the quantity sold.

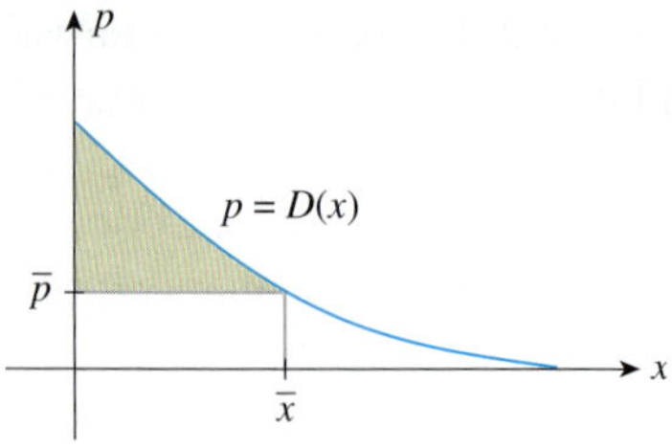

FIGURE 40
Consumers' surplus

The consumer's surplus is given by the area of the region bounded above by the demand curve $p = D(x)$ and below by the straight line $p = \bar{p}$ from $x = 0$ to $x = \bar{x}$ (Figure 40). We can also see this if we rewrite Equation (16) in the form

$$\int_0^{\bar{x}} [D(x) - \bar{p}]\, dx$$

and interpret the result geometrically.

Analogously, we can derive a formula for computing the producers' surplus. Suppose $p = S(x)$ is the supply equation that relates the unit price p of a certain commodity to the quantity x that the supplier will make available in the market at that price.

Again, suppose a fixed market price $\bar{p}$ has been established for the commodity and, corresponding to this unit price, a quantity of $\bar{x}$ units will be made available in the market by the supplier (Figure 41). Then, the suppliers who would be willing to make the commodity available at a lower price stand to gain from the fact that the market price is set as such. The difference between what the suppliers actually receive and what they would be willing to receive is called the **producers' surplus.** Proceeding in a manner similar to the derivation of the equation for computing the consumers' surplus, we find that the producers' surplus PS is defined as follows:

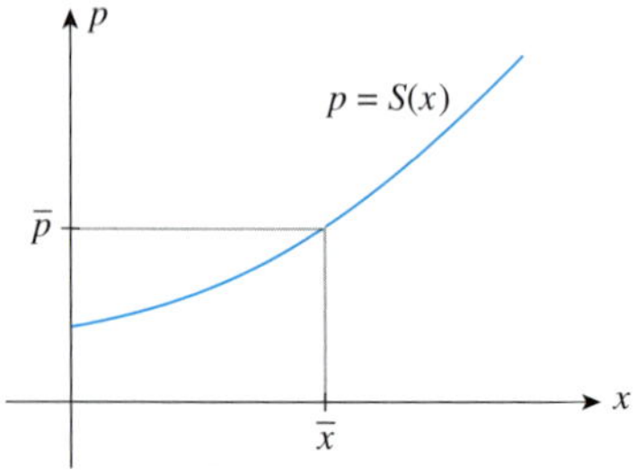

FIGURE 41
$S(x)$ is a supply function.

Producers' Surplus

The producers' surplus is given by

$$PS = \bar{p}\,\bar{x} - \int_0^{\bar{x}} S(x)\, dx \qquad \textbf{(17)}$$

where $S(x)$ is the supply function, $\bar{p}$ is the unit market price, and $\bar{x}$ is the quantity supplied.

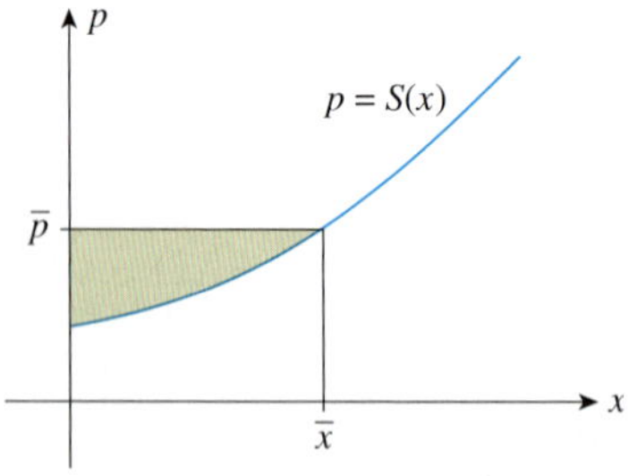

FIGURE 42
Producers' surplus

Geometrically, the producers' surplus is given by the area of the region bounded above by the straight line $p = \bar{p}$ and below by the supply curve $p = S(x)$ from $x = 0$ to $x = \bar{x}$ (Figure 42).

We can also show that the last statement is true by converting Equation (17) to the form

$$\int_0^{\bar{x}} [\bar{p} - S(x)]\, dx$$

and interpreting the definite integral geometrically.

VIDEO

EXAMPLE 1 The demand function for a certain make of 10-speed bicycle is given by

$$p = D(x) = -0.001x^2 + 250$$

where p is the unit price in dollars and x is the quantity demanded in units of a thousand. The supply function for these bicycles is given by

$$p = S(x) = 0.0006x^2 + 0.02x + 100$$

where p stands for the unit price in dollars and x stands for the number of bicycles that the supplier will put on the market, in units of a thousand. Determine the consumers' surplus and the producers' surplus if the market price of a bicycle is set at the equilibrium price.

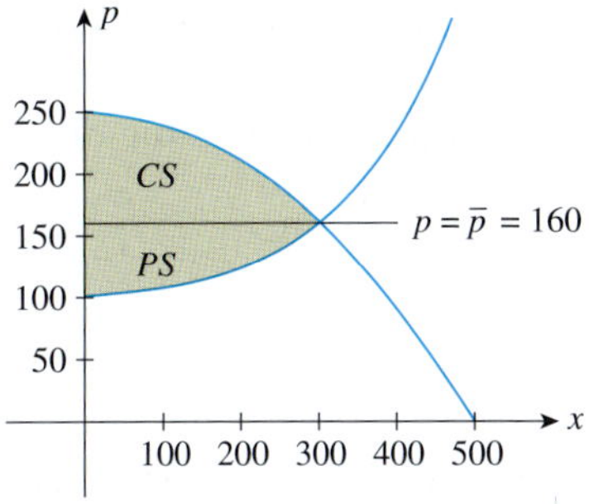

FIGURE 43
Consumers' surplus and producers' surplus when market price = equilibrium price

Solution Recall that the equilibrium price is the unit price of the commodity when market equilibrium occurs. We determine the equilibrium price by solving for the point of intersection of the demand curve and the supply curve (Figure 43). To solve the system of equations

$$p = -0.001x^2 \qquad\qquad + 250$$
$$p = 0.0006x^2 + 0.02x + 100$$

we simply substitute the first equation into the second, obtaining

$$\begin{aligned} 0.0006x^2 + 0.02x + 100 &= -0.001x^2 + 250 \\ 0.0016x^2 + 0.02x - 150 &= 0 \\ 16x^2 + 200x - 1{,}500{,}000 &= 0 \\ 2x^2 + 25x - 187{,}500 &= 0 \end{aligned}$$

Factoring this last equation, we obtain

$$(2x + 625)(x - 300) = 0$$

Thus, $x = -625/2$ or $x = 300$. The first number lies outside the interval of interest, so we are left with the solution $x = 300$, with a corresponding value of

$$p = -0.001(300)^2 + 250 = 160$$

Thus, the equilibrium point is (300, 160); that is, the equilibrium quantity is 300,000, and the equilibrium price is \$160. Setting the market price at \$160 per unit and using Formula (16) with $\bar{p} = 160$ and $\bar{x} = 300$, we find that the consumers' surplus is given by

$$\begin{aligned} CS &= \int_0^{300} (-0.001x^2 + 250)\,dx - (160)(300) \\ &= \left(-\frac{1}{3000}x^3 + 250x\right)\Bigg|_0^{300} - 48{,}000 \\ &= -\frac{300^3}{3000} + (250)(300) - 48{,}000 \\ &= 18{,}000 \end{aligned}$$

or \$18,000,000.
(Recall that x is measured in units of a thousand.) Next, using (17), we find that the producers' surplus is given by

$$\begin{aligned} PS &= (160)(300) - \int_0^{300} (0.0006x^2 + 0.02x + 100)\,dx \\ &= 48{,}000 - (0.0002x^3 + 0.01x^2 + 100x)\Big|_0^{300} \\ &= 48{,}000 - [(0.0002)(300)^3 + (0.01)(300)^2 + 100(300)] \\ &= 11{,}700 \end{aligned}$$

or \$11,700,000. ■

The Future and Present Value of an Income Stream

Suppose a firm generates a stream of income over a period of time—for example, the revenue generated by a large chain of retail stores over a 5-year period. As the income is realized, it is reinvested and earns interest at a fixed rate. The **accumulated future income stream** over the 5-year period is the amount of money the firm ends up with at the end of that period.

The definite integral can be used to determine this accumulated, or total, future income stream over a period of time. The total future value of an income stream gives us a way to measure the value of such a stream. To find the **total future value of an income stream,** suppose

$$R(t) = \text{Rate of income generation at any time } t \qquad \text{Dollars per year}$$
$$r = \text{Interest rate compounded continuously}$$
$$T = \text{Term} \qquad \text{In years}$$

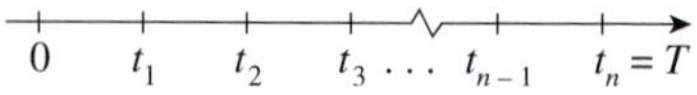

FIGURE **44**
The time interval [0, T] is partitioned into n subintervals.

Let's divide the time interval [0, T] into n subintervals of equal length $\Delta t = T/n$ and denote the right endpoints of these intervals by $t_1, t_2, \ldots, t_n = T$, as shown in Figure 44.

If R is a continuous function on [0, T], then $R(t)$ will not differ by much from $R(t_1)$ in the subinterval [0, t_1] provided that the subinterval is small (which is true if n is large). Therefore, the income generated over the time interval [0, t_1] is approximately

$$R(t_1)\Delta t \qquad \text{Constant rate of income} \times \text{Length of time}$$

dollars. The future value of this amount, T years from now, calculated as if it were earned at time t_1, is

$$[R(t_1)\Delta t]e^{r(T-t_1)} \qquad \text{Equation (10), Section 5.3}$$

dollars. Similarly, the income generated over the time interval [t_1, t_2] is approximately $R(t_2)\Delta t$ dollars and has a future value, T years from now, of approximately

$$[R(t_2)\Delta t]e^{r(T-t_2)}$$

dollars. Therefore, the sum of the future values of the income stream generated over the time interval [0, T] is approximately

$$R(t_1)e^{r(T-t_1)}\Delta t + R(t_2)e^{r(T-t_2)}\Delta t + \cdots + R(t_n)e^{r(T-t_n)}\Delta t$$
$$= e^{rT}[R(t_1)e^{-rt_1}\Delta t + R(t_2)e^{-rt_2}\Delta t + \cdots + R(t_n)e^{-rt_n}\Delta t]$$

dollars. But this sum is just the Riemann sum of the function $e^{rT}R(t)e^{-rt}$ over the interval [0, T] with representative points $t_1, t_2, \ldots, t_n$. Letting n approach infinity, we obtain the following result.

Accumulated or Total Future Value of an Income Stream

The accumulated, or total, future value after T years of an income stream of $R(t)$ dollars per year, earning interest at the rate of r per year compounded continuously, is given by

$$A = e^{rT}\int_0^T R(t)e^{-rt}\,dt \qquad \textbf{(18)}$$

APPLIED EXAMPLE 2 Income Stream Crystal Car Wash recently bought an automatic car-washing machine that is expected to generate \$40,000 in revenue per year, t years from now, for the next 5 years. If the income is reinvested in a business earning interest at the rate of 12% per year compounded continuously, find the total accumulated value of this income stream at the end of 5 years.

Solution We are required to find the total future value of the given income stream after 5 years. Using Equation (18) with $R(t) = 40{,}000$, $r = 0.12$, and $T = 5$, we see that the required value is given by

$$\begin{aligned}
e^{0.12(5)}\int_0^5 40{,}000e^{-0.12t}\,dt &= e^{0.6}\left[-\frac{40{,}000}{0.12}e^{-0.12t}\right]\Bigg|_0^5 \quad \text{Integrate using the substitution } u = -0.12t.\\
&= -\frac{40{,}000e^{0.6}}{0.12}(e^{-0.6}-1) \approx 274{,}039.60
\end{aligned}$$

or approximately $274,040.

Another way of measuring the value of an income stream is by considering its present value. The **present value of an income stream** of $R(t)$ dollars per year over a term of T years, earning interest at the rate of r per year compounded continuously, is the principal P that will yield the same accumulated value as the income stream itself when P is invested today for a period of T years at the same rate of interest. In other words,

$$Pe^{rT} = e^{rT}\int_0^T R(t)e^{-rt}\,dt$$

Dividing both sides of the equation by e^{rT} gives the following result.

Present Value of an Income Stream

The present value of an income stream of $R(t)$ dollars per year, earning interest at the rate of r per year compounded continuously, is given by

$$PV = \int_0^T R(t)e^{-rt}\,dt \tag{19}$$

VIDEO

APPLIED EXAMPLE 3 Investment Analysis The owner of a local cinema is considering two alternative plans for renovating and improving the theater. Plan A calls for an immediate cash outlay of $250,000, whereas plan B requires an immediate cash outlay of $180,000. It has been estimated that adopting plan A would result in a net income stream generated at the rate of

$$f(t) = 630{,}000$$

dollars per year, whereas adopting plan B would result in a net income stream generated at the rate of

$$g(t) = 580{,}000$$

dollars per year for the next 3 years. If the prevailing interest rate for the next 5 years is 10% per year, which plan will generate a higher net income by the end of 3 years?

Solution Since the initial outlay is $250,000, we find—using Equation (19) with $R(t) = 630{,}000$, $r = 0.1$, and $T = 3$—that the present value of the net income under plan A is given by

$$\begin{aligned}
\int_0^3 630{,}000e^{-0.1t}\,dt - 250{,}000 &= \frac{630{,}000}{-0.1}e^{-0.1t}\Bigg|_0^3 - 250{,}000 \quad \text{Integrate using the substitution } u = -0.1t.\\
&= -6{,}300{,}000e^{-0.3} + 6{,}300{,}000 - 250{,}000\\
&\approx 1{,}382{,}845
\end{aligned}$$

or approximately $1,382,845.

To find the present value of the net income under plan B, we use (19) with $R(t) = 580{,}000$, $r = 0.1$, and $T = 3$, obtaining

$$\int_0^3 580{,}000e^{-0.1t}\,dt - 180{,}000$$

dollars. Proceeding as in the previous computation, we see that the required value is \$1,323,254 (see Exercise 8, page 474).

Comparing the present value of each plan, we conclude that plan A would generate a higher net income by the end of 3 years. ■

Note The function R in Example 3 is a constant function. If R is not a constant function, then we may need more sophisticated techniques of integration to evaluate the integral in (19). Exercises 7.1 and 7.2 contain problems of this type. ■

The Amount and Present Value of an Annuity

An annuity is a sequence of payments made at regular time intervals. The time period in which these payments are made is called the *term* of the annuity. Although the payments need not be equal in size, they are equal in many important applications, and we will assume that they are equal in our discussion. Examples of annuities are regular deposits to a savings account, monthly home mortgage payments, and monthly insurance payments.

The **amount of an annuity** is the sum of the payments plus the interest earned. A formula for computing the amount of an annuity A can be derived with the help of (18). Let

P = Size of each payment in the annuity
r = Interest rate compounded continuously
T = Term of the annuity (in years)
m = Number of payments per year

The payments into the annuity constitute a constant income stream of $R(t) = mP$ dollars per year. With this value of $R(t)$, (18) yields

$$\begin{aligned} A &= e^{rT}\int_0^T R(t)e^{-rt}\,dt = e^{rT}\int_0^T mPe^{-rt}\,dt \\ &= mPe^{rT}\left[-\frac{e^{-rt}}{r}\right]\Bigg|_0^T = mPe^{rT}\left[-\frac{e^{-rT}}{r} + \frac{1}{r}\right] \\ &= \frac{mP}{r}(e^{rT} - 1) \qquad \text{Since } e^{rT}\cdot e^{-rT} = 1 \end{aligned}$$

This leads us to the following formula.

Amount of an Annuity

The amount of an annuity is

$$A = \frac{mP}{r}(e^{rT} - 1) \tag{20}$$

where P, r, T, and m are as defined earlier.

APPLIED EXAMPLE 4 IRAs On January 1, 1994, Marcus Chapman deposited \$2000 into an Individual Retirement Account (IRA) paying interest at the rate of 5% per year compounded continuously. Assuming that he deposited \$2000 annually into the account, how much did he have in his IRA at the beginning of 2010?

Solution We use (20), with $P = 2000$, $r = 0.05$, $T = 16$, and $m = 1$, obtaining

$$\begin{aligned} A &= \frac{2000}{0.05}(e^{0.8} - 1) \\ &\approx 49{,}021.64 \end{aligned}$$

Thus, Marcus had approximately \$49,022 in his account at the beginning of 2008. ■

Exploring with TECHNOLOGY

Refer to Example 4. Suppose Marcus wished to know how much he would have in his IRA at any time in the future, not just at the beginning of 2010, as you were asked to compute in the example.

1. Using Formula (18) and the relevant data from Example 4, show that the required amount at any time x (x measured in years, $x > 0$) is given by

$$A = f(x) = 40{,}000(e^{0.05x} - 1)$$

2. Use a graphing utility to plot the graph of f, using the viewing window $[0, 30] \times [2000, 400{,}000]$.
3. Using ZOOM and TRACE, or using the function evaluation capability of your graphing utility, use the result of part 2 to verify the result obtained in Example 4. Comment on the advantage of the mathematical model found in part 1.

Using (19), we can derive the following formula for the present value of an annuity.

Present Value of an Annuity

The present value of an annuity is given by

$$PV = \frac{mP}{r}(1 - e^{-rT}) \tag{21}$$

where P, r, T, and m are as defined earlier.

APPLIED EXAMPLE 5 Sinking Funds Tomas Perez, the proprietor of a hardware store, wants to establish a fund from which he will withdraw \$1000 per month for the next 10 years. If the fund earns interest at the rate of 6% per year compounded continuously, how much money does he need to establish the fund?

Solution We want to find the present value of an annuity with $P = 1000$, $r = 0.06$, $T = 10$, and $m = 12$. Using Equation (21), we find

$$\begin{aligned} PV &= \frac{12{,}000}{0.06}(1 - e^{-(0.06)(10)}) \\ &\approx 90{,}237.70 \end{aligned}$$

Thus, Tomas needs approximately \$90,238 to establish the fund. ■

Lorentz Curves and Income Distributions

One method used by economists to study the distribution of income in a society is based on the **Lorentz curve,** named after American statistician M.D. Lorentz. To describe the Lorentz curve, let $f(x)$ denote the proportion of the total income received by the poorest $100x\%$ of the population for $0 \le x \le 1$. Using this terminology, $f(0.3) = 0.1$ simply states that the lowest 30% of the income recipients receive 10% of the total income.

The function f has the following properties:

1. The domain of f is $[0, 1]$.
2. The range of f is $[0, 1]$.
3. $f(0) = 0$ and $f(1) = 1$.
4. $f(x) \le x$ for every x in $[0, 1]$.
5. f is increasing on $[0, 1]$.

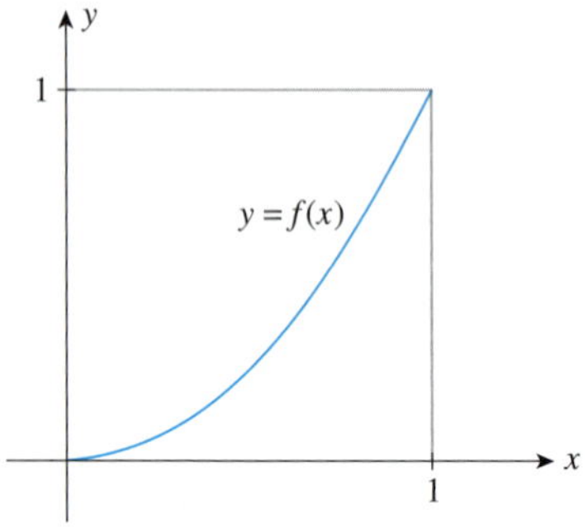

FIGURE 45
A Lorentz curve

The first two properties follow from the fact that both x and $f(x)$ are fractions of a whole. Property 3 is a statement that 0% of the income recipients receive 0% of the total income and 100% of the income recipients receive 100% of the total income. Property 4 follows from the fact that the lowest $100x\%$ of the income recipients cannot receive more than $100x\%$ of the total income. A typical Lorentz curve is shown in Figure 45.

APPLIED EXAMPLE 6 Lorentz Curves A developing country's income distribution is described by the function

$$f(x) = \frac{19}{20}x^2 + \frac{1}{20}x$$

a. Sketch the Lorentz curve for the given function.
b. Compute $f(0.2)$ and $f(0.8)$ and interpret your results.

Solution

a. The Lorentz curve is shown in Figure 46.

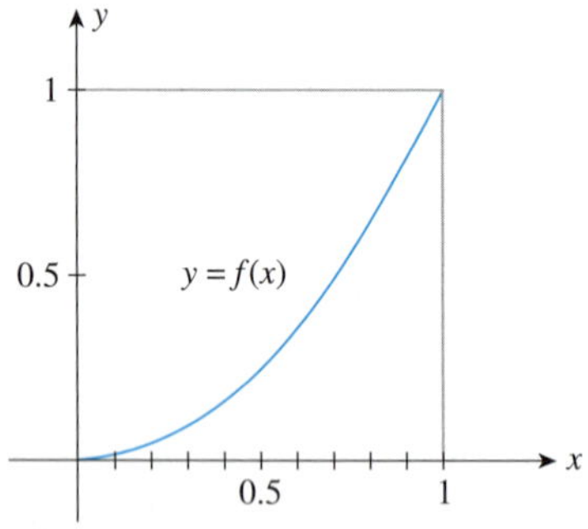

FIGURE 46
The Lorentz curve $f(x) = \frac{19}{20}x^2 + \frac{1}{20}x$

b.
$$f(0.2) = \frac{19}{20}(0.2)^2 + \frac{1}{20}(0.2) = 0.048$$

Thus, the lowest 20% of the people receive 4.8% of the total income.

$$f(0.8) = \frac{19}{20}(0.8)^2 + \frac{1}{20}(0.8) = 0.648$$

Thus, the lowest 80% of the people receive 64.8% of the total income. ■

Next, let's consider the Lorentz curve described by the function $y = f(x) = x$. Since exactly $100x\%$ of the total income is received by the lowest $100x\%$ of income recipients, the line $y = x$ is called the **line of complete equality.** For example, 10% of the total income is received by the lowest 10% of income recipients, 20% of the total income is received by the lowest 20% of income recipients, and so on. Now, it is evident that the closer a Lorentz curve is to this line, the more equitable the income distribution is among the income recipients. But the proximity of a Lorentz curve to the line of complete equality is reflected by the area between the Lorentz curve and the line $y = x$ (Figure 47). The closer the curve is to the line, the smaller the enclosed area.

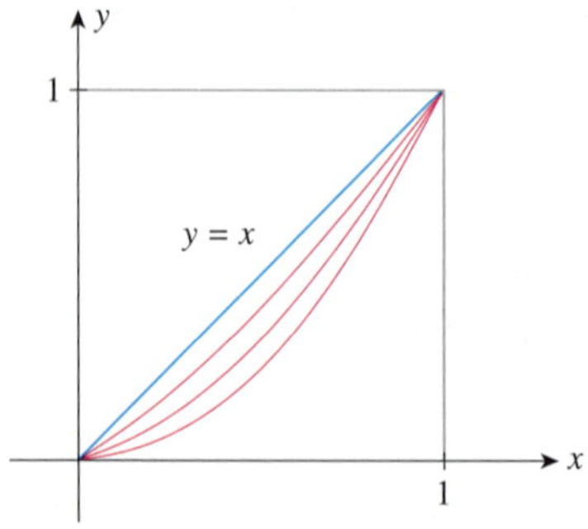

FIGURE 47
The closer the Lorentz curve is to the line, the more equitable the income distribution.

This observation suggests that we may define a number, called the coefficient of inequality of a Lorentz curve, as the ratio of the area between the line of complete equality and the Lorentz curve to the area under the line of complete equality. Since the area under the line of complete equality is $\frac{1}{2}$, we see that the coefficient of inequality is given by the following formula.

Coefficient of Inequality of a Lorentz Curve

The coefficient of inequality, or **Gini index,** of a Lorentz curve is

$$L = 2\int_0^1 [x - f(x)]\, dx \qquad \textbf{(22)}$$

The coefficient of inequality is a number between 0 and 1. For example, a coefficient of zero implies that the income distribution is perfectly uniform.

APPLIED EXAMPLE 7 Income Distributions In a study conducted by a certain country's Economic Development Board with regard to the income distribution of certain segments of the country's workforce, it was found that the Lorentz curves for the distribution of income of medical doctors and of movie actors are described by the functions

$$f(x) = \frac{14}{15}x^2 + \frac{1}{15}x \quad \text{and} \quad g(x) = \frac{5}{8}x^4 + \frac{3}{8}x$$

respectively. Compute the coefficient of inequality for each Lorentz curve. Which profession has a more equitable income distribution?

Solution The required coefficients of inequality are, respectively,

$$\begin{aligned} L_1 &= 2\int_0^1 \left[x - \left(\frac{14}{15}x^2 + \frac{1}{15}x\right)\right] dx = 2\int_0^1 \left(\frac{14}{15}x - \frac{14}{15}x^2\right) dx \\ &= \frac{28}{15}\int_0^1 (x - x^2)\, dx = \frac{28}{15}\left(\frac{1}{2}x^2 - \frac{1}{3}x^3\right)\Bigg|_0^1 \\ &= \frac{14}{45} \approx 0.311 \\ L_2 &= 2\int_0^1 \left[x - \left(\frac{5}{8}x^4 + \frac{3}{8}x\right)\right] dx = 2\int_0^1 \left(\frac{5}{8}x - \frac{5}{8}x^4\right) dx \\ &= \frac{5}{4}\int_0^1 (x - x^4)\, dx = \frac{5}{4}\left(\frac{1}{2}x^2 - \frac{1}{5}x^5\right)\Bigg|_0^1 \\ &= \frac{15}{40} = 0.375 \end{aligned}$$

We conclude that in this country the incomes of medical doctors are more evenly distributed than the incomes of movie actors. ■

6.7 Self-Check Exercises

The demand function for a certain make of exercise bicycle that is sold exclusively through cable television is

$$p = d(x) = \sqrt{9 - 0.02x}$$

where p is the unit price in hundreds of dollars and x is the quantity demanded/week. The corresponding supply function is given by

$$p = s(x) = \sqrt{1 + 0.02x}$$

where p has the same meaning as before and x is the number of exercise bicycles the supplier will make available at price p. Determine the consumers' surplus and the producers' surplus if the unit price is set at the equilibrium price.

The solution to Self-Check Exercise 6.7 can be found on page 476.

6.7 Concept Questions

1. a. Define consumers' surplus. Give a formula for computing it.
   b. Define producers' surplus. Give a formula for computing it.
2. a. Define the accumulated (future) value of an income stream. Give a formula for computing it.
   b. Define the present value of an income stream. Give a formula for computing it.
3. Define the amount of an annuity. Give a formula for computing it.
4. Explain the following terms: (a) Lorentz curve (b) Coefficient of inequality of a Lorentz curve.

6.7 Exercises

1. **Consumers' Surplus** The demand function for a certain make of replacement cartridges for a water purifier is given by

$$p = -0.01x^2 - 0.1x + 6$$

where p is the unit price in dollars and x is the quantity demanded each week, measured in units of a thousand. Determine the consumers' surplus if the market price is set at \$4/cartridge.

2. **Consumers' Surplus** The demand function for a certain brand of CD is given by

$$p = -0.01x^2 - 0.2x + 8$$

where p is the wholesale unit price in dollars and x is the quantity demanded each week, measured in units of a thousand. Determine the consumers' surplus if the wholesale market price is set at \$5/disc.

3. **Consumers' Surplus** It is known that the quantity demanded of a certain make of portable hair dryer is x hundred units/week and the corresponding wholesale unit price is

$$p = \sqrt{225 - 5x}$$

dollars. Determine the consumers' surplus if the wholesale market price is set at \$10/unit.

4. **Producers' Surplus** The supplier of the portable hair dryers in Exercise 3 will make x hundred units of hair dryers available in the market when the wholesale unit price is

$$p = \sqrt{36 + 1.8x}$$

dollars. Determine the producers' surplus if the wholesale market price is set at \$9/unit.

5. **Producers' Surplus** The supply function for the CDs of Exercise 2 is given by

$$p = 0.01x^2 + 0.1x + 3$$

where p is the unit wholesale price in dollars and x stands for the quantity that will be made available in the market by the supplier, measured in units of a thousand. Determine the producers' surplus if the wholesale market price is set at the equilibrium price.

6. **Consumers' and Producers' Surplus** The management of the Titan Tire Company has determined that the quantity demanded x of their Super Titan tires/week is related to the unit price p by the relation

$$p = 144 - x^2$$

where p is measured in dollars and x is measured in units of a thousand. Titan will make x units of the tires available in the market if the unit price is

$$p = 48 + \frac{1}{2}x^2$$

dollars. Determine the consumers' surplus and the producers' surplus when the market unit price is set at the equilibrium price.

7. **Consumers' and Producers' Surplus** The quantity demanded x (in units of a hundred) of the Mikado miniature cameras/week is related to the unit price p (in dollars) by

$$p = -0.2x^2 + 80$$

and the quantity x (in units of a hundred) that the supplier is willing to make available in the market is related to the unit price p (in dollars) by

$$p = 0.1x^2 + x + 40$$

If the market price is set at the equilibrium price, find the consumers' surplus and the producers' surplus.

8. Refer to Example 3, page 470. Verify that

$$\int_0^3 580{,}000e^{-0.1t}\,dt - 180{,}000 \approx 1{,}323{,}254$$

9. **Present Value of an Investment** Suppose an investment is expected to generate income at the rate of

$$R(t) = 200{,}000$$

dollars/year for the next 5 yr. Find the present value of this investment if the prevailing interest rate is 8%/year compounded continuously.

10. **Franchises** Camille purchased a 15-yr franchise for a computer outlet store that is expected to generate income at the rate of

$$R(t) = 400{,}000$$

dollars/year. If the prevailing interest rate is 8%/year compounded continuously, find the present value of the franchise.

11. **The Amount of an Annuity** Find the amount of an annuity if \$250/month is paid into it for a period of 20 yr, earning interest at the rate of 8%/year compounded continuously.

12. **The Amount of an Annuity** Find the amount of an annuity if \$400/month is paid into it for a period of 20 yr, earning interest at the rate of 6%/year compounded continuously.

13. **The Amount of an Annuity** Aiso deposits \$150/month in a savings account paying 6%/year compounded continuously. Estimate the amount that will be in his account after 15 yr.

14. **Custodial Accounts** The Armstrongs wish to establish a custodial account to finance their children's education. If they deposit \$200 monthly for 10 yr in a savings account paying 6%/year compounded continuously, how much will their savings account be worth at the end of this period?

15. **IRA Accounts** Refer to Example 4, page 470. Suppose Marcus made his IRA payment on April 1, 1994, and annually thereafter. If interest is paid at the same initial rate, approximately how much did Marcus have in his account at the beginning of 2010?

16. **Present Value of an Annuity** Estimate the present value of an annuity if payments are \$800 monthly for 12 yr and the account earns interest at the rate of 5%/year compounded continuously.

17. **Present Value of an Annuity** Estimate the present value of an annuity if payments are \$1200 monthly for 15 yr and the account earns interest at the rate of 6%/year compounded continuously.

18. **Lottery Payments** A state lottery commission pays the winner of the "Million Dollar" lottery 20 annual installments of \$50,000 each. If the prevailing interest rate is 6%/year compounded continuously, find the present value of the winning ticket.

19. **Reverse Annuity Mortgages** Sinclair wishes to supplement his retirement income by \$300/month for the next 10 yr. He plans to obtain a reverse annuity mortgage (RAM) on his home to meet this need. Estimate the amount of the mortgage he will require if the prevailing interest rate is 8%/year compounded continuously.

20. **Reverse Annuity Mortgage** Refer to Exercise 19. Leah wishes to supplement her retirement income by \$400/month for the next 15 yr by obtaining a RAM. Estimate the amount of the mortgage she will require if the prevailing interest rate is 6%/year compounded continuously.

21. **Lorentz Curves** A certain country's income distribution is described by the function

$$f(x) = \frac{15}{16}x^2 + \frac{1}{16}x$$

a. Sketch the Lorentz curve for this function.
b. Compute $f(0.4)$ and $f(0.9)$ and interpret your results.

22. **Lorentz Curves** In a study conducted by a certain country's Economic Development Board, it was found that the Lorentz curve for the distribution of income of college teachers was described by the function

$$f(x) = \frac{13}{14}x^2 + \frac{1}{14}x$$

and that of lawyers by the function

$$g(x) = \frac{9}{11}x^4 + \frac{2}{11}x$$

a. Compute the coefficient of inequality for each Lorentz curve.
b. Which profession has a more equitable income distribution?

23. **Lorentz Curves** A certain country's income distribution is described by the function

$$f(x) = \frac{14}{15}x^2 + \frac{1}{15}x$$

a. Sketch the Lorentz curve for this function.
b. Compute $f(0.3)$ and $f(0.7)$.

24. **Lorentz Curves** In a study conducted by a certain country's Economic Development Board, it was found that the Lorentz curve for the distribution of income of stockbrokers was described by the function

$$f(x) = \frac{11}{12}x^2 + \frac{1}{12}x$$

and that of high school teachers by the function

$$g(x) = \frac{5}{6}x^2 + \frac{1}{6}x$$

a. Compute the coefficient of inequality for each Lorentz curve.
b. Which profession has a more equitable income distribution?

6.7 Solutions to Self-Check Exercises

We find the equilibrium price and equilibrium quantity by solving the system of equations

$$p = \sqrt{9 - 0.02x}$$
$$p = \sqrt{1 + 0.02x}$$

simultaneously. Substituting the first equation into the second, we have

$$\sqrt{9 - 0.02x} = \sqrt{1 + 0.02x}$$

Squaring both sides of the equation then leads to

$$9 - 0.02x = 1 + 0.02x$$
$$x = 200$$

Therefore,

$$p = \sqrt{9 - 0.02(200)}$$
$$= \sqrt{5} \approx 2.24$$

The equilibrium price is \$224, and the equilibrium quantity is 200. The consumers' surplus is given by

$$CS = \int_0^{200} \sqrt{9 - 0.02x}\,dx - (2.24)(200)$$
$$= \int_0^{200} (9 - 0.02x)^{1/2}\,dx - 448$$
$$= -\frac{1}{0.02}\left(\frac{2}{3}\right)(9 - 0.02x)^{3/2}\Big|_0^{200} - 448 \quad \text{Integrate by substitution.}$$
$$= -\frac{1}{0.03}(5^{3/2} - 9^{3/2}) - 448$$
$$\approx 79.32$$

or approximately \$7932.

Next, the producers' surplus is given by

$$PS = (2.24)(200) - \int_0^{200} \sqrt{1 + 0.02x}\,dx$$
$$= 448 - \int_0^{200} (1 + 0.02x)^{1/2}\,dx$$
$$= 448 - \frac{1}{0.02}\left(\frac{2}{3}\right)(1 + 0.02x)^{3/2}\Big|_0^{200}$$
$$= 448 - \frac{1}{0.03}(5^{3/2} - 1)$$
$$\approx 108.66$$

or approximately \$10,866.

USING TECHNOLOGY

Business and Economic Applications/Technology Exercises

1. Re-solve Example 1, Section 6.7, using a graphing utility.
   Hint: Use the intersection operation to find the equilibrium quantity and the equilibrium price. Use the numerical integral operation to evaluate the definite integral.
2. Re-solve Exercise 7, Section 6.7, using a graphing utility.
   Hint: See Exercise 1.
3. The demand function for a certain brand of travel alarm clocks is given by

$$p = -0.01x^2 - 0.3x + 10$$

   where p is the wholesale unit price in dollars and x is the quantity demanded each month, measured in units of a thousand. The supply function for this brand of clocks is given by

$$p = -0.01x^2 + 0.2x + 4$$

   where p has the same meaning as before and x is the quantity, in thousands, the supplier will make available in the marketplace per month. Determine the consumers' surplus and the producers' surplus when the market unit price is set at the equilibrium price.
4. The quantity demanded of a certain make of compact disc organizer is x thousand units per week, and the corresponding wholesale unit price is

$$p = \sqrt{400 - 8x}$$

dollars. The supplier of the organizers will make x thousand units available in the market when the unit wholesale price is

$$p = 0.02x^2 + 0.04x + 5$$

dollars. Determine the consumers' surplus and the producers' surplus when the market unit price is set at the equilibrium price.

5. Investment A is expected to generate income at the rate of

$$R_1(t) = 50{,}000 + 10{,}000\sqrt{t}$$

dollars/year for the next 5 years and investment B is expected to generate income at the rate of

$$R_2(t) = 50{,}000 + 6000t$$

dollars/year over the same period of time. If the prevailing interest rate for the next 5 years is 10%/year, which investment will generate a higher net income by the end of 5 years?

6. Investment A is expected to generate income at the rate of

$$R_1(t) = 40{,}000 + 5000t + 100t^2$$

dollars/year for the next 10 years and investment B is expected to generate income at the rate of

$$R_2(t) = 60{,}000 + 2000t$$

dollars/year over the same period of time. If the prevailing interest rate for the next 10 years is 8%/year, which investment will generate a higher net income by the end of 10 years?

CHAPTER 6 Summary of Principal Formulas and Terms

FORMULAS

1. Indefinite integral of a constant	$\int k\,du = ku + C$
2. Power rule	$\int u^n\,du = \dfrac{u^{n+1}}{n+1} + C \quad (n \neq -1)$
3. Constant multiple rule	$\int kf(u)\,du = k\int f(u)\,du$ (k, a constant)
4. Sum rule	$\int [f(u) \pm g(u)]\,du = \int f(u)\,du \pm \int g(u)\,du$
5. Indefinite integral of the exponential function	$\int e^u\,du = e^u + C$
6. Indefinite integral of $f(u) = \dfrac{1}{u}$	$\int \dfrac{du}{u} = \ln\lvert u\rvert + C$
7. Method of substitution	$\int f'(g(x))g'(x)\,dx = \int f'(u)\,du$

8. Definite integral as the limit of a sum	$\int_a^b f(x)\,dx = \lim_{n\to\infty} S_n$, where S_n is a Riemann sum
9. Fundamental theorem of calculus	$\int_a^b f(x)\,dx = F(b) - F(a), F'(x) = f(x)$
10. Average value of f over $[a, b]$	$\frac{1}{b-a}\int_a^b f(x)\,dx$
11. Area between two curves	$\int_a^b [f(x) - g(x)]\,dx, f(x) \geq g(x)$
12. Consumers' surplus	$CS = \int_0^{\bar{x}} D(x)\,dx - \bar{p}\,\bar{x}$
13. Producers' surplus	$PS = \bar{p}\,\bar{x} - \int_0^{\bar{x}} S(x)\,dx$
14. Accumulated (future) value of an income stream	$A = e^{rT}\int_0^T R(t)e^{-rt}\,dt$
15. Present value of an income stream	$PV = \int_0^T R(t)e^{-rt}\,dt$
16. Amount of an annuity	$A = \frac{mP}{r}(e^{rT} - 1)$
17. Present value of an annuity	$PV = \frac{mP}{r}(1 - e^{-rT})$
18. Coefficient of inequality of a Lorentz curve	$L = 2\int_0^1 [x - f(x)]\,dx$

TERMS

antiderivative (398)
antidifferentiation (400)
integration (400)
indefinite integral (400)
integrand (400)
constant of integration (400)
differential equation (404)
initial value problem (404)
Riemann sum (425)
definite integral (425)
lower limit of integration (425)
upper limit of integration (425)
Lorentz curve (472)
line of complete equality (472)

CHAPTER 6 Concept Review Questions

Fill in the blanks.

1. a. A function F is an antiderivative of f on an interval, if _____ for all x in I.

b. If F is an antiderivative of f on an interval I, then every antiderivative of f on I has the form _____.

2. a. $\int cf(x)\,dx =$ _____

b. $\int [f(x) \pm g(x)]\,dx =$ _____

3. a. A differential equation is an equation that involves the derivative or differential of a/an _____ function.

b. A solution of a differential equation on an interval I is any _____ that satisfies the differential equation.

4. If we let $u = g(x)$, then $du =$ _____, and the substitution transforms the integral $\int f(g(x))g'(x)\,dx$ into the integral _____ involving only u.

5. a. If f is continuous and nonnegative on an interval $[a, b]$, then the area of the region under the graph of f on $[a, b]$ is given by _____.

b. If f is continuous on an interval $[a, b]$, then $\int_a^b f(x)\,dx$ is equal to the area(s) of the regions lying above the x-axis and bounded by the graph of f on $[a, b]$ _____ the area(s) of the regions lying below the x-axis and bounded by the graph of f on $[a, b]$.

6. a. The fundamental theorem of calculus states that if f is continuous on $[a, b]$, then $\int_a^b f(x)\,dx =$ _____, where F is a/an _____ of f.
   b. The net change in a function f over an interval $[a, b]$ is given by $f(b) - f(a) =$ _____, provided f' is continuous on $[a, b]$.

7. a. If f is continuous on $[a, b]$, then the average value of f over $[a, b]$ is the number _____.
   b. If f is a continuous and nonnegative function on $[a, b]$, then the average value of f over $[a, b]$ may be thought of as the _____ of the rectangle with base lying on the interval $[a, b]$ and having the same _____ as the region under the graph of f on $[a, b]$.

8. If f and g are continuous on $[a, b]$ and $f(x) \geq g(x)$ for all x in $[a, b]$, then the area of the region between the graphs of f and g and the vertical lines $x = a$ and $x = b$ is $A =$ _____.

9. a. The consumers' surplus is given by $CS =$ _____.
   b. The producers' surplus is given by $PS =$ _____.

10. a. The accumulated value after T years of an income stream of $R(t)$ dollars/year, earning interest of r/year compounded continuously, is given by $A =$ _____.
    b. The present value of an income stream is given by $PV =$ _____.

11. The amount of an annuity is $A =$ _____.

12. The coefficient of inequality of a Lorentz curve is $L =$ _____.

CHAPTER 6 Review Exercises

In Exercises 1–20, find each indefinite integral.

1. $\int (x^3 + 2x^2 - x)\,dx$
2. $\int \left(\frac{1}{3}x^3 - 2x^2 + 8\right) dx$
3. $\int \left(x^4 - 2x^3 + \frac{1}{x^2}\right) dx$
4. $\int (x^{1/3} - \sqrt{x} + 4)\,dx$
5. $\int x(2x^2 + x^{1/2})\,dx$
6. $\int (x^2 + 1)(\sqrt{x} - 1)\,dx$
7. $\int \left(x^2 - x + \frac{2}{x} + 5\right) dx$
8. $\int \sqrt{2x + 1}\,dx$
9. $\int (3x - 1)(3x^2 - 2x + 1)^{1/3}\,dx$
10. $\int x^2(x^3 + 2)^{10}\,dx$
11. $\int \frac{x - 1}{x^2 - 2x + 5}\,dx$
12. $\int 2e^{-2x}\,dx$
13. $\int \left(x + \frac{1}{2}\right)e^{x^2+x+1}\,dx$
14. $\int \frac{e^{-x} - 1}{(e^{-x} + x)^2}\,dx$
15. $\int \frac{(\ln x)^5}{x}\,dx$
16. $\int \frac{\ln x^2}{x}\,dx$
17. $\int x^3(x^2 + 1)^{10}\,dx$
18. $\int x\sqrt{x + 1}\,dx$
19. $\int \frac{x}{\sqrt{x - 2}}\,dx$
20. $\int \frac{3x}{\sqrt{x + 1}}\,dx$

In Exercises 21–32, evaluate each definite integral.

21. $\int_0^1 (2x^3 - 3x^2 + 1)\,dx$
22. $\int_0^2 (4x^3 - 9x^2 + 2x - 1)\,dx$
23. $\int_1^4 (\sqrt{x} + x^{-3/2})\,dx$
24. $\int_0^1 20x(2x^2 + 1)^4\,dx$
25. $\int_{-1}^0 12(x^2 - 2x)(x^3 - 3x^2 + 1)^3\,dx$
26. $\int_4^7 x\sqrt{x - 3}\,dx$
27. $\int_0^2 \frac{x}{x^2 + 1}\,dx$
28. $\int_0^1 \frac{dx}{(5 - 2x)^2}$
29. $\int_0^2 \frac{4x}{\sqrt{1 + 2x^2}}\,dx$
30. $\int_0^2 xe^{(-1/2)x^2}\,dx$
31. $\int_{-1}^0 \frac{e^{-x}}{(1 + e^{-x})^2}\,dx$
32. $\int_1^e \frac{\ln x}{x}\,dx$

In Exercises 33–36, find the function f given that the slope of the tangent line to the graph at any point $(x, f(x))$ is $f'(x)$ and that the graph of f passes through the given point.

33. $f'(x) = 3x^2 - 4x + 1$; (1, 1)
34. $f'(x) = \frac{x}{\sqrt{x^2 + 1}}$; (0, 1)
35. $f'(x) = 1 - e^{-x}$; (0, 2)
36. $f'(x) = \frac{\ln x}{x}$; (1, −2)

37. Let $f(x) = -2x^2 + 1$ and compute the Riemann sum of f over the interval [1, 2] by partitioning the interval into five subintervals of the same length ($n = 5$), where the points p_i ($1 \leq i \leq 5$) are taken to be the *right* endpoints of the respective subintervals.

38. **Marginal Cost Functions** The management of National Electric has determined that the daily marginal cost function associated with producing their automatic drip coffeemakers is given by

$$C'(x) = 0.00003x^2 - 0.03x + 20$$

where $C'(x)$ is measured in dollars/unit and x denotes the number of units produced. Management has also determined that the daily fixed cost incurred in producing these coffeemakers is \$500. What is the total cost incurred by National in producing the first 400 coffeemakers/day?

39. **Marginal Revenue Functions** Refer to Exercise 38. Management has also determined that the daily marginal revenue function associated with producing and selling their coffeemakers is given by

$$R'(x) = -0.03x + 60$$

where x denotes the number of units produced and sold and $R'(x)$ is measured in dollars/unit.

a. Determine the revenue function $R(x)$ associated with producing and selling these coffeemakers.
b. What is the demand equation relating the wholesale unit price to the quantity of coffeemakers demanded?

40. **Computer Resale Value** Franklin National Life Insurance Company purchased new computers for \$200,000. If the rate at which the computers' resale value changes is given by the function

$$V'(t) = 3800(t - 10)$$

where t is the length of time since the purchase date and $V'(t)$ is measured in dollars/year, find an expression $V(t)$ that gives the resale value of the computers after t yr. How much would the computers cost after 6 yr?

41. **Measuring Temperature** The temperature on a certain day as measured at the airport of a city is changing at the rate of

$$T'(t) = 0.15t^2 - 3.6t + 14.4 \qquad (0 \le t \le 4)$$

°F/hour, where t is measured in hours, with $t = 0$ corresponding to 6 a.m. The temperature at 6 a.m. was 24°F.

a. Find an expression giving the temperature T at the airport at any time between 6 a.m. and 10 a.m.
b. What was the temperature at 10 a.m.?

42. **DVD Sales** The total number of DVDs sold to U.S. dealers for rental and sale from 1999 through 2003 grew at the rate of approximately

$$R(t) = -0.03t^2 + 0.218t - 0.032 \qquad (0 \le t \le 4)$$

billion units/year, where t is measured in years, with $t = 0$ corresponding to 1999. The total number of DVDs sold as of 1999 was 0.1 billion units.

a. Find an expression giving the total number of DVDs sold by year t $(0 \le t \le 4)$.
b. How many DVDs were sold in 2003?

Source: Adams Media

43. **Air Pollution** On an average summer day, the level of carbon monoxide (CO) in a city's air is 2 parts per million (ppm). An environmental protection agency's study predicts that, unless more stringent measures are taken to protect the city's atmosphere, the CO concentration present in the air will increase at the rate of

$$0.003t^2 + 0.06t + 0.1$$

ppm/year, t yr from now. If no further pollution-control efforts are made, what will be the CO concentration on an average summer day 5 yr from now?

44. **Projection TV Sales** The marketing department of Vista Vision forecasts that sales of their new line of projection television systems will grow at the rate of

$$3000 - 2000e^{-0.04t} \qquad (0 \le t \le 24)$$

units/month once they are introduced into the market. Find an expression giving the total number of the projection television systems that Vista may expect to sell t mo from the time they are put on the market. How many units of the television systems can Vista expect to sell during the first year?

45. **Commuter Trends** Due to the increasing cost of fuel, the manager of the City Transit Authority estimates that the number of commuters using the city subway system will increase at the rate of

$$3000(1 + 0.4t)^{-1/2} \qquad (0 \le t \le 36)$$

per month, t mo from now. If 100,000 commuters are currently using the system, find an expression giving the total number of commuters who will be using the subway t mo from now. How many commuters will be using the subway 6 mo from now?

46. **Sales: Loudspeakers** Sales of the Acrosonic model F loudspeaker systems have been growing at the rate of

$$f'(t) = 2000(3 - 2e^{-t})$$

units/year, where t denotes the number of years these loudspeaker systems have been on the market. Determine the number of loudspeaker systems that were sold in the first 5 yr after they appeared on the market.

47. **Supply: Women's Boots** The rate of change of the unit price p (in dollars) of Apex women's boots is given by

$$p'(x) = \frac{240}{(5 - x)^2}$$

where x is the number of pairs in units of a hundred that the supplier will make available in the market daily when the unit price is \$$p$/pair. Find the supply equation for these boots if the quantity the supplier is willing to make available is 200 pairs daily $(x = 2)$ when the unit price is \$50/pair.

48. **Marginal Cost Functions** The management of a division of Ditton Industries has determined that the daily marginal cost function associated with producing their hot-air corn poppers is given by

$$C'(x) = 0.00003x^2 - 0.03x + 10$$

where $C'(x)$ is measured in dollars/unit and x denotes the number of units manufactured. Management has also determined that the daily fixed cost incurred in producing these corn poppers is \$600. Find the total cost incurred by Ditton in producing the first 500 corn poppers.

49. **U.S. Census** The number of Americans aged 45–54 yr (which stood at 25 million at the beginning of 1990) grew at the rate of

$$R(t) = 0.00933t^3 + 0.019t^2 - 0.10833t + 1.3467$$

million people/year, t yr from the beginning of 1990. How many Americans aged 45 to 54 were added to the population between 1990 and the year 2000?
Source: U.S. Census Bureau

50. **Online Retail Sales** Since the inception of the Web, online commerce has enjoyed phenomenal growth. But growth, led by such major sectors as books, tickets, and office supplies, is expected to slow in the coming years. The projected growth of online retail sales is given by

$$R(t) = 15.82e^{-0.176t} \qquad (0 \le t \le 4)$$

where t is measured in years with $t = 0$ corresponding to 2007 and $R(t)$ is measured in billions of dollars per year. Online retail sales in 2007 were \$116 billion.

a. Find an expression for online retail sales in year t.
b. If the projection holds true, what will be online retail sales in 2011?

Source: Jupiter Research

51. Find the area of the region under the curve $y = 3x^2 + 2x + 1$ from $x = -1$ to $x = 2$.

52. Find the area of the region under the curve $y = e^{2x}$ from $x = 0$ to $x = 2$.

53. Find the area of the region bounded by the graph of the function $y = 1/x^2$, the x-axis, and the lines $x = 1$ and $x = 3$.

54. Find the area of the region bounded by the curve $y = -x^2 - x + 2$ and the x-axis.

55. Find the area of the region bounded by the graphs of the functions $f(x) = e^x$ and $g(x) = x$ and the vertical lines $x = 0$ and $x = 2$.

56. Find the area of the region that is completely enclosed by the graphs of $f(x) = x^4$ and $g(x) = x$.

57. Find the area of the region between the curve $y = x(x - 1)(x - 2)$ and the x-axis.

58. **Oil Production** Based on current production techniques, the rate of oil production from a certain oil well t yr from now is estimated to be

$$R_1(t) = 100e^{0.05t}$$

thousand barrels/year. Based on a new production technique, however, it is estimated that the rate of oil production from that oil well t yr from now will be

$$R_2(t) = 100e^{0.08t}$$

thousand barrels/year. Determine how much additional oil will be produced over the next 10 yr if the new technique is adopted.

59. Find the average value of the function

$$f(x) = \frac{x}{\sqrt{x^2 + 16}}$$

over the interval [0, 3].

60. **Average Temperature** The temperature (in °F) in Boston over a 12-hr period on a certain December day was given by

$$T = -0.05t^3 + 0.4t^2 + 3.8t + 5.6 \qquad (0 \le t \le 12)$$

where t is measured in hours, with $t = 0$ corresponding to 6 a.m. Determine the average temperature on that day over the 12-hr period from 6 a.m. to 6 p.m.

61. **Average Velocity of a Truck** A truck traveling along a straight road has a velocity (in feet/second) at time t (in seconds) given by

$$v(t) = \frac{1}{12}t^2 + 2t + 44 \qquad (0 \le t \le 5)$$

What is the average velocity of the truck over the time interval from $t = 0$ to $t = 5$?

62. **Membership in Credit Unions** Credit unions in Massachusetts have grown remarkably in recent years. Their tax-exempt status allows them to offer deposit and loan rates that are often more favorable than those offered by banks. The membership in Massachusetts credit unions grew at the rate of

$$R(t) = -0.0039t^2 + 0.0374t + 0.0046 \qquad (0 \le t \le 9)$$

million members/year between 1994 ($t = 0$) and 2003 ($t = 9$). Find the average rate of growth of membership in Massachusetts credit unions over the period in question.
Source: Massachusetts Credit Union League

63. **Demand for Digital Camcorder Tapes** The demand function for a brand of blank digital camcorder tapes is given by

$$p = -0.01x^2 - 0.2x + 23$$

where p is the wholesale unit price in dollars and x is the quantity demanded each week, measured in units of a thousand. Determine the consumers' surplus if the wholesale unit price is \$8/tape.

64. Consumers' and Producers' Surplus The quantity demanded x (in units of a hundred) of the Sportsman 5×7 tents, per week, is related to the unit price p (in dollars) by the relation

$$p = -0.1x^2 - x + 40$$

The quantity x (in units of a hundred) that the supplier is willing to make available in the market is related to the unit price by the relation

$$p = 0.1x^2 + 2x + 20$$

If the market price is set at the equilibrium price, find the consumers' surplus and the producers' surplus.

65. Retirement Account Savings Chi-Tai plans to deposit \$4000/year in his Keogh Retirement Account. If interest is compounded continuously at the rate of 8%/year, how much will he have in his retirement account after 20 yr?

66. Installment Contracts Glenda sold her house under an installment contract whereby the buyer gave her a down payment of \$20,000 and agreed to make monthly payments of \$925/month for 30 yr. If the prevailing interest rate is 6%/year compounded continuously, find the present value of the purchase price of the house.

67. Present Value of a Franchise Alicia purchased a 10-yr franchise for a health spa that is expected to generate income at the rate of

$$P(t) = 80{,}000$$

dollars/year. If the prevailing interest rate is 10%/year compounded continuously, find the present value of the franchise.

68. Income Distribution of a Country A certain country's income distribution is described by the function

$$f(x) = \frac{17}{18}x^2 + \frac{1}{18}x$$

a. Sketch the Lorentz curve for this function.
b. Compute $f(0.3)$ and $f(0.6)$ and interpret your results.
c. Compute the coefficient of inequality for this Lorentz curve.

69. Population Growth The population of a certain Sunbelt city, currently 80,000, is expected to grow exponentially in the next 5 yr with a growth constant of 0.05. If the prediction comes true, what will be the average population of the city over the next 5 yr?

CHAPTER 6 Before Moving On . . .

1. Find $\int \left(2x^3 + \sqrt{x} + \frac{2}{x} - \frac{2}{\sqrt{x}}\right) dx$.

2. Find f if $f'(x) = e^x + x$ and $f(0) = 2$.

3. Find $\int \frac{x}{\sqrt{x^2 + 1}}\, dx$.

4. Evaluate $\int_0^1 x\sqrt{2 - x^2}\, dx$

5. Find the area of the region completely enclosed by the graphs of $y = x^2 - 1$ and $y = 1 - x$.

APPENDIX

A.1 The Inverse of a Function

Consider the position function

$$s = f(t) = 4t^2 \qquad (0 \le t \le 30) \tag{1}$$

giving the position of the maglev at any time t in its domain [0, 30]. The graph of f is shown in Figure 1. Equation (1) enables us to compute algebraically the position of the maglev at any time t. Geometrically, we can find the position of the maglev at any given time t by following the path indicated in Figure 1.

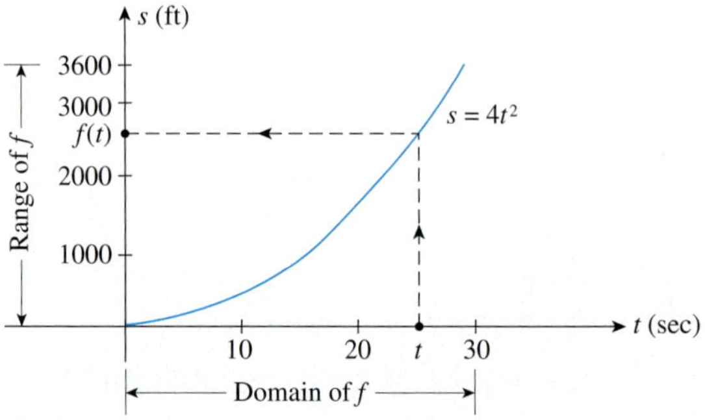

FIGURE 1
Each t in the domain of f is associated with the (unique) position $s = f(t)$ of the maglev.

Now consider the reverse problem: Knowing the position function of the maglev, can we find some way of obtaining the time it takes for the maglev to reach a given position? Geometrically, this problem is easily solved: Locate the point on the s-axis corresponding to the given position. Follow the path considered earlier but traced in the *opposite* direction. This path associates the given position s with the desired time t.

Algebraically, we can obtain a formula for the time t it takes for the maglev to get to the position s by solving (1) for t in terms of s. Thus,

$$t = \frac{1}{2}\sqrt{s}$$

(We reject the negative root because t lies in [0, 30].) Observe that the function g defined by

$$t = g(s) = \frac{1}{2}\sqrt{s}$$

has domain [0, 3600] (the range of f) and range [0, 30] (the domain of f) (Figure 2).

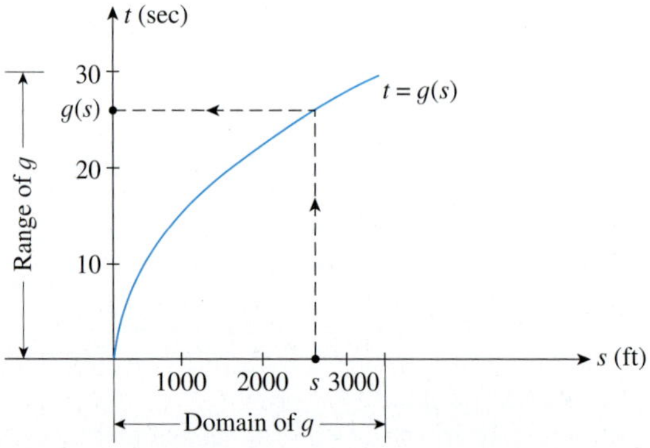

FIGURE 2
Each s in the domain of g is associated with the (unique) time $t = g(s)$.

The functions f and g have the following properties:

1. The domain of g is the range of f and vice versa.
2. $(g \circ f)t = g[f(t)] = \frac{1}{2}\sqrt{f(t)} = \frac{1}{2}\sqrt{4t^2} = t$

   and

   $$(f \circ g)(t) = f[g(t)] = 4[g(t)]^2 = 4\left(\frac{1}{2}\sqrt{t}\right)^2 = t$$

In other words, one undoes what the other does. This is to be expected because f maps t onto $s = f(t)$ and g maps $s = f(t)$ back onto t.

The functions f and g are said to be *inverses* of each other. More generally, we have the following definition.

Inverse Functions

A function g is the inverse of the function f if

$$f[g(tx)] = x \text{ for every } x \text{ in the domain of } g$$
$$g[f(x)] = x \text{ for every } x \text{ in the domain of } f$$

Equivalently, g is the inverse of f if the following condition is satisfied:

$$y = f(x) \quad \text{if and only if} \quad x = g(y)$$

for every x in the domain of f and for every y in its range.

Note The inverse of f is normally denoted by f^{-1} (read "f inverse"). ■

Do not confuse $f^{-1}(x)$ with $[f(x)]^{-1} = \dfrac{1}{f(x)}$!

EXAMPLE 1 Show that the functions $f(x) = x^{1/3}$ and $g(x) = x^3$ are inverses of each other.

Solution First, observe that the domain and range of both f and g are $(-\infty, \infty)$. Therefore, both composite functions $f \circ g$ and $g \circ f$ are defined. Next, we compute

$$(f \circ g)(x) = f[g(x)] = [g(x)]^{1/3} = [x^3]^{1/3} = x$$

and

$$(g \circ f)(x) = g[f(x)] = [f(x)]^3 = (x^{1/3})^3 = x$$

Since $f[g(x)] = g[f(x)] = x$, we conclude that f and g are inverses of each other. In short, $f^{-1}(x) = x^3$. ■

Interpreting Our Results We can view f as a cube root extracting machine and g as a "cubing" machine. In this light, it is easy to see that one function does undo what the other does. So f and g are indeed inverses of each other.

A.2 The Graphs of Inverse Functions

The graphs of $f(x) = x^{1/3}$ and $f^{-1}(x) = x^3$ are shown in Figure 3.

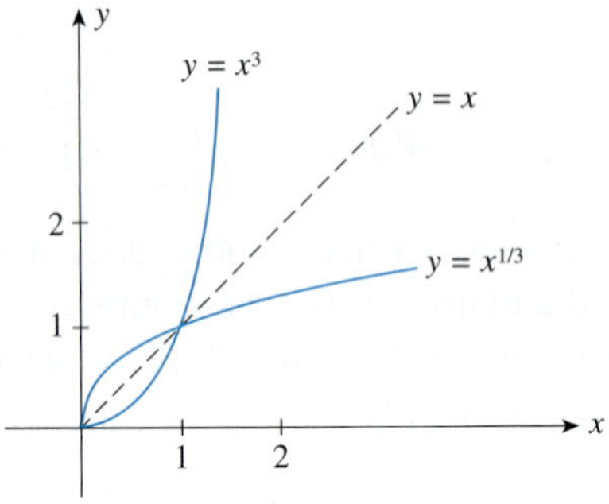

FIGURE 3
The functions $y = x^{1/3}$ and $f^{-1}(x) = x^3$ are inverses of each other. Their graphs are symmetric with respect to the line $y = x$.

They seem to suggest that the graphs of inverse functions are mirror images of each other with respect to the line $y = x$. This is true in general, a fact that we will not prove.

> **The Graphs of Inverse Functions**
> The graph of f^{-1} is the reflection of the graph of f with respect to the line $y = x$ and vice versa.

A.3 Functions That Have Inverses

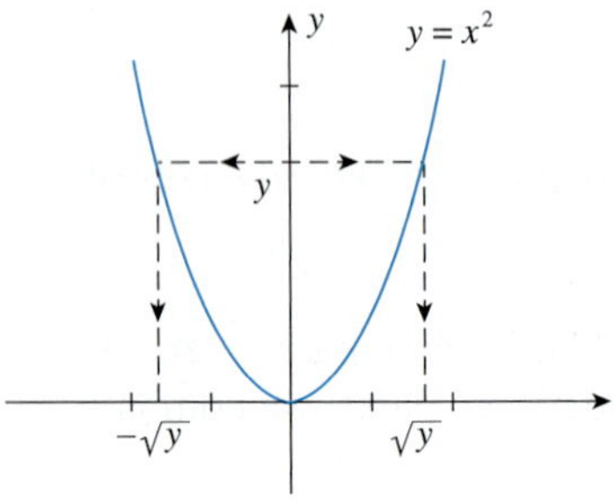

FIGURE 4
Each value of y is associated with two values of x.

Not every function has an inverse. Consider, for example, the function f defined by $y = x^2$ with domain $(-\infty, \infty)$ and range $[0, \infty)$. From the graph of f shown in Figure 4, you can see that each value of y in the range $[0, \infty)$ of f is associated with exactly *two* numbers $x = \pm\sqrt{y}$ (except for $y = 0$) in the domain $(-\infty, \infty)$ of f. This implies that f does not have an inverse because the uniqueness requirement of a function cannot be satisfied in this case. Observe that any horizontal line $y = c$ $(c > 0)$ intersects the graph of f at more than one point.

Next, consider the function g defined by the same rule as that of f, namely $y = x^2$, but with domain restricted to $[0, \infty)$. From the graph of g shown in Figure 5, you can see that each value of y in the range $[0, \infty)$ of g is mapped onto exactly *one* number $x = \sqrt{y}$ in the domain $[0, \infty)$ of g.

Thus, in this case, we can define the inverse function of g, from the range $[0, \infty)$ of g onto the domain $[0, \infty)$ of g. To find the rule for g^{-1}, we solve the equation $y = x^2$ for x in terms of y. Thus, $x = \sqrt{y}$ (because $x \geq 0$) and so $g^{-1}(y) = \sqrt{y}$, or, since y is a dummy variable, we can write $g^{-1}(x) = \sqrt{x}$. Also, observe that every horizontal line intersects the graph of g at no more than one point.

Why does g have an inverse but f does not? Observe that f takes on the same value twice; that is, there are two values of x that are mapped onto each value of y (except $y = 0$). On the other hand, g never takes on the same value more than once; that is, any two values of x have different images. The function g is said to be *one-to-one*.

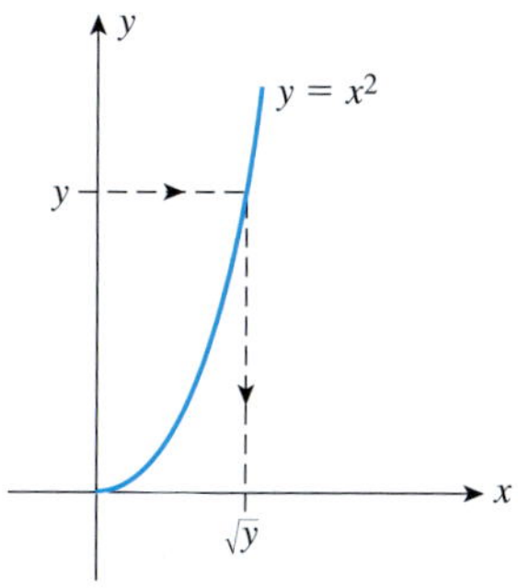

FIGURE 5
Each value of y is associated with exactly one value of x.

One-to-One Function

A function f with domain D is **one-to-one** if no two points in D have the same image: that is, $f(x_1) \neq f(x_2)$ whenever $x_1 \neq x_2$.

Geometrically, a function is one-to-one if every horizontal line intersects its graph at no more than one point. This is called the **horizontal-line test.**

The next theorem tells us when an inverse function exists.

THEOREM 1

The Existence of an Inverse Function

A function has an inverse if and only if it is one-to-one.

A.4 Finding the Inverse of a Function

Here is a summary of the steps for finding the inverse of a function (if it exists).

Guidelines for Finding the Inverse of a Function

1. Write $y = f(x)$.
2. Solve for x in terms of y (if possible).
3. Interchange x and y to obtain $y = f^{-1}(x)$.

EXAMPLE 1 Find the inverse of the function defined by $f(x) = \dfrac{1}{\sqrt{2x-3}}$.

Solution To find the rule for this inverse, write

$$y = \frac{1}{\sqrt{2x-3}} \qquad (y > 0)$$

and then solve the equation for x:

$$y^2 = \frac{1}{2x-3} \qquad \text{Square both sides.}$$

$$2x - 3 = \frac{1}{y^2} \qquad \text{Take reciprocals.}$$

$$2x = \frac{1}{y^2} + 3 = \frac{3y^2+1}{y^2}$$

$$x = \frac{3y^2+1}{2y^2}$$

Finally, interchanging x and y, we obtain

$$y = \frac{3x^2+1}{2x^2}$$

giving the rule for f^{-1} as

$$f^{-1}(x) = \frac{3x^2+1}{2x^2}$$

The graphs of both f and f^{-1} are shown in Figure 6.

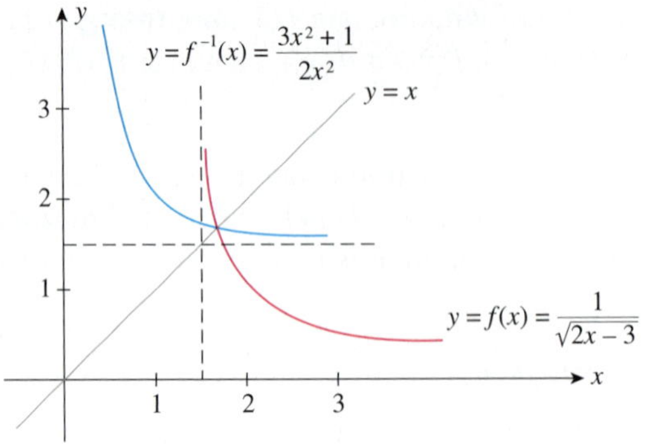

FIGURE 6
The graphs of f and f^{-1}. Notice that they are reflections of each other about the line $y = x$.

Exercises

In Exercises 1–6, show that f and g are inverses of each other by showing that $f[g(x)] = x$ and $g[f(x)] = x$.

1. $f(x) = \frac{1}{3}x^3$; $g(x) = \sqrt[3]{3x}$ **2.** $f(x) = \frac{1}{x}$; $g(x) = \frac{1}{x}$

3. $f(x) = 2x + 3$; $g(x) = \frac{x - 3}{2}$

4. $f(x) = x^2 + 1, (x \leq 0)$; $g(x) = -\sqrt{x - 1}$

5. $f(x) = 4(x + 1)^{2/3}, (x \geq -1)$; $g(x) = \frac{1}{8}(x^{3/2} - 8), (x \geq 0)$

6. $f(x) = \frac{1 + x}{1 - x}$; $g(x) = \frac{x - 1}{x + 1}$

In Exercises 7–12, you are given the graph of a function f. Determine whether f is one-to-one.

7.

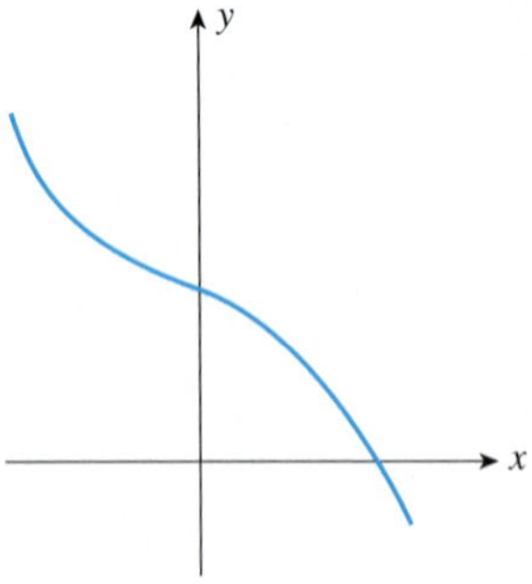

8.

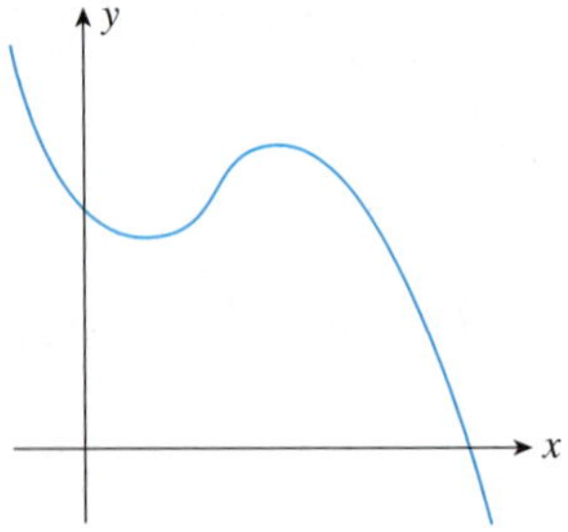

9.

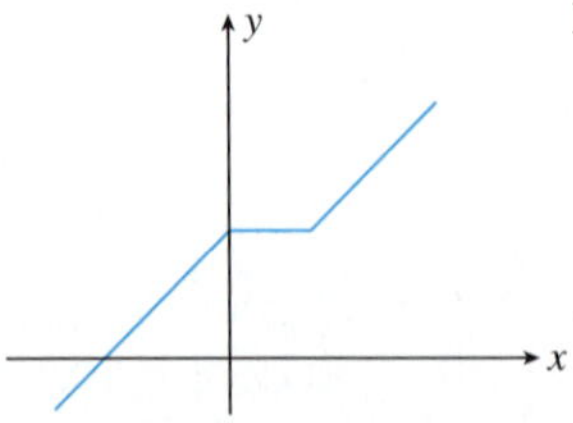

10.

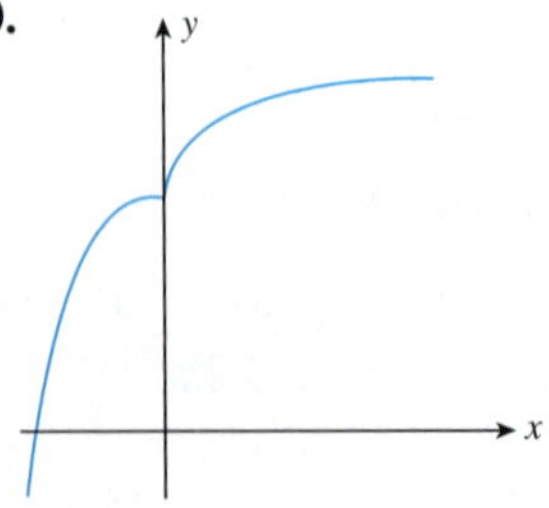

11.

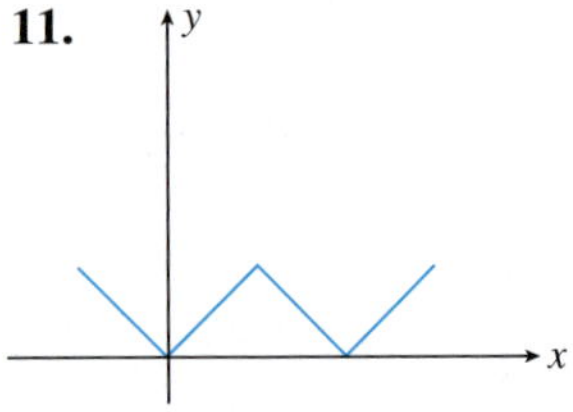

12.

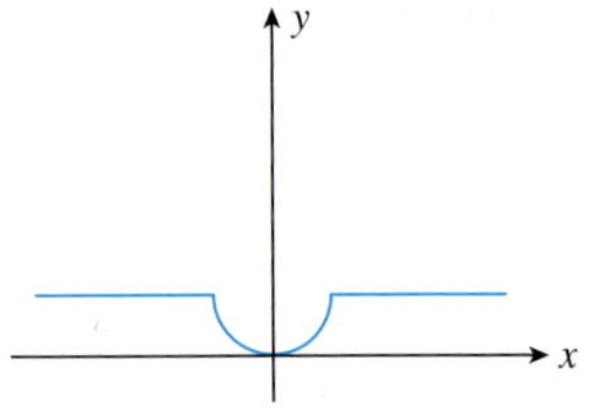

In Exercises 13–18, find the inverse of f. Then sketch the graphs of f and f^{-1} on the same set of axes.

13. $f(x) = 3x - 2$ **14.** $f(x) = x^2, x \leq 0$

15. $f(x) = x^3 + 1$ **16.** $f(x) = 2\sqrt{x + 3}$

17. $f(x) = \sqrt{9 - x^2}, x \geq 0$ **18.** $f(x) = x^{3/5} + 1$

19. A hot-air balloon rises vertically from the ground so that its height after t sec is $h = \frac{1}{2}t^2 + \frac{1}{2}t$ ft $(0 \leq t \leq 60)$.
a. Find the inverse of the function $f(t) = \frac{1}{2}t^2 + \frac{1}{2}t$ and explain what it represents.
b. Use the result of part (a) to find the time when the balloon is between an altitude of 120 and 210 ft.

20. **Aging Population** The population of Americans age 55 yr and older as a percentage of the total population is approximated by the function

$$f(t) = 10.72(0.9t + 10)^{0.3} \qquad (0 \leq t \leq 20)$$

where t is measured in years and $t = 0$ corresponds to the year 2000.
a. Find the rule for f^{-1}.
b. Evaluate $f^{-1}(25)$ and interpret your result.
Source: U.S. Census Bureau

B.1 Indeterminate Forms and l'Hôpital's Rule

In Section 2.4, we encountered the limit

$$\lim_{t\to 2}\frac{4(t^2-4)}{t-2} \tag{1}$$

when we attempted to find the velocity of the maglev at time $t = 2$, and in Section 12.3, we studied the limit

$$\lim_{x\to 0}\frac{\sin x}{x} \tag{2}$$

Observe that both the numerator and the denominator of each of these expressions approach zero as t approaches zero.

More generally, if $\lim_{x\to a} f(x) = 0$ and $\lim_{x\to a} g(x) = 0$, then the limit

$$\lim_{x\to a}\frac{f(x)}{g(x)}$$

is called an **indeterminate form of the type 0/0.** As the name implies, the undefined expression 0/0 does not provide us with a definitive answer concerning the existence of the limit or its value, if the limit exists.

Recall that we evaluated the limit in (1) through algebraic sleight of hand. Thus,

$$\lim_{t\to 2}\frac{4(t^2-4)}{t-2} = \lim_{t\to 2}\frac{4(t+2)(t-2)}{t-2} = \lim_{t\to 2}4(t+2) = 16$$

This method, however, will not work in evaluating the limit in (2).

These examples raise the following question: Given an indeterminate form of the type 0/0, is there a more general and efficient method for resolving whether the limit

$$\lim_{x\to a}\frac{f(x)}{g(x)}$$

exists, and if so, what is the limit?

B.2 The Indeterminate Forms 0/0 and ∞/∞

To gain insight into the nature of an indeterminate form of the type 0/0, let's consider the following limits:

a. $\lim_{x\to 0^+}\frac{x^2}{x}$ **b.** $\lim_{x\to 0^+}\frac{2x}{3x}$ **c.** $\lim_{x\to 0^+}\frac{x}{x^2}$

Each of these limits is an indeterminate form of the type 0/0. We can evaluate each limit as follows:

a. $\lim_{x\to 0^+}\frac{x^2}{x} = \lim_{x\to 0^+} x = 0$ **b.** $\lim_{x\to 0^+}\frac{2x}{3x} = \lim_{x\to 0^+}\frac{2}{3} = \frac{2}{3}$ **c.** $\lim_{x\to 0^+}\frac{x}{x^2} = \lim_{x\to 0^+}\frac{1}{x} = \infty$

Let's examine each limit in greater detail. In part (a), the numerator $f_1(x) = x^2$ goes to zero faster than the denominator $g_1(x) = x$, when x is close to zero. So it is plausible that the ratio $f_1(x)/g_1(x)$ should approach zero as x approaches zero. In part (b), the numerator $f_2(x) = 2x$ goes to zero at $(2x)/(3x) = 2/3$, or two-thirds of the rate that $g_2(x) = 3x$ goes to zero, and so the answer seems reasonable. Finally, in part (c), the denominator $g_3(x) = x^2$ goes to zero faster than the numerator $f_3(x) = x$, and, consequently, we expect the ratio to "blow up."

These three examples show that the existence or nonexistence of the limit, as well as the value of the limit, depends on how fast the numerator $f(x)$ and the denominator $g(x)$ go to zero. This observation suggests the following technique for evaluating these indeterminate forms: Because both $f(x)$ and $g(x)$ go to zero as x approaches zero, we cannot determine the limit of the quotient by using the quotient rule for limits. So, we might consider the limit of the ratio of their *derivatives*, $f'(x)$ and $g'(x)$, since the derivatives measure how fast $f(x)$ and $g(x)$ change. In other words, it might be plausible that if both $f(x) \to 0$ and $g(x) \to 0$ as $x \to 0$, then

$$\lim_{x\to 0}\frac{f(x)}{g(x)} = \lim_{x\to 0}\frac{f'(x)}{g'(x)}$$

Let's try this on the limits in (1) and (2). For the limit in (1), we have

$$\lim_{t\to 2}\frac{4(t^2-4)}{t-2} = \lim_{t\to 2}\frac{\frac{d}{dt}[4(t^2-4)]}{\frac{d}{dt}(t-2)} = \lim_{t\to 2}\frac{8t}{1} = 16$$

which is the value we obtained before! For the limit in (2), we find

$$\lim_{x\to 0}\frac{\sin x}{x} = \lim_{x\to 0}\frac{\frac{d}{dx}(\sin x)}{\frac{d}{dx}(x)} = \lim_{x\to 0}\frac{\cos x}{1} = 1$$

a result we used in Section 12.3.

This method, which we have arrived at intuitively, is given validity by the theorem known as l'Hôpital's rule. The theorem is named after the French mathematician Guillaume Francois Antoine de l'Hôpital (1661–1704), who published the first calculus text in 1696. But, before stating l'Hôpital's rule, we need to define another type of indeterminate form.

If $\lim_{x\to a} f(x) = \pm\infty$ and $\lim_{x\to\infty} g(x) = \pm\infty$ then the limit

$$\lim_{x\to a}\frac{f(x)}{g(x)}$$

is also an indeterminate form of the type ∞/∞, $-\infty/\infty$, $\infty/-\infty$, or $-\infty/-\infty$. To see why this limit is an indeterminate form, we simply write

$$\lim_{x\to a}\frac{f(x)}{g(x)} = \lim_{x\to a}\frac{\frac{1}{g(x)}}{\frac{1}{f(x)}}$$

which has the form 0/0 and is therefore indeterminate. We refer to each of these limits as an **indeterminate form of the type ∞/∞** since the sign provides no useful information.

THEOREM 1

l'Hôpital's Rule

Suppose f and g are differentiable on an open interval I that contains a, with the possible exception of a itself and $g'(x) \neq 0$ for all x in I. If $\lim_{x\to a}\frac{f(x)}{g(x)}$ has an indeterminate form of the type 0/0 or ∞/∞, then

$$\lim_{x\to 0}\frac{f(x)}{g(x)} = \lim_{x\to 0}\frac{f'(x)}{g'(x)}$$

provided that the limit on the right exists or is infinite.

The expression $f'(x)/g'(x)$ is the *ratio* of the derivatives of $f(x)$ and $g(x)$—it is *not* obtained from f/g using the quotient rule.

Notes **1.** l'Hôpital's rule is also valid for one-sided limits as well as limits at infinity or negative infinity; that is, we can replace $x \to a$ by any of the symbols $x \to a^+$, $x \to a^-$, $x \to -\infty$, and $x \to \infty$.

2. Before applying l'Hôpital's rule, check to see that the limit has one of the indeterminate forms. For example, $\cos x \to 1$, as $x \to 0^+$, and so

$$\lim_{x \to 0^+} \frac{\cos x}{x} = \infty$$

If we had applied l'Hôpital's rule to evaluate the limit without first ascertaining that it had an indeterminate form, we would have obtained the erroneous result

$$\lim_{x \to 0^+} \frac{\cos x}{x} = \lim_{x \to 0^+} \frac{-\sin x}{1} = 0$$

■

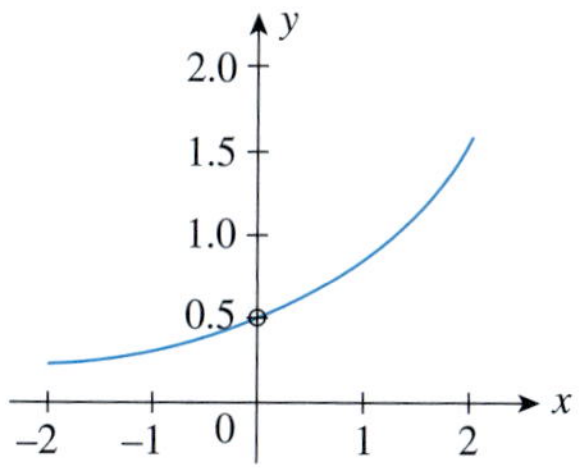

FIGURE 1
The graph of $y = (e^x - 1)/2x$ gives a visual confirmation of the result of Example 1.

EXAMPLE 1 Evaluate $\lim_{x \to 0} \dfrac{2^x - 1}{2x}$.

Solution We have an indeterminate form of the type 0/0. Applying l'Hôpital's rule, we obtain

$$\lim_{x \to 0} \frac{e^x - 1}{2x} = \lim_{x \to 0} \frac{\frac{d}{dx}(e^x - 1)}{\frac{d}{dx}(2x)} = \lim_{x \to 0} \frac{e^x}{2} = \frac{1}{2}$$

(See Figure 1.) ■

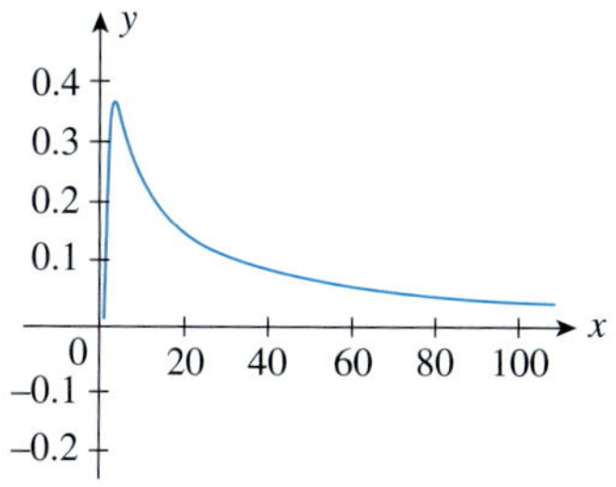

FIGURE 2
The graph of $y = (\ln x)/x$ shows that $y \to 0$ as $x \to \infty$.

EXAMPLE 2 Evaluate $\lim_{x \to \infty} \dfrac{\ln x}{x}$.

Solution We have an indeterminate form of the type ∞/∞. Applying l'Hôpital's rule, we obtain

$$\lim_{x \to \infty} \frac{\ln x}{x} = \lim_{x \to \infty} \frac{\frac{d}{dx}(\ln x)}{\frac{d}{dx}(x)} = \lim_{x \to \infty} \frac{1}{x} = 0$$

(See Figure 2.) ■

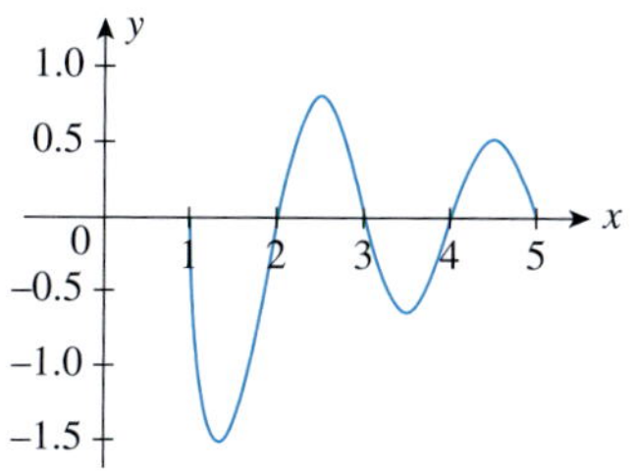

FIGURE 3
The graph of $y = (\sin \pi x)/(x - 1)^{1/2}$ shows that $\lim_{x \to 1^+} (\sin \pi x)/(\sqrt{x - 1}) = 0$.

EXAMPLE 3 Evaluate $\lim_{x \to 1^+} \dfrac{\sin \pi x}{\sqrt{x - 1}}$.

Solution We have an indeterminate form of the type 0/0. Applying l'Hôpital's rule, we obtain

$$\begin{aligned}\lim_{x \to 1^+} \frac{\sin \pi x}{(x - 1)^{1/2}} &= \lim_{x \to 1^+} \frac{\pi \cos \pi x}{\frac{1}{2}(x - 1)^{-1/2}} \\ &= \lim_{x \to 1^+} 2\pi(\cos \pi x)\sqrt{x - 1} \\ &= 0\end{aligned}$$

(See Figure 3.) ■

To resolve a limit involving an indeterminate form, we sometimes need to apply l'Hôpital's rule more than once. This is illustrated in the next two examples.

EXAMPLE 4 Evaluate $\lim_{x\to\infty} \frac{x^3}{e^{2x}}$.

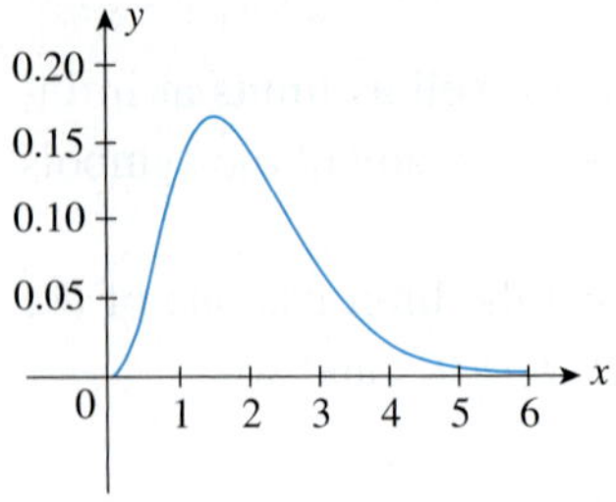

FIGURE 4
The graph of $y = x^3/e^{2x}$ shows that $y \to 0$ as $x \to \infty$.

Solution Applying l'Hôpital's rule (three times), we obtain

$$\begin{aligned}\lim_{x\to\infty} \frac{x^3}{e^{2x}} &= \lim_{x\to\infty} \frac{3x^2}{2e^{2x}} && \text{Type: } \infty/\infty \\ &= \lim_{x\to\infty} \frac{6x}{4e^{2x}} && \text{Type: } \infty/\infty \\ &= \lim_{x\to\infty} \frac{6}{8e^{2x}} = 0\end{aligned}$$

(See Figure 4.) ■

EXAMPLE 5 Evaluate $\lim_{x\to 0} \frac{x^3}{x - \tan x}$.

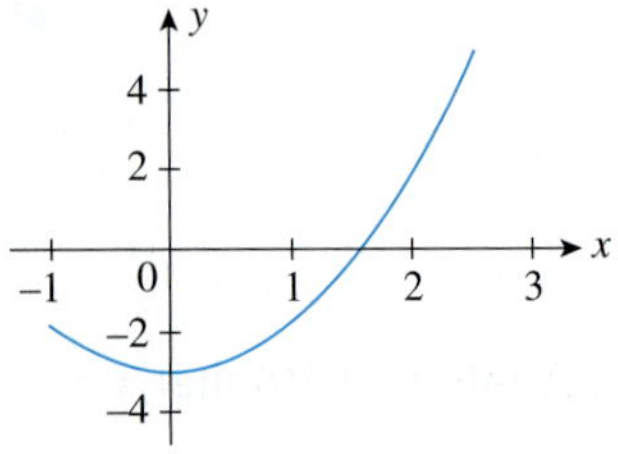

FIGURE 5
The graph of $y = x^3/(x - \tan x)$ shows that $y \to -3$ as $x \to 0$. Note that y is not defined at $x = 0$.

Solution We have an indeterminate form of the type 0/0. Using l'Hôpital's rule repeatedly, we obtain

$$\begin{aligned}\lim_{x\to 0} \frac{x^3}{x - \tan x} &= \lim_{x\to 0} \frac{3x^2}{1 - \sec^2 x} && \text{Type: } 0/0 \\ &= \lim_{x\to 0} \frac{6x}{-2\sec^2 x \tan x} && \text{Type: } 0/0 \\ &= \lim_{x\to 0} \frac{6}{-4\sec^2 x \tan^2 x - 2\sec^4 x} = \frac{6}{-2} = -3\end{aligned}$$

(See Figure 5.) ■

Exercises

In Exercises 1–26, evaluate the limit using l'Hôpital's rule if appropriate.

1. $\lim_{x\to 1} \frac{x - 1}{x^2 - 1}$
2. $\lim_{x\to -1} \frac{x^2 - 2x - 3}{x + 1}$
3. $\lim_{x\to 2} \frac{x^3 - 8}{x - 2}$
4. $\lim_{x\to 1} \frac{x^7 - 1}{x^4 - 1}$
5. $\lim_{x\to 0} \frac{e^x - 1}{x^2 + x}$
6. $\lim_{x\to 1} \frac{\ln x}{x - 1}$
7. $\lim_{t\to \pi} \frac{\sin t}{\pi - t}$
8. $\lim_{x\to 0} \frac{e^x - 1}{x + \sin x}$
9. $\lim_{\theta\to 0} \frac{\tan 2\theta}{\theta}$
10. $\lim_{x\to 0} \frac{\sin 2x}{x}$
11. $\lim_{x\to \infty} \frac{x + \cos x}{2x + 1}$
12. $\lim_{\theta\to 0} \frac{\theta + \sin\theta}{\tan\theta}$
13. $\lim_{x\to 0} \frac{\sin x - x\cos x}{\tan^3 x}$
14. $\lim_{u\to \pi} \frac{2\sin^2 u}{1 + \cos u}$
15. $\lim_{x\to \infty} \frac{\sqrt{x}}{\ln x}$
16. $\lim_{x\to \infty} \frac{e^x}{x^4}$
17. $\lim_{x\to \infty} \frac{(\ln x)^3}{x^2}$
18. $\lim_{x\to 1} \frac{x^{1/2} - x^{1/3}}{x - 1}$
19. $\lim_{x\to \infty} \frac{\ln(1 + e^x)}{x^2}$
20. $\lim_{x\to 1} \frac{a^{\ln x} - x}{\ln x}$
21. $\lim_{x\to -1} \frac{\sqrt{x + 2} + x}{\sqrt[3]{2x + 1} + 1}$
22. $\lim_{x\to 0} \frac{\ln(x^2 + 1)}{\cos x - 1}$
23. $\lim_{x\to 0^+} \frac{e^{x^2} + x - 1}{1 - \sqrt{1 - x^2}}$
24. $\lim_{x\to 0} \frac{\ln(1 + x) - \tan x}{x^2}$
25. $\lim_{x\to 0} \frac{\sin x - x}{e^x - e^{-x} - 2x}$
26. $\lim_{x\to 0} \frac{e^{x^2} - 1}{1 - \cos x}$

If $\lim_{x\to\infty} f(x) = \infty$ and $\lim_{x\to\infty} g(x) = \infty$, then $\lim_{x\to a} [f(x) - g(x)]$ is said to be an *indeterminate form of the type* $\infty - \infty$. An indeterminate form of this type can be expressed as one of the type 0/0 or ∞/∞ by algebraic manipulation. In Exercises 27–30, use this observation to evaluate the limits.

27. $\lim_{x\to 0^+} \left(\frac{1}{x} - \frac{1}{e^x - 1}\right)$

28. $\lim_{x\to 0^+} \left(\frac{1}{x} - \frac{1}{1 - \cos x}\right)$

29. $\lim_{t\to \pi/2} (\tan t - \sec t)$

30. $\lim_{x\to 1} \left(\frac{1}{\ln x} - \frac{1}{x - 1}\right)$

31. Evaluate $\lim_{x\to\infty} \left(1 + \frac{1}{x}\right)^x$.

(*Note:* This is the definition of the number e used in Section 5.1.)

32. Evaluate $\lim_{m\to\infty} P\left(1 + \frac{r}{m}\right)^{mt}$.

(*Note:* This gives the continuous compound interest formula of Section 5.3.)

In Exercises 33–36, determine whether the statement is true or false. If it is true, explain why it is true. If it is false, explain why, or give an example to show why it is false.

33. $\lim_{x\to 0} \frac{x}{e^x} = \lim_{x\to 0} \frac{1}{e^x} = 1$

34. Suppose f and g satisfy the hypothesis for l'Hôpital's rule. If $\lim_{x\to a} f(x) = \lim_{x\to a} g(x) = 0$, then

$$\lim_{x\to a} \frac{f(x)}{g(x)} = \lim_{x\to a} \frac{d}{dx}\left[\frac{f(x)}{g(x)}\right]$$

35. Suppose f and g satisfy the hypothesis for l'Hôpital's rule. If $\lim_{x\to a} f(x) = \lim_{x\to a} g(x) = 0$, then $\lim_{x\to a} \frac{f(x)}{g(x)}$ does not exist.

36. Suppose f and g satisfy the hypothesis for l'Hôpital's rule. If $\lim_{x\to a} f(x) = \lim_{x\to a} g(x) = 0$, $\lim_{x\to a} f'(x) = L$, and $\lim_{x\to a} g'(x) = M \neq 0$, then $\lim_{x\to a} \frac{f(x)}{g(x)} = \frac{L}{M}$.

C.1 The Standard Normal Distribution

TABLE C.1

The Standard Normal Distribution

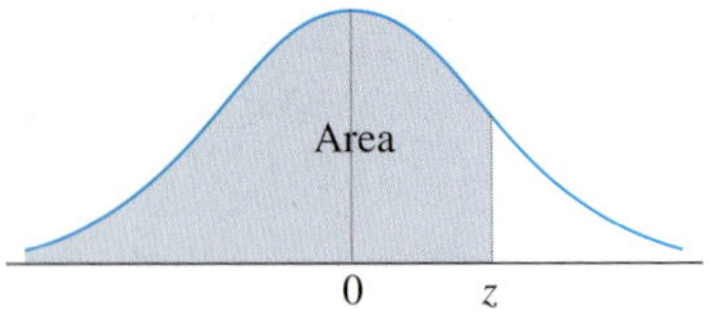

$F_z(z) = P(Z \leq z)$

z	0.00	0.01	0.02	0.03	0.04	0.05	0.06	0.07	0.08	0.09
−3.4	0.0003	0.0003	0.0003	0.0003	0.0003	0.0003	0.0003	0.0003	0.0003	0.0002
−3.3	0.0005	0.0005	0.0005	0.0004	0.0004	0.0004	0.0004	0.0004	0.0004	0.0003
−3.2	0.0007	0.0007	0.0006	0.0006	0.0006	0.0006	0.0006	0.0005	0.0005	0.0005
−3.1	0.0010	0.0009	0.0009	0.0009	0.0008	0.0008	0.0008	0.0008	0.0007	0.0007
−3.0	0.0013	0.0013	0.0013	0.0012	0.0012	0.0011	0.0011	0.0011	0.0010	0.0010
−2.9	0.0019	0.0018	0.0017	0.0017	0.0016	0.0016	0.0015	0.0015	0.0014	0.0014
−2.8	0.0026	0.0025	0.0024	0.0023	0.0023	0.0022	0.0021	0.0021	0.0020	0.0019
−2.7	0.0035	0.0034	0.0033	0.0032	0.0031	0.0030	0.0029	0.0028	0.0027	0.0026
−2.6	0.0047	0.0045	0.0044	0.0043	0.0041	0.0040	0.0039	0.0038	0.0037	0.0036
−2.5	0.0062	0.0060	0.0059	0.0057	0.0055	0.0054	0.0052	0.0051	0.0049	0.0048
−2.4	0.0082	0.0080	0.0078	0.0075	0.0073	0.0071	0.0069	0.0068	0.0066	0.0064
−2.3	0.0107	0.0104	0.0102	0.0099	0.0096	0.0094	0.0091	0.0089	0.0087	0.0084
−2.2	0.0139	0.0136	0.0132	0.0129	0.0125	0.0122	0.0119	0.0116	0.0113	0.0110
−2.1	0.0179	0.0174	0.0170	0.0166	0.0162	0.0158	0.0154	0.0150	0.0146	0.0143
−2.0	0.0228	0.0222	0.0217	0.0212	0.0207	0.0202	0.0197	0.0192	0.0188	0.0183
−1.9	0.0287	0.0281	0.0274	0.0268	0.0262	0.0256	0.0250	0.0244	0.0239	0.0233
−1.8	0.0359	0.0352	0.0344	0.0336	0.0329	0.0322	0.0314	0.0307	0.0301	0.0294
−1.7	0.0446	0.0436	0.0427	0.0418	0.0409	0.0401	0.0392	0.0384	0.0375	0.0367
−1.6	0.0548	0.0537	0.0526	0.0516	0.0505	0.0495	0.0485	0.0475	0.0465	0.0455
−1.5	0.0668	0.0655	0.0643	0.0630	0.0618	0.0606	0.0594	0.0582	0.0571	0.0559
−1.4	0.0808	0.0793	0.0778	0.0764	0.0749	0.0735	0.0722	0.0708	0.0694	0.0681
−1.3	0.0968	0.0951	0.0934	0.0918	0.0901	0.0885	0.0869	0.0853	0.0838	0.0823
−1.2	0.1151	0.1131	0.1112	0.1093	0.1075	0.1056	0.1038	0.1020	0.1003	0.0985
−1.1	0.1357	0.1335	0.1314	0.1292	0.1271	0.1251	0.1230	0.1210	0.1190	0.1170
−1.0	0.1587	0.1562	0.1539	0.1515	0.1492	0.1469	0.1446	0.1423	0.1401	0.1379
−0.9	0.1841	0.1814	0.1788	0.1762	0.1736	0.1711	0.1685	0.1660	0.1635	0.1611
−0.8	0.2119	0.2090	0.2061	0.2033	0.2005	0.1977	0.1949	0.1922	0.1894	0.1867
−0.7	0.2420	0.2389	0.2358	0.2327	0.2296	0.2266	0.2236	0.2206	0.2177	0.2148
−0.6	0.2743	0.2709	0.2676	0.2643	0.2611	0.2578	0.2546	0.2514	0.2483	0.2451
−0.5	0.3085	0.3050	0.3015	0.2981	0.2946	0.2912	0.2877	0.2843	0.2810	0.2776
−0.4	0.3446	0.3409	0.3372	0.3336	0.3300	0.3264	0.3228	0.3192	0.3156	0.3121
−0.3	0.3821	0.3783	0.3745	0.3707	0.3669	0.3632	0.3594	0.3557	0.3520	0.3483
−0.2	0.4207	0.4168	0.4129	0.4090	0.4052	0.4013	0.3974	0.3936	0.3897	0.3859
−0.1	0.4602	0.4562	0.4522	0.4483	0.4443	0.4404	0.4364	0.4325	0.4286	0.4247
−0.0	0.5000	0.4960	0.4920	0.4880	0.4840	0.4801	0.4761	0.4721	0.4681	0.4641

TABLE C.1

The Standard Normal Distribution (*continued*)

$F_z(z) = P(Z \leq z)$

z	**0.00**	**0.01**	**0.02**	**0.03**	**0.04**	**0.05**	**0.06**	**0.07**	**0.08**	**0.09**
0.0	0.5000	0.5040	0.5080	0.5120	0.5160	0.5199	0.5239	0.5279	0.5319	0.5359
0.1	0.5398	0.5438	0.5478	0.5517	0.5557	0.5596	0.5636	0.5675	0.5714	0.5753
0.2	0.5793	0.5832	0.5871	0.5910	0.5948	0.5987	0.6026	0.6064	0.6103	0.6141
0.3	0.6179	0.6217	0.6255	0.6293	0.6331	0.6368	0.6406	0.6443	0.6480	0.6517
0.4	0.6554	0.6591	0.6628	0.6664	0.6700	0.6736	0.6772	0.6808	0.6844	0.6879
0.5	0.6915	0.6950	0.6985	0.7019	0.7054	0.7088	0.7123	0.7157	0.7190	0.7224
0.6	0.7257	0.7291	0.7324	0.7357	0.7389	0.7422	0.7454	0.7486	0.7517	0.7549
0.7	0.7580	0.7611	0.7642	0.7673	0.7704	0.7734	0.7764	0.7794	0.7823	0.7852
0.8	0.7881	0.7910	0.7939	0.7967	0.7995	0.8023	0.8051	0.8078	0.8106	0.8133
0.9	0.8159	0.8186	0.8212	0.8238	0.8264	0.8289	0.8315	0.8340	0.8365	0.8389
1.0	0.8413	0.8438	0.8461	0.8485	0.8508	0.8531	0.8554	0.8577	0.8599	0.8621
1.1	0.8643	0.8665	0.8686	0.8708	0.8729	0.8749	0.8770	0.8790	0.8810	0.8830
1.2	0.8849	0.8869	0.8888	0.8907	0.8925	0.8944	0.8962	0.8980	0.8997	0.9015
1.3	0.9032	0.9049	0.9066	0.9082	0.9099	0.9115	0.9131	0.9147	0.9162	0.9177
1.4	0.9192	0.9207	0.9222	0.9236	0.9251	0.9265	0.9278	0.9292	0.9306	0.9319
1.5	0.9332	0.9345	0.9357	0.9370	0.9382	0.9394	0.9406	0.9418	0.9429	0.9441
1.6	0.9452	0.9463	0.9474	0.9484	0.9495	0.9505	0.9515	0.9525	0.9535	0.9545
1.7	0.9554	0.9564	0.9573	0.9582	0.9591	0.9599	0.9608	0.9616	0.9625	0.9633
1.8	0.9641	0.9649	0.9656	0.9664	0.9671	0.9678	0.9686	0.9693	0.9699	0.9706
1.9	0.9713	0.9719	0.9726	0.9732	0.9738	0.9744	0.9750	0.9756	0.9761	0.9767
2.0	0.9772	0.9778	0.9783	0.9788	0.9793	0.9798	0.9803	0.9808	0.9812	0.9817
2.1	0.9821	0.9826	0.9830	0.9834	0.9838	0.9842	0.9846	0.9850	0.9854	0.9857
2.2	0.9861	0.9864	0.9868	0.9871	0.9875	0.9878	0.9881	0.9884	0.9887	0.9890
2.3	0.9893	0.9896	0.9898	0.9901	0.9904	0.9906	0.9909	0.9911	0.9913	0.9916
2.4	0.9918	0.9920	0.9922	0.9925	0.9927	0.9929	0.9931	0.9932	0.9934	0.9936
2.5	0.9938	0.9940	0.9951	0.9943	0.9945	0.9946	0.9948	0.9949	0.9951	0.9952
2.6	0.9953	0.9955	0.9956	0.9957	0.9959	0.9960	0.9961	0.9962	0.9963	0.9964
2.7	0.9965	0.9966	0.9967	0.9968	0.9969	0.9970	0.9971	0.9972	0.9973	0.9974
2.8	0.9974	0.9975	0.9976	0.9977	0.9977	0.9978	0.9979	0.9979	0.9980	0.9981
2.9	0.9981	0.9982	0.9982	0.9983	0.9984	0.9984	0.9985	0.9985	0.9986	0.9986
3.0	0.9987	0.9987	0.9987	0.9988	0.9988	0.9989	0.9989	0.9989	0.9990	0.9990
3.1	0.9990	0.9991	0.9991	0.9991	0.9992	0.9992	0.9992	0.9992	0.9993	0.9993
3.2	0.9993	0.9993	0.9994	0.9994	0.9994	0.9994	0.9994	0.9995	0.9995	0.9995
3.3	0.9995	0.9995	0.9995	0.9996	0.9996	0.9996	0.9996	0.9996	0.9996	0.9997
3.4	0.9997	0.9997	0.9997	0.9997	0.9997	0.9997	0.9997	0.9997	0.9997	0.9998

Answers to Odd-Numbered Exercises

CHAPTER 1

Exercises 1.1, page 13

1. 

3. 

5. 

7. 9 **9.** 1 **11.** 4 **13.** 7 **15.** $\frac{1}{5}$ **17.** 2

19. 2 **21.** 1 **23.** True **25.** False **27.** False

29. False **31.** False **33.** $\dfrac{1}{(xy)^2}$ **35.** $\dfrac{1}{x^{5/6}}$ **37.** $\dfrac{1}{(s+t)^3}$

39. $x^{13/3}$ **41.** $\dfrac{1}{x^3}$ **43.** x **45.** $\dfrac{9}{x^2y^4}$ **47.** $\dfrac{y^8}{x^{10}}$

49. $2x^{11/6}$ **51.** $-2xy^2$ **53.** $2x^{4/3}y^{1/2}$ **55.** 2.828

57. 5.196 **59.** 31.62 **61.** 316.2 **63.** $\dfrac{3\sqrt{x}}{2x}$

65. $\dfrac{2\sqrt{3y}}{3}$ **67.** $\dfrac{\sqrt[3]{x^2}}{x}$ **69.** $\dfrac{2x}{3\sqrt{x}}$ **71.** $\dfrac{2y}{\sqrt{2xy}}$

73. $\dfrac{xz}{y\sqrt[3]{xz^2}}$ **75.** $9x^2+3x+1$ **77.** $4y^2+y+8$

79. $-x-1$ **81.** $\frac{2}{3}+e-e^{-1}$ **83.** $6\sqrt{2}+8+\frac{1}{2}\sqrt{x}-\frac{11}{4}\sqrt{y}$

85. $x^2+6x-16$ **87.** $a^2+10a+25$ **89.** $x^2+4xy+4y^2$

91. $4x^2-y^2$ **93.** $-2x$ **95.** $2t(2\sqrt{t}+1)$

97. $2x^3(2x^2-6x-3)$ **99.** $7a^2(a^2+7ab-6b^2)$

101. $e^{-x}(1-x)$ **103.** $\frac{1}{2}x^{-5/2}(4-3x)$ **105.** $(2a+b)(3c-2d)$

107. $(2a+b)(2a-b)$ **109.** $-2(3x+5)(2x-1)$

111. $3(x-4)(x+2)$ **113.** $2(3x-5)(2x+3)$

115. $(3x-4y)(3x+4y)$ **117.** $(x^2+5)(x^4-5x^2+25)$

119. x^3-xy^2 **121.** $4(x-1)(3x-1)(2x+2)^3$

123. $4(x-1)(3x-1)(2x+2)^3$

125. $2x(x^2+2)^2(5x^4+20x^2+17)$

127. -4 and 3 **129.** -1 and $\frac{1}{2}$ **131.** 2 and 2

133. -2 and $\frac{3}{4}$ **135.** $\frac{1}{2}+\frac{1}{4}\sqrt{10}$ and $\frac{1}{2}-\frac{1}{4}\sqrt{10}$

137. $-1+\frac{1}{2}\sqrt{10}$ and $-1-\frac{1}{2}\sqrt{10}$

139. **a.** 53,886 **b.** 19,052 **c.** 4820 **141.** True

Exercises 1.2, page 23

1. $\dfrac{x-1}{x-2}$ **3.** $\dfrac{3(2t+1)}{2t-1}$ **5.** $-\dfrac{7}{(4x-1)^2}$ **7.** -8

9. $\dfrac{3x-1}{2}$ **11.** $\dfrac{t+20}{3t+2}$ **13.** $-\dfrac{x(2x-13)}{(2x-1)(2x+5)}$

15. $-\dfrac{x+27}{(x-3)^2(x+3)}$ **17.** $\dfrac{x+1}{x-1}$ **19.** $\dfrac{4x^2+7}{\sqrt{2x^2+7}}$

21. $\dfrac{x-1}{x^2\sqrt{x+1}}$ **23.** $\dfrac{x-1}{(2x+1)^{3/2}}$ **25.** $\dfrac{\sqrt{3}+1}{2}$

27. $\dfrac{\sqrt{x}+\sqrt{y}}{x-y}$ **29.** $\dfrac{(\sqrt{a}+\sqrt{b})^2}{a-b}$ **31.** $\dfrac{x}{3\sqrt{x}}$

33. $-\dfrac{2}{3(1+\sqrt{3})}$ **35.** $-\dfrac{x+1}{\sqrt{x+2}(1-\sqrt{x+2})}$

37. False **39.** False **41.** $(-\infty, 2)$ **43.** $(-\infty, -5]$

45. $(-4, 6)$ **47.** $(-\infty, -3) \cup (3, \infty)$ **49.** $(-2, 3)$

51. $[-3, 5]$ **53.** $(-\infty, 1] \cup [\frac{3}{2}, \infty)$ **55.** $(-\infty, -3] \cup (2, \infty)$

57. $(-\infty, 0] \cup (1, \infty)$ **59.** 4 **61.** 2 **63.** $5\sqrt{3}$ **65.** $\pi+1$

67. 2 **69.** False **71.** False **73.** True **75.** False

77. True **79.** False **81.** [362, 488.7] **83.** \$12,300

85. \$32,000 **87.** $|x-0.5| \leq 0.01$

89. Between 1000 and 4000 units

91. Between 98.04 and 98.36% of the toxic pollutants

93. Between 10:18 a.m. and 12:42 a.m. **95.** False **97.** True

Exercises 1.3, page 30

1. (3, 3); Quadrant I **3.** (2, −2); Quadrant IV

5. (−4, −6); Quadrant III **7.** *A* **9.** *E*, *F*, and *G*

11. *F* **13–19.** See the accompanying figure.

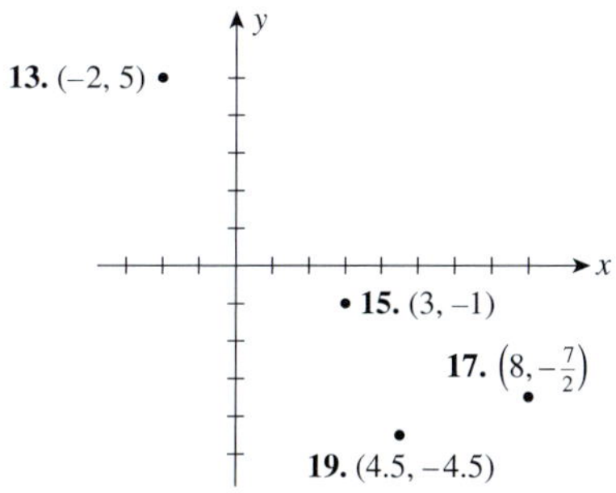

21. 5 **23.** $\sqrt{61}$ **25.** $(-8, -6)$ and $(8, -6)$

29. $(x - 2)^2 + (y + 3)^2 = 25$ **31.** $x^2 + y^2 = 25$

33. $(x - 2)^2 + (y + 3)^2 = 34$ **35.** 3400 mi **37.** Route 1

39. Model C **41.** $10\sqrt{13}t$; 72.1 mi

43. a. $\sqrt{16{,}000{,}000 + x^2}$ **b.** 20,396 ft **45.** True

47. True **49. b.** $(\frac{1}{2}, -\frac{3}{2})$

Exercises 1.4, page 41

1. e **3.** a **5.** f **7.** $\frac{1}{2}$ **9.** Not defined **11.** 5

13. $\frac{5}{6}$ **15.** $\dfrac{d - b}{c - a}$ $(a \neq c)$ **17. a.** 4 **b.** -8

19. Parallel **21.** Perpendicular **23.** -5 **25.** $y = -3$

27. $y = 2x - 10$ **29.** $y = 2$ **31.** $y = 3x - 2$

33. $y = x + 1$ **35.** $y = 3x + 4$ **37.** $y = 5$

39. $y = \frac{1}{2}x$; $m = \frac{1}{2}$; $b = 0$ **41.** $y = \frac{2}{3}x - 3$; $m = \frac{2}{3}$; $b = -3$

43. $y = -\frac{1}{2}x + \frac{7}{2}$; $m = -\frac{1}{2}$; $b = \frac{7}{2}$ **45.** $y = \frac{1}{2}x + 3$

47. $y = -6$ **49.** $y = b$ **51.** $y = \frac{2}{3}x - \frac{2}{3}$ **53.** $k = 8$

55.

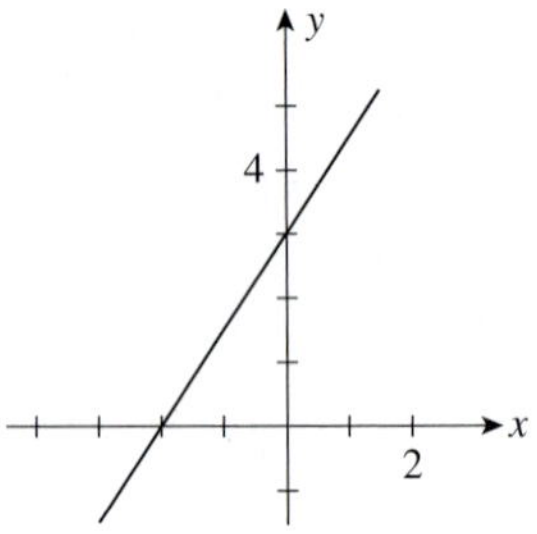

57.

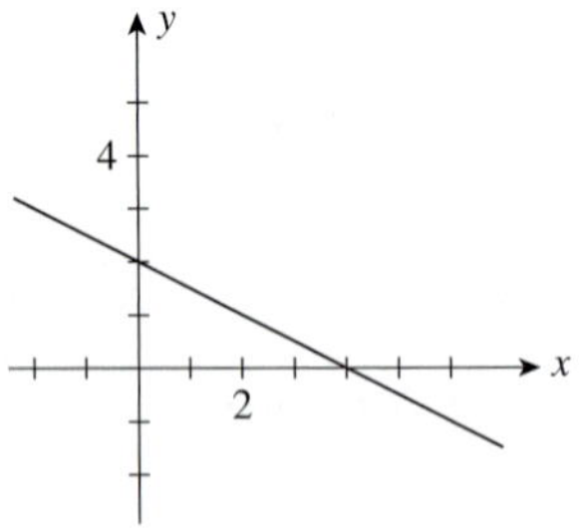

59.

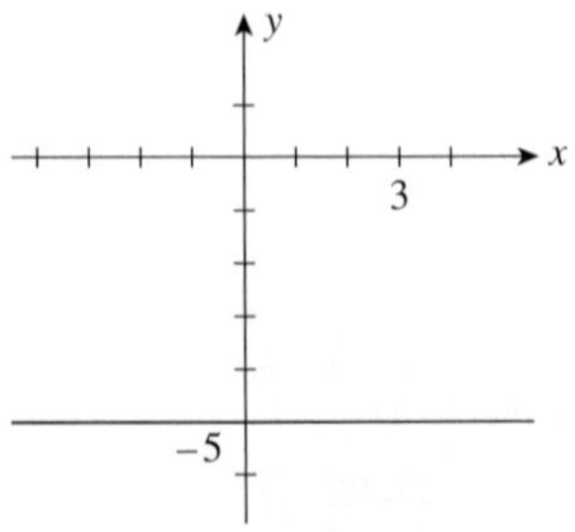

63. $y = -2x - 4$ **65.** $y = \frac{1}{8}x - \frac{1}{2}$ **67.** Yes

69. a.

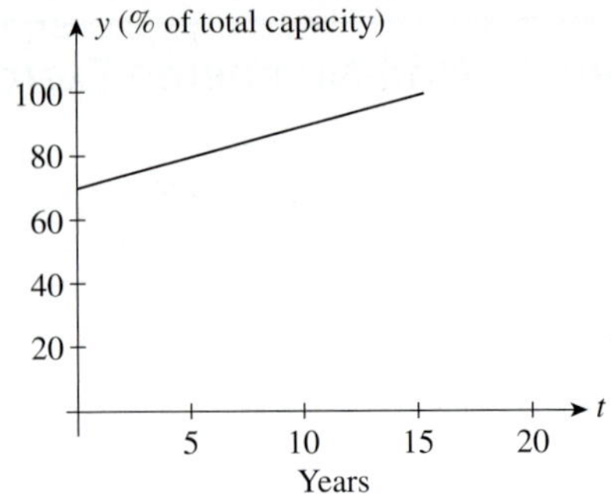

b. 1.9467; 70.082
c. The capacity utilization has been increasing by 1.9467% each year since 1990 when it stood at 70.082%.
d. Shortly after 2005

71. a. $y = 0.55x$ **b.** 2000 **73.** 89.6% of men's wages

75. a. and **b.**

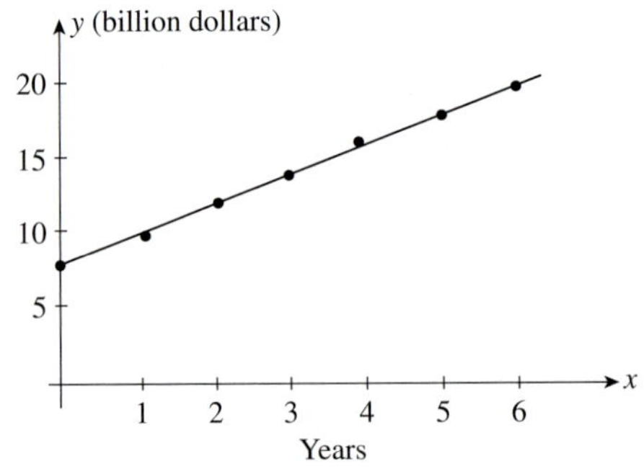

c. $y = 1.82x + 7.9$ **d.** \$17 billion; same

77. True **79.** False **81.** True

Chapter 1 Concept Review, page 46

1. Ordered; abscissa (x-coordinate); ordinate (y-coordinate)

2. a. x-; y-; **b.** Third **3.** $\sqrt{(c - a)^2 + (d - b)^2}$

4. $(x - a)^2 + (y - b)^2 = r^2$

5. a. $\dfrac{y_2 - y_1}{x_2 - x_1}$ **b.** Undefined **c.** 0 **d.** Positive

6. $m_1 = m_2$; $m_1 = -\dfrac{1}{m_2}$

7. a. $y - y_1 = m(x - x_1)$; point-slope form
b. $y = mx + b$; slope-intercept

8. a. $Ax + By + C = 0$ (A, B, not both zero) **b.** $-\dfrac{a}{b}$

Chapter 1 Review Exercises, page 47

1. $[-2, \infty)$ **2.** $[-1, 2]$ **3.** $(-\infty, -4) \cup (5, \infty)$

4. $(-\infty, -5) \cup (5, \infty)$ **5.** 4 **6.** 1 **7.** $\pi - 6$

8. $8 - 3\sqrt{3}$ **9.** $\frac{27}{8}$ **10.** 25 **11.** $\frac{1}{144}$ **12.** -32

13. $\frac{1}{4}$ **14.** $3\sqrt[3]{3}$ **15.** $4(x^2 + y)^2$ **16.** $\dfrac{a^{15}}{b^{11}}$ **17.** $\dfrac{2x}{3z}$

18. $-x^{1/2}$ **19.** $6xy^7$ **20.** $9x^2y^4$ **21.** $-2\pi r^2(\pi r - 50)$

22. $2vw(v^2 + w^2 + u^2)$ **23.** $(4 - x)(4 + x)$

24. $6t(2t - 3)(t + 1)$ **25.** $-\frac{3}{4}$ and $\frac{1}{2}$ **26.** -2 and $\frac{1}{3}$

27. $0, -3, 1$ **28.** $\dfrac{\sqrt{2}}{2}$ and $-\dfrac{\sqrt{2}}{2}$ **29.** $[-2, \frac{1}{2}]$ **30.** $(-2, -\frac{3}{2})$

31. $(-1, 4)$ **32.** $\frac{3}{2}; \frac{2}{3}$ **33.** $1 + \sqrt{6}, 1 - \sqrt{6}$

34. $-2 + \dfrac{\sqrt{2}}{2}; -2 - \dfrac{\sqrt{2}}{2}$ **35.** $\dfrac{180}{(t + 6)^2}$

36. $\dfrac{15x^2 + 24x + 2}{4(x + 2)(3x^2 + 2)}$ **37.** $\dfrac{78x^2 - 8x - 27}{3(2x^2 - 1)(3x - 1)}$

38. $\dfrac{2(x + 2)}{\sqrt{x + 1}}$ **39.** $\dfrac{1}{\sqrt{x} + 1}$ **40.** $\dfrac{x - \sqrt{x}}{2x}$ **41.** 5

42. 2 **43.** $x = -2$ **44.** $y = 4$ **45.** $y = -\frac{1}{10}x + \frac{19}{5}$

46. $y = -\frac{4}{5}x + \frac{12}{5}$ **47.** $y = \frac{5}{2}x + 9$ **48.** $y = \frac{3}{4}x + \frac{11}{2}$

49. $y = -\frac{3}{4}x + \frac{9}{2}$ **50.** $y = -\frac{3}{5}x + \frac{12}{5}$ **51.** $y = 3x + 7$

52. $y = -\frac{3}{2}x - 7$

53.

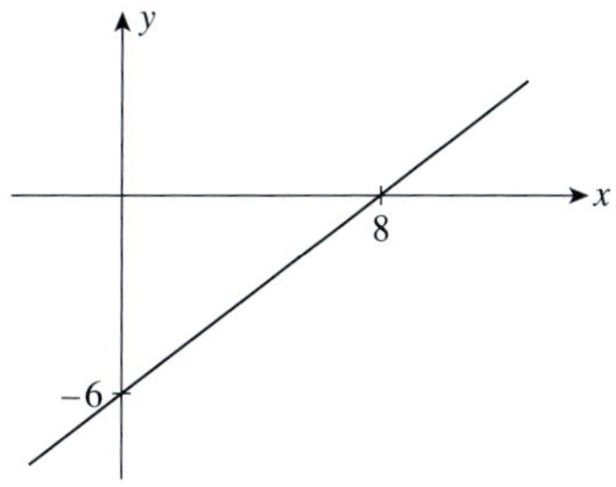

54.

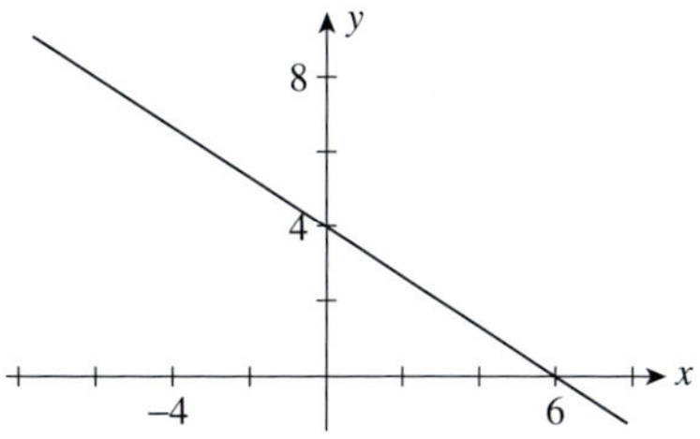

55. \$100 **56.** \$400 **57.** Between 1 sec and 3 sec

58. **a.** and **b.**

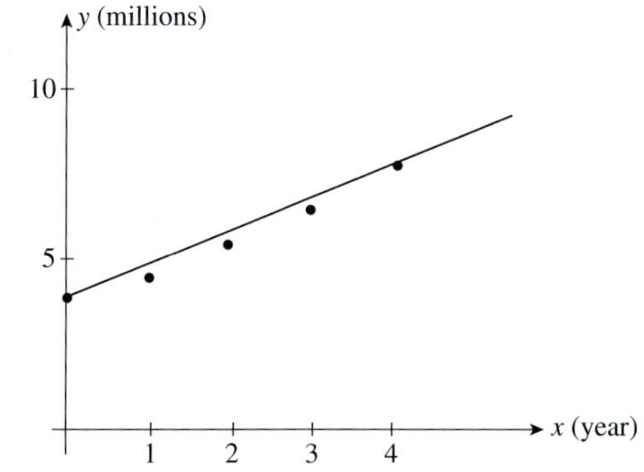

c. $y = 0.975x + 3.9$ **d.** 6,825,000; differ by 25,000

Chapter 1 Before Moving On, page 48

1. **a.** $\sqrt{3} + \sqrt{2} - \pi$ **b.** -3 **2.** **a.** $12x^5y$ **b.** $\dfrac{b^5}{a^3}$

3. **a.** $\dfrac{2x\sqrt{y}}{3y}$ **b.** $\dfrac{x(\sqrt{x} + 4)}{x - 16}$

4. **a.** $\dfrac{1 - 3x^2}{2\sqrt{x}(x^2 + 1)^2}$ **b.** $\dfrac{6\sqrt{x + 2}}{x + 2}$ **5.** $\dfrac{x - y}{(\sqrt{x} - \sqrt{y})^2}$

6. **a.** $2x(3x + 2)(2x - 3)$ **b.** $(2b + 3c)(x - y)$

7. **a.** $x = -\frac{1}{4}$, or 1 **b.** $\dfrac{5 \pm \sqrt{13}}{6}$ **8.** $4\sqrt{5}$

9. $y = \frac{7}{5}x - \frac{3}{5}$ **10.** $y = -\frac{1}{3}x + \frac{4}{3}$

CHAPTER 2

Exercises 2.1, page 57

1. $21, -9, 5a + 6, -5a + 6, 5a + 21$

3. $-3, 6, 3a^2 - 6a - 3, 3a^2 + 6a - 3, 3x^2 - 6$

5. $2a + 2h + 5, -2a + 5, 2a^2 + 5, 2a - 4h + 5, 4a - 2h + 5$

7. $\dfrac{8}{15}, 0, \dfrac{2a}{a^2 - 1}, \dfrac{2(2 + a)}{a^2 + 4a + 3}, \dfrac{2(t + 1)}{t(t + 2)}$

9. $8, \dfrac{2a^2}{\sqrt{a - 1}}, \dfrac{2(x + 1)^2}{\sqrt{x}}, \dfrac{2(x - 1)^2}{\sqrt{x - 2}}$ **11.** 5, 1, 1 **13.** $\frac{5}{2}$, 3, 3, 9

15. **a.** -2 **b.** (i) $x = 2$; (ii) $x = 1$ **c.** $[0, 6]$ **d.** $[-2, 6]$

17. Yes **19.** Yes **21.** 7 **23.** $(-\infty, \infty)$ **25.** $(-\infty, 0) \cup (0, \infty)$

27. $(-\infty, \infty)$ **29.** $(-\infty, 5]$ **31.** $(-\infty, -1) \cup (-1, 1) \cup (1, \infty)$

33. $[-3, \infty)$ **35.** $(-\infty, -2) \cup (-2, 1]$

37. **a.** $(-\infty, \infty)$
b. $6, 0, -4, -6, -\frac{25}{4}, -6, -4, 0$
c.

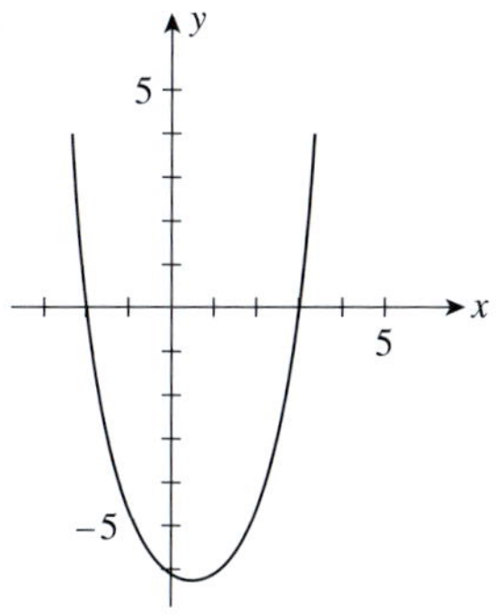

39.

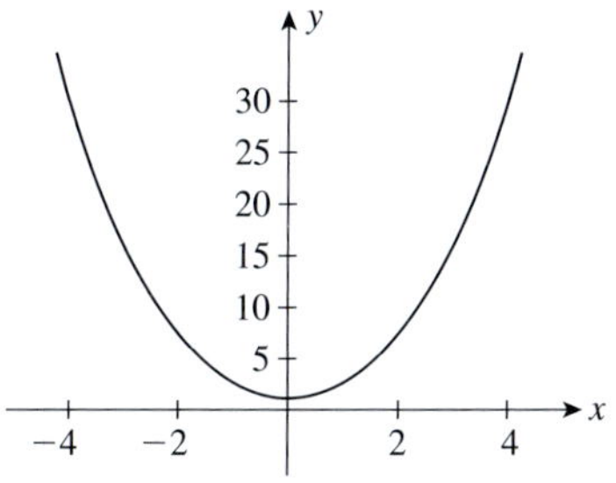

$(-\infty, \infty)$; $[1, \infty)$

41.

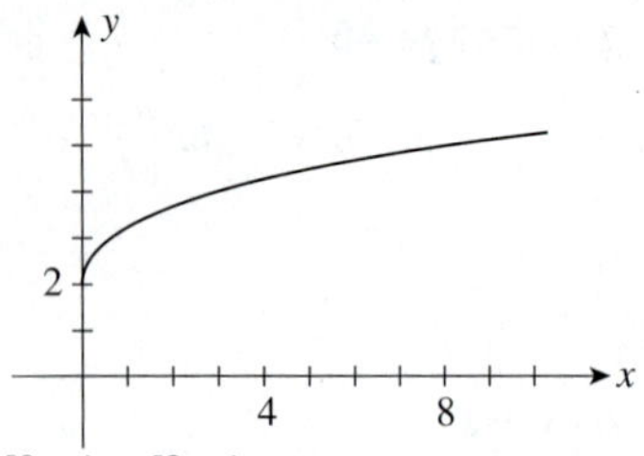

$[0, \infty)$; $[2, \infty)$

43.

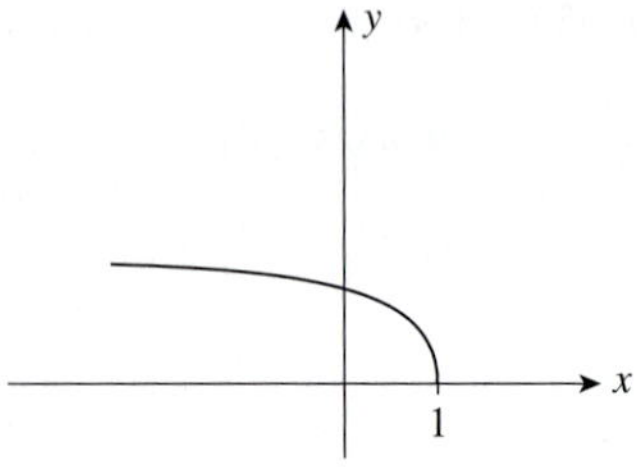

$(-\infty, 1]$; $[0, \infty)$

45.

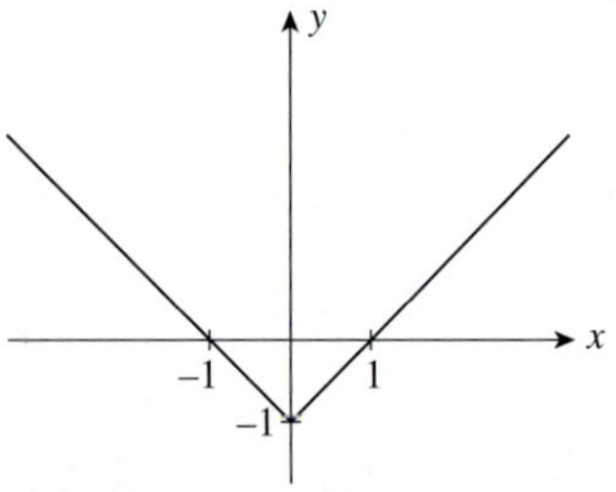

$(-\infty, \infty)$; $[-1, \infty)$

47.

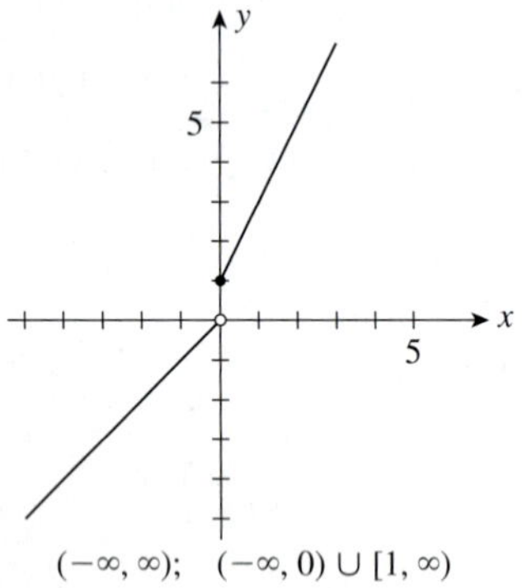

$(-\infty, \infty)$; $(-\infty, 0) \cup [1, \infty)$

49.

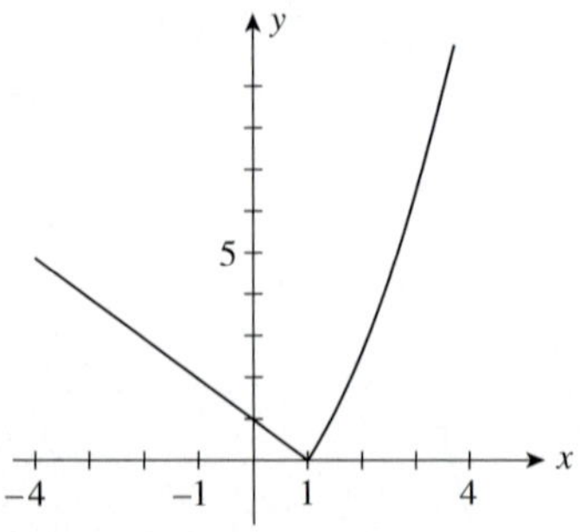

$(-\infty, \infty)$; $[0, \infty)$

51. Yes **53.** No **55.** Yes **57.** Yes

59. 10π in. **61.** 8

63. **a.** From 1985 to 1990 **b.** From 1990 on
c. 1990; \$3.5 billion

65. **a.** $f(t) = \begin{cases} 0.0185t + 0.58 & \text{if } 0 \le t \le 20 \\ 0.015t + 0.65 & \text{if } 20 < t \le 30 \end{cases}$

b. 0.0185/yr from 1960 through 1980; 0.015/yr from 1980 through 1990

c. 1983

67. **a.** $0.06x$ **b.** \$12.00; \$0.34 **69.** 160 mg

71. **a.** $f(t) = 7.5t + 20$ **b.** 65 million

73. **a.** $V = -12{,}000n + 120{,}000$

b.

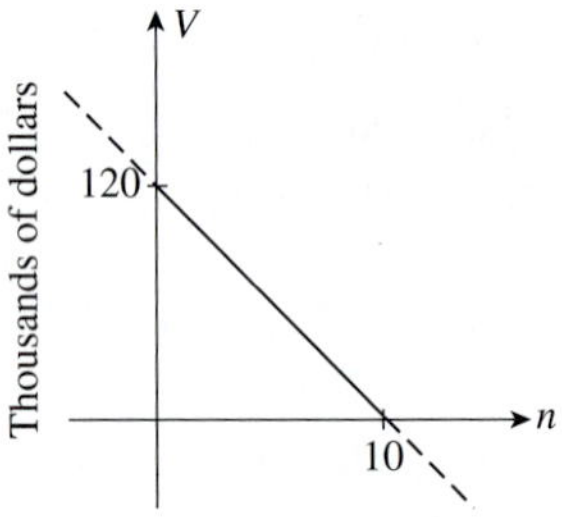

c. \$48,000 **d.** \$12,000/yr

75. $(0, \infty)$

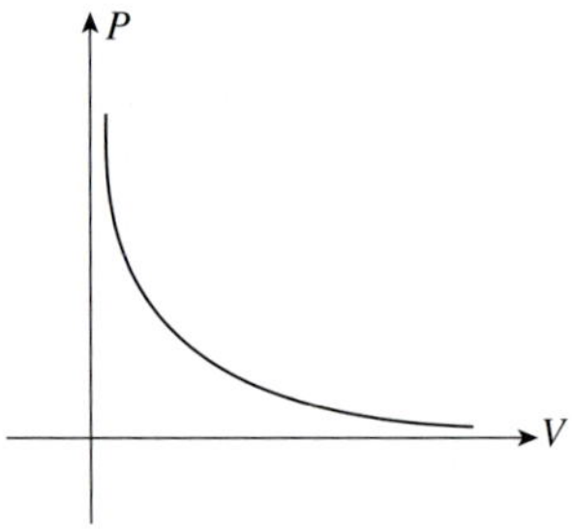

77. **a.** 3.6 million; 9.5 million **b.** 11.2 million **79.** 20; 26

81. \$5.6 billion; \$7.8 billion

83. **a.** \$0.6 trillion; \$0.6 trillion **b.** \$0.96 trillion; \$1.2 trillion

85. **a.** 130 tons/day; 100 tons/day; 40 tons/day

b.

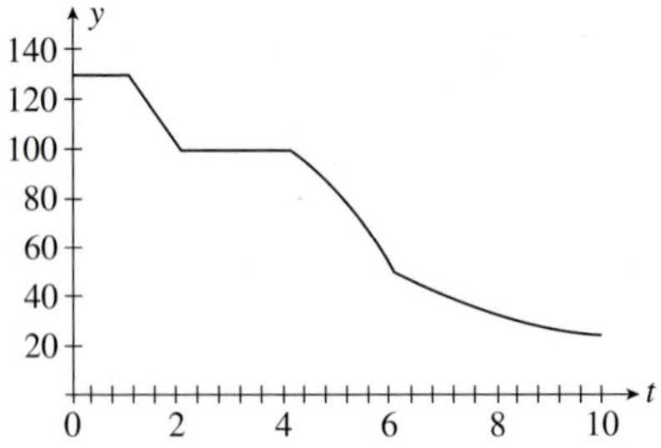

87. **a.** $D(t) = \begin{cases} 14t & \text{if } 0 \le t \le 2 \\ 2\sqrt{74t^2 - 100t + 100} & \text{if } 2 < t \end{cases}$

b. 76.16 miles

89. False **91.** False **93.** False

Using Technology Exercises 2.1, page 66

1. **a.**

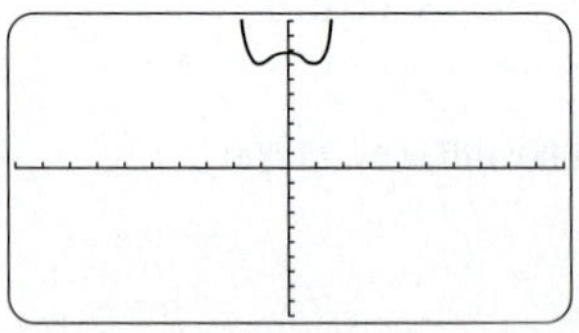

b.

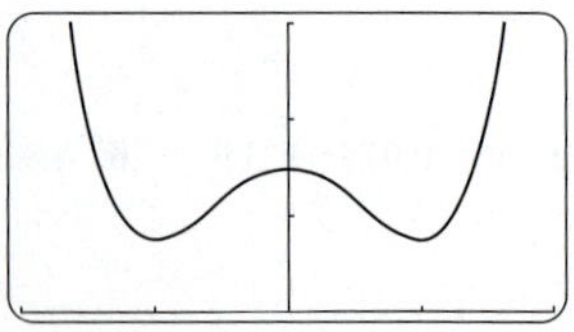

3. a.

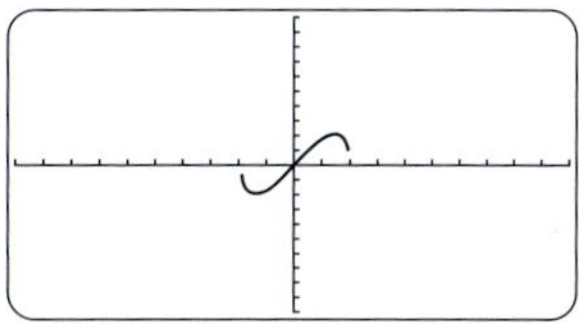

b.

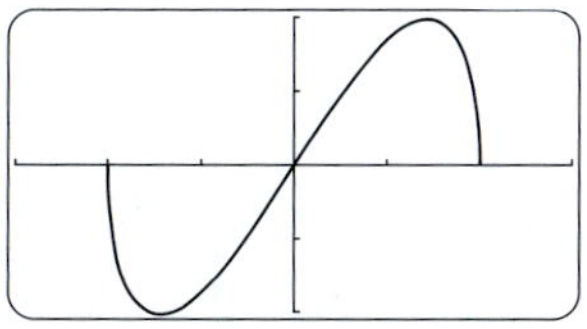

5.

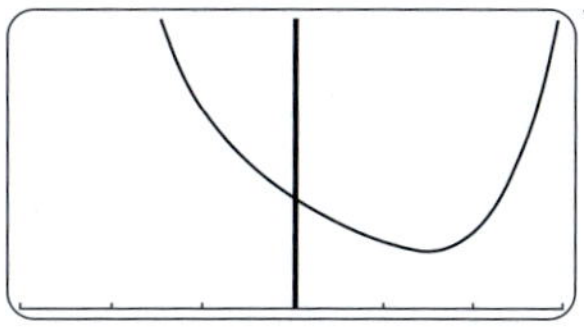

7.

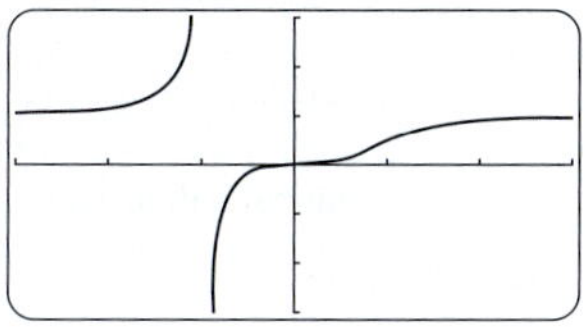

9. 18.5505 **11.** 4.1616

13. a.

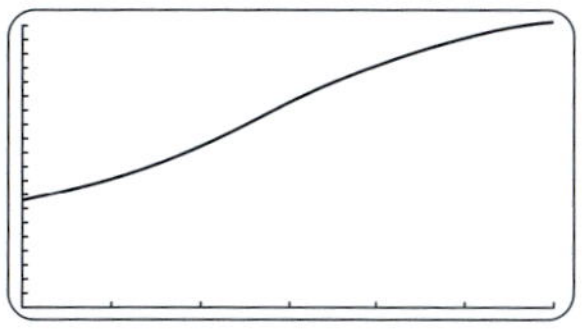

b. \$62.96 million; \$107.66 million

15. a.

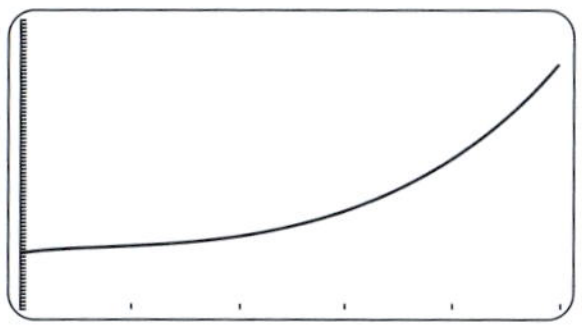

b. 44.7; 52.7; 129.2

Exercises 2.2, page 72

1. $f(x) + g(x) = x^3 + x^2 + 3$ **3.** $f(x)g(x) = x^5 - 2x^3 + 5x^2 - 10$

5. $\dfrac{f(x)}{g(x)} = \dfrac{x^3 + 5}{x^2 - 2}$ **7.** $\dfrac{f(x)g(x)}{h(x)} = \dfrac{x^5 - 2x^3 + 5x^2 - 10}{2x + 4}$

9. $f(x) + g(x) = x - 1 + \sqrt{x + 1}$

11. $f(x)g(x) = (x - 1)\sqrt{x + 1}$

13. $\dfrac{g(x)}{h(x)} = \dfrac{\sqrt{x + 1}}{2x^3 - 1}$ **15.** $\dfrac{f(x)g(x)}{h(x)} = \dfrac{(x - 1)\sqrt{x + 1}}{2x^3 - 1}$

17. $\dfrac{f(x) - h(x)}{g(x)} = \dfrac{x - 2x^3}{\sqrt{x + 1}}$

19. $f(x) + g(x) = x^2 + \sqrt{x} + 3$;
$f(x) - g(x) = x^2 - \sqrt{x} + 7$;
$f(x)g(x) = (x^2 + 5)(\sqrt{x} - 2)$; $\dfrac{f(x)}{g(x)} = \dfrac{x^2 + 5}{\sqrt{x} - 2}$

21. $f(x) + g(x) = \dfrac{(x - 1)\sqrt{x + 3} + 1}{x - 1}$;
$f(x) - g(x) = \dfrac{(x - 1)\sqrt{x + 3} - 1}{x - 1}$;
$f(x)g(x) = \dfrac{\sqrt{x + 3}}{x - 1}$; $\dfrac{f(x)}{g(x)} = (x - 1)\sqrt{x + 3}$

23. $f(x) + g(x) = \dfrac{2(x^2 - 2)}{(x - 1)(x - 2)}$;
$f(x) - g(x) = \dfrac{-2x}{(x - 1)(x - 2)}$;
$f(x)g(x) = \dfrac{(x + 1)(x + 2)}{(x - 1)(x - 2)}$; $\dfrac{f(x)}{g(x)} = \dfrac{(x + 1)(x - 2)}{(x - 1)(x + 2)}$

25. $f(g(x)) = x^4 + x^2 + 1$; $g(f(x)) = (x^2 + x + 1)^2$

27. $f(g(x)) = \sqrt{x^2 - 1} + 1$; $g(f(x)) = x + 2\sqrt{x}$

29. $f(g(x)) = \dfrac{x}{x^2 + 1}$; $g(f(x)) = \dfrac{x^2 + 1}{x}$ **31.** 49

33. $\dfrac{\sqrt{5}}{5}$ **35.** $f(x) = 2x^3 + x^2 + 1$ and $g(x) = x^5$

37. $f(x) = x^2 - 1$ and $g(x) = \sqrt{x}$

39. $f(x) = x^2 - 1$ and $g(x) = \dfrac{1}{x}$

41. $f(x) = 3x^2 + 2$ and $g(x) = \dfrac{1}{x^{3/2}}$ **43.** $3h$ **45.** $-h(2a + h)$

47. $2a + h$ **49.** $3a^2 + 3ah + h^2 - 1$ **51.** $-\dfrac{1}{a(a + h)}$

53. The total revenue in dollars from both restaurants at time t

55. The value in dollars of Nancy's shares of IBM at time t

57. The carbon monoxide pollution in parts per million at time t

59. $C(x) = 0.6x + 12{,}100$

61. a. $f(t) = 267$; $g(t) = 2t^2 + 46t + 733$
b. $f(t) + g(t) = 2t^2 + 46t + 1000$ **c.** 1936 tons

63. a. 23; In 2002, 23% of reported serious crimes ended in the arrests or in the identification of the suspects.
b. 18; In 2007, 18% of reported serious crimes ended in the arrests or in the identification of the suspects.

65. a. $P(x) = -0.000003x^3 - 0.07x^2 + 300x - 100{,}000$
b. \$182,375

67. a. $3.5t^2 + 2.4t + 71.2$ **b.** 71,200; 109,900

69. a. 55%; 98.2% **b.** \$444,700; \$1,167,600

71. a. $s(x) = f(x) + g(x) + h(x)$

73. True **75.** False

Exercises 2.3, page 85

1. Yes; $y = -\frac{2}{3}x + 2$ **3.** Yes; $y = \frac{1}{2}x + 2$

5. Yes; $y = \frac{1}{2}x + \frac{9}{4}$ **7.** No **9.** Polynomial function; degree 6

11. Polynomial function; degree 6

13. Some other function **15.** $m = -1$; $b = 2$

17. a. $C(x) = 8x + 40{,}000$ **b.** $R(x) = 12x$
c. $P(x) = 4x - 40{,}000$ **d.** A loss of \$8000; a profit of \$8000

19. \$43,200 **21.** 104 mg **23.** \$128,000

25. 123,780,000 kWh; 175,820,000 kWh

27. **a.** \$3.25 billion **b.** \$3.88 billion; \$4.39 billion; \$4.78 billion

c.

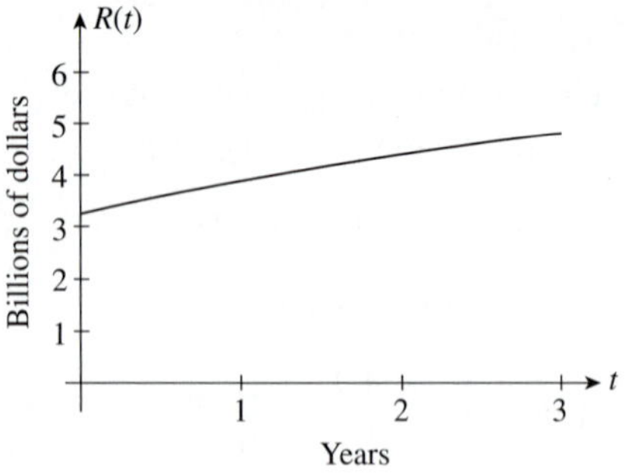

29. \$751.50/yr; \$1772.38/yr

31. **a.** 320,000 **b.** 3,923,200

33. **a.**

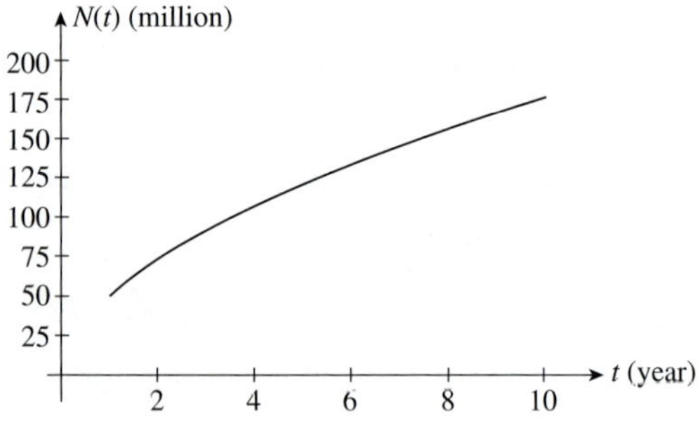

b. 176,000,000

35. 582,650; 1,605,590 **37.** \$699; \$130 **39.** **b.** 2003

41. $\dfrac{110}{\frac{1}{2}t + 1} - 26\left(\dfrac{1}{4}t^2 - 1\right)^2 - 52$; \$32, \$6.71, \$3; the gap was closing.

45. \$4770; \$6400; \$7560

47. **a.**

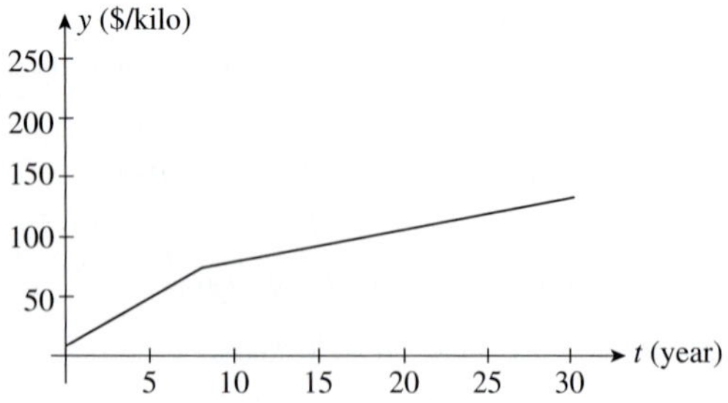

b. \$7.44/kilo; \$108.48/kilo

49. **a.**

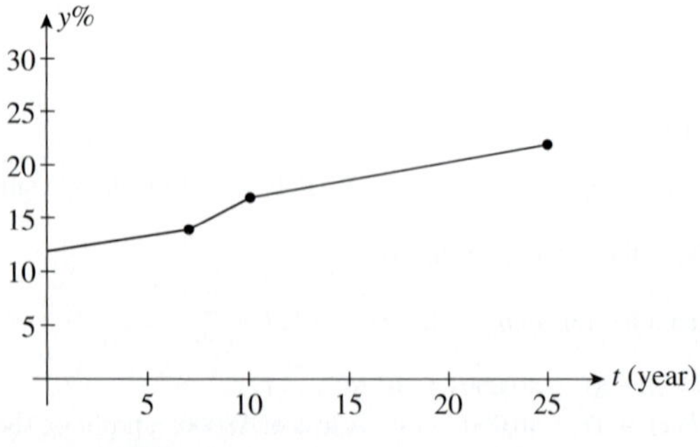

b. 13.43%; $18\frac{2}{3}\%$

51. **a.**

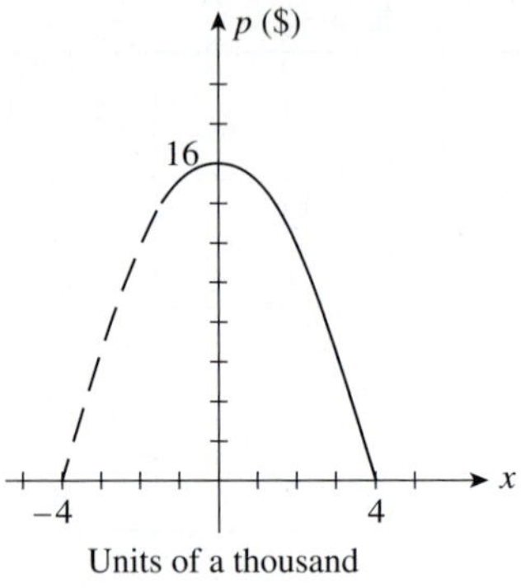

b. 3000 units

53. **a.**

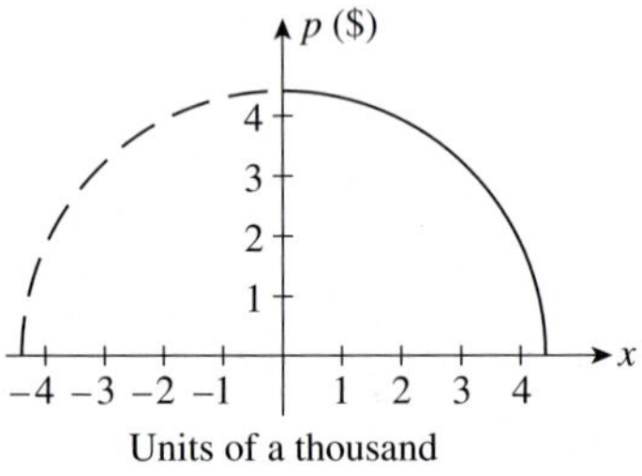

b. 3000

55. **a.**

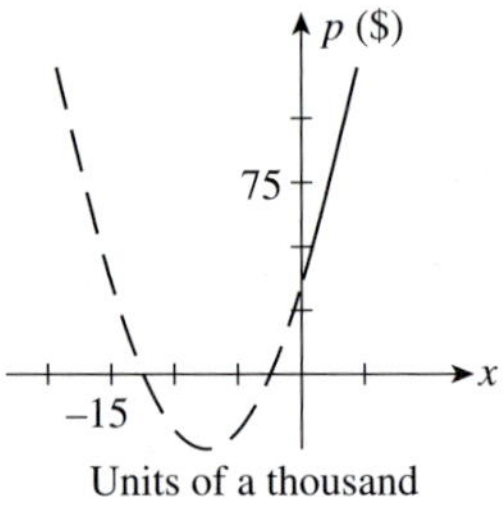

b. \$76

57. **a.**

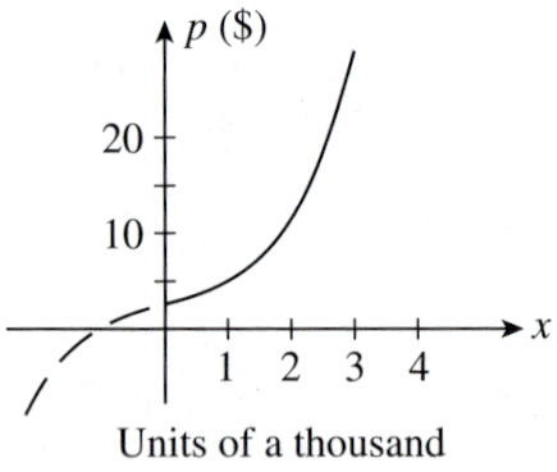

b. \$15

59. L_2; for each dollar decrease in the price of a clock radio, the additional quantity demanded of model B clock radios will be greater than that of model A clock radios.

61. **a.**

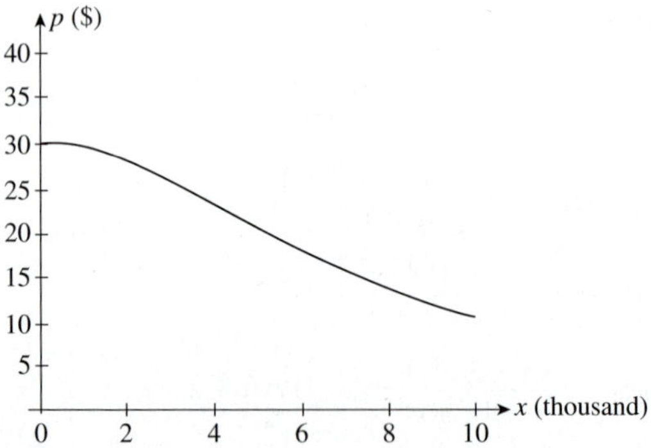

b. \$10

63. **a.**

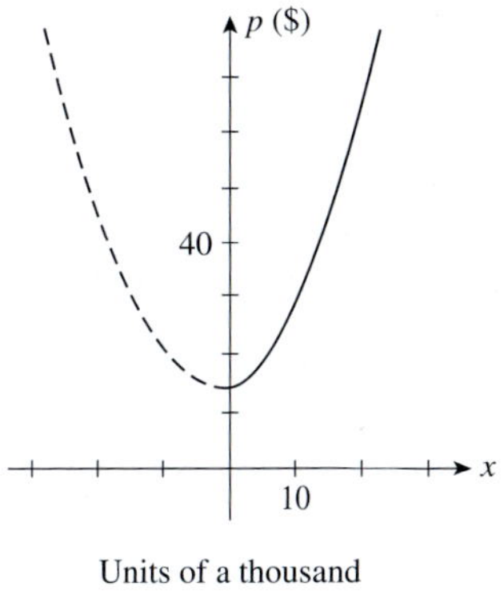

b. \$20

65. **a.** $\dfrac{b-d}{c-a}; \dfrac{bc-ad}{c-a}$

b. If the unit price is increased, then the equilibrium quantity decreases while the equilibrium price increases.

c. If the upper bound for the unit price of a commodity is lowered, then both the equilibrium quantity and the equilibrium price drop.

67. 2500; \$67.50 **69.** 11,000; \$3 **71.** 8000; \$80

73. $f(x) = 2x + \dfrac{500}{x}; x > 0$ **75.** $f(x) = 0.5x^2 + \dfrac{8}{x}$

77. $f(x) = (22 + x)(36 - 2x)$ bushels/acre

79. **a.** $P(x) = (10{,}000 + x)(5 - 0.0002x)$ **b.** \$60,800

81. False **83.** False

Using Technology Exercises 2.3, page 94

1. $(-3.0414, 0.1503)$; $(3.0414, 7.4497)$

3. $(-2.3371, 2.4117)$; $(6.0514, -2.5015)$

5. $(-1.0219, -6.3461)$; $(1.2414, -1.5931)$; $(5.7805, 7.9391)$

7. **a.**

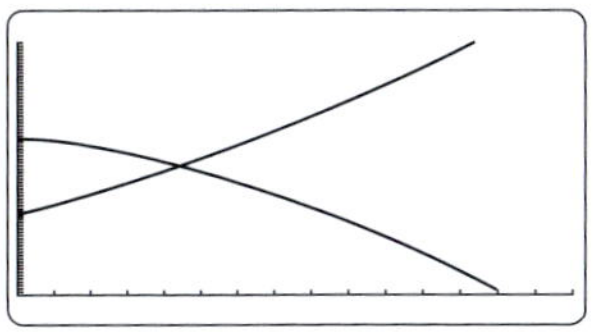

b. 438 wall clocks; \$40.92

9. **a.** $f(t) = 1.85t + 16.9$

b.

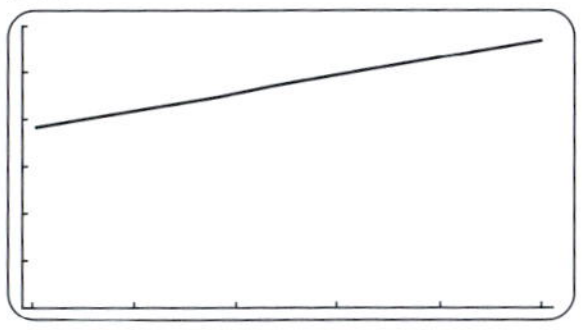

c.

t	1	2	3	4	5	6
y	18.8	20.6	22.5	24.3	26.2	28.0

d. 31.7 gallons

11. **a.** $f(t) = -0.221t^2 + 4.14t + 64.8$

b.

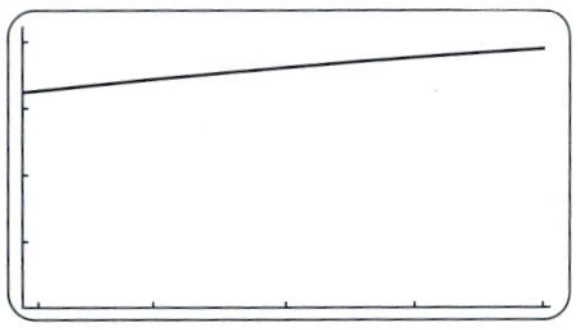

c. 77.8 million

13. **a.**

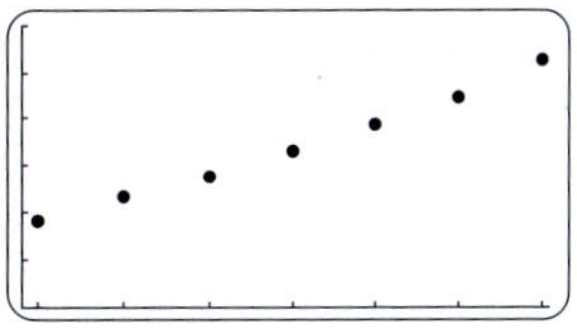

b. $f(t) = 2.94t^2 + 38.75t + 188.5$ **c.** \$604 billion

15. **a.** $f(t) = -0.00081t^3 + 0.0206t^2 + 0.125t + 1.69$

b.

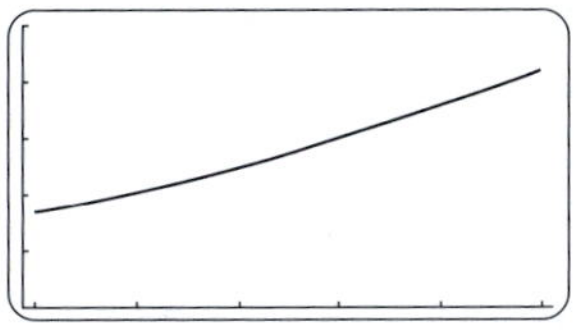

c. \$1.8 trillion; \$2.7 trillion; \$4.2 trillion

17. **a.** $f(t) = -0.425t^3 + 3.6571t^2 + 4.018t + 43.7$

b.

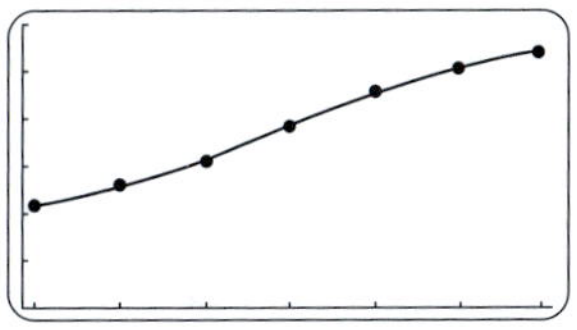

c. \$43.7 million; \$77.2 million; \$107.7 million

19. **a.** $f(t) = -2.4167t^3 + 24.5t^2 - 123.33t + 506$

b.

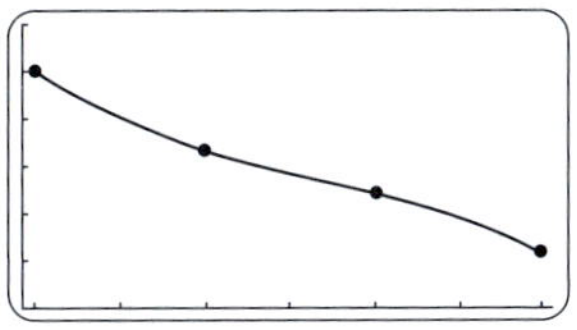

c. 506,000; 338,000; 126,000 **d.** Approximately 200,000

21. **a.** $f(t) = 0.000133t^4 + 0.00353t^3 - 0.04487t^2 + 0.143t + 1.71$

b.

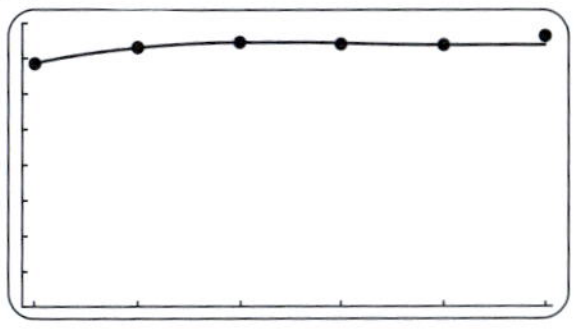

c. 1.71 mg; 1.81 mg; 1.85 mg; 1.84 mg; 1.82 mg; 1.83 mg

d. 1.9 mg/cigarette

Exercises 2.4, page 111

1. $\lim_{x \to -2} f(x) = 3$ **3.** $\lim_{x \to 3} f(x) = 3$ **5.** $\lim_{x \to -2} f(x) = 3$

7. The limit does not exist.

9.

x	1.9	1.99	1.999
$f(x)$	4.61	4.9601	4.9960

x	2.001	2.01	2.1
$f(x)$	5.004	5.0401	5.41

$\lim_{x\to 2}(x^2 + 1) = 5$

11.

x	−0.1	−0.01	−0.001
$f(x)$	−1	−1	−1

x	0.001	0.01	0.1
$f(x)$	1	1	1

The limit does not exist.

13.

x	0.9	0.99	0.999
$f(x)$	100	10,000	1,000,000

x	1.001	1.01	1.1
$f(x)$	1,000,000	10,000	100

The limit does not exist.

15.

x	0.9	0.99	0.999	1.001	1.01	1.1
$f(x)$	2.9	2.99	2.999	3.001	3.01	3.1

$\lim_{x\to 1}\dfrac{x^2 + x - 2}{x - 1} = 3$

17.

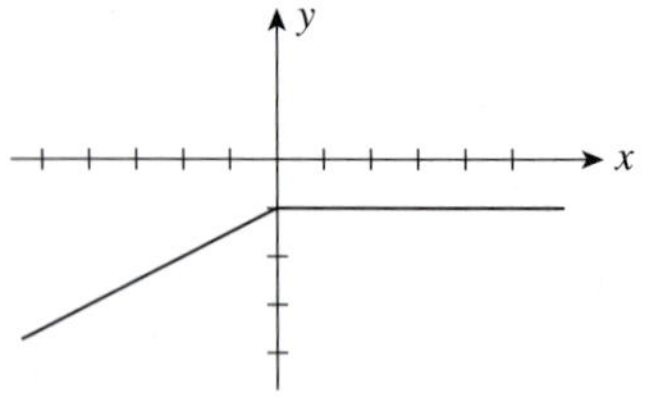

$\lim_{x\to 0} f(x) = -1$

19.

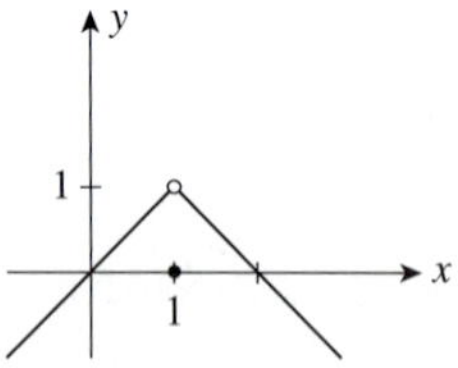

$\lim_{x\to 1} f(x) = 1$

21.

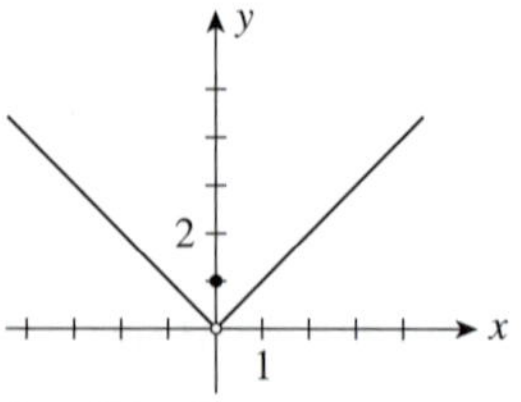

$\lim_{x\to 0} f(x) = 0$

23. 3 **25.** 3 **27.** −1 **29.** 2 **31.** −4 **33.** $\frac{5}{4}$

35. 2 **37.** $\sqrt{171} = 3\sqrt{19}$ **39.** $\frac{3}{2}$ **41.** −1 **43.** −6

45. 2 **47.** $\frac{1}{6}$ **49.** 2 **51.** −1 **53.** −10

55. The limit does not exist. **57.** $\frac{5}{3}$ **59.** $\frac{1}{2}$ **61.** $\frac{1}{3}$

63. $\lim_{x\to\infty} f(x) = \infty$; $\lim_{x\to-\infty} f(x) = \infty$ **65.** 0; 0

67. $\lim_{x\to\infty} f(x) = -\infty$; $\lim_{x\to-\infty} f(x) = -\infty$

69.

x	1	10	100	1000
$f(x)$	0.5	0.009901	0.0001	0.000001

x	−1	−10	−100	−1000
$f(x)$	0.5	0.009901	0.0001	0.000001

$\lim_{x\to\infty} f(x) = 0$ and $\lim_{x\to-\infty} f(x) = 0$

71.

x	1	5	10	100
$f(x)$	12	360	2910	2.99×10^6

x	1000	−1	−5
$f(x)$	2.999×10^9	6	−390

x	−10	−100	−1000
$f(x)$	−3090	-3.01×10^6	-3.0×10^9

$\lim_{x\to\infty} f(x) = \infty$ and $\lim_{x\to-\infty} f(x) = -\infty$

73. 3 **75.** 3 **77.** $\lim_{x\to-\infty} f(x) = -\infty$ **79.** 0

81. **a.** \$0.5 million; \$0.75 million; \$1,166,667; \$2 million; \$4.5 million; \$9.5 million
b. The limit does not exist; as the percent of pollutant to be removed approaches 100, the cost becomes astronomical.

83. \$2.20; the average cost of producing x DVDs will approach \$2.20/disc in the long run.

85. **a.** \$24 million; \$60 million; \$83.1 million **b.** \$120 million

87. **a.** 76.1¢/mi; 30.5¢/mi; 23¢/mi; 20.6¢/mi; 19.5¢/mi
b.

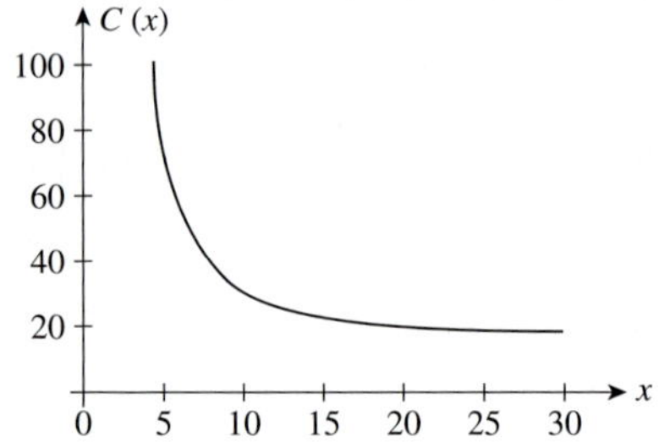

c. It approaches 17.8¢/mi.

89. False **91.** True **93.** True

95. a moles/liter/second **97.** No

Using Technology Exercises 2.4, page 117

1. 5 **3.** 3 **5.** $\frac{2}{3}$ **7.** e^2

11. **a.**

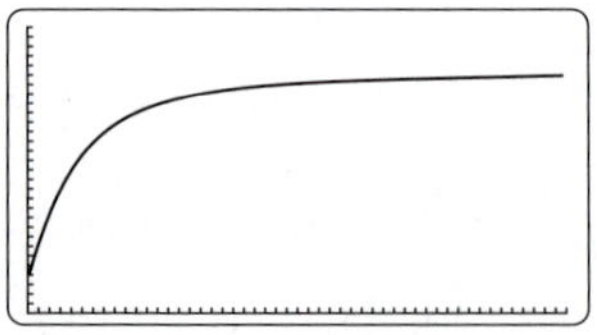

b. 25,000

Exercises 2.5, page 126

1. 3; 2; the limit does not exist.

3. The limit does not exist; 2; the limit does not exist.

5. 0; 2; the limit does not exist.

7. −2; 2; the limit does not exist. **9.** True **11.** True

13. False **15.** True **17.** False **19.** True **21.** 6

23. $-\frac{1}{4}$ **25.** The limit does not exist. **27.** −1 **29.** 0

31. −4 **33.** The limit does not exist. **35.** 4 **37.** 0; 0

39. $x = 0$; conditions 2 and 3 **41.** Continuous everywhere

43. $x = 0$; condition 3 **45.** $(-\infty, \infty)$ **47.** $(-\infty, \infty)$

49. $(-\infty, \frac{1}{2}) \cup (\frac{1}{2}, \infty)$ **51.** $(-\infty, -2) \cup (-2, 1) \cup (1, \infty)$

53. $(-\infty, \infty)$ **55.** $(-\infty, \infty)$ **57.** −1 and 1 **59.** 1 and 2

61. f is discontinuous at $x = 1, 2, \ldots, 12$.

63. Michael makes progress toward solving the problem until $x = x_1$. Between $x = x_1$ and $x = x_2$, he makes no further progress. But at $x = x_2$ he suddenly achieves a breakthrough, and at $x = x_3$ he proceeds to complete the problem.

65. Conditions 2 and 3 are not satisfied at each of these points.

67.

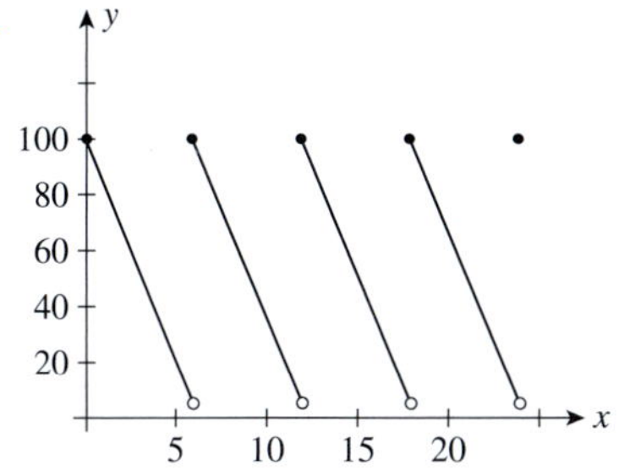

69.

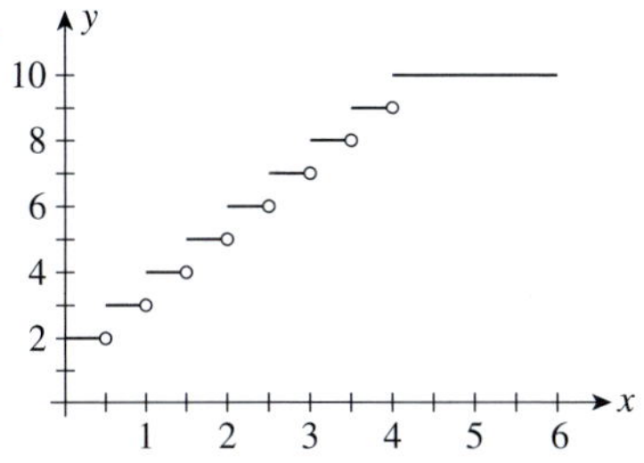

f is discontinuous at $x = \frac{1}{2}, 1, 1\frac{1}{2}, \ldots, 4$.

71. **a.** ∞; As the time taken to excite the tissue is made shorter and shorter, the strength of the electric current gets stronger and stronger.
b. b; As the time taken to excite the tissue is made longer and longer, the strength of the electric current gets weaker and weaker and approaches b.

73. 3 **75.** **a.** Yes **b.** No

77. **a.** f is a polynomial of degree 2. **b.** $f(1) = 3$ and $f(3) = -1$

79. **a.** f is a polynomial of degree 3. **b.** $f(-1) = -4$ and $f(1) = 4$

81. $x \approx 0.59$ **83.** ≈ 1.34

85. **c.** $\frac{1}{2}$; $\frac{7}{2}$; Joan sees the ball on its way up $\frac{1}{2}$ sec after it was thrown and again $3\frac{1}{2}$ sec later.

87. False **89.** False **91.** False **93.** False **95.** False

97. No **99.** **c.** $\pm\frac{\sqrt{2}}{2}$

Using Technology Exercises 2.5, page 132

1. $x = 0, 1$ **3.** $x = 0, \frac{1}{2}$ **5.** $x = -\frac{1}{2}, 2$ **7.** $x = -2, 1$

9.

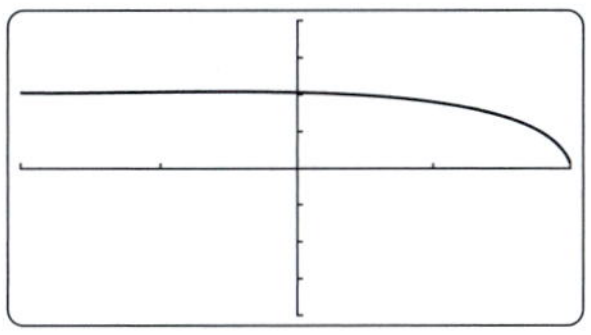

11.

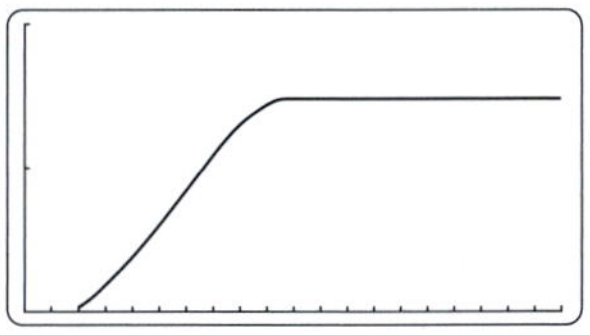

Exercises 2.6, page 145

1. 1.5 lb/mo; 0.58 lb/mo; 1.25 lb/mo **3.** 3.1%/hr; −21.2%/hr

5. **a.** Car A **b.** They are traveling at the same speed.
c. Car B **d.** Both cars covered the same distance.

7. **a.** P_2 **b.** P_1 **c.** Bactericide B; bactericide A

9. 0 **11.** 2 **13.** $6x$ **15.** $-2x + 3$ **17.** 2; $y = 2x + 7$

19. 6; $y = 6x - 3$ **21.** $\frac{1}{9}$; $y = \frac{1}{9}x - \frac{2}{3}$

23. **a.** $4x$ **b.** $y = 4x - 1$
c.

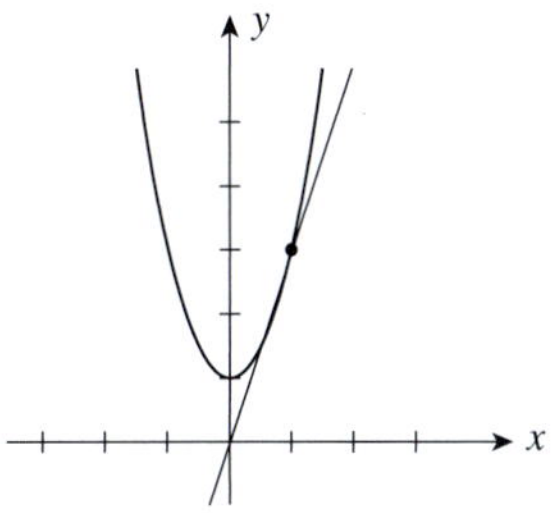

25. **a.** $2x - 2$ **b.** (1, 0)
c.

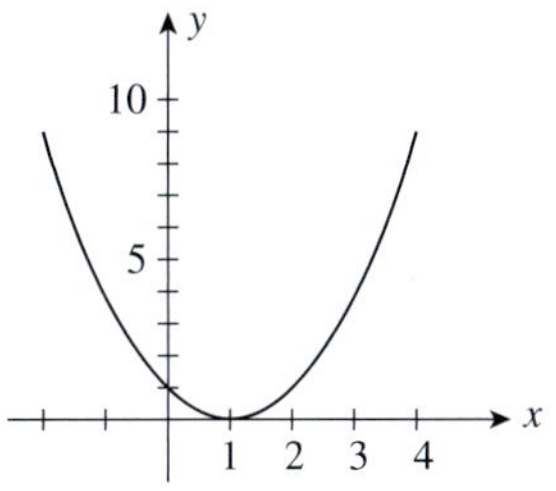

d. 0

27. **a.** 6; 5.5; 5.1 **b.** 5
c. The computations in part (a) show that as h approaches zero, the average velocity approaches the instantaneous velocity.

29. **a.** 130 ft/sec; 128.2 ft/sec; 128.02 ft/sec **b.** 128 ft/sec
c. The computations in part (a) show that as the time intervals over which the average velocity are computed become smaller and smaller, the average velocity approaches the instantaneous velocity of the car at $t = 20$.

31. a. 5 sec **b.** 80 ft/sec **c.** 160 ft/sec

33. a. $-\frac{1}{6}$ liter/atmosphere **b.** $-\frac{1}{4}$ liter/atmosphere

35. a. $-\frac{2}{3}x + 7$ **b.** \$333/quarter; −\$13,000/quarter

37. \$6 billion/yr; \$10 billion/yr

39. a. $f'(h)$ gives the instantaneous rate of change of the temperature at a given height h.
b. Negative **c.** ≈ -0.05°F

41. Average rate of change of the seal population over $[a, a + h]$; instantaneous rate of change of the seal population at $x = a$

43. Average rate of change of the country's industrial production over $[a, a + h]$; instantaneous rate of change of the country's industrial production at $x = a$

45. Average rate of change of atmospheric pressure over $[a, a + h]$; instantaneous rate of change of atmospheric pressure at $x = a$

47. a. Yes **b.** No **c.** No

49. a. Yes **b.** Yes **c.** No

51. a. No **b.** No **c.** No

53. 32.1, 30.939, 30.814, 30.8014, 30.8001, 30.8000; 30.8 ft/sec

55. False

57.

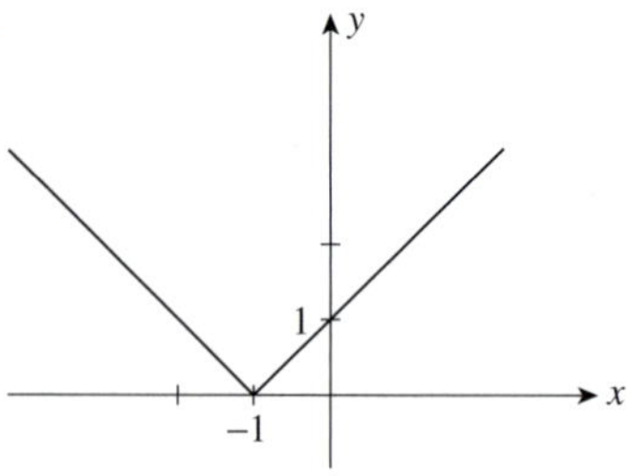

59. $a = 2, b = -1$

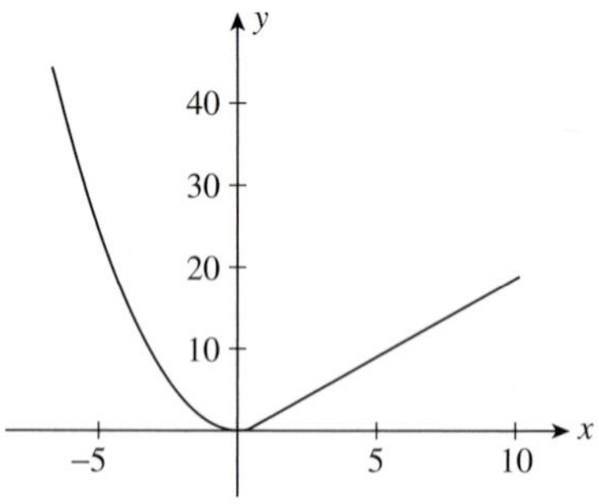

Using Technology Exercises 2.6, page 151

1. a. $y = 9x - 11$
b.

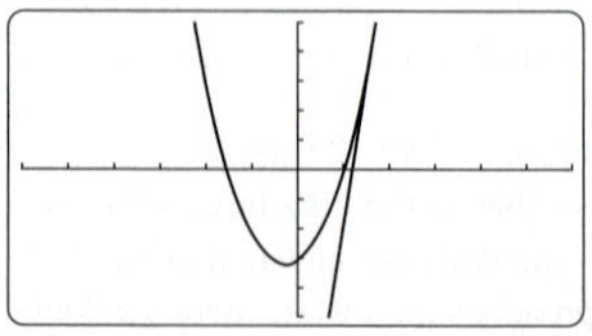

3. a. $y = \frac{1}{4}x + 1$
b.

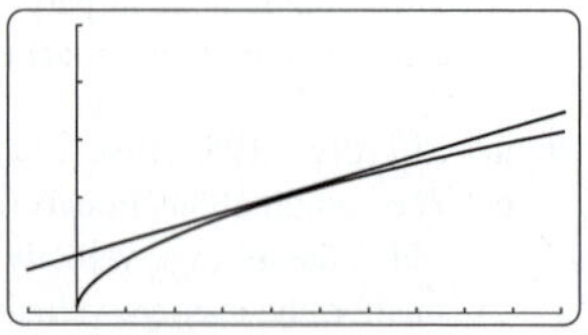

5. a. 4
b. $y = 4x - 1$
c.

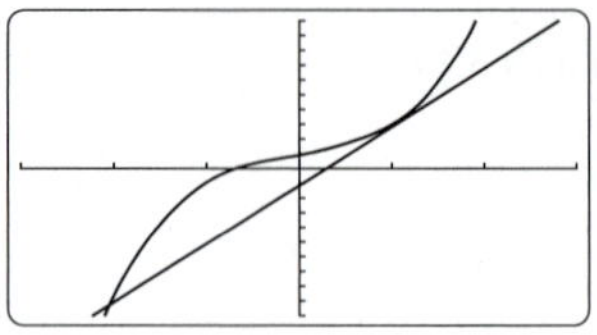

7. a. 4.02
b. $y = 4.02x - 3.57$
c.

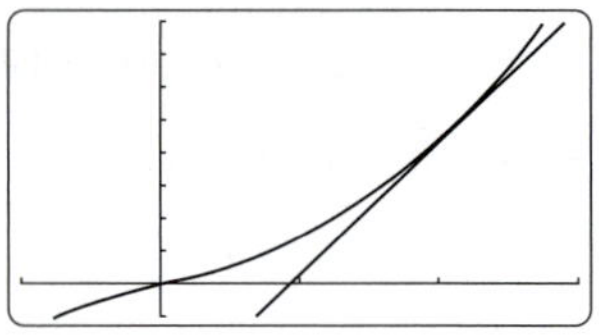

9. a.

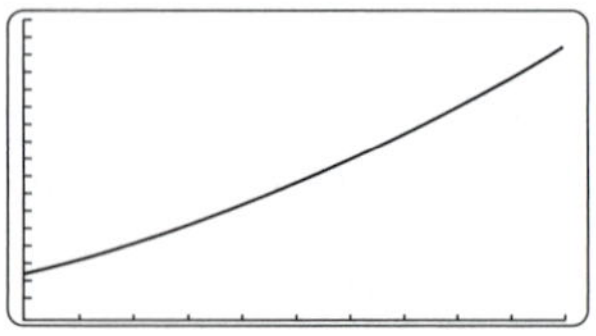

b. 41.22¢/mi **c.** 1.22¢/mi/yr

Chapter 2 Concept Review, page 152

1. Domain; range; B

2. Domain, $f(x)$; vertical, point

3. $f(x) \pm g(x)$; $f(x)g(x)$; $\dfrac{f(x)}{g(x)}$; $A \cap B$; $A \cap B$; 0

4. $g[f(x)]$; f; $f(x)$; g

5. a. $P(x) = a_nx^n + a_{n-1}x^{n-1} + \cdots + a_1x + a_0$ ($a_n \neq 0$, n, a positive integer)
b. Linear; quadratic; cubic
c. Quotient; polynomials
d. x^r (r, a real number)

6. L; $f(x)$; L; a

7. a. L^r **b.** $L \pm M$
c. LM **d.** $\dfrac{L}{M}$; $M \neq 0$

8. a. L; x **b.** M; negative; absolute

9. a. Right **b.** Left **c.** L; L

10. a. Continuous **b.** Discontinuous **c.** Every

11. a. a; a; $g(a)$; **b.** Everywhere **c.** $Q(x)$

12. a. $[a, b]$; $f(c) = M$ **b.** $f(x) = 0$; (a, b)

13. a. $f'(a)$ **b.** $y = f(a) + m(x - a)$

14. a. $\dfrac{f(a + h) - f(a)}{h}$ **b.** $\lim_{h \to 0} \dfrac{f(a + h) - f(a)}{h}$

Chapter 2 Review Exercises, page 153

1. a. $(-\infty, 9]$ **b.** $(-\infty, -1) \cup (-1, \frac{3}{2}) \cup (\frac{3}{2}, \infty)$

2. a. $(-\infty, -3) \cup (-3, 2]$ **b.** $(-\infty, \infty)$

3. a. 0 **b.** $3a^2 + 17a + 20$ **c.** $12a^2 + 10a - 2$
d. $3a^2 + 6ah + 3h^2 + 5a + 5h - 2$

4. a. $4x^2 - 2x + 6$ **b.** $2x^2 + 8xh + 8h^2 - x - 2h + 1$

5. **a.**

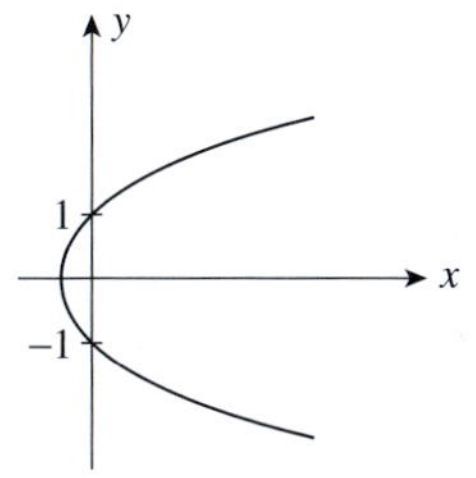

b. No **c.** Yes

6.

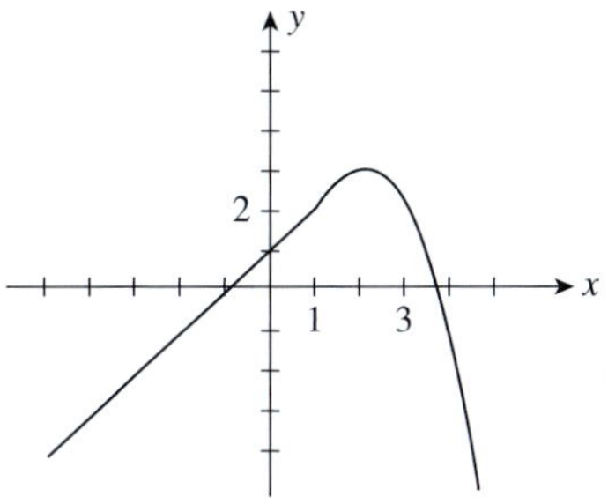

7. **a.** $\frac{2x+3}{x}$ **b.** $\frac{1}{x(2x+3)}$ **c.** $\frac{1}{2x+3}$ **d.** $\frac{2}{x}+3$

8. **a.** $2x^2+7$; $4x^2-4x+5$ **b.** $\frac{3(x+1)}{3x+4}$; $\frac{1}{7-3x}$

c. $\frac{1}{\sqrt{x+1}}-3$; $\frac{1}{\sqrt{x-2}}$

9. **a.** $f(x)=2x^2+x+1$; $g(x)=\frac{1}{x^3}$

b. $f(x)=x^2+x+4$; $g(x)=\sqrt{x}$

10. $-\frac{3}{8}$ **11.** -3 **12.** 2 **13.** -21 **14.** 0 **15.** -1

16. The limit does not exist. **17.** 7 **18.** $\frac{9}{2}$ **19.** 1 **20.** $\frac{1}{2}$

21. 1 **22.** 1 **23.** $\frac{3}{2}$ **24.** The limit does not exist.

25.

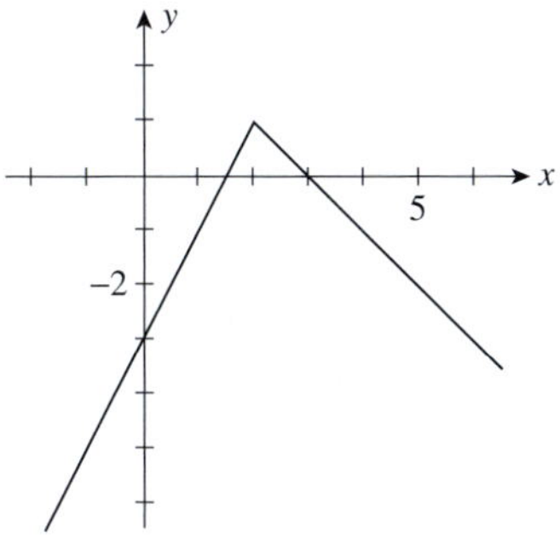

1; 1; 1

26.

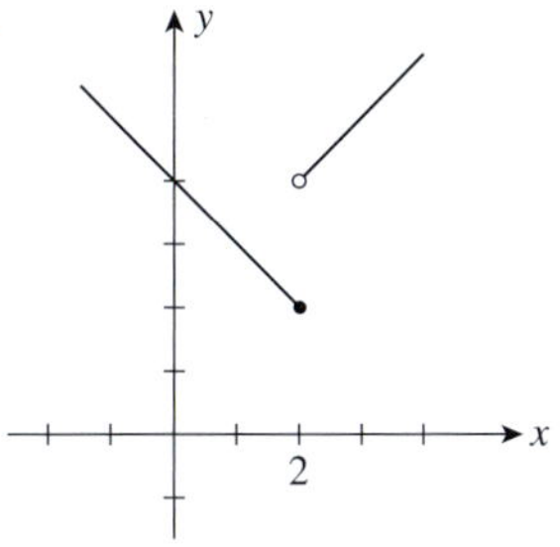

4; 2; the limit does not exist.

27. $x=2$ **28.** $x=-\frac{1}{2}, 1$ **29.** $x=-1$ **30.** $x=0$

31. **a.** 3; 2.5; 2.1 **b.** 2 **32.** 4 **33.** $\frac{1}{x^2}$

34. $\frac{3}{2}$; $y=\frac{3}{2}x+5$ **35.** -4; $y=-4x+4$

36. **a.** Yes **b.** No **37.** 54,000

38. **a.** $S(t)=t+2.4$ **b.** \$5.4 million

39. **a.** $C(x)=6x+30{,}000$ **b.** $R(x)=10x$
c. $P(x)=4x-30{,}000$ **d.** (\$6000); \$2000; \$18,000

40. $\left(6, \frac{21}{2}\right)$ **41.** $P(x)=8x-20{,}000$ **42.** 6000; \$22

43. 117 mg **44.** \$400,000 **45.** \$45,000

46. 400; 800 **47.** 990; 2240

48.

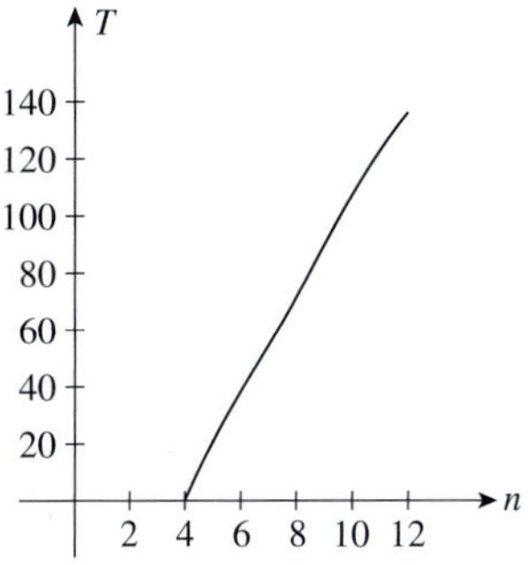

As the length of the list increases, the time taken to learn the list increases by a very large amount.

49. After $5\frac{1}{2}$ years **50.** 5000; \$20

51. **a.** 714,300 **b.** 8,330,000

52. 648,000; 902,000; 1,345,200; 1,762,800

53. **a.** \$16.4 billion; \$17.6 billion; \$18.3 billion; \$18.8 billion; \$19.3 billion

b.

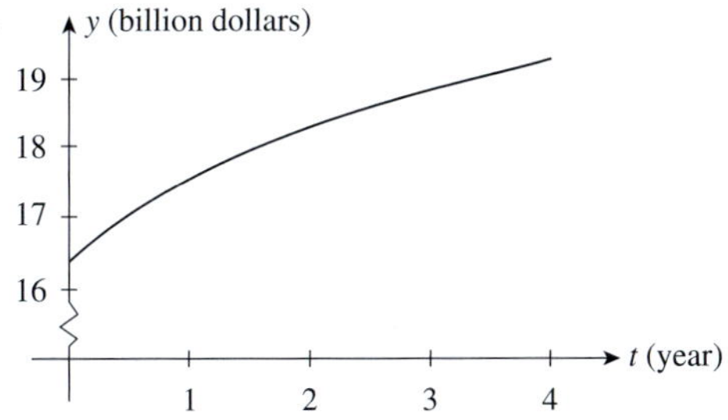

54. **a.** 59.8%; 58.9%; 59.2%; 60.7%, 61.7%

b.

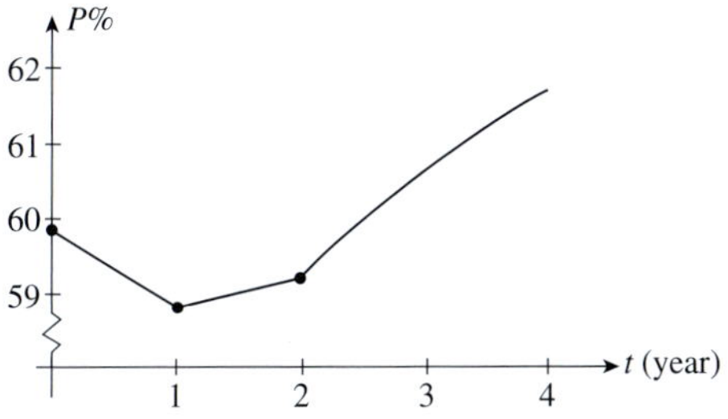

c. 60.66%

55. **a.** πr^2 **b.** $2t$ **c.** $4\pi t^2$ **d.** 3600π ft^2

56. $x(20 - 2x)^2$

57. $100x^2 + \dfrac{1350}{x}$

58. $C(x) = \begin{cases} 5 & \text{if } 1 \le x \le 100 \\ 9 & \text{if } 100 < x \le 200 \\ 12.50 & \text{if } 200 < x \le 300 \\ 15.00 & \text{if } 300 < x \le 400 \\ 7 + 0.02x & \text{if } x > 400 \end{cases}$

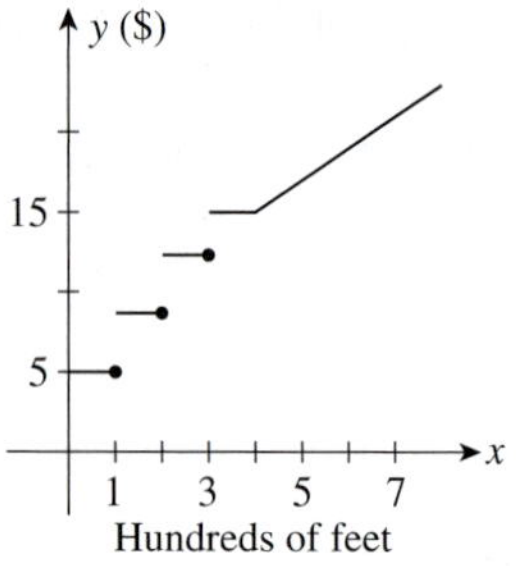

The function is discontinuous at $x = 100$, 200, and 300.

59. 20

60. **a.** $C'(x)$ gives the instantaneous rate of change of the total manufacturing cost C in dollars when x units of a certain product are produced.
b. Positive
c. $\approx$\$20

Chapter 2 Before Moving On, page 156

1. **a.** 3 **b.** 2 **c.** $\frac{17}{4}$

2. **a.** $\dfrac{1}{x+1} + x^2 + 1$ **b.** $\dfrac{x^2+1}{x+1}$ **c.** $\dfrac{1}{x^2+2}$
d. $\dfrac{1}{(x+1)^2} + 1$

3. $108x^2 - 4x^3$ 4. 2 5. **a.** 0 **b.** 1; no

6. -1; $y = -x$

CHAPTER 3

Exercises 3.1, page 164

1. 0 3. $5x^4$ 5. $2.1x^{1.1}$ 7. $6x$ 9. $2\pi r$ 11. $\dfrac{3}{x^{2/3}}$

13. $\dfrac{3}{2\sqrt{x}}$ 15. $-84x^{-13}$ 17. $10x - 3$ 19. $-3x^2 + 4x$

21. $0.06x - 0.4$ 23. $2x - 4 - \dfrac{3}{x^2}$ 25. $16x^3 - 7.5x^{3/2}$

27. $-\dfrac{3}{x^2} - \dfrac{8}{x^3}$ 29. $-\dfrac{16}{t^5} + \dfrac{9}{t^4} - \dfrac{2}{t^2}$ 31. $2 - \dfrac{5}{2\sqrt{x}}$

33. $-\dfrac{4}{x^3} + \dfrac{1}{x^{4/3}}$ 35. **a.** 20 **b.** -4 **c.** 20 37. 3

39. 11 41. $m = 5$; $y = 5x - 4$ 43. $m = -2$; $y = -2x + 2$

45. **a.** $(0, 0)$
b.

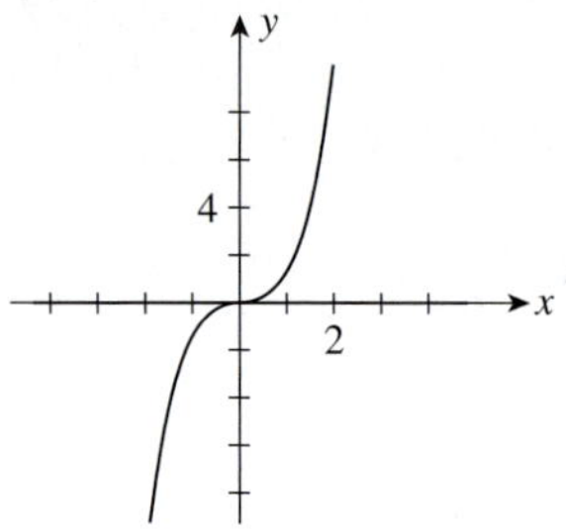

47. **a.** $(-2, -7)$, $(2, 9)$
b. $y = 12x + 17$ and $y = 12x - 15$
c.

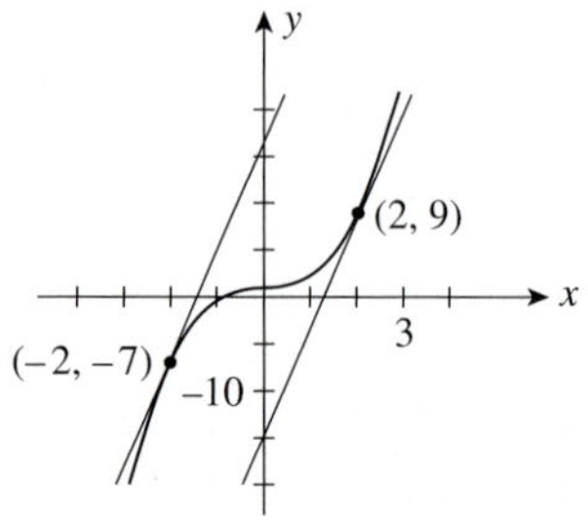

49. **a.** $(0, 0)$; $(1, -\frac{13}{12})$ **b.** $(0, 0)$; $(2, -\frac{8}{3})$; $(-1, -\frac{5}{12})$
c. $(0, 0)$; $(4, \frac{80}{3})$; $(-3, \frac{81}{4})$

51. **a.** $\dfrac{16\pi}{9}$ cm^3/cm **b.** $\dfrac{25\pi}{4}$ cm^3/cm

53. **a.** 16.3 million **b.** 14.3 million/yr **c.** 66.8 million
d. 11.7 million/yr

55. **a.** 49.6%; 41.13%; 36.87%; 34.11%
b. -5.55%/yr; -3.32%/yr

57. **a.** 157 million **b.** 10.4 million/yr

59. **a.** $120 - 30t$ **b.** 120 ft/sec **c.** 240 ft

61. **a.** 5%; 11.3%; 15.5% **b.** 0.63%/yr; 0.525%/yr

63. **a.** -0.9 thousand metric tons/yr; 20.3 thousand metric tons/yr
b. Yes

65. **a.** 15 pts/yr; 12.6 pts/yr; 0 pts/yr **b.** 10 pts/yr

67. **a.** $(0.0001)(\frac{5}{4})x^{1/4}$ **b.** \$0.00125/radio

69. **a.** $20\left(1 - \dfrac{1}{\sqrt{t}}\right)$ **b.** 50 mph; 30 mph; 33.43 mph
c. -8.28; 0; 5.86; at 6:30 a.m., the average velocity is decreasing at the rate of 8.28 mph/hr; at 7 a.m., it is unchanged; and at 8 a.m., it is increasing at the rate of 5.86 mph.

71. 32 turtles/yr; 428 turtles/yr; 3260 turtles

73. **a.** 12%; 23.9% **b.** 0.8%/yr; 1.1%/yr

75. **a.** The total population, including the population of the developed countries and that of the underdeveloped/emerging countries
b. $0.92t + 3.727$; $\approx$13 million people/yr

77. True

Using Technology Exercises 3.1, page 170

1. 1 **3.** 0.4226 **5.** 0.1613

7. a.

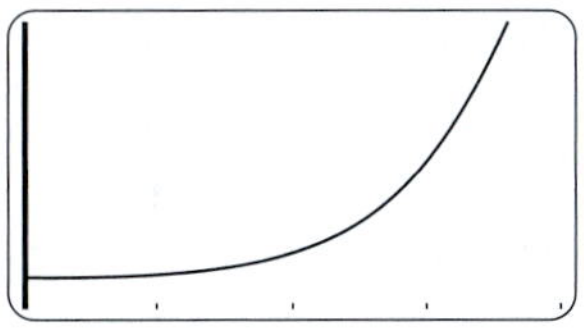

b. 3.4295 ppm/40 yr; 105.4332 ppm/40 yr

9. a. $f(t) = 0.611t^3 + 9.702t^2 + 32.544t + 473.5$

b.

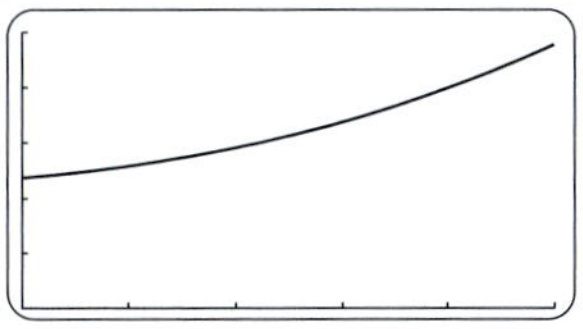

c. At the beginning of 2000, the assets of the hedge funds were increasing at the rate of \$53.781 billion/yr, and at the beginning of 2003, they were increasing at the rate of \$139.488 billion/yr.

Exercises 3.2, page 177

1. $2x(2x) + (x^2 + 1)(2)$, or $6x^2 + 2$

3. $(t - 1)(2) + (2t + 1)(1)$, or $4t - 1$

5. $(3x + 1)(2x) + (x^2 - 2)(3)$, or $9x^2 + 2x - 6$

7. $(x^3 - 1)(1) + (x + 1)(3x^2)$, or $4x^3 + 3x^2 - 1$

9. $(w^3 - w^2 + w - 1)(2w) + (w^2 + 2)(3w^2 - 2w + 1)$, or $5w^4 - 4w^3 + 9w^2 - 6w + 2$

11. $(5x^2 + 1)(x^{-1/2}) + (2x^{1/2} - 1)(10x)$, or $\dfrac{25x^2 - 10x\sqrt{x} + 1}{\sqrt{x}}$

13. $\dfrac{(x^2 - 5x + 2)(x^2 + 2)}{x^2} + \dfrac{(x^2 - 2)(2x - 5)}{x}$, or $\dfrac{3x^4 - 10x^3 + 4}{x^2}$

15. $\dfrac{-1}{(x - 2)^2}$ **17.** $\dfrac{2x + 1 - (x - 1)(2)}{(2x + 1)^2}$, or $\dfrac{3}{(2x + 1)^2}$

19. $-\dfrac{2x}{(x^2 + 1)^2}$ **21.** $\dfrac{s^2 + 2s + 4}{(s + 1)^2}$

23. $\dfrac{(\frac{1}{2}x^{-1/2})[x^2 + 1 - 4x^{3/2}(x^{1/2} + 1)]}{(x^2 + 1)^2}$, or $\dfrac{-3x^2 - 4x^{3/2} + 1}{2\sqrt{x}(x^2 + 1)^2}$

25. $\dfrac{2x^3 + 2x^2 + 2x - 2x^3 - x^2 - 4x - 2}{(x^2 + x + 1)^2}$, or $\dfrac{x^2 - 2x - 2}{(x^2 + x + 1)^2}$

27. $\dfrac{(x - 2)(3x^2 + 2x + 1) - (x^3 + x^2 + x + 1)}{(x - 2)^2}$, or $\dfrac{2x^3 - 5x^2 - 4x - 3}{(x - 2)^2}$

29. $\dfrac{(x^2 - 4)(x^2 + 4)(2x + 8) - (x^2 + 8x - 4)(4x^3)}{(x^2 - 4)^2(x^2 + 4)^2}$, or $\dfrac{-2x^5 - 24x^4 + 16x^3 - 32x - 128}{(x^2 - 4)^2(x^2 + 4)^2}$

31. 8 **33.** -9 **35.** $2(3x^2 - x + 3)$; 10

37. $\dfrac{-3x^4 + 2x^2 - 1}{(x^4 - 2x^2 - 1)^2}$; $-\dfrac{1}{2}$ **39.** 60; $y = 60x - 102$

41. $-\frac{1}{2}$; $y = -\frac{1}{2}x + \frac{3}{2}$ **43.** 8 **45.** $y = 7x - 5$

47. $(\frac{1}{3}, \frac{50}{27})$; $(1, 2)$ **49.** $(\frac{4}{3}, -\frac{770}{27})$; $(2, -30)$

51. $y = -\frac{1}{2}x + 1$; $y = 2x - \frac{3}{2}$

53. 0.125, 0.5, 2, 50; the cost of removing all of the pollutant is prohibitively high.

55. -5000/min; -1600/min; 7000; 4000

57. a. $\dfrac{180}{(t + 6)^2}$ **b.** 3.7; 2.2; 1.8; 1.1

c.

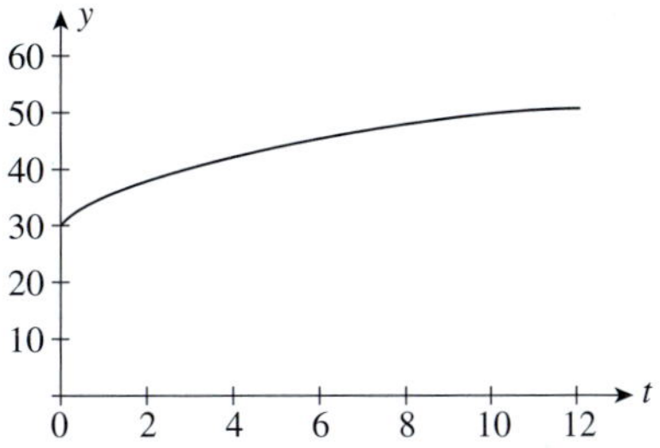

Yes

d. 50 words/min

59. Dropping at the rate of 0.0375 ppm/yr; dropping at the rate of 0.006 ppm/yr

61. False **63.** False

Using Technology Exercises 3.2, page 181

1. 0.8750 **3.** 0.0774

5. -0.5000 **7.** 87,322/yr

Exercises 3.3, page 189

1. $8(2x - 1)^3$ **3.** $10x(x^2 + 2)^4$

5. $3(2x - x^2)^2(2 - 2x)$, or $6x^2(1 - x)(2 - x)^2$ **7.** $\dfrac{-4}{(2x + 1)^3}$

9. $3x\sqrt{x^2 - 4}$ **11.** $\dfrac{3}{2\sqrt{3x - 2}}$ **13.** $\dfrac{-2x}{3(1 - x^2)^{2/3}}$

15. $-\dfrac{6}{(2x + 3)^4}$ **17.** $\dfrac{-1}{(2t - 3)^{3/2}}$ **19.** $-\dfrac{3(16x^3 + 1)}{2(4x^4 + x)^{5/2}}$

21. $-2(3x^2 + 2x + 1)^{-3}(6x + 2) = -4(3x + 1)(3x^2 + 2x + 1)^{-3}$

23. $3(x^2 + 1)^2(2x) - 2(x^3 + 1)(3x^2)$, or $6x(2x^2 - x + 1)$

25. $3(t^{-1} - t^{-2})^2(-t^{-2} + 2t^{-3})$

27. $\dfrac{1}{2\sqrt{x - 1}} + \dfrac{1}{2\sqrt{x + 1}}$

29. $2x^2(4)(3 - 4x)^3(-4) + (3 - 4x)^4(4x)$, or $(-12x)(4x - 1)(3 - 4x)^3$

31. $8(x - 1)^2(2x + 1)^3 + 2(x - 1)(2x + 1)^4$, or $6(x - 1)(2x - 1)(2x + 1)^3$

33. $3\left(\frac{x+3}{x-2}\right)^2\left[\frac{(x-2)(1)-(x+3)(1)}{(x-2)^2}\right]$, or $-\frac{15(x+3)^2}{(x-2)^4}$

35. $\frac{3}{2}\left(\frac{t}{2t+1}\right)^{1/2}\left[\frac{(2t+1)(1)-t(2)}{(2t+1)^2}\right]$, or $\frac{3t^{1/2}}{2(2t+1)^{5/2}}$

37. $\frac{1}{2}\left(\frac{u+1}{3u+2}\right)^{-1/2}\left[\frac{(3u+2)(1)-(u+1)(3)}{(3u+2)^2}\right]$, or $-\frac{1}{2\sqrt{u+1}\,(3u+2)^{3/2}}$

39. $\frac{(x^2-1)^4(2x)-x^2(4)(x^2-1)^3(2x)}{(x^2-1)^8}$, or $\frac{(-2x)(3x^2+1)}{(x^2-1)^5}$

41. $\frac{2x(x^2-1)^3(3x^2+1)^2[9(x^2-1)-4(3x^2+1)]}{(x^2-1)^8}$, or $-\frac{2x(3x^2+13)(3x^2+1)^2}{(x^2-1)^5}$

43. $\frac{(2x+1)^{-1/2}[(x^2-1)-(2x+1)(2x)]}{(x^2-1)^2}$, or $-\frac{3x^2+2x+1}{\sqrt{2x+1}(x^2-1)^2}$

45. $\frac{(t^2+1)^{1/2}(\frac{1}{2})(t+1)^{-1/2}(1)-(t+1)^{1/2}(\frac{1}{2})(t^2+1)^{-1/2}(2t)}{t^2+1}$, or $-\frac{t^2+2t-1}{2\sqrt{t+1}(t^2+1)^{3/2}}$

47. $4(3x+1)^3(3)(x^2-x+1)^3+(3x+1)^4(3)(x^2-x+1)^2(2x-1)$, or $3(3x+1)^3(x^2-x+1)^2(10x^2-5x+3)$

49. $\frac{4}{3}u^{1/3}$; $6x$; $8x(3x^2-1)^{1/3}$

51. $-\frac{2}{3u^{5/3}}$; $6x^2-1$; $-\frac{2(6x^2-1)}{3(2x^3-x+1)^{5/3}}$

53. $\frac{1}{2}u^{-1/2}-\frac{1}{2}u^{-3/2}$; $3x^2-1$; $\frac{(3x^2-1)(x^3-x-1)}{2(x^3-x)^{3/2}}$

55. -12 57. 6 59. No 61. $y=-33x+57$

63. $y=\frac{43}{5}x-\frac{54}{5}$

65. 0.333 million/wk; 0.305 million/wk; 16 million; 22.7 million

67. $\frac{6.87775}{(5+t)^{0.795}}$; 0.53%/yr; 64.9%

69. **a.** \$8.7 billion/yr **b.** \$92.3 billion

71. **a.** $0.027(0.2t^2+4t+64)^{-1/3}(0.1t+1)$ **b.** 0.0091 ppm/yr

73. **a.** $0.03[3t^2(t-7)^4+t^3(4)(t-7)^3]$, or $0.21t^2(t-3)(t-7)^3$
b. 90.72; 0; -90.72; at 8 a.m. the level of nitrogen dioxide is increasing; at 10 a.m. the level stops increasing; at 11 a.m. the level is decreasing.

75. $300\left[\frac{(t+25)\frac{1}{2}(\frac{1}{2}t^2+2t+25)^{-1/2}(t+2)-(\frac{1}{2}t^2+2t+25)^{1/2}(1)}{(t+25)^2}\right]$, or $\frac{3450t}{(t+25)^2\sqrt{\frac{1}{2}t^2+2t+25}}$; 2.9 beats/min², 0.7 beats/min², 0.2 beats/min², 179 beats/min

77. 160π ft²/sec 79. -27 mph/decade; 19 mph

81. $(1.42)\left[\frac{(3t^2+80t+550)(14t+140)-(7t^2+140t+700)(6t+80)}{(3t^2+80t+550)^2}\right]$, or $\frac{1.42(140t^2+3500t+21{,}000)}{(3t^2+80t+550)^2}$; 31,312 jobs/yr

83. -400 wristwatches/(dollar price increase)

85. True 87. True

Using Technology Exercises 3.3, page 193

1. 0.5774 3. 0.9390 5. -4.9498

7. 10,146,200/decade; 7,810,520/decade

Exercises 3.4, page 204

1. **a.** $C(x)$ is always increasing because as the number of units x produced increases, the amount of money that must be spent on production also increases.
b. 4000

3. **a.** \$1.80; \$1.60 **b.** \$1.80; \$1.60

5. **a.** $100+\frac{200{,}000}{x}$ **b.** $-\frac{200{,}000}{x^2}$
c. $\overline{C}(x)$ approaches \$100 if the production level is very high.

7. $\frac{2000}{x}+2-0.0001x$; $-\frac{2000}{x^2}-0.0001$

9. **a.** $8000-200x$ **b.** 200, 0, -200 **c.** \$40

11. **a.** $-0.04x^2+600x-300{,}000$
b. $-0.08x+600$ **c.** 200; -40
d. The profit increases as production increases, peaking at 7500 units; beyond this level, profit falls.

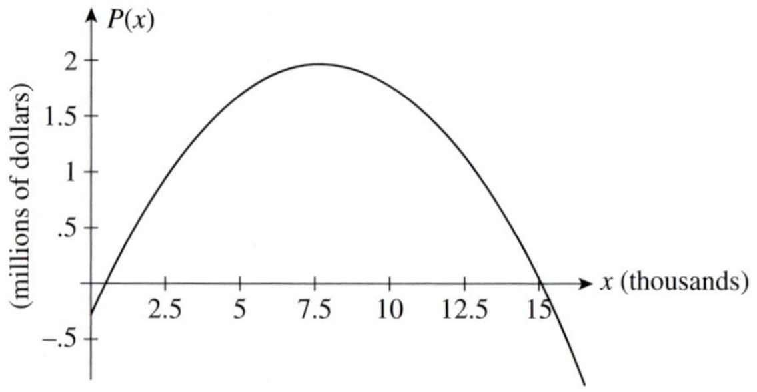

13. **a.** $600x-0.05x^2$; $-0.000002x^3-0.02x^2+200x-80{,}000$
b. $0.000006x^2-0.06x+400$; $600-0.1x$; $-0.000006x^2-0.04x+200$
c. 304; 400; 96

d.

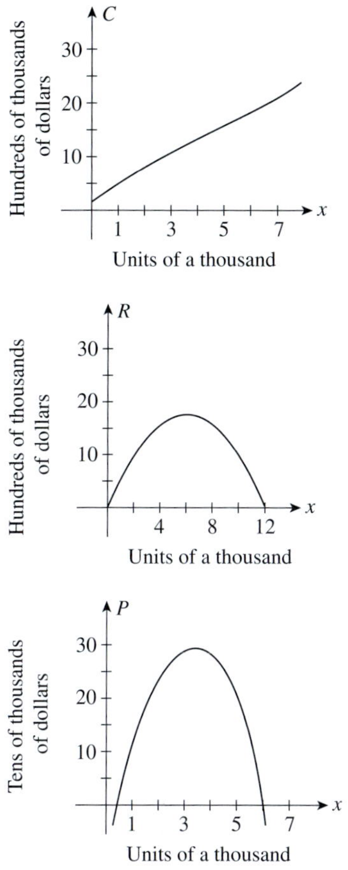

15. $0.000002x^2 - 0.03x + 400 + \dfrac{80{,}000}{x}$

a. $0.000004x - 0.03 - \dfrac{80{,}000}{x^2}$

b. -0.0132; 0.0092; the marginal average cost is negative (average cost is decreasing) when 5000 units are produced and positive (average cost is increasing) when 10,000 units are produced.

c.

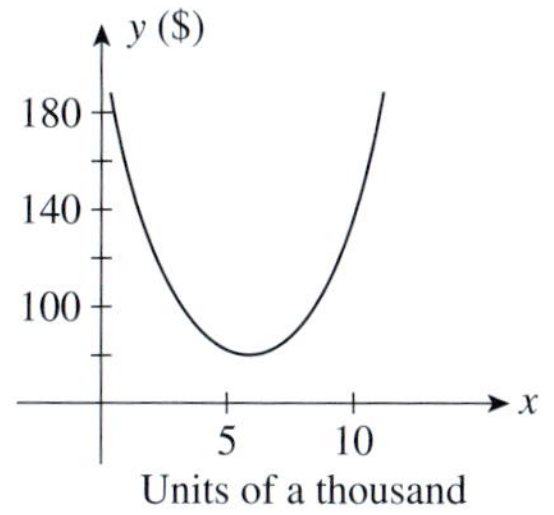

17. **a.** $\dfrac{50x}{0.01x^2 + 1}$ **b.** $\dfrac{50 - 0.5x^2}{(0.01x^2 + 1)^2}$

c. \$44,380; when the level of production is 2000 units, the revenue increases at the rate of \$44,380 per additional 1000 units produced.

19. \$1.21 billion/billion dollars **21.** \$0.288 billion/billion dollars

23. $\frac{5}{3}$; elastic **25.** 1; unitary **27.** 0.104; inelastic

29. **a.** Inelastic; elastic **b.** When $p = 8.66$
c. Increase **d.** Increase

31. **a.** Inelastic **b.** Increase

33. $\dfrac{2p^2}{9 - p^2}$; for $p < \sqrt{3}$, demand is inelastic; for $p = \sqrt{3}$, demand is unitary; and for $p > \sqrt{3}$, demand is elastic.

35. True

Exercises 3.5, page 212

1. $8x - 2$; 8 **3.** $6x^2 - 6x$; $6(2x - 1)$

5. $4t^3 - 6t^2 + 12t - 3$; $12(t^2 - t + 1)$

7. $10x(x^2 + 2)^4$; $10(x^2 + 2)^3(9x^2 + 2)$

9. $6t(2t^2 - 1)(6t^2 - 1)$; $6(60t^4 - 24t^2 + 1)$

11. $14x(2x^2 + 2)^{5/2}$; $28(2x^2 + 2)^{3/2}(6x^2 + 1)$

13. $(x^2 + 1)(5x^2 + 1)$; $4x(5x^2 + 3)$

15. $\dfrac{1}{(2x + 1)^2}$; $-\dfrac{4}{(2x + 1)^3}$ **17.** $\dfrac{2}{(s + 1)^2}$; $-\dfrac{4}{(s + 1)^3}$

19. $-\dfrac{3}{2(4 - 3u)^{1/2}}$; $-\dfrac{9}{4(4 - 3u)^{3/2}}$ **21.** $72x - 24$ **23.** $-\dfrac{6}{x^4}$

25. $\frac{81}{8}(3s - 2)^{-5/2}$ **27.** $192(2x - 3)$ **29.** 128 ft/sec; 32 ft/sec^2

31. **a.** and **b.**

t	0	1	2	3	4	5	6	7
$N'(t)$	0	2.7	4.8	6.3	7.2	7.5	7.2	6.3
$N''(t)$					0.6	0	-0.6	-1.2

33. 8.1 million; 0.204 million/yr; -0.03 million/yr^2. At the beginning of 1998, there were 8.1 million people receiving disability benefits; the number was increasing at the rate of 0.2 million/yr; the rate of the rate of change of the number of people was decreasing at the rate of 0.03 million people/yr^2.

35. **a.** $\frac{1}{4}t^3 - 3t^2 + 8t$ **b.** 0 ft/sec; 0 ft/sec; 0 ft/sec
c. $\frac{3}{4}t^2 - 6t + 8$ **d.** 8 ft/sec^2; -4 ft/sec^2; 8 ft/sec^2
e. 0 ft; 16 ft; 0 ft

37. -0.01%/yr^2. The rate of the rate of change of the percent of Americans aged 55 and over decreases at the rate of 0.01%/yr^2.

39. False **41.** True **43.** True **45.** $f(x) = x^{n+1/2}$

Using Technology Exercises 3.5, page 215

1. -18 **3.** 15.2762 **5.** -0.6255 **7.** 0.1973

9. -68.46214; at the beginning of 1988, the rate of the rate of the rate at which banks were failing was 68 banks/yr/yr/yr.

Exercises 3.6, page 223

1. **a.** $-\frac{1}{2}$ **b.** $-\frac{1}{2}$ **3.** **a.** $-\dfrac{1}{x^2}$ **b.** $-\dfrac{y}{x}$

5. **a.** $2x - 1 + \dfrac{4}{x^2}$ **b.** $3x - 2 - \dfrac{y}{x}$

7. **a.** $\dfrac{1 - x^2}{(1 + x^2)^2}$ **b.** $-2y^2 + \dfrac{y}{x}$ **9.** $-\dfrac{x}{y}$ **11.** $\dfrac{x}{2y}$

13. $1 - \frac{y}{x}$ **15.** $-\frac{y}{x}$ **17.** $-\frac{\sqrt{y}}{\sqrt{x}}$ **19.** $2\sqrt{x+y} - 1$

21. $-\frac{y^3}{x^3}$ **23.** $\frac{2\sqrt{xy} - y}{x - 2\sqrt{xy}}$ **25.** $\frac{6x - 3y - 1}{3x + 1}$

27. $\frac{2(2x - y^{3/2})}{3x\sqrt{y} - 4y}$ **29.** $-\frac{2x^2 + 2xy + y^2}{x^2 + 2xy + 2y^2}$ **31.** $y = 2$

33. $y = -\frac{3}{2}x + \frac{5}{2}$ **35.** $\frac{2y}{x^2}$ **37.** $\frac{2y(y - x)}{(2y - x)^3}$

39. **a.** $\frac{dV}{dt} = \pi r\left(r\frac{dh}{dt} + 2h\frac{dr}{dt}\right)$ **b.** 3.6π cu in./sec

41. Dropping at the rate of 111 tires/wk

43. Increasing at the rate of 44 headphones/wk

45. Dropping at the rate of 3.7¢/carton/wk **47.** 0.37; inelastic

49. 160π ft²/sec **51.** 188.5 ft²/sec **53.** 13.44 ft/sec

55. 7.69 ft/sec **59.** 196.8 ft/sec **61.** 9 ft/sec

63. 19.2 ft/sec **65.** $\frac{1}{7}$ L/sec **67.** False **69.** True

Exercises 3.7, page 233

1. $4x\,dx$ **3.** $(3x^2 - 1)\,dx$ **5.** $\frac{dx}{2\sqrt{x+1}}$ **7.** $\frac{6x+1}{2\sqrt{x}}\,dx$

9. $\frac{x^2 - 2}{x^2}\,dx$ **11.** $\frac{-x^2 + 2x + 1}{(x^2+1)^2}\,dx$ **13.** $\frac{6x - 1}{2\sqrt{3x^2 - x}}\,dx$

15. **a.** $2x\,dx$ **b.** 0.04 **c.** 0.0404

17. **a.** $-\frac{dx}{x^2}$ **b.** -0.05 **c.** -0.05263

19. 3.167 **21.** 7.0357 **23.** 1.983 **25.** 0.298

27. 2.50375 **29.** ± 8.64 cm³ **31.** 18.85 ft³

33. It will drop by 40%. **35.** 274 sec **37.** 283,937

39. Decrease of \$1.33 **41.** $\pm$\$64,800

43. Decrease of 11 crimes/yr

45. **a.** $100{,}000\left(1 + \frac{r}{12}\right)^{119} dr$ **b.** \$220.50; \$440.99; \$661.48

47. True

Using Technology Exercises 3.7, page 237

1. 7.5787 **3.** 0.03122 **5.** -0.01988

7. \$48.35/mo; \$64.47/mo; \$80.59/mo **9.** 625

Chapter 3 Concept Review, page 238

1. **a.** 0 **b.** nx^{n-1} **c.** $cf'(x)$ **d.** $f'(x) \pm g'(x)$

2. **a.** $f(x)g'(x) + g(x)f'(x)$ **b.** $\frac{g(x)f'(x) - f(x)g'(x)}{[g(x)]^2}$

3. **a.** $g'[f(x)]f'(x)$ **b.** $n[f(x)]^{n-1}f'(x)$

4. Marginal cost; marginal revenue; marginal profit; marginal average cost

5. **a.** $-p\frac{f'(p)}{f(p)}$ **b.** Elastic; unitary; inelastic

6. Both sides; $\frac{dy}{dx}$ **7.** y; $\frac{dy}{dt}$; a **8.** $\frac{-f(t)f'(t)}{g(t)}$; $\frac{-f(t)g'(t)}{g(t)}$

9. **a.** $x_2 - x_1$ **b.** $f(x + \Delta x) - f(x)$ **10.** Δx; Δx; x; $f'(x)\,dx$

Chapter 3 Review Exercises, page 239

1. $15x^4 - 8x^3 + 6x - 2$ **2.** $24x^5 + 8x^3 + 6x$

3. $\frac{6}{x^4} - \frac{3}{x^2}$ **4.** $4t - 9t^2 + \frac{1}{2}t^{-3/2}$ **5.** $-\frac{1}{t^{3/2}} - \frac{6}{t^{5/2}}$

6. $2x - \frac{2}{x^2}$ **7.** $1 - \frac{2}{t^2} - \frac{6}{t^3}$ **8.** $4s + \frac{4}{s^2} - \frac{1}{s^{3/2}}$

9. $2x + \frac{3}{x^{5/2}}$ **10.** $\frac{(2x-1)(1) - (x+1)(2)}{(2x-1)^2}$, or $-\frac{3}{(2x-1)^2}$

11. $\frac{(2t^2+1)(2t) - t^2(4t)}{(2t^2+1)^2}$, or $\frac{2t}{(2t^2+1)^2}$

12. $\frac{(t^{1/2}+1)\frac{1}{2}t^{-1/2} - t^{1/2}(\frac{1}{2}t^{-1/2})}{(t^{1/2}+1)^2}$, or $\frac{1}{2\sqrt{t}(\sqrt{t}+1)^2}$

13. $\frac{(x^{1/2}+1)(\frac{1}{2}x^{-1/2}) - (x^{1/2}-1)(\frac{1}{2}x^{-1/2})}{(x^{1/2}+1)^2}$ or $\frac{1}{\sqrt{x}(\sqrt{x}+1)^2}$

14. $\frac{(2t^2+1)(1) - t(4t)}{(2t^2+1)^2}$, or $\frac{1 - 2t^2}{(2t^2+1)^2}$

15. $\frac{(x^2-1)(4x^3+2x) - (x^4+x^2)(2x)}{(x^2-1)^2}$, or $\frac{2x(x^4 - 2x^2 - 1)}{(x^2-1)^2}$

16. $3(4x+1)(2x^2+x)^2$ **17.** $8(3x^3-2)^7(9x^2)$, or $72x^2(3x^3-2)^7$

18. $5(x^{1/2}+2)^4 \cdot \frac{1}{2}x^{-1/2}$, or $\frac{5(\sqrt{x}+2)^4}{2\sqrt{x}}$

19. $\frac{1}{2}(2t^2+1)^{-1/2}(4t)$, or $\frac{2t}{\sqrt{2t^2+1}}$

20. $\frac{1}{3}(1 - 2t^3)^{-2/3}(-6t^2)$, or $-2t^2(1 - 2t^3)^{-2/3}$

21. $-4(3t^2 - 2t + 5)^{-3}(3t - 1)$, or $-\frac{4(3t-1)}{(3t^2 - 2t + 5)^3}$

22. $-\frac{3}{2}(2x^3 - 3x^2 + 1)^{-5/2}(6x^2 - 6x)$, or $-9x(x-1)(2x^3 - 3x^2 + 1)^{-5/2}$

23. $2\left(x + \frac{1}{x}\right)\left(1 - \frac{1}{x^2}\right)$, or $\frac{2(x^2+1)(x^2-1)}{x^3}$

24. $\frac{(2x^2+1)^2(1) - (1+x)2(2x^2+1)(4x)}{(2x^2+1)^4}$, or $-\frac{6x^2 + 8x - 1}{(2x^2+1)^3}$

25. $(t^2+t)^4(4t) + 2t^2 \cdot 4(t^2+t)^3(2t+1)$, or $4t^2(5t+3)(t^2+t)^3$

26. $(2x+1)^3 \cdot 2(x^2+x)(2x+1) + (x^2+x)^2\, 3(2x+1)^2(2)$, or $2(2x+1)^2(x^2+x)(7x^2+7x+1)$

27. $x^{1/2} \cdot 3(x^2 - 1)^2(2x) + (x^2 - 1)^3 \cdot \frac{1}{2}x^{-1/2}$, or $\frac{(13x^2 - 1)(x^2 - 1)^2}{2\sqrt{x}}$

28. $\frac{(x^3 + 2)^{1/2}(1) - x \cdot \frac{1}{2}(x^3 + 2)^{-1/2} \cdot 3x^2}{x^3 + 2}$, or $\frac{4 - x^3}{2(x^3 + 2)^{3/2}}$

29. $\frac{(4x - 3)\frac{1}{2}(3x + 2)^{-1/2}(3) - (3x + 2)^{1/2}(4)}{(4x - 3)^2}$, or $-\frac{12x + 25}{2\sqrt{3x + 2}(4x - 3)^2}$

30. $\frac{(t + 1)^3 \frac{1}{2}(2t + 1)^{-1/2}(2) - (2t + 1)^{1/2} \cdot 3(t + 1)^2(1)}{(t + 1)^6}$, or $-\frac{5t + 2}{\sqrt{2t + 1}(t + 1)^4}$

31. $2(12x^2 - 9x + 2)$ 32. $-\frac{1}{4x^{3/2}} + \frac{3}{4x^{5/2}}$

33. $\frac{(t^2 + 4)^2(-2t) - (4 - t^2)2(t^2 + 4)(2t)}{(t^2 + 4)^4}$, or $\frac{2t(t^2 - 12)}{(t^2 + 4)^3}$

34. $2(15x^4 + 12x^2 + 6x + 1)$

35. $2(2x^2 + 1)^{-1/2} + 2x\left(-\frac{1}{2}\right)(2x^2 + 1)^{-3/2}(4x)$, or $\frac{2}{(2x^2 + 1)^{3/2}}$

36. $(t^2 + 1)(14t) + (7t^2 + 1)(2)(t^2 + 1)(2t)$, or $6t(t^2 + 1)(7t^2 + 3)$

37. $\frac{2x}{y}$ 38. $\frac{2x^2 - y}{x}$ 39. $-\frac{2x}{y^2 - 1}$ 40. $-\frac{x(1 + 2y^2)}{y(2x^2 + 1)}$

41. $\frac{x - 2y}{2x + y}$ 42. $\frac{4y - 6xy - 1}{3x^2 - 4x - 2}$ 43. $\frac{2(x^4 - 1)}{x^3}\,dx$

44. $-\frac{3x^2}{2(x^3 + 1)^{3/2}}\,dx$

45. **a.** $\frac{2x}{\sqrt{2(x^2 + 2)}}\,dx$ **b.** 0.1333 **c.** 0.1335; differ by 0.0002

46. 2.9926

47. **a.** $(2, -25)$ and $(-1, 14)$ **b.** $y = -4x - 17$; $y = -4x + 10$

48. **a.** $(-2, \frac{25}{3})$ and $(1, -\frac{13}{6})$ **b.** $y = -2x + \frac{13}{3}$; $y = -2x - \frac{1}{6}$

49. $y = -\frac{\sqrt{3}}{3}x + \frac{4}{3}\sqrt{3}$ 50. $y = 112x - 80$

51. $-\frac{48}{(2x - 1)^4}$; $(-\infty; \frac{1}{2}) \cup (\frac{1}{2}, \infty)$

52. **a.** $\frac{1}{3}$; inelastic **b.** 1; unitary **c.** 3; elastic

53. $\frac{25}{2(25 - \sqrt{p})}$; for $p > 156.25$, demand is elastic; for $p = 156.25$, demand is unitary; and for $p < 156.25$, demand is inelastic.

54. **a.** Inelastic **b.** Increase 55. **a.** Elastic **b.** Decrease

56. **a.** \$33.2 billion **b.** \$7.7 billion/yr

57. **a.** 51.5% **b.** 1.95%/yr 58. **a.** 20,430 **b.** 225 cameras/yr

59. **a.** \$3.1 billion; \$10 billion **b.** \$0.68 billion/yr; \$2.08 billion/yr

60. **a.** 15%; 31.99% **b.** 0.51%/yr; 1.04%/yr

61. **a.** 4445 **b.** 494 people/yr

62. **a.** 1,024,000 **b.** 133,000 copies/wk

63. 200 subscribers/wk

64. $-14.346(1 + t)^{-1.45}$; -2.92¢/min/yr; 19.45¢/min

65. ≈ 75 yr; 0.07 yr/yr

66. **a.** \$2.20; \$2.20 **b.** $\frac{2500}{x} + 2.2$; $-\frac{2500}{x^2}$
c. $\lim_{x \to \infty} \left(\frac{2500}{x} + 2.2\right) = 2.2$

67. \$0.9487/1000 radios

68. A decrease of \$15/1000 units

69. **a.** \$39.07 **b.** \$39

70. **a.** $-0.02x^2 + 600x$ **b.** $-0.04x + 600$
c. 200; the sale of the 10,001st phone will bring a revenue of \$200.

71. **a.** $2000x - 0.04x^2$; $-0.000002x^3 - 0.02x^2 + 1000x - 120{,}000$; $0.000002x^2 - 0.02x + 1000 + \frac{120{,}000}{x}$
b. $0.000006x^2 - 0.04x + 1000$; $2000 - 0.08x$; $-0.000006x^2 - 0.04x + 1000$; $0.0000004x - 0.02 - \frac{120{,}000}{x^2}$
c. 934; 1760; 826
d. -0.0048; 0.010125; at a production level of 5000, the average cost is decreasing by 0.48¢/unit; at a production level of 8000, the average cost is increasing by 1.0125¢/unit.

72. **a.** $80 + \frac{150{,}000}{x}$ **b.** $-\frac{150{,}000}{x^2}$
c. If the production level is very high, then the unit-cost approaches \$80/desk.

73. 1.2; -1.2; the GDP is increasing at the rate of \$1.2 billion/yr; the rate of the rate of change of the GDP is decreasing at a rate of \$1.2 billion/yr/yr.

74. $\frac{17}{3}$ (ft/sec); $\frac{76}{27}$ (ft/sec^2)

Chapter 3 Before Moving On, page 242

1. $6x^2 - \frac{1}{x^{2/3}} - \frac{10}{3x^{5/3}}$ 2. $\frac{4x^2 - 1}{\sqrt{2x^2 - 1}}$ 3. $-\frac{2x^2 + 2x - 1}{(x^2 + x + 1)^2}$

4. $-\frac{1}{2(x + 1)^{3/2}}$; $\frac{3}{4(x + 1)^{5/2}}$; $-\frac{15}{8(x + 1)^{7/2}}$ 5. $\frac{-y^2 + 2xy - 3x^2}{2xy - x^2}$

6. **a.** $\frac{2x^2 + 5}{\sqrt{x^2 + 5}}\,dx$ **b.** 0.0433

CHAPTER 4

Exercises 4.1, page 255

1. Decreasing on $(-\infty, 0)$ and increasing on $(0, \infty)$

3. Increasing on $(-\infty, -1) \cup (1, \infty)$ and decreasing on $(-1, 1)$

5. Decreasing on $(-\infty, 0) \cup (2, \infty)$ and increasing on $(0, 2)$

7. Decreasing on $(-\infty, -1) \cup (1, \infty)$ and increasing on $(-1, 1)$

9. Increasing on $(20.2, 20.6) \cup (21.7, 21.8)$, constant on $(19.6, 20.2) \cup (20.6, 21.1)$, and decreasing on $(21.1, 21.7) \cup (21.8, 22.7)$

11. **a.** Positive **b.** Positive **c.** Zero **d.** Zero **e.** Negative **f.** Negative **g.** Positive

13. Increasing on $(-\infty, \infty)$

15. Decreasing on $(-\infty, \frac{3}{2})$ and increasing on $(\frac{3}{2}, \infty)$

17. Decreasing on $(-\infty, -\sqrt{3}/3) \cup (\sqrt{3}/3, \infty)$ and increasing on $(-\sqrt{3}/3, \sqrt{3}/3)$

19. Increasing on $(-\infty, -2) \cup (0, \infty)$ and decreasing on $(-2, 0)$

21. Increasing on $(-\infty, 3) \cup (3, \infty)$

23. Decreasing on $(-\infty, 0) \cup (0, 3)$ and increasing on $(3, \infty)$

25. Decreasing on $(-\infty, 2) \cup (2, \infty)$

27. Decreasing on $(-\infty, 1) \cup (1, \infty)$

29. Increasing on $(-\infty, 0) \cup (0, \infty)$ 31. Increasing on $(-1, \infty)$

33. Increasing on $(-4, 0)$; decreasing on $(0, 4)$

35. Increasing on $(-\infty, 0) \cup (0, \infty)$

37. Relative maximum: $f(0) = 1$; relative minima: $f(-1) = 0$ and $f(1) = 0$

39. Relative maximum: $f(-1) = 2$; relative minimum: $f(1) = -2$

41. Relative maximum: $f(1) = 3$; relative minimum: $f(2) = 2$

43. Relative minimum: $f(0) = 2$ 45. a 47. d

49. Relative minimum: $f(2) = -4$

51. Relative maximum: $f(3) = 15$ 53. None

55. Relative maximum: $g(0) = 4$; relative minimum: $g(2) = 0$

57. Relative maximum: $f(0) = 0$; relative minima: $f(-1) = -\frac{1}{2}$ and $f(1) = -\frac{1}{2}$

59. Relative minimum: $F(3) = -5$; relative maximum: $F(-1) = \frac{17}{3}$

61. Relative minimum: $g(3) = -19$ 63. None

65. Relative maximum: $f(-3) = -4$; relative minimum: $f(3) = 8$

67. Relative maximum: $f(1) = \frac{1}{2}$; relative minimum: $f(-1) = -\frac{1}{2}$

69. Relative minimum: $f(1) = 0$

71.

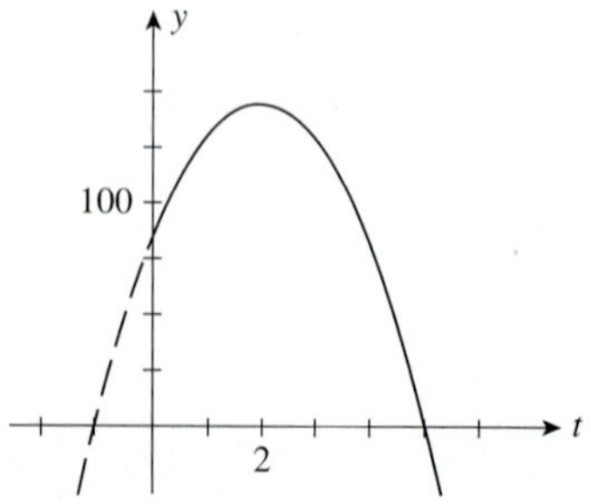

Rising in the time interval $(0, 2)$; falling in the time interval $(2, 5)$; when $t = 5$ sec

73. The percent of the U.S. population age 65 and over afflicted by the disease increases with age.

75. Rising on $(0, 33)$ and descending on $(33, T)$ for some positive number T.

77. f is decreasing on $(0, 1)$ and increasing on $(1, 4)$. The average speed decreases from 6 a.m. to 7 a.m. and then picks up from 7 a.m. to 10 a.m.

79. **a.** Increasing on $(0, 6)$ **b.** Sales will be increasing.

83. Spending was increasing from 2001 to 2006.

85. Increasing on $(0, 1)$ and decreasing on $(1, 4)$

87. Increasing on $(0, 4.5)$ and decreasing on $(4.5, 11)$; the pollution is increasing from 7 a.m. to 11:30 a.m. and decreasing from 11:30 a.m. to 6 p.m.

89. **a.** $0.0021t^2 - 0.0061t + 0.1$
    b. Decreasing on $(0, 1.5)$ and increasing on $(1.5, 15)$. The gap (shortage of nurses) was decreasing from 2000 to mid-2001 and is expected to be increasing from mid-2001 to 2015.
    c. $(1.5, 0.096)$. The gap was smallest ($\approx$ 96,000) in mid-2001.

91. True 93. True 95. False 99. $a = -4$; $b = 24$

101. **a.** $-2x$ if $x \neq 0$ **b.** No

Using Technology Exercises 4.1, page 263

1. **a.** f is decreasing on $(-\infty, -0.2934)$ and increasing on $(-0.2934, \infty)$.
   b. Relative minimum: $f(-0.2934) = -2.5435$

3. **a.** f is increasing on $(-\infty, -1.6144) \cup (0.2390, \infty)$ and decreasing on $(-1.6144, 0.2390)$.
   b. Relative maximum: $f(-1.6144) = 26.7991$; relative minimum: $f(0.2390) = 1.6733$

5. **a.** f is decreasing on $(-\infty, -1) \cup (0.33, \infty)$ and increasing on $(-1, 0.33)$.
   b. Relative maximum: $f(0.33) = 1.11$; relative minimum: $f(-1) = -0.63$

7. **a.** f is decreasing on $(-1, -0.71)$ and increasing on $(-0.71, 1)$.
   b. Relative minimum: $f(-0.71) = -1.41$

9. **a.**

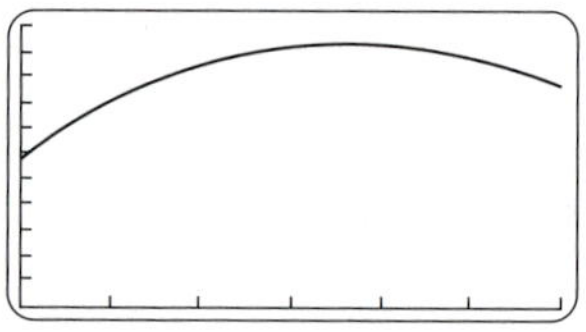

   b. Increasing on $(0, 3.6676)$ and decreasing on $(3.6676, 6)$

11. Increasing on $(0, 4.5)$ and decreasing on $(4.5, 11)$; 11:30 a.m.; 164 PSI

Exercises 4.2, page 274

1. Concave downward on $(-\infty, 0)$ and concave upward on $(0, \infty)$; inflection point: $(0, 0)$

3. Concave downward on $(-\infty, 0) \cup (0, \infty)$

5. Concave upward on $(-\infty, 0) \cup (1, \infty)$ and concave downward on $(0, 1)$; inflection points: $(0, 0)$ and $(1, -1)$

7. Concave downward on $(-\infty, -2) \cup (-2, 2) \cup (2, \infty)$

9. **a.** Concave upward on $(0, 2) \cup (4, 6) \cup (7, 9) \cup (9, 12)$ and concave downward on $(2, 4) \cup (6, 7)$
   b. $(2, \frac{5}{2})$, $(4, 2)$, $(6, 2)$, and $(7, 3)$

11. a **13.** b

15. **a.** $D_1'(t) > 0$, $D_2'(t) > 0$, $D_1''(t) > 0$, and $D_2''(t) < 0$ on $(0, 12)$
b. With or without the proposed promotional campaign, the deposits will increase; with the promotion, the deposits will increase at an increasing rate; without the promotion, the deposits will increase at a decreasing rate.

17. At the time t_0, corresponding to its t-coordinate, the restoration process is working at its peak.

23. Concave upward on $(-\infty, \infty)$

25. Concave downward on $(-\infty, 0)$; concave upward on $(0, \infty)$

27. Concave upward on $(-\infty, 0) \cup (3, \infty)$; concave downward on $(0, 3)$

29. Concave downward on $(-\infty, 0) \cup (0, \infty)$

31. Concave downward on $(-\infty, 4)$

33. Concave downward on $(-\infty, 2)$; concave upward on $(2, \infty)$

35. Concave upward on $(-\infty, -\sqrt{6}/3) \cup (\sqrt{6}/3, \infty)$; concave downward on $(-\sqrt{6}/3, \sqrt{6}/3)$

37. Concave downward on $(-\infty, 1)$; concave upward on $(1, \infty)$

39. Concave upward on $(-\infty, 0) \cup (0, \infty)$

41. Concave upward on $(-\infty, 2)$; concave downward on $(2, \infty)$

43. $(0, -2)$ **45.** $(1, -15)$ **47.** $(0, 1)$ and $(\frac{2}{3}, \frac{11}{27})$

49. $(0, 0)$ **51.** $(1, 2)$ **53.** $(-\sqrt{3}/3, 3/2)$ and $(\sqrt{3}/3, 3/2)$

55. Relative maximum: $f(1) = 5$ **57.** None

59. Relative maximum: $f(-1) = -\frac{22}{3}$, relative minimum: $f(5) = -\frac{130}{3}$

61. Relative maximum: $g(-3) = -6$; relative minimum: $g(3) = 6$

63. None **65.** Relative minimum: $f(-2) = 12$

67. Relative maximum: $g(1) = \frac{1}{2}$; relative minimum: $g(-1) = -\frac{1}{2}$

69. Relative maximum: $f(0) = 0$; relative minimum: $f(\frac{4}{3}) = \frac{256}{27}$

71.

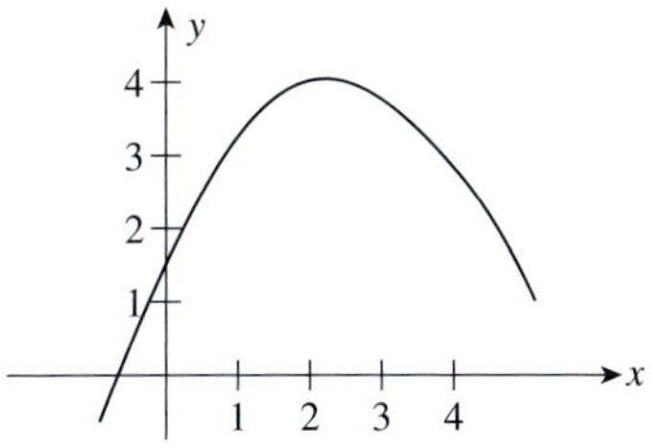

73.

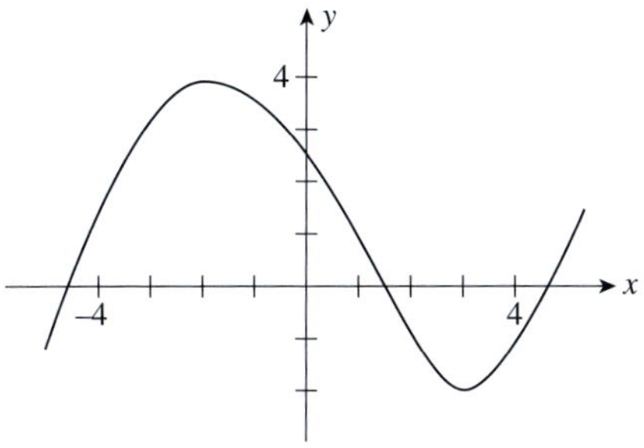

75.

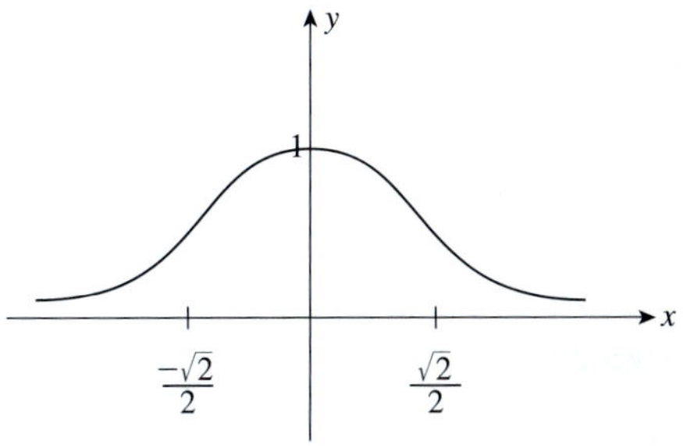

77. **a.** N is increasing on $(0, 12)$.
b. $N''(t) < 0$ on $(0, 6)$ and $N''(t) > 0$ on $(6, 12)$
c. The rate of growth of the number of help-wanted advertisements was decreasing over the first 6 mo of the year and increasing over the last 6 mo.

79. $f(t)$ increases at an increasing rate until the water level reaches the middle of the vase at which time (corresponding to the inflection point) $f(t)$ is increasing at the fastest rate. After that, $f(t)$ increases at a decreasing rate until the vase is filled.

81. **b.** The rate of increase of the average state cigarette tax was decreasing from 2001 to 2008.

83. **b.** The rate was increasing. **85.** 10 a.m.

87. The rate of business spending on technology was increasing from 2000 through 2005.

89. **a.** Concave upward on $(0, 150)$; concave downward on $(150, 400)$; $(150, 28{,}550)$
b. \$140,000

93. $(1.9, 784.9)$; the rate of annual pharmacy spending slowed down near the end of 2000.

95. **a.** $74.925t^2 - 99.62t + 41.25$; $149.85t - 99.62$
c. $(0.66, 12.91)$; the rate was increasing least rapidly around August 1999.

97. **a.** 506,000; 125,480
b. The number of measles deaths was dropping from 1999 through 2005.
c. May 2002; approximately -41 deaths/yr

103. True **105.** True

Using Technology Exercises 4.2, page 283

1. **a.** f is concave upward on $(-\infty, 0) \cup (1.1667, \infty)$ and concave downward on $(0, 1.1667)$.
b. $(1.1667, 1.1153)$; $(0, 2)$

3. **a.** f is concave downward on $(-\infty, 0)$ and concave upward on $(0, \infty)$.
b. $(0, 2)$

5. **a.** f is concave downward on $(-\infty, 0)$ and concave upward on $(0, \infty)$.
b. $(0, 0)$

7. **a.** f is concave downward on $(-\infty, -2.4495) \cup (0, 2.4495)$ and concave upward on $(-2.4495, 0) \cup (2.4495, \infty)$.
b. $(2.4495, 0.3402)$; $(-2.4495, -0.3402)$

9. **a.**

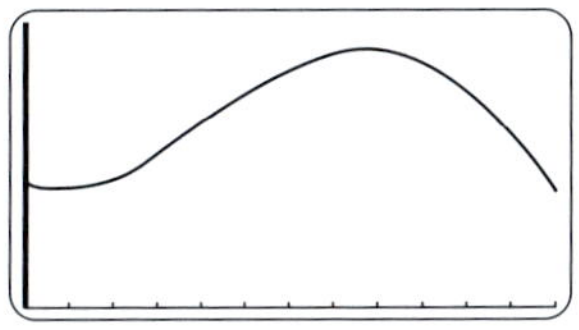

b. $(3.9024, 77.0919)$

11. a.

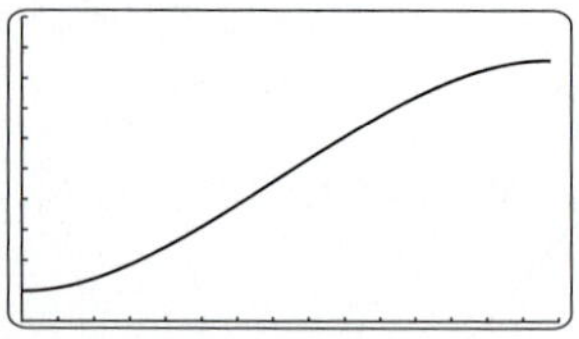

b. May 1993 ($t = 7.36$)

Exercises 4.3, page 291

1. Horizontal asymptote: $y = 0$
3. Horizontal asymptote: $y = 0$; vertical asymptote: $x = 0$
5. Horizontal asymptote: $y = 0$; vertical asymptotes: $x = -1$ and $x = 1$
7. Horizontal asymptote: $y = 3$; vertical asymptote: $x = 0$
9. Horizontal asymptotes: $y = 1$ and $y = -1$
11. Horizontal asymptote: $y = 0$; vertical asymptote: $x = 0$
13. Horizontal asymptote: $y = 0$; vertical asymptote: $x = 0$
15. Horizontal asymptote: $y = 1$; vertical asymptote: $x = -1$
17. None
19. Horizontal asymptote: $y = 1$; vertical asymptotes: $t = -3$ and $t = 3$
21. Horizontal asymptote: $y = 0$; vertical asymptotes: $x = -2$ and $x = 3$
23. Horizontal asymptote: $y = 2$; vertical asymptote: $t = 2$
25. Horizontal asymptote: $y = 1$; vertical asymptotes: $x = -2$ and $x = 2$
27. None

29. f is the derivative function of the function g.

31.

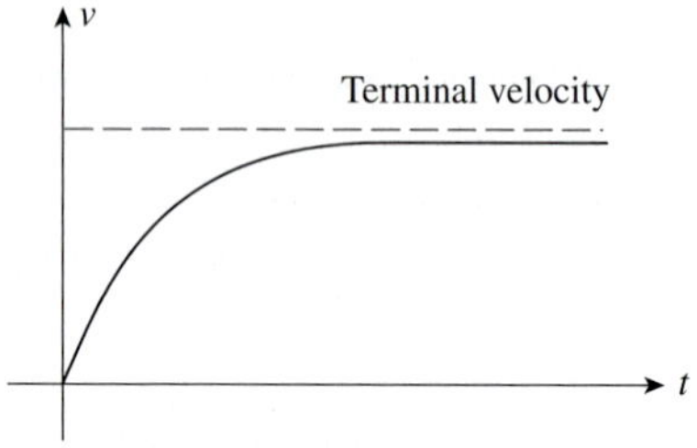

33.

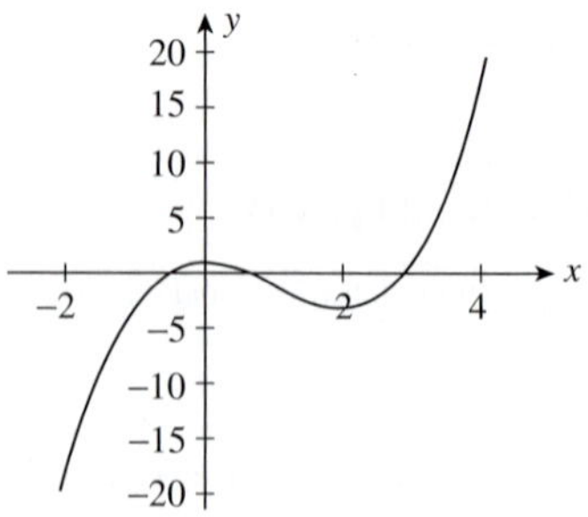

35.

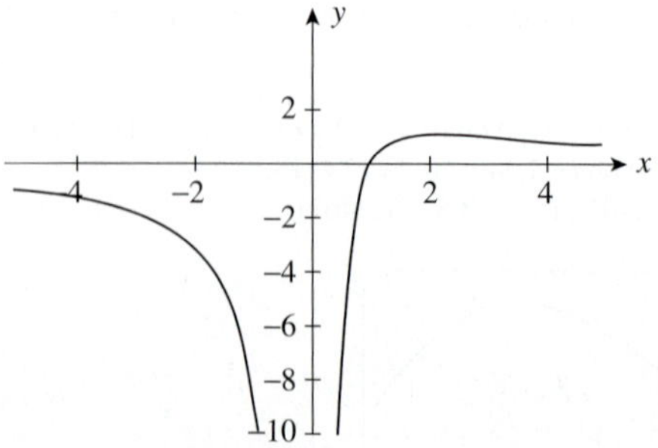

37.

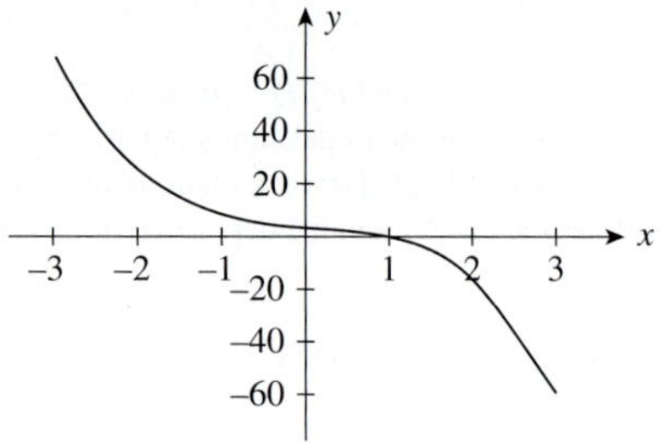

39.

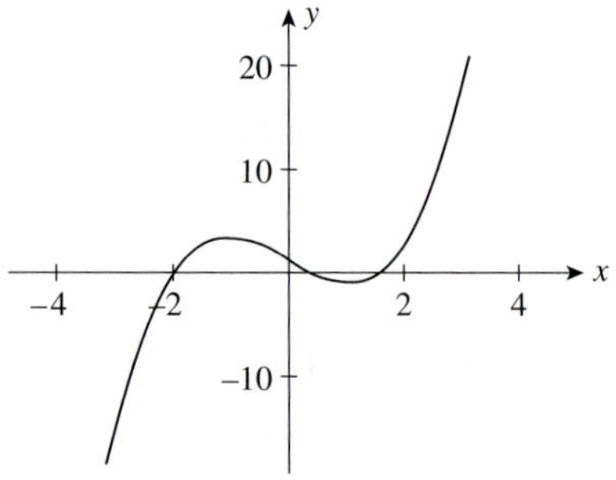

41.

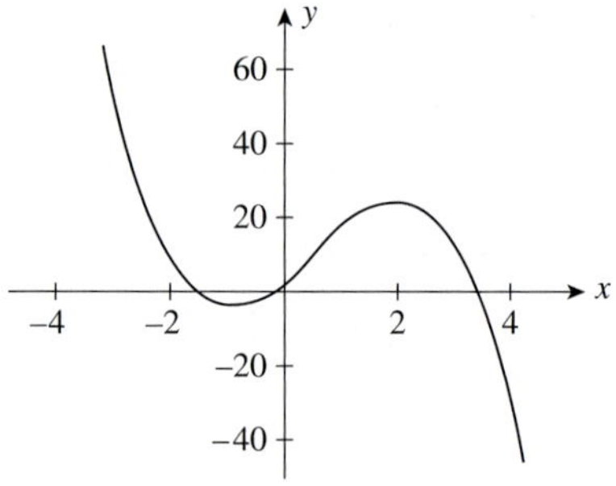

43.

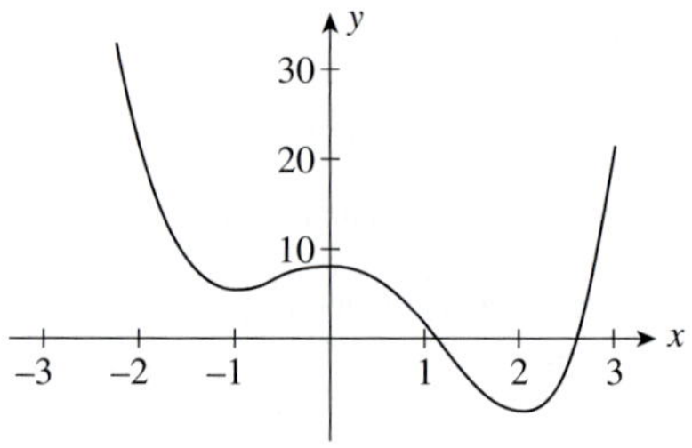

45.

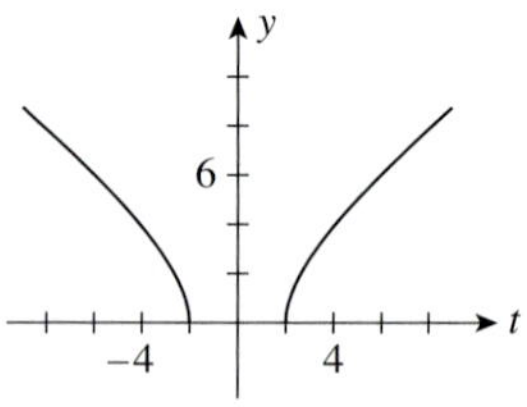

47.

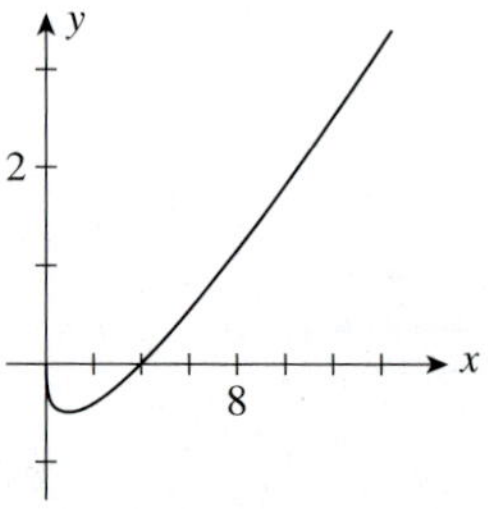

49.

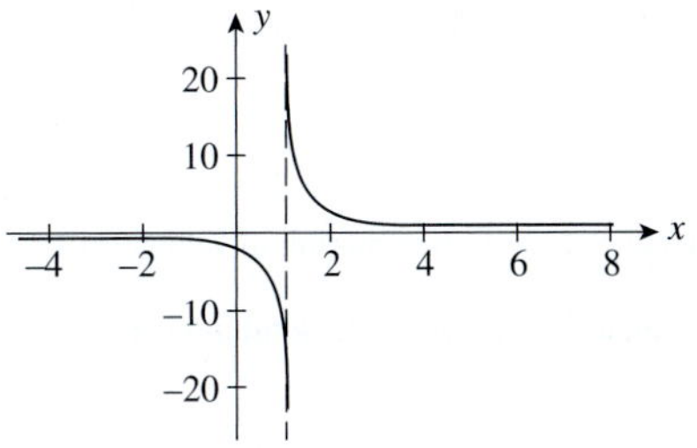

51.

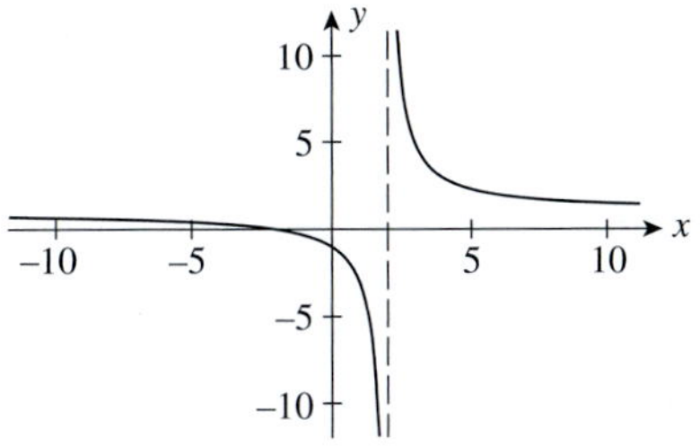

53.

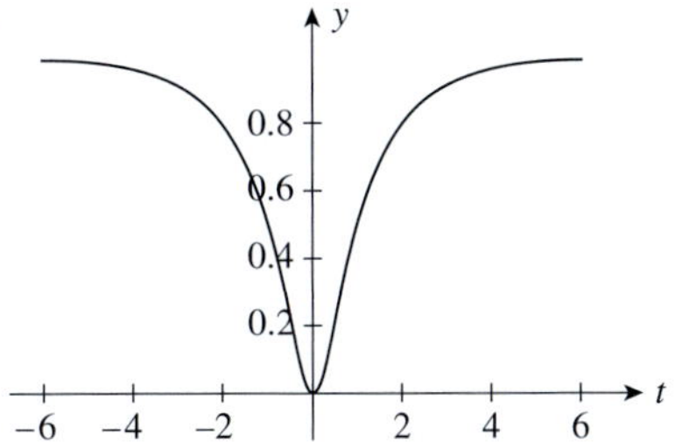

55.

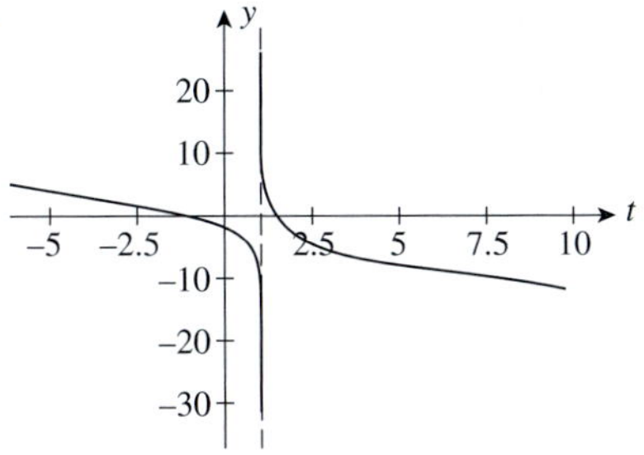

57.

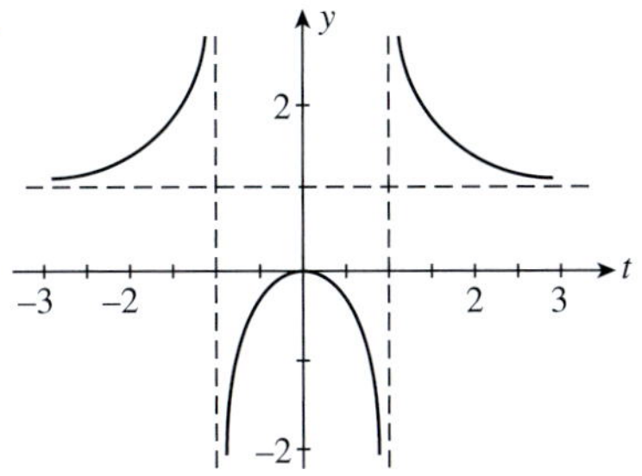

59.

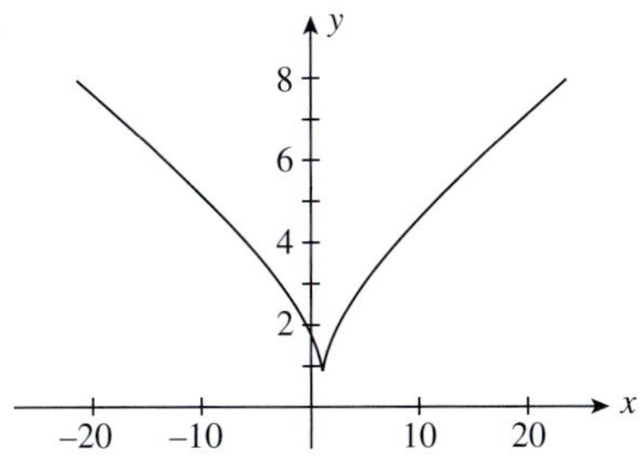

61. **a.** $x = 100$ **b.** No

63. **a.** $y = 0$
b. As time passes, the concentration of the drug decreases and approaches zero.

65.

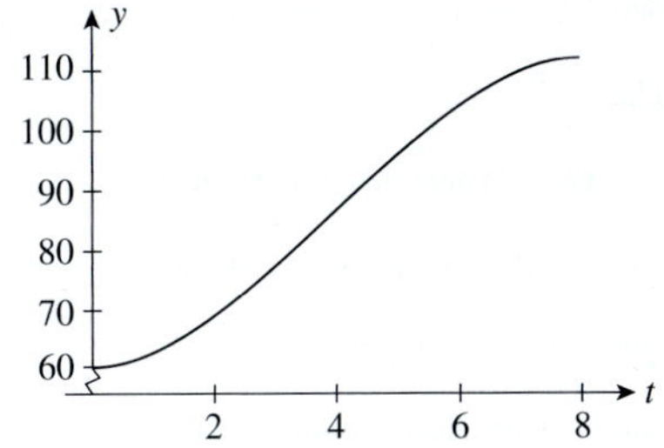

67.

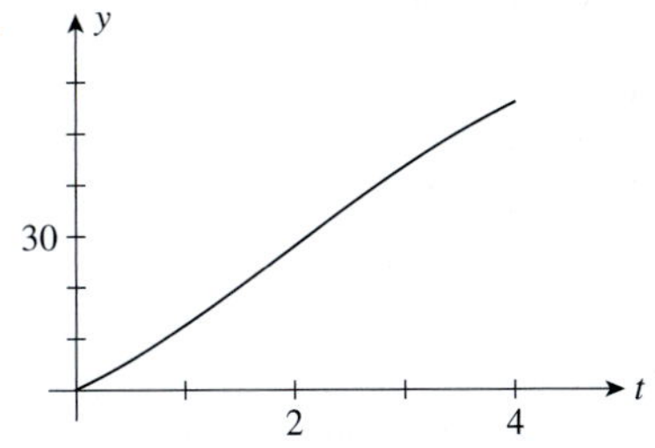

69.

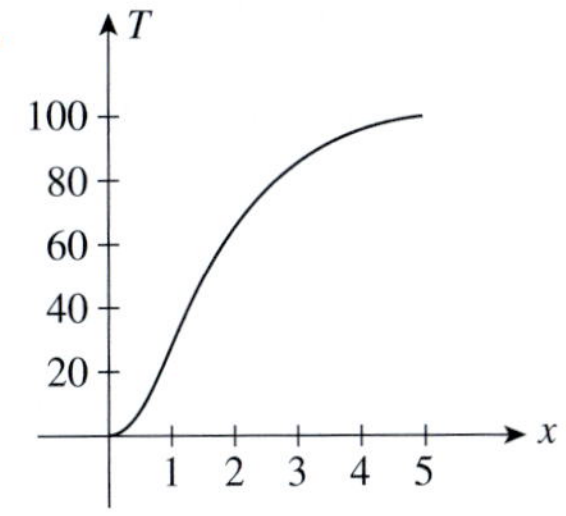

71.

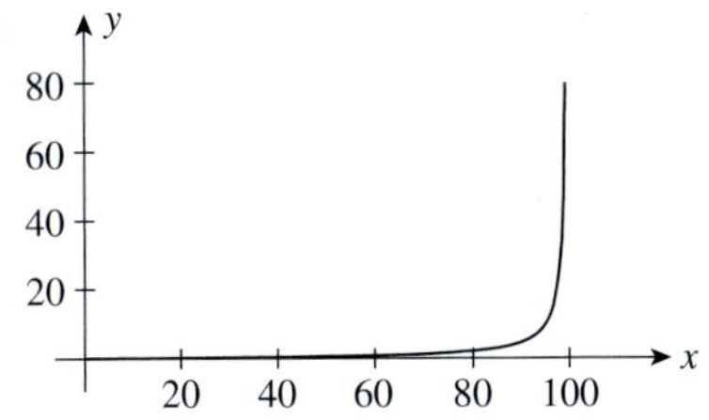

Using Technology Exercises 4.3, page 297

1.

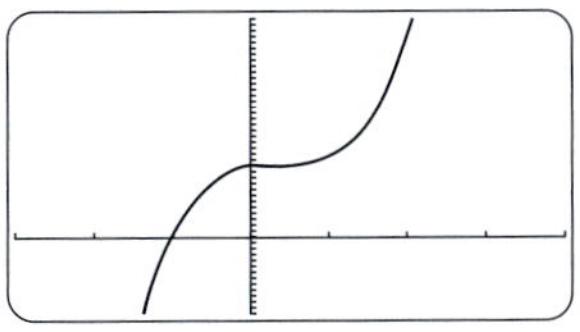

3.

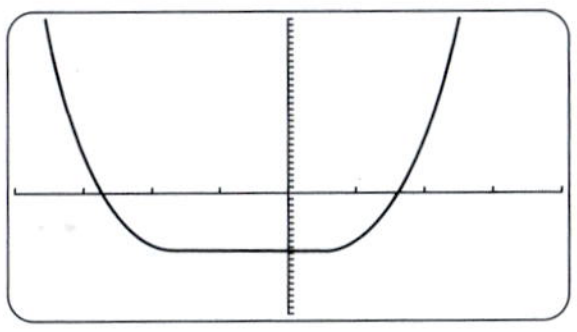

5. −0.9733; 2.3165, 4.6569 **7.** 1.5142

9.

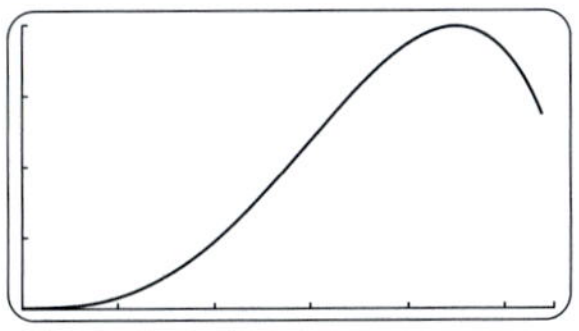

Exercises 4.4, page 305

1. None 3. Absolute minimum value: 0

5. Absolute maximum value: 3; absolute minimum value: -2

7. Absolute maximum value: 3; absolute minimum value: $-\frac{27}{16}$

9. Absolute minimum value: $-\frac{41}{8}$

11. No absolute extrema 13. Absolute maximum value: 1

15. Absolute maximum value: 5; absolute minimum value: -4

17. Absolute maximum value: 10; absolute minimum value: 1

19. Absolute maximum value: 19; absolute minimum value: -1

21. Absolute maximum value: 16; absolute minimum value: -1

23. Absolute maximum value: 3; absolute minimum value: $\frac{5}{3}$

25. Absolute maximum value: $\frac{37}{3}$; absolute minimum value: 5

27. Absolute maximum value ≈ 1.04; absolute minimum value: -1.5

29. No absolute extrema

31. Absolute maximum value: 1; absolute minimum value: 0

33. Absolute maximum value: 0; absolute minimum value: -3

35. Absolute maximum value: $\sqrt{2}/4 \approx 0.35$; absolute minimum value: $-\frac{1}{3}$

37. Absolute maximum value: $\sqrt{2}/2$; absolute minimum value: $-\sqrt{2}/2$

39. 144 ft 41. 17.72%

43. $f(6) = 3.60, f(0.5) = 1.13$; the number of nonfarm, full-time, self-employed women over the time interval from 1963 to 1993 reached its highest level, 3.6 million, in 1993.

45. \$3600 47. 6000 49. 3333

51. **a.** $0.0025x + 80 + \dfrac{10{,}000}{x}$ **b.** 2000 **c.** 2000 **d.** Same

53. 533

55. **a.** 2 days after the organic waste was dumped into the pond
b. 3.5 days after the organic waste was dumped into the pond

65. \$52.79/sq ft

67. **a.** 2000; \$105.8 billion **b.** 1995; \$7.6 billion

71. $R = r$; $\dfrac{E^2}{4r}$ watts 75. False 77. False

81. **c.**

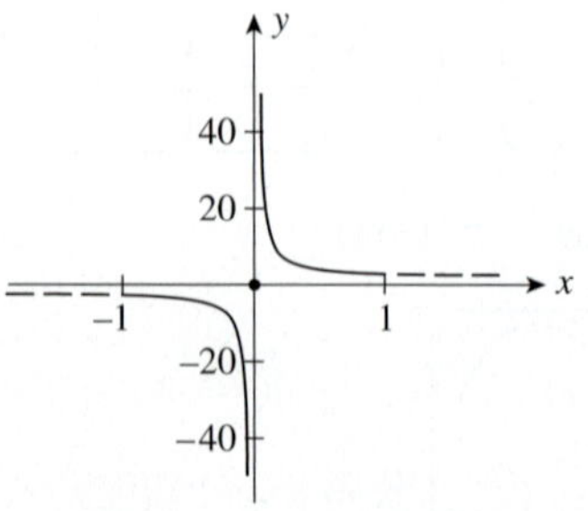

Using Technology Exercises 4.4, page 312

1. Absolute maximum value: 145.8985; absolute minimum value: -4.3834

3. Absolute maximum value: 16; absolute minimum value: -0.1257

5. Absolute maximum value: 2.8889; absolute minimum value: 0

7. **a.**

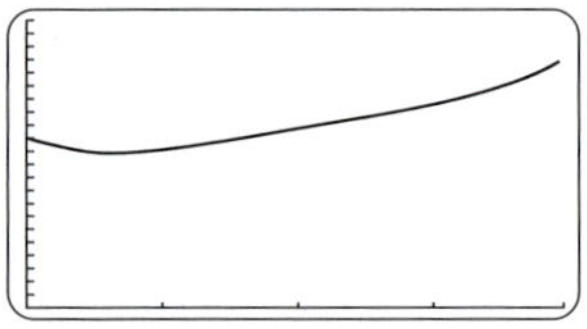

b. 21.51%

9. **b.** 1145

Exercises 4.5, page 319

1. 25 ft × 25 ft

3. 750 yd × 1500 yd; 1,125,000 yd^2

5. $10\sqrt{2}$ ft × $40\sqrt{2}$ ft

7. $\frac{16}{3}$ in. × $\frac{16}{3}$ in. × $\frac{4}{3}$ in.

9. 5.04 in. × 5.04 in. × 5.04 in.

11. 18 in. × 18 in. × 36 in.; 11,664 in.^{3}

13. $r = \dfrac{36}{\pi}$ in.; $l = 36$ in.; $\dfrac{46{,}656}{\pi}$ in.^{3}

15. $\frac{2}{3}\sqrt[3]{9}$ ft × $\sqrt[3]{9}$ ft × $\frac{2}{5}\sqrt[3]{9}$ ft

17. 250; \$62,500; \$250

19. 85; \$28,900; \$340 21. 60 miles/hr

23. $w \approx 13.86$ in.; $h \approx 19.60$ in.

25. $x = 2250$ ft 27. $x \approx 2.68$

29. 440 ft; 140 ft; 184,874 sq ft 31. 45, 44,445

Chapter 4 Concept Review, page 324

1. **a.** $f(x_1) < f(x_2)$ **b.** $f(x_1) > f(x_2)$

2. **a.** Increasing **b.** $f'(x) < 0$ **c.** Constant

3. **a.** $f(x) \le f(c)$ **b.** $f(x) \ge f(c)$

4. **a.** Domain; $= 0$; exist **b.** Critical number
c. Relative extremum

5. **a.** $f'(x)$ **b.** > 0 **c.** Concavity
d. Relative maximum; relative extremum

6. $\pm\infty$; $\pm\infty$ 7. 0; 0 8. b; b

9. **a.** $f(x) \le f(c)$; absolute maximum value
b. $f(x) \ge f(c)$; open interval

10. Continuous; absolute; absolute

Chapter 4 Review Exercises, page 324

1. **a.** f is increasing on $(-\infty, 1) \cup (1, \infty)$ **b.** No relative extrema
c. Concave down on $(-\infty, 1)$; concave up on $(1, \infty)$
d. $(1, -\frac{17}{3})$

2. **a.** f is increasing on $(-\infty, 2) \cup (2, \infty)$ **b.** No relative extrema
c. Concave down on $(-\infty, 2)$; concave up on $(2, \infty)$ **d.** $(2, 0)$

3. **a.** f is increasing on $(-1, 0) \cup (1, \infty)$ and decreasing on $(-\infty, -1) \cup (0, 1)$
b. Relative maximum value: 0; relative minimum value: -1
c. Concave up on $\left(-\infty, -\frac{\sqrt{3}}{3}\right) \cup \left(\frac{\sqrt{3}}{3}, \infty\right)$; concave down on $\left(-\frac{\sqrt{3}}{3}, \frac{\sqrt{3}}{3}\right)$
d. $\left(-\frac{\sqrt{3}}{3}, -\frac{5}{9}\right)$; $\left(\frac{\sqrt{3}}{3}, -\frac{5}{9}\right)$

4. **a.** f is increasing on $(-\infty, -2) \cup (2, \infty)$ and decreasing on $(-2, 0) \cup (0, 2)$
b. Relative maximum value: -4; relative minimum value: 4
c. Concave down on $(-\infty, 0)$; concave up on $(0, \infty)$ **d.** None

5. **a.** f is increasing on $(-\infty, 0) \cup (2, \infty)$; decreasing on $(0, 1) \cup (1, 2)$
b. Relative maximum value: 0; relative minimum value: 4
c. Concave up on $(1, \infty)$; concave down on $(-\infty, 1)$ **d.** None

6. **a.** f is increasing on $(1, \infty)$ **b.** No relative extrema
c. Concave down on $(1, \infty)$ **d.** None

7. **a.** f is decreasing on $(-\infty, 1) \cup (1, \infty)$ **b.** No relative extrema
c. Concave down on $(-\infty, 1)$; concave up on $(1, \infty)$ **d.** $(1, 0)$

8. **a.** f is increasing on $(1, \infty)$ **b.** No relative extrema
c. Concave down on $(1, \frac{4}{3})$; concave up on $(\frac{4}{3}, \infty)$
d. $\left(\frac{4}{3}, \frac{4\sqrt{3}}{9}\right)$

9. **a.** f is increasing on $(-\infty, -1) \cup (-1, \infty)$
b. No relative extrema
c. Concave down on $(-1, \infty)$; concave up on $(-\infty, -1)$
d. None

10. **a.** f is decreasing on $(-\infty, 0)$ and increasing on $(0, \infty)$
b. Relative minimum value: -1
c. Concave down on $\left(-\infty, -\frac{1}{\sqrt{3}}\right) \cup \left(\frac{1}{\sqrt{3}}, \infty\right)$; concave up on $\left(-\frac{1}{\sqrt{3}}, \frac{1}{\sqrt{3}}\right)$
d. $\left(-\frac{1}{\sqrt{3}}, -\frac{3}{4}\right)$; $\left(\frac{1}{\sqrt{3}}, -\frac{3}{4}\right)$

11.

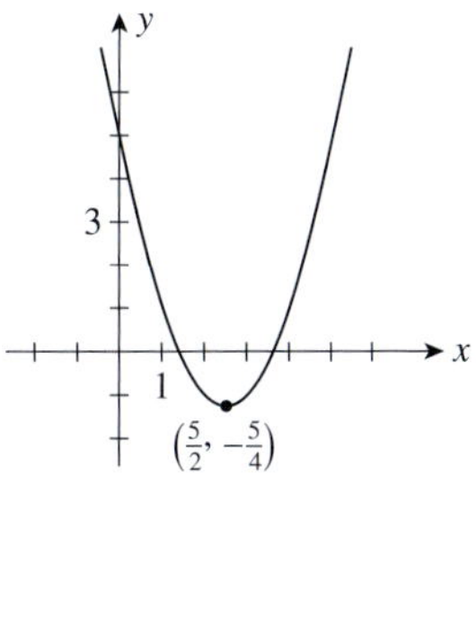

12.

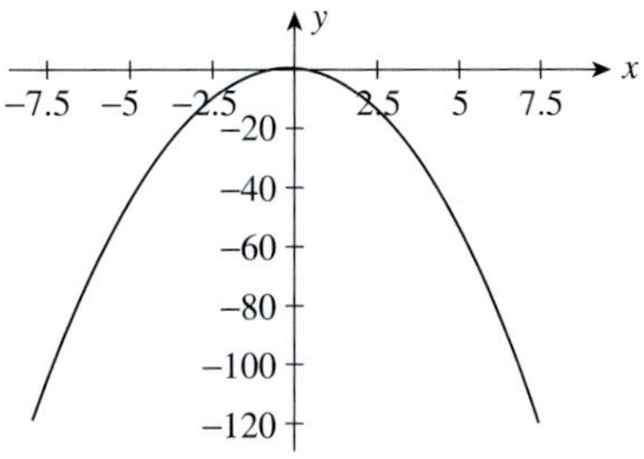

13.

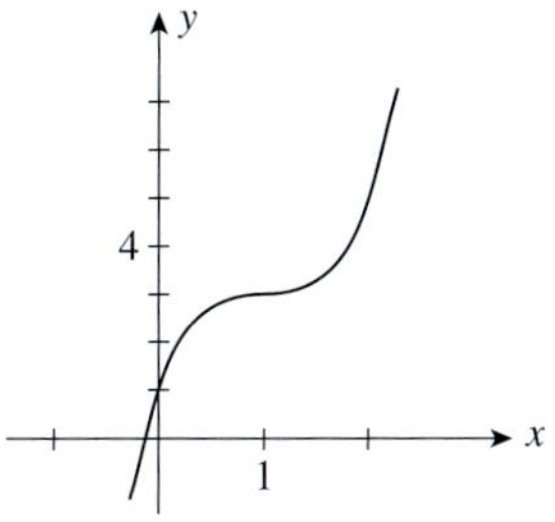

14.

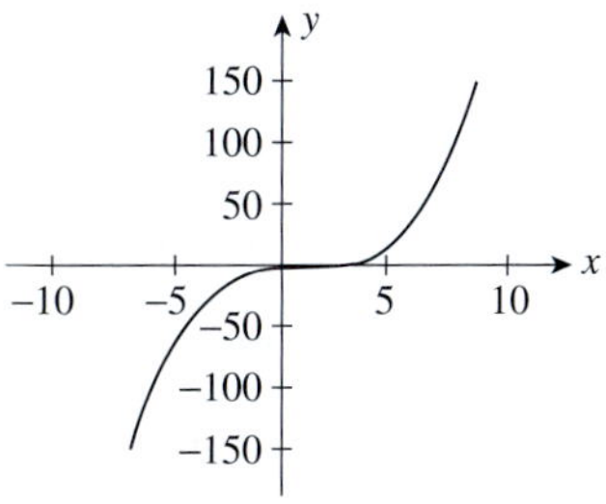

15.

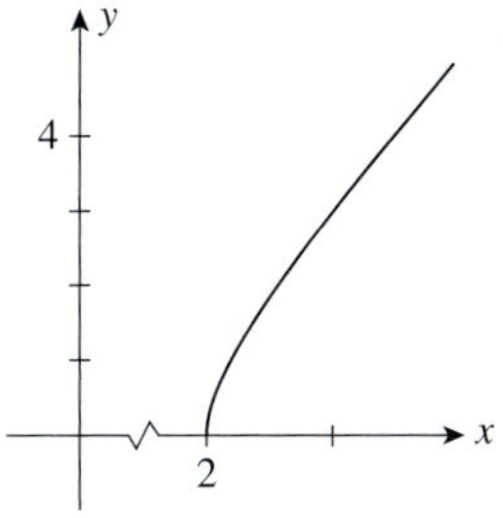

16.

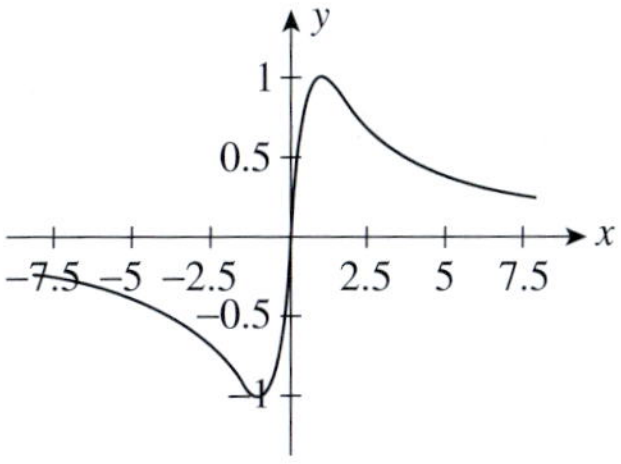

17.

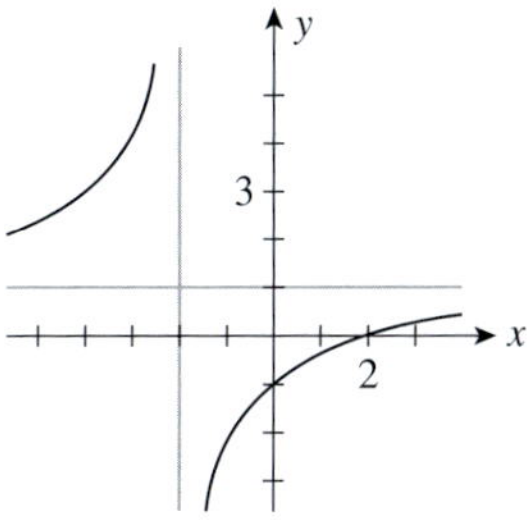

18.

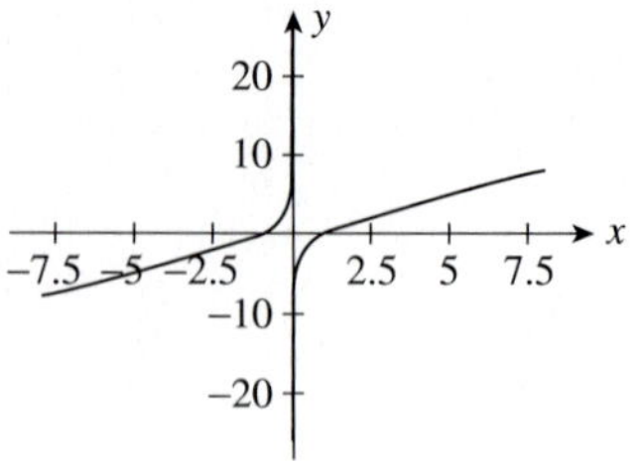

19. Vertical asymptote: $x = -\frac{3}{2}$; horizontal asymptote: $y = 0$

20. Horizontal asymptote: $y = 2$; vertical asymptote: $x = -1$

21. Vertical asymptotes: $x = -2, x = 4$, horizontal asymptote: $y = 0$

22. Horizontal asymptote: $y = 1$; vertical asymptote: $x = 1$

23. Absolute minimum value: $-\frac{25}{8}$

24. Absolute minimum value: 0

25. Absolute maximum value: 5; absolute minimum value: 0

26. Absolute maximum value: $\frac{5}{3}$; absolute minimum value: 1

27. Absolute maximum value: -16; absolute minimum value: -32

28. Absolute maximum value: $\frac{1}{2}$; absolute minimum value: 0

29. Absolute maximum value: $\frac{8}{3}$; absolute minimum value: 0

30. Absolute maximum value: $\frac{215}{9}$; absolute minimum value: 7

31. Absolute maximum value: $\frac{1}{2}$; absolute minimum value: $-\frac{1}{2}$

32. No absolute extrema **33.** \$4000

34. **c.** Online travel spending is expected to increase at an increasing rate over that period of time.

35. **a.** R is increasing on (0, 6).
b. The revenue is always increasing from 1997 through 2003.

36. **a.** Decreasing on (0, 21.4); increasing on (21.4, 30)
b. The percentage of men 65 yr and older in the workforce was decreasing from 1970 until mid-1991 and increasing from mid-1991 through 2000.

37. **a.** Concave downward on (0, 0.67) and concave upward on (0.67, 4).
b. (0.67,138.1); the rate of increase of shipments is slowest at $t = 0.67$; that is a little after mid-2001.

38. **a.** $16.25t + 24.625$; sales were increasing.
b. 16.25; the rate of sales was increasing from 2002 to 2005.

39. (100, 4600); sales increase rapidly until \$100,000 is spent on advertising; after that, any additional expenditure results in increased sales but at a slower rate of increase.

40. (266.67, 11,874.08); the rate of increase is lowest when 267 calculators are produced.

41. **a.** $I'(t) = -\dfrac{200t}{(t^2 + 10)^2}$

b. $I''(t) = \dfrac{-200(10 - 3t^2)}{(t^2 + 10)^3}$; concave up on $(\sqrt{10/3}, \infty)$; concave down on $(0, \sqrt{10/3})$

c.

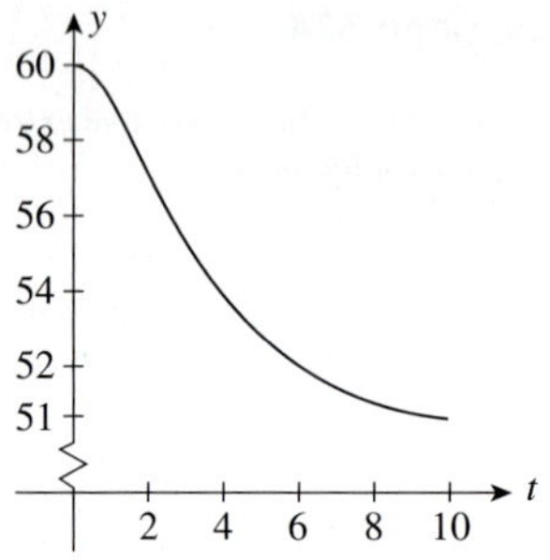

d. The rate of decline in the environmental quality of the wildlife was increasing the first 1.8 yr. After that time the rate of decline decreased.

42. 168 **43.** 3000

44. **a.** $0.001x + 100 + \dfrac{4000}{x}$ **b.** 2000

45. 10 a.m.

46. **a.** Decreasing on (0, 12.7); increasing on (12.7, 30)
b. (12.7, 7.9)
c. The percent of women 65 years and older in the workforce was decreasing from 1970 to Sept. 1982 and increasing from Sept. 1982 to 2000. It reached a minimum value of 7.9% in Sept. 1982.

48. 74.07 in.^{3} **49.** Radius: 2 ft; height: 8 ft

50. 1 ft × 2 ft × 2 ft **51.** 20,000 cases

52. If $a > 0$, f is decreasing on $(-\infty, -\frac{b}{2a})$ and increasing on $(-\frac{b}{2a}, \infty)$; if $a < 0$, f is increasing on $(-\infty, -\frac{b}{2a})$ and decreasing on $(-\frac{b}{2a}, \infty)$.

53. $a = -4$; $b = 11$ **54.** $c \geq \frac{3}{2}$

56. **a.** $f'(x) = 3x^2$ if $x \neq 0$ **b.** No

Chapter 4 Before Moving On, page 327

1. Decreasing on $(-\infty, 0) \cup (2, \infty)$; increasing on $(0, 1) \cup (1, 2)$

2. Rel. min: $(1, -10)$

3. Concave downward on $(-\infty, \frac{1}{4})$; concave upward on $(\frac{1}{4}, \infty)$; $(\frac{1}{4}, \frac{83}{96})$

4.

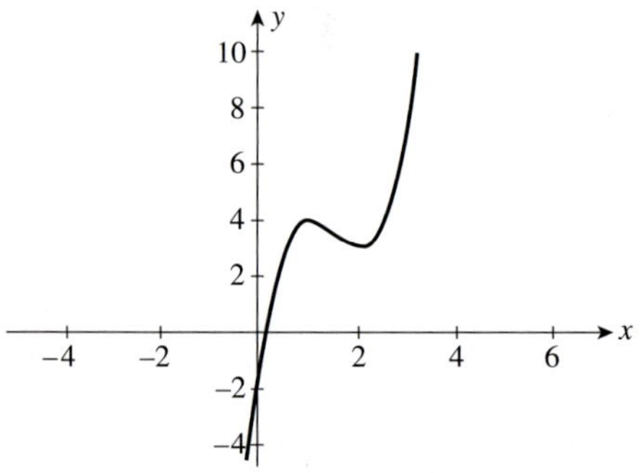

5. Abs. min. value: -5; abs. max. value: 80

6. $r = h = \dfrac{1}{\sqrt[3]{\pi}}$ (ft)

CHAPTER 5

Exercises 5.1, page 334

1. a. 16 **b.** 27 **3. a.** 3 **b.** $\sqrt{5}$

5. a. -3 **b.** 8 **7. a.** 25 **b.** $4^{1.8}$

9. a. $4x^3$ **b.** $5xy^2\sqrt{x}$ **11. a.** $\dfrac{2}{a^2}$ **b.** $\frac{1}{3}b^2$

13. a. $8x^9y^6$ **b.** $16x^4y^4z^6$ **15. a.** $\dfrac{64x^6}{y^4}$ **b.** $(x-y)(x+y)$

17. 2 **19.** 3 **21.** 3 **23.** $\frac{5}{4}$ **25.** 1 or 2

27.

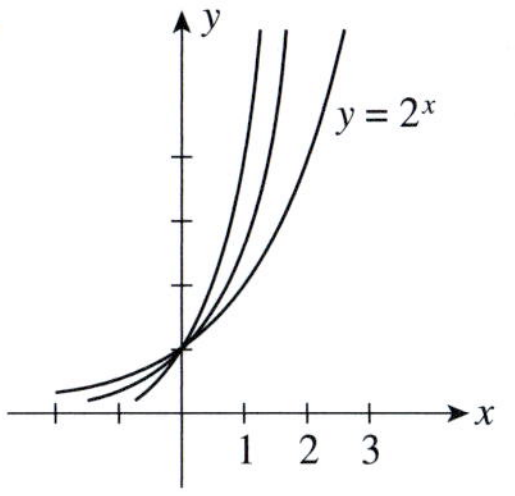

29.

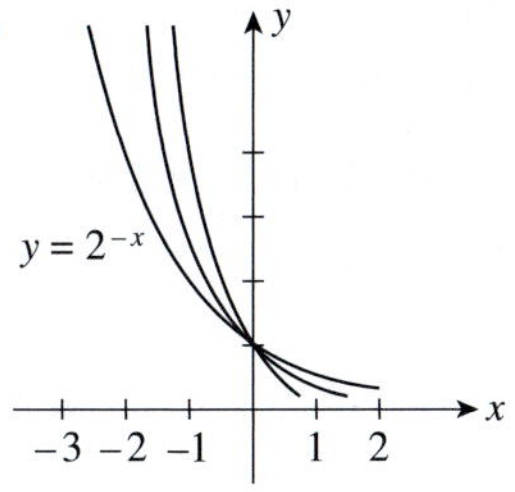

31.

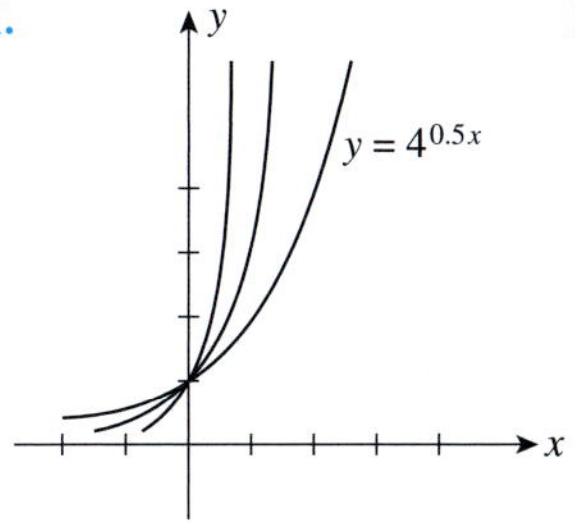

33.

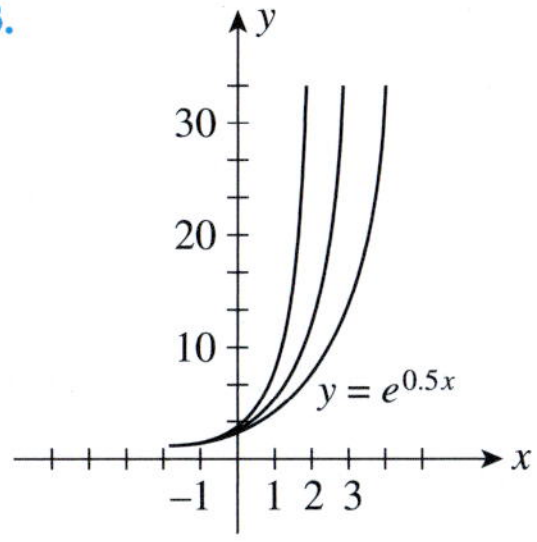

35.

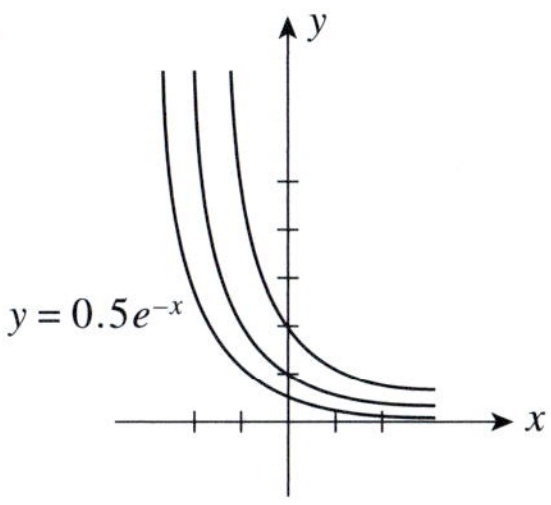

37. $f(x) = 100(\frac{6}{5})^x$ **39.** 54.56

41. a. 26.3%; 24.67%; 21.71%; 19.72%

b.

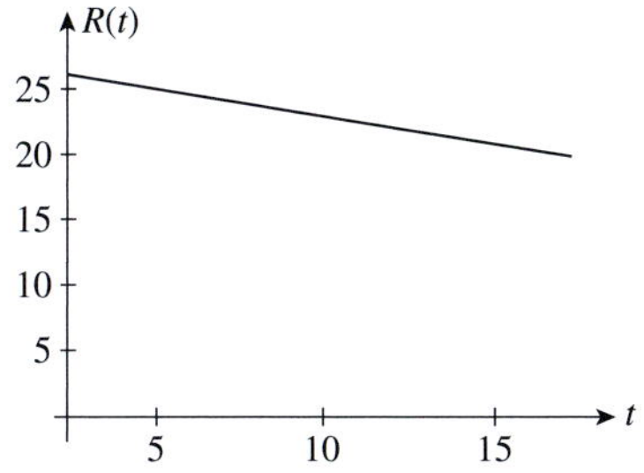

43. a.

Year	0	1	2	3	4	5
Web Addresses (billions)	0.45	0.80	1.41	2.49	4.39	7.76

b.

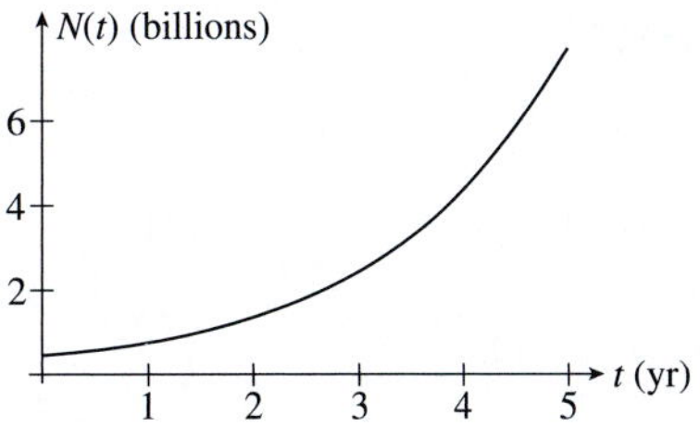

45. 34,210,000

47. a. 0.08 g/cm³ **b.** 0.12 g/cm³ **c.** 0.2 g/cm³

d.

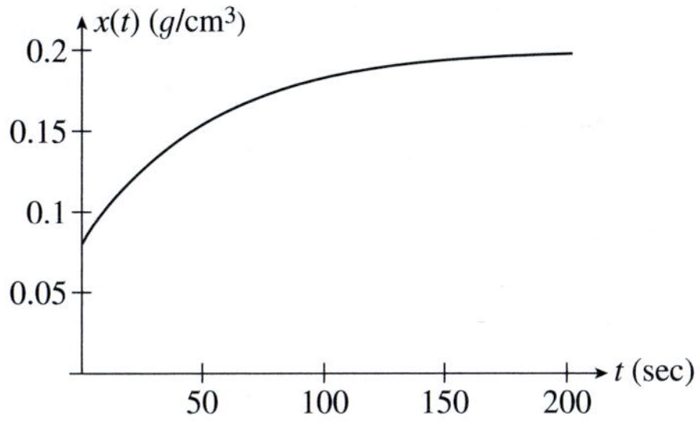

49. False **51.** True

Using Technology Exercises 5.1, page 336

1.

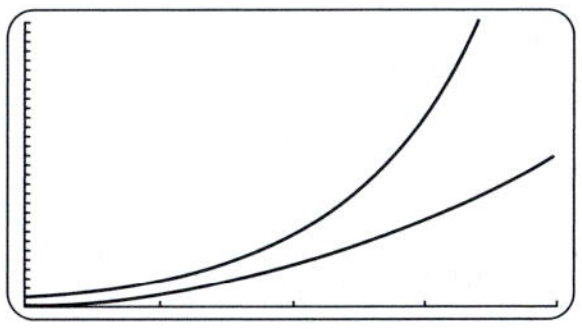

3.

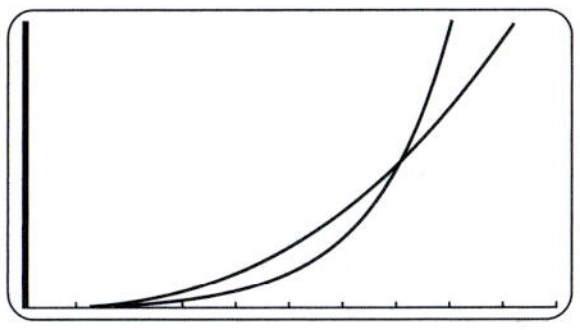

5.

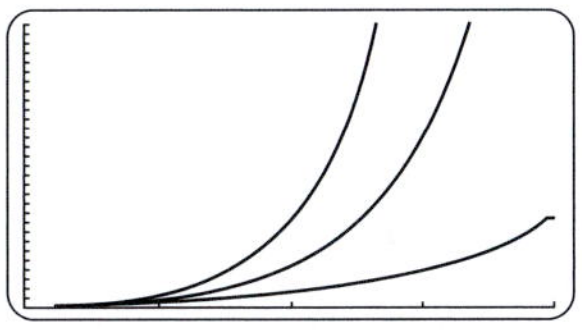

7.

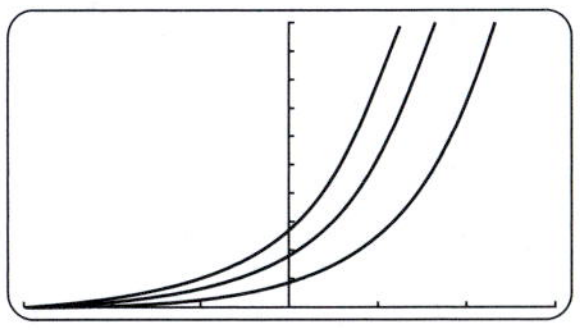

9.

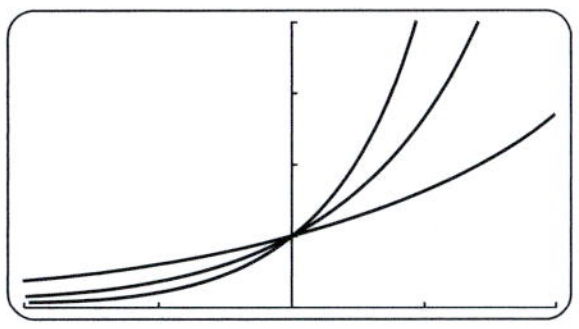

11. a.

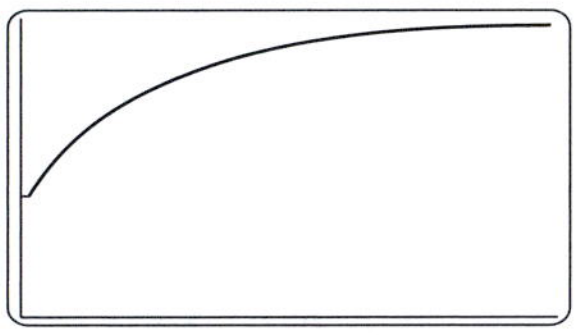

b. 0.08 g/cm³ **c.** 0.12 g/cm³ **d.** 0.2 g/cm³

13. a.

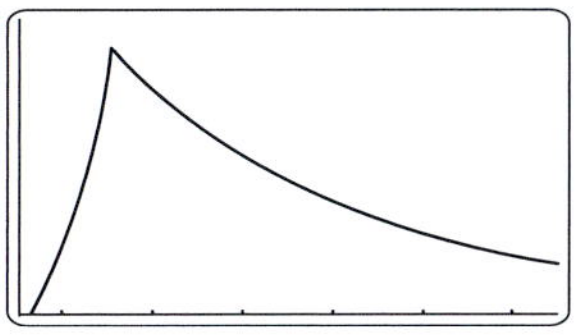

b. 20 sec **c.** 35.1 sec

Exercises 5.2, page 343

1. $\log_2 64 = 6$ 3. $\log_3 \frac{1}{9} = -2$ 5. $\log_{1/3} \frac{1}{3} = 1$

7. $\log_{32} 8 = \frac{3}{5}$ 9. $\log_{10} 0.001 = -3$ 11. 1.0792

13. 1.2042 15. 1.6813 17. $\ln a^2b^3$ 19. $\ln \dfrac{3\sqrt{xy}}{\sqrt[3]{z}}$

21. $\log x + 4\log(x+1)$ 23. $\frac{1}{2}\log(x+1) - \log(x^2+1)$

25. $\ln x - x^2$ 27. $-\frac{3}{2}\ln x - \frac{1}{2}\ln(1+x^2)$

29.

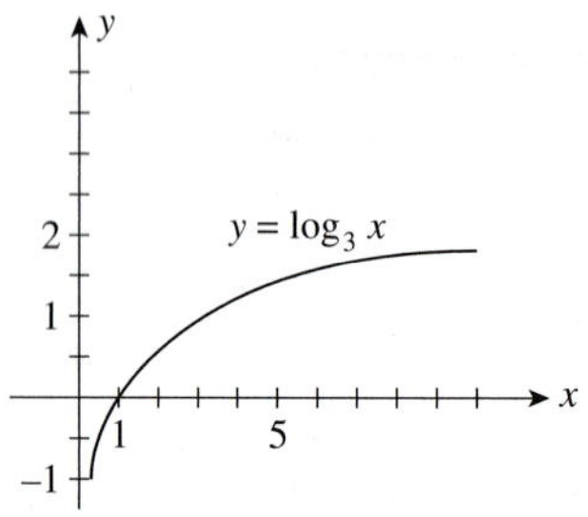

31.

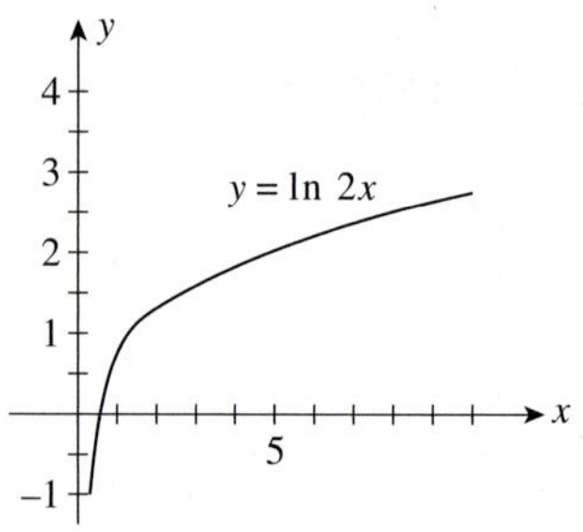

33.

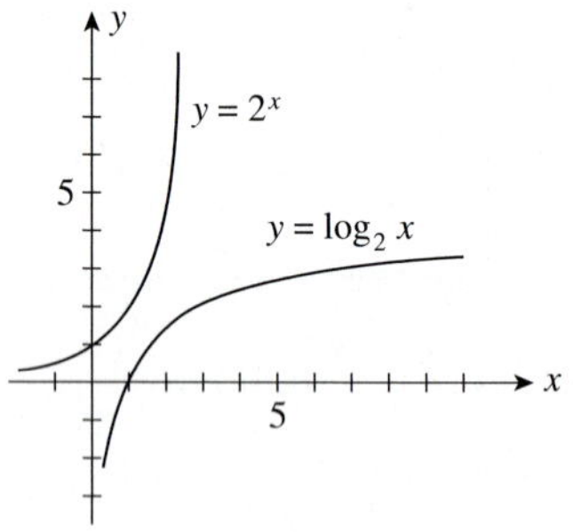

35. 5.1986 37. -0.0912 39. -8.0472 41. -4.9041

43. $-2\ln\left(\dfrac{A}{B}\right)$ 45. $2 + 2.8854\ln x$ 47. 105.7 mm

49. **a.** $10^3 I_0$ **b.** 100,000 times greater
c. 10,000,000 times greater

51. 27.40 yr 53. 6.44 yr 55. **a.** 9.12 sec **b.** 20.27 sec

57. False 59. True 61. **a.** ln 2

Exercises 5.3, page 356

1. \$4974.47 3. \$223,403.11

5. **a.** 10.25%/yr **b.** 9.31%/yr

7. **a.** \$29,227.61 **b.** \$29,137.83

9. \$6885.64 11. \$8693.70 13. \$16,705.40

15. 13.59%/yr 17. 12.1%/yr 19. 14.87%/yr

21. 2.2 yr 23. 7.7 yr 25. 6.08%/yr 27. 2.06 yr

29. \$254,084.69 31. \$2.844 million 33. \$23,329.48

35. \$731,250 37. 91¢ 39. 12.75% 41. \$40,000

43. **a.** \$33,885.14 **b.** \$33,565.38

45. **a.** \$16,262.79 **b.** \$12,047.77 **c.** \$6611.96

49. Bank B 51. 9.531%

Using Technology Exercises 5.3, page 360

1. \$5872.78 3. 8.95%/yr 5. \$29,743.30

Exercises 5.4, page 366

1. $3e^{3x}$ 3. $-e^{-t}$ 5. $e^x + 1$ 7. $x^2e^x(x+3)$

9. $\dfrac{2e^x(x-1)}{x^2}$ 11. $3(e^x - e^{-x})$ 13. $-\dfrac{1}{e^w}$ 15. $6e^{3x-1}$

17. $-2xe^{-x^2}$ 19. $\dfrac{3e^{-1/x}}{x^2}$ 21. $25e^x(e^x+1)^{24}$ 23. $\dfrac{e^{\sqrt{x}}}{2\sqrt{x}}$

25. $e^{3x+2}(3x-2)$ 27. $\dfrac{2e^x}{(e^x+1)^2}$ 29. $2(8e^{-4x} + 9e^{3x})$

31. $6e^{3x}(3x+2)$ 33. $y = 2x - 2$

35. f is increasing on $(-\infty, 0)$ and decreasing on $(0, \infty)$.

37. Concave downward on $(-\infty, 0)$; concave upward on $(0, \infty)$

39. $(1, e^{-2})$ 41. $y = e^{-1/2}(-\sqrt{2}x + 2)$; $y = e^{-1/2}(\sqrt{2}x + 2)$

43. Absolute maximum value: 1; absolute minimum value: e^{-1}

45. Absolute minimum value: -1; absolute maximum value: $2e^{-3/2}$

47.

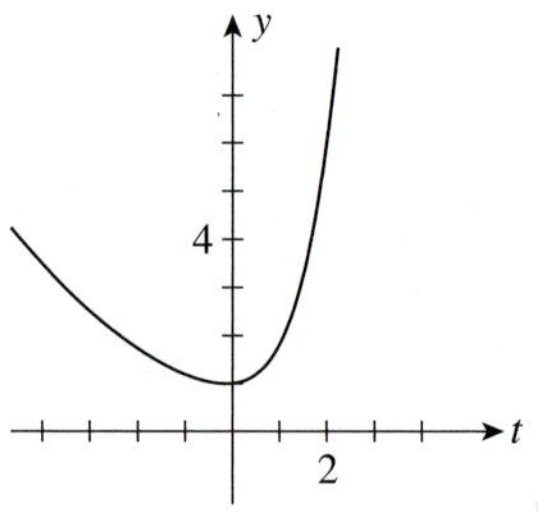

49.

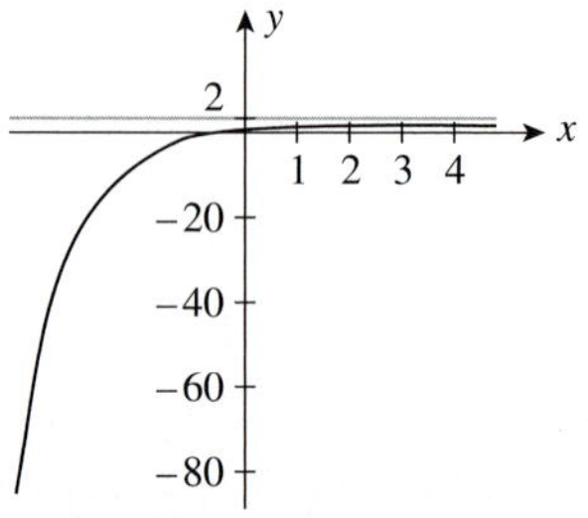

51. $-0.1694, -0.1549, -0.1415$; the percentage of the total population relocating was decreasing at the rate of 0.17%/yr in 1970, 0.15%/yr in 1980, and 0.14%/yr in 1990.

53. **a.** 70,000; 353,716 **b.** 37,800/decade; 191,000/decade

55. **a.** \$4.6 trillion; \$3.62 trillion
b. \$184 billion/yr; \$145 billion/yr

57. **a.** −\$6065/day; −\$3679/day; −\$2231/day; −\$1353/day
b. 2 days

59. **b.** 4505/yr; 273 cases/yr

61. **a.** 181/100,000
b. 0/100,000 people; −27/100,000 people; −38/100,000 people; −32/100,000 people
c. Around 1970 **d.** 52/100,000

63. 10,000; \$367,879

65. **a.** −1.68¢/case **b.** \$40.36/case

67. **a.** 45.6 kg **b.** 4.2 kg 69. 7.72 yr; \$160,207.69

71. $2\frac{1}{2}$ hours after drinking; 0.2115%

73. **a.** 70°F **b.** −14.7°F **c.** 30°F

75. **a.** $\dfrac{\ln(\frac{a}{b})}{a-b}$ sec **b.** $\dfrac{2\ln(\frac{a}{b})}{a-b}$ sec

79. **a.** 0.05 g/cm³/sec **b.** −0.01 g/cm³/sec **c.** 20 sec
d. 0.90 g/cm³

81. False 83. True

Using Technology Exercises 5.4, page 370

1. 5.4366 3. 12.3929 5. 0.1861

7. **a.** 50 **c.**

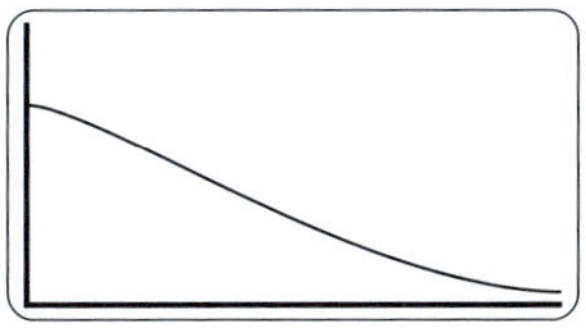

9. **a.**

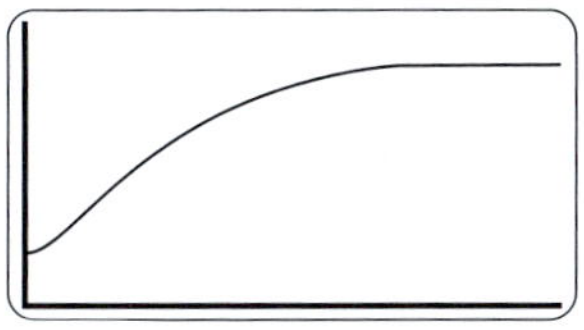

b. 4.2720 billion/half century

11. **a.** 153,024; 235,181 **b.** −634; 18,401

13. **a.** 69.63% **b.** 5.094%/decade

Exercises 5.5, page 377

1. $\dfrac{5}{x}$ 3. $\dfrac{1}{x+1}$ 5. $\dfrac{8}{x}$ 7. $\dfrac{1}{2x}$ 9. $\dfrac{-2}{x}$

11. $\dfrac{2(4x-3)}{4x^2-6x+3}$ 13. $\dfrac{1}{x(x+1)}$ 15. $x(1+2\ln x)$

17. $\dfrac{2(1-\ln x)}{x^2}$ 19. $\dfrac{3}{u-2}$ 21. $\dfrac{1}{2x\sqrt{\ln x}}$

23. $\dfrac{3(\ln x)^2}{x}$ 25. $\dfrac{3x^2}{x^3+1}$ 27. $\dfrac{(x\ln x+1)e^x}{x}$

29. $\dfrac{e^{2t}[2(t+1)\ln(t+1)+1]}{t+1}$ 31. $\dfrac{1-2\ln x}{x^3}$ 33. $\dfrac{1}{x\ln x}$

35. $-\dfrac{1}{x^2}$ 37. $\dfrac{2(2-x^2)}{(x^2+2)^2}$ 39. $3+2\ln x$

41. $(x+1)(5x+7)(x+2)^2$

43. $(x-1)(x+1)^2(x+3)^3(9x^2+14x-7)$

45. $\dfrac{(2x^2-1)^4(38x^2+40x+1)}{2(x+1)^{3/2}}$ 47. $3^x\ln 3$

49. $(x^2+1)^{x-1}[2x^2+(x^2+1)\ln(x^2+1)]$ 51. $(\ln x+1)y$

53. $y = x - 1$

55. f is decreasing on $(-\infty, 0)$ and increasing on $(0, \infty)$.

57. Concave up: $(-\infty, -1) \cup (1, \infty)$; concave down: $(-1, 0) \cup (0, 1)$

59. $(-1, \ln 2)$ and $(1, \ln 2)$ 61. $y = 4x - 3$

63. Absolute minimum value: 1; absolute maximum value: $3 - \ln 3$

65. 0.0580%/kg; 0.0133%/kg

67. **a.** 78.82 million **b.** 3.95 million/yr

69. **a.** \$38,400 **b.** −22.3%

71. **b.** W is concave downward on (1, 6).

73. **b.**

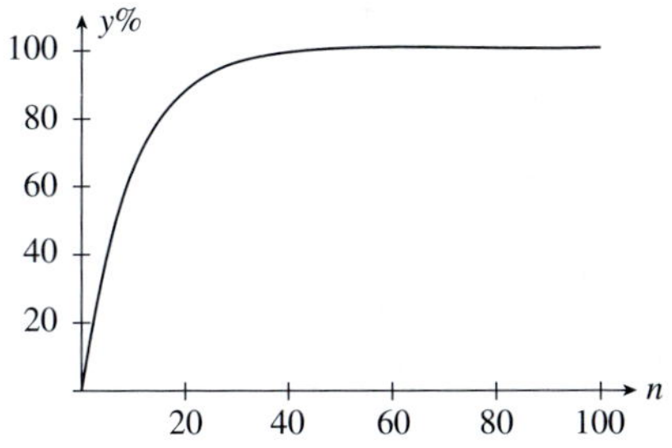

c. 100

77. $\dfrac{V(x_0-p)}{x_0-p+k}$

79.

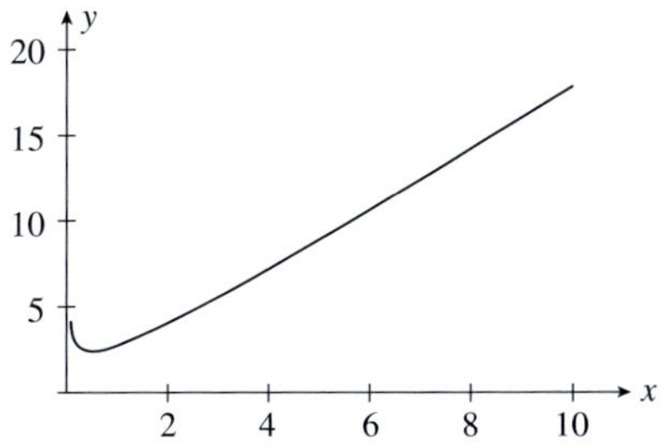

81. **b.** $\dfrac{1}{(\ln 10)x}$

83. $\dfrac{[(x+1)(\ln[10])-1]10^x}{(x+1)^2}$

85. $2x\left[(\ln 3)3^{x^2} + \dfrac{1}{(\ln 2)(x^2+1)}\right]$

87. True

Exercises 5.6, page 386

1. **a.** 0.05 **b.** 400
c.

t	0	10	20	100	1000
Q	400	660	1087	59,365	2.07×10^{24}

3. **a.** $Q(t) = 100e^{0.035t}$ **b.** 266 min **c.** $Q(t) = 1000e^{0.035t}$

5. **a.** 54.93 yr **b.** 14.25 billion 7. 8.7 lb/in.^{2}; -0.0004 lb/in.^{2}/ft

9. $Q(t) = 100e^{-0.049t}$; 70.7 g; 3.451 g/day 11. 13,412 yr ago

13.
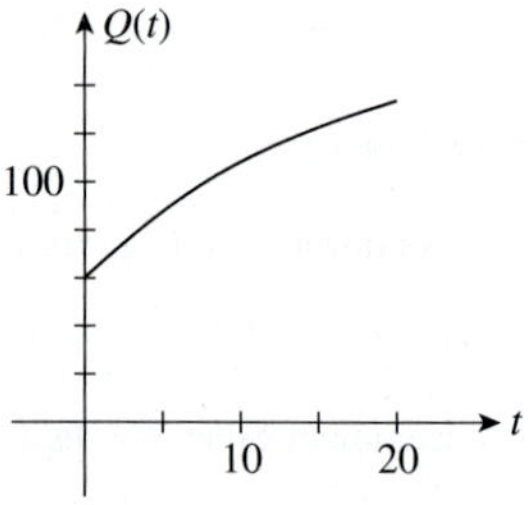

a. 60 words/min **b.** 107 words/min **c.** 136 words/min

15. **a.** \$5.806 trillion; \$8.575 trillion
b. \$0.45 trillion/yr; \$0.67 trillion/yr

17.
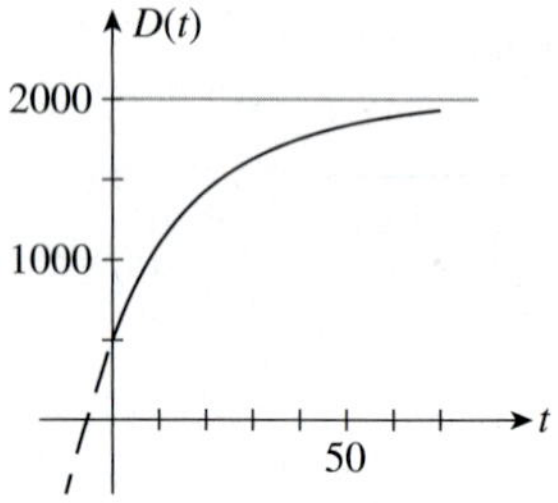

a. 573 computers; 1177 computers; 1548 computers; 1925 computers
b. 2000 computers **c.** 46 computers/mo

19. **a.** 122.3 cm **b.** 14 cm/yr **c.** 200 cm

21. **a.** 86.1% **b.** 10.44%/yr **c.** 1970

23. 76.4 million 25. 1080; 280/hr 27. 15 lb

29. **a.** $\dfrac{\ln \frac{b}{a}}{b-a}$ min **b.** 0 g/cm^3 31. **b.** 5599 yr

33. **b.** Q increases most rapidly at $t = \dfrac{\ln B}{k}$.

35. 0.14

Using Technology Exercises 5.6, page 391

1. **a.**
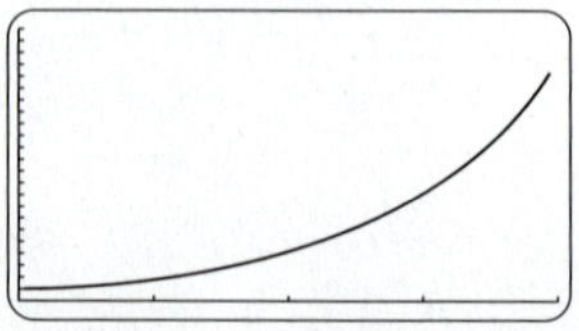

b. 12.146%/yr **c.** 9.474%/yr/yr

3. **a.**
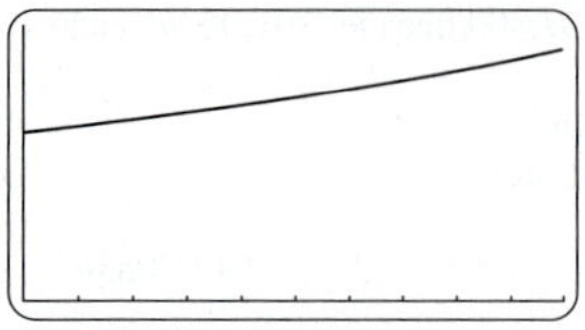

b. 666 million, 926.8 million **c.** 38.3 million/yr

5. **a.**
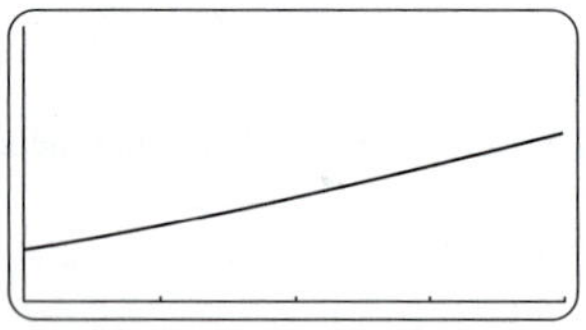

b. 325 million **c.** 76.84 million/30 yr

7. **a.**
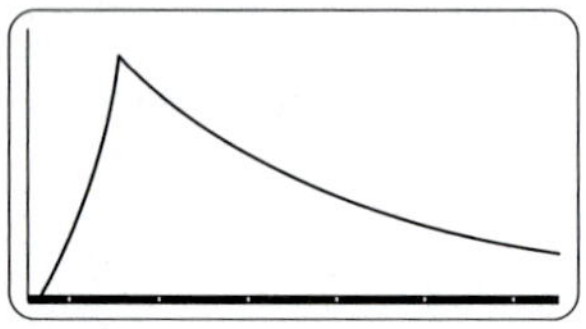

b. 0 **c.** 0.237 g/cm^3
d. 0.760 g/cm^3 **e.** 0

9. **a.**
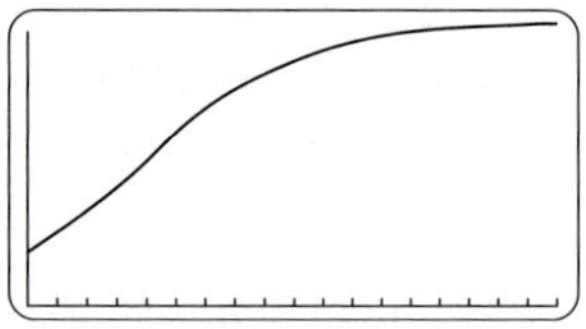

b. 36.6%; 84.2% **c.** 7.9%/yr; 1%/yr
d. The end of 1984

11. **a.** $f(t) = \dfrac{544.65}{1 + 1.65e^{-0.1846t}}$
b.
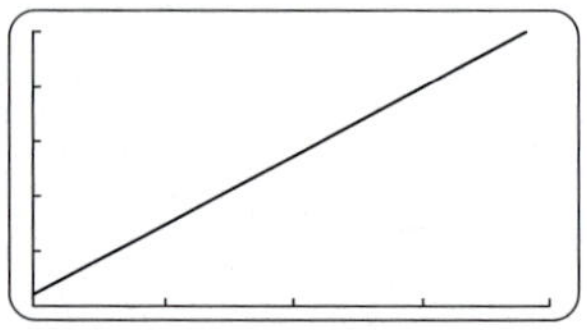

c. 24.5 million/yr; 25.1 million/yr

Chapter 5 Concept Review, page 393

1. Power; 0; 1; exponential

2. **a.** $(-\infty, \infty)$; $(0, \infty)$ **b.** $(0, 1)$; $(-\infty, \infty)$

3. **a.** $(0, \infty)$; $(-\infty, \infty)$; $(1, 0)$ **b.** <1; >1

4. **a.** x **b.** x

5. Accumulated amount; principal; nominal interest rate; number of conversion periods; term

6. $(1 + \frac{r}{m})^m - 1$ 7. Pe^{rt}

8. **a.** $e^{f(x)}f'(x)$ **b.** $\dfrac{f'(x)}{f(x)}$

9. **a.** Initially; growth **b.** Decay **c.** Time; one-half

10. **a.** Horizontal asymptote; C
b. Horizontal asymptote; A, carrying capacity

Chapter 5 Review Exercises, page 394

1. **a.** and **b.**

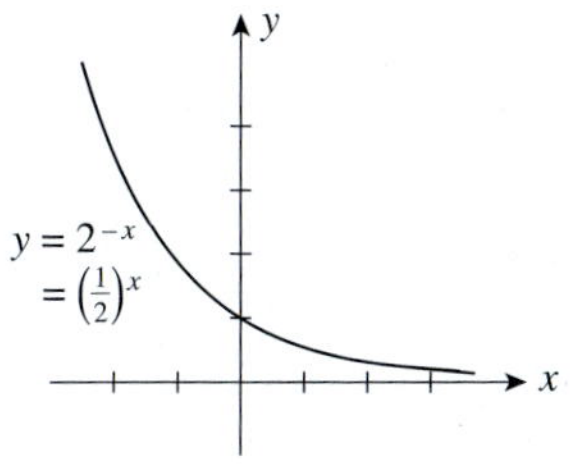

2. $\log_{2/3}\left(\frac{27}{8}\right) = -3$ **3.** $\log_{16} 0.125 = -\frac{3}{4}$ **4.** $\frac{15}{2}$ **5.** $x = 2$

6. $x + y + z$ **7.** $x + 2y - z$ **8.** $y + 2z$

9.

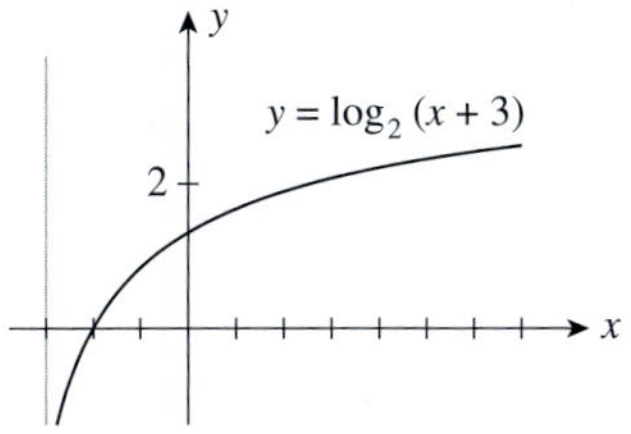

10.

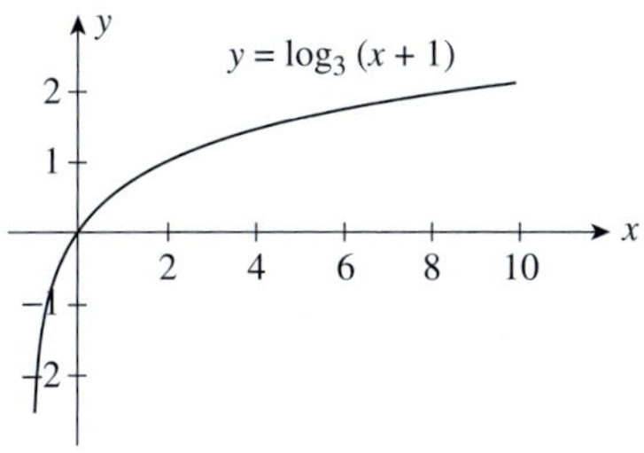

11. **a.** \$11,274.86 **b.** \$11,274.97 **12.** 6.12%/yr

13. 6.8 yr **14.** 7.77%/yr

15. $(2x + 1)e^{2x}$ **16.** $\dfrac{e^t}{2\sqrt{t}} + \sqrt{t}e^t + 1$ **17.** $\dfrac{1 - 4t}{2\sqrt{te^{2t}}}$

18. $\dfrac{e^x(x^2 + x + 1)}{\sqrt{1 + x^2}}$ **19.** $\dfrac{2(e^{2x} + 2)}{(1 + e^{-2x})^2}$ **20.** $4xe^{2x^2-1}$

21. $(1 - 2x^2)e^{-x^2}$ **22.** $3e^{2x}(1 + e^{2x})^{1/2}$ **23.** $(x + 1)^2e^x$

24. $\ln t + 1$ **25.** $\dfrac{2xe^{x^2}}{e^{x^2} + 1}$ **26.** $\dfrac{\ln x - 1}{(\ln x)^2}$

27. $\dfrac{x - x\ln x + 1}{x(x + 1)^2}$ **28.** $(x + 2)e^x$ **29.** $\dfrac{4e^{4x}}{e^{4x} + 3}$

30. $\dfrac{(r^3 - r^2 + r + 1)e^r}{(1 + r^2)^2}$ **31.** $\dfrac{1 + e^x(1 - x\ln x)}{x(1 + e^x)^2}$

32. $\dfrac{(2x^2 + 2x^2 \cdot \ln x - 1)e^{x^2}}{x(1 + \ln x)^2}$ **33.** $-\dfrac{9}{(3x + 1)^2}$ **34.** $\dfrac{1}{x}$

35. 0 **36.** −2 **37.** $6x(x^2 + 2)^2(3x^3 + 2x + 1)$

38. $\dfrac{(4x^3 - 5x^2 + 2)(x^2 - 2)}{(x - 1)^2}$ **39.** $y = -(2x - 3)e^{-2}$

40. $y = \dfrac{1}{e}$

41.

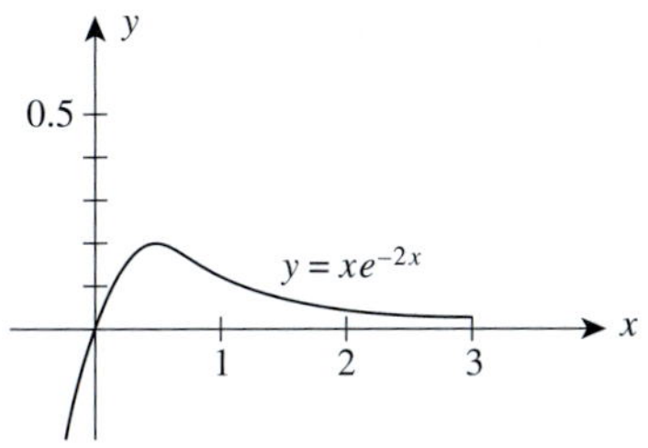

42.

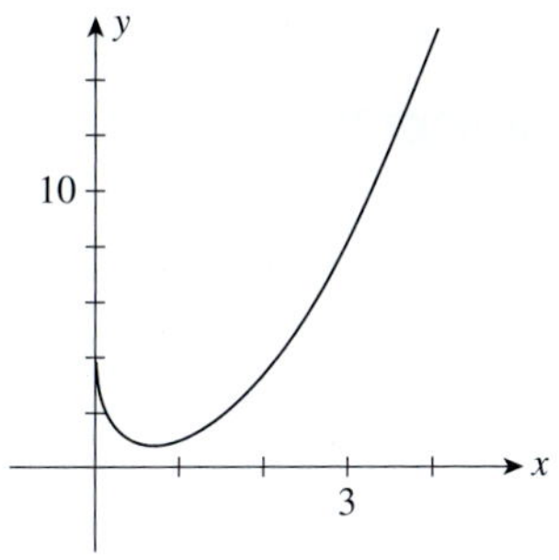

43. Absolute maximum value: $\dfrac{1}{e}$

44. Absolute maximum value: $\frac{\ln 2}{2}$; absolute minimum value: 0

45. 12%/yr **46.** \$20,136.31 **47.** \$80,000 **48.** 9.58 yr

49. **a.** $Q(t) = 2000e^{0.01831t}$ **b.** 161,992 **50.** 0.0004332

51.

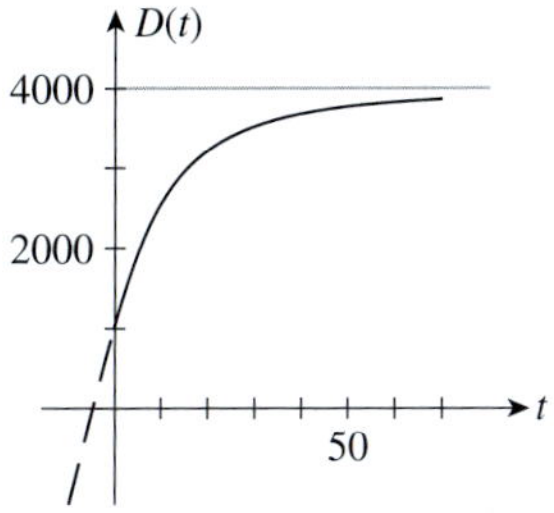

a. 1175, 2540, 3289 **b.** 4000

52. 200 g; −21.2 g/yr

53. 1.8; −0.11; −0.23; −0.13; the rate of change of the amount of oil used is 1.8 barrels per \$1000 of output per decade in 1965; it is decreasing at the rate of 0.11 barrels per \$1000 of output per decade in 1966, and so on.

54. **a.** \$9/unit **b.** \$8/week **c.** \$18/unit

55. 970

56. **a.** 12.5/1000 live births; 9.3/1000 live births; 6.9/1000 live births

b.

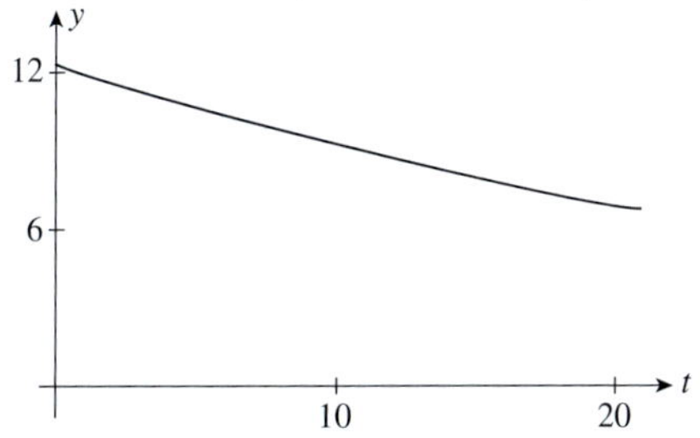

57. 5000; \$36,787.90

58. **a.** 0 g/cm³ **b.** 0.0361 g/cm³ **c.** 0.08 g/cm³
d.

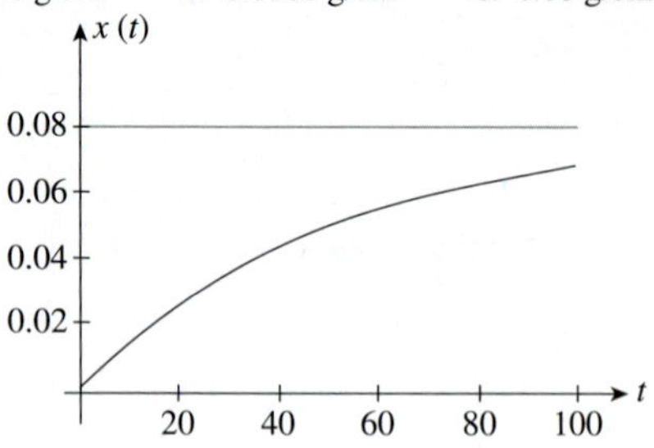

Chapter 5 Before Moving On, page 396

1. −0.9589 **2.** \$4130.37 **3.** $\dfrac{e^{\sqrt{x}}}{2\sqrt{x}}$ **4.** $1 + \ln 2$

5. $e^{2x}\left(\dfrac{4x^2 \ln 3x + 4x - 1}{x^2}\right)$ **6.** 8.7 min

CHAPTER 6

Exercises 6.1, page 406

5. **b.** $y = 2x + C$

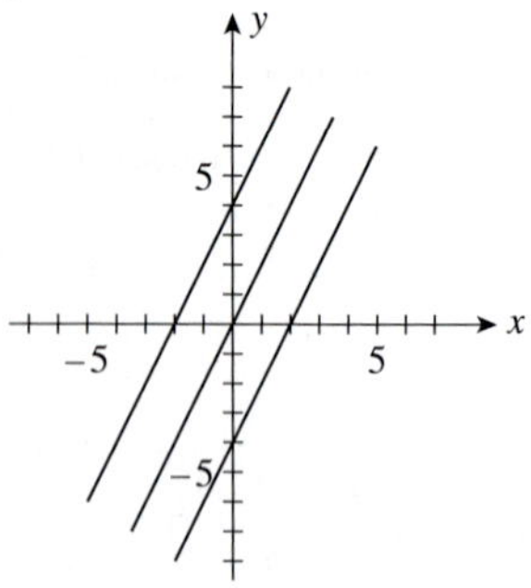

7. **b.** $y = \frac{1}{3}x^3 + C$

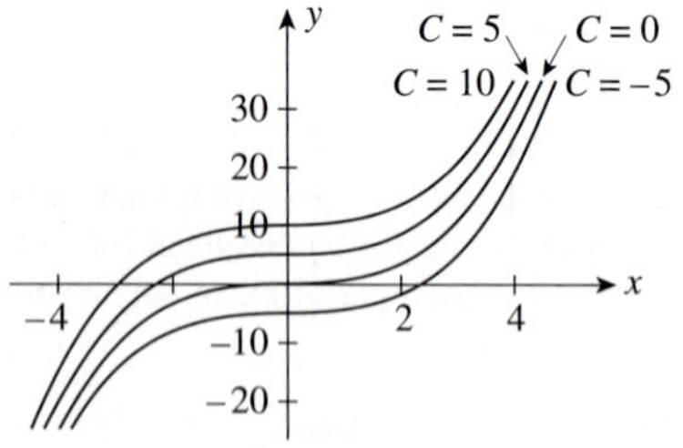

9. $6x + C$ **11.** $\frac{1}{4}x^4 + C$ **13.** $-\dfrac{1}{3x^3} + C$

15. $\frac{3}{5}x^{5/3} + C$ **17.** $-\dfrac{4}{x^{1/4}} + C$ **19.** $-\dfrac{2}{x} + C$

21. $\frac{2}{3}\pi t^{3/2} + C$ **23.** $3x - x^2 + C$

25. $\dfrac{1}{3}x^3 + \dfrac{1}{2}x^2 - \dfrac{1}{2x^2} + C$ **27.** $4e^x + C$

29. $x + \frac{1}{2}x^2 + e^x + C$ **31.** $x^4 + \dfrac{2}{x} - x + C$

33. $\frac{2}{7}x^{7/2} + \frac{4}{5}x^{5/2} - \frac{1}{2}x^2 + C$ **35.** $\frac{2}{3}x^{3/2} + 6\sqrt{x} + C$

37. $\frac{1}{9}u^3 + \frac{1}{3}u^2 - \frac{1}{3}u + C$ **39.** $\frac{2}{3}t^3 - \frac{3}{2}t^2 - 2t + C$

41. $\frac{1}{3}x^3 - 2x - \dfrac{1}{x} + C$ **43.** $\frac{1}{3}s^3 + s^2 + s + C$

45. $e^t + \dfrac{t^{e+1}}{e+1} + C$ **47.** $\frac{1}{2}x^2 + x - \ln|x| - \dfrac{1}{x} + C$

49. $\ln|x| + \dfrac{4}{\sqrt{x}} - \dfrac{1}{x} + C$ **51.** $x^2 + x + 1$

53. $x^3 + 2x^2 - x - 5$ **55.** $x - \dfrac{1}{x} + 2$ **57.** $x + \ln|x|$

59. $\sqrt{x}$ **61.** $e^x + \frac{1}{2}x^2 + 2$ **63.** Branch A

65. $s(t) = \frac{4}{3}t^{3/2}$ **67.** \$3370 **69.** 5000 units; \$34,000

71. **a.** $0.0029t^2 + 0.159t + 1.6$ **b.** \$4.16 trillion

73. **a.** $-0.125t^3 + 1.05t^2 + 2.45t + 1.5$ **b.** 24.375 million

75. **a.** $-1.493t^3 + 34.9t^2 + 279.5t + 2917$ **b.** \$9168

77. **a.** $3.133t^3 - 6.7t^2 + 14.07t + 36.7$ **b.** 103,201

79. **a.** $y = 4.096t^3 - 75.2797t^2 + 695.23t + 3142$ **b.** \$4264.11

81. 21,960 **83.** $-t^3 + 96t^2 + 120t$; 63,000 ft

85. **a.** $0.75t^4 - 5.9815t^3 + 14.3611t^2 + 26.632t + 108$
b. \$321.25 million

87. **a.** $9.3e^{-0.02t}$ **b.** 7030 **c.** 6619

89. $\frac{1}{2}k(R^2 - r^2)$ **91.** $9\frac{7}{9}$ ft/sec²; 396 ft **93.** 0.924 ft/sec²

95. **a.** $\dfrac{2t}{t+4}$ **b.** $\frac{2}{5}$ in.; $\frac{2}{3}$ in. **97.** True **99.** True

Exercises 6.2, page 418

1. $\frac{1}{5}(4x + 3)^5 + C$ **3.** $\frac{1}{3}(x^3 - 2x)^3 + C$

5. $-\dfrac{1}{2(2x^2 + 3)^2} + C$ **7.** $\frac{2}{3}(t^3 + 2)^{3/2} + C$

9. $\frac{1}{20}(x^2 - 1)^{10} + C$ **11.** $-\frac{1}{5}\ln|1 - x^5| + C$

13. $\ln(x - 2)^2 + C$ **15.** $\frac{1}{2}\ln(0.3x^2 - 0.4x + 2) + C$

17. $\frac{1}{6}\ln|3x^2 - 1| + C$ **19.** $-\frac{1}{2}e^{-2x} + C$ **21.** $-e^{2-x} + C$

23. $-\frac{1}{2}e^{-x^2} + C$ **25.** $e^x + e^{-x} + C$ **27.** $\ln(1 + e^x) + C$

29. $2e^{\sqrt{x}} + C$ **31.** $-\dfrac{1}{6(e^{3x} + x^3)^2} + C$ **33.** $\frac{1}{8}(e^{2x} + 1)^4 + C$

35. $\frac{1}{2}(\ln 5x^2) + C$ **37.** $\ln|\ln x| + C$ **39.** $\frac{2}{3}(\ln x)^{3/2} + C$

41. $\frac{1}{2}e^{x^2} - \frac{1}{2}\ln(x^2 + 2) + C$

43. $\frac{2}{3}(\sqrt{x} - 1)^3 + 3(\sqrt{x} - 1)^2 + 8(\sqrt{x} - 1) + 4\ln|\sqrt{x} - 1| + C$

45. $\dfrac{(6x + 1)(x - 1)^6}{42} + C$

47. $5 + 4\sqrt{x} - x - 4\ln(1 + \sqrt{x}) + C$

49. $-\frac{1}{252}(1 - v)^7(28v^2 + 7v + 1) + C$

51. $\frac{1}{2}[(2x - 1)^5 + 5]$ **53.** $e^{-x^2+1} - 1$ **55.** 17,341,000

57. $21{,}000 - \dfrac{20{,}000}{\sqrt{1 + 0.2t}}$; 6858 **59.** $\dfrac{250}{\sqrt{16 + x^2}}$

61. $30(\sqrt{2t + 4} - 2)$; $14{,}400\,\pi$ ft^2

63. $\dfrac{65.8794}{1 + 2.449e^{-0.3277t}} + 0.3$; 56.22 in. **65.** $\frac{r}{a}(1 - e^{-at})$ **67.** True

Exercises 6.3, page 428

1. 4.27

3. a. 6

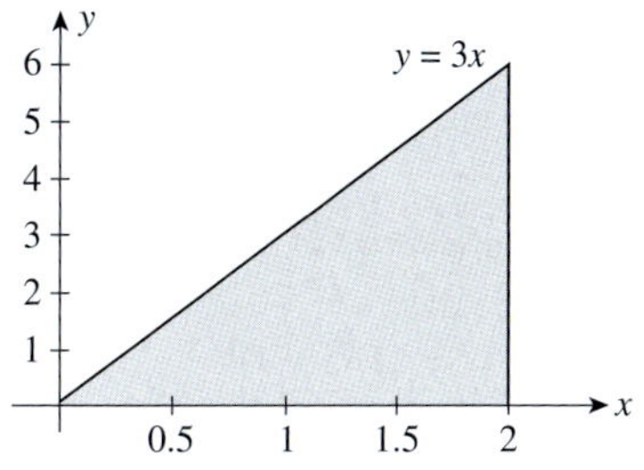

b. 4.5 **c.** 5.25 **d.** Yes

5. a. 4

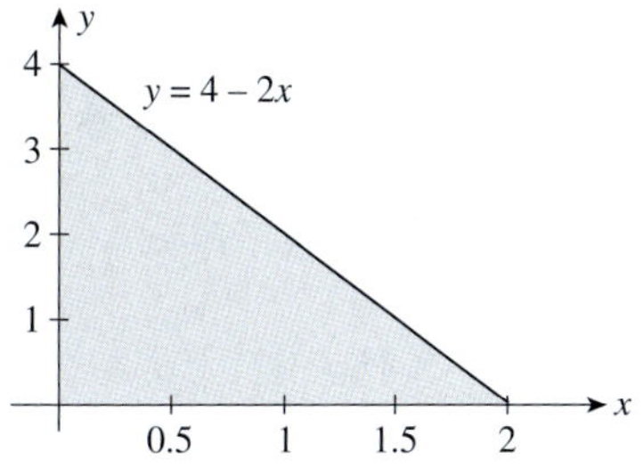

b. 4.8 **c.** 4.4 **d.** Yes

7. a. 18.5 **b.** 18.64 **c.** 18.66 **d.** $\approx$18.7

9. a. 25 **b.** 21.12 **c.** 19.88 **d.** $\approx$19.9

11. a. 0.0625 **b.** 0.16 **c.** 0.2025 **d.** $\approx$0.2

13. 4.64 **15.** 0.95 **17.** 9400 sq ft

Exercises 6.4, page 438

1. 6 **3.** 8 **5.** 12 **7.** 9 **9.** ln 2 **11.** $17\frac{1}{3}$ **13.** $18\frac{1}{4}$

15. $(e^2 - 1)$ **17.** 6 **19.** 14 **21.** $18\frac{2}{3}$ **23.** $\frac{4}{3}$ **25.** 45

27. $\frac{7}{12}$ **29.** ln 2 **31.** 56 **33.** $\frac{256}{15}$ **35.** $\frac{2}{3}$ **37.** $2\frac{2}{3}$

39. $19\frac{1}{2}$ **41. a.** \$4100 **b.** \$900 **43. a.** \$2800 **b.** \$219.20

45. a. $0.86t^{0.96} + 0.04$ **b.** \$4.84 billion **47.** $10{,}133\frac{1}{3}$ ft

49. a. $0.2833t^3 - 1.936t^2 + 5t + 5.6$ **b.** 12.8% **c.** 5.2%

51. 695.5 million **53.** 149.14 million **55.** $\frac{23}{15}$

57. False **59.** False

Using Technology Exercises 6.4, page 441

1. 6.1787 **3.** 0.7873 **5.** −0.5888 **7.** 2.7044

9. 3.9973 **11.** 46%; 24% **13.** 333,209 **15.** 903,213

Exercises 6.5, page 448

1. 10 **3.** $\frac{19}{15}$ **5.** $32\frac{4}{15}$ **7.** $\sqrt{3} - 1$ **9.** $24\frac{1}{5}$

11. $\frac{32}{15}$ **13.** $18\frac{2}{15}$ **15.** $\frac{1}{2}(e^4 - 1)$ **17.** $\frac{1}{2}e^2 + \frac{5}{6}$ **19.** 0

21. 2 ln 4 **23.** $\frac{1}{3}(\ln 19 - \ln 3)$ **25.** $2e^4 - 2e^2 - \ln 2$

27. $\frac{1}{2}(e^{-4} - e^{-8} - 1)$ **29.** 6 **31.** $\frac{1}{2}$ **33.** $2(\sqrt{e} - \frac{1}{e})$ **35.** 5

37. $\frac{17}{3}$ **39.** −1 **41.** $\frac{13}{6}$ **43.** $\frac{1}{4}(e^4 - 1)$

45. 120.3 billion metric tons

47. $\approx$\$2.24 million **49.** \$40,339.50 **51.** \$3.24 billion/yr

53. a. 160.7 billion gal/yr **b.** 150.1 billion gal/yr/yr

55. 9.8% **57.** 16,863 **59.** \$14.78 **61.** 80.7%

69. Property 5 **71.** 0 **73. a.** −1 **b.** 5 **c.** −13

75. True **77.** False **79.** True

Using Technology Exercises 6.5, page 452

1. 7.71667 **3.** 17.56487 **5.** 10,140 **7.** 60.45 mg/day

Exercises 6.6, page 459

1. 108 **3.** $\frac{2}{3}$ **5.** $2\frac{2}{3}$ **7.** $1\frac{1}{2}$ **9.** 3 **11.** $3\frac{1}{3}$

13. 27 **15.** $2(e^2 - e^{-1})$ **17.** $12\frac{2}{3}$ **19.** $3\frac{1}{3}$ **21.** $4\frac{3}{4}$

23. $12 - \ln 4$ **25.** $e^2 - e - \ln 2$ **27.** $2\frac{1}{2}$ **29.** $7\frac{1}{3}$ **31.** $\frac{3}{2}$

33. $e^3 - 4 + \frac{1}{e}$ **35.** $20\frac{5}{6}$ **37.** $\frac{1}{12}$ **39.** $\frac{71}{6}$ **41.** 18

43. S is the additional revenue that Odyssey Travel could realize by switching to the new agency; $S = \int_0^b [g(x) - f(x)]\,dx$

45. Shortfall $= \int_{2010}^{2050} [f(t) - g(t)]\,dt$

47. a. $A_2 - A_1$ **b.** The distance car 2 is ahead of car 1 after T sec

49. 30 ft/sec faster **51.** 21,850 **53.** True **55.** False

Using Technology Exercises 6.6, page 464

1. a.

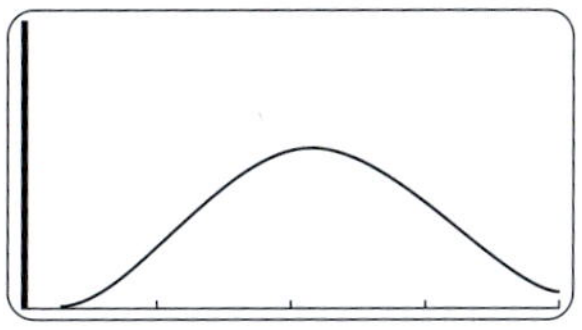

b. 1074.2857

3. a.

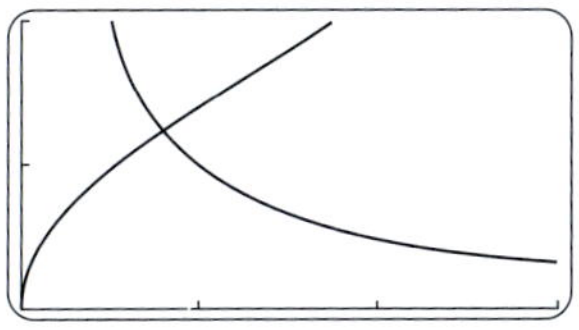

b. 0.9961

5. a.

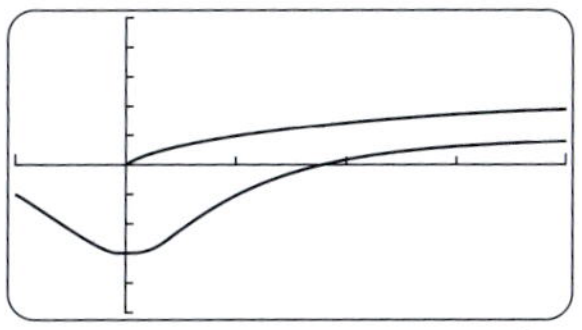

b. 5.4603

7. a.

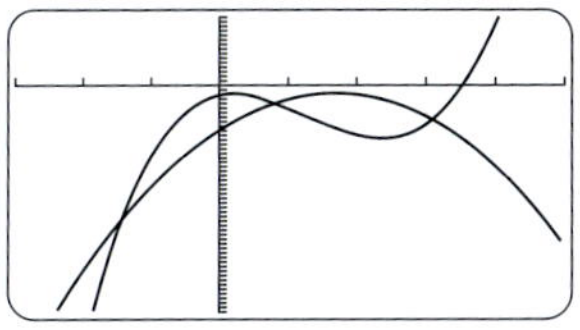

b. 25.8549

9. a.

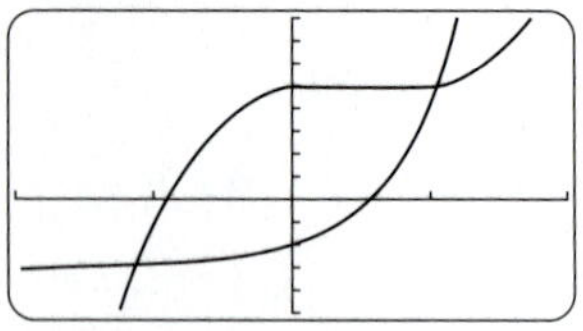

b. 10.5144

11. a.

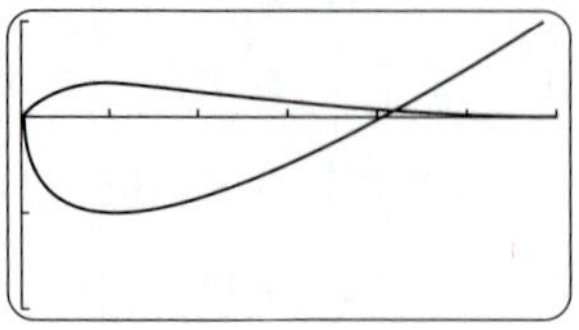

b. 3.5799

13. 207.43

Exercises 6.7, page 474

1. \$11,667 3. \$6667 5. \$11,667

7. Consumers' surplus: \$13,333; producers' surplus: \$11,667

9. \$824,200 11. \$148,239 13. \$43,788

15. \$47,916 17. \$142,423 19. \$24,780

21. a.

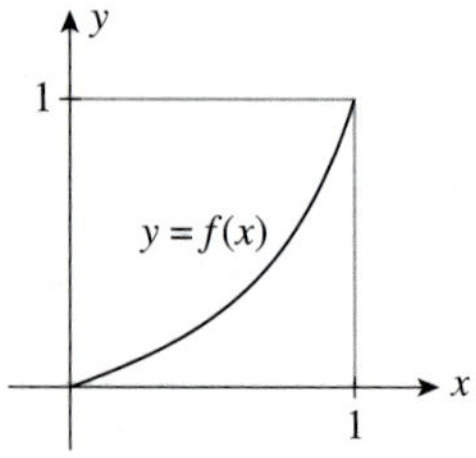

b. 0.175; 0.816

23. a.

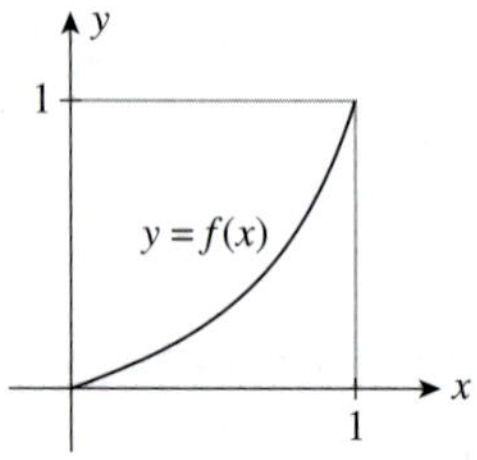

b. 0.104; 0.504

Using Technology Exercises 6.7, page 476

1. Consumers' surplus: \$18,000,000; producers' surplus: \$11,700,000

3. Consumers' surplus: \$33,120; producers' surplus: \$2880

5. Investment A

Chapter 6 Concept Review, page 478

1. a. $F'(x) = f(x)$ b. $F(x) + C$

2. a. $c\int f(x)\,dx$ b. $\int f(x)\,dx \pm \int g(x)\,dx$

3. a. Unknown b. Function 4. $g'(x)\,dx$; $\int f(u)\,du$

5. a. $\int_a^b f(x)\,dx$ b. Minus

6. a. $F(b) - F(a)$; antiderivative b. $\int_a^b f'(x)\,dx$

7. a. $\dfrac{1}{b-a}\displaystyle\int_a^b f(x)\,dx$ b. Area; area 8. $\int_a^b [f(x) - g(x)]\,dx$

9. a. $\int_0^{\bar{x}} D(x)\,dx - \bar{p}\,\bar{x}$ b. $\bar{p}\,\bar{x} - \int_0^{\bar{x}} S(x)\,dx$

10. a. $e^{rT}\int_0^T R(t)e^{-rt}\,dt$ b. $\int_0^T R(t)e^{-rt}\,dt$

11. $A = \dfrac{mP}{r}(e^{rT} - 1)$ 12. $L = 2\int_0^1 [x - f(x)]\,dx$

Chapter 6 Review Exercises, page 479

1. $\frac{1}{4}x^4 + \frac{2}{3}x^3 - \frac{1}{2}x^2 + C$ 2. $\frac{1}{12}x^4 - \frac{2}{3}x^3 + 8x + C$

3. $\frac{1}{5}x^5 - \frac{1}{2}x^4 - \dfrac{1}{x} + C$ 4. $\frac{3}{4}x^{4/3} - \frac{2}{3}x^{3/2} + 4x + C$

5. $\frac{1}{2}x^4 + \frac{2}{5}x^{5/2} + C$ 6. $\frac{2}{7}x^{7/2} - \frac{1}{3}x^3 + \frac{2}{3}x^{3/2} - x + C$

7. $\frac{1}{3}x^3 - \frac{1}{2}x^2 + 2\ln|x| + 5x + C$ 8. $\frac{1}{3}(2x + 1)^{3/2} + C$

9. $\frac{3}{8}(3x^2 - 2x + 1)^{4/3} + C$ 10. $\dfrac{(x^3 + 2)^{11}}{33} + C$

11. $\frac{1}{2}\ln(x^2 - 2x + 5) + C$ 12. $-e^{-2x} + C$

13. $\frac{1}{2}e^{x^2+x+1} + C$ 14. $\dfrac{1}{e^{-x} + x} + C$ 15. $\frac{1}{6}(\ln x)^6 + C$

16. $(\ln x)^2 + C$ 17. $\dfrac{(11x^2 - 1)(x^2 + 1)^{11}}{264} + C$

18. $\frac{2}{15}(3x - 2)(x + 1)^{3/2} + C$ 19. $\frac{2}{3}(x + 4)\sqrt{x - 2} + C$

20. $2(x - 2)\sqrt{x + 1} + C$ 21. $\frac{1}{2}$ 22. -6 23. $5\frac{2}{3}$

24. 242 25. -80 26. $\frac{132}{5}$ 27. $\frac{1}{2}\ln 5$ 28. $\frac{1}{15}$

29. 4 30. $1 - \dfrac{1}{e^2}$ 31. $\dfrac{e - 1}{2(1 + e)}$ 32. $\frac{1}{2}$

33. $f(x) = x^3 - 2x^2 + x + 1$ 34. $f(x) = \sqrt{x^2 + 1}$

35. $f(x) = x + e^{-x} + 1$ 36. $f(x) = \frac{1}{2}(\ln x)^2 - 2$

37. -4.28 38. \$6740

39. a. $-0.015x^2 + 60x$; b. $p = -0.015x + 60$

40. $V(t) = 1900(t - 10)^2 + 10{,}000$; \$40,400

41. a. $0.05t^3 - 1.8t^2 + 14.4t + 24$ b. 56°F

42. a. $-0.01t^3 + 0.109t^2 - 0.032t + 0.1$ b. 1.076 billion

43. 3.375 ppm 44. $3000t - 50{,}000(1 - e^{-0.04t})$; 16,939

45. $N(t) = 15{,}000\sqrt{1 + 0.4t} + 85{,}000$; 112,659

46. 26,027 47. $\dfrac{240}{5 - x} - 30$

48. \$3100 49. 37.7 million

50. a. $S(t) = 205.89 - 89.89e^{-0.176t}$ b. \$161.43 billion

51. 15 52. $\frac{1}{2}(e^4 - 1)$

53. $\frac{2}{3}$ 54. $\frac{9}{2}$

55. $(e^2 - 3)$ 56. $\frac{3}{10}$ 57. $\frac{1}{2}$

58. 234,500 barrels 59. $\frac{1}{3}$ 60. 26°F

61. 49.7 ft/sec 62. 67,600/yr 63. \$270,000

64. Consumers' surplus: \$2083; producers' surplus: \$3333

65. \$197,652 **66.** \$174,420 **67.** \$505,696

68. a.

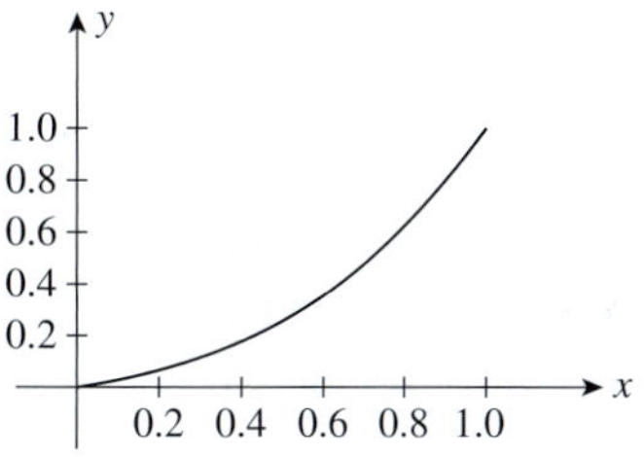

b. 0.1017; 0.3733 **c.** 0.315

69. 90,888

Chapter 6 Before Moving On, page 482

1. $\frac{1}{2}x^4 + \frac{2}{3}x^{3/2} + 2\ln|x| - 4\sqrt{x} + C$ **2.** $e^x + \frac{1}{2}x^2 + 1$

3. $\sqrt{x^2+1} + C$ **4.** $\frac{1}{3}(2\sqrt{2} - 1)$ **5.** $\frac{9}{2}$

CHAPTER 7

Exercises 7.1, page 488

1. $\frac{1}{4}e^{2x}(2x-1) + C$ **3.** $4(x-4)e^{x/4} + C$

5. $\frac{1}{2}e^{2x} - 2(x-1)e^x + \frac{1}{3}x^3 + C$ **7.** $xe^x + C$

9. $\dfrac{2(x+2)}{\sqrt{x+1}} + C$ **11.** $\frac{2}{3}x(x-5)^{3/2} - \frac{4}{15}(x-5)^{5/2} + C$

13. $\dfrac{x^2}{4}(2\ln 2x - 1) + C$ **15.** $\dfrac{x^4}{16}(4\ln x - 1) + C$

17. $\frac{2}{9}x^{3/2}(3\ln\sqrt{x} - 1) + C$ **19.** $-\dfrac{1}{x}(\ln x + 1) + C$

21. $x(\ln x - 1) + C$ **23.** $-(x^2 + 2x + 2)e^{-x} + C$

25. $\frac{1}{4}x^2[2(\ln x)^2 - 2\ln x + 1] + C$ **27.** $2\ln 2 - 1$

29. $4\ln 4 - 3$ **31.** $\frac{1}{4}(3e^4 + 1)$ **33.** $-\frac{1}{2}xe^{-2x} - \frac{1}{4}e^{-2x} + \frac{13}{4}$

35. $5\ln 5 - 4$ **37.** 1485 ft **39.** 2.04 mg/mL

41. $-20e^{-0.1t}(t+10) + 200$ **43.** \$131,324 **45.** 101,606

47. $(c_2 - c_1)\left[\dfrac{r_1}{r_2 - r_1} + \dfrac{1}{\ln r_1 - \ln r_2}\right] + c_2$ moles/L **49.** True

Exercises 7.2, page 496

1. $\frac{2}{9}[2 + 3x - 2\ln|2 + 3x|] + C$

3. $\frac{3}{32}[(1+2x)^2 - 4(1+2x) + 2\ln|1+2x|] + C$

5. $2\left[\dfrac{x}{8}\left(\dfrac{9}{4} + 2x^2\right)\sqrt{\dfrac{9}{4} + x^2} - \dfrac{81}{128}\ln\left(x + \sqrt{\dfrac{9}{4} + x^2}\right)\right] + C$

7. $\ln\left(\dfrac{\sqrt{1+4x} - 1}{\sqrt{1+4x} + 1}\right) + C$ **9.** $\frac{1}{2}\ln 3$ **11.** $\dfrac{x}{9\sqrt{9 - x^2}} + C$

13. $\dfrac{x}{8}(2x^2 - 4)\sqrt{x^2 - 4} - 2\ln|x + \sqrt{x^2 - 4}| + C$

15. $\sqrt{4 - x^2} - 2\ln\left|\dfrac{2 + \sqrt{4 - x^2}}{x}\right| + C$

17. $\frac{1}{4}(2x - 1)e^{2x} + C$ **19.** $\ln|\ln(1+x)| + C$

21. $\dfrac{1}{9}\left[\dfrac{1}{1 + 3e^x} + \ln(1 + 3e^x)\right] + C$

23. $6[e^{(1/2)x} - \ln(1 + e^{(1/2)x})] + C$

25. $\frac{1}{9}(2 + 3\ln x - 2\ln|2 + 3\ln x|) + C$

27. $e - 2$ **29.** $\dfrac{x^3}{9}(3\ln x - 1) + C$

31. $x[(\ln x)^3 - 3(\ln x)^2 + 6\ln x - 6] + C$ **33.** $\approx$\$2329

35. 27,136 **37.** 44; 49 **39.** 26,157 **41.** \$326,222

Exercises 7.3, page 508

1. 2.7037; 2.6667; $2\frac{2}{3}$ **3.** 0.2656; 0.2500; $\frac{1}{4}$

5. 0.6970; 0.6933; $\approx$0.6931 **7.** 0.5090; 0.5004; $\frac{1}{2}$

9. 5.2650; 5.3046; $\frac{16}{3}$ **11.** 0.6336; 0.6321; $\approx$0.6321

13. 0.3837; 0.3863; $\approx$0.3863 **15.** 1.1170; 1.1114

17. 1.3973; 1.4052 **19.** 0.8806; 0.8818 **21.** 3.7757; 3.7625

23. a. 3.6 **b.** 0.0324 **25. a.** 0.013 **b.** 0.00043

27. a. 0.0078125 **b.** 0.0002848 **29.** 52.84 mi

31. 21.65 mpg **33.** 17.1 million barrels **35.** 1922.4 ft^3/sec

37. a. \$51,558 **b.** \$51,708 **39.** 474.77 million barrels

41. $\approx$50% **43.** 6.42 L/min **45.** False **47.** True

Exercises 7.4, page 518

1. $\frac{2}{3}$ **3.** 1 **5.** 2 **7.** $\frac{2}{3}$ **9.** $\frac{1}{2}e^4$ **11.** 1 **13. a.** $\frac{2}{3}b^{3/2}$

15. 1 **17.** 2 **19.** Divergent **21.** $-\frac{1}{8}$ **23.** 1 **25.** 1

27. $\frac{1}{2}$ **29.** Divergent **31.** -1 **33.** Divergent **35.** 0

37. 0 **39.** Divergent **41.** $\frac{1}{2}$ **43.** \$18,750

45. $\dfrac{10{,}000r + 4000}{r^2}$ dollars **47.** True **49.** False

51. b. \$83,333

Exercises 7.5, page 526

1. $\dfrac{2\pi}{3}$ **3.** $\dfrac{153\pi}{5}$ **5.** 3π **7.** $\dfrac{15\pi}{2}$

9. $\dfrac{4\pi}{3}$ **11.** $\dfrac{16\pi}{15}$ **13.** $\dfrac{\pi}{2}(e^2 - 1)$

15. $\dfrac{2\pi}{15}$ **17.** $\dfrac{136\pi}{15}$ **19.** $\dfrac{64\sqrt{2}\pi}{3}$

21. $\dfrac{\pi}{2}(e^2 - 2 + e^{-2})$ **23.** $\dfrac{\pi}{6}$

25. $\dfrac{6517\pi}{240}$ **27.** $\dfrac{64\sqrt{2}\pi}{3}$

29. $\dfrac{8(\sqrt{2} - 1)\pi}{3}$ **31.** $\dfrac{4\pi}{3}r^3$

33. $50{,}000\pi$ cu ft

Chapter 7 Concept Review, page 528

1. Product: $uv - \int v\,du$; u; easy to integrate **2.** $x^2 + 1$; $2x\,dx$; (27)

3. $\dfrac{\Delta x}{2}[f(x_0) + 2f(x_1) + \cdots + 2f(x_{n-1}) + f(x_n)]$; $\dfrac{M(b-a)^3}{12n^2}$

4. $\dfrac{\Delta x}{3}[f(x_0) + 4f(x_1) + 2f(x_2) + 4f(x_3) + 2f(x_4) + \cdots + 4f(x_{n-1}) + f(x_n)]$; even; $\dfrac{M(b-a)^5}{180n^4}$

5. $\lim\limits_{a\to-\infty} \int_a^b f(x)\,dx$; $\lim\limits_{b\to\infty} \int_a^b f(x)\,dx$; $\int_{-\infty}^c f(x)\,dx + \int_c^\infty f(x)\,dx$

6. region; $x = a$; $x = b$; $V = \pi \int_a^b [f(x)]^2\,dx$

Chapter 7 Review Exercises, page 529

1. $-2(1 + x)e^{-x} + C$ **2.** $\frac{1}{16}(4x - 1)e^{4x} + C$

3. $x(\ln 5x - 1) + C$ **4.** $4 \ln 8 - \ln 2 - 3$ **5.** $\frac{1}{4}(1 - 3e^{-2})$

6. $\frac{1}{4}(1 + 3e^4)$ **7.** $2\sqrt{x}(\ln x - 2) + 2$

8. $-\frac{1}{3}xe^{-3x} - \frac{1}{9}e^{-3x} + \frac{1}{9}$

9. $\dfrac{1}{8}\left[3 + 2x - \dfrac{9}{3 + 2x} - 6 \ln|3 + 2x|\right] + C$

10. $\frac{2}{3}(x - 3)\sqrt{2x + 3} + C$ **11.** $\frac{1}{32}e^{4x}(8x^2 - 4x + 1) + C$

12. $-\dfrac{x}{25\sqrt{x^2 - 25}} + C$ **13.** $\dfrac{1}{4}\dfrac{\sqrt{x^2 - 4}}{x} + C$

14. $\frac{1}{2}x^4(4 \ln 2x - 1) + C$ **15.** $\frac{1}{2}$ **16.** $\frac{1}{3}$

17. Divergent **18.** 1 **19.** $\frac{1}{10}$ **20.** 3

21. 0.8421; 0.8404 **22.** 1.491; 1.464 **23.** 2.2379; 2.1791

24. 8.1310; 8.041 **25. a.** 0.002604 **b.** 0.000033 **26.** \$1,157,641

27. 48,761 **28.** \$41,100 **29.** 274,000 ft²; 278,667 ft²

30. 7850 sq ft **31.** \$111,111 **32.** $\frac{2\pi}{3}$ **33.** $\frac{3\pi}{10}$

Chapter 7 Before Moving On, page 530

1. $\frac{1}{3}x^3 \ln x - \frac{1}{9}x^3 + C$ **2.** $-\dfrac{\sqrt{8 + 2x^2}}{8x} + C$ **3.** 6.3367

4. 3.00358 **5.** $\dfrac{1}{2e^2}$ **6. b.** 2π

CHAPTER 8

Exercises 8.1, page 538

1. $f(0, 0) = -4$; $f(1, 0) = -2$; $f(0, 1) = -1$; $f(1, 2) = 4$; $f(2, -1) = -3$

3. $f(1, 2) = 7$; $f(2, 1) = 9$; $f(-1, 2) = 1$; $f(2, -1) = 1$

5. $g(1, 2) = 4 + 3\sqrt{2}$; $g(2, 1) = 8 + \sqrt{2}$; $g(0, 4) = 2$; $g(4, 9) = 56$

7. $h(1, e) = 1$; $h(e, 1) = -1$; $h(e, e) = 0$

9. $g(1, 1, 1) = e$; $g(1, 0, 1) = 1$; $g(-1, -1, -1) = -e$

11. All real values of x and y

13. All real values of u and v except those satisfying the equation $u = v$

15. All real values of r and s satisfying $rs \geq 0$

17. All real values of x and y satisfying $x + y > 5$

19.

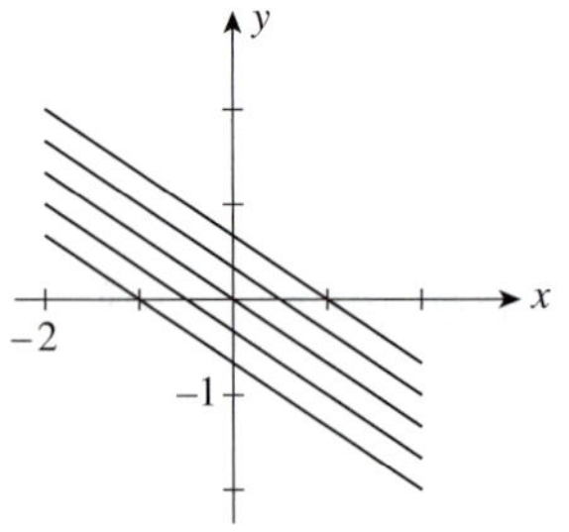

21.

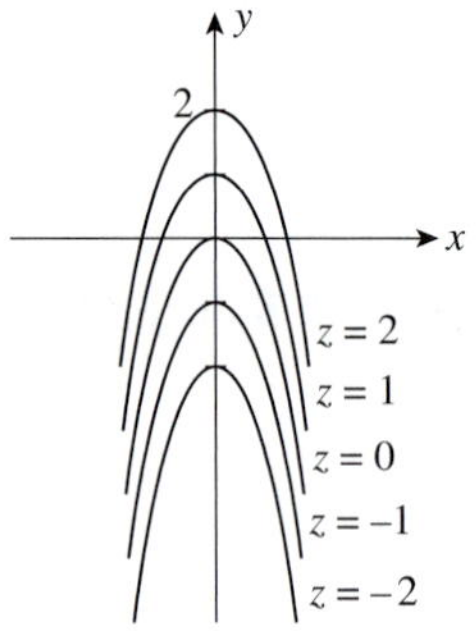

23.

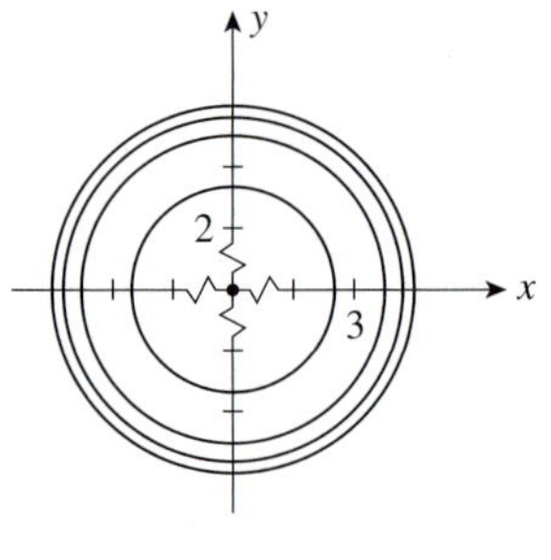

25. $\sqrt{x^2 + y^2} = 5$ **27.** 9π ft³ **29. a.** 24.69 **b.** 81 kg

31. a. $-\frac{1}{5}x^2 - \frac{1}{4}y^2 - \frac{1}{5}xy + 200x + 160y$
b. The set of all points (x, y) satisfying $200 - \frac{1}{5}x - \frac{1}{10}y \geq 0$, $160 - \frac{1}{10}x - \frac{1}{4}y \geq 0$, $x \geq 0$, $y \geq 0$

33. a. $-0.005x^2 - 0.003y^2 - 0.002xy + 20x + 15y$
b. The set of all ordered pairs (x, y) for which
$20 - 0.005x - 0.001y \geq 0$
$15 - 0.001x - 0.003y \geq 0$, $x \geq 0$, $y \geq 0$

35. a. The set of all ordered pairs (P, T), where P and T are positive numbers
b. 11.10 L

37. \$7200 billion **39.** 103

41. a. \$1798.65; \$2201.29 **b.** \$2509.32

43. 40.28 times gravity

45. The level curves of V have equation $\dfrac{kT}{P} = C$ (C, a positive constant). The level curves are a family of straight lines $T = \left(\dfrac{C}{k}\right)P$ lying in the first quadrant since k, T, and P are positive. Every point on the level curve $V = C$ gives the same volume C.

47. False **49.** False **51.** False

Exercises 8.2, page 550

1. a. 4; 4

b. $f_x(2, 1) = 4$ says that the slope of the tangent line to the curve of intersection of the surface $z = x^2 + 2y^2$ and the plane $y = 1$ at the point (2, 1, 6) is 4. $f_y(2, 1) = 4$ says that the slope of the tangent line to the curve of intersection of the surface $z = x^2 + 2y^2$ and the plane $x = 2$ at the point (2, 1, 6) is 4.

c. $f_x(2, 1) = 4$ says that the rate of change of $f(x, y)$ with respect to x with y held fixed with a value of 1 is 4 units/unit change in x. $f_y(2, 1) = 4$ says that the rate of change of $f(x, y)$ with respect to y with x held fixed with a value of 2 is 4 units/unit change in y.

3. 2; 3 **5.** $4x$; 4 **7.** $-\dfrac{4y}{x^3}; \dfrac{2}{x^2}$ **9.** $\dfrac{2v}{(u+v)^2}; -\dfrac{2u}{(u+v)^2}$

11. $3(2s - t)(s^2 - st + t^2)^2; 3(2t - s)(s^2 - st + t^2)^2$

13. $\dfrac{4x}{3(x^2 + y^2)^{1/3}}; \dfrac{4y}{3(x^2 + y^2)^{1/3}}$ **15.** $ye^{xy+1}; xe^{xy+1}$

17. $\ln y + \dfrac{y}{x}; \dfrac{x}{y} + \ln x$ **19.** $e^u \ln v; \dfrac{e^u}{v}$

21. $yz + y^2 + 2xz; xz + 2xy + z^2; xy + 2yz + x^2$

23. $ste^{rst}; rte^{rst}; rse^{rst}$ **25.** $f_x(1, 2) = 8; f_y(1, 2) = 5$

27. $f_x(2, 1) = 1; f_y(2, 1) = 3$ **29.** $f_x(1, 2) = \frac{1}{2}; f_y(1, 2) = -\frac{1}{4}$

31. $f_x(1, 1) = e; f_y(1, 1) = e$

33. $f_x(1, 0, 2) = 0; f_y(1, 0, 2) = 8; f_z(1, 0, 2) = 0$

35. $f_{xx} = 2y; f_{xy} = 2x + 3y^2 = f_{yx}; f_{yy} = 6xy$

37. $f_{xx} = 2; f_{xy} = f_{yx} = -2; f_{yy} = 4$

39. $f_{xx} = \dfrac{y^2}{(x^2 + y^2)^{3/2}}; f_{xy} = f_{yx} = -\dfrac{xy}{(x^2 + y^2)^{3/2}};$
$f_{yy} = \dfrac{x^2}{(x^2 + y^2)^{3/2}}$

41. $f_{xx} = \dfrac{1}{y^2} e^{-x/y}; f_{xy} = \dfrac{y - x}{y^3} e^{-x/y} = f_{yx};$
$f_{yy} = \dfrac{x}{y^3}\left(\dfrac{x}{y} - 2\right)e^{-x/y}$

43. **a.** $f_x = 7.5; f_y = 40$ **b.** Yes

45. $p_x = 10$—at (0, 1), the price of land is changing at the rate of \$10/ft²/mile change to the right; $p_y = 0$—at (0, 1), the price of land is constant/mile change upward.

47. Complementary commodities

49. \$30/unit change in finished desks; −\$25/unit change in unfinished desks. The weekly revenue increases by \$30/unit for each additional finished desk produced (beyond 300) when the level of production of unfinished desks remains fixed at 250; the revenue decreases by \$25/unit when each additional unfinished desk (beyond 250) is produced and the level of production of finished desks remains fixed at 300.

51. **a.** ≈20°F **b.** ≈ −0.3°F

53. 0.039 L/degree; −0.015 L/mm of mercury. The volume increases by 0.039 L when the temperature increases by 1 degree (beyond 300 K) and the pressure is fixed at 800 mm of mercury. The volume decreases by 0.015 L when the pressure increases by 1 mm of mercury (beyond 800 mm) and the temperature is fixed at 300 K.

57. False **59.** True

Using Technology Exercises 8.2, page 553

1. 1.3124; 0.4038 **3.** −1.8889; 0.7778 **5.** −0.3863; −0.8497

Exercises 8.3, page 561

1. (0, 0); relative maximum value: $f(0, 0) = 1$

3. (1, 2); saddle point: $f(1, 2) = 4$

5. (8, −6); relative minimum value: $f(8, -6) = -41$

7. (1, 2) and (2, 2); saddle point: $f(1, 2) = -1$; relative minimum value: $f(2, 2) = -2$

9. $(-\frac{1}{3}, \frac{11}{3})$ and (1, 5); saddle point: $f(-\frac{1}{3}, \frac{11}{3}) = -\frac{319}{27}$; relative minimum value: $f(1, 5) = -13$

11. (0, 0) and (1, 1); saddle point: $f(0, 0) = -2$; relative minimum value: $f(1, 1) = -3$

13. (2, 1); relative minimum value: $f(2, 1) = 6$

15. (0, 0); saddle point: $f(0, 0) = -1$

17. (0, 0); relative minimum value: $f(0, 0) = 1$

19. (0, 0); relative minimum value: $f(0, 0) = 0$

21. 200 finished units and 100 unfinished units; \$10,500

23. Price of land (\$200/ft²) is highest at $(\frac{1}{2}, 1)$

25. (0, 1) gives desired location. **27.** 10″ × 10″ × 5″; 500 in.³

29. 30″ × 40″ × 10″; \$7200 **31.** False **33.** False **35.** True

Exercises 8.4, page 570

1. **a.** $y = 2.3x + 1.5$

b.

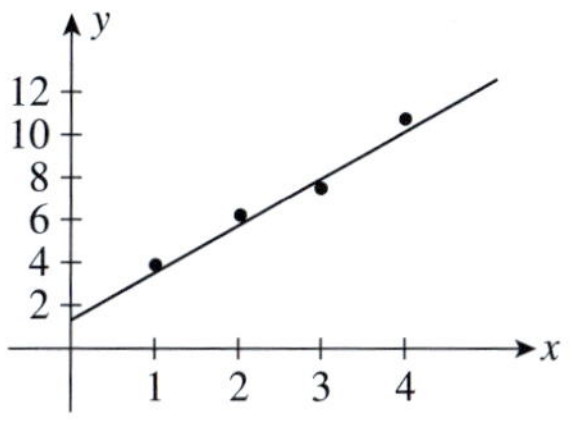

3. **a.** $y = -0.77x + 5.74$

b.

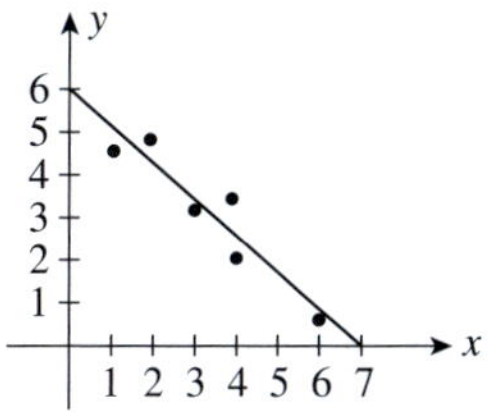

5. **a.** $y = 1.2x + 2$

b.

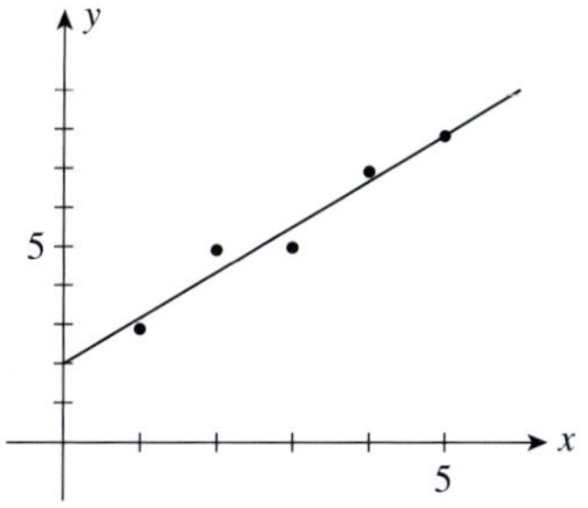

7. a. $y = 0.34x - 0.9$
b.
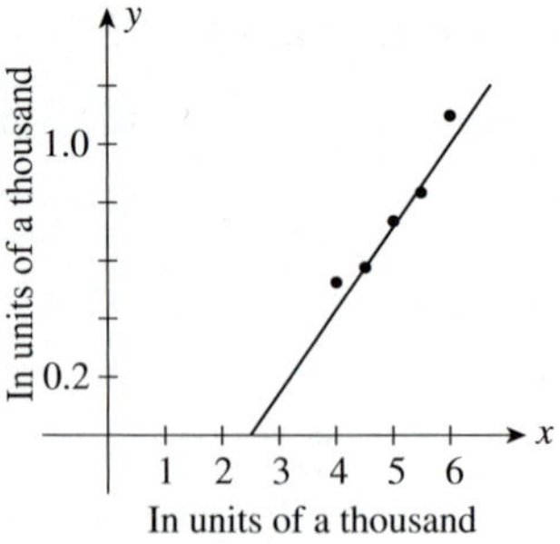

c. 1276 applications

9. a. $y = -2.8x + 440$
b.
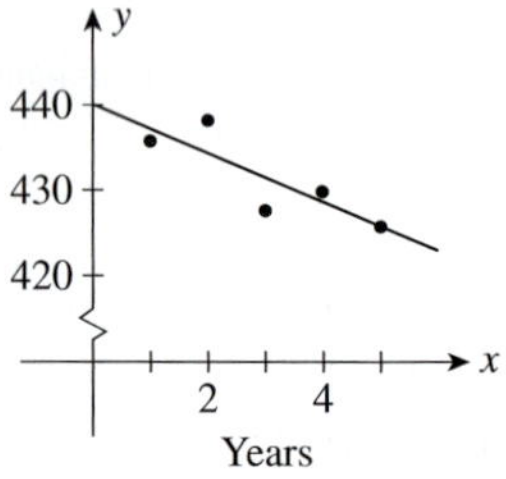

c. 420

11. a. $y = 2.8x + 17.6$ b. \$40,000,000

13. a. $y = 22.7x + 124.2$ b. \$260.4 billion

15. a. $y = 25.6x + 74$ b. 330 million

17. a. $y = 0.436x + 3.58$ b. \$436 million/yr

19. a. $y = 0.305x + 0.19$ b. \$0.305 billion/yr c. \$3.24 billion

21. a. $y = 0.4x + 4.42$ b. \$0.4 billion/yr

23. a. $y = 3.17x + 82.1$ b. \$113,800

25. a. $y = 0.087x + 15.90$ b. 19.38 yr c. 18.5 yr

27. a. $y = 0.23x + 1.16$ b. 2.8 billion bushels

29. False 31. True

Using Technology Exercises 8.4, page 574

1. $y = 2.3596x + 3.8639$ 3. $y = -1.1948x + 3.5525$

5. a. $y = 1.03x + 2.33$ b. \$10.57 billion

7. a. $y = 13.321x + 72.571$ b. 192 million tons

9. a. $y = 1.95x + 12.19$ b. \$29.74 billion

Exercises 8.5, page 583

1. Min. of $\frac{3}{4}$ at $(\frac{3}{4}, \frac{1}{4})$ 3. Max. of $-\frac{7}{4}$ at $(2, \frac{7}{2})$

5. Min. of 4 at $(\sqrt{2}, \sqrt{2}/2)$ and $(-\sqrt{2}, -\sqrt{2}/2)$

7. Max. of $-\frac{3}{4}$ at $(\frac{3}{2}, 1)$

9. Max. of $2\sqrt{3}$ at $(\sqrt{3}/3, -\sqrt{6})$ and $(\sqrt{3}/3, \sqrt{6})$

11. Max. of 8 at $(2\sqrt{2}, 2\sqrt{2})$ and $(-2\sqrt{2}, -2\sqrt{2})$; min. of -8 at $(2\sqrt{2}, -2\sqrt{2})$ and $(-2\sqrt{2}, 2\sqrt{2})$

13. Max.: $\frac{2\sqrt{3}}{9}$; min.: $-\frac{2\sqrt{3}}{9}$ 15. Min. of $\frac{18}{7}$ at $(\frac{9}{7}, \frac{6}{7}, \frac{3}{7})$

17. 140 finished and 60 unfinished units

19. $10\sqrt{2}$ ft $\times$ $40\sqrt{2}$ ft 21. $r = \frac{4}{3}\sqrt[3]{\frac{18}{\pi}}$ in.; $h = 2\sqrt[3]{\frac{18}{\pi}}$ in.

23. $\frac{2}{3}\sqrt[3]{9} \times \frac{2}{3}\sqrt[3]{9} \times \sqrt[3]{9}$

25. 1500 units on labor and 250 units of capital

27. False

29. True

Exercises 8.6, page 590

1. $2x\ dx + 2\ dy$ 3. $(4x - 3y + 4)\ dx - 3x\ dy$

5. $\frac{x}{\sqrt{x^2 + y^2}}\ dx + \frac{y}{\sqrt{x^2 + y^2}}\ dy$

7. $-\frac{5y}{(x - y)^2}\ dx + \frac{5x}{(x - y)^2}\ dy$

9. $(10x^4 + 3ye^{-3x})\ dx - e^{-3x}\ dy$

11. $\left(2xe^y + \frac{y}{x}\right) dx + (x^2e^y + \ln x)\ dy$

13. $y^2z^3\ dx + 2xyz^3\ dy + 3xy^2z^2\ dz$

15. $\frac{1}{y + z}\ dx - \frac{x}{(y + z)^2}\ dy - \frac{x}{(y + z)^2}\ dz$

17. $(yz + e^{yz})\ dx + xz(1 + e^{yz})\ dy + xy(1 + e^{yz})\ dz$

19. 0.04 21. -0.01 23. -0.10 25. -0.18

27. 0.06 29. -0.1401 31. An increase of \$19,250/mo

33. An increase of \$5000/mo 35. \$1.25 37. 38.4π cu cm

39. 104 sq in. 41. 3.3256 Pa 43. 0.55 ohm

Exercises 8.7, page 597

1. $\frac{7}{2}$ 3. 0 5. $4\frac{1}{2}$ 7. $(e^2 - 1)(1 - e^{-2})$ 9. 1

11. $\frac{2}{3}$ 13. $\frac{188}{3}$ 15. $\frac{84}{5}$ 17. $2\frac{2}{3}$ 19. 1 21. $\frac{1}{2}(3 - e)$

23. $\frac{1}{4}(e^4 - 1)$ 25. $\frac{2}{3}(e - 1)$ 27. False

Exercises 8.8, page 604

1. 42 3. 20 5. $\frac{64}{3}$ 7. $\frac{1}{3}$ 9. 4 11. $3\frac{1}{3}$

13. $2(e^2 - 1)$ 15. $\frac{2}{35}$ 17. 54 19. $\frac{1}{3}$ 21. 1

23. 43,329 25. 312,455 27. \$194/sq ft 29. True

Chapter 8 Concept Review, page 606

1. xy; ordered pair; real number; $f(x, y)$

2. Independent; dependent; value

3. $z = f(x, y)$; f; surface

4. $f(x, y) = c$; level curve; level curves; c

5. Constant; x 6. Slope; $(a, b, f(a, b))$; x; b

7. $\leq$; (a, b); $\leq$; domain

8. Domain; $f_x(a, b) = 0$ and $f_y(a, b) = 0$; exist; candidate

9. Scatter; minimizing; least-squares; normal

10. $g(x, y) = 0$; $f(x, y) + \lambda g(x, y)$; $F_x = 0$; $F_y = 0$; $F_\lambda = 0$; extrema

11. Independent, $dx = \Delta x$, $dy = \Delta y$, $\dfrac{\partial f}{\partial x}dx + \dfrac{\partial f}{\partial y}dy$

12. Volume; solid 13. Iterated; $\int_3^5 \int_0^1 (2x + y^2)\, dx\, dy$

Chapter 8 Review Exercises, page 607

1. 0, 0, $\frac{1}{2}$; no 2. $e, \dfrac{e^2}{1 + \ln 2}, \dfrac{2e}{1 + \ln 2}$; no

3. 2, $-(e + 1)$, $-(e + 1)$

4. The set of all ordered pairs (u, v) such that $u \neq v$

5. The set of all ordered pairs (x, y) such that $y \neq -x$

6. The set of all ordered pairs (x, y) such that $x \leq 1$ and $y \geq 0$

7. The set of all triplets (x, y, z) such that $z \geq 0$ and $x \neq 1$, $y \neq 1$, and $z \neq 1$

8. $2x + 3y = z$

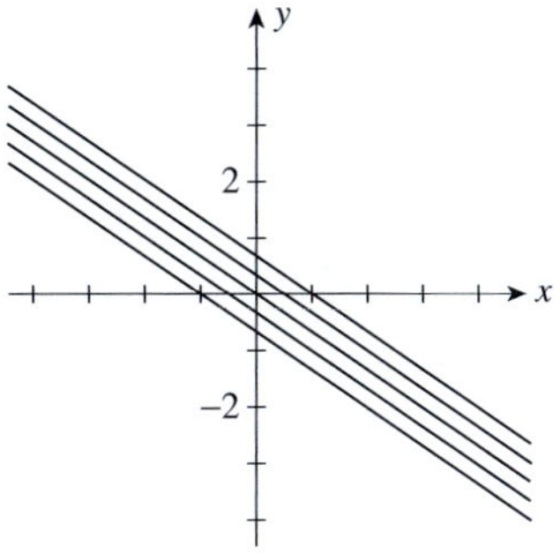

9. $z = y - x^2$

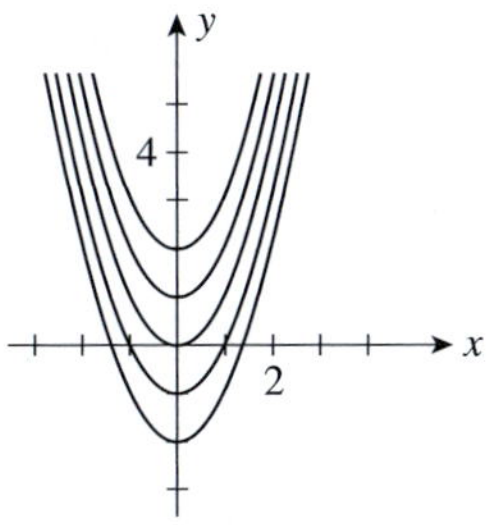

10. $z = \sqrt{x^2 + y^2}$

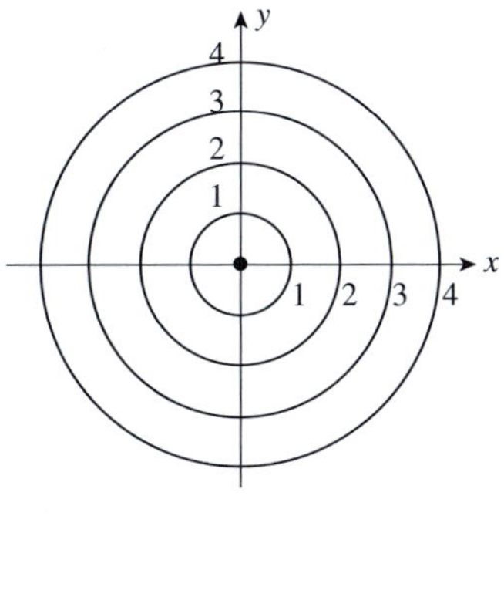

11. $z = e^{xy}$

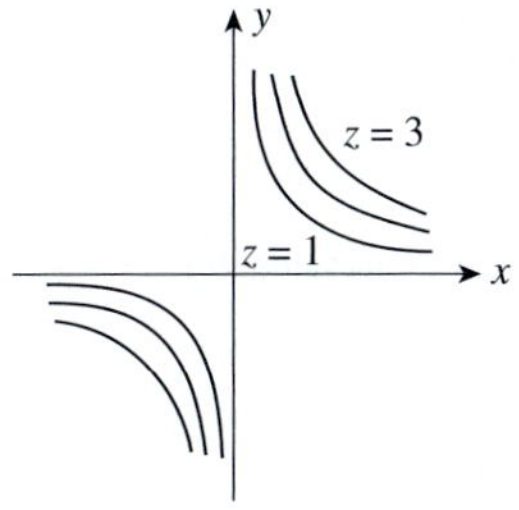

12. $f_x = 2xy^3 + 3y^2 + \dfrac{1}{y}$; $f_y = 3x^2y^2 + 6xy - \dfrac{x}{y^2}$

13. $f_x = \sqrt{y} + \dfrac{y}{2\sqrt{x}}$; $f_y = \dfrac{x}{2\sqrt{y}} + \sqrt{x}$

14. $f_u = \dfrac{v^2 - 2}{2\sqrt{uv^2 - 2u}}$; $f_v = \dfrac{uv}{\sqrt{uv^2 - 2u}}$

15. $f_x = \dfrac{3y}{(y + 2x)^2}$; $f_y = -\dfrac{3x}{(y + 2x)^2}$

16. $g_x = \dfrac{y(y^2 - x^2)}{(x^2 + y^2)^2}$; $g_y = \dfrac{x(x^2 - y^2)}{(x^2 + y^2)^2}$

17. $h_x = 10y(2xy + 3y^2)^4$; $h_y = 10(x + 3y)(2xy + 3y^2)^4$

18. $f_x = \dfrac{e^y}{2(xe^y + 1)^{1/2}}$; $f_y = \dfrac{xe^y}{2(xe^y + 1)^{1/2}}$

19. $f_x = 2x(1 + x^2 + y^2)e^{x^2+y^2}$; $f_y = 2y(1 + x^2 + y^2)e^{x^2+y^2}$

20. $f_x = \dfrac{4x}{1 + 2x^2 + 4y^4}$; $f_y = \dfrac{16y^3}{1 + 2x^2 + 4y^4}$

21. $f_x = \dfrac{2x}{x^2 + y^2}$; $f_y = -\dfrac{2x^2}{y(x^2 + y^2)}$

22. $f_{xx} = 6x - 4y$; $f_{xy} = -4x = f_{yx}$; $f_{yy} = 2$

23. $f_{xx} = 12x^2 + 4y^2$; $f_{xy} = 8xy = f_{yx}$; $f_{yy} = 4x^2 - 12y^2$

24. $f_{xx} = 12(2x^2 + 3y^2)(10x^2 + 3y^2)$;
$f_{xy} = 144xy(2x^2 + 3y^2) = f_{yx}$;
$f_{yy} = 18(2x^2 + 3y^2)(2x^2 + 15y^2)$

25. $g_{xx} = \dfrac{-2y^2}{(x + y^2)^3}$; $g_{xy} = \dfrac{2y(x - y^2)}{(x + y^2)^3} = g_{yx}$;
$g_{yy} = \dfrac{2x(3y^2 - x)}{(x + y^2)^3}$

26. $g_{xx} = 2(1 + 2x^2)e^{x^2+y^2}$; $g_{xy} = 4xye^{x^2+y^2} = g_{yx}$;
$g_{yy} = 2(1 + 2y^2)e^{x^2+y^2}$

27. $h_{ss} = -\dfrac{1}{s^2}$; $h_{st} = h_{ts} = 0$; $h_{tt} = \dfrac{1}{t^2}$

28. $f_x(1, 1, 0) = 3$; $f_y(1, 1, 0) = 3$; $f_z(1, 1, 0) = -2$

29. (2, 3); relative minimum value: $f(2, 3) = -13$

30. (8, −2); saddle point at $f(8, -2) = -8$

31. (0, 0) and $(\frac{3}{2}, \frac{9}{4})$; saddle point at $f(0, 0) = 0$; relative minimum value: $f(\frac{3}{2}, \frac{9}{4}) = -\frac{27}{16}$

32. $(-\frac{1}{3}, \frac{13}{3})$, $(3, 11)$; saddle point at $f(-\frac{1}{3}, \frac{13}{3}) = (-\frac{445}{27})$; relative minimum value: $f(3, 11) = -35$

33. $(0, 0)$; relative minimum value: $f(0, 0) = 1$

34. $(1, 1)$; relative minimum value: $f(1, 1) = \ln 2$

35. $f(\frac{12}{11}, \frac{20}{11}) = -\frac{32}{11}$

36. $f(-\frac{1}{22}, \frac{23}{22}) = \frac{175}{44}$

37. Relative maximum value: $f(5, -5) = 26$; relative minimum value: $f(-5, 5) = -24$

38. Relative maximum value: $f\left(\frac{\sqrt{2}}{2}, -\frac{\sqrt{2}}{2}\right) = e^{\sqrt{2}}$; relative minimum value: $f\left(-\frac{\sqrt{2}}{2}, \frac{\sqrt{2}}{2}\right) = e^{-\sqrt{2}}$

39. $45\,dx + 240\,dy$ **40.** $2\,dx$ **41.** 0.04 **42.** $\frac{227}{54}$

43. 48 **44.** $\frac{1}{2}(e^{-2} - 1)^2$ **45.** $\frac{2}{63}$ **46.** $\frac{1}{4}(3 - 2\ln 2)$

47. $\frac{34}{3}$ **48.** $10\frac{4}{5}$ **49.** 3

50. $k = \frac{100\,m}{c}$

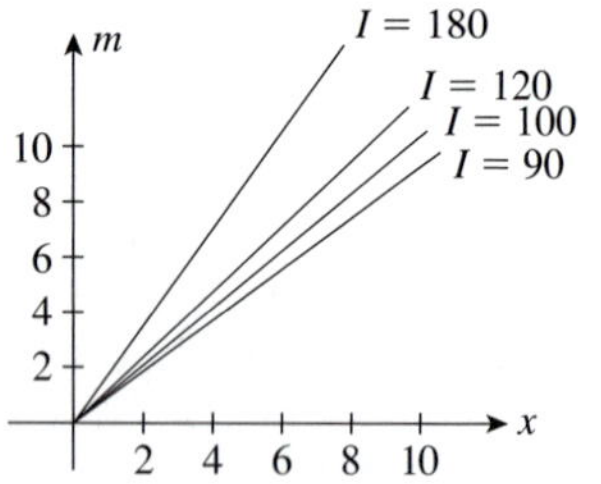

51. **a.** $R(x, y) = -0.02x^2 - 0.2xy - 0.05y^2 + 80x + 60y$
b. The set of all points satisfying $0.02x + 0.1y \leq 80$, $0.1x + 0.05y \leq 60$, $x \geq 0$, $y \geq 0$
c. 15,300; the revenue realized from the sale of 100 16-speed and 300 10-speed electric blenders is \$15,300.

52. Complementary **53.** \$490

54. **a.** $y = 8.2x + 361.2$ **b.** 7 hr 31 min

55. **a.** $y = 0.059x + 19.45$
b. 21.8 yr; same **c.** 21.2 yr

56. **a.** $y = 14.1x + 114.4$
b. 14.1 million/yr **c.** 184.9 million

57. The company should spend \$11,000 on advertising and employ 14 agents in order to maximize its revenue.

58. 337.5 yd × 900 yd **59.** 75 units on labor; 25 units on capital

Chapter 8 Before Moving On, page 610

1. All real values of x and y satisfying $x \geq 0$, $x \neq 1$, $y \geq 0$, $y \neq 2$

2. $f_x = 2xy + ye^{xy}$; $f_{xx} = 2y + y^2e^{xy}$; $f_{xy} = 2x + (xy + 1)e^{xy} = f_{yx}$; $f_y = x^2 + xe^{xy}$; $f_{yy} = x^2e^{xy}$

3. Rel. min.: $(1, 1, -7)$ **4.** $y = 2.04x + 2.88$ **5.** $f(\frac{1}{2}, \frac{1}{2}) = \frac{5}{2}$

6. **a.** $(4x - y)\,dx - x\,dy$ **b.** -0.09 **7.** $\frac{1}{8}$

CHAPTER 9

Exercises 9.1, page 617

13. $y = 12x^2 - 2x$ **15.** $y = \frac{1}{x}$ **17.** $y = -\frac{e^x}{x} + \frac{1}{2}xe^x$

19. $\frac{dQ}{dt} = -kQ$; $Q(0) = Q_0$ **21.** $\frac{dA}{dt} = k(C - A)$

23. $\frac{dC}{dt} = -kC$; $C(0) = C_0$ **29.** True **31.** False

33. False

Exercises 9.2, page 623

1. $\frac{1}{3}y^3 = \frac{1}{2}x^2 + x + C$ **3.** $\frac{1}{3}y^3 = e^x + C$ **5.** $y = Ce^{2x}$

7. $y = -\frac{2}{x^2 + 2C}$ **9.** $y = -\frac{4}{3} + Ce^{-6x}$

11. $y^3 = \frac{1}{3}x^3 + x + C$ **13.** $y^{1/2} - x^{1/2} = C$

15. $y = Ce^{\frac{1}{2}(\ln x)^2}$ **17.** $y = -\sqrt{2x^2 + 2}$

19. $y = 2 + e^{-x}$ **21.** $y = \frac{2}{3} + \frac{1}{3}e^{(3/2)x^2}$ **23.** $y = \sqrt{x^2 + 1}$

25. $y = e^{(x-1)e^x}$ **27.** $y = \ln(x^3 + e)$

29. $y = \sqrt{x^3 + 8}$ **31.** $Q(t) = Q_0e^{-kt}$ **33.** $A = \frac{C}{k} - de^{-kt}$

35. $S(t) = D - (D - S_0)e^{-kt}$ **37.** True **39.** True

41. False

Exercises 9.3, page 632

1. $y = y_0e^{-kt}$ **3.** $Q(t) = 4.5e^{0.02t}$; 8.5 billion **5.** $\frac{1}{2}$ in.

7. $Q(t) = \frac{50}{4t + 1}$; 5.56 g **9.** 3.6 min **11.** 23

13. 310.8 million **15.** 395

17. **a.** $\left(\sqrt{H} - \frac{B}{A}t\sqrt{\frac{g}{2}}\right)^2$ **b.** $\frac{A}{B}\sqrt{\frac{2H}{g}}$ **c.** 9.6 min

19. **a.** $Le^{-\ln(L/P_0)e^{-ct}}$ **b.** L

23. **a.** $L - (L - x_0)e^{-kt}$ **b.** $L - (L - 0.4)\left(\frac{L - 10}{L - 0.4}\right)^t$

c.

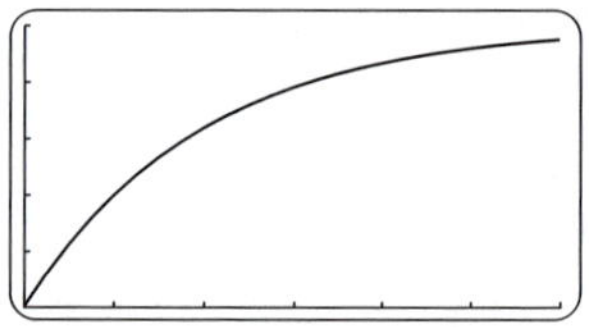

d. Between 5 and 9 years old

25. $x(t) = 40(1 - e^{-(3/20)t})$; 38 lb; 40 lb

Exercises 9.4, page 640

1. **a.** $y(1) = \frac{369}{128} \approx 2.8828$ **b.** $y(1) = \frac{70{,}993}{23{,}328} \approx 3.043$

3. **a.** $y(2) = \frac{51}{16} \approx 3.1875$ **b.** $y(2) = \frac{793}{243} \approx 3.2634$

5. **a.** $y(0.5) \approx 0.8324$ **b.** $y(0.5) \approx 0.8207$

7. **a.** $y(1.5) \approx 1.7831$ **b.** $y(1.5) \approx 1.7920$

9. **a.** $y(1) \approx 1.3390$ **b.** $y(1) \approx 1.3654$

11.

x	0.0	0.2	0.4	0.6	0.8	1
y_n	1	1	1.02	1.0608	1.1245	1.2144

13.

x	0.0	0.2	0.4	0.6	0.8	1
y_n	2	1.8	1.72	1.736	1.8288	1.9830

15.

x	0.0	0.1	0.2	0.3	0.4	0.5
y_n	1	1.1	1.211	1.3361	1.4787	1.6426

Chapter 9 Concept Review, page 641

1. **a.** differential equation **b.** satisfies

2. **a.** general **b.** particular

3. **a.** highest; unknown **b.** $f(x)g(y)$; not separable; separable **c.** separate; variable

4. **a.** kQ **b.** Q_0e^{kt} **c.** restricted

5. approximate; actual; polygonal

Chapter 9 Review Exercises, page 642

4. $y = \dfrac{1}{x^2 + 1}$ 5. $y = (9x + 8)^{-1/3}$ 6. $\frac{1}{3}y^3 = \frac{1}{4}x^4 + x + C$

7. $y = 4 - Ce^{-2t}$ 8. $y = Ce^{(1/2)(\ln x)^2}$ 9. $y = -\dfrac{2}{2x^3 + 2x + 1}$

10. $y = 1 - 3e^{-(1/3)x^3}$ 11. $y = 3e^{-x^3/2}$ 12. $y = \dfrac{4}{(x^2 + 1)^2}$

13. **a.** $y(1) \approx 0.3849$ **b.** $y(1) \approx 0.4361$

14. **a.** $y(1) \approx 0.2219$ **b.** $y(1) = 0.2628$

15. **a.** $y(1) \approx 1.3258$ **b.** $y(1) \approx 1.4570$

16. **a.** $y(1) \approx 1.9083$ **b.** $y(1) = 2.1722$

17.

x	0.0	0.2	0.4	0.6	0.8	1
y_n	1	1	1.08	1.2528	1.5535	2.0506

18.

x	0	0.2	0.4	0.6	0.8	1
y_n	1	1.2	1.496	1.9756	2.8282	4.5560

19. **a.** $S = 50{,}000(0.8)^t$ **b.** \$16,384 20. $R = CS^k$

21. $A = \dfrac{P}{r}(e^{rt} - 1)$; \$342,549.50 22. $S = D - (D - S_0)e^{-kt}$

23. ≈7:30 P.M. 24. 105 words/min 25. **a.** 183

26. $x = 120(1 - e^{-0.1t})$; 120 lb

Chapter 9 Before Moving On, page 643

1. **b.** $y = 2x^2$ 2. $y = \dfrac{1}{xe^{x-1}}$ 3. $5000e^{0.1386t}$; 6.3 years

4. ≈ 1.7639

CHAPTER 10

Exercises 10.1, page 654

13. $k = \frac{1}{3}$ 15. $k = \frac{1}{8}$ 17. $k = \frac{3}{16}$ 19. $k = 2$

21. **a.** $\frac{1}{2}$ **b.** $\frac{5}{8}$ **c.** $\frac{7}{8}$ **d.** 0

23. **a.** $\frac{11}{16}$ **b.** $\frac{1}{2}$ **c.** $\frac{27}{32}$ **d.** 0

25. **a.** $\frac{1}{2}$ **b.** $\frac{1}{2}(2\sqrt{2} - 1)$ **c.** 0 **d.** $\frac{1}{2}$

27. **a.** ≈1 **b.** ≈.14 29. **a.** .375 **b.** .75 **c.** .5 **d.** .875

35. $k = 2$ 37. $k = 12e^2$ 39. **a.** $\frac{1}{4}$ **b.** $\frac{1}{96}$

41. **a.** $\frac{1}{56}(2\sqrt{2} - 1)$ **b.** ≈.2575

43. **a.** ≈.63 **b.** ≈.30 **c.** ≈.30 45. **a.** ≈.10 **b.** ≈.30

47. ≈.9355 49. ≈.4815; ≈.7407 51. $\frac{1}{5}$ 53. ≈ .041

55. False

Using Technology Exercises 10.1, page 657

1.

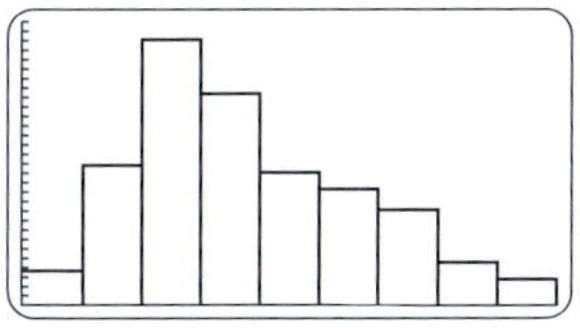

3. **a.**

x	0	1	2	3	4
$P(X = x)$	0.2	0.7	0.3	.12	.23

x	5	6	7	8	9	10
$P(X = x)$	.13	.17	.10	.05	.07	.02

b.

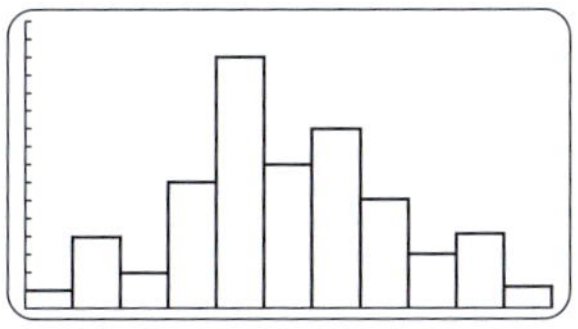

Exercises 10.2, page 670

1. $\mu = \frac{9}{2}$; $\text{Var}(X) = \frac{3}{4}$; $\sigma \approx 0.8660$

3. $\mu = \frac{15}{4}$; $\text{Var}(X) = \frac{15}{16}$; $\sigma \approx 0.9682$

5. $\mu = 3$; $\text{Var}(X) = 0.8$; $\sigma \approx 0.8944$

7. $\mu \approx 2.3765$; $\text{Var}(X) \approx 2.3522$; $\sigma \approx 1.5337$

9. $\mu = \frac{93}{35}$; $\text{Var}(X) \approx 0.7151$; $\sigma \approx 0.846$

11. $\mu = \frac{3}{2}$; $\text{Var}(X) = \frac{3}{4}$; $\sigma = \frac{1}{2}\sqrt{3}$ 13. $\mu = 4$; $\text{Var}(X) = 16$; $\sigma = 4$

15. 100 days 17. $3\frac{1}{3}$ min; $\frac{25}{18}$ 19. 1.5 ft; $\frac{9}{20}$ 21. 2500 lb/wk

23. $\dfrac{b + a}{2}$; $\dfrac{(b - a)^2}{12}$ 25. $a \approx 3.3$; $b \approx -5.7$

27. **a.** $\frac{1}{4}$ **b.** 6:30 a.m. 29. $m = 5$ 31. $m \approx 2.52$

33. $m = \frac{1}{2}$ 35. True

Using Technology Exercises 10.2, page 672

1. a.
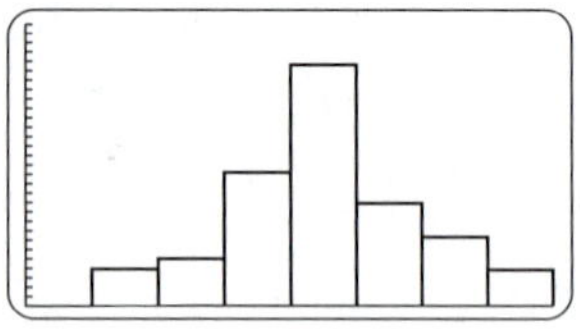

b. $\mu = 4$ and $\sigma \approx 1.40$

3. a. X gives the minimum age requirement for a regular driver's license.

b.

x	15	16	17	18	19	21
$P(X = x)$	.02	.30	.08	.56	.02	.02

c.
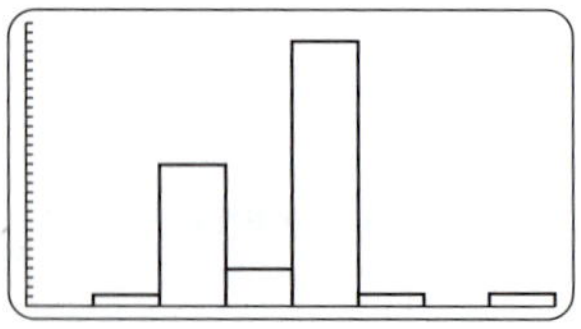

d. $\mu = 17.34$ and $\sigma = 1.11$

5. a. Let X denote the random variable that gives the weight of a carton of sugar.

b.

x	4.96	4.97	4.98	4.99	5.00	5.01
$P(X = x)$	$\frac{3}{30}$	$\frac{4}{30}$	$\frac{4}{30}$	$\frac{1}{30}$	$\frac{1}{30}$	$\frac{5}{30}$

x	5.02	5.03	5.04	5.05	5.06
$P(X = x)$	$\frac{3}{30}$	$\frac{3}{30}$	$\frac{4}{30}$	$\frac{1}{30}$	$\frac{1}{30}$

c. $\mu \approx 5.00$; $\text{Var}(X) \approx 0.0009$; $\sigma \approx 0.03$

Exercises 10.3, page 682

1. .9265 3. .0401 5. .8657

7. a.
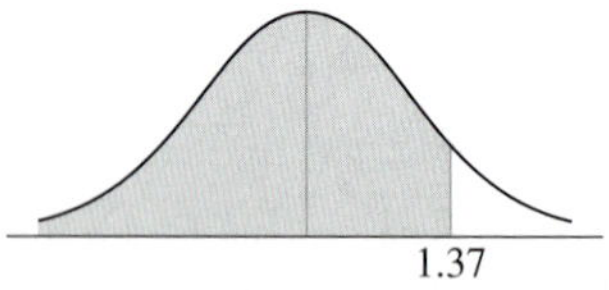

b. 0.9147

9. a.
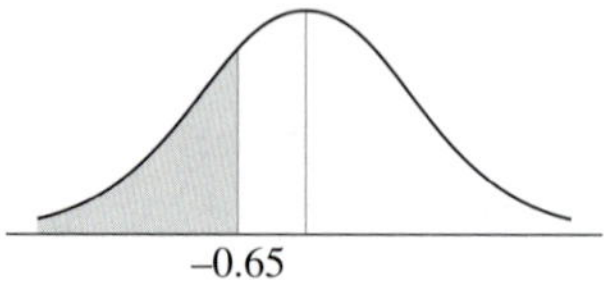

b. .2578

11. a.
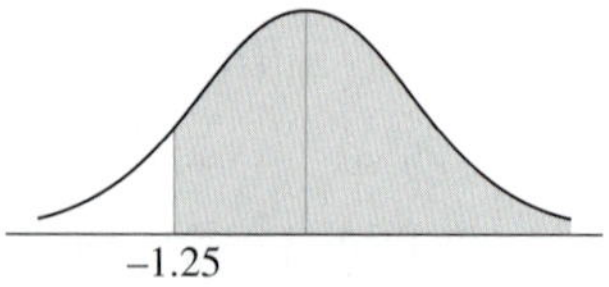

b. .8944

13. a.
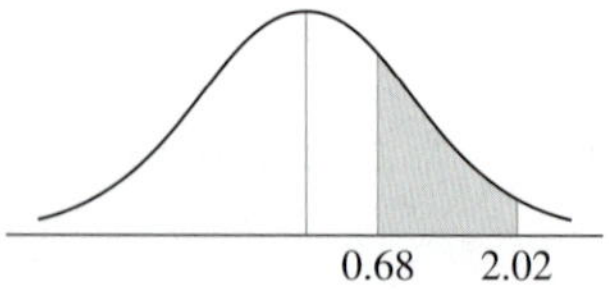

b. .2266

15. a. 1.23 b. −0.81 17. a. 1.9 b. −1.9

19. a. .9772 b. .9192 c. .7333

21. a. .2206 b. .2206 c. .2960

23. a. .0228 b. .0228 c. .4772 d. .7258

25. a. .0038 b. .0918 c. .4082 d. .2514

27. .6247 29. 0.62% 31. A: 80; B: 73; C: 62; D: 54

Chapter 10 Concept Review, page 685

1. a. experiment b. outcomes, sample points c. sample space d. 0, 1

2. a. random b. finite discrete c. 0, 1, 1

3. a. continuous b. nonnegative, 1, $\int_a^b f(x)\,dx$

4. nonnegative, 1, $\iint_R f(x, y)\,dA$

5. a. $(x_1 + x_2 + \cdots + x_n)/n$ b. central tendency, X
c. $x_1p_1 + x_2p_2 + \cdots + x_np_n$, X, probabilities
d. $\int_a^b x f(x)\,dx$, probability density, $[a, b]$

6. a. dispersion
b. $p_1(x_1 - \mu)^2 + p_2(x_2 - \mu)^2 + \cdots + p_n(x_n - \mu)^2$
c. $\sqrt{\text{Var}(X)}$

7. $\int_a^b (x - \mu)^2 f(x)\,dx$, $\sqrt{\text{Var}(X)}$

8. a. $x = \mu$ b. $x = \mu$ c. above, the x-axis d. 1
e. standard deviation, 2, 3

Chapter 10 Review Exercises, page 686

5. $\frac{1}{243}$ 6. $\frac{1}{6}$ 7. $\frac{3}{2}$ 8. 1.5

9. a. $\frac{4}{7}$ b. 0 c. $\frac{1}{3}$ 10. a. $\frac{1}{2}$ b. $\frac{1}{2}$ c. $\frac{3}{4}$

11. a. ≈.52 b. ≈.65 c. 0 12. a. $\frac{9}{10}$ b. $\frac{1}{4}$ c. $\frac{1}{2}$

13. $\mu = \frac{9}{2}$; $\text{Var}(X) \approx 2.083$; $\sigma \approx 1.44$

14. $\mu = \frac{50}{21}$; $\text{Var}(X) = 1.1882$; $\sigma \approx 1.0900$

15. $\mu = 0$; $\text{Var}(X) = \frac{7}{15}$; $\sigma \approx 0.6831$

16. $\mu = \frac{4}{3}$; $\text{Var}(X) = \frac{2}{9}$; $\sigma = 0.4714$ 17. .9875

18. .8925 19. .3049 20. .8294 22. a. $\frac{3}{16}$ b. $\frac{5}{6}$

23. a. .6915 b. .8944 c. .4681

24. a. .9525 b. .9525 c. .9050

25. a. .2231 b. .3935 c. 4 days

26. a. .5507 b. .3012 c. .2210

Chapter 10 Before Moving On, page 687

1. $\frac{6}{7}$ 2. .2387 3. $\frac{13}{3}$, 5.42, 2.33

4. a. .9938 b. .8944 c. .9876

CHAPTER 11

Exercises 11.1, page 698

1. $P_1(x) = 1 - x$; $P_2(x) = 1 - x + \frac{1}{2}x^2$; $P_3(x) = 1 - x + \frac{1}{2}x^2 - \frac{1}{6}x^3$

3. $P_1(x) = 1 - x$; $P_2(x) = 1 - x + x^2$; $P_3(x) = 1 - x + x^2 - x^3$

5. $P_1(x) = 1 - (x - 1)$; $P_2(x) = 1 - (x - 1) + (x - 1)^2$;
$P_3(x) = 1 - (x - 1) + (x - 1)^2 - (x - 1)^3$

7. $P_1(x) = 1 - \frac{1}{2}x$; $P_2(x) = 1 - \frac{1}{2}x - \frac{1}{8}x^2$;
$P_3(x) = 1 - \frac{1}{2}x - \frac{1}{8}x^2 - \frac{1}{16}x^3$

9. $P_1(x) = -x$; $P_2(x) = -x - \frac{1}{2}x^2$; $P_3(x) = -x - \frac{1}{2}x^2 - \frac{1}{3}x^3$

11. $P_2(x) = 16 + 32(x-2) + 24(x-2)^2$

13. $P_4(x) = (x-1) - \frac{1}{2}(x-1)^2 + \frac{1}{3}(x-1)^3 - \frac{1}{4}(x-1)^4$

15. $P_4(x) = e + e(x-1) + \frac{1}{2}e(x-1)^2 + \frac{1}{6}e(x-1)^3 + \frac{1}{24}e(x-1)^4$

17. $P_3(x) = 1 - \frac{1}{3}x - \frac{1}{9}x^2 - \frac{5}{81}x^3$

19. $P_3(x) = \frac{1}{3} - \frac{2}{9}x + \frac{4}{27}x^2 - \frac{8}{81}x^3$

21. $P_n(x) = 1 - x + x^2 - x^3 + \cdots + (-1)^n x^n$; 0.9091; 0.909090. . .

23. $P_4(x) = 1 - \frac{1}{2}x + \frac{1}{8}x^2 - \frac{1}{48}x^3 + \frac{1}{384}x^4$; 0.90484

25. $P_2(x) = 4 + \frac{1}{8}(x-16) - \frac{1}{512}(x-16)^2$; 3.9496875

27. $P_3(x) = x - \frac{1}{2}x^2 + \frac{1}{3}x^3$; 0.109; 0.108

29. 2.01494375; 0.00000042

31. 1.248; 0.0048828125; 1.25 **33.** 0.095; 0.00033

35. **a.** 0.04167 **b.** 0.625 **c.** 0.04167 **d.** 0.007121

37. 0.47995 **39.** 48% **41.** 2600 **43.** False **45.** True

Exercises 11.2, page 707

1. 1, 2, 4, 8, 16 **3.** $0, \frac{1}{3}, \frac{2}{4}, \frac{3}{5}, \frac{4}{6}$ **5.** $1, 1, \frac{4}{6}, \frac{8}{24}, \frac{16}{120}$

7. $e, \frac{e^2}{8}, \frac{e^3}{27}, \frac{e^4}{64}, \frac{e^5}{125}$ **9.** $1, \frac{11}{9}, \frac{25}{19}, \frac{45}{33}, \frac{71}{51}$ **11.** $a_n = 3n - 2$

13. $a_n = \dfrac{1}{n^3}$ **15.** $a_n = \dfrac{2^{2n-1}}{5^{n-1}}$ **17.** $a_n = \dfrac{(-1)^{n+1}}{2^{n-1}} = \left(-\dfrac{1}{2}\right)^{n-1}$

19. $a_n = \dfrac{n}{(n+1)(n+2)}$ **21.** $a_n = \dfrac{e^{n-1}}{(n-1)!}$

23.

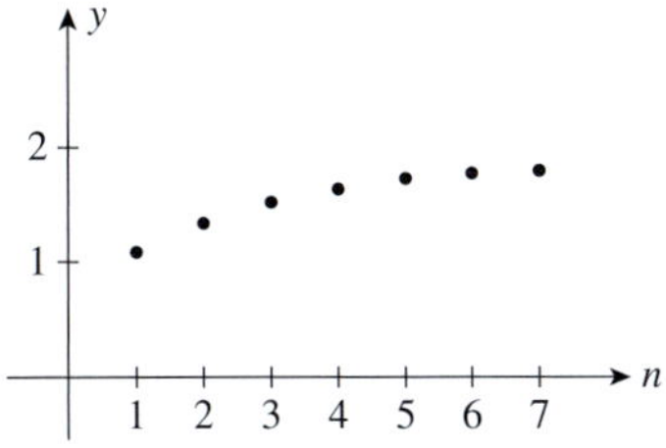

25.

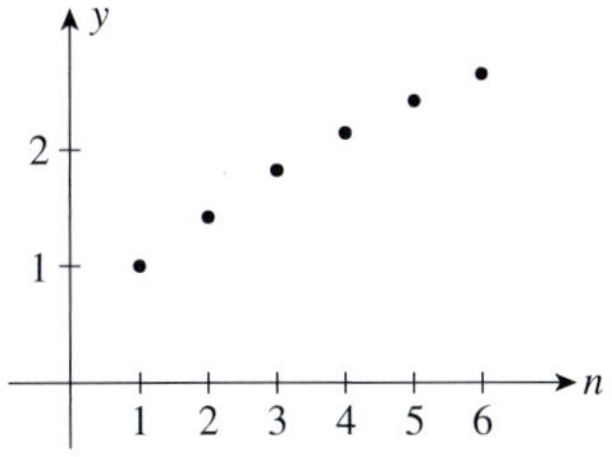

27.

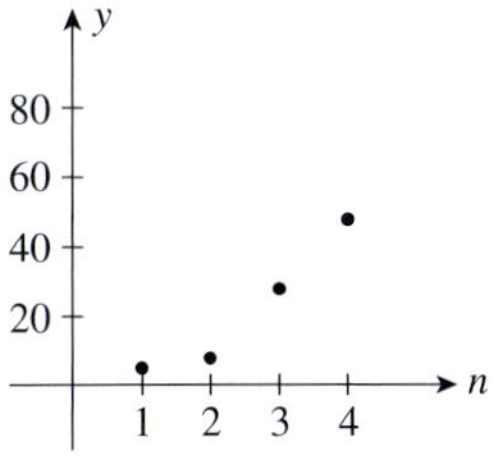

29.

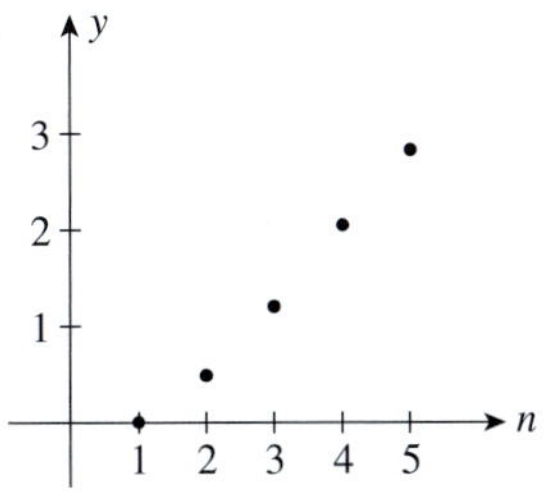

31. Converges; $\frac{1}{2}$ **33.** Converges; 0 **35.** Converges; 1

37. Diverges **39.** Converges; 2 **41.** Converges; 0

43. Converges; $\dfrac{\sqrt{2}}{2}$

45. **a.** $a_1 = 0.015$, $a_{10} \approx 0.140$, $a_{100} \approx 0.77939$, $a_{1000} \approx 0.999999727$
b. 1

47. **b.** $a_n = 100(1.005)^n$
c. $a_{24} \approx 112.72$. The accumulated amount at the end of 2 yr is \$112.72.

49. True **51.** False

Exercises 11.3, page 718

1. $\lim_{N\to\infty} S_N$ does not exist; divergent

3. $S_N = \dfrac{1}{2} - \dfrac{1}{N+2}; \dfrac{1}{2}$ **5.** Converges; $\frac{3}{2}$ **7.** Diverges

9. Converges; $\frac{3}{5}$ **11.** Converges; $\frac{1}{3}$ **13.** Converges; ≈ 5.52

15. Converges; $\dfrac{\pi}{\pi + 3}$ **17.** Converges; 13

19. Converges; $2\frac{3}{4}$ **21.** Converges; $7\frac{1}{2}$ **23.** Diverges

25. $\dfrac{e^3 - 2\pi e + \pi^2}{(\pi - e)(e^2 - \pi)}$ **27.** $\frac{1}{3}$ **29.** $\frac{404}{333}$

31. $-1 < x < 1; \dfrac{1}{1+x}$ **33.** $\dfrac{1}{2} < x < \dfrac{3}{2}; \dfrac{2(x-1)}{3-2x}$

35. $\approx$\$303 billion **37.** $\frac{6}{11}$ **41.** $\dfrac{P}{e^r - 1}$ **43.** **b.** $\dfrac{1}{k}\ln\dfrac{S}{S-C}$ hr

45. False **47.** True

Exercises 11.4, page 729

11. Diverges **13.** Diverges **15.** Converges

17. Converges **19.** Converges **21.** Converges

23. Converges **25.** Converges **27.** Converges

29. Diverges **31.** Diverges **33.** Converges **35.** Diverges

37. Diverges **39.** Converges **41.** Converges

43. Diverges **45.** Converges **47.** Diverges **49.** $p > 1$

51. $a = 1$ **55.** True **57.** False **59.** False

Exercises 11.5, page 738

1. $R = 1$; $(0, 2)$ **3.** $R = 1$; $(-1, 1)$ **5.** $R = 4$; $(-4, 4)$

7. $R = \infty$; $(-\infty, \infty)$ **9.** $R = 0$; $x = -2$ **11.** $R = 1$; $(-4, -2)$

13. $R = \infty$; $(-\infty, \infty)$ **15.** $R = \frac{1}{2}$; $(-\frac{1}{2}, \frac{1}{2})$ **17.** $R = 0$; $x = -1$

19. $R = 3$; $(0, 6)$ **21.** $\sum_{n=0}^{\infty} (-1)^n(x-1)^n$; $R = 1$; $(0, 2)$

23. $\sum_{n=0}^{\infty} (-1)^n \frac{(x-2)^n}{3^{n+1}}$; $R = 3$; $(-1, 5)$

25. $\sum_{n=0}^{\infty} (-1)^{n+1}(x-2)^n$; $R = 1$; $(1, 3)$

27. $1 + \frac{1}{2}(x-1) + \sum_{n=2}^{\infty} (-1)^{n+1} \frac{1 \cdot 3 \cdot 5 \cdots (2n-3)}{n!\, 2^n}(x-1)^n$;

$R = 1$; $(0, 2)$

29. $\sum_{n=0}^{\infty} \frac{2^n}{n!} x^n$; $R = \infty$; $(-\infty, \infty)$

31. $\sum_{n=0}^{\infty} (-1)^n \frac{1 \cdot 3 \cdot 5 \cdots (2n-1)}{n!\, 2^n} x^n$; $R = 1$; $(-1, 1)$ **35.** True

Exercises 11.6, page 745

1. $\sum_{n=0}^{\infty} (-1)^{n+1}(x-2)^n$; $(1, 3)$ **3.** $\sum_{n=0}^{\infty} (-1)^n 3^n x^n$; $(-\frac{1}{3}, \frac{1}{3})$

5. $\sum_{n=0}^{\infty} \frac{3^n}{4^{n+1}} x^n$; $(-\frac{4}{3}, \frac{4}{3})$ **7.** $\sum_{n=0}^{\infty} x^{2n}$; $(-1, 1)$

9. $\sum_{n=0}^{\infty} (-1)^n \frac{x^n}{n!}$; $(-\infty, \infty)$ **11.** $\sum_{n=0}^{\infty} (-1)^n \frac{x^{2n+2}}{n!}$; $(-\infty, \infty)$

13. $\sum_{n=0}^{\infty} \frac{x^{2n}}{(2n)!}$; $(-\infty, \infty)$

15. $\sum_{n=1}^{\infty} (-1)^{n-1} \frac{2^n x^n}{n}$; $(-\frac{1}{2}, \frac{1}{2}]$ **17.** $\sum_{n=1}^{\infty} (-1)^{n+1} \frac{x^{2n}}{n}$; $(-1, 1)$

19. $(\ln 2)(x-2) + \sum_{n=1}^{\infty} (-1)^{n-1} \left(\frac{1}{n2^n}\right)(x-2)^{n+1}$; $(0, 4]$

21. $1 - x + x^2 + \cdots + (-1)^n x^n + \cdots$

23. $f(x) = x - \frac{1}{2}x^2 + \frac{1}{3}x^3 + \cdots + \frac{(-1)^{n+1}}{n} x^n + \cdots$

25. 0.4812 **27.** 0.7475 **29.** 3.34 **31.** 15.85%

Exercises 11.7, page 753

1. 1.732051 **3.** 2.645751 **5.** 2.410142 **7.** 0.7709

9. 1.1219; 2.5745 **11.** 2.30278 **13.** 1.11634

15. 1.61803 **17.** 0.5671 **19. b.** 0.19356 **21. b.** 1.87939

23. 2.9365 **25.** 1.2785 **27.** 0.4263 **29.** 294 units/day

31. 8:39 P.M. **33.** 26.82%/yr **35.** 0.01 **37.** 10%/yr

39. 11,671 units; \$21.17/unit **41.** 2.546

Chapter 11 Concept Review, page 756

1. a. $f(a) + f'(a)(x-a) + \cdots + \frac{f^{(n)}(a)}{n!}(x-a)^n$

b. $f(a), f'(a), \frac{f^{(n)}(a)}{n!}$

2. a. $I, a, x, \frac{f^{(n+1)}(c)}{(n+1)!}(x-a)^{n+1}$

3. a. function, integers, $f(n)$, nth term **b.** large, converge

4. $cA, A + B, AB, \frac{A}{B}, B \neq 0$

5. a. partial sums **b.** geometric, 1, 1

6. a. 0, diverges **b.** $cA + B$

7. a. convergent, divergent **b.** $\sum_{n=1}^{\infty} \frac{1}{n^p}, p > 1, p \leq 1$

8. converges, $a_n \geq b_n$

9. a. $\sum_{n=0}^{\infty} a_n(x-a)^n$ **b.** $x = a, x, (a - R, a + R)$, outside

10. a. $\frac{f^{(n)}(a)}{n!}$ **b.** $\lim_{N \to \infty} R_N(x) = 0$, the Taylor series

Chapter 11 Review Exercises, page 757

1. $f(x) = 1 - (x+1) + (x+1)^2 - (x+1)^3 + (x+1)^4$

2. $f(x) = e^{-1}\left[1 - (x-1) + \frac{(x-1)^2}{2!} - \frac{(x-1)^3}{3!} + \frac{(x-1)^4}{4!}\right]$

3. $f(x) = x^2 - \frac{1}{2}x^4$ **4.** $f(x) = 1 - 2x + 3x^2 - 4x^3 + 5x^4$

5. $f(x) = 2 + \frac{1}{12}(x-8) - \frac{1}{288}(x-8)^2$; 1.983

6. $f(x) = 1 - x^2 + \frac{x^4}{2} - \frac{x^6}{6}$; ≈ 0.7429

7. 2.9992591; 8×10^{-11} **8.** 0.909; 0.0001; 0.9090909. . .

9. 0.37 **10.** 0.199 **11.** Converges; $\frac{1}{3}$ **12.** Converges; 0

13. Converges; $\frac{2}{3}$ **14.** Diverges **15.** Converges; 1

16. Converges; -1 **17.** 2 **18.** $\frac{3}{5}$ **19.** $\frac{1}{\sqrt{2}+1}$ **20.** $\frac{1}{e-1}$

21. $\frac{7}{6}$ **22.** $\frac{1}{2}$ **23.** $\frac{141}{99}$ **24.** $\frac{3139}{999}$ **25.** Diverges

26. Diverges **27.** Converges **28.** Converges

29. Converges **30.** Converges **31.** Converges **32.** Converges

33. $R = 1$; $(-1, 1)$ **34.** $R = 1$; $(-1, 1)$ **35.** $R = 1$; $(0, 2)$

36. $R = \frac{1}{e}$; $\left(2 - \frac{1}{e}, 2 + \frac{1}{e}\right)$

37. $f(x) = -1 - 2x - 4x^2 - 8x^3 - \cdots - 2^n x^n - \cdots$; $(-\frac{1}{2}, \frac{1}{2})$

38. $f(x) = \frac{1}{e} - \frac{1}{e}(x-1) + \frac{1}{2!e}(x-1)^2 - \cdots + \frac{(-1)^n}{n!e}(x-1)^n + \cdots$; $(-\infty, \infty)$

39. $f(x) = 2x - 2x^2 + \frac{8}{3}x^3 - \cdots + \frac{(-1)^{n+1}2^n}{n} x^n + \cdots$; $(-\frac{1}{2}, \frac{1}{2}]$

40. $f(x) = x^2 - 2x^3 + 2x^4 - \cdots + \frac{(-1)^n 2^n x^{n+2}}{n!} + \cdots$; $(-\infty, \infty)$

41. $(-\infty, \infty)$ **42.** $\ln|x| + \sum_{n=1}^{\infty} (-1)^n \frac{x^n}{n!n} + C$ **43.** $\sum_{n=1}^{\infty} \frac{x^n}{n!n} + C$

44. 2.28943 **45.** $\approx$0.7549 **46.** $\approx$(0.35173, 0.70346)

47. 2.65 sec **48.** \$323,433.33

49. a. \$115 billion **b.** \$90 billion **50.** 31.08%

Chapter 11 Before Moving On, page 758

1. $P_1(x) = x$, $P_2(x) = x - x^2$, and $P_3(x) = x - x^2 + \frac{x^3}{2}$

2. a. $\frac{2}{3}$ **b.** 2 **3. a.** $\frac{25}{12}$ **b.** $\frac{1}{3}$

4. a. convergent **b.** divergent **5.** $-\sum_{n=0}^{\infty} (x+2)^n$; 1

CHAPTER 12

Exercises 12.1, page 763

1. $\frac{5\pi}{2}$ radians **3.** $-\frac{3\pi}{2}$ radians

5. a. III **b.** III **c.** II **d.** I

7. $\frac{5\pi}{12}$ radians **9.** $\frac{8\pi}{9}$ radians **11.** $\frac{7\pi}{2}$ radians

13. 120° **15.** −270° **17.** 220°

19.

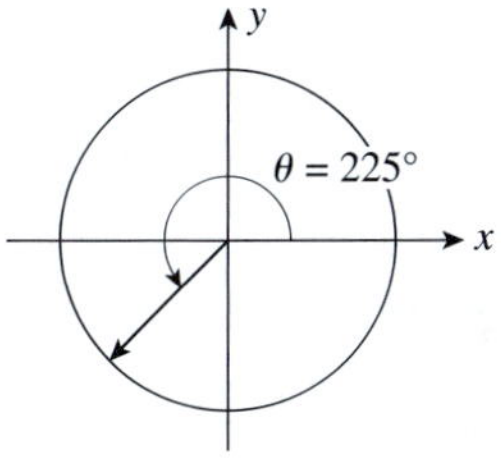

21.

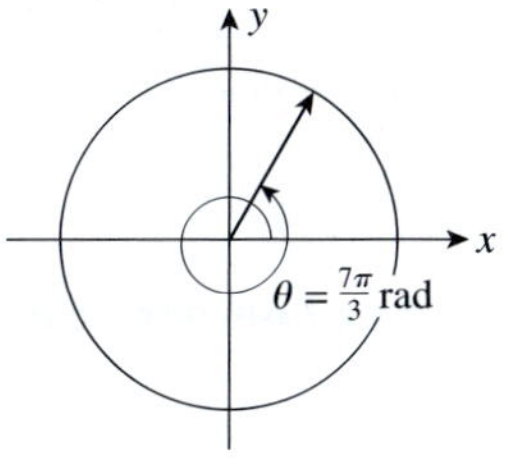

23. 150°, −210° **25.** 315°, −45° **27.** True **29.** True

Exercises 12.2, page 771

1. 0 **3.** 1 **5.** $\frac{\sqrt{3}}{2}$ **7.** $\frac{\sqrt{3}}{3}$ **9.** −2.6131

11. $\sin\frac{\pi}{2} = 1$, $\cos\frac{\pi}{2} = 0$, $\tan\frac{\pi}{2}$ is undefined, $\csc\frac{\pi}{2} = 1$, $\sec\frac{\pi}{2}$ is undefined, $\cot\frac{\pi}{2} = 0$

13. $\sin\frac{5\pi}{3} = -\frac{\sqrt{3}}{2}$, $\cos\frac{5\pi}{3} = \frac{1}{2}$, $\tan\frac{5\pi}{3} = -\sqrt{3}$, $\csc\frac{5\pi}{3} = -\frac{2\sqrt{3}}{3}$, $\sec\frac{5\pi}{3} = 2$, $\cot\frac{5\pi}{3} = -\frac{\sqrt{3}}{3}$

15. $\frac{7\pi}{6}, \frac{11\pi}{6}$ **17.** $\frac{5\pi}{6}, \frac{11\pi}{6}$ **19.** π **21.** $\frac{\pi}{3}, \frac{2\pi}{3}$

23.

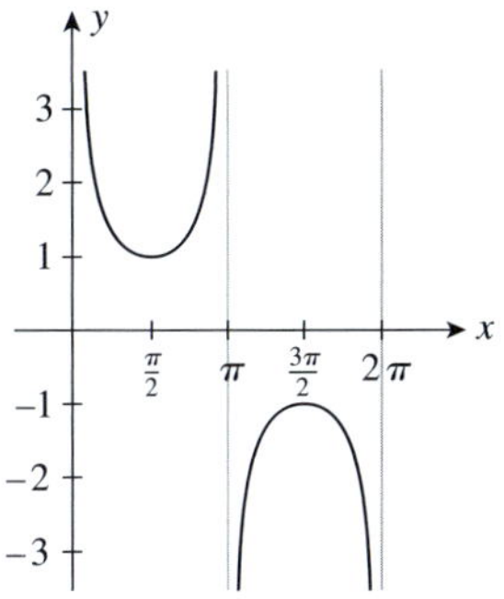

25.

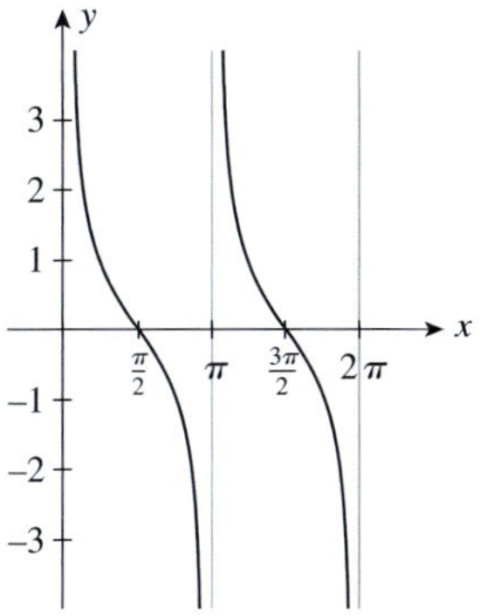

27.

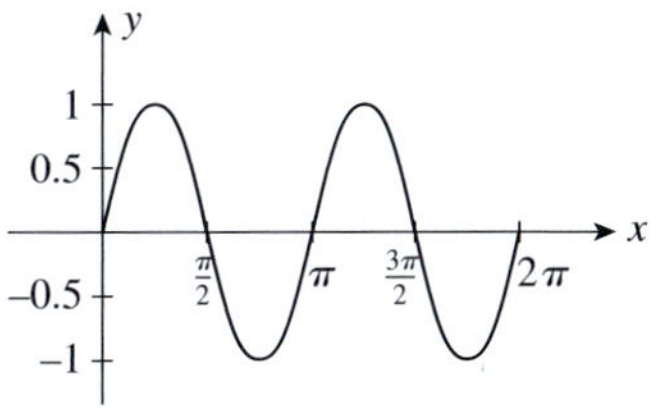

29.

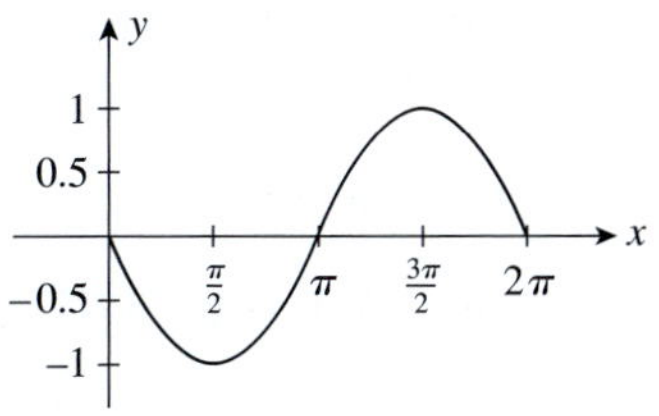

41. $\sin\theta = \frac{5}{13}$, $\cos\theta = \frac{12}{13}$, $\tan\theta = \frac{5}{12}$, $\csc\theta = \frac{13}{5}$, $\sec\theta = \frac{13}{12}$, $\cot\theta = \frac{12}{5}$

43. b. $\frac{\pi(4n+1)}{12}$ $(n = 0, 1, 2, \ldots)$; $\frac{\pi(4n+3)}{12}$ $(n = 0, 1, 2, \ldots)$

45. False **47.** True

Exercises 12.3, page 781

1. $-3\sin 3x$ **3.** $-2\pi\sin\pi x$ **5.** $2x\cos(x^2+1)$

7. $4x\sec^2 2x^2$ **9.** $x\cos x + \sin x$ **11.** $6(\cos 3x - \sin 2x)$

13. $2x(\cos 2x - x\sin 2x)$ **15.** $\frac{x\cos\sqrt{x^2-1}}{\sqrt{x^2-1}}$

17. $e^x\sec x(1+\tan x)$ **19.** $\cos\frac{1}{x} + \frac{1}{x}\sin\frac{1}{x}$

21. $\frac{x\sin x}{(1+\cos x)^2}$ **23.** $\frac{\sec^2 x}{2\sqrt{\tan x}}$ **25.** $\frac{x\cos x - \sin x}{x^2}$

27. $2\tan x\sec^2 x$ **29.** $-\csc^2 x\cdot e^{\cot x}$ **31.** $y = -2x + \frac{\pi}{2}$

33. Increasing on $\left(0, \frac{\pi}{4}\right) \cup \left(\frac{5\pi}{4}, 2\pi\right)$; decreasing on $\left(\frac{\pi}{4}, \frac{5\pi}{4}\right)$

35. $f(x) = \sin x + \cos x$

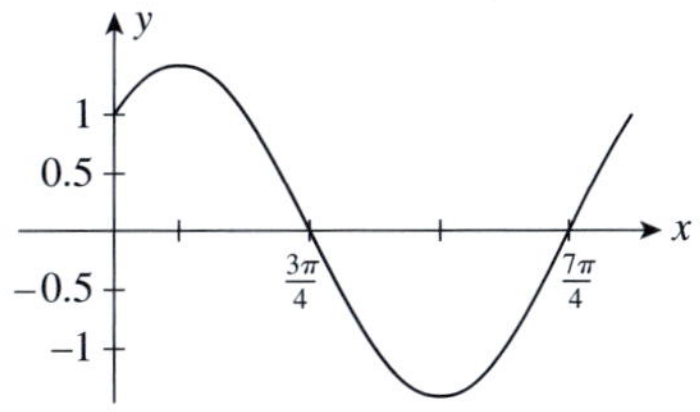

37. $f(x) = 2\sin x + \sin 2x$

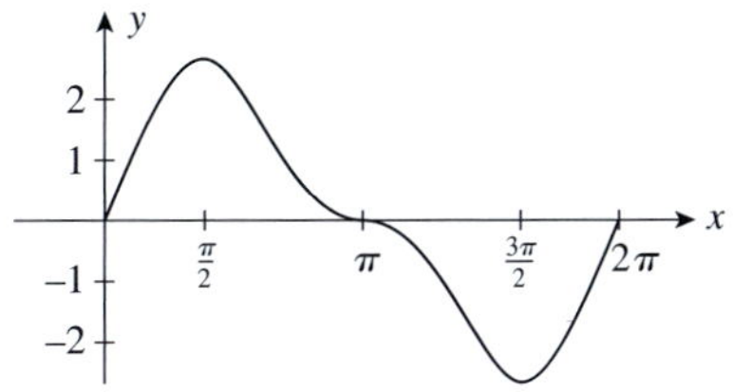

39. a. $f(x) = \sin x = x - \frac{x^3}{3!} + \frac{x^5}{5!} - \frac{x^7}{7!} + \cdots + (-1)^n\frac{x^{2n+1}}{(2n+1)!} + \cdots$

41. Zero wolves/mo; −1571 caribou/mo **43.** 0.05

45. Warmest day is July 25; coldest day is January 23.

47. Sixth week

49. Maximum when $t = 1, 5, 9, 13, \ldots$; minimum when $t = 3, 7, 11, 15, \ldots$

51. 70.7 ft/sec **53.** 60° **55.** 1.4987 radians **57.** True

59. False

Using Technology Exercises 12.3, page 785

1. 1.2038 **3.** 0.7762 **5.** -0.2368 **7.** 0.8415; -0.2172

9. 1.1271; 0.2013

11. a.

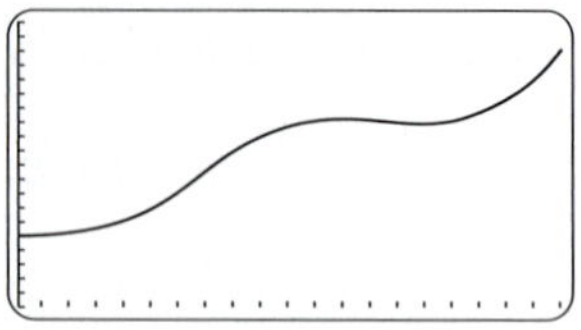

b. $\approx$\$0.90 **c.** $\approx$\$37.86

13. a.

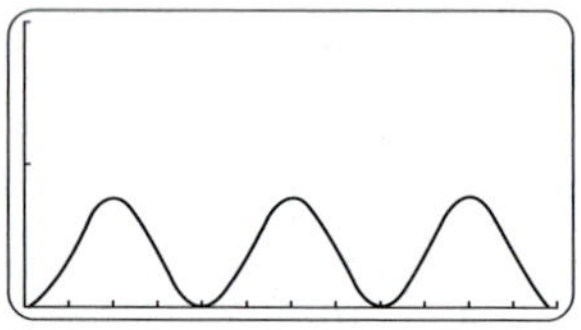

15. $\approx$0.006 ft

Exercises 12.4, page 789

1. $-\frac{1}{3}\cos 3x + C$ **3.** $-3\cos x + 4\sin x + C$

5. $\frac{1}{2}\tan 2x + C$ **7.** $\frac{1}{2}\sin x^2 + C$ **9.** $-\frac{1}{\pi}\csc \pi x + C$

11. 2 **13.** $-\frac{1}{2}\ln\frac{1}{2}$ **15.** $\frac{1}{4}\sin^4 x + C$

17. $\frac{1}{\pi}\ln|\sec \pi x + \tan \pi x| + C$ **19.** $\frac{1}{3}\ln(1 + \sqrt{2})$

21. $-\frac{2}{3}(\cos x)^{3/2} + C$ **23.** $-\frac{1}{9}(1 - 2\sin 3x)^{3/2} + C$

25. $\frac{1}{4}\tan^4 x + C$ **27.** $-\frac{1}{4}(\cot x - 1)^4 + C$ **29.** 2

31. $\frac{1}{2}x[\sin(\ln x) - \cos(\ln x)] + C$ **33.** $2\sqrt{2}$

35. $\frac{1}{2}\ln 2$ **37.** $\frac{1}{2}(\pi^2 - 4)$ **39.** $\approx$\$85

41. \$120,000 **43.** $\dfrac{1.2}{\pi}\left(1 - \cos\dfrac{\pi t}{2}\right)$ **45.** 162 fruit flies

47. $\approx$0.9045 **49.** True **51.** True

Using Technology Exercises 12.4, page 792

1. 0.5419 **3.** 0.7544 **5.** 0.2231 **7.** 0.6587

9. -0.2032 **11.** 0.9045

13. a.

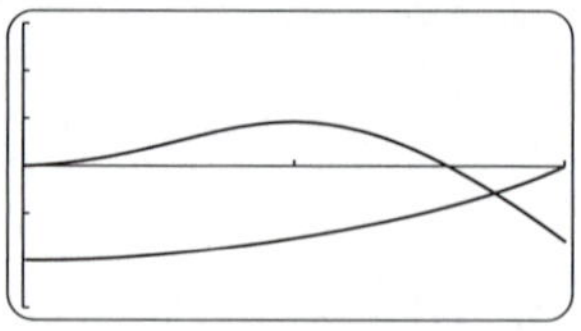

b. 2.2687

15. a.

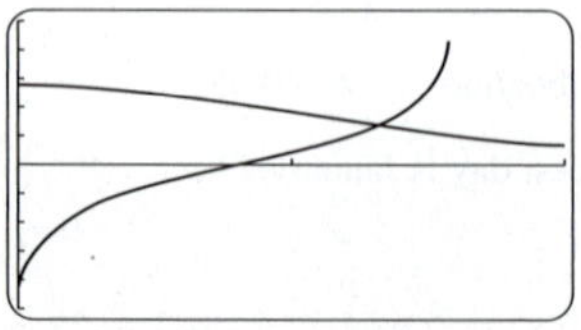

b. 1.8239

17. a.

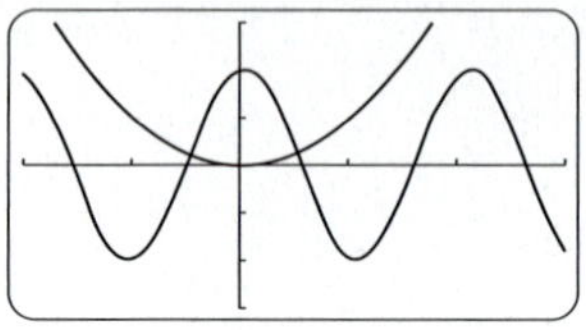

b. 1.2484

19. a.

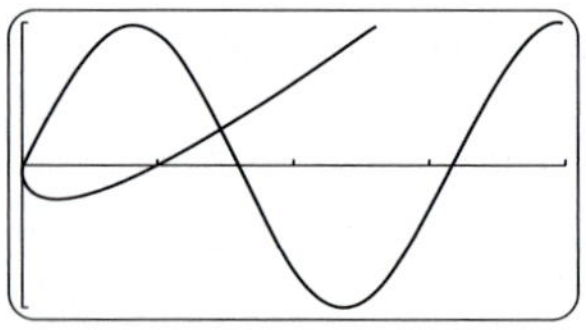

b. 1.0983

21. 7.6 ft

Chapter 12 Concept Review, page 793

1. a. radian **b.** 2π, 360 **2. a.** $\frac{\pi}{180}x$ **b.** $\frac{180}{\pi}x$

3. $x, y, \dfrac{y}{x}, \dfrac{1}{y}, \dfrac{1}{x}, \dfrac{x}{y}$ **4. a.** $\cos^2\theta$ **b.** 1 **c.** 1 **d.** $\cos 2\theta$

5. a. $\cos 2\theta$ **b.** + **c.** $\sin A \cos A$ **d.** $\sin^2 A$

6. a. $[\cos f(x)]f'(x)$ **b.** $-[\sin f(x)]f'(x)$
c. $[\sec^2 f(x)]f'(x)$ **d.** $-[\csc f(x)][\cot f(x)]f'(x)$
e. $[\sec f(x)][\tan f(x)]f'(x)$ **f.** $-[\csc^2 f(x)]f'(x)$

7. a. $-\cos x + C$ **b.** $\sin x + C$ **c.** $\tan x + C$
d. $-\cot x + C$ **e.** $\sec x + C$ **f.** $-\csc x + C$

Chapter 12 Review Exercises, page 794

1. $\frac{2\pi}{3}$ radians **2.** $\frac{5\pi}{2}$ radians **3.** $-\frac{5\pi}{4}$ radians **4.** 330°

5. $-450°$ **6.** $-315°$ **7.** $\frac{\pi}{3}$ or $\frac{5\pi}{3}$ **8.** 150° or 330°

9. $3\cos 3x$ **10.** $-\sin\frac{x}{2}$ **11.** $2\cos x + 6\sin 2x$

12. $\dfrac{\sec^2\sqrt{x}\tan\sqrt{x}}{\sqrt{x}}$ **13.** $e^{-x}(3\sec^2 3x - \tan 3x)$

14. $4(1 - \csc 2x)(\csc 2x \cot 2x)$ **15.** $4\cos 2x$ or $4(\cos^2 x - \sin^2 x)$

16. $-\dfrac{\sin x}{(1 - \cos x)^2}$

17. $\dfrac{(\cot x - 1)\sec^2 x - (1 - \tan x)\csc^2 x}{(1 - \cot x)^2} = -\sec^2 x$

18. $-2\tan x$ **19.** $\cos(\sin x)\cos x$ **20.** $e^{\sin x}(-\sin x + \cos^2 x)$

21. $y = 4x + 1 - \pi$

22.

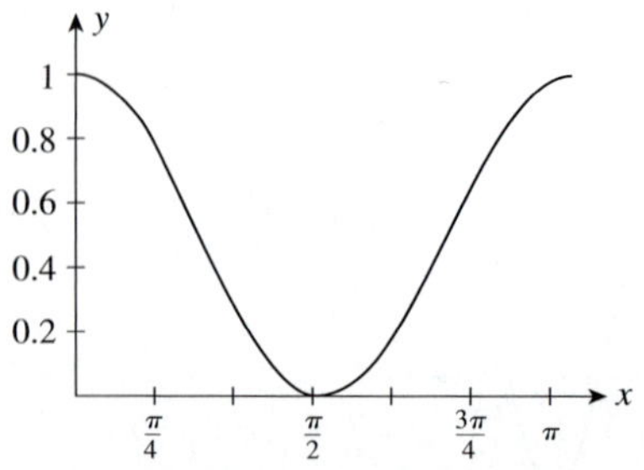

a. ↗ on $(\frac{\pi}{2}, \pi)$ and ↘ on $(0, \frac{\pi}{2})$ **b.** Relative minimum: $f(\frac{\pi}{2}) = 0$
c. Concave up on $(\frac{\pi}{4}, \frac{3\pi}{4})$ and concave down on $(0, \frac{\pi}{4})$ and $(\frac{3\pi}{4}, \pi)$
d. $(\frac{\pi}{4}, \frac{1}{2})$ and $(\frac{3\pi}{4}, \frac{1}{2})$

23. $\frac{3}{2}\sin\frac{2}{3}x + C$ 24. $\frac{1}{2}x - \frac{1}{8}\sin 4x + C$ 25. $-\frac{1}{2}\csc x^2 + C$

26. $x\sin x + \cos x + C$ 27. $\frac{1}{3}\sin^3 x + C$ 28. $-\cos e^x + C$

29. $-\csc x + C$ 30. $\frac{1}{2}\ln 2$ 31. 1 32. 2

33. $2\sqrt{2}$

34. Fox population ↗ at the rate of ≈6.5 foxes/mo and the rabbit population ↘ at the rate of ≈113.4 rabbits/mo

35. December 1; September 1 36. 78.5%

Chapter 12 Before Moving On, page 794

1.

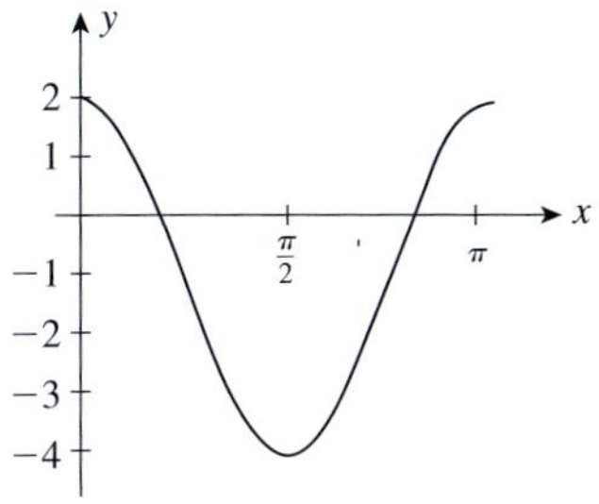

3. $(3\cos 3x - 2\sin 3x)e^{-2x}$ 4. $\frac{\pi + 2}{8}$ 5. $\frac{\sqrt{2}}{12}$

APPENDIX A

Exercises, page 800

7. Yes 9. No 11. No

13. $\frac{1}{3}(x + 2)$

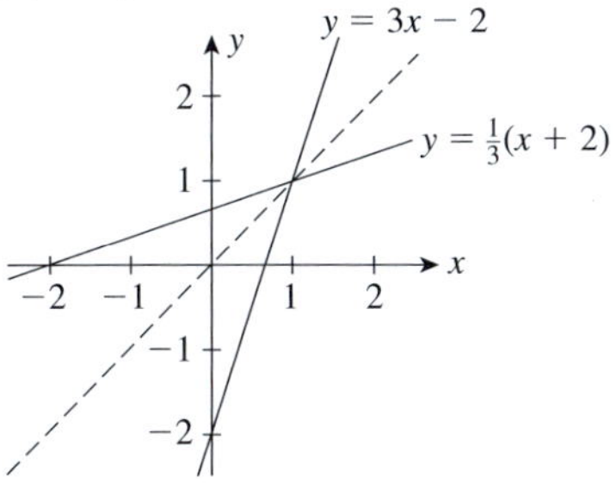

15. $\sqrt[3]{x - 1}$

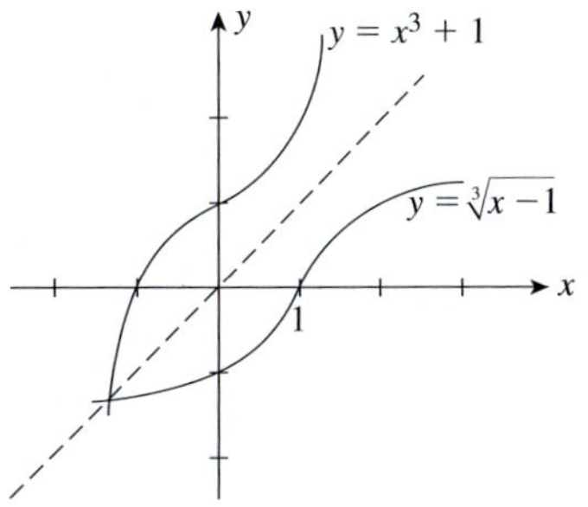

17. $\sqrt{9 - x^2}\ (x \geq 0)$

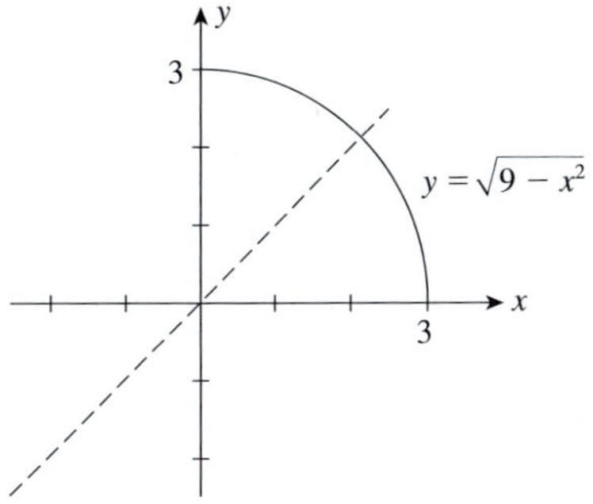

19. **a.** $-\frac{1}{2} + \frac{1}{2}\sqrt{1 + 8h}$; f^{-1} represents the time at which the balloon is at height h.
b. Between 15 and 20 sec

APPENDIX B

Exercises, page 804

1. $\frac{1}{2}$ 3. 12 5. 1 7. 1 9. 2 11. $\frac{1}{2}$ 13. $\frac{1}{3}$

15. ∞ 17. 0 19. 0 21. $\frac{9}{4}$ 23. ∞ 25. $-\frac{1}{2}$

27. $\frac{1}{2}$ 29. 0 31. e 33. False 35. False

INDEX

How-To Technology Index